Probability Theory and Stochastic Modelling

Volume 101

The Probability Theory and Stochastic Modelling series is a merger and continuation of Springer's two well established series Stochastic Modelling and Applied Probability and Probability and Its Applications. It publishes research monographs that make a significant contribution to probability theory or an applications domain in which advanced probability methods are fundamental. Books in this series are expected to follow rigorous mathematical standards, while also displaying the expository quality necessary to make them useful and accessible to advanced students as well as researchers. The series covers all aspects of modern probability theory including:

- Gaussian processes
- Markov processes
- Random Fields, point processes and random sets
- Random matrices
- Statistical mechanics and random media
- Stochastic analysis

as well as applications that include (but are not restricted to):

- Branching processes and other models of population growth
- Communications and processing networks
- Computational methods in probability and stochastic processes, including simulation
- Genetics and other stochastic models in biology and the life sciences
- Information theory, signal processing, and image synthesis
- Mathematical economics and finance
- Statistical methods (e.g. empirical processes, MCMC)
- Statistics for stochastic processes
- Stochastic control
- Stochastic models in operations research and stochastic optimization
- Stochastic models in the physical sciences

All titles in this series are peer-reviewed to the usual standards of mathematics and its applications.

More information about this series at http://www.springer.com/series/13205

Qi Lü · Xu Zhang

Mathematical Control Theory for Stochastic Partial Differential Equations

Springer

Qi Lü
School of Mathematics
Sichuan University
Chengdu, China

Xu Zhang
School of Mathematics
Sichuan University
Chengdu, China

ISSN 2199-3130 ISSN 2199-3149 (electronic)
Probability Theory and Stochastic Modelling
ISBN 978-3-030-82330-6 ISBN 978-3-030-82331-3 (eBook)
https://doi.org/10.1007/978-3-030-82331-3

Mathematics Subject Classification: 93E03, 93B05, 93B07, 93E20, 60H15, 60H30

This Springer imprint is published by the registered company Springer Nature Switzerland AG
The registered company address is: Gewerbestrasse 11, 6330 Cham, Switzerland

Preface

It is well-known that Control Theory was founded by N. Wiener in 1948 ([349]). After that, this theory was greatly extended to various complicated setting and widely used in sciences and technologies. Particularly, the rapid development of Control Theory began in the mid 1950's, partially in response to practical problems in many branches of engineering and economics.

Roughly speaking, Control Theory can be divided into two parts. The first part is control theory for deterministic systems, and the second one is that for stochastic systems. Of course, these two parts are not completely separated but rather they are inextricably linked each other.

Control theory for deterministic systems can be again divided into two parts. The first part is control theory for finite dimensional systems, mainly governed by ordinal differential equations, and the second one is that for (deterministic) distributed parameter systems, mainly described by differential equations in infinite dimensional spaces, typically by partial differential equations.

Control theory for finite dimensional systems is by now relatively mature. Classics in the field include, among others, the three milestones of modern control theory, i.e., L. S. Pontryagin's maximum principle ([281]), R. Bellman's dynamic programming method ([20]), and R. E. Kalman's optimal linear regulator theory ([163]). We refer to [1, 28, 37, 45, 59, 102, 193], just to mention a few, for some of the subsequent progresses.

There exist a huge list of works on control theory for distributed parameter systems, and it is still quite active. Pioneers in this field include A. V. Balakrishnan ([9]), A. G. Butkovskiĭ ([44]), Yu V. Egorov ([85]), H. O. Fattorini ([91]), C. C. Kwan and K. N. Wang ([183]), J.-L. Lions ([205]), D. L. Russell ([292]), P. K. C. Wang ([343]) and so on. For early books on this topic, we mention [44, 66, 205]. For the extensive works in the recent three decades, we refer to [7, 10, 61, 67, 88, 92, 101, 110, 116, 117, 123, 144, 168, 170, 172, 186, 199, 202, 220, 305, 320, 323, 325, 338, 388, 395, 396] and the rich references cited therein.

Likewise, control theory for stochastic systems can be divided into two parts. The first part is control theory for stochastic finite dimensional systems, governed by stochastic (ordinary) differential equations, and the second one is that for stochastic distributed parameter systems, described by stochastic differential equations in infinite dimensions, typically by stochastic partial differential equations.

One can also find a huge list of publications on control theory for stochastic finite dimensional systems and its applications (say, in mathematical finance). Pioneers in this field include R. Bellman ([21]), W. H. Fleming and M. Nisio ([97]), J. J. Florentin ([100]), H. J. Kushner ([177]) and so on. For early books on this topic, we mention [6], [29] and [179]. For the extensive (and/or closely related) works in this respect in the recent decades, we refer to [5, 27, 48, 52, 53, 98, 124, 173, 181, 204, 251, 272, 279, 322, 340, 363, 364, 371, 383] and the rich references cited therein. Nevertheless, most of the existing works in this respect are mainly addressed/related to optimal control problems. As we shall see in Chapter 6 (of this book), so far controllability theory for stochastic finite dimensional systems is NOT well-developed.

By contrast, control theory for stochastic distributed parameter systems, is, in our opinion, still at its very beginning stage. Indeed, though one can find a considerably number of papers (See [24, 30, 90, 96, 178, 302, 326] for the early ones) and five related books [15, 89, 167, 242, 259] [1] (See also the books [66, 270] for several related chapters) on this topic, as far as we know, there exist no any monographs addressing systematically to this rather new branch of mathematical control theory, which is actually the main concern of this book.

One of the most essential difficulties in the study of control theory for stochastic distributed parameter systems is that, compared to the deterministic setting, people know very little about stochastic evolution equation (and particularly, about stochastic partial differential equations) although many progresses have been made there (See [68, 136, 137, 145, 157, 329, 335], etc.). On the other hand, as we shall see in the context of this book, both the formulation of stochastic control problems in infinite dimensions and the tools to solve them may differ considerably from their deterministic/finite-dimensional counterparts. Because of this, one has to develop new mathematical tools, say, *the stochastic transposition method* introduced in our works [241, 242, 244, 245, 248] (See also Chapters 4 and 12–13, and particularly Section 1.4 of Chapter 1 for more details and explanations), to solve some problems in this field.

In some sense, the stochastic distributed parameter control system is the most general control system in the framework of classical physics, and therefore the study of this field may provide some useful hints for that of quantum control systems.

---

[1] Actually, the books [15, 167, 259] addressed mainly to some slightly different topics.

This book is an introduction to (mathematical) control theory for stochastic distributed parameter systems. Our aim is to provide a basic reference for both beginners and experts who are interested in this challenging, vigorous and fascinating field. It is designed for a reader having some basic knowledge of functional analysis, partial differential equations and control theory for deterministic systems but not being familiar with probability theory and, particularly, stochastic analysis.

The authors would like to take this opportunity to thank Professors Shuping Chen, Jean-Michel Coron, Hélène Frankowska, Xiaoyu Fu, Andrei Fursikov, Zechun Hu, Oleg Yu. Imanuvilov, Gilles Lebeau, Xu Liu, Shige Peng, Luc Robbiano, Shanjian Tang, Emmanuel Trélat, Jan van Neerven, Gengsheng Wang, Tianxiao Wang, Zhongqi Yin, Jiongmin Yong, Bingyu Zhang, Haisen Zhang and Enrique Zuazua for collaborations/comments/discussions/encouragement/supports during the preparation of this book. This work is partially supported by NSF of China under grants 12025105, 11971334, 11931011 and 11221101.

Chengdu, China *Qi Lü*
Chengdu, China *Xu Zhang*
June 30, 2020

Contents

1

Introduction

In this chapter, we expound the necessity, novelty, importance and challenge to study stochastic distributed parameter control systems. We also illustrate pedagogically the main problems to be considered and the main methods to be employed in this book.

Throughout this book, $\mathbb{N}$, $\mathbb{Z}$, $\mathbb{Q}$, $\mathbb{R}$ and $\mathbb{C}$ stand for respectively the set of positive integers, integers, rational numbers, real numbers and complex numbers. Also, we shall denote by $\mathcal{C}$ a generic positive constant which may change from line to line (unless other stated).

1.1 Why Stochastic Distributed Parameter Control Systems?

After N. Wiener's classical work ([349]), owing to the great effort of numerous mathematicians and engineers, Control Theory has grown into a new discipline—Control Science. Mathematical control theory, an active branch of applied mathematics, which aims to provide the mathematical foundations for Control Science, refers to mathematical theory and methods for the analysis and design of control systems. Stimulated by the seminal works [20, 163, 281], rapid development of mathematical control theory (for both deterministic and stochastic systems) began in the 1960s. Usually, in terms of the state space method (e.g., [163]), people describe the control system under consideration as a suitable state equation.

The main concern of this book belongs to a new branch of mathematical control theory, i.e., control theory for stochastic distributed parameter systems, that is, systems governed by stochastic differential equations in infinite dimensions, including stochastic differential equations with delays, stochastic partial differential equations, random partial differential equations (i.e., partial differential equations with random parameters), and so on.

One may ask a natural question, i.e., *why should people study stochastic distributed parameter control systems?* Our answer is, this is because, for

Q. Lü and X. Zhang, *Mathematical Control Theory for Stochastic Partial Differential Equations*, Probability Theory and Stochastic Modelling 101, https://doi.org/10.1007/978-3-030-82331-3_1

Control Theory, it is now stochastic distributed parameter control systems' turn! Let us justify it below by recalling rapidly the history of modern control theory (particularly mathematical control theory, of course).

Since the middle of 1950s, the rapid development of space technology requests urgently to solve the control problems such as launching the rockets, satellites, etc. into the pre-selected orbit accurately with the minimum fuel consumption or in the shortest flight time. These control problems are too complex to be solved by the classical control theory developed in [349]. By the early 1960s, a set of new principles and methods, which are based on the state space method to analyze and design the control system, has been established. This marks the formation of modern control theory (and also mathematical control theory). Three milestones therein are achieved by L. Pontryagin (Pontryagin's maximum principle, [281]), R. Bellman (Bellman's dynamic programming method, [20]) and R. Kalman (Kalman's filtering and linear quadratic optimal control, [162, 163]). Note that these classics are addressed mainly to finite dimensional control systems, which are governed by ordinary differential (or difference) equations. Inspired by Pontryagin, Bellman and Kalman's works, starting from 1960s there are numerous publications addressing to this topic. Now, control theory for finite dimensional systems is relatively mature though many problems in this field are still remained to be done.

Fundamentally, all physical systems are intrinsically distributed in nature. Actually, for practical problems, finite dimensional systems are only approximations of distributed parameter systems (i.e., infinite dimensional systems) to a certain extend. Therefore, theoretically distributed parameter systems are more realistic models. As for concrete applications, although it seems that approximate models have more significance in view of engineering (especially by considering computer implementation), the study on more realistic models contributes to giving reasonable approximate models, and particularly giving reasonable control laws, from the point of view of Control Theory. Distributed parameter control systems are control systems described by differential equations in infinite dimensional spaces, for example, partial differential equations, delay differential equations, etc.

Control theory for distributed parameter systems originated in the 1960s, and the early works on this field were mainly done by A. G. Butkovskiĭ ([44]), Yu V. Egorov ([85]), H. O. Fattorini ([91]), J.-L. Lions ([205]), D. L. Russell ([292]), P. K. C. Wang ([343]) and so on. In the past nearly 60 years, this theory has been developed very rapidly, for which more than one hundred monographs and thousands of research papers have been published one after another. So far, there exists no sign of slowing growth of this field. In the recent two decades, many new tools are introduced/refined, and lots of breakthroughs are obtained in this respect. We refer the readers to [45, 46, 49, 62, 65, 88, 95, 110, 116, 123, 133, 144, 168, 170, 172, 186, 199, 200, 220, 288, 305, 323, 325, 333, 338, 388, 395, 396] and the references therein for the development of control theory for distributed parameter systems in the new century.

The world is full of uncertainties, and therefore it is impossible to predict accurately the future, just like "a storm may arise from a clear sky". Therefore, studying random phenomena from the viewpoint of Control Theory, without doubt, has basic, important scientific significance and practical values. Strictly speaking, any real system contains various random factors, but these factors may be ignored in many situations. However, when these factors cannot be ignored, the behavior of the control system, if designed according to the deterministic control theory, would deviate from the expected design requirements, and sometimes it may even happen that, as a Chinese proverb said, "Cha Zhi Hao Li, Shi Zhi Qian Li", i.e., "an error, the breadth of a single hair, can lead a thousand li astray".

Control theory for stochastic finite dimensional systems originated in the late 1950s, which has developed very rapidly for the past (more than) 60 years. The early works on this field were mainly accomplished by R. Bellman, W. H. Fleming, H. J. Kushner, M. Nisio, etc. ([21, 97, 177]). Thanks to J.-M. Bismut ([33]), W. H. Fleming ([96]), P.-L. Lions ([210, 211, 212]), S. Peng ([273]) and so on, deep studies in this respect have been done (see [98, 173, 371]). An important byproduct in the study of this field is the appearance of a new theory, i.e., backward (or more generally, forward-backward) stochastic differential equations (see [33, 251, 271]), which is by now widely used in financial mathematics, stochastic analysis, partial differential equations, etc. (e.g., [56, 73, 86, 251, 274, 321]).

Now, we are in a position to return to the main topic of this book, i.e., control theory for stochastic distributed parameter systems. Theoretically, motivated by the success of control theories for finite dimensional systems, distributed parameter systems and stochastic finite dimensional systems, as a natural development, people hope to study stochastic distributed parameter control systems. Indeed, soon after the publication of early works on stochastic finite dimensional control systems, some pioneers including A. Bensoussan [24], M. Kh. Bikchantayev ([30]), P. L. Falb ([90]), W. H. Fleming ([96]), H. J. Kushner ([178]), T. K. Sirazetdinov ([302]), G. Tzafestas and J. M. Nightingale ([326]) and so on began the study of stochastic distributed parameter control systems. On the other hand, in the application aspects, because of the inherent complexity of the underlying physical processing, many control systems in reality (such as that in the microelectronics industry, in the atmospheric motion, in communications and transportation, and so on) exhibit very complex dynamics, including substantial model uncertainty, actuator and state constraints, and high dimensionality (usually infinite). These systems are often best described by stochastic partial differential equations or even more complicated stochastic equations in infinite dimensions (e.g., [263]).

In our opinion, the field of stochastic distributed parameter control systems is far from mature till now, in which many aspects are much less-understood or even still blank, compared to the deterministic setting and/or stochastic finite dimensional problems. More precisely, we mention the following issues.

- Compared to the deterministic cases (for both finite dimensional and distributed parameter control systems), the study of stochastic controllability/observability is quite unsatisfactory, even for the setting of stochastic finite dimensional control systems (see Chapter 6 for a more detailed analysis). Indeed, so far there exists no nontrivial controllability result for stochastic nonlinear systems. Also, for most of positive controllability results (in the existing literatures, e.g., [228, 231, 247, 315]) for stochastic partial differential equations, controls are introduced into the diffusion terms and are assumed to be active everywhere.
- In order to derive Pontryagin-type maximum principle for general nonlinear stochastic distributed parameter control systems, the key in [242] is to introduce the notion of relaxed transposition solutions to an operator-valued backward stochastic Lyapunov equation, and via which the well-posednss of such an equation can be established. Although this notion of solutions and its modification are useful and enough to tackle several different problems (say, the first and second order necessary conditions for optimal controls of general nonlinear stochastic evolution equations in infinite dimensions ([103, 104, 240, 242, 244, 245]), characterization of optimal feedback operators for general stochastic linear quadratic control problems (with random coefficients) in infinite dimensions ([235, 248]), etc.), it is much better to show that the underlying operator-valued backward stochastic evolution equations are well-posed in a stronger sense, i.e., in the sense of transposition solutions rather than relaxed transposition solutions (See Chapters 12–13 for more details), which remains an unsolved problem.
- The extension of Bellman's dynamic programming method to the stochastic setting, though some very interesting and important progresses were made ([89, 213, 214]), the problem is far from well-understood for the systems with random coefficients, even in the setting of stochastic finite dimensional control systems. Indeed, as far as we know, in this case, one has to face a challenging problem to establish the well-posednss of backward stochastic Hamilton-Jacobi-Bellman equations.
- In order to find optimal feedback operators for stochastic linear quadratic control problems (with random coefficients) in infinite dimensions, one has to solve suitable operator-valued backward stochastic Riccati equations. However, so far the well-posedness of this sort of Riccati equations is known only for some very special case, again even in the setting of stochastic control systems in finite dimensions ([248]).

Consequently, control theory for stochastic distributed parameter systems is still at its very beginning stage, though after the development of more than 50 years. Compared with other directions in mathematical control theory (including control theory for finite dimensional systems, that for distributed parameter systems and that for stochastic finite dimensional systems), at this moment it is an "*ugly duckling*"!

One will meet many substantially difficulties in the study of control problems for stochastic distributed parameter systems, for example:

- Unlike the deterministic setting, the solution to a stochastic evolution equation (even in finite dimensions) is usually non-differentiable w.r.t. the variables with noises.
- The usual compactness embedding result fails to be true for solution spaces related to the stochastic evolution equations under consideration.
- The "time" in the stochastic setting is not reversible, even for stochastic hyperbolic equations.
- Generally speaking, stochastic control problems cannot be reduced to deterministic ones.

Actually, the main difficulty in this field lies in the fact that, as we shall see in the rest of this book, one has to develop new mathematical tools even for some very simple stochastic distributed parameter control systems. Indeed, many tools and methods, which are effective in the deterministic case, do not work anymore in the stochastic setting. Technically, the most essential difficulty in the field of stochastic distributed parameter control systems, is the very fact that the theory of stochastic partial differential equations itself is far from mature (though significant progress has been made there in recent years, e. g., [136, 137]). This can be seen even from the title of J. Glimm's review article ([122]): *Nonlinear and stochastic phenomena: the grand challenge for partial differential equations*.

We think, control theory for stochastic distributed parameter systems is a direction which deserves to be developed with great efforts in the whole Control Science, especially in mathematical control theory. Clearly, in the framework of classical physics, the stochastic distributed parameter system is very likely the most general control systems. A deep study on this topic may provide some useful hints for the development of mathematical theory for quantum control systems.

In conclusion, the field of stochastic distributed parameter control systems is full of challenging problems, which offers a rare opportunity for the new generations in Control Theory. Definitely, we believe that control theory for stochastic distributed parameter systems will become a "*white swan*" in the near future!

1.2 Two Fundamental Issues in Control Theory

Roughly speaking, "control" means one can change the dynamics of the involved system, by means of a suitable way. By our understanding, there are at least two fundamental issues in Control Theory[1], i.e., feasibility and optimality, which will be explained more below (c.f. [220]).

[1] We refer the readers to [389] for more fundamental issues in Control Theory.

The first fundamental issue in Control Theory is *feasibility*, or in the terminology of Control Theory, *controllability*, which means that, one can find at least one way to achieve a goal. More precisely, for simplicity, let us consider the following controlled system governed by a linear ordinary differential equation:

$$\begin{cases} y_t(t) = Ay(t) + Bu(t), & \text{a.e. } t \in [0,T], \\ y(0) = y_0. \end{cases} \tag{1.1}$$

In (1.1), $A \in \mathbb{R}^{n\times n}$, $B \in \mathbb{R}^{n\times m}$ $(n, m \in \mathbb{N})$, $T > 0$, $y(\cdot)$ is the state variable, $u(\cdot)$ is the control variable, and $\mathbb{R}^n$ and $\mathbb{R}^m$ are respectively the state and control spaces. We begin with the following notion (More notations/notions, which may be used below, will be given in Chapter 2):

Definition 1.1. *The system (1.1) is called exactly controllable (at time T) if for any initial state $y_0 \in \mathbb{R}^n$ and final state $y_1 \in \mathbb{R}^n$, there is a control $u(\cdot) \in L^1(0,T;\mathbb{R}^m)$ such that the solution $y(\cdot)$ to (1.1) satisfies*

$$y(T) = y_1, \tag{1.2}$$

i.e., the control $u(\cdot)$ transfers the state of (1.1) from y_0 to y_1 at time T.

One has the following result:

Theorem 1.2. ([163]) *The system (1.1) is exactly controllable at time T if and only if the following Kalman rank condition holds*

$$\operatorname{rank}[B, AB, \cdots, A^{n-1}B] = n. \tag{1.3}$$

In the sequel, denote by $B^\top$ the transpose of B, and write

$$G_T = \int_0^T e^{At}BB^\top e^{A^\top t}dt. \tag{1.4}$$

One can show the following further result:

Theorem 1.3. ([163]) *If the system (1.1) is exactly controllable at time T, then $\det G_T \neq 0$. Moreover, for any $y_0, y_1 \in \mathbb{R}^n$, the following control of explicit form:*

$$u(t) = -B^\top e^{A^\top(T-t)}G_T^{-1}(e^{AT}y_0 - y_1) \tag{1.5}$$

transfers the state of (1.1) from y_0 to y_1 at time T.

The proof of Theorems 1.2–1.3 will be given in that of Theorem 1.13 (in Section 1.3).

Remark 1.4. Unlike most of the previous literatures, in Definition 1.1 we deliberately choose the control class to be $L^1(0,T;\mathbb{R}^m)$. From Theorem 1.3, it is easy to see that, if (1.1) is exactly controllable at time T (by means of

L^1-(in time) controls), then the same controllability can be achieved by using analytic-(in time) controls. Actually the same can be said for the case that the control class $L^1(0,T;\mathbb{R}^m)$ in Definition 1.1 is replaced by $L^p(0,T;\mathbb{R}^m)$ for any $p \in [1,\infty]$. However, in Chapter 6, we shall see a completely different phenomenon even in the simplest stochastic situation.

The above definition of controllability can be easily extended to both deterministic and stochastic systems (See Definition 5.6 in Chapter 5 for a general formulation of controllability). In the general setting, it may happen that the requirement (1.2) has to be relaxed in one way or another. This leads to the notions of approximate controllability, null controllability, partial controllability, and so on. Also, the above B can be unbounded (in a suitable way) for general controlled systems.

Controllability is strongly related to (or in some situation, even equivalent to) other important issues in Control Theory, say observability, stabilization and so on. One can find numerous literatures on these topics (e.g., [4, 14, 41, 42, 61, 62, 117, 154, 163, 189, 190, 199, 207, 288, 295, 388, 395, 396] and rich references therein).

One of the main concerns in this book is to study controllability problems for stochastic evolution systems, in particularly that for several typical stochastic partial differential equations.

The second fundamental issue in Control Theory is *optimality*, or in the terminology of Control Theory, *optimal control*, which means that people are expected to find the best way, in some sense, to achieve their goals. As an example, we fix $y_0 \in \mathbb{R}^n$, a subset S of $\mathbb{R}^n$, a suitable nonlinear function $f : [0,T] \times \mathbb{R}^n \times \mathbb{R}^m \to \mathbb{R}^n$ and a given nonempty subset U of $\mathbb{R}^m$, and consider the following controlled system:

$$\begin{cases} y_t(t) = f(t, y(t), u(t)), \quad \text{a.e. } t \in [0,T], \\ y(0) = y_0, \end{cases} \tag{1.6}$$

where $y(\cdot)$ is the state variable and $u(\cdot)$ is the control variable, valued in $\mathbb{R}^n$ and U, respectively. It is easy to see that, if there exists a control $u(\cdot) \in \mathcal{U} \overset{\Delta}{=} \{v(\cdot) : [0,T] \to U \mid v(\cdot) \text{ is (Lebesgue) measurable}\}$ such that the solution $y(\cdot) \in C([0,T];\mathbb{R}^n)$ to (1.6) satisfies

$$y(T) \in S, \tag{1.7}$$

then, very likely, one may find another control verifying the same conditions. Naturally, one hopes to find the "best" control fulfilling these conditions. To be more precisely, we fix two suitable functions $g(\cdot,\cdot,\cdot) : [0,T] \times \mathbb{R}^n \times \mathbb{R}^m \to \mathbb{R}$ and $h : \mathbb{R}^n \to \mathbb{R}$, and denote by $\mathcal{U}_{ad}$ the set of controls $u(\cdot) \in \mathcal{U}$ so that the solution $y(\cdot)$ to (1.6) satisfies (1.7) and $g(\cdot, y(\cdot), u(\cdot)) \in L^1(0,T)$. Suppose that

$$\mathcal{U}_{ad} \neq \emptyset, \tag{1.8}$$

and introduce the following cost functional:

$$\mathcal{J}(u(\cdot)) = \int_0^T g(t, y(t), u(t))dt + h(y(T)), \quad u(\cdot) \in \mathcal{U}_{ad}. \tag{1.9}$$

A typical optimal control problem for the system (1.6) is to find a $\bar{u}(\cdot) \in \mathcal{U}_{ad}$, called an *optimal control*, which minimizes the cost functional (1.9), i.e.,

$$\mathcal{J}(\bar{u}(\cdot)) = \inf_{u(\cdot) \in \mathcal{U}_{ad}} \mathcal{J}(u(\cdot)). \tag{1.10}$$

The corresponding solution to (1.6) is denoted by $\bar{y}(\cdot)$, and $(\bar{y}(\cdot), \bar{u}(\cdot))$ is called an *optimal pair*.

The simplest case is $S = \mathbb{R}^m$, which means that there is no constraint on the final states of (1.6). For this case, people introduce the *Hamiltonian* as follows:

$$\begin{aligned} H(t, y, u, p) &\stackrel{\Delta}{=} \langle p, f(t, y, u) \rangle_{\mathbb{R}^n} - g(t, y, u), \\ &\qquad (t, y, u, p) \in [0, T] \times \mathbb{R}^n \times U \times \mathbb{R}^n. \end{aligned}$$

Then, one can show the following Pontryagin Maximum Principle ([281]):

Theorem 1.5. *Let $S = \mathbb{R}^m$ and $(\bar{y}(\cdot), \bar{u}(\cdot))$ be an optimal pair. Then*

$$H(t, \bar{y}(t), \bar{u}(t), p(t)) = \max_{u \in U} H(t, \bar{y}(t), u, p(t)), \qquad \text{a.e. } t \in [0, T], \tag{1.11}$$

where $p(\cdot) : [0, T] \to \mathbb{R}^n$ solves

$$\begin{cases} p_t(t) = -f_y(t, \bar{y}(t), \bar{u}(t))^\top p(t) + g_y(t, \bar{y}(t), \bar{u}(t)), & \text{a.e. } t \in [0, T], \\ p(T) = -h_y(\bar{y}(T)). \end{cases} \tag{1.12}$$

Remark 1.6. Thanks to Pontryagin's Maximum Principle, the optimization problem (1.10) (which is usually infinite dimensional) is reduced to a finite dimensional optimization problem (1.11) (in the pointwise sense). Since Theorem 1.5 can be viewed as a special case of Theorems 12.2 or 12.17 in Chapter 12, we will not prove it in here. Instead, in the next section we shall prove a linear quadratic version of this theorem, and via which we emphasize the difficulty to formulate the stochastic version of the "backward" equations (1.12) and (1.41) (in Section 1.3).

The above optimal control problem can be easily extended to more general setting (See Problem (SOP) in Section 5.3 of Chapter 5 for a general formulation of stochastic optimal control problems), particularly for the case that the controls take values in a more general set (instead of a subset of $\mathbb{R}^m$), say any metric space, which does not need to enjoy any linearity or convexity structure.

Optimal control problems are strongly related to the classical calculus of variations and optimization theory. Nevertheless, since the control class may be quite general, the classical variation technique cannot be applied to optimal control problems directly. Various optimal control problems are extensively studied in the literatures (e.g., [1, 5, 20, 27, 44, 59, 64, 98, 102, 158, 163, 179, 186, 202, 206, 242, 281, 323, 371] and rich references cited therein).

Another main concern in this book is to study optimal control problems for stochastic evolution systems in infinite dimensions, including Pontryagin-type Maximum Principle for the general setting and characterization of optimal feedback operators for linear quadratic optimal control problems.

It is easy to see that, the study of controllability problems is a basis to investigate further optimal control problems. Indeed, the usual nonempty assumption on the set of feasible/admissible control-state pairs for optimal control problems (say, $\mathcal{U}_{ad} \neq \emptyset$ in (1.8)) is actually a controllability condition. Nevertheless, in the previous literatures, it seems that the studies of controllability and optimal control problems are almost independent. A few typical exceptions that we know are the following:

1) In [154], some techniques from optimal control theory are employed to derive the observability estimate and null controllability for parabolic type equations.
2) In [338], some techniques developed in the study of controllability and observability problems are adopted to solve several time optimal control problems.
3) Very recently, in [220], a new link between (finite codimensional exact) controllability and optimal control problems for general evolution equations with endpoint state constraints is presented.

In our opinion, it is now the time to solve controllability and optimal control problems as a whole, at least in some sense and to some extent, though they are two quite different issues in Control Theory. This is by no means an easy task. Actually, for many concrete problems, it is highly nontrivial to verify the above mentioned assumption that the set of feasible/admissible control-state pairs is nonempty, even in the setting of finite dimensional control systems.

In this book, we shall try to present the existing theory on controllability and optimal control for stochastic distributed parameter systems in a unified manner, at least from the perspective of methodology. Indeed, as we shall explain in the next section, throughout this book we shall systematically use the duality argument.

1.3 Range Inclusion and the Duality Argument

Clearly, any controllability problem (formulated in Definition 1.1 or more generally, in Definition 5.6 in Chapter 5) can be viewed as an equation problem, in which both the state $x(\cdot)$ and the control $u(\cdot)$ variables are unknowns. Namely, instead of viewing $u(\cdot)$ as a control variable, we may simply regard it as another unknown variable[2]. One of the main concerns in this book is to study the controllability problems for linear stochastic evolution equations. As we shall see later, this is far from an easy task.

It is easy to see that, in many cases solving linear equations are equivalent to showing range inclusion for suitable linear operators. Because of this, we shall present below two known range inclusion theorems (i.e., Theorems 1.7 and 1.10 below) in an abstract setting.

Throughout this section, X, Y and Z are Banach spaces. Denote by $\mathcal{L}(X;Y)$ the Banach space of all bounded linear operators from X to Y, with the usual operator norm. When $X = Y$, we simply write $\mathcal{L}(X)$ instead of $\mathcal{L}(X;X)$. For any $L \in \mathcal{L}(X;Y)$, denote by $\mathcal{R}(L)$ the range of L. We begin with the following result.

Theorem 1.7. *Let $F \in \mathcal{L}(X;Z)$ and $G \in \mathcal{L}(Y;Z)$. The following assertions hold:*

1) *If $\mathcal{R}(F) \supseteq \mathcal{R}(G)$, then there is a constant $\mathcal{C} > 0$ such that*

$$|G^* z^*|_{Y^*} \leq \mathcal{C}|F^* z^*|_{X^*}, \quad \forall\, z^* \in Z^*. \tag{1.13}$$

2) *If X is reflexive and* (1.13) *holds for some constant $\mathcal{C} > 0$, then $\mathcal{R}(F) \supseteq \mathcal{R}(G)$.*

Proof: 1) We use the contradiction argument. Suppose (1.13) did not hold. Then, for each $n \in \mathbb{N}$, we could find $z_n^* \in Z^*$ such that

$$|F^* z_n^*|_{X^*} < \frac{1}{n}|G^* z_n^*|_{Y^*}. \tag{1.14}$$

From (1.14), $G^* z_n^* \neq 0$. Define

$$\tilde{z}_n^* = \frac{\sqrt{n}}{|G^* z_n^*|_{Y^*}} z_n^*, \quad n \geq 1. \tag{1.15}$$

Then,

$$\begin{cases} |F^* \tilde{z}_n^*|_{X^*} = \dfrac{\sqrt{n}}{|G^* z_n^*|_{Y^*}}|F^* z_n^*|_{X^*} < \dfrac{1}{\sqrt{n}} \to 0, \\ |G^* \tilde{z}_n^*|_{Y^*} = \sqrt{n} \to \infty, \end{cases} \quad \text{as } n \to \infty. \tag{1.16}$$

On the other hand, for any $y \in Y$, there exists $x \in X$ such that $Gy = Fx$. Thus,

$$\begin{aligned} \langle G^* \tilde{z}_n^*, y\rangle_{Y^*,Y} &= \langle \tilde{z}_n^*, Gy\rangle_{Z^*,Z} = \langle \tilde{z}_n^*, Fx\rangle_{Z^*,Z} \\ &= \langle F^* \tilde{z}_n^*, x\rangle_{X^*,X} \to 0, \qquad n \to \infty. \end{aligned} \tag{1.17}$$

[2] Nevertheless, the resulting equation is NOT well-posed, or even highly ill-posed. Indeed, generally speaking, none of the existence, uniqueness and continuous dependence for such an equation can be guaranteed.

This means that, for each $y \in Y$, $(G^*\tilde{z}_n^*)(y)$ tends to 0 as $n \to +\infty$. Hence, by the Principle of Uniform Boundedness, the sequence $\{G^*\tilde{z}_n^*\}_{n\in\mathbb{N}}$ must be uniformly bounded, contradicting the second assertion in (1.16).

2) Fix a $y \in Y$, define a functional f on $\mathcal{R}(F^*)(\subset X^*)$ as follows:

$$f(x^*) = \langle y, G^*z^*\rangle_{Y,Y^*}, \qquad \forall\, z^* \in Z^*, \text{ with } x^* = F^*z^*. \tag{1.18}$$

For any $z_1^*, z_2^* \in Z^*$ with $F^*z_1^* = F^*z_2^*$, by (1.13),

$$\big|G^*(z_1^* - z_2^*)\big|_{Y^*} \le \mathcal{C}\big|F^*(z_1^* - z_2^*)\big|_{X^*} = 0,$$

which implies that f is well-defined. Clearly, f is linear and, by (1.13) again,

$$\begin{aligned}|f(x^*)| &= |\langle y, G^*z^*\rangle_{Y,Y^*}| \le |y|_Y\, |G^*z^*|_{Y^*}\\ &\le \mathcal{C}|y|_Y\, |F^*z^*|_{X^*} = \mathcal{C}|y|_Y\, |x^*|_{X^*}.\end{aligned}$$

Hence, f is a linear bounded functional on $\mathcal{R}(F^*)$, with the norm inherited from X^*. By the Hahn-Banach Theorem and the reflexivity of X, we can find $x \in X^{**} = X$ such that

$$f(x^*) = \langle x^*, x\rangle_{X^*,X}, \qquad \forall\, x^* \in \mathcal{R}(F^*). \tag{1.19}$$

Consequently, by (1.18) and (1.19),

$$\begin{aligned}\langle z^*, Gy\rangle_{Z^*,Z} &= \langle y, G^*z^*\rangle_{Y,Y^*} = f(x^*) = \langle x^*, x\rangle_{X^*,X}\\ &= \langle F^*z^*, x\rangle_{X^*,X} = \langle z^*, Fx\rangle_{Z^*,Z}, \qquad \forall\, z^* \in Z^*.\end{aligned}$$

Hence, $Gy = Fx$ and therefore, $\mathcal{R}(G) \subseteq \mathcal{R}(F)$. This completes the proof of Theorem 1.7. □

Remark 1.8. By Theorem 1.7, it is easy to see that, if X is reflexive, then, for some constant $\mathcal{C} > 0$,

$$|G^*z^*|_{Y^*} \le \mathcal{C}|F^*z^*|_{X^*}, \quad \forall\, z^* \in Z^* \iff \mathcal{R}(G) \subseteq \mathcal{R}(F). \tag{1.20}$$

As shown in [36], the equivalence (1.20) may fail whenever X is not reflexive.

In this book, we shall need a range inclusion theorem in which X is NOT reflexive. In this case, instead of the reflexivity assumption for X, people assume the surjection condition for G, and then show that the equivalence (1.20) remains true. In order to establish such an equivalence, we need the following result, which is an easy consequence of [291, Theorem 3.4, p. 58].

Lemma 1.9. *Suppose that $\mathcal{O}_1$ and $\mathcal{O}_2$ are two disjoint closed convex sets in X, and $\mathcal{O}_1$ is compact. Then, there exists a real-valued functional $f \in X^*$ and two constants $\alpha, \beta \in \mathbb{R}$, such that*

$$\sup_{x\in\mathcal{O}_1} \langle x, f\rangle_{X,X^*} < \alpha < \beta < \sup_{x\in\mathcal{O}_2} \langle x, f\rangle_{X,X^*}.$$

The following result holds:

Theorem 1.10. *If* $F \in \mathcal{L}(X;Z)$, $G \in \mathcal{L}(Y;Z)$ *is surjective and* (1.13) *holds for some constant* $\mathcal{C} > 0$, *then*

$$G(B_Y) \subset \mathcal{C}F(B_X) \equiv \{\mathcal{C}Fx \mid x \in B_X\}, \tag{1.21}$$

where B_X *and* B_Y *are respectively the open unit balls in* X *and* Y, *and hence* $\mathcal{R}(F) \supseteq \mathcal{R}(G)$.

Proof: We first claim that

$$\overline{F(B_X)} \supseteq \frac{1}{\mathcal{C}}G(B_Y). \tag{1.22}$$

If the above was not true, then we could find some $y_0 \in Y$ with $|y_0|_Y \le \frac{1}{\mathcal{C}}$ such that $Gy_0 \notin \overline{F(B_X)}$. By Lemma 1.9, one can find a real-valued functional $\zeta^* \in Z^*$ and two constants $\alpha, \beta \in \mathbb{R}$, such that

$$\langle Gy_0, \zeta^* \rangle_{Z,Z^*} < \alpha < \beta < \langle z, \zeta^* \rangle_{Z,Z^*}, \quad \forall\, z \in \overline{F(B_X)}.$$

This implies that

$$\langle z, -\zeta^* \rangle_{Z,Z^*} < -\beta < -\alpha < \langle Gy_0, -\zeta^* \rangle_{Z,Z^*}, \quad \forall\, z \in \overline{F(B_X)}.$$

Noting $0 \in \overline{F(B_X)}$, we then see that $-\beta > 0$. Hence, after suitable scaling, for some $z^* \in Z^*$, we have

$$\langle z, z^* \rangle_{Z,Z^*} \le 1 < \langle Gy_0, z^* \rangle_{Z,Z^*}, \quad \forall\, z \in \overline{F(B_X)}.$$

Further, $\overline{F(B_X)}$ is symmetric w.r.t. the origin. Thus, the above also implies

$$|\langle z, z^* \rangle_{Z,Z^*}| \le 1 < \langle Gy_0, z^* \rangle_{Z,Z^*}, \quad \forall\, z \in \overline{F(B_X)}.$$

Consequently,

$$|\langle x, F^*z^* \rangle_{X,X^*}| = |\langle Fx, z^* \rangle_{Z,Z^*}| \le 1, \quad \forall\, x \in B_X,$$

which yields $|F^*z^*|_{X^*} \le 1$. Hence, by (1.13), we arrive at

$$\begin{aligned}\frac{1}{\mathcal{C}} &< \frac{1}{\mathcal{C}}|\langle Gy_0, z^* \rangle_{Z,Z^*}| = \frac{1}{\mathcal{C}}|\langle y_0, G^*z^* \rangle_{Y,Y^*}| \\ &\le \frac{1}{\mathcal{C}}\,|y_0|_Y\,|G^*z^*|_{Y^*} \le |y_0|_Y\,|F^*z^*|_{X^*} \le |y_0|_Y\,,\end{aligned}$$

which is a contradiction. This proves (1.22).

Further, we claim that

$$F(B_X) \supseteq \frac{1}{\mathcal{C}} G(B_Y). \tag{1.23}$$

To show this, for each $y \in Y$ and $\varepsilon > 0$, let us find $x \in X$ such that

$$|x|_X \le \mathcal{C}\,|y|_Y\,, \qquad |Gy - Fx|_Z < \varepsilon. \tag{1.24}$$

This can be achieved as follows: If $y = 0$, we simply choose $x = 0$. Otherwise, consider $v = \frac{y}{|y|_Y}$. Clearly, $|v|_Y = 1$ and $v \in \overline{B_Y}$. Note that, by (1.22), we have

$$\overline{F(B_X)} \supseteq \frac{1}{\mathcal{C}}\overline{G(B_Y)} \supseteq \frac{1}{\mathcal{C}} G(\overline{B_Y}).$$

Hence, there is a $u \in \mathcal{C}B_X$ such that $|Gv - Fu|_Z < \frac{\varepsilon}{|y|_Y}$. Now, let us put $x = |y|_Y\, u$. We then have (1.24).

Clearly, $\ker G = \{y \in Y \mid Gy = 0\}$ is a closed subspace of Y. Denote by $\widetilde{Y}$ the quotient space $Y/\ker G$, which is a Banach space with the usual quotient norm. Since G is surjective, the operator $\mathcal{G} : \widetilde{Y} \to Z$ defined by

$$\mathcal{G}(y + \ker G) = G(y), \quad \forall\, y \in Y,$$

is a bijective bounded linear operator from $\widetilde{Y}$ onto Z. Now, fix any $y_1 \in B_Y$. Pick $\varepsilon_n > 0$ so that

$$\sum_{n=1}^{\infty} \varepsilon_n < 1 - |y_1|_Y\,.$$

Assume for $n \ge 1$, y_n is picked. By (1.24), there is x_n such that $|x_n|_X \le \mathcal{C}\,|y_n|_Y$ and $|Gy_n - Fx_n|_Z < \frac{\varepsilon_n}{2||\mathcal{G}^{-1}||_{\mathcal{L}(Z;\widetilde{Y})}}$. Let us define $\tilde{y}_{n+1} \in \widetilde{Y}$ by

$$\tilde{y}_{n+1} = \mathcal{G}^{-1}(Gy_n - Fx_n).$$

Then, one may find $y_{n+1} \in Y$ so that

$$\begin{cases} Gy_{n+1} = Gy_n - Fx_n, \\ |y_{n+1}|_Y \le 2|\tilde{y}_{n+1}|_{\widetilde{Y}} \le 2||\mathcal{G}^{-1}||_{\mathcal{L}(Z;\widetilde{Y})}\,|Gy_n - Fx_n|_Z < \varepsilon_n. \end{cases}$$

By induction, this process defines two sequences $\{x_n\} \subset X$ and $\{y_n\} \subset Y$. Note that

$$|x_{n+1}|_X \le \mathcal{C}\,|y_{n+1}|_Y < \mathcal{C}\varepsilon_n.$$

Hence,

$$\sum_{n=1}^{\infty} |x_n|_X \le |x_1|_X + \sum_{n=1}^{\infty} \varepsilon_n \le \mathcal{C}\,|y_1|_Y + \mathcal{C}\sum_{n=1}^{\infty} \varepsilon_n < \mathcal{C}.$$

It follows that $x = \sum_{n=1}^{\infty} x_n$ is in $\mathcal{C}B_X$ and that

$$Fx = \lim_{N\to\infty} \sum_{n=1}^{N} Fx_n = \lim_{N\to\infty} \sum_{n=1}^{N} (Gy_n - Gy_{n+1}) = Gy_1,$$

since $y_{N+1} \to 0$ as $N \to \infty$. Thus $Gy_1 = Fx \in \mathcal{C}F(B_X)$, which proves (1.23).

Finally, $\mathcal{R}(F) \supseteq \mathcal{R}(G)$ follows immediately from (1.23). This completes the proof of Theorem 1.10. □

Remark 1.11. Clearly, by Theorem 1.10 and the first assertion in Theorem 1.7, we see that the equivalence (1.20) holds whenever G is surjective (even without the reflexivity assumption for X). Also, the inclusion relationship (1.21) in Theorem 1.10 provides a "quantitative" characterization on $\mathcal{R}(F) \supseteq \mathcal{R}(G)$. Such a characterization is sometimes useful (e.g., [239]).

We shall also need the following result.

Theorem 1.12. *Let $F \in \mathcal{L}(X;Z)$. Then $\overline{\mathcal{R}(F)} = Z$ if and only if*

$$F^*z^* = 0 \textit{ for some } z^* \in Z^* \Rightarrow z^* = 0. \tag{1.25}$$

Proof: The **"if" part**. We use the contradiction argument. If $\overline{\mathcal{R}(F)} = Z$ was not true, then we could find some nonzero $z_0 \in Z$ such that $z_0 \notin \overline{\mathcal{R}(F)}$. By Lemma 1.9, similarly to that in the proof of Theorem 1.10, one can find a nonzero $z^* \in Z^*$ with $\langle z_0, z^* \rangle_{Z,Z^*} = 1$ such that

$$\langle z, z^* \rangle_{Z,Z^*} = 0, \quad \forall\, z \in \overline{\mathcal{R}(F)}.$$

Consequently,

$$\langle Fx, z^* \rangle_{Z,Z^*} = 0, \quad \forall\, x \in X, \tag{1.26}$$

which implies that $F^*z^* = 0$. But this, by (1.25), leads to $z^* = 0$, a contradiction.

The **"only if" part**. Suppose that $F^*z^* = 0$ for some $z^* \in Z^*$. Then, it is easy to see that (1.26) holds. This, combined with $\overline{\mathcal{R}(F)} = Z$, indicates that $z^* = 0$. □

The key in Theorems 1.7 and 1.10 is to reduce the problem of range inclusion (i.e., $\mathcal{R}(F) \supseteq \mathcal{R}(G)$) to the majorization of dual operators, i.e., the estimate (1.13) (See also Theorem 1.12). Such a method, i.e., solving the problem under consideration via its dual one, is classical, called the *duality argument.*

Inspired by the study of deterministic control problems (especially these for deterministic partial differential equations), in this book we shall employ the above range inclusion theorems and the duality argument to handle the controllability problems and optimal control problems for stochastic evolution equations

To explain the idea, in the rest of this section, we shall focus on the simplest control system, i.e., the system (1.1) (governed by a linear ordinary differential equation), and show how to apply the above mentioned method to solve the exact controllability problem and an optimal control problem for such a system by means of its dual problem.

To analyze the exact controllability for (1.1), people introduce the following dual equation (of (1.1)), which is a (backward) ordinary differential equation:

$$\begin{cases} z_t = -A^\top z & \text{in } [0,T], \\ z(T) = z_1 \in \mathbb{R}^n. \end{cases} \tag{1.27}$$

We have the following result (which implies Theorems 1.2–1.3):

Theorem 1.13. *The following five statements are equivalent:*

1) *The system* (1.1) *is exactly controllable;*
2) *Solutions to* (1.27) *satisfy the following observability estimate:*

$$|z_1|_{\mathbb{R}^n} \le \mathcal{C} \max_{t\in[0,T]} |B^\top z(t)|_{\mathbb{R}^m}, \qquad \forall\, z_1 \in \mathbb{R}^n; \tag{1.28}$$

3) *Solutions to* (1.27) *enjoy the following observability:*

$$B^\top z(\cdot) \equiv 0 \textit{ in } (0,T) \Rightarrow z_1 = 0; \tag{1.29}$$

4) *The rank condition* (1.3) *holds;*
5) G_T *defined by* (1.4) *is non-degenerate, i.e.,* $\det G_T \neq 0$.

Moreover, under any one of the above conditions, for any $y_0, y_1 \in \mathbb{R}^n$*, the control* $u(\cdot)$ *given by* (1.5) *transfers the state of (1.1) from* y_0 *to* y_1 *at time* T.

Proof: "1)$\Longleftrightarrow$2)". Clearly, it suffices to consider the special case that the initial datum x_0 in the system (1.1) is equal to 0. We define an operator $G : L^1(0,T;\mathbb{R}^m) \to \mathbb{R}^n$ by

$$G(u(\cdot)) = y(T), \quad \forall\, u(\cdot) \in L^1(0,T;\mathbb{R}^m), \tag{1.30}$$

where $y(\cdot)$ is the corresponding solution to (1.1) with $y_0 = 0$. Then, by (1.1) (with $y_0 = 0$) and (1.27), we obtain that

$$\langle\, y(T), z_1 \,\rangle_{\mathbb{R}^n} = \int_0^T \langle\, Bu(t), z(t) \,\rangle_{\mathbb{R}^n} dt, \quad \forall\, z_1 \in \mathbb{R}^n. \tag{1.31}$$

where $z(\cdot)$ solves (1.27). By (1.30)–(1.31), it is clear that,

$$(G^* z_1)(t) = B^\top z(t), \quad \text{a.e. } t \in [0,T]. \tag{1.32}$$

Now, the exact controllability of (1.1) is equivalent to that $\mathcal{R}(G) = \mathbb{R}^n$; while the latter, by (1.32) and in view of Theorem 1.10 and the first assertion in Theorem 1.7, is equivalent to the estimate (1.28).

"2)$\Longleftrightarrow$3)". By (1.29), one can define a norm $||\cdot||_1$ in $\mathbb{R}^n$ as follows:

$$||z_1||_1 = \max_{t\in[0,T]} |B^\top z(t)|_{\mathbb{R}^m},$$

where z solves (1.27). By the equivalence of all norms in $\mathbb{R}^n$, we obtain that "2)$\Longleftarrow$3)". "2)$\Longrightarrow$3)" is obvious.

"4)$\Longleftrightarrow$3)". Let us assume that $B^\top z(\cdot) \equiv 0$ in $(0,T)$ for some $z_1 \in \mathbb{R}^n$. Then, by

$$z(t) = e^{A^\top(T-t)} z_1, \tag{1.33}$$

this is equivalent to that

$$B^\top (A^k)^\top z_1 = 0, \quad k \in \mathbb{N} \cup \{0\}.$$

Therefore, by the Hamilton-Cayley Theory (in Matrix Theory), the above is equivalent to that

$$z_1^\top [B,\ AB,\ A^2B,\ \cdots\ A^{n-1}B] = 0. \tag{1.34}$$

By (1.34), it follows that 4)$\Longleftrightarrow$3).

"5)$\Longleftrightarrow$3)". Using again the expression (1.33) for solutions to (1.27), we obtain that 5)$\Longleftrightarrow$3).

Finally, by (1.4), the corresponding solution $y(\cdot)$ to (1.1) with $u(\cdot)$ given by (1.5) satisfies

$$\begin{aligned} y(T) &= e^{AT}y_0 + \int_0^T e^{A(T-t)}Bu(t)dt \\ &= e^{AT}y_0 - \int_0^T e^{A(T-t)}BB^\top e^{A^\top(T-t)}dt G_T^{-1}(e^{AT}y_0 - y_1) \\ &= e^{AT}y_0 - G_T G_T^{-1}(e^{AT}y_0 - y_1) = y_1. \end{aligned}$$

This completes the proof of Theorem 1.13. □

Let us analyze a little more the dual equation (1.27). Write

$$\begin{cases} w(t) = z(T-t), \\ w_0 = z_1, \end{cases} \tag{1.35}$$

i.e., we simply reverse the time in (1.27). Then, (1.27) is reduced to the following:

$$\begin{cases} w_t = A^\top w, \quad t \in [0,T], \\ w(0) = w_0. \end{cases} \tag{1.36}$$

By Theorem 1.13, it is easy to see that the system (1.1) is exactly controllable if and only if solutions to (1.36) enjoy the following observability:

$$B^\top w(\cdot) \equiv 0 \text{ in } (0,T) \Rightarrow w_0 = 0.$$

Things will be much more complicated to handle the controllability problems for stochastic equations, even in finite dimensions. Indeed, in the stochastic setting, as we will explain in Chapter 4, one cannot use a similar transformation as that in (1.35) to reduce the dual equation (1.27) to a forward one, and therefore, in order to solve the stochastic controllability problem under consideration, generally speaking we will have to handle the observability for the corresponding (genuinely) backward stochastic equation. On the other hand, since one may put controls on both drift and diffusion terms in the stochastic setting (as we shall see in Chapter 6, sometimes it is also necessary to introduce controls in such a way), even the choice of control class (and therefore the definition of controllability) is a delicate problem. Indeed, in order that the drift term makes sense, one needs the control to be L^1-integrable in time; while for the diffusion one, people need the control to be L^2-integrable in time. The problem is then how to balance the time integrability of controls and minimize the number of controls. As far as we know, such a problem is actually unclear even for controllability of stochastic differential equations in finite dimensions (See some progress in this respect in Chapter 6).

We now use the duality argument to analyze a quadratic optimal control problem for the linear system (1.1) (with a given $y_0 \in \mathbb{R}^n$), for which the cost functional is given by

$$\mathcal{J}(u(\cdot)) = \frac{1}{2}\int_0^T (\langle My(t), y(t) \rangle_{\mathbb{R}^n} + \langle Ru(t), u(t) \rangle_{\mathbb{R}^m})\, dt + \frac{1}{2}\langle Gy(T), y(T) \rangle_{\mathbb{R}^n}, \tag{1.37}$$

where $M = M^\top \in \mathbb{R}^{n\times n}$, $R = R^\top \in \mathbb{R}^{m\times m}$, $G = G^\top \in \mathbb{R}^{n\times n}$, $u(\cdot) \in L^2(0,T;\mathbb{R}^m)$. Similarly to that for (1.10), $\bar{u}(\cdot) \in L^2(0,T;\mathbb{R}^m)$ is called an *optimal control* for the control system (1.1) with the cost functional (1.37)), if

$$\mathcal{J}(\bar{u}(\cdot)) = \inf_{u(\cdot)\in L^2(0,T;\mathbb{R}^m)} \mathcal{J}(u(\cdot)).$$

The corresponding solution $\bar{y}(\cdot) \overset{\Delta}{=} y(\cdot)$ to (1.1) (with $u(\cdot)$ replaced by $\bar{u}(\cdot)$) is called an *optimal trajectory*, and $(\bar{u}(\cdot), \bar{y}(\cdot))$ is called an *optimal pair.*

In order to give a necessary condition for which $\bar{u}(\cdot)$ and $\bar{y}(\cdot)$ satisfy, by the classical variational method, we take any $\varepsilon \in \mathbb{R}$ and $u(\cdot) \in L^2(0,T;\mathbb{R}^m)$, and set

$$u^\varepsilon(\cdot) = \bar{u}(\cdot) + \varepsilon u(\cdot). \tag{1.38}$$

Denote by $y^\varepsilon(\cdot)$ the corresponding solutio to (1.1) (with $u(\cdot)$ replaced by $u^\varepsilon(\cdot)$). Then

$$y^\varepsilon(\cdot) = \bar{y}(\cdot) + \varepsilon r(\cdot),$$

where $r(\cdot)$ satisfies the following variational equation:

$$\begin{cases} r_t(t) = Ar(t) + Bu(t), & t \in [0,T], \\ r(0) = 0. \end{cases} \tag{1.39}$$

Clearly,

$$
\begin{aligned}
\mathcal{J}(y^{\varepsilon}(\cdot)) &= \mathcal{J}(\bar{y}(\cdot)) + 2\varepsilon\Big[\int_0^T (\langle M\bar{y}(t), r(t)\rangle_{\mathbb{R}^n} + \langle R\bar{u}(t), u(t)\rangle_{\mathbb{R}^m})\,dt \\
&\quad + \langle G\bar{y}(T), r(T)\rangle_{\mathbb{R}^n}\Big] + O(\varepsilon^2) \geq \mathcal{J}(\bar{y}(\cdot)), \qquad \text{as } \varepsilon \to 0.
\end{aligned}
$$

Hence,

$$
\int_0^T (\langle M\bar{y}(t), r(t)\rangle_{\mathbb{R}^n} + \langle R\bar{u}(t), u(t)\rangle_{\mathbb{R}^n})\,dt + \langle G\bar{y}(T), r(T)\rangle_{\mathbb{R}^n} = 0. \tag{1.40}
$$

Obviously, (1.40) is a necessary condition for the optimal pair $(\bar{y}(\cdot), \bar{u}(\cdot))$. In order to re-write (1.40) in a better form, similarly to (1.27), people introduce a "backward" ordinary differential equation as follows:

$$
\begin{cases}
\psi_t(t) = -A^{\top}\psi(t) + M\bar{y}(t), \\
\psi(T) = -G\bar{y}(T).
\end{cases} \tag{1.41}
$$

Computing $\frac{d}{dt}\langle r(t), \psi(t)\rangle_{\mathbb{R}^n}$, from (1.39) and (1.41), we obtain that

$$
\begin{aligned}
&\int_0^T \big(\langle M\bar{y}(t), r(t)\rangle_{\mathbb{R}^n} + \langle B^{\top}\psi(t), u(t)\rangle_{\mathbb{R}^m}\big)\,dt = -\langle G\bar{y}(T), r(T)\rangle_{\mathbb{R}^n}, \\
&\qquad \forall\, u(\cdot) \in L^2(0, T; \mathbb{R}^m).
\end{aligned} \tag{1.42}
$$

Finally, combining (1.40) and (1.42), we conclude the following necessary condition for $(\bar{y}(\cdot), \bar{u}(\cdot))$:

$$
R\bar{u}(t) = B^{\top}\psi(t). \tag{1.43}
$$

Using a similar method, one can prove the maximum condition (1.11) except that the details are more complicated than that to prove (1.43), and more importantly, since the control set U is any given subset of $\mathbb{R}^m$ (without assuming any linearity or convexity conditions on U), the classical variation (1.38) has to be replaced by the following spike variation:

$$
u^{\varepsilon}(t) = \begin{cases}
\bar{u}(t), & t \in [0, T] \setminus E_{\varepsilon}, \\
u(t), & t \in E_{\varepsilon},
\end{cases} \tag{1.44}
$$

where $u(\cdot) \in \mathcal{U}_{ad}$, $\varepsilon \in (0, T]$, $E_{\varepsilon} (\subseteq [0, T])$ is a measurable set with the Lebesgue measure $\mathbf{m}(E_{\varepsilon}) = \varepsilon$.

Nevertheless, there are some essential difficulties to extend the results for optimal controls in deterministic problems to the stochastic situation. Firstly, similarly to the dual equation (1.27) for the controllability problem, in the study of stochastic optimal control problems, as a counterpart of the dual equation (1.12) (also (1.41)), people need to introduce backward stochastic differential equations ([31]). Secondly, again similarly to the controllability problems, since one may put controls on both drift and diffusion terms in the stochastic setting, the corresponding necessary conditions for optimal controls (even in finite dimensions) are quite different from the deterministic problems, for example the first order necessary condition when the diffusion term may contain the control variable and the set of controls is allowed to be nonconvex ([273]) and the second order necessary condition when the diffusion term contains the control variable even if the set of controls is convex ([378, 379, 380]). Thirdly and more challengingly, since we are mainly concerned with optimal controls for stochastic evolution equations in infinite dimensions, we have to handle operator-valued backward stochastic evolution equations, for which there exists no any satisfactory theory (in the previous literatures) on stochastic integration/evolution equations that can be employed to derive the well-posedness of such equations in the usual sense. In order to overcome this difficulty, we introduce a new concept of solution, i.e., transposition solutions to the desired operator valued backward stochastic evolution equation, and prove the corresponding well-posedness (See [242, 248] and Chapters 12 and 13 for details). Note that, as we shall explain in the next section, our stochastic transposition method can also be viewed as a variant of the duality argument.

1.4 Two Basic Methods in This Book

In this section, we shall present two basic methods (via illuminating examples) that will be systematically used throughout this book.

The main method that we employ in this book to deal with the analysis of the structure of stochastic distributed parameter systems is the *global Carleman type estimate*. This method was introduced by T. Carleman ([47]) in 1939 to prove the uniqueness of solutions to second order elliptic partial differential equations with two variables. The key in [47] is an elementary energy estimate with some exponential weight. This type of weighted energy estimates, now referred to as *Carleman estimates*, have become one of the major tools in the study of unique continuation property, inverse problems and control problems for many partial differential equations. However, it is only in the last ten plus years that the power of the global Carleman estimate in the context of controllability of stochastic partial differential equations came to be realized. For the readers' convenience, we explain the main idea of Carleman estimate by the following very simple example:

Example 1.14. Consider the following ordinary differential equation in $\mathbb{R}^n$:

$$\begin{cases} y_t(t) = a(t)y(t) & \text{in } [0,T], \\ y(0) = y_0. \end{cases} \tag{1.45}$$

It is well-known that if $a \in L^\infty(0,T)$, then there is a constant $C_T > 0$ such that for all solutions of (1.45), it holds that

$$\max_{t\in[0,T]} |y(t)|_{\mathbb{R}^n} \le \mathcal{C}_T |y_0|_{\mathbb{R}^n}, \qquad \forall\, y_0 \in \mathbb{R}^n. \tag{1.46}$$

Now we give a slightly different proof of this result via Carleman-type estimate:

For any $\lambda \in \mathbb{R}$, it is easy to see that

$$\begin{aligned}&\frac{d}{dt}\big(e^{-2\lambda t}|y(t)|^2_{\mathbb{R}^n}\big)\\ &= -2\lambda e^{-2\lambda t}|y(t)|^2_{\mathbb{R}^n} + 2e^{-2\lambda t}\langle\, y_t(t), y(t)\,\rangle_{\mathbb{R}^n} = 2\big(a(t)-\lambda\big)e^{-2\lambda t}|y(t)|^2_{\mathbb{R}^n}.\end{aligned} \tag{1.47}$$

Choosing $\lambda = |a|_{L^\infty(0,T)}$, we find that

$$|y(t)|_{\mathbb{R}^n} \le e^{\lambda T}|y_0|_{\mathbb{R}^n}, \quad t \in [0,T],$$

which proves (1.46).

It is easy to see that, the equality (1.47) can be re-written as the following pointwise identity:

$$2e^{-2\lambda t}\langle\, y_t(t), y(t)\,\rangle_{\mathbb{R}^n} = \frac{d}{dt}\big(e^{-2\lambda t}|y(t)|^2_{\mathbb{R}^n}\big) + 2\lambda e^{-2\lambda t}|y(t)|^2_{\mathbb{R}^n}. \tag{1.48}$$

Note that $\frac{d}{dt}$ is the principal operator of the first equation in (1.45). The main idea of (1.48) is to establish a pointwise identity (and/or estimate) on the principal operator $\frac{d}{dt}$ in terms of the sum of a "divergence" term $\frac{d}{dt}(e^{-2\lambda t}|y(t)|^2_{\mathbb{R}^n})$ and an "energy" term $2\lambda e^{-2\lambda t}|y(t)|^2_{\mathbb{R}^n}$. One then chooses λ being large enough to absorb the undesired terms.

In the above example, the weight function depends only on the time variable t. By suitably choosing time and spatial dependent weight functions, one can apply the method of Carleman-type estimates to handle many problems in both deterministic and stochastic partial differential equations. Nevertheless, usually for many control problems for stochastic partial differential equations, people have to establish Carleman-type estimates for backward stochastic partial differential equations (in which one has two unknowns in one equation), and therefore the results that can be obtained so far are not so satisfactory. It seems that people need to introduce some Carleman-type estimates with the weight functions being suitable stochastic process (i.e., which should depend on the sample point as well) but this remains to be done.

The main method that we employ in this book to deal with optimal control problems for stochastic distributed parameter systems is the *stochastic transposition method* (See Chapters 12 and 13 for details). More precisely, we solve the corresponding dual equations, including operator-valued backward stochastic evolution equations in particular, which are key tools to establish the Pontryagin type maximum principle, in the sense of transposition solution. In the stochastic setting in finite dimensions, based on a new and delicate

Riesz-type Representation Theorem (to characterize the dual of the Banach space of some vector-valued stochastic processes having different integrability with respect to the time variable and the probability measure) (See [239] and Theorem 2.73 in Chapter 2), the stochastic transposition method (for solving backward stochastic differential equations) was first introduced in our paper [241] (See [371, pp. 353–354] for a rudiment of this method). Our method was further developed in [242] (and also [244, 245]) for the stochastic setting in infinite dimensions (See also Chapters 4 and 12). Recently, in [248] we employ the stochastic transposition method to characterize the optimal feedbacks for general stochastic linear quadratic control problems (with random coefficients) in infinite dimensions, in which the point is to introduce a suitable notion of solutions (i.e., transposition solutions) to the corresponding operator-valued backward stochastic Riccati equations (See also Chapter 13).

Although our stochastic transposition method was developed to handle backward stochastic evolution equations, it is stimulated by the classical transposition method to solve the non-homogeneous boundary value problems for deterministic partial differential equations (See [208] for a systematic introduction to this topic). For the readers' convenience, as in [244], let us recall the main idea of the classical transposition method to solve the following wave equation with non-homogeneous Dirichelt boundary conditions:

$$\begin{cases} y_{tt} - \Delta y = 0 & \text{in } Q, \\ y = u & \text{on } \Sigma, \\ y(0) = y_0, \quad y_t(0) = y_1 & \text{in } G. \end{cases} \tag{1.49}$$

Here $G \subset \mathbb{R}^n$ is a nonempty open bounded domain with a C^2 boundary Γ, $Q \equiv (0,T) \times G$, $\Sigma \equiv (0,T) \times \Gamma$, $(y_0, y_1) \in L^2(G) \times H^{-1}(G)$ and $u \in L^2((0,T) \times \Gamma)$ are given, and y is the unknown.

When $u \equiv 0$, one can use the standard semigroup theory to prove the well-posedness of (1.49) in the space $C([0,T]; L^2(G)) \bigcap C^1([0,T]; H^{-1}(G))$.

When $u \not\equiv 0$, $y|_\Sigma = u$ does NOT make sense any more by the usual trace theorem in theory of Sobolv spaces. One needs to use the transposition method. To this end, for any $f \in L^1(0,T; L^2(G))$ and $g \in L^1(0,T; H_0^1(G))$, let us consider the following test equation:

$$\begin{cases} \zeta_{tt} - \Delta \zeta = f + g_t, & \text{in } Q, \\ \zeta = 0, & \text{on } \Sigma, \\ \zeta(T) = \zeta_t(T) = 0, & \text{in } G. \end{cases} \tag{1.50}$$

(1.50) admits a unique solution $\zeta \in C([0,T]; H_0^1(G)) \bigcap C^1([0,T]; L^2(G))$, which enjoys a hidden regularity $\frac{\partial \zeta}{\partial \nu} \in L^2(\Sigma)$ (e.g. [207]). Here and henceforth, $\nu(x) = (\nu^1(x), \cdots, \nu^n(x))$ stands for the *unit outward normal vector* of G at $x \in \Gamma$.

To give a reasonable definition for solutions to (1.49), we first consider the case that y is smooth. Assume that $f \in L^2(Q)$, $g \in C_0^\infty(0,T; H_0^1(G))$,

$y_1 \in L^2(G)$, and that $y \in H^2(Q)$. Then, multiplying the first equation in (1.49) by ζ, integrating it in Q, and using integration by parts, we find that

$$\int_Q (fy - gy_t)dxdt = \int_G \big(\zeta(0)y_1 - \zeta_t(0)y_0\big)dx - \int_\Sigma \frac{\partial \zeta}{\partial \nu} u d\Sigma. \tag{1.51}$$

Clearly, (1.51) still makes sense even if the regularity of y is relaxed as $y \in C([0,T]; L^2(G)) \bigcap C^1([0,T]; H^{-1}(G))$. This motivates the following notion:

Definition 1.15. *We call $y \in C([0,T]; L^2(G)) \bigcap C^1([0,T]; H^{-1}(G))$ a transposition solution to* (1.49)*, if $y(0) = y_0$, $y_t(0) = y_1$, and for any $f \in L^1(0,T; L^2(G))$ and $g \in L^1(0,T; H_0^1(G))$, it holds that*

$$\begin{aligned} &\int_Q fydxdt - \int_0^T \langle g, y_t\rangle_{H_0^1(G),H^{-1}(G)}dt \\ &= \langle \zeta(0), y_1\rangle_{H_0^1(G),H^{-1}(G)} - \int_G \zeta_t(0)y_0dx - \int_\Sigma \frac{\partial \zeta}{\partial \nu} u d\Sigma, \end{aligned}$$

where ζ is the unique solution to (1.50).

One can show the well-posedness of (1.49) in the sense of Definition 1.15, by means of the classical transposition method ([208]). The point of this method is to interpret the solution to the forward wave equation (1.49) (with non-homogeneous Dirichlet boundary conditions) by means of the test equation (1.50), which is a backward wave equation with non-homogeneous source terms. In the deterministic setting, since the wave equation is time-reversible, there exists no essential difference between the forward equation and the backward one. Nevertheless, this reminds us to interpret backward stochastic differential/evolution equations in terms of forward stochastic differential/evolution equations, as we have done in [241, 242] and so on.

Clearly, the transposition method is a variant of the classical duality argument, and in some sense it provides a way to see something which is not easy to be detected directly. Indeed, it is easy to see that, the main idea of this method is to interpret the solution to a less understood equation by means of another well understood one. Nevertheless, the philosophy of our stochastic transposition method is quite different. Indeed, the main purpose to introduce stochastic transposition method in our works is not to solve the non-homogeneous boundary value problems for stochastic partial differential equations (though it does work for this sort of problems, see Proposition 8.8 of Chapter 8 in this book, for example) but to solve the backward stochastic equations (especially the operator-valued backward stochastic evolution equations mentioned above), in which there exist no boundary conditions explicitly (or even no boundary conditions at all)!

1.5 Notes and Comments

Sections 1.1 and 1.2 are partially based on [389] and [220], respectively.

Theorem 1.7 can be found in [202, Lemma 2.4 in Chapter 7]), for example. Theorem 1.10 can be found in [291, Lemma 4.13, pp. 94–95 and Theorem 4.15, p. 97] and [330, Theorem 1.2 and Remark 1.3]). Theorem 1.12 is well-known (e.g., [377, Theorem 2.1 in Chapter 2 of Part IV, p. 207]).

It seems that Example 1.14 was firstly introduced in [201] to explain the main idea of Carleman estimates (See [188] and also [110] for much more general pointwise weighted identities/inequalities). The material in Section 1.4 is partially based on [201, 244].

2

Some Preliminaries in Stochastic Calculus

This chapter is a concise introduction to stochastic calculus (in infinite dimensions in particular), which will play a fundamental role throughout this book. Especially, we collect the most relevant preliminaries for studying control problems in stochastic distributed parameter systems. Also, we will provide some unified notations (which may differ from one paper/book to another) to be used in later chapters.

To present the proofs for all the results introduced in this chapter would considerably increase both the size and scope of this book. Thus, we distinguish two different cases. When we think that the proof is important in understanding the subsequent materials and/or when there is no immediate reference available, we will provide the details. Otherwise, we omit the proofs but giving standard and easily accessible references. Knowledgeable readers may skip this chapter or regard it as a quick reference.

2.1 Measures and Probability, Measurable Functions and Random Variables

Fix a nonempty set Ω and a family $\mathcal{F}$ of subsets of Ω. For any $E \in \mathcal{F}$, denote by $\chi_E(\cdot)$ the characteristic function of E, defined on Ω. $\mathcal{F}$ is called a σ-field on Ω if i) $\Omega \in \mathcal{F}$; ii) $\Omega \setminus A \in \mathcal{F}$ for any $A \in \mathcal{F}$; and iii) $\bigcup_{i=1}^{\infty} A_i \in \mathcal{F}$ whenever each $A_i \in \mathcal{F}$.

If $\mathcal{F}$ is a σ-field on Ω, then $(\Omega, \mathcal{F})$ is called a measurable space. An element $A \in \mathcal{F}$ is called a measurable set on $(\Omega, \mathcal{F})$, or simply a measurable set. A set function $\mu : \mathcal{F} \to [0, +\infty]$ is called a measure on $(\Omega, \mathcal{F})$ if $\mu(\emptyset) = 0$ and μ is countably additive, i.e., $\mu(\bigcup_{i=1}^{\infty} A_i) = \sum_{i=1}^{\infty} \mu(A_i)$ whenever $\{A_i\}_{i=1}^{\infty}$ are a sequence of mutually disjoint sets from $\mathcal{F}$. The triple $(\Omega, \mathcal{F}, \mu)$ is called a measure space. The measure μ is called finite (*resp.* σ-finite) if $\mu(\Omega) < \infty$ (*resp.* there exists a sequence $\{A_i\}_{i=1}^{\infty} \subset \mathcal{F}$ so that $\Omega = \bigcup_{i=1}^{\infty} A_i$ and $\mu(A_i) < \infty$ for each $i \in \mathbb{N}$). We call A a μ-null (measurable) set if $\mu(A) = 0$.

Q. Lü and X. Zhang, *Mathematical Control Theory for Stochastic Partial Differential Equations*, Probability Theory and Stochastic Modelling 101, https://doi.org/10.1007/978-3-030-82331-3_2

A probability space is a finite measure space $(\Omega, \mathcal{F}, \mathbb{P})$ for which $\mathbb{P}(\Omega) = 1$. In this case, $\mathbb{P}$ is also called a probability measure; any $\omega \in \Omega$ is called a sample point; any $A \in \mathcal{F}$ is called an event, and $\mathbb{P}(A)$ represents the probability of the event A. If an event $A \in \mathcal{F}$ is such that $\mathbb{P}(A) = 1$, then we may alternatively say that A holds, $\mathbb{P}$-a.s., or simply A holds a.s. (if the probability $\mathbb{P}$ is clear from the context).

For any measure space $(\Omega, \mathcal{F}, \mu)$, write $\mathcal{N} = \{B \subset \Omega \mid B \subset A$ for some μ-null set $A\}$. $(\Omega, \mathcal{F}, \mu)$ is said to be complete if $\mathcal{N} \subset \mathcal{F}$. Especially, one may define complete probability space $(\Omega, \mathcal{F}, \mathbb{P})$. If $(\Omega, \mathcal{F}, \mu)$ is incomplete, then the class $\overline{\mathcal{F}}$ of all sets of the form $(E \setminus N) \cup (N \setminus E)$, with $E \in \mathcal{F}$ and $N \in \mathcal{N}$, is a σ-field which contains $\mathcal{F}$ as a proper sub-class, and the set function $\bar{\mu}$ defined by $\bar{\mu}((E \setminus N) \cup (N \setminus E)) = \mu(E)$ is a complete measure on $\overline{\mathcal{F}}$. The measure $\bar{\mu}$ is called the completion of μ.

Let $(\Omega_1, \mathcal{F}_1), \cdots, (\Omega_n, \mathcal{F}_n)$ be measurable spaces, $n \in \mathbb{N}$. Denote by $\mathcal{F}_1 \times \cdots \times \mathcal{F}_n$ the σ-field[1] (on the Cartesian product $\Omega_1 \times \cdots \times \Omega_n$) generated by the subsets of the form $A_1 \times \cdots \times A_n$, where $A_i \in \mathcal{F}_i$, $1 \leq i \leq n$.

Let μ_i be a measure on $(\Omega_i, \mathcal{F}_i)$. We call μ a product measure on $(\Omega_1 \times \cdots \times \Omega_n, \mathcal{F}_1 \times \cdots \times \mathcal{F}_n)$ induced by $\mu_1, \cdots, \mu_n$ if

$$\mu(A_1 \times \cdots \times A_n) = \prod_{i=1}^{n} \mu_i(A_i), \qquad \forall\, A_i \in \mathcal{F}_i.$$

Theorem 2.1. *Let $(\Omega_i, \mathcal{F}_i, \mu_i)$ be a σ-finite measure space, $1 \leq i \leq n$. Then there is one and only one product measure μ (denoted by $\mu_1 \times \cdots \times \mu_n$) on $(\Omega_1 \times \cdots \times \Omega_n, \mathcal{F}_1 \times \cdots \times \mathcal{F}_n)$ induced by $\{\mu_i\}_{i=1}^n$.*

Further, we consider the product spaces and product measures for countably many measurable spaces and measures. In this case, for simplicity, we only present the definition and related result for probability spaces, which is enough for the rest of this book. Let $\{(\Omega_i, \mathcal{F}_i)\}_{i=1}^\infty$ be a sequence of measurable spaces. Denote by $\prod_{i=1}^\infty \mathcal{F}_i$ the σ-field on the Cartesian product $\prod_{i=1}^\infty \Omega_i$ generated by the subsets of the form

$$\prod_{i=1}^\infty A_i, \quad \text{where } A_i \in \mathcal{F}_i \text{ for all } i \in \mathbb{N}, \text{ and } A_i = \Omega_i \text{ for all but a finite number of values of } i. \tag{2.1}$$

Let $\mathbb{P}_i$ be a probability measure on $(\Omega_i, \mathcal{F}_i)$, $i \in \mathbb{N}$. We call $\mathbb{P}$ a product probability measure on $(\prod_{i=1}^\infty \Omega_i, \prod_{i=1}^\infty \mathcal{F}_i)$ if for any $\prod_{i=1}^\infty A_i$ of the form (2.1),

$$\mathbb{P}\Big(\prod_{i=1}^{\infty} A_i\Big) = \prod_{i=1}^{\infty} \mathbb{P}_i(A_i).$$

[1] Note that here and henceforth the σ-field $\mathcal{F}_1 \times \cdots \times \mathcal{F}_n$ does not stand for the Cartesian product of $\mathcal{F}_1, \cdots, \mathcal{F}_n$.

Theorem 2.2. *Let $\{(\Omega_i, \mathcal{F}_i, \mathbb{P}_i)\}_{i=1}^{\infty}$ be a sequence of probability spaces. Then, there is a unique product probability measure $\mathbb{P}$ (denoted by $\prod_{i=1}^{\infty} \mathbb{P}_i$) on $(\prod_{i=1}^{\infty} \Omega_i, \prod_{i=1}^{\infty} \mathcal{F}_i)$ induced by $\{\mathbb{P}_i\}_{i=1}^{\infty}$.*

If $(G, \mathcal{J})$ is a topological space (on a given nonempty set G), then the smallest σ-field containing all open sets of G is called the Borel σ-field of G, denoted by $\mathcal{B}(G)$. Any set $A \in \mathcal{B}(G)$ is called a Borel set (in $(G, \mathcal{J})$). Let H be a metric space with metric $\mathbf{d}$. We call any $A \in \mathcal{B}(H)$ a Borel set in H.

Let $(\Omega, \mathcal{F})$ and $(\Omega', \mathcal{F}')$ be two measurable spaces and $f : \Omega \to \Omega'$ be a map. The map f is said to be $\mathcal{F}/\mathcal{F}'$-measurable or simply $\mathcal{F}$-measurable or even measurable (in the case that no confusion would occur) if $f^{-1}(\mathcal{F}') \subset \mathcal{F}$. Particularly if $(\Omega', \mathcal{F}') = (H, \mathcal{B}(H))$, then f is said to be an (H-valued) $\mathcal{F}$-measurable (or simply measurable) function. In the context of probability theory, f is called an (H-valued) $\mathcal{F}$-random variable or simply a random variable (if the meaning is clear from the context). Note that measurable map or random variable is defined without measures. Also, it is clear that random variable is a special case of measurable map.

For a measurable map $f : (\Omega, \mathcal{F}) \to (\Omega', \mathcal{F}')$, it is easy to show that $f^{-1}(\mathcal{F}')$ is a sub-σ-field of $\mathcal{F}$. We call it the σ-field generated by f, and denote it by $\sigma(f)$. Further, for a given index set Λ and a family of measurable maps $\{f_\lambda\}_{\lambda \in \Lambda}$ (defined on $(\Omega, \mathcal{F})$, with possibly different ranges), we denote by $\sigma(f_\lambda; \lambda \in \Lambda)$ the σ-field generated by $\cup_{\lambda \in \Lambda} \sigma(f_\lambda)$.

Now, let us introduce an important notation, independence, which distinguishes probability theory from the usual measure theory.

Definition 2.3. *Let $(\Omega, \mathcal{F}, \mathbb{P})$ be a probability space.*

1) *We say that two events A and B are independent if $\mathbb{P}(A \cap B) = \mathbb{P}(A)\mathbb{P}(B)$;*

2) *Let $\mathcal{J}_1$ and $\mathcal{J}_2$ be two subsets of $\mathcal{F}$. We say that $\mathcal{J}_1$ and $\mathcal{J}_2$ are independent if $\mathbb{P}(A \cap B) = \mathbb{P}(A)\mathbb{P}(B)$ for any $A \in \mathcal{J}_1$ and $B \in \mathcal{J}_2$;*

3) *Let f and g be two random variables defined on $(\Omega, \mathcal{F})$ (with possibly different ranges), and $\mathcal{J} \subset \mathcal{F}$. We say that f and g are independent if $\sigma(f)$ and $\sigma(g)$ are independent. We say that f and $\mathcal{J}$ are independent if $\sigma(f)$ and $\mathcal{J}$ are independent;*

4) *Let $\{A_\lambda\}_{\lambda \in \Lambda} \subset \mathcal{F}$. We say that $\{A_\lambda\}_{\lambda \in \Lambda}$ is a class of mutually independent events if $\mathbb{P}(A_{\lambda_1} \cap \cdots \cap A_{\lambda_k}) = \mathbb{P}(A_{\lambda_1}) \cdots \mathbb{P}(A_{\lambda_k})$ for any $k \in \mathbb{N}$, $\lambda_1, \cdots, \lambda_k \in \Lambda$ satisying $\lambda_i \neq \lambda_j$ whenever $i \neq j$, $i, j = 1, \cdots, k$.*

The following result is known as the Borel-Cantelli lemma.

Theorem 2.4. 1) *Suppose that $\{A_i\}_{i=1}^{\infty}$ is a sequence of measurable sets in a measure space $(\Omega, \mathcal{F}, \mu)$. If $\sum_{i=1}^{\infty} \mu(A_i) < +\infty$, then $\mu(\overline{\lim_{i\to\infty}} A_i) = 0$. Here and henceforth,*

$$\overline{\lim_{i\to\infty}} A_i \overset{\Delta}{=} \bigcap_{k=1}^{\infty} \bigcup_{i=k}^{\infty} A_i.$$

2) *If $(\Omega, \mathcal{F}, \mathbb{P})$ is a probability space and $\{A_i\}_{i=1}^{\infty}$ is a sequence of mutually independent events, then the condition $\sum_{i=1}^{\infty} \mathbb{P}(A_i) = +\infty$ implies that $\mathbb{P}(\overline{\lim_{i\to\infty}} A_i) = 1$.*

Remark 2.5. In the probability context, the first conclusion in Theorem 2.4 states a condition that (almost surely) an event occurs only finitely often; and the second one in this theorem states that, in the independence case, if the condition is not satisfied, the event is almost sure to occur infinitely times.

Let f be a map from Ω to Ω' and $\mathfrak{P}$ be a property concerning the map f at some elements in Ω. In the rest of this book, we shall simply denote by $\{\mathfrak{P}\}$ the subset $\{\omega \in \Omega \mid \mathfrak{P} \text{ holds for } f(\omega)\}$. For example, when $\Omega' = \mathbb{R}$ we use $\{f \geq 0\}$ to denote $\{\omega \in \Omega \mid f(\omega) \geq 0\}$.

The following two notions of convergence are used frequently.

Definition 2.6. *Let $(\Omega, \mathcal{F}, \mu)$ be a measure space and $\{f_i\}_{i=0}^{\infty}$ be a sequence of H-valued functions defined on Ω.*

1) *The sequence $\{f_k\}_{k=1}^{\infty}$ is said to converge to f_0 in measure, denoted by $f_0 = \mu\text{-}\lim_{k\to\infty} f_k$, if $\mathbf{d}(f_k, f_0)$ is $\mathcal{F}$-measurable and for every $\varepsilon > 0$,*

$$\lim_{k\to\infty} \mu(\{\mathbf{d}(f_k, f_0) \geq \varepsilon\}) = 0.$$

(In the context of probability theory, $\{f_k\}_{k=1}^{\infty}$ is said to converge to f_0 in probability).

2) *The sequence $\{f_k\}_{k=1}^{\infty}$ is said to converge to f_0 almost everywhere, denoted by $\lim_{k\to\infty} f_k = f_0$, μ-a.e., if $\{\lim_{k\to\infty} f_k \neq f_0\} \in \mathcal{F}$, and*

$$\mu(\{\lim_{k\to\infty} f_k \neq f_0\}) = 0.$$

(Clearly, in the context of probability theory, $\{f_k\}_{k=1}^{\infty}$ converges to f_0 a.s.).

The relationship between convergence in measure and convergence almost everywhere is as follows.

Theorem 2.7. *Suppose that $(\Omega, \mathcal{F}, \mu)$ is a measure space and $\{f_i\}_{i=0}^{\infty}$ is a sequence of H-valued functions defined on Ω.*

1) *If $\{f_k\}_{k=1}^{\infty}$ converges to f_0 in measure, then there is a subsequence $\{f_{n_k}\}_{k=1}^{\infty}$ of $\{f_k\}_{k=1}^{\infty}$ such that $f_{n_k} \to f_0$, μ-a.e. as $k \to \infty$.*

2) *Suppose in addition that μ is finite, and $\mathbf{d}(f_k, f_0)$ is $\mathcal{F}$-measurable for each $k \in \mathbb{N}$. Then the μ-a.e. convergence of $\{f_k\}_{k=1}^{\infty}$ to f_0 implies that $\{f_k\}_{k=1}^{\infty}$ converges to f_0 in measure.*

Clearly, the H-valued measurable functions/random variables introduced in the above are natural generalization of the usual real valued measurable functions/random variables. However, as pointed in [264], when H is a non-separable Banach space, the sum of two H-valued measurable functions is not necessary to be measurable. Because of this, one needs to introduce other notions for measurable functions (or random variables in the context of probability) valued in metric spaces.

Definition 2.8. *Let $f(\cdot)$ be an H-valued function defined on a measurable space $(\Omega, \mathcal{F})$.*

1) The function $f(\cdot)$ is called weakly $\mathcal{F}$-measurable (or simply, weakly measurable) if H is a Banach space[2]*, and for any $\varphi \in H^*$, the (scalar) function $\varphi(f(\cdot))$ is measurable;*

2) We call $f(\cdot)$ an $\mathcal{F}$-simple function if[3]

$$f(\cdot) = \sum_{i=1}^{k} \chi_{E_i}(\cdot) h_i, \tag{2.2}$$

for some $k \in \mathbb{N}$, $h_i \in H$, and mutually disjoint sets $E_1, \cdots, E_k \in \mathcal{F}$ satisfying $\bigcup_{i=1}^{k} E_i = \Omega$;

3) We call $f(\cdot)$ an $\mathcal{F}$-countably valued function if $f(\cdot) = \sum_{i=1}^{\infty} \chi_{E_i}(\cdot) h_i$, for some $h_1, h_2, \cdots \in H$, and mutually disjoint sets $E_1, E_2, \cdots \in \mathcal{F}$ satisfying $\bigcup_{i=1}^{\infty} E_i = \Omega$;

4) The function $f(\cdot)$ is said to be strongly $\mathcal{F}$-measurable or simply strongly measurable if there exists a sequence of (H-valued) $\mathcal{F}$-simple functions $\{f_k\}_{k=1}^{\infty}$ converging pointwise to f;

5) The function $f(\cdot)$ is said to be separably valued if its range $\{f(\omega) \mid \omega \in \Omega\}$ is separable.

The following result indicates that when H is separable, the above defined strong $\mathcal{F}$-measurability, $\mathcal{F}$-measurability and weak $\mathcal{F}$-measurability are equivalent.

Theorem 2.9. *Let $f(\cdot)$ be an H-valued function defined on a measurable space $(\Omega, \mathcal{F})$. Then, the following two statements are equivalent:*

1) $f(\cdot)$ is strongly $\mathcal{F}$-measurable;

2) $f(\cdot)$ is $\mathcal{F}$-measurable and separably valued.

Further, if H is a Banach space, then the above two statements are equivalent to:

3) $f(\cdot)$ is weakly $\mathcal{F}$-measurable and separably valued.

For vector-valued functions defined on measure spaces, one may slightly change/weaken the requirement in Definition 2.8 4)–5) as follows.

Definition 2.10. *Let $f(\cdot)$ be an H-valued function defined on a σ-finite measure space $(\Omega, \mathcal{F}, \mu)$.*

1) The function $f(\cdot)$ is said to be strongly $\mathcal{F}$-measurable or simply strongly measurable (w.r.t. μ) if there exists a sequence of (H-valued) $\mathcal{F}$-simple functions $\{f_k\}_{k=1}^{\infty}$ such that $\lim_{k \to \infty} f_k = f$, μ-a.e.;

2) The function $f(\cdot)$ is called μ-almost separably valued if there exists a $E_0 \in \mathcal{F}$ such that $\mu(E_0) = 0$ and $\{f(\omega) \mid \omega \in \Omega \setminus E_0\}$ is separable.

---

[2] When H is a Banach space, we denote by H^* the dual space of H.

[3] Since $E_1, \cdots, E_k$ are mutually disjoint, the function $f(\cdot)$ in (2.2) is simply defined by $f(x) = h_i$ when $x \in E_i$ for some $i = 1, \cdots, k$.

Note that, for any measurable space $(\Omega, \mathcal{F})$, one may define a trivial measure μ_0 by $\mu_0(\emptyset) = 0$ and $\mu_0(A) = \infty$ for any nonempty $A \in \mathcal{F}$. Hence, the strongly $\mathcal{F}$-measurable functions and separably valued functions in Definition 2.8 are special cases of that in Definition 2.10. Clearly, the norm of (H-valued) strongly $\mathcal{F}$-measurable function (w.r.t. μ) is measurable, and any linear combination of two strongly $\mathcal{F}$-measurable functions (w.r.t. μ) is strongly measurable whenever H is a Banach space; the (μ-almost everywhere) strong limit of a sequence of strongly $\mathcal{F}$-measurable functions is strongly measurable.

Similar to Theorem 2.9, one has the following result.

Theorem 2.11. *Let $f(\cdot)$ be an H-valued function defined on a complete σ-finite measure space $(\Omega, \mathcal{F}, \mu)$. Then, the following two statements are equivalent:*

1) $f(\cdot)$ is strongly $\mathcal{F}$-measurable w.r.t. μ;

2) $f(\cdot)$ is $\mathcal{F}$-measurable and μ-almost separably valued.

Further, if H is a Banach space, then the above two statements are equivalent to:

3) $f(\cdot)$ is weakly $\mathcal{F}$-measurable and μ-almost separably valued.

By Theorem 2.11, it is clear that when $(\Omega, \mathcal{F}, \mu)$ is a complete σ-finite measure space and H is a separable Banach space, the strong $\mathcal{F}$-measurability w.r.t. μ, the $\mathcal{F}$-measurability and the weak $\mathcal{F}$-measurability are equivalent. Also, we remark that, the completeness assumption on $(\Omega, \mathcal{F}, \mu)$ in the second part of Theorem 2.11 cannot be dropped (Indeed, without this assumption, the strong $\mathcal{F}$-measurability w.r.t. μ does not imply the weak $\mathcal{F}$-measurability, even in finite dimensions).

Also, one has the following characterization on the strong $\mathcal{F}$-measurability w.r.t. μ.

Proposition 2.12. *Let $(\Omega, \mathcal{F}, \mu)$ be a σ-finite measure space. Then, $f(\cdot)$ is an H-valued, strongly $\mathcal{F}$-measurable (w.r.t. μ) function on $(\Omega, \mathcal{F}, \mu)$ if and only if there exist a μ-null set Ω_0 and a sequence of $\mathcal{F}$-countably (H-)valued functions $\{f_i(\cdot)\}_{i=1}^{\infty}$ such that $\{f_i(\cdot)\}$ converges to $f(\cdot)$ uniformly in $\Omega \setminus \Omega_0$.*

We shall use the following result, known as the Doob-Dynkin lemma.

Theorem 2.13. *Let $(\Omega, \mathcal{F})$ be a measurable space, H_1 and H_2 be two separable Banach spaces, and $f_k : \Omega \to H_k$ $(k = 1, 2)$ be $\mathcal{F}$-measurable functions. Then $f_2 = g(f_1)$ for some $\mathcal{B}(H_1)$-measurable function $g : H_1 \to H_2$ if and only if $\sigma(f_2) \subset \sigma(f_1)$.*

Proof: We need only to prove the "if" part. Put

$$\mathcal{H} \stackrel{\Delta}{=} \big\{\eta(f_1(\cdot)) \bigm| \eta : H_1 \to H_2 \text{ is } \mathcal{B}(H_1)\text{-measurable}\big\}.$$

It suffices to show that

$$\mathcal{H} \text{ contains all } H_2\text{-valued, } \sigma(f_1)\text{-measurable functions.} \tag{2.3}$$

Now we prove (2.3). Clearly, $\mathcal{H}$ is a linear space. Let

$$\mathcal{F}' \stackrel{\Delta}{=} \{A \subset \Omega \mid \chi_A(\cdot)h \in \mathcal{H}, \ \forall\, h \in H_2\}.$$

If $A \in \sigma(f_1)$, then $A = f_1^{-1}(B)$ for some $B \in \mathcal{B}(H_1)$, which leads to $\chi_A(\cdot)h = \chi_B(f_1(\cdot))h \in \mathcal{H}$ for any $h \in H_2$. Hence, $\sigma(f_1) \subset \mathcal{F}'$. For any H_2-valued, $\sigma(f_1)$-measurable function ζ, we select a sequence of H_2-valued, $\sigma(f_1)$-simple functions $\{\zeta_k\}_{k=1}^\infty$ such that $\lim_{k\to\infty} \zeta_k(\omega) = \zeta(\omega)$ for any $\omega \in \Omega$. Without loss of generality, we assume that $\zeta_k(\cdot) = \sum_{i=1}^{\ell_k} \chi_{f_1^{-1}(B_i)}(\cdot)h_i$, for some $\ell_k \in \mathbb{N}$, $\ell_k \to \infty$ as $k \to \infty$, $h_i \in H_2$, $B_i \in \mathcal{B}(H_1)$ and $B_i \cap B_j = \emptyset$ when $i \neq j$ $(j = 1, \cdots, \ell_k)$. Write

$$\eta_k(\cdot) = \sum_{i=1}^{\ell_k} \chi_{B_i}(\cdot)h_i, \qquad B = \bigcup_{i=1}^{\infty} B_i.$$

It is easy to see that $\eta_k(f_1(\cdot)) = \zeta_k(\cdot)$. Define

$$\eta(x) = \begin{cases} \lim\limits_{k\to\infty} \eta_k(x), & x \in B, \\ 0, & x \in H_1 \setminus B. \end{cases}$$

Clearly, $\eta: H_1 \to H_2$ is $\mathcal{B}(H_1)$-measurable and $\zeta(\cdot) = \eta(f_1(\cdot))$. Thus, $\zeta(\cdot) \in \mathcal{H}$. This completes the proof of (2.3). □

2.2 Integrals and Expectation

In this section, we recall the definitions and some basic results for the Bochner integral and the Pettis integral. We omit the proofs and refer the readers to [74, 143]. Let us fix a σ-finite measure space $(\Omega, \mathcal{F}, \mu)$, a probability space $(\Omega, \mathcal{F}, \mathbb{P})$ and a Banach space H.

Let $f(\cdot)$ be an (H-valued) $\mathcal{F}$-simple function in the form (2.2). We call $f(\cdot)$ Bochner integrable if $\mu(E_i) < \infty$ for each $i = 1, \cdots, k$. In this case, for any $E \in \mathcal{F}$, the Bochner integral of $f(\cdot)$ over E is defined by

$$\int_E f(s)d\mu = \sum_{i=1}^{k} \mu(E \cap E_i)h_i.$$

In general, we have the following notion.

Definition 2.14. *A strongly $\mathcal{F}$-measurable function $f(\cdot): \Omega \to H$ is said to be Bochner integrable (w.r.t. μ) if there exists a sequence of Bochner integrable $\mathcal{F}$-simple functions $\{f_i(\cdot)\}_{i=1}^\infty$ converging strongly to $f(\cdot)$, μ-a.e. in Ω, so that*

$$\lim_{i,j\to\infty} \int_\Omega |f_i(s) - f_j(s)|_H d\mu = 0.$$

For any $E \in \mathcal{F}$, the Bochner integral of $f(\cdot)$ over E is defined by

$$\int_E f(s)d\mu = \lim_{i\to\infty} \int_\Omega \chi_E(s) f_i(s)d\mu(s) \quad \text{in } H. \tag{2.4}$$

It is easy to verify that the limit in the right hand side of (2.4) exists and its value is independent of the choice of the sequence $\{f_i(\cdot)\}_{i=1}^\infty$. Clearly, when $H = \mathbb{R}^n$ (for some $n \in \mathbb{N}$), the above Bochner integral coincides the usual integral for $\mathbb{R}^n$-valued functions.

Particularly, if $(\Omega, \mathcal{F}, \mathbb{P})$ ia a probability space and $f : \Omega \to H$ is Bochner integrable (w.r.t. $\mathbb{P}$), then we say that f has a mean, denoted by

$$\mathbb{E}f = \int_\Omega f d\mathbb{P}. \tag{2.5}$$

We also call $\mathbb{E}f$ the (mathematical) expectation of f.

The following result reveals the relationship between the Bochner integral (for vector-valued functions) and the usual integral (for scalar functions).

Theorem 2.15. *Let $f(\cdot) : \Omega \to H$ be a strongly $\mathcal{F}$-measurable w.r.t. μ. Then, $f(\cdot)$ is Bochner integrable w.r.t. μ if and only if the scalar function $|f(\cdot)|_H : \Omega \to \mathbb{R}$ is integrable w.r.t. μ.*

Further properties for Bochner integral are collected as follows.

Theorem 2.16. *Let $f(\cdot),\ g(\cdot) : (\Omega, \mathcal{F}, \mu) \to H$ be Bochner integrable. Then,*

1) *For any $a, b \in \mathbb{C}$, the function $af(\cdot) + bg(\cdot)$ is Bochner integrable, and*

$$\int_E (af(s) + bg(s))d\mu = a\int_E f(s)d\mu + b\int_E g(s)d\mu, \qquad \forall\, E \in \mathcal{F}.$$

2) *For any $E \in \mathcal{F}$,*

$$\Big|\int_E f(s)d\mu\Big|_H \le \int_E |f(s)|_H d\mu.$$

3) *The Bochner integral is μ-absolutely continuous, that is,*

$$\lim_{E\in\mathcal{F},\ \mu(E)\to 0} \int_E f(s)d\mu = 0 \quad \text{in } H.$$

4) *If $F \in \mathcal{L}(H; \widetilde{H})$, then $Ff(\cdot)$ is an $\widetilde{H}$-valued Bochner integrable function, and for any $E \in \mathcal{F}$,*

$$\int_E Ff(s)d\mu = F\int_E f(s)d\mu.$$

It is easy to show the following result.

Proposition 2.17. *Let H be a Hilbert space. If f and g are independent, Bochner integrable random variables on $(\Omega, \mathcal{F}, \mathbb{P})$, valued in H, then $(f,g)_H$ is integrable, and*

$$\mathbb{E}(f,g)_H = (\mathbb{E}f, \mathbb{E}g)_H. \tag{2.6}$$

The following result, known as Dominated Convergence Theorem, will be used frequently in the sequel.

Theorem 2.18. *Let $f : (\Omega, \mathcal{F}, \mu) \to H$ be strongly measurable, and let $g : (\Omega, \mathcal{F}, \mu) \to \mathbb{R}$ be a real valued nonnegative integrable function. Assume that, $f_i : (\Omega, \mathcal{F}, \mu) \to H$ is Bochner integrable so that $|f_i|_H \leq g$, μ-a.e. for each $i \in \mathbb{N}$ and $\lim_{i\to\infty} f_i = f$, μ-a.e. Then, f is Bochner integrable, and*

$$\lim_{i\to\infty} \int_E f_i(s)d\mu = \int_E f(s)d\mu, \qquad \forall\, E \in \mathcal{F}.$$

The following result, known as Monotone Convergence Theorem (also called Levi's lemma), will be used in the sequel.

Theorem 2.19. *Let $f : \Omega \to H$ be strongly measurable. Assume that, $f_i : \Omega \to H$ is Bochner integrable w.r.t. μ so that $|f_i|_H \leq |f_{i+1}|_H$, μ-a.e. for each $i \in \mathbb{N}$, $\sup_{i\in\mathbb{N}} \int_\Omega |f_i|_H d\mu < \infty$ and $\lim_{i\to\infty} f_i = f$, μ-a.e. Then the same conclusion as that in Theorem 2.18 holds.*

The following result, known as Fatou's lemma, will be useful in the sequel.

Theorem 2.20. *Let $f, f_i : (\Omega, \mathcal{F}, \mu) \to \mathbb{R}$ be real valued integrable functions ($i \in \mathbb{N}$). Assume that $f \leq f_i$, μ-a.e. for each $i \in \mathbb{N}$. Then,*

$$\int_E \varliminf_{i\to\infty} f_i(s)d\mu \leq \varliminf_{i\to\infty} \int_E f_i(s)d\mu, \qquad \forall\, E \in \mathcal{F}.$$

Also, one has the following Fubini Theorem (on Bochner integrals).

Theorem 2.21. *Let $(\Omega_1, \mathcal{F}_1, \mu_1)$ and $(\Omega_2, \mathcal{F}_2, \mu_2)$ be σ-finite measure spaces. If $f(\cdot,\cdot) : \Omega_1 \times \Omega_2 \to H$ is Bochner integrable, then the functions $y(\cdot) \equiv \int_{\Omega_1} f(t,\cdot)d\mu_1(t)$ and $z(\cdot) \equiv \int_{\Omega_2} f(\cdot,s)d\mu_2(s)$ are a.e. defined and Bochner integrable on Ω_2 and Ω_1, respectively. Moreover,*

$$\int_{\Omega_1\times\Omega_2} f(t,s)d\mu_1 \times \mu_2 = \int_{\Omega_1} z(t)d\mu_1(t) = \int_{\Omega_2} y(s)d\mu_2(s).$$

For any $p \in (0,\infty)$, denote by $L^p_{\mathcal{F}}(\Omega; H) \overset{\Delta}{=} L^p(\Omega, \mathcal{F}, \mu; H)$ the set of all (equivalence classes of) strongly measurable functions $f : (\Omega, \mathcal{F}, \mu) \to H$ such that $\int_\Omega |f|^p_H d\mu < \infty$. When $p \in [1,\infty)$, this is a Banach space with the norm

$$|f|_{L^p_{\mathcal{F}}(\Omega;H)} = \left(\int_\Omega |f|^p_H d\mu\right)^{1/p}. \tag{2.7}$$

(Particularly, when H is a Hilbert space, so is $L^2_{\mathcal{F}}(\Omega; H)$). We denote by $L^\infty_{\mathcal{F}}(\Omega; H) \overset{\Delta}{=} L^\infty(\Omega, \mathcal{F}, \mu; H)$ the set of all (equivalence classes of) strongly measurable (H-valued) functions f such that $\operatorname{ess\,sup}_{\omega\in\Omega}|f(\omega)|_H < \infty$. This is also a Banach space with the norm

$$|f|_{L^\infty_{\mathcal{F}}(\Omega;H)} = \operatorname{ess\,sup}_{\omega\in\Omega}|f(\omega)|_H. \tag{2.8}$$

For $1 \le p \le \infty$ and $T > 0$, we shall simply denote $L^p((0,T), \mathcal{L}, \mathbf{m}; H)$ by $L^p(0,T;H)$, where $\mathcal{L}$ is the family of Lebesgue measurable sets in $(0,T)$, and $\mathbf{m}$ is the Lebesgue measure on $(0,T)$. Also, we simply denote $L^p_{\mathcal{F}}(\Omega;\mathbb{R})$ and $L^p(0,T;\mathbb{R})$ by $L^p_{\mathcal{F}}(\Omega)$ and $L^p(0,T)$, respectively.

The following easy result is known as Chebyshev's inequality.

Theorem 2.22. *For any $p \in [1, +\infty)$, $f \in L^p_{\mathcal{F}}(\Omega; H)$ and $\lambda > 0$, it holds that*

$$\mu(\{|f|_H \ge \lambda\}) \le \frac{1}{\lambda^p}\int_\Omega |f|^p_H d\mu. \tag{2.9}$$

The following simple result is sometimes useful.

Lemma 2.23. *If $p \in (0, \infty)$ and $f \in L^p_{\mathcal{F}}(\Omega; H)$, then*

$$\int_\Omega |f|^p_H d\mu = p\int_0^{+\infty} \lambda^{p-1}\mu(\{|f|_H \ge \lambda\})d\lambda. \tag{2.10}$$

Proof: Since $p \in (0, +\infty)$, we have

$$\begin{aligned}\int_\Omega |f|^p_H d\mu &= p\int_\Omega d\mu \int_0^{|f|_H} \lambda^{p-1}d\lambda = p\int_\Omega d\mu \int_0^{+\infty} \chi_{\{|f|_H\ge\lambda\}}\lambda^{p-1}d\lambda \\ &= p\int_0^{+\infty} \lambda^{p-1}\mu(\{|f|_H \ge \lambda\})d\lambda,\end{aligned}$$

which gives (2.10). □

In the sequel, we will need the following concept of uniform integrability.

Definition 2.24. *Suppose $\mathcal{K} \subset L^1_{\mathcal{F}}(\Omega; H)$. We call $\mathcal{K}$ a uniformly integrable subset (of $L^1_{\mathcal{F}}(\Omega; H)$) if $\int_{\{|f|_H\ge s\}} |f|_H d\mu$ converges to 0 uniformly (for any $f \in \mathcal{K}$) as $s \to +\infty$.*

Obviously, if $\mu(\Omega) < \infty$ and $\mathcal{K}$ is a uniformly integrable subset of $L^1_{\mathcal{F}}(\Omega; H)$, then it is also bounded in $L^1_{\mathcal{F}}(\Omega; H)$.

The following result characterizes the L^1-convergence in terms of the uniform integrability.

Theorem 2.25. *Let $\{f_i\}_{i=0}^\infty \subset L^1_{\mathcal{F}}(\Omega; H)$. Then $f_i \to f_0$ in $L^1_{\mathcal{F}}(\Omega; H)$ as $i \to \infty$ if and only if $\{f_i\}_{i=1}^\infty$ converges to f_0 in measure and $\{f_i\}_{i=1}^\infty$ is uniformly integrable.*

For nonnegative measurable functions, one has the following simple characterization of L^1-convergence:

Theorem 2.26. *Let $\{f_i\}_{i=0}^{\infty}$ be a sequence of nonnegative integrable functions on $(\Omega, \mathcal{F}, \mu)$. Then $f_i \to f_0$ in $L^1_{\mathcal{F}}(\Omega)$ as $i \to \infty$ if and only if $\{f_i\}_{i=1}^{\infty}$ converges to f_0 in measure and $\lim_{i\to\infty} \int_\Omega f_i d\mu = \int_\Omega f_0 d\mu$.*

Let $(\Omega', \mathcal{F}')$ be another measurable space, and $\Phi : (\Omega, \mathcal{F}) \to (\Omega', \mathcal{F}')$ be a measurable map. Then Φ induces a measure ν on $(\Omega', \mathcal{F}')$ via

$$\nu(A') \stackrel{\Delta}{=} \mu(\Phi^{-1}(A')), \qquad \forall\, A' \in \mathcal{F}'. \tag{2.11}$$

The following is a change-of-variable formula:

Theorem 2.27. *A function $f(\cdot) : \Omega' \to H$ is Bochner integrable w.r.t. ν if and only if $f(\Phi(\cdot))$ (defined on $(\Omega, \mathcal{F})$) is Bochner integrable w.r.t. μ. Furthermore,*

$$\int_{\Omega'} f(\omega')d\nu(\omega') = \int_\Omega f(\Phi(\omega))d\mu(\omega). \tag{2.12}$$

2.3 Signed/Vector Measures, Conditional Expectation

We fix a measurable space $(\Omega, \mathcal{F})$ and a Banach space H.

2.3.1 Signed Measures

Let us begin with the following notion.

Definition 2.28. *A function $\mu : \mathcal{F} \to [-\infty, +\infty]$ is called a signed measure on $(\Omega, \mathcal{F})$ if*

1) $\mu(\emptyset) = 0$;

2) $\mu(\bigcup_{j=1}^{\infty} A_j) = \sum_{j=1}^{\infty} \mu(A_j)$ for any sequence $\{A_j\}$ of mutually disjoint sets from $\mathcal{F}$; and

3) μ assumes at most one of the values $+\infty$ and $-\infty$.

Example 2.29. Let ν be a measure on $(\Omega, \mathcal{F})$ and f be a real valued integrable function defined on $(\Omega, \mathcal{F}, \nu)$. Then

$$\mu(A) = \int_A f d\nu, \qquad \forall\, A \in \mathcal{F},$$

defines a signed measure in $(\Omega, \mathcal{F})$. More generally, the above μ is still a signed measure if f is a measurable function on $(\Omega, \mathcal{F})$, and one of f^+ and f^-, the positive and negative parts of f, is integrable on $(\Omega, \mathcal{F}, \nu)$.

If μ is a signed measure on $(\Omega, \mathcal{F})$, we call a set $E \subset \Omega$ positive (*resp.* negative) (w.r.t. μ) if for every $F \in \mathcal{F}$, $E \cap F$ is measurable, and $\mu(E \cap F) \geq 0$ (*resp.* $\mu(E \cap F) \leq 0$).

Theorem 2.30. *If μ is a signed measure on $(\Omega, \mathcal{F})$, then there exist two disjoint sets A and B, whose union is Ω, such that A is positive and B is negative w.r.t. μ.*

The sets A and B in Theorem 2.30 are said a *Hahn decomposition of Ω w.r.t. μ*. It is not difficult to construct examples to show that Hahn decomposition is not unique. However, if

$$\Omega = A_1 \cup B_1 \quad \text{and} \quad \Omega = A_2 \cup B_2$$

are two Hahn decomposition of Ω, then it is easy to show that, for every measurable set E, it holds

$$\mu(E \cap A_1) = \mu(E \cap A_2) \quad \text{and} \quad \mu(E \cap B_1) = \mu(E \cap B_2).$$

From this fact, we may define unambiguously two set functions μ^+ and μ^- on $(\Omega, \mathcal{F})$ as follows:

$$\mu^+(E) = \mu(E \cap A), \quad \mu^-(E) = -\mu(E \cap B), \qquad \forall\, E \in \mathcal{F}.$$

We call μ^+ and μ^- respectively the upper variation and the lower variation of μ. The set function $|\mu|$, defined for every $E \in \mathcal{F}$ by

$$|\mu|(E) = \mu^+(E) + \mu^-(E)$$

is called the total variation of μ. Obviously

$$\mu(E) = \mu^+(E) - \mu^-(E), \qquad |\mu|(E) \geq |\mu(E)|, \quad \forall E \in \mathcal{F}.$$

Also, it is easy to show that the upper, lower, and the total variations of a signed measure μ are measures.

Definition 2.31. *A signed measure μ on $(\Omega, \mathcal{F})$ is said to be finite if $|\mu(E)| < +\infty$ for any $E \in \mathcal{F}$; μ is said to be σ-finite if for any $E \in \mathcal{F}$, there exists a sequence $\{E_i\}_{i=1}^{\infty} \subset \mathcal{F}$ such that $|\mu(E_i)| < +\infty$ for every $i \in \mathbb{N}$ and $E \subset \bigcup_{i=1}^{\infty} E_i$.*

Let μ be a σ-finite signed measure on $(\Omega, \mathcal{F})$. If $f : (\Omega, \mathcal{F}) \to H$ is Bochner integrable w.r.t. $|\mu| = \mu^+ + \mu^-$, then f is said to be Bochner integrable w.r.t. μ. The Bochner integral of f in Ω w.r.t. μ is defined by

$$\int_{\Omega} f d\mu \stackrel{\Delta}{=} \int_{\Omega} f d\mu^+ - \int_{\Omega} f d\mu^-.$$

For any $E \in \mathcal{F}$, we define

$$\int_E f d\mu = \int_{\Omega} \chi_E f d\mu.$$

The properties of the Bochner integral w.r.t. the above signed measure μ is similarly to the usual (Bochner) integral except that all conditions such as "null measure sets" or "almost everywhere" w.r.t. μ should be changed to that w.r.t. $|\mu|$.

Definition 2.32. *Let μ and ν be two signed measures on $(\Omega, \mathcal{F})$. We say that μ is absolutely continuous w.r.t. ν, denoted by $\mu \ll \nu$, if $\mu(E) = 0$ for every $E \in \mathcal{F}$ with $|\nu|(E) = 0$.*

We recall the following fundamental result (known as the Radon-Nikodým Theorem) concerning absolute continuity.

Theorem 2.33. *Let μ and ν be two σ-finite signed measures on $(\Omega, \mathcal{F})$ and $\mu \ll \nu$. Then there is a real valued measurable function f on Ω such that*

$$\mu(E) = \int_E f d\nu, \qquad \forall\, E \in \mathcal{F}.$$

The function f is unique in the sense that if also $\mu(E) = \int_E f_1 d\nu$ for another real valued measurable function f_1 on Ω and for all $E \in \mathcal{F}$, then $f = f_1$, $|\nu|$-a.e.

The function f in Theorem 2.33 is called the Radon-Nikodým derivative (of μ w.r.t. ν), and is denoted by

$$f = \frac{d\mu}{d\nu}.$$

By Theorem 2.33, it is easy to conclude the following result.

Theorem 2.34. *Let the assumptions in Theorem 2.33 hold and $g : (\Omega, \mathcal{F}) \to H$ be strongly $\mathcal{F}$-measurable. Then g is Bochner integrable w.r.t. μ if and only if $g\frac{d\mu}{d\nu}$ is Bochner integrable w.r.t. ν. Furthermore,*

$$\int_E g d\mu = \int_E g \frac{d\mu}{d\nu} d\nu, \qquad \forall\, E \in \mathcal{F}.$$

2.3.2 Distribution, Density and Characteristic Functions

Let $\mathbb{P}$ be a probability measure on $(\Omega, \mathcal{F})$, and $X : \Omega \to H$ be a strongly measurable random variable. Then, as a special case of (2.11), X induces a probability measure $\mathbb{P}_X$ on $(H, \mathcal{B}(H))$ via

$$\mathbb{P}_X(A) \overset{\Delta}{=} \mathbb{P}(X^{-1}(A)), \qquad \forall\, A \in \mathcal{B}(H). \tag{2.13}$$

We call $\mathbb{P}_X$ the distribution of X. If X is Bochner integrable w.r.t. $\mathbb{P}$, then by (2.5) and using (2.12) in Theorem 2.27, we see that

$$\mathbb{E}X = \int_H x d\mathbb{P}_X(x).$$

In the case of $H = \mathbb{R}^m$ (for some $m \in \mathbb{N}$), $\mathbb{P}_X$ can be uniquely determined by the following function:

$$F(x) \overset{\Delta}{=} F(x_1, \cdots, x_m) \overset{\Delta}{=} \mathbb{P}\{X_i \le x_i,\ 1 \le i \le m\}, \tag{2.14}$$

where $x = (x_1, \cdots, x_m)$ and $(X_1, \cdots, X_m) = X$. We call $F(x)$ the distribution function of X. If $\mathbb{P}_X$ is absolutely continuous w.r.t. the Lebesgue measure in $\mathbb{R}^m$, then by the Radon-Nikodým theorem (i.e., Theorem 2.33), there exists a (nonnegative) function $f \in L^1(\mathbb{R}^m)$ such that

$$\mathbb{P}_X(A) = \int_A f(x)dx, \qquad \forall\, A \in \mathcal{B}(\mathbb{R}^m). \tag{2.15}$$

Particularly,

$$F(x) = \int_{-\infty}^{x_1} \cdots \int_{-\infty}^{x_m} f(\xi_1, \cdots, \xi_m) d\xi_1 \cdots d\xi_m. \tag{2.16}$$

The function $f(x)$ is called the density of X. As a special case, if $f(x)$ is of the following form:

$$f(x) = [(2\pi)^m \det Q]^{-1/2} \exp\Big\{-\frac{1}{2}(x-\lambda)Q^{-1}(x-\lambda)^\top\Big\}, \quad x \in \mathbb{R}^m, \tag{2.17}$$

for some $\lambda \in \mathbb{R}^m$, $Q \in \mathbb{R}^{m\times m}$ with $Q^\top = Q > 0$, then we say that X has a normal distribution with parameter (λ, Q), denoted by $X \sim \mathcal{N}(\lambda, Q)$. We call X a Gaussian random variable (valued in $\mathbb{R}^m$) if X has a normal distribution or X is constant.

Remark 2.35. Gaussian random variable appears in many physical models. Indeed, under some mild conditions, the mean of many independent and identically-distributed random variables is close to a Gaussian random variable (We refer to [38] for more details).

For Gaussian random variables, one has the following result.

Theorem 2.36. *Let $X = (X_1, \cdots, X_m)$ be a Gaussian random variable valued in $\mathbb{R}^m$. Then $X_1, \cdots, X_m$ are independent if and only if the matrix Q (in* (2.17)*) is diagonal.*

In the rest of this subsection, we assume that H and G are two Hilbert spaces. For any $x \in H$ and $y \in G$, the tensor product $x \otimes y$ (of x and y) is a bounded linear operator from H to G, defined by

$$(x \otimes y)h = \langle h, x \rangle_H y, \qquad \forall\, h \in H. \tag{2.18}$$

Let $Q \in \mathcal{L}(H;G)$. We call Q a trace-class operator from H to G if

$$trQ \overset{\Delta}{=} \sup\Big\{ \sum_{\mu\in\Lambda} |\langle Qh_\mu, g_\mu \rangle_G| \;\Big|\; \{h_\mu\}_{\mu\in\Lambda} \text{ and } \{g_\mu\}_{\mu\in\Lambda} \text{ are respectively}$$
$$\text{orthonormal basis in } H \text{ and } G \Big\} < \infty.$$

Denote by $\mathcal{L}_1(H;G)$ (or simply $\mathcal{L}_1(H)$ if $H = G$) the space of trace-class operators, which is a subset of the space of all compact operators (from H to G). It is well known that $\mathcal{L}_1(H;G)$ equipped with the trace norm trQ is a Banach space, and the following result holds (e.g. [299, Chapter III] and [355, Chapter 6]):

Proposition 2.37. *Let H be a separable Hilbert space. Then, a nonnegative operator $Q \in \mathcal{L}(H)$ is of trace-class if and only if for an orthonormal basis $\{h_i\}_{i=1}^{\infty}$ on H,*

$$\sum_{i=1}^{\infty} \langle Qh_i, h_i \rangle_H < \infty.$$

Moreover, in this case $trQ = \sum_{i=1}^{\infty} \langle Qh_i, h_i \rangle_H$.

For any $X, Y \in L^2_{\mathcal{F}}(\Omega; H)$, we define the covariance operator of X and Y by

$$\mathrm{Cov}\,(X, Y) \overset{\Delta}{=} \mathbb{E}\big[(X - \mathbb{E}X) \otimes (Y - \mathbb{E}Y)\big]. \tag{2.19}$$

It is easy to show that $\mathrm{Cov}\,(X, Y) \in \mathcal{L}_1(H)$. Particularly,

$$\mathrm{Var}\, X \overset{\Delta}{=} \mathrm{Cov}\,(X, X) \tag{2.20}$$

is called the variance operator of X. It is not difficult to show that for a normal distribution $\mathcal{N}(\lambda, Q)$ (defined above), λ is the mean and Q is the variance operator. When H is finite dimensional, the covariance/variance operator is usually called the covariance/variance matrix.

For any H-valued, strongly measurable random variable X and $\xi \in H$, when $e^{i\langle \xi, X(\cdot)\rangle_H}$ is integrable w.r.t. $\mathbb{P}$ (It is always the case provided that H is a real Hilbert space), write

$$\varphi_X(\xi) = \mathbb{E}e^{i\langle \xi, X\rangle_H} = \int_H e^{i\langle \xi, x\rangle_H} d\,\mathbb{P}_X(x). \tag{2.21}$$

We call $\varphi_X(\cdot)$ the characteristic function of X. It can be regarded as the Fourier transform of the probability measure induced by $X(\cdot)$.

The characteristic function is unique in the following sense.

Theorem 2.38. *Assume that X and Y are two H-valued, strongly measurable random variables. If $\varphi_X(\cdot) = \varphi_Y(\cdot)$, then $\mathbb{P}_X = \mathbb{P}_Y$.*

We will need the following notion.

Definition 2.39. *An H-valued random variable X is said to be Gaussian, denoted by $X \sim \mathcal{N}(\lambda, Q)$ if for some $\lambda \in H$ and nonnegative symmetric operator $Q \in \mathcal{L}_1(H)$,*

$$\varphi_X(\xi) = \exp\left\{ i\langle \lambda, \xi\rangle_H - \frac{1}{2}\langle Q\xi, \xi\rangle_H \right\}, \qquad \forall\, \xi \in H. \tag{2.22}$$

If an H-valued random variable $X \sim \mathcal{N}(\lambda, Q)$, then it is easy to check that $\mathbb{E}X = \lambda$ and $\mathrm{Var}\, X = Q$.

2.3.3 Vector Measures

The results of this subsection can be found in [74]. In this subsection, we fix a finite measure ν on $(\Omega, \mathcal{F})$

We begin with the following definition.

Definition 2.40. *An H-valued function $\mu : \mathcal{F} \to H$ is called a vector measure (on $(\Omega, \mathcal{F})$) if $\mu(\emptyset) = 0$, and μ is countably additive, i.e., for any sequence $\{A_k\}_{k=1}^{\infty}$ of mutually disjoint sets in $\mathcal{F}$,*

$$\mu\Big(\bigcup_{k=1}^{\infty} A_k\Big) = \sum_{k=1}^{\infty} \mu(A_k), \quad \text{in } H.$$

An example of vector measures is as follows.

Example 2.41. For $p \in [1, \infty)$, define $\mu : \mathcal{F} \to L^p(\Omega, \mathcal{F}, \nu)$ by $\mu(A) = \chi_A(\cdot)$, $\forall\, A \in \mathcal{F}$. It is easy to see that μ is a vector measure on $(\Omega, \mathcal{F})$.

Definition 2.42. *Let μ be an (H-valued) vector measure on $(\Omega, \mathcal{F})$. The variation of μ is the nonnegative function $|\mu|(\cdot)$ on $\mathcal{F}$, defined by*

$$|\mu|(A) = \sup\Big\{ \sum_{i=1}^{k} |\mu(A_i)|_H \;\Big|\; A = \bigcup_{i=1}^{k} A_i \text{ for some } k \in \mathbb{N} \text{ and mutually disjoint sets } A_1, \cdots, A_k \in \mathcal{F} \Big\}, \qquad \forall\, A \in \mathcal{F}.$$

If $|\mu|(\Omega) < +\infty$, then we call μ a vector measure of bounded variation.

It is easy to check that the vector measure μ given in Example 2.41 is of bounded variation.

Theorem 2.43. *Let μ be a vector measure of bounded variation on $(\Omega, \mathcal{F})$. Then the variation $|\mu|$ is a finite measure on $(\Omega, \mathcal{F})$.*

The following result is sometimes useful.

Proposition 2.44. *If $g \in L^1(\Omega, \mathcal{F}, \nu; H)$, $\mu(A) = \int_A g d\nu$ for all $A \in \mathcal{F}$, then μ is a vector measure of bounded variation on $(\Omega, \mathcal{F})$, and*

$$|\mu|(A) = \int_A |g|_H d\nu, \quad \forall A \in \mathcal{F}.$$

Definition 2.45. *An (H-valued) vector measure μ on $(\Omega, \mathcal{F})$ is called ν-continuous, denoted by $\mu \ll \nu$, if for any $\varepsilon > 0$, there is a $\delta > 0$ such that $|\mu(A)|_H < \varepsilon$ whenever $A \in \mathcal{F}$ satisfying $\nu(A) < \delta$.*

The result below gives a criterion for the ν-continuous vector measure.

Theorem 2.46. *An (H-valued) vector measure μ on $(\Omega, \mathcal{F})$ is ν-continuous if and only if $\mu(A) = 0$ for any $A \in \mathcal{F}$ with $\nu(A) = 0$.*

We shall need the following notion.

Definition 2.47. *We call H to have the Radon-Nikodým property w.r.t. $(\Omega, \mathcal{F}, \nu)$, if for each ν-continuous (H-valued) vector measure μ of bounded variation on $(\Omega, \mathcal{F})$, there exists a $g \in L^1(\Omega, \mathcal{F}, \nu; H)$ such that*

$$\mu(A) = \int_A g d\nu, \qquad \forall\, A \in \mathcal{F}.$$

H is said to have the Radon-Nikodým property, if it has the Radon-Nikodým property w.r.t. any finite measure space.

It is well-known that there does exist a Banach space for which the Radon-Nikodým property fails. The following result is due to R. S. Phillips.

Theorem 2.48. *Reflexive Banach spaces (and hence Hilbert spaces) have the Radon-Nikodým property.*

2.3.4 Conditional Expectation

In this subsection, we fix a probability measure $\mathbb{P}$ on $(\Omega, \mathcal{F})$, and a function $f \in L^1_{\mathcal{F}}(\Omega; H) \overset{\Delta}{=} L^1(\Omega, \mathcal{F}, \mathbb{P}; H)$.

Definition 2.49. *Let $B \in \mathcal{F}$ with $\mathbb{P}(B) > 0$. For any event $A \in \mathcal{F}$, put*

$$\mathbb{P}(A \mid B) = \frac{\mathbb{P}(A \cap B)}{\mathbb{P}(B)}.$$

Then $\mathbb{P}(\cdot \mid B)$ is a probability measure on $(\Omega, \mathcal{F})$, called the conditional probability given the event B, and denoted by $\mathbb{P}_B(\cdot)$. For any given $A \in \mathcal{F}$, $\mathbb{P}(A \mid B)$ is called the conditional probability of A given B. The conditional expectation of f given the event B is defined by

$$\mathbb{E}(f \mid B) = \int_\Omega f d\mathbb{P}_B = \frac{1}{\mathbb{P}(B)} \int_B f d\mathbb{P}.$$

Clearly, the conditional expectation of f given the event B represents the average value of f on B.

In many concrete problems, it is not enough to consider the conditional expectation given only one event. Instead, it is quite useful to define the conditional expectation to be a suitable random variable. For example, when consider two conditional expectations $\mathbb{E}(f \mid B)$ and $\mathbb{E}(f \mid B^c)$ simultaneously, we simply define it as a function $\mathbb{E}(f \mid B)\chi_B(\omega) + \mathbb{E}(f \mid B^c)\chi_{B^c}(\omega)$ rather than regarding it as two numbers. Before considering the general setting, we begin with the following special case.

Definition 2.50. *Let* $\{B_i\}_{i=1}^{\infty} \subset \mathcal{F}$ *be a sequence of mutually disjoint sets in* Ω *such that* $\mathbb{P}(B_i) > 0$ *for all* $i = 1, 2, \cdots$ *and* $\Omega = \bigcup_{i=1}^{\infty} B_i$. *Put* $\mathcal{J} = \sigma\{B_1, B_2, \cdots\}$. *Then the following* $\mathcal{J}$*-measurable function*

$$\mathbb{E}(f \mid \mathcal{J})(\omega) \stackrel{\Delta}{=} \sum_{k=1}^{\infty} \mathbb{E}(f \mid B_k)\chi_{B_k}(\omega)$$

is called the conditional expectation of f *given the* σ*-field* $\mathcal{J}$.

Clearly, when $\mathcal{J} = \sigma\{B_1, B_2, \cdots\}$, the conditional expectation $\mathbb{E}(f \mid \mathcal{J})$ of f takes its average value in every B_k. Also, it is easy to check that

$$\int_{B_k} \mathbb{E}(f \mid \mathcal{J})d\mathbb{P} = \int_{B_k} fd\mathbb{P}, \qquad k = 1, 2, \cdots. \tag{2.23}$$

In the rest of this subsection, we assume that H has the Radon-Nikodým property.

Now, let us consider the more general case that $\mathcal{J}$ is a given sub-σ-field of $\mathcal{F}$. Stimulated by (2.23), one defines a set function on $\mathcal{J}$ by

$$\mu(B) \stackrel{\Delta}{=} \int_B fd\mathbb{P}, \quad \forall\, B \in \mathcal{J}.$$

It is easy to see that μ is an (H-valued) vector measure of bounded variation on $(\Omega, \mathcal{J})$, and $\mu \ll \mathbb{P}$. Hence, one may find a (unique) function in $L^1_{\mathcal{J}}(\Omega; H) \stackrel{\Delta}{=} L^1(\Omega, \mathcal{J}, \mathbb{P}; H)$, denoted by $\mathbb{E}(f \mid \mathcal{J})$, such that

$$\int_B \mathbb{E}(f \mid \mathcal{J})d\mathbb{P} = \int_B fd\mathbb{P}, \qquad \forall\, B \in \mathcal{J}. \tag{2.24}$$

This leads to the following notion:

Definition 2.51. *The (*H*-valued) function* $\mathbb{E}(f \mid \mathcal{J})$*, determined by* (2.24)*, is called the conditional expectation of* f *given the* σ*-field* $\mathcal{J}$.

We collect some basic properties of the conditional expectation as follows.

Theorem 2.52. *Let* $\mathcal{J}$ *be a sub-*σ*-field of* $\mathcal{F}$. *It holds that:*

1) *The map* $\mathbb{E}(\cdot \mid \mathcal{J}) : L^1_{\mathcal{F}}(\Omega; H) \to L^1_{\mathcal{J}}(\Omega; H)$ *is linear and continuous;*

2) $\mathbb{E}(a \mid \mathcal{J}) = a$, $\mathbb{P}|_{\mathcal{J}}$-a.s. , $\forall\, a \in H$;

3) *If* $f_1, f_2 \in L^1_{\mathcal{F}}(\Omega)$ *with* $f_1 \geq f_2$, *then*

$$\mathbb{E}(f_1 \mid \mathcal{J}) \geq \mathbb{E}(f_2 \mid \mathcal{J}), \qquad \mathbb{P}|_{\mathcal{J}}\text{-a.s.};$$

4) *Let* $m, n, k \in \mathbb{N}$, $f_1 \in L^1_{\mathcal{J}}(\Omega; \mathbb{R}^{m\times n})$ *and* $f_2 \in L^1_{\mathcal{F}}(\Omega; \mathbb{R}^{n\times k})$ *such that* $f_1 f_2 \in L^1_{\mathcal{F}}(\Omega; \mathbb{R}^{m\times k})$. *Then*

$$\mathbb{E}(f_1 f_2 \mid \mathcal{J}) = f_1\mathbb{E}(f_2 \mid \mathcal{J}), \qquad \mathbb{P}|_{\mathcal{J}}\text{-a.s.}$$

Particularly, $\mathbb{E}(f_1 \mid \mathcal{J}) = f_1$, $\mathbb{P}|_{\mathcal{J}}$ – a.s. *Also, for any* $f_3 \in L^1_{\mathcal{F}}(\Omega; \mathbb{R}^{m\times n})$,

$$\mathbb{E}(\mathbb{E}(f_3 \mid \mathcal{J})f_2 \mid \mathcal{J}) = \mathbb{E}(f_3 \mid \mathcal{J})\mathbb{E}(f_2 \mid \mathcal{J}), \qquad \mathbb{P}|_{\mathcal{J}}\text{-a.s.};$$

5) *If* f *is independent of* $\mathcal{J}$, *then*

$$\mathbb{E}(f \mid \mathcal{J}) = \mathbb{E}f, \qquad \mathbb{P}|_{\mathcal{J}}\text{-a.s.};$$

6) *Let* $\mathcal{J}'$ *be a sub-*σ*-field of* $\mathcal{J}$. *Then*

$$\mathbb{E}(\mathbb{E}(f \mid \mathcal{J}) \mid \mathcal{J}') = \mathbb{E}(\mathbb{E}(f \mid \mathcal{J}') \mid \mathcal{J}) = \mathbb{E}(f \mid \mathcal{J}'), \qquad \mathbb{P}|_{\mathcal{J}'}\text{-a.s.};$$

7) *(Jensen's inequality) Let* $\phi : H \to \mathbb{R}$ *be a convex function such that* $\phi(f) \in L^1_{\mathcal{F}}(\Omega)$. *Then*

$$\phi(\mathbb{E}(f \mid \mathcal{J})) \le \mathbb{E}(\phi(f) \mid \mathcal{J}), \qquad \mathbb{P}|_{\mathcal{J}}\text{-a.s.}$$

Particularly, for any $p \ge 1$, *we have*

$$\big|\mathbb{E}(f \mid \mathcal{J})\big|_H^p \le \mathbb{E}(|f|_H^p \mid \mathcal{J}), \qquad \mathbb{P}|_{\mathcal{J}}\text{-a.s.}$$

provided that $\mathbb{E}|f|_H^p$ *exists.*

Remark 2.53. Given two sub-σ-fields $\mathcal{J}^k$ $(k = 1, 2)$ of $\mathcal{F}$, generally

$$\mathbb{E}(\mathbb{E}(f|\mathcal{J}^1)|\mathcal{J}^2) \ne \mathbb{E}(\mathbb{E}(f|\mathcal{J}^2)|\mathcal{J}^1) \ne \mathbb{E}(f|\mathcal{J}^1 \cap \mathcal{J}^2), \qquad \mathbb{P}|_{\mathcal{J}^1\cap\mathcal{J}^2}\text{-a.s.}$$

Proposition 2.54. *Let* $\{\mathcal{F}_\alpha\}_{\alpha\in\Lambda}$ *be a family of sub-*σ*-fields of* $\mathcal{F}$, *where* Λ *is a given index set. Then* $\{\mathbb{E}(f \mid \mathcal{F}_\alpha)\}_{\alpha\in\Lambda}$ *is uniformly integrable.*

Proof: By Theorem 2.22, for any $s > 0$,

$$\mathbb{P}(|\mathbb{E}(f \mid \mathcal{F}_\alpha)|_H \ge s) \le s^{-1}\mathbb{E}|\mathbb{E}(f \mid \mathcal{F}_\alpha)|_H \le s^{-1}\mathbb{E}|f|_H, \qquad \forall\, \alpha \in \Lambda.$$

Hence

$$\begin{aligned}
&\int_{(|\mathbb{E}(f \mid \mathcal{F}_\alpha)|_H \ge s)} |\mathbb{E}(f \mid \mathcal{F}_\alpha)|_H d\mathbb{P} \le \int_{(|\mathbb{E}(f \mid \mathcal{F}_\alpha)|_H \ge s)} |f|_H d\mathbb{P}\\
&\le \sqrt{s}\mathbb{P}(|\mathbb{E}(f \mid \mathcal{F}_\alpha)|_H \ge s) + \int_{(|f|_H \ge \sqrt{s})} |f|_H d\mathbb{P}\\
&\le \frac{1}{\sqrt{s}}\mathbb{E}|f|_H + \int_{(|f|_H \ge \sqrt{s})} |f|_H d\mathbb{P},
\end{aligned}$$

which yields the uniform integrability of $\{\mathbb{E}(f \mid \mathcal{F}_\alpha)\}_{\alpha\in\Lambda}$. ☐

2.4 A Riesz-Type Representation Theorem

In this section, we shall prove a Riesz-type representation theorem, which will play important roles in the study of both controllability and optimal control problems for stochastic evolution equations.

Let $(X_1, \mathcal{M}_1, \mu_1)$ and $(X_2, \mathcal{M}_2, \mu_2)$ be two finite measure spaces, and let H be a Banach space. Let $\mathcal{M}$ be a sub-σ-field of $\mathcal{M}_1 \times \mathcal{M}_2$, and for any $1 \le p, q < \infty$, let

$$\begin{aligned} L^p_{\mathcal{M}}(X_1; L^q(X_2; H)) = \Big\{ \varphi : X_1 \times X_2 \to H \;\Big|\; &\varphi(\cdot) \text{ is strongly } \mathcal{M}\text{-measurable} \\ \text{w.r.t. } \mu_1 \times \mu_2 \text{ and } &\int_{X_1} \Big(\int_{X_2} |\varphi(x_1, x_2)|^q_H d\mu_2 \Big)^{\frac{p}{q}} d\mu_1 < \infty \Big\}. \end{aligned} \tag{2.25}$$

Likewise, let

$$\begin{aligned} L^\infty_{\mathcal{M}}(X_1; L^q(X_2; H)) = \Big\{ \varphi : X_1 \times X_2 \to H \;\Big|\; &\varphi(\cdot) \text{ is strongly } \mathcal{M}\text{-measurable} \\ \text{w.r.t. } \mu_1 \times \mu_2 \text{ and } \operatorname{ess\,sup}_{x_1 \in X_1} &\Big(\int_{X_2} |\varphi(x_1, x_2)|^q_H d\mu_2 \Big)^{\frac{1}{q}} < \infty \Big\}, \end{aligned} \tag{2.26}$$

$$\begin{aligned} L^p_{\mathcal{M}}(X_1; L^\infty(X_2; H)) = \Big\{ \varphi : X_1 \times X_2 \to H \;\Big|\; &\varphi(\cdot) \text{ is strongly } \mathcal{M}\text{-measurable} \\ \text{w.r.t. } \mu_1 \times \mu_2 \text{ and } &\int_{X_1} \big(\operatorname{ess\,sup}_{x_2 \in X_2} |\varphi(x_1, x_2)|^p_H \big) d\mu_1 < \infty \Big\}, \end{aligned} \tag{2.27}$$

$$\begin{aligned} L^\infty_{\mathcal{M}}(X_1; L^\infty(X_2; H)) = \Big\{ \varphi : X_1 \times X_2 \to H \;\Big|\; &\varphi(\cdot) \text{ is strongly } \mathcal{M}\text{-measurable} \\ \text{w.r.t. } \mu_1 \times \mu_2 \text{ and } \operatorname{ess\,sup}_{(x_1,x_2) \in X_1 \times X_2} &|\varphi(x_1, x_2)|_H < \infty \Big\}. \end{aligned} \tag{2.28}$$

For any $f \in L^p_{\mathcal{M}}(X_1; L^q(X_2; H))$, $1 \le p, q < \infty$, we denote

$$|f|_{p,q;H} \equiv |f|_{L^p_{\mathcal{M}}(X_1; L^q(X_2; H))} \overset{\Delta}{=} \Big[\int_{X_1} \Big(\int_{X_2} |f(x_1, x_2)|^q_H d\mu_2 \Big)^{\frac{p}{q}} d\mu_1 \Big]^{\frac{1}{p}}.$$

The definition of $|f|_{\infty,q;H}$, $|f|_{p,\infty;H}$ and $|f|_{\infty,\infty;H}$ are similar. In a similar way, we may define $L^q_{\mathcal{M}}(X_2; L^p(X_1; H))$ $(1 \le p, q \le \infty)$. One can show that both $L^p_{\mathcal{M}}(X_1; L^q(X_2; H))$ and $L^q_{\mathcal{M}}(X_2; L^p(X_1; H))$ are Banach spaces. We shall simply denote $L^p_{\mathcal{M}}(X_1; L^q(X_2; \mathbb{R}))$ and $|\cdot|_{p,q;\mathbb{R}}$ by $L^p_{\mathcal{M}}(X_1; L^q(X_2))$ and $|\cdot|_{p,q}$, respectively. When $1 \le p < \infty$, by the Fubini Theorem (i.e., Theorem 2.21), one can identify $L^p_{\mathcal{M}}(X_1; L^p(X_2; H))$ and $L^p_{\mathcal{M}}(X_2; L^p(X_1; H))$ by $L^p_{\mathcal{M}}(X_1 \times X_2; H)$.

By Hölder's inequality and Minkovski's inequality, we have the following inclusions:

$$\begin{aligned} & L^p_{\mathcal{M}}(X_1; L^q(X_2; H)) \subseteq L^r_{\mathcal{M}}(X_1; L^s(X_2; H)), \\ & \qquad\qquad\qquad 1 \le r \le p \le \infty, \quad 1 \le s \le q \le \infty, \end{aligned} \tag{2.29}$$

and,

$$L^p_{\mathcal{M}}(X_1; L^q(X_2; H)) \subseteq L^q_{\mathcal{M}}(X_2; L^p(X_1; H)), \quad 1 \le p \le q \le \infty. \tag{2.30}$$

Next, for any $p \in [1, \infty]$, denote by p' its Hölder conjugate, i.e.,

$$p' = \begin{cases} p/(p-1), & 1 < p < \infty, \\ 1, & p = \infty, \\ \infty, & p = 1. \end{cases}$$

The definition of $q' \in [1, \infty]$ for $q \in [1, \infty]$ is similar.

Let us introduce the following assumption.

Condition 2.1 *For any $\mathcal{M}$-measurable, nonnegative and bounded function ξ, the following function Ξ (defined on $X_1 \times X_2$) is $\mathcal{M}$-measurable:*

$$\Xi(x_1, x_2) = \Big(\int_{X_2} \xi(x_1, s) d\mu_2(s) \Big)^{\frac{p'}{q'} - 1}, \qquad (x_1, x_2) \in X_1 \times X_2.$$

We have the following Riesz-type representation theorem for the dual of $L^p_{\mathcal{M}}(X_1; L^q(X_2; H))$.

Theorem 2.55. *Let $1 \le p, q < \infty$, and let Condition 2.1 hold. Then, H^* has the Radon-Nikodým property w.r.t. $(X_1 \times X_2, \mathcal{M}, \mu_1 \times \mu_2)$ if and only if for any $F \in L^p_{\mathcal{M}}(X_1; L^q(X_2; H))^*$, there exists a unique $g \in L^{p'}_{\mathcal{M}}(X_1; L^{q'}(X_2; H^*))$ such that for any $f \in L^p_{\mathcal{M}}(X_1; L^q(X_2; H))$,*

$$F(f) = \int_{X_1 \times X_2} \langle f(x_1, x_2), g(x_1, x_2) \rangle_{H, H^*} d\mu_1(x_1) d\mu_2(x_2), \tag{2.31}$$

and

$$|F|_{L^p_{\mathcal{M}}(X_1; L^q(X_2; H))^*} = |g|_{L^{p'}_{\mathcal{M}}(X_1; L^{q'}(X_2; H^*))}. \tag{2.32}$$

Remark 2.56. Due to Theorem 2.55, we may make the following identification (for the case that H^* has the Radon-Nikodým property w.r.t. $(X_1 \times X_2, \mathcal{M}, \mu_1 \times \mu_2)$ and Condition 2.1 holds):

$$L^p_{\mathcal{M}}(X_1; L^q(X_2; H))^* = L^{p'}_{\mathcal{M}}(X_1; L^{q'}(X_2; H^*)), \qquad 1 \le p, q < \infty. \tag{2.33}$$

This indicates that the dual of $L^p_{\mathcal{M}}(X_1; L^q(X_2; H))$ can be identified as $L^{p'}_{\mathcal{M}}(X_1; L^{q'}(X_2; H^*))$.

Remark 2.57. Surprisingly, when Condition 2.1 is dropped, the "only if" part of Theorem 2.55 does not hold (e.g. [234]).

Remark 2.58. From Subsection 2.4.3, it is easy to see that Condition 2.1 is not needed in the "if" part of Theorem 2.55.

Remark 2.59. When $p, q \in (1, \infty)$, a different proof of Theorem 2.55 was given in a recent very interesting and important monograph [151].

Remark 2.60. Theorem 2.55 is of fundamental importance in the sequel of this book. Indeed, this theorem and in particular its consequence (i.e., Theorem 2.73 in Section 2.6) consist in the theoretic basis of our stochastic transposition method.

The proof of Theorem 2.55 will be given in the next three subsections.

2.4.1 Proof of the Necessity for a Special Case

In this subsection, we prove the "only if" part of Theorem 2.55 for the case that $H = \mathbb{R}$.

For any $g \in L^{p'}_{\mathcal{M}}(X_1; L^{q'}(X_2))$, define $F_g : L^p_{\mathcal{M}}(X_1; L^q(X_2)) \to \mathbb{R}$ by

$$F_g(f) = \int_{X_1 \times X_2} f(x_1, x_2) g(x_1, x_2) d\mu_1 d\mu_2, \qquad \forall\, f \in L^p_{\mathcal{M}}(X_1; L^q(X_2)).$$

Clearly, $g \mapsto F_g$ is a linear map. It follows from Hölder's inequality that

$$|F_g(f)| \leq |f|_{L^p_{\mathcal{M}}(X_1; L^q(X_2))} |g|_{L^{p'}_{\mathcal{M}}(X_1; L^{q'}(X_2))}, \quad \forall\, f \in L^p_{\mathcal{M}}(X_1; L^q(X_2)).$$

Hence $F_g \in L^p_{\mathcal{M}}(X_1; L^q(X_2))^*$ and

$$|F_g|_{L^p_{\mathcal{M}}(X_1; L^q(X_2))^*} \leq |g|_{p', q'}. \tag{2.34}$$

Therefore, $g \mapsto F_g$ is a linear non-expanding map. Now, we show that this map is surjective and is an isometry.

To show the surjectivity of $g \mapsto F_g$, take any $F \in L^p_{\mathcal{M}}(X_1; L^q(X_2))^*$. Since $\chi_A \in L^p_{\mathcal{M}}(X_1; L^q(X_2))$ for any $A \in \mathcal{M}$, we may define

$$\nu(A) = F(\chi_A), \qquad \forall\, A \in \mathcal{M}.$$

Then ν is a finite signed measure on $(X_1 \times X_2, \mathcal{M})$, and $\nu \ll \mu_1 \times \mu_2$. By Theorem 2.33, there is a $g \in L^1_{\mathcal{M}}(X_1 \times X_2)$ such that

$$\nu(A) = \int_A g d\mu_1 d\mu_2, \qquad \forall\, A \in \mathcal{M},$$

i.e.,

$$F(\chi_A) = \int_{X_1 \times X_2} g\chi_A d\mu_1 d\mu_2, \qquad \forall\, A \in \mathcal{M}.$$

Consequently, for any $\mathcal{M}$-measurable simple function f,

$$F(f) = \int_{X_1 \times X_2} f(x_1, x_2) g(x_1, x_2) d\mu_1 d\mu_2.$$

Select a sequence $\{A_j\}_{j=1}^{\infty} \subset \mathcal{M}$ such that

$$A_j \subset A_{j+1}, \quad j = 1, 2, \cdots, \quad (\mu_1 \times \mu_2)\Big((X_1 \times X_2) \setminus \bigcup_{j=1}^{\infty} A_j\Big) = 0,$$

and g is bounded on each A_j. For any $j \geq 1$, the following linear functional

$$f \mapsto \int_{X_1 \times X_2} f(x_1, x_2) g(x_1, x_2) \chi_{A_j}(x_1, x_2) d\mu_1 d\mu_2$$

is bounded on $L^p_{\mathcal{M}}(X_1; L^q(X_2))$. This functional agrees with F on all $\mathcal{M}$-measurable simple functions which vanishes off A_j. It follows that

$$F(f\chi_{A_j}) = \int_{X_1 \times X_2} fg\chi_{A_j} d\mu_1 d\mu_2, \qquad \forall\, f \in L^p_{\mathcal{M}}(X_1; L^q(X_2)). \tag{2.35}$$

Since $g\chi_{A_j}$ is bounded, one has $g\chi_{A_j} \in L^{p'}_{\mathcal{M}}(X_1; L^{q'}(X_2))$. We claim that $g \in L^{p'}_{\mathcal{M}}(X_1; L^{q'}(X_2))$, and

$$|g|_{p',q'} \leq |F|_{L^p_{\mathcal{M}}(X_1; L^q(X_2))^*}. \tag{2.36}$$

To show this, without loss of generality, we assume that $g \neq 0$, and distinguish four cases.

Case 1: $p, q \in (1, \infty)$. Choose

$$f = \begin{cases} a\Big(\displaystyle\int_{X_2} |g|^{q'} \chi_{A_j} d\mu_2\Big)^{\frac{p'}{q'}-1} |g|^{q'-1} (\operatorname{sgn} g) \chi_{A_j}, & \text{if } \displaystyle\int_{X_2} |g|^{q'} \chi_{A_j} d\mu_2 \neq 0, \\ 0, & \text{if } \displaystyle\int_{X_2} |g|^{q'} \chi_{A_j} d\mu_2 = 0, \end{cases}$$

where

$$a = \Big[\int_{X_1} \Big(\int_{X_2} |g|^{q'} \chi_{A_j} d\mu_2\Big)^{\frac{p'}{q'}} d\mu_1\Big]^{\frac{1}{p'}-1}.$$

By Condition 2.1, it is easy to see that f is $\mathcal{M}$-measurable. Moreover, $f = f\chi_{A_j}$ and

$$
\begin{aligned}
|f|_{p,q} &= \left[\int_{X_1}\Big(\int_{X_2}|f|^q d\mu_2\Big)^{\frac{p}{q}} d\mu_1\right]^{\frac{1}{p}} \\
&= \left\{\int_{X_1}\left[\int_{X_2} a^q\Big(\int_{X_2}|g|^{q'}\chi_{A_j} d\mu_2\Big)^{\left(\frac{p'}{q'}-1\right)q} |g|^{(q'-1)q}\chi_{A_j} d\mu_2\right]^{\frac{p}{q}} d\mu_1\right\}^{\frac{1}{p}} \\
&= a\left\{\int_{X_1}\left[\Big(\int_{X_2}|g|^{q'}\chi_{A_j} d\mu_2\Big)^{\left(\frac{p'}{q'}-1\right)p}\Big(\int_{X_2}|g|^{q'}\chi_{A_j} d\mu_2\Big)^{\frac{p}{q}}\right] d\mu_1\right\}^{\frac{1}{p}} \\
&= a\left\{\int_{X_1}\Big(\int_{X_2}|g|^{q'}\chi_{A_j} d\mu_2\Big)^{\frac{p'}{q'}} d\mu_1\right\}^{\frac{1}{p}} = 1.
\end{aligned}
$$

Substituting the above f into (2.35), we find that

$$
\begin{aligned}
F(f) &= \int_{X_1}\int_{X_2} fg\chi_{A_j} d\mu_2 d\mu_1 \\
&= a\int_{X_1}\left[\int_{X_2}\Big(\int_{X_2}|g|^{q'}\chi_{A_j} d\mu_2\Big)^{\frac{p'}{q'}-1}|g|^{q'}\chi_{A_j} d\mu_2\right] d\mu_1 \\
&= a\int_{X_1}\Big(\int_{X_2}|g|^{q'}\chi_{A_j} d\mu_2\Big)^{\frac{p'}{q'}} d\mu_1 = \left[\int_{X_1}\Big(\int_{X_2}|g|^{q'}\chi_{A_j} d\mu_2\Big)^{\frac{p'}{q'}} d\mu_1\right]^{\frac{1}{p'}} \\
&= |g\chi_{A_j}|_{p',q'},
\end{aligned}
\tag{2.37}
$$

which gives

$$
|g\chi_{A_j}|_{p',q'} \le |F|_{L^p_{\mathcal{M}}(X_1;L^q(X_2))^*}. \tag{2.38}
$$

Letting $j \to \infty$ in (2.38), and by making use of Fatou's lemma (or using the Monotone Convergence Theorem), one concludes (2.36).

Case 2: $p = 1$, $1 < q < \infty$. In this case, we first take $p \in (1, \infty)$, and take f as in Case 1. Then

$$
\begin{aligned}
|f|_{1,q} &= \int_{X_1}\Big(\int_{X_2}|f|^q d\mu_2\Big)^{\frac{1}{q}} d\mu_1 \le \left[\int_{X_1}\Big(\int_{X_2}|f|^q d\mu_2\Big)^{\frac{p}{q}} d\mu_1\right]^{\frac{1}{p}} \mu_1(X_1)^{\frac{1}{p'}} \\
&= \mu_1(X_1)^{\frac{1}{p'}}.
\end{aligned}
$$

Consequently, by (2.37),

$$
\begin{aligned}
|g\chi_{A_j}|_{p',q'} = F(f) &\le |F|_{L^1_{\mathcal{M}}(X_1;L^q(X_2))^*}|f|_{1,q} \\
&\le |F|_{L^1_{\mathcal{M}}(X_1;L^q(X_2))^*}\mu_1(X_1)^{\frac{1}{p'}}.
\end{aligned}
\tag{2.39}
$$

Letting $j \to \infty$ and then letting $p \to 1$ (which means $p' \to \infty$) in (2.39), we obtain

$$
|g|_{\infty,q'} \le |F|_{L^1_{\mathcal{M}}(X_1;L^q(X_2))^*},
$$

which is (2.36) for the case $p = 1$.

Case 3: $1 < p < \infty$, $q = 1$. In this case, we first take $q \in (1, \infty)$, and take f as in Case 1. Then

$$\begin{aligned}|f|_{p,1} &= \left[\int_{X_1}\left(\int_{X_2}|f|d\mu_2\right)^p d\mu_1\right]^{\frac{1}{p}}\\ &\le \left[\int_{X_1}\left(\int_{X_2}|f|^q d\mu_2\right)^{\frac{p}{q}} d\mu_1\right]^{\frac{1}{p}} \mu_2(X_2)^{\frac{1}{q'}} = \mu_2(X_2)^{\frac{1}{q'}}.\end{aligned}$$

Hence, by (2.37),

$$\begin{aligned}|g\chi_{A_j}|_{p',q'} &= F(f) \le |F|_{L^p_{\mathcal{M}}(X_1;L^1(X_2))^*}|f|_{p,1}\\ &\le |F|_{L^p_{\mathcal{M}}(X_1;L^1(X_2))^*}\mu_2(X_1)^{\frac{1}{q'}}\end{aligned} \tag{2.40}$$

Letting $j \to \infty$ and then letting $q \to 1$ (which means $q' \to \infty$) in (2.40), we obtain

$$|g|_{p',\infty} \le |F|_{L^p_{\mathcal{M}}(X_1;L^1(X_2))^*},$$

which is (2.36) for the case $q = 1$.

Case 4: $p = q = 1$. In this case, we still first let $p, q \in (1, \infty)$, and take f as in Case 1. Then

$$\begin{aligned}|f|_{1,1} &= \int_{X_1}\int_{X_2}|f|d\mu_2 d\mu_1 \le \left[\int_{X_1}\left(\int_{X_2}|f|^q d\mu_2\right)^{\frac{p}{q}} d\mu_1\right]^{\frac{1}{p}} \mu_1(X_1)^{\frac{1}{p'}}\mu_2(X_2)^{\frac{1}{q'}}\\ &= \mu_1(X_1)^{\frac{1}{p'}}\mu_2(X_2)^{\frac{1}{q'}}.\end{aligned}$$

Consequently, by (2.37),

$$\begin{aligned}|g\chi_{A_j}|_{p',q'} &= F(f) \le |F|_{L^1_{\mathcal{M}}(X_1;L^1(X_2))^*}|f|_{1,1}\\ &\le |F|_{L^1_{\mathcal{M}}(X_1;L^1(X_2))^*}\mu_1(X_1)^{\frac{1}{p'}}\mu_2(X_1)^{\frac{1}{q'}}.\end{aligned} \tag{2.41}$$

Letting $j \to \infty$ and then letting $p, q \to 1$ (which means $p', q' \to \infty$) in (2.41), we obtain

$$|g|_{\infty,\infty} \le |F|_{L^1_{\mathcal{M}}(X_1;L^1(X_2))^*},$$

which is (2.36) for the case $p, q = 1$.

Finally, (2.34) means that $F_g \in (L^p_{\mathcal{M}}(X_1; L^q(X_2)))^*$ and since F and F_g coincide on a dense subset of $L^p_{\mathcal{M}}(X_1; L^q(X_2))$, we see that $F = F_g$. Also, (2.32) follows easily from (2.34) and (2.36).

2.4.2 Proof of the Necessity for the General Case

In this subsection, we prove the "only if" part of Theorem 2.55 for the general case. The proof is divided into two steps.

Step 1. We show that $L^{p'}_{\mathcal{M}}(X_1; L^{q'}(X_2; H^*))$ is isometrically isomorphic to a subspace $\mathcal{H}$ of $L^p_{\mathcal{M}}(X_1; L^q(X_2; H))^*$.

For any given $g \in L^{p'}_{\mathcal{M}}(X_1; L^{q'}(X_2; H^*))$, define a linear functional F_g on $L^p_{\mathcal{M}}(X_1; L^q(X_2; H))$ as follows:

$$F_g(f) = \int_{X_1 \times X_2} \langle f(x_1, x_2), g(x_1, x_2) \rangle_{H,H^*} d\mu_1 d\mu_2,$$
$$\forall\, f \in L^p_{\mathcal{M}}(X_1; L^q(X_2; H)).$$

Then, by means of the Hölder inequality and similarly to (2.34), we conclude that F_g belongs to $L^p_{\mathcal{M}}(X_1; L^q(X_2; H))^*$, and

$$|F_g|_{L^p_{\mathcal{M}}(X_1; L^q(X_2; H))^*} \leq |g|_{p',q';H^*}. \tag{2.42}$$

Therefore the norm of F_g is not greater than $|g|_{p',q';H^*}$. Define

$$\mathcal{H} \equiv \big\{ F_g \mid g \in L^{p'}_{\mathcal{M}}(X_1; L^{q'}(X_2; H^*)) \big\}. \tag{2.43}$$

It remains to prove the reverse of inequality (2.42). Clearly, without loss of generality, we may assume that $g \neq 0$.

Suppose first that $g = \sum_{i=1}^{\infty} h_i^* \chi_{E_i}$ where h_i^* is a sequence in H^* and $\{E_i\}_{i=1}^{\infty}$ is a sequences of mutually disjoint sets from $\mathcal{M}$ such that $(\mu_1 \times \mu_2)(E_i) > 0$ for all i and $X_1 \times X_2 = \bigcup_{i=1}^{\infty} E_i$. Since it has been shown that $L^p_{\mathcal{M}}(X_1; L^q(X_2))^* = L^{p'}_{\mathcal{M}}(X_1; L^{q'}(X_2))$ (in the last subsection) and noting that $0 < |g|_{H^*} \in L^{p'}_{\mathcal{M}}(X_1; L^{q'}(X_2))$, we see that for any $\varepsilon > 0$, there exists a nonnegative function $\varphi \in L^p_{\mathcal{M}}(X_1; L^q(X_2))$ such that

$$0 < |\varphi|_{p,q} \leq 1, \qquad |g|_{p',q';H^*} - \varepsilon \leq \int_{X_1 \times X_2} |g|_{H^*} \varphi d\mu_1 d\mu_2.$$

Further, choose an $h_i \in H$ with $|h_i|_H = 1$ such that $\langle h_i, h_i^* \rangle_{H,H^*}$ is nonnegative and

$$|h_i^*|_{H^*} - \frac{\varepsilon}{|\varphi|_{1,1}} \leq \langle h_i, h_i^* \rangle_{H,H^*},$$

and define

$$f = \sum_{i=1}^{\infty} \chi_{E_i} \varphi h_i \in L^p_{\mathcal{M}}(X_1; L^q(X_2; H)).$$

It is easy to check that $|f(\cdot)|_H = \varphi(\cdot)$ in $X_1 \times X_2$. Hence, $|f|_{p,q;H} = |\varphi|_{p,q} \leq 1$, and

$$\begin{aligned}
&\int_{X_1\times X_2} \langle f(x_1,x_2), g(x_1,x_2)\rangle_{H,H^*} d\mu_1 d\mu_2 \\
&= \int_{X_1\times X_2} \varphi \sum_{i=1}^{\infty} \langle h_i, h_i^* \rangle_{H,H^*} \chi_{E_i} d\mu_1 d\mu_2 \\
&\geq \int_{X_1\times X_2} \varphi \sum_{i=1}^{\infty} \Big(|h_i^*|_{H^*} - \frac{\varepsilon}{|\varphi|_{1,1}} \Big) \chi_{E_i} d\mu_1 d\mu_2 \\
&\geq \int_{X_1\times X_2} |g|_{H^*} \varphi d\mu_1 d\mu_2 - \frac{\varepsilon}{|\varphi|_{1,1}} \int_{X_1\times X_2} \varphi d\mu_1 d\mu_2 \geq |g|_{p',q';H^*} - 2\varepsilon.
\end{aligned}$$

This gives

$$|F_g|_{L^p_{\mathcal{M}}(X_1;L^q(X_2;H))^*} \geq |g|_{p',q';H^*},$$

and therefore

$$|F_g|_{L^p_{\mathcal{M}}(X_1;L^q(X_2;H))^*} = |g|_{p',q';H^*},$$

whenever $g \in L^{p'}_{\mathcal{M}}(X_1; L^{q'}(X_2; H^*))$ is countably valued.

For the general case, by Proposition 2.12, we may choose a sequence $\{g_j\}_{j=1}^{\infty} \subset L^{p'}_{\mathcal{M}}(X_1; L^{q'}(X_2; H^*))$ such that each g_j is countably valued and

$$\lim_{j\to\infty} |g_j - g|_{p',q';H^*} = 0. \tag{2.44}$$

We have obtained that

$$|F_{g_j}|_{L^p_{\mathcal{M}}(X_1;L^q(X_2;H))^*} = |g_j|_{p',q';H^*}, \tag{2.45}$$

and by virtue of (2.42),

$$|F_{g_j} - F_g|_{L^p_{\mathcal{M}}(X_1;L^q(X_2;H))^*} = |F_{g_j-g}|_{L^p_{\mathcal{M}}(X_1;L^q(X_2;H))^*} \leq |g_j - g|_{p',q';H^*}.$$

Therefore, noting (2.44) and (2.45), we end up with

$$\begin{aligned}
\Big|F_g\Big|_{L^p_{\mathcal{M}}(X_1;L^q(X_2;H))^*} &= \lim_{j\to\infty} |F_{g_j}|_{L^p_{\mathcal{M}}(X_1;L^q(X_2;H))^*} = \lim_{j\to\infty} |g_j|_{p',q';H^*} \\
&= |g|_{p',q';H^*}.
\end{aligned}$$

Hence, $L^{p'}_{\mathcal{M}}(X_1; L^{q'}(X_2; H^*))$ is isometrically isomorphic to $\mathcal{H}$.

Step 2. We show that the subspace $\mathcal{H}$ (defined in (2.43)) is equal to $L^p_{\mathcal{M}}(X_1; L^q(X_2; H))^*$.

To this end, for $F \in L^p_{\mathcal{M}}(X_1; L^q(X_2; H))^*$, we define

$$G(E)(h) = F(h\chi_E), \qquad \forall\, E \in \mathcal{M},\ h \in H. \tag{2.46}$$

By

$$\begin{aligned}|F(h\chi_E)| &\leq |F|_{L^p_{\mathcal{M}}(X_1;L^q(X_2;H))^*}|h\chi_E|_{p,q;H}\\ &\leq |F|_{L^p_{\mathcal{M}}(X_1;L^q(X_2;H))^*}|h|_H|\chi_E|_{p,q},\end{aligned}$$

we see that $G : \mathcal{M} \to H^*$ and it is countably additive. Let $E_1, \cdots, E_k$ ($k \in \mathbb{N}$) be mutually disjoint sets from $\mathcal{M}$ such that $X_1 \times X_2 = \bigcup_{i=1}^k E_i$. Then $G(E_i) \in H^*$. For any $\varepsilon > 0$, one can find an $h_i \in H$ with $|h_i|_H = 1$ such that $G(E_i)(h_i)$ is nonnegative and

$$|G(E_i)|_{H^*} - \frac{\varepsilon}{k} < G(E_i)(h_i).$$

It follows that

$$\begin{aligned}&\sum_{i=1}^k |G(E_i)|_{H^*} - \varepsilon < \sum_{i=1}^k G(E_i)(h_i) = F\Big(\sum_{i=1}^k h_i\chi_{E_i}\Big)\\ &\leq |F|_{L^p_{\mathcal{M}}(X_1;L^q(X_2;H))^*}\Big|\sum_{i=1}^k h_i\chi_{E_i}\Big|_{p,q;H}\\ &\leq |F|_{L^p_{\mathcal{M}}(X_1;L^q(X_2;H))^*}\mu_1(X_1)^{\frac{1}{p}}\mu_2(X_2)^{\frac{1}{q}}.\end{aligned}$$

Hence $|G|_{H^*}(X_1 \times X_2) < \infty$ and G is a $\mu_1 \times \mu_2$-continuous vector-valued measure of bounded variation. Since H^* has the Radon-Nikodým property w.r.t. $(X_1 \times X_2, \mathcal{M}, \mu_1 \times \mu_2)$, there exists a Bochner integrable function $g : X_1 \times X_2 \to H^*$ such that

$$G(E) = \int_E g d\mu_1 d\mu_2, \qquad \forall\, E \in \mathcal{M}. \tag{2.47}$$

Clearly, if $f \in L^p_{\mathcal{M}}(X_1; L^q(X_2; H))$ is a simple function, then

$$F(f) = \int_{X_1\times X_2} \langle f(x_1, x_2), g(x_1, x_2)\rangle_{H,H^*} d\mu_1 d\mu_2.$$

Select an expanding sequence $\{E_j\}_{j=1}^\infty$ in $\mathcal{M}$ such that $\bigcup_{j=1}^\infty E_j = X_1 \times X_2$ and such that g is bounded on each E_j. Fixing an arbitrary $j_0 \in \mathbb{N}$ and noting that $\int_{E_{j_0}} \langle \cdot, g(x_1, x_2)\rangle_{H,H^*} d\mu_1 d\mu_2$ is a bounded linear functional (on $L^p_{\mathcal{M}}(X_1; L^q(X_2; H))$), which agrees with F on all simple functions supported on E_{j_0}, we deduce that

$$\begin{aligned}F(f\chi_{E_{j_0}}) = \int_{X_1\times X_2} \langle f(x_1, x_2), g(x_1, x_2)\chi_{E_{j_0}}\rangle_{H,H^*} d\mu_1 d\mu_2,\\ \forall\, f \in L^p_{\mathcal{M}}(X_1; L^q(X_2; H)).\end{aligned} \tag{2.48}$$

Further, since $g\chi_{E_{j_0}}$ is bounded, one has $g\chi_{E_{j_0}} \in L^{p'}_{\mathcal{M}}(X_1; L^{q'}(X_2; H^*))$. By the result proved in Step 1 and (2.48), it follows that

$$|g\chi_{E_{j_0}}|_{p',q';H^*} = |F_{g\chi_{E_{j_0}}}|_{L^p_{\mathcal{M}}(X_1;L^q(X_2;H))^*} \leq |F|_{L^p_{\mathcal{M}}(X_1;L^q(X_2;H))^*}. \qquad (2.49)$$

Since the inequality (2.49) holds for each j_0, by the Monotone Convergence Theorem, we conclude that $g \in L^{p'}_{\mathcal{M}}(X_1; L^{q'}(X_2; H^*))$.

Finally, for any $f \in L^p_{\mathcal{M}}(X_1; L^q(X_2; H))$, it follows from (2.48) that

$$\begin{aligned} F(f) &= \lim_{j\to\infty} \int_{X_1\times X_2} \langle f(x_1,x_2), g(x_1,x_2)\chi_{E_j}\rangle_{H,H^*} d\mu_1 d\mu_2 \\ &= \int_{X_1\times X_2} \langle f(x_1,x_2), g(x_1,x_2)\rangle_{H,H^*} d\mu_1 d\mu_2 = F_g(f). \end{aligned}$$

This means that $F = F_g$. Hence $L^p_{\mathcal{M}}(X_1; L^q(X_2; H))^*$ coincides with the space $L^{p'}_{\mathcal{M}}(X_1; L^{q'}(X_2; H^*))$.

2.4.3 Proof of the Sufficiency

In this subsection, we prove the "if" part of Theorem 2.55.

Let $G : \mathcal{M} \to H^*$ be a $\mu_1 \times \mu_2$-continuous vector measure of bounded variation. We shall show that there exists a $\tilde{g} \in L^1_{\mathcal{M}}(X_1 \times X_2; H^*)$ such that

$$G(E) = \int_E \tilde{g} d\mu_1 d\mu_2, \qquad \forall\ E \in \mathcal{M}. \qquad (2.50)$$

By Theorem 2.43, the variation $|G|$ of G is a finite measure on $(X_1 \times X_2, \mathcal{M})$. It is easy to see that $|G|$ is a $\mu_1\times\mu_2$-continuous ($\mathbb{R}^+$-valued) measure. Applying the Radon–Nikodým Theorem, i.e. Theorem 2.33 (to $|G|$ and $\mu_1 \times \mu_2$), we can find a nonnegative function $k(\cdot) \in L^1_{\mathcal{M}}(X_1 \times X_2)$ such that

$$|G|(A) = \int_A k(x_1,x_2) d\mu_1 d\mu_2, \qquad \forall\ A \in \mathcal{M}. \qquad (2.51)$$

Clearly, there exists a sequence $\{A_j\}_{j=1}^{\infty}$ of mutually disjoint members from $\mathcal{M}$ so that $\bigcup_{j=1}^{\infty} A_j = X_1 \times X_2$, and $k(\cdot)$ is bounded by a constant $k_j > 0$ on each A_j. It follows from (2.51) that

$$|G|(A) \leq k_j(\mu_1 \times \mu_2)(A), \qquad \forall\ A \in \mathcal{M} \text{ with } A \subset A_j. \qquad (2.52)$$

For each $j \in \mathbb{N}$, define a linear functional ℓ_j on the subspace $\mathcal{S}$ of $\mathcal{M}$-simple functions in $L^p_{\mathcal{M}}(X_1; L^q(X_2; H))$ as follows:

$$\ell_j(f) = \sum_{i=1}^{i_0} G(E_i \cap A_j)(x_i),$$

where $i_0 \in \mathbb{N}$, and $f = \sum_{i=1}^{i_0} x_i\chi_{E_i}$ for some $x_i \in H$ and mutually disjoint sets $E_1, \cdots E_{i_0}$ from $\mathcal{M}$ so that $\bigcup_{i=1}^{i_0} E_i = X_1 \times X_2$. Then, by (2.52),

$$
\begin{aligned}
|\ell_j(f)| &= \Big|\sum_{i=1}^{i_0} G(E_i \cap A_j)(x_i)\Big| \\
&\le \sum_{i=1}^{i_0} |G(E_i \cap A_j)|_{H^*}|x_i|_H \le \sum_{i=1}^{i_0} |G|(E_i \cap A_j)|x_i|_H \\
&\le k_j \sum_{i=1}^{i_0} (\mu_1 \times \mu_2)(E_i \cap A_j)|x_i|_H \le k_j |f|_{L^1(X_1\times X_2;H)} \\
&\le k_j \mu_1(X_1)^{\frac{1}{p}} \mu_2(X_2)^{\frac{1}{q}} |f|_{L^p_{\mathcal{M}}(X_1;L^q(X_2;H))}.
\end{aligned}
$$

Thus, ℓ_j is a bounded linear functional on $\mathcal{S}$. By the Hahn-Banach Theorem, ℓ_j can be extended to a bounded linear functional on $L^p_{\mathcal{M}}(X_1; L^q(X_2; H))$ (The extension is still denoted by ℓ_j). Hence there exists a $g_j \in L^{p'}_{\mathcal{M}}(X_1; L^{q'}(X_2; H^*))$ such that

$$
\ell_j(f) = \int_{X_1\times X_2} \langle f, g_j\rangle_{H,H^*} d\mu_1 d\mu_2, \quad \forall\, f \in L^p_{\mathcal{M}}(X_1; L^q(X_2; H)).
$$

We have

$$
G(E \cap A_j)(x) = \ell_j(x\chi_E) = \int_E \langle x, g_j\rangle_{H,H^*} d\mu_1 d\mu_2, \qquad \forall\, x \in H,\ E \in \mathcal{M}.
$$

Since $g_j \in L^{p'}_{\mathcal{M}}(X_1; L^{q'}(X_2; H^*))$ is Bochner integrable, we see that

$$
G(E \cap A_j)(x) = \Big(\int_E g_j d\mu_1 d\mu_2\Big)(x), \qquad \forall\, x \in H,\ E \in \mathcal{M}.
$$

Consequently,

$$
G(E \cap A_j) = \int_E g_j d\mu_1 d\mu_2, \qquad \forall\, E \in \mathcal{M}. \tag{2.53}
$$

Noting that $A_j \in \mathcal{M}$, and therefore replacing E in (2.53) by $E \cap A_j$, we see that

$$
G(E \cap A_j) = \int_{E\cap A_j} g_j d\mu_1 d\mu_2, \qquad \forall\, E \in \mathcal{M},\ j \in \mathbb{N}.
$$

Define $\tilde{g} : X_1 \times X_2 \to H^*$ by $\tilde{g}(x_1, x_2) = g_j(x_1, x_2)$ if $(x_1, x_2) \in A_j$. It is obvious that $\tilde{g}$ is strongly $\mathcal{M}$-measurable w.r.t. $\mu_1 \times \mu_2$. Moreover, for each $E \in \mathcal{M}$ and all $k \in \mathbb{N}$, it holds

$$
G\Big(E \bigcap \Big(\bigcup_{j=1}^{k} A_j\Big)\Big) = \int_{E\cap(\cup_{j=1}^k A_j)} \tilde{g} d\mu_1 d\mu_2 = \int_E \tilde{g}\chi_{\cup_{j=1}^k A_j} d\mu_1 d\mu_2. \tag{2.54}
$$

Consequently,

$$G(E) = \lim_{k\to\infty} \int_E \tilde{g}\chi_{\cup_{j=1}^k A_j} d\mu_1 d\mu_2, \qquad \forall\, E \in \mathcal{M}. \tag{2.55}$$

It remains to show that $\tilde{g} \in L^1_{\mathcal{M}}(X_1 \times X_2; H^*)$. Since $|G|(X_1 \times X_2)$ is finite, by (2.54) and using Proposition 2.44, we see that

$$\int_{X_1\times X_2} |\tilde{g}|_{H^*}\chi_{\cup_{j=1}^k A_j} d\mu_1 d\mu_2 \le |G|(X_1 \times X_2), \qquad \forall\, k \in \mathbb{N}.$$

By the Monotone Convergence Theorem, $|\tilde{g}|_{H^*} \in L^1_{\mathcal{M}}(X_1 \times X_2)$. Hence $\tilde{g}$ is Bochner integrable. By (2.55) and Theorem 2.18, we obtain (2.50), which proves the Radon-Nikodým property of H^* w.r.t. $(X_1 \times X_2, \mathcal{M}, \mu_1 \times \mu_2)$.

2.5 A Sequential Banach-Alaoglu-Type Theorem in the Operator Version

The classical Banach-Alaoglu Theorem (e.g. [60, p. 130]) states that the closed unit ball of the dual space of a normed vector space is compact in the weak* topology. This theorem has an important special (sequential) version, asserting that the closed unit ball of the dual space of a separable normed vector space (*resp.*, the closed unit ball of a reflexive Banach space) is sequentially compact in the weak* topology (*resp.*, the weak topology). In this section, we shall present a sequential Banach-Alaoglu-type theorem for uniformly bounded linear operators (between suitable Banach spaces). As we shall see later, this result will play crucial roles in the study of the optimal control problems for stochastic evolution equations.

Let X and Y be two Banach spaces. Let $\{y_n\}_{n=1}^\infty \subset Y$, $y \in Y$, $\{z_n\}_{n=1}^\infty \subset Y^*$ and $z \in Y^*$. In the sequel, we denote by

$$\text{(w)-}\lim_{n\to\infty} y_n = y \quad \text{in } Y$$

when $\{y_n\}_{n=1}^\infty$ weakly converges to y in Y; and by

$$\text{(w*)-}\lim_{n\to\infty} z_n = z \quad \text{in } Y^*$$

when $\{z_n\}_{n=1}^\infty$ weakly* converges to z in Y^*. Let us show first the following simple result:

Lemma 2.61. *Let X be a separable Banach space and let Y be a reflexive Banach space. Assume that $\{G_n\}_{n=1}^\infty \subset \mathcal{L}(X;Y)$ is a sequence of bounded linear operators such that $\{G_n x\}_{n=1}^\infty$ is bounded in Y for any given $x \in X$. Then, there exist a subsequence $\{G_{n_k}\}_{k=1}^\infty$ and a $G \in \mathcal{L}(X;Y)$ satisfying that*

$$\text{(w)-}\lim_{k\to\infty} G_{n_k} x = Gx \quad \textit{in } Y, \qquad \forall\, x \in X,$$

$$(\text{w}^*)\text{-}\lim_{k\to\infty} G^*_{n_k} y^* = G^* y^* \quad in \ X^*, \qquad \forall\, y^* \in Y^*,$$

and

$$|G|_{\mathcal{L}(X;Y)} \le \sup_{n\in\mathbb{N}} |G_n|_{\mathcal{L}(X;Y)} (< \infty). \tag{2.56}$$

Proof: Noting that X is separable, we can find a countable subset $\{x_i\}_{i=1}^\infty$ of X such that $\{x_1, x_2, \cdots\}$ is dense in X. Since $\{G_n x_1\}_{n=1}^\infty$ is bounded in Y and Y is reflexive, there exists a subsequence $\{n_k^{(1)}\}_{k=1}^\infty \subset \{n\}_{n=1}^\infty$ such that (w)-$\lim_{k\to\infty} G_{n_k^{(1)}} x_1 = y_1$. Now, the sequence $\{G_{n_k^{(1)}} x_2\}_{k=1}^\infty$ is still bounded in Y, one can find a subsequence $\{n_k^{(2)}\}_{k=1}^\infty \subset \{n_k^{(1)}\}_{k=1}^\infty$ such that (w)-$\lim_{k\to\infty} G_{n_k^{(2)}} x_2 = y_2$. Generally, for each $m \in \mathbb{N}$, we can find a subsequence $\{n_k^{(m+1)}\}_{k=1}^\infty \subset \{n_k^{(m)}\}_{k=1}^\infty$ such that (w)-$\lim_{k\to\infty} G_{n_k^{(m+1)}} x_{m+1} = y_{m+1}$. We now use the classical diagonalisation argument. Write $n_m = n_m^{(m)}$, $m = 1, 2, \cdots$. Then, $\{G_{n_m} x_i\}_{m=1}^\infty$ converges weakly to y_i in Y.

Let us define an operator G (from X to Y) as follows: For any $x \in X$,

$$Gx = \lim_{k\to\infty} y_{i_k} = \lim_{k\to\infty} \Big((\text{w})\text{-}\lim_{m\to\infty} G_{n_m} x_{i_k} \Big),$$

where $\{x_{i_k}\}_{k=1}^\infty$ is any subsequence of $\{x_i\}_{i=1}^\infty$ such that $\lim_{k\to\infty} x_{i_k} = x$ in X. We shall show below that $G \in \mathcal{L}(X;Y)$.

First, we show that G is well-defined. By the Principle of Uniform Boundedness, it is clear that $\{G_n\}_{n=1}^\infty$ is uniformly bounded in $\mathcal{L}(X;Y)$. We choose an $M > 0$ such that $|G_n|_{\mathcal{L}(X;Y)} \le M$ for all $n \in \mathbb{N}$. Since $\{x_{i_k}\}_{k=1}^\infty$ is a Cauchy sequence in X, for any $\varepsilon > 0$, there is an $N > 0$ such that $|x_{i_{k_1}} - x_{i_{k_2}}|_X < \frac{\varepsilon}{M}$ whenever $k_1, k_2 > N$. Hence, $|G_n(x_{i_{k_1}} - x_{i_{k_2}})|_Y < \varepsilon$ for any $n \in \mathbb{N}$. Then, by the weakly sequentially lower semicontinuity (on the norms of Banach spaces), we deduce that

$$|y_{i_{k_1}} - y_{i_{k_2}}|_Y \le \varliminf_{m\to\infty} |G_{n_m}(x_{i_{k_1}} - x_{i_{k_2}})|_Y < \varepsilon,$$

which implies that $\{y_{i_k}\}_{k=1}^\infty$ is a Cauchy sequence in Y. Therefore, $\lim_{k\to\infty} y_{i_k}$ exists in Y. On the other hand, assume that there is another subsequence $\{x'_{i_k}\}_{k=1}^\infty \subset \{x_i\}_{i=1}^\infty$ such that $\lim_{k\to\infty} x'_{i_k} = x$. Let y'_{i_k} be the corresponding weak limit of $G_{n_m} x'_{i_k}$ in Y for $m \to \infty$. Then,

$$\begin{aligned}
&\Big| \lim_{k\to\infty} y_{i_k} - \lim_{k\to\infty} y'_{i_k} \Big|_Y \\
&\le \lim_{k\to\infty} \varliminf_{m\to\infty} |G_{n_m}(x_{i_k} - x'_{i_k})|_Y \le M \lim_{k\to\infty} |x_{i_k} - x'_{i_k}|_X \\
&\le M \lim_{k\to\infty} |x_{i_k} - x|_X + M \lim_{k\to\infty} |x - x'_{i_k}|_X = 0.
\end{aligned}$$

Hence, G is well-defined.

Next, we prove that G is a bounded linear operator. For any $x \in X$ and the above sequence $\{x_{i_k}\}_{k=1}^{\infty}$, it follows that

$$|Gx|_Y = \lim_{k\to\infty} |y_{i_k}|_Y \le \lim_{k\to\infty} \varliminf_{m\to\infty} |G_{n_m} x_{i_k}|_Y \le M \lim_{k\to\infty} |x_{i_k}|_X \le M|x|_X.$$

Hence, G is a bounded operator. Further, it is easy to see that G is linear. Therefore, $G \in \mathcal{L}(X;Y)$.

Also, for any $x \in X$ and $y^* \in Y^*$, it holds that

$$(x, G^*y^*)_{X,X^*} = (Gx, y^*)_{Y,Y^*} = \lim_{k\to\infty} (G_{n_k}x, y^*)_{Y,Y^*} = \lim_{k\to\infty} (x, G^*_{n_k}y^*)_{X,X^*}.$$

Hence,

$$(\text{w}^*)\text{-}\lim_{k\to\infty} G^*_{n_k}y^* = G^*y^* \text{ in } X^*.$$

Finally, from the above proof, (2.56) is obvious. This completes the proof of Lemma 2.61. □

Remark 2.62. Lemma 2.61 is not a direct consequence of the classical sequential Banach-Alaoglu Theorem. Indeed, the Banach space $\mathcal{L}(X;Y)$ is neither reflexive nor separable even if both X and Y are (infinite dimensional) separable Hilbert spaces (See Problem 99 in [135]).

Let $(\Omega_1, \mathcal{M}_1, \mu_1)$ and $(\Omega_2, \mathcal{M}_2, \mu_2)$ be two finite measure spaces. Let $\mathcal{M}$ be a sub-σ-field of $\mathcal{M}_1 \times \mathcal{M}_2$. For any $1 \le p, q \le \infty$, we define a Banach space $L^p_{\mathcal{M}}(\Omega_1; L^q(\Omega_2; X))$ as that in (2.25)–(2.28). Fix any $r_1, r_2, r_3, r_4 \in [1, \infty]$, we denote by $\mathcal{L}_{pd}\big(L^{r_1}_{\mathcal{M}}(\Omega_1; L^{r_2}(\Omega_2; X)); L^{r_3}_{\mathcal{M}}(\Omega_1; L^{r_4}(\Omega_2;\ Y))\big)$ (*resp.* $\mathcal{L}_{pd}\big(X; L^{r_3}_{\mathcal{M}}(\Omega_1; L^{r_4}(\Omega; Y))\big)$) the vector space of all bounded, pointwise defined linear operators $\mathcal{L}$ from $L^{r_1}_{\mathcal{M}}(\Omega_1; L^{r_2}(\Omega_2; X))$ (*resp.* X) to $L^{r_3}_{\mathcal{M}}(\Omega_1; L^{r_4}(\Omega_2; Y))$, i.e., for a.e. $(\omega_1, \omega_2) \in \Omega_1 \times \Omega_2$, there exists an $L(\omega_1, \omega_2) \in \mathcal{L}(X;Y)$ satisfying that $\big(\mathcal{L}u(\cdot)\big)(\omega_1, \omega_2) = L(\omega_1, \omega_2)u(\omega_1, \omega_2), \forall\, u(\cdot) \in L^{r_1}_{\mathcal{M}}(\Omega_1; L^{r_2}(\Omega_2; X))$ (*resp.* $(\mathcal{L}x)(\omega_1, \omega_2) = L(\omega_1, \omega_2)x, \ \forall\ x \in X$). In a similar way, one can define the spaces $\mathcal{L}_{pd}\big(L^{r_1}_{\mathcal{M}_1}(\Omega_1; X); L^{r_2}_{\mathcal{M}}(\Omega_1; L^{r_3}(\Omega_2; Y))\big)$ and $\mathcal{L}_{pd}\big(L^{r_1}_{\mathcal{M}_1}(\Omega_1; X); L^{r_2}_{\mathcal{M}_2}(\Omega_2; Y)\big)$, etc. In the sequel, in the case that no confusion would occur, sometimes we identify $\mathcal{L}$ with $L(\cdot, \cdot)$.

We now show the following sequential Banach-Alaoglu-type theorem in the operator version.

Theorem 2.63. *Let X be a separable Banach space, and let Y be a reflexive Banach space. Let $1 \le p_1, p_2 < \infty$ and $1 < q_1, q_2 < \infty$, and let $L^{p_1}_{\mathcal{M}}(\Omega_1; L^{p_2}(\Omega_2; \mathbb{C}))$ be separable. Let Condition 2.1 hold with X_1, X_2, p and q being replaced respectively by Ω_1, Ω_2, q_1 and q_2. Assume that $\{\mathcal{G}_n\}_{n=1}^{\infty} \subset \mathcal{L}_{pd}\big(L^{p_1}_{\mathcal{M}}(\Omega_1;\ L^{p_2}(\Omega_2; X)); L^{q_1}_{\mathcal{M}}(\Omega_1; L^{q_2}(\Omega_2; Y))\big)$ are uniformly bounded. Then, there exist a subsequence $\{\mathcal{G}_{n_k}\}_{k=1}^{\infty} \subset \{\mathcal{G}_n\}_{n=1}^{\infty}$ and a $\mathcal{G} \in \mathcal{L}_{pd}\big(L^{p_1}_{\mathcal{M}}(\Omega_1; L^{p_2}(\Omega_2; X)); L^{q_1}_{\mathcal{M}}(\Omega_1; L^{q_2}(\Omega_2; Y))\big)$ such that, for all $u(\cdot) \in L^{p_1}_{\mathcal{M}}(\Omega_1; L^{p_2}(\Omega_2; X))$,*

$$\mathcal{G}u(\cdot) = (\mathrm{w})\text{-}\lim_{k\to\infty} \mathcal{G}_{n_k} u(\cdot) \quad \textit{in } L^{q_1}_{\mathcal{M}}(\Omega_1; L^{q_2}(\Omega_2; Y)).$$

Moreover,

$$\begin{aligned}&|\mathcal{G}|_{\mathcal{L}(L^{p_1}_{\mathcal{M}}(\Omega_1;L^{p_2}(\Omega_2;X));L^{q_1}_{\mathcal{M}}(\Omega_1;L^{q_2}(\Omega_2;Y)))}\\&\leq \sup_{n\in\mathbb{N}} |\mathcal{G}_n|_{\mathcal{L}(L^{p_1}_{\mathcal{M}}(\Omega_1;L^{p_2}(\Omega_2;X));L^{q_1}_{\mathcal{M}}(\Omega_1;L^{q_2}(\Omega_2;Y)))}.\end{aligned}$$

Proof: We divide the proof into several steps.

Step 1. Since $\mathcal{G}_n \in \mathcal{L}_{pd}\big(L^{p_1}_{\mathcal{M}}(\Omega_1; L^{p_2}(\Omega_2; X)); L^{q_1}_{\mathcal{M}}(\Omega_1; L^{q_2}(\Omega_2; Y))\big)$ (for each $n \in \mathbb{N}$), for a.e. $(\omega_1, \omega_2) \in \Omega_1 \times \Omega_2$, there exists an $G_n(\omega_1, \omega_2) \in \mathcal{L}(X; Y)$ satisfying that, for all $u(\cdot) \in L^{p_1}_{\mathcal{M}}(\Omega_1; L^{p_2}(\Omega_2; X))$,

$$\big(\mathcal{G}_n u(\cdot)\big)(\omega_1, \omega_2) = G_n(\omega_1, \omega_2) u(\omega_1, \omega_2). \tag{2.57}$$

Write

$$M = \sup_{n\in\mathbb{N}} |\mathcal{G}_n|_{\mathcal{L}(L^{p_1}_{\mathcal{M}}(\Omega_1;L^{p_2}(\Omega_2;X));L^{q_1}_{\mathcal{M}}(\Omega_1;L^{q_2}(\Omega_2;Y)))}.$$

By Lemma 2.61, we conclude that there exist a bounded linear operator $\mathcal{G}$ from $L^{p_1}_{\mathcal{M}}(\Omega_1; L^{p_2}(\Omega_2; X))$ to $L^{q_1}_{\mathcal{M}}(\Omega_1; L^{q_2}(\Omega_2; Y))$ and a subsequence $\{\mathcal{G}_{n_k}\}_{k=1}^{\infty} \subset \{\mathcal{G}_n\}_{n=1}^{\infty}$ such that, for all $u(\cdot) \in L^{p_1}_{\mathcal{M}}(\Omega_1; L^{p_2}(\Omega_2; X))$,

$$\mathcal{G}u(\cdot) = (\mathrm{w})\text{-}\lim_{k\to\infty} \mathcal{G}_{n_k} u(\cdot) \quad \text{in } L^{q_1}_{\mathcal{M}}(\Omega_1; L^{q_2}(\Omega_2; Y)), \tag{2.58}$$

and

$$|\mathcal{G}u(\cdot)|_{L^{q_1}_{\mathcal{M}}(\Omega_1;L^{q_2}(\Omega_2;Y))} \leq M |u(\cdot)|_{L^{p_1}_{\mathcal{M}}(\Omega_1;L^{p_2}(\Omega_2;X))}. \tag{2.59}$$

We claim that

$$\begin{aligned}\sum_{i=1}^{m} f_i \mathcal{G} u_i &= \mathcal{G}\Big(\sum_{i=1}^{m} f_i u_i\Big)\\&= (\mathrm{w})\text{-}\lim_{k\to\infty} \sum_{i=1}^{m} f_i \mathcal{G}_{n_k} u_i \qquad \text{in } L^{q_1}_{\mathcal{M}}(\Omega_1; L^{q_2}(\Omega_2; Y))\end{aligned} \tag{2.60}$$

and

$$\Big|\sum_{i=1}^{m} f_i \mathcal{G} u_i\Big|_{L^{q_1}_{\mathcal{M}}(\Omega_1;L^{q_2}(\Omega_2;Y))} \leq M \Big|\sum_{i=1}^{m} f_i u_i\Big|_{L^{p_1}_{\mathcal{M}}(\Omega_1;L^{p_2}(\Omega_2;X))}, \tag{2.61}$$

where $m \in \mathbb{N}$, $f_i \in L^{\infty}_{\mathcal{M}}(\Omega_1 \times \Omega_2)$ and $u_i \in L^{p_1}_{\mathcal{M}}(\Omega_1; L^{p_2}(\Omega_2; X))$, $i = 1, 2, \cdots, m$. To show this, write $q_1' = \frac{q_1}{q_1 - 1}$ and $q_2' = \frac{q_2}{q_2 - 1}$. Denote by μ the product measure of μ_1 and μ_2. By Theorems 2.48 and 2.55, from (2.57)–(2.58), we conclude that for any $v(\cdot) \in L^{q_1'}_{\mathcal{M}}(\Omega_1; L^{q_2'}(\Omega_2; Y^*))$,

$$
\begin{aligned}
&\int_{\Omega_1\times\Omega_2}\Big\langle\sum_{i=1}^m f_i(\omega_1,\omega_2)(\mathcal{G}u_i)(\omega_1,\omega_2),v(\omega_1,\omega_2)\Big\rangle_{Y,Y^*}d\mu\\
&=\int_{\Omega_1\times\Omega_2}\sum_{i=1}^m\big\langle(\mathcal{G}u_i)(\omega_1,\omega_2),f_i(\omega_1,\omega_2)v(\omega_1,\omega_2)\big\rangle_{Y,Y^*}d\mu\\
&=\lim_{k\to\infty}\int_{\Omega_1\times\Omega_2}\sum_{i=1}^m\big\langle\big(\mathcal{G}_{n_k}u_i\big)(\omega_1,\omega_2),f_i(\omega_1,\omega_2)v(\omega_1,\omega_2)\big\rangle_{Y,Y^*}d\mu\\
&=\lim_{k\to\infty}\int_{\Omega_1\times\Omega_2}\sum_{i=1}^m\big\langle(f_i\mathcal{G}_{n_k}u_i)(\omega_1,\omega_2),v(\omega_1,\omega_2)\big\rangle_{Y,Y^*}d\mu,
\end{aligned}
\tag{2.62}
$$

and

$$
\begin{aligned}
&\lim_{k\to\infty}\int_{\Omega_1\times\Omega_2}\sum_{i=1}^m\big\langle\big(\mathcal{G}_{n_k}u_i\big)(\omega_1,\omega_2),f_i(\omega_1,\omega_2)v(\omega_1,\omega_2)\big\rangle_{Y,Y^*}d\mu\\
&=\lim_{k\to\infty}\int_{\Omega_1\times\Omega_2}\sum_{i=1}^m\big\langle G_{n_k}(\omega_1,\omega_2)u_i(\omega_1,\omega_2),f_i(\omega_1,\omega_2)v(\omega_1,\omega_2)\big\rangle_{Y,Y^*}d\mu\\
&=\lim_{k\to\infty}\int_{\Omega_1\times\Omega_2}\Big\langle G_{n_k}(\omega_1,\omega_2)\Big(\sum_{i=1}^m f_i(\omega_1,\omega_2)u_i(\omega_1,\omega_2)\Big),v(\omega_1,\omega_2)\Big\rangle_{Y,Y^*}d\mu\\
&=\lim_{k\to\infty}\int_{\Omega_1\times\Omega_2}\Big\langle\Big(\mathcal{G}_{n_k}\Big(\sum_{i=1}^m f_iu_i\Big)\Big)(\omega_1,\omega_2),v(\omega_1,\omega_2)\Big\rangle_{Y,Y^*}d\mu\\
&=\int_{\Omega_1\times\Omega_2}\Big\langle\Big(\mathcal{G}\Big(\sum_{i=1}^m f_iu_i\Big)\Big)(\omega_1,\omega_2),v(\omega_1,\omega_2)\Big\rangle_{Y,Y^*}d\mu.
\end{aligned}
\tag{2.63}
$$

By (2.62)–(2.63) and noting (2.59), we obtain (2.60)–(2.61).

Step 2. It is easy to see that, each $x\in X$ can be regarded as an element (i.e., $\chi_{\Omega_1\times\Omega_2}(\cdot)x$) in $L^{p_1}_{\mathcal{M}}(\Omega_1;L^{p_2}(\Omega_2;X))$. Hence, $\mathcal{G}x$ makes sense and belongs to $L^{q_1}_{\mathcal{M}}(\Omega_1;L^{q_2}(\Omega_2;Y))$, and $\mathcal{G}$ is a bounded linear operator from X to $L^{q_1}_{\mathcal{M}}(\Omega_1;L^{q_2}(\Omega_2;Y))$.

Write $B_X=\big\{x\in X\ \big|\ |x|_X\le 1\big\}$. By the separability of X, it is easy to see that the real valued function $\sup\limits_{x\in B_X}\big|(\mathcal{G}x)(\cdot)\big|_Y$ is $\mathcal{M}$-measurable. We claim that

$$
\sup_{x\in B_X}\big|(\mathcal{G}x)(\omega_1,\omega_2)\big|_Y<\infty,\qquad \text{a.e. } (\omega_1,\omega_2)\in\Omega_1\times\Omega_2. \tag{2.64}
$$

In the rest of this step, we shall prove (2.64) by the contradiction argument.

Assume that (2.64) was not true. By the measurability of the function $\sup\limits_{x\in B_X}\big|(\mathcal{G}x)(\cdot)\big|_Y$ w. r. t. $\mathcal{M}$, there would exist a set $A\in\mathcal{M}$ such that $\mu(A)>0$ and that

$$\sup_{x\in B_X} \big|(\mathcal{G}x)(\omega_1,\omega_2)\big|_Y = \infty, \text{ for } (\omega_1,\omega_2)\in A.$$

Let $\{x_i\}_{i=1}^\infty$ be a sequence in B_X such that it is dense in B_X. Then

$$\sup_{i\in\mathbb{N}} \big|(\mathcal{G}x_i)(\omega_1,\omega_2)\big|_Y = \sup_{x\in B_X} \big|(\mathcal{G}x)(\omega_1,\omega_2)\big|_Y = \infty, \qquad \text{for } (\omega_1,\omega_2)\in A.$$

For any $n\in\mathbb{N}$ and $i\in\mathbb{N}\setminus\{1\}$, we define a sequence of subsets of $\Omega_1\times\Omega_2$ in the following way:

$$\begin{cases} A_1^{(n)} = \Big\{(\omega_1,\omega_2)\in\Omega_1\times\Omega_2 \;\Big|\; \big|(\mathcal{G}x_1)(\omega_1,\omega_2)\big|_Y \geq n\Big\}, \\ A_i^{(n)} = \Big\{(\omega_1,\omega_2)\in(\Omega_1\times\Omega_2)\setminus\big(\bigcup_{k=1}^{i-1} A_k^{(n)}\big) \;\Big|\; \big|(\mathcal{G}x_i)(\omega_1,\omega_2)\big|_Y \geq n\Big\}. \end{cases} \tag{2.65}$$

It follows from the measurability of $\big|(\mathcal{G}x)(\cdot)\big|_Y$ that $A_i^{(n)}\in\mathcal{M}$ for every $i\in\mathbb{N}$ and $n\in\mathbb{N}$. It is clear that $A\subset\bigcup_{i=1}^\infty A_i^{(n)}$ for any $n\in\mathbb{N}$ and $A_i^{(n)}\cap A_j^{(n)}=\emptyset$ for $i\neq j$. Hence, we see that

$$\sum_{i=1}^\infty \mu(A_i^{(n)}) = \mu\big(\bigcup_{i=1}^\infty A_i^{(n)}\big) \geq \mu(A) > 0, \qquad \forall\, n\in\mathbb{N}.$$

Thus, for each $n\in\mathbb{N}$, there is an $N_n\in\mathbb{N}$ such that

$$\sum_{i=1}^{N_n} \mu(A_i^{(n)}) = \mu\big(\bigcup_{i=1}^{N_n} A_i^{(n)}\big) \geq \frac{\mu(A)}{2} > 0. \tag{2.66}$$

Write

$$x^{(n)}(\omega_1,\omega_2) = \sum_{i=1}^{N_n} \chi_{A_i^{(n)}}(\omega_1,\omega_2)x_i. \tag{2.67}$$

Obviously, $x^{(n)}(\cdot)$ is an X-valued, strongly $\mathcal{M}$-measurable function. By $\mathcal{G}\in\mathcal{L}(L^{p_1}_{\mathcal{M}}(\Omega_1;L^{p_2}(\Omega_2;X));L^{q_1}_{\mathcal{M}}(\Omega_1;L^{q_2}(\Omega_2;Y)))$ and $|x^{(n)}(\omega_1,\omega_2)|_X \leq 1$ for a.e. $(\omega_1,\omega_2)\in\Omega_1\times\Omega_2$, and noting (2.60)–(2.61), we find that

$$\begin{aligned} &|\mathcal{G}x^{(n)}|_{L^{q_1}_{\mathcal{M}}(\Omega_1;L^{q_2}(\Omega_2;Y))} \\ &\leq M\Big\{\int_{\Omega_1}\Big[\int_{\Omega_2}|x^{(n)}(\omega_1,\omega_2)|_X^{p_1}d\mu_2(\omega_2)\Big]^{\frac{p_2}{p_1}}d\mu_1(\omega_1)\Big\}^{\frac{1}{p_2}} \\ &\leq M\big(\mu_1(\Omega_1)\big)^{1/p_2}\big(\mu_2(\Omega_2)\big)^{1/p_1}, \qquad \forall\, n\in\mathbb{N}. \end{aligned} \tag{2.68}$$

Now, let us choose an integer $n > \frac{2M}{\mu(A)}\big(\mu_1(\Omega_1)\big)^{\frac{1}{p_2}+\frac{1}{q_1'}}\big(\mu_2(\Omega_2)\big)^{\frac{1}{p_1}+\frac{1}{q_2'}}$. By (2.65)–(2.67), it follows that

$$
\begin{aligned}
&|\mathcal{G}x^{(n)}|_{L^{q_1}_{\mathcal{M}}(\Omega_1;L^{q_2}(\Omega_2;Y))}\\
&\geq \big(\mu_1(\Omega_1)\big)^{-\frac{1}{q_1'}}\big(\mu_2(\Omega_2)\big)^{-\frac{1}{q_2'}}|\mathcal{G}x^{(n)}|_{L^{1}_{\mathcal{M}}(\Omega_1;L^{1}(\Omega_2;Y))}\\
&= \big(\mu_1(\Omega_1)\big)^{-\frac{1}{q_1'}}\big(\mu_2(\Omega_2)\big)^{-\frac{1}{q_2'}}\sum_{i=1}^{N_n}\int_{A_i^{(n)}}\big|\mathcal{G}x_i\big|_Y d\mu\\
&\geq \big(\mu_1(\Omega_1)\big)^{-\frac{1}{q_1'}}\big(\mu_2(\Omega_2)\big)^{-\frac{1}{q_2'}} n\sum_{i=1}^{N_n}\mu(A_i^{(n)})\\
&\geq \frac{\mu(A)}{2}\big(\mu_1(\Omega_1)\big)^{-\frac{1}{q_1'}}\big(\mu_2(\Omega_2)\big)^{-\frac{1}{q_2'}} n > M\big(\mu_1(\Omega_1)\big)^{1/p_2}\big(\mu_2(\Omega_2)\big)^{1/p_1},
\end{aligned}
$$

which contradicts the inequality (2.68). Therefore, (2.64) holds.

Step 3. By (2.64), for a.e. $(\omega_1,\omega_2)\in\Omega_1\times\Omega_2$, we may define an operator $G(\omega_1,\omega_2)\in\mathcal{L}(X;Y)$ by

$$
X\ni x\mapsto G(\omega_1,\omega_2)x=\big(\mathcal{G}x\big)(\omega_1,\omega_2). \tag{2.69}
$$

Further, we introduce the following subspace of $L^{p_1}_{\mathcal{M}}(\Omega_1;L^{p_2}(\Omega_2;X))$:

$$
\mathcal{X}=\Big\{u(\cdot)=\sum_{i=1}^{m}\chi_{A_i}(\cdot)h_i\;\Big|\;m\in\mathbb{N},\ A_i\in\mathcal{M},h_i\in X\Big\}.
$$

It is clear that $\mathcal{X}$ is dense in $L^{p_1}_{\mathcal{M}}(0,T;L^{p_2}(\Omega;X))$. We now define a linear operator $\widetilde{\mathcal{G}}$ from $\mathcal{X}$ to $L^{q_1}_{\mathcal{M}}(\Omega_1;L^{q_2}(\Omega_2;Y))$ by

$$
\mathcal{X}\ni u(\cdot)=\sum_{i=1}^{m}\chi_{A_i}(\cdot)h_i\mapsto(\widetilde{\mathcal{G}}u)(\omega_1,\omega_2)=\sum_{i=1}^{m}\chi_{A_i}(\omega_1,\omega_2)G(\omega_1,\omega_2)h_i. \tag{2.70}
$$

We claim that

$$
(\widetilde{\mathcal{G}}u)(\cdot)=(\mathcal{G}u)(\cdot),\qquad\forall\,u(\cdot)\in\mathcal{X}. \tag{2.71}
$$

Indeed, it follows from (2.57)–(2.58) that for any $v(\cdot)\in L^{q_1'}_{\mathcal{M}}(\Omega_1;L^{q_2'}(\Omega_2;Y^*))$, and $u(\cdot)$ to be of the form in (2.70),

$$
\begin{aligned}
&\int_{\Omega_1\times\Omega_2}\big\langle(\widetilde{\mathcal{G}}u)(\omega_1,\omega_2),v(\omega_1,\omega_2)\big\rangle_{Y,Y^*}d\mu\\
&=\int_{\Omega_1\times\Omega_2}\big\langle\sum_{i=1}^m\chi_{A_i}(\omega_1,\omega_2)G(\omega_1,\omega_2)h_i,v(\omega_1,\omega_2)\big\rangle_{Y,Y^*}d\mu\\
&=\int_{\Omega_1\times\Omega_2}\big\langle\sum_{i=1}^m\chi_{A_i}(\omega_1,\omega_2)\big(\mathcal{G}h_i\big)(\omega_1,\omega_2),v(\omega_1,\omega_2)\big\rangle_{Y,Y^*}d\mu\\
&=\sum_{i=1}^m\int_{\Omega_1\times\Omega_2}\big\langle\big(\mathcal{G}h_i\big)(\omega_1,\omega_2),\chi_{A_i}(\omega_1,\omega_2)v(\omega_1,\omega_2)\big\rangle_{Y,Y^*}d\mu\\
&=\sum_{i=1}^m\lim_{k\to\infty}\int_{\Omega_1\times\Omega_2}\big\langle G_{n_k}(\omega_1,\omega_2)h_i,\chi_{A_i}(\omega_1,\omega_2)v(\omega_1,\omega_2)\big\rangle_{Y,Y^*}d\mu\\
&=\lim_{k\to\infty}\int_{\Omega_1\times\Omega_2}\big\langle G_{n_k}(\omega_1,\omega_2)\Big(\sum_{i=1}^m\chi_{A_i}(\omega_1,\omega_2)h_i\Big),v(\omega_1,\omega_2)\big\rangle_{Y,Y^*}d\mu\\
&=\int_{\Omega_1\times\Omega_2}\big\langle(\mathcal{G}u)(\omega_1,\omega_2),v(\omega_1,\omega_2)\big\rangle_{Y,Y^*}d\mu.
\end{aligned}
$$

This gives (2.71).

Recall that $\mathcal{G}$ is a bounded linear operator from $L^{p_1}_{\mathcal{M}}(\Omega_1;L^{p_2}(\Omega_2;X))$ to $L^{q_1}_{\mathcal{M}}(\Omega_1;L^{q_2}(\Omega_2;Y))$. Hence, it is also a bounded linear operator from $\mathcal{X}$ to $L^{q_1}_{\mathcal{M}}(\Omega_1;L^{q_2}(\Omega_2;Y))$. By (2.71), we see that $\widetilde{\mathcal{G}}$ is a bounded linear operator from $\mathcal{X}$ to $L^{q_1}_{\mathcal{M}}(\Omega_1;L^{q_2}(\Omega_2;Y))$. Since $\mathcal{X}$ is dense in $L^{p_1}_{\mathcal{M}}(\Omega_1;L^{p_2}(\Omega_2;X))$, it is clear that $\widetilde{\mathcal{G}}$ can be uniquely extended as a bounded linear operator from $L^{p_1}_{\mathcal{M}}(\Omega_1;L^{p_2}(\Omega_2;X))$ to $L^{q_1}_{\mathcal{M}}(\Omega_1;L^{q_2}(\Omega_2;Y))$ (We still denote by $\widetilde{\mathcal{G}}$ its extension). By (2.71) again, we conclude that

$$
\widetilde{\mathcal{G}}=\mathcal{G}. \tag{2.72}
$$

It remains to show that

$$
\big(\widetilde{\mathcal{G}}u(\cdot)\big)(\omega_1,\omega_2)=G(\omega_1,\omega_2)u(\omega_1,\omega_2),\qquad \text{a.e. }(\omega_1,\omega_2)\in\Omega_1\times\Omega_2, \tag{2.73}
$$

for all $u\in L^{p_1}_{\mathcal{M}}(\Omega_1;L^{p_2}(\Omega_2;X))$. For this purpose, by the fact that $\mathcal{X}$ is dense in $L^{p_1}_{\mathcal{M}}(\Omega_1;L^{p_2}(\Omega_2;X))$, we may assume that

$$
u(\cdot)=\sum_{i=1}^\infty\chi_{A_i}(\cdot)h_i, \tag{2.74}
$$

for some $A_i\in\mathcal{M}$ and $h_i\in X$, $i=1,2,\cdots$ (Note that here we assume neither $A_i\bigcap A_j=\emptyset$ nor $h_i\neq h_j$ for $i,j=1,2,\cdots$). For each $n\in\mathbb{N}$, write $u^n(\cdot)=\sum_{i=1}^n\chi_{A_i}(\cdot)h_i$. From (2.74), it is clear that

$$u(\cdot) = \lim_{n\to\infty} u^n(\cdot) \qquad \text{in } L^{p_1}_{\mathcal{M}}(\Omega_1; L^{p_2}(\Omega_2; X)). \tag{2.75}$$

By (2.61), (2.69)–(2.70) and (2.75), it is easy to see that

$$\big(\widetilde{\mathcal{G}} u^n(\cdot)\big)(\omega_1,\omega_2) = \sum_{i=1}^{n} \chi_{A_i}(\omega_1,\omega_2) G(\omega_1,\omega_2) h_i \tag{2.76}$$

is a Cauchy sequence in $L^{q_1}_{\mathcal{M}}(\Omega_1; L^{q_2}(\Omega_2; Y))$. Hence, by (2.76) and recalling that $\widetilde{\mathcal{G}}$ is a bounded linear operator from $L^{p_1}_{\mathcal{M}}(\Omega_1; L^{p_2}(\Omega_2; X))$ to $L^{q_1}_{\mathcal{M}}(\Omega_1; L^{q_2}(\Omega_2; Y))$, we conclude that

$$\big(\widetilde{\mathcal{G}} u(\cdot)\big)(\omega_1,\omega_2) = \sum_{i=1}^{\infty} \chi_{A_i}(\omega_1,\omega_2) G(\omega_1,\omega_2) h_i. \tag{2.77}$$

Combining (2.74) and (2.77), we obtain (2.73).

Finally, by (2.72) and (2.73), the desired result follows. This completes the proof of Theorem 2.63. □

Remark 2.64. i) Clearly, the most difficult part in the proof of Theorem 2.63 is to show that the weak limit operator $\mathcal{G} \in \mathcal{L}_{pd}\big(L^{p_1}_{\mathcal{M}}(\Omega_1; L^{p_2}(\Omega_2; X)); L^{q_1}_{\mathcal{M}}(\Omega_1; L^{q_2}(\Omega_2; Y))\big)$. Note that, a simple application of Lemma 2.61 to the operators $\{\mathcal{G}_n\}_{n=1}^{\infty}$ does not guarantee this point but only that $\mathcal{G} \in \mathcal{L}\big(L^{p_1}_{\mathcal{M}}(\Omega_1; L^{p_2}(\Omega_2; X)); L^{q_1}_{\mathcal{M}}(\Omega_1; L^{q_2}(\Omega_2; Y))\big)$.

ii) By Theorem 2.63, it is easy to deduce that $\mathcal{L}_{pd}\big(L^{p_1}_{\mathcal{M}}(\Omega_1; L^{p_2}(\Omega_2; X)); L^{q_1}_{\mathcal{M}}(\Omega_1; L^{q_2}(\Omega_2; Y))\big)$ is a closed linear subspace of $\mathcal{L}\big(L^{p_1}_{\mathcal{M}}(\Omega_1; L^{p_2}(\Omega_2; X)); L^{q_1}_{\mathcal{M}}(\Omega_1; L^{q_2}(\Omega_2; Y))\big)$.

2.6 Stochastic Processes

In this section, we recall some elements on stochastic processes. We fix a probability space $(\Omega, \mathcal{F}, \mathbb{P})$, a Banach space H and a nonempty index set $\mathcal{I}$.

We begin with the following definition:

Definition 2.65. *A family of (strongly $\mathcal{F}$-measurable) random variables $X \stackrel{\Delta}{=} X(\cdot) \stackrel{\Delta}{=} \{X(t)\}_{t\in\mathcal{I}}$ from $(\Omega, \mathcal{F}) \to (H, \mathcal{B}(H))$ is called an (H-valued) stochastic process. For any $\omega \in \Omega$, the map $t \mapsto X(t,\omega)$ is called a sample path (of X).*

In what follows, we will choose $\mathcal{I} = [0,T]$ with $T > 0$, or $\mathcal{I} = [0,+\infty)$ unless otherwise stated. Also, a stochastic process will be simply called a process if no ambiguity. An (H-valued) process $X(\cdot)$ is said to be continuous (*resp.*, cádlàg, i.e., right-continuous with left limits) if there is a $\mathbb{P}$-null set $N \in \mathcal{F}$, such that for any $\omega \in \Omega \setminus N$, the sample path $X(\cdot,\omega)$ is continuous (*resp.*, cádlàg) in H. In a similar way, one can define right-continuous stochastic processes, etc.

In the case that H is $\mathbb{R}^m$ (for some $m \in \mathbb{N}$) with the standard topology, for a given stochastic process $\{X(t)\}_{t\in\mathcal{I}}$, we set

$$F_{t_1,\cdots,t_j}(x_1,\cdots,x_j) \stackrel{\Delta}{=} \mathbb{P}(\{X(t_1) \le x_1,\cdots,X(t_j) \le x_j\}), \tag{2.78}$$

where $j \in \mathbb{N}$, $t_i \in \mathcal{I}$, $x_i \in \mathbb{R}^m$ and $X(t_i) \le x_i$ stands for componentwise inequalities ($i = 1,\cdots,j$). Functions defined in (2.78) are called the finite dimensional distributions of X.

Similar to the distribution function of random variable, the finite dimensional distributions $F_{t_1,\cdots,t_j}(x_1,\cdots,x_j)$ of X include the main probability properties of the process.

When H is a Hilbert space, an (H-valued) stochastic process $\{X(t)\}_{t\in\mathcal{I}}$ is called Gaussian, if any finite linear combination $\sum_{i=1}^k a_i X(t_i)$ (with $a_i \in \mathbb{R}$ and $t_i \in \mathcal{I}$ ($i = 1,\cdots,k$) and $k \in \mathbb{N}$) is an H-valued Gaussian random variable.

We shall use the following two notions in the sequel.

Definition 2.66. *A class of stochastic processes $\{X_\lambda(\cdot)\}_{\lambda\in\Lambda}$ are called independent, if for any $n,\ell \in \mathbb{N}$, $\{\lambda_1,\cdots,\lambda_n\} \subset \Lambda$ and $\{t_1,\cdots,t_\ell\} \subset \mathcal{I}$, the random variables $(X_{\lambda_1}(t_1),\cdots,X_{\lambda_1}(t_\ell)),\cdots,(X_{\lambda_n}(t_1),\cdots,X_{\lambda_n}(t_\ell))$ are independent.*

Definition 2.67. *Two (H-valued) processes $X(\cdot)$ and $\overline{X}(\cdot)$ are said to be stochastically equivalent if $\mathbb{P}(\{X(t) = \overline{X}(t)\}) = 1$ for any $t \in \mathcal{I}$. In this case, one is said to be a modification of the other.*

Obviously, in Definition 2.67, $X(\cdot)$'s and $\overline{X}(\cdot)$'s finite dimensional distributions are the same, provided that they are stochastically equivalent, However, generally, the $\mathbb{P}$-null set $N_t \stackrel{\Delta}{=} \{X(t) \neq \overline{X}(t)\}$ depends on t. Therefore, the sample paths of $X(\cdot)$ and $\overline{X}(\cdot)$ can differ significantly. The following is a simple example.

Example 2.68. Let $\Omega = [0,1]$, $\mathcal{I} = [0,1]$, $\mathbb{P}$ be the Lebesgue measure $\mathbf{m}$ on $[0,1]$, $X(t,\omega) \equiv 0$, and

$$\overline{X}(t,\omega) = \begin{cases} 0, & t \neq \omega, \\ 1, & t = \omega. \end{cases}$$

Then $X(\cdot)$ and $\overline{X}(\cdot)$ are stochastically equivalent. But, each sample path $X(\cdot,\omega)$ is continuous and none of the sample paths $\overline{X}(\cdot,\omega)$ is continuous. In the present case, we actually have $\bigcup_{t\in\mathcal{I}} N_t = \Omega$.

We call a family of sub-σ-fields $\{\mathcal{F}_t\}_{t\in\mathcal{I}}$ in $\mathcal{F}$ a filtration if $\mathcal{F}_{t_1} \subset \mathcal{F}_{t_2}$ for all $t_1, t_2 \in \mathcal{I}$ with $t_1 \le t_2$. For any $t \in \mathcal{I}$, we put

$$\mathcal{F}_{t+} \stackrel{\Delta}{=} \bigcap_{s\in(t,+\infty)\cap\mathcal{I}} \mathcal{F}_s, \qquad \mathcal{F}_{t-} \stackrel{\Delta}{=} \bigcup_{s\in[0,t)\cap\mathcal{I}} \mathcal{F}_s.$$

If $\mathcal{F}_{t+} = \mathcal{F}_t$ (*resp.* $\mathcal{F}_{t-} = \mathcal{F}_t$), then $\{\mathcal{F}_t\}_{t\in\mathcal{I}}$ is said to be right (*resp.* left) continuous. In the sequel, for simplicity, we write $\mathbf{F} = \{\mathcal{F}_t\}_{t\in\mathcal{I}}$ unless we want to emphasize what $\mathcal{F}_t$ or I exactly is. We call $(\Omega, \mathcal{F}, \mathbf{F})$ a filtered measurable space and $(\Omega, \mathcal{F}, \mathbf{F}, \mathbb{P})$ a filtered probability space. We say that $(\Omega, \mathcal{F}, \mathbf{F}, \mathbb{P})$ satisfies the usual condition if $(\Omega, \mathcal{F}, \mathbb{P})$ is complete, $\mathcal{F}_0$ contains all $\mathbb{P}$-null sets in $\mathcal{F}$, and $\mathbf{F}$ is right continuous.

Remark 2.69. For each $t \in \mathcal{I}$, the σ-algebra $\mathcal{F}_t$ can be interpreted as the set of information that one has collected up to time t.

In the rest of this section, we fix a filtration $\mathbf{F}$ on $(\Omega, \mathcal{F})$.

Definition 2.70. *Let $X(\cdot)$ be an H-valued process.*

1) *$X(\cdot)$ is said to be measurable if the map $(t, \omega) \mapsto X(t, \omega)$ is strongly $(\mathcal{B}(\mathcal{I}) \times \mathcal{F})/\mathcal{B}(H)$-measurable;*

2) *$X(\cdot)$ is said to be $\mathbf{F}$-adapted if it is measurable, and for each $t \in \mathcal{I}$, the map $\omega \mapsto X(t, \omega)$ is strongly $\mathcal{F}_t/\mathcal{B}(H)$-measurable;*

3) *$X(\cdot)$ is said to be $\mathbf{F}$-progressively measurable if for each $t \in \mathcal{I}$, the map $(s, \omega) \mapsto X(s, \omega)$ from $[0, t] \times \Omega$ to H is strongly $(\mathcal{B}([0, t]) \times \mathcal{F}_t)/\mathcal{B}(H)$-measurable.*

Definition 2.71. *A set $A \in \mathcal{I} \times \Omega$ is called progressively measurable w.r.t. $\mathbf{F}$ if the process $\chi_A(\cdot)$ is progressive. The class of all progressively measurable sets is a σ-algebra, called the progressive σ-field w.r.t. $\mathbf{F}$, denoted by $\mathbb{F}$.*

The following result is sometimes useful.

Lemma 2.72. *An (H-valued) process $X(\cdot) : [0, T] \times \Omega \to H$ is $\mathbf{F}$-progressively measurable if and only if it is strongly $\mathbb{F}$-measurable.*

It is clear that if $X(\cdot)$ is $\mathbf{F}$-progressively measurable, it must be $\mathbf{F}$-adapted. Conversely, it can be proved that, for any $\mathbf{F}$-adapted process $X(\cdot)$, there is a $\mathbf{F}$-progressively measurable process $\widetilde{X}(\cdot)$ which is stochastically equivalent to $X(\cdot)$ (e.g. [261, pp. 68] for the case $\dim H < \infty$. The case when $\dim H = \infty$ can be considered similarly). For this reason, unless otherwise indicated, in the sequel, by saying that a process is $\mathbf{F}$-adapted, we mean that it is $\mathbf{F}$-progressively measurable.

The notations to be given below will be used in the rest of this book.

For any $p, q \in [1, \infty)$, write $L^p_{\mathcal{F}_t}(\Omega; H) \overset{\Delta}{=} L^p(\Omega, \mathcal{F}_t, \mathbb{P}; H)$ ($t \in [0, T]$), and define

$$
\begin{aligned}
L^p_{\mathbb{F}}(\Omega; L^q(0, T; H)) \overset{\Delta}{=} \Big\{ \varphi : (0, T) \times \Omega \to H \ \Big| \ & \varphi(\cdot) \text{ is } \mathbf{F}\text{-adapted} \\
& \text{and } \mathbb{E}\Big(\int_0^T |\varphi(t)|_H^q dt \Big)^{\frac{p}{q}} < \infty \Big\}, \\
L^q_{\mathbb{F}}(0, T; L^p(\Omega; H)) \overset{\Delta}{=} \Big\{ \varphi : (0, T) \times \Omega \to H \ \Big| \ & \varphi(\cdot) \text{ is } \mathbf{F}\text{-adapted} \\
& \text{and } \int_0^T \Big(\mathbb{E}|\varphi(t)|_H^p \Big)^{\frac{q}{p}} dt < \infty \Big\}.
\end{aligned}
$$

Similarly, we may define (for $1 \le p, q < \infty$)

$$\begin{cases} L^\infty_{\mathbb{F}}(\Omega; L^q(0,T;H)), & L^p_{\mathbb{F}}(\Omega; L^\infty(0,T;H)), & L^\infty_{\mathbb{F}}(\Omega; L^\infty(0,T;H)), \\ L^\infty_{\mathbb{F}}(0,T; L^p(\Omega;H)), & L^q_{\mathbb{F}}(0,T; L^\infty(\Omega;H)), & L^\infty_{\mathbb{F}}(0,T; L^\infty(\Omega;H)). \end{cases}$$

Clearly, for $1 \le p,q \le \infty$, $L^p_{\mathbb{F}}(\Omega; L^q(0,T;H))$ and $L^q_{\mathbb{F}}(0,T; L^p(\Omega;H))$ are respectively special cases of $L^p_{\mathcal{M}}(X_1; L^q(X_2;H))$ and $L^q_{\mathcal{M}}(X_2; L^p(X_1;H))$ defined in Section 2.4, and therefore, they are Banach spaces (with the canonical norms). In the sequel, we shall simply denote $L^p_{\mathbb{F}}(\Omega; L^p(0,T;H)) \equiv L^p_{\mathbb{F}}(0,T; L^p(\Omega;H))$ by $L^p_{\mathbb{F}}(0,T;H)$; and further simply denote $L^p_{\mathbb{F}}(0,T;\mathbb{R})$ by $L^p_{\mathbb{F}}(0,T)$.

As a consequence of Theorem 2.55 and Lemma 2.72, we have the following result (which is a Riesz-type Representation Theorem for the dual of the space $L^p_{\mathbb{F}}(0,T; L^q(\Omega;H))$).

Theorem 2.73. *Suppose* $1 \le p, q < \infty$, *and that* H^* *has the Radon-Nikodým property. Then*

$$L^p_{\mathbb{F}}(0,T; L^q(\Omega;H))^* = L^{p'}_{\mathbb{F}}(0,T; L^{q'}(\Omega;H^*)), \tag{2.79}$$

where p' *and* q' *are the Hölder conjugates of* p *and* q, *respectively.*

Remark 2.74. 1) As we shall see later, Theorem 2.73 is the theoretic basis of our stochastic transposition method.

2) Surprisingly, as shown in [234], a quite "natural" variant of (2.79), i.e.,

$$L^p_{\mathbb{F}}(\Omega; L^q(0,T;H))^* = L^{p'}_{\mathbb{F}}(\Omega; L^{q'}(0,T;H^*))$$

is NOT correct! Note that, Condition 2.1 fails for the latter.

For any $p \in [1,\infty)$, set

$$L^p_{\mathbb{F}}(\Omega; C([0,T];H)) \stackrel{\Delta}{=} \Big\{ \varphi : [0,T] \times \Omega \to H \ \Big|\ \varphi(\cdot) \text{ is continuous, } \mathbf{F}\text{-adapted and } \mathbb{E}\big(|\varphi(\cdot)|^p_{C([0,T];H)}\big) < \infty \Big\}$$

and

$$C_{\mathbb{F}}([0,T]; L^p(\Omega;H)) \stackrel{\Delta}{=} \Big\{ \varphi : [0,T] \times \Omega \to H \ \Big|\ \varphi(\cdot) \text{ is } \mathbf{F}\text{-adapted and } \varphi(\cdot) : [0,T] \to L^p_{\mathcal{F}_T}(\Omega;H) \text{ is continuous} \Big\}.$$

One can show that both $L^p_{\mathbb{F}}(\Omega; C([0,T];H))$ and $C_{\mathbb{F}}([0,T]; L^p(\Omega;H))$ are Banach spaces with norms given by $|\varphi(\cdot)|_{L^p_{\mathbb{F}}(\Omega;C([0,T];H))} = \big(\mathbb{E}(|\varphi(\cdot)|^p_{C([0,T];H)})\big)^{1/p}$ and $|\varphi(\cdot)|_{C_{\mathbb{F}}([0,T];L^p(\Omega;H))} = \max_{t\in[0,T]} \big(\mathbb{E}(|\varphi(t)|^p_H)\big)^{1/p}$, respectively. Similarly, for any $k \in \mathbb{N}$, one can define the Banach spaces $L^p_{\mathbb{F}}(\Omega; C^k([0,T];H))$ and

$C_{\mathbb{F}}^k([0,T];L^p(\Omega;H))$. Also, we denote by $L_{\mathbb{F}}^p(\Omega;D([0,T];H))$ the Banach space of all H-valued, $\mathbf{F}$-adapted, càdlàg stochastic processes $X(\cdot)$ such that $\mathbb{E}(|X(\cdot)|^p_{L^\infty_{\mathbb{F}}(0,T;H)}) < \infty$, with the canonical norm. Similarly, one can define $D_{\mathbb{F}}([0,T];L^p(\Omega;H))$. For any $X(\cdot) \in L_{\mathbb{F}}^p(\Omega;D([0,T];H))$, by the Dominated Convergence Theorem, it is easy to see that $X(t)$ is càdlàg in $L^p_{\mathcal{F}_T}(\Omega;H)$, w.r.t. $t \in [0,T]$. Hence,

$$L_{\mathbb{F}}^p(\Omega;D([0,T];H)) \subset D_{\mathbb{F}}([0,T];L^p(\Omega;H)).$$

In the sequel, we simply denote $L_{\mathbb{F}}^p(\Omega;C^k([0,T];\mathbb{R}))$ and $L_{\mathbb{F}}^p(\Omega;D([0,T];\mathbb{R}))$ by $L_{\mathbb{F}}^p(\Omega;C([0,T]))$ and $L_{\mathbb{F}}^p(\Omega;D([0,T]))$, respectively.

Next, let $(G,\mathcal{G},\mu)$ be a finite measure space. For any $p,q,r \in [1,\infty)$, let $L^p(G;L^q_{\mathbb{F}}(\Omega;L^r(0,T;H)))$ be the set of all strongly $\mathcal{G}\times\mathcal{B}([0,T])\times\mathcal{F}_T$-measurable functions $\Phi : G\times[0,T]\times\Omega\to H$ such that for any $x\in G$, the process $\Phi(x,\cdot,\cdot)$ is $\mathbf{F}$-adapted and

$$\int_G \Big[\mathbb{E}\Big(\int_0^T |\Phi(x,t)|_H^r dt\Big)^{\frac{q}{r}}\Big]^{\frac{p}{q}} d\mu(x) < \infty,$$

and let $L^p(G;L^r_{\mathbb{F}}(0,T;L^q(\Omega;H)))$ be the set of all strongly $\mathcal{G}\times\mathcal{B}([0,T])\times\mathcal{F}_T$-measurable functions $\Phi : G\times[0,T]\times\Omega\to H$ such that for any $x\in G$, the process $\Phi(x,\cdot,\cdot)$ is $\mathbf{F}$-adapted and

$$\int_G \Big[\int_0^T \Big(\mathbb{E}|\Phi(x,t)|_H^q dt\Big)^{\frac{r}{q}}\Big]^{\frac{p}{r}} d\mu(x) < \infty.$$

In a similar way, we also define

$$\begin{cases} L^\infty(G;L^q_{\mathbb{F}}(\Omega;L^r(0,T;H))), & L^p(G;L^q_{\mathbb{F}}(\Omega;L^\infty(0,T;H))),\\ L^\infty(G;L^q_{\mathbb{F}}(\Omega;L^\infty(0,T;H))), & L^p(G;L^q_{\mathbb{F}}(\Omega;C([0,T];H))),\\ L^\infty(G;L^r_{\mathbb{F}}(0,T;L^q(\Omega;H))), & L^p(G;L^\infty_{\mathbb{F}}(0,T;L^q(\Omega;H))),\\ L^\infty(G;L^\infty_{\mathbb{F}}(0,T;L^q(\Omega;H))), & L^p(G;C_{\mathbb{F}}([0,T];L^q(\Omega;H))). \end{cases}$$

One can show that, for any $p,q,r\in[1,\infty]$, both $L^p(G;L^q_{\mathbb{F}}(\Omega;L^r(0,T;H)))$ and $L^p(G;L^r_{\mathbb{F}}(0,T;L^q(\Omega;H)))$ are Banach spaces (with the canonical norms).

Denote by $L_{S,\mathbb{F}}(0,T;H)$ the set of all H-valued, $\mathbf{F}$-adapted, simple processes of the following form:

$$f(t,\omega) = \sum_{i=0}^{\ell-1} \chi_{[t_i,t_{i+1})}(t)\xi_i(\omega), \tag{2.80}$$

where $\ell\in\mathbb{N}$, $0=t_0<t_1<\cdots<t_\ell=T$, $\xi_i\in L^\infty_{\mathcal{F}_{t_i}}(\Omega;H)$. We simply denote $L_{S,\mathbb{F}}(0,T;\mathbb{R})$ by $L_{S,\mathbb{F}}(0,T)$. One can show the following result.

Lemma 2.75. *For $r\in[1,\infty)$ and $s\in[1,\infty]$, the set $L_{S,\mathbb{F}}(0,T;H)$ is dense in $L^r_{\mathbb{F}}(0,T;L^s(\Omega;H))$ and $L^s_{\mathbb{F}}(\Omega;L^r(0,T;H))$.*

For any given $s \in [1,\infty]$, we denote by $C^\infty_{0,\mathbb{F}}((0,T);L^s(\Omega;H))$ the set of all H-valued, s-th power integrable (w.r.t. the sample point), $\mathbf{F}$-adapted processes $\phi(\cdot)$ such that $\phi(\cdot):(0,T)\to L^s_{\mathcal{F}_T}(\Omega;H)$ is an infinitely differentiable (vector-valued) function and has a compact support in $(0,T)$. We have the following result.

Lemma 2.76. *For any $r\in[1,\infty)$ and $s\in[1,\infty]$, the set $C^\infty_{0,\mathbb{F}}((0,T);L^\infty(\Omega;H))$ is dense in $L^r_{\mathbb{F}}(0,T;L^s(\Omega;H))$.*

Proof: It suffices to show that for any given $f\in L^r_{\mathbb{F}}(0,T;L^s(\Omega;H))$ and $\varepsilon>0$, there is a $g\in C^\infty_{0,\mathbb{F}}((0,T);L^\infty(\Omega;H))$ such that $|f-g|_{L^r_{\mathbb{F}}(0,T;L^s(\Omega;H))}<\varepsilon$. By Lemma 2.75, we can find an $f_n=\sum_{i=1}^n\chi_{[t_i,t_{i+1})}(t)\xi_i$, where $n\in\mathbb{N}$, $0=t_1<t_2<\cdots<t_n<t_{n+1}=T$ and $\xi_i\in L^\infty_{\mathcal{F}_{t_i}}(\Omega;H)$, such that

$$|f-f_n|_{L^r_{\mathbb{F}}(0,T;L^s(\Omega;H))}<\frac{\varepsilon}{2}.$$

On the other hand, for each $\chi_{[t_i,t_{i+1})}$, we can find a $g_i\in C^\infty_0(t_i,t_{i+1})$ such that

$$|\chi_{[t_i,t_{i+1})}-g_i|_{L^r(0,T)}\le\frac{\varepsilon}{2n(1+|\xi_i|_{L^s(\Omega;H)})}.$$

Write $g=\sum_{i=1}^n g_i(t)\xi_i$. Then, $g\in C^\infty_{0,\mathbb{F}}((0,T);L^\infty(\Omega;H))$, and

$$\begin{aligned}&|f-g|_{L^r_{\mathbb{F}}(0,T;L^s(\Omega;H))}\le|f-f_n|_{L^r_{\mathbb{F}}(0,T;L^s(\Omega;H))}+|f_n-g|_{L^r_{\mathbb{F}}(0,T;L^s(\Omega;H))}\\&<\frac{\varepsilon}{2}+\sum_{i=1}^n|\chi_{[t_i,t_{i+1})}\xi_i-g_i\xi_i|_{L^r_{\mathbb{F}}(0,T;L^s(\Omega;H))}<\varepsilon.\end{aligned}$$

This completes the proof of Lemma 2.76. □

We shall present below several sequential Banach-Alaoglu-type theorems (for operator-valued stochastic processes/random variables), which will be useful later.

In the rest of this section, we assume that Y_1 is a separable Banach space, Y_2 is a reflexive Banach space, and $L^p(\Omega,\mathcal{F}_T,\mathbb{P})$ (with $1\le p<\infty$) is separable. By means of Theorems 2.63 and 2.73, it is easy to show the following result.

Theorem 2.77. *Let $1\le p_1,p_2<\infty$ and $1<q_1,q_2<\infty$. Assume that $\{\mathcal{G}_n\}_{n=1}^\infty$ is a sequence of uniformly bounded, pointwise defined linear operators from $L^{p_1}_{\mathbb{F}}(0,T;L^{p_2}(\Omega;Y_1))$ to $L^{q_1}_{\mathbb{F}}(0,T;L^{q_2}(\Omega;Y_2))$. Then, there exist a subsequence $\{\mathcal{G}_{n_k}\}_{k=1}^\infty\subset\{\mathcal{G}_n\}_{n=1}^\infty$ and a $\mathcal{G}\in\mathcal{L}_{pd}\big(L^{p_1}_{\mathbb{F}}(0,T;L^{p_2}(\Omega;Y_1));L^{q_1}_{\mathbb{F}}(0,T;L^{q_2}(\Omega;Y_2))\big)$ such that, for any $u(\cdot)\in L^{p_1}_{\mathbb{F}}(0,T;L^{p_2}(\Omega;Y_1))$,*

$$\mathcal{G}u(\cdot)=(\mathrm{w})\text{-}\lim_{k\to\infty}\mathcal{G}_{n_k}u(\cdot)\quad\text{in } L^{q_1}_{\mathbb{F}}(0,T;L^{q_2}(\Omega;Y_2)).$$

Moreover,

$$|\mathcal{G}|_{\mathcal{L}(L^{p_1}_{\mathbb{F}}(0,T;L^{p_2}(\Omega;Y_1));L^{q_1}_{\mathbb{F}}(0,T;L^{q_2}(\Omega;Y_2)))}$$
$$\leq \sup_{n\in\mathbb{N}} |\mathcal{G}_n|_{\mathcal{L}(L^{p_1}_{\mathbb{F}}(0,T;L^{p_2}(\Omega;Y_1));L^{q_1}_{\mathbb{F}}(0,T;L^{q_2}(\Omega;Y_2))}.$$

From the proof of Theorem 2.63, similarly to Theorem 2.77, it is easy to deduce the following result.

Corollary 2.78. *Assume that* $1 < q_1, q_2 < \infty$, *and* $\{\mathcal{G}_n\}_{n=1}^{\infty}$ *is uniformly bounded in* $\mathcal{L}_{pd}\big(Y_1; L^{q_1}_{\mathbb{F}}(0,T;L^{q_2}(\Omega;Y_2))\big)$. *Then, there exist a subsequence* $\{\mathcal{G}_{n_k}\}_{k=1}^{\infty} \subset \{\mathcal{G}_n\}_{n=1}^{\infty}$ *and a* $\mathcal{G} \in \mathcal{L}_{pd}\big(Y_1; L^{q_1}_{\mathbb{F}}(0,T;L^{q_2}(\Omega;Y_2))\big)$ *such that*

$$\mathcal{G}y = (\mathrm{w})\text{-}\lim_{k\to\infty} \mathcal{G}_{n_k} y \quad \textit{in } L^{q_1}_{\mathbb{F}}(0,T;L^{q_2}(\Omega;Y_2)), \qquad \forall\, y \in Y_1.$$

Moreover, $|\mathcal{G}|_{\mathcal{L}(Y_1;L^{q_1}_{\mathbb{F}}(0,T;L^{q_2}(\Omega;Y_2)))} \leq \sup_{n\in\mathbb{N}} |\mathcal{G}_n|_{\mathcal{L}(Y_1;L^{q_1}_{\mathbb{F}}(0,T;L^{q_2}(\Omega;Y_2))}$.

Proceeding exactly as in the proof of Theorem 2.63, similarly to Theorem 2.77, we can show the following two results (Hence, the detailed proof will be omitted).

Theorem 2.79. *Assume that* $1 \leq p_1 < \infty$, $1 < q_1 < \infty$, *and* $\{\mathcal{G}_n\}_{n=1}^{\infty}$ *is uniformly bounded in* $\mathcal{L}_{pd}\big(L^{p_1}_{\mathcal{F}_T}(\Omega;Y_1); L^{q_1}_{\mathcal{F}_T}(\Omega;Y_2)\big)$. *Then, there exist a subsequence* $\{\mathcal{G}_{n_k}\}_{k=1}^{\infty} \subset \{\mathcal{G}_n\}_{n=1}^{\infty}$ *and an* $\mathcal{G} \in \mathcal{L}_{pd}\big(L^{p_1}_{\mathcal{F}_T}(\Omega;Y_1); L^{q_1}_{\mathcal{F}_T}(\Omega;Y_2)\big)$ *such that*

$$\mathcal{G}u(\cdot) = (\mathrm{w})\text{-}\lim_{k\to\infty} \mathcal{G}_{n_k} u(\cdot) \quad \textit{in } L^{q_1}_{\mathcal{F}_T}(\Omega;Y_2), \qquad \forall\, u(\cdot) \in L^{p_1}_{\mathcal{F}_T}(\Omega;Y_1).$$

Moreover, $|\mathcal{G}|_{\mathcal{L}(L^{p_1}_{\mathcal{F}_T}(\Omega;Y_1);L^{q_1}_{\mathcal{F}_T}(\Omega;Y_2))} \leq \sup_{n\in\mathbb{N}} |\mathcal{G}_n|_{\mathcal{L}(L^{p_1}_{\mathcal{F}_T}(\Omega;Y_1);L^{q_1}_{\mathcal{F}_T}(\Omega;Y_2))}$.

Theorem 2.80. *Assume that* $1 \leq p_1 < \infty$, $1 < q_1, q_2 < \infty$, $0 \leq t_0 \leq T$, *and* $\{\mathcal{G}_n\}_{n=1}^{\infty}$ *is uniformly bounded in* $\mathcal{L}_{pd}\big(L^{p_1}_{\mathcal{F}_{t_0}}(\Omega;Y_1); L^{q_1}_{\mathbb{F}}(t_0,T;L^{q_2}(\Omega;Y_2))\big)$. *Then, there exist a subsequence* $\{\mathcal{G}_{n_k}\}_{k=1}^{\infty} \subset \{\mathcal{G}_n\}_{n=1}^{\infty}$ *and a* $\mathcal{G} \in \mathcal{L}_{pd}\big(L^{p_1}_{\mathcal{F}_{t_0}}(\Omega;Y_1); L^{q_1}_{\mathbb{F}}(t_0,T;L^{q_2}(\Omega;Y_2))\big)$ *such that*

$$\mathcal{G}u(\cdot) = (\mathrm{w})\text{-}\lim_{k\to\infty} \mathcal{G}_{n_k} u(\cdot) \quad \textit{in } L^{q_1}_{\mathbb{F}}(t_0,T;L^{q_2}(\Omega;Y_2)), \quad \forall\, u(\cdot) \in L^{p_1}_{\mathcal{F}_{t_0}}(\Omega;Y_1).$$

Moreover,

$$|\mathcal{G}|_{\mathcal{L}(L^{p_1}_{\mathcal{F}_{t_0}}(\Omega;Y_1);L^{q_1}_{\mathbb{F}}(t_0,T;L^{q_2}(\Omega;Y_2)))} \leq \sup_{n\in\mathbb{N}} |\mathcal{G}_n|_{\mathcal{L}(L^{p_1}_{\mathcal{F}_{t_0}}(\Omega;Y_1);L^{q_1}_{\mathbb{F}}(t_0,T;L^{q_2}(\Omega;Y_2))}.$$

2.7 Stopping Times

In this section, we shall introduce a special class of random variables, i.e., stopping times. We fix a filtered probability space $(\Omega, \mathcal{F}, \mathbf{F}, \mathbb{P})$ with right continuous filtration $\mathbf{F} = \{\mathcal{F}_t\}_{t\in[0,\infty)}$ and a Banach space H.

Definition 2.81. *A map $\tau : \Omega \to [0, \infty]$ is called an $\mathbf{F}$-stopping time, or simply a stopping time in the case that the filtration $\mathbf{F}$ is clear from the context, if $\{\tau \le t\} \in \mathcal{F}_t$ for any $t \ge 0$.*

Obviously, when τ is a stopping time, it is a nonnegative random variable taking possibly infinite value. Define

$$\mathcal{F}_\tau = \Big\{A \in \mathcal{F} \ \big|\ A \cap \{\tau \le t\} \in \mathcal{F}_t,\ \forall\, t \ge 0\Big\}.$$

It is clear that $\mathcal{F}_\tau$ is a sub-σ-field of $\mathcal{F}$. The sets in $\mathcal{F}_\tau$ can be thought of as events which may occur before the time τ. The constants, i.e., $\tau(\omega) \equiv s$ ($s \ge 0$) for every ω, are stopping times and in that case $\mathcal{F}_\tau = \mathcal{F}_s$ (Recall that $\mathbf{F}$ can be regarded as a set of information describing the history of some phenomenon and $\mathcal{F}_s$ is the σ-field of events prior to s). Stopping times thus appear as generalizations of constant times for each of which one can define a "past" which is consistent with the "past" of constant times. A stopping time τ is a random variable such that the event "τ has occurred up to t" depends only on the history up to t, and not on any further information in the future.

By means of the right-continuity of the filtration, one has

Proposition 2.82. 1) *A map $\tau : \Omega \to [0, \infty]$ is a stopping time if and only if $\{\tau < t\} \in \mathcal{F}_t$ for all $t > 0$.*

2) *Let τ be a stopping time. Then $A \in \mathcal{F}_\tau$ if and only if $A \cap \{\tau < t\} \in \mathcal{F}_t$ for all $t > 0$.*

A stopping time may be thought of as the first time when some physical event occurs.

Example 2.83. Let $X(\cdot)$ be an $\mathbf{F}$-adapted and continuous process with values in H. Let $D \subset H$ be an open set. Then, both the first entrance time of the process $X(\cdot)$ to D, i.e.,

$$\sigma_D(\omega) \stackrel{\Delta}{=} \inf\Big\{t \ge 0 \ \big|\ X(t, \omega) \in D\Big\},$$

and the first exit time of the process $X(\cdot)$ from D, i.e.,

$$\tau_D(\omega) \stackrel{\Delta}{=} \inf\Big\{t \ge 0 \ \big|\ X(t, \omega) \notin D\Big\}$$

are stopping times. (Here, we agree that $\inf \emptyset = +\infty$).

Some basic properties of stopping times are listed in the following proposition.

Proposition 2.84. *Let σ, τ and σ_i $(i = 1, 2, \cdots)$ are stopping times. Then,*

1) The random variables $\sigma + \tau$, $\sup_{i\in\mathbb{N}} \sigma_i$, $\inf_{i\in\mathbb{N}}\sigma_i$, $\overline{\lim_{i\to\infty}}\, \sigma_i$ and $\underline{\lim}_{i\to\infty}\, \sigma_i$ are stopping times. Also, the events $\{\sigma > \tau\}$, $\{\sigma \geq \tau\}$ and $\{\sigma = \tau\}$ belong to $\mathcal{F}_{\sigma\wedge\tau}$;

2) The process $Y(\cdot) \overset{\Delta}{=} \tau \wedge \cdot$ is $\mathbf{F}$-progressively measurable;

3) For any $A \in \mathcal{F}_\sigma$, it holds $A \cap \{\sigma \leq \tau\} \in \mathcal{F}_\tau$. Particularly, if $\sigma \leq \tau$ a.s., then $\mathcal{F}_\sigma \subset \mathcal{F}_\tau$;

4) Let $\hat{\sigma} = \inf_{i\in\mathbb{N}} \sigma_i$. Then $\bigcap_{i=1}^{\infty} \mathcal{F}_{\sigma_i} = \mathcal{F}_{\hat{\sigma}}$; Particularly, $\mathcal{F}_{\sigma_1} \cap \mathcal{F}_{\sigma_2} = \mathcal{F}_{\sigma_1\wedge\sigma_2}$.

The following result will be sometimes technically useful.

Proposition 2.85. *Every stopping time is the decreasing limit of a sequence of stopping times taking only finitely many values.*

Proof: For a stopping time τ, write

$$\tau_k = \sum_{\ell=1}^{k2^k} \frac{\ell}{2^k} \chi_{\{\frac{\ell-1}{2^k} \leq \tau < \frac{\ell}{2^k}\}} + (+\infty)\chi_{\{\tau \geq k\}}, \quad k = 1, 2, \cdots .$$

It is easy to check that τ_k is a stopping time and that $\{\tau_k\}_{k=1}^{\infty}$ decreases to τ. □

The following result provides a characterization of $\mathcal{F}_\tau$-random variable.

Proposition 2.86. *Let τ be a stopping time and ξ be a random variable with values in H. Then ξ is $\mathcal{F}_\tau$-measurable if and only if $\xi\chi_{\{\tau\leq t\}}$ is $\mathcal{F}_t$-measurable for all $t \geq 0$.*

Proof: For simplicity, we only consider the case $H = \mathbb{R}$. If ξ is $\mathcal{F}_\tau$-measurable, then there is a sequence of $\mathcal{F}_\tau$-measurable simple functions

$$\xi_j \overset{\Delta}{=} \sum_{i=1}^{n_j} \xi_j^i \chi_{A_j^i} \to \xi, \qquad \text{as } j \to \infty, \ \mathbb{P}\text{-a.s.},$$

where $\xi_j^i \in \mathbb{R}$, $n_j \in \mathbb{N}$, $A_j^i \in \mathcal{F}_\tau$ and $j = 1, 2, \cdots$. Obviously, $\xi_j\chi_{\{\tau\leq t\}} = \sum_{i=1}^{n_j} \xi_j^i \chi_{A_j^i\cap\{\tau\leq t\}}$ is $\mathcal{F}_t$-measurable. Letting $j \to \infty$, we see that $\xi\chi_{\{\tau\leq t\}}$ is also $\mathcal{F}_t$-measurable for all $t \geq 0$.

Conversely, if $\xi\chi_{\{\tau\leq t\}}$ is $\mathcal{F}_t$-measurable for all $t \geq 0$, then

$$\begin{aligned}
&\{\xi \leq a\} \cap \{\tau \leq t\} = \{\xi\chi_{\{\tau\leq t\}} \leq a\} \in \mathcal{F}_t, \qquad \forall\, a < 0, \\
&\{\xi > a\} \cap \{\tau \leq t\} = \{\xi\chi_{\{\tau\leq t\}} > a\} = \{\xi\chi_{\{\tau\leq t\}} \leq a\}^c \in \mathcal{F}_t, \qquad \forall\, a \geq 0.
\end{aligned}$$

Therefore, $\{\xi \leq a\} \in \mathcal{F}_\tau$ for all $a \in \mathbb{R}$, which implies that ξ is $\mathcal{F}_\tau$-measurable. □

Proposition 2.87. *Let σ and τ be two stopping times, and X be an integrable random variable with values in H. Then*

$$\begin{cases} \chi_{\{\sigma>\tau\}}\mathbb{E}(X \mid \mathcal{F}_\tau) = \mathbb{E}(\chi_{\{\sigma>\tau\}}X \mid \mathcal{F}_\tau) = \chi_{\{\sigma>\tau\}}\mathbb{E}(X \mid \mathcal{F}_{\sigma\wedge\tau}), \\ \chi_{\{\sigma\geq\tau\}}\mathbb{E}(X \mid \mathcal{F}_\tau) = \mathbb{E}(\chi_{\{\sigma\geq\tau\}}X \mid \mathcal{F}_\tau) = \chi_{\{\sigma\geq\tau\}}\mathbb{E}(X \mid \mathcal{F}_{\sigma\wedge\tau}), \\ \mathbb{E}(\mathbb{E}(X \mid \mathcal{F}_\tau) \mid \mathcal{F}_\sigma) = \mathbb{E}(X \mid \mathcal{F}_{\sigma\wedge\tau}). \end{cases}$$

Proof: The first equalities in the first two assertions are obvious.

To prove the second equality in the first assertion, we note that

$$\chi_{\{\sigma>\tau\}}\mathbb{E}(X \mid \mathcal{F}_\tau)\chi_{\{\sigma\wedge\tau\leq t\}} = \mathbb{E}(X \mid \mathcal{F}_\tau)\chi_{\{\tau\leq t\}}\chi_{\{\sigma>\tau,\,\sigma\wedge\tau\leq t\}}.$$

Since $\mathbb{E}(X \mid \mathcal{F}_\tau)$ is $\mathcal{F}_\tau$-measurable, by Proposition 2.86, $\mathbb{E}(X \mid \mathcal{F}_\tau)\chi_{(\tau\leq t)}$ is $\mathcal{F}_t$-measurable. Recall also that $\{\sigma > \tau\} \in \mathcal{F}_{\sigma\wedge\tau}$. Thus $\chi_{\{\sigma>\tau,\,\sigma\wedge\tau\leq t\}}$ is $\mathcal{F}_t$-measurable. Hence $\chi_{\{\sigma>\tau\}}\mathbb{E}(X \mid \mathcal{F}_\tau)\chi_{\{\sigma\wedge\tau\leq t\}}$ is $\mathcal{F}_t$-measurable for all $t \geq 0$. Hence, by Proposition 2.86 again, $\chi_{\{\sigma>\tau\}}\mathbb{E}(X \mid \mathcal{F}_\tau)$ is $\mathcal{F}_{\sigma\wedge\tau}$-measurable. Then,

$$\begin{aligned} \chi_{\{\sigma>\tau\}}\mathbb{E}(X \mid \mathcal{F}_\tau) &= \mathbb{E}(\chi_{\{\sigma>\tau\}}\mathbb{E}(X \mid \mathcal{F}_\tau) \mid \mathcal{F}_{\sigma\wedge\tau}) \\ &= \chi_{\{\sigma>\tau\}}\mathbb{E}(\mathbb{E}(X \mid \mathcal{F}_\tau) \mid \mathcal{F}_{\sigma\wedge\tau}) = \chi_{\{\sigma>\tau\}}\mathbb{E}(X \mid \mathcal{F}_{\sigma\wedge\tau}), \end{aligned}$$

which proves the first assertion. The second one can be proved similarly. Finally,

$$\begin{aligned} &\mathbb{E}(\mathbb{E}(X \mid \mathcal{F}_\tau) \mid \mathcal{F}_\sigma) = \mathbb{E}(\chi_{\{\sigma>\tau\}}\mathbb{E}(X \mid \mathcal{F}_\tau) \mid \mathcal{F}_\sigma) + \mathbb{E}(\chi_{\{\tau\geq\sigma\}}\mathbb{E}(X \mid \mathcal{F}_\tau) \mid \mathcal{F}_\sigma) \\ &= \mathbb{E}(\chi_{\{\sigma>\tau\}}\mathbb{E}(X \mid \mathcal{F}_{\sigma\wedge\tau}) \mid \mathcal{F}_\sigma) + \chi_{\{\tau\geq\sigma\}}\mathbb{E}(\mathbb{E}(X \mid \mathcal{F}_\tau) \mid \mathcal{F}_{\tau\wedge\sigma}) \\ &= \chi_{\{\sigma>\tau\}}\mathbb{E}(X \mid \mathcal{F}_{\sigma\wedge\tau}) + \chi_{\{\tau\geq\sigma\}}\mathbb{E}(X \mid \mathcal{F}_{\tau\wedge\sigma}) = \mathbb{E}(X \mid \mathcal{F}_{\tau\wedge\sigma}), \end{aligned}$$

which gives the third assertion. □

Finally, we show the following result.

Proposition 2.88. *Let $X(\cdot)$ be an $\mathbf{F}$-adapted process with values in H, and τ be a stopping time. Then the random variable $X(\tau)$ is $\mathcal{F}_\tau$-measurable and the process $X(\tau \wedge \cdot)$ is $\mathbf{F}$-adapted.*

Proof: We first prove that $X(\tau \wedge \cdot)$ is $\mathbf{F}$-adapted. By the second conclusion in Proposition 2.84, the process $\tau \wedge \cdot$ is $\mathbf{F}$-progressively measurable. Thus for each $t \geq 0$, the map $(s, \omega) \mapsto (\tau(\omega) \wedge s, \omega)$ is measurable from $([0, t] \times \Omega, \mathcal{B}([0, t]) \times \mathcal{F}_t)$ into itself. On the other hand, by the progressively measurability of $X(t)$, the map $(s, \omega) \mapsto X(s, \omega)$ is measurable from $([0, t] \times \Omega, \mathcal{B}([0, t]) \times \mathcal{F}_t)$ into $(H, \mathcal{B}(H))$. Hence, the map $(s, \omega) \mapsto X(\tau(\omega) \wedge s, \omega)$ is measurable from $([0, t] \times \Omega, \mathcal{B}([0, t]) \times \mathcal{F}_t)$ into $(H, \mathcal{B}(H))$, which yields the $\mathbf{F}$-progressive measurability of $X(\tau\wedge t)$. Particularly, $X(\tau\wedge t)$ is $\mathcal{F}_t$-measurable for all $t \geq 0$.

Next, for any $B \in \mathcal{B}(H)$,

$$\{X(\tau) \in B\} \cap \{\tau \leq t\} = \{X(\tau \wedge t) \in B\} \cap \{\tau \leq t\} \in \mathcal{F}_t, \quad \forall\, t \geq 0.$$

Therefore, $\{X(\tau) \in B\} \in \mathcal{F}_\tau$. Thus $X(\tau)$ is $\mathcal{F}_\tau$-measurable. □

2.8 Martingales

Let $(\Omega, \mathcal{F}, \mathbf{F}, \mathbb{P})$ (with $\mathbf{F} = \{\mathcal{F}_t\}_{t\in\mathcal{I}}$) be a filtered probability space satisfying the usual condition, and denote by $\mathbb{F}$ the progressive σ-field w.r.t. $\mathbf{F}$. We first consider the case of real valued martingales. Then we address to the vector-valued martingales.

2.8.1 Real Valued Martingales

Let $\mathcal{I}$ be a time parameter set: $\mathcal{I} = \{0, 1, 2, \cdots\}$ ($\hat{\mathcal{I}} = \{0, 1, 2, \cdots, \infty\}$) in the discrete time case or $\mathcal{I} = [0, \infty)$ ($\hat{\mathcal{I}} = [0, \infty]$) in the continuous time case. We begin with the following notions.

Definition 2.89. *A real valued process $X = \{X(t)\}_{t\in\mathcal{I}}$ is called an $\{\mathcal{F}_t\}_{t\in\mathcal{I}}$-martingale, or simply a martingale (resp. supermartingale, submartingale) in the case that the filtration $\{\mathcal{F}_t\}_{t\in\mathcal{I}}$ is clear from the context, if*

1) $X(t)$ is $\mathcal{F}_t$-measurable and integrable (w.r.t. the probability measure $\mathbb{P}$) for each $t \in \mathcal{I}$; and

2) $\mathbb{E}(X(t) \mid \mathcal{F}_s) = X(s)$ (resp. $\leq X(s)$, $\geq X(s)$), a.s. *for any $t, s \in \mathcal{I}$ with $s < t$.*

In the sequel, for any real numbers b and c, write $b \vee c \equiv \max(b, c)$ and $b \wedge c \equiv \min(b, c)$.

Example 2.90. 1) Suppose that X is an integrable random variable. Then $\{\mathbb{E}(X \mid \mathcal{F}_t)\}_{t\in\mathcal{I}}$ is a martingale.

2) Let $X(\cdot)$ be a martingale (*resp.* submartingale) and f be a convex function (*resp.* non-decreasing convex function) so that $f(X(t))$ is integrable for each $t \in \mathcal{I}$. Then $f(X(\cdot))$ is a submartingale. Particularly, for any $a \in \mathbb{R}$, the processes $X(\cdot)^+$ and $X(\cdot) \vee a$ are submartingales whenever so is $X(\cdot)$.

Obviously, $X(\cdot)$ is a martingale if and only if it is both sub- and suppermartingale. Martingales are a class of important stochastic processes, which are easily computable and estimable.

First, we consider the discrete time case. In this case, we write X_n instead of $X(n)$, and all stopping times are assumed to be valued in $\{0, 1, 2, \cdots, \infty\}$.

The following result, called Doob's stopping theorem, is the basis to show the inequalities for martingales in the sequel.

Theorem 2.91. *Let $\{X_n\}_{n\in\mathcal{I}}$ be a martingale (resp. supermartingale, submartingale), σ and τ be two bounded stopping times with $\sigma \leq \tau$,* a.s. *Then*

$$\mathbb{E}(X_\tau \mid \mathcal{F}_\sigma) = X_\sigma \ (\textit{resp.} \leq X_\sigma,\ \geq X_\sigma), \quad \text{a.s.}$$

Proof: It suffices to consider the case of supermartingale.

Suppose $\sigma \vee \tau \leq M$, a.s. We need to show that for every $A \in \mathcal{F}_\sigma$, it holds

$$\int_A X_\sigma d\mathbb{P} \geq \int_A X_\tau d\mathbb{P}.$$

Suppose first that $\sigma \leq \tau \leq \sigma + 1$. Put

$$B_n \stackrel{\Delta}{=} A \cap \{\sigma = n\} \cap \{\tau > \sigma\} = A \cap \{\sigma = n\} \cap \{\tau > n\} \in \mathcal{F}_n.$$

It is clear that $A \cap \{\tau > \sigma\} = \bigcup_{n=0}^{\infty} B_n$. Therefore

$$\int_A (X_\sigma - X_\tau) d\mathbb{P} = \int_{A\cap\{\tau\geq\sigma\}} (X_\sigma - X_\tau) d\mathbb{P} = \int_{A\cap\{\tau>\sigma\}} (X_\sigma - X_\tau) d\mathbb{P}$$

$$= \sum_{n=0}^{\infty} \int_{B_n} (X_\sigma - X_\tau) d\mathbb{P} = \sum_{n=0}^{\infty} \int_{B_n} (X_n - X_{n+1}) d\mathbb{P} \geq 0.$$

In the general case, write $\gamma_k = \tau \wedge (\sigma + k)$, for $k = 0, 1, 2, \cdots, M$. Then γ_k are $\{\mathcal{F}_n\}_{n=1}^{\infty}$-stopping times, and

$$0 \leq \gamma_{j+1} - \gamma_j \leq 1, \qquad j = 0, 1, 2, \cdots, M-1.$$

Note that $\gamma_0 = \sigma$ and $\gamma_M = \tau$. Thus from the case discussed above,

$$\int_A X_\sigma d\mathbb{P} = \int_A X_{\gamma_0} d\mathbb{P} \geq \int_A X_{\gamma_1} d\mathbb{P} \geq \cdots \geq \int_A X_{\gamma_M} d\mathbb{P} = \int_A X_\tau d\mathbb{P}.$$

□

The following result is called Doob's inequality.

Theorem 2.92. *Let $\{X_n\}_{n\in\mathcal{I}}$ be a submartingale. Then for every $\lambda > 0$ and $m \in \mathcal{I}$,*

$$\lambda\mathbb{P}(\{\max_{0\leq n\leq m} X_n \geq \lambda\}) \leq \int_{\{\max_{0\leq n\leq m} X_n \geq \lambda\}} X_m d\mathbb{P} \leq \mathbb{E}|X_m|,$$

and

$$\lambda\mathbb{P}(\{\min_{0\leq n\leq m} X_n \leq -\lambda\}) \leq \mathbb{E}(|X_0| + |X_m|).$$

Proof: We define a stopping time as follows

$$\sigma = \begin{cases} \min\{n \leq m \mid X_n \geq \lambda\}, & \text{if } \{n \leq m \mid X_n \geq \lambda\} \neq \emptyset, \\ m, & \text{if } \{n \leq m \mid X_n \geq \lambda\} = \emptyset. \end{cases}$$

Obviously, $\sigma \leq m$. Therefore

$$\mathbb{E}X_m \geq \mathbb{E}X_\sigma = \int_{\{\max_{0\leq n\leq m} X_n \geq \lambda\}} X_\sigma d\mathbb{P} + \int_{\{\max_{0\leq n\leq m} X_n < \lambda\}} X_\sigma d\mathbb{P}$$

$$\geq \lambda\mathbb{P}(\{\max_{0\leq n\leq m} X_n \geq \lambda\}) + \int_{\{\max_{0\leq n\leq m} X_n < \lambda\}} X_m d\mathbb{P},$$

which yields the first inequality.

The second one is obtained from $\mathbb{E}X_0 \leq \mathbb{E}X_\tau$, where

$$\tau = \begin{cases} \min\{n \leq m \mid X_n \leq -\lambda\}, & \text{if } \{n \leq m \mid X_n \leq -\lambda\} \neq \emptyset, \\ m, & \text{if } \{n \leq m \mid X_n \leq -\lambda\} = \emptyset. \end{cases}$$

This completes the proof of Theorem 2.92. □

As a consequence of Theorem 2.92, one has the following result.

Corollary 2.93. *Let $\{X_n\}_{n\in\mathcal{I}}$ be a martingale or nonnegative submartingale such that $\mathbb{E}|X_n|^p < \infty$ for some $p \geq 1$ and all $n \in \mathcal{I}$. Then, for every $m \in \mathcal{I}$ and $\lambda > 0$,*

$$\mathbb{P}(\{\max_{0\leq n\leq m} |X_n| \geq \lambda\}) \leq \lambda^{-p}\mathbb{E}|X_m|^p,$$

and if $p > 1$,

$$\mathbb{E}(\max_{0\leq n\leq m} |X_n|^p) \leq \Big(\frac{p}{p-1}\Big)^p \mathbb{E}|X_m|^p.$$

Proof: Obviously, $\{|X_n|^p\}_{n\in\mathcal{I}}$ is a submartingale and the first assertion follows from Theorem 2.92.

As for the second one, we set $Y = \max\limits_{0\leq n\leq m} |X_n|$. Then, by Theorem 2.92, we have

$$\lambda\mathbb{P}(\{Y \geq \lambda\}) \leq \int_{\{Y\geq\lambda\}} |X_m| d\mathbb{P}. \tag{2.81}$$

Hence, by (2.81) and Lemma 2.23,

$$\begin{aligned} \mathbb{E}Y^p &\leq p\int_0^\infty \lambda^{p-2}\int_\Omega \chi_{\{Y\geq\lambda\}}|X_m|d\mathbb{P}d\lambda \\ &= p\int_\Omega |X_m|\int_0^Y \lambda^{p-2}d\lambda d\mathbb{P} = \frac{p}{p-1}\int_\Omega |X_m|Y^{p-1}d\mathbb{P}, \end{aligned}$$

which yields the desired result. □

For a real valued, **F**-adapted process $X = \{X_n\}_{n\in\mathcal{I}}$, and an interval $[a,b]$, with $-\infty < a < b < \infty$, we set

$$\left.\begin{aligned} &\tau_0 = 0, \\ &\tau_1 = \min\{n \mid X_n \leq a\}, \\ &\tau_2 = \min\{n \geq \tau_1 \mid X_n \geq b\}, \\ &\cdots \\ &\tau_{2k+1} = \min\{n \geq \tau_{2k} \mid X_n \leq a\}, \\ &\tau_{2k+2} = \min\{n \geq \tau_{2k+1} \mid X_n \geq b\}, \\ &\cdots. \end{aligned}\right\} \tag{2.82}$$

($\min\emptyset = +\infty$ unless otherwise stated). It is clear that $\{\tau_n\}$ is an increasing sequence of stopping times. For any $m \in \mathcal{I}$, set

$$U_m^X(a,b)(\omega) \stackrel{\Delta}{=} \max\{k \mid \tau_{2k}(\omega) \le m\}.$$

Obviously, $U_m^X(a,b)$ is the number of upcrossings of $[a,b]$ by $\{X_n\}_{n=0}^m$. The following result is called Doob's upcrossing inequality, which is used to derive convergence results for martingales.

Theorem 2.94. *Let $X = \{X_n\}_{n\in\mathcal{I}}$ be a submartingale. Then,*

$$\mathbb{E}(U_m^X(a,b)) \le \frac{1}{b-a}\mathbb{E}[(X_m-a)^+ - (X_0-a)^+].$$

Proof: Write $Y_n = (X_n-a)^+$ and $Y = \{Y_n\}_{n\in\mathcal{I}}$. Obviously, Y is a submartingale and $U_m^X(a,b) = U_m^Y(0,b-a)$. Let τ_i be defined as in (2.82) with X, a and b replaced by Y, 0 and $b-a$, respectively. Set $\tau_n' = \tau_n \wedge m$. Then, if $2j > m$,

$$Y_m - Y_0 = \sum_{n=1}^{2j}(Y_{\tau_n'} - Y_{\tau_{n-1}'}) = \sum_{n=1}^{j}(Y_{\tau_{2n}'} - Y_{\tau_{2n-1}'}) + \sum_{n=0}^{j-1}(Y_{\tau_{2n+1}'} - Y_{\tau_{2n}'}).$$

It is easy to see that the first term in the right-hand side is greater than or equal to $(b-a)U_m^Y(0,b-a)$. Also, $EY_{\tau_{2n+1}'} \ge EY_{\tau_{2n}'}$. These two facts yield the desired results. □

The following result characterizes the convergence of submartingales.

Theorem 2.95. *If $X = \{X_n\}_{n\in\mathcal{I}}$ is a submartingale such that $\sup_{n\in\mathcal{I}}\mathbb{E}X_n^+ < \infty$, or equivalently $\sup_{n\in\mathcal{I}}\mathbb{E}|X_n| < \infty$, then $X_\infty = \lim_{n\to\infty}X_n$ exists* a.s., *and X_∞ is integrable. In order that $\hat{X} = \{X_n\}_{n\in\hat{\mathcal{I}}}$ is a submartingale, i.e.,*

$$X_n \le \mathbb{E}(X_m \mid \mathcal{F}_n), \qquad \forall\, 0 \le n < m \le \infty,$$

it is necessary and sufficient that $\{X_n^+\}_{n\in\mathcal{I}}$ is uniformly integrable.

Proof: Since X is a submartingale, the equivalence of $\sup_{n\in\mathcal{I}}\mathbb{E}X_n^+ < \infty$ and $\sup_{n\in\mathcal{I}}\mathbb{E}|X_n| < \infty$ follows from

$$\sup_{n\in\mathcal{I}}\mathbb{E}X_n^+ \le \sup_{n\in\mathcal{I}}\mathbb{E}|X_n| = \sup_{n\in\mathcal{I}}(2\mathbb{E}X_n^+ - \mathbb{E}X_n) \le -\mathbb{E}X_0 + 2\sup_{n\in\mathcal{I}}\mathbb{E}X_n^+.$$

Set $U_\infty^X(a,b) = \lim_{m\to\infty}U_m^X(a,b)$. Clearly,

$$\{\underline{\lim}_{n\to\infty} X_n < \overline{\lim}_{n\to\infty} X_n\} \subset \bigcup_{r,r'\in\mathbb{Q},r<r'} \{U_\infty^X(r,r') = \infty\}.$$

By Theorem 2.94, for any $r, r' \in \mathbb{Q}$ with $r < r'$, noting $(X_m - r)^+ \le X_m^+ + |r|$, we get

$$\begin{aligned}\mathbb{E}(U_\infty^X(r,r')) &\le \lim_{m\to\infty}\mathbb{E}(U_m^X(r,r')) \le \frac{1}{r'-r}\overline{\lim_{m\to\infty}}\,\mathbb{E}((X_m-r)^+ - (X_0-r)^+)\\ &\le \frac{1}{r'-r}\overline{\lim_{m\to\infty}}\,\mathbb{E}(X_m^+ + |r| - (X_0-r)^+),\end{aligned}$$

and this limit is finite. Consequently,

$$\mathbb{P}(\{\varliminf_{n\to\infty} X_n < \varlimsup_{n\to\infty} X_n\}) = 0,$$

which proves that $X_\infty = \lim_{n\to\infty} X_n$ exists a.s. By Fatou's lemma, it follows that

$$\int_\Omega |X_\infty| d\mathbb{P} \le \varliminf_{n\to\infty} \int_\Omega |X_n| d\mathbb{P} \le \sup_{n\in\mathcal{I}} \mathbb{E}|X_n| < \infty,$$

which shows the integrability of X_∞.

Now, if $X_n \le \mathbb{E}(X_\infty \mid \mathcal{F}_n)$ for $n = 0, 1, 2, \cdots$, then by the property of conditional expectation, we have

$$X_n^+ \le \mathbb{E}(X_\infty^+ \mid \mathcal{F}_n), \quad \text{a.s.}$$

Clearly, $\{\mathbb{E}(X_\infty^+ \mid \mathcal{F}_n)\}_{n\in\mathcal{I}}$ is uniformly integrable. Hence, so is $\{X_n^+\}_{n\in\mathcal{I}}$.

Conversely, if $\{X_n^+\}_{n\in\mathcal{I}}$ is uniformly integrable, then, by

$$|X_n \vee (-a)| \le a + X_n^+, \qquad \forall a > 0,$$

one concludes that $\{X_n \vee (-a)\}_{n\in\mathcal{I}}$ is uniformly integrable. Thus, by Theorem 2.25, we see that

$$\lim_{n\to\infty} X_n \vee (-a) = X_\infty \vee (-a) \text{ in } L^1_{\mathcal{F}}(\Omega).$$

Since $\{X_n \vee (-a)\}_{n\in\mathcal{I}}$ is a submartingale,

$$\mathbb{E}(X_\infty \vee (-a) \mid \mathcal{F}_n) = \lim_{m\to\infty} \mathbb{E}(X_m \vee (-a) \mid \mathcal{F}_n) \ge X_n \vee (-a),$$

and so, letting $a \to +\infty$, one gets $\mathbb{E}(X_\infty \mid \mathcal{F}_n) \ge X_n$. □

For the martingales, one has the following further result.

Theorem 2.96. *Let $X = \{X_n\}_{n\in\mathcal{I}}$ be a martingale. Then the following conditions are equivalent.*

1) *$\lim_{n\to\infty} X_n = X_\infty$ exists a.s. and in $L^1_{\mathcal{F}}(\Omega)$;*

2) *$X_n = E(X_\infty \mid \mathcal{F}_n)$ for all $n \in \mathcal{I}$,* a.s.*;*

3) *$\{X_n\}_{n\in\mathcal{I}}$ is uniformly integrable.*

Furthermore, under one of the above conditions, $\hat{X} = \{X_n\}_{n\in\hat{\mathcal{I}}}$ is a martingale.

Proof: "1)⇒2)". For any $m \in \mathcal{I}$, we have $\mathbb{E}(X_n \mid \mathcal{F}_m) = X_m$ for all $n \ge m$. Letting $n \to \infty$, we obtain that $X_m = \mathbb{E}(X_\infty \mid \mathcal{F}_m)$, a.s.

"2)⇒3)". Obviously.

"3)⇒1)". The uniform integrability of $\{X_n\}_{n\in\mathcal{I}}$ implies the boundedness of $\{X_n\}_{n\in\mathcal{I}}$ in $L^1_{\mathcal{F}}(\Omega)$ and the uniform integrability of $\{X_n^+\}_{n\in\mathcal{I}}$. Now, the desired result follows from Theorems 2.95 and 2.25. □

Next, we consider, for a moment, martingales with the "reversed" time. Let $\{\mathcal{F}_n\}_{n=-\infty}^0$ be a family of sub-σ-fields of $\mathcal{F}$ such that $\mathcal{F}_0 \supset \mathcal{F}_{-1} \supset \mathcal{F}_{-2} \supset \cdots \supset$

$\mathcal{F}_{-\infty}$. We say $X = \{X_n\}_{n=-\infty}^0$ to be a martingale (*resp.* supermartingale, submartingale) if X_n is $\mathcal{F}_n$-adapted integrable random variable such that

$$\mathbb{E}(X_n \mid \mathcal{F}_m) = X_m \quad (\textit{resp.} \ \leq X_m, \ \geq X_m)$$

for every $n, m \in \{0, -1, -2, \cdots\}$ with $n > m$. One has the following convergence result.

Theorem 2.97. *Let* $X = \{X_n\}_{n=-\infty}^0$ *be a submartingale. Then*

1) $X_{-\infty} = \lim_{n\to-\infty} X_n$ *exists* a.s.;

2) X *is uniformly integrable and* $\lim_{n\to-\infty} X_n = X_{-\infty}$ *in* $L^1_{\mathcal{F}}(\Omega)$ *provided that* $\lim_{n\to-\infty} \mathbb{E}X_n > -\infty$.

Proof: The first assertion can be proved similarly to that of Theorem 2.95.

For the second one, in view of Theorem 2.25, it suffices to show the uniform integrability of X. For this, we fix any $\varepsilon > 0$ and choose an integer k such that

$$|\mathbb{E}X_{k_1} - \mathbb{E}X_{k_2}| < \varepsilon, \qquad \forall \text{ integers } k_1, k_2 \leq k.$$

Then, for any $\lambda > 0$ and integer $n \leq k$,

$$\begin{aligned}\int_{\{|X_n|>\lambda\}} |X_n| d\mathbb{P} &= \int_{\{X_n>\lambda\}} X_n d\mathbb{P} + \int_{\{X_n\geq-\lambda\}} X_n d\mathbb{P} - \mathbb{E}X_n \\ &\leq \int_{\{X_n>\lambda\}} X_k d\mathbb{P} + \int_{\{X_n\geq-\lambda\}} X_k d\mathbb{P} - \mathbb{E}X_k + \varepsilon \leq 2\int_{\{|X_n|>\lambda\}} |X_k| d\mathbb{P} + \varepsilon.\end{aligned}$$

Also,

$$\mathbb{P}(\{|X_n| > \lambda\}) \leq \frac{1}{\lambda}\mathbb{E}|X_n| = \frac{1}{\lambda}(2\mathbb{E}X_n^+ - \mathbb{E}X_n) \leq \frac{1}{\lambda}(2\mathbb{E}X_0^+ - \lim_{n\to-\infty} \mathbb{E}X_n).$$

Therefore, the uniform integrability of $\{X_n\}_{n=-\infty}^0$ follows. □

Now, we consider the continuous time case, i.e., $\mathcal{I} = [0, \infty)$. We begin with the following result.

Proposition 2.98. *Let* $X = \{X(t)\}_{t\in\mathcal{I}}$ *be a submartingale. Then,* $\mathbb{Q} \cap \mathcal{I} \ni t \mapsto X(t)$ *is finitely valued a.s. and possesses*

$$\lim_{\mathbb{Q}\cap\mathcal{I}\ni s\to t+} X(s) \ \ \textit{and} \ \ \lim_{\mathbb{Q}\cap\mathcal{I}\ni s\to t-} X(s), \qquad \textit{a.s.}, \forall\, t \geq 0.$$

Proof: Let $T > 0$ be given and $\{r_1, r_2, \cdots\}$ be an enumeration of the set $\mathbb{Q} \cap [0, T]$. For every $n \in \mathbb{N}$, if $\{s_1, s_2, \cdots, s_n\}$ is the rearrangement of the set $\{r_1, r_2, \cdots, r_n\}$ according to the natural order, then

$$Y_0 = X(0), \ Y_1 = X(s_1), \cdots, Y_n = X(s_n), \ Y_{n+1} = X(T)$$

defines a submartingale. Write $Y = \{Y_i\}_{i\in\{0,1,\cdots,n+1\}}$. By Theorems 2.92 and 2.94, for each $\lambda > 0$, we get

$$\mathbb{P}(\{\max_{0\le i\le n+1}|Y_i|>\lambda\})\le\frac{2}{\lambda}\big(\mathbb{E}|X(0)|+\mathbb{E}|X(T)|\big),$$

and for any $a,b\in\mathbb{Q}$ with $a<b$,

$$\mathbb{E}(U_n^Y(a,b))\le\frac{1}{b-a}\mathbb{E}(X(T)-a)^+.$$

Since this holds for every n, we have

$$\mathbb{P}(\{\sup_{t\in\mathbb{Q}\cap[0,T]}|X(t)|>\lambda\})\le\frac{2}{\lambda}\big(\mathbb{E}|X(0)|+\mathbb{E}|X(T)|\big),$$

and

$$\mathbb{E}\big(U_\infty^{X|_{\mathbb{Q}\cap[0,T]}}(a,b)\big)\le\frac{1}{b-a}\mathbb{E}\big(X(T)-a\big)^+,$$

where $X|_{\mathbb{Q}\cap[0,T]}$ denotes the restriction of X on the set $\mathbb{Q}\cap[0,T]$. By letting λ and $a<b$ run over respectively positive integers and pairs of rational, the assertion of the theorem follows. □

The following result is quite useful.

Theorem 2.99. *Let $X=\{X(t)\}_{t\in\mathcal{I}}$ be a submartingale. Then,*

1) $\widetilde{X}(t)\overset{\Delta}{=}\lim_{\mathbb{Q}\cap\mathcal{I}\ni r\to t+}X(r)$ *exists* a.s. *for any $t\in\mathcal{I}$, and the stochastic process $\widetilde{X}=\{\widetilde{X}(t)\}_{t\in\mathcal{I}}$ is a cádlàg submartingale;*

2) $X(t)\le\widetilde{X}(t)$ a.s. , $\quad\forall\, t\in\mathcal{I}$;

3) $\mathbb{P}(\{X(t)=\widetilde{X}(t)\})=1$ *for every $t\in\mathcal{I}$ if and only if $\mathbb{E}X(\cdot)$ is right-continuous.*

Proof: 1) By Proposition 2.98, we see that $\widetilde{X}(t)$ is well-defined and **F**-adapted.

To prove that $\{\widetilde{X}(t)\}_{t\in\mathcal{I}}$ is a submartingale, we fix any $s>t$ and choose arbitrarily a sequence $\{\varepsilon_n\}_{n=1}^\infty$ decreasing to 0 so that $t+\varepsilon_n\in\mathbb{Q}\cap\mathcal{I}$ for all $n\in\mathbb{N}$. Then, by Theorem 2.97, we see that

$$\lim_{n\to\infty}|X(t+\varepsilon_n)-\widetilde{X}(t)|_{L^1_{\mathcal{F}}(\Omega)}=0.$$

Similarly, we choose a sequence $\{\varepsilon_n'\}_{n=1}^\infty$ decreasing to 0 such that $s+\varepsilon_n'\in\mathbb{Q}\cap\mathcal{I}$ for all $n\in\mathbb{N}$. Then

$$\lim_{n\to\infty}|X(s+\varepsilon_n')-\widetilde{X}(s)|_{L^1_{\mathcal{F}}(\Omega)}=0.$$

Hence, for any $B\in\mathcal{F}_t$,

$$\int_B\widetilde{X}(t)d\mathbb{P}=\lim_{n\to\infty}\int_B X(t+\varepsilon_n)d\mathbb{P}\le\lim_{n\to\infty}\int_B X(s+\varepsilon_n')d\mathbb{P}=\int_B\widetilde{X}(s)d\mathbb{P}.$$

This implies that $\{\widetilde{X}(t)\}_{t\in\mathcal{I}}$ is a submartingale.

Now, using Proposition 2.98 again, we see that, for any $t_0 \in I$ and $\varepsilon > 0$, there is a $\delta > 0$ such that for any $s \in (t_0, t_0 + \delta) \cap \mathbb{Q}$,

$$|\widetilde{X}(t_0) - X(s)| < \varepsilon, \text{ a.s.}$$

Therefore, for any $r \in (t_0, t_0 + \delta)$,

$$|\widetilde{X}(t_0) - \widetilde{X}(r)| = \lim_{\mathbb{Q}\cap\mathcal{I}\ni s\to r+} |\widetilde{X}(t_0) - X(s)| \leq \varepsilon, \text{ a.s.}$$

Hence, $\lim_{r\to t_0+} \widetilde{X}(r) = \widetilde{X}(t_0)$, a.s. This yields the right-continuity of $\widetilde{X}$.

To show the existence of $\lim_{t\to t_0-} \widetilde{X}(t)$, we use Proposition 2.98 (to the submartingale $\{\widetilde{X}(t)\}_{t\in\mathcal{I}}$) to conclude that

$$\lim_{\mathbb{Q}\cap\mathcal{I}\ni t\to t_0-} \widetilde{X}(t), \quad \text{exists, a.s.}$$

Thus, for any $\varepsilon > 0$, there is a $\delta > 0$ such that

$$|\widetilde{X}(t_1) - \widetilde{X}(t_2)| < \varepsilon, \qquad \forall\, t_1, t_2 \in (t_0 - \delta, t_0) \cap \mathbb{Q} \cap \mathcal{I}.$$

By the right-continuity of $\widetilde{X}(t)$, the above inequality can be strengthened as

$$|\widetilde{X}(t_1) - \widetilde{X}(t_2)| < \varepsilon, \qquad \forall\, t_1, t_2 \in (t_0 - \delta, t_0) \cap \mathcal{I},$$

which implies the existence of the desired left limit.

2) It is easy to see that

$$\int_B X(t)d\mathbb{P} \leq \int_B \widetilde{X}(t)d\mathbb{P}, \qquad \forall\, B \in \mathcal{F}_t$$

and hence $X(t) \leq \widetilde{X}(t)$, a.s.

3) The **"if" part**. By the right-continuity of $\mathbb{E}X(\cdot)$, we get

$$\mathbb{E}X(t) = \lim_{\mathbb{Q}\cap\mathcal{I}\ni s\to t+} \mathbb{E}X(s) = \mathbb{E}\widetilde{X}(t).$$

Note that, $\widetilde{X}(t) \geq X(t)$, a.s. Hence $\mathbb{P}(\widetilde{X}(t) = X(t)) = 1$.

The **"only if" part**. Choose any sequence $\{s_n\}$ decreasing to t. Then, from the proof of Proposition 2.98, we see that,

$$\lim_{s_n\to t+} X(s_n) = \lim_{\mathbb{Q}\cap\mathcal{I}\ni s\to t+} X(s) = \widetilde{X}(t), \qquad \text{a.s. } \forall\, t \in \mathcal{I}.$$

Thus, by Theorem 2.97,

$$\lim_{s_n\to t+} \mathbb{E}X(s_n) = \mathbb{E}\widetilde{X}(t) = \mathbb{E}X(t),$$

which gives the right-continuity of $\mathbb{E}X(\cdot)$, □

An immediate consequence of Theorem 2.99 is as follows.

Corollary 2.100. *Let $X = \{X(t)\}_{t\in\mathcal{I}}$ be a submartingale such that $\mathbb{E}X(\cdot)$ is right-continuous. Then the process $\widetilde{X} = \{\widetilde{X}(t)\}_{t\in\mathcal{I}}$ in Theorem 2.99 is a cádlàg modification of X.*

The following result is a continuous counterpart of Corollary 2.93.

Theorem 2.101. *If $X = \{X(t)\}_{t\in\mathcal{I}}$ is a right-continuous martingale or non-negative submartingale so that $\mathbb{E}|X(T)|^p < \infty$ for some $p \geq 1$ and $T > 0$, then for each $\lambda > 0$,*

$$\begin{aligned}&\mathbb{P}\big(\{\sup_{t\in[0,T]}|X(t)| \geq \lambda\}\big) \leq \lambda^{-p}\mathbb{E}|X(T)|^p, \qquad p \geq 1,\\ &\mathbb{E}\big(\sup_{t\in[0,T]}|X(t)|^p\big) \leq \Big(\frac{p}{p-1}\Big)^p\mathbb{E}|X(T)|^p, \qquad p > 1.\end{aligned} \tag{2.83}$$

Proof: Let $\{\Delta_n\}_{n=1}^\infty$ be a family of subsets of $(\mathbb{Q}\cap[0,T])\cup\{T\}$ such that $\bigcup_{n=1}^\infty \Delta_n = (\mathbb{Q}\cap[0,T])\cup\{T\}$ and Δ_n has n elements. From Corollary 2.93, we find that

$$\mathbb{P}\big(\{\sup_{t\in\Delta_n}|X(t)| \geq \lambda\}\big) \leq \lambda^{-p}\mathbb{E}|X(T)|^p, \qquad p \geq 1.$$

Hence,

$$\mathbb{P}\big(\{\sup_{t\in(\mathbb{Q}\cap[0,T])\cup\{T\}}|X(t)| \geq \lambda\}\big) \leq \lambda^{-p}\mathbb{E}|X(T)|^p, \qquad p \geq 1.$$

This, together with the right continuous of $|X(t)|$, implies the first inequality in (2.83). The proof for the second inequality of (2.83) is similar. □

Corollary 2.102. *Let $T > 0$, $p \in (1,\infty]$ and $X(\cdot) \in L^p_{\mathbb{F}}(0,T)$ be a martingale. Then, there is an $\widetilde{X}(\cdot) \in L^p_{\mathbb{F}}(\Omega; D([0,T]))$ such that $\widetilde{X}(\cdot)$ is a modification of $X(\cdot)$.*

Proof: By Corollary 2.100, there is a cádlàg martingale $\widetilde{X}(\cdot)$ such that $\mathbb{P}(\{X(t) = \widetilde{X}(t)\}) = 1$ for all $t \in [0,T]$. Thus, $\widetilde{X}(\cdot) \in L^p_{\mathbb{F}}(0,T)$. This, together with the fact that the stochastic process $\widetilde{X}(\cdot)$ is cádlàg, implies $\widetilde{X}(T) \in L^p_{\mathcal{F}_T}(\Omega)$. By Theorem 2.101, we conclude that $\mathbb{E}(\sup_{t\in[0,T]}|\widetilde{X}(t)|^p) < \infty$. Hence, $\widetilde{X}(\cdot) \in L^p_{\mathbb{F}}(\Omega; D([0,T]))$. □

Similar to the proof of Theorem 2.101, by Theorems 2.95–2.96, one can show the following two results.

Theorem 2.103. *Let $X = \{X(t)\}_{t\in\mathcal{I}}$ be a right-continuous submartingale so that $\sup_{t\in\mathcal{I}}\mathbb{E}X(t)^+ < \infty$. Then,*

1) $\lim_{t\to\infty}X(t) = X_\infty$ a.s., *for some* $X_\infty \in L^1_{\mathcal{F}}(\Omega)$;

2) *If $\{X(t)^+\}_{t\in\mathcal{I}}$ is uniformly integrable, then $\hat{X} = \{X(t)\}_{t\in\hat{\mathcal{I}}}$ is a submartingale;*

3) *If $\{X(t)\}_{t\in\mathcal{I}}$ is uniformly integrable, then $\lim_{t\to\infty}X(t) = X_\infty$ in $L^1_{\mathcal{F}}(\Omega)$.*

Theorem 2.104. *Let $X = \{X(t)\}_{t\in\mathcal{I}}$ be a martingale. Then the following conditions are equivalent:*

1) $\lim_{t\to\infty} X(t) = X_\infty$ *in* $L^1_{\mathcal{F}}(\Omega)$;
2) $X(t) = E(X_\infty \mid \mathcal{F}_t)$ *a.s.,* $\forall\, t \in \mathcal{I}$;
3) $\{X(t)\}_{t\in\mathcal{I}}$ *is uniformly integrable.*

Furthermore, under one of the above conditions, $\hat{X} = \{X(t)\}_{t\in\hat{\mathcal{I}}}$ *is a martingale.*

The following result is a continuous version of Theorem 2.91.

Theorem 2.105. (Doob's stopping theorem) *Let $X = \{X(t)\}_{t\in\mathcal{I}}$ be a right-continuous submartingale, σ and τ be two bounded stopping times such that $\mathbb{P}(\{\sigma \le \tau\}) = 1$. Then $X(\sigma)$ and $X(\tau)$ are integrable, and*

$$\mathbb{E}(X(\tau) \mid \mathcal{F}_\sigma) \ge X_\sigma, \quad \text{a.s.}$$

Proof: Since σ and τ are bounded, one can find a positive integer N such that $\sigma \vee \tau \le N$. Put

$$\sigma_n = \sum_{k=1}^{2^n N} \frac{k}{2^n} \chi_{\{\frac{k-1}{2^n} \le \sigma < \frac{k}{2^n}\}}, \quad \tau_n = \sum_{k=1}^{2^n N} \frac{k}{2^n} \chi_{\{\{\frac{k-1}{2^n} \le \tau < \frac{k}{2^n}\}}.$$

Then $\sigma_n \le \tau_n \le N$, a.s. By Theorem 2.91, we see that $\mathbb{E}(X(\tau_n) \mid \mathcal{F}_{\sigma_n}) \ge X(\sigma_n)$, a.s. This means that, for any $A \in \mathcal{F}_\sigma$,

$$\int_A X(\tau_n) d\mathbb{P} \ge \int_A X(\sigma_n) d\mathbb{P}, \qquad \forall\, n.$$

Hence, by Theorem 2.97, we get

$$\lim_{n\to\infty} X(\tau_n) = X(\tau), \quad \lim_{n\to\infty} X(\sigma_n) = X(\sigma), \quad \text{in } L^1_{\mathcal{F}}(\Omega).$$

Consequently, $\int_A X(\tau) d\mathbb{P} \ge \int_A X(\sigma) d\mathbb{P}$, which gives the desired result. $\square$

Some consequences of Theorem 2.105 are in order.

Corollary 2.106. *Let $X(\cdot)$ be a right-continuous, **F**-submartingale, and let $\{\sigma_t\}_{t\in\mathcal{I}}$ be a family of right-continuous, bounded stopping times such that $\mathbb{P}(\sigma_t \le \sigma_s) = 1$ when $t < s$. Set*

$$\overline{X}(t) = X(\sigma_t), \quad \overline{\mathcal{F}}_t = \mathcal{F}_{\sigma_t}, \quad \forall\, t \in \mathcal{I}.$$

Then $\overline{X} = \{\overline{X}(t)\}_{t\in\mathcal{I}}$ be a right-continuous $\{\overline{\mathcal{F}}_t\}_{t\in I}$-submartingale.

Corollary 2.107. *Let $X = \{X(t)\}_{t\in\mathcal{I}}$ be a martingale, and $\sigma \le \tau$ be two stopping times satisfying $\sigma \le \tau$, a.s. Then*

$$\mathbb{E}(X(t\wedge\tau) - X(t\wedge\sigma) \mid \mathcal{F}_\sigma) = 0, \quad \forall\, t \in \mathcal{I}.$$

Proof: By Proposition 2.88, $X(t\wedge\tau)$ is $\mathcal{F}_t$-measurable. Hence, by Corollary 2.106, for any $t \ge 0$,

$$X(t\wedge\sigma) = \mathbb{E}(X(t\wedge\tau) \mid \mathcal{F}_{t\wedge\sigma}) = \mathbb{E}(\mathbb{E}(X(t\wedge\tau) \mid \mathcal{F}_t) \mid \mathcal{F}_\sigma) = \mathbb{E}(X(t\wedge\tau) \mid \mathcal{F}_\sigma).$$

$\square$

2.8.2 Vector-Valued Martingales

Let $T > 0$ and H be a separable Hilbert space.

Definition 2.108. *An H-valued, $\mathbf{F}$-adapted process $X = \{X(t)\}_{t\in[0,T]}$ is called an $\mathbf{F}$-martingale, or simply a martingale when the filtration $\mathbf{F}$ is clear from the context, if*

1) *$X(t)$ is Bochner integrable for each $t \in [0,T]$; and*
2) *$\mathbb{E}(X(t) \mid \mathcal{F}_s) = X(s)$ a.s., for every $t, s \in [0,T]$ with $s < t$.*

We have the following result.

Theorem 2.109. *If $X(\cdot)$ is an H-valued martingale such that $|X(\cdot)|_H$ is right-continuous and $X(T) \in L^p_{\mathcal{F}_T}(\Omega; H)$ for some $p \geq 1$, then,*

$$\begin{aligned}
&\mathbb{P}\big(\{\sup_{t\in[0,T]} |X(t)|_H > \lambda\}\big) \leq \frac{1}{\lambda^p}\mathbb{E}|X(T)|^p_H, \quad p \geq 1,\ \lambda > 0,\\
&\mathbb{E}\big(\sup_{t\in[0,T]} |X(t)|^p_H\big) \leq \Big(\frac{p}{p-1}\Big)^p \mathbb{E}|X(T)|^p_H, \quad p > 1.
\end{aligned} \tag{2.84}$$

Proof: Noting that the process $|X(\cdot)|_H$ is a real valued submartingale, by Theorem 2.101, we obtain the desired results. □

As an immediate corollary of Corollary 2.107, we have the following result.

Corollary 2.110. *Let $X = \{X(t)\}_{t\in\mathcal{I}}$ be an H-valued martingale, and $\sigma \leq \tau$ be two stopping times satisfying $\sigma \leq \tau$, a.s. Then*

$$\mathbb{E}(X(t \wedge \tau) - X(t \wedge \sigma) \mid \mathcal{F}_\sigma) = 0, \quad \forall\, t \in \mathcal{I}.$$

We give below a generalization of Corollary 2.102 to the case of H-valued martingales. This result will play a key role in the sequel.

Theorem 2.111. *Let $X(\cdot)$ be an H-valued martingale such that $X(T) \in L^p_{\mathcal{F}_T}(\Omega; H)$ for some $p \in (1, \infty]$. Then there is an $\widetilde{X}(\cdot) \in L^p_{\mathbb{F}}(\Omega; D([0,T]; H))$ such that $\widetilde{X}(\cdot)$ is a modification of $X(\cdot)$.*

Proof: We consider only the case that $p \in (1, \infty)$. Assume that $X(T) = \sum_{i=1}^\infty \xi_i e_i$, where $\{e_i\}_{i=1}^\infty$ is an orthonormal basis of H. Obviously,

$$X(t) = \mathbb{E}(X(T) \mid \mathcal{F}_t) = \sum_{i=1}^\infty \mathbb{E}(\xi_i \mid \mathcal{F}_t)e_i, \qquad \forall\, t \in [0,T].$$

Put $X_i(t) = \mathbb{E}(\xi_i \mid \mathcal{F}_t)$. Since $X_i(\cdot)$ is a $\mathbb{C}$-valued martingale for each $i \in \mathbb{N}$, by Corollary 2.100, one can find a $\mathbb{C}$-valued cádlág process $\widetilde{X}_i(\cdot)$ such that it is a modification of $X_i(\cdot)$. Write

$$\widetilde{X}(t) = \sum_{i=1}^\infty \widetilde{X}_i(t)e_i, \qquad \forall\, t \in [0,T].$$

Then $\widetilde{X}(\cdot)$ is an H-valued martingale and $\widetilde{X}(T) \in L^p_{\mathcal{F}_T}(\Omega; H)$.

Let us prove that $\widetilde{X}(\cdot)$ is cádlàg. For each $n \in \mathbb{N}$, put $\widetilde{X}_n(\cdot) = \sum_{i=1}^n \widetilde{X}_i(\cdot)e_i$. Clearly, $\widetilde{X}_n(\cdot)$ is cádlàg. Since $\widetilde{X}(\cdot)$ and $\widetilde{X}_n(\cdot)$ are martingales, $\widetilde{X}(\cdot) - \widetilde{X}_n(\cdot)$ is also a martingale. From the definition of $\widetilde{X}_n(\cdot)$, there is a subsequence $\{n_k\}_{k=1}^\infty \subset \{n\}_{n=1}^\infty$ such that for any $k \in \mathbb{N}$,

$$|\widetilde{X}(T) - \widetilde{X}_{n_k}(T)|_{L^p_{\mathcal{F}_T}(\Omega;H)} \le \frac{1}{k^3}. \tag{2.85}$$

Noting that $\big|\widetilde{X}(\cdot) - \widetilde{X}_{n_k}(\cdot)\big|_H$ is a non-negative submartingale, by Theorem 2.109, we find that

$$\begin{aligned}&\mathbb{P}\Big(\Big\{\sup_{t\in[0,T]} \big|\widetilde{X}(t) - \widetilde{X}_{n_k}(t)\big|_H \ge \frac{1}{k}\Big\}\Big)\\&\le k\mathbb{E}\big|\widetilde{X}(T) - \widetilde{X}_{n_k}(T)\big|_H \le k\big|\widetilde{X}(T) - \widetilde{X}_{n_k}(T)\big|_{L^p_{\mathcal{F}_T}(\Omega;H)} \le \frac{1}{k^2}.\end{aligned}$$

Since $\sum_{k=1}^\infty \frac{1}{k^2} < +\infty$, by the Borel-Cantelli lemma (Theorem 2.4), we conclude that

$$\mathbb{P}\Big(\overline{\lim_{k\to\infty}}\Big\{\sup_{t\in[0,T]} \big|\widetilde{X}(t) - \widetilde{X}_{n_k}(t)\big|_H \ge \frac{1}{k}\Big\}\Big) = 0. \tag{2.86}$$

This implies that

$$\mathbb{P}\Big(\underline{\lim}_{k\to\infty}\Big\{\sup_{t\in[0,T]} \big|\widetilde{X}(t) - \widetilde{X}_{n_k}(t)\big|_H < \frac{1}{k}\Big\}\Big) = 1. \tag{2.87}$$

Hence there exists an event Ω_0 such that $\mathbb{P}(\Omega_0) = 1$ and for each $\omega \in \Omega_0$, there exists a positive integer $N(\omega)$ such that

$$\sup_{t\in[0,T]} \big|\widetilde{X}(t) - \widetilde{X}_{n_k}(t)\big|_H < \frac{1}{k}, \qquad \forall\, k \ge N(\omega). \tag{2.88}$$

Thus for $\omega \in \Omega_0$, the sequence of functions $\{\widetilde{X}_{n_k}(\cdot,\omega)\}_{k=1}^\infty$ converges uniformly to $\widetilde{X}(\cdot,\omega)$ on $[0,T]$. Since for each $k \in \mathbb{N}$, the stochastic process $\widetilde{X}_{n_k}(\cdot)$ is cádlàg, there exists an event Ω_k with $\mathbb{P}(\Omega_k) = 1$ and for any $\omega \in \Omega_k$, the function $\widetilde{X}_{n_k}(\cdot,\omega)$ is cádlàg.

Put $\widetilde{\Omega} = \bigcap_{k=0}^\infty \Omega_k$. Then $\mathbb{P}(\widetilde{\Omega}) = 1$ and for each $\omega \in \widetilde{\Omega}$, the sequence $\widetilde{X}_{n_k}(\cdot,\omega)$ is a sequence of cádlàg functions that converges uniformly to $\widetilde{X}(\cdot,\omega)$ on $[0,T]$. It follows that $\widetilde{X}(\cdot,\omega)$ is a cádlàg function for each $\omega \in \widetilde{\Omega}$. Hence $\widetilde{X}(\cdot)$ is an H-valued, cádlàg stochastic process.

Now, by Theorem 2.109, one can show that $\widetilde{X}(\cdot) \in L^p_{\mathbb{F}}(\Omega; D([0,T]; H))$. This completes the proof of Theorem 2.111. □

Finally, we introduce some Hilbert spaces which will be useful later.

$$\begin{cases} \mathcal{M}^2([0,T];H) = \{X \in L^2_{\mathbb{F}}(0,T;H) \mid X \text{ is a right-continuous} \\ \qquad\qquad \mathbf{F}\text{-martingale with } X(0)=0, \text{ a.s.}\}, \\ \mathcal{M}^2_c([0,T];H) = \{X \in \mathcal{M}^2([0,T];H) \mid X \text{ is continuous}\}. \end{cases}$$

We identify $X, Y \in \mathcal{M}^2([0,T];H)$ if there exists a set $N \in \mathcal{F}$ with $\mathbb{P}(N) = 0$, such that $X(t,\omega) = Y(t,\omega)$, for all $t \geq 0$ and $\omega \notin N$. Define

$$|X|_{\mathcal{M}^2([0,T];H)} = \sqrt{\mathbb{E}|X(T)|^2_H}, \qquad \forall\, X \in \mathcal{M}^2([0,T];H).$$

We shall simply denote $\mathcal{M}^2([0,T];\mathbb{R})$ and $\mathcal{M}^2_c([0,T];\mathbb{R})$ by $\mathcal{M}^2[0,T]$ and $\mathcal{M}^2_c[0,T]$, respectively.

We have the following result.

Lemma 2.112. *$(\mathcal{M}^2([0,T];H), |\cdot|_{\mathcal{M}^2([0,T];H)})$ is a Hilbert space, and $\mathcal{M}^2_c([0,T];H)$ is a closed subspace of $\mathcal{M}^2([0,T];H)$.*

Proof: First we note that if $|X-Y|_{\mathcal{M}^2([0,T];H)} = 0$, then $X = Y$. Indeed, $|X-Y|_{\mathcal{M}^2([0,T];H)} = 0$ implies $X(T) = Y(T)$ a.s. Therefore, $X(t) = \mathbb{E}(X(T) \mid \mathcal{F}_t) = \mathbb{E}(Y(T) \mid \mathcal{F}_t) = Y(t)$ for $t \in [0,T]$, a.s. By the right-continuity of $X(\cdot)$ and $Y(\cdot)$, we conclude that $X = Y$.

Next, suppose that $\{X^n\}_{n=1}^{\infty}$ is a Cauchy sequence in $\mathcal{M}^2([0,T];H)$, i.e., $\lim_{n,m\to\infty} |X^n - X^m|_{\mathcal{M}^2([0,T];H)} = 0$. Then, by the second assertion in Theorem 2.109, we conclude that

$$\mathbb{E}\big(\sup_{0\leq t\leq T} |X^n(t) - X^m(t)|^2_H\big) \leq 4\mathbb{E}|X^n(T) - X^m(T)|^2_H.$$

Hence, there is an $X = \{X(t)\}_{t\in[0,T]}$ such that $\mathbb{E}|X^n(t) - X(t)|^2_H \to 0$ as $n \to \infty$ uniformly for $t \in [0,T]$, and we see from this that $X \in \mathcal{M}^2([0,T];H)$ and $|X^n(t) - X(t)|_{\mathcal{M}^2([0,T];H)} \to 0$ as $n \to \infty$.

Finally, it is also clear from this proof that if $X^n \in \mathcal{M}^2_c([0,T];H)$, then $X \in \mathcal{M}^2_c([0,T];H)$. □

Finally, we give below the notion of local martingale.

Definition 2.113. *An H-valued, $\mathbf{F}$-adapted process $X = \{X(t)\}_{t\in[0,T]}$ is called an $\mathbf{F}$-local martingale, or simply a local martingale when the filtration $\mathbf{F}$ is clear from the context, if there exists a sequence of stopping times $\{\tau_k\}_{k=1}^{\infty}$ such that τ_k increases monotonically to T a.s. as $k \to +\infty$, and for each $k \in \mathbb{N}$, $X(\cdot \wedge \tau_k)$ is an $\mathbf{F}$-martingale.*

2.9 Brownian Motions

Let $(\Omega, \mathcal{F}, \mathbf{F}, \mathbb{P})$ (with $\mathbf{F} = \{\mathcal{F}_t\}_{t\in[0,\infty)}$) be a filtered probability space satisfying the usual condition. In this section, we present a class of important examples of martingales, i.e., Brownian motions.

2.9.1 Brownian Motions in Finite Dimensions

The Brownian motion of pollen particles in a liquid, which owes the name to its discovery by the English botanist R. Brown in 1827, is due to the incessant hitting of pollen by the much small molecules of the liquid. The hits occur a large number of times in any small interval of time, independently of each other and the effect of a particular hit is so small compared to the total effect. This motion has the following properties:

1) The displacement of a pollen particle over disjoint time intervals are independent random variables;
2) The displacements are Gaussian random variables;
3) The motion is continuous.

In this subsection, we address to Brownian motions in $\mathbb{R}^m$ ($m \in \mathbb{N}$). Denote by I_m the identity matrix of order $m \times m$. We begin with the following notion.

Definition 2.114. *A continuous, $\mathbb{R}^m$-valued, $\mathbf{F}$-adapted process $\{W(t)\}_{t\geq 0}$ is called an m-dimensional (standard) Brownian motion, if*

1) $\mathbb{P}(\{W(0)=0\})=1$*; and*

2) *For any $s,t\in[0,\infty)$ with $0\leq s<t<\infty$, the random variable $W(t)-W(s)$ is independent of $\mathcal{F}_s$, and $W(t)-W(s)\sim\mathcal{N}(0,(t-s)I_m)$.*

Similarly, one can define $\mathbb{R}^m$-valued Brownian motions over any time interval $[a,b]$ or $[a,b)$ with $0\leq a<b\leq\infty$.

If $W(\cdot)$ is a Brownian motion on $(\Omega,\mathcal{F},\mathbf{F},\mathbb{P})$, we may define

$$\mathcal{F}_t^W \overset{\Delta}{=} \sigma(W(s);\ s\in[0,t])\subset\mathcal{F}_t,\qquad \forall\, t\geq 0. \tag{2.89}$$

Generally, the filtration $\{\mathcal{F}_t^W\}_{t\geq 0}$ is left-continuous, but not necessarily right-continuous. Nevertheless, the augmentation $\{\hat{\mathcal{F}}_t^W\}_{t\geq 0}$ of $\{\mathcal{F}_t^W\}_{t\geq 0}$ by adding all $\mathbb{P}$-null sets is continuous, and $W(\cdot)$ is still a Brownian motion on the (augmented) filtered probability space $(\Omega,\mathcal{F},\{\hat{\mathcal{F}}_t^W\}_{t\geq 0},\mathbb{P})$ (*see* [164, pp. 89 and 122] for detailed discussion). In the sequel, by saying that $\mathbf{F}$ is the natural filtration generated by (the Brownian motion) $W(\cdot)$, we mean that $\mathbf{F}$ is generated as (2.89) with the above augmentation, and hence in this case $\mathbf{F}$ is continuous.

The following result shows that the sample path of a Brownian motion is highly irregular, almost surely.

Proposition 2.115. *Let $W(\cdot)$ be a real valued Brownian motion. Then, for almost all $\omega\in\Omega$, the map $t\mapsto W(t,\omega)$ is nowhere differentiable.*

Proof: It suffices to show that

$$\mathbb{P}\Big(\Big\{\ \overline{\lim_{s\to 0+}}\ \frac{|W(t+s)-W(t)|}{s}=+\infty\Big\}\Big)=1,\qquad \forall\, t\geq 0.$$

For this purpose, for any $t \geq 0$, $s > 0$ and positive integer n, denote

$$A_{t,n,s} \stackrel{\Delta}{=} \{|W(t+s) - W(t)| < ns\}.$$

Noting that $W(t+s) - W(t) \sim \mathcal{N}(0, s)$, we get

$$P(A_{t,n,s}) = \frac{1}{\sqrt{2\pi s}} \int_{-ns}^{ns} e^{-\frac{x^2}{2s}} dx \leq C\sqrt{s},$$

where $C = C(n) > 0$ is a generic constant. We choose a sequence $\{s_k\}_{k=1}^{\infty}$ so that $\lim_{k\to\infty} s_k = 0$ and $\sum_{k=1}^{\infty} \sqrt{s_k} < \infty$. Then $\sum_{k=1}^{\infty} P(A_{t,n,s_k}) < \infty$.

Denote

$$A_{t,n} = \varliminf_{k\to\infty} (A_{t,n,s_k})^c.$$

By Theorem 2.4, we conclude that $\mathbb{P}(A_{t,n}) = 1$. Then for any $\omega \in A_{t,n}$, there is a $k(\omega)$, such that

$$\left| \frac{W(t+s_k, \omega) - W(t, \omega)}{s_k} \right| \geq n, \qquad \forall\, k \geq k(\omega).$$

Therefore,

$$\varlimsup_{s\to 0+} \frac{|W(t+s, \omega) - W(t, \omega)|}{s} \geq n, \qquad \forall\, \omega \in A_{t,n}.$$

Put $B = \bigcap_{n\geq 1} A_{t,n}$. Then, $\mathbb{P}(B) = 1$ and

$$\varlimsup_{s\to 0+} \frac{|W(t+s, \omega) - W(t, \omega)|}{s} = +\infty, \qquad \forall\, \omega \in B.$$

This completes the proof. □

2.9.2 Construction of Brownian Motions in one Dimension

In this subsection, we present the Lévy construction of real valued Brownian motions. For a systematic introduction to the construction of Brownian motions in finite dimensions, we refer the readers to [164, pp. 49–66].

We begin with the following assumption.

(H) *There exists a sequence of independent random variables* $\{X_k\}_{k=1}^{\infty}$ *on the probability space* $(\Omega, \mathcal{F}, \mathbb{P})$, *and* $X_k \sim \mathcal{N}(0,1)$ *for each* $k \in \mathbb{N}$.

We now present a probability space $(\Omega, \mathcal{F}, \mathbb{P})$ satisfying the assumption (H). For $i \in \mathbb{N}$, let $(\Omega_i, \mathcal{F}_i, \mathbb{P}_i)$ be a probability space on which there is a random variable $Y_i \sim \mathcal{N}(0,1)$. Let $(\prod_{i=1}^{\infty} \Omega_i, \prod_{i=1}^{\infty} \mathcal{F}_i)$ be the product measurable space of $\{(\Omega_i, \mathcal{F}_i)\}_{i=1}^{\infty}$ and $\mathbb{P}$ the produce measure induced by $\{\mathbb{P}_i\}_{i=1}^{\infty}$. We define a sequence of random variables $\{X_k\}_{k=1}^{\infty}$ on $(\prod_{i=1}^{\infty} \Omega_i, \prod_{i=1}^{\infty} \mathcal{F}_i, \mathbb{P})$ as follows:

$$X_k(\omega_1, \cdots, \omega_k, \cdots) = Y_k(\omega_k), \qquad \text{for } (\omega_1, \cdots, \omega_k, \cdots) \in \prod_{i=1}^{\infty} \Omega_i.$$

One can check that $\{X_k\}_{k=1}^{\infty}$ are independent, and $X_k \sim \mathcal{N}(0,1)$.

We first construct a Brownian motion on $[0,1]$.

Let us recall the Haar functions $\{h_k(\cdot)\}_{k=1}^{\infty}$ on $[0,1]$: $h_1(\cdot) \equiv 1$ in $[0,1]$;

$$h_2(t) = \begin{cases} 1, & \text{if } t \in \left[0, \frac{1}{2}\right), \\ -1, & \text{if } t \in \left[\frac{1}{2}, 1\right]; \end{cases}$$

while for $n \in \mathbb{N}$, $k \in I(n) \overset{\Delta}{=} [2^n + 1, 2^{n+1} + 1) \cap \mathbb{N}$,

$$h_k(t) = \begin{cases} 2^{n/2}, & \text{if } t \in \left[\frac{k-2^n-1}{2^n}, \frac{k-2^n-1/2}{2^n}\right), \\ -2^{n/2}, & \text{if } t \in \left[\frac{k-2^n-1/2}{2^n}, \frac{k-2^n}{2^n}\right), \\ 0, & \text{otherwise.} \end{cases}$$

It is well-known that $\{h_k\}_{k=1}^{\infty}$ forms a complete, orthonormal basis for the Hilbert space $L^2(0,1)$ (with the standard inner product $\langle \cdot, \cdot \rangle$). Hence,

$$\langle f, g \rangle = \sum_{k=1}^{\infty} \langle f, h_k \rangle \langle g, h_k \rangle, \qquad f, g \in L^2(0,1), \tag{2.90}$$

where $\langle \cdot, \cdot \rangle$ stands for the usual inner product in $L^2(0,1)$. Let

$$s_k(t) = \int_0^t h_k(\tau) d\tau, \qquad t \in [0,1], \quad \text{for } k = 1, 2, \cdots.$$

Applying (2.90) to $f(\cdot) = \chi_{[0,t]}(\cdot)$ and $g(\cdot) = \chi_{[0,s]}(\cdot)$ with $t, s \in [0,1]$, we find

$$\sum_{k=1}^{\infty} s_k(t) s_k(s) = \min(t, s). \tag{2.91}$$

For $n \in \mathbb{N}$ and the sequence $\{X_k\}_{k=1}^{\infty}$ given in the assumption (H), put

$$W_n(t, \omega) = \sum_{k=1}^{n} s_k(t) X_k(\omega), \qquad (t, \omega) \in [0,1] \times \Omega. \tag{2.92}$$

One has the following result.

Lemma 2.116. *The sequence $\{W_n(\cdot)\}_{n=1}^{\infty}$ (given by (2.92)) converges uniformly w.r.t. $t \in [0,1]$, a.s.*

Proof: For any $x > 0$ and $k \geq 2$,

$$\mathbb{P}(\{|X_k| > x\}) = \frac{2}{\sqrt{2\pi}} \int_x^\infty e^{-\frac{s^2}{2}} ds \leq \frac{2}{\sqrt{2\pi}} e^{-\frac{x^2}{4}} \int_x^\infty e^{-\frac{s^2}{4}} ds \leq 2e^{-\frac{x^2}{4}}. \tag{2.93}$$

Let us choose $x = 4\sqrt{\ln k}$ in (2.93). Then,

$$\mathbb{P}(\{|X_k| \geq 4\sqrt{\ln k}\}) \leq 2e^{-4\ln k} = \frac{2}{k^4}.$$

From $\sum_{k=2}^\infty \frac{1}{k^4} < \infty$ and the Borel-Cantelli lemma (i.e., Theorem 2.4), we find that

$$\mathbb{P}\big(\{\overline{\lim_{k\to\infty}} |X_k| \geq 4\sqrt{\ln k}\}\big) = 0.$$

Thus, for a.e. $\omega \in \Omega$, there is a constant $K = K(\omega)$ such that

$$|X_k(\omega)| \leq 4\sqrt{\ln k}, \qquad \text{for } k \geq K(\omega). \tag{2.94}$$

Put $Y_n \overset{\Delta}{=} \max_{k\in I(n)} |X_k|$. It is clear that $Y_n \leq \mathcal{C}\sqrt{n+1}$. For any $\varepsilon > 0$, from the definition of $s_k(\cdot)$, one can find an $m \in \mathbb{N}$ large enough such that, for any $t \in [0,1]$,

$$\sum_{k=2^m}^\infty |X_k||s_k(t)| \leq \sum_{n=m}^\infty Y_n \max_{k\in I(n),t\in[0,1]} |s_k(t)| \leq \mathcal{C} \sum_{n=m}^\infty \frac{\sqrt{n+1}}{2^{n/2-1}} < \varepsilon.$$

This completes the proof of Lemma 2.116. □

For $t \in [0,1]$, write

$$W(t) = \sum_{k=1}^\infty s_k(t) X_k, \quad \mathcal{F}_t = \sigma(W(s);\ s \in [0,t]). \tag{2.95}$$

We have the following result.

Theorem 2.117. *Let the assumption* (H) *hold. Then, the real valued stochastic process* W *(given by* (2.95)*) is a Brownian motion on* $(\Omega, \mathcal{F}, \{\mathcal{F}_t\}_{t\in[0,1]}, \mathbb{P})$.

Proof: Clearly, $W(0) = 0$ a.s. By Lemma 2.116, it is easy to see that $W(\cdot)$ is continuous on $[0,1]$, a.s.

For any $s,t \in [0,1]$ satisfying $0 \leq s \leq t \leq 1$, in view of (2.91), we have

$$\begin{aligned}
\mathbb{E}(e^{i\xi(W(t)-W(s))}) &= \mathbb{E}(e^{i\xi\sum_{k=1}^\infty [s_k(t)-s_k(s)]X_k}) = \prod_{k=1}^\infty \mathbb{E}(e^{i\xi[s_k(t)-s_k(s)]X_k}) \\
&= \prod_{k=1}^\infty e^{-\frac{\xi^2}{2}[s_k(t)-s_k(s)]^2} = e^{-\frac{\xi^2}{2}\sum_{k=1}^\infty [s_k(t)-s_k(s)]^2} \\
&= e^{-\frac{\xi^2}{2}\sum_{k=1}^\infty [s_k^2(t)-2s_k(t)s_k(s)+s_k^2(s)]} \\
&= e^{-\frac{\xi^2}{2}(t-s)}, \qquad \forall\, \xi \in \mathbb{R}.
\end{aligned}$$

Hence, by Definition 2.39, $W(t)-W(s)\sim\mathcal{N}(0,t-s)$.

We now prove that for any $m\in\mathbb{N}$ and for $\{t_i\}_{i=0}^m$ satisfying $0=t_0<t_1<\cdots<t_m\le 1$, the random variables $W(t_1)$, $W(t_2)-W(t_1),\cdots,W(t_m)-W(t_{m-1})$ are independent. For simplicity, we only consider the case that $m=2$. Using (2.91) again, we obtain that

$$
\begin{aligned}
&\mathbb{E}(e^{i[\xi_1 W(t_1)+\xi_2(W(t_2)-W(t_1))]})=\mathbb{E}(e^{i[(\xi_1-\xi_2)W(t_1)+\xi_2 W(t_2)]})\\
&=\mathbb{E}(e^{i[(\xi_1-\xi_2)\sum_{k=1}^{\infty}s_k(t_1)X_k+\xi_2\sum_{k=1}^{\infty}s_k(t_2)X_k]})=\prod_{k=1}^{\infty}\mathbb{E}(e^{i[(\xi_1-\xi_2)s_k(t_1)+\xi_2 s_k(t_2)]X_k})\\
&=\prod_{k=1}^{\infty}e^{-\frac{1}{2}[(\xi_1-\xi_2)s_k(t_1)+\xi_2 s_k(t_2)]^2}=e^{-\frac{1}{2}[(\xi_1-\xi_2)^2t_1+2(\xi_1-\xi_2)\xi_2 t_1+\xi_2^2 t_2]}\\
&=e^{-\frac{1}{2}[\xi_1^2 t_1+\xi_2^2(t_2-t_1)]},\qquad\forall\,\xi_1,\xi_2\in\mathbb{R}.
\end{aligned}
$$

This, together with Definition 2.39 and Theorem 2.36, implies that $W(t_1)$ and $W(t_2)-W(t_1)$ are independent. Consequently, $W(t)-W(s)$ is independent of $\mathcal{F}_s$. □

One can patch the Brownian motion (on $[0,1]$) constructed in Theorem 2.117 to get a Brownian motion on $[0,\infty)$.

2.9.3 Vector-Valued Brownian Motions

Let V be a separable, real Hilbert space. In this subsection, we address to the V-valued Brownian motions.

Let $Q\in\mathcal{L}_1(V)$ be positive definite. Then, there is an orthonormal basis $\{e_j\}_{j=1}^{\infty}$ in V and a sequence $\{\lambda_j\}_{j=1}^{\infty}$ of positive numbers satisfying $\sum_{j=1}^{\infty}\lambda_j<\infty$ and $Qe_j=\lambda_j e_j$ for $j=1,2,\cdots$ (Hence, λ_j and e_j are respectively eigenvalue and eigenvector of Q). Here, when V is an m-dimensional Hilbert space (for some $m\in\mathbb{N}$), the sequence $\{\lambda_j\}_{j=1}^{\infty}$ is reduced to $\{\lambda_j\}_{j=1}^{m}$.

Similar to Definition 2.114, one introduces the following notion.

Definition 2.118. *A continuous, V-valued, $\mathbf{F}$-adapted process $\{W(t)\}_{t\ge 0}$ is called a (standard) Q-Brownian motion if*

1) $\mathbb{P}(\{W(0)=0\})=1$; and

2) For any $s,t\in[0,\infty)$ satisfying $0\le s<t<\infty$, the random variable $W(t)-W(s)$ is independent of $\mathcal{F}_s$, and $W(t)-W(s)\sim\mathcal{N}(0,(t-s)Q)$.

In the sequel, we call $\Delta_n\stackrel{\Delta}{=}\{t_0,t_1,\cdots,t_n\}$ (for $n\in\mathbb{N}$) a partition of $[0,T]$ if $0=t_0<t_1<\cdots<t_{n-1}<t_n=T$, and write $|\Delta_n|\stackrel{\Delta}{=}\max_{0\le i\le n-1}\{t_{i+1}-t_i\}$.

The following result reveals the relationship between a Q-Brownian Motion and its finite dimensional projections.

Proposition 2.119. *Assume that* $\{W(t)\}_{t\geq 0}$ *is a* Q*-Brownian motion. Then, for any* $t \geq 0$,

$$W(t) = \sum_{j=1}^{\infty} \sqrt{\lambda_j} W_j(t) e_j, \tag{2.96}$$

where $W_j(t) = \frac{1}{\sqrt{\lambda_j}} \langle W(t), e_j \rangle_V$*. Furthermore,* $W_1(\cdot), W_2(\cdot), \cdots$ *are independent real valued Brownian motions, and the series in* (2.96) *is convergent in* $L^2_{\mathbb{F}}(\Omega; C([0,T]; V))$ *for each* $T > 0$.

Proof: We only prove that $W_1(\cdot), W_2(\cdot), \cdots$ are independent because the other conclusions in Proposition 2.119 can be easily verified. Assume that $n \in \mathbb{N}$ and $j_1, \cdots, j_n \in \mathbb{N}$ satisfying $j_i \neq j_k$ for $i, k \in \{1, 2, \cdots, n\}$ with $i \neq k$. Let $\Delta_\ell = \{t_0, t_1, \cdots, t_\ell\}$, $\ell \in \mathbb{N}$ be a partition of $[0, T]$. By Definition 2.66, it remains to prove that

$$\sigma(W_{j_1}(t_1), \cdots, W_{j_1}(t_\ell)), \cdots, \sigma(W_{j_n}(t_1), \cdots, W_{j_n}(t_\ell)) \text{ are independent.} \tag{2.97}$$

We use the induction argument (w.r.t. ℓ). For the case $\ell = 1$, it is clear that $(W_{j_1}(t_1), \cdots, W_{j_n}(t_1))$ is a n-dimensional Gaussian random variable. Further,

$$\begin{aligned} &\mathbb{E}(W_{j_i}(t_1) W_{j_k}(t_1)) = \frac{1}{\sqrt{\lambda_{j_i}\lambda_{j_k}}} \mathbb{E}\big(\langle W(t_1), e_{j_i}\rangle_V \langle W(t_1), e_{j_k}\rangle_V\big) \\ &= \frac{1}{\sqrt{\lambda_{j_i}\lambda_{j_k}}} t_1 (Q e_{j_i}, e_{j_k})_V = \sqrt{\frac{\lambda_{j_i}}{\lambda_{j_k}}} t_1 \delta_{j_i j_k}. \end{aligned}$$

Hence, the covariance matrix of $(W_{j_1}(t_1), \cdots, W_{j_n}(t_1))$ is diagonal. By Theorem 2.36, it follows that $W_{j_1}(t_1), \cdots, W_{j_n}(t_1)$ are independent.

We now assume that (2.97) holds for ℓ and prove that it also holds for $\ell + 1$. Similar to the above argument, one can show that $W_{j_1}(t_{\ell+1}) - W_{j_1}(t_\ell), \cdots, W_{j_n}(t_{\ell+1}) - W_{j_n}(t_\ell)$ are independent. Further, for $1 \leq i \leq n$, it holds that

$$\begin{aligned} &\sigma\big(W_{j_i}(t_1) \cdots, W_{j_i}(t_\ell), W_{j_i}(t_{\ell+1})\big) \\ &= \sigma\big(W_{j_i}(t_1) \cdots, W_{j_i}(t_\ell), W_{j_i}(t_{\ell+1}) - W_{j_i}(t_\ell)\big), \end{aligned}$$

Let $B_{i,k} \in \mathcal{B}(\mathbb{R})$, $1 \leq i \leq n$, $1 \leq k \leq \ell + 1$. Utilizing the fact that $\sigma(W(s); s \in [0, t_\ell])$ and $\sigma(W(t_{\ell+1}) - W(t_\ell))$ are independent, we obtain that

$$\begin{aligned}
&\mathbb{P}\Big(\bigcap_{i=1}^{n}\{W_{j_i}(t_1)\in B_{i,1},\cdots,W_{j_i}(t_\ell)\in B_{i,\ell},W_{j_i}(t_{\ell+1})-W_{j_i}(t_\ell)\in B_{i,\ell+1}\}\Big)\\
&=\mathbb{P}\Big(\Big[\bigcap_{i=1}^{n}\bigcap_{k=1}^{\ell}\{W_{j_i}(t_k)\in B_{i,k}\}\Big]\bigcap\Big[\bigcap_{i=1}^{n}\{W_{j_i}(t_{\ell+1})-W_{j_i}(t_\ell)\in B_{i,\ell+1}\}\Big]\Big)\\
&=\mathbb{P}\Big(\bigcap_{i=1}^{n}\bigcap_{k=1}^{\ell}\{W_{j_i}(t_k)\in B_{i,k}\}\Big)\mathbb{P}\Big(\bigcap_{i=1}^{n}\{W_{j_i}(t_{\ell+1})-W_{j_i}(t_\ell)\in B_{i,\ell+1}\}\Big)\\
&=\Big[\prod_{i=1}^{n}\mathbb{P}\Big(\bigcap_{k=1}^{\ell}\{W_{j_i}(t_k)\in B_{i,k}\}\Big)\Big]\Big[\prod_{i=1}^{n}\mathbb{P}\Big(\{W_{j_i}(t_{\ell+1})-W_{j_i}(t_\ell)\in B_{i,\ell+1}\}\Big)\Big]\\
&=\prod_{i=1}^{n}\mathbb{P}\Big(\Big[\bigcap_{k=1}^{\ell}\{W_{j_i}(t_k)\in B_{i,k}\}\Big]\bigcap\{W_{j_i}(t_{\ell+1})-W_{j_i}(t_\ell)\in B_{i,\ell+1}\}\Big).
\end{aligned}$$

Hence the independence of $W_1(\cdot),W_2(\cdot),\cdots$ follows. □

In the rest of this subsection, we consider another kind of vector-valued Brownian motions.

Let $\{w_j(\cdot)\}_{j=1}^{\infty}$ be a sequence of independent real valued, standard Brownian motions. We need the following notion.

Definition 2.120. *The following formal series (in V)*

$$W(t)=\sum_{j=1}^{\infty}w_j(t)e_j,\quad t\geq 0,\tag{2.98}$$

is called a (V-valued) cylindrical Brownian motion.

Clearly, the series in (2.98) does not converge in V. Nevertheless, this series does converge in a larger space, as we shall see below.

Let H be another separable Hilbert spaces. An operator $F\in\mathcal{L}(V;H)$ is said to be a Hilbert-Schmidt operator if $\sum_{j=1}^{\infty}|Fe_j|_H^2<\infty$. It is easy to show that the number $\sum_{j=1}^{\infty}|Fe_j|_H^2$ is independent of the choice of orthonormal basis $\{e_j\}_{j=1}^{\infty}$ in V. Denote by $\mathcal{L}_2(V;H)$ the space of all Hilbert-Schmidt operators from V into H. When V and H are clear from the context, we simply write $\mathcal{L}_2^0$ for $\mathcal{L}_2(V;H)$. One can show that, $\mathcal{L}_2^0$ equipped with the inner product

$$\langle F,G\rangle_{\mathcal{L}_2^0}=\sum_{j=1}^{\infty}\langle Fe_j,Ge_j\rangle_H,\qquad\forall\,F,G\in\mathcal{L}_2^0,\tag{2.99}$$

is a separable Hilbert space. For any $v\in V$ and $h\in H$, it is easy to show that the tensor product $v\otimes h\in\mathcal{L}_2^0$ (Recall (2.18) for the definition of $v\otimes h$).

Generally, for any bounded bilinear functional $\psi(\cdot,\cdot)$ on H, one can define an indefinite inner product on $\mathcal{L}_2^0$ as follows:

$$\psi(F,G)_{\mathcal{L}_2^0} = \sum_{j=1}^{\infty} \psi(Fe_j, Ge_j), \qquad \forall\, F,G \in \mathcal{L}_2^0. \tag{2.100}$$

It is easy to see that

$$|\psi(F,G)_{\mathcal{L}_2^0}| \le ||\psi|| |F|_{\mathcal{L}_2^0} |G|_{\mathcal{L}_2^0}, \tag{2.101}$$

where

$$||\psi|| = \sup\big\{|\psi(x,y)| \;\big|\; x,y \in H, |x|_H \le 1, |y|_H \le 1\big\}. \tag{2.102}$$

For more details on the Hilbert-Schmidt operators, we refer to [299] for example.

Fix any sequence $\{\mu_j\}_{j=1}^{\infty}$ of positive numbers such that $\sum_{j=1}^{\infty} \mu_j^2 < \infty$. Let V_1 be the completion of V w.r.t. the following norm:

$$|f|_{V_1} = \sqrt{\sum_{j=1}^{\infty} \mu_j^2 |\langle f, e_j\rangle_V|^2}, \qquad \forall\, f \in V.$$

Then, V_1 is a separable Hilbert space, $V \subset V_1$ and the embedding map $J : V \to V_1$ is a Hilbert-Schmidt operator. Let $Q_1 = JJ^*$. Then $Q_1 \in \mathcal{L}_1(V_1)$ is positive definite and symmetric.

Proposition 2.121. *The series*

$$W(t) = \sum_{j=1}^{\infty} w_j(t) Je_j, \quad t \ge 0, \tag{2.103}$$

converges in $L^2_{\mathbb{F}}(\Omega; C([0,T]; V_1))$ for any $T > 0$ and defines a V_1-valued, standard Q_1-Brownian motion.

Proof: We first prove that (2.103) is a V_1-valued, standard Q_1-Brownian motion.

Let $W_n(\cdot) = \sum_{j=1}^{n} w_j(\cdot) Je_j$ for $n \in \mathbb{N}$. Then $W_n(\cdot)$ is a continuous V_1-valued martingale. By Theorem 2.109, for $m \in \mathbb{N}$ with $n < m$, we have that

$$\mathbb{E} \sup_{t\in[0,T]} |W_m(t) - W_n(t)|_{V_1}^2 \le 4\mathbb{E}\Big|\sum_{j=n+1}^{m} w_j(T) Je_j\Big|_{V_1}^2 = 4T \sum_{j=n+1}^{m} |Je_j|_{V_1}^2.$$

This, together with the fact that $|J|^2_{\mathcal{L}_2(V;V_1)} = \sum_{j=1}^{\infty} |Je_j|^2_{V_1} < \infty$, implies that $\{W_n(\cdot)\}_{n=1}^{\infty}$ is a Cauchy sequence in $L^2_{\mathbb{F}}(\Omega; C([0,T]; V_1))$. Hence, its limit $W(\cdot) \in L^2_{\mathbb{F}}(\Omega; C([0,T]; V_1))$ for any $T > 0$.

Next, for any $f \in V_1$, it is easy to see that $\langle W(t) - W(s), f\rangle_{V_1}$ is a Gaussian random variable for any $s,t \in [0,T]$ with $s < t$. Further,

$$\mathbb{E}\langle W(t) - W(s), f\rangle_{V_1} = \Big\langle \sum_{j=1}^{\infty} \mathbb{E}\big[w_j(t) - w_j(s)\big] Je_j, f \Big\rangle_{V_1} = 0$$

and

$$\begin{aligned}
&\mathbb{E}\big(\langle W(t) - W(s), f_1\rangle_{V_1} \langle W(t) - W(s), f_2\rangle_{V_1}\big) \\
&= \sum_{j=1}^{\infty} (t-s)\langle Je_j, f_1\rangle_{V_1} \langle Je_j, f_2\rangle_{V_1} = (t-s)\sum_{j=1}^{\infty} \langle e_j, J^* f_1\rangle_V \langle e_j, J^* f_2\rangle_V \\
&= (t-s)\langle J^* f_1, J^* f_2\rangle_V = (t-s)\langle JJ^* f_1, f_2\rangle_{V_1} = (t-s)\langle Q_1 f_1, f_2\rangle_{V_1}.
\end{aligned}$$

Hence, $W(t) - W(s) \sim \mathcal{N}(0, (t-s)Q_1)$. Also, it is easy to show that the random variable $W(t) - W(s)$ is independent of $\mathcal{F}_s$. Thus, $W(\cdot)$ is a V_1-valued, standard Q_1-Brownian motion. □

2.10 Stochastic Integrals

Let $T > 0$ and $(\Omega, \mathcal{F}, \mathbf{F}, \mathbb{P})$ (with $\mathbf{F} = \{\mathcal{F}_t\}_{t\in[0,T]}$) be a fixed filtered probability space satisfying the usual condition, and denote by $\mathbb{F}$ the progressive σ-field w.r.t. $\mathbf{F}$.

In this section, we shall define the integral

$$\int_0^T X(t)dW(t) \tag{2.104}$$

of a stochastic process $X(\cdot)$ w.r.t. a Brownian motion $W(t)$. Such an integral will play an essential role in the sequel. Note that if for $\omega \in \Omega$, the map $t \mapsto W(t, \omega)$ was of bounded variation, then a natural definition of (2.104) would be a Lebesgue-Stieltjes type integral, simply regarding ω as a parameter. Unfortunately, as shown in Proposition 2.115, the map $t \mapsto W(t, \omega)$ is not of bounded variation for a.e. $\omega \in \Omega$ even for the simplest case of real valued Brownian motions. Thus, the integral $\int_0^T f(s, \omega)dW(s, \omega)$ cannot be defined pointwise. However, one can define the integral for a large class of processes by means of the martingale property of Brownian motion. This was first done by K. Itô ([155]) and is now known as Itô's integral.

2.10.1 Itô's Integrals w.r.t. Brownian Motions in Finite Dimensions

To begin with, we assume that $W(\cdot)$ is a one dimensional standard Brownian motion.

We shall define the Itô's integral as a mapping $f \in L^2_{\mathbb{F}}(0, T) \mapsto I(f) \in \mathcal{M}^2_c[0, T]$. As the situation for the Lebesgue integral, we will first define the Itô's integral for simple integrands.

For a moment, assume that $f \in L_{S,\mathbb{F}}(0,T)$ takes the form

$$f(t) = \sum_{j=0}^{n} \chi_{[t_j,t_{j+1})}(t) f_j, \quad t \in [0,T], \tag{2.105}$$

where $n \in \mathbb{N}$, $0 = t_0 < t_1 < \cdots < t_{n+1} = T$, f_j is $\mathcal{F}_{t_j}$-measurable with $\sup\{|f_j(\omega)| \mid j \in \{0,\cdots,n\}, \omega \in \Omega\} < \infty$. We set

$$I(f)(t,\omega) = \sum_{j=0}^{n} f_j(\omega)[W(t \wedge t_{j+1},\omega) - W(t \wedge t_j,\omega)]. \tag{2.106}$$

We have the following key preliminary result.

Proposition 2.122. *Let $f \in L_{S,\mathbb{F}}(0,T)$. Then*
1) $I(f) \in \mathcal{M}_c^2[0,T]$;
2) *For any $t \in [0,T]$, $\mathbb{E}[I(f)(t)] = 0$;*
3) *For any $t \in (0,T]$, $|I(f)|_{\mathcal{M}_c^2[0,t]} = |f|_{L^2_{\mathbb{F}}(0,t)}$; and*
4) *$\big(I(f)(t)\big)^2 - \int_0^t f^2(s)ds$ is a martingale.*

Proof: Suppose that f takes the form of (2.105). From (2.106), it is clear that $I(f)$ is continuous. Further, $0 \le s \le t \le T$,

$$\begin{aligned}
&\mathbb{E}(I(f)(t) \mid \mathcal{F}_s) = \mathbb{E}\big(I(f)(s) + I(f)(t) - I(f)(s) \mid \mathcal{F}_s\big)\\
&= I(f)(s) + \mathbb{E}\Big(\sum_{j=0}^{n} f_j(\omega)[W(t \wedge t_{j+1},\omega) - W(t \wedge t_j,\omega)]\\
&\quad - \sum_{j=0}^{n} f_j(\omega)[W(s \wedge t_{j+1},\omega) - W(s \wedge t_j,\omega)] \,\Big|\, \mathcal{F}_s\Big)\\
&= I(f)(s), \quad \text{a.s.}
\end{aligned}$$

Hence, the conclusion 1) holds.

For each $j \in \{0,1,\cdots,n\}$,

$$\begin{aligned}
&\mathbb{E}\big\{f_j[W(t \wedge t_{j+1}) - W(t \wedge t_j)]\big\} = \mathbb{E}\big\{\mathbb{E}\big(f_j[W(t \wedge t_{j+1}) - W(t \wedge t_j)] \mid \mathcal{F}_{t_j}\big)\big\}\\
&= \mathbb{E}\big\{f_j\mathbb{E}\big([W(t \wedge t_{j+1}) - W(t \wedge t_j)] \mid \mathcal{F}_{t_j}\big)\big\}\\
&= \mathbb{E}\big\{f_j\mathbb{E}[W(t \wedge t_{j+1}) - W(t \wedge t_j)]\big\} = 0.
\end{aligned}$$

Hence, $\mathbb{E}I(f) = 0$, which proves the second conclusion.

Clearly,

$$|I(f)(t)|^2 = \sum_{i,j=0}^{n} f_i f_j [W(t \wedge t_{i+1}) - W(t \wedge t_i)][W(t \wedge t_{j+1}) - W(t \wedge t_j)].$$

For $i < j$,

$$
\begin{aligned}
&\mathbb{E}\{f_i f_j [W(t\wedge t_{i+1}) - W(t\wedge t_i)][W(t\wedge t_{j+1}) - W(t\wedge t_j)]\}\\
&= \mathbb{E}\{\mathbb{E}(f_i f_j [W(t\wedge t_{i+1}) - W(t\wedge t_i)][W(t\wedge t_{j+1}) - W(t\wedge t_j)] \mid \mathcal{F}_{t_j})\}\\
&= \mathbb{E}\{f_i f_j [W(t\wedge t_{i+1}) - W(t\wedge t_i)]\mathbb{E}([W(t\wedge t_{j+1}) - W(t\wedge t_j)] \mid \mathcal{F}_{t_j})\} = 0
\end{aligned}
$$

For $i = j$,

$$
\begin{aligned}
&\mathbb{E}\{f_j^2 [W(t\wedge t_{j+1}) - W(t\wedge t_j)]^2\}\\
&= \mathbb{E}\{\mathbb{E}(f_j^2 [W(t\wedge t_{j+1}) - W(t\wedge t_j)]^2 \mid \mathcal{F}_{t_j})\}\\
&= \mathbb{E}\{f_j^2 \mathbb{E}([W(t\wedge t_{j+1}) - W(t\wedge t_j)]^2 \mid \mathcal{F}_{t_j})\} = (t\wedge t_{j+1} - t\wedge t_j)\mathbb{E} f_j^2.
\end{aligned}
$$

Hence, the conclusion 3) holds.

For any $0 \le s \le t$,

$$
\begin{aligned}
&\mathbb{E}\Big(\big[I(f)(t)\big]^2 - \int_0^t f^2(r)dr \,\Big|\, \mathcal{F}_s\Big)\\
&= \mathbb{E}\Big(\big[I(f)(s)\big]^2 - \int_0^s f^2(r)dr \,\Big|\, \mathcal{F}_s\Big)\\
&\quad + \mathbb{E}\Big(\big[I(f)(t) - I(f)(s) + I(f)(s)\big]^2 - \big[I(f)(s)\big]^2 - \int_s^t f^2(r)dr \,\Big|\, \mathcal{F}_s\Big)\\
&= \big(I(f)(s)\big)^2 - \int_0^s f^2(r)dr + \mathbb{E}\Big\{\big[I(f)(t) - I(f)(s)\big]^2\\
&\quad + 2I(f)(s)\big[I(f)(t) - I(f)(s)\big] \,\Big|\, \mathcal{F}_s\Big\} - \mathbb{E}\Big(\int_s^t f^2(r)dr \,\Big|\, \mathcal{F}_s\Big)\\
&= \big(I(f)(s)\big)^2 - \int_0^s f^2(r)dr.
\end{aligned}
$$

This proves the last conclusion. □

Next, for any given $f \in L^2_{\mathbb{F}}(0,T)$, by Lemma 2.75, we can find a sequence of $\{f_n\}_{n=1}^\infty \subset L_{S,\mathbb{F}}(0,T)$ such that $|f_n - f|_{L^2_{\mathbb{F}}(0,T)} \to 0$ as $n \to \infty$. Since $|I(f_n) - I(f_m)|_{\mathcal{M}^2_c[0,T]} = |f_n - f_m|_{L^2_{\mathbb{F}}(0,T)}$ for $m, n \in \mathbb{N}$, one deduces that $\{I(f_n)\}_{n=1}^\infty$ is a Cauchy sequence in $\mathcal{M}^2_c[0,T]$ and therefore, it converges to a unique element $X \in \mathcal{M}^2_c[0,T]$. Clearly, the process X is determined uniquely by f and it is independent of the particular choice of $\{f_n\}_{n=1}^\infty$. This process is called Itô's integral of $f \in L^2_{\mathbb{F}}(0,T)$ (w.r.t. the Brownian motion $W(\cdot)$). We shall denote it by

$$
\int_0^t f(s,\omega)dW(s,\omega) \quad \text{or simply} \quad \int_0^t f(s)dW(s) \quad \text{or even} \quad \int_0^t f dW.
$$

It is easy to show the following result.

Theorem 2.123. *For each $f \in L^2(a,b)$ (for which f is a deterministic function, i.e., it does not depend on $\omega \in \Omega$) with $0 \le a < b \le T$, the Itô integral $\int_a^b f(t)dW(t)$ is a Gaussian random variable with mean 0 and variance $\int_a^b |f(t)|^2 dt$.*

Now, fix $m_1, m_2 \in \mathbb{N}$. For an $f = (f^{ij})_{1\le i\le m_1,\, 1\le j\le m_2} \in L^2_{\mathbb{F}}(0,T;\mathbb{R}^{m_1\times m_2})$ and an $\mathbb{R}^{d_2}$-valued Brownian motion $W(\cdot) = (W^1(\cdot),\cdots,W^{m_2}(\cdot))^\top$, we define

$$\int_0^t f(s)dW(s) \triangleq \Big(\sum_{j=1}^{m_2}\int_0^t f^{1j}(s)dW^j(s),\cdots,\sum_{j=1}^{m_2}\int_0^t f^{m_1 j}(s)dW^j(s)\Big)^\top.$$

Further, for $0 < s < t$, we define

$$\int_s^t f(r)dW(r) = \int_0^t f(r)dW(r) - \int_0^s f(r)dW(r).$$

Itô's integral (w.r.t. the $\mathbb{R}^{m_2}$-valued Brownian motion $W(\cdot)$) has the properties:

Theorem 2.124. *Let $f, g \in L^2_{\mathbb{F}}(0,T;\mathbb{R}^{m_1\times m_2})$, $a, b \in \mathbb{R}$ and $T \ge t > s \ge 0$. Then*

1)

$$\int_0^t (af+bg)dW = a\int_0^t fdW + b\int_0^t gdW, \quad \text{a.s.};$$

2)

$$\mathbb{E}\Big(\int_s^t fdW \,\Big|\, \mathcal{F}_s\Big) = 0, \quad \text{a.s.};$$

3)

$$\mathbb{E}\Big(\Big\langle \int_s^t fdW, \int_s^t gdW\Big\rangle_{\mathbb{R}^{m_1}} \,\Big|\, \mathcal{F}_s\Big) = \mathbb{E}\Big(\int_s^t \langle f(r,\cdot), g(r,\cdot)\rangle_{\mathbb{R}^{m_1\times m_2}} dr \,\Big|\, \mathcal{F}_s\Big), \text{ a.s.};$$

4) *The stochastic process $\int_0^\cdot fdW \in \mathcal{M}^2_c([0,T];\mathbb{R}^{m_1})$.*

Proof: It is easy to show that the conclusions in Theorem 2.124 hold for simple processes. The proof for the general f and g is based on these facts and the definition of stochastic integral, and hence we omit it here. □

The following result shows that Itô's integral enjoys some local property though it is not defined pointwise w.r.t. the sample point.

Lemma 2.125. *Let $f, g \in L^2_{\mathbb{F}}(0,T;\mathbb{R}^{m_1\times m_2})$ and $W(\cdot)$ be an $\mathbb{R}^{m_2}$-valued Brownian motion. Let $\Omega_1 = \{\omega \in \Omega \mid f(t,\omega) = g(t,\omega)$ for a.e. $t \in [0,T]\}$. Then,*

$$\int_0^T f(t)dW(t) = \int_0^T g(t)dW(t), \quad a.e.\ \omega \in \Omega_1.$$

Proof: Without loss of generality, we may assume that $g(t,\omega)=0$ for a.e. $(t,\omega)\in[0,T]\times\Omega$. Let

$$\tau(\omega)\overset{\Delta}{=}\inf\Big\{t\in[0,T)\ \Big|\ \int_0^t|f(s,\omega)|^2ds\neq 0\Big\}.$$

Here we agree that $\inf\emptyset=T$. Then, $\int_0^T\chi_{[0,\tau]}(t)|f(t)|^2dt=0$, a.s. For any $t\in(0,T]$,

$$\{\tau<t\}=\bigcup_{r\in\mathbb{Q}\cap(0,t)}\Big\{\int_0^r|f(s)|^2ds>0\Big\}\in\mathcal{F}_t.$$

For $t=0$, we have that $\{\tau\le 0\}=\emptyset\in\mathcal{F}_0$. Hence, τ is a stopping time.

Define a random variable

$$Y(\tau)=\int_0^\tau f(s)dW(s)=\int_0^T\chi_{[0,\tau]}(s)f(s)dW(s).$$

Since $\chi_{[0,\tau]}f\in L^2_{\mathbb{F}}(0,T)$, we see that $Y(\tau)$ is well defined and

$$\mathbb{E}|Y(\tau)|^2=\mathbb{E}\int_0^T\chi_{[0,\tau]}(t)|f(t)|^2dt=0,$$

which implies that $Y(\tau)=0$, a.s.

If $\omega\in\Omega_1$, then $\tau(\omega)=T$. Hence, $Y(\tau(\omega))=\int_0^T f(s)dW(s)$. This show that $\int_0^T f(s)dW(s)=0$, a.e. $\omega\in\Omega_1$. □

The following result will play a crucial role later.

Lemma 2.126. *Let $f\in L^2_{\mathbb{F}}(0,T;\mathbb{R}^{m_1\times m_2})$ and $W(\cdot)$ be an $\mathbb{R}^{m_2}$-valued Brownian motion. Then, for any positive numbers ε and λ,*

$$\mathbb{P}\Big(\Big\{\Big|\int_0^T f(t)dW(t)\Big|_{\mathbb{R}^{m_1}}\ge\varepsilon\Big\}\Big)\le\frac{\lambda}{\varepsilon^2}+\mathbb{P}\Big(\Big\{\int_0^T|f(t)|^2_{\mathbb{R}^{m_1\times m_2}}dt>\lambda\Big\}\Big).\tag{2.107}$$

Proof: For simplicity, we only consider the case that $m_1=m_2=1$. Define a stochastic process $f_\lambda(t,\omega)$ by

$$f_\lambda(t,\omega)=\begin{cases}f(t,\omega), & \text{if }\int_0^t|f(s,\omega)|^2ds\le\lambda,\\ 0, & \text{otherwise.}\end{cases}\tag{2.108}$$

Clearly,

$$\begin{aligned}&\Big\{\Big|\int_0^T f(t)dW(t)\Big|\ge\varepsilon\Big\}\\&\subset\Big\{\Big|\int_0^T f_\lambda(t)dW(t)\Big|\ge\varepsilon\Big\}\bigcup\Big\{\int_0^T f(t)dW(t)\neq\int_0^T f_\lambda(t)dW(t)\Big\}.\end{aligned}\tag{2.109}$$

Further, by Lemma 2.125 and the definition of $f_\lambda(\cdot)$, we see that

$$\Big\{\int_0^T f(t)dW(t) \neq \int_0^T f_\lambda(t)dW(t)\Big\} \subset \Big\{\int_0^T |f(t)|^2 dt > \lambda\Big\}. \qquad (2.110)$$

From (2.109) and (2.110), we see that

$$\mathbb{P}\Big\{\Big|\int_0^T f(t)dW(t)\Big| \geq \varepsilon\Big\} \leq \mathbb{P}\Big\{\Big|\int_0^T f_\lambda(t)dW(t)\Big| \geq \varepsilon\Big\} + \mathbb{P}\Big\{\int_0^T |f(t)|^2 dt > \lambda\Big\}. \qquad (2.111)$$

By (2.108), we have $\int_0^T |f_\lambda(t)|^2 dt \leq \lambda$, a.s. Hence, $\mathbb{E}\int_0^T |f_\lambda(t)|^2 dt \leq \lambda$. This, together with Chebyshev's inequality (in Theorem 2.22) and (2.111), implies that

$$\begin{aligned}
&\mathbb{P}\Big\{\Big|\int_0^T f(t)dW(t)\Big| \geq \varepsilon\Big\} \leq \frac{1}{\varepsilon^2}\mathbb{E}\Big|\int_0^T f_\lambda(t)dW(t)\Big|^2 + \mathbb{P}\Big\{\int_0^T |f(t)|^2 dt > \lambda\Big\} \\
&= \frac{1}{\varepsilon^2}\mathbb{E}\int_0^T |f_\lambda(t)|^2 dt + \mathbb{P}\Big\{\int_0^T |f(t)|^2 dt > \lambda\Big\} \leq \frac{\lambda}{\varepsilon^2} + \mathbb{P}\Big\{\int_0^T |f(t)|^2 dt > \lambda\Big\},
\end{aligned}$$

which gives (2.107). This completes the proof of Lemma 2.126. □

The above Itô integral $\int_0^T f(t)dW(t)$ is defined for **F**-adapted stochastic processes $f(t)$ satisfying the condition $\mathbb{E}\int_0^T |f(t)|^2_{\mathbb{R}^{m_1\times m_2}} dt < \infty$. Next, we will extend the Itô integral $\int_0^T f(t)dW(t)$ to more general **F**-adapted stochastic processes $f(\cdot)$.

Let Y be a Banach space. For any $p \in [1, \infty)$, denote by $L^{p,loc}_{\mathbb{F}}(0, T; Y)$ the set of all Y-valued, **F**-adapted stochastic processes $f(\cdot)$ satisfying $\int_0^T |f(t)|^p_Y dt < \infty$, a.s.

Now, we extend the Itô integral $\int_0^T f(t)dW(t)$ to the case that the integrand $f(\cdot) \in L^{2,loc}_{\mathbb{F}}(0, T; \mathbb{R}^{m_1\times m_2})$. For simplicity, we consider only the case that $m_1 = m_2 = 1$, and simply denote $L^{p,loc}_{\mathbb{F}}(0, T; \mathbb{R})$ by $L^{p,loc}_{\mathbb{F}}(0, T)$. The reasons for this generalization are as follows. First, generally speaking, it is very hard to check whether an **F**-adapted stochastic process $f(\cdot)$ belongs to $L^2_{\mathbb{F}}(0, T)$ and it is easier to see whether $f \in L^{2,loc}_{\mathbb{F}}(0, T)$. For example, when $f(\cdot)$ is a continuous, **F**-adapted process, then it belongs to $L^{2,loc}_{\mathbb{F}}(0, T)$. Second, we need to consider the integral of processes $f \in L^r_{\mathbb{F}}(\Omega; L^2(0, T))$ for $r \in [1, 2)$ in the study of stochastic differential/evolution equations.

We need the following technical result.

Lemma 2.127. *Let* $f \in L^{2,loc}_{\mathbb{F}}(0, T)$. *Then there exists a sequence* $\{f_n\}_{n=1}^\infty \subset L^2_{\mathbb{F}}(0, T)$ *such that*

$$\lim_{n\to\infty}\int_0^T |f_n(t) - f(t)|^2 dt = 0, \quad \text{a.s.} \qquad (2.112)$$

Proof: For each $n \in \mathbb{N}$, let us set

$$f_n(t,\omega) = \begin{cases} f(t,\omega), & \text{if } \int_0^t |f(s,\omega)|^2 ds \leq n, \\ 0, & \text{otherwise.} \end{cases} \tag{2.113}$$

Clearly, f_n is $\mathbf{F}$-adapted and $\int_0^T |f_n(t,\omega)|^2 dt \leq n$, a.s., which implies that $\mathbb{E}\int_0^T |f_n(t)|^2 dt \leq n$. Hence, $f_n \in L^2_{\mathbb{F}}(0,T)$.

Fix an $\omega \in \Omega$. When $n > \int_0^T |f(t,\omega)|^2 dt$, by the definition of f_n, we have $f_n(t,\omega) = f(t,\omega)$ for all $t \in [0,T]$. Thus, (2.112) holds. □

Now, for any given $f \in L^{2,loc}_{\mathbb{F}}(0,T)$, by Lemma 2.127, one can find a sequence $\{f_n\}_{n=1}^{\infty} \subset L^2_{\mathbb{F}}(0,T)$ such that (2.112) holds. Apply Lemma 2.126 to $f = f_n - f_m$ (for $m,n \in \mathbb{N}$) with $\varepsilon > 0$ and $\lambda = \frac{\varepsilon^3}{2}$, we obtain that

$$\begin{aligned} &\mathbb{P}\Big\{\Big|\int_0^T f_n(t)dW(t) - \int_0^T f_m(t)dW(t)\Big| > \varepsilon\Big\} \\ &\leq \frac{\varepsilon}{2} + \mathbb{P}\Big\{\int_0^T |f_n(t) - f_m(t)|^2 dt > \frac{\varepsilon^3}{2}\Big\}. \end{aligned} \tag{2.114}$$

Hence, $\{\int_0^T f_n(t)dW(t)\}_{n=1}^{\infty}$ is a Cauchy sequence in probability. This indicates that the sequence $\{\int_0^T f_n(t)dW(t)\}_{n=1}^{\infty}$ converges in probability. Therefore, we can define

$$\int_0^T f(t)dW(t) = \mathbb{P}\text{-}\lim_{n\to\infty} \int_0^T f_n(t)dW(t). \tag{2.115}$$

One can easily see that $\int_0^T f(t)dW(t)$ is independent of the choice of $\{f_n\}_{n=1}^{\infty}$. Hence, the integral "$\int_0^T f(t)dW(t)$" is well-defined. If $f \in L^p_{\mathbb{F}}(\Omega; L^2(0,T))$ for $p \geq 1$, then one can show that $\int_0^T f(t)dW(t) \in L^p(\Omega)$.

Remark 2.128. If $f \in L^2_{\mathbb{F}}(0,T)$, then we can take $f_n = f$ for all $n \in \mathbb{N}$. In this case, the stochastic integral defined by (2.115) is obviously the same as the one defined for $f \in L^2_{\mathbb{F}}(0,T)$. This shows that the new stochastic integral defined by (2.115) reduces to the old stochastic integral when $f \in L^2_{\mathbb{F}}(0,T)$.

We list some basic properties of the Itô's integral of $L^{2,loc}_{\mathbb{F}}(0,T;\mathbb{R}^{m_1\times m_2})$-processes as follows.

Theorem 2.129. *Let $f, g \in L^{2,loc}_{\mathbb{F}}(0,T;\mathbb{R}^{m_1\times m_2})$ and $W(\cdot)$ be an $\mathbb{R}^{m_2}$-valued Brownian motion. Then:*

1) *For any $\mathbf{F}$-stopping time σ, it holds that*

$$\int_0^{t\wedge\sigma} f dW = \int_0^t f\chi_{[0,\sigma]}(s)dW(s);$$

2) *For any bounded* $\mathbf{F}$*-stopping times* τ *and* σ *with* $\tau \leq \sigma$ *a.s., and any bounded* $\mathcal{F}_\tau$*-measurable random variable* ξ_1 *and* ξ_2*, it holds*

$$\int_\tau^\sigma (\xi_1 f + \xi_2 g)dW = \xi_1 \int_\tau^\sigma f dW + \xi_2 \int_\tau^\sigma g dW;$$

3) *The stochastic process* $X(\cdot) \stackrel{\Delta}{=} \int_0^\cdot f(s)dW(s)$ *is a continuous, local martingale;*

4) *For any* $\varepsilon > 0$ *and* $\lambda > 0$*,*

$$\mathbb{P}\Big(\Big\{ \sup_{s\in[0,T]} \Big| \int_0^s f dW \Big|_{\mathbb{R}^{m_1}} \geq \varepsilon \Big\}\Big) \leq \frac{\lambda}{\varepsilon^2} + \mathbb{P}\Big(\Big\{ \int_0^T |f(t)|^2_{\mathbb{R}^{m_1\times m_2}} dt > \lambda \Big\}\Big). \tag{2.116}$$

Proof: Let us prove the last two conclusions 3) and 4). We only consider the case that $m_1 = m_2 = 1$.

To prove 3), for any $n \in \mathbb{N}$, put

$$\tau_n(\omega) \stackrel{\Delta}{=} \inf\Big\{ t \in [0,T) \Bigm| \int_0^t |f(s,\omega)|^2 ds > n \Big\}. \tag{2.117}$$

Then, τ_n is a stopping time and $\lim_{n\to\infty} \tau_n = T$, a.s. Clearly, $\chi_{[0,\tau_n]} f \in L^2_{\mathbb{F}}(0,T)$ and

$$X(t\wedge\tau_n) = \int_0^{t\wedge\tau_n} f(s)dW(s) = \int_0^t \chi_{[0,\tau_n]}(s) f(s) dW(s).$$

Thus, $\{X(t\wedge\tau_n)\}_{t\in[0,T]}$ is a martingale, which implies that $\{X(t)\}_{t\in[0,T]}$ is a local martingale.

For each $n \in \mathbb{N}$, let us set

$$f_n(t,\omega) = \begin{cases} f(t,\omega), & \text{if } \displaystyle\int_0^t |f(s,\omega)|^2 ds \leq n, \\ 0, & \text{otherwise.} \end{cases} \tag{2.118}$$

It is easy to see that $f_n \in L^2_{\mathbb{F}}(0,T)$. Let $X^{(n)}(t) \stackrel{\Delta}{=} \int_0^t f_n(s)dW(s)$. By the fourth conclusion in Theorem 2.124, $X^{(n)}(\cdot)$ is continuous. Let $\Omega_n \stackrel{\Delta}{=} \{\int_0^T |f(t)|^2 dt \leq n\}$. The sequence $\{\Omega_n\}_{n=1}^\infty$ is increasing. Let $\widetilde{\Omega} \stackrel{\Delta}{=} \bigcup_{n=1}^\infty \Omega_n$. Since $\int_0^T |f(t)|^2 dt < \infty$, a.s., we find that $\mathbb{P}(\widetilde{\Omega}) = 1$. If $\omega \in \Omega_n$, then $f_n(t,\omega) = f_m(t,\omega)$ for any $t \in [0,T]$ and integer $m \geq n$. Thus, by Lemma 2.125, for almost all $\omega \in \Omega_n$,

$$X^{(m)}(t,\omega) = X^{(n)}(t,\omega), \quad \forall\, m \geq n \text{ and } t \in [0,T]. \tag{2.119}$$

This implies that for a.e. $\omega \in \widetilde{\Omega}$, $\lim_{m\to\infty} X^{(m)}(t,\omega)$ exists for all $t \in [0,T]$. We define a stochastic process $Y(t,\omega)$ by

$$Y(t,\omega) \triangleq \begin{cases} \lim_{m\to\infty} X^{(m)}(t,\omega), & \text{if } \omega \in \widetilde{\Omega}, \\ 0, & \text{if } \omega \notin \widetilde{\Omega}. \end{cases}$$

Then $Y(\cdot)$ is a continuous stochastic process. On the other hand, from the definition of the stochastic integral, we have that $X(t) = \mathbb{P}\text{-}\lim_{m\to\infty} X^{(m)}(t)$. Therefore, for every $t \in [0,T]$, $X(t) = Y(t)$, a.s. Hence $Y(\cdot)$ is a continuous modification of $X(\cdot)$.

We now prove 4). We choose $f^\lambda(\cdot)(\in L^2_{\mathbb{F}}(0,T))$ as that in (2.108). Then, thanks to Chebyshev's inequality (in Theorem 2.22) and Doob's inequality (in Theorem 2.101), we have

$$\begin{aligned}
&\mathbb{P}\Big(\Big\{\sup_{s\in[0,T]}\Big|\int_0^s fdW\Big| \geq \varepsilon\Big\}\Big)\\
&\leq \mathbb{P}\Big(\Big\{\sup_{s\in[0,T]}\Big|\int_0^s (f-f^\lambda)dW\Big| > 0\Big\}\Big) + \mathbb{P}\Big(\Big\{\sup_{s\in[0,T]}\Big|\int_0^s f^\lambda dW\Big| \geq \varepsilon\Big\}\Big)\\
&\leq \mathbb{P}\Big(\Big\{\int_0^T |f(r)|^2 dr \geq \lambda\Big\}\Big) + \varepsilon^{-2}\mathbb{E}\Big(\int_0^T f^\lambda dW\Big)^2\\
&= \mathbb{P}\Big(\Big\{\int_0^T |f(r)|^2 dr \geq \lambda\Big\}\Big) + \varepsilon^{-2}\mathbb{E}\int_0^T |f^\lambda(r)|^2 dr\\
&\leq \mathbb{P}\Big(\Big\{\int_0^T |f(r)|^2 dr \geq \lambda\Big\}\Big) + \frac{\lambda}{\varepsilon^2},
\end{aligned} \tag{2.120}$$

which gives (2.116). □

2.10.2 Itô's Integrals w.r.t. Vector-Valued Brownian Motions

Let H be a separable Hilbert space with an orthonormal basis $\{h_k\}_{k=1}^\infty$. Let V be another separable Hilbert space and $\mathcal{L}_2^0 \equiv \mathcal{L}_2(V;H)$.

We consider first the case of Q-Brownian motion $W(\cdot)$, where $Q \in \mathcal{L}_1(V)$ is given as that in Subsection 2.9.3 (Recall that $\{\lambda_k\}_{k=1}^\infty$ and $\{e_k\}_{k=1}^\infty$ are respectively sequences of eigenvalues and eigenvectors of Q). By Proposition 2.119, for any t, $W(t)$ has the expansion $W(t) = \sum_{j=1}^\infty \sqrt{\lambda_j} W_j(t) e_j$, where

$$W_j = \frac{1}{\sqrt{\lambda_j}} \langle W(t), e_j\rangle_V, \qquad j = 1, 2, \cdots,$$

are mutually independent real valued Brownian motions.

For each $N \in \mathbb{N}$, we define $\widetilde{W}_N(t) = \sum_{j=1}^N \sqrt{\lambda_j} W_j(t) e_j$. Let $\Phi \in L^2_{\mathbb{F}}(0,T;\mathcal{L}(V;H))$. Put

$$X_{n,N} = \sum_{k=1}^n \sum_{j=1}^N \sqrt{\lambda_j} h_k \int_0^T \langle \Phi(s)e_j, h_k\rangle_H dW_j(s).$$

Then, for any $m, n \in \mathbb{N}$ with $m < n$, we have that

$$
\begin{aligned}
\mathbb{E}|X_{n,N} - X_{m,N}|_H^2 &= \mathbb{E}\Big|\sum_{k=m+1}^{n}\sum_{j=1}^{N}\sqrt{\lambda_j}h_k\int_0^T\langle\Phi(s)e_j, h_k\rangle_H dW_j(s)\Big|_H^2 \\
&= \sum_{k=m+1}^{n}\mathbb{E}\Big|\sum_{j=1}^{N}\sqrt{\lambda_j}\int_0^T\langle\Phi(s)e_j, h_k\rangle_H dW_j(s)\Big|^2 \\
&= \sum_{k=m+1}^{n}\sum_{j=1}^{N}\lambda_j\mathbb{E}\int_0^T\big|\langle\Phi(s)e_j, h_k\rangle_H\big|^2 ds,
\end{aligned}
\tag{2.121}
$$

which implies that $\{X_{n,N}\}_{n=1}^{\infty}$ is a Cauchy sequence in $L^2_{\mathcal{F}_T}(\Omega; H)$. We define

$$
\int_0^T\Phi(s)d\widetilde{W}_N(s) = \sum_{k=1}^{\infty}\sum_{j=1}^{N}\sqrt{\lambda_j}h_k\int_0^T\langle\Phi(s)e_j, h_k\rangle_H dW_j(s) \in L^2_{\mathcal{F}_T}(\Omega; H).
$$

For $t \in [0, T]$, put $Y_N(t) = \int_0^t\Phi(s)d\widetilde{W}_N(s)$. Then, for any $M, N \in \mathbb{N}$ with $M > N$, we have that

$$
\begin{aligned}
&\mathbb{E}|Y_N(T) - Y_M(T)|_H^2 = \mathbb{E}\Big|\sum_{k=1}^{\infty}\sum_{j=N+1}^{M}\sqrt{\lambda_j}h_k\int_0^T\langle\Phi(s)e_j, h_k\rangle_H dW_j(s)\Big|_H^2 \\
&= \sum_{k=1}^{\infty}\mathbb{E}\Big|\sum_{j=N+1}^{M}\sqrt{\lambda_j}\int_0^T\langle\Phi(s)e_j, h_k\rangle_H dW_j(s)\Big|^2 \\
&= \sum_{j=N+1}^{M}\lambda_j\mathbb{E}\int_0^T\sum_{k=1}^{\infty}\big|\langle\Phi(s)e_j, h_k\rangle_H\big|^2 ds \\
&= \mathbb{E}\sum_{j=N+1}^{M}\lambda_j\int_0^T\big|\Phi(s)e_j\big|_H^2 ds \le |\Phi|^2_{L^2_{\mathbb{F}}(0,T;\mathcal{L}(V;H))}\sum_{j=N+1}^{M}\lambda_j.
\end{aligned}
\tag{2.122}
$$

This shows that $\{Y_N(T)\}_{N=1}^{\infty}$ is a Cauchy sequence in $L^2_{\mathcal{F}_T}(\Omega; H)$. Hence, we may define the integral of Φ w.r.t. $W(\cdot)$ as follows:

$$
\int_0^T\Phi(s)dW(s) = \lim_{N\to\infty}\int_0^T\Phi(s)d\widetilde{W}_N(s), \quad \text{in } L^2_{\mathcal{F}_T}(\Omega; H). \tag{2.123}
$$

Further, we define

$$
\int_0^t\Phi(s)dW(s) = \int_0^T\chi_{[0,t]}\Phi(s)dW(s) \quad \text{for } t \in [0, T],
$$

and

$$\int_s^t \Phi(s)dW(s) = \int_0^t \Phi(s)dW(s) - \int_0^s \Phi(s)dW(s) \quad \text{for } 0 \le s < t \le T.$$

Similar to the case of finite dimensions, for any $p \in [1, \infty)$, one may introduce the space $L_{\mathbb{F}}^{p,loc}(0,T;\mathcal{L}(V;H))$ and define the stochastic integral of $\Phi \in L_{\mathbb{F}}^{2,loc}(0,T;\mathcal{L}(V;H))$ (especially $\Phi \in L_{\mathbb{F}}^p(\Omega;L^2(0,T;\mathcal{L}(V;H))))$ w.r.t. $W(\cdot)$.

Similar to Theorems 2.124 and 2.129, one can show that the Itô's integral (with respect to a Q-Brownian motion $W(\cdot)$ valued in V) has the following elementary properties:

Theorem 2.130. *Let $f, g \in L_{\mathbb{F}}^{2,loc}(0,T;\mathcal{L}(V;H))$, $a, b \in \mathbb{R}$ and $T \ge t > s \ge 0$. The following results hold:*

1)

$$\int_0^t (af + bg)dW = a\int_0^t fdW + b\int_0^t gdW, \quad \text{a.s.};$$

2) *The stochastic process $\int_0^\cdot fdW$ is an H-valued, continuous, local martingale;*

3) *If $f \in L_{\mathbb{F}}^p(\Omega;L^2(0,T;\mathcal{L}(V;H)))$, then*

$$\mathbb{E}\Big(\int_s^t fdW \;\Big|\; \mathcal{F}_s\Big) = 0, \quad \text{a.s.};$$

4) *If $f, g \in L_{\mathbb{F}}^2(0,T;\mathcal{L}(V;H))$, then $\int_0^\cdot fdW \in \mathcal{M}_c^2([0,T];H)$, and*

$$\begin{aligned}&\mathbb{E}\Big(\Big\langle \int_s^t fdW, \int_s^t gdW\Big\rangle_H \;\Big|\; \mathcal{F}_s\Big)\\ &= \mathbb{E}\Big(\int_s^t \langle f(r,\cdot)Q^{\frac{1}{2}}, g(r,\cdot)Q^{\frac{1}{2}}\rangle_{\mathcal{L}_2^0} dr \;\Big|\; \mathcal{F}_s\Big), \quad \text{a.s.}\end{aligned}$$

As a consequence of the conclusion 4) in Theorem 2.130, we have the following result (recall (2.100) for the notation $\psi(\cdot,\cdot)_{\mathcal{L}_2^0}$).

Corollary 2.131. *For any bounded bilinear functional $\psi(\cdot,\cdot)$ on H and $f, g \in L_{\mathbb{F}}^2(0,T;\mathcal{L}(V;H))$, it holds that*

$$\begin{aligned}&\mathbb{E}\Big(\psi\Big(\int_s^t fdW(r), \int_s^t gdW(r)\Big)_H \;\Big|\; \mathcal{F}_s\Big)\\ &= \mathbb{E}\Big(\int_s^t \psi(f(r,\cdot)Q^{\frac{1}{2}}, g(r,\cdot)Q^{\frac{1}{2}})_{\mathcal{L}_2^0} dr \;\Big|\; \mathcal{F}_s\Big), \quad \text{a.s.}\end{aligned}$$

Proof: Since $\psi(\cdot,\cdot)$ is a bounded bilinear functional on H, there is a $P \in \mathcal{L}(H)$ such that $\psi(x,y) = \langle Px, y\rangle_H$ for any $x, y \in H$. Hence, by (2.99), (2.100) and the third conclusion in Theorem 2.130, it follows that

$$
\begin{aligned}
&\mathbb{E}\Big(\psi\Big(\int_s^t fdW(r), \int_s^t gdW(r)\Big)_H \Bigm| \mathcal{F}_s\Big) \\
&= \mathbb{E}\Big(\Big\langle \int_s^t PfdW(r), \int_s^t gdW(r)\Big\rangle_H \Bigm| \mathcal{F}_s\Big) \\
&= \mathbb{E}\Big(\int_s^t \langle Pf(r,\cdot)Q^{\frac{1}{2}}, g(r,\cdot)Q^{\frac{1}{2}}\rangle_{\mathcal{L}_2^0} dr \Bigm| \mathcal{F}_s\Big) \\
&= \mathbb{E}\Big(\int_s^t \psi(f(r,\cdot)Q^{\frac{1}{2}}, g(r,\cdot)Q^{\frac{1}{2}})_{\mathcal{L}_2^0} dr \Bigm| \mathcal{F}_s\Big), \text{ a.s.}
\end{aligned}
$$

This completes the proof of Corollary 2.131. □

We now proceed with the definition of the stochastic integral w.r.t. a cylindrical Wiener process.

Let $\{w_k(\cdot)\}_{k=1}^\infty$ be a sequence of independent real valued standard Brownian motions, and $W(\cdot)$ be the corresponding V-valued cylindrical Brownian motion (given in Definition 2.120). Recall that $L^2_{\mathbb{F}}(0,T;\mathcal{L}_2^0)$ is the Hilbert space consisting of all $\mathcal{L}_2^0$-valued, $\mathbf{F}$-adapted processes Φ such that

$$
|\Phi|^2_{L^2_{\mathbb{F}}(0,T;\mathcal{L}_2^0)} \overset{\Delta}{=} \mathbb{E}\int_0^T |\Phi(t)|^2_{\mathcal{L}_2^0} dt = \sum_{k=1}^\infty \mathbb{E}\int_0^T |\Phi(t)e_k|_H^2 dt < \infty. \tag{2.124}
$$

For each $N \in \mathbb{N}$, write $\widehat{W}_N(t) = \sum_{j=1}^N w_j(t)e_j$. For any $\Phi \in L^2_{\mathbb{F}}(0,T;\mathcal{L}_2^0)$, put

$$
X_{n,N} = \sum_{k=1}^n \sum_{j=1}^N h_k \int_0^T \langle \Phi(s)e_j, h_k\rangle_H dw_j(s).
$$

Then, similarly to (2.121), for any $m, n \in \mathbb{N}$ with $m < n$, we have that

$$
\mathbb{E}|X_{n,N} - X_{m,N}|_H^2 = \sum_{k=m+1}^n \sum_{j=1}^N \mathbb{E}\int_0^T \big|\langle \Phi(s)e_j, h_k\rangle_H\big|^2 ds,
$$

which implies that $\{X_{n,N}\}_{n=1}^\infty$ is a Cauchy sequence in $L^2_{\mathcal{F}_T}(\Omega;H)$. Hence, we define

$$
\int_0^T \Phi(s)d\widehat{W}_N(s) = \sum_{k=1}^\infty \sum_{j=1}^N h_k \int_0^T \langle \Phi(s)e_j, h_k\rangle_H dw_j(s) \in L^2_{\mathcal{F}_T}(\Omega;H).
$$

For any $t \in [0,T]$, put $Y_N(t) = \int_0^t \Phi(s)d\widehat{W}_N(s)$. Then, similarly to (2.122), for any $M, N \in \mathbb{N}$ with $M > N$, we have that

$$
\mathbb{E}|Y_N(T) - Y_M(T)|_H^2 = \mathbb{E}\sum_{j=N+1}^M \int_0^T \big|\Phi(s)e_j\big|_H^2 ds. \tag{2.125}
$$

By (2.124) and (2.125), $\{Y_N(T)\}_{N=1}^{\infty}$ is a Cauchy sequence in $L^2_{\mathcal{F}_T}(\Omega;H)$. Hence, we may define the integral of Φ w.r.t. $W(\cdot)$ as follows:

$$\int_0^T \Phi(s)dW(s) = \lim_{N\to\infty}\int_0^T \Phi(s)d\widehat{W}_N(s), \quad \text{in } L^2_{\mathcal{F}_T}(\Omega;H).$$

Similarly as in the case of finite dimensions, for any $p\in[1,\infty)$, one may introduce the space $L^{p,loc}_{\mathbb{F}}(0,T;\mathcal{L}_2^0))$ and define the stochastic integral of $\Phi\in L^{2,loc}_{\mathbb{F}}(0,T;\mathcal{L}_2^0))$ (especially $\Phi\in L^p_{\mathbb{F}}(\Omega;L^2(0,T;\mathcal{L}_2^0)))$ w.r.t. $W(\cdot)$. Further, we define

$$\int_0^t \Phi(s)dW(s) = \int_0^T \chi_{[0,t]}\Phi(s)dW(s), \quad \text{for } t\in[0,T],$$

and

$$\int_s^t \Phi(s)dW(s) = \int_0^t \Phi(s)dW(s) - \int_0^s \Phi(s)dW(s), \quad \text{for } 0\le s<t\le T.$$

Remark 2.132. As we mentioned before, the series (for the definition of cylindrical Brownian motion $W(\cdot)$) in (2.98) does not converge in V. Nevertheless, for any operator $\Phi_0\in\mathcal{L}_2^0$, one can show that $\Phi_0W(t)\in H$ for any $t\in[0,T]$, and

$$\int_0^T \Phi_0 dW(s) = \Phi_0 W(T). \tag{2.126}$$

Similar to the proof of Theorems 2.124 and 2.129, we can show that the Itô's integral (w.r.t. the cylindrical Brownian motion $W(\cdot)$) has the following properties:

Theorem 2.133. *Let $f,g\in L^{2,loc}_{\mathbb{F}}(0,T;\mathcal{L}_2^0)$, $a,b\in\mathbb{R}$ and $T\ge t>s\ge 0$. The following results hold:*

1)
$$\int_0^t (af+bg)dW = a\int_0^t fdW + b\int_0^t gdW, \quad \text{a.s.};$$

2) *The stochastic process $\int_0^\cdot fdW$ is an H-valued, continuous, local martingale;*

3) *If $f\in L^p_{\mathbb{F}}(\Omega;L^2(0,T;\mathcal{L}_2^0))$, then*

$$\mathbb{E}\Big(\int_s^t fdW \,\Big|\, \mathcal{F}_s\Big) = 0, \quad \text{a.s.};$$

4) *If $f,g\in L^2_{\mathbb{F}}(0,T;\mathcal{L}_2^0)$, then $\int_0^\cdot fdW\in\mathcal{M}_c^2([0,T];H)$, and*

$$\mathbb{E}\Big(\Big\langle \int_s^t fdW, \int_s^t gdW\Big\rangle_H \,\Big|\, \mathcal{F}_s\Big) = \mathbb{E}\Big(\int_s^t \langle f(r,\cdot),g(r,\cdot)\rangle_{\mathcal{L}_2^0}dr \,\Big|\, \mathcal{F}_s\Big), \text{ a.s.}$$

Similar to Corollary 2.131, we have the following result.

Corollary 2.134. *For any bounded bilinear functional $\psi(\cdot,\cdot)$ on H and $f, g \in L^2_{\mathbb{F}}(0,T;\mathcal{L}_2^0)$, it holds that*

$$\begin{aligned}&\mathbb{E}\Big(\psi\Big(\int_s^t fdW(\tau), \int_s^t gdW(\tau)\Big)_H \;\Big|\; \mathcal{F}_s\Big)\\&= \mathbb{E}\Big(\int_s^t \psi(f(\tau,\cdot), g(\tau,\cdot))_{\mathcal{L}_2^0} d\tau \;\Big|\; \mathcal{F}_s\Big), \quad \text{a.s.}\end{aligned}$$

As an easy consequence of Theorem 2.109 and Corollary 2.110, one has the following Doob inequalities for stochastic integrals.

Theorem 2.135. *Let $W(\cdot)$ be a V-valued, Q-Brownian motion (resp. cylindrical Brownian motion) and $f \in L^2_{\mathbb{F}}(0,T;\mathcal{L}(V,H))$ (resp. $f \in L^2_{\mathbb{F}}(0,T;\mathcal{L}_2^0)$). Let τ be a stopping time. Then, for any $p \geq 1$ and $\lambda > 0$,*

$$\mathbb{P}\Big(\sup_{t\in[0,T]}\Big|\int_0^{t\wedge\tau} f(s)dW(s)\Big|_H \geq \lambda\Big) \leq \frac{1}{\lambda^p}\mathbb{E}\Big|\int_0^{T\wedge\tau} f(s)dW(s)\Big|_H^p,$$

and for any $p > 1$,

$$\mathbb{E}\Big(\sup_{t\in[0,T]}\Big|\int_0^{t\wedge\tau} f(s)dW(s)\Big|_H^p\Big) \leq \Big(\frac{p}{p-1}\Big)^p\mathbb{E}\Big|\int_0^{T\wedge\tau} f(s)dW(s)\Big|_H^p.$$

The following result provides a useful link between Itô's and Lebesgue's integrals.

Lemma 2.136. *Let $W(\cdot)$ be a V-valued, Q-Brownian motion (resp. cylindrical Brownian motion) and $f \in L^{2,loc}_{\mathbb{F}}(0,T;\mathcal{L}(V,H))$ (resp. $f \in L^{2,loc}_{\mathbb{F}}(0,T;\mathcal{L}_2^0)$). Then, for any positive numbers ε and λ,*

$$\begin{aligned}&\mathbb{P}\Big(\Big\{\sup_{s\in[0,T]}\Big|\int_0^s fdW\Big|_H \geq \varepsilon\Big\}\Big)\\&\leq \frac{1}{\varepsilon^2}\mathbb{E}\Big(\lambda\wedge\int_0^T |f(t)Q^{\frac{1}{2}}|^2_{\mathcal{L}_2^0}dt\Big) + \mathbb{P}\Big(\Big\{\int_0^T |f(t)Q^{\frac{1}{2}}|^2_{\mathcal{L}_2^0}dt > \lambda\Big\}\Big)\end{aligned} \tag{2.127}$$

(resp.

$$\begin{aligned}&\mathbb{P}\Big(\Big\{\sup_{s\in[0,T]}\Big|\int_0^s fdW\Big|_H \geq \varepsilon\Big\}\Big)\\&\leq \frac{1}{\varepsilon^2}\mathbb{E}\Big(\lambda\wedge\int_0^T |f(t)|^2_{\mathcal{L}_2^0}dt\Big) + \mathbb{P}\Big(\Big\{\int_0^T |f(t)|^2_{\mathcal{L}_2^0}dt > \lambda\Big\}\Big)).\end{aligned} \tag{2.128}$$

Proof: We only consider the case that $W(\cdot)$ is a (V-valued) Q-Brownian motion. Similar to the proof of Lemma 2.126, we define a stochastic process $f_\lambda(t,\omega)$ by

$$f_\lambda(t,\omega)=\begin{cases} f(t,\omega), & \text{if } \int_0^t |f(s,\omega)Q^{\frac{1}{2}}|^2_{\mathcal{L}_2^0}ds \le \lambda, \\ 0, & \text{otherwise.} \end{cases} \tag{2.129}$$

By Chebyshev's inequality (in Theorem 2.22) and Doob's inequality (in Theorem 2.135), similarly to (2.120), we have that

$$\begin{aligned} &\mathbb{P}\Big(\Big\{\sup_{s\in[0,T]}\Big|\int_0^s f dW\Big|_H \ge \varepsilon\Big\}\Big) \\ &\le \mathbb{P}\Big(\Big\{\sup_{s\in[0,T]}\Big|\int_0^s f^\lambda dW\Big|_H \ge \varepsilon\Big\}\Big) + \mathbb{P}\Big(\Big\{\sup_{s\in[0,T]}\Big|\int_0^s (f-f^\lambda) dW\Big|_H > 0\Big\}\Big) \\ &\le \varepsilon^{-2}\mathbb{E}\Big|\int_0^T f^\lambda dW\Big|_H^2 + \mathbb{P}\Big(\Big\{\int_0^T |f(r)Q^{\frac{1}{2}}|_H^2 dr > \lambda\Big\}\Big) \\ &= \varepsilon^{-2}\mathbb{E}\int_0^T |f^\lambda(r)Q^{\frac{1}{2}}|^2_{\mathcal{L}_2^0} dr + \mathbb{P}\Big(\Big\{\int_0^T |f(r)Q^{\frac{1}{2}}|_H^2 dr > \lambda\Big\}\Big) \\ &\le \frac{1}{\varepsilon^2}\mathbb{E}\Big(\lambda\wedge\int_0^T |fQ^{\frac{1}{2}}|^2_{\mathcal{L}_2^0}ds\Big) + \mathbb{P}\Big(\Big\{\int_0^T |f(r)Q^{\frac{1}{2}}|_H^2 dr > \lambda\Big\}\Big), \end{aligned} \tag{2.130}$$

which gives (2.127). □

Remark 2.137. In principle, one should introduce the stochastic integral w.r.t. a Brownian sheet. This integral plays an important role in the study of the stochastic partial differential equations perturbed by white noise depending on both the time and spatial variables (See [335]). One can show that such kind of integral equals the integral w.r.t. a cylindrical Brownian motion for the same integrand (e. g., [68]). Therefore, we omit the details here.

2.11 Properties of Stochastic Integrals

Let $T>0$ and $(\Omega,\mathcal{F},\mathbf{F},\mathbb{P})$ (with $\mathbf{F}=\{\mathcal{F}_t\}_{t\in[0,T]}$) be a fixed filtered probability space satisfying the usual condition, and denote by $\mathbb{F}$ the progressive σ-field w.r.t. $\mathbf{F}$. Let H and V be two separable Hilbert spaces, $Q\in\mathcal{L}_1(V)$ be given as that in Subsection 2.9.3 and write $\mathcal{L}_2^0 \overset{\Delta}{=} \mathcal{L}_2(V;H)$. Denote by I the identity operator on H.

2.11.1 Itô's Formula for Itô's Processes (in a Strong Form)

In this subsection, we present a stochastic version of the *chain rule*, called *Itô's formula*, which plays a fundamental role in stochastic calculus.

We first give the notion of (H-valued) Itô process (in a strong form).

Definition 2.138. *For any $b(\cdot) \in L_{\mathbb{F}}^{1,loc}(0,T;H)$, V-valued Q-Brownian motion (resp. cylindrical Brownian motion) $W(\cdot)$ and $\sigma(\cdot) \in L_{\mathbb{F}}^{2,loc}(0,T;\mathcal{L}(V;H))$ (resp. $\sigma(\cdot) \in L_{\mathbb{F}}^{2,loc}(0,T;\mathcal{L}_2^0)$), the following form of $\mathbf{F}$-adapted process*

$$X(t) = X(0) + \int_0^t b(s)ds + \int_0^t \sigma(s)dW(s), \qquad t \in [0,T], \tag{2.131}$$

is called an H-valued Itô process, or simply an Itô process (if the meaning is clear from the context).

For any $h \in H$, $h^* \in H^*$ and $f \in \mathcal{L}(V;H)$, we define two maps $\langle\!\langle f,h \rangle\!\rangle_H$, $\langle\!\langle h^*,f \rangle\!\rangle_H : V \to \mathbb{R}$ as follows:

$$\langle\!\langle f,h \rangle\!\rangle_H(v) = \langle f(v),h \rangle_H, \quad \langle\!\langle h^*,f \rangle\!\rangle_H(v) = \langle h^*,f(v) \rangle_{H^*,H}, \qquad \forall\, v \in V. \tag{2.132}$$

Clearly, $\langle\!\langle f,h \rangle\!\rangle_H$, $\langle\!\langle h^*,f \rangle\!\rangle_H \in \mathcal{L}(V;\mathbb{R})$ (It is easy to check that $\mathcal{L}(V;\mathbb{R}) \equiv \mathcal{L}_2(V;\mathbb{R})$)).

Let $\mathcal{X}$, $\mathcal{Y}$ and $\mathcal{Z}$ be three Banach spaces, and O and O' be respectively nonempty open subsets of $\mathcal{X}$ and $\mathcal{Y}$. Let us recall that, a function $f : O \to \mathcal{Y}$ is called Fréchet differentiable at $x_0 \in O$ if there exists an $A(x_0) \in \mathcal{L}(\mathcal{X};\mathcal{Y})$ such that

$$\lim_{h \to 0} \frac{|f(x_0+h) - f(x_0) - A(x_0)h|_{\mathcal{Y}}}{|h|_{\mathcal{X}}} = 0.$$

In this case, we write $f_x(x_0) = A(x_0)$ and call it the Fréchet derivative of f at x_0. A function f that is Fréchet differentiable at any point of O is said to be C^1 in O, denoted by $f \in C^1(O;\mathcal{Y})$, if the function

$$f_x : O \to \mathcal{L}(\mathcal{X};\mathcal{Y}), \quad x \mapsto f_x(x)$$

is continuous. Similarly, one can define the Fréchet derivatives of higher orders and the space $C^m(O;\mathcal{Y})$ ($m \in \mathbb{N}$). This is done by induction so $f \in C^m(O;\mathcal{Y})$ if $f \in C^1(O;\mathcal{Y})$ and $f_x \in C^{m-1}(O;\mathcal{L}(\mathcal{X};\mathcal{Y}))$. Especially, the Fréchet derivative f_{xx} of f_x is called the second derivative of f, and $f_{xx}(x_0) \in \mathcal{L}(\mathcal{X};\mathcal{L}(\mathcal{X};\mathcal{Y}))$. Note that $\mathcal{L}(\mathcal{X};\ \mathcal{L}(\mathcal{X};\ \mathcal{Y}))$ is isomorphic (as a Banach space) to the Banach space $\mathcal{L}(\mathcal{X},\mathcal{X};\mathcal{Y})$ of all bilinear maps from $\mathcal{X}$ to $\mathcal{Y}$, with the canonic norm. Similarly, for a function $g : O \times O' \to \mathcal{Z}$ and $y_0 \in O'$, one can define the (second) mixed Fréchet derivative of g, i.e., $g_{xy}(x_0,y_0) \in \mathcal{L}(\mathcal{X},\mathcal{Y};\mathcal{Z})$, where $\mathcal{L}(\mathcal{X},\mathcal{Y};\mathcal{Z})$ stands for the Banach space of all maps M from $\mathcal{X} \times \mathcal{Y}$ to $\mathcal{Z}$ so that, for any $a_1,a_2 \in \mathbb{C}$, $x,x_1,x_2 \in \mathcal{X}$ and $y,y_1,y_2 \in \mathcal{Y}$,

$$\begin{cases} M(a_1x_1 + a_2x_2, y) = a_1M(x_1,y) + a_2M(x_2,y), \\ M(x, a_1y_1 + a_2y_2) = a_1M(x,y_1) + a_2M(x,y_2), \end{cases}$$

with the following norm:

$$|M|_{\mathcal{L}(\mathcal{X},\mathcal{Y};\mathcal{Z})} = \sup\big\{|M(x,y)|_{\mathcal{Z}} \bigm| x \in \mathcal{X}, y \in \mathcal{Y}, |x|_{\mathcal{X}} \le 1, |y|_{\mathcal{Y}} \le 1\big\}.$$

The following result is known as Itô's formula.

Theorem 2.139. *Let $X(\cdot)$ be given by (2.131). Let $F : [0,T] \times H \to \mathbb{R}$ be a function such that its partial (Fréchet) derivatives F_t, F_x and F_{xx} are uniformly continuous on any bounded subsets of $[0,T] \times H$. Then, a.s., for all $t \in [0,T]$ (Recall* (2.100) *for the notation $F_{xx}(s,X(s))(\cdot,\cdot)_{\mathcal{L}_2^0}$),*

$$
\begin{aligned}
F(t,X(t)) = F(0,X(0)) &+ \int_0^t \langle\!\langle\, F_x(s,X(s)),\sigma(s)\,\rangle\!\rangle_H dW(s)\\
&+ \int_0^t \Big[F_t(s,X(s)) + \big\langle F_x(s,X(s)),b(s)\big\rangle_{H^*,H}\\
&+\frac{1}{2}F_{xx}(s,X(s))\big(\sigma(s)Q^{\frac{1}{2}},\sigma(s)Q^{\frac{1}{2}}\big)_{\mathcal{L}_2^0}\Big]ds
\end{aligned}
\tag{2.133}
$$

for the case that $\{W(t)\}_{t\in[0,T]}$ is a Q-Brownian motion, and

$$
\begin{aligned}
F(t,X(t)) = F(0,X(0)) &+ \int_0^t \langle\!\langle\, F_x(s,X(s)),\sigma(s)\,\rangle\!\rangle_H dW(s)\\
&+ \int_0^t \Big[F_t(s,X(s)) + \big\langle F_x(s,X(s)),b(s)\big\rangle_{H^*,H}\\
&+\frac{1}{2}F_{xx}(s,X(s))\big(\sigma(s),\sigma(s)\big)_{\mathcal{L}_2^0}\Big]ds
\end{aligned}
\tag{2.134}
$$

for the case that $\{W(t)\}_{t\in[0,T]}$ is a cylindrical Brownian motion.

Proof: We only consider the case that $W(\cdot)$ is a (V-valued) cylindrical Brownian motion. We divide the proof into several steps. The main idea is to use Taylor's formula. We fix any $t \in [0,T]$.

Step 1. We claim that, it suffices to prove (2.134) under the additional assumptions that $|X(\cdot)|_H$, $|\int_0^\cdot \sigma dW|_H$, $\int_0^\cdot |\sigma(r)|^2_{\mathcal{L}_2^0}dr$ and $\int_0^\cdot |b(r)|_H dr$ are uniformly bounded in $[0,t] \times \Omega$, a.s.

Indeed, for the general case, for $n \in \mathbb{N}$, we set

$$
\tau^n \overset{\Delta}{=} \inf\left\{ s\in[0,t] \;\Big|\; \Big|\int_0^s \sigma dW\Big|_H \vee \Big|\int_0^s |\sigma(s)|^2_{\mathcal{L}_2^0}ds\Big| \vee \Big|\int_0^s |b(s)|_H ds\Big| \geq n \right\},
$$

($\inf\emptyset = t$). It is easy to see that $\{\tau^n\}_{n=1}^\infty$ is a sequence of $\mathbf{F}$-stopping times.

Put

$$
X^n(t) \overset{\Delta}{=} X_0^n + \int_0^t \sigma^n dW + \int_0^t b^n ds,
$$

where

$$
X_0^n = X(0)\chi_{(|X(0)|_H \leq n)},\quad \sigma^n(s) = \sigma(s)\chi_{[0,\tau^n]}(s),\quad b^n(s) = b(s)\chi_{[0,\tau^n]}(s).
$$

Hence, $|X_0^n|_H$, $|\int_0^\cdot \sigma^n dW|_H$, $\int_0^\cdot |\sigma^n(s)|^2_{\mathcal{L}_2^0}ds$ and $\int_0^\cdot |b^n(s)|_H ds$ are uniformly bounded almost surely. If the formula (2.134) holds for X^n, then we have

$$
\begin{aligned}
F(t,X^n(t)) = F(0,X_0^n)) &+ \int_0^t \langle\!\langle\, F_x(s,X^n(s)),\sigma^n(s)\,\rangle\!\rangle_H dW(s) \\
&+ \int_0^t \Big[F_t(s,X^n(s)) + \big\langle F_x(s,X^n(s)),b^n(s)\big\rangle_{H^*,H} \qquad (2.135)\\
&+\frac{1}{2}F_{xx}(s,X^n(s))\big(\sigma^n(s),\sigma^n(s)\big)_{\mathcal{L}_2^0}\Big]ds, \quad \text{a.s.}
\end{aligned}
$$

Clearly,

$$
F(t,X^n(t)) \to F(t,X(t)), \quad F(0,X_0^n) \to F(0,X(0)), \quad \text{a.s.} \quad \text{as } n\to\infty.
$$

Now, for any fixed $\omega \in \Omega$, $F_x(s,X(s))$ is continuous w.r.t. $s \in [0,t]$ and therefore it is bounded by a constant $\mathcal{C}$. Hence

$$
\begin{aligned}
&|\langle\!\langle\, F_x(s,X^n(s)),\sigma^n(s)\,\rangle\!\rangle_H - \langle\!\langle\, F_x(s,X(s)),\sigma(s)\,\rangle\!\rangle_H|^2_{\mathcal{L}_2(V;\mathbb{R})}\\
&\le \mathcal{C}|\sigma(s)|^2_{\mathcal{L}_2^0} \in L^{1,loc}_{\mathbb{F}}(0,T),
\end{aligned}
$$

which implies that

$$
\int_0^t |\langle\!\langle\, F_x(s,X^n(s)),\sigma^n(s)\,\rangle\!\rangle_H - \langle\!\langle\, F_x(s,X(s)),\sigma(s)\,\rangle\!\rangle_H|^2_{\mathcal{L}_2(V;\mathbb{R})}ds \to 0, \text{ a.s.}
$$

as $n \to \infty$. Thus, by the third conclusion in Theorem 2.133,

$$
\begin{aligned}
&\mathbb{P}\text{-}\lim_{n\to+\infty} \int_0^t \langle\!\langle\, F_x(s,X^n(s)),\sigma^n(s)\,\rangle\!\rangle_H dW(s)\\
&= \int_0^t \langle\!\langle\, F_x(s,X(s)),\sigma(s)\,\rangle\!\rangle_H dW(s).
\end{aligned}
\tag{2.136}
$$

Similarly,

$$
\int_0^t |\langle F_x(s,X^n(s)),b^n(s)\rangle_{H^*,H} - \langle F_x(s,X(s)),b(s)\rangle_{H^*,H}|ds \to 0, \text{ a.s.}
$$

as $n\to\infty$. Also, using the boundedness for any fixed ω, we obtain that

$$
\begin{aligned}
&\int_0^t \Big[F_s(s,X^n(s)) + \frac{1}{2}F_{xx}(s,X^n(s))\big(\sigma^n(s),\sigma^n(s)\big)_{\mathcal{L}_2^0}\Big]ds\\
&\to \int_0^t \Big[F_s(s,X(s)) + \frac{1}{2}F_{xx}(s,X(s))\big(\sigma(s),\sigma(s)\big)_{\mathcal{L}_2^0}\Big]ds, \text{ a.s. as } n\to\infty.
\end{aligned}
$$

From (2.136), we conclude that there is a subsequence $\{n_k\}_{k=1}^\infty$ such that

$$
\begin{aligned}
&\lim_{k\to+\infty} \int_0^t \langle\!\langle\, F_x(s,X^{n_k}(s)),\sigma^{n_k}(s)\,\rangle\!\rangle_H dW(s)\\
&= \int_0^t \langle\!\langle\, F_x(s,X(s)),\sigma(s)\,\rangle\!\rangle_H dW(s), \quad \text{a.s.}
\end{aligned}
$$

Now, replacing n in (2.135) by n_k and then letting $k \to \infty$, one gets (2.134).

Step 2. Now, let us show (2.134) under the additional conditions that $|X(\cdot)|_H$, $|\int_0^\cdot \sigma dW|_H$, $\int_0^\cdot |\sigma(r)|^2_{\mathcal{L}_2^0} dr$ and $\int_0^\cdot |b(r)|_H dr$ are uniformly bounded in $[0,t] \times \Omega$ by some positive constant K. Denote by $\mathcal{C}$ the upper bound of $F_t(t,x)$, $F_x(t,x)$ and $F_{xx}(t,x)$ over $L \stackrel{\Delta}{=} [0,t] \times \{x \in H \mid |x|_H \le K\}$.

Using Taylor's expansion, we see that there is a (scalar) function $\varepsilon(r)$ (defined on $[0,\infty)$), decreasing to 0 as $r \to 0$, such that

$$\begin{aligned}&|F(s,x_2) - F(s,x_1) - \langle F_x(s,x_1), x_2 - x_1 \rangle_{H^*,H} \\ &\quad -\frac{1}{2} F_{xx}(s,x_1)(x_2 - x_1, x_2 - x_1)| \\ &\le \varepsilon(|x_2 - x_1|_H)|x_2 - x_1|_H^2, \quad \forall\, (s,x_1),(s,x_2) \in L,\end{aligned}$$

and

$$\begin{aligned}&|F(s_2,x) - F(s_1,x) - F_s(s_1,x)(s_2 - s_1)| \\ &\le \varepsilon(|s_2 - s_1|)|s_2 - s_1|, \quad \forall\, (s_1,x),(s_2,x) \in L.\end{aligned}$$

For any $h > 0$ and $k = 0,1,2\cdots$, put

$$\begin{cases}\tau_0 = 0, \\ \hat\tau_{k+1} = \inf\Big\{s \in [\tau_k, t] \;\Big|\; \Big|\int_{\tau_k}^s \sigma dW\Big|_H \vee \int_{\tau_k}^s |\sigma(r)|^2_{\mathcal{L}_2^0} dr \vee \int_{\tau_k}^s |b(r)|_H dr \ge h\Big\}, \\ \tau_{k+1} = \hat\tau_{k+1} \wedge (\tau_k + h) \wedge t,\end{cases}$$

($\inf \emptyset = t$). One can show that $\{\tau_k\}$ is a sequence of **F**-stopping times, and

$$\tau_k \le \tau_{k+1} \le \tau_k + h, \quad \Big|\int_{\tau_k}^{\tau_{k+1}} \sigma dW\Big|_H \le h, \quad \int_{\tau_k}^{\tau_{k+1}} |b(s)|_H ds \le h.$$

It is easy to see that

$$\begin{aligned}\mathcal{I} &\stackrel{\Delta}{=} F(t,X(t)) - F(0,X(0)) \\ &= \sum_{k=0}^\infty [F(\tau_{k+1}, X(\tau_{k+1})) - F(\tau_k, X(\tau_{k+1}))] \\ &\quad + \sum_{k=0}^\infty [F(\tau_k, X(\tau_{k+1})) - F(\tau_k, X(\tau_k))].\end{aligned}$$

Put $\Delta\tau_k = \tau_{k+1} - \tau_k$, $\Delta X(\tau_k) = X(\tau_{k+1}) - X(\tau_k)$, and

$$\begin{aligned}\mathcal{I}_h = \sum_{k=0}^\infty \Big[&F_s(\tau_k, X(\tau_{k+1}))\Delta\tau_k + \langle F_x(\tau_k, X(\tau_k)), \Delta X(\tau_k) \rangle_{H^*,H} \\ &+ \frac{1}{2} F_{xx}(\tau_k, X(\tau_k))(\Delta X(\tau_k), \Delta X(\tau_k))\Big].\end{aligned}$$

Then

$$
\begin{aligned}
|\mathcal{I}-\mathcal{I}_h| &\le \sum_{k=0}^{\infty}\Big(\varepsilon(h)|\Delta\tau_k|+\varepsilon(2h)|\Delta X(\tau_k)|_H^2\Big)\\
&\le \varepsilon(2h)\Big[t+2\sum_{k=0}^{\infty}\Big(\Big|\int_{\tau_k}^{\tau_{k+1}}\sigma dW\Big|_H^2+\Big|\int_{\tau_k}^{\tau_{k+1}}b(u)du\Big|_H^2\Big)\Big].
\end{aligned}
$$

Note that

$$
\mathbb{E}\sum_{k=0}^{\infty}\Big|\int_{\tau_k}^{\tau_{k+1}}\sigma dW\Big|_H^2=\mathbb{E}\sum_{k=0}^{\infty}\int_{\tau_k}^{\tau_{k+1}}|\sigma(s)|_{\mathcal{L}_2^0}^2ds=\mathbb{E}\int_0^t|\sigma(s)|_{\mathcal{L}_2^0}^2ds.
$$

Hence,

$$
\varepsilon(2h)\mathbb{E}\sum_{k=0}^{\infty}\Big|\int_{\tau_k}^{\tau_{k+1}}\sigma dW\Big|_H^2\to 0 \quad \text{as } h\to 0.
$$

On the other hand,

$$
\sum_{k=0}^{\infty}\Big|\int_{\tau_k}^{\tau_{k+1}}b(r)dr\Big|_H^2\le h\sum_{k=0}^{\infty}\int_{\tau_k}^{\tau_{k+1}}|b(r)|_H dr\le h\int_0^t|b(r)|_H dr.
$$

Hence,

$$
\lim_{h\to 0}\mathbb{E}|\mathcal{I}_h-\mathcal{I}|=0.
$$

Step 3. In what follows, we will analyze the limit of $\mathcal{I}_h$ in probability as $h\to 0$. Clearly, the first term in $\mathcal{I}_h$ tends to $\int_0^t F_s(s,X(s))ds$ a.s. as $h\to 0$.

Since

$$
\Delta X(\tau_k)=\int_{\tau_k}^{\tau_{k+1}}\sigma dW+\int_{\tau_k}^{\tau_{k+1}}b(u)du,
$$

the term "$\sum_{k=0}^{\infty}\langle F_x(\tau_k,X(\tau_k)),\Delta X(\tau_k)\rangle_{H^*,H}$" in $\mathcal{I}_h$ can be split into two terms. The last term tends to $\int_0^t\langle F_x(s,X(s)),b(s)\rangle_{H^*,H}ds$ a.s. as $h\to 0$; while the previous term reads

$$
\begin{aligned}
&\sum_{k=0}^{\infty}\langle F_x(\tau_k,X(\tau_k)),\int_{\tau_k}^{\tau_{k+1}}\sigma dW\rangle_{H^*,H}\\
&=\sum_{k=0}^{\infty}\int_{\tau_k}^{\tau_{k+1}}\langle\!\langle F_x(\tau_k,X(\tau_k)),\sigma(s)\rangle\!\rangle_H dW(s)\\
&=\int_0^t\sum_{k=0}^{\infty}\langle\!\langle F_x(\tau_k,X(\tau_k))\chi_{(\tau_k,\tau_{k+1}]}(s),\sigma(s)\rangle\!\rangle_H dW(s).
\end{aligned}
$$

However, it follows from the dominated convergence theorem that

$$\mathbb{E}\int_0^t \Big|\sum_{k=0}^{\infty} F_x(\tau_k, X(\tau_k))\chi_{(\tau_k,\tau_{k+1}]}(s) - F_x(s, X(s))\Big|_H^2 |\sigma(s)|^2_{\mathcal{L}_2^0} ds \to 0$$

as $h \to 0$. Hence,

$$\sum_{k=0}^{\infty} \langle F_x(\tau_k, X(\tau_k)), \int_{\tau_k}^{\tau_{k+1}} \sigma dW \rangle_{H^*,H} \to \int_0^t \langle\!\langle F_x(s, X(s)), \sigma(s) \rangle\!\rangle_H dW(s)$$

in probability as $h \to 0$.

Similarly, the term "$\frac{1}{2}\sum_{k=0}^{\infty} F_{xx}(\tau_k, X(\tau_k))(\Delta X(\tau_k), \Delta X(\tau_k))$" in $\mathcal{I}_h$ can be split into three terms. The last two terms tend to 0 a.s. as $h \to 0$; while the first one reads

$$\frac{1}{2}\sum_{k=0}^{\infty} F_{xx}(\tau_k, X(\tau_k))\Big(\int_{\tau_k}^{\tau_{k+1}} \sigma dW, \int_{\tau_k}^{\tau_{k+1}} \sigma dW\Big) \equiv \mathcal{I}_1 + \mathcal{I}_2,$$

where

$$\mathcal{I}_1 = \frac{1}{2}\sum_{k=0}^{\infty} \int_{\tau_k}^{\tau_{k+1}} F_{xx}(\tau_k, X(\tau_k))\big(\sigma(s), \sigma(s)\big)_{\mathcal{L}_2^0} ds,$$

$$\begin{aligned}\mathcal{I}_2 = \frac{1}{2}\sum_{k=0}^{\infty} \Big[&F_{xx}(\tau_k, X(\tau_k))\Big(\int_{\tau_k}^{\tau_{k+1}} \sigma dW, \int_{\tau_k}^{\tau_{k+1}} \sigma dW\Big) \\ &- \int_{\tau_k}^{\tau_{k+1}} F_{xx}(\tau_k, X(\tau_k))\big(\sigma(s), \sigma(s)\big)_{\mathcal{L}_2^0} ds\Big].\end{aligned}$$

Obviously, $\mathcal{I}_1$ tends to $\frac{1}{2}\int_0^t F_{xx}(s, X(s))\big(\sigma(s), \sigma(s)\big)_{\mathcal{L}_2^0} ds$ a.s. as $h \to 0$. On the other hand, by Corollary 2.134 and using (2.101), we have

$$\begin{aligned}\mathbb{E}(\mathcal{I}_2)^2 &= \frac{1}{4}\mathbb{E}\sum_{k=0}^{\infty} \Big[F_{xx}(\tau_k, X(\tau_k))\Big(\int_{\tau_k}^{\tau_{k+1}} \sigma dW, \int_{\tau_k}^{\tau_{k+1}} \sigma dW\Big) \\ &\qquad - \int_{\tau_k}^{\tau_{k+1}} F_{xx}(\tau_k, X(\tau_k))\big(\sigma(s), \sigma(s)\big)_{\mathcal{L}_2^0} ds\Big]^2 \\ &\leq \frac{\mathcal{C}^2}{2}\mathbb{E}\sum_{k=0}^{\infty} \Big[\Big|\int_{\tau_k}^{\tau_{k+1}} \sigma dW\Big|_H^4 + \Big(\int_{\tau_k}^{\tau_{k+1}} |\sigma(s)|^2_{\mathcal{L}_2^0} ds\Big)^2\Big] \\ &\leq \frac{\mathcal{C}^2 h}{2}\Big(h\mathbb{E}\Big|\int_0^t \sigma dW\Big|_H^2 + \mathbb{E}\int_0^t |\sigma(s)|^2_{\mathcal{L}_2^0} ds\Big) \to 0, \qquad \text{as } h \to 0.\end{aligned}$$

Hence, $\lim_{h\to 0}\mathcal{I}_2 = 0$ in probability.

Combining the above analysis, we conclude that

$$\begin{aligned}\mathcal{I}_h \to &\int_0^t \Big\{F_s(s, X(s)) + \langle F_x(s, X(s)), b(s) \rangle_{H^*,H} \\ &+ \frac{1}{2}F_{xx}(s, X(s))\big(\sigma(s), \sigma(s)\big)_{\mathcal{L}_2^0}\Big\} ds + \int_0^t \langle\!\langle F_x(s, X(s)), \sigma(s) \rangle\!\rangle_H dW(s)\end{aligned}$$

in probability as $h \to 0$, which gives the desired result. $\square$

2.11.2 Burkholder-Davis-Gundy Inequality

In this subsection, we shall prove the following Burkholder-Davis-Gundy inequality (for stochastic integrals), which will be quite useful in the sequel (Compared to Lemma 2.136, this inequality can be regarded as a further link between Itô's and Lebesgue's integrals).

Theorem 2.140. *For any $p > 0$, there exists a constant $\mathcal{C}_p > 0$ such that for any $T > 0$, V-valued Q-Brownian motion (resp. cylindrical Brownian motion) $W(\cdot)$ and $f \in L^p_{\mathbb{F}}(\Omega; L^2(0,T;\mathcal{L}(V;H)))$ (resp. $f \in L^p_{\mathbb{F}}(\Omega; L^2(0,T;\mathcal{L}_2^0)))$,*

$$\mathbb{E}\Big(\sup_{t\in[0,T]}\Big|\int_0^t f(s)dW(s)\Big|_H^p\Big) \le \mathcal{C}_p\mathbb{E}\Big(\int_0^T |f(s)Q^{\frac{1}{2}}|^2_{\mathcal{L}_2^0}ds\Big)^{\frac{p}{2}} \tag{2.137}$$

(resp.

$$\mathbb{E}\Big(\sup_{t\in[0,T]}\Big|\int_0^t f(s)dW(s)\Big|_H^p\Big) \le \mathcal{C}_p\mathbb{E}\Big(\int_0^T |f(s)|^2_{\mathcal{L}_2^0}ds\Big)^{\frac{p}{2}}\ ). \tag{2.138}$$

Proof: We shall only deal with the case that $W(\cdot)$ is cylindrical Brownian motion. The case that $W(\cdot)$ is Q-Brownian motion can be handled similarly. If $p = 2$, then (2.138) follows from Theorem 2.135. Hence, we only need to deal with the case that $p \neq 2$.

Put

$$X(t) \overset{\Delta}{=} \int_0^t f(s)dW(s), \qquad t \in [0,T] \tag{2.139}$$

We first consider the case that X is bounded.

Step 1. In this step, we prove (2.138) for $p > 2$. Let $g(x) = |x|_H^p$ for $x \in H$. Since

$$g_{xx}(x) = \begin{cases} p(p-2)|x|_H^{p-4}x\otimes x + p|x|_H^{p-2}I, & x \neq 0, \\ 0, & x = 0, \end{cases}$$

and recalling the definition of $||\cdot||$ in (2.102), we have that

$$||g_{xx}(x)|| \le p(p-1)|x|_H^{p-2}, \qquad \forall\, x \in H.$$

By Itô's formula,

$$\begin{aligned}
\mathbb{E}|X(t)|_H^p &= \frac{1}{2}\mathbb{E}\int_0^t g_{xx}(X)(f,f)_{\mathcal{L}_2^0}ds \le \frac{1}{2}p(p-1)\mathbb{E}\int_0^t |X|_H^{p-2}|f|^2_{\mathcal{L}_2^0}ds \\
&\le \frac{1}{2}p(p-1)\mathbb{E}\Big(\sup_{s\in[0,t]}|X(s)|_H^{p-2}\int_0^t |f(s)|^2_{\mathcal{L}_2^0}ds\Big) \\
&\le \frac{1}{2}p(p-1)\Big(\mathbb{E}\sup_{s\in[0,t]}|X(s)|_H^p\Big)^{\frac{p-2}{p}}\Big[\mathbb{E}\Big(\int_0^t |f(s)|^2_{\mathcal{L}_2^0}ds\Big)^{\frac{p}{2}}\Big]^{\frac{2}{p}}.
\end{aligned}$$

This, together with Theorem 2.135, implies that

$$\mathbb{E}|X(t)|_H^p \leq \frac{1}{2}p(p-1)\Big[\Big(\frac{p}{p-1}\Big)^p \mathbb{E}|X(t)|_H^p\Big]^{\frac{p-2}{p}} \Big[\mathbb{E}\Big(\int_0^t |f(s)|^2_{\mathcal{L}_2^0} ds\Big)^{\frac{p}{2}}\Big]^{\frac{2}{p}}.$$

Hence,

$$\big(\mathbb{E}|X(t)|_H^p\big)^{\frac{2}{p}} \leq \frac{1}{2}p(p-1)\Big(\frac{p}{p-1}\Big)^{p-2} \Big[\mathbb{E}\Big(\int_0^t |f(s)|^2_{\mathcal{L}_2^0} ds\Big)^{\frac{p}{2}}\Big]^{\frac{2}{p}}. \tag{2.140}$$

It follows from (2.140) and Theorem 2.135 again that

$$\mathbb{E} \sup_{s\in[0,t]} |X(s)|_H^p \leq \Big(\frac{p}{p-1}\Big)^p \mathbb{E}|X(t)|_H^p \leq C_p\Big[\mathbb{E}\Big(\int_0^t |f(s)|^2_{\mathcal{L}_2^0} ds\Big)^{\frac{p}{2}}\Big]. \tag{2.141}$$

Step 2. In this step, we consider the case that $p \in (0,2)$.
By Lemma 2.23, it is easy to see that

$$\mathbb{E}\big(\sup_{s\in[0,t]} |X(s)|_H^p\big) = p\int_0^\infty \lambda^{p-1}\mathbb{P}\big(\sup_{s\in[0,t]} |X(s)|_H \geq \lambda\big)d\lambda. \tag{2.142}$$

By (2.142) and Lemma 2.136, we get that

$$\begin{aligned}
&\mathbb{E}\big(\sup_{s\in[0,t]} |X(s)|_H^p\big)\\
&\leq p\int_0^\infty \lambda^{p-1}\mathbb{P}\Big(\int_0^t |f|^2_{\mathcal{L}_2^0} ds > \lambda^2\Big)d\lambda + p\int_0^\infty \lambda^{p-3}\mathbb{E}\Big(\int_0^t |f|^2_{\mathcal{L}_2^0} ds \wedge \lambda^2\Big)d\lambda.
\end{aligned} \tag{2.143}$$

Write $a = \sqrt{\int_0^t |f|^2_{\mathcal{L}_2^0} ds}$. A direct computation yields that

$$p\int_0^\infty \lambda^{p-1}\mathbb{P}\Big(\int_0^t |f|^2_{\mathcal{L}_2^0} ds > \lambda^2\Big)d\lambda = \mathbb{E}\Big(\int_0^t |f|^2_{\mathcal{L}_2^0} ds\Big)^{\frac{p}{2}} \tag{2.144}$$

and

$$\begin{aligned}
&p\int_0^\infty \lambda^{p-3}\mathbb{E}\Big(\int_0^t |f|^2_{\mathcal{L}_2^0} ds \wedge \lambda^2\Big)d\lambda = p\mathbb{E}\int_0^\infty \lambda^{p-3}\Big(\int_0^t |f|^2_{\mathcal{L}_2^0} ds \wedge \lambda^2\Big)d\lambda\\
&= p\mathbb{E}\int_0^a \lambda^{p-1}d\lambda + p\mathbb{E}\Big(a^2\int_a^{+\infty} \lambda^{p-3}d\lambda\Big)\\
&= \mathbb{E}(a^p) + \frac{p}{2-p}\mathbb{E}(a^p) = \frac{2}{2-p}\mathbb{E}\Big(\int_0^t |f|^2_{\mathcal{L}_2^0} ds\Big)^{\frac{p}{2}}.
\end{aligned} \tag{2.145}$$

From (2.143), (2.144) and (2.145), we get that

$$\mathbb{E}\big(\sup_{s\in[0,t]} |X(s)|_H^p\big) \leq \mathbb{E}\Big(\int_0^t |f|^2_{\mathcal{L}_2^0} ds\Big)^{\frac{p}{2}} + \frac{2}{2-p}\mathbb{E}\Big(\int_0^t |f|^2_{\mathcal{L}_2^0} ds\Big)^{\frac{p}{2}}. \tag{2.146}$$

This concludes that (2.138) holds for $p \in (0,2)$.

Step 3. In this step, we consider the general X. For each $k \in \mathbb{N}$, put

$$\tau_k \triangleq \begin{cases} \inf\{t \in [0,T] \mid |X(t)|_H \geq k\}, & \text{if } \{t \in [0,T] \mid |X(t)|_H \geq k\} \neq \emptyset, \\ T, & \text{otherwise.} \end{cases}$$

Write $f_k = \chi_{[0,\tau_k]} f$. Then by the result in Step 2, we have that

$$\mathbb{E}\Big(\sup_{t\in[0,T]}\Big|\int_0^t f_k(s)dW(s)\Big|_H^p\Big) \leq \mathcal{C}_p \mathbb{E}\Big(\int_0^T |f_k(s)|^2_{\mathcal{L}_2^0} ds\Big)^{\frac{p}{2}}. \tag{2.147}$$

Letting $k \to +\infty$ in both sides of (2.147), by Fatou's lemma (See Theorem 2.20), we arrive at the desired inequality (2.138). This completes the proof of Theorem 2.140. □

2.11.3 Stochastic Fubini Theorem

The following result is a stochastic version of Fubini's theorem.

Theorem 2.141. *Let $W(\cdot)$ be a V-valued, Q-Brownian motion (resp. cylindrical Brownian motion). Let $(G, \mathcal{G}, \mu)$ be a finite measure space and $\Phi : (G \times (0,T) \times \Omega, \mathcal{G} \times \mathbb{F}) \to (\mathcal{L}(V;H), \mathcal{B}(\mathcal{L}(V;H)))$ (resp. $\Phi : (G \times (0,T) \times \Omega, \mathcal{G} \times \mathbb{F}) \to (\mathcal{L}_2^0, \mathcal{B}(\mathcal{L}_2^0))$) be a measurable mapping. If $\int_G (\int_0^T |\Phi(x,t,\cdot)|^2_{\mathcal{L}(V;H)} dt)^{1/2} \mu(dx) < \infty$ (resp. $\int_G (\int_0^T |\Phi(x,t,\cdot)|^2_{\mathcal{L}_2^0} dt)^{1/2} \mu(dx) < \infty$), a.s., then $\int_G \Phi(x,\cdot,\cdot)\mu(dx) \in L^{2,loc}_{\mathbb{F}}(0,T;\mathcal{L}(V;H))$ (resp. $\int_G \Phi(x,\cdot,\cdot)\mu(dx) \in L^{2,loc}_{\mathbb{F}}(0,T;\mathcal{L}_2^0)$), and a.s., $\int_G |\int_0^T \Phi(x,t,\cdot)dW(t)|_H \mu(dx) < \infty$ and*

$$\int_G \int_0^T \Phi(x,t,\cdot)dW(t)\mu(dx) = \int_0^T \int_G \Phi(x,t,\cdot)\mu(dx)dW(t). \tag{2.148}$$

Proof: We shall only deal with the case that $W(\cdot)$ is cylindrical Brownian motion. Let us first show that (2.148) holds for the functions in the following set

$$\mathcal{S} \triangleq \Big\{ \Phi_n = \sum_{j=1}^n \sum_{k=1}^n \beta_j^n \chi_{(s_k^n, s_{k+1}^n]} \alpha_k^n \;\Big|\; n \in \mathbb{N},\ \alpha_k^n \in L^2_{\mathcal{F}_{s_k^n}}(\Omega;\mathcal{L}_2^0),$$
$$\beta_j^n \text{ are real valued, bounded } \mathcal{G}\text{-measurable functions, and}$$
$$\{s_k^n, \cdots, s_{k+1}^n\} \text{ is a partition of } [0,T] \Big\}.$$

By Remark 2.132, for any Φ_n in the above form,

$$
\begin{aligned}
&\int_G \int_0^T \Phi_n(x,t,\cdot)dW(t)d\mu(x) \\
&= \int_G \sum_{j=1}^n \sum_{k=1}^n \beta_j^n(x)\chi_{(s_k^n,s_{k+1}^n]}\alpha_k^n\big[W(s_{k+1}^n)-W(s_k^n)\big]d\mu(x) \\
&= \sum_{k=1}^n \Big[\sum_{j=1}^n \chi_{(s_k^n,s_{k+1}^n]}\int_G \beta_j^n(x)d\mu(x)\Big]\alpha_k^n\big[W(s_{k+1}^n)-W(s_k^n)\big] \\
&= \int_0^T \int_G \Phi_n(x,t,\cdot)d\mu(x)dW(t).
\end{aligned}
\tag{2.149}
$$

Next, we consider the general case. We claim that there exist $\{\Phi_n\}_{n=1}^\infty \subset \mathcal{S}$ such that

$$
\mathbb{P}\text{-}\lim_{n\to\infty}\int_G \Big(\int_0^T |\Phi_n(x,t,\cdot)-\Phi(x,t,\cdot)|_{\mathcal{L}_2^0}^2 dt\Big)^{1/2}\mu(dx) = 0. \tag{2.150}
$$

To see this, for each $n \in \mathbb{N}$, let us set

$$
\Lambda_n(\cdot,\cdot,\omega) = \begin{cases} \Phi(\cdot,\cdot,\omega) & \text{if } \int_G \Big(\int_0^T |\Phi(x,t,\omega)|_{\mathcal{L}_2^0}^2 dt\Big)^{1/2}\mu(dx) \le n, \\ 0 & \text{otherwise.} \end{cases} \tag{2.151}
$$

It is easy to see that $\int_G \Big(\int_0^T |\Lambda_n(x,t,\omega)|_{\mathcal{L}_2^0}^2 dt\Big)^{1/2}\mu(dx) \le n$, a.s. Hence, $\Lambda_n \in L^1_{\mathcal{G}}(G; L^1_{\mathbb{F}}(\Omega; L^2(0,T;\mathcal{L}_2^0)))$. Fix an $\omega \in \Omega$. By (2.151), it is easy to see that

$$
\lim_{n\to\infty}\int_G \Big(\int_0^T |\Lambda_n(x,t,\cdot)-\Phi(x,t,\cdot)|_{\mathcal{L}_2^0}^2 dt\Big)^{1/2}\mu(dx) = 0, \quad \text{a.s.} \tag{2.152}
$$

By the definition of Bochner's integral, for each $m \in \mathbb{N}$, one can find $q_i(\cdot,\cdot) \in L^1_{\mathbb{F}}(\Omega; L^2(0,T;\mathcal{L}_2^0))$ $(i = 1,2,\cdots,m)$ and $E_i \in \mathcal{G}$ such that

$$
\lim_{m\to\infty}\int_G \Big|\Lambda_n(\cdot,\cdot,x) - \sum_{i=1}^m q_i(\cdot,\cdot)\chi_{E_i}(x)\Big|_{L^1_{\mathbb{F}}(\Omega;L^2(0,T;\mathcal{L}_2^0))}\mu(dx) = 0. \tag{2.153}
$$

By Lemma 2.75, the set $L_{\mathcal{S},\mathbb{F}}(0,T;\mathcal{L}_2^0)$ is dense in $L^1_{\mathbb{F}}(\Omega; L^2(0,T;\mathcal{L}_2^0))$. This fact, together with (2.152)–(2.153), gives (2.150) for suitably chosen $\Phi_n \in \mathcal{S}$.

We claim that $\int_G \big|\int_0^T \Phi(x,t,\cdot)dW(t)\big|_H \mu(dx) < \infty$, a.s. Indeed, for any $n \in \mathbb{N}$, by the definition of Λ_n in (2.151), using the Burkholder-Davis-Gundy inequality (in Theorem 2.140), we obtain that

$$
\begin{aligned}
&\mathbb{P}\Big(\Big\{\int_G\Big|\int_0^T \Phi(x,t,\cdot)dW(t)\Big|_H\mu(dx)\geq n^2\Big\}\Big)\\
&\leq \mathbb{P}\Big(\Big\{\int_G\Big|\int_0^T (\Phi(x,t,\cdot)-\Lambda_n(x,t,\cdot))dW(t)\Big|_H\mu(dx)>0\Big\}\Big)\\
&\quad+\mathbb{P}\Big(\Big\{\int_G\Big|\int_0^T \Lambda_n(x,t,\cdot)dW(t)\Big|_H\mu(dx)\geq n^2\Big\}\Big)\\
&\leq \mathbb{P}\Big(\Big\{\int_G\Big(\int_0^T |\Phi(x,t,\cdot)|^2_{\mathcal{L}_2^0}dt\Big)^{1/2}\mu(dx)\geq n\Big\}\Big)\\
&\quad+n^{-2}\int_G\mathbb{E}\Big|\int_0^T \Lambda_n(x,t,\cdot)dW(t)\Big|_H\mu(dx)\\
&\leq \mathbb{P}\Big(\Big\{\int_G\Big(\int_0^T |\Phi(x,t,\cdot)|^2_{\mathcal{L}_2^0}dt\Big)^{1/2}\mu(dx)\geq n\Big\}\Big)\\
&\quad+\mathcal{C}n^{-2}\int_G\mathbb{E}\Big(\int_0^T |\Lambda_n(x,t,\cdot)|^2_{\mathcal{L}_2^0}dt\Big)^{1/2}\mu(dx)\\
&\leq \mathbb{P}\Big(\Big\{\int_G\Big(\int_0^T |\Phi(x,t,\cdot)|^2_{\mathcal{L}_2^0}dt\Big)^{1/2}\mu(dx)\geq n\Big\}\Big)+\frac{\mathcal{C}}{n}.
\end{aligned}
\tag{2.154}
$$

Hence, $\mathbb{P}\big(\{\int_G|\int_0^T\Phi(x,t,\cdot)dW(t)|_H\mu(dx)=\infty\}\big)=0$.

By (2.30) and $\int_G(\int_0^T|\Phi(x,t,\cdot)|^2_{\mathcal{L}_2^0}dt)^{1/2}\mu(dx)<\infty$, a.s., it is easy to show that $\int_G\Phi(x,\cdot,\cdot)\mu(dx)\in L^{2,loc}_{\mathbb{F}}(0,T;\mathcal{L}_2^0))$, a.s.

Similar to (2.154), one can show that, for any $\varepsilon>0$ and $\delta>0$,

$$
\begin{aligned}
&\mathbb{P}\Big(\Big\{\Big|\int_G\int_0^T \Phi_n(x,t,\cdot)dW(t)\mu(dx)-\int_G\int_0^T \Phi(x,t,\cdot)dW(t)\mu(dx)\Big|_H\geq\varepsilon\Big\}\Big)\\
&\leq \mathbb{P}\Big(\Big\{\int_G\Big|\int_0^T (\Phi_n(x,t,\cdot)-\Phi(x,t,\cdot))dW(t)\Big|_H\mu(dx)\geq\varepsilon\Big\}\Big)\\
&\leq \mathbb{P}\Big(\Big\{\int_G\Big(\int_0^T |\Phi_n(x,t,\cdot)-\Phi(x,t,\cdot)|^2_{\mathcal{L}_2^0}dt\Big)^{1/2}\mu(dx)\geq\delta\Big\}\Big)+\varepsilon^{-1}\delta.
\end{aligned}
\tag{2.155}
$$

This, together with (2.150), yields that

$$
\mathbb{P}\text{-}\lim_{n\to\infty}\int_G\int_0^T \Phi_n(x,t,\cdot)dW(t)\mu(dx)=\int_G\int_0^T \Phi(x,t,\cdot)dW(t)\mu(dx). \tag{2.156}
$$

On the other hand, similarly to (2.155), one can show that,

$$
\begin{aligned}
&\mathbb{P}\Big(\Big\{\Big|\int_0^T\int_G \Phi_n(x,t,\cdot)\mu(dx)dW(t)-\int_0^T\int_G \Phi(x,t,\cdot)\mu(dx)dW(t)\Big|_H\geq\varepsilon\Big\}\Big)\\
&=\mathbb{P}\Big(\Big\{\Big|\int_0^T\Big(\int_G \Phi_n(x,t,\cdot)\mu(dx)-\int_G \Phi(x,t,\cdot)\mu(dx)\Big)dW(t)\Big|_H\geq\varepsilon\Big\}\Big)\\
&\leq\mathbb{P}\Big(\Big\{\int_0^T\Big|\int_G (\Phi_n(x,t,\cdot)-\Phi(x,t,\cdot))\mu(dx)\Big|^2_{\mathcal{L}_2^0}dt\geq\delta\Big\}\Big)+\varepsilon^{-1}\delta\\
&\leq\mathbb{P}\Big(\Big\{\int_G\Big(\int_0^T\Big|\Phi_n(x,t,\cdot)-\Phi(x,t,\cdot)\Big|^2_{\mathcal{L}_2^0}dt\Big)^{1/2}\mu(dx)\geq\delta^{1/2}\Big\}\Big)+\varepsilon^{-1}\delta.
\end{aligned}
$$

This, together with (2.150), yields that

$$
\mathbb{P}\text{-}\lim_{n\to\infty}\int_0^T\int_G \Phi_n(x,t,\cdot)\mu(dx)dW(t)=\int_0^T\int_G \Phi(x,t,\cdot)\mu(dx)dW(t). \quad (2.157)
$$

Finally, by (2.156) and (2.157), and noting (2.149), we obtain the desired equality (2.148). □

2.11.4 Itô's Formula for Itô's processes in a Weak Form

Theorem 2.139 works well for Itô's processes in the (strong) form (2.131). However, usually this is too restrictive in the study of stochastic differential equations in infinite dimensions. Indeed, in the infinite dimensional setting sometimes one has to handle Itô's processes in a weaker form, to be presented below.

Let $\mathcal{V}$ be a Hilbert space such that the embedding $\mathcal{V}\subset H$ is continuous and dense. Denote by $\mathcal{V}^*$ the dual space of $\mathcal{V}$ with respect to the pivot space H. Hence, $\mathcal{V}\subset H=H^*\subset\mathcal{V}^*$, continuously and densely and

$$
\langle z,v\rangle_{\mathcal{V},\mathcal{V}^*}=\langle z,v\rangle_H,\qquad \forall\, v\in H,\ z\in\mathcal{V}.
$$

We have the following Itô's formula for a weak form of Itô process (Recall (2.132) for the notation $\langle\!\langle\cdot,\cdot\rangle\!\rangle_H$).

Theorem 2.142. *Suppose that $X_0\in L^2_{\mathcal{F}_0}(\Omega;H)$, $W(\cdot)$ is a V-valued, Q-Brownian motion (resp. cylindrical Brownian motion), $\phi(\cdot)\in L^2_{\mathbb{F}}(0,T;\mathcal{V}^*)$, and $\Phi(\cdot)\in L^p_{\mathbb{F}}(\Omega;L^2(0,T;\mathcal{L}(V;H)))$ (resp. $\Phi(\cdot)\in L^p_{\mathbb{F}}(\Omega;\ L^2(0,T;\mathcal{L}_2^0)))$ for some $p\geq 1$. Let*

$$
X(t)=X_0+\int_0^t\phi(s)ds+\int_0^t\Phi(s)dW(s),\quad t\in[0,T]. \quad (2.158)
$$

If $X\in L^2_{\mathbb{F}}(0,T;\mathcal{V})$, then $X(\cdot)\in C([0,T];H)$, a.s., and for any $t\in[0,T]$,

$$
\begin{aligned}
|X(t)|_H^2=&\,|X_0|_H^2+2\int_0^t\langle\phi(s),X(s)\rangle_{\mathcal{V}^*,\mathcal{V}}ds\\
&+2\int_0^t\langle\!\langle\Phi(s),X(s)\rangle\!\rangle_H dW(s)+\int_0^t|\Phi(s)Q^{\frac12}|^2_{\mathcal{L}_2^0}ds,\quad a.s.
\end{aligned}
\quad (2.159)
$$

(resp.

$$
\begin{aligned}
|X(t)|_H^2 = |X_0|_H^2 + 2\int_0^t \langle\, \phi(s), X(s)\,\rangle_{\mathcal{V}^*,\mathcal{V}} ds \\
+2\int_0^t \langle\!\langle\, \Phi(s), X(s)\,\rangle\!\rangle_H dW(s) + \int_0^t |\Phi(s)|_{\mathcal{L}_2^0}^2 ds, \quad a.s.\ ).
\end{aligned}
\tag{2.160}
$$

Proof: We only consider the case that $W(\cdot)$ is a V-valued, cylindrical Brownian motion.

Let us introduce the infinitesimal generators of some C_0-semigroups. Denote by A the Riesz isometry from $\mathcal{V}$ to $\mathcal{V}^*$, defined by

$$
\langle\, u, Av\,\rangle_{\mathcal{V},\mathcal{V}^*} = \langle\, u, v\,\rangle_{\mathcal{V}}, \qquad \forall\, u, v \in \mathcal{V}. \tag{2.161}
$$

From (2.161), we deduce that

$$
\langle\, u, Av\,\rangle_H = \langle\, u, v\,\rangle_{\mathcal{V}}, \qquad \forall\, u, v \in \mathcal{V} \text{ such that } Av \in H. \tag{2.162}
$$

Write $\mathcal{D} = \{v \in \mathcal{V} \mid Av \in H\}$. Clearly, $A : \mathcal{D}(\subset H) \to H$ is a densely defined, closed operator. Also, one can check that A is self-adjoint, and $\langle\, v, Av\,\rangle_H = |v|_{\mathcal{V}}^2$ for any $v \in \mathcal{D}$. Hence, by the Lumer-Phillips theorem, $-A$ generates a contractive C_0-semigroup on H. Similarly, $-A$ also generates contractive C_0-semigroups on V and V' (Here, to simplify the notation, we denote by $-A$ the "same" infinitesimal generator of three different C_0-semigroups). For the identity operator I on H (also on V and V') and any $\lambda > 0$, put $I_\lambda = \lambda(\lambda I + A)^{-1}$. It is clear that $I_\lambda \to I$ strongly on H (also on V and V') as $\lambda \to +\infty$. Write $X_\lambda(\cdot) = I_\lambda X(\cdot)$, $X_{0,\lambda} = I_\lambda X_0$, $\phi_\lambda(\cdot) = I_\lambda \phi(\cdot)$ and $\Phi_\lambda(\cdot) = I_\lambda \Phi(\cdot)$. Clearly, $\phi_\lambda(\cdot) \in L^2_{\mathbb{F}}(0,T;H)$ and $\Phi_\lambda(\cdot) \in L^p_{\mathbb{F}}(\Omega; L^2(0,T;\mathcal{L}_2^0))$. We claim that

$$
\begin{cases}
\lim\limits_{\lambda\to\infty} X_{0,\lambda} = X_0 \ \text{ in } L^2_{\mathcal{F}_0}(\Omega; H),\\
\lim\limits_{\lambda\to\infty} \phi_\lambda(\cdot) = \phi(\cdot) \text{ in } L^2_{\mathbb{F}}(0,T;\mathcal{V}^*),\\
\lim\limits_{\lambda\to\infty} \Phi_\lambda(\cdot) = \Phi(\cdot) \text{ in } L^p_{\mathbb{F}}(\Omega; L^2(0,T;\mathcal{L}_2^0)).
\end{cases}
\tag{2.163}
$$

Indeed, by $A \geq 0$, we see that for any $\lambda \geq 0$, $|\lambda(\lambda + A)^{-1}|_{\mathcal{L}(H)} \leq 1$. Hence, $\big|\lambda(\lambda + A)^{-1}X_0 - X_0\big|_H \leq 2|X_0|_H$. This, together with Lebesgue's dominated convergence theorem, implies that $\lim_{\lambda\to\infty} \mathbb{E}\big|\lambda(\lambda+A)^{-1}X_0 - X_0\big|_H^2 = 0$, which gives the first equality in (2.163). Similarly, we can prove that the second conclusion in (2.163) is true. Now, for any $\lambda > 0$, $v \in V$ and $K \in \mathcal{L}_2^0$, $|\lambda(\lambda + A)^{-1}Kv|_H \leq |Kv|_H$. Thus, $|\lambda(\lambda + A)^{-1}K|_{\mathcal{L}_2^0} \leq |K|_{\mathcal{L}_2^0}$. Therefore, $|\Phi_\lambda - \Phi|_{\mathcal{L}_2^0} \leq 2|\Phi|_{\mathcal{L}_2^0}$. Hence, using Lebesgue's dominated convergence theorem again, we obtain the third equality in (2.163).

It follows from (2.158) that

$$
X_\lambda(t) = X_{0,\lambda} + \int_0^t \phi_\lambda(s) ds + \int_0^t \Phi_\lambda(s) dW(s), \quad t \in [0,T]. \tag{2.164}
$$

According to (2.163), we get that $X_\lambda(\cdot)$ is an Itô process in H. Similar to the proof of (2.163), we can show that

$$\lim_{\lambda\to\infty} X_\lambda(\cdot) = X(\cdot) \ \text{ in } L^2_{\mathbb{F}}(0,T;\mathcal{V}). \tag{2.165}$$

Applying Theorem 2.139 to $X_\lambda(\cdot)$, we find that, for any $t\in[0,T]$, a.s.,

$$\begin{aligned} |X_\lambda(t)|_H^2 = |X_{0,\lambda}|_H^2 &+ 2\int_0^t \langle\, \phi_\lambda(s), X_\lambda(s)\,\rangle_{\mathcal{V}^*,\mathcal{V}} ds \\ &+2\int_0^t \langle\!\langle\, \varPhi_\lambda(s), X_\lambda(s)\,\rangle\!\rangle_H dW(s) + \int_0^t |\varPhi_\lambda(s)|^2_{\mathcal{L}_2^0} ds \end{aligned} \tag{2.166}$$

By Burkholder-Davis-Gundy's inequality (in Theorem 2.140), for any $\varepsilon>0$, we see that

$$\begin{aligned} &\mathbb{E}\sup_{t\in[0,T]}\Big|\int_0^t \langle\!\langle\, \varPhi_\lambda(s), X_\lambda(s)\,\rangle\!\rangle_H dW(s)\Big| \le \mathcal{C}\mathbb{E}\Big(\int_0^T |\varPhi_\lambda(s)|^2_{\mathcal{L}_2^0}|X_\lambda(s)|_H^2 ds\Big)^{\frac12} \\ &\le \mathcal{C}\mathbb{E}\Big[\sup_{t\in[0,T]}|X_\lambda(t)|_H\Big(\int_0^T |\varPhi_\lambda(s)|^2_{\mathcal{L}_2^0} ds\Big)^{\frac12}\Big] \\ &\le \varepsilon\mathbb{E}\sup_{t\in[0,T]}|X_\lambda(t)|_H^2 + \frac{\mathcal{C}}{\varepsilon}\mathbb{E}\Big(\int_0^T |\varPhi_\lambda(s)|^2_{\mathcal{L}_2^0} ds\Big). \end{aligned} \tag{2.167}$$

Further,

$$\begin{aligned} &\mathbb{E}\sup_{t\in[0,T]}\Big|\int_0^t \langle\, \phi_\lambda(s), X_\lambda(s)\,\rangle_{\mathcal{V}^*,\mathcal{V}} ds\Big|_H \le \mathbb{E}\int_0^T |\phi_\lambda(s)|_{\mathcal{V}^*}|X_\lambda(s)|_{\mathcal{V}} ds \\ &\le |\phi_\lambda(\cdot)|_{L^2_{\mathbb{F}}(0,T;\mathcal{V}^*)}|X_\lambda(\cdot)|_{L^2_{\mathbb{F}}(0,T;\mathcal{V})} \le \mathcal{C}. \end{aligned} \tag{2.168}$$

By (2.166)–(2.168), we find that $\mathbb{E}\sup_{t\in[0,T]}|X_\lambda(t)|_H^2 \le \mathcal{C}$, where the constant $\mathcal{C}$ is independent of λ. Hence, there is a subsequence $\{\lambda_n\}_{n=1}^\infty \subset (0,+\infty)$ and a $\hat{X}\in L^2_{\mathbb{F}}(\Omega;L^\infty(0,T;H))$ such that $\lim_{n\to\infty} X_{\lambda_n} = \hat{X}$ weakly star in $L^2_{\mathbb{F}}(\Omega;L^\infty(0,T;H))$. On the other hand, since $X(\cdot)\in L^2_{\mathbb{F}}(0,T;V)$, we see that $\lim_{\lambda\to\infty} X_\lambda = X$ in $L^2_{\mathbb{F}}(0,T;V)\subset L^2_{\mathbb{F}}(0,T;H)$. Thus, $X=\hat{X}\in L^2_{\mathbb{F}}(\Omega;L^\infty(0,T;H))$.

Letting $\lambda\to+\infty$ in (2.166) and noting (2.165), we obtain (2.160). Hence, $|X(\cdot)|_H$ is continuous. This, together with that $X(\cdot)$ is continuous in $\mathcal{V}^*$, implies that $X(\cdot)$ is continuous in H. □

Remark 2.143. When $\mathcal{V}$ is only assumed to be a reflexive Banach space such that the embedding $\mathcal{V}\subset H$ is continuous and dense, one can prove a similar result as that in Theorem 2.142 provide that there is a C_0-semigroup $\{S(t)\}_{t\ge0}$ on H such that the restriction of $\{S(t)\}_{t\ge0}$ on $\mathcal{V}$ is a C_0-semigroup on $\mathcal{V}$.

2.11.5 Martingale Representation Theorem

In this subsection, we always assume that $\mathbf{F} = \{\mathcal{F}_t\}_{t\in[0,T]}$ is the natural filtration generated by the underlying Brownian motion $W(\cdot)$, which can be an $\mathbb{R}^m$-valued (for some $m \in \mathbb{N}$), or a V-valued, Q-Brownian motion, or a V-valued, cylindrical Brownian motion but it can be clarified by the context. The goal of this subsection is to prove that any $\eta \in L^2_{\mathcal{F}_T}(\Omega; H)$ with mean 0 can be represented as an Itô integral of a suitable process w.r.t. $W(\cdot)$.

We begin with some auxiliary results.

Lemma 2.144. *Let $W(\cdot) = (W_1(\cdot), \cdots, W_m(\cdot))^\top$ be an $\mathbb{R}^m$-valued Brownian motion. Then, the following set*

$$\Big\{\phi(W_1(t_1), \cdots, W_1(t_n), \cdots, W_m(t_1), \cdots, W_m(t_n)) \;\Big|\; 0 < t_1 < \cdots < t_n \le T, \\ \phi \in C_0^\infty(\mathbb{R}^{mn}),\ n \in \mathbb{N}\Big\}$$

is dense in $L^2_{\mathcal{F}_T}(\Omega)$.

Proof: Without loss of generality, we only consider the case that $m = 1$. The proof for the general case is similar.

Let $\{s_i\}_{i=1}^\infty$ be a dense subset of $(0, T]$ and for each $n = 1, 2, \cdots$, let $\{s_i\}_{i=1}^n = \{t_i\}_{i=1}^n$ with $0 < t_1 < \cdots < t_n \le T$, and $\mathcal{F}^n = \sigma(W(t_1), W(t_2) - W(t_1), \cdots, W(t_n) - W(t_{n-1}))$. Then, $\mathcal{F}^n \subset \mathcal{F}^{n+1}$ and $\mathcal{F}_T$ is the smallest σ-algebra containing all of the $\mathcal{F}^n$'s. Therefore, $\mathcal{F}_T = \sigma\Big(\bigcup_{n=1}^\infty \mathcal{F}^n\Big)$.

Fix any $g \in L^2_{\mathcal{F}_T}(\Omega)$. By Theorem 2.96, we have that

$$g = \mathbb{E}(g \mid \mathcal{F}_T) = \lim_{n\to\infty} \mathbb{E}(g \mid \mathcal{F}^n), \quad \text{in } L^2_{\mathcal{F}_T}(\Omega). \tag{2.169}$$

By the Doob-Dynkin lemma (i.e., Theorem 2.13), for each n, we have that

$$\mathbb{E}(g \mid \mathcal{F}^n) = g_n(W(t_1), W(t_2) - W(t_1), \cdots, W(t_n) - W(t_{n-1}))$$

for some Borel measurable function $g_n : \mathbb{R}^n \to \mathbb{R}$.

It suffices to prove that there is a sequence of functions $\{\phi_{n_k}\}_{k=1}^\infty \subset C_0^\infty(\mathbb{R}^n)$ such that $\phi_{n_k}(W(t_1), W(t_2) - W(t_1), \cdots, W(t_n) - W(t_{n-1})) \to g_n(W(t_1), W(t_2) - W(t_1), \cdots, W(t_n) - W(t_{n-1}))$ in $L^2_{\mathcal{F}_T}(\Omega)$ as $k \to \infty$. To prove this, we fix an $n \in \mathbb{N}$ and an $\varepsilon > 0$. Write $t_0 = 0$. By

$$\begin{aligned} &|g_n(W(t_1), W(t_2) - W(t_1), \cdots, W(t_n) - W(t_{n-1}))|^2_{L^2_{\mathcal{F}_T}(\Omega)} \\ &= \int_{\mathbb{R}^n} g_n^2(x_1, \cdots, x_n) \prod_{i=1}^n \frac{e^{-\frac{x_i^2}{2(t_i - t_{i-1})}}}{\sqrt{2\pi(t_i - t_{i-1})}} dx_1 \cdots dx_n < \infty, \end{aligned}$$

we conclude that there exists a bounded, open subset $G \subset \mathbb{R}^n$ such that

$$\int_{\mathbb{R}^n\setminus G} g_n^2(x_1,\cdots,x_n)\prod_{i=1}^{n}\frac{e^{-\frac{x_i^2}{2(t_i-t_{i-1})}}}{\sqrt{2\pi(t_i-t_{i-1})}}dx_1\cdots dx_n<\frac{\varepsilon^2}{4}. \tag{2.170}$$

On the other hand, one can find a $\varphi\in C_0^\infty(G)$ such that

$$\int_G |g_n(x_1,\cdots,x_n)-\varphi(x_1,\cdots,x_n)|^2\prod_{i=1}^{n}\frac{e^{-\frac{x_i^2}{2(t_i-t_{i-1})}}}{\sqrt{2\pi(t_i-t_{i-1})}}dx_1\cdots dx_n<\frac{\varepsilon^2}{4}. \tag{2.171}$$

From (2.170) and (2.171), we obtain that $|g_n-\varphi|_{L^2_{\mathcal{F}_T}(\Omega)}<\varepsilon$. This completes the proof of Lemma 2.144. $\square$

Lemma 2.145. *Let $F:\mathbb{C}^m\to\mathbb{C}$ be an analytic function. If*

$$F(z_1,\cdots,z_m)=0$$

on the set $\{(z_1,\cdots,z_m)\in\mathbb{C}^m\bigm|\operatorname{Im} z_1=\cdots\operatorname{Im} z_m=0\}$, then $F\equiv 0$ in $\mathbb{C}^m$.

Proof: For simplicity, we assume that $m=1$. The proof for the general l is similar. Write $z_1=x+iy$ for $x,y\in\mathbb{R}$, and denote by u and v respectively the real and imaginary parts of F. Clearly, both u and v are real analytic functions. Since F is analytic, it follows that

$$\frac{\partial u(x,0)}{\partial x}=\frac{\partial v(x,0)}{\partial y}=0,\quad -\frac{\partial u(x,0)}{\partial y}=\frac{\partial v(x,0)}{\partial x}=0,\quad \forall\, x\in\mathbb{R}. \tag{2.172}$$

Thus,

$$\frac{\partial^2 u(x,0)}{\partial x^2}=\frac{\partial^2 v(x,0)}{\partial y\partial x}=0,\quad -\frac{\partial^2 u(x,0)}{\partial y\partial x}=\frac{\partial^2 v(x,0)}{\partial x^2}=0,\quad \forall\, x\in\mathbb{R}. \tag{2.173}$$

From (2.172) and (2.173), we get that

$$\frac{\partial^2 u(x,0)}{\partial y^2}=-\frac{\partial^2 v(x,0)}{\partial y\partial x}=0,\quad \frac{\partial^2 v(x,0)}{\partial y^2}=\frac{\partial^2 u(x,0)}{\partial y\partial x}=0,\quad \forall\, x\in\mathbb{R}.$$

Inductively, for any $j,k\in\mathbb{N}$,

$$\frac{\partial^{j+k} u(x,0)}{\partial x^j\partial y^k}=\frac{\partial^{j+k} v(x,0)}{\partial x^j\partial y^k}=0,\ \forall\, x\in\mathbb{R}.$$

Therefore, $u=v\equiv 0$. This completes the proof of Lemma 2.145. $\square$

Lemma 2.146. *Let $W(\cdot)=(W_1(\cdot),\cdots,W_m(\cdot))^\top$ be an $\mathbb{R}^m$-valued Brownian motion. Then the following set*

$$\mathcal{X}\overset{\Delta}{=}\operatorname{span}\Big\{\exp\Big\{\int_0^T h(s)^\top dW(s)-\frac{1}{2}\int_0^T|h(s)|^2_{\mathbb{R}^m}ds\Big\}\Bigm| h\in L^2(0,T;\mathbb{R}^m)\Big\} \tag{2.174}$$

is dense in $L^2_{\mathcal{F}_T}(\Omega)$.

Proof: Suppose that $g \in L^2_{\mathcal{F}_T}(\Omega) \setminus \{0\}$ is orthogonal (in $L^2_{\mathcal{F}_T}(\Omega)$) to all functions in $\mathcal{X}$. For any given $t_1, \cdots, t_n \in [0,T]$ with $0 < t_1 < \cdots < t_n \le T$, we define an analytic function in the complex space $\mathbb{C}^{mn}$ as follows:

$$\begin{aligned} &F(z) = \int_\Omega \exp\Big\{\sum_{i=1}^n \sum_{j=1}^m z_{ij} W_j(t_i)\Big\} g d\mathbb{P}, \\ &\qquad\qquad \forall\, z = (z_{11}, \cdots, z_{1n}, \cdots, z_{m1}, \cdots, z_{mn}) \in \mathbb{C}^{mn}. \end{aligned} \tag{2.175}$$

For any given $\lambda_1^0 \in \mathbb{R}^m, \cdots, \lambda_n^0 \in \mathbb{R}^m$, let us choose

$$h = \sum_{i=1}^n \lambda_i^0 \chi_{[0,t_i]} \in L^2(0,T;\mathbb{R}^m). \tag{2.176}$$

By a suitable choice of $(\lambda_1^0, \cdots, \lambda_n^0)$ in (2.176), we see that

$$F(\lambda) = \int_\Omega \exp\{\lambda_1^\top W(t_1) + \cdots + \lambda_n^\top W(t_n)\} g d\mathbb{P} = 0 \tag{2.177}$$

for all $\lambda = (\lambda_1, \cdots, \lambda_n)$ with $\lambda_1 \in \mathbb{R}^m, \cdots, \lambda_n \in \mathbb{R}^m$ and $t_1, \cdots, t_n \in [0,T]$ with $0 < t_1 < \cdots < t_n \le T$. Since $F = 0$ on $\mathbb{R}^{mn}$ and F is analytic, by Lemma 2.145, we conclude that $F \equiv 0$ in $\mathbb{C}^{mn}$. Particularly, $F(iy) = 0$ for all $y = (y_{11}, \cdots, y_{1n}, \cdots, y_{m1}, \cdots, y_{mn}) \in \mathbb{R}^{mn}$. But then, for any $\phi \in C_0^\infty(\mathbb{R}^{mn})$,

$$\begin{aligned} &\int_\Omega \phi\big(W_1(t_1), \cdots, W_1(t_n), \cdots, W_m(t_1), \cdots, W_m(t_n)\big) g d\mathbb{P} \\ &= (2\pi)^{-mn} \int_\Omega \Big(\int_{\mathbb{R}^{mn}} \hat\phi(y) \exp\Big\{\sum_{i=1}^n \sum_{j=1}^m i y_{ji} W_j(t_i)\Big\} dy\Big) g d\mathbb{P} \\ &= (2\pi)^{-mn} \int_{\mathbb{R}^{mn}} \hat\phi(y) \Big(\int_\Omega \exp\Big\{\sum_{i=1}^n \sum_{j=1}^m i y_{ji} W_j(t_i)\Big\} g d\mathbb{P}\Big) dy \\ &= (2\pi)^{-mn} \int_{\mathbb{R}^{mn}} \hat\phi(y) F(iy) dy = 0, \end{aligned} \tag{2.178}$$

where $\hat\phi(y) = (2\pi)^{-mn} \int_{\mathbb{R}^{mn}} \phi(x) e^{-i\langle x,y\rangle_{\mathbb{R}^{mn}}} dx$ is the Fourier transform of ϕ. By (2.178) and Lemma 2.144, g is orthogonal to a dense subset of $L^2_{\mathcal{F}_T}(\Omega)$ and hence $g = 0$. Therefore $\mathcal{X}$ is dense in $L^2_{\mathcal{F}_T}(\Omega)$. □

The following result is known as the Martingale Representation Theorem.

Theorem 2.147. *Let $W(\cdot)$ be a V-valued, Q-Brownian motion (resp. cylindrical Brownian motion). If $X \in L^2_{\mathcal{F}_T}(\Omega; H)$, then there exists one and only one* **F***-adapted process $\Phi(\cdot) \in L^2_{\mathbb{F}}(0,T;\mathcal{L}(V;H))$ (resp. $\Phi(\cdot) \in L^2_{\mathbb{F}}(0,T;\mathcal{L}_2^0)$) such that*

$$X = \mathbb{E}X + \int_0^T \Phi(s) dW(s). \tag{2.179}$$

Proof: We only consider the case of cylindrical Brownian motion. We divide the proof into three steps.

Step 1. We consider the case that $V = \mathbb{R}^m$, $H = \mathbb{R}$ and $W(\cdot)$ is an $\mathbb{R}^m$-valued, standard Brownian motion.

Assume first that $X = \exp\Big\{ \int_0^T h(s)^\top dW(s) - \frac{1}{2}\int_0^T |h(s)|_{\mathbb{R}^m}^2 ds\Big\}$ for some $h(\cdot) \in L^2(0,T;\mathbb{R}^m)$. Define

$$Y(t) = \exp\Big\{ \int_0^t h(s)^\top dW(s) - \frac{1}{2}\int_0^t |h(s)|_{\mathbb{R}^m}^2 ds\Big\}, \quad t \in [0,T].$$

Then, by Itô's formula, it follows that

$$\begin{aligned} dY(t) &= Y(t)\Big[h(t)^\top dW(t) - \frac{1}{2}|h(t)|_{\mathbb{R}^m}^2 dt\Big] + \frac{1}{2}Y(t)|h(t)|_{\mathbb{R}^m}^2 dt \\ &= Y(t)h(t)^\top dW(t), \end{aligned}$$

which gives $Y(t) = 1 + \int_0^t Y(s)h(s)^\top dW(s)$. This leads to

$$X = Y(T) = 1 + \int_0^T Y(s)h(s)dW(s)$$

and $\mathbb{E}X = 1$. Hence, (2.179) holds for such kind of X's.

By the linearity, (2.179) also holds for any element of $\mathcal{X}$ given by (2.174). Now for any $X \in L^2_{\mathcal{F}_T}(\Omega)$, by Lemma 2.146, we may approximate it in $L^2_{\mathcal{F}_T}(\Omega)$ by a sequence $\{X_n\}_{n=1}^\infty \subset \mathcal{X}$. Then for each $n \in \mathbb{N}$, we have

$$X_n = \mathbb{E}X_n + \int_0^T \Phi_n(s)^\top dW(s), \qquad \text{for some } \Phi_n \in L^2_{\mathbb{F}}(0,T;\mathbb{R}^m).$$

By the third conclusion in Theorem 2.124, for any $k \in \mathbb{N}$,

$$\begin{aligned} \mathbb{E}(X_n - X_k)^2 &= \mathbb{E}\Big[\mathbb{E}(X_n - X_k) + \int_0^T (\Phi_n - \Phi_k)(s)^\top dW(s)\Big]^2 \\ &= [\mathbb{E}(X_n - X_k)]^2 + \int_0^T \mathbb{E}|\Phi_n - \Phi_k|_{\mathbb{R}^m}^2 ds \to 0, \end{aligned}$$

as $k, n \to \infty$. Hence, $\{\Phi_n\}_{n=1}^\infty$ is a Cauchy sequence in $L^2_{\mathbb{F}}(0,T;\mathbb{R}^m)$ and hence it converges to some $\Phi \in L^2_{\mathbb{F}}(0,T;\mathbb{R}^m)$. Hence,

$$X = \lim_{n\to\infty} X_n = \lim_{n\to\infty} \Big[\mathbb{E}X_n + \int_0^T \Phi_n(s)^\top dW(s)\Big] = \mathbb{E}X + \int_0^T \Phi(s)^\top dW(s),$$

the limit being taken in $L^2_{\mathcal{F}_T}(\Omega)$. Hence the representation (2.179) holds for all $X \in L^2_{\mathcal{F}_T}(\Omega)$. The uniqueness of Φ is obvious.

Step 2. In this step, we consider the case that $H = \mathbb{R}$ and $W(\cdot)$ is a V-valued, cylindrical Brownian motion. By Definition 2.120, $W(\cdot)$ can be formally written as $W(\cdot) = \sum_{i=1}^{\infty} w_i(\cdot)e_i$, where $\{w_i(\cdot)\}_{i=1}^{\infty}$ is a sequence of independent real valued, standard Brownian motions, and $\{e_i\}_{i=1}^{\infty}$ is an orthonormal basis of V.

Clearly, ℓ^2 can be regarded as a subspace of $\mathcal{L}_2(V;\mathbb{R}) \equiv \mathcal{L}(V;\mathbb{R})$. Indeed, for any $a = (a_1, a_2, \cdots) \in \ell^2$, it can be defined as an element of $\mathcal{L}_2(V;\mathbb{R})$ by $a(v) = \sum_{i=1}^{\infty} a_i v_i$ for any $v = \sum_{i=1}^{\infty} v_i e_i$.

For each $k \in \mathbb{N}$, denoted by $\{\mathcal{F}_t^k\}_{t\in[0,T]}$ the natural filtration generated by $\{w_i(\cdot)\}_{i=1}^{k}$. For any $X \in L^2_{\mathcal{F}_T}(\Omega)$, similarly to (2.169), there exists an $X_k \in L^2_{\mathcal{F}_T^k}(\Omega)$ such that

$$\lim_{k\to\infty} X_k = X \quad \text{in } L^2_{\mathcal{F}_T}(\Omega). \tag{2.180}$$

For every $k \in \mathbb{N}$, by Step 1, there is a $\Phi_k \in L^2_{\mathbb{F}}(0,T;\mathbb{R}^k) \subset L^2_{\mathbb{F}}(0,T;\ell^2) \subset L^2_{\mathbb{F}}(0,T;\mathcal{L}_2(V;\mathbb{R}))$ such that

$$X_k = \mathbb{E}X_k + \int_0^T \Phi_k(s)^\top dW(s) = \mathbb{E}X_k + \int_0^T \Phi_k(s)dW(s). \tag{2.181}$$

In the last term of the equality (2.181), we have regarded $\Phi_k(\cdot)$ as an element of $L^2_{\mathbb{F}}(0,T;\mathcal{L}_2(V;\mathbb{R}))$. From (2.181), using the conclusion 4) in Theorem 2.133, we find that, for any $k, n \in \mathbb{N}$,

$$\begin{aligned} &|\Phi_k - \Phi_n|^2_{L^2_{\mathbb{F}}(0,T;\mathcal{L}_2(V;\mathbb{R}))} = \mathbb{E}\Big|\int_0^T \big(\Phi_k(s) - \Phi_n(s)\big)dW(s)\Big|^2 \\ &= \mathbb{E}\Big|\int_0^T \Phi_k(t)^\top dW(s) - \int_0^T \Phi_n(s)^\top dW(s)\Big|^2 \le \mathbb{E}|X_k - X_n|^2. \end{aligned} \tag{2.182}$$

This, together with (2.180), indicates that $\{\Phi_k\}_{k=1}^{\infty}$ is a Cauchy sequence in $L^2_{\mathbb{F}}(0,T;\mathcal{L}_2(V;\mathbb{R}))$. Denote by Φ the limit of $\{\Phi_k\}_{k=1}^{\infty}$. Then,

$$\begin{aligned} X &= \lim_{k\to\infty} X_k = \lim_{k\to\infty} \Big[\mathbb{E}X_k + \int_0^T \Phi_k(s)dW(s)\Big] \\ &= \mathbb{E}X + \int_0^T \Phi(s)dW(s) \quad \text{in } L^2_{\mathcal{F}_T}(\Omega). \end{aligned}$$

Step 3. In this step, we consider the general separable Hilbert space H. Let $\{h_j\}_{j=1}^{\infty}$ be an orthogonal normal basis of H. Then, $\langle X, h_j\rangle_H \in L^2_{\mathcal{F}_T}(\Omega)$ for $j \in \mathbb{N}$. Thus, we can find a $\phi_j \in L^2_{\mathbb{F}}(0,T;\mathcal{L}_2(V;\mathbb{R}))$ such that

$$\langle X, h_j\rangle_H = \mathbb{E}\langle X, h_j\rangle_H + \int_0^T \phi_j(s)dW(s). \tag{2.183}$$

Similar to (2.182), it holds that

$$|\phi_j|^2_{L^2_{\mathbb{F}}(0,T;\mathcal{L}_2(V;\mathbb{R}))} = \mathbb{E}\Big|\int_0^T \phi_j(s)dW(s)\Big|^2 \le \mathbb{E}|\langle X, h_j\rangle_H|^2. \tag{2.184}$$

Therefore,

$$X = \mathbb{E}X + \sum_{j=1}^{\infty}\int_0^T \phi_j(s)dW(s)h_j. \tag{2.185}$$

Hence, we only need to prove that there is a $\Phi \in L^2_{\mathbb{F}}(0,T;\mathcal{L}_2^0)$ such that

$$\int_0^T \Phi(s)dW(s) = \sum_{j=1}^{\infty}\int_0^T \phi_j(s)dW(s)h_j. \tag{2.186}$$

First, by (2.184), we have that

$$\mathbb{E}\Big|\sum_{j=1}^{\infty}\int_0^T \phi_j(s)dW(s)h_j\Big|_H^2 = \sum_{j=1}^{\infty}\mathbb{E}\int_0^T |\phi_j(s)|^2_{\mathcal{L}_2(V;\mathbb{R})}ds \le \mathbb{E}|X|_H^2.$$

Let us define $\Phi(\cdot)$ by $\Phi(\cdot)v = \sum_{j=1}^{\infty}(\phi_j v)h_j$ for any $v \in V$. Then,

$$\begin{aligned}
|\Phi(\cdot)|^2_{L^2_{\mathbb{F}}(0,T;\mathcal{L}_2^0)} &= \mathbb{E}\int_0^T \sum_{i=1}^{\infty}\Big|\sum_{j=1}^{\infty}(\phi_j e_i)h_j\Big|_H^2 dt\\
&= \mathbb{E}\int_0^T \sum_{j=1}^{\infty}\sum_{i=1}^{\infty}|(\phi_j e_i)|_H^2 dt = \sum_{j=1}^{\infty}|\phi_j(s)|^2_{L^2_{\mathbb{F}}(0,T;\mathcal{L}_2(V;\mathbb{R}))}ds < +\infty,
\end{aligned}$$

and the equality (2.186) holds. This completes the proof of Theorem 2.147. □

Remark 2.148. By (2.179), it is easy to see that

$$\mathbb{E}(X \mid \mathcal{F}_t) = \mathbb{E}X + \int_0^t \Phi(s)dW(s) \tag{2.187}$$

and

$$\mathbb{E}|X - \mathbb{E}X|_H^2 = \mathbb{E}\int_0^T |\Phi(s)Q^{\frac{1}{2}}|^2_{\mathcal{L}_2^0}ds \tag{2.188}$$

(*resp.*

$$\mathbb{E}|X - \mathbb{E}X|_H^2 = \mathbb{E}\int_0^T |\Phi(s)|^2_{\mathcal{L}_2^0}ds\ \ ). \tag{2.189}$$

As a consequence of Theorem 2.147, we have the following result, which will play a crucial role in the study of backward stochastic evolution equations.

Corollary 2.149. *Let $W(\cdot)$ be a V-valued, Q-Brownian motion (resp. cylindrical Brownian motion). Then, for any $f \in L^1_{\mathbb{F}}(0,T;L^2(\Omega;H))$, there is a unique $K(\cdot,\cdot,\cdot) \in L^1(0,T;L^2_{\mathbb{F}}(0,T;\mathcal{L}(V;H)))$ (resp. $K(\cdot,\cdot,\cdot) \in L^1(0,T;L^2_{\mathbb{F}}(0,T;\mathcal{L}^0_2))$) satisfying the following conditions:*

1) *$K(s,\sigma,\cdot) = 0$ for $\sigma > s$;*

2) *For a.e. $s \in [0,T]$,*

$$f(s) = \mathbb{E}f(s) + \int_0^s K(s,\sigma)dW(\sigma), \quad a.s.; \tag{2.190}$$

3)

$$\begin{aligned} &|K(\cdot,\cdot,\cdot)|_{L^1(0,T;L^2_{\mathbb{F}}(0,T;\mathcal{L}(V;H)))} \leq |f|_{L^1_{\mathbb{F}}(0,T;L^2(\Omega;H))} \\ &(resp.\ |K(\cdot,\cdot,\cdot)|_{L^1(0,T;L^2_{\mathbb{F}}(0,T;\mathcal{L}^0_2))} \leq |f|_{L^1_{\mathbb{F}}(0,T;L^2(\Omega;H))}). \end{aligned} \tag{2.191}$$

Proof: We only consider the case that $W(\cdot)$ is a cylindrical Brownian motion here. For $f \in L^1_{\mathbb{F}}(0,T;L^2(\Omega;H))$, we can find a sequence $\{f_n\}_{n=1}^\infty$ of simple processes such that f_n tends to f in $L^1_{\mathbb{F}}(0,T;L^2(\Omega;H))$ as $n \to \infty$, where $f_n(s,\omega) = \sum_{i=0}^{n-1} \chi_{[t_{n,i},t_{n,i+1})}(s)\xi_{n,i}(\omega)$ for some partition $\{t_{n,0},t_{n,1},\cdots,t_{n,n}\}$ of $[0,T]$ and $\xi_{n,i} \in L^2_{\mathcal{F}_{t_{n,i}}}(\Omega;H)$.

By Theorem 2.147 and Remark 2.148, for every $\xi_{n,i}$, there is a $k_{n,i} \in L^2_{\mathbb{F}}(0,T;\mathcal{L}^0_2)$ such that

$$\xi_{n,i} = \mathbb{E}\xi_{n,i} + \int_0^{t_{n,i}} k_{n,i}(\sigma)dW(\sigma), \quad k_{n,i}(\sigma) = 0 \text{ for } \sigma > t_{n,i}. \tag{2.192}$$

Put $K_n(s,\sigma) = \sum_{i=0}^{n-1} \chi_{[t_{n,i},t_{n,i+1})}(s)\chi_{[0,s]}(\sigma)k_{n,i}(\sigma)$. Then,

$$f_n(s) = \mathbb{E}f_n(s) + \int_0^s K_n(s,\sigma)dW(\sigma). \tag{2.193}$$

From (2.193), we see that, for any $n,k \in \mathbb{N}$,

$$\begin{aligned} &\int_0^T \Big(\mathbb{E}\int_0^T |K_n(s,\sigma) - K_k(s,\sigma)|^2_{\mathcal{L}^0_2} d\sigma\Big)^{1/2} ds \\ &= \int_0^T \Big(\mathbb{E}\int_0^s |K_n(s,\sigma) - K_k(s,\sigma)|^2_{\mathcal{L}^0_2} d\sigma\Big)^{1/2} ds \\ &= \int_0^T \Big(\mathbb{E}\Big|\int_0^s K_n(s,\sigma)dW(\sigma) - \int_0^s K_k(s,\sigma)dW(\sigma)\Big|^2_H\Big)^{1/2} ds \\ &= \int_0^T \Big(\mathbb{E}\Big|f_n(s) - f_k(s) - \big(\mathbb{E}f_n(s) - \mathbb{E}f_k(s)\big)\Big|^2_H\Big)^{1/2} ds \\ &\leq |f_n - f_k|_{L^1_{\mathbb{F}}(0,T;L^2(\Omega;H))}. \end{aligned} \tag{2.194}$$

Since $\{f_n\}_{n=1}^\infty$ is a Cauchy sequence in $L^1_{\mathbb{F}}(0,T;L^2(\Omega;H))$, by (2.194), we find that $\{K_n\}_{n=1}^\infty$ is a Cauchy sequence in $L^1(0,T;L^2_{\mathbb{F}}(0,T;\mathcal{L}^0_2))$. Thus, there is a

$K \in L^1(0,T;L^2_{\mathbb{F}}(0,T;\mathcal{L}_2^0))$ so that $\lim_{n\to\infty} K_n = K$ in $L^1(0,T;L^2_{\mathbb{F}}(0,T;\mathcal{L}_2^0))$. This, together with (2.193), implies (2.190).

Similar to (2.194), one has

$$\int_0^T \Big(\mathbb{E}\int_0^T |K_n(s,\sigma)|^2_{\mathcal{L}_2^0} d\sigma\Big)^{1/2} ds \le |f_n|_{L^1_{\mathbb{F}}(0,T;L^2(\Omega;H))}. \tag{2.195}$$

Taking $n \to \infty$ in (2.195), we obtain (2.191). □

2.12 Notes and Comments

Stochastic Analysis is a huge subject which is still growing up and flourishing. In this chapter, we only present a brief introduction to the relevant results will be used in the rest of the book, and we omit the proofs of some classical results, which can be found in almost all standard books (e.g., [57, 164, 287]). Interested readers can find more materials in the references such as [58, 68, 129, 140]. Theorem 2.55 was proved in [239]. Section 2.5 was based on [242]. The proof of Theorem 2.140 (Burkholder-Davis-Gundy inequality) can be found in [68], for example. Note that the inverse inequality of (2.137) (*resp.* (2.138)) in Theorem 2.140 is also true, i.e., one can prove that

$$\mathbb{E}\Big(\sup_{t\in[0,T]}\Big|\int_0^t f(s)dW(s)\Big|_H^p\Big) \ge \frac{1}{\mathcal{C}_p}\mathbb{E}\Big(\int_0^T |f(s)Q^{\frac{1}{2}}|^2_{\mathcal{L}_2^0} ds\Big)^{\frac{p}{2}}$$

(*resp.*

$$\mathbb{E}\Big(\sup_{t\in[0,T]}\Big|\int_0^t f(s)dW(s)\Big|_H^p\Big) \ge \frac{1}{\mathcal{C}_p}\mathbb{E}\Big(\int_0^T |f(s)|^2_{\mathcal{L}_2^0} ds\Big)^{\frac{p}{2}}\ ),$$

but we will not use it in this book. The readers are referred to [327] and the references therein for the general stochastic Fubini theorem. The proof of Theorem 2.142 can be found in [282, 286]. In this book, we present an elementary but long proof of the Martingale Representation Theorem (There exist some shorter proofs which need more preliminaries on stochastic analysis, e.g., [160, 175]).

In this book, we use the Itô integral, which is the most common and useful way to define stochastic integral w.r.t. a Brownian motion. There are two other typical ways to define stochastic integral: Stratonovich integral (e.g., [176]) and rough path integral (e.g., [105, 250]). Stochastic integral (including multiple stochastic integral) w.r.t. more general stochastic processes can be found in [129, 176, 182].

3

Stochastic Evolution Equations

This book is mainly addressed to studying the control problems governed by stochastic evolution equations. In this chapter, we shall present a short introduction to the well-posedness and regularity of solutions to this sort of equations, valued in Hilbert spaces.

Throughout this chapter, $T > 0$, $(\Omega, \mathcal{F}, \mathbf{F}, \mathbb{P})$ (with $\mathbf{F} = \{\mathcal{F}_t\}_{t\in[0,T]}$) is a fixed filtered probability space satisfying the usual condition, and we denote by $\mathbb{F}$ the progressive σ-field w.r.t. $\mathbf{F}$; H and V are two separable Hilbert spaces; we denote by I the identity operator on H, and write $\mathcal{L}_2^0 \overset{\Delta}{=} \mathcal{L}_2(V; H)$.

3.1 Stochastic Evolution Equations in Finite Dimensions

In this section, we shall consider the well-posedness of stochastic evolution equations in finite dimensions, i.e., stochastic (ordinary) differential equations.

Fix $d, n \in \mathbb{N}$ and $p \geq 1$. Let $W(\cdot) = (W^1(\cdot), \cdots, W^d(\cdot))^\top$ be an d-dimensional standard Brownian motion. We consider the following stochastic differential equation:

$$\begin{cases} dX(t) = F(t, X)dt + \widetilde{F}(t, X)dW(t) \quad & \text{in } [0, T], \\ X(0) = X_0. \end{cases} \tag{3.1}$$

In (3.1), X is an unknown, and the initial datum X_0 is a given $\mathcal{F}_0$-measurable $\mathbb{R}^n$-valued function; $F(\cdot,\cdot,\cdot) : [0, T] \times \Omega \times \mathbb{R}^n \to \mathbb{R}^n$ and $\widetilde{F}(\cdot,\cdot,\cdot) : [0, T] \times \Omega \times \mathbb{R}^n \to \mathbb{R}^{n\times d}$ are two given functions. Here and in what follows, when the context is clear, we omit the $\omega(\in \Omega)$ argument in the defined functions. In the sequel, we call $F(t, X)dt$ and $\widetilde{F}(t, X)dW(t)$ the drift and diffusion terms of (3.1), respectively.

To begin with, as for ordinal differential equations, one needs to define the solution to (3.1). The following is the most natural one.

Q. Lü and X. Zhang, *Mathematical Control Theory for Stochastic Partial Differential Equations*, Probability Theory and Stochastic Modelling 101, https://doi.org/10.1007/978-3-030-82331-3_3

Definition 3.1. *An $\mathbb{R}^n$-valued, $\mathbf{F}$-adapted, continuous process $X(\cdot)$ is called a solution to (3.1) if $F(\cdot, X(\cdot)) \in L^1(0,T;\mathbb{R}^n)$ a.s., $\widetilde{F}(\cdot, X(\cdot)) \in L^{2,loc}_{\mathbb{F}}(0,T;\mathbb{R}^{n\times d})$, and for each $t \in [0,T]$,*

$$X(t) = X_0 + \int_0^t F(s, X(s))ds + \int_0^t \widetilde{F}(s, X(s))dW(s), \quad \text{a.s.} \tag{3.2}$$

The solution to (3.1) is said to be unique, if for any other solution $Y(\cdot)$ to (3.1), one has $\mathbb{P}(\{X(t) = Y(t),\ t \in [0,T]\}) = 1$.

In the rest of this section, we assume that

1) Both $F(\cdot, x)$ and $\widetilde{F}(\cdot, x)$ are $\mathbf{F}$-adapted for each $x \in \mathbb{R}^n$; and

2) $F(\cdot, 0) \in L^p_{\mathbb{F}}(\Omega; L^1(0,T;\mathbb{R}^n))$ and $\widetilde{F}(\cdot, 0) \in L^p_{\mathbb{F}}(\Omega; L^2(0,T;\mathbb{R}^{n\times d}))$, and there exist two nonnegative functions $L_1(\cdot) \in L^1(0,T)$ and $L_2(\cdot) \in L^2(0,T)$ such that for any $x, y \in \mathbb{R}^n$ and a.e. $t \in [0,T]$,

$$\begin{cases} |F(t,x) - F(t,y)|_{\mathbb{R}^n} \le L_1(t)|x-y|_{\mathbb{R}^n}, \\ |\widetilde{F}(t,x) - \widetilde{F}(t,y)|_{\mathbb{R}^{n\times d}} \le L_2(t)|x-y|_{\mathbb{R}^n}, \end{cases} \quad \text{a.s.}$$

The fundamental existence and uniqueness result for the equation (3.1) is as follows.

Theorem 3.2. *For any $X_0 \in L^p_{\mathcal{F}_0}(\Omega;\mathbb{R}^n)$, the equation (3.1) admits a unique solution $X(\cdot) \in L^p_{\mathbb{F}}(\Omega; C([0,T];\mathbb{R}^n))$ satisfying*

$$\begin{aligned} |X(\cdot)|_{L^p_{\mathbb{F}}(\Omega;C([0,T];\mathbb{R}^n))} \le \mathcal{C}\big(&|X_0|_{L^p_{\mathcal{F}_0}(\Omega;\mathbb{R}^n)} + |F(\cdot,0)|_{L^p_{\mathbb{F}}(\Omega;L^1(0,T;\mathbb{R}^n))} \\ &+ |\widetilde{F}(\cdot,0)|_{L^p_{\mathbb{F}}(\Omega;L^2(0,T;\mathbb{R}^{n\times d}))}\big). \end{aligned} \tag{3.3}$$

Moreover, if $\widetilde{X}_0 \in L^p_{\mathcal{F}_0}(\Omega;\mathbb{R}^n)$ is another initial datum and $\widetilde{X}(\cdot)$ is the corresponding solution to (3.1), then

$$|X(\cdot) - \widetilde{X}(\cdot)|_{L^p_{\mathbb{F}}(\Omega;C([0,T];\mathbb{R}^n))} \le \mathcal{C}|X_0 - \widetilde{X}_0|_{L^p_{\mathcal{F}_0}(\Omega;\mathbb{R}^n)}. \tag{3.4}$$

Proof: For any $T_1 \in (0,T)$ and $x(\cdot) \in L^p_{\mathbb{F}}(\Omega; C([0,T_1];\mathbb{R}^n))$, we define

$$\mathcal{J}(x(\cdot))(t) = X_0 + \int_0^t F(s, x(s))ds + \int_0^t \widetilde{F}(s, x(s))dW(s), \quad t \in [0,T_1]. \tag{3.5}$$

By Hölder's inequality, it follows that

$$\begin{aligned} &\mathbb{E}\sup_{0\le t\le T_1}\Big|\int_0^t F(s,x(s))ds\Big|^p_{\mathbb{R}^n} \\ &\le \mathcal{C}\Big(\mathbb{E}\sup_{0\le t\le T_1}\Big|\int_0^t F(s,0)ds\Big|^p_{\mathbb{R}^n} + \mathbb{E}\sup_{0\le t\le T_1}\Big|\int_0^t L_1(s)|x(s)|_{\mathbb{R}^n}ds\Big|^p\Big) \\ &\le \mathcal{C}\Big[\mathbb{E}\Big(\int_0^{T_1}|F(s,0)|_{\mathbb{R}^n}ds\Big)^p + |L_1(\cdot)|^p_{L^1(0,T_1)}|x(\cdot)|^p_{L^p_{\mathbb{F}}(\Omega;C([0,T_1];\mathbb{R}^n))}\Big], \end{aligned} \tag{3.6}$$

where $\mathcal{C}$ is independent of T_1. Similarly, by the Burkholder-Davis-Gundy inequality (i.e., Theorem 2.140), we have

$$
\begin{aligned}
&\mathbb{E}\sup_{0\le t\le T_1}\Big|\int_0^t \widetilde F(s,x(s))dW(s)\Big|_{\mathbb{R}^n}^p \le \mathcal{C}\mathbb{E}\Big(\int_0^{T_1}|\widetilde F(s,x(s))|_{\mathbb{R}^{n\times d}}^2 ds\Big)^{p/2}\\
&\le \mathcal{C}\Big[\mathbb{E}\Big(\int_0^{T_1}|\widetilde F(s,x(s))-\widetilde F(s,0)|_{\mathbb{R}^{n\times d}}^2 ds\Big)^{p/2}+\mathbb{E}\Big(\int_0^{T_1}|\widetilde F(s,0)|_{\mathbb{R}^{n\times d}}^2 ds\Big)^{p/2}\Big]\\
&\le \mathcal{C}\Big[\mathbb{E}\Big(\int_0^{T_1}|L_2(s)|^2|x(s)|_{\mathbb{R}^n}^2 ds\Big)^{p/2}+\mathbb{E}\Big(\int_0^{T_1}|\widetilde F(s,0)|_{\mathbb{R}^{n\times d}}^2 ds\Big)^{p/2}\Big]\\
&\le \mathcal{C}\Big[|L_2(\cdot)|_{L^2(0,T_1)}^p|x(\cdot)|_{L^p_{\mathbb{F}}(\Omega;C([0,T_1];\mathbb{R}^n))}^p+\mathbb{E}\Big(\int_0^{T_1}|\widetilde F(s,0)|_{\mathbb{R}^{n\times d}}^2 ds\Big)^{p/2}\Big].
\end{aligned}
\tag{3.7}
$$

Hence, by (3.5)–(3.7), we arrive at

$$
\begin{aligned}
&\mathbb{E}\sup_{0\le t\le T_1}|\mathcal{J}(x(\cdot)(t)|_{\mathbb{R}^n}^p\\
&\le \mathcal{C}\Big[\Big(|L_1(\cdot)|_{L^1(0,T_1)}^p+|L_2(\cdot)|_{L^2(0,T_1)}^p\Big)|x(\cdot)|_{L^p_{\mathbb{F}}(\Omega;C([0,T_1];\mathbb{R}^n))}^p\\
&\quad+\mathbb{E}|X_0|_{\mathbb{R}^n}^p+\mathbb{E}\Big(\int_0^{T_1}|F(s,0)|_{\mathbb{R}^n}ds\Big)^p+\mathbb{E}\Big(\int_0^{T_1}|\widetilde F(s,0)|_{\mathbb{R}^{n\times d}}^2 ds\Big)^{p/2}\Big].
\end{aligned}
\tag{3.8}
$$

By (3.8), it is easy to see that $\mathcal{J}(x(\cdot))\in L^p_{\mathbb{F}}(\Omega;C([0,T_1];\mathbb{R}^n))$. Further, from the proof of (3.8), we see that

$$
\begin{aligned}
&|\mathcal{J}(x(\cdot))-\mathcal{J}(y(\cdot))|_{L^p_{\mathbb{F}}(\Omega;C([0,T_1];\mathbb{R}^n))}\\
&\le \mathcal{C}\Big(|L_1(\cdot)|_{L^1(0,T_1)}+|L_2(\cdot)|_{L^2(0,T_1)}\Big)|x(\cdot)-y(\cdot)|_{L^p_{\mathbb{F}}(\Omega;C([0,T_1];\mathbb{R}^n))},
\end{aligned}
\tag{3.9}
$$

for any $x(\cdot),y(\cdot)\in L^p_{\mathbb{F}}(\Omega;C([0,T_1];\mathbb{R}^n))$. We choose a $T_1\in(0,T)$ such that

$$
\mathcal{C}\Big(|L_1(\cdot)|_{L^1(0,T_1)}+|L_2(\cdot)|_{L^2(0,T_1)}\Big)<1. \tag{3.10}
$$

By (3.9), the map $\mathcal{J}:\ L^p_{\mathbb{F}}(\Omega;C([0,T_1];\mathbb{R}^n))\to L^p_{\mathbb{F}}(\Omega;C([0,T_1];\mathbb{R}^n))$ is contractive. Thus, $\mathcal{J}$ admits a unique fixed point, which gives a solution $X(\cdot)$ to (3.1) on $[0,T_1]$. Further, by (3.8) and noting (3.10), we see that $X(\cdot)$ satisfies the estimate (3.3) with T replaced by T_1. Repeating this procedure, we obtain a solution $X(\cdot)$ to (3.1) on $[0,T]$ and it satisfies (3.3). The other conclusions in Theorem 3.2 follow also from the above argument. □

Now, we consider the following linear equation:

$$
\begin{cases}
dX(t)=\big(A(t)X(t)+f(t)\big)dt+\displaystyle\sum_{i=1}^d\big(C_i(t)X(t)+g_i(t)\big)dW^i(t) \ \text{ in } [s,T],\\
X(s)=\xi.
\end{cases}
\tag{3.11}
$$

Here $s \in [0,T)$, $\xi \in L^2_{\mathcal{F}_s}(\Omega;\mathbb{R}^n)$, and

$$\begin{cases} A(\cdot) \in L^1_{\mathbb{F}}(0,T;L^\infty(\Omega;\mathbb{R}^{n\times n})), \quad C_i(\cdot) \in L^2_{\mathbb{F}}(0,T;L^\infty(\Omega;\mathbb{R}^{n\times n})), \\ f(\cdot) \in L^2_{\mathbb{F}}(\Omega;L^1(0,T;\mathbb{R}^n)), \quad g_i(\cdot) \in L^2_{\mathbb{F}}(0,T;\mathbb{R}^n). \end{cases} \tag{3.12}$$

In the rest of this section, we shall derive a representation formula for solutions to the equation (3.11). For this purpose, let us consider the following two ($\mathbb{R}^{n\times n}$-valued) stochastic differential equations:

$$\begin{cases} d\Phi(t) = A(t)\Phi(t)dt + \sum_{i=1}^{d} C_i(t)\Phi(t)dW^i(t) \ \text{ in } [s,T], \\ \Phi(0) = I_n, \end{cases} \tag{3.13}$$

and

$$\begin{cases} d\Psi(t) = \Psi(t)\big(- A(t) + \sum_{i=1}^{d} C_i(t)^2\big)dt - \sum_{i=1}^{d} \Psi(t)C_i(t)dW^i(t) \ \text{ in } [s,T], \\ \Psi(0) = I_n, \end{cases} \tag{3.14}$$

where I_n is the identity matrix in $\mathbb{R}^{n\times n}$. By Itô's formula, we have that

$$\begin{aligned} & d\big(\Psi(t)\Phi(t)\big) \\ & = d\Psi(t)\Phi(t) + \Psi(t)d\Phi(t) + d\Psi(t)d\Phi(t) \\ & = -\Psi(t)A(t)\Phi(t)dt + \Psi(t)\sum_{i=1}^{d} C_i(t)^2\Phi(t)dt - \sum_{i=1}^{d}\Psi(t)C_i(t)\Phi(t)dW^i(t) \\ & \quad +\Psi(t)A(t)\Phi(t)dt + \sum_{i=1}^{d}\Psi(t)C_i(t)\Phi(t)dW^i(t) - \Psi(t)\sum_{i=1}^{d} C_i(t)^2\Phi(t)dt \\ & = 0. \end{aligned}$$

This, together with $\Psi(0)\Phi(0) = I_n$ implies that $\Psi(\cdot)\Phi(\cdot) = I_n$ in $[0,T]$. Hence, $\Psi(t) = \Phi(t)^{-1}$ for all $t \in [0,T]$. We have the following *variation of constants formula* for solutions to (3.11).

Theorem 3.3. *Let the assumption* (3.12) *hold. Then, for any* $\xi \in L^2_{\mathcal{F}_0}(\Omega;\mathbb{R}^n)$, *the equation* (3.11) *admits a unique solution* $X(\cdot)$, *which can be represented by the following formula:*

$$\begin{aligned} X(t) = \Phi(t)\xi + \Phi(t)\int_s^t \Phi(r)^{-1}\big(f(r) - \sum_{i=1}^{d} C_i(r)g_i(r)\big)dr \\ + \sum_{i=1}^{d}\Phi(t)\int_s^t \Phi(r)^{-1}g_i(r)dW^i(r), \qquad \forall\, t \in [s,T]. \end{aligned} \tag{3.15}$$

Proof: Write

$$Y(t)=\xi+\int_s^t \Phi(r)^{-1}\Big(f(r)-\sum_{i=1}^d C_i(r)g_i(r)\Big)dr+\sum_{i=1}^d\int_s^t \Phi(r)^{-1}g_i(r)dW^i(r).$$

By (3.13) and Itô's formula, it follows that

$$d(\Phi(t)Y(t))=\big(A(t)\Phi(t)Y(t)+f(t)\big)dt+\sum_{i=1}^d\big(C_i(t)\Phi(t)Y(t)+g_i(t)\big)dW^i(t).$$

Also, $\Phi(0)Y(0)=\xi$. By the uniqueness of solutions to the equation (3.11), we conclude that $X(t)=\Phi(t)Y(t)$, which gives the desired result. $\square$

Remark 3.4. As we shall see later (in Theorem 3.20), the conditions $A(\cdot)\in L^1_{\mathbb{F}}(0,T;L^\infty(\Omega;\mathbb{R}^{n\times n}))$ and $C_i(\cdot)\in L^2_{\mathbb{F}}(0,T;L^\infty(\Omega;\mathbb{R}^{n\times n}))$ in (3.12) (for Theorem 3.3) can be weakened as $A(\cdot)\in L^\infty_{\mathbb{F}}(\Omega;L^1(0,T;\mathbb{R}^{n\times n}))$ and $C_i(\cdot)\in L^\infty_{\mathbb{F}}(\Omega;L^2(0,T;\mathbb{R}^{n\times n}))$, respectively.

3.2 Well-Posedness of Stochastic Evolution Equations

In the rest of this chapter, $W(\cdot)$ may be a V-valued, Q-Brownian motion or cylindrical Brownian motion but we only consider the case of cylindrical Brownian motion.

We consider the following stochastic evolution equation:

$$\begin{cases} dX(t)=\big(AX(t)+F(t,X(t))\big)dt+\widetilde{F}(t,X(t))dW(t) & \text{in } (0,T],\\ X(0)=X_0. \end{cases} \tag{3.16}$$

Here $X_0:\Omega\to H$ is an $\mathcal{F}_0$-random variable, A generates a C_0-semigroup $\{S(t)\}_{t\geq 0}$ on H, and $F(\cdot,\cdot):\ [0,T]\times\Omega\times H\to H$ and $\widetilde{F}(\cdot,\cdot):\ [0,T]\times\Omega\times H\to\mathcal{L}_2^0$ are two given functions. Similarly to the case of finite dimensions, in the sequel, we call $(AX(t)+F(t,X(t)))\,dt$ and $\widetilde{F}(t,X)dW(t)$ the drift and diffusion terms of (3.16), respectively.

It is easy to show the following result (and hence we omit the details).

Proposition 3.5. *For any $g\in L^{2,loc}_{\mathbb{F}}(0,T;\mathcal{L}_2^0)$, the H-valued, stochastic process $\int_0^\cdot S(\cdot-s)g(s)dW(s)$ has a continuous modification*[1].

3.2.1 Notions of Solutions

Similar to the case of deterministic evolution equations, one has several ways to define solutions to (3.16). The most natural way seems to be the following one.

[1] In the sequel, we shall always use such a modification for $\int_0^\cdot S(\cdot-s)g(s)dW(s)$.

Definition 3.6. *An H-valued, $\mathbf{F}$-adapted, continuous stochastic process $X(\cdot)$ is called a strong solution to (3.16) if*

1) *$X(t) \in D(A)$ for a.e. $(t,\omega) \in [0,T] \times \Omega$ and $AX(\cdot) \in L^1(0,T;H)$ a.s.;*
2) *$F(\cdot,X(\cdot)) \in L^1(0,T;H)$ a.s., $\widetilde{F}(\cdot,X(\cdot)) \in L^{2,loc}_{\mathbb{F}}(0,T;\mathcal{L}_2^0)$; and*
3) *For all $t \in [0,T]$,*

$$X(t) = X_0 + \int_0^t \big(AX(s) + F(s,X(s))\big)ds + \int_0^t \widetilde{F}(s,X(s))dW(s), \quad \text{a.s.}$$

Example 3.7. Let $V = \mathbb{R}$, $\{T(t)\}_{t\geq 0}$ be another C_0-semigroups on H with the infinitesimal generator B so that $D(A) \subset D(B^2)$ and $S(\cdot)T(\cdot) = T(\cdot)S(\cdot)$ (Particularly, we may take $B = I$). Then, by Itô's formula, a strong solution to the following equation

$$\begin{cases} dX(t) = AX(t)dt + \dfrac{1}{2}B^2X(t)dt + BX(t)dW & \text{in } [0,\infty), \\ X(0) = X_0 \in D(A), \end{cases}$$

is given by $X(t) = S(t)\big(T(W(t))X_0\big)$.

Unfortunately, generally the conditions which guarantee the well-posedness of (3.16) in the sense of Definition 3.6 are very restrictive. Therefore, following the idea of weak and mild solutions to deterministic evolution equations, people also introduce similar notions for its stochastic counterpart (Recall (2.132) for the notation $\langle\!\langle \cdot,\cdot \rangle\!\rangle_H$).

Definition 3.8. *An H-valued, $\mathbf{F}$-adapted, continuous stochastic process $X(\cdot)$ is called a weak solution to (3.16) if $F(\cdot,X(\cdot)) \in L^1(0,T;H)$ a.s., $\widetilde{F}(\cdot,X(\cdot)) \in L^{2,loc}_{\mathbb{F}}(0,T;\mathcal{L}_2^0)$, and for any $t \in [0,T]$ and $\xi \in D(A^*)$,*

$$\begin{aligned} \langle X(t),\xi\rangle_H &= \langle X_0,\xi\rangle_H + \int_0^t \big(\langle X(s),A^*\xi\rangle_H + \langle F(s,X(s)),\xi\rangle_H\big)ds \\ &\quad + \int_0^t \langle\!\langle \widetilde{F}(s,X(s)),\xi \rangle\!\rangle_H dW(s), \qquad \text{a.s.} \end{aligned} \tag{3.17}$$

Definition 3.9. *An H-valued, $\mathbf{F}$-adapted, continuous stochastic process $X(\cdot)$ is called a mild solution to (3.16) if $F(\cdot,X(\cdot)) \in L^1(0,T;H)$ a.s., $\widetilde{F}(\cdot,X(\cdot)) \in L^{2,loc}_{\mathbb{F}}(0,T;\mathcal{L}_2^0)$, and for any $t \in [0,T]$,*

$$\begin{aligned} X(t) &= S(t)X_0 + \int_0^t S(t-s)F(s,X(s))ds \\ &\quad + \int_0^t S(t-s)\widetilde{F}(s,X(s))dW(s), \qquad \text{a.s.} \end{aligned} \tag{3.18}$$

It is easy to see that, if an H-valued, $\mathbf{F}$-adapted, continuous stochastic process $X(\cdot)$ is a strong solution to (3.16), then it is also a weak solution to the same equation. The following result shows the equivalence between weak and mild solutions to (3.16).

Theorem 3.10. *An H-valued, $\mathbf{F}$-adapted, continuous stochastic process $X(\cdot)$ is a weak solution to (3.16) if and only if it is a mild solution to the same equation.*

Proof: **The "only if" part**. Since $X(\cdot)$ is a weak solution to (3.16), for any fixed $\phi \in D((A^*)^2)$, $t \in [0,T]$ and $r \in [t,T]$, we choose $\xi = S^*(r-t)A^*\phi$ in (3.17), and obtain that

$$
\begin{aligned}
&\big\langle X(t), S^*(r-t)A^*\phi\big\rangle_H \\
&= \big\langle X_0, S^*(r-t)A^*\phi\big\rangle_H + \int_0^t \big\langle X(s), A^*S^*(r-t)A^*\phi\big\rangle_H ds \\
&\quad + \int_0^t \big\langle F(s,X(s)), S^*(r-t)A^*\phi\big\rangle_H ds \\
&\quad + \int_0^t \langle\!\langle \widetilde{F}(s,X(s)), S^*(r-t)A^*\phi\rangle\!\rangle_H dW(s).
\end{aligned}
\tag{3.19}
$$

Integrating (3.19) w.r.t. t from 0 to r, we obtain that

$$
\begin{aligned}
&\int_0^r \big\langle X(t), S^*(r-t)A^*\phi\big\rangle_H dt \\
&= \int_0^r \big\langle X_0, S^*(r-t)A^*\phi\big\rangle_H dt + \int_0^r\int_0^t \big\langle X(s), A^*S^*(r-t)A^*\phi\big\rangle_H dsdt \\
&\quad + \int_0^r\int_0^t \big\langle F(s,X(s)), S^*(r-t)A^*\phi\big\rangle_H dsdt \\
&\quad + \int_0^r\int_0^t \langle\!\langle \widetilde{F}(s,X(s)), S^*(r-t)A^*\phi\rangle\!\rangle_H dW(s)dt.
\end{aligned}
\tag{3.20}
$$

A direct computation shows that

$$
\begin{aligned}
\int_0^r \big\langle X_0, S^*(r-t)A^*\phi\big\rangle_H dt &= \int_0^r \big\langle AS(r-t)X_0, \phi\big\rangle_H dt \\
&= \big\langle S(r)X_0, \phi\big\rangle_H - \big\langle X_0, \phi\big\rangle_H.
\end{aligned}
\tag{3.21}
$$

By Fubini's theorem (i.e., Theorem 2.21), we find that

$$
\begin{aligned}
&\int_0^r \int_0^t \langle X(s), A^* S^*(r-t) A^* \phi \rangle_H ds dt \\
&= \int_0^r \int_s^r \langle X(s), A^* S^*(r-t) A^* \phi \rangle_H dt ds \\
&= \int_0^r \langle X(s), S^*(r-s) A^* \phi \rangle_H ds - \int_0^r \langle X(s), A^* \phi \rangle_H ds
\end{aligned}
\tag{3.22}
$$

and

$$
\begin{aligned}
&\int_0^r \int_0^t \langle F(s, X(s)), S^*(r-t) A^* \phi \rangle_H ds dt \\
&= \int_0^r \langle F(s, X(s)), S^*(r-s) \phi \rangle_H ds - \int_0^r \langle F(s, X(s)), \phi \rangle_H ds.
\end{aligned}
\tag{3.23}
$$

By the stochastic Fubini theorem (i.e., Theorem 2.141), we see that

$$
\begin{aligned}
&\int_0^r \int_0^t \langle\!\langle \widetilde{F}(s, X(s)), S^*(r-t) A^* \phi \rangle\!\rangle_H dW(s) dt \\
&= \int_0^r \int_s^r \langle\!\langle \widetilde{F}(s, X(s)), S^*(r-t) A^* \phi \rangle\!\rangle_H dt dW(s) \\
&= \int_0^r \langle\!\langle \widetilde{F}(s, X(s)), S^*(r-s) \phi \rangle\!\rangle_H dW(s) - \int_0^r \langle\!\langle \widetilde{F}(s, X(s)), \phi \rangle\!\rangle_H dW(s) \\
&= \Big\langle \int_0^r S(r-s) \widetilde{F}(s, X(s)) dW(s), \phi \Big\rangle_H - \int_0^r \langle\!\langle \widetilde{F}(s, X(s)), \phi \rangle\!\rangle_H dW(s).
\end{aligned}
\tag{3.24}
$$

From (3.20)–(3.24) and using (3.17) again, we end up with

$$
\begin{aligned}
&\Big\langle X(r) - S(r) X_0 - \int_0^r S(r-s) F(s, X(s)) ds \\
&\quad - \int_0^r S(r-s) \widetilde{F}(s, X(s)) dW(s),\ \phi \Big\rangle_H = 0.
\end{aligned}
\tag{3.25}
$$

Since $D((A^*)^2)$ is dense in H, (3.25) also holds for any $\phi \in H$, which indicates that $X(\cdot)$ is a mild solution to (3.16).

The "if" part. Assume that $X(\cdot)$ is a mild solution to (3.16). Then for any $\xi \in D(A^*)$ and $r \in [0, T]$, we have that

$$
\begin{aligned}
&\Big\langle X(r) - S(r) X_0 - \int_0^r S(r-s) F(s, X(s)) ds \\
&\quad - \int_0^r S(r-s) \widetilde{F}(s, X(s)) dW(s),\ A^* \xi \Big\rangle_H = 0.
\end{aligned}
\tag{3.26}
$$

Fix $t \in [0, T]$. Integrating (3.26) from 0 to t w.r.t. r, we find that

$$
\begin{aligned}
&\int_0^t \langle X(r), A^*\xi\rangle_H \\
&= \int_0^t \langle S(r)X_0, A^*\xi\rangle_H dr + \int_0^t \int_0^r \langle S(r-s)F(s,X(s)), A^*\xi\rangle_H dsdr \qquad (3.27)\\
&\quad + \int_0^t \int_0^r \langle\!\langle\, S(r-s)\widetilde{F}(s,X(s)), A^*\xi \,\rangle\!\rangle_H dW(s)dr.
\end{aligned}
$$

Clearly,

$$
\int_0^t \langle S(r)X_0, A^*\xi\rangle_H dr = \langle S(t)X_0, \xi\rangle_H - \langle X_0, \xi\rangle_H. \qquad (3.28)
$$

By Fubini's theorem, we find that

$$
\begin{aligned}
&\int_0^t \int_0^r \langle S(r-s)F(s,X(s)), A^*\xi\rangle_H dsdr \\
&= \int_0^t \int_s^t \langle AS(r-s)F(s,X(s)), \xi\rangle_H drds \qquad (3.29)\\
&= \int_0^t \langle S(t-s)F(s,X(s)), \xi\rangle_H ds - \int_0^t \langle F(s,X(s)), \xi\rangle_H ds.
\end{aligned}
$$

Thanks to the stochastic Fubini theorem, we see that

$$
\begin{aligned}
&\int_0^t \int_0^r \langle\!\langle\, S(r-s)\widetilde{F}(s,X(s)), A^*\xi \,\rangle\!\rangle_H dW(s)dr \\
&= \int_0^t \int_s^t \langle\!\langle\, AS(r-s)\widetilde{F}(s,X(s)), \xi \,\rangle\!\rangle_H drdW(s) \\
&= \int_0^t \langle\!\langle\, S(t-s)\widetilde{F}(s,X(s)), \xi \,\rangle\!\rangle_H dW(s) - \int_0^t \langle\!\langle\, \widetilde{F}(s,X(s)), \xi \,\rangle\!\rangle_H dW(s).
\end{aligned}
\qquad (3.30)
$$

From (3.27)–(3.30) and noting (3.26), we see that $X(\cdot)$ satisfies (3.17). Hence, $X(\cdot)$ is a weak solution to (3.16). □

The following result provides a sufficient condition for a mild solution to be a strong solution to (3.16).

Lemma 3.11. *A mild solution $X(\cdot)$ to (3.16) is a strong solution (to the same equation) if the following three conditions hold for all $x \in H$ and $0 \le s \le t \le T$, a.s.,*

1) $X_0 \in D(A)$, $S(t-s)F(s,x) \in D(A)$ and $S(t-s)\widetilde{F}(s,x) \in \mathcal{L}_2(V; D(A))$;

2) $|AS(t-s)F(s,x)|_H \le \alpha(t-s)|x|_H$ for some real-valued stochastic process $\alpha(\cdot)$ with $\alpha(\cdot) \in L^1(0,T)$ a.s.; and

3) $|AS(t-s)\widetilde{F}(s,x)|_{\mathcal{L}_2^0} \le \beta(t-s)|x|_H$ for some real-valued stochastic process $\beta(\cdot) \in L^{2,loc}_{\mathbb{F}}(0,T)$.

Proof: Since $X(\cdot)$ is an H-valued, continuous stochastic process, by the conditions in Lemma 3.11, we have

$$\int_0^T \int_0^t \big|AS(t-r)F(r,X(r))\big|_H drdt < \infty, \quad \text{a.s.}$$

and

$$\int_0^T \int_0^t |AS(T-r)\widetilde{F}(r,X(r))|^2_{\mathcal{L}_2^0} drdt < \infty, \quad \text{a.s.}$$

By Fubini's theorem, we have

$$\begin{aligned}&\int_0^t \int_0^s AS(s-r)F(r,X(r))drds = \int_0^t \int_r^t AS(s-r)F(r,X(r))dsdr\\&= \int_0^t S(t-r)F(r,X(r))dr - \int_0^t F(r,X(r))dr.\end{aligned}$$

On the other hand, by Theorem 2.141, we have that

$$\begin{aligned}&\int_0^t \int_0^s AS(s-r)\widetilde{F}(r,X(r))dW(r)ds = \int_0^t \int_r^t AS(s-r)\widetilde{F}(r,X(r))dsdW(r)\\&= \int_0^t S(t-r)\widetilde{F}(r,X(r))dW(r) - \int_0^t \widetilde{F}(r,X(r))dW(r).\end{aligned}$$

Therefore, by (3.18), $AX(\cdot)$ is integrable a.s. and

$$\begin{aligned}\int_0^t AX(s)ds &= S(t)X_0 - X_0 + \int_0^t S(t-r)F(r,X(r))dr - \int_0^t F(r,X(r))dr\\&\quad + \int_0^t S(t-r)\widetilde{F}(r,X(r))dW(r) - \int_0^t \widetilde{F}(r,X(r))dW(r)\\&= X(t) - X_0 - \int_0^t F(r,X(r))dr - \int_0^t \widetilde{F}(r,X(r))dW(r).\end{aligned}$$

This indicated that $X(\cdot)$ is a strong solution to (3.16). □

3.2.2 Well-Posedness in the Sense of Mild Solution

In the rest of this section, we will focus on the well-posedness of stochastic evolution equations in the sense of mild solution.

Let us begin with the following result, which is an easy consequence of the Burkholder-Davis-Gundy inequality (in Theorem 2.140).

Proposition 3.12. *For any $p \geq 1$, there exists a constant $\mathcal{C}_p > 0$ such that for any $g \in L^p_{\mathbb{F}}(\Omega; L^2(0,T;\mathcal{L}_2^0))$ and $t \in [0,T]$,*

$$\mathbb{E}\Big|\int_0^t S(t-s)g(s)dW(s)\Big|_H^p \leq \mathcal{C}_p \mathbb{E}\Big(\int_0^t |g(s)|^2_{\mathcal{L}_2^0} ds\Big)^{\frac{p}{2}}. \qquad (3.31)$$

We consider first the following linear equation:

$$\begin{cases} dX(t) = \big(AX(t) + \mathcal{A}X(t) + f(t)\big)dt + \big(\mathcal{B}X(t) + g(t)\big)dW(t) & \text{in } (0,T], \\ X(0) = X_0. \end{cases} \tag{3.32}$$

Here X_0, $\mathcal{A}$, $\mathcal{B}$, $f(\cdot)$ and $g(\cdot)$ are suitable vector/operator valued functions to be given later. We have the following well-posedness result for (3.32) with $\mathcal{A} = 0$ and $\mathcal{B} = 0$.

Theorem 3.13. *Let $X_0 \in L^p_{\mathcal{F}_0}(\Omega; H)$ for some $p \geq 1$, $\mathcal{A} = 0$, $\mathcal{B} = 0$, $f(\cdot) \in L^p_{\mathbb{F}}(\Omega; L^1(0,T;H))$ and $g(\cdot) \in L^p(\Omega; L^2(0,T;\mathcal{L}_2^0))$. Then, the equation (3.32) admits a unique mild solution $X(\cdot) \in C_{\mathbb{F}}([0,T]; L^p(\Omega;H))$. Moreover,*

$$\begin{aligned} & |X(\cdot)|_{C_{\mathbb{F}}([0,T];L^p(\Omega;H))} \\ & \leq \mathcal{C}\big(|X_0|_{L^p_{\mathcal{F}_0}(\Omega;H)} + |f|_{L^p_{\mathbb{F}}(\Omega;L^1(0,T;H))} + |g|_{L^p_{\mathbb{F}}(\Omega;L^2(0,T;\mathcal{L}_2^0))}\big). \end{aligned} \tag{3.33}$$

Proof: Clearly,

$$X(t) = S(t)X_0 + \int_0^t S(t-s)f(s)ds + \int_0^t S(t-s)g(s)dW(s) \tag{3.34}$$

is a mild solution to (3.32). Now, we prove that $X(\cdot) \in C_{\mathbb{F}}([0,T]; L^p(\Omega;H))$.

First, by Proposition 3.12, for any $t \in [0,T]$,

$$\begin{aligned} & |X(t)|_{L^p_{\mathcal{F}_t}(\Omega;H)} \\ & = \Big|S(t)X_0 + \int_0^t S(t-s)f(s)ds + \int_0^t S(t-s)g(s)dW(s)\Big|_{L^p_{\mathcal{F}_t}(\Omega;H)} \\ & \leq \big|S(\cdot)X_0\big|_{L^p_{\mathcal{F}_0}(\Omega;H)} + \Big|\int_0^t S(t-s)f(s)ds\Big|_{L^p_{\mathcal{F}_t}(\Omega;H)} \\ & \quad + \Big|\int_0^t S(t-s)g(s)dW(s)\Big|_{L^p_{\mathcal{F}_t}(\Omega;H)} \\ & \leq \mathcal{C}\big(|X_0|_{L^p_{\mathcal{F}_0}(\Omega;H)} + |f|_{L^p_{\mathbb{F}}(\Omega;L^1(0,t;H))} + |g|_{L^p_{\mathbb{F}}(\Omega;L^2(0,t;\mathcal{L}_2^0))}\big). \end{aligned} \tag{3.35}$$

Hence,

$$\begin{aligned} & |X(\cdot)|_{L^\infty_{\mathbb{F}}(0,T;L^p(\Omega;H))} \\ & \leq \mathcal{C}\big(|X_0|_{L^p_{\mathcal{F}_0}(\Omega;H)} + |f|_{L^p_{\mathbb{F}}(\Omega;L^1(0,T;H))} + |g|_{L^p_{\mathbb{F}}(\Omega;L^2(0,T;\mathcal{L}_2^0))}\big). \end{aligned} \tag{3.36}$$

Obvious, $S(\cdot)X_0 + \int_0^\cdot S(\cdot - s)f(s)ds \in C_{\mathbb{F}}([0,T]; L^p(\Omega;H))$. It remains to show that $\int_0^\cdot S(\cdot - s)g(s)dW(s) \in C_{\mathbb{F}}([0,T]; L^p(\Omega;H))$. For this purpose, for any t_1 and t_2 satisfying $0 \leq t_1 \leq t_2 \leq T$, we have

$$
\begin{aligned}
&\Big|\int_0^{t_2} S(t_2-s)g(s)dW(s) - \int_0^{t_1} S(t_1-s)g(s)dW(s)\Big|_{L^p_{\mathcal{F}_{t_2}}(\Omega;H)}\\
&\leq \Big|\int_{t_1}^{t_2} S(t_2-s)g(s)dW(s)\Big|_{L^p_{\mathcal{F}_{t_2}}(\Omega;H)}\\
&\quad+\Big|\int_0^{t_1} S(t_1-s)\big[S(t_2-t_1)-I\big]g(s)dW(s)\Big|_{L^p_{\mathcal{F}_{t_2}}(\Omega;H)}.
\end{aligned}
$$

By Proposition 3.12, it follows that

$$
\Big|\int_{t_1}^{t_2} S(t_2-s)g(s)dW(s)\Big|_{L^p_{\mathcal{F}_{t_2}}(\Omega;H)} \leq \mathcal{C}|g|_{L^p_{\mathbb{F}}(\Omega;L^2(t_1,t_2;\mathcal{L}_2^0))}.
$$

Since $\int_{t_1}^{t_2}|g(s)|^2_{\mathcal{L}_2^0}ds \leq \int_0^{t_2}|g(s)|^2_{\mathcal{L}_2^0}ds$, by Lebesgue's dominated convergence theorem, we find that

$$
\begin{aligned}
&\lim_{t_1\to t_2^-}\Big|\int_{t_1}^{t_2} S(t_2-s)g(s)dW(s)\Big|_{L^p_{\mathcal{F}_{t_2}}(\Omega;H)}\\
&\leq \mathcal{C}\lim_{t_1\to t_2^-}\Big[\mathbb{E}\Big(\int_{t_1}^{t_2}|g(s)|^2_{\mathcal{L}_2^0}ds\Big)^{\frac{p}{2}}\Big]^{\frac{1}{p}} = 0.
\end{aligned}
\tag{3.37}
$$

Likewise,

$$
\lim_{t_1\to t_2^-}\Big|\int_0^{t_1} S(t_1-s)\big[S(t_2-t_1)-I\big]g(s)dW(s)\Big|_{L^p_{\mathcal{F}_{t_2}}(\Omega;H)} = 0. \tag{3.38}
$$

By (3.34) and (3.37)–(3.38), we conclude that $\lim_{t\to t_0^-}|X(t)-X(t_0)|_{L^p_{\mathcal{F}_{t_2}}(\Omega;H)} = 0$ for any $t_0\in[0,T]$. Similarly, $\lim_{t\to t_0^+}|X(t)-X(t_0)|_{L^p_{\mathcal{F}_{t_2}}(\Omega;H)} = 0$. By (3.36), we obtain (3.33). This completes the proof of Theorem 3.13. □

In the sequel, we need the following assumption on the nonlinearities in (3.16).

Condition 3.1 $F(\cdot,\cdot):[0,T]\times\Omega\times H\to H$ *and* $\widetilde F(\cdot,\cdot):[0,T]\times\Omega\times H\to\mathcal{L}_2^0$ *are two given functions satisfying*

1) *Both* $F(\cdot,x)$ *and* $\widetilde F(\cdot,x)$ *are* $\mathbf{F}$*-adapted for each* $x\in H$*; and*

2) *There exist two nonnegative (real-valued) functions* $L_1(\cdot)\in L^1(0,T)$ *and* $L_2(\cdot)\in L^2(0,T)$ *such that, for any* $x,y\in H$ *and a.e.* $t\in[0,T]$,

$$
\begin{cases}
|F(t,x)-F(t,y)|_H \leq L_1(t)|x-y|_H,\\
|\widetilde F(t,x)-\widetilde F(t,y)|_{\mathcal{L}_2^0} \leq L_2(t)|x-y|_H,
\end{cases}
\quad \text{a.s.} \tag{3.39}
$$

We have the following well-posedness result for (3.16).

Theorem 3.14. *Let Condition 3.1 hold, and $F(\cdot,0)\in L^p_{\mathbb{F}}(\Omega;L^1(0,T;H))$ and $\widetilde F(\cdot,0)\in L^p_{\mathbb{F}}(\Omega;L^2(0,T;\mathcal{L}_2^0))$ for some $p\geq 2$. Then, for any $X_0\in L^p_{\mathcal{F}_0}(\Omega;H)$, the equation (3.16) admits a unique mild solution $X(\cdot)\in C_{\mathbb{F}}([0,T];L^p(\Omega;H))$. Moreover,*

$$
\begin{aligned}
&|X(\cdot)|_{C_{\mathbb{F}}([0,T];L^p(\Omega;H))}\\
&\leq \mathcal{C}\big(|X_0|_{L^p_{\mathcal{F}_0}(\Omega;H)}+|F(\cdot,0)|_{L^p_{\mathbb{F}}(\Omega;L^1(0,T;H))}+|\widetilde F(\cdot,0)|_{L^p_{\mathbb{F}}(\Omega;L^2(0,T;\mathcal{L}_2^0))}\big).
\end{aligned}
\tag{3.40}
$$

Proof: Fix any $T_1\in[0,T]$. For any $Y\in C_{\mathbb{F}}([0,T_1];L^p(\Omega;H))$, we consider the following equation

$$
\begin{cases}
dX(t)=\big(AX(t)+F(t,Y(t))\big)dt+\widetilde F(t,Y(t))dW(t) & \text{in } (0,T_1],\\
X(0)=X_0.
\end{cases}
\tag{3.41}
$$

By Condition 3.1 and noting $L^1_{\mathbb{F}}(0,T_1;L^p(\Omega;H))\subset L^p_{\mathbb{F}}(\Omega;L^1(0,T_1;H))$, we see that

$$
\begin{aligned}
&|F(\cdot,Y(\cdot))|_{L^p_{\mathbb{F}}(\Omega;L^1(0,T_1;H))}\\
&\leq |F(\cdot,0)|_{L^p_{\mathbb{F}}(\Omega;L^1(0,T_1;\mathcal{L}_2^0))}+\mathcal{C}|L_1(\cdot)|Y(\cdot)|_H|_{L^1_{\mathbb{F}}(0,T_1;L^p(\Omega))}\\
&\leq |F(\cdot,0)|_{L^p_{\mathbb{F}}(\Omega;L^1(0,T_1;\mathcal{L}_2^0))}+\mathcal{C}|L_1(\cdot)|_{L^1(0,T_1)}|Y(\cdot)|_{C_{\mathbb{F}}([0,T_1];L^p(\Omega;H))}.
\end{aligned}
\tag{3.42}
$$

Using Condition 3.1 again and noting $p\geq 2$ (and hence $L^2_{\mathbb{F}}(0,T_1;L^p(\Omega;\mathcal{L}_2^0))\subset L^p_{\mathbb{F}}(\Omega;L^2(0,T_1;\mathcal{L}_2^0))$, we see that

$$
\begin{aligned}
&|\widetilde F(\cdot,Y(\cdot))|_{L^p_{\mathbb{F}}(\Omega;L^2(0,T_1;\mathcal{L}_2^0))}\\
&\leq |\widetilde F(\cdot,0)|_{L^p_{\mathbb{F}}(\Omega;L^2(0,T_1;\mathcal{L}_2^0))}+\mathcal{C}|L_2(\cdot)|Y(\cdot)|_H|_{L^2_{\mathbb{F}}(0,T_1;L^p(\Omega))}\\
&\leq |\widetilde F(\cdot,0)|_{L^p_{\mathbb{F}}(\Omega;L^2(0,T_1;\mathcal{L}_2^0))}+\mathcal{C}|L_2(\cdot)|_{L^2(0,T_1)}|Y(\cdot)|_{C_{\mathbb{F}}([0,T_1];L^p(\Omega;H))}.
\end{aligned}
\tag{3.43}
$$

By (3.42)–(3.43) and Theorem 3.13, the equation (3.41) admits a unique mild solution $X(\cdot)\in C_{\mathbb{F}}([0,T_1];L^p(\Omega;H))$. We define a map $\mathcal{J}:C_{\mathbb{F}}([0,T_1];L^p(\Omega;H))\to C_{\mathbb{F}}([0,T_1];L^p(\Omega;H))$ by $\mathcal{J}(Y)=X$.

Now we show that the map $\mathcal{J}$ is contractive provided that T_1 is small enough. Indeed, for another $\widehat Y\in C_{\mathbb{F}}([0,T_1];L^p(\Omega;H))$, we define $\widehat X=\mathcal{J}(\widehat Y)$. Write $\widetilde X(\cdot)=X(\cdot)-\widehat X(\cdot)$, $F_1(\cdot)=F(\cdot,Y(\cdot))-F(\cdot,\widehat Y(\cdot))$ and $\widetilde F_1(\cdot)=\widetilde F(\cdot,Y(\cdot))-\widetilde F(\cdot,\widehat Y(\cdot))$. Clearly, $\widetilde X(\cdot)$ solves the following equation

$$
\begin{cases}
d\widetilde X(t)=\big(A\widetilde X(t)+F_1(t)\big)dt+\widetilde F_1(t)dW(t) & \text{in } (0,T_1],\\
\widetilde X(0)=0.
\end{cases}
\tag{3.44}
$$

Similar to (3.42)–(3.43), we have

$$
|F_1(\cdot)|_{L^p_{\mathbb{F}}(\Omega;L^1(0,T_1;H))}\leq \mathcal{C}|L_1(\cdot)|_{L^1(0,T_1)}|Y(\cdot)-\widehat Y(\cdot)|_{C_{\mathbb{F}}([0,T_1];L^p(\Omega;H))}
\tag{3.45}
$$

and

$$|\widetilde{F}_1(\cdot)|_{L^p_{\mathbb{F}}(\Omega;L^2(0,T_1;\mathcal{L}_2^0))} \leq \mathcal{C}|L_2(\cdot)|_{L^2(0,T_1)}|Y(\cdot)-\widehat{Y}(\cdot)|_{C_{\mathbb{F}}([0,T_1];L^p(\Omega;H))}. \quad (3.46)$$

Applying Theorem 3.13 to (3.44) and noting (3.45)–(3.46), we find that

$$\begin{aligned}
&|\widetilde{X}(\cdot)|_{C_{\mathbb{F}}([0,T_1];L^p(\Omega;H))}\\
&\leq \mathcal{C}\big(|F_1(\cdot)|_{L^p_{\mathbb{F}}(\Omega;L^1(0,T_1;H))} + |\widetilde{F}_1(\cdot)|_{L^p_{\mathbb{F}}(\Omega;L^2(0,T_1;\mathcal{L}_2^0))}\big)\\
&\leq \mathcal{C}\big(|L_1(\cdot)|_{L^1(0,T_1)} + |L_2(\cdot)|_{L^2(0,T_1)}\big)|Y(\cdot) - \widehat{Y}(\cdot)|_{C_{\mathbb{F}}([0,T_1];L^p(\Omega;H))}.
\end{aligned}$$

Choose T_1 so that $\mathcal{C}\big(|L_1(\cdot)|_{L^1(0,T_1)} + |L_2(\cdot)|_{L^2(0,T_1)}\big) < 1$. Then, $\mathcal{J}$ is contractive. By means of the Banach fixed point theorem, $\mathcal{J}$ enjoys a unique fixed point $X(\cdot) \in C_{\mathbb{F}}([0,T_1];L^p(\Omega;H))$. It is clear that $X(\cdot)$ is a mild solution to the following equation:

$$\begin{cases} dX(t) = \big(AX(t) + F(t,X(t))\big)dt + \widetilde{F}(t,X(t))dW(t) & \text{in } (0,T_1],\\ X(0) = X_0. \end{cases} \quad (3.47)$$

By (3.42)–(3.43), we see that $F(\cdot,X(\cdot)) \in L^p_{\mathbb{F}}(\Omega;L^1(0,T_1;H))$, $\widetilde{F}(\cdot,X(\cdot)) \in L^p_{\mathbb{F}}(\Omega;L^2(0,T_1;\mathcal{L}_2^0))$, and

$$\begin{aligned}
&|F(\cdot,X(\cdot))|_{L^p_{\mathbb{F}}(\Omega;L^1(0,T_1;H))} + |\widetilde{F}(\cdot,X(\cdot))|_{L^p_{\mathbb{F}}(\Omega;L^2(0,T_1;\mathcal{L}_2^0))}\\
&\leq |F(\cdot,0)|_{L^p_{\mathbb{F}}(\Omega;L^1(0,T_1;H))} + |\widetilde{F}(\cdot,0)|_{L^p_{\mathbb{F}}(\Omega;L^2(0,T_1;\mathcal{L}_2^0))}\\
&\quad +\mathcal{C}\big(|L_1(\cdot)|_{L^1(0,T_1)} + |L_2(\cdot)|_{L^2(0,T_1)}\big)|X(\cdot)|_{C_{\mathbb{F}}([0,T_1];L^p(\Omega;H))}.
\end{aligned}$$

Therefore, using Theorem 3.13 again, we find that

$$\begin{aligned}
&|X(\cdot)|_{C_{\mathbb{F}}([0,T_1];L^p(\Omega;H))}\\
&\leq \mathcal{C}\Big(|X_0|_{L^p_{\mathcal{F}_0}(\Omega;H)}+ |F(\cdot,X(\cdot))|_{L^p_{\mathbb{F}}(\Omega;L^1(0,T_1;H))}+ |\widetilde{F}(\cdot,X(\cdot))|_{L^p_{\mathbb{F}}(\Omega;L^2(0,T_1;\mathcal{L}_2^0))}\Big)\\
&\leq \mathcal{C}\Big[|X_0|_{L^p_{\mathcal{F}_0}(\Omega;H)} + |F(\cdot,0)|_{L^p_{\mathbb{F}}(\Omega;L^1(0,T_1;H))} + |\widetilde{F}(\cdot,0)|_{L^p_{\mathbb{F}}(\Omega;L^2(0,T_1;\mathcal{L}_2^0))}\\
&\quad +\big(|L_1(\cdot)|_{L^1(0,T_1)} + |L_2(\cdot)|_{L^2(0,T_1)}\big)|X(\cdot)|_{C_{\mathbb{F}}([0,T_1];L^p(\Omega;H))}\Big].
\end{aligned} \quad (3.48)$$

Since $\mathcal{C}\big(|L_1(\cdot)|_{L^1(0,T_1)} + |L_2(\cdot)|_{L^2(0,T_1)}\big) < 1$, it follows from (3.48) that

$$\begin{aligned}
&\big|X(\cdot)\big|_{C_{\mathbb{F}}([0,T_1];L^p(\Omega;H))}\\
&\leq\mathcal{C}\big(|X_0|_{L^p_{\mathcal{F}_0}(\Omega;H)}+|F(\cdot,0)|_{L^p_{\mathbb{F}}(\Omega;L^1(0,T_1;H))}+|\widetilde{F}(\cdot,0)|_{L^p_{\mathbb{F}}(\Omega;L^2(0,T_1;\mathcal{L}_2^0))}\big).
\end{aligned} \quad (3.49)$$

Repeating the above argument, we obtain a mild solution to the equation (3.16). The uniqueness of such solution to (3.16) is obvious. The desired estimate (3.40) follows from (3.49). This completes the proof of Theorem 3.14.

□

Remark 3.15. In the above proof (of Theorem 3.14), $p \geq 2$ is a key assumption. Otherwise, we cannot show that

$$\widetilde{F}(\cdot, X(\cdot)) \in L^p_{\mathbb{F}}(\Omega; L^2(0,T; \mathcal{L}_2^0)), \qquad \forall\, X(\cdot) \in C_{\mathbb{F}}([0,T]; L^p(\Omega; H)).$$

As far as we know, it is an unsolved problem to establish the well-posedness result (as that in Theorem 3.14) for the general stochastic evolution equation (3.16) without the assumption $p \geq 2$.

3.3 Regularity of Mild Solutions to Stochastic Evolution Equations

In this section, we shall consider the time/space regularities of mild solutions to the equation (3.16).

3.3.1 Burkholder-Davis-Gundy Type Inequality and Time Regularity

By comparing Theorems 3.2 and 3.14, we see that the time regularity in the solution space $L^p_{\mathbb{F}}(\Omega; C([0,T]; \mathbb{R}^n))$ for finite dimensions is a little better than that in $C_{\mathbb{F}}([0,T]; L^p(\Omega; H))$ for the general stochastic evolution equation (3.16). It is easy to see that, the difference comes from the use of Burkholder-Davis-Gundy type inequality for the stochastic involution "$\int_0^\cdot S(\cdot - s)g(s)dW(s)$".

We consider first the linear equation (3.32). For this, we introduce some notations. Fix $p \geq 1$ and $q \geq 1$. Let H_1 and H_2 be two Hilbert spaces. Write

$$\begin{aligned}\Upsilon_{p,q}(H_1; H_2) \triangleq \big\{ J(\cdot,\cdot) \in \mathcal{L}_{pd}(L^p_{\mathbb{F}}(\Omega; L^\infty(0,T; H_1)); L^p_{\mathbb{F}}(\Omega; L^q(0,T; H_2))) \;\big|\; \\ |J(\cdot,\cdot)|_{\mathcal{L}(H_1;H_2)} \in L^\infty_{\mathbb{F}}(\Omega; L^q(0,T)) \big\}.\end{aligned} \tag{3.50}$$

In the sequel, we shall simply denote $\Upsilon_{p,p}(H_1; H_2)$ (*resp.* $\Upsilon_{p,p}(H; H)$) by $\Upsilon_p(H_1; H_2)$ (*resp.* $\Upsilon_p(H)$).

Remark 3.16. Note that, for any $J(\cdot,\cdot) \in \Upsilon_{p,q}(H_1; H_2)$, one does not need to have $J(\cdot,\cdot) \in L^\infty_{\mathbb{F}}(\Omega; L^p(0,T; \mathcal{L}(H_1; H_2)))$. Nevertheless, in some sense $\Upsilon_{p,q}(H_1; H_2)$ is a nice "replacement" of the space $L^\infty_{\mathbb{F}}(\Omega; L^q(0,T; \mathcal{L}(H_1; H_2)))$.

Also, we introduce the following assumption:

Condition 3.2 *There is a constant $\mathcal{C}_p > 0$ such that the following Burkholder-Davis-Gundy type inequality holds (for "$\int_0^\cdot S(\cdot - s)g(s)dW(s)$"):*

$$\begin{aligned}\mathbb{E}\Big(\sup_{t\in[0,T]} \Big| \int_0^t S(t-s)g(s)dW(s) \Big|_H^p \Big) \leq \mathcal{C}_p |g|^p_{L^p_{\mathbb{F}}(\Omega; L^2(0,T; \mathcal{L}_2^0))}, \\ \forall\, g \in L^p_{\mathbb{F}}(\Omega; L^2(0,T; \mathcal{L}_2^0)).\end{aligned} \tag{3.51}$$

Remark 3.17. For any two stopping times σ and τ with $0 \le \sigma \le \tau \le T$ a.s., similarly to $L^p_{\mathbb{F}}(\Omega; C([0,T]; H))$ and $L^p_{\mathbb{F}}(\Omega; L^q(0,T;H))$, one can define the Banach spaces $L^p_{\mathbb{F}}(\Omega; C([\sigma,\tau]; H))$ and $L^p_{\mathbb{F}}(\Omega; L^q(\sigma,\tau;H))$. Under Condition 3.2, one can show that

$$\mathbb{E}\Big(\sup_{t\in[\sigma,\tau]}\Big|\int_\sigma^t S(t-s)g(s)dW(s)\Big|_H^p\Big) \le \mathcal{C}_p|g|^p_{L^p_{\mathbb{F}}(\Omega;L^2(\sigma,\tau;\mathcal{L}_2^0))}, \qquad \forall\, g\in L^p_{\mathbb{F}}(\Omega;L^2(\sigma,\tau;\mathcal{L}_2^0)). \tag{3.52}$$

We believe that Condition 3.2 holds for any C_0-semigroup $S(\cdot)$ and $p \ge 1$. However, as far as we know, this is a longstanding unsolved problem. So far, the best known result is as follows:

Theorem 3.18. *Condition 3.2 holds whenever one of the following conditions is satisfied.*

1) *$S(\cdot)$ is a C_0-group on H;*

2) *$S(\cdot)$ is a generalized contractive C_0-semigroup, i.e., $|S(s)|_{\mathcal{L}(H)} \le e^{cs}$ for some constant $c \in \mathbb{R}$ and any $s \ge 0$.*

Proof: The first case is obvious.

We now consider the second case. Without loss of generality, we assume $c = 0$, i.e., $S(\cdot)$ is a contractive semigroup. For this case, by Sz.-Nagy's theorem on unitary dilations (e.g. [312, p. 29, Theorem 8.1 in Chapter I]), there exist a Hilbert space $\mathcal{H}$ and a unitary C_0-group $\{U(t)\}_{t\in\mathbb{R}}$ on $\mathcal{H}$ such that H embeds isometrically into $\mathcal{H}$ and $\Gamma U(t) = S(t)$ on H for all $t \ge 0$, where Γ is the orthogonal projection from $\mathcal{H}$ onto H. Therefore, taking any $g \in L^p_{\mathbb{F}}(\Omega; L^2(0,T;\mathcal{L}_2^0))$, we get

$$\begin{aligned}
&\mathbb{E}\sup_{t\in[0,T]}\Big|\int_0^t S(t-s)g(s)dW(s)\Big|_H^p = \mathbb{E}\sup_{t\in[0,T]}\Big|\int_0^t \Gamma U(t-s)g(s)dW(s)\Big|_{\mathcal{H}}^p\\
&= \mathbb{E}\sup_{t\in[0,T]}\Big|\Gamma U(t)\int_0^t U(-s)g(s)dW(s)\Big|_{\mathcal{H}}^p \le \mathbb{E}\sup_{t\in[0,T]}\Big|\int_0^t U(-s)g(s)dW(s)\Big|_{\mathcal{H}}^p\\
&\le \mathcal{C}_p\mathbb{E}\Big(\int_0^T |U(-s)g(s)|^2_{\mathcal{L}_2(V;\mathcal{H})}\,ds\Big)^{p/2} \le \mathcal{C}_p\mathbb{E}\Big(\int_0^T |g(s)|^2_{\mathcal{L}_2(V;\mathcal{H})}ds\Big)^{p/2}\\
&= \mathcal{C}_p\mathbb{E}\Big(\int_0^T |g(s)|^2_{\mathcal{L}_2^0}ds\Big)^{p/2},
\end{aligned}$$

where we have used the fact that $\mathcal{L}_2(V;\mathcal{H})$- and $(\mathcal{L}_2(V;H) =)\mathcal{L}_2^0$-norms of $g(\cdot)$ coincide, because $g(\cdot)$ is $\mathcal{L}_2^0$-valued. □

Remark 3.19. It is easy to see that, when H is finite dimensional, $S(\cdot)$ is always a generalized contractive C_0-semigroup. Hence, Condition 3.2 is always satisfied when $\dim H < \infty$.

We have the following time-regularity results for mild solutions to the equation (3.32).

Theorem 3.20. *Let Condition 3.2 hold, $X_0 \in L^p_{\mathcal{F}_0}(\Omega; H)$, $\mathcal{A} \in \Upsilon_1(H)$, $\mathcal{B} \in \Upsilon_2(H; \mathcal{L}_2^0)$, $f(\cdot) \in L^p_{\mathbb{F}}(\Omega; L^1(0,T;H))$ and $g(\cdot) \in L^p(\Omega; L^2(0,T;\mathcal{L}_2^0))$. Then, the equation (3.32) admits a unique mild solution $X(\cdot) \in L^p_{\mathbb{F}}(\Omega; C([0,T];H))$. Moreover,*

$$
\begin{aligned}
&|X(\cdot)|_{L^p_{\mathbb{F}}(\Omega;C([0,T];H))}\\
&\leq \mathcal{C}\big(|X_0|_{L^p_{\mathcal{F}_0}(\Omega;H)} + |f|_{L^p_{\mathbb{F}}(\Omega;L^1(0,T;H))} + |g|_{L^p_{\mathbb{F}}(\Omega;L^2(0,T;\mathcal{L}_2^0))}\big).
\end{aligned}
\tag{3.53}
$$

Proof: We first consider the case that $\mathcal{A} = 0$ and $\mathcal{B} = 0$. For such a case, the mild solution $X(\cdot)$ to (3.32) is still given by (3.34). Using Condition 3.2 (instead of Proposition 3.12), similarly to (3.35)–(3.36), we have

$$
\begin{aligned}
&|X(\cdot)|_{L^p_{\mathbb{F}}(\Omega;L^\infty(0,T;H))}\\
&\leq \mathcal{C}\big(|X_0|_{L^p_{\mathcal{F}_0}(\Omega;H)} + |f|_{L^p_{\mathbb{F}}(\Omega;L^1(0,T;H))} + |g|_{L^p_{\mathbb{F}}(\Omega;L^2(0,T;\mathcal{L}_2^0))}\big).
\end{aligned}
\tag{3.54}
$$

By Proposition 3.5, it is easy to see that $X(\cdot) \in L^p_{\mathbb{F}}(\Omega; C([0,T];H))$.

Now we handle the general case. For any $a \in \mathbb{R}$, denote by $\lceil a \rceil$ the biggest integer less than or equal to a. Write

$$
N = \Big\lceil \frac{1}{\varepsilon}\Big(\big||\mathcal{A}|_{\mathcal{L}(H)}\big|^p_{L^\infty_{\mathbb{F}}(\Omega;L^1(0,T))} + \big||\mathcal{B}|_{\mathcal{L}(H;\mathcal{L}_2^0)}\big|^p_{L^\infty_{\mathbb{F}}(\Omega;L^2(0,T))}\Big)\Big\rceil + 1,
$$

where $\varepsilon > 0$ is a constant to be determined later. Define a sequence of stopping times $\{\tau_{j,\varepsilon}\}_{j=1}^N$ as follows:

$$
\left\{
\begin{aligned}
&\tau_{1,\varepsilon}(\omega) = \inf\Big\{t \in [0,T] \;\Big|\; \Big(\int_0^t |\mathcal{A}(s,\omega)|_{\mathcal{L}(H)} ds\Big)^p\\
&\qquad\qquad\qquad + \Big(\int_0^t |\mathcal{B}(s,\omega)|^2_{\mathcal{L}(H;\mathcal{L}_2^0)} ds\Big)^{\frac{p}{2}} = \varepsilon\Big\},\\
&\tau_{k,\varepsilon}(\omega) = \inf\Big\{t \in [\tau_{k-1,\varepsilon},T] \;\Big|\; \Big(\int_{\tau_{k-1,\varepsilon}}^t |\mathcal{A}(s,\omega)|_{\mathcal{L}(H)} ds\Big)^p\\
&\qquad\qquad\qquad + \Big(\int_{\tau_{k-1,\varepsilon}}^t |\mathcal{B}(s,\omega)|^2_{\mathcal{L}(H;\mathcal{L}_2^0)} ds\Big)^{\frac{p}{2}} = \varepsilon\Big\},\\
&\qquad\qquad\qquad k = 2, \cdots, N.
\end{aligned}
\right.
$$

Here, we agree that $\inf\emptyset = T$.

Consider the following stochastic evolution equation:

$$
\left\{
\begin{aligned}
&dX = (AX + \hat{f})dt + \hat{g}dW(t) \quad \text{in } (0,T],\\
&X(0) = X_0.
\end{aligned}
\right.
\tag{3.55}
$$

Here $\hat{f} \in L^p_{\mathbb{F}}(\Omega; L^1(0,T;H))$ and $\hat{g} \in L^p_{\mathbb{F}}(\Omega; L^2(0,T;\mathcal{L}_2^0))$. Clearly, (3.55) admits a unique mild solution $X \in L^p_{\mathbb{F}}(\Omega; C([0,T];H))$. Define a map $\mathcal{J} : L^p_{\mathbb{F}}(\Omega; C([0,\tau_{1,\varepsilon}];H)) \to L^p_{\mathbb{F}}(\Omega; C([0,\tau_{1,\varepsilon}];H))$ as follows:

$$L^p_{\mathbb{F}}(\Omega; C([0,\tau_{1,\varepsilon}];H)) \ni \widetilde{X} \mapsto X = \mathcal{J}(\widetilde{X}),$$

where X is the solution to (3.55) with $\hat{f}$ and $\hat{g}$ being replaced by $\mathcal{A}\widetilde{X} + f$ and $\mathcal{B}\widetilde{X} + g$, respectively. We claim that $\mathcal{J}$ is contractive. Indeed, by means of (3.52), for any $\widetilde{X}_1, \widetilde{X}_2 \in L^p_{\mathbb{F}}(\Omega; C([0,\tau_{1,\varepsilon}];H))$,

$$\begin{aligned}
&|\mathcal{J}(\widetilde{X}_1) - \mathcal{J}(\widetilde{X}_2)|^p_{L^p_{\mathbb{F}}(\Omega;C([0,\tau_{1,\varepsilon}];H))}\\
&\leq \mathbb{E}\Big(\sup_{t\in[0,\tau_{1,\varepsilon}]}\Big|\int_0^t S(t-r)\mathcal{A}(\widetilde{X}_1-\widetilde{X}_2)dr+\int_0^t S(t-r)\mathcal{B}(\widetilde{X}_1-\widetilde{X}_2)dW(r)\Big|_H\Big)^p\\
&\leq \mathcal{C}\mathbb{E}\Big(\sup_{t\in[0,\tau_{1,\varepsilon}]}\Big|\int_0^t S(t-r)\mathcal{A}(\widetilde{X}_1-\widetilde{X}_2)dr\Big|_H\Big)^p\\
&\quad+\mathcal{C}\mathbb{E}\sup_{t\in[0,\tau_{1,\varepsilon}]}\Big(\Big|\int_0^t S(t-r)\mathcal{B}(\widetilde{X}_1-\widetilde{X}_2)dW(r)\Big|_H\Big)^p\\
&\leq \mathcal{C}\Big(\sup_{t\in[0,T]}|S(t)|^p_{\mathcal{L}(H)}\Big)\big||\mathcal{A}|_{\mathcal{L}(H)}\big|^p_{L^\infty_{\mathbb{F}}(\Omega;L^1(0,\tau_{1,\varepsilon}))}|\widetilde{X}_1-\widetilde{X}_2|^p_{L^p_{\mathbb{F}}(\Omega;C([0,\tau_{1,\varepsilon}];H))}\\
&\quad+\mathcal{C}\Big(\sup_{t\in[0,T]}|S(t)|^p_{\mathcal{L}(H)}\Big)\big||\mathcal{B}|_{\mathcal{L}(H;\mathcal{L}_2^0)}\big|^p_{L^\infty_{\mathbb{F}}(\Omega;L^2(0,\tau_{1,\varepsilon}))}|\widetilde{X}_1-\widetilde{X}_2|^p_{L^p_{\mathbb{F}}(\Omega;C([0,\tau_{1,\varepsilon}];H))}.
\end{aligned}\tag{3.56}$$

Let us choose $\varepsilon = \frac{1}{4\mathcal{C}\sup_{t\in[0,T]}|S(t)|^p_{\mathcal{L}(H)}}$. Then, from (3.56), we find that

$$|\mathcal{J}(\widetilde{X}_1) - \mathcal{J}(\widetilde{X}_2)|^p_{L^p_{\mathbb{F}}(\Omega;C([0,\tau_{1,\varepsilon}];H))} \leq \frac{1}{2}|\widetilde{X}_1-\widetilde{X}_2|^p_{L^p_{\mathbb{F}}(\Omega;C([0,\tau_{1,\varepsilon}];H))}. \tag{3.57}$$

This implies that $\mathcal{J}$ is contractive. Thus, it has a unique fixed point $X \in L^p_{\mathbb{F}}(\Omega; C([0,\tau_{1,\varepsilon}];H))$, which solves (3.32) in $[0,\tau_{1,\varepsilon}]$ (in the sense of mild solution). Inductively, we conclude that (3.32) admits a mild solution X in $[\tau_{k-1,\varepsilon}, \tau_{k,\varepsilon}]$ for $k = 2, \cdots, N$. Furthermore,

$$
\begin{aligned}
&|X|^p_{L^p_{\mathbb{F}}(\Omega;C([0,\tau_{1,\varepsilon}];H))}\\
&\leq \mathbb{E}\sup_{t\in[0,\tau_{1,\varepsilon}]}\Big(\Big|S(t)X_0+\int_0^t S(t-r)(\mathcal{A}X+f)dr\\
&\qquad\qquad +\int_0^t S(t-r)(\mathcal{B}X+g)dW(r)\Big|_H\Big)^p\\
&\leq \mathcal{C}\mathbb{E}\sup_{t\in[0,\tau_{1,\varepsilon}]}\Big(\Big|\int_0^t S(t-r)\mathcal{A}Xdr\Big|_H\Big)^p\\
&\quad +\mathcal{C}\mathbb{E}\sup_{t\in[0,\tau_{1,\varepsilon}]}\Big(\Big|\int_0^t S(t-r)\mathcal{B}xdW(r)\Big|_H\Big)^p\\
&\quad +\mathcal{C}\mathbb{E}\sup_{t\in[0,T]}\Big(\Big|S(t)X_0+\int_0^t S(t-r)fdr+\int_0^t S(t-r)gdW(r)\Big|_H\Big)^p\\
&\leq \mathcal{C}\Big(\sup_{t\in[0,T]}|S(t)|^p_{\mathcal{L}(H)}\Big)\big||\mathcal{A}|_{\mathcal{L}(H)}\big|^p_{L^\infty_{\mathbb{F}}(\Omega;L^1(0,\tau_{1,\varepsilon}))}|X|^p_{L^p_{\mathbb{F}}(\Omega;C([0,\tau_{1,\varepsilon}];H))}\\
&\quad +\mathcal{C}\Big(\sup_{t\in[0,T]}|S(t)|^p_{\mathcal{L}(H)}\Big)\big||\mathcal{B}|_{\mathcal{L}(H;\mathcal{L}_2^0)}\big|^p_{L^\infty_{\mathbb{F}}(\Omega;L^2(0,\tau_{1,\varepsilon}))}|X|^p_{L^p_{\mathbb{F}}(\Omega;C([0,\tau_{1,\varepsilon}];H))}\\
&\quad +\mathcal{C}\big(\mathbb{E}|X_0|^p_H+|f|^p_{L^p_{\mathbb{F}}(\Omega;L^1(0,T;H))}+|g|^p_{L^p_{\mathbb{F}}(\Omega;L^2(0,T;\mathcal{L}_2^0))}\big).
\end{aligned}
\tag{3.58}
$$

This, together with the choice of $\tau_{1,\varepsilon}$, implies that

$$
\begin{aligned}
&|X|_{L^p_{\mathbb{F}}(\Omega;C([0,\tau_{1,\varepsilon}];H))}\\
&\leq \mathcal{C}\big(|X_0|_{L^p_{\mathcal{F}_0}(\Omega;H)}+|f|_{L^p_{\mathbb{F}}(\Omega;L^1(0,T;H))}+|g|_{L^p_{\mathbb{F}}(\Omega;L^2(0,T;\mathcal{L}_2^0))}\big).
\end{aligned}
\tag{3.59}
$$

Repeating the above argument, we obtain (3.53). The uniqueness of solutions is obvious. □

Theorem 3.21. *Let Conditions 3.1 and 3.2 hold,* $F(\cdot,0)\in L^p_{\mathbb{F}}(\Omega;L^1(0,T;H))$ *and* $\widetilde{F}(\cdot,0)\in L^p_{\mathbb{F}}(\Omega;L^2(0,T;\mathcal{L}_2^0))$*. Then, for any* $X_0\in L^p_{\mathcal{F}_0}(\Omega;H)$*, the equation* (3.16) *admits a unique mild solution* $X(\cdot)\in L^p_{\mathbb{F}}(\Omega;C([0,T];H))$*, and*

$$
\begin{aligned}
&|X(\cdot)|_{L^p_{\mathbb{F}}(\Omega;C([0,T];H))}\\
&\leq \mathcal{C}\big(|X_0|_{L^p_{\mathcal{F}_0}(\Omega;H)}+|F(\cdot,0)|_{L^p_{\mathbb{F}}(\Omega;L^1(0,T;H))}+|\widetilde{F}(\cdot,0)|_{L^p_{\mathbb{F}}(\Omega;L^2(0,T;\mathcal{L}_2^0))}\big).
\end{aligned}
\tag{3.60}
$$

Proof: The proof is very similarly to that of Theorem 3.14.

Fix any $T_1\in[0,T]$. Similar to the proof of Theorem 3.14, we define a map $\mathcal{J}:L^p_{\mathbb{F}}(\Omega;C([0,T_1];H))\to L^p_{\mathbb{F}}(\Omega;C([0,T_1];H))$ by $\mathcal{J}(Y)=X$ for any $Y\in L^p_{\mathbb{F}}(\Omega;C([0,T_1];H))$, where X solves the equation (3.41). To show that the map $\mathcal{J}$ is contractive for T_1 small enough, we choose arbitrarily another $\widehat{Y}\in L^p_{\mathbb{F}}(\Omega;C([0,T_1];H))$, and define $\widehat{X}=\mathcal{J}(\widehat{Y})$. Write $\widetilde{X}(\cdot)$, $F_1(\cdot)$ and $\widetilde{F}_1(\cdot)$ as that in the proof of Theorem 3.14. Then $\widetilde{X}(\cdot)$ solves (3.44). By Conditions 3.1–3.2 and $p\geq 1$, similarly to (3.45)–(3.46), we see that

$$|F_1(\cdot)|_{L^p_{\mathbb{F}}(\Omega;L^1(0,T_1;H))} \leq \mathcal{C}|L_1(\cdot)|_{L^1(0,T_1)}|Y(\cdot)-\widehat{Y}(\cdot)|_{L^p_{\mathbb{F}}(\Omega;C([0,T_1];H))} \tag{3.61}$$

and

$$|\widetilde{F}_1(\cdot)|_{L^p_{\mathbb{F}}(\Omega;L^2(0,T_1;\mathcal{L}_2^0))} \leq \mathcal{C}|L_2(\cdot)|_{L^2(0,T_1)}|Y(\cdot)-\widehat{Y}(\cdot)|_{L^p_{\mathbb{F}}(\Omega;C([0,T_1];H))}. \tag{3.62}$$

Applying the estimate (3.54) (for the linear problem) to the equation (3.44) (for $\widetilde{X}(\cdot)$) and noting (3.61)–(3.62), we find that

$$\begin{aligned}
&|\widetilde{X}(\cdot)|_{L^p_{\mathbb{F}}(\Omega;C([0,T_1];H))}\\
&\leq \mathcal{C}\big(|F_1(\cdot)|_{L^p_{\mathbb{F}}(\Omega;L^1(0,T_1;H))}+|\widetilde{F}_1(\cdot)|_{L^p_{\mathbb{F}}(\Omega;L^2(0,T_1;\mathcal{L}_2^0))}\big)\\
&\leq \mathcal{C}\big(|L_1(\cdot)|_{L^1(0,T_1)}+|L_2(\cdot)|_{L^2(0,T_1)}\big)|Y(\cdot)-\widehat{Y}(\cdot)|_{L^p_{\mathbb{F}}(\Omega;C([0,T_1];H))}.
\end{aligned} \tag{3.63}$$

Choose T_1 so that $\mathcal{C}\big(|L_1(\cdot)|_{L^1(0,T_1)}+|L_2(\cdot)|_{L^2(0,T_1)}\big)<1$. Then, $\mathcal{J}$ is contractive. By means of the Banach fixed point theorem, $\mathcal{J}$ enjoys a unique fixed point $X(\cdot)\in L^p_{\mathbb{F}}(\Omega;C([0,T_1];H))$. Clearly, $X(\cdot)$ is a mild solution to the following equation:

$$\begin{cases}
dX(t)=\big(AX(t)+F(t,X(t))\big)dt+\widetilde{F}(t,X(t))dW(t) & \text{in } (0,T_1],\\
X(0)=X_0.
\end{cases} \tag{3.64}$$

Using again Condition 3.1 and by $X(\cdot)\in L^p_{\mathbb{F}}(\Omega;C([0,T_1];H))$, similarly to (3.61)–(3.62), we see that $F(\cdot,X(\cdot))\in L^p_{\mathbb{F}}(\Omega;L^1(0,T_1;H))$, $\widetilde{F}(\cdot,X(\cdot))\in L^p_{\mathbb{F}}(\Omega;L^2(0,T_1;\mathcal{L}_2^0))$, and

$$\begin{aligned}
&|F(\cdot,X(\cdot))|_{L^p_{\mathbb{F}}(\Omega;L^1(0,T_1;H))}+|\widetilde{F}(\cdot,X(\cdot))|_{L^p_{\mathbb{F}}(\Omega;L^2(0,T_1;\mathcal{L}_2^0))}\\
&\leq |F(\cdot,0)|_{L^p_{\mathbb{F}}(\Omega;L^1(0,T_1;H))}+|\widetilde{F}(\cdot,0)|_{L^p_{\mathbb{F}}(\Omega;L^2(0,T_1;\mathcal{L}_2^0))}\\
&\quad+\mathcal{C}\big(|L_1(\cdot)|_{L^1(0,T_1)}+|L_2(\cdot)|_{L^2(0,T_1)}\big)|X(\cdot)|_{L^p_{\mathbb{F}}(\Omega;C([0,T_1];H))}.
\end{aligned}$$

Therefore, similarly to (3.54), we have

$$\begin{aligned}
&|X(\cdot)|_{L^p_{\mathbb{F}}(\Omega;C([0,T_1];H))}\\
&\leq \mathcal{C}\Big(|X_0|_{L^p_{\mathcal{F}_0}(\Omega;H)}+|F(\cdot,X(\cdot))|_{L^p_{\mathbb{F}}(\Omega;L^1(0,T_1;H))}+|\widetilde{F}(\cdot,X(\cdot))|_{L^p_{\mathbb{F}}(\Omega;L^2(0,T_1;\mathcal{L}_2^0))}\Big)\\
&\leq \mathcal{C}\Big[|X_0|_{L^p_{\mathcal{F}_0}(\Omega;H)}+|F(\cdot,0)|_{L^p_{\mathbb{F}}(\Omega;L^1(0,T_1;H))}+|\widetilde{F}(\cdot,0)|_{L^p_{\mathbb{F}}(\Omega;L^2(0,T_1;\mathcal{L}_2^0))}\\
&\quad+\big(|L_1(\cdot)|_{L^1(0,T_1)}+|L_2(\cdot)|_{L^2(0,T_1)}\big)|X(\cdot)|_{L^p_{\mathbb{F}}(\Omega;C([0,T_1];H))}\Big].
\end{aligned} \tag{3.65}$$

Since $\mathcal{C}\big(|L_1(\cdot)|_{L^1(0,T_1)}+|L_2(\cdot)|_{L^2(0,T_1)}\big)<1$, it follows from (3.65) that

$$\begin{aligned}
&|X(\cdot)|_{L^p_{\mathbb{F}}(\Omega;C([0,T_1];H))}\\
&\leq \mathcal{C}\big(|X_0|_{L^p_{\mathcal{F}_0}(\Omega;H)}+|F(\cdot,0)|_{L^p_{\mathbb{F}}(\Omega;L^1(0,T_1;H))}+|\widetilde{F}(\cdot,0)|_{L^p_{\mathbb{F}}(\Omega;L^2(0,T_1;\mathcal{L}_2^0))}\big).
\end{aligned} \tag{3.66}$$

Repeating the above argument, we obtain a mild solution to (3.16). The uniqueness of mild solutions to (3.16) is obvious. The desired estimate (3.60) follows from (3.66). This completes the proof of Theorem 3.21. □

3.3.2 Space Regularity

Sometimes we will need the space regularity of mild/weak solutions to (3.16).

Generally speaking, the stochastic convolution $\int_0^t S(t-s)\widetilde{F}(s,X(s))dW(s)$ is no longer a martingale. Hence, one cannot apply Itô's formula directly to mild solutions to (3.16). For example, when establishing the pointwise identities (for Carleman estimates) for stochastic partial differentia equations of second order, we need the function to be second differentiable in the sense of weak derivative w.r.t the space variables. Nevertheless, this sort of problems can be solved by the following strategy:

1. Introduce suitable approximating equations with strong solutions such that the limit of these strong solutions is the mild/weak solution to the original equation;
2. Obtain the desired properties for the above strong solutions;
3. Utilize the density argument to establish the desired properties for the mild/weak solutions.

There are many methods to implement the above three steps in the setting of deterministic partial differential equations. Roughly speaking, any of these methods, which does not destroy the adaptedness of solutions, can be applied to stochastic partial differential equations. Here we only present an approach based on Lemma 3.11. To this end, we introduce the following approximating equations of (3.16):

$$\begin{cases} dX_\lambda(t) = \big(AX_\lambda(t) + R(\lambda)F(t,X_\lambda(t))\big)dt + R(\lambda)\widetilde{F}(t,X_\lambda(t))dW(t) & \text{in } (0,T],\\ X_\lambda(0) = R(\lambda)X_0. \end{cases} \tag{3.67}$$

Here, λ belongs to $\rho(A)$, the resolvent set of A, and $R(\lambda) \stackrel{\Delta}{=} \lambda(\lambda I - A)^{-1}$.

We have the following result.

Theorem 3.22. *Let Condition 3.1 hold and $p \geq 2$ (resp. Conditions 3.1 and 3.2 hold for some $p \geq 1$), $F(\cdot,0) \in L^p_{\mathbb{F}}(\Omega; L^1(0,T;H))$ and $\widetilde{F}(\cdot,0) \in L^p_{\mathbb{F}}(\Omega; L^2(0,T;\mathcal{L}_2^0))$. Then, for each $X_0 \in L^p_{\mathcal{F}_0}(\Omega;H)$ and $\lambda \in \rho(A)$, the equation (3.67) admits a unique strong solution $X_\lambda(\cdot) \in C_{\mathbb{F}}([0,T];L^p(\Omega;D(A)))$ (resp. $X_\lambda(\cdot) \in L^p_{\mathbb{F}}(\Omega;C([0,T];D(A)))$). Moreover, as $\lambda \to \infty$, the solution $X_\lambda(\cdot)$ converges to $X(\cdot)$ in $C_{\mathbb{F}}([0,T];L^p(\Omega;H))$ (resp. $L^p_{\mathbb{F}}(\Omega;C([0,T];H))$), where $X(\cdot)$ solves (3.16) in the sense of the mild solution.*

Proof: We only consider the case that Condition 3.1 holds and $p \geq 2$ (Another case can be considered similarly).

Since $AR(\lambda) = \lambda^2(\lambda I - A)^{-1} - \lambda I$ is a bounded operator, by Lemma 3.11, we obtain the existence of unique strong solution $X_\lambda(t)$ to the equation (3.67).

Now we prove the convergence conclusion in Theorem 3.22. For any $t \geq 0$,

$$\begin{aligned}
&X(t) - X_\lambda(t) \\
&= S(t)\big(X_0 - R(\lambda)X_0\big) + \int_0^t S(t-s)\big(F(s,X(s)) - R(\lambda)F(s,X_\lambda(s))\big)ds \\
&\quad + \int_0^t S(t-s)\big(\widetilde{F}(s,X(s)) - R(\lambda)\widetilde{F}(s,X_\lambda(s))\big)dW(s).
\end{aligned}$$

For any $r \in (0,T]$, we have that,

$$\sup_{t\in[0,r]} \mathbb{E}\big|X(t) - X_\lambda(t)\big|_H^p \leq 3^p(I_1 + I_2 + I_3),$$

where

$$\begin{aligned}
I_1 &= \sup_{t\in[0,r]} \mathbb{E}\Big|\int_0^t S(t-s)R(\lambda)\big(F(s,X(s)) - F(s,X_\lambda(s))\big)ds\Big|_H^p, \\
I_2 &= \sup_{t\in[0,r]} \mathbb{E}\Big|\int_0^t S(t-s)R(\lambda)\big(\widetilde{F}(s,X(s)) - \widetilde{F}(s,X_\lambda(s))\big)dW(s)\Big|_H^p, \\
I_3 &= \sup_{t\in[0,r]} \mathbb{E}\Big|S(t)\big(X_0 - R(\lambda)X_0\big) + \int_0^t S(t-s)\big(I - R(\lambda)\big)F(s,X(s))ds \\
&\qquad + \int_0^t S(t-s)\big(I - R(\lambda)\big)\widetilde{F}(s,X(s))dW(s)\Big|_H^p.
\end{aligned}$$

Since $|R(\lambda)|_{\mathcal{L}(H)} \leq 2$ for $\lambda > 0$ large enough, by (2.30) and Condition 3.1, similarly to (3.45), it follows that

$$\begin{aligned}
I_1 &\leq \sup_{t\in[0,r]} \mathbb{E}\Big[\int_0^t \big|S(t-s)R(\lambda)\big(F(s,X(s)) - F(s,X_\lambda(s))\big)\big|_H ds\Big]^p \\
&\leq \mathcal{C} \sup_{t\in[0,r]} \mathbb{E}\Big(\int_0^t \big|F(s,X(s)) - F(s,X_\lambda(s))\big|_H ds\Big)^p \\
&\leq \mathcal{C} \sup_{t\in[0,r]} \Big[\int_0^t \Big(\mathbb{E}\big|F(x,X(s)) - F(s,X_\lambda(s))\big|_H^p\Big)^{1/p} ds\Big]^p \\
&\leq \mathcal{C}|L_1(\cdot)|_{L^1(0,r)}^p \sup_{t\in[0,r]} \mathbb{E}\big|X(t) - X_\lambda(t)\big|_H^p.
\end{aligned}$$

Further, by means of the Burkholder-Davis-Gundy inequality (i.e., Proposition 3.12), similarly to (3.46), we have that,

$$I_2 \leq \mathcal{C}|L_2(\cdot)|_{L^2(0,r)}^p \sup_{t\in[0,r]} \mathbb{E}\big|X(t) - X_\lambda(t)\big|_H^p.$$

Hence,

$$\sup_{t\in[0,r]} \mathbb{E}\big|X(t) - X_\lambda(t)\big|_H^p$$
$$\leq \mathcal{C}\Big[\big(|L_1(\cdot)|^p_{L^1(0,r)} + |L_2(\cdot)|^p_{L^2(0,r)}\big) \sup_{t\in[0,r]} \mathbb{E}\big|X(t) - X_\lambda(t)\big|_H^p + I_3\Big].$$

Choose r so that $\mathcal{C}\big(|L_1(\cdot)|^p_{L^1(0,r)} + |L_2(\cdot)|^p_{L^2(0,r)}\big) < 1$. Then,

$$\sup_{t\in[0,r]} \mathbb{E}\big|X(t) - X_\lambda(t)\big|_H^p \leq \mathcal{C} I_3. \tag{3.68}$$

On the other hand,

$$\begin{aligned} I_3 \leq 3^p \Big[&\sup_{t\in[0,r]} \mathbb{E}\big|S(t)(X_0 - R(\lambda)X_0)\big|_H^p \\ &+ \sup_{t\in[0,r]} \mathbb{E}\Big|\int_0^t S(t-s)\big(I - R(\lambda)\big)F(s,X(s))ds\Big|_H^p \\ &+ \sup_{t\in[0,r]} \mathbb{E}\Big|\int_0^t S(t-s)\big(I - R(\lambda)\big)\widetilde{F}(s,X(s))dW(s)\Big|_H^p\Big]. \end{aligned} \tag{3.69}$$

We now estimate each term in (3.69). First,

$$\sup_{t\in[0,r]} \mathbb{E}\big|S(t)(X_0 - R(\lambda)X_0)\big|_H^p \leq \mathcal{C}\mathbb{E}\big|X_0 - R(\lambda)X_0\big|_H^p \to 0, \qquad \text{as } \lambda \to \infty.$$

Next, by Condition 3.1 again and Theorem 3.14, we obtain that

$$\begin{aligned} &\sup_{t\in[0,r]} \mathbb{E}\Big|\int_0^t S(t-s)\big(I - R(\lambda)\big)F(s,X(s))ds\Big|_H^p \\ &\leq \mathcal{C}\mathbb{E}\Big|\int_0^r \big|\big(I - R(\lambda)\big)F(s,X(s))\big|_H ds\Big|^p \to 0, \qquad \text{as } \lambda \to \infty. \end{aligned}$$

Similarly, by means of Proposition 3.12 again, we have that

$$\begin{aligned} &\sup_{t\in[0,r]} \mathbb{E}\Big|\int_0^t S(t-s)\big(I - R(\lambda)\big)\widetilde{F}(x,X(s))dW(s)\Big|_H^p \\ &\leq \mathcal{C}\mathbb{E}\Big|\int_0^r \big|\big(I - R(\lambda)\big)\widetilde{F}(s,X(s))\big|^2_{\mathcal{L}_2^0} ds\Big|^{p/2} \to 0, \qquad \text{as } \lambda \to \infty. \end{aligned}$$

Hence, by (3.68), it follows that

$$\sup_{t\in[0,r]} \mathbb{E}\big|X(t) - X_\lambda(t)\big|_H^p \leq o(1),$$

where $o(1) \to 0$ as $\lambda \to \infty$. Repeating the above argument, we deduce that

$$\sup_{t\in[0,T]} \mathbb{E}\big|X(t)-X_\lambda(t)\big|_H^p \le o(1)\to 0, \qquad \text{as } \lambda\to\infty,$$

which gives the desired convergence. □

Likewise, for any $\lambda \in \rho(A)$, we introduce the following approximating equation of (3.32) with $\mathcal{A}=0$ and $\mathcal{B}=0$:

$$\begin{cases} dX_\lambda(t) = \big(AX_\lambda(t) + R(\lambda)f(t)\big)dt + R(\lambda)\tilde f(t)dW(t) & \text{in } (0,T], \\ X_\lambda(0) = R(\lambda)X_0. \end{cases} \tag{3.70}$$

By Theorem 3.13, from the proof of Theorem 3.22, it is easy to show the following result.

Corollary 3.23. *For each $p\ge 1$, $X_0 \in L^p_{\mathcal{F}_0}(\Omega;H)$, $f(\cdot)\in L^p_{\mathbb{F}}(\Omega;L^1(0,T;H))$ and $\tilde f(\cdot)\in L^p(\Omega;L^2(0,T;\mathcal{L}_2^0))$, the equation (3.70) admits a unique strong solution $X_\lambda(\cdot)\in C_{\mathbb{F}}([0,T];L^p(\Omega;D(A)))$. Moreover, as $\lambda\to\infty$, the solution $X_\lambda(\cdot)$ converges to $X(\cdot)$ in $C_{\mathbb{F}}([0,T];L^p(\Omega;H))$, where $X(\cdot)$ solves (3.32) with $\mathcal{A}=0$ and $\mathcal{B}=0$ in the sense of mild solution.*

The following result shows the space smoothing effect of mild solutions to a class of stochastic evolutions equations, for example, the stochastic parabolic equation.

Theorem 3.24. *Let Condition 3.1 hold, $F(\cdot,0)\in L^p_{\mathbb{F}}(\Omega;L^1(0,T;H))$ and $\widetilde F(\cdot,0)\in L^p_{\mathbb{F}}(\Omega;L^2(0,T;\mathcal{L}_2^0))$ for some $p\ge 1$, and A be a self-adjoint, negative definite (unbounded linear) operator on H. Then for any $X_0\in L^p_{\mathcal{F}_0}(\Omega;H)$, the equation (3.16) admits a unique mild solution $X(\cdot)\in L^p_{\mathbb{F}}(\Omega;C([0,T];H))\cap L^p_{\mathbb{F}}(\Omega;L^2(0,T;D((-A)^{\frac12})))$. Moreover,*

$$\begin{aligned} &|X(\cdot)|_{L^p_{\mathbb{F}}(\Omega;C([0,T];H))} + |X(\cdot)|_{L^p_{\mathbb{F}}(\Omega;L^2(0,T;D((-A)^{\frac12})))} \\ &\le \mathcal{C}\big(|X_0|_{L^p_{\mathcal{F}_0}(\Omega;H)} + |F(\cdot,0)|_{L^p_{\mathbb{F}}(\Omega;L^1(0,T;H))} + |\widetilde F(\cdot,0)|_{L^p_{\mathbb{F}}(\Omega;L^2(0,T;\mathcal{L}_2^0))}\big). \end{aligned} \tag{3.71}$$

Proof: Clearly, A generates a contractive C_0-semigroup. Hence, Condition 3.2 hold.

By Theorem 3.22, for each $\lambda\in\rho(A)$, the equation (3.67) admits a unique strong solution $X_\lambda(\cdot)\in L^p_{\mathbb{F}}(\Omega;C([0,T];D(A)))$, and $\lim\limits_{\lambda\to\infty} X_\lambda(\cdot) = X(\cdot)$ in $L^p_{\mathbb{F}}(\Omega;C([0,T];H))$. Moreover,

$$\begin{aligned} &|X_\lambda(\cdot)|_{L^p_{\mathbb{F}}(\Omega;C([0,T];H))} \\ &\le \mathcal{C}\big(|X_0|_{L^p_{\mathcal{F}_0}(\Omega;H)} + |F(\cdot,0)|_{L^p_{\mathbb{F}}(\Omega;L^1(0,T;H))} + |\widetilde F(\cdot,0)|_{L^p_{\mathbb{F}}(\Omega;L^2(0,T;\mathcal{L}_2^0))}\big), \end{aligned} \tag{3.72}$$

for some constant $\mathcal{C}>0$, independent of λ.

Applying Itô's formula to $|X_\lambda(\cdot)|_H^2$ and using (3.67), we obtain that

$$
\begin{aligned}
&|X_\lambda(T)|_H^2 - |X_\lambda(0)|_H^2 \\
&= 2\int_0^T \big\langle AX_\lambda(s), X_\lambda(s)\big\rangle_H ds + 2\int_0^T \big\langle R(\lambda)F(s,X_\lambda(s)), X_\lambda(s)\big\rangle_H ds \\
&\quad +2\int_0^T \langle\!\langle\, R(\lambda)\widetilde F(s,X_\lambda(s)), X_\lambda(s)\,\rangle\!\rangle_H dW(s) \\
&\quad + \int_0^T \big|R(\lambda)\widetilde F(s,X_\lambda(s))\big|_{\mathcal{L}_2^0}^2 ds.
\end{aligned}
\tag{3.73}
$$

By (3.73), we see that

$$
\begin{aligned}
&\mathbb{E}\Big|\int_0^T |(-A)^{\frac12}X_\lambda(s)|_H^2 ds\Big|^{\frac p2} \\
&\le \mathcal{C}\Big[\mathbb{E}|X_\lambda(0)|_H^p + \mathbb{E}\Big|\int_0^T \big\langle R(\lambda)F(s,X_\lambda(s)), X_\lambda(s)\big\rangle_H ds\Big|^{\frac p2} \\
&\quad +\mathbb{E}\Big|\int_0^T \langle\!\langle\, R(\lambda)\widetilde F(s,X_\lambda(s)), X_\lambda(s)\,\rangle\!\rangle_H dW(s)\Big|^{\frac p2} \\
&\quad +\mathbb{E}\Big(\int_0^T \big|R(\lambda)\widetilde F(s,X_\lambda(s))\big|_{\mathcal{L}_2^0}^2 ds\Big)^{\frac p2}\Big].
\end{aligned}
\tag{3.74}
$$

It follows from Burkholder-Davis-Gundy inequality that

$$
\begin{aligned}
&\mathbb{E}\Big|\int_0^T \langle\!\langle\, R(\lambda)\widetilde F(s,X_\lambda(s)), X_\lambda(s)\,\rangle\!\rangle_H dW(s)\Big|^{\frac p2} \\
&\le \mathcal{C}\mathbb{E}\Big(\int_0^T \big|\widetilde F(s,X_\lambda(s))\big|_{\mathcal{L}_2^0}^2 |X_\lambda(s)|_H^2 ds\Big)^{\frac p4} \\
&\le \mathcal{C}\mathbb{E}\Big[\Big(\int_0^T \big|\widetilde F(s,X_\lambda(s))\big|_{\mathcal{L}_2^0}^2 ds\Big)^{\frac p4} \sup_{t\in[0,T]} |X_\lambda(t)|_H^{\frac p2}\Big] \\
&\le \mathcal{C}\mathbb{E}\Big[\Big(\int_0^T \big|\widetilde F(s,X_\lambda(s))\big|_{\mathcal{L}_2^0}^2 ds\Big)^{\frac p2} + \sup_{t\in[0,T]} |X_\lambda(t)|_H^p\Big].
\end{aligned}
\tag{3.75}
$$

By Condition 3.1, it holds that

$$
\begin{aligned}
&\mathbb{E}\Big(\int_0^T \big|\widetilde F(s,X_\lambda(s))\big|_{\mathcal{L}_2^0}^2 ds\Big)^{\frac p2} \\
&\le \mathcal{C}\mathbb{E}\Big[\Big(\int_0^T \big|\widetilde F(s,0)\big|_{\mathcal{L}_2^0}^2 ds\Big)^{\frac p2} + \Big(\int_0^T L_2^2(s)\big|X_\lambda(s)\big|_H^2 ds\Big)^{\frac p2}\Big] \\
&\le \mathcal{C}\Big[|\widetilde F(\cdot,0)|^p_{L^p_{\mathbb{F}}(\Omega;L^2(0,T;\mathcal{L}_2^0))} + |L_2(\cdot)|^p_{L^2(0,T)} \sup_{t\in[0,T]} \mathbb{E}|X_\lambda(t)|_H^p\Big].
\end{aligned}
\tag{3.76}
$$

Combining (3.72), (3.74) and (3.75)–(3.76), we end up with

$$\begin{aligned}&\mathbb{E}\Big|\int_0^T|(-A)^{\frac{1}{2}}X_\lambda(s)|_H^2ds\Big|^{\frac{p}{2}}\\&\leq\mathcal{C}\big(|X_0|_{L^p_{\mathcal{F}_0}(\Omega;H)}+|F(\cdot,0)|_{L^p_{\mathbb{F}}(\Omega;L^1(0,T;H))}+|\widetilde{F}(\cdot,0)|_{L^p_{\mathbb{F}}(\Omega;L^2(0,T;\mathcal{L}_2^0))}\big).\end{aligned}\tag{3.77}$$

Similarly, one can further show that

$$\lim_{\lambda,\mu\to\infty}|(-A)^{\frac{1}{2}}X_\lambda(\cdot)-(-A)^{\frac{1}{2}}X_\mu(\cdot)|_{L^p_{\mathbb{F}}(\Omega;L^2(0,T;H))}=0.\tag{3.78}$$

By (3.78) and noting that $(-A)^{\frac{1}{2}}$ is a closed operator, we conclude that $X(\cdot)\in L^p_{\mathbb{F}}(\Omega;L^2(0,T;D((-A)^{\frac{1}{2}})))$. By (3.77), it is easy to see that the estimate (3.71) holds. □

3.4 Notes and Comments

Since the seminal work in [156], stochastic differential/evolution equations were studied extensively in the literature and attracted a great interest even today. Detailed introduction to stochastic differential equations (*resp.* stochastic evolution equations) can be found in [267, 284] (*resp.* [68, 217, 223]), etc.

Throughout this book, stochastic integrals appeared in stochastic differential/evolution equations are understood in Itô's sense. For different definitions of stochastic integral, one has different types of stochastic differential/evolution equations. By a simple transformation, a stochastic differential/evolution equation in Stratonovich's sense can be reduced to that in Itô's sense. On the other hand, we refer the interested readers to [105, 136, 249] for the study of stochastic differential/evolution equations when the Brownian motion is regarded as a rough path.

In this chapter, we only consider stochastic evolution equations with the nonlinearities satisfying (3.39). Such kind of assumptions are essential in the proofs of Theorems 3.14 and 3.21. To drop it, generally speaking, one has to assume that the nonlinearities have "good" signs (e.g., [68]).

When studying the stochastic evolution equations driven by cylindrical Brownian motions, we assume that the integrand is a Hilbert-Schmidt operator-values process. Without this assumption, people usually need some conditions on the operator A, for example some smoothing effect of the C_0-semigroup generated by A, to show the well-posedness of the corresponding stochastic evolution equation (e.g., [72]).

Almost all results presented in this chapter are either well-known (e.g., [68]) or small modification of the known ones. The very simple proof of the second case in Theorem 3.18 is due to E. Hausenblas and J. Seidler ([138]). The space $\Upsilon_{p,q}(H_1;H_2)$ introduced in (3.50) seems to be new; while Theorem 3.20 was taken from [248], which can be viewed as an infinite dimensional counterpart of the related result appeared in [284, Chapter V, Section 3].

4

Backward Stochastic Evolution Equations

In this chapter, we present an introduction to backward stochastic evolution equations (valued in Hilbert spaces), which appear naturally in the study of control problems for stochastic distributed parameter systems. In the case of natural filtration, by means of the Martingale Representation Theorem, these equations are proved to be well-posed in the sense of mild solutions; while for the general filtration, using our stochastic transposition method, we also establish their well-posedness.

Throughout this chapter, $T > 0$, $(\Omega, \mathcal{F}, \mathbf{F}, \mathbb{P})$ (with $\mathbf{F} \stackrel{\Delta}{=} \{\mathcal{F}_t\}_{t\in[0,T]}$) is a fixed filtered probability space satisfying the usual condition, and we denote by $\mathbb{F}$ the progressive σ-field w.r.t. $\mathbf{F}$; H and V are two separable Hilbert spaces; we denote by I the identity operator on H, and write $\mathcal{L}_2^0 \stackrel{\Delta}{=} \mathcal{L}_2(V; H)$.

4.1 The Case of Finite Dimensions and Natural filtration

In this section, we consider backward stochastic differential equations, i.e., backward stochastic evolution equations in $\mathbb{R}^n$ (for some $n \in \mathbb{N}$). It is well-known that, for a deterministic ordinary differential equation, the terminal value problem on $[0, T]$ is equivalent to an initial value problem on $[0, T]$ under a simple time transformation, i.e., $t \mapsto T - t$. However, things are completely different in the stochastic setting. Indeed, for stochastic problems one cannot simply reverse the "time" when looking for solutions that are adapted to the given filtration $\mathbf{F}$. As a result, one has to introduce some new idea to treat the terminal value problem for stochastic differential equations. This leads to a new class of differential equations in the stochastic setting, i.e., backward stochastic differential equations.

To begin with, let us analyze how to formulate correctly backward stochastic differential equations.

Let $W(\cdot) = (W^1(\cdot), \cdots, W^m(\cdot))^\top$ (for some $m \in \mathbb{N}$) be an m-dimensional standard Brownian motion, and $\mathbf{F} = \{\mathcal{F}_t\}_{t\in[0,T]}$ be the natural filtration

Q. Lü and X. Zhang, *Mathematical Control Theory for Stochastic Partial Differential Equations*, Probability Theory and Stochastic Modelling 101, https://doi.org/10.1007/978-3-030-82331-3_4

generated by $W(\cdot)$. Consider the following simple terminal value problem for stochastic differential equation:

$$\begin{cases} dy(t) = 0, & t \in [0,T], \\ y(T) = y_T \in L^2_{\mathcal{F}_T}(\Omega). \end{cases} \tag{4.1}$$

We want to find an $\mathbf{F}$-adapted solution $y(\cdot)$ to (4.1). However, this is impossible since the only possible solution to (4.1) is $y(\cdot) \equiv y_T$, which is not necessarily $\mathbf{F}$-adapted (unless y_T itself is $\mathcal{F}_0$-measurable). Namely, the equation (4.1) is not well-formulated if one expects to find an $\mathbf{F}$-adapted solution.

In what follows, we will see that the following modified version of (4.1):

$$\begin{cases} dy(t) = Y(t)dW(t), & t \in [0,T], \\ y(T) = y_T \end{cases} \tag{4.2}$$

is an appropriate reformation of (4.1). Comparing (4.2) with (4.1), we see that an additional term, i.e., $Y(t)dW(t)$, has been added. The process $Y(\cdot)$ is not *a priori* known but a part of the solution! As a matter of fact, the presence of the term $Y(t)dW(t)$ "corrects" the "non-adaptiveness" of the original $y(\cdot)$ in (4.1).

Generally, we consider the following backward stochastic differential equation:

$$\begin{cases} dy(t) = h(t, y(t), Y(t))dt + Y(t)dW(t), & t \in [0,T], \\ y(T) = y_T. \end{cases} \tag{4.3}$$

Here $y_T \in L^2_{\mathcal{F}_T}(\Omega; \mathbb{R}^n)$ and $h : [0,T] \times \mathbb{R}^n \times \mathbb{R}^{n\times m} \times \Omega \to \mathbb{R}^n$ is a given function. The main goal of this section is to find a pair of $\mathbf{F}$-adapted processes $y : [0,T] \times \Omega \to \mathbb{R}^n$ and $Y : (0,T) \times \Omega \to \mathbb{R}^{n\times m}$ satisfying (4.3) in a suitable sense.

Similarly to Definition 3.1, the following notion of solution to (4.3) seems to be the most natural one.

Definition 4.1. *A pair of stochastic processes $(y(\cdot), Y(\cdot))$ is called a solution to (4.3) if $y(\cdot)$ is an $\mathbb{R}^n$-valued, $\mathbf{F}$-adapted, continuous process, $Y(\cdot)) \in L^{2,loc}_{\mathbb{F}}(0,T; \mathbb{R}^{n\times m})$, $h(\cdot, y(\cdot), Y(\cdot)) \in L^1(0,T; \mathbb{R}^n)$ a.s., and for each $t \in [0,T]$,*

$$y(t) = y_T - \int_t^T h(s, y(s), Y(s))ds - \int_t^T Y(s)dW(s), \quad \text{a.s.} \tag{4.4}$$

Equation (4.3) is said to admit a unique solution if for any two solutions $(y(\cdot), Y(\cdot))$ and $(\tilde{y}(\cdot), \widetilde{Y}(\cdot))$ to (4.3), it holds that

$$\mathbb{P}(\{y(t) = \tilde{y}(t),\ \forall\, t \in [0,T]\}) = 1, \quad \mathbb{P}(\{Y(t) = \widetilde{Y}(t)\}) = 1, \text{ a.e. } t \in [0,T].$$

In the rest of this section, we assume that:

1) $h(\cdot, y, z)$ is $\mathbf{F}$-adapted for each $y \in \mathbb{R}^n$ and $z \in \mathbb{R}^{n\times m}$; and

2) $h(\cdot,0,0)\in L^2_{\mathbb{F}}(\Omega;\, L^1(0,T;\mathbb{R}^n))$, and there exist two nonnegative functions $L_1(\cdot)\in L^1(0,T)$ and $L_2(\cdot)\in L^2(0,T)$ such that

$$
\begin{aligned}
&|h(t,y,z)-h(t,\hat y,\hat z)|_{\mathbb{R}^n}\le L_1(t)|y-\hat y|_{\mathbb{R}^n}+L_2(t)|z-\hat z|_{\mathbb{R}^{n\times m}},\\
&\qquad\forall\ y,\hat y\in\mathbb{R}^n,\ z,\hat z\in\mathbb{R}^{n\times m},\ a.e.t\in[0,T],\ \text{a.s.}
\end{aligned}
\tag{4.5}
$$

One has the following well-posedness result for the equation (4.3).

Theorem 4.2. *For any given $y_T\in L^2_{\mathcal{F}_T}(\Omega;\mathbb{R}^n)$, the equation (4.3) admits a unique solution $(y(\cdot),Y(\cdot))\in L^2_{\mathbb{F}}(\Omega;C([0,T];\mathbb{R}^n))\times L^2_{\mathbb{F}}(0,T;\mathbb{R}^{n\times m})$, and*

$$
\begin{aligned}
&|(y(\cdot),Y(\cdot))|_{L^2_{\mathbb{F}}(\Omega;C([0,T];\mathbb{R}^n))\times L^2_{\mathbb{F}}(0,T;\mathbb{R}^{n\times m})}\\
&\le \mathcal{C}\big(|y_T|_{L^2_{\mathcal{F}_T}(\Omega;\mathbb{R}^n)}+|h(\cdot,0,0)|_{L^2_{\mathbb{F}}(\Omega;L^1(0,T;\mathbb{R}^n))}\big).
\end{aligned}
\tag{4.6}
$$

Proof: For $s\in[0,T)$, put

$$
\mathcal{V}(s,T)\overset{\Delta}{=}L^2_{\mathbb{F}}(\Omega;C([s,T];\mathbb{R}^n))\times L^2_{\mathbb{F}}(s,T;\mathbb{R}^{n\times m}).
$$

We divide the proof into two steps.

Step 1. In this step, we shall show that, for any $h(\cdot)\in L^2_{\mathbb{F}}(\Omega;L^1(0,T;\mathbb{R}^n))$ and $y_T\in L^2_{\mathcal{F}_T}(\Omega;\mathbb{R}^n)$, there is a pair $\big(y(\cdot),Y(\cdot)\big)\in\mathcal{V}(0,T)$ such that

$$
y(t)=y_T-\int_t^T h(s)ds-\int_t^T Y(s)dW(s),\quad \forall\, t\in[0,T],\ \text{a.s.} \tag{4.7}
$$

Indeed, put

$$
M(t)=\mathbb{E}\Big(y_T-\int_0^T h(s)ds\,\Big|\,\mathcal{F}_t\Big),\quad y(t)=\mathbb{E}\Big(y_T-\int_t^T h(s)ds\,\Big|\,\mathcal{F}_t\Big). \tag{4.8}
$$

Then, $M(0)=y(0)=\mathbb{E}M(T)$. By the Martingale Representation Theorem (i.e., Theorem 2.147 and Remark 2.148), one can find a $Y(\cdot)\in L^2_{\mathbb{F}}(0,T;\mathbb{R}^{n\times m})$ such that

$$
M(t)=M(0)+\int_0^t Y(s)dW(s). \tag{4.9}
$$

Hence,

$$
\mathbb{E}\int_0^T|Y(t)|^2_{\mathbb{R}^{n\times m}}dt\le\mathbb{E}|M(T)|^2_{\mathbb{R}^n}\le 2\mathbb{E}\Big[|y_T|^2_{\mathbb{R}^n}+\Big(\int_0^T|h(t)|_{\mathbb{R}^n}dt\Big)^2\Big]. \tag{4.10}
$$

From (4.9), we see that

$$
y_T-\int_0^T h(s)ds=M(T)=M(0)+\int_0^T Y(s)dW(s)=y(0)+\int_0^T Y(s)dW(s),
$$

and

$$\begin{aligned}y(t) &= M(t) + \int_0^t h(s)ds = y(0) + \int_0^t Y(s)dW(s) + \int_0^t h(s)ds \\ &= y_T - \int_t^T h(s)ds - \int_t^T Y(s)dW(s).\end{aligned}$$

Hence, $y(\cdot) = y(0) + \int_0^{\cdot} Y(s)dW(s) + \int_0^{\cdot} h(s)ds$. By the fourth conclusion in Theorem 2.124, it is easy to see that $y(\cdot) \in L^2_{\mathbb{F}}(\Omega; C([0,T];\mathbb{R}^n))$. From (4.8), it follows that

$$|y(\cdot)|^2_{L^2_{\mathbb{F}}(\Omega;C([0,T];\mathbb{R}^n))} \leq \mathcal{C}\mathbb{E}\Big[|y_T|^2_{\mathbb{R}^n} + \Big(\int_0^T |h(t)|_{\mathbb{R}^n} dt\Big)^2\Big]. \tag{4.11}$$

By (4.10) and (4.11), we see that $(y(\cdot), Y(\cdot))$ satisfies

$$|(y(\cdot), Y(\cdot))|^2_{\mathcal{V}(0,T)} \leq \mathcal{C}\mathbb{E}\Big[|y_T|^2 + \Big(\int_0^T |h(t)|dt\Big)^2\Big]. \tag{4.12}$$

Step 2. Fix any $T_1 \in [0,T)$. For any fixed $\big(z(\cdot), Z(\cdot)\big) \in \mathcal{V}(T_1,T)$, it is easy to see that $h(\cdot) \equiv h(\cdot, z(\cdot), Z(\cdot)) \in L^2_{\mathbb{F}}(\Omega; L^1(T_1,T;\mathbb{R}^n))$. Consider the following backward stochastic differential equation:

$$\begin{cases} dy(t) = h(t, z(t), Z(t))dt + Y(t)dW(t), & t \in [T_1,T), \\ y(T) = y_T. \end{cases} \tag{4.13}$$

By the result in Step 1, the equation (4.13) admits a unique solution $(y(\cdot), Y(\cdot)) \in \mathcal{V}(T_1,T)$. This defines a map $\mathcal{J} : \mathcal{V}(T_1,T) \to \mathcal{V}(T_1,T)$ by $(z(\cdot), Z(\cdot)) \mapsto (y(\cdot), Y(\cdot))$.

We claim that, for T_1 being sufficiently close to T,

$$\begin{aligned}\big|\mathcal{J}(z,Z) - \mathcal{J}(\tilde{z},\widetilde{Z})\big|_{\mathcal{V}(T_1,T)} &\leq \frac{1}{2}\big|(z,Z) - (\tilde{z},\widetilde{Z})\big|_{\mathcal{V}(T_1,T)}, \\ &\qquad \forall\, (z,Z),\ \ (\tilde{z},\widetilde{Z}) \in \mathcal{V}(T_1,T).\end{aligned} \tag{4.14}$$

To show (4.14), put $(\hat{y}(\cdot), \hat{Y}(\cdot)) = \mathcal{J}(z,Z) - \mathcal{J}(\tilde{z},\widetilde{Z})$ and $\hat{h}(\cdot) = h(\cdot, z(\cdot), Z(\cdot)) - h(\cdot, \tilde{z}(\cdot), \widetilde{Z}(\cdot))$. Then, $(\hat{y}(\cdot), \widehat{Y}(\cdot))$ solves

$$\begin{cases} d\hat{y}(t) = \hat{h}(t)dt + \hat{Y}(t)dW(t), & t \in [T_1,T), \\ \hat{y}(T) = 0. \end{cases} \tag{4.15}$$

From (4.12) and using (4.5), we find that

$$\begin{aligned}&|(\hat{y}(\cdot), \hat{Y}(\cdot))|^2_{\mathcal{V}(T_1,T)} \leq \mathcal{C}\mathbb{E}\Big(\int_{T_1}^T |\hat{h}(t)|_{\mathbb{R}^n} dt\Big)^2 \\ &= \mathcal{C}\mathbb{E}\Big(\int_{T_1}^T |h(t, z(t), Z(t)) - h(t, \tilde{z}(t), \widetilde{Z}(t))|_{\mathbb{R}^n} dt\Big)^2 \\ &\leq \mathcal{C}\big(|L_1(\cdot)|^2_{L^1(T_1,T)} + |L_2(\cdot)|^2_{L^2(T_1,T)}\big)|(z(\cdot) - \hat{z}(\cdot), Z(\cdot) - \hat{Z}(\cdot))|^2_{\mathcal{V}(T_1,T)}\end{aligned} \tag{4.16}$$

and

$$
\begin{aligned}
&|(y(\cdot),Y(\cdot))|^2_{\mathcal{V}(T_1,T)} \le \mathcal{C}\mathbb{E}\Big[|y_T|^2_{\mathbb{R}^n} + \Big(\int_{T_1}^T |h(t,z(t),Z(t))|_{\mathbb{R}^n}dt\Big)^2\Big]\\
&\le \mathcal{C}\big(|y_T|^2_{L^2_{\mathcal{F}_T}(\Omega;\mathbb{R}^n)} + |h(\cdot,0,0)|^2_{L^2_{\mathbb{F}}(\Omega;L^1(0,T;\mathbb{R}^n))}\big)\\
&\quad +\mathcal{C}\big(|L_1(\cdot)|^2_{L^1(T_1,T)} + |L_2(\cdot)|^2_{L^2(T_1,T)}\big)|(z(\cdot),Z(\cdot))|^2_{\mathcal{V}(T_1,T)}.
\end{aligned}
\tag{4.17}
$$

Let us choose T_1 such that

$$
\mathcal{C}\big(|L_1(\cdot)|^2_{L^1(T_1,T)} + |L_2(\cdot)|^2_{L^2(T_1,T)}\big) \le \frac{1}{4}. \tag{4.18}
$$

Then, by (4.16), we get (4.14). This implies that the map $\mathcal{J}$ is contractive. Hence, it admits a unique fixed point, which is a solution to (4.3) on $[T_1,T]$. Moreover, by (4.17)–(4.18), it follows that

$$
|(y(\cdot),Y(\cdot))|^2_{\mathcal{V}(T_1,T)} \le \mathcal{C}\big(|y_T|^2_{L^2_{\mathcal{F}_T}(\Omega;\mathbb{R}^n)} + |h(\cdot,0,0)|^2_{L^2_{\mathbb{F}}(\Omega;L^1(0,T;\mathbb{R}^n))}\big). \tag{4.19}
$$

Repeating the above argument, we obtain the existence of solutions to (4.3). The uniqueness and the estimate (4.6) follow from (4.19). □

4.2 The Case of Infinite Dimensions

In the rest of this chapter, $W(\cdot)$ is a V-valued, Q-Brownian motion or cylindrical Brownian motion but we only consider the case of cylindrical Brownian motion.

We consider the following backward stochastic evolution equation:

$$
\begin{cases}
dy(t) = -Ay(t)dt + F(t,y(t),Y(t))dt + Y(t)dW(t) & \text{in } [0,T),\\
y(T) = y_T.
\end{cases}
\tag{4.20}
$$

Here and in the rest of this chapter, $y_T : \Omega \to H$ is an $\mathcal{F}_T$-random variable, A generates a C_0-semigroup $\{S(t)\}_{t\ge 0}$ on H, and $F : [0,T]\times\Omega\times H\times\mathcal{L}_2^0 \to H$ is a given function.

Remark 4.3. The diffusion term "$Y(t)dW(t)$" in (4.20) may be replaced by a more general form "$\big(\widetilde{F}(t,y(t))+Y(t)\big)dW(t)$" (for a function $\widetilde{F} : [0,T]\times\Omega\times H\to\mathcal{L}_2^0$). Clearly, the later can be reduced to the former one by means of a simple transformation:

$$
\tilde{y}(\cdot) = y(\cdot), \qquad \widetilde{Y}(\cdot) = \widetilde{F}(\cdot,y(\cdot)) + Y(\cdot),
$$

under suitable assumptions on the function $\widetilde{F}$.

It is easy to see that, for any $f(\cdot) \in L^{1,loc}_{\mathbb{F}}(0,T;H)$, the following equation

$$\begin{cases} dy(t) = -Ay(t)dt + f(t)dt + Y(t)dW(t) \quad \text{in } [0,T), \\ y(T) = y_T \end{cases} \tag{4.21}$$

is a special case of (4.20), in which the function $F(\cdot,\cdot,\cdot)$ is independent of y and Y.

4.2.1 Notions of Solutions

Similar to the case of stochastic evolution equations considered in the last chapter, we introduce below notions of strong/weak/mild solutions to the equation (4.20) (Recall (2.132) for the notation $\langle\!\langle \cdot,\cdot \rangle\!\rangle_H$).

Definition 4.4. *An $H \times \mathcal{L}_2^0$-valued process $(y(\cdot), Y(\cdot))$ is called a strong solution to (4.20) if*

1) *$y(\cdot)$ is $\mathbf{F}$-adapted and continuous, $Y(\cdot) \in L^{2,loc}_{\mathbb{F}}(0,T;\mathcal{L}_2^0)$;*

2) *$y(t) \in D(A)$ for a.e. $(t,\omega) \in [0,T] \times \Omega$, and $Ay(\cdot) \in L^1(0,T;H)$ and $F(\cdot, y(\cdot), Y(\cdot)) \in L^1(0,T;H)$ a.s.; and*

3) *For any $t \in [0,T]$,*

$$y(t) = y_T + \int_t^T \big(Ay(s) - F(s,y(s),Y(s))\big)ds - \int_t^T Y(s)dW(s), \quad \text{a.s.} \tag{4.22}$$

Definition 4.5. *An $H \times \mathcal{L}_2^0$-valued process $(y(\cdot), Y(\cdot))$ is called a weak solution to (4.20) if*

1) *$y(\cdot)$ is $\mathbf{F}$-adapted and continuous, $Y(\cdot) \in L^{2,loc}_{\mathbb{F}}(0,T;\mathcal{L}_2^0)$, and $F(\cdot, y(\cdot), Y(\cdot)) \in L^1(0,T;H)$ a.s.; and*

2 *For any $t \in [0,T]$ and $\eta \in D(A^*)$,*

$$\begin{aligned} \langle y(t), \eta \rangle_H &= \langle y_T, \eta \rangle_H + \int_t^T \langle y(s), A^*\eta \rangle_H ds \\ &\quad - \int_t^T \langle F(s,y(s),Y(s)), \eta \rangle_H ds - \int_t^T \langle\!\langle Y(s), \eta \rangle\!\rangle_H dW(s), \quad \text{a.s.} \end{aligned} \tag{4.23}$$

Definition 4.6. *An $H \times \mathcal{L}_2^0$-valued process $(y(\cdot), Y(\cdot))$ is called a mild solution to (4.20) if*

1) *$y(\cdot)$ is $\mathbf{F}$-adapted and continuous, $Y(\cdot) \in L^{2,loc}_{\mathbb{F}}(0,T;\mathcal{L}_2^0)$, and $F(\cdot, y(\cdot), Y(\cdot)) \in L^1(0,T;H)$ a.s.; and*

2 *For any $t \in [0,T]$,*

$$\begin{aligned} y(t) = S(T-t)y_T &- \int_t^T S(s-t)F(s,y(s),Y(s))ds \\ &- \int_t^T S(s-t)Y(s)dW(s), \qquad \text{a.s.} \end{aligned} \tag{4.24}$$

It is easy to show the following result.

Proposition 4.7. *If $(y(\cdot),Y(\cdot))$ is a strong solution to (4.20), then it is also a weak solution to the same equation. Conversely, if a weak solution $(y(\cdot),Y(\cdot))$ to (4.20) satisfies that $y(t)\in D(A)$ for a.e. $(t,\omega)\in[0,T]\times\Omega$, and $Ay(\cdot)\in L^1(0,T;H)$ a.s., then $(y(\cdot),Y(\cdot))$ is also a strong solution to (4.20).*

Similarly to Theorem 3.10, we have the following result concerning the equivalence between weak and mild solutions to the equation (4.20).

Theorem 4.8. *An $H\times\mathcal{L}_2^0$-valued process $(y(\cdot),Y(\cdot))$ is a weak solution to (4.20) if and only if it is a mild solution to the same equation.*

Proof: **The "only if" part**. Let $(y(\cdot),Y(\cdot))$ be a weak solution to (4.20). For any $\phi\in D((A^*)^2)$, $t\in[0,T]$ and $r\in[t,T]$, by (4.23), we obtain that

$$
\begin{aligned}
&\big\langle y(r),S(r-t)^*A^*\phi\big\rangle_H\\
&=\big\langle y_T,S(r-t)^*A^*\phi\big\rangle_H+\int_r^T\big\langle y(s),A^*S(r-t)^*A^*\phi\big\rangle_H ds\\
&\quad-\int_r^T\big\langle F(s,y(s),Y(s)),S(r-t)^*A^*\phi\big\rangle_H ds\\
&\quad-\int_r^T\langle\!\langle Y(s),S(r-t)^*A^*\phi\rangle\!\rangle_H dW(s).
\end{aligned}
$$

Integrating the above equality w.r.t. r from t to T, we obtain that

$$
\int_t^T\big\langle y(r),S(r-t)^*A^*\phi\big\rangle_H dr=I_1+I_2-I_3-I_4, \tag{4.25}
$$

where

$$
\begin{aligned}
I_1&\overset{\Delta}{=}\int_t^T\big\langle y_T,S(r-t)^*A^*\phi\big\rangle_H dr,\\
I_2&\overset{\Delta}{=}\int_t^T\int_r^T\big\langle y(s),A^*S(r-t)^*A^*\phi\big\rangle_H dsdr,\\
I_3&\overset{\Delta}{=}\int_t^T\int_r^T\big\langle F(s,y(s),Y(s)),S(r-t)^*A^*\phi\big\rangle_H dsdr,\\
I_4&\overset{\Delta}{=}\int_t^T\int_r^T\langle\!\langle Y(s),S(r-t)^*A^*\phi\rangle\!\rangle_H dW(s)dr.
\end{aligned}
$$

We have

$$
I_1=\Big\langle y_T,\int_t^T S(r-t)^*A^*\phi dr\Big\rangle_H=\big\langle y_T,S(T-t)^*\phi-\phi\big\rangle_H. \tag{4.26}
$$

By Fubini's theorem (i.e., Theorem 2.21), we obtain that

$$\begin{aligned} I_2 &= \int_t^T \int_t^s \big\langle y(s), A^* S(r-t)^* A^* \phi \big\rangle_H drds \\ &= \int_t^T \big\langle y(s), S(s-t)^* A^* \phi \big\rangle_H ds - \int_t^T \big\langle y(s), A^* \phi \big\rangle_H ds \end{aligned} \tag{4.27}$$

and

$$\begin{aligned} I_3 &= \int_t^T \int_t^s \big\langle F(s, y(s), Y(s)), S(r-t)^* A^* \phi \big\rangle_H drds \\ &= \int_t^T \big\langle F(s, y(s), Y(s)), S(s-t)^* \phi \big\rangle_H ds \\ &\quad - \int_t^T \big\langle F(s, y(s), Y(s)), \phi \big\rangle_H ds. \end{aligned} \tag{4.28}$$

Further, by the stochastic Fubini theorem (i.e., Theorem 2.141), we conclude that

$$\begin{aligned} I_4 &= \int_t^T \int_t^s \langle\!\langle Y(s), S(r-t)^* A^* \phi \rangle\!\rangle_H drdW(s) \\ &= \int_t^T \langle\!\langle Y(s), S(s-t)^* \phi \rangle\!\rangle_H dW(s) - \int_t^T \langle\!\langle Y(s), \phi \rangle\!\rangle_H dW(s). \end{aligned} \tag{4.29}$$

From (4.25)–(4.29) and using (4.23) again, we obtain that

$$\begin{aligned} \big\langle y(t), \phi \big\rangle_H &= \big\langle y_T, S(T-t)^* \phi \big\rangle_H - \int_t^T \big\langle F(s, y(s), Y(s)), S(s-t)^* \phi \big\rangle_H ds \\ &\quad - \int_t^T \langle\!\langle Y(s), S(s-t)^* \phi \rangle\!\rangle_H dW(s). \end{aligned}$$

which implies that

$$\begin{aligned} \big\langle y(t), \phi \big\rangle_H &= \big\langle S(T-t) y_T, \phi \big\rangle_H - \int_t^T \big\langle S(s-t) F(s, y(s), Y(s)), \phi \big\rangle_H ds \\ &\quad - \int_t^T \langle\!\langle S(s-t) Y(s), \phi \rangle\!\rangle_H dW(s). \end{aligned} \tag{4.30}$$

Since $D((A^*)^2)$ is dense in H, we see that (4.30) holds for any $\phi \in H$. Therefore, $(y(\cdot), Y(\cdot))$ is a mild solution to (4.20).

The "if" part. Assume that $(y(\cdot), Y(\cdot))$ is a mild solution to (4.20). By (4.24), for any $\eta \in D(A^*)$, $t \in [0, T]$ and $s \in [t, T)$, it holds

$$\begin{aligned} &\big\langle y(s), A^* \eta \big\rangle_H \\ &= \big\langle y_T, S(T-s)^* A^* \eta \big\rangle_H - \int_s^T \big\langle F(r, y(r), Y(r)), S(r-s)^* A^* \eta \big\rangle_H dr \\ &\quad - \int_s^T \langle\!\langle Y(r), S(r-s)^* A^* \eta \rangle\!\rangle_H dW(r). \end{aligned}$$

Integrating the above equality w.r.t. s from t to T, we find that

$$\int_t^T \langle y(s), A^*\eta \rangle_H ds = I_5 - I_6 - I_7, \tag{4.31}$$

where

$$\begin{aligned}
I_5 &\triangleq \int_t^T \langle y_T, S(T-s)^*A^*\eta \rangle_H ds, \\
I_6 &\triangleq \int_t^T \int_s^T \langle F(r, y(r), Y(r)), S(r-s)^*A^*\eta \rangle_H dr ds, \\
I_7 &\triangleq \int_t^T \int_s^T \langle\!\langle Y(r), S(r-s)^*A^*\eta \rangle\!\rangle_H dW(r) ds.
\end{aligned}$$

Clearly,

$$I_5 = \Big\langle y_T, \int_t^T S(T-s)^*A^*\eta ds \Big\rangle_H = \langle y_T, S(T-t)^*\eta \rangle_H - \langle y_T, \eta \rangle_H. \tag{4.32}$$

By means of Fubini's theorem, we get that

$$\begin{aligned}
I_6 &= \int_t^T \int_t^r \langle F(r, y(r), Y(r)), S(r-s)^*A^*\eta \rangle_H ds dr \\
&= \int_t^T \langle F(r, y(r), Y(r)), S(r-t)^*\eta \rangle_H dr \\
&\quad - \int_t^T \langle F(r, y(r), Y(r)), \eta \rangle_H dr.
\end{aligned} \tag{4.33}$$

By virtue of the stochastic Fubini theorem, we obtain that

$$\begin{aligned}
I_7 &= \int_t^T \langle\!\!\int_t^r \langle Y(r), S(r-s)^*A^*\eta \rangle\!\rangle_H ds dW(r) \\
&= \int_t^T \langle\!\langle Y(r), S(r-t)^*\eta \rangle\!\rangle_H dW(r) - \int_t^T \langle\!\langle Y(r), \eta \rangle\!\rangle_H dW(r).
\end{aligned} \tag{4.34}$$

From (4.31)–(4.34) and noting (4.24), we find that $(y(\cdot), Y(\cdot))$ satisfies (4.23). This implies that $(y(\cdot), Y(\cdot))$ is a weak solution to (4.20). $\square$

Similar to Lemma 3.11, we have the following result.

Lemma 4.9. *A mild solution $(y(\cdot), Y(\cdot))$ to (4.20) is a strong solution (to the same equation) if the following three conditions hold,*

1) $y_T \in D(A)$, $S(s-t)F(s, y(s), Y(s)) \in D(A)$ and $S(s-t)Y(s) \in \mathcal{L}_2(V; D(A))$ for any $t \in [0,T]$ and $s \in [t,T]$, a.s.;

2) $\int_0^T \int_t^T |AS(s-t)F(s, y(s), Y(s))|_H ds dt < \infty$; and

3) $\int_0^T \int_t^T |AS(s-t)Y(s)|^2_{\mathcal{L}_2^0} ds dt < \infty$.

Proof: Recall that for every $t \in [0,T]$, $y(t)$ satisfies (4.24). By the conditions in Lemma 4.9, we see that, for any $t \in [0,T]$ and a.s., $y(t) \in D(A)$,

$$A \int_t^T S(s-t)F(s,y(s),Y(s))ds = \int_t^T AS(s-t)F(s,y(s),Y(s))ds, \quad (4.35)$$

and

$$A \int_t^T S(s-t)Y(s)dW(s) = \int_t^T AS(s-t)Y(s)dW(s). \quad (4.36)$$

From (4.24) and (4.35)–(4.36), we find that for any $t \in [0,T]$,

$$\begin{aligned} Ay(t) = AS(T-t)y_T - \int_t^T AS(s-t)F(s,y(s),Y(s))ds \\ - \int_t^T AS(s-t)Y(s)dW(s). \end{aligned} \quad (4.37)$$

Hence, $Ay(\cdot) \in L^1(0,T;H)$ a.s. This, together with Proposition 4.7 and Theorem 4.8, implies that $(y(\cdot),Y(\cdot))$ is a strong solution to (4.20). □

In the sequel, we will need the following assumption on the nonlinearity in (4.20).

Condition 4.1 $F : [0,T] \times \Omega \times H \times \mathcal{L}_2^0 \to H$ *is a given function satisfying*

1) $F(\cdot,y,z)$ *is* $\mathbf{F}$*-adapted for each* $(y,z) \in H \times \mathcal{L}_2^0$*; and*

2) *There exist two nonnegative functions* $L_1(\cdot) \in L^1(0,T)$ *and* $L_2(\cdot) \in L^2(0,T)$ *such that, for any* $y_1,y_2 \in H$, $z_1,z_2 \in \mathcal{L}_2^0$, *a.e.* $t \in [0,T]$,

$$|F(t,y_1,z_1)-F(t,y_2,z_2)|_H \le L_1(t)|y_1-y_2|_H + L_2(t)|z_1-z_2|_{\mathcal{L}_2^0}, \quad \text{a.s.} \quad (4.38)$$

Generally speaking, it is impossible to find mild solutions to (4.20) under Condition 4.1 except for the case of natural filtration (See the next subsection). Because of this, in Section 4.3 we shall introduce a weaker concept of solution to (4.20) for the case of general filtration.

4.2.2 Well-Posedness in the Sense of Mild Solution for the Case of Natural Filtration

In this subsection, we assume that $\mathbf{F}$ is the natural filtration generated by the underlying Brownian motion $W(\cdot)$.

Similar to Theorem 3.14, we have the following well-posedness of (4.20) in the sense of mild solution.

Theorem 4.10. *Let Condition 4.1 hold and* $F(\cdot,0,0) \in L^1_{\mathbb{F}}(0,T;L^2(\Omega;H))$. *Then, for any* $y_T \in L^2_{\mathcal{F}_T}(\Omega;H)$, *the equation* (4.20) *admits a unique mild solution* $(y(\cdot),Y(\cdot)) \in L^2_{\mathbb{F}}(\Omega;C([0,T];H)) \times L^2_{\mathbb{F}}(0,T;\mathcal{L}_2^0)$. *Further,*

$$\begin{aligned} &|(y(\cdot),Y(\cdot))|_{L^2_{\mathbb{F}}(\Omega;C([0,T];H)) \times L^2_{\mathbb{F}}(0,T;\mathcal{L}_2^0)} \\ &\le \mathcal{C}\big(|y_T|_{L^2_{\mathcal{F}_T}(\Omega;H)} + |F(\cdot,0,0)|_{L^1_{\mathbb{F}}(0,T;L^2(\Omega;H))}\big). \end{aligned} \quad (4.39)$$

Proof: For any $s \in [0,T)$, let $\mathcal{X}[s,T] \stackrel{\Delta}{=} L^2_{\mathbb{F}}(\Omega; C([s,T];H)) \times L^2_{\mathbb{F}}(s,T;\mathcal{L}_2^0)$. In what follows, we divide the proof into two steps.

Step 1. In this step, we consider the linear equation (4.21) with $f(\cdot) \in L^1_{\mathbb{F}}(0,T;L^2(\Omega;H))$ and $y_T \in L^2_{\mathcal{F}_T}(\Omega;H)$.

By the Martingale Representation Theorem (Theorem 2.147), there is an $l(\cdot) \in L^2_{\mathbb{F}}(0,T;\mathcal{L}_2^0)$ such that

$$\mathbb{E}(y_T \mid \mathcal{F}_t) = \mathbb{E}y_T + \int_0^t l(\sigma)dW(\sigma). \tag{4.40}$$

By Corollary 2.149, there is an $h(\cdot,\cdot) \in L^1(0,T;L^2_{\mathbb{F}}(0,T;\mathcal{L}_2^0))$ such that for a.e. $s \in [0,T]$,

$$f(s) = \mathbb{E}f(s) + \int_0^s h(s,\sigma)dW(\sigma) \tag{4.41}$$

and

$$|h(\cdot,\cdot)|_{L^1(0,T;L^2_{\mathbb{F}}(0,T;\mathcal{L}_2^0))} \le |f(\cdot)|_{L^1_{\mathbb{F}}(0,T;L^2(\Omega;H))}. \tag{4.42}$$

Put

$$y(t) = \mathbb{E}\Big(S(T-t)y_T - \int_t^T S(s-t)f(s)ds \;\Big|\; \mathcal{F}_t\Big). \tag{4.43}$$

This, together with (4.40) and (4.41), implies that

$$\begin{aligned} y(t) &= S(T-t)\Big(y_T - \int_t^T l(\sigma)dW(\sigma)\Big) \\ &\quad - \int_t^T S(s-t)\Big(f(s) + \int_t^s h(s,\sigma)dW(\sigma)\Big)ds. \end{aligned} \tag{4.44}$$

From the stochastic Fubini Theorem, we get that

$$\begin{aligned} y(t) &= S(T-t)y_T - \int_t^T S(s-t)f(s)ds - \int_t^T S(T-t)l(\sigma)dW(\sigma) \\ &\quad - \int_t^T \int_\sigma^T S(s-t)h(s,\sigma)dsdW(\sigma) \\ &= S(T-t)y_T - \int_t^T S(s-t)f(s)ds - \int_t^T S(s-t)Y(s)dW(s), \end{aligned} \tag{4.45}$$

where

$$Y(s) \stackrel{\Delta}{=} S(T-s)l(s) + \int_s^T S(\eta-s)h(\eta,s)d\eta. \tag{4.46}$$

Hence, $(y(\cdot),Y(\cdot))$ is a mild solution to (4.21).

From (4.43) and (4.46), and noting (4.42), we find that

$$|(y(\cdot),Y(\cdot))|_{\mathcal{X}[0,T]} \le \mathcal{C}\big(|y_T|_{L^2_{\mathcal{F}_T}(\Omega;H)} + |f(\cdot)|_{L^1_{\mathbb{F}}(0,T;L^2(\Omega;H))}\big). \tag{4.47}$$

Step 2. Fix any $T_1 \in [0,T)$. For any fixed $\big(z(\cdot), Z(\cdot)\big) \in \mathcal{X}(T_1,T)$, it is easy to see that $F(\cdot, z(\cdot), Z(\cdot)) \in L^1_{\mathbb{F}}(T_1,T; L^2(\Omega;H))$. Consider the following equation

$$
\begin{cases} dy(t) = -Ay(t)dt + F(t,z(t),Z(t))dt + Y(t)dW(t), & t \in [T_1,T), \\ y(T) = y_T. \end{cases} \tag{4.48}
$$

By the result in Step 1, the equation (4.48) admits a unique mild solution $(y(\cdot), Y(\cdot)) \in \mathcal{X}(T_1,T)$. This defines a map $\mathcal{J}: \mathcal{X}(T_1,T) \to \mathcal{X}(T_1,T)$ by $(z(\cdot), Z(\cdot)) \mapsto (y(\cdot), Y(\cdot))$.

We claim that, for T_1 being sufficiently close to T,

$$
\begin{array}{r} \big|\mathcal{J}(z,Z) - \mathcal{J}(\tilde z, \widetilde Z)\big|_{\mathcal{X}(T_1,T)} \leq \dfrac{1}{2}\big|(z,Z) - (\tilde z, \widetilde Z)\big|_{\mathcal{X}(T_1,T)}, \\ \forall\, (z,Z),\ \ (\tilde z, \widetilde Z) \in \mathcal{X}(T_1,T). \end{array} \tag{4.49}
$$

To show (4.49), put $\hat f(\cdot) = F(\cdot, z(\cdot), Z(\cdot)) - F(\cdot, \tilde z(\cdot), \widetilde Z(\cdot))$ and $(\hat y(\cdot), \widehat Y(\cdot)) = \mathcal{J}(z,Z) - \mathcal{J}(\tilde z, \widetilde Z)$. Then, $(\hat y(\cdot), \widehat Y(\cdot))$ satisfies

$$
\begin{cases} d\hat y(t) = -A\hat y(t)dt + \hat f(t)dt + \widehat Y(t)dW(t), & t \in [T_1,T), \\ \hat y(T) = 0. \end{cases} \tag{4.50}
$$

From (4.47), we find that

$$
\begin{array}{ll} & |(\hat y(\cdot), \widehat Y(\cdot))|^2_{\mathcal{X}(T_1,T)} \leq \mathcal{C}|F(\cdot, z(\cdot), Z(\cdot)) - F(\cdot, \tilde z(\cdot), \widetilde Z(\cdot))|^2_{L^1_{\mathbb{F}}(T_1,T;L^2(\Omega;H))} \\ & \leq \mathcal{C}\big(|L_1(\cdot)|^2_{L^1(T_1,T)} + |L_2(\cdot)|^2_{L^2(T_1,T)}\big)|(z(\cdot) - \tilde z(\cdot), Z(\cdot) - \widetilde Z(\cdot))|^2_{\mathcal{X}(T_1,T)} \end{array} \tag{4.51}
$$

and

$$
\begin{array}{ll} & |(y(\cdot), Y(\cdot))|^2_{\mathcal{X}(T_1,T)} \leq \mathcal{C}\Big(|y_T|^2_{L^2_{\mathcal{F}_T}(\Omega;H)} + |F(\cdot, z(\cdot), Z(\cdot))|^2_{L^1_{\mathbb{F}}(T_1,T;L^2(\Omega;H))}\Big) \\ & \leq \mathcal{C}\big(|y_T|^2_{L^2_{\mathcal{F}_T}(\Omega;H)} + |F(\cdot,0,0)|^2_{L^1_{\mathbb{F}}(T_1,T;L^2(\Omega;H))}\big) \\ & \quad + \mathcal{C}\big(|L_1(\cdot)|^2_{L^1(T_1,T)} + |L_2(\cdot)|^2_{L^2(T_1,T)}\big)|(z(\cdot), Z(\cdot))|^2_{\mathcal{X}(T_1,T)}. \end{array} \tag{4.52}
$$

Let us choose T_1 such that

$$
\mathcal{C}\big(|L_1(\cdot)|^2_{L^1(T_1,T)} + |L_2(\cdot)|^2_{L^2(T_1,T)}\big) \leq \frac{1}{4}. \tag{4.53}
$$

Then, by (4.51), we get (4.49). This implies that the map $\mathcal{J}$ is contractive. Hence, it admits a unique fixed point, which is a mild solution to (4.20) on $[T_1,T]$. Moreover, by (4.52)–(4.53), it follows that

$$
|(y(\cdot), Y(\cdot))|^2_{\mathcal{X}(T_1,T)} \leq \mathcal{C}\big(|y_T|^2_{L^2_{\mathcal{F}_T}(\Omega;H)} + |F(\cdot,0,0)|^2_{L^1_{\mathbb{F}}(0,T;L^2(\Omega;H))}\big). \tag{4.54}
$$

Repeating the above argument, we obtain the existence of mild solutions to (4.20). The uniqueness and the estimate (4.39) follow from (4.54). □

Similar to the proof of Theorem 3.24, we can show the following result, which describes the smoothing effect of mild solutions to a class of backward stochastic evolution equations.

Theorem 4.11. *Let Condition 4.1 hold, $F(\cdot,0,0) \in L^1_{\mathbb{F}}(0,T;L^2(\Omega;H))$, and A be a self-adjoint, negative definite (unbounded linear) operator on H. Then, for any $y_T \in L^2_{\mathcal{F}_T}(\Omega;H)$, the equation (4.20) admits a unique mild solution $(y(\cdot),Y(\cdot)) \in \big(L^2_{\mathbb{F}}(\Omega;C([0,T];H)) \cap L^2_{\mathbb{F}}(0,T;D((-A)^{\frac{1}{2}}))\big) \times L^2_{\mathbb{F}}(0,T;\mathcal{L}_2^0)$. Moreover,*

$$\begin{aligned}&|y(\cdot)|_{L^2_{\mathbb{F}}(\Omega;C([0,T];H))} + |y(\cdot)|_{L^2_{\mathbb{F}}(0,T;D((-A)^{\frac{1}{2}}))} + |Y(\cdot)|_{L^2_{\mathbb{F}}(0,T;\mathcal{L}_2^0)}\\&\le \mathcal{C}\big(|Y_T|_{L^2_{\mathcal{F}_T}(\Omega;H)} + |F(\cdot,0,0)|_{L^1_{\mathbb{F}}(0,T;L^2(\Omega;H))}\big).\end{aligned} \tag{4.55}$$

As the case of stochastic evolution equations in infinite dimensions, we cannot employ Itô's formula for mild solutions to backward stochastic evolution equations directly on most occasions. Indeed, usually the mild solution does not have enough regularity. Nevertheless, similarly to the last chapter, the problem can also be solved by the following strategy.

1. Introduce suitable approximating (backward stochastic evolution) equations with strong solutions such that the limit of these strong solutions is the mild or weak solution to the equation (4.20).
2. Obtain the desired properties for the above strong solutions.
3. Utilize the density argument to obtain the desired properties for mild/weak solutions to (4.20).

Similar to the case of stochastic evolution equations, there are many mays to implement the above three steps for backward stochastic evolution equations. For example, for each $\lambda \in \rho(A)$, we may introduce an approximating equation of (4.20) as follows:

$$\begin{cases} dy_\lambda(t) = -Ay_\lambda(t)dt + R(\lambda)F(t,y_\lambda(t),Y_\lambda(t))dt + Y_\lambda(t)dW(t) & \text{in } (0,T],\\ y_\lambda(T) = R(\lambda)y_T, \end{cases} \tag{4.56}$$

where $R(\lambda) = \lambda(\lambda I - A)^{-1}$.

By Lemma 4.9, similarly to the proof of Theorem 3.22, one can show the following result.

Theorem 4.12. *Let Condition 4.1 hold and $F(\cdot,0,0) \in L^1_{\mathbb{F}}(0,T;L^2(\Omega;H))$. Then, for each $y_T \in L^2_{\mathcal{F}_T}(\Omega;H)$ and $\lambda \in \rho(A)$, (4.56) admits a unique strong solution $(y_\lambda(\cdot),Y_\lambda(\cdot)) \in L^2_{\mathbb{F}}(\Omega;C([0,T];D(A))) \times L^2_{\mathbb{F}}(0,T;\mathcal{L}_2^0)$. Moreover, as $\lambda \to \infty$, $(y_\lambda(\cdot),Y_\lambda(\cdot))$ converges to $(y(\cdot),Y(\cdot))$ (in $L^2_{\mathbb{F}}(\Omega;C([0,T];H)) \times L^2_{\mathbb{F}}(0,T;\mathcal{L}_2^0)$), the mild solution to (4.20).*

4.3 The Case of General Filtration

In this section, we shall employ the stochastic transposition method (developed first in our paper [241] for backward stochastic differential equations) to introduce another concept of solution, i.e., the transposition solution, to the equation (4.20). This sort of equation will play an important role in establishing the Pontryagin type maximum principle for optimal control problems of stochastic evolution equations in infinite dimensions. It is notable that we do not assume the condition that **F** is the natural filtration, and therefore we cannot use the Martingale Representation Theorem (Theorem 2.147) anymore. Instead, as we shall see later, the Riesz-type representation theorem presented in Chapter 2 (i.e., Theorem 2.55 or more precisely Theorem 2.73, established in [239]) will play a fundamental role to derive the well-posedness of (4.20) in the sense of transposition solution.

Throughout this section, we assume that $y_T \in L^p_{\mathcal{F}_T}(\Omega; H))$ for a given $p \in (1,2]$ unless other stated.

In order to define the transposition solution to (4.20), we introduce the following (forward) stochastic evolution equation:

$$
\begin{cases} dz = (A^* z + v_1)ds + v_2 dW(s) \ \text{ in } (t,T], \\ z(t) = \eta, \end{cases} \tag{4.57}
$$

where $t \in [0,T]$, $v_1 \in L^1_{\mathbb{F}}(t,T;L^q(\Omega;H))$, $v_2 \in L^q_{\mathbb{F}}(\Omega;L^2(t,T;\mathcal{L}_2^0))$, $\eta \in L^q_{\mathcal{F}_t}(\Omega;H)$, and $q = \frac{p}{p-1}$. By Theorem 3.13 and noting (2.30), the equation (4.57) admits a unique mild solution $z \in C_{\mathbb{F}}([t,T];L^q(\Omega;H))$, and

$$
\begin{aligned}
&|z|_{C_{\mathbb{F}}([t,T];L^q(\Omega;H))} \\
&\le \mathcal{C}\big(|\eta|_{L^q_{\mathcal{F}_t}(\Omega;H)} + |v_1|_{L^1_{\mathbb{F}}(t,T;L^q(\Omega;H))} + |v_2|_{L^q_{\mathbb{F}}(\Omega;L^2(t,T;\mathcal{L}_2^0))}\big).
\end{aligned} \tag{4.58}
$$

We now introduce the following notion.

Definition 4.13. *We call* $(y(\cdot),Y(\cdot)) \in D_{\mathbb{F}}([0,T];L^p(\Omega;H)) \times L^2_{\mathbb{F}}(0,T;L^p(\Omega;\mathcal{L}_2^0))$ *a transposition solution to the equation (4.20) if for any* $t \in [0,T]$, $v_1(\cdot) \in L^1_{\mathbb{F}}(t,T;L^q(\Omega;H))$, $v_2(\cdot) \in L^q_{\mathbb{F}}(\Omega;L^2(t,T;\mathcal{L}_2^0))$, $\eta \in L^q_{\mathcal{F}_t}(\Omega;H)$ *and the corresponding solution* $z \in C_{\mathbb{F}}([t,T];L^q(\Omega;H))$ *to (4.57), it holds that* $\big\langle z(\cdot), F(\cdot,y(\cdot),Y(\cdot))\big\rangle_H \in L^1_{\mathbb{F}}(\Omega;L^1(t,T))$, *and*

$$
\begin{aligned}
&\mathbb{E}\big\langle z(T), y_T\big\rangle_H - \mathbb{E}\int_t^T \big\langle z(s), F(s,y(s),Y(s))\big\rangle_H ds \\
&= \mathbb{E}\big\langle \eta, y(t)\big\rangle_H + \mathbb{E}\int_t^T \big\langle v_1(s), y(s)\big\rangle_H ds + \mathbb{E}\int_t^T \big\langle v_2(s), Y(s)\big\rangle_{\mathcal{L}_2^0} ds.
\end{aligned} \tag{4.59}
$$

Before showing the well-posedness of (4.20) in the sense of transposition solution, we collect some preliminaries.

First, we need the following technical result, which can be viewed as a variant of the classical Lebesgue Theorem (on the Lebesgue points).

Lemma 4.14. *Assume that* $\alpha \in [1,\infty)$, $\alpha' = \begin{cases} \frac{\alpha}{\alpha-1}, & \text{if } \alpha \in (1,\infty), \\ \infty, & \text{if } \alpha = 1, \end{cases}$ $r \in (1,\infty)$, $r' = \frac{r}{r-1}$, $f_1 \in L^r_{\mathbb{F}}(0,T;L^{\alpha}(\Omega;H))$ *and* $f_2 \in L^{r'}_{\mathbb{F}}(0,T;L^{\alpha'}(\Omega;H))$. *Then there exists a monotonic sequence* $\{h_n\}_{n=1}^{\infty}$ *of positive numbers such that* $\lim_{n\to\infty} h_n = 0$, *and*

$$\lim_{n\to\infty}\frac{1}{h_n}\int_t^{t+h_n}\mathbb{E}\langle f_1(t), f_2(\tau)\rangle_H d\tau = \mathbb{E}\langle f_1(t), f_2(t)\rangle_H, \quad \text{a.e. } t\in[0,T].$$

Proof: Write $\tilde f_2 = \begin{cases} f_2, & t\in[0,T], \\ 0, & t\in(T,2T]. \end{cases}$ Then, $\tilde f_2 \in L^{r'}_{\mathbb{F}}(0,2T;L^{\alpha'}(\Omega;H))$ and $|\tilde f_2|_{L^{r'}_{\mathbb{F}}(0,2T;L^{\alpha'}(\Omega;H))} = |\tilde f_2|_{L^{r'}_{\mathbb{F}}(0,T;L^{\alpha'}(\Omega;H))} = |f_2|_{L^{r'}_{\mathbb{F}}(0,T;L^{\alpha'}(\Omega;H))}$. By Lemma 2.76, for any $\varepsilon > 0$, one can find an $f_2^0 \in C_{\mathbb{F}}([0,2T];L^{\alpha'}(\Omega;H))$ such that

$$|\tilde f_2 - f_2^0|_{L^{r'}_{\mathbb{F}}(0,2T;L^{\alpha'}(\Omega;H))} \le \varepsilon. \tag{4.60}$$

Thanks to (4.60), we find that

$$\begin{aligned}
&\int_0^T \Big|\mathbb{E}\langle f_1(t), \tilde f_2(t)\rangle_H - \mathbb{E}\big(f_1(t), f_2^0(t)\rangle_H\Big|\, dt \\
&\le |f_1|_{L^r_{\mathbb{F}}(0,T;L^{\alpha}(\Omega;H))}|\tilde f_2 - f_2^0|_{L^{r'}_{\mathbb{F}}(0,2T;L^{\alpha'}(\Omega;H))} \le \varepsilon |f_1|_{L^r_{\mathbb{F}}(0,T;L^{\alpha}(\Omega;H))}.
\end{aligned} \tag{4.61}$$

Utilizing (4.60) again, for $h\in(0,T)$, we see that

$$\begin{aligned}
&\int_0^T \Big|\frac{1}{h}\int_t^{t+h}\mathbb{E}\langle f_1(t), \tilde f_2(\tau)\rangle_H d\tau - \frac{1}{h}\int_t^{t+h}\mathbb{E}\langle f_1(t), f_2^0(\tau)\rangle_H d\tau\Big| dt \\
&= \frac{1}{h}\int_0^T \Big|\int_t^{t+h}\mathbb{E}\langle f_1(t), \tilde f_2(\tau) - f_2^0(\tau)\rangle_H d\tau\Big| dt \\
&\le \frac{1}{h}\int_0^T\int_t^{t+h} |f_1(t)|_{L^{\alpha}_{\mathcal{F}_T}(\Omega;H)}|\tilde f_2(\tau) - f_2^0(\tau)|_{L^{\alpha'}_{\mathcal{F}_T}(\Omega;H)} d\tau dt \\
&\le \frac{1}{h}\Big(\int_0^T\int_t^{t+h}|f_1(t)|^r_{L^{\alpha}_{\mathcal{F}_T}(\Omega;H)}d\tau dt\Big)^{\frac{1}{r}}\Big(\int_0^T\int_t^{t+h}|\tilde f_2(\tau)-f_2^0(\tau)|^{r'}_{L^{\alpha'}_{\mathcal{F}_T}(\Omega;H)}d\tau dt\Big)^{\frac{1}{r'}} \\
&= |f_1|_{L^r_{\mathbb{F}}(0,T;L^{\alpha}(\Omega;H))}\Big(\frac{1}{h}\int_0^T\int_0^h |\tilde f_2(t+\tau) - f_2^0(t+\tau)|^{r'}_{L^{\alpha'}_{\mathcal{F}_T}(\Omega;H)} d\tau dt\Big)^{1/r'} \\
&= |f_1|_{L^r_{\mathbb{F}}(0,T;L^{\alpha}(\Omega;H))}\Big(\frac{1}{h}\int_0^h\int_{\tau}^{T+\tau} |\tilde f_2(t) - f_2^0(t)|^{r'}_{L^{\alpha'}_{\mathcal{F}_T}(\Omega;H)} dt d\tau\Big)^{1/r'} \\
&\le |f_1|_{L^r_{\mathbb{F}}(0,T;L^{\alpha}(\Omega;H))}\Big(\frac{1}{h}\int_0^h\int_0^{2T} |\tilde f_2(t) - f_2^0(t)|^{r'}_{L^{\alpha'}_{\mathcal{F}_T}(\Omega;H)} dt d\tau\Big)^{1/r'} \\
&\le \varepsilon |f_1|_{L^r_{\mathbb{F}}(0,T;L^{\alpha}(\Omega;H))}.
\end{aligned} \tag{4.62}$$

On the other hand, by the uniform continuity of $f_2^0(\cdot)$ in $L^{\alpha'}_{\mathcal{F}_T}(\Omega;H)$, one can find a $\delta = \delta(\varepsilon) > 0$ such that

$$|f_2^0(s_1) - f_2^0(s_2)|_{L^{\alpha'}_{\mathcal{F}_T}(\Omega;H)} \leq \varepsilon, \quad \forall\, s_1, s_2 \in [0, 2T] \text{ satisfying } |s_1 - s_2| \leq \delta.$$

Hence, for each $h \in (0, \delta]$, we have

$$\begin{aligned}
&\int_0^T \Big| \frac{1}{h} \int_t^{t+h} \mathbb{E}\langle f_1(t), f_2^0(\tau)\rangle_H d\tau - \mathbb{E}\langle f_1(t), f_2^0(t)\rangle_H \Big| dt \\
&= \frac{1}{h} \int_0^T \Big| \int_t^{t+h} \mathbb{E}\langle f_1(t), f_2^0(\tau) - f_2^0(t)\rangle_H d\tau \Big| dt \\
&\leq \frac{1}{h} \int_0^T \int_t^{t+h} |f_1(t)|_{L^{\alpha}_{\mathcal{F}_T}(\Omega;H)} |f_2^0(\tau) - f_2^0(t)|_{L^{\alpha'}_{\mathcal{F}_T}(\Omega;H)} d\tau dt \\
&\leq \varepsilon \int_0^T |f_1(t)|_{L^{\alpha}_{\mathcal{F}_T}(\Omega;H)} dt \leq C\varepsilon |f_1|_{L^r_{\mathbb{F}}(0,T;L^\alpha(\Omega;H))}.
\end{aligned} \tag{4.63}$$

By (4.61), (4.62) and (4.63), we conclude that, for $h \in (0, \delta]$

$$\int_0^T \left| \frac{1}{h} \int_t^{t+h} \mathbb{E}\langle f_1(t), \tilde{f}_2(\tau)\rangle_H d\tau - \mathbb{E}\langle f_1(t), \tilde{f}_2(t)\rangle_H \right| dt \leq C\varepsilon |f_1|_{L^r_{\mathbb{F}}(0,T;L^\alpha(\Omega;H))}.$$

This leads to

$$\lim_{h\to 0} \int_0^T \left| \frac{1}{h} \int_t^{t+h} \mathbb{E}\langle f_1(t), \tilde{f}_2(\tau)\rangle_H d\tau - \mathbb{E}\langle f_1(t), \tilde{f}_2(t)\rangle_H \right| dt = 0.$$

Hence, one can find a monotonic sequence $\{h_n\}_{n=1}^\infty$ of positive numbers with $\lim_{n\to\infty} h_n = 0$ such that

$$\lim_{n\to\infty} \frac{1}{h_n} \int_t^{t+h_n} \mathbb{E}\langle f_1(t), \tilde{f}_2(\tau)\rangle_H d\tau = \mathbb{E}\langle f_1(t), \tilde{f}_2(t)\rangle_H, \quad \text{a.e. } t \in [0, T].$$

By this and the definition of $\tilde{f}_2(\cdot)$, we conclude that

$$\begin{aligned}
&\lim_{n\to\infty} \frac{1}{h_n} \int_t^{t+h_n} \mathbb{E}\langle f_1(t), f_2(\tau)\rangle_H d\tau = \lim_{n\to\infty} \frac{1}{h_n} \int_t^{t+h_n} \mathbb{E}\langle f_1(t), \tilde{f}_2(\tau)\rangle_H d\tau \\
&= \mathbb{E}\langle f_1(t), \tilde{f}_2(t)\rangle_H = \mathbb{E}\langle f_1(t), f_2(t)\rangle_H, \quad \text{a.e. } t \in [0, T].
\end{aligned}$$

This completes the proof of Lemma 4.14. □

We also need the following technical result.

Lemma 4.15. *If $p \in [1, \infty]$, $y_T \in L^p_{\mathcal{F}_T}(\Omega; H)$ and $f(\cdot) \in L^p_{\mathbb{F}}(\Omega; L^1(0, T; H))$, then the process*

$$\zeta^{\cdot} \stackrel{\Delta}{=} \mathbb{E}\Big(S(T - \cdot) y_T - \int_{\cdot}^T S(s - \cdot) f(s) ds \,\Big|\, \mathcal{F}_{\cdot} \Big) \tag{4.64}$$

has a càdlàg modification in $L^p_{\mathbb{F}}(\Omega; D([0, T]; H))$.

Proof: We consider only the case $p \in [1, \infty)$. Recall that for any $\lambda \in \rho(A)$, the Yosida approximation $A_\lambda \stackrel{\Delta}{=} \lambda^2(\lambda I - A)^{-1} - \lambda I$ of A generates a C_0-group $\{S_\lambda(t)\}_{t\in\mathbb{R}}$ on H (e.g., [372]).

For $t \in [0, T]$, put

$$\zeta_\lambda^t \stackrel{\Delta}{=} \mathbb{E}\Big(S_\lambda(T-t)y_T - \int_t^T S_\lambda(s-t)f(s)ds \,\Big|\, \mathcal{F}_t\Big) \tag{4.65}$$

and

$$X_\lambda(t) \stackrel{\Delta}{=} S_\lambda(t)\zeta_\lambda^t - \int_0^t S_\lambda(s)f(s)ds. \tag{4.66}$$

It is easy to see that $X_\lambda(\cdot) = \mathbb{E}\Big(S_\lambda(T)y_T - \int_0^T S_\lambda(s)f(s)ds \,\Big|\, \mathcal{F}_\cdot\Big)$ is an H-valued $\mathbb{F}$-martingale. By Theorem 2.111, it enjoys a càdlàg modification in $L^p_{\mathbb{F}}(\Omega; D([0,T]; H))$, and hence so does the following process

$$\zeta_\lambda^t = S_\lambda(-t)\Big[X_\lambda(t) + \int_0^t S_\lambda(s)f(s)ds\Big].$$

Here we have used the fact that $\{S_\lambda(t)\}_{t\in\mathbb{R}}$ is a C_0-group on H. We still use $\zeta_\lambda^\cdot$ to stand for its càdlàg modification.

From (4.64) and (4.65), it follows that

$$\begin{aligned}
&\lim_{\lambda\to\infty} |\zeta^\cdot - \zeta_\lambda^\cdot|_{L^p_{\mathbb{F}}(\Omega;L^\infty(0,T;H))}\\
&\le \lim_{\lambda\to\infty} \Big|S(T-\cdot)y_T - S_\lambda(T-\cdot)y_T\Big|_{L^p_{\mathbb{F}}(\Omega;L^\infty(0,T;H))}\\
&\quad + \lim_{\lambda\to\infty} \Big|\int_\cdot^T S(s-\cdot)f(s)ds - \int_\cdot^T S_\lambda(s-\cdot)f(s)ds\Big|_{L^p_{\mathbb{F}}(\Omega;L^\infty(0,T;H))}.
\end{aligned} \tag{4.67}$$

Let us prove the right hand side of (4.67) equals zero. For this purpose, noting that the set of simple functions is dense in $L^p_{\mathcal{F}_T}(\Omega; H)$, we conclude that there exists a sequence $\{y^m\}_{m=1}^\infty \subset L^p_{\mathcal{F}_T}(\Omega; H)$ satisfying the following two conditions:

1) $y^m = \sum_{k=1}^{N_m} \alpha_k^m \chi_{\Omega_k^m}(\omega)$, where $N_m \in \mathbb{N}$, $\alpha_k^m \in H$ and $\Omega_k^m \in \mathcal{F}_T$ with $\{\Omega_k^m\}_{k=1}^{N_m}$ to be a partition of Ω; and

2) $\lim_{m\to\infty} |y^m - y_T|_{L^p_{\mathcal{F}_T}(\Omega;H)} = 0$.

Likewise, since the set of simple adapted processes is dense in $L^p_{\mathbb{F}}(\Omega; L^1(0,T; H))$, there exists a sequence $\{f^m\}_{m=1}^\infty \subset L^p_{\mathbb{F}}(\Omega; L^1(0,T;H))$ satisfying the following two conditions:

i) $f^m = \sum_{j=1}^{L_m} \sum_{k=1}^{M_j^m} \alpha_{j,k}^m \chi_{\Omega_{j,k}^m}(\omega)\chi_{[t_j^m, t_{j+1}^m)}(t)$, where $L_m \in \mathbb{N}$, $M_j^m \in \mathbb{N}$, $\alpha_{j,k}^m \in H$, $\Omega_{j,k}^m \in \mathcal{F}_{t_j^m}$ with $\{\Omega_{j,k}^m\}_{k=1}^{M_j^m}$ being a partition of Ω, and $0 = t_1^m < t_2^m \cdots < t_{J_m}^m < t_{J_m+1}^m = T$; and

ii) $\lim_{m\to\infty} |f^m - f|_{L^p_{\mathbb{F}}(\Omega;L^1(0,T;H))} = 0$.

Now, let us prove that

$$\lim_{\lambda\to\infty}\Big|S(T-\cdot)y_T - S_\lambda(T-\cdot)y_T\Big|_{L^p_{\mathbb{F}}(\Omega;L^\infty(0,T;H))} = 0. \tag{4.68}$$

Since $\{S(t)\}_{t\geq 0}$ is a C_0-semigroup, for any $\varepsilon > 0$, there is an $M > 0$ such that for any $m > M$, it holds that

$$|S(T-\cdot)y_T - S(T-\cdot)y^m - S_\lambda(T-\cdot)y_T + S_\lambda(T-\cdot)y^m|_{L^p_{\mathbb{F}}(\Omega;L^\infty(0,T;H))} < \frac{\varepsilon}{2}.$$

On the other hand, by the property of Yosida approximations, we deduce that for any $\alpha \in H$, it holds that $\lim_{\lambda\to\infty}|S(T-\cdot)\alpha - S_\lambda(T-\cdot)\alpha|_{L^\infty(0,T;H)} = 0$. Thus, there is a $\Lambda = \Lambda(m) > 0$ such that for any $\lambda > \Lambda$, it holds that

$$|S(T-\cdot)\alpha_k^m - S_\lambda(T-\cdot)\alpha_k^m|_{L^\infty(0,T;H)} < \frac{\varepsilon}{2N_m}, \qquad k = 1, 2, \cdots, N_m,$$

which implies that

$$\begin{aligned}&|S(T-\cdot)y^m - S_\lambda(T-\cdot)y^m|_{L^p_{\mathbb{F}}(\Omega;L^\infty(0,T;H))}\\&\leq \sum_{k=1}^{N_m}|S(T-\cdot)\alpha_k^m - S_\lambda(T-\cdot)\alpha_k^m|_{L^\infty(0,T;H)} < \frac{\varepsilon}{2}.\end{aligned}$$

Therefore, for each $m > M$, there is a $\Lambda = \Lambda(m)$ such that when $\lambda > \Lambda(m)$, it holds that

$$\begin{aligned}&\Big|S(T-\cdot)y_T - S_\lambda(T-\cdot)y_T\Big|_{L^p_{\mathbb{F}}(\Omega;L^\infty(0,T;H))}\\&\leq |S(T-\cdot)y_T - S_\lambda(T-\cdot)y_T - S(T-\cdot)y^m + S_\lambda(T-\cdot)y^m|_{L^p_{\mathbb{F}}(\Omega;L^\infty(0,T;H))}\\&\quad+|S(T-\cdot)y^m - S_\lambda(T-\cdot)y^m|_{L^p_{\mathbb{F}}(\Omega;L^\infty(0,T;H))}\\&< \frac{\varepsilon}{2} + \frac{\varepsilon}{2} = \varepsilon.\end{aligned}$$

This gives (4.68).

Further, we show that

$$\lim_{\lambda\to\infty}\Big|\int_\cdot^T S(s-\cdot)f(s)ds - \int_\cdot^T S_\lambda(s-\cdot)f(s)ds\Big|_{L^p_{\mathbb{F}}(\Omega;L^\infty(0,T;H))} = 0. \tag{4.69}$$

For any $\varepsilon > 0$, there is a $M^* > 0$ such that for any $m > M^*$,

$$\begin{aligned}&\Big|\int_\cdot^T S(s-\cdot)f(s)ds - \int_\cdot^T S(s-\cdot)f^m(s)ds\Big|_{L^p_{\mathbb{F}}(\Omega;L^\infty(0,T;H))}\\&\quad+\Big|\int_\cdot^T S_\lambda(s-\cdot)f(s)ds - \int_\cdot^T S_\lambda(s-\cdot)f^m(s)ds\Big|_{L^p_{\mathbb{F}}(\Omega;L^\infty(0,T;H))}\\&< \frac{\varepsilon}{2}.\end{aligned}$$

By the property of Yosida approximations again, for any $\alpha \in H$, it holds that

$$\lim_{\lambda\to\infty}\Big|\int_\cdot^T S(s-\cdot)\alpha ds-\int_\cdot^T S_\lambda(s-\cdot)\alpha ds\Big|_{L^\infty(0,T;H)}=0.$$

Hence, there is a $\Lambda^*=\Lambda^*(m)>0$ such that for any $\lambda>\Lambda^*$,

$$\begin{aligned}&\Big|\int_\cdot^T S(s-\cdot)\alpha_{j,k}^m ds-\int_\cdot^T S_\lambda(s-\cdot)\alpha_{j,k}^m ds\Big|_{L^\infty(0,T;H)}\\&<\frac{\varepsilon}{2J_m\max(M_1^m,M_2^m,\cdots,M_{J_m}^m)},\qquad j=1,2,\cdots,L_m;\ k=1,2,\cdots,M_j^m.\end{aligned}$$

This implies that

$$\begin{aligned}&\Big|\int_\cdot^T S(s-\cdot)f^m(s)ds-\int_\cdot^T S_\lambda(s-\cdot)f^m(s)ds\Big|_{L^p_{\mathbb F}(\Omega;L^\infty(0,T;H))}\\&\le\sum_{j=1}^{L_m}\sum_{k=1}^{M_j^m}\Big|\int_\cdot^T S(s-\cdot)\alpha_{j,k}^m ds-\int_\cdot^T S_\lambda(s-\cdot)\alpha_{j,k}^m ds\Big|_{L^\infty(0,T;H)}<\frac{\varepsilon}{2}.\end{aligned}$$

Therefore, for any $m>M^*$ and $\lambda>\Lambda^*=\Lambda^*(m)$, we have

$$\Big|\int_\cdot^T S(s-\cdot)f(s)ds-\int_\cdot^T S_\lambda(s-\cdot)f(s)ds\Big|_{L^p_{\mathbb F}(\Omega;L^\infty(0,T;H))}<\frac{\varepsilon}{2}+\frac{\varepsilon}{2}=\varepsilon,$$

which gives (4.69).

By (4.67), (4.68) and (4.69), we obtain that

$$\lim_{\lambda\to\infty}|\zeta^\cdot-\zeta^\cdot_\lambda|_{L^p_{\mathbb F}(\Omega;L^\infty(0,T;H))}=0.$$

Since $\zeta^\cdot_\lambda\in L^p_{\mathbb F}(\Omega;D([0,T];H))$, we see that $\zeta^\cdot$ enjoys a cádlág modification in $L^p_{\mathbb F}(\Omega;D([0,T];H))$. □

We have the following well-posedness result for a special case of (4.20), i.e., the linear equation (4.21), in the sense of transposition solution.

Theorem 4.16. *For any $y_T\in L^p_{\mathcal F_T}(\Omega;H)$ and $f(\cdot)\in L^1_{\mathbb F}(0,T;L^p(\Omega;H))$, the equation* (4.21) *admits one and only one transposition solution $(y(\cdot),Y(\cdot))\in D_{\mathbb F}([0,T];L^p(\Omega;H))\times L^2_{\mathbb F}(0,T;L^p(\Omega;\mathcal L_2^0))$. Furthermore,*

$$\begin{aligned}&|(y(\cdot),Y(\cdot))|_{D_{\mathbb F}([0,T];L^p(\Omega;H))\times L^2_{\mathbb F}(0,T;L^p(\Omega;\mathcal L_2^0))}\\&\le\mathcal C\big(|y_T|_{L^p_{\mathcal F_T}(\Omega;H)}+|f(\cdot)|_{L^1_{\mathbb F}(0,T;L^p(\Omega;H))}\big).\end{aligned}\tag{4.70}$$

Proof: The proof is divided into four steps.

Step 1. For any $t\in[0,T]$, we define a linear functional ℓ (depending on t) on the Banach space $L^1_{\mathbb F}(t,T;L^q(\Omega;H))\times L^2_{\mathbb F}(t,T;L^q(\Omega;\mathcal L_2^0))\times$

$L^q_{\mathcal{F}_t}(\Omega;H)$ as follows: First, for any $\big(v_1(\cdot),v_2(\cdot),\eta\big)\in L^1_{\mathbb{F}}(t,T;L^q(\Omega;H))\times L^q_{\mathbb{F}}(\Omega;L^2(t,T;\mathcal{L}_2^0))\times L^q_{\mathcal{F}_t}(\Omega;H)$,

$$\ell\big(v_1(\cdot),v_2(\cdot),\eta\big)\stackrel{\Delta}{=}\mathbb{E}\big\langle z(T),y_T\big\rangle_H-\mathbb{E}\int_t^T\big\langle z(s),f(s)\big\rangle_H ds, \tag{4.71}$$

where $z(\cdot)\in C_{\mathbb{F}}([t,T];L^q(\Omega;H))$ solves the equation (4.57). By the Hölder inequality and noting (4.58), we see that

$$\begin{aligned}
&\big|\ell\big(v_1(\cdot),v_2(\cdot),\eta\big)\big|\\
&\leq|z(T)|_{L^q_{\mathcal{F}_T}(\Omega;H)}|y_T|_{L^p_{\mathcal{F}_T}(\Omega;H)}+|z(\cdot)|_{C_{\mathbb{F}}([t,T];L^q(\Omega;H))}|f|_{L^1_{\mathbb{F}}(t,T;L^p(\Omega;H))}\\
&\leq\mathcal{C}\Big(|y_T|_{L^p_{\mathcal{F}_T}(\Omega;H)}+|f(\cdot)|_{L^1_{\mathbb{F}}(t,T;L^p(\Omega;H))}\Big)\\
&\qquad\times\big|\big(v_1(\cdot),v_2(\cdot),\eta\big)\big|_{L^1_{\mathbb{F}}(t,T;L^q(\Omega;H))\times L^q_{\mathbb{F}}(\Omega;L^2(t,T;\mathcal{L}_2^0))\times L^q_{\mathcal{F}_t}(\Omega;H)},\quad\forall\, t\in[0,T],
\end{aligned} \tag{4.72}$$

where the positive constant $\mathcal{C}=\mathcal{C}(T,A)$ is independent of t. From (4.72), it follows that ℓ is a bounded linear functional on $L^1_{\mathbb{F}}(t,T;L^q(\Omega;H))\times L^q_{\mathbb{F}}(\Omega;L^2(t,T;\mathcal{L}_2^0))\times L^q_{\mathcal{F}_t}(\Omega;H)$. Since $q\in[2,\infty)$, by (2.30), it is easy to see that $L^q_{\mathbb{F}}(\Omega;L^2(t,T;\mathcal{L}_2^0))\supset L^2_{\mathbb{F}}(t,T;L^q(\Omega;\mathcal{L}_2^0))$. Then, ℓ is also a bounded linear functional on $L^1_{\mathbb{F}}(t,T;L^q(\Omega;H))\times L^2_{\mathbb{F}}(t,T;L^q(\Omega;\mathcal{L}_2^0))\times L^q_{\mathcal{F}_t}(\Omega;H)$.

Thanks to Theorem 2.73, there is a triple $\big(y^t(\cdot),Y^t(\cdot),\xi^t\big)\in L^\infty_{\mathbb{F}}(t,T;L^p(\Omega;H))\times L^2_{\mathbb{F}}(t,T;L^p(\Omega;\mathcal{L}_2^0))\times L^p_{\mathcal{F}_t}(\Omega;H)$ such that

$$\begin{aligned}
&\mathbb{E}\big\langle z(T),y_T\big\rangle_H-\mathbb{E}\int_t^T\big\langle z(\tau),f(\tau)\big\rangle_H d\tau\\
&=\mathbb{E}\int_t^T\big\langle v_1(\tau),y^t(\tau)\big\rangle_H d\tau+\mathbb{E}\int_t^T\big\langle v_2(\tau),Y^t(\tau)\big\rangle_{\mathcal{L}_2^0}d\tau+\mathbb{E}\big\langle\eta,\xi^t\big\rangle_H,\\
&\quad\forall\,\big(v_1(\cdot),v_2(\cdot),\eta\big)\in L^1_{\mathbb{F}}(t,T;L^q(\Omega;H))\times L^q_{\mathbb{F}}(\Omega;L^2(t,T;\mathcal{L}_2^0))\times L^q_{\mathcal{F}_t}(\Omega;H).
\end{aligned} \tag{4.73}$$

It is clear that $\xi^T=y_T$. Furthermore, there is a positive constant $\mathcal{C}=\mathcal{C}(T,A)$, independent of t, such that

$$\begin{aligned}
&|(y^t(\cdot),Y^t(\cdot),\xi^t)|_{L^\infty_{\mathbb{F}}(t,T;L^p(\Omega;H))\times L^2_{\mathbb{F}}(t,T;L^p(\Omega;\mathcal{L}_2^0))\times L^p_{\mathcal{F}_t}(\Omega;H)}\\
&\leq\mathcal{C}\Big(|y_T|_{L^p_{\mathcal{F}_T}(\Omega;H)}+|f(\cdot)|_{L^1_{\mathbb{F}}(t,T;L^p(\Omega;H))}\Big),\qquad\forall\, t\in[0,T].
\end{aligned} \tag{4.74}$$

Step 2. Note that the function $(y^t(\cdot),Y^t(\cdot))$ obtained in Step 1 may depend on t. In this step, we show the time consistency of $(y^t(\cdot),Y^t(\cdot))$, that is, for any t_1 and t_2 with $0\leq t_2\leq t_1\leq T$ and a.e. $(\tau,\omega)\in[t_1,T]\times\Omega$, it holds that

$$\big(y^{t_2}(\tau,\omega),Y^{t_2}(\tau,\omega)\big)=\big(y^{t_1}(\tau,\omega),Y^{t_1}(\tau,\omega)\big). \tag{4.75}$$

Since the solution $z(\cdot)$ to (4.57) depends on t, we denote it by $z^t(\cdot)$ whenever there exists a possible confusion. To show (4.75), we fix arbitrarily

$\varrho(\cdot) \in L^1_{\mathbb{F}}(t_1, T; L^q(\Omega; H))$ and $\varsigma(\cdot) \in L^q_{\mathbb{F}}(\Omega; L^2(t_1, T; \mathcal{L}_2^0))$, and choose first $t = t_1$, $\eta = 0$, $v_1(\cdot) = \varrho(\cdot)$ and $v_2(\cdot) = \varsigma(\cdot)$ in (4.57). From (4.73), we obtain that

$$\begin{aligned} &\mathbb{E}\big\langle z^{t_1}(T), y_T\big\rangle_H - \mathbb{E}\int_{t_1}^T \big\langle z^{t_1}(\tau), f(\tau)\big\rangle_H d\tau \\ &= \mathbb{E}\int_{t_1}^T \big\langle \varrho(\tau), y^{t_1}(\tau)\big\rangle_H d\tau + \mathbb{E}\int_{t_1}^T \big\langle \varsigma(\tau), Y^{t_1}(\tau)\big\rangle_{\mathcal{L}_2^0} d\tau. \end{aligned} \tag{4.76}$$

Then, choose $t = t_2$, $\eta = 0$, $v_1(t,\omega) = \chi_{[t_1,T]}(t)\varrho(t,\omega)$ and $v_2(t,\omega) = \chi_{[t_1,T]}(t)\varsigma(t,\omega)$ in (4.57). It is clear that $z^{t_2}(\cdot) = \begin{cases} z^{t_1}(\cdot), & t \in [t_1, T], \\ 0, & t \in [t_2, t_1). \end{cases}$ From the equality (4.73), it follows that

$$\begin{aligned} &\mathbb{E}\big\langle z^{t_1}(T), y_T\big\rangle_H - \mathbb{E}\int_{t_1}^T \big\langle z^{t_1}(\tau), f(\tau)\big\rangle_H d\tau \\ &= \mathbb{E}\int_{t_1}^T \big\langle \varrho(\tau), y^{t_2}(\tau)\big\rangle_H d\tau + \mathbb{E}\int_{t_1}^T \big\langle \varsigma(\tau), Y^{t_2}(\tau)\big\rangle_{\mathcal{L}_2^0} d\tau. \end{aligned} \tag{4.77}$$

Combining (4.76) and (4.77), we obtain that

$$\begin{aligned} &\mathbb{E}\int_{t_1}^T \big\langle \varrho(\tau), y^{t_1}(\tau) - y^{t_2}(\tau)\big\rangle_H d\tau = \mathbb{E}\int_{t_1}^T \big\langle \varsigma(\tau), Y^{t_1}(\tau) - Y^{t_2}(\tau)\big\rangle_{\mathcal{L}_2^0} d\tau, \\ &\qquad\qquad \forall\ \varrho(\cdot) \in L^1_{\mathbb{F}}(t_1, T; L^q(\Omega; H)), \quad \varsigma(\cdot) \in L^q_{\mathbb{F}}(\Omega; L^2(t_1, T; \mathcal{L}_2^0)). \end{aligned}$$

This yields the desired equality (4.75).

Put

$$y(t,\omega) = y^0(t,\omega), \quad Y(t,\omega) = Y^0(t,\omega), \quad \forall\ (t,\omega) \in [0,T] \times \Omega. \tag{4.78}$$

From (4.75), it follows that

$$\big(y^t(\tau,\omega), Y^t(\tau,\omega)\big) = \big(y(\tau,\omega), Y(\tau,\omega)\big), \quad \text{a.e. } (\tau,\omega) \in [t,T] \times \Omega. \tag{4.79}$$

Combining (4.73) and (4.79), we end up with

$$\begin{aligned} &\mathbb{E}\big\langle z(T), y_T\big\rangle_H - \mathbb{E}\int_t^T \big\langle z(\tau), f(\tau)\big\rangle_H d\tau \\ &= \mathbb{E}\big\langle \eta, \xi^t\big\rangle_H + \mathbb{E}\int_t^T \big\langle v_1(\tau), y(\tau)\big\rangle_H d\tau + \mathbb{E}\int_t^T \big\langle v_2(\tau), Y(\tau)\big\rangle_{\mathcal{L}_2^0} d\tau, \\ &\quad \forall\ \big(v_1(\cdot), v_2(\cdot), \eta\big) \in L^1_{\mathbb{F}}(t, T; L^q(\Omega; H)) \times L^q_{\mathbb{F}}(\Omega; L^2(t, T; \mathcal{L}_2^0)) \times L^q_{\mathcal{F}_t}(\Omega; H). \end{aligned} \tag{4.80}$$

Step 3. We show in this step that $\xi^\cdot$ has a càdlàg modification in the space $L^p_{\mathbb{F}}(\Omega; D([0,T]; H))$.

We claim that, for each $t \in [0,T]$,

$$\mathbb{E}\Big(S(T-t)y_T - \int_t^T S(s-t)f(s)ds \,\Big|\, \mathcal{F}_t\Big) = \xi^t, \quad \text{a.s.} \tag{4.81}$$

To prove this, we note that for any $\eta \in L^q_{\mathcal{F}_t}(\Omega; H)$, $v_1 = 0$ and $v_2 = 0$, the corresponding solution to (4.57) is given by $z(s) = S^*(s-t)\eta$ for $s \in [t,T]$. Hence, by (4.80), we obtain that

$$\mathbb{E}\big\langle S^*(T-t)\eta, y_T\big\rangle_H - \mathbb{E}\langle \eta, \xi^t\rangle_H = \mathbb{E}\int_t^T \big\langle S^*(s-t)\eta, f(s)\big\rangle_H ds. \tag{4.82}$$

Noting that

$$\mathbb{E}\big\langle S^*(T-t)\eta, y_T\big\rangle_H = \mathbb{E}\big\langle \eta, S(T-t)y_T\big\rangle_H = \mathbb{E}\big\langle \eta, \mathbb{E}(S(T-t)y_T \mid \mathcal{F}_t)\big\rangle_H$$

and

$$\begin{aligned}\mathbb{E}\int_t^T \big\langle S^*(s-t)\eta, f(s)\big\rangle_H ds &= \mathbb{E}\Big\langle \eta, \int_t^T S(s-t)f(s)ds\Big\rangle_H \\ &= \mathbb{E}\Big\langle \eta, \mathbb{E}\Big(\int_t^T S(s-t)f(s)ds \,\Big|\, \mathcal{F}_t\Big)\Big\rangle_H,\end{aligned}$$

by (4.82), we conclude that, for all $\eta \in L^q_{\mathcal{F}_t}(\Omega; H)$,

$$\mathbb{E}\Big\langle \eta, \mathbb{E}\Big(S(T-t)y_T - \int_t^T S(s-t)f(s)ds \,\Big|\, \mathcal{F}_t\Big) - \xi^t\Big\rangle_H = 0. \tag{4.83}$$

Clearly, (4.81) follows from (4.83) immediately.

Since the process $\mathbb{E}\big(S(T-\cdot)y_T - \int_\cdot^T S(s-\cdot)f(s)ds \mid \mathcal{F}_\cdot\big)$ has a càdlàg modification in $L^p_{\mathbb{F}}(\Omega; D([0,T]; H)) \subset D_{\mathbb{F}}([0,T]; L^p(\Omega; H))$, by Lemma 4.15, we deduce that so does $\xi^\cdot$.

Step 4. In this step, we show that, for a.e. $t \in [0,T]$,

$$\xi^t = y(t), \quad \text{a.s.} \tag{4.84}$$

Fix any $\gamma \in L^q_{\mathcal{F}_{t_2}}(\Omega; H)$. Choosing $t = t_2$, $v_1(\cdot) = 0$, $v_2(\cdot) = 0$ and $\eta = (t_1 - t_2)\gamma$ in (4.57), utilizing (4.80), we obtain that

$$\begin{aligned}&\mathbb{E}\big\langle S^*(T-t_2)((t_1-t_2)\gamma), y_T\big\rangle_H - \mathbb{E}\big\langle (t_1-t_2)\gamma, \xi^{t_2}\big\rangle_H \\ &= \mathbb{E}\int_{t_2}^T \big\langle S^*(\tau-t_2)((t_1-t_2)\gamma), f(\tau)\big\rangle_H d\tau.\end{aligned} \tag{4.85}$$

Choosing $t = t_2$, $v_1(\tau,\omega) = \chi_{[t_2,t_1]}(\tau)\gamma(\omega)$, $v_2(\cdot) = 0$ and $\eta = 0$ in (4.57), utilizing (4.80) again, we find that

$$
\begin{aligned}
&\mathbb{E}\Big\langle \int_{t_2}^{T} S^*(T-s)(\chi_{[t_2,t_1]}(s)\gamma)ds, y_T\Big\rangle_H \\
&= \mathbb{E}\int_{t_2}^{t_1} \Big\langle \int_{t_2}^{\tau} S^*(\tau-s)\gamma ds, f(\tau)\Big\rangle_H d\tau \\
&\quad +\mathbb{E}\int_{t_1}^{T} \Big\langle S^*(\tau-t_1)\int_{t_2}^{t_1} S^*(t_1-s)\gamma ds, f(\tau)\Big\rangle_H d\tau + \mathbb{E}\int_{t_2}^{t_1} \langle \gamma, y(\tau)\rangle_H d\tau.
\end{aligned}
\tag{4.86}
$$

From (4.85) and (4.86), we find that

$$
\begin{aligned}
&\mathbb{E}\langle \gamma, \xi^{t_2}\rangle_H \\
&= \frac{1}{t_1-t_2}\int_{t_2}^{t_1} \mathbb{E}\,\langle \gamma, y(\tau)\rangle_H d\tau + \mathbb{E}\big\langle S^*(T-t_2)\gamma, y_T\big\rangle_H \\
&\quad -\frac{1}{t_1-t_2}\mathbb{E}\Big\langle \int_{t_2}^{T} S^*(T-\tau)(\chi_{[t_2,t_1]}(\tau)\gamma)d\tau, y_T\Big\rangle_H \\
&\quad -\mathbb{E}\int_{t_2}^{T} \langle S^*(\tau-t_2)\gamma, f(\tau)\rangle_H d\tau + \frac{1}{t_1-t_2}\mathbb{E}\int_{t_2}^{t_1}\Big\langle \int_{t_2}^{\tau} S^*(\tau-s)\gamma ds, f(\tau)\Big\rangle_H d\tau \\
&\quad +\frac{1}{t_1-t_2}\mathbb{E}\int_{t_1}^{T} \Big\langle S^*(\tau-t_1)\int_{t_2}^{t_1} S^*(t_1-s)\gamma ds, f(\tau)\Big\rangle_H d\tau.
\end{aligned}
\tag{4.87}
$$

Now we analyze the terms in the right hand side of (4.87) one by one. First, it is easy to show that

$$
\lim_{t_1\to t_2+0}\frac{1}{t_1-t_2}\mathbb{E}\int_{t_2}^{t_1}\Big\langle \int_{t_2}^{\tau} S^*(\tau-s)\gamma ds, f(\tau)\Big\rangle_H d\tau = 0, \quad \forall\, \gamma\in L^q_{\mathcal{F}_{t_2}}(\Omega;H).
\tag{4.88}
$$

Further,

$$
\begin{aligned}
&\lim_{t_1\to t_2+0}\frac{1}{t_1-t_2}\mathbb{E}\Big\langle \int_{t_2}^{T} S^*(T-\tau)(\chi_{[t_2,t_1]}(\tau)\gamma)d\tau, y_T\Big\rangle_H \\
&= \lim_{t_1\to t_2+0}\frac{1}{t_1-t_2}\mathbb{E}\Big\langle \int_{t_2}^{t_1} S^*(T-\tau)\gamma d\tau, y_T\Big\rangle_H \\
&= \mathbb{E}\big\langle S^*(T-t_2)\gamma, y_T\big\rangle_H.
\end{aligned}
\tag{4.89}
$$

Utilizing the semigroup property of $\{S^*(t)\}_{t\geq 0}$, we have

$$
\begin{aligned}
&\lim_{t_1\to t_2+0}\frac{1}{t_1-t_2}\mathbb{E}\int_{t_1}^{T}\Big\langle S^*(\tau-t_1)\int_{t_2}^{t_1} S^*(t_1-s)\gamma ds, f(\tau)\Big\rangle_H d\tau \\
&= \mathbb{E}\int_{t_2}^{T} \langle S^*(\tau-t_2)\gamma, f(\tau)\rangle_H d\tau.
\end{aligned}
\tag{4.90}
$$

From (4.87), (4.88), (4.89) and (4.90), we arrive at

$$\lim_{t_1 \to t_2+0} \frac{1}{t_1 - t_2} \int_{t_2}^{t_1} \mathbb{E}\langle \gamma, y(\tau) \rangle_H d\tau = \mathbb{E}\big\langle \gamma, \xi^{t_2} \big\rangle_H, \tag{4.91}$$

for arbitrary $t_2 \in [0,T)$ and $\gamma \in L^q_{\mathcal{F}_{t_2}}(\Omega; H)$.

By Lemma 4.14, we can find a monotonic sequence $\{h_n\}_{n=1}^\infty$ of positive numbers with $\lim_{n\to\infty} h_n = 0$, such that

$$\lim_{n\to\infty} \frac{1}{h_n} \int_{t_2}^{t_2+h_n} \mathbb{E}\langle \gamma, y(\tau) \rangle_H d\tau = \mathbb{E}\langle \gamma, y(t_2) \rangle_H, \qquad \text{a.e. } t_2 \in [0,T). \tag{4.92}$$

By (4.91) and (4.92), we arrive at

$$\mathbb{E}\langle \gamma, \xi^{t_2} \rangle_H = \mathbb{E}\langle \gamma, y(t_2) \rangle_H, \qquad \text{a.e. } t_2 \in [0,T], \tag{4.93}$$

which implies (4.84) immediately (by choosing $\gamma = \xi^{t_2} - y(t_2)$).

Now, by (4.84) and recalling that ξ^t has a càdlàg modification, we see that there is a càdlàg H-valued process $\tilde{y}(\cdot)$ such that $y(\cdot) = \tilde{y}(\cdot)$ in $[0,T]\times\Omega$ a.e. It is easy to check that $(\tilde{y}(\cdot), Y(\cdot))$ is a transposition solution to the equation (4.21). To simplify the notation, we still use y (instead of $\tilde{y}$) to denote the first component of this solution. Clearly, $(y(\cdot), Y(\cdot)) \in D_{\mathbb{F}}([0,T]; L^p(\Omega; H)) \times L^2_{\mathbb{F}}(0,T; L^p(\Omega; \mathcal{L}_2^0))$ and satisfies (4.70). Also, the uniqueness of the transposition solution to (4.21) is obvious. □

Remark 4.17. In the proof of Theorem 4.16, we need to use the assumption $p \in (1,2]$ to guarantee that $L^2_{\mathbb{F}}(t,T; L^q(\Omega; \mathcal{L}_2^0)) \subset L^q_{\mathbb{F}}(\Omega; L^2(t,T; \mathcal{L}_2^0))$. How to drop this restrictive assumption on p is an open problem.

Similar to Theorem 4.16, we have the following result, in which the assumption for $f(\cdot)$ is slightly relaxed to be $f(\cdot) \in L^p(\Omega; L^1_{\mathbb{F}}(0,T;H))$ while a stronger requirement on $S(\cdot)$ is imposed.

Theorem 4.18. *Let $S(\cdot)$ be a generalized contractive C_0-semigroup. Then, for any $y_T \in L^p_{\mathcal{F}_T}(\Omega; H)$ and $f(\cdot) \in L^p_{\mathbb{F}}(\Omega; L^1(0,T;H))$, the equation (4.21) admits one and only one transposition solution $(y(\cdot), Y(\cdot)) \in D_{\mathbb{F}}([0,T]; L^p(\Omega; H)) \times L^2_{\mathbb{F}}(0,T; L^p(\Omega; \mathcal{L}_2^0))$. Furthermore,*

$$\begin{aligned}
&|(y(\cdot), Y(\cdot))|_{D_{\mathbb{F}}([0,T];L^p(\Omega;H))\times L^2_{\mathbb{F}}(0,T;L^p(\Omega;\mathcal{L}_2^0))} \\
&\le \mathcal{C}\left(|y_T|_{L^p_{\mathcal{F}_T}(\Omega;H)} + |f(\cdot)|_{L^p_{\mathbb{F}}(\Omega;L^1(0,T;H))} \right).
\end{aligned}$$

Proof: When $S(\cdot)$ is a generalized contractive C_0-semigroup, by Theorems 3.21 and 3.18, the equation (4.57) admits a unique mild solution $z \in L^q_{\mathbb{F}}(\Omega; C([t,T]; H))$, and

$$\begin{aligned}
&|z|_{L^q_{\mathbb{F}}(\Omega;C([t,T];H))} \\
&\le \mathcal{C}\big(|\eta|_{L^q_{\mathcal{F}_t}(\Omega;H)} + |v_1|_{L^q_{\mathbb{F}}(\Omega;L^1(t,T;H))} + |v_2|_{L^q_{\mathbb{F}}(\Omega;L^2(t,T;\mathcal{L}_2^0))} \big).
\end{aligned} \tag{4.94}$$

By (4.94) and proceeding as in the proof of Theorem 4.16, we can prove Theorem 4.18 (The only difference is that in (4.72) we use (4.94) instead of (4.58)), and hence we omit the details. □

Based on Theorem 4.16, we have the following well-posedness result for (4.20) in the sense of transposition solution.

Theorem 4.19. *Let Condition 4.1 hold and* $F(\cdot,0,0)\in L^1_{\mathbb{F}}(0,T;L^p(\Omega;H))$. *Then, for any* $y_T\in L^p_{\mathcal{F}_T}(\Omega;H)$, *the equation* (4.20) *admits one and only one transposition solution* $(y(\cdot),Y(\cdot))\in D_{\mathbb{F}}([0,T];L^p(\Omega;\,H))\times L^2_{\mathbb{F}}(0,T;L^p(\Omega;\mathcal{L}_2^0))$. *Furthermore,*

$$
\begin{aligned}
&|(y(\cdot),Y(\cdot))|_{D_{\mathbb{F}}([0,T];L^p(\Omega;H))\times L^2_{\mathbb{F}}(0,T;L^p(\Omega;\mathcal{L}_2^0))}\\
&\le \mathcal{C}\big(|y_T|_{L^p_{\mathcal{F}_T}(\Omega;H)}+|F(\cdot,0,0)|_{L^1_{\mathbb{F}}(0,T;L^p(\Omega;H))}\big).
\end{aligned}
\tag{4.95}
$$

Proof: We proceed as that in Step 2 in the proof of Theorem 4.10. Fix any $T_1\in[0,T)$. For any fixed $\big(z(\cdot),Z(\cdot)\big)\in D_{\mathbb{F}}([T_1,T];L^p(\Omega;H))\times L^2_{\mathbb{F}}(T_1,T;L^p(\Omega;\mathcal{L}_2^0))$, by Condition 4.1 and $F(\cdot,0,0)\in L^1_{\mathbb{F}}(0,T;L^p(\Omega;H))$, it is easy to see that $F(\cdot,z(\cdot),Z(\cdot))\in L^1_{\mathbb{F}}(T_1,T;L^p(\Omega;H))$. Consider the following equation

$$
\begin{cases}
dy(t)=-Ay(t)dt+F(t,z(t),Z(t))dt+Y(t)dW(t), & t\in[T_1,T),\\
y(T)=y_T.
\end{cases}
\tag{4.96}
$$

By Theorem 4.16, the equation (4.96) admits a unique transposition solution $(y(\cdot),Y(\cdot))\in D_{\mathbb{F}}([T_1,T];L^p(\Omega;H))\times L^2_{\mathbb{F}}(T_1,T;L^p(\Omega;\mathcal{L}_2^0))$. This defines a map $\mathcal{J}$ from $D_{\mathbb{F}}([T_1,T];L^p(\Omega;H))\times L^2_{\mathbb{F}}(T_1,T;L^p(\Omega;\mathcal{L}_2^0))$ into itself by $(z(\cdot),Z(\cdot))\mapsto(y(\cdot),Y(\cdot))$.

We claim that, for T_1 being sufficiently close to T,

$$
\begin{aligned}
&\big|\mathcal{J}(z,Z)-\mathcal{J}(\tilde z,\widetilde Z)\big|_{D_{\mathbb{F}}([T_1,T];L^p(\Omega;H))\times L^2_{\mathbb{F}}(T_1,T;L^p(\Omega;\mathcal{L}_2^0))}\\
&\le\frac12\big|(z,Z)-(\tilde z,\widetilde Z)\big|_{D_{\mathbb{F}}([T_1,T];L^p(\Omega;H))\times L^2_{\mathbb{F}}(T_1,T;L^p(\Omega;\mathcal{L}_2^0))},\\
&\qquad\forall\,(z,Z),\ \ (\tilde z,\widetilde Z)\in D_{\mathbb{F}}([T_1,T];L^p(\Omega;H))\times L^2_{\mathbb{F}}(T_1,T;L^p(\Omega;\mathcal{L}_2^0)).
\end{aligned}
\tag{4.97}
$$

To show (4.97), put $\hat f(\cdot)=F(\cdot,z(\cdot),Z(\cdot))-F(\cdot,\tilde z(\cdot),\widetilde Z(\cdot))$ and $(\hat y(\cdot),\widehat Y(\cdot))=\mathcal{J}(z,Z)-\mathcal{J}(\tilde z,\widetilde Z)$. Then, $(\hat y(\cdot),\widehat Y(\cdot))$ satisfies

$$
\begin{cases}
d\hat y(t)=-A\hat y(t)dt+\hat f(t)dt+\widehat Y(t)dW(t), & t\in[T_1,T),\\
\hat y(T)=0.
\end{cases}
\tag{4.98}
$$

Applying the estimate (4.70) respectively to the equations (4.98) and (4.96), we find that

$$
\begin{aligned}
&|(\hat{y}(\cdot),\widehat{Y}(\cdot))|_{D_{\mathbb{F}}([T_1,T];L^p(\Omega;H))\times L^2_{\mathbb{F}}(T_1,T;L^p(\Omega;\mathcal{L}_2^0))}\\
&\le \mathcal{C}|F(\cdot,z(\cdot),Z(\cdot))-F(\cdot,\tilde{z}(\cdot),\widetilde{Z}(\cdot))|_{L^1_{\mathbb{F}}(T_1,T;L^p(\Omega;H))}\\
&\le \mathcal{C}\big(|L_1(\cdot)|_{L^1(T_1,T)}+|L_2(\cdot)|_{L^2(T_1,T)}\big)\\
&\quad\times|(z(\cdot)-\tilde{z}(\cdot),Z(\cdot)-\widetilde{Z}(\cdot))|_{D_{\mathbb{F}}([T_1,T];L^p(\Omega;H))\times L^2_{\mathbb{F}}(T_1,T;L^p(\Omega;\mathcal{L}_2^0))}
\end{aligned}
\tag{4.99}
$$

and

$$
\begin{aligned}
&|(y(\cdot),Y(\cdot))|_{D_{\mathbb{F}}([T_1,T];L^p(\Omega;H))\times L^2_{\mathbb{F}}(T_1,T;L^p(\Omega;\mathcal{L}_2^0))}\\
&\le \mathcal{C}\big(|y_T|_{L^p_{\mathcal{F}_T}(\Omega;H)}+|F(\cdot,z(\cdot),Z(\cdot))|_{L^1_{\mathbb{F}}(T_1,T;L^p(\Omega;H))}\big)\\
&\le \mathcal{C}\big(|y_T|_{L^p_{\mathcal{F}_T}(\Omega;H)}+|F(\cdot,0,0)|_{L^1_{\mathbb{F}}(T_1,T;L^2(\Omega;H))}\big)\\
&\quad+\mathcal{C}\big(|L_1(\cdot)|_{L^1(T_1,T)}+|L_2(\cdot)|_{L^2(T_1,T)}\big)\\
&\qquad\times|(z(\cdot),Z(\cdot))|_{D_{\mathbb{F}}([T_1,T];L^p(\Omega;H))\times L^2_{\mathbb{F}}(T_1,T;L^p(\Omega;\mathcal{L}_2^0))}.
\end{aligned}
\tag{4.100}
$$

Let us choose T_1 such that

$$
\mathcal{C}\big(|L_1(\cdot)|_{L^1(T_1,T)}+|L_2(\cdot)|_{L^2(T_1,T)}\big)\le\frac{1}{2}.
\tag{4.101}
$$

Then, by (4.99), we get (4.97). This implies that the map $\mathscr{J}$ is contractive. Hence, it admits a unique fixed point, which is a transposition solution to (4.20) on $[T_1,T]$. Moreover, by (4.100)–(4.101), it follows that

$$
\begin{aligned}
&|(y(\cdot),Y(\cdot))|_{D_{\mathbb{F}}([T_1,T];L^p(\Omega;H))\times L^2_{\mathbb{F}}(T_1,T;L^p(\Omega;\mathcal{L}_2^0))}\\
&\le \mathcal{C}\big(|y_T|_{L^p_{\mathcal{F}_T}(\Omega;H)}+|F(\cdot,0,0)|_{L^1_{\mathbb{F}}(0,T;L^p(\Omega;H))}\big).
\end{aligned}
\tag{4.102}
$$

Repeating the above argument, we obtain the existence of transposition solutions to (4.20). The uniqueness and the estimate (4.95) follow from the estimate (4.102). □

Proceeding exactly as that in the proof of Theorem 4.19, based on Theorem 4.18, one can show the following well-posedness result for (4.20).

Theorem 4.20. *Let Condition 4.1 hold, $S(\cdot)$ be a generalized contractive C_0-semigroup and $F(\cdot,0,0)\in L^p_{\mathbb{F}}(\Omega;L^1(0,T;H))$. Then, for any $y_T\in L^p_{\mathcal{F}_T}(\Omega;H)$, the equation (4.20) admits one and only one transposition solution $(y(\cdot),Y(\cdot))\in D_{\mathbb{F}}([0,T];L^p(\Omega;H))\times L^2_{\mathbb{F}}(0,T;L^p(\Omega;\mathcal{L}_2^0))$. Furthermore,*

$$
\begin{aligned}
&|(y(\cdot),Y(\cdot))|_{D_{\mathbb{F}}([0,T];L^p(\Omega;H))\times L^2_{\mathbb{F}}(0,T;L^p(\Omega;\mathcal{L}_2^0))}\\
&\le \mathcal{C}\big(|y_T|_{L^p_{\mathcal{F}_T}(\Omega;H)}+|F(\cdot,0,0)|_{L^p_{\mathbb{F}}(\Omega;L^1(0,T;H))}\big).
\end{aligned}
$$

Remark 4.21. At the very beginning of this chapter, we assume that H is a separable Hilbert space. Similarly to [242] (in which V is assumed to be $\mathbb{R}$), we believe that the separability assumption on H can be dropped but this remains to be done.

4.4 The Case of Natural Filtration Revisited

In this section, we assume that $\mathbf{F}$ is the natural filtration condition, and revisit the equation (4.20), and its special case, the equation (4.21), by means of the transposition method.

We begin with the following result.

Proposition 4.22. *Let Condition 4.1 hold and* $F(\cdot,0,0)\in L^1_{\mathbb{F}}(0,T;L^2(\Omega;H))$. *Then, for each* $y_T\in L^2_{\mathcal{F}_T}(\Omega;H)$, *the corresponding mild solution* $(y(\cdot),Y(\cdot))\in L^2_{\mathbb{F}}(\Omega;C([0,T];H))\times L^2_{\mathbb{F}}(0,T;\mathcal{L}_2^0)$ *to the equation (4.20) is also a transposition solution to the same equation.*

Proof: By Theorem 4.12, for each $\lambda\in\rho(A)\cap\rho(A^*)$, the equation (4.56) admits a unique strong solution $(y_\lambda(\cdot),Y_\lambda(\cdot))\in L^2_{\mathbb{F}}(\Omega;C([0,T];D(A)))\times L^2_{\mathbb{F}}(0,T;\mathcal{L}_2^0)$, and

$$\lim_{\lambda\to\infty}|(y_\lambda(\cdot),Y_\lambda(\cdot))-(y(\cdot),Y(\cdot))|_{L^2_{\mathbb{F}}(\Omega;C([0,T];H))\times L^2_{\mathbb{F}}(0,T;\mathcal{L}_2^0)}=0. \tag{4.103}$$

On the other hand, by Corollary 3.23, for any $t\in[0,T]$, $v_1(\cdot)\in L^2_{\mathbb{F}}(\Omega;L^1(t,T;H))$, $v_2(\cdot)\in L^2_{\mathbb{F}}(t,T;\mathcal{L}_2^0)$ and $\eta\in L^2_{\mathcal{F}_t}(\Omega;H)$, the following equation

$$\begin{cases} dz_\lambda=(A^*z_\lambda+R(\lambda)^*v_1)ds+R(\lambda)^*v_2dW(s) & \text{in } (t,T],\\ z_\lambda(t)=R(\lambda)^*\eta \end{cases} \tag{4.104}$$

admits a unique strong solution $z_\lambda(\cdot)\in C_{\mathbb{F}}([t,T];L^2(\Omega;D(A^*)))$, and

$$\lim_{\lambda\to\infty}|z_\lambda(\cdot)-z(\cdot)|_{C_{\mathbb{F}}([t,T];L^2(\Omega;H))}=0, \tag{4.105}$$

where $z(\cdot)$ solves (4.57) in the sense of mild solution.

Applying Itô's formula to $\langle z_\lambda(\cdot),y_\lambda(\cdot)\rangle_H$ and noting the equations (4.104) and (4.56), we obtain that

$$\begin{aligned}
&\mathbb{E}\big\langle z_\lambda(T),R(\lambda)y_T\big\rangle_H-\mathbb{E}\int_t^T\big\langle z_\lambda(s),R(\lambda)F(s,y_\lambda(s),Y_\lambda(s))\big\rangle_H ds\\
&=\mathbb{E}\big\langle R(\lambda)^*\eta,y_\lambda(t)\big\rangle_H+\mathbb{E}\int_t^T\big\langle R(\lambda)^*v_1(s),y_\lambda(s)\big\rangle_H ds\\
&\quad+\mathbb{E}\int_t^T\big\langle R(\lambda)^*v_2(s),Y_\lambda(s)\big\rangle_{\mathcal{L}_2^0}ds.
\end{aligned} \tag{4.106}$$

Letting $\lambda\to\infty$ in (4.106), and noting (4.103) and (4.105), we find that $(y(\cdot),Y(\cdot))$ satisfies (4.59), and therefore it is a transposition solution to the equation (4.20). □

We have the following result for the linear equation (4.21).

Theorem 4.23. *For any* $p \in (1,2]$, $y_T \in L^p_{\mathcal{F}_T}(\Omega;H)$ *and* $f(\cdot) \in L^1_{\mathbb{F}}(0,T;L^p(\Omega;H))$ *(resp.* $f(\cdot) \in L^p_{\mathbb{F}}(\Omega;L^1(0,T;H))$ *and* $S(\cdot)$ *is a generalized contractive* C_0*-semigroup), the equation* (4.21) *admits one and only one mild solution* $(y(\cdot),Y(\cdot)) \in L^p_{\mathbb{F}}(\Omega;C([0,T];\, H)) \times L^2_{\mathbb{F}}(0,T;L^p(\Omega;\mathcal{L}_2^0))$*. Furthermore,*

$$
\begin{aligned}
&|(y(\cdot),Y(\cdot))|_{L^p_{\mathbb{F}}(\Omega;C([0,T];H))\times L^2_{\mathbb{F}}(0,T;L^p(\Omega;\mathcal{L}_2^0))}\\
&\leq \mathcal{C}\big(|y_T|_{L^p_{\mathcal{F}_T}(\Omega;H)} + |f(\cdot)|_{L^1_{\mathbb{F}}(0,T;L^p(\Omega;H))}\big)\\
&\qquad (resp.\ \ \mathcal{C}\big(|y_T|_{L^p_{\mathcal{F}_T}(\Omega;H)} + |f(\cdot)|_{L^p_{\mathbb{F}}(\Omega;L^1(0,T;H))}\big)).
\end{aligned}
\tag{4.107}
$$

Proof: We consider only the case that $f(\cdot) \in L^1_{\mathbb{F}}(0,T;L^p(\Omega;H))$.

By Theorem 4.16, the equation (4.21) admits one and only one transposition solution $(y(\cdot),Y(\cdot)) \in D_{\mathbb{F}}([0,T];L^p(\Omega;H)) \times L^2_{\mathbb{F}}(0,T;L^p(\Omega;\mathcal{L}_2^0))$, and it satisfies (4.107). It remains to show that $(y(\cdot),Y(\cdot))$ is also a mild solution to (4.21).

We choose two sequences $\{y_T^k\}_{k=1}^\infty \subset L^\infty_{\mathcal{F}_T}(\Omega;H)$ and $\{f^k(\cdot)\}_{k=1}^\infty \subset L^1_{\mathbb{F}}(0,T;\, L^\infty(\Omega;H))$ such that

$$
\lim_{k\to\infty} |y_T^k - y_T|_{L^p_{\mathcal{F}_T}(\Omega;H)} = 0, \qquad \lim_{k\to\infty} |f^k(\cdot) - f(\cdot)|_{L^1_{\mathbb{F}}(0,T;L^p(\Omega;H))} = 0. \tag{4.108}
$$

Since $L^\infty_{\mathcal{F}_T}(\Omega;H) \subset L^2_{\mathcal{F}_T}(\Omega;H)$ and $L^1_{\mathbb{F}}(0,T;L^\infty(\Omega;H)) \subset L^1_{\mathbb{F}}(0,T;L^2(\Omega;H))$, by Theorem 4.10, the following equation

$$
\begin{cases}
dy^k(t) = -Ay^k(t)dt + f^k(t)dt + Y^k(t)dW(t) \ \text{ in } [0,T),\\
y^k(T) = y_T^k
\end{cases}
\tag{4.109}
$$

admits a unique mild solution $(y^k(\cdot),Y^k(\cdot)) \in L^2_{\mathbb{F}}(\Omega;C([0,T];H)) \times L^2_{\mathbb{F}}(0,T;\mathcal{L}_2^0)$. Hence, for any $t \in [0,T]$,

$$
\begin{aligned}
y^k(t) = S(T-t)y_T^k &- \int_t^T S(s-t)f^k(s)ds\\
&- \int_t^T S(s-t)Y^k(s)dW(s), \qquad \text{a.s.}
\end{aligned}
\tag{4.110}
$$

By Proposition 4.22, $(y^k(\cdot),Y^k(\cdot))$ is also a transposition solution to (4.109). Then, applying Theorem 4.16 to the following equation:

$$
\begin{cases}
d\big(y(t)-y^k(t)\big) = -A\big(y(t)-y^k(t)\big)dt + \big(f(t)-f^k(t)\big)dt\\
\qquad\qquad\qquad\qquad +\big(Y(t)-Y^k(t)\big)dW(t) \quad \text{in } [0,T),\\
y(T)-y^k(T) = y_T - y_T^k,
\end{cases}
$$

we conclude that

$$
\begin{aligned}
&\big|\big(y^k(\cdot)-y(\cdot),Y^k(\cdot)-Y(\cdot)\big)\big|_{D_{\mathbb{F}}([0,T];L^p(\Omega;H))\times L^2_{\mathbb{F}}(0,T;L^p(\Omega;\mathcal{L}_2^0))}\\
&\leq \mathcal{C}\left(|y_T^k - y_T|_{L^p_{\mathcal{F}_T}(\Omega;H)} + |f^k(\cdot)-f(\cdot)|_{L^1_{\mathbb{F}}(0,T;L^p(\Omega;H))}\right).
\end{aligned}
\tag{4.111}
$$

Taking $k \to \infty$ in (4.110), and noting (4.111) and (4.108), we find that $(y(\cdot), Y(\cdot))$ verifies (4.24), and hence it is a mild solution to (4.21). Noting also that $\{y^k\}_{k=1}^{\infty}$ is a Cauchy sequence in $L^p_{\mathbb{F}}(\Omega; C([0,T]; H))$, we then obtain (4.107). $\square$

By Theorem 4.23 and the Banach fixed point theorem, similar to Theorem 4.19, it is easy to derive the following result for the equation (4.20) (Hence we omit the proof).

Theorem 4.24. *Let Condition 4.1 hold and $p \in (1,2]$. Assume that $F(\cdot,0,0) \in L^1_{\mathbb{F}}(0,T; L^p(\Omega; H))$ (resp. $F(\cdot,0,0) \in L^p_{\mathbb{F}}(\Omega; L^1(0,T;H))$ and $S(\cdot)$ is a generalized contractive C_0-semigroup). Then, for any $y_T \in L^p_{\mathcal{F}_T}(\Omega; H)$, the equation (4.20) admits a unique mild solution $(y(\cdot), Y(\cdot)) \in L^p_{\mathbb{F}}(\Omega; C([0,T]; H)) \times L^2_{\mathbb{F}}(0,T; L^p(\Omega; \mathcal{L}_2^0))$, which is also a transposition solution to the same equation. Furthermore,*

$$
\begin{aligned}
&|(y(\cdot), Y(\cdot))|_{L^p_{\mathbb{F}}(\Omega;C([0,T];H))\times L^2_{\mathbb{F}}(0,T;L^p(\Omega;\mathcal{L}_2^0))}\\
&\leq \mathcal{C}\big(|y_T|_{L^p_{\mathcal{F}_T}(\Omega;H)} + |F(\cdot,0,0)|_{L^1_{\mathbb{F}}(0,T;L^p(\Omega;H))}\big) \qquad (4.112)\\
&\quad \big(resp.\ \ \mathcal{C}\big(|y_T|_{L^p_{\mathcal{F}_T}(\Omega;H)} + |F(\cdot,0,0)|_{L^p_{\mathbb{F}}(\Omega;L^1(0,T;H))}\big)\big).
\end{aligned}
$$

4.5 Notes and Comments

The theory of backward stochastic evolution equations remains a new and relatively unexplored area. This chapter only gives a very short introduction to this important topic.

People have studied two different kinds of backward stochastic differential/evolution equations in the literature. The main difference between them is whether their solutions are adapted. We refer the readers to [174, 266] for the classical works for backward stochastic differential equations with non-adapted solutions. In this chapter, we only consider backward stochastic differential/evolution equations with adapted solutions, which are quite naturally in the study of stochastic control problems. Linear backward stochastic differential equation with adapted solutions was introduced in [31], for the purpose of studying the Pontryagin-type maximum principle for stochastic optimal control problems. The well-posedness of nonlinear backward stochastic differential equations was established in [271]. A little later, this sort of nonlinear equations was also independently introduced in [81] for the study of the so-called *recursive utility* in finance. Detailed introduction to backward stochastic differential equations can be found in [48, 272, 322, 381].

In order to use the Martingale Representation Theorem, a key assumption in the theory of backward stochastic differential equations is to assume that the filtration under consideration is generated by the underlying Brownian motion. This restriction was dropped in several works (e.g., [203, 241]) by generalizing the definition of solutions.

As far as we know, backward stochastic evolution equation was first introduced in [26] to derive maximum principles for some stochastic distributed parameter control systems (See also [148, 149, 319] and so on for some early results in this respect). Unlike stochastic evolution equations or backward stochastic differential equation (i.e., backward stochastic evolution equations in finite dimensions), there are no published books addressing exclusively to backward stochastic evolution equations (though some results on this topic can be found in the monograph [242]).

The main result in Section 4.1 is due to [271]. The results presented in Section 4.2 are some small modification of that from [3, 148, 149, 256, 258]. Sections 4.3-4.4 are some extensions of the related results appeared in [241, 242, 244], obtained by means of our stochastic transposition method.

Compared to backward stochastic differential equations, there are many things for backward stochastic evolution equations which remain to be done, for example:

- In finite dimensions, people also studied forward-backward stochastic differential equations ([251]). It would be quite interesting to study their infinite dimensional counterpart, i.e., forward-backward stochastic evolution equations. It seems that there are almost no publications in this direction (See [84] for the only one reference that we know).
- In the setting of finite dimensions, people also studied backward stochastic differential equations driven by G-Brownian motions ([147, 276, 277], forward-backward doubly stochastic differential equations ([278]) and backward stochastic Volterra integral equations ([367, 368]). It would be quite interesting to study their infinite dimensional counterpart, i.e., backward stochastic evolution equations driven by G-Brownian motions, forward-backward doubly stochastic evolution equations and backward stochastic Volterra integral equation in infinite dimensions. As far as we know, there exists no publication in this respect.
- It is quite interesting to study backward stochastic evolution equations driven by general martingales but this seems far from well-understood (See [344] for a result in this respect).
- As mentioned before, besides applications in control problems, backward (or more generally, forward-backward) stochastic differential equations are also widely used in financial mathematics, stochastic analysis, partial differential equations, etc. (e.g., [56, 73, 86, 251, 274, 321]). It is quite interesting to consider similar applications of backward stochastic evolution equations but it seems that almost nothing has been done, yet.
- It is almost blank about the numerical analysis on backward stochastic evolution equations. As far as we know, the earliest work in this respect (for backward stochastic parabolic differential equations) is the Ph D thesis [344] by Y. Wang (See [84, 283, 345, 346] for a few pioneer works).

Several more comments on our stochastic transposition method and transposition solutions to the backward stochastic evolution equation (4.20) are as follows:

- The usual duality analysis is contained in the definition of transposition solutions, and therefore, one does not need to utilize any deep result in martingale theory to deduce this sort of duality relationship any more. Moreover, when $H = V = \mathbb{R}$, it is easy to deduce a comparison theorem (See [241, Theorem 5.1]) for transposition solutions to (4.20) by using almost the same method as that in the case of natural filtration ([86]). It is notable that, by means of this comparison theorem, some interesting new results were obtained in a recent paper [354].
- As we shall see in Sections 12.2 and 12.3 in Chapter 12, by our stochastic transposition method, it is even easier to establish Pontryagin-type maximum principles for stochastic optimal control problems (for both finite and infinite dimensions) than to solve the same problem with the natural filtration ([273, 371]) because, again, the desired duality analysis is already contained in the definition of transposition solution.
- Our method does NOT need to use the Martingale Representation Theorem, and therefore we can study the problem with a general filtration. Because of this, as we have done in [242, 248] (See also Chapters 12 and 13), the stochastic transposition method can be employed to solve some very difficult well-posedness problems for operator-valued backward stochastic evolution equations. Note that, when dealing with backward stochastic evolution equations with operator-valued unknowns, as far as we know, there exists no suitable Martingale Representation Theorem (for the $\mathcal{L}(H)$-valued martingale) even if $\mathbf{F}$ is the natural filtration.
- By our method, when $p = 2$, the correction part $Y(\cdot)$ in the transposition solution to (4.20) can be obtained by the classical Riesz Representation Theorem for Hilbert spaces, and therefore, as mentioned in [241, Remark 3.2], one can even give a numerical scheme (at least conceptually) for $Y(\cdot)$ although the detailed analysis is beyond the scope of this book. Indeed, let $\{H_m\}_{m=1}^{\infty}$ be a sequence of subspaces of $L^2_{\mathbb{F}}(0,T;\mathcal{L}_2^0)$ such that for any $g(\cdot) \in L^2_{\mathbb{F}}(0,T;\mathcal{L}_2^0)$, there exists a sequence $\{g_m(\cdot)\}_{m=1}^{\infty}$ satisfying that

$$g_m(\cdot) \in H_m \text{ and } \lim_{m\to\infty} |g_m - g|_{L^2_{\mathbb{F}}(0,T;\mathcal{L}_2^0)} = 0.$$

Now, for any fixed $m \in \mathbb{N}$, choosing $t = 0$, $\eta = 0$, $v_1(\cdot) = 0$ and $v_2(\cdot) = v_m(\cdot) \in H_m$ in (4.57) (and denote by $z_m(\cdot)$ the corresponding solution). From (4.73), we see that

$$\mathbb{E}\langle z_m(T), y_T \rangle_H - \mathbb{E}\int_0^T \langle z_m(\tau), f(\tau) \rangle_H d\tau = \mathbb{E}\int_0^T \langle v_m(\tau), Y(\tau) \rangle_{\mathcal{L}_2^0} d\tau. \tag{4.113}$$

On the other hand, using the same argument to obtain $Y(\cdot)$ (by Riesz's Representation Theorem), we can find $Y_m(\cdot) \in H_m$ such that

$$\mathbb{E}\langle z_m(T), y_T \rangle_H - \mathbb{E}\int_0^T \langle z_m(\tau), f(\tau) \rangle_H d\tau = \mathbb{E}\int_0^T \langle v_m(\tau), Y_m(\tau) \rangle_{\mathcal{L}_2^0} d\tau. \tag{4.114}$$

From (4.113)–(4.114), it follows that $Y_m(\cdot)$ is the orthogonal projection of $Y(\cdot)$ to H_m. By the definition of H_m, we find that

$$\lim_{m\to\infty} |Y_m(\cdot) - Y(\cdot)|_{L^2_{\mathbb{F}}(0,T;\mathcal{L}_2^0)} = 0.$$

Therefore, one may obtain a "good" approximation of $Y(\cdot)$ if one can choose a suitable sequence $\{H_m\}_{m=1}^\infty$ such that $Y_m(\cdot)$ (say, belongs to a finite dimensional space) can be computed efficiently and that $Y_m(\cdot)$ converges to $Y(\cdot)$ in $L^2_{\mathbb{F}}(0,T;\mathcal{L}_2^0)$ in some sense. Whenever $Y(\cdot)$ is computed, it seems that another part $y(\cdot)$ in the transposition solution to (4.20) can be obtained (numerically) even easier. On the other hand, similar to the classical finite element method solving deterministic partial differential equations, our stochastic transposition method leads naturally numerical schemes to solve both vector-valued and operator-valued backward stochastic evolution equations. Indeed, our Definition 4.13 can be viewed as a variational formulation of the equation (4.20), even if the underlying filtration **F** is the natural one. These ideas were partially adopted in the works [342] and [69, 159, 341], which gave approximate solutions to backward stochastic evolution equations and forward-backward stochastic evolution equations (both in finite dimensions) based on our stochastic transposition method. We expect that the method in [69, 159, 341, 342] may be extended to the setting of infinite dimensions but a careful study will be needed.

- Clearly, in some sense, the transposition solution $(y(\cdot), Y(\cdot))$ to the backward stochastic evolution equation (4.20) is, in spirit, very close to the distribution solution to partial differential equations. It is then very natural to study the further regularity of $(y(\cdot), Y(\cdot))$ but this is beyond the scope of this book.

5

Control Problems for Stochastic Distributed Parameter Systems

In this chapter, we shall present several typical controlled stochastic evolution equations, including a new refined controlled stochastic wave equation in particular. Then, we shall give a quite general formulation of some typical control problems, such as controllability/observability problems and optimal control problems for stochastic distributed parameter systems.

5.1 An Example of Controlled Stochastic Differential Equations

Throughout this section and the next one, $T > 0$, $(\Omega, \mathcal{F}, \mathbf{F}, \mathbb{P})$ (with $\mathbf{F} = \{\mathcal{F}_t\}_{t\in[0,T]}$) is a fixed filtered probability space (satisfying the usual condition), on which a one dimensional standard Brownian motion $W(\cdot)$ (unless other stated) is defined.

Before introducing control problems for stochastic distributed parameter systems, we first present some control problems for a linear stochastic differential equation appeared in the financial market (e.g., [371]).

Consider a market where a bond and n stocks are traded continuously, where $n \in \mathbb{N}$. The price process $P_0(t)$ of the bond is governed by the following ordinary differential equation:

$$\begin{cases} dP_0(t) = r(t)P_0(t)dt \quad \text{in } [0,T], \\ P_0(0) = p_0 > 0, \end{cases} \tag{5.1}$$

where $r(t) > 0$ is called the *interest rate* of the bond, p_0 is its initial price. The price processes $P_1(t), \cdots P_n(t)$ of the stocks satisfy the following stochastic differential equation:

$$\begin{cases} dP_j(t) = a_j(t)P_j(t)dt + b_j(t)P_j(t)dW(t) \quad \text{in } [0,T], \\ P_j(0) = p_j > 0, \end{cases} \tag{5.2}$$

Q. Lü and X. Zhang, *Mathematical Control Theory for Stochastic Partial Differential Equations*, Probability Theory and Stochastic Modelling 101, https://doi.org/10.1007/978-3-030-82331-3_5

where, $a_j : [0,T] \times \Omega \to \mathbb{R}$ is called the *appreciation rate*, and $b_j : [0,T] \times \Omega \to \mathbb{R}$ is called the *volatility* of the j-th stock, p_j is its initial price, $j = 1, \cdots, n$. All a_j and b_j are assumed to be $\mathbf{F}$-adapted. The diffusion term in (5.2) reflects the fluctuation of the stock prices.

Denote by $x(t)$ the total wealth of an investor at time $t \geq 0$. Suppose that he/she holds $N_j(t)$ shares of the j-th asset ($j = 0, 1, \cdots, n$) at time t. Clearly,

$$x(t) = \sum_{j=0}^{n} N_j(t) P_j(t), \qquad t \geq 0. \tag{5.3}$$

Assume that the trading of shares and the payment of dividends (at the rate $\mu_j(t)$ per unit time and per unit of money invested in the j-th stock) takes place continuously, while for the bond, the rate $\mu_0(t) \equiv 0$. Write $c(t)$ for the rate of withdrawal for consumption. Then, with time increment Δt, the change of this investor's total wealth is given by

$$\begin{aligned} & x(t+\Delta t) - x(t) \\ & = \sum_{j=0}^{n} N_j(t)\big(P_j(t+\Delta t) - P_j(t)\big) + \sum_{j=1}^{n} \mu_j(t) N_j(t) P_j(t) \Delta t - c(t)\Delta t. \end{aligned} \tag{5.4}$$

Put

$$u_j(t) \overset{\Delta}{=} N_j(t) P_j(t), \qquad j = 0, 1, 2 \cdots, n,$$

which stands for the market value of the investor's wealth in the j-th bond/stock. By letting $\Delta t \to 0$, we obtain that

$$\begin{aligned} dx(t) &= \sum_{j=0}^{n} N_j(t) dP_j(t) + \sum_{j=1}^{n} \mu_j(t) N_j(t) P_j(t) dt - c(t) dt \\ &= \Big[r(t) N_0(t) P_0(t) + \sum_{j=1}^{n} \big(a_j(t) + \mu_j(t)\big) N_j(t) P_j(t) - c(t) \Big] dt \\ &\quad + \sum_{j=1}^{n} N_j(t) P_j(t) b_j(t) dW(t) \\ &= \Big[r(t) x(t) + \sum_{j=1}^{n} \big(a_j(t) + \mu_j(t) - r(t)\big) u_j(t) - c(t) \Big] dt \\ &\quad + \sum_{j=1}^{n} b_j(t) u_j(t) dW(t). \end{aligned}$$

Hence, $x(\cdot)$ satisfies the following stochastic differential equation:

$$\begin{cases} dx(t) = \Big[r(t)x(t) + \sum_{j=1}^{n}\big(a_j(t) + \mu_j(t) - r(t)\big)u_j(t) - c(t)\Big]dt \\ \qquad\qquad + \sum_{j=1}^{n} b_j(t)u_j(t)dW(t), \\ x(0) = x_0, \end{cases} \tag{5.5}$$

where x_0 represents the investor's initial wealth.

Remark 5.1. For simplicity, we assume that the investor can do short-selling and borrow money. In this case, each $u_j(\cdot)$ can be negative. When $u_j(t) < 0$ (for some $j = 1, 2, \cdots, n$), the investor is *short-selling* the j-th stock; when $u_0(t) < 0$, the investor borrows money of the amount $|u_0(t)|$ with an interest rate $r(t)$. On the other hand, it is clear that $c(\cdot) \geq 0$.

The investor can change the "allocation" of his/her wealth in these assets by changing $u(\cdot) = (u_0(\cdot), u_1(\cdot), \cdots, u_n(\cdot))^\top$ and/or $c(\cdot)$, both of which can be viewed as controls.

Clearly, there is an implicit constraint on the portfolio $u(\cdot)$ due to the stochastic environment, that is, at any time the investor can only utilize the past information. Therefore, the controls must be nonanticipative, i.e., adapted to the filtration $\mathbf{F}$.

For a given initial wealth $x_0 > 0$ and an expected return $x_T \in L^2_{\mathcal{F}_T}(\Omega)$, a natural objective of the investor is to choose an investment portfolio $u(\cdot)$ and a consumption plan $c(\cdot)$ such that

$$x(T) = x_T, \qquad \text{a.s.} \tag{5.6}$$

This is typically a controllability problem. Usually, the requirement (5.6) is too strong, particularly in the stochastic setting. A more realistic requirement is to relax (5.6) as follows:

$$x(T) \in \Lambda, \qquad \text{a.s.}, \tag{5.7}$$

for some given subset $\Lambda \in L^2_{\mathcal{F}_T}(\Omega)$. This is also a sort of controllability problem.

On the other hand, given an initial wealth $x_0 > 0$, in order to prohibit bankruptcy and guarantee a rewarding investment, the investor needs to choose $u(\cdot)$ and $c(\cdot)$ so that

$$x(t) \geq 0, \quad \forall\, t \in [0, T], \qquad \text{a.s.}, \tag{5.8}$$

$$\mathbb{E}\, x(T) \geq \theta x_0, \qquad \text{a.s.} \tag{5.9}$$

for some given $\theta > 1$, and maximizes the stream of discounted *utility*:

$$\mathcal{J}(u(\cdot), c(\cdot)) = \mathbb{E}\Big(\int_0^T e^{-\gamma t}\varphi(c(t))dt + e^{-\gamma T}h(x(T))\Big), \tag{5.10}$$

where $\gamma > 0$ is the discount rate, $\varphi(\cdot)$ is the instantaneous utility from consumption and $h(\cdot)$ is the utility that comes from bequests. This is typically an optimal control problem. Obviously, in order that this optimal control problem makes sense, one needs the nonempty-ness of the set of all $u(\cdot)$ and $c(\cdot)$ so that both (5.8) and (5.9) hold and that the integration in (5.10) is meaningful. Such a nonempty-ness condition can be viewed also as a controllability problem.

In summary, the above investment problem has the following features:

(1) The system is governed by a stochastic evolution equation;
(2) There are usually alternative decisions that can affect the dynamics of the system;
(3) The aim is to drive the solutions to the system to a given state (in some sense) by selecting a nonanticipative decision;
(4) There are some *constraints* that the decisions and/or the state are subject to;
(5) There exists a *criterion* which measures the performance of the decisions, and the goal is to *optimize* (maximize or minimize) the criterion by selecting a *nonanticipative* decision among the ones satisfying all the constraints.

In the next section, we shall present some control problems governed by stochastic evolution equations in infinite dimensions.

5.2 Control Systems Governed by Stochastic Partial Differential Equations

Generally speaking, any ordinary/partial differential equation can be regarded as a stochastic ordinary/partial differential equation provided that its coefficients, forcing terms, initial and boundary conditions, or at least one of them is random. The terminology "stochastic partial differential equation" is a little misused, which may mean different types of equations in different places. The analysis of equations with random coefficients differs dramatically from that of equations with random noises. In this book, we focus on the latter ones.

The study of stochastic partial differential equations was motivated by two aspects:

- One is due to the rapid recent development of stochastic analysis. The topic of stochastic partial differential equations is an interdisciplinary area involved both stochastic processes (random fields) and partial differential equations, which has quickly become a discipline in itself. In the last two decades, stochastic partial differential equations have been one of the most active areas in stochastic analysis.
- Another is the requirement from some branches of sciences. In many phenomena in physics, biology, economics and Control Theory (including filter

theory in particular), stochastic effects play a central role. Thus, stochastic corrections to the deterministic models are indispensably needed. These backgrounds have been influencing the development of stochastic partial differential equations remarkably.

In this section, we shall present some physical/biological models described by stochastic partial differential equations, including particularly the derivation of a refined stochastic wave equation in one space dimension, motivated by the study of some control problems. The readers are referred to [70, 96, 145, 161, 169, 171, 257, 290, 316, 332] for typical references on systems governed by stochastic equations in finite/infinite dimensions.

Example 5.2. **Stochastic parabolic equations**

The following equation was introduced to describe the evolution of the density of a bacteria population (e.g., [71]):

$$\begin{cases} dy = \nu \partial_{xx} y dt + \alpha \sqrt{y} dW(t) & \text{in } (0,T) \times (0,L), \\ y_x = 0 & \text{on } (0,T) \times \{0,L\}, \\ y(0) = y_0 & \text{in } (0,L), \end{cases} \tag{5.11}$$

where $\nu > 0$, $\alpha > 0$ and $L > 0$ are given constants, and $y_0 \in L^2(0,L)$. The derivation of (5.11) is as follows:

Suppose that such a bacteria population is distributed in the interval $[0,L]$. Denote by $y(t,x)$ the density of this population at time $t \in [0,T]$ and position $x \in [0,L]$. If there is no large-scale migration, the variation of the density is governed by the following

$$dy(t,x) = \nu y_{xx}(t,x)dt + d\xi(t,x,y), \tag{5.12}$$

where $\nu y_{xx}(t,x)$ describes the population's diffusion from the high density place to the low one, while $\xi(t,x,y)$ is a random perturbation caused by lots of small independent random disturbances. Suppose that the random perturbation $\xi(t,x,y)$ at time t and position x can be approximated by a Gaussian stochastic process whose variance Var ξ is monotone w.r.t. y. When L is small, which is usual for the study of bacteria, we may assume that ξ is independent of x. Further, when y is small, the variance of ξ can be approximated by $\alpha^2 y$, where α^2 is the derivative of Var ξ at $y = 0$. Under these assumptions, it follows that

$$d\xi(t,x,y) = \alpha \sqrt{y} dW(t). \tag{5.13}$$

Combining (5.12) and (5.13), we arrive at the first equation of (5.11). If there is no bacteria entering or leaving $[0,L]$ from its boundary, we have $y_x(t,0) = y_x(t,1) = 0$. According to these, we obtain the second equation of (5.11).

To control the population's density, people can put in and/or draw out some species. Under such sort of actions, the equation (5.11) becomes the following controlled stochastic parabolic equation:

$$
\begin{cases}
dy = (\nu\partial_{xx}y + u)dt + (\alpha\sqrt{y} + v)dW(t) & \text{in } (0,T)\times(0,L),\\
y_x = 0 & \text{on } (0,T)\times\{0,L\},\\
y(0) = y_0 & \text{in } (0,L),
\end{cases}
\tag{5.14}
$$

where u and v denote the way of putting in and/or drawing out species.

The following equation has been proposed in [335] to study the propagation of an electric potential in a neuron:

$$
\begin{cases}
dy = (\nu\partial_{xx}y - y)dt + g(y)dW(t) & \text{in } (0,T)\times(0,L),\\
y_x = 0 & \text{on } (0,T)\times\{0,L\},\\
y(0) = y_0 & \text{in } (0,L),
\end{cases}
\tag{5.15}
$$

in which $g(\cdot)$ is a suitable operator-valued function, and $\{W(t)\}_{t\geq 0}$ is a Q-Brownian motion. To control the electric potential, one puts some charge in the system. In this way, the equation (5.15) becomes a controlled stochastic parabolic equation as follows:

$$
\begin{cases}
dy = (\nu\partial_{xx}y - y + u)dt + (g(y) + v)dW(t) & \text{in } (0,T)\times(0,L),\\
y_x = 0 & \text{on } (0,T)\times\{0,L\},\\
y(0) = y_0 & \text{in } (0,L).
\end{cases}
\tag{5.16}
$$

where u and v stand for the way of putting charges.

Similarly to Section 5.1, one may pose the controllability and optimal control problems for (5.14) and/or (5.16) but we omit the details here.

Some other models governed by stochastic parabolic equations can be found in [68, 356], for example.

Example 5.3. **A refined stochastic wave equation**

To study the vibration of thin string/membrane perturbed by the random force, people introduced the following stochastic wave equation (e.g. [115]):

$$
\begin{cases}
dy_t = \Delta y dt + \alpha(t) y dW(t) & \text{in } (0,T)\times G,\\
y = 0 & \text{on } (0,T)\times\partial G,\\
y(0) = y_0,\ y_t(0) = y_1 & \text{in } G.
\end{cases}
\tag{5.17}
$$

Here $G \subset \mathbb{R}^m$ $(m \in \mathbb{N})$ is a bounded domain, $\alpha(\cdot)$ is a suitable function and $(y_0, y_1) \in H_0^1(G)\times L^2(G)$. Let us recall below a derivation of (5.17) for $m = 1$ by studying the motion of a DNA molecule in a fluid (e.g., [115, 300]).

Compared with its length, the diameter of such a DNA molecule is very small, and therefore, it can be viewed as a thin and long elastic string. One can describe its position by using an $\mathbb{R}^3$-valued function $y = (y_1, y_2, y_3)$ defined on $[0,L]\times[0,+\infty)$ for some $L > 0$. Usually, a DNA molecule floats in a fluid. Thus, it is always struck by the fluid molecules, just as a particle of pollen floating in a fluid.

Let the mass of this string per unit length be equal to 1. The acceleration at position $x \in [0, L]$ along the string at time $t \in [0, +\infty)$ is $y_{tt}(t, x)$. There are mainly four kinds of forces acting on the string: the elastic force $F_1(t, x)$, the friction $F_2(t, x)$ due to viscosity of the fluid, the impact force $F_3(t, x)$ from the flowing of the fluid and the random impulse $F_4(t, x)$ from the impacts of the fluid molecules. By Newton's second law, it follows that

$$y_{tt}(t, x) = F_1(t, x) + F_2(t, x) + F_3(t, x) + F_4(t, x). \tag{5.18}$$

Similar to the derivation of deterministic wave equations, the elastic force $F_1(t, x) = y_{xx}(t, x)$. The fiction depends on the property of the fluid and the velocity and shape of the string. When y, y_x and y_t are small, $F_2(t, x)$ approximately depends on them linearly, that is, $F_2 = a_1 y_x + a_2 y_t + a_3 y$ for some suitable functions a_1, a_2 and a_3. From the classical theory of Statistical Mechanics (e.g., [306, Chapter VI]), the random impulse $F_4(t, x)$ at time t and position x can be approximated by a Gaussian stochastic process with a covariance $k(\cdot, y)$, which also depends on the property of the fluid. More precisely,

$$F_4(t, x) = \int_0^t k(x, y(s, x)) dW(s).$$

Thus, the equation (5.18) can be rewritten as

$$dy_t(t, x) = y_{xx}(t, x)dt + F_2(t, x)dt + F_3(t, x)dt + k(x, y(t, x))dW(t). \tag{5.19}$$

When y is small, we may assume that $k(\cdot, y)$ is linear w.r.t. y. In this case, $k(x, y(t, x)) = k_1(x)y(t, x)$ for a suitable $k_1(\cdot)$. More generally, one may assume that k_1 depends on t (and even on the sample point $\omega \in \Omega$). Thus, (5.19) is reduced to the following:

$$dy_t(t, x) = y_{xx}(t, x)dt + F_2(t, x)dt + F_3(t, x)dt + k_1(t, x)y(t, x)dW(t). \tag{5.20}$$

Many biological behaviors are related to the motions of DNA molecules. Hence, there is a strong motivation for controlling these motions. In (5.20), $F_3(\cdot, \cdot)$ in the drift term and a part of the diffusion term can be designed as controls on the fluid. Furthermore, one can act some forces on the endpoints of the molecule. In this way, we arrive at the following controlled stochastic wave equation:

$$\begin{cases} dy_t = (y_{xx} + a_1 y_x + a_2 y_t + a_3 y + f)dt \\ \qquad\quad +(a_4 y + g)dW(t) & \text{in } (0, T) \times (0, L), \\ y = h_1 & \text{on } (0, T) \times \{0\}, \\ y = h_2 & \text{on } (0, T) \times \{L\}, \\ y(0) = y_0, \quad y_t(0) = y_1 & \text{in } (0, L). \end{cases} \tag{5.21}$$

In (5.21), (f, g, h_1, h_2) are controls which belong to some suitable spaces, a_4 is a suitable function, while (y_0, y_1) is an initial datum.

A natural question is as follows:

Can we drive the DNA molecule (governed by (5.21)) from any (y_0, y_1) *to a given configuration by choosing suitable controls* (f, g, h_1, h_2)*?*

Clearly, this is a controllability problem for the stochastic wave equation (5.21). Although there are four controls in the system (5.21), the answer to this question is negative (see Theorem 10.2 in Chapter 10). This differs completely from the well-known controllability property for the deterministic version of (5.21). Indeed, for the latter, under some mild assumptions, one can prove the corresponding exact controllability (e.g., [110, 207, 388, 396]). Since (5.20) is a generalization of the classical wave equation to the stochastic setting, as remarked in our paper [247], from the viewpoint of Control Theory, some key feature has been ignored in the derivation of this equation.

Motivated by the above-mentioned negative controllability result, in what follows, we shall propose a refined model to describe the DNA molecule. For this purpose, we partially employ a dynamical theory of Brownian motions, developed in [265], to describe the motion of a particle perturbed by random forces. In our opinion, the essence of the theory in [265] is a stochastic Newton's law, at least in some sense.

According to [265, Chapter 11], we may suppose that

$$y(t,x) = \int_0^t \tilde{y}(s,x)ds + \int_0^t F(s,x,y(s))dW(s). \tag{5.22}$$

Here $\tilde{y}(\cdot,\cdot)$ is the expected velocity, $F(\cdot,\cdot,\cdot)$ is the random perturbation from the fluid molecule. When y is small, one can assume that $F(\cdot,\cdot,\cdot)$ is linear in the third argument, i.e.,

$$F(s,x,y(t,x)) = b_1(t,x)y(t,x) \tag{5.23}$$

for a suitable $b_1(\cdot,\cdot)$.

Formally, the acceleration at position x along the string at time t is $\tilde{y}_t(t,x)$. By Newton's second law, it follows that

$$\tilde{y}_t(t,x) = F_1(t,x) + F_2(t,x) + F_3(t,x) + F_4(t,x).$$

Similar to (5.20), we obtain that

$$d\tilde{y}(t,x) = y_{xx}(t,x)dt + F_2(t,x)dt + F_3(t,x)dt + k_1(t,x)y(t,x)dW(t). \tag{5.24}$$

Combining (5.22), (5.23) and (5.24), we arrive at the following modified version of (5.20):

$$\begin{cases} dy = \tilde{y}dt + b_1 y dW(t) & \text{in } (0,T)\times(0,L), \\ d\tilde{y} = (y_{xx} + a_1 y_x + a_2 y_t + a_3 y)dt + a_4 y dW(t) & \text{in } (0,T)\times(0,L). \end{cases} \tag{5.25}$$

Then, similar to (5.21), we obtain the following control system:

$$
\begin{cases}
dy = \tilde{y}dt + (b_1 y + f)dW(t) & \text{in } (0,T)\times(0,L),\\
d\tilde{y} = (y_{xx} + a_1 y_x + a_2 y_t + a_3 y)dt & \\
\qquad +(a_4 y + g)dW(t) & \text{in } (0,T)\times(0,L),\\
y = h_1 & \text{on } (0,T)\times\{0\},\\
y = h_2 & \text{on } (0,T)\times\{L\},\\
y(0) = y_0, \quad \tilde{y}(0) = y_1 & \text{in } (0,L),
\end{cases}
\tag{5.26}
$$

where (f, g, h_1, h_2) are controls which belong to some suitable spaces. Under some assumptions, we shall show the exact controllability of the above system (See Theorem 10.12 in Chapter 10). This, in turn, justifies our modification (5.25).

Example 5.4. **Stochastic Schrödinger equation**

To investigate open quantum systems, people introduced the following stochastic Schrödinger equation (e.g., [169]):

$$
\begin{cases}
dy(t,x) = \Big(\frac{i}{2}\Delta y(t,x) - iV(x)y(t,x) - \frac{\alpha}{2}|x|^2 y(t,x)\Big)dt & \\
\qquad\qquad +\sqrt{\alpha}y(t,x)x dW(t) & \text{in } (0,T)\times G,\\
y = 0 & \text{on } (0,T)\times\partial G,\\
y(0) = y_0 & \text{in } G.
\end{cases}
\tag{5.27}
$$

Here G is a domain in $\mathbb{R}^m$ ($m \in \mathbb{N}$), $y(t,x)$ stands for the wave function at time $t \in [0,T]$ and $x \in \overline{G}$, $W(\cdot)$ is an m-dimensional Brownian motion, α is a positive constant, V is a suitable real function and $y_0 \in L^2(G;\mathbb{C})$ is the initial state. A derivation of (5.27) is given below.

For simplicity, we consider the continuous measurement of the position of a one-dimensional quantum particle ($m = 1$) governed by the standard Hamiltonian

$$
H = iV(x) - \frac{i}{2}\partial_{xx}.
$$

Let us first recall the following postulates in Quantum Mechanics (e.g., [184]).

1. **Non-ideal measurement principle.**
   After a measurement, the wave function $y(t,x)$ is transformed into a new one which, up to a normalization, reads
   $$
   y(t,x)e^{-\alpha(x-q(t))^2},
   $$
   where $q(t) \in \mathbb{R}$ is the result of the measurement at time t and α characterizes the accuracy of the measurement.
2. **Continuous limit of discrete observations.**
   Fix a time t and denote by n the times of the measurements. Take the above non-ideal measurements of the position of the quantum particle at

moments $t_k = k\delta$, where $\delta = t/n$, $n \in \mathbb{N}$. As $n \to +\infty$, the accuracy of each measurement will be proportional to the time duration δ between successive measurements, i.e. $\alpha = \lambda\delta$, where λ reflects the properties of the measuring apparatus.

The above two postulates are standard in Quantum Mechanics. Next, we introduce a postulate about the errors of the measurement.

3. **Gaussian model of errors.** Suppose that $q(t)$ obeys the following relation:
$$\int_0^s q(s)ds = \frac{1}{2\sqrt{\lambda}} W(t) \quad \text{for a constant } \lambda > 0. \tag{5.28}$$

Note that the evolution of the quantum system between measurements is governed by the law of free Hamiltonian. Hence, after n times measurements, the resulting wave function reads

$$y(n\delta) = \prod_{k=1}^{n} \exp\{-\delta\lambda(x - q(k\delta))^2\} \exp(-\delta H) y_0. \tag{5.29}$$

Letting $n \to +\infty$ in (5.29), we obtain that

$$y = e^{-(tH + \int_0^t \lambda(x-q(s))^2 ds)} y_0.$$

Hence,

$$y_t = -(H + \lambda(x - q(t))^2)y. \tag{5.30}$$

The term $\lambda q(t)^2 y$ can be dropped in (5.30) because solutions to the equation with and without the term $\lambda q(t)^2 y$ are proportional and this difference is irrelevant in Quantum Mechanics (in which people are only interested in normalized states). In this way, we obtain the following equation:

$$y_t + (H + \lambda x^2)y = 2\lambda x q y. \tag{5.31}$$

Noting (5.28), we then replace the equation (5.31) by the following one:

$$dy + (H + \lambda x^2)y dt = 2\sqrt{\lambda} x y dW(t). \tag{5.32}$$

This, after some obvious scaling, gives the first equation of (5.27) for one space dimension.

People have two possible ways to put controls on (5.32). One is absorbing/delivering the wave on the boundary. This leads to a boundary control in the drift term. The other is to put some external force when doing the measurement. This leads to an internal control in the diffusion term. With such controls, the equation (5.27) becomes the following control system:

$$\begin{cases} dy = \Big(\frac{i}{2}\Delta y - iV(x)y - \frac{\alpha}{2}|x|^2 y\Big)dt \\ \qquad\quad +(\sqrt{\alpha} y + v)x dW(t) & \text{in } (0,T)\times G, \\ y = f & \text{on } (0,T)\times \partial G, \\ y(0) = y_0 & \text{in } G. \end{cases} \tag{5.33}$$

Example 5.5. **Stochastic Navier-Stokes equation**

To study turbulence, people introduced the statistical approach (e.g. [257, 262, 296, 331]). In this framework, the velocity field of an incompressible fluid is described by a space-time vector-valued random field, which is governed by the following Navier-Stokes equation with some random force:

$$\begin{cases} u_t = \nu\Delta u - u\nabla u - \nabla p + f & \text{in } (0,T)\times G, \\ \operatorname{div} u = 0 & \text{in } (0,T)\times G, \\ u(0) = u_0 & \text{in } G. \end{cases} \tag{5.34}$$

Here $G \subset \mathbb{R}^m$ ($m = 2$ or 3) is a domain, ν is a positive real number, $u(t,x) = (u^1(t,x), \cdots, u^m(t,x))$ is the velocity of the fluid at time t and point x, u_0 is the initial datum of u, $p(t,x)$ is the pressure of the fluid, $f = f(t,x,u)$ is the external force acted to the fluid, which is caused by lots of small independent random disturbances. In many situations, it is reasonable to assume that the random force $f(t,x,u)$ at time t and position x can be approximated by a Gaussian stochastic process as follows:

$$df(t,x,u) = g(u)dW(t),$$

where g is a suitable process depending on u, while $W(\cdot)$ is a cylindrical Brownian motion. Consequently, instead of (5.34), we obtain the following stochastic Navier-Stokes equation:

$$\begin{cases} du = (\nu\Delta u - u\nabla u - \nabla p)dt + g(u)dW(t) & \text{in } (0,T)\times G, \\ \operatorname{div} u = 0 & \text{in } (0,T)\times G, \\ u(0) = u_0 & \text{in } G. \end{cases} \tag{5.35}$$

To control the velocity of the fluid, one can put a force h_1 in the drift term and a force h_2 in the diffusion term. Then the equation (5.35) becomes

$$\begin{cases} du = (\nu\Delta u - u\nabla u - \nabla p + h_1)dt + (g(u) + h_2)dW(t) & \text{in } (0,T)\times G, \\ \operatorname{div} u = 0 & \text{in } (0,T)\times G, \\ u(0) = u_0 & \text{in } G. \end{cases}$$

Although the control problems for stochastic Navier-Stokes equations are very important both theoretically and in applications, due to the space limitation, we will not consider these problems in this book (See [128, 132, 301, 304] for some interesting works on this topic).

5.3 Some Control Problems for Stochastic Distributed Parameter Systems

Let H and V be two separable Hilbert spaces, and write $\mathcal{L}_2^0 \overset{\Delta}{=} \mathcal{L}_2(V;H)$. Let $\{W(t)\}_{t\in[0,T]}$ be a V-valued, $\mathbf{F}$-adapted, standard Q-Brownian motion (with

Q being a given positive definite, trace-class operator on V) or a cylindrical Brownian motion on $(\Omega, \mathcal{F}, \mathbf{F}, \mathbb{P})$. Denote by $\mathbb{F}$ the progressive σ-field (in $[0,T]\times\Omega$) w.r.t. $\mathbf{F}$. In the sequel, to simplify the presentation, we only consider the case of cylindrical Brownian motion.

Suppose that A generates a C_0-semigroup $\{S(t)\}_{t\geq 0}$ on H, U is a given separable metric space and $a(\cdot): [0,T]\times\Omega\times H\times U\to H$ and $b(\cdot): [0,T]\times\Omega\times H\times U\to\mathcal{L}_2^0$ are two suitably given functions. Let us consider the following controlled stochastic evolution equation:

$$
\begin{cases}
dx(t) = \big(Ax(t) + a(t, x(t), u(t))\big)dt \\
\qquad\qquad +b(t, x(t), u(t))dW(t) \qquad \text{in } (0,T], \\
x(0) = x_0.
\end{cases}
\tag{5.36}
$$

Here x_0 is the initial state (in a suitable set $\mathcal{H}_0$ consisting of some H-valued, $\mathcal{F}_0$-measurable random variables), $x(\cdot)$ is the state variable (valued in H), $u(\cdot)$ is the control variable (valued in U), and hence H and U serve as the state and the control spaces, respectively.

Write

$$
\mathcal{U}[0,T] \overset{\Delta}{=} \big\{u : [0,T]\times\Omega\to U \ \big|\ u(\cdot) \text{ is } \mathbf{F}\text{-adapted}\big\}.
$$

This is the (biggest) control class that we shall use. Such a choice of controls means that one can only use the information (as specified by the filtration $\mathbf{F}$) of what has happened up to the present moment, rather than what is going to happen afterwards due to the uncertainty of the system under consideration (as a consequence, for any $t\in[0,T]$ one cannot perform his/her action $u(t)$ before the time t really comes). This nonanticipative restriction is the reason for us to choose $u(\cdot)$ to be $\mathbf{F}$-adapted.

Let $\mathcal{U}^0[0,T]$ be a suitable subset of $\mathcal{U}[0,T]$, and $\mathcal{H}_T$ a suitable set consisting of some H-valued, $\mathcal{F}_T$-measurable random variables. We suppose that the choices of the functions $a(\cdot)$ and $b(\cdot)$ and the sets $\mathcal{H}_0$ and the control class $\mathcal{U}^0[0,T]$ are such that, for any $x_0\in\mathcal{H}_0$ and $u(\cdot)\in\mathcal{U}^0[0,T]$, the system (5.36) admits a mild solution[1] $x(\cdot)$ in some solution space $\mathcal{S}[0,T]$ (in which solutions are understood in the sense of Definition 3.9), say $C_{\mathbb{F}}([0,T];L^p(\Omega;H))$ (for some $p\geq 2$) as that in Theorem 3.14.

Let Z be a nonempty set, and Γ be a nonempty subset of Z. Suppose $F: \mathcal{S}[0,T]\times\mathcal{U}^0[0,T]\to Z$ is a given map. Motivated by [370, p. 14] and [50], we introduce the following notions of controllability:

Definition 5.6. 1) *The system* (5.36) *is said to be* (F,Γ)*-exactly controllable (or simply exactly controllable in the case that* (F,Γ) *is clear from the context)*

[1] Here, we do not need to assume the solution is unique. Solutions to (5.36) can also be understood in other sense, say transpositions (for the case of unbounded control operators) which will be introduced in Definition 7.11 of Chapter 7.

if for any $x_0 \in \mathcal{H}_0$ *and* $x_1 \in \Gamma$, *there is a control* $u(\cdot) \in \mathcal{U}^0[0,T]$ *such that* (5.36) *admits a mild solution* $x(\cdot)$ *satisfying*

$$F(x(\cdot), u(\cdot)) = x_1; \tag{5.37}$$

2) *The system* (5.36) *is said to be* (F,Γ)*-approximately controllable (or simply approximately controllable in the case that* (F,Γ) *is clear from the context) if* Z *is another metric space with a metric* $d_Z(\cdot,\cdot)$, *and for any* $x_0 \in \mathcal{H}_0$, $x_1 \in \Gamma$ *and* $\varepsilon > 0$, *there is a control* $u(\cdot) \in \mathcal{U}^0[0,T]$ *such that* (5.36) *admits a mild solution* $x(\cdot)$ *satisfying*

$$d_Z(F(x(\cdot), u(\cdot)), x_1) < \varepsilon;$$

3) *The system* (5.36) *is said to be* (F,Γ)*-controllable (or simply controllable in the case that* (F,Γ) *is clear from the context) if for any* $x_0 \in \mathcal{H}_0$, *there is a control* $u(\cdot) \in \mathcal{U}^0[0,T]$ *such that* (5.36) *admits a mild solution* $x(\cdot)$ *satisfying*

$$F(x(\cdot), u(\cdot)) \in \Gamma. \tag{5.38}$$

Clearly, the above controllability notions are quite general, which include the usual exact/null/approximate controllability studied in the literature. For example, let us choose $\mathcal{H}_1$ to be a metric space consisting of suitably H-valued, $\mathcal{F}_T$-measurable random variables so that $0 \in \mathcal{H}_1$. For $Z = \mathcal{H}_1$, $\Gamma = Z$ and $F(x(\cdot), u(\cdot)) = x(T)$, we get the exact/approximate controllability; for $Z = \mathcal{H}_1$, $\Gamma = \{0\}$ and $F(x(\cdot), u(\cdot)) = x(T)$, we get the null controllability. Further, for

$$Z = \mathcal{H}_1 \times \mathcal{H}_1, \quad \Gamma = \{0\} \times \{0\}$$

and

$$F(x(\cdot), u(\cdot)) = \begin{pmatrix} x(T) \\ \displaystyle\int_0^T M(T-s)x(s)ds \end{pmatrix},$$

in which $M(\cdot) : [0,T] \to \mathbb{R}$ is a given function, we get the memory-type null controllability of (5.36) (See [50] for examples in the deterministic setting). Clearly, the (F,Γ)-controllability concept is too general to obtain meaningful results, and therefore in the rest of this book we will focus mainly on exact/null controllability. Nevertheless, as we shall see later, (F,Γ)-controllability is a basis for the study of optimal control problems for (5.36). This is quite natural because, as we mentioned in the first chapter, controllability actually means "feasibility" for many activities (at least in some broad sense).

Clearly, the map $F : \mathcal{S}[0,T] \times \mathcal{U}^0[0,T] \to Z$ can also be viewed an input-output operator, while Γ can be regarded as a ("measurable") observation. Similarly to Definition 5.6, we introduce the following notions of observability:

Definition 5.7. 1) *The system* (5.36) *is said to be* (F,Γ)*-exactly observable (or simply exactly observable in the case that* (F,Γ) *is clear from the context)* *if for any control* $u(\cdot) \in \mathcal{U}^0[0,T]$ *and observation* $\gamma \in \Gamma$*, there is a unique initial datum* $x_0 \in \mathcal{H}_0$ *such that the corresponding mild solution* $x(\cdot)$ *to* (5.36) *is unique and satisfies*

$$F(x(\cdot), u(\cdot)) = \gamma; \tag{5.39}$$

2) *The system* (5.36) *is said to be* (F,Γ)*-continuously observable (or simply continuously observable in the case that* (F,Γ) *is clear from the context) if* $\mathcal{S}[0,T]$*,* $\mathcal{U}^0[0,T]$ *and* Γ *are metric spaces, and for any control* $\hat{u}(\cdot) \in \mathcal{U}^0[0,T]$*, observation* $\hat{\gamma} \in \Gamma$ *and* $\varepsilon > 0$*, there exists* $\delta = \delta(\varepsilon) > 0$ *and a unique initial datum* $\hat{x}_0 \in \mathcal{H}_0$ *such that the corresponding mild solution* $\hat{x}(\cdot)$ *to* (5.36) *(with* (x_0, u) *replaced by* $(\hat{x}_0, \hat{u})$*) is unique and satisfies*

$$F(\hat{x}(\cdot), \hat{u}(\cdot)) = \hat{\gamma}, \tag{5.40}$$

and for any control $u(\cdot) \in \mathcal{U}^0[0,T]$ *and observation* $\gamma \in \Gamma$ *satisfying*

$$d_{\mathcal{U}^0[0,T]}(u(\cdot), \hat{u}(\cdot)) + d_\Gamma(\gamma, \hat{\gamma}) < \varepsilon, \tag{5.41}$$

there is a unique initial datum $x_0 \in \mathcal{H}_0$ *such that the corresponding mild solution* $x(\cdot)$ *to* (5.36) *is unique, and satisfies* (5.39) *and*

$$d_{\mathcal{S}[0,T]}(x(\cdot), \hat{x}(\cdot)) \leq \delta;$$

3) *The system* (5.36) *is said to be* (F,Γ)*-continuously initially observable (or simply continuously initially observable in the case that* (F,Γ) *is clear from the context) if* $\mathcal{H}_0$*,* $\mathcal{U}^0[0,T]$ *and* Γ *are metric spaces, and for any control* $\hat{u}(\cdot) \in \mathcal{U}^0[0,T]$*, observation* $\hat{\gamma} \in \Gamma$ *and* $\varepsilon > 0$*, there exists* $\delta = \delta(\varepsilon) > 0$ *and a unique initial datum* $\hat{x}_0 \in \mathcal{H}_0$ *such that the corresponding mild solution* $\hat{x}(\cdot)$ *to* (5.36) *(with* (x_0, u) *replaced by* $(\hat{x}_0, \hat{u})$*) is unique and satisfies* (5.40)*, and for any control* $u(\cdot) \in \mathcal{U}^0[0,T]$ *and observation* $\gamma \in \Gamma$ *satisfying* (5.41)*, there is a unique initial datum* $x_0 \in \mathcal{H}_0$ *such that the corresponding mild solution* $x(\cdot)$ *to* (5.36) *is unique, and satisfies* (5.39) *and*

$$d_{\mathcal{H}_0}(x_0, \hat{x}_0) \leq \delta;$$

4) *The system* (5.36) *is said to be* (F,Γ)*-continuously finally observable (or simply continuously finally observable in the case that* (F,Γ) *is clear from the context) if* $\mathcal{H}_T$*,* $\mathcal{U}^0[0,T]$ *and* Γ *are metric spaces, and for any control* $\hat{u}(\cdot) \in \mathcal{U}^0[0,T]$*, observation* $\hat{\gamma} \in \Gamma$ *and* $\varepsilon > 0$*, there exists* $\delta = \delta(\varepsilon) > 0$ *and a unique initial datum* $\hat{x}_0 \in \mathcal{H}_0$ *such that the corresponding mild solution* $\hat{x}(\cdot)$ *to* (5.36) *(with* (x_0, u) *replaced by* $(\hat{x}_0, \hat{u})$*) is unique and satisfies* (5.40)*, and for any control* $u(\cdot) \in \mathcal{U}^0[0,T]$ *and observation* $\gamma \in \Gamma$ *satisfying* (5.41)*, there is a unique initial datum* $x_0 \in \mathcal{H}_0$ *such that the corresponding mild solution* $x(\cdot)$ *to* (5.36) *is unique, and satisfies* (5.39) *and*

$$d_{\mathcal{H}_T}(x(T), \hat{x}(T)) \leq \delta.$$

The above observability notions are also quite general, which include the usual exact/internal/boundary observability studied in the literature. Particularly, the above continuous observability and continuous initial observability include various observability estimates which are extensively studied for many partial differential equations. Roughly speaking, observability means that people may recognize the whole state under consideration through the observation part. In the deterministic setting, the observation problems are strongly related to the controllability problems, which are actually dual to each other at least for linear systems. However, in the stochastic situation, as we shall see in the next two chapters, the relationship between the controllability problems and the observation problems are much more complicated.

Note also that, the observation problems make sense even if there is no control variable in the system (5.36) (The corresponding problems can be easily formulated as special cases as that in Definition 5.7). In this way, the observation problems may contain also various inverse problems which are also extensively studied in the literature for deterministic differential equations. Nevertheless, in this book we shall focus more on control problems, and therefore, the observation problems are considered only as auxiliary tools for studying the controllability problems.

More importantly, in order to solve the controllability problem for the forward system (5.36), as we shall see in Section 7.3 of Chapter 7, stimulated also by [275], people need to analyze suitable controllability and/or observability problems for some backward stochastic evolution equations. Because of this, we shall give below a general formulation of backward controllability/observability problems in the stochastic setting.

Suppose that $a(\cdot):\ [0,T]\times\Omega\times H\times\mathcal{L}_2^0\times U\to H$ and $b(\cdot):\ [0,T]\times\Omega\times H\times U\to\mathcal{L}_2^0$ are two suitably given functions. Let us consider the following backward stochastic equation evolved in H:

$$\begin{cases} dy(t)=\big(-A^*y(t)+a(t,y(t),u(t),Y(t))\big)dt\\ \qquad\qquad +\big(b(t,u(t),y(t))+Y(t)\big)dW(t) & \text{in } [0,T),\\ y(T)=y_T. \end{cases}\tag{5.42}$$

Here $u(\cdot)\in\mathcal{U}^0[0,T]$, $y_T\in\mathcal{H}_T$ is the final state. We suppose that the choices of the functions $a(\cdot)$ and $b(\cdot)$ and the sets $\mathcal{U}^0[0,T]$ and $\mathcal{H}_T$ are such that, for any $y_T\in\mathcal{H}_T$ and $u(\cdot)\in\mathcal{U}^0[0,T]$, the system (5.42) admits a (mild/transposition) solution[2] $(y(\cdot),Y(\cdot))$ in some solution space $\mathcal{S}[0,T]\times\mathcal{T}[0,T]$ (in which solutions are understood in the sense of Definition 4.6 or Definition 4.13), say $(y(\cdot),Y(\cdot))\in L^p_{\mathbb{F}}(\Omega;C([0,T];H))\times L^2_{\mathbb{F}}(0,T;L^p(\Omega;\mathcal{L}_2^0))$ (for some $p\in(1,2]$) as that in Theorem 4.24.

[2] Here, similarly to that for the system (5.36), we do not need to assume the solution to (5.42) is unique. Also, solutions to (5.42) can be understood in other sense, say transpositions (for the case of unbounded control operators, even if the filtration **F** is natural) which will be introduced in Definition 7.13 of Chapter 7.

Suppose $G: \mathcal{S}[0,T]\times\mathcal{T}[0,T]\times\mathcal{U}^0[0,T]\to Z$ is a given map. Similarly to Definition 5.6, we introduce the following notions of controllability:

Definition 5.8. 1) *The system* (5.42) *is said to be* (G,Γ)*-exactly controllable (or simply exactly controllable in the case that* (G,Γ) *is clear from the context) if for any* $y_T\in\mathcal{H}_T$ *and* $y_0\in\Gamma$*, there is a control* $u(\cdot)\in\mathcal{U}^0[0,T]$ *such that* (5.42) *admits a mild/transposition solution* $(y(\cdot),Y(\cdot))$ *satisfying*

$$G(y(\cdot),Y(\cdot),u(\cdot))=y_0;$$

2) *The system* (5.42) *is said to be* (G,Γ)*-approximately controllable (or simply approximately controllable in the case that* (G,Γ) *is clear from the context) if* Z *is another metric space, and for any* $y_T\in\mathcal{H}_T$*,* $y_0\in\Gamma$ *and* $\varepsilon>0$*, there is a control* $u(\cdot)\in\mathcal{U}^0[0,T]$ *such that* (5.42) *admits a mild/transposition solution* $(y(\cdot),Y(\cdot))$ *satisfying*

$$d_Z(G(y(\cdot),Y(\cdot),u(\cdot)),y_0)<\varepsilon;$$

3) *The system* (5.42) *is said to be* (G,Γ)*-controllable (or simply controllable in the case that* (G,Γ) *is clear from the context) if for any* $y_T\in\mathcal{H}_T$*, there is a control* $u(\cdot)\in\mathcal{U}^0[0,T]$ *such that* (5.42) *admits a mild/transposition solution* $(y(\cdot),Y(\cdot))$ *satisfying*

$$G(y(\cdot),Y(\cdot),u(\cdot))\in\Gamma. \tag{5.43}$$

Also, similarly to Definition 5.7, we introduce the following notions of observability:

Definition 5.9. 1) *The system* (5.42) *is said to be* (G,Γ)*-exactly observable (or simply exactly observable in the case that* (G,Γ) *is clear from the context) if for any control* $u(\cdot)\in\mathcal{U}^0[0,T]$ *and observation* $\gamma\in\Gamma$*, there is a unique final initial datum* $y_T\in\mathcal{H}_T$ *such that the corresponding solution* $(y(\cdot),Y(\cdot))$ *to* (5.42) *is unique and satisfies*

$$G(y(\cdot),Y(\cdot),u(\cdot))=\gamma; \tag{5.44}$$

2) *The system* (5.42) *is said to be* (G,Γ)*-continuously observable (or simply continuously observable in the case that* (G,Γ) *is clear from the context) if* $\mathcal{S}[0,T]$*,* $\mathcal{T}[0,T]$*,* $\mathcal{U}^0[0,T]$ *and* Γ *are metric spaces, and for any control* $\hat u(\cdot)\in\mathcal{U}^0[0,T]$*, observation* $\hat\gamma\in\Gamma$ *and* $\varepsilon>0$*, there exists* $\delta=\delta(\varepsilon)>0$ *and a unique final datum* $\hat y_T\in\mathcal{H}_T$ *such that the corresponding solution* $(\hat y(\cdot),\widehat Y(\cdot))$ *to* (5.42) *(with* (y_T,u) *replaced by* $(\hat y_T,\hat u)$*) is unique and satisfies*

$$G(\hat y(\cdot),\widehat Y(\cdot),\hat u(\cdot))=\hat\gamma, \tag{5.45}$$

and for any control $u(\cdot)\in\mathcal{U}^0[0,T]$ *and observation* $\gamma\in\Gamma$ *satisfying*

$$d_{\mathcal{U}^0[0,T]}(u(\cdot),\hat u(\cdot))+d_\Gamma(\gamma,\hat\gamma)<\varepsilon, \tag{5.46}$$

there is a unique final datum $y_T \in \mathcal{H}_T$ *such that the corresponding solution* $(y(\cdot), Y(\cdot))$ *to* (5.42) *is unique, and satisfies* (5.44) *and*

$$d_{\mathcal{S}[0,T]}(y(\cdot), \hat{y}(\cdot)) + d_{\mathcal{T}[0,T]}(Y(\cdot), \widehat{Y}(\cdot)) \le \delta;$$

3) *The system* (5.42) *is said to be* (G, Γ)*-continuously finally observable (or simply continuously finally observable in the case that* (G, Γ) *is clear from the context) if* $\mathcal{H}_T$, $\mathcal{U}^0[0,T]$ *and* Γ *are metric spaces, and for any control* $\hat{u}(\cdot) \in \mathcal{U}^0[0,T]$, *observation* $\hat{\gamma} \in \Gamma$ *and* $\varepsilon > 0$, *there exists* $\delta = \delta(\varepsilon) > 0$ *and a unique final datum* $\hat{y}_T \in \mathcal{H}_T$ *such that the corresponding solution* $(\hat{y}(\cdot), \widehat{Y}(\cdot))$ *to* (5.42) *(with* (y_T, u) *replaced by* $(\hat{y}_T, \hat{u})$*) is unique and satisfies* (5.45), *and for any control* $u(\cdot) \in \mathcal{U}^0[0,T]$ *and observation* $\gamma \in \Gamma$ *satisfying* (5.46), *there is a unique final datum* $y_T \in \mathcal{H}_T$ *such that the corresponding solution* $(y(\cdot), Y(\cdot))$ *to* (5.42) *is unique, and satisfies* (5.44) *and*

$$d_{\mathcal{H}_T}(y_T, \hat{y}_T) \le \delta;$$

4) *The system* (5.42) *is said to be* (G, Γ)*-continuously initially observable (or simply continuously initially observable in the case that* (G, Γ) *is clear from the context) if* $\mathcal{H}_0$, $\mathcal{U}^0[0,T]$ *and* Γ *are metric spaces, and for any control* $\hat{u}(\cdot) \in \mathcal{U}^0[0,T]$, *observation* $\hat{\gamma} \in \Gamma$ *and* $\varepsilon > 0$, *there exists* $\delta = \delta(\varepsilon) > 0$ *and a unique final datum* $\hat{y}_T \in \mathcal{H}_T$ *such that the corresponding solution* $(\hat{y}(\cdot), \widehat{Y}(\cdot))$ *to* (5.42) *(with* (y_T, u) *replaced by* $(\hat{y}_T, \hat{u})$*) is unique and satisfies* (5.45), *and for any control* $u(\cdot) \in \mathcal{U}^0[0,T]$ *and observation* $\gamma \in \Gamma$ *satisfying* (5.46), *there is a unique final datum* $y_T \in \mathcal{H}_T$ *such that the corresponding solution* $(y(\cdot), Y(\cdot))$ *to* (5.42) *is unique, and satisfies* (5.44) *and*

$$d_{\mathcal{H}_0}(y(0), \hat{y}(0)) \le \delta.$$

In this book, as a key step to solve the stochastic controllability problems, we shall employ mainly the global Carleman estimates to derive observability estimates for some backward/forward stochastic evolution equations in infinite dimensions.

In the rest of this section, we fix a nonempty set Z, a nonempty subset Γ of Z and a map $F: \mathcal{S}[0,T] \times \mathcal{U}^0[0,T] \to Z$, and we suppose that the system (5.36) is (F, Γ)-controllable.

For any fixed functions $f(\cdot,\cdot,\cdot): [0,T] \times \mathcal{S}[0,T] \times \mathcal{U}^0[0,T] \to \mathbb{R}$ and $h(\cdot): H \to \mathbb{R}$, we introduce a cost function $\mathcal{J}(\cdot,\cdot): \mathcal{S}[0,T] \times \mathcal{U}^0[0,T] \to \mathbb{R}$ for the control system (5.36) as follows:

$$\begin{aligned} \mathcal{J}(x(\cdot), u(\cdot)) = \mathbb{E}\left(\int_0^T f(t, x(\cdot), u(\cdot))dt + h(x(T))\right), \\ \forall\, (x(\cdot), u(\cdot)) \in \mathcal{S}[0,T] \times \mathcal{U}^0[0,T]. \end{aligned} \tag{5.47}$$

Denote by $\mathcal{U}_{ad}[0,T]$ the set of all controls $u(\cdot) \in \mathcal{U}^0[0,T]$ such that for some $x_0 \in \mathcal{H}_0$, (5.36) admits a mild solution $x(\cdot)$ satisfying (5.38) and

$$f(\cdot, x(\cdot), u(\cdot)) \in L^1_{\mathbb{F}}(0,T) \quad \text{and} \quad h(x(T)) \in L^1_{\mathcal{F}_T}(\Omega). \tag{5.48}$$

Also, denote by $\mathcal{P}_{ad}[0,T]$ the set of all pairs $(x(\cdot), u(\cdot)) \in \mathcal{S}[0,T] \times \mathcal{U}^0[0,T]$ such that for some $x_0 \in \mathcal{H}_0$, $x(\cdot)$ is a mild solution to (5.36) satisfying (5.38) and (5.48). Clearly, under our (F, Γ)-controllability assumption on the system (5.36), both $\mathcal{U}_{ad}[0,T]$ and $\mathcal{P}_{ad}[0,T]$ are non-empty. Any control $u(\cdot) \in \mathcal{U}_{ad}[0,T]$ (*resp.* pair $(x(\cdot), u(\cdot)) \in \mathcal{P}_{ad}[0,T]$) is called an admissible control (*resp.* pair).

The optimal control problem for the system (5.36) with the cost functional (5.47) is formulated as follows:

Problem (SOP) *Find a* $(\bar{x}(\cdot), \bar{u}(\cdot)) \in \mathcal{P}_{ad}[0,T]$ *such that*

$$\mathcal{J}(\bar{x}(\cdot), \bar{u}(\cdot)) = \inf_{(x(\cdot),u(\cdot)) \in \mathcal{P}_{ad}[0,T]} \mathcal{J}(x(\cdot), u(\cdot)). \tag{5.49}$$

Any $(\bar{x}(\cdot), \bar{u}(\cdot)) \in \mathcal{P}_{ad}[0,T]$ satisfying (5.49) is called an *optimal pair* (for Problem (SOP)), and $\bar{x}(\cdot)$ and $\bar{u}(\cdot)$ are called respectively an *optimal state* and an *optimal control.*

Similarly, we fix a map $G: \ \mathcal{S}[0,T] \times \mathcal{T}[0,T] \times \mathcal{U}^0[0,T] \to Z$, and we suppose that the system (5.42) is (G, Γ)-controllable.

For any fixed functions $g(\cdot,\cdot,\cdot): \ [0,T] \times \mathcal{S}[0,T] \times \mathcal{T}[0,T] \times \mathcal{U}^0[0,T] \to \mathbb{R}$ and $h(\cdot): \ H \to \mathbb{R}$, we introduce a cost function $\mathcal{J}(\cdot,\cdot,\cdot): \ \mathcal{S}[0,T] \times \mathcal{T}[0,T] \times \mathcal{U}^0[0,T] \to \mathbb{R}$ for the control system (5.42) as follows:

$$\begin{aligned} &\mathcal{J}(y(\cdot), Y(\cdot), u(\cdot)) = \mathbb{E}\int_0^T g(t, y(\cdot), Y(\cdot), u(\cdot))dt + h(y(0)), \\ &\qquad\qquad \forall\, (y(\cdot), Y(\cdot), u(\cdot)) \in \mathcal{S}[0,T] \times \mathcal{T}[0,T] \times \mathcal{U}^0[0,T]. \end{aligned} \tag{5.50}$$

Denote by $\mathcal{U}_{ad}[0,T]$ the set of all controls $u(\cdot) \in \mathcal{U}^0[0,T]$ such that for some $y_T \in \mathcal{H}_T$, (5.42) admits a solution $(y(\cdot), Y(\cdot))$ satisfying (5.43) and

$$g(\cdot, y(\cdot), Y(\cdot), u(\cdot)) \in L^1_{\mathbb{F}}(0,T). \tag{5.51}$$

Also, denote by $\mathcal{P}_{ad}[0,T]$ the set of all triples $(y(\cdot), Y(\cdot), u(\cdot)) \in \mathcal{S}[0,T] \times \mathcal{T}[0,T] \times \mathcal{U}^0[0,T]$ such that for some $y_T \in \mathcal{H}_T$, $(y(\cdot), Y(\cdot))$ is a solution to (5.42) satisfying (5.43) and (5.51). Clearly, under our (G, Γ)-controllability assumption on the system (5.42), both $\mathcal{U}_{ad}[0,T]$ and $\mathcal{P}_{ad}[0,T]$ are non-empty. Any control $u(\cdot) \in \mathcal{U}_{ad}[0,T]$ is called an admissible control.

The optimal control problem for the system (5.42) with the cost functional (5.50) is formulated as follows:

Problem (BSOP) *Find a triple* $(\bar{y}(\cdot), \overline{Y}(\cdot), \bar{u}(\cdot)) \in \mathcal{P}_{ad}[0,T]$ *such that*

$$\mathcal{J}(\bar{y}(\cdot), \overline{Y}(\cdot), \bar{u}(\cdot)) = \inf_{(y(\cdot),Y(\cdot),u(\cdot)) \in \mathcal{P}_{ad}[0,T]} \mathcal{J}(y(\cdot), Y(\cdot), u(\cdot)). \tag{5.52}$$

Any $(\bar{y}(\cdot), \overline{Y}(\cdot), \bar{u}(\cdot)) \in \mathcal{P}_{ad}[0,T]$ satisfying (5.52) is called an *optimal triple* (for Problem (BSOP)), and $(\bar{y}(\cdot), \overline{Y}(\cdot))$ and $\bar{u}(\cdot)$ are called respectively an *optimal state* and an *optimal control.*

In this book, we shall address only Problem (SOP) (Actually, Problem (BSOP) is much easier to handle than Problem (SOP)). We shall derive the first and second order necessary conditions for optimal controls for Problem (SOP), and also characterize the optimal feedback operators for a special version of this problem. The main difficulty to do all of these is how to establish the well-posedness of the operator-valued backward stochastic evolution equations (including operator-valued backward stochastic Lyapunov/Riccati equations in infinite dimensions in particular). To do this, we shall systematically employ our stochastic transposition method.

5.4 Notes and Comments

The stochastic control systems presented in this chapter are well-known except the system (5.26) (and its high dimensional counterpart (10.3) in Chapter (10)) introduced in [247].

Our formulation of (F,Γ)- (*resp.* (G,Γ)-) controllability/observability problems for the system (5.36) (*resp.* (5.42)) are quite general, even if the system is reduced to deterministic equations in finite dimensions. The same can be said for optimal control problems for the systems (5.36) and (5.42), i.e. Problems (SOP) and (BSOP). Of course, in this book (as any of other works on Control Theory), we have to focus on more specific controllability and optimal control problems. Also, we emphasize again that controllability is the basis for optimal control problems though the controllability for the stochastic optimal control problems to be studied in this book is obvious.

There are many other control problems for stochastic distributed parameter systems which deserve careful studies (but they are not addressed to in this book), for example:

- In the field of control theory for stochastic finite dimensional systems, stochastic approximation, system identification and adaptive control ([52, 53, 181, 340]) are important topics. It would be quite interesting to study their infinite dimensional counterparts but it seems that not too much are known in this respect (See however [83] and the two books [15, 167] for some very interesting related works).
- There exist numerous works on inverse problems for deterministic equations. However, very little are known for the relatively unexplored but important subject of inverse problems for stochastic partial differential equations. By means of the tools that we developed for solving the s- tochastic controllability problems, in [226, 229, 243] we initiated the s- tudy of inverse problems for stochastic partial differential equations (See [198, 238, 353, 358, 366, 373, 374, 375] for further progress), in which the

point is that, unlike most of the previous works in this topic, the problem is genuinely stochastic and therefore it cannot be reduced to any deterministic one. Especially, it was found in [243] that both the formulation of stochastic inverse problems and the tools to solve them differ considerably from their deterministic counterpart. Indeed, as the counterexample in [243, Remark 2.7] shows, the inverse problem considered therein makes sense only in the stochastic setting!

- There exist numerous works on the stability/stabilization problems for deterministic equations. However, very little are known for the same problems but for stochastic partial differential equations (We refer to [216] for some results on this topic). Further, we remark that, it would be quite interesting to extend the famous backstepping method for handling the stabilization problems of some deterministic partial differential equations ([172]) to the stochastic setting but this remains to be done.
- It is well-known that Game Theory is strongly related to Control Theory. In recent years, there are many studies on stochastic differential games (e.g., [39, 48, 99]). It would be quite interesting to develop game theory for stochastic distributed parameter systems but, as far as we know, nothing has been done in this direction.
- In recent years, there are many works on the time-inconsistent optimal control problems for stochastic differential equations (See [369] and the references cited therein). It would be quite interesting to study the same problems but for stochastic evolution equations in infinite dimensions. As far as we know, [76] is the only publication on this topic.
- In recent years, there are some works on control problems for stochastic systems with fractional Brownian motions (e.g., [82]). It would be quite interesting to study carefully this sort of problems (especially those in infinite dimensions).
- It seems that numerical methods for solving control problems of stochastic distributed parameter systems are almost blank (See [283] for the only one work that we know). Very likely, people need to develop completely new algorithms for these problems.
- The control problems presented in this chapter and mentioned above make sense also for forward-backward stochastic evolution equations, stochastic evolution equations driven by G-Brownian motions, forward-backward doubly stochastic evolution equations, stochastic Volterra integral equation in infinite dimensions or stochastic evolution equations driven by general martingales (even with incomplete information). Also, it seems interesting to study recursive optimal control problems for stochastic evolution equations and/or the above mentioned stochastic equations. As far as we know, all of these problems remain to be done, and very likely people might need to develop new tools to obtain interesting new results.

6

Controllability for Stochastic Differential Equations in Finite Dimensions

In this chapter, we are concerned with the controllability problem for the simplest stochastic linear evolution equations, i.e., stochastic linear differential equations (with constant coefficients) in finite dimensions. A Kalman-type rank condition to characterize the exact controllability is given for the special case that the control matrix in the diffusion term is of full rank. Nevertheless, generally speaking, the controllability of stochastic linear differential equations in finite dimensions is surprisingly different from its deterministic counterpart. Indeed, it will be shown further that

- There is a very simple stochastic control system (in one dimension) which is exactly controllable by means of L^1-in time controls but the same system is NOT exactly controllable anymore provided that one chooses L^p-in time controls for any $p \in (1, \infty]$;
- One can construct a class of very simple stochastic control systems (in two dimensions) for which neither null nor approximate controllability is robust w.r.t. small perturbations.

6.1 The Control Systems With Controls in Both Drift and Diffusion Terms

Let $T > 0$ and $(\Omega, \mathcal{F}, \mathbf{F}, \mathbb{P})$ be a complete filtered probability space on which a one dimensional Brownian motion $W(\cdot)$ is defined and $\mathbf{F} = \{\mathcal{F}_t\}_{t\in[0,T]}$ is the natural filtration generated by $W(\cdot)$. Denote by $\mathbb{F}$ the progressive σ-field w.r.t. $\mathbf{F}$.

Fix $m, n \in \mathbb{N}$. Let us consider the following linear controlled stochastic differential equation in $\mathbb{R}^n$:

$$\begin{cases} dy = (Ay + Bu)dt + (Cy + Du)dW(t) \ \text{ in } [0, T], \\ y(0) = y_0, \end{cases} \tag{6.1}$$

Q. Lü and X. Zhang, *Mathematical Control Theory for Stochastic Partial Differential Equations*, Probability Theory and Stochastic Modelling 101, https://doi.org/10.1007/978-3-030-82331-3_6

where $A \in \mathbb{R}^{n\times n}$, $B \in \mathbb{R}^{n\times m}$, $C \in \mathbb{R}^{n\times n}$ and $D \in \mathbb{R}^{n\times m}$, $u(\cdot) \in L^2_{\mathbb{F}}(0,T;\mathbb{R}^m)$ is the control and $y(\cdot) \in L^2_{\mathbb{F}}(\Omega; C([0,T];\mathbb{R}^n))$ is the state.

We introduce the following notion of exact controllability for (10.2).

Definition 6.1. *The system (6.1) is called exactly controllable (at time T) if for any $y_0 \in \mathbb{R}^n$ and $y_1 \in L^2_{\mathcal{F}_T}(\Omega;\mathbb{R}^n)$, one can find a control $u(\cdot) \in L^2_{\mathbb{F}}(0,T;\mathbb{R}^m)$ such that the corresponding solution y to (6.1) satisfies that $y(T) = y_1$.*

Define a (deterministic) function $\eta(\cdot)$ on $[0,T]$ by

$$\eta(t) = \begin{cases} 1, & \text{for } t \in \Big[\Big(1-\dfrac{1}{2^{2j}}\Big)T, \Big(1-\dfrac{1}{2^{2j+1}}\Big)T\Big), \quad j = 0,1,2,\cdots, \\ -1, & \text{otherwise}. \end{cases} \tag{6.2}$$

One can check that ([275]) there exists a constant $\beta > 0$ such that

$$\int_t^T \big|\eta(s) - c\big|^2 ds \geq 4\beta(T-t), \quad \forall\, (c,t) \in \mathbb{R}\times[0,T]. \tag{6.3}$$

The following result is useful later.

Proposition 6.2. *Let $\xi = \displaystyle\int_0^T \eta(t)dW(t)$, where $\eta(\cdot)$ is given by (6.2). Then, it is impossible to find $(\varrho_1,\varrho_2) \in L^2_{\mathbb{F}}(0,T)\times C_{\mathbb{F}}([0,T];L^2(\Omega))$ and $y_0 \in \mathbb{R}$ such that*

$$\xi = y_0 + \int_0^T \varrho_1(t)dt + \int_0^T \varrho_2(t)dW(t). \tag{6.4}$$

Proof: We use the contradiction argument. Assume that (6.4) held for some $(\varrho_1,\varrho_2) \in L^2_{\mathbb{F}}(0,T)\times C_{\mathbb{F}}([0,T];L^2(\Omega))$ and $y_0 \in \mathbb{R}$. Then,

$$\int_0^T \big(\eta(t) - \varrho_2(t)\big)dW(t) = y_0 + \int_0^T \varrho_1(t)dt.$$

Hence,

$$\int_0^t \big(\eta(s) - \varrho_2(s)\big)dW(s) = y_0 + \int_0^t \varrho_1(s)ds + \mathbb{E}\Big(\int_t^T \varrho_1(s)ds \;\Big|\; \mathcal{F}_t\Big).$$

This gives

$$\int_t^T \big(\eta(s) - \varrho_2(s)\big)dW(s) = \int_t^T \varrho_1(s)ds - \mathbb{E}\Big(\int_t^T \varrho_1(s)ds \;\Big|\; \mathcal{F}_t\Big),$$

which yields that

$$\begin{aligned} &\mathbb{E}\int_t^T \big|\eta(s) - \varrho_2(s)\big|^2 ds = \mathbb{E}\Big[\int_t^T \varrho_1(s)ds - \mathbb{E}\Big(\int_t^T \varrho_1(s)ds \;\Big|\; \mathcal{F}_t\Big)\Big]^2 \\ &\leq \mathbb{E}\Big(\int_t^T \varrho_1(s)ds\Big)^2 \leq (T-t)\int_t^T \big|\varrho_1(s)\big|^2 ds. \end{aligned} \tag{6.5}$$

On the other hand, it follows from the inequality (6.3) that

$$\begin{aligned}
&\mathbb{E}\int_t^T \left|\eta(s)-\varrho_2(s)\right|^2 ds \\
&\geq \frac{1}{2}\mathbb{E}\int_t^T \left|\eta(s)-\varrho_2(T)\right|^2 ds - \mathbb{E}\int_t^T \left|\varrho_2(T)-\varrho_2(s)\right|^2 ds \\
&\geq 2\beta(T-t) - \mathbb{E}\int_t^T \left|\varrho_2(T)-\varrho_2(s)\right|^2 ds.
\end{aligned} \tag{6.6}$$

Since $\varrho_2 \in C_{\mathbb{F}}([0,T];L^2(\Omega))$, there exists $\tilde{t} \in [0,T)$ such that

$$\mathbb{E}\left|\varrho_2(T)-\varrho_2(s)\right|^2 \leq \beta, \quad \forall\, s \in [\tilde{t},T].$$

This, together with (6.6) implies that

$$\mathbb{E}\int_t^T \left|\eta(s)-\varrho_2(s)\right|^2 ds \geq \beta(T-t), \quad \forall\, t \in [\tilde{t},T]. \tag{6.7}$$

It follows from (6.5) and (6.7) that

$$\beta \leq \int_t^T \left|\eta(s)-\varrho_2(s\right|^2 ds, \quad \forall\, t \in [\tilde{t},T],$$

a contradiction. □

The following result gives some necessary conditions for the exact controllability of (6.1).

Proposition 6.3. *If the system* (6.1) *is exactly controllable, then*
1) $\operatorname{rank} D = n$;
2) (A,B) *fulfills the Kalman rank condition* (1.3).

Proof: We use the contradiction argument to prove the conclusion 1). Assume that the system (6.1) was exactly controllable for some matrix D with $\operatorname{rank} D < n$. Then, we could find a vector $v \in \mathbb{R}^n$ with $|v|_{\mathbb{R}^n} = 1$ such that $v^\top D y_0 = 0$ for any $y_0 \in \mathbb{R}^m$.

Let $y_1 = \xi v$, where ξ is given in Proposition 6.2. By the exact controllability of (6.1), one can find a control $u \in L^2_{\mathbb{F}}(0,T;\mathbb{R}^m)$ such that

$$y_1 = y_0 + \int_0^T \big(Ay(t)+Bu(t)\big)dt + \int_0^T \big(Cy(t)+Du(t)\big)dW(t),$$

which implies that

$$\int_0^T \eta(t)dW(t) = v^\top y_0 + \int_0^T v^\top\big(Ay(t)+Bu(t)\big)dt + \int_0^T v^\top Cy(t)dW(t).$$

This contradicts Proposition 6.2.

In order to prove the conclusion 2), write $\tilde{y} = \mathbb{E}y$, where y is a solution to (6.1) for some $y_0 \in \mathbb{R}^n$ and $u(\cdot) \in L^2_{\mathbb{F}}(0,T;\mathbb{R}^m)$. Then $\tilde{y}$ solves

$$\begin{cases} \dfrac{d\tilde{y}}{dt} = A\tilde{y} + B\mathbb{E}u \quad \text{in } [0,T], \\ \tilde{y}(0) = y_0. \end{cases} \tag{6.8}$$

For any $y_1 \in \mathbb{R}^n$, since (6.1) is exactly controllable at time T, we can find a control $u \in L^2_{\mathbb{F}}(0,T;\mathbb{R}^m)$ such that the corresponding solution to (6.1) satisfies that $y(T) = y_1$. Then we have that $\tilde{y}(T) = \mathbb{E}(y(T)) = \mathbb{E}y_1 = y_1$. This implies that (6.8) is exactly controllable at time T. Thus, in view of Theorem 1.2, (A, B) fulfills the Kalman rank condition. □

By means of Proposition 6.3, we should assume that $\operatorname{rank} D = n$ and (A, B) fulfills the Kalman rank condition if we expect the exact controllability of the system (6.1). In the rest of this section, we keep this assumption.

Since $\operatorname{rank} D = n$, we have $n \leq m$. Thus, there exist two matrices $K_1 \in \mathbb{R}^{m\times m}$ and $K_2 \in \mathbb{R}^{m\times n}$ such that

$$DK_1 = (I_n, 0), \qquad DK_2 = -C, \tag{6.9}$$

where I_n is the identity matrix in $\mathbb{R}^{n\times n}$. Introduce a simple linear transformation as

$$u = K_1 \begin{pmatrix} Y \\ v \end{pmatrix} + K_2 y, \tag{6.10}$$

where $Y \in L^2_{\mathbb{F}}(0,T;\mathbb{R}^n)$ and $v \in L^2_{\mathbb{F}}(0,T;\mathbb{R}^{m-n})$. Then the system (6.1) is reduced to

$$\begin{cases} dy = (A_1 y + A_2 Y + B_1 v)dt + Y dW(t) \quad \text{in } [0,T], \\ y(0) = y_0, \end{cases} \tag{6.11}$$

where

$$\begin{aligned} &A_1 = A + BK_2, \quad A_2 \in \mathbb{R}^{n\times n}, \quad B_1 \in \mathbb{R}^{n\times(m-n)}, \\ &A_2 Y + B_1 v = BK_1 \begin{pmatrix} Y \\ v \end{pmatrix}. \end{aligned} \tag{6.12}$$

Clearly, (6.11) can be viewed as a controlled system, in which (Y, v) is the control variable, and y is still the state variable. Similarly to Definition 6.1, one can define the exact controllability (at time T) of (6.11). It is easy to show the following result.

Proposition 6.4. *If* $\operatorname{rank} D = n$ *and* (6.11) *is exactly controllable, then so is the system* (6.1).

Remark 6.5. At this moment, it seems unclear whether the reverse of Proposition 6.4 is true or not. Indeed, suppose that the system (6.1) is exactly controllable, then by Proposition 6.3, $\operatorname{rank} D = n$; moreover, for any $y_0 \in \mathbb{R}^n$ and $y_1 \in L^2_{\mathcal{F}_T}(\Omega; \mathbb{R}^n)$, there is a control $u(\cdot) \in L^2_{\mathbb{F}}(0, T; \mathbb{R}^m)$ so that the corresponding solution y to (6.1) verifies that $y(T) = y_1$. Write

$$\begin{cases} Y = Du - DK_2 y, \\ B_1 v = B(u - K_2 y) - A_2 Y = (B - A_2 D)(u - K_2 y). \end{cases}$$

Then, by (6.9)–(6.12), it is easy to see that y, Y and $B_1 v$ satisfy (6.11) and $y(T) = y_1$. However, we do not know whether it is possible to find v from the algebraic equation $B_1 v = (B - A_2 D)(u - K_2 y)$. If v can be found in this way, then (6.11) is exactly controllable at time T, as well.

In order to deal with the exact controllability problem for (6.11), we consider the following controlled backward stochastic differential system:

$$\begin{cases} dy = (A_1 y + A_2 Y + B_1 v)dt + Y dW(t) \quad \text{in } [0, T], \\ y(T) = y_1, \end{cases} \tag{6.13}$$

where $y_1 \in L^2_{\mathcal{F}_T}(\Omega; \mathbb{R}^n)$, $v \in L^2_{\mathbb{F}}(0, T; \mathbb{R}^{m-n})$ is the control variable.

Definition 6.6. *The system* (6.13) *is called exactly controllable (at time* 0*) if for any $y_1 \in L^2_{\mathcal{F}_T}(\Omega; \mathbb{R}^n)$ and $y_0 \in \mathbb{R}^n$, there is a control $v \in L^2_{\mathbb{F}}(0, T; \mathbb{R}^{n \times (m-n)})$ such that the corresponding solution $(y(\cdot), Y(\cdot)) \in L^2_{\mathbb{F}}(\Omega; C([0, T]; \mathbb{R}^n)) \times L^2_{\mathbb{F}}(0, T; \mathbb{R}^n)$ to* (6.13) *satisfies $y(0) = y_0$.*

It is easy to show the following result:

Proposition 6.7. *The system* (6.11) *is exactly controllable if and only if so is the system* (6.13).

The dual equation of the system (6.13) is the following (forward) stochastic ordinary differential equation:

$$\begin{cases} dz = -A_1^\top z dt - A_2^\top z dW(t) \quad \text{in } [0, T], \\ z(0) = z_0 \in \mathbb{R}^n. \end{cases} \tag{6.14}$$

Similar to Theorem 1.2, one can show the following result:

Theorem 6.8. *The following statements are equivalent:*

1) *The system* (6.13) *is exactly controllable;*
2) *Solutions to* (6.14) *satisfy the following observability estimate:*

$$|z_0|^2 \leq C\mathbb{E} \int_0^T |B_1^\top z(t)|^2 dt, \qquad \forall\, z_0 \in \mathbb{R}^n; \tag{6.15}$$

3) *Solutions to* (6.14) *enjoy the following observability:*

$$B_1^\top z(\cdot) \equiv 0 \text{ in } (0,T), \text{ a.s. } \Rightarrow z_0 = 0; \tag{6.16}$$

4) *The following rank condition holds:*

$$\text{rank } [B_1,\ A_1B_1,\ A_2B_1,\ A_1^2B_1,\ A_1A_2B_1,\ A_2^2B,\ A_2A_1B_1, \cdots] = n. \tag{6.17}$$

Proof: "1)$\Longleftrightarrow$2)". Similarly to the deterministic setting, it suffices to consider the special case that the final datum y_1 in the system (6.13) is equal to 0. We define an operator $G : L^2_{\mathbb{F}}(0,T;\mathbb{R}^{m-n}) \to \mathbb{R}^n$ by

$$G(v(\cdot)) = y(0), \quad \forall\, v(\cdot) \in L^2_{\mathbb{F}}(0,T;\mathbb{R}^{m-n}), \tag{6.18}$$

where $(y(\cdot), Y(\cdot))$ is the corresponding solution to (6.13) with $y_1 = 0$. Then, applying Itô's formula to (6.13) (with $y_1 = 0$) and (6.14), we obtain that

$$\langle\, y(0), z_0 \,\rangle_{\mathbb{R}^n} = -\mathbb{E}\int_0^T \langle\, B_1 v(t), z(t) \,\rangle_{\mathbb{R}^n} dt, \quad \forall\, z_0 \in \mathbb{R}^n. \tag{6.19}$$

where $z(\cdot)$ solves (6.14). By (6.18)–(6.19), it is clear that,

$$(G^* z_0)(t) = -B_1^\top z(t), \quad \text{a.e. } t \in [0,T]. \tag{6.20}$$

Now, the exact controllability of (6.13) is equivalent to that $\mathcal{R}(G) = \mathbb{R}^n$; while the latter, by (6.20) and using Theorem 1.7 (with G defined by (6.18) and F being the identity operator in $\mathbb{R}^n$), is equivalent to the estimate (6.15).

The proof of "2)$\Longleftrightarrow$3)" is easy as that in Theorem 1.13 (because $\mathbb{R}^n$ is finite dimensional).

"4)$\Longrightarrow$3)". We use an idea from the proof of [275, Theorem 3.2]. Let us assume that $B_1^\top z(\cdot) \equiv 0$ in $(0,T)$, a.s. for some $z_0 \in \mathbb{R}^n$. Then,

$$B_1^\top z(t) = B_1^\top z_0 + \int_0^t B_1^\top A_1^\top z(s)ds + \int_0^t B_1^\top A_2^\top z(s)dW(s) = 0, \quad \forall\, t \in (0,T).$$

Therefore,

$$B_1^\top z_0 = 0, \quad B_1^\top A_1^\top z \equiv 0, \quad B_1^\top A_2^\top z \equiv 0. \tag{6.21}$$

Hence $B_1^\top A_1^\top z_0 = B_1^\top A_2^\top z_0 = 0$.

Noticing that $z(\cdot)$ solves (6.14), by (6.21), we have

$$B_1^\top A_1^\top z = B_1^\top A_1^\top z_0 + \int_0^t B_1^\top A_1^\top A_1^\top z(s)ds + \int_0^t B_1^\top A_1^\top A_2^\top z(s)dW(s) = 0$$

and

$$B_1^\top A_2^\top z = B_1^\top A_2^\top z_0 + \int_0^t B_1^\top A_2^\top A_1^\top z(s)ds + \int_0^t B_1^\top A_2^\top A_2^\top z(s)dW(s) = 0.$$

Hence,

$$B_1^\top A_1^\top A_1^\top z \equiv B_1^\top A_1^\top A_2^\top z \equiv B_1^\top A_2^\top A_1^\top z \equiv B_1^\top A_2^\top A_2^\top z \equiv 0,$$

which implies

$$B_1^\top A_1^\top A_2^\top z_0 = B_1^\top A_1^\top A_1^\top z_0 = B_1^\top A_2^\top A_1^\top z_0 = B_1^\top A_2^\top A_2^\top z_0 = 0.$$

Repeating the above argument, we conclude that

$$z_0^\top [B_1,\ A_1B_1,\ A_2B_1,\ A_1^2B_1,\ A_1A_2B_1,\ A_2^2B,\ A_2A_1B_1, \cdots] = 0. \tag{6.22}$$

By (6.17) and (6.22), it follows that $z_0 = 0$.

"3)$\Longrightarrow$4)". We use the contradiction argument. Assume that (6.17) was false. Then, we could find a nonzero $z_0 \in \mathbb{R}^n$ satisfying (6.22). For this z_0, denote by $z(\cdot)$ the corresponding solution to (6.14). Clearly, $z(\cdot)$ can be approximated (in $L^2_{\mathbb{F}}(\Omega; C([0,T];\mathbb{R}^n))$) by the Picard sequence $\{z_k(\cdot)\}_{k=0}^\infty$ defined as follows:

$$\begin{cases} z_0(t) = z_0, \\ z_k(t) = z_0 + \displaystyle\int_0^t A_1^\top z_{k-1}(s)ds + \int_0^t A_2^\top z_{k-1}(s)dW(s), \quad k \in \mathbb{N}, \end{cases} \tag{6.23}$$

for any $t \in [0,T]$. By (6.22) and (6.23), via a direct computation, one can show that

$$B_1^\top z_k(\cdot) = 0, \qquad k = 0, 1, 2, \cdots. \tag{6.24}$$

By (6.24), we deduce that $B_1^\top z(\cdot) \equiv 0$ in $(0,T)$. Hence, by (6.16), it follows that $z_0 = 0$, which is a contradiction. □

As a consequence of Theorem 6.8, we have the following characterization on the exact controllability of (6.11).

Corollary 6.9. ([275]) *The system* (6.11) *is exactly controllable at time* T *if and only if the rank condition* (6.17) *holds.*

Note that the rank condition (6.17) is not easy to verify because the matrix $(B_1,\ A_1B_1,\ A_2B_1,\ A_1^2B_1,\ A_1A_2B_1,\ A_2^2B,\ A_2A_1B_1, \cdots)$ has infinite columns. To simplify this condition, we introduce two sequences of matrices $\{M_{1,k}\}_{k=1}^\infty$ and $\{M_{2,k}\}_{k=1}^\infty$ inductively as follows:

$$M_{1,1} = A_1B_1, \quad M_{2,1} = A_2B_1, \tag{6.25}$$

and for $k \in \mathbb{N}$,

$$M_{1,k+1} = (A_1M_{1,k}, A_1M_{2,k}), \quad M_{2,k+1} = (A_2M_{1,k}, A_2M_{2,k}). \tag{6.26}$$

We have the following Kalman-type rank condition to characterize the exact controllability of (6.11):

Theorem 6.10. *The rank condition* (6.17) *holds if and only if*

$$\mathrm{rank}\,(B_1,\ M_{1,1},\ M_{2,1},\ M_{1,2},\ M_{2,2},\cdots,M_{1,n-1},\ M_{2,n-1}) = n. \tag{6.27}$$

Proof: The **"if" part** is obvious. It suffices to prove the **"only if" part**. For any matrix M, we denote by span $\{M\}$ the vector space spanned by the column vectors of M.

If for each $k \in \{1,\cdots,n-1\}$, there exists an $\eta_k \in \mathrm{span}\,\{M_{1,k}, M_{2,k}\}$ such that

$$\eta_k \notin \mathrm{span}\,\{B_1, M_{1,1}, M_{2,1}, M_{1,2}, M_{2,2},\cdots,M_{1,k-1}, M_{2,k-1}\}.$$

then, $\{B_1, \eta_1,\cdots,\eta_{n-1}\}$ is linearly independent, and therefore (6.27) holds.

Now, we suppose that there is a $k \in \{1,\cdots,n-1\}$ such that

$$\begin{aligned}&\mathrm{span}\,\{M_{1,k}, M_{2,k}\}\\&\subset \mathrm{span}\,\{B_1, M_{1,1}, M_{2,1}, M_{1,2}, M_{2,2},\cdots,M_{1,k-1}, M_{2,k-1}\}.\end{aligned} \tag{6.28}$$

Then, we claim that for any $\ell > k$,

$$\begin{aligned}&\mathrm{span}\,\{M_{1,\ell}, M_{2,\ell}\}\\&\subset \mathrm{span}\,\{B_1, M_{1,1}, M_{2,1}, M_{1,2}, M_{2,2},\cdots,M_{1,k-1}, M_{2,k-1}\}.\end{aligned} \tag{6.29}$$

We first consider the case that $\ell = k+1$. Let $\eta \in \mathrm{span}\,\{M_{1,k+1}\}$. Then $\eta = A_1\eta_1$, where $\eta_1 \in \mathrm{span}\,\{M_{1,k}, M_{2,k}\}$. By (6.28), we find that

$$\eta_1 \in \mathrm{span}\,\{B_1, M_{1,1}, M_{2,1}, M_{1,2}, M_{2,2},\cdots,M_{1,k-1}, M_{2,k-1}\}.$$

Thus,

$$\begin{aligned}A_1\eta_1 &\in \mathrm{span}\,\{A_1B_1, A_1M_{1,1}, A_1M_{2,1}, A_1M_{1,2}, A_1M_{2,2},\cdots,\\&\qquad\qquad A_1M_{1,k-1}, A_1M_{2,k-1}\}\\&\subset \mathrm{span}\,\{B_1, M_{1,1}, M_{2,1}, M_{1,2}, M_{2,2},\cdots,M_{1,k-1}, M_{2,k-1}, M_{1,k}, M_{2,k}\}\\&\subset \mathrm{span}\,\{B_1, M_{1,1}, M_{2,1}, M_{1,2}, M_{2,2},\cdots,M_{1,k-1}, M_{2,k-1}\}.\end{aligned}$$

This implies that

$$\mathrm{span}\,\{M_{1,k+1}\} \subset \mathrm{span}\,\{B_1, M_{1,1}, M_{2,1}, M_{1,2}, M_{2,2},\cdots,M_{1,k-1}, M_{2,k-1}\}.$$

Similarly,

$$\mathrm{span}\,\{M_{2,k+1}\} \subset \mathrm{span}\,\{B_1, M_{1,1}, M_{2,1}, M_{1,2}, M_{2,2},\cdots,M_{1,k-1}, M_{2,k-1}\}.$$

Hence, (6.29) holds for $\ell = k+1$. By the induction argument, we see that (6.29) holds for all $\ell > k$. Hence,

$$\begin{aligned}&\operatorname{span}\{B_1, A_1B_1, A_2B_1, A_1^2B_1, A_1A_2B_1, A_2^2B, A_2A_1B_1, \cdots\}\\&\subset \operatorname{span}\{B_1, M_{1,1}, M_{2,1}, M_{1,2}, M_{2,2}, \cdots, M_{1,k-1}, M_{2,k-1}\}.\end{aligned}$$

It follows from (6.17) that the dimension of $\operatorname{span}\{B_1, A_1B_1, A_2B_1, A_1^2B_1, A_1A_2B_1, A_2^2B, A_2A_1B_1, \cdots\}$ is n. Thus, the dimension of $\operatorname{span}\{B_1, M_{1,1}, M_{2,1}, M_{1,2}, M_{2,2}, \cdots, M_{1,k-1}, M_{2,k-1}\}$ is also equal to n. This completes the proof of Theorem 6.10. □

Remark 6.11. By Remark 6.5, it is easy to see that the rank condition (6.17) or (6.27) is sufficient for the exact controllability of (6.1) but maybe not necessary. It would be quite interesting to derive a rank condition to characterize the exact controllability of this system directly (rather than through that of the reduced system (6.11) or (6.13)). However, as far as we know, this is an unsolved problem.

By Proposition 6.3, in order that the system (6.1) is exactly controllable, one needs $\operatorname{rank} D = n$, which means that the control has to be active everywhere in the diffusion term of this system. Clearly, this is unsatisfactory for many practice problems. How to avoid this? One possibility is to relax the class of control functions in the drift term of (6.1), which is the main concern in the next section.

6.2 Control System With a Control in the Drift Term

In this subsection, we study a special case for which the control is only acted in the drift term.

Let us consider the following special case of the system (6.1):

$$\begin{cases} dy = (Ay + Bu)dt + CydW(t) \text{ in } [0, T], \\ y(0) = y_0. \end{cases} \tag{6.30}$$

In (6.30), $u(\cdot) \in L^1_{\mathbb{F}}(0, T; L^2(\Omega; \mathbb{R}^m))$ is the control variable and $y(\cdot)$ is the state. Clearly, $y(\cdot) \in L^2_{\mathbb{F}}(\Omega; C([0, T]; \mathbb{R}^n))$. The exact controllability for (6.30) is defined as follows:

Definition 6.12. *The system (6.30) is called exactly controllable (at time T) if for any $y_0 \in \mathbb{R}^n$ and $y_1 \in L^2_{\mathcal{F}_T}(\Omega; \mathbb{R}^n)$, there exists a control $u(\cdot) \in L^1_{\mathbb{F}}(0, T; L^2(\Omega; \mathbb{R}^m))$ so that the corresponding solution y to (6.30) verifies that $y(T) = y_1$, a.s.*

Let us first show the following preliminary result.

Theorem 6.13. *Let H be a Hilbert space and $p \in [1, \infty)$. For any $\xi \in L^p_{\mathcal{F}_T}(\Omega; H)$, there is an $f \in L^1_{\mathbb{F}}(0, T; L^p(\Omega; H))$ such that*

$$\xi = \int_0^T f(t)dt \tag{6.31}$$

and

$$|f(\cdot)|_{L^1_{\mathbb{F}}(0,T;L^p(\Omega;H))} \le |\xi|_{L^p_{\mathcal{F}_T}(\Omega;H))}. \tag{6.32}$$

Proof: For any $p \in [1,\infty)$, define an operator $\mathbb{L}_T : L^1_{\mathbb{F}}(0,T;L^p(\Omega;H)) \to L^p_{\mathcal{F}_T}(\Omega;H)$ by

$$\mathbb{L}_T\big(f(\cdot)\big) = \int_0^T f(t)dt, \qquad \forall\, f(\cdot) \in L^1_{\mathbb{F}}(0,T;L^p(\Omega;H)).$$

Clearly, to prove (6.31), we only need to prove that

$$\mathcal{R}(\mathbb{L}_T) = L^p_{\mathcal{F}_T}(\Omega;H). \tag{6.33}$$

Note that $\big(L^p_{\mathcal{F}_T}(\Omega;H)\big)' = L^{p'}_{\mathcal{F}_T}(\Omega;H)$, where p' denotes the Hölder conjugates of p, and, by Theorem 2.73,

$$\big(L^1_{\mathbb{F}}(0,T;L^p(\Omega;H))\big)' = L^\infty_{\mathbb{F}}(0,T;L^{p'}(\Omega;H)).$$

Hence, in order to prove (6.33) and (6.32), by Theorem 1.10, it suffices to derive the following inequality:

$$|\mathbb{L}_T^*\eta|_{L^\infty_{\mathbb{F}}(0,T;L^{p'}(\Omega;H))} \ge |\eta|_{L^{p'}_{\mathcal{F}_T}(\Omega;H)}, \quad \forall\, \eta \in L^{p'}_{\mathcal{F}_T}(\Omega;H). \tag{6.34}$$

In order to prove (6.34), let us first find the dual operator $\mathbb{L}_T^*$ of $\mathbb{L}_T$. For any $f(\cdot) \in L^1_{\mathbb{F}}(0,T;L^p(\Omega;H))$, and $\eta \in L^{p'}_{\mathcal{F}_T}(\Omega;H)$, we have

$$\begin{aligned}&\langle \mathbb{L}_T f, \eta\rangle_{L^p_{\mathcal{F}_T}(\Omega;H),L^{p'}_{\mathcal{F}_T}(\Omega;H)}\\ &= \mathbb{E}\Big\langle \int_0^T f(t)dt, \eta\Big\rangle_H = \int_0^T \mathbb{E}\langle f(t),\eta\rangle_H dt = \int_0^T \mathbb{E}\langle f(t), \mathbb{E}(\eta \mid \mathcal{F}_t)\rangle_H dt.\end{aligned}$$

Consequently, $\mathbb{L}_T^* : L^{p'}_{\mathcal{F}_T}(\Omega;H) \to L^\infty_{\mathbb{F}}(0,T;L^{p'}(\Omega;H))$ is given by

$$(\mathbb{L}_T^*\eta)(t) = \mathbb{E}(\eta \mid \mathcal{F}_t), \quad \text{a.e. } t \in [0,T],\ \forall\, \eta \in L^{p'}_{\mathcal{F}_T}(\Omega;H). \tag{6.35}$$

For $p > 1$, making use of (6.35), we find that

$$\begin{aligned}|\mathbb{L}_T^*\eta|_{L^\infty_{\mathbb{F}}(0,T;L^{p'}(\Omega;H))} &= \Big(\sup_{t\in[0,T]} \mathbb{E}\big|\mathbb{E}(\eta \mid \mathcal{F}_t)\big|_H^{p'}\Big)^{\frac{1}{p'}}\\ &\ge \Big(\mathbb{E}\big|\mathbb{E}(\eta \mid \mathcal{F}_T)\big|_H^{p'}\Big)^{\frac{1}{p'}} = \big(\mathbb{E}|\eta|^{p'}\big)^{\frac{1}{p'}} = |\eta|_{L^{p'}_{\mathcal{F}_T}(\Omega;H)}.\end{aligned}$$

Therefore, (6.34) holds for $p > 1$.

Next, for $p = 1$, we have that

$$\begin{aligned}|\mathbb{L}_T^*\eta|_{L_{\mathbb{F}}^\infty(0,T;L^\infty(\Omega;H))} &= \sup_{t\in[0,T]}\Big(\operatorname{ess\,sup}_{\omega\in\Omega}|\mathbb{E}(\eta \mid \mathcal{F}_t)|_H\Big)\\ &\geq \operatorname{ess\,sup}_{\omega\in\Omega}\big(|\mathbb{E}(\eta \mid \mathcal{F}_T)|_H\big) = \operatorname{ess\,sup}_{\omega\in\Omega}|\eta(\omega)|_H = |\eta|_{L^\infty_{\mathcal{F}_T}(\Omega;H)}.\end{aligned}$$

This implies that (6.34) also holds for $p = 1$. □

Remark 6.14. Theorem 6.13 is also true when H is a Banach space having the Radon-Nikodým property. We refer the readers to [239, Theorem 3.1] for the proof.

By Theorem 6.13, it is easy to show that the system (6.30) is exactly controllable whenever $C = 0$ rank $B = n$. Generally, we have the following result.

Theorem 6.15. *System* (6.30) *is exactly controllable if and only if*

$$\operatorname{rank} B = n.$$

Proof: The **"only if" part**. We use the contradiction argument. If $\operatorname{rank} B < n$, then there would exist an $\eta \in \mathbb{R}^n$ with $|\eta|_{\mathbb{R}^n} = 1$ such that $\eta^\top B = 0$. Let $\xi \in L^2_{\mathcal{F}_T}(\Omega)$ be the random variable given in Proposition 6.2 and put $y_1 = \xi\eta$. Since the system (6.30) is exactly controllable, one can find a control $u \in L^1_{\mathbb{F}}(0,T;L^2(\Omega;\mathbb{R}^m))$ such that

$$\xi\eta = y_0 + \int_0^T \big(Ay(t) + Bu(t)\big)dt + \int_0^T Cy(t)dW(t). \tag{6.36}$$

Multiplying both sides of (6.36) by $\eta^\top$, we obtain that

$$\xi = \eta^\top y_0 + \int_0^T \eta^\top Ay(t)dt + \int_0^T \eta^\top Cy(t)dW(t),$$

which contradicts Proposition 6.2.

The **"if" part**. Similarly to the proof of Theorem 1.13 (for "1)$\Longleftrightarrow$2)"), it suffices to consider the special case that the initial datum y_0 in the system (6.30) is equal to 0. We define an operator $G : L^1_{\mathbb{F}}(0,T;L^2(\Omega;\mathbb{R}^m)) \to L^2_{\mathcal{F}_T}(\Omega;\mathbb{R}^n)$ by

$$G(u(\cdot)) = y(T), \quad \forall\, u(\cdot) \in L^1_{\mathbb{F}}(0,T;L^2(\Omega;\mathbb{R}^m)), \tag{6.37}$$

where $y(\cdot)$ is the corresponding solution to (6.30) with $y_0 = 0$. We need to give an explicit form of the dual operator $G^* : L^2_{\mathcal{F}_T}(\Omega;\mathbb{R}^n) \to L^\infty_{\mathbb{F}}(0,T;L^2(\Omega;\mathbb{R}^m))$. For this purpose, we introduce the following backward stochastic differential equation:

$$\begin{cases} dz(t) = -\big(A^\top z(t) + C^\top Z(t)\big)dt + Z(t)dW(t) & \text{in } [0,T), \\ z(T) = z_1, \end{cases} \tag{6.38}$$

where $z_1 \in L^2_{\mathcal{F}_T}(\Omega;\mathbb{R}^n)$. Clearly, the equation (6.38) admits a unique mild solution $(z(\cdot),Z(\cdot)) \in L^2_{\mathbb{F}}(\Omega;C([0,T];\mathbb{R}^n)) \times L^2_{\mathbb{F}}(0,T;\mathbb{R}^n)$. Applying Itô's formula to (6.30) (with $y_0 = 0$) and (6.38), we obtain that

$$\mathbb{E}\langle y(T), z_1 \rangle_{\mathbb{R}^n} = \mathbb{E}\int_0^T \langle Bu(t), z(t) \rangle_{\mathbb{R}^n} dt, \quad \forall\, z_1 \in L^2_{\mathcal{F}_T}(\Omega;\mathbb{R}^n). \tag{6.39}$$

where $(z(\cdot),Z(\cdot))$ solves (6.38). By (6.37) and (6.39), it is clear that,

$$(G^* z_1)(t) = B^\top z(t), \quad \text{a.e. } t \in [0,T]. \tag{6.40}$$

Now, the exact controllability of (6.30) is equivalent to that

$$\mathcal{R}(G) = L^2_{\mathcal{F}_T}(\Omega;\mathbb{R}^n);$$

while the latter, by (6.40) and using Theorem 1.10 (with G defined by (6.18) and F being the identity operator in $L^2_{\mathcal{F}_T}(\Omega;\mathbb{R}^n)$), is equivalent to that all solutions $(z(\cdot),Z(\cdot)) \in L^2_{\mathbb{F}}(\Omega;C([0,T];\mathbb{R}^n)) \times L^2_{\mathbb{F}}(0,T;\mathbb{R}^n)$ to (6.38) satisfy the following estimate:

$$|z_1|_{L^2_{\mathcal{F}_T}(\Omega;\mathbb{R}^n)} \le \mathcal{C}|B^\top z|_{L^\infty_{\mathbb{F}}(0,T;L^2(\Omega;\mathbb{R}^m))}, \quad \forall\, z_1 \in L^2_{\mathcal{F}_T}(\Omega;\mathbb{R}^n). \tag{6.41}$$

Now, by $\operatorname{rank} B = n$, $z(\cdot) \in L^2_{\mathbb{F}}(\Omega;C([0,T];\mathbb{R}^n))$ and the second equation in (6.38), the desired estimate (6.41) is obvious. This completes the proof of Theorem 6.15. □

Remark 6.16. Fix any $q \in [1,2)$. By means of Theorem 6.13, we may give another "proof" of the "if" part in of Theorem 6.15. Indeed, without loss of generality, we assume that $B \in \mathbb{R}^{n\times n}$. Fix any $y_1 \in L^2_{\mathcal{F}_T}(\Omega;\mathbb{R}^n)$. Since $\Phi(T)^{-1}\big(y_1 - \Phi(T)y_0\big) \in L^{\frac{2+q}{2}}_{\mathcal{F}_T}(\Omega;\mathbb{R}^n)$, it follows from Theorem 6.13 that there is an $f \in L^1_{\mathbb{F}}(0,T;L^{\frac{2+q}{2}}(\Omega;\mathbb{R}^n))$ such that

$$\Phi(T)^{-1}\big(y_1 - \Phi(T)y_0\big) = \int_0^T f(s)ds. \tag{6.42}$$

Here $\Phi(\cdot)$ is the solution to (3.13) with $A(t) = A$, $d = 1$ and $C_1(t) = C$.

Let $u(\cdot) = B^{-1}\Phi(\cdot)f(\cdot)$. It is easy to see that $u(\cdot) \in L^1_{\mathbb{F}}(0,T;L^q(\Omega;\mathbb{R}^m))$. By Theorem 3.3, we have that

$$y(T) = \Phi(T)y_0 + \Phi(T)\int_0^T \Phi(s)^{-1}Bu(s)ds = y_1. \tag{6.43}$$

Hence, we conclude the exact controllability of the system (6.30) provided that the control class is chosen as follows:

$$
\mathcal{U} \stackrel{\Delta}{=} \big\{u(\cdot) \in L^1_{\mathbb{F}}(0,T;L^q(\Omega;\mathbb{R}^m)) \bigm| \text{The corresponding solution to } (6.30) \text{ with } y_0 = 0 \text{ satisfies that } y(T) \in L^2_{\mathcal{F}_T}(\Omega;\mathbb{R}^n)\big\}. \tag{6.44}
$$

In this case, $u(\cdot) \in \mathcal{U}$ and $y(\cdot)$ are respectively the control and the state variables for the system (6.30). Then, clearly, $y(\cdot) \in L^q_{\mathbb{F}}(\Omega;C([0,T];\mathbb{R}^n))$. Note that, in the above "proof", q cannot be chosen to be 2, and therefore, the control class $\mathcal{U}$ is a proper subset of $L^1_{\mathbb{F}}(0,T;L^2(\Omega;\mathbb{R}^m))$.

Next, we show that, for any $r \in (1,\infty)$ and $p \in (1,\infty)$, the system (6.30) with $C = 0$ and $m = n$ is not exactly controllable if the control class is chosen to be $L^r_{\mathbb{F}}(0,T;L^p(\Omega;\mathbb{R}^m))$. For simplicity, we only consider the case that $m = n = 1$.

Theorem 6.17. *For any $r,p \in (1,\infty)$, there exists an $\eta \in L^p_{\mathcal{F}_T}(\Omega)$ such that for any $f \in L^r_{\mathbb{F}}(0,T;L^p(\Omega))$,*

$$
\eta \neq \int_0^T f(t)dt. \tag{6.45}
$$

Proof: For any $r,p \in (1,\infty)$, define an operator $\mathbb{L}_T : L^r_{\mathbb{F}}(0,T;L^p(\Omega)) \to L^p_{\mathcal{F}_T}(\Omega)$ by

$$
\mathbb{L}_T\big(f(\cdot)\big) = \int_0^T f(t)dt, \qquad \forall\, f(\cdot) \in L^r_{\mathbb{F}}(0,T;L^p(\Omega)).
$$

Denote by r' and p' the Hölder conjugates of r and p, respectively. Similarly to (6.35), by Theorem 2.73, $\mathbb{L}_T^* : L^{p'}_{\mathcal{F}_T}(\Omega) \to L^{r'}_{\mathbb{F}}(0,T;L^{p'}(\Omega))$ is given by

$$
(\mathbb{L}_T^*\eta)(t) = \mathbb{E}(\eta \mid \mathcal{F}_t), \quad \text{a.e. } t \in [0,T],\ \forall\, \eta \in L^{p'}_{\mathcal{F}_T}(\Omega). \tag{6.46}
$$

Let us use the contradiction argument to prove Theorem 6.17. If there was an $r \in (1,\infty)$ so that $\mathcal{R}(\mathbb{L}_T) = L^p_{\mathcal{F}_T}(\Omega)$, then, in view of Theorem 1.7, we could find a constant $\mathcal{C} > 0$ such that

$$
|\eta|_{L^{p'}_{\mathcal{F}_T}(\Omega)} \leq \mathcal{C}|\mathbb{L}_T^*\eta|_{L^{r'}_{\mathbb{F}}(0,T;L^{p'}(\Omega))}, \quad \forall\, \eta \in L^{p'}_{\mathcal{F}_T}(\Omega). \tag{6.47}
$$

Consider a sequence of random variables $\{\eta_n\}_{n=1}^{\infty}$ defined by

$$
\eta_n = \int_0^T e^{nt}dW(t), \qquad n \in \mathbb{N}.
$$

It is obvious that $\eta_n \in L^{p'}_{\mathcal{F}_T}(\Omega)$ for any $n \in \mathbb{N}$. By Theorem 2.123, the integral $\int_0^T e^{nt}dW(t)$ is a Gaussian random variable with mean 0 and variance $\frac{e^{2nT}-1}{2n}$. Hence,

$$
\begin{aligned}
\Big(\mathbb{E}\Big|\int_0^T e^{nt}dW(t)\Big|^{p'}\Big)^{\frac{1}{p'}} &= \Big[\int_{-\infty}^{\infty}\frac{\sqrt{n}\,|x|^{p'}}{\sqrt{(e^{2nT}-1)\pi}}e^{-\frac{nx^2}{e^{2nT}-1}}dx\Big]^{\frac{1}{p'}}\\
&= \Big[\int_{-\infty}^{\infty}\Big(\frac{e^{2nT}-1}{n}\Big)^{p'/2}\frac{|x|^{p'}}{\sqrt{\pi}}e^{-x^2}dx\Big]^{\frac{1}{p'}} \qquad (6.48)\\
&= \Big(\frac{1}{\sqrt{\pi}}\int_{-\infty}^{\infty}|x|^{p'}\,e^{-x^2}dx\Big)^{\frac{1}{p'}}\sqrt{\frac{e^{2nT}-1}{n}}.
\end{aligned}
$$

Now, by (6.48), it is easy to see that

$$
\begin{aligned}
|\eta_n|_{L^{p'}_{\mathcal{F}_T}(\Omega)} &= \Big(\mathbb{E}\Big|\int_0^T e^{nt}dW(t)\Big|^{p'}\Big)^{\frac{1}{p'}}\\
&= \Big(\frac{1}{\sqrt{\pi}}\int_{-\infty}^{\infty}|x|^{p'}\,e^{-x^2}dx\Big)^{\frac{1}{p'}}\sqrt{\frac{e^{2nT}-1}{n}}.
\end{aligned} \qquad (6.49)
$$

Using (6.48) again, we have

$$
\begin{aligned}
&\big|\mathbb{E}(\eta_n \mid \mathcal{F}_t)\big|_{L^{r'}_{\mathbb{F}}(0,T;L^{p'}(\Omega))} = \Big[\int_0^T\Big(\mathbb{E}\Big|\int_0^t e^{n\tau}dW(\tau)\Big|^{p'}\Big)^{\frac{r'}{p'}}dt\Big]^{\frac{1}{r'}}\\
&= \Big\{\int_0^T\Big[\Big(\frac{1}{\sqrt{\pi}}\int_{-\infty}^{\infty}|x|^{p'}\,e^{-x^2}dx\Big)^{\frac{1}{p'}}\sqrt{\frac{e^{2nt}-1}{n}}\Big]^{r'}dt\Big\}^{\frac{1}{r'}} \qquad (6.50)\\
&\le \frac{1}{\sqrt{n}}\Big(\frac{1}{\sqrt{\pi}}\int_{-\infty}^{\infty}|x|^{p'}\,e^{-x^2}dx\Big)^{\frac{1}{p'}}\Big(\int_0^T e^{nr't}dt\Big)^{\frac{1}{r'}}\\
&\le \frac{1}{\sqrt{n}}\Big(\frac{1}{\sqrt{\pi}}\int_{-\infty}^{\infty}|x|^{p'}\,e^{-x^2}dx\Big)^{\frac{1}{p'}}\frac{e^{nT}}{(nr')^{\frac{1}{r'}}}.
\end{aligned}
$$

From (6.49) and (6.50), it follows that

$$
\lim_{n\to\infty}\frac{\big|\mathbb{E}(\eta_n \mid \mathcal{F}_t)\big|_{L^{r'}_{\mathbb{F}}(0,T;L^{p'}(\Omega))}}{|\eta_n|_{L^{p'}_{\mathcal{F}_T}(\Omega)}} \le \lim_{n\to\infty}\frac{e^{nT}}{(nr')^{\frac{1}{r'}}\sqrt{e^{2nT}-1}} = 0.
$$

This, combined with (6.46), gives

$$
\lim_{n\to\infty}\frac{|\mathbb{L}_T^*\eta_n|_{L^{r'}_{\mathbb{F}}(0,T;L^{p'}(\Omega))}}{|\eta_n|_{L^{p'}_{\mathcal{F}_T}(\Omega)}} = 0,
$$

which contradicts the inequality (6.47). This completes the proof of Theorem 6.17. □

Remark 6.18. At this moment, it is unclear whether the same result in Theorem 6.17 holds for $p = 1$ or not.

For any $r \in [1, \infty)$, let us "embed" the controlled ordinary equation (1.1) with $m = n$ into our stochastic setting, i.e., we consider the following controlled stochastic differential equation:

$$\begin{cases} dy = (Ay + Bu)dt & \text{in } [0, T], \\ y(0) = y_0, \end{cases} \tag{6.51}$$

where $A \in \mathbb{R}^{n\times n}$, $B \in \mathbb{R}^{n\times n}$, $u(\cdot) \in L^r_{\mathbb{F}}(0, T; L^2(\Omega; \mathbb{R}^n))$ is the control variable and $y(\cdot) \in L^2_{\mathbb{F}}(\Omega; C([0, T]; \mathbb{R}^n))$ is the state variable. Similarly to Definition 6.12, the system (6.51) is called exactly controllable (at time T) if for any $y_0 \in \mathbb{R}^n$ and $y_1 \in L^2_{\mathcal{F}_T}(\Omega; \mathbb{R}^n)$, one can find a control $u(\cdot) \in L^r_{\mathbb{F}}(0, T; L^2(\Omega; \mathbb{R}^n))$ such that the corresponding solution y to (6.51) satisfies $y(T) = y_1$. Then, as an easy consequence of Theorems 6.15 and 6.17, the following result holds:

Corollary 6.19. *The system* (6.51) *is exactly controllable if and only if* $\operatorname{rank} B = n$ *and* $r = 1$.

To the best of our knowledge, unlike the deterministic case, so far there exists no universally accepted notion for stochastic controllability, even for stochastic differential equations in finite dimensions. Motivated by Theorem 6.15 and Corollary 6.19, we introduced a corrected formulation for the exact controllability of the system (6.1) as follows:

Definition 6.20. *The system* (6.1) *is called exactly controllable if for any* $y_0 \in \mathbb{R}^n$ *and* $y_1 \in L^2_{\mathcal{F}_T}(\Omega; \mathbb{R}^n)$, *one can find a control* $u(\cdot) \in L^1_{\mathbb{F}}(0, T; \mathbb{R}^m)$ *so that* $Du \in L^{2,loc}_{\mathbb{F}}(0, T; \mathbb{R}^n)$ *and the corresponding solution* $y(\cdot)$ *to* (6.1) *satisfies* $y(T) = y_1$, *a.s.*

Remark 6.21. Clearly, the control class in Definition 6.20 is the biggest one so that the solutions to (6.1) make sense.

The above definition seems to be a reasonable notion for exact controllability of stochastic differential equations. Nevertheless, a complete study on this problem is still under consideration and it does not seem to be easy, even for the very simple case when $n = 2$.

On the other hand, from Corollary 6.19, it is easy to see that exact controllability seems a too strong controllability requirement for stochastic evolution equations. Naturally, people hope to see what happens if such a controllability requirement is weakened, say to be null/approximate controllability, which is the main concern in the next section.

6.3 Lack of Robustness for Null/Approximate Controllability

In this section we shall show the lack of robustness for null/approximate controllability for stochastic differential equations.

We begin with the following two notions of controllability for the system (6.1).

Definition 6.22. *The system (6.1) is called null controllable (at time T) if for any $y_0 \in \mathbb{R}^n$, there exists a control $u(\cdot) \in L^1_{\mathbb{F}}(0,T;L^2(\Omega;\mathbb{R}^m))$ so that $Du \in L^{2,loc}_{\mathbb{F}}(0,T;\mathbb{R}^n)$ and the corresponding solution $y(\cdot)$ to (6.1) satisfies $y(T)=0$, a.s.*

Definition 6.23. *The system (6.1) is called approximately controllable (at time T) if for any $y_0 \in \mathbb{R}^n$, $y_1 \in L^2_{\mathcal{F}_T}(\Omega;\mathbb{R}^n)$ and $\varepsilon > 0$, there exists a control $u(\cdot) \in L^1_{\mathbb{F}}(0,T;L^2(\Omega;\mathbb{R}^m))$ so that $Du \in L^{2,loc}_{\mathbb{F}}(0,T;\mathbb{R}^n)$ and the corresponding solution $y(\cdot)$ to (6.1) satisfies that $y(T) \in L^2_{\mathcal{F}_T}(\Omega;\mathbb{R}^n)$ and that $|y(T)-y_1|_{L^2_{\mathcal{F}_T}(\Omega;\mathbb{R}^n)} < \varepsilon$.*

To show the lack of robustness for null/approximate controllability for (6.1), we consider the following very simple stochastic control system (in two dimensions):

$$\begin{cases} dy_1 = y_2 dt + \varepsilon y_2 dW(t) & \text{in } [0,T], \\ dy_2 = u dt & \text{in } [0,T], \\ y_1(0) = y_{10}, \ y_2(0) = y_{20}, \end{cases} \tag{6.52}$$

where $(y_{10}, y_{20}) \in \mathbb{R}^2$, $u(\cdot) \in L^1_{\mathbb{F}}(0,T;L^2(\Omega))$ is the control variable, ε is a parameter.

Clearly, if $\varepsilon = 0$, then (6.52) is null controllable for any $T > 0$. However, this system is NOT null controllable anymore whenever $\varepsilon \neq 0$. To see this, we use the contradiction argument. Assume that there was an $\varepsilon_0 > 0$ such that for all $\varepsilon \in [-\varepsilon_0, \varepsilon_0]$ and $T > 0$, (6.52) would be null controllable. Let us take $y_{10} = 0$, $y_{20} = 1$, $\varepsilon = \varepsilon_0$ and $T = \frac{\varepsilon_0^2}{2}$. By the null controllability of (6.52) at $T = \frac{\varepsilon_0^2}{2}$, we have

$$y_1\Big(\frac{\varepsilon_0^2}{2}\Big) = \int_0^{\frac{\varepsilon_0^2}{2}} y_2 dt + \varepsilon_0 \int_0^{\frac{\varepsilon_0^2}{2}} y_2 dW(t) = 0.$$

Thus,

$$\mathbb{E}\Big|\int_0^{\frac{\varepsilon_0^2}{2}} y_2 dt\Big|^2 = \mathbb{E}\Big|\varepsilon_0 \int_0^{\frac{\varepsilon_0^2}{2}} y_2 dW(t)\Big|^2 = \varepsilon_0^2 \int_0^{\frac{\varepsilon_0^2}{2}} \mathbb{E}|y_2|^2 dt. \tag{6.53}$$

On the other hand,

$$\mathbb{E}\Big|\int_0^{\frac{\varepsilon_0^2}{2}} y_2 dt\Big|^2 \leq \mathbb{E}\Big|\Big(\int_0^{\frac{\varepsilon_0^2}{2}} 1 dt\Big)\Big(\int_0^{\frac{\varepsilon_0^2}{2}} |y_2|^2 dt\Big)\Big| \leq \frac{\varepsilon_0^2}{2} \int_0^{\frac{\varepsilon_0^2}{2}} \mathbb{E}|y_2|^2 dt. \tag{6.54}$$

It follows from (6.53) and (6.54) that $\int_0^{\frac{\varepsilon_0^2}{2}} \mathbb{E}|y_2|^2 dt = 0$, which contradicts the choice of $y_2(0)$.

Next, we consider the approximate controllability of (6.52). For this purpose, we introduce the following backward stochastic differential equation:

$$\begin{cases} dz_1 = Z_1 dW(t) & \text{in } [0,T], \\ dz_2 = -(z_1 + \varepsilon Z_1)dt + Z_2 dW(t) & \text{in } [0,T], \\ z_1(T) = z_{1T},\ z_2(T) = z_{2T}, \end{cases} \tag{6.55}$$

where $(z_{1T}, z_{2T}) \in L^2_{\mathcal{F}_T}(\Omega;\mathbb{R}^2)$. We define an operator $\mathcal{L} : L^1_{\mathbb{F}}(0,T;L^2(\Omega)) \to L^2_{\mathcal{F}_T}(\Omega;\mathbb{R}^2)$ by

$$\mathcal{L}(u(\cdot)) = (y_1(T), y_2(T)), \quad \forall\, u(\cdot) \in L^1_{\mathbb{F}}(0,T;L^2(\Omega)), \tag{6.56}$$

where $(y_1(\cdot), y_2(\cdot))$ is the corresponding solution to (6.52) with $y_{10} = y_{20} = 0$. Then, applying Itô's formula to (6.52) (with $y_{10} = y_{20} = 0$) and (6.55), we obtain that

$$\begin{aligned} \langle (y_1(T), y_2(T)), (z_{1T}, z_{2T}) \rangle_{L^2_{\mathcal{F}_T}(\Omega;\mathbb{R}^2)} &= \mathbb{E}\int_0^T u(t) z_2(t) dt, \\ &\forall\, (z_{1T}, z_{2T}) \in L^2_{\mathcal{F}_T}(\Omega;\mathbb{R}^2). \end{aligned} \tag{6.57}$$

By (6.56)–(6.57), it is clear that, $\mathcal{L}^* : L^2_{\mathcal{F}_T}(\Omega;\mathbb{R}^2) \to L^\infty_{\mathbb{F}}(0,T;L^2(\Omega))$ is given as follows:

$$(\mathcal{L}^*(z_{1T}, z_{2T}))(t) = z_2(t), \quad \text{a.e. } t \in [0,T]. \tag{6.58}$$

Obviously, the system (6.52) is approximately controllable at time T if and only if $\overline{\mathcal{R}(\mathcal{L})} = L^2_{\mathcal{F}_T}(\Omega;\mathbb{R}^2)$. Hence, by Theorem 1.12, the approximate controllability of (6.52) is equivalent to the following observability property for (solutions to) (6.55): *If $z_2(\cdot) = 0$ in $(0,T)$, then $z_{1T} = z_{2T} = 0$, a.s.*

If $\varepsilon = 0$ and $z_2(\cdot) = 0$, then by the second equation in (6.55), we obtain that

$$z_2(t) = -\int_0^t z_1(s)ds + \int_0^t Z_2(s)dW(s) = 0, \quad \forall\, t \in [0,T]. \tag{6.59}$$

By Itô's formula, we see that

$$\mathbb{E}|z_2(T)|^2 - \mathbb{E}|z_2(0)|^2 = \int_0^T |Z_2(s)|^2 ds = 0, \tag{6.60}$$

which implies that $Z_2(t) = 0$ for a.e. $(t,\omega) \in [0,T] \times \Omega$. Thus, it follows from (6.59) that $\int_0^t z_1(s)ds = 0$ for any $t \in [0,T]$. This concludes that $z_1(t) = 0$ for a.e. $(t,\omega) \in [0,T]\times\Omega$. Then, by $(z_1(\cdot), z_2(\cdot)) \in L^2_{\mathbb{F}}(\Omega; C([0,T];\mathbb{R}^2))$, we deduce that $z_{1T} = z_{2T} = 0$, a.s. Therefore, we conclude that (6.52) is approximately controllable if $\varepsilon = 0$.

However, if $\varepsilon \neq 0$, then it is easy to check that

$$
\begin{aligned}
&(z_1(t), z_2(t), Z_1(t), Z_2(t))\\
&= \Big(\exp\Big\{ -\frac{W(t)}{\varepsilon} - \frac{t}{2\varepsilon^2}\Big\}, 0, -\frac{1}{\varepsilon}\exp\Big\{ -\frac{W(t)}{\varepsilon} - \frac{t}{2\varepsilon^2}\Big\}, 0\Big)
\end{aligned}
$$

is a nonzero solution to (6.55) with the final datum

$$
(z_{1T}, z_{2T}) = \Big(\exp\Big\{ -\frac{W(T)}{\varepsilon} - \frac{T}{2\varepsilon^2}\Big\}, -\frac{1}{\varepsilon}\exp\Big\{ -\frac{W(T)}{\varepsilon} - \frac{T}{2\varepsilon^2}\Big\}\Big).
$$

Hence, the above observability property for (6.55) does not hold. Therefore, (6.52) is not approximately controllable whenever $\varepsilon \neq 0$.

Remark 6.24. By the above two examples, we deduce that both null and approximate controllability for stochastic differential equations in finite dimensions are unstable under small perturbations. This differs significantly from the controllability properties for deterministic problems. Note that, a similar phenomenon was firstly observed in [219] for control systems governed by stochastic parabolic equations (See also Section 9.3).

6.4 Notes and Comments

Except Theorem 6.10 (which was communicated privately to us by Qida Xie), the results in Section 6.1 were first proven in [275] (See also [246, Section 3] for a slightly different presentation). The results in Section 6.2 are modification of that in [239]. Section 6.3 is taken from [246]. We refer to [13, 40, 51, 125, 127, 141, 347, 348, 376, 382] for some other controllability results on stochastic differential equations in finite dimensions.

Compared to stochastic optimal control theory, the controllability theory for stochastic differential equations is surprisingly far from satisfactory, even for the case in finite dimensions. Actually, in our opinion (as remarked in [246]), compared to the deterministic case, the controllability/observability for stochastic differential equations (even in finite dimensions) is still at its enfant stage. There are numerous works which remain to be done, for example:

- Generally speaking, when $n > 1$, the exact/null/approximate controllability for the linear system (6.1) (in the sense of Definitions 6.20, 6.22 and 6.23) are far from well-understood;
- In the deterministic setting, at least for the case of finite dimensions, controllability is closely related to another important notion in Control Theory, i.e., stabilization. It seems that so far there exist no studies on stochastic stabilization via controllability;
- It is well-known that, the classical controllability theory for the deterministic setting is the basis for linear filtering and prediction problems

([162, 163]. Now, since the stochastic controllability theory is completely unsatisfactory (as mentioned above), it seems quite interesting to revisit the corresponding filtering and prediction problems;

- Similarly to the deterministic setting ([197, 350]), it would be quite interesting to give topological classification of stochastic linear control systems, at least for the relatively simple system (6.11) but, as far as we know, no results have been published for the stochastic problems;
- To the best of our knowledge, there exist no nontrivial results about the controllability of stochastic nonlinear control systems. It would be quite interesting to give some Lie bracket conditions (as people did for the deterministic setting, e.g., [61, 311]) for stochastic controllability problems, again at least for the nonlinear version of the relatively simple system (6.11).

7

Controllability for Stochastic Linear Evolution Equations

This chapter is devoted to presenting some results on controllability for forward and backward stochastic linear evolution equations with possibly unbounded control operators. By the duality argument, these controllability problems can be reduced to suitable observability for the dual equations. Explicit forms of controls for the controllability problems are also provided. Finally, the controllability of some forward stochastic evolution equations are shown to be equivalent to that of suitably chosen backward stochastic evolution equations.

7.1 Formulation of the Problems

The (F, Γ)-controllability concept introduced in Definition 5.6 (in Section 5.3 of Chapter 5) is quite general, which is very hard to be studied (even in the deterministic setting and in finite dimensions). In this chapter, we shall focus mainly on exact/null/approximate controllability problems for stochastic linear evolution equations in the abstract setting.

Throughout this chapter, H, V and U are three Hilbert spaces which are identified with their dual spaces, $\mathcal{L}_2^0 \stackrel{\Delta}{=} \mathcal{L}_2(V; H)$, and A is the generator of a C_0-semigroup $\{S(t)\}_{t\geq 0}$ on H. In this chapter, H and U will serve as respectively the state and the control spaces for the systems under consideration unless other stated.

For any fixed $T > 0$ and a control operator $B \in \mathcal{L}(U; H)$, we begin with the following deterministic controlled evolution equation:

$$\begin{cases} y_t(t) = Ay(t) + Bu(t) & \text{in } (0, T], \\ y(0) = y_0, \end{cases} \tag{7.1}$$

where $y_0 \in H$, y is the state variable, and $u(\in L^2(0, T; U))$ is the control variable. The exact/null/approximate controllability of (7.1) can be defined

Q. Lü and X. Zhang, *Mathematical Control Theory for Stochastic Partial Differential Equations*, Probability Theory and Stochastic Modelling 101, https://doi.org/10.1007/978-3-030-82331-3_7

as that in Definition 5.6. For example, the system (7.1) is called exactly (*resp.* null) controllable at time T if for any $y_0, y_T \in H$ (*resp.* $y_0 \in H$), one can find a control $u \in L^2(0,T;U)$ so that the corresponding solution to (7.1) verifies $y(T) = y_T$ (*resp.* $y(T) = 0$).

Similarly to the equation (1.27) for the exact controllability of (1.1), people introduce the following dual equation (of (7.1)), which is a (backward) equation evolved in H:

$$\begin{cases} z_t = -A^* z \quad \text{in } [0,T), \\ z(T) = z_T. \end{cases} \tag{7.2}$$

One can show the following result.

Theorem 7.1. *The following statements hold:*

1) *The system* (7.1) *is exactly controllable at time* T *if and only if solutions of* (7.2) *satisfy*

$$|z_T|_H \le \mathcal{C} \int_0^T |B^* z(t)|^2 dt, \quad \forall\, z_T \in H; \tag{7.3}$$

2) *The system* (7.1) *is null controllable at time* T *if and only if solutions of* (7.2) *satisfy*

$$|z(0)|_H \le \mathcal{C} \int_0^T |B^* z(t)|^2 dt, \quad \forall\, z_T \in H; \tag{7.4}$$

3) *The system* (7.1) *is approximately controllable at time* T *if and only if solutions to* (7.2) *enjoy the following property:*

$$B^* z(\cdot) \equiv 0 \ \text{in } (0,T) \Rightarrow z_T = 0. \tag{7.5}$$

The proof of Theorem 7.1 can be found, say, in that of [377, Theorems 2.4–2.6 of Chapter 2].

Remark 7.2. 1) It is easy to see that, when A generates a C_0-group $\{S(t)\}_{t\in\mathbb{R}}$ on H (hence both (7.1) and (7.2) are time reversible), the estimates (7.3) and (7.4) are equivalent, and therefore in this case the exact controllability and null controllability for the deterministic system (7.1) are equivalent. At the end of this section, we shall see that this is NOT true anymore in the stochastic setting.

2) From (7.4) and (7.5), it is clear that, if the dual equation (7.2) enjoys the following property:

$$z(0) = 0 \Rightarrow z_T = 0,$$

then, the null controllability of (7.1) implies the approximate controllability of the same system. This is the case, for example, for deterministic parabolic equations. However, as we shall see in Chapter 9, this is NOT true for stochastic parabolic equations (See Theorems 9.7 and 9.9 for more details).

A main goal in this chapter is to extend Theorem 7.1 to the stochastic setting. For this purpose, suppose that $(\Omega, \mathcal{F}, \mathbf{F}, \mathbb{P})$ (with $\mathbf{F} = \{\mathcal{F}_t\}_{t\in[0,T]}$) is a fixed filtered probability space satisfying the usual condition. Let $\{W(t)\}_{t\in[0,T]}$ be a V-valued, $\mathbf{F}$-adapted, standard Q-Brownian motion (with Q being a given positive definite, trace-class operator on V) or a cylindrical Brownian motion on $(\Omega, \mathcal{F}, \mathbf{F}, \mathbb{P})$. Denote by $\mathbb{F}$ the progressive σ-field (in $[0,T]\times\Omega$) w.r.t. $\mathbf{F}$. In the sequel, to simplify the presentation, we only consider the case of cylindrical Brownian motion and $\mathbb{F}$ is the corresponding natural filtration (generated by $W(\cdot)$). Also, in the rest of this chapter, to simplify the notation, we write $\mathcal{U}_T \overset{\Delta}{=} L^2_{\mathbb{F}}(0,T;U)$.

We consider the following forward stochastic control system evolved in H:

$$
\begin{cases}
dy(t) = \big(Ay(t) + F(t)y(t) + f(t) + Bu(t)\big)dt \\
\qquad\qquad + \big(G(t)y(t) + Du(t)\big)dW(t) \qquad \text{in } (0,T], \\
y(0) = y_0,
\end{cases}
\tag{7.6}
$$

where $F \in L^\infty_{\mathbb{F}}(0,T;\mathcal{L}(H))$, $f \in L^2_{\mathbb{F}}(\Omega;L^1(0,T;H))$, $G \in L^\infty_{\mathbb{F}}(0,T;\mathcal{L}(H;\mathcal{L}_2^0))$ and $D \in \mathcal{L}(U;\mathcal{L}_2^0)$. In (7.6), $y_0 (\in H)$ is the initial state, y is the state variable, and $u(\in \mathcal{U}_T)$ is the control variable. By Theorem 3.14, the system (7.6) admits a unique mild solution $y(\cdot) \in C_{\mathbb{F}}([0,T];L^2(\Omega;H))$.

Some notions of controllability of (7.6) are given as follows:

Definition 7.3. 1) *System (7.6) is called exactly controllable at time T if for any $y_0 \in H$ and $y_T \in L^2_{\mathcal{F}_T}(\Omega;H)$, there exists $u \in \mathcal{U}_T$ such that the corresponding mild solution to (7.6) satisfies $y(T) = y_T$ a.s.*

2) *System (7.6) is called null controllable at time T if for any $y_0 \in H$, there exists $u \in \mathcal{U}_T$ such that the corresponding mild solution to (7.6) satisfies $y(T) = 0$ a.s.*

3) *System (7.6) is called approximately controllable at time T if for any $y_0 \in H$, $y_T \in L^2_{\mathcal{F}_T}(\Omega;H)$ and $\varepsilon > 0$, there exists $u \in \mathcal{U}_T$ such that the corresponding mild solution to (7.6) satisfies $|y(T) - y_T|_{L^2_{\mathcal{F}_T}(\Omega;H)} \le \varepsilon$.*

In view of the conclusion 1) in Proposition 6.3 and also motivated by the controlled backward stochastic differential equation (6.13), we introduce the following controlled backward stochastic evolution equation:

$$
\begin{cases}
dy(t) = -A^*y(t)dt + \big(J(t)y(t) + K(t)Y(t) + g(t) + Bu(t)\big)dt \\
\qquad\qquad + Y(t)dW(t) \qquad\qquad \text{in } [0,T), \\
y(T) = y_T,
\end{cases}
\tag{7.7}
$$

where $J \in L^\infty_{\mathbb{F}}(0,T;\mathcal{L}(H))$, $K \in L^\infty_{\mathbb{F}}(0,T;\mathcal{L}(\mathcal{L}_2^0;H))$ and $g \in L^2_{\mathbb{F}}(\Omega;L^1(0,T;H))$. In (7.7), $y_T (\in L^2_{\mathcal{F}_T}(\Omega;H))$ is the final state, y and $u(\in \mathcal{U}_T)$ are viewed as the state and control variables, respectively. By Theorem 4.10, the system (7.7) admits a unique mild solution $(y(\cdot), Y(\cdot)) \in L^2_{\mathbb{F}}(\Omega;C([0,T];H)) \times L^2_{\mathbb{F}}(0,T;\mathcal{L}_2^0)$.

Similarly to Definition 7.3, some notions of controllability of (7.7) are introduced as follows:

Definition 7.4. 1) *System* (7.7) *is called exactly controllable at time* 0 *if for any* $y_T \in L^2_{\mathcal{F}_T}(\Omega; H)$ *and* $y_0 \in H$, *there exists* $u \in \mathcal{U}_T$ *such that the corresponding mild solution to* (7.7) *satisfies* $y(0) = y_0$.

2) *System* (7.7) *is called null controllable at time* 0 *if for any* $y_T \in L^2_{\mathcal{F}_T}(\Omega; H)$, *there exists* $u \in \mathcal{U}_T$ *such that the corresponding mild solution to* (7.7) *satisfies* $y(0) = 0$.

3) *System* (7.7) *is called approximately controllable at time* 0 *if for any* $y_T \in L^2_{\mathcal{F}_T}(\Omega; H)$, $y_0 \in H$ *and* $\varepsilon > 0$, *there exists* $u \in \mathcal{U}_T$ *such that the corresponding mild solution to* (7.7) *satisfies* $|y(0) - y_0|_H \leq \varepsilon$.

As that in the deterministic setting, for control problems of stochastic partial differential equations, there are many cases that the control operators are unbounded, say the Dirichlet boundary controls for stochastic wave/heat/Schrödinger equations. For such kind of problems, one cannot obtain the existence of the H-valued mild solutions to (7.6) or (7.7). Nevertheless, as we shall see in the next section that, when the control operators have some further property, one can show the well-posedness of the corresponding control systems in the sense of transposition solutions.

In the rest of this chapter, O is another Hilbert space which is dense in H and denote by O' its dual space w.r.t. the pivot space H. Hence, $O \subset H \equiv H' \subset O'$. Let A generate also a C_0-semigroup $\{S(t)\}_{t\geq 0}$ on O and O', respectively. Denote by H_{-1} the completion of H w.r.t. the norm

$$|x|_{H_{-1}} \stackrel{\Delta}{=} |(\beta I - A)^{-1} x|_H,$$

where β is a fixed real number which belongs to the resolvent set of A. We fix two unbounded operators $\mathcal{B} \in \mathcal{L}(U; H_{-1})$ and $\mathcal{D} \in \mathcal{L}(U; \mathcal{L}_2(V; H_{-1}))$.

Similarly to (7.6), we consider the following forward stochastic control system evolved in the state space O':

$$\begin{cases} dy(t) = \big(Ay(t) + \mathcal{F}(t)y(t) + \mathbf{f}(t) + \mathcal{B}u(t)\big)dt \\ \qquad\qquad +\big(\mathcal{G}(t)y(t) + \mathcal{D}u(t)\big)dW(t) \qquad \text{in } (0,T], \\ y(0) = y_0, \end{cases} \tag{7.8}$$

where $\mathcal{F} \in L^\infty_{\mathbb{F}}(0,T;\mathcal{L}(O'))$, $\mathbf{f} \in L^2_{\mathbb{F}}(\Omega; L^1(0,T;O'))$ and $\mathcal{G} \in L^\infty_{\mathbb{F}}(0,T;\mathcal{L}(O';\mathcal{L}_2(V;O')))$. In (7.8), $y_0 \in O'$ is the initial state, y is the state variable, while $u (\in \mathcal{U}_T)$ is the control variable.

Also, similarly to (7.7), we consider the following backward stochastic control system evolved in the state space O':

$$\begin{cases} dy(t) = -A^* y(t)dt + \big(\mathcal{J}(t)y(t) + \mathcal{K}(t)Y(t) + \mathbf{g}(t) + \mathcal{B}u(t)\big)dt \\ \qquad\qquad +Y(t)dW(t) \qquad \text{in } [0,T), \\ y(T) = y_T, \end{cases} \tag{7.9}$$

where $\mathcal{J} \in L^\infty_{\mathbb{F}}(0,T;\mathcal{L}(O'))$, $\mathcal{K} \in L^\infty_{\mathbb{F}}(0,T;\mathcal{L}(\mathcal{L}_2(V;O');O'))$, $\mathbf{g} \in L^1_{\mathbb{F}}(0,T;L^2(\Omega;O'))$. In (7.9), $y_T(\in L^2_{\mathcal{F}_T}(\Omega;O'))$ is the final state, y and $u(\in \mathcal{U}_T)$ are viewed as the state and control variables, respectively.

Clearly, unlike that in (7.6) and (7.7), the control operator $\mathcal{B}$ in both (7.8) and (7.9) and the control operator $\mathcal{D}$ in (7.8) are allowed unbounded. As we shall see in the next section, under some assumptions, the systems (7.8) and (7.9) are well-posed in the sense of transposition solutions, in the solution spaces $C_{\mathbb{F}}([0,T];L^2(\Omega;O'))$ and $C_{\mathbb{F}}([0,T];L^2(\Omega;O')) \times L^2_{\mathbb{F}}(0,T;\mathcal{L}_2(V;O'))$, respectively.

The notions of controllability for the systems (7.8) and (7.9) can be defined very similarly to that for (7.6) and (7.7) (The only two differences are to replace the Hilbert space H and mild solutions used Definitions 7.3 and 7.4 by the Hilbert space O' and transposition solutions, respectively), and therefore we omit the details.

Remark 7.5. 1) In the system (7.6) (and also (7.8)), the control u enters in both the drift and diffusion terms. This is natural in the stochastic setting. Indeed, as we shall see in the Chapters 8–11 (see also that in the last chapter), for many stochastic controllability problems, people have to introduce controls in this way. On the other hand, even if in some case it is enough to have only one control in the drift (or diffusion) term, this control may affect the diffusion (or drift) term one way or another.

2) In view of Definition 6.20, it is more natural to replace the control class $\mathcal{U}_T$ employed in the definition of controllability for the system (7.6) (*resp.* (7.8)) by the set $\{u \in L^1_{\mathbb{F}}(0,T;U) \mid Du \in L^{2,loc}_{\mathbb{F}}(0,T;\mathcal{L}^0_2)\}$ (*resp.* $\{u \in L^1_{\mathbb{F}}(0,T;U) \mid \mathcal{D}u \in L^{2,loc}_{\mathbb{F}}(0,T;\mathcal{L}_2(V;O'))\}$). Nevertheless, so far it seems difficult to obtain useful results using such a control class.

3) The control operator $\mathcal{D}$ in (7.8) is also allowed to be unbounded. Compared to the deterministic problems, this is completely new for the stochastic setting though the analysis for concrete problems remains to be done and there might exist some essential difficulties.

In the sequel, the controllability for the forward (*resp.* backward) stochastic evolution equations (7.6) and (7.8) (*resp.* (7.7) and (7.9)) are simply referred to the forward (*resp.* backward) controllability. In the deterministic setting, it is easy to see that any backward controllability problem can be reduced to a forward one by simply reversing the time variable. However, such a technique does not work anymore for stochastic problems because of our adapted-ness requirement on stochastic processes, and therefore one has to analyze the forward and the backward controllability separately.

In the rest of this section, we shall present more new phenomena for the controllability problems in the stochastic setting. For simplicity, we only consider the case of bounded control operators and the special case that $V = \mathbb{R}$, i.e., $W(\cdot)$ is a one dimensional standard Brownian motion.

First, as a simple generalization of the conclusion 1) in Proposition 6.3, we have the following necessary condition for the exact controllability of (7.6).

Proposition 7.6. *If $f(\cdot) \in L^2_{\mathbb{F}}(0,T;H)$ and $G(\cdot) \in C_{\mathbb{F}}([0,T];L^\infty(\Omega;\mathcal{L}(H)))$ and the system (7.6) is exactly controllable at time T, then*

$$D(A^*) \cap \overline{\mathcal{R}(D)}^{\perp} = \emptyset.$$

Proof: We use the contradiction argument. Assume that $D(A^*) \cap \overline{\mathcal{R}(D)}^{\perp} \neq \emptyset$. Then, we could find $\varphi \in D(A^*)$ with $|\varphi|_H = 1$, such that $\langle Dv, \varphi\rangle_H = 0$ for any $v \in U$. Hence, for any $u \in \mathcal{U}_T$, by (7.6),

$$\begin{aligned} &\langle y(T), \varphi\rangle_H - \langle y_0, \varphi\rangle_H \\ &= \int_0^T \big(\langle y(t), A^*\varphi\rangle_H + \langle F(t)y(t) + f(t) + Bu(t), \varphi\rangle_H\big)dt \\ &\quad + \int_0^T \langle G(t)y(t), \varphi\rangle_H dW(t). \end{aligned} \tag{7.10}$$

Let us choose $y_0 = 0$ and $y_T = \Big(\int_0^T \eta(t)dW(t)\Big)\varphi$, where $\eta(\cdot)$ is given by (6.2). Then, by the exact controllability of the system (7.6) at time T, it follows from (7.10) that, for some $u \in \mathcal{U}_T$,

$$\begin{aligned} \int_0^T \eta(t)dW(t) &= \int_0^T \big(\langle y(t), A^*\varphi\rangle_H + \langle F(t)y(t) + f(t) + Bu(t), \varphi\rangle_H\big)dt \\ &\quad + \int_0^T \langle G(t)y(t), \varphi\rangle_H dW(t). \end{aligned} \tag{7.11}$$

Since $y(\cdot) \in C_{\mathbb{F}}([0,T];L^2(\Omega;H))$ and $u(\cdot) \in \mathcal{U}_T$, we see that

$$\langle y(\cdot), A^*\varphi\rangle_H + \langle F(\cdot)y(\cdot) + f(\cdot) + Bu(\cdot), \varphi\rangle_H \in L^2_{\mathbb{F}}(0,T)$$

and $\langle G(\cdot)y(\cdot), \varphi\rangle_H \in C_{\mathbb{F}}([0,T];L^2(\Omega))$. Hence, it is clear that (7.11) contradicts Proposition 6.2. □

Remark 7.7. By Proposition 7.6 (See also the conclusion 1) in Proposition 6.3), we see that, in order that the system (7.6) is exactly controllable at time T, one has to choose the control operator D in the diffusion term of (7.6) to be sufficiently "effective" (or even active everywhere in some sense). This is because, in the stochastic setting, the "target" set (i.e., $L^2_{\mathcal{F}_T}(\Omega;H)$) is too large, compared with the set of initial state (i.e., H).

Next, we consider the following control system with deterministic coefficients:

$$\begin{cases} dy(t) = \big(Ay(t) + F(t)y(t) + f(t) + Bu(t)\big)dt + G(t)y(t)dW(t) & \text{in } (0,T], \\ y(0) = y_0 \end{cases} \tag{7.12}$$

and its deterministic counterpart:

$$\begin{cases} \dfrac{d\hat{y}(t)}{dt} = A\hat{y}(t) + F(t)\hat{y}(t) + f(t) + B\hat{u}(t) & \text{in } (0,T], \\ \hat{y}(0) = y_0, \end{cases} \tag{7.13}$$

where $y_0 \in H$, $F \in L^\infty(0,T;\mathcal{L}(H))$, $f \in L^2(0,T;H)$ and $G \in L^\infty_{\mathbb{F}}(0,T)$ (Clearly, any $G \in L^\infty_{\mathbb{F}}(0,T)$ can be viewed as an element in $L^\infty_{\mathbb{F}}(0,T;\mathcal{L}(H))$). In (7.12) (*resp.* (7.13)), y (*resp.* $\hat{y}$) is the state variable and $u \in \mathcal{U}_T$ (*resp.* $\hat{u} \in L^2(0,T;U)$) is the control variable.

We have the following simple result:

Proposition 7.8. *The system* (7.12) *is null controllable at time T if and only if so is the system* (7.13).

Proof: The **"only if" part**. For arbitrary $y_0 \in H$, let $u(\cdot) \in \mathcal{U}_T$ be a control which transfers the state $y(\cdot)$ of (7.12) from y_0 to zero at time T. Then, clearly, $\mathbb{E}y$ is the solution to (7.13) with control $\mathbb{E}u$. Since $y(T) = 0$, a.s., we see that $\mathbb{E}y(T) = 0$, which implies that (7.13) is null controllable.

The **"if" part**. For arbitrary $y_0 \in H$, let $\hat{u}(\cdot) \in L^2(0,T;U)$ be a control which transfers the state $\hat{y}(\cdot)$ of (7.13) from y_0 to zero at time T. Let

$$y(t) = e^{\int_0^t G(s)dW(s) - \frac{1}{2}\int_0^t G(s)^2 ds}\hat{y}(t).$$

Then, it is easy to see that $y(\cdot)$ is the unique solution to (7.12) with the control $u(\cdot) = e^{\int_0^\cdot G(s)dW(s) - \frac{1}{2}\int_0^\cdot G(s)^2 ds}\hat{u}(\cdot)$. Since $\hat{y}(T) = 0$, we see that $y(T) = e^{\int_0^T G(s)dW(s) - \frac{1}{2}\int_0^T G(s)^2 ds}\hat{y}(T) = 0$. Further,

$$\begin{aligned} \mathbb{E}\int_0^T |u(t)|_U^2 dt &= \int_0^T |\hat{u}(t)|_U^2 \mathbb{E}e^{2\int_0^t G(s)dW(s) - \int_0^t G(s)^2 ds} dt \\ &\le \mathcal{C}\int_0^T |\hat{u}(t)|_U^2 dt < \infty. \end{aligned}$$

This complete the proof of Proposition 7.8. □

Remark 7.9. The same technique to prove Proposition 7.8 cannot be applied to get the exact controllability or approximate controllability of the system (7.12). Indeed, let us assume that (7.13) is exactly controllable. Put

$$\begin{aligned} \mathcal{A}_T \overset{\Delta}{=} \big\{ y(T) \bigm| & y(\cdot) \text{ solves (7.12) for some } y_0 \in H \text{ and} \\ & u(\cdot) = e^{\int_0^\cdot G(s)dW(s) - \frac{1}{2}\int_0^\cdot G(s)^2 ds}\hat{u}(\cdot) \text{ for some } \hat{u}(\cdot) \in L^2(0,T;U) \big\}. \end{aligned}$$

Then, by the exact controllability of (7.13), it is easy to see that

$$\mathcal{A}_T = \big\{ e^{\int_0^T G(s)dW(s) - \frac{1}{2}\int_0^T G(s)^2 ds} y_1 \bigm| y_1 \in H \big\}.$$

Clearly, $\mathcal{A}_T \neq L^2_{\mathcal{F}_T}(\Omega;H)$ and $\mathcal{A}_T$ is not dense in $L^2_{\mathcal{F}_T}(\Omega;H)$, which means that the system (7.12) is neither exactly controllable nor approximately controllable. By Remark 7.2, it is clear that this is significantly different from the deterministic setting, at least when A generates a C_0-group on H.

7.2 Well-Posedness of Stochastic Systems With Unbounded Control Operators

In this section, we shall use the transposition method to establish the well-posedness of stochastic control systems (7.8) and (7.9).

In order to define transposition solutions to (7.8), we introduce the following (backward stochastic) test equation evolved in O:

$$\begin{cases} d\xi(t) = -\big(A^*\xi(t) + \mathcal{F}(t)^*\xi(t) + \mathcal{G}(t)^*\Xi(t)\big)dt + \Xi(t)dW(t) & \text{in } [0,\tau), \\ \xi(\tau) = \xi_\tau. \end{cases} \tag{7.14}$$

In (7.14), $\tau \in [0,T]$ and $\xi_\tau \in L^2_{\mathcal{F}_\tau}(\Omega; O)$.

Recall that we assume $\mathbb{F}$ is the natural filtration. Hence, by Theorem 4.10, the equation (7.14) admits a unique mild solution $(\xi, \Xi) \in L^2_{\mathbb{F}}(\Omega; C([0,\tau]; O)) \times L^2_{\mathbb{F}}(0,\tau; \mathcal{L}_2(V; O))$ such that

$$\big|(\xi, \Xi)\big|_{L^2_{\mathbb{F}}(\Omega;C([0,\tau];O))\times L^2_{\mathbb{F}}(0,\tau;\mathcal{L}_2(V;O))} \leq \mathcal{C}(\mathcal{F},\mathcal{G})|\xi_\tau|_{L^2_{\mathcal{F}_\tau}(\Omega;O)}. \tag{7.15}$$

We make the following additional assumptions.

Condition 7.1 *There exists a sequence* $\{u_n\}_{n=1}^\infty \subset \mathcal{U}_T$ *such that* $\mathcal{B}u_n \in L^2_{\mathbb{F}}(0,T;O')$ *and* $\mathcal{D}u_n \in L^2_{\mathbb{F}}(0,T;\mathcal{L}_2(V;O'))$ *for each* $n \in \mathbb{N}$ *and*

$$\lim_{n\to\infty} u_n = u \qquad \text{in } \mathcal{U}_T. \tag{7.16}$$

Condition 7.2 *There exists a constant* $\mathcal{C} > 0$ *such that for any* $\tau \in [0,T]$ *and* $\xi_\tau \in L^2_{\mathcal{F}_\tau}(\Omega;O)$, *the solution* (ξ,Ξ) *to (7.14) satisfies that*

$$\big|\mathcal{B}^*\xi + \mathcal{D}^*\Xi\big|_{L^2_{\mathbb{F}}(0,\tau;U)} \leq \mathcal{C}\big|\xi_\tau\big|_{L^2_{\mathcal{F}_\tau}(\Omega;O)}. \tag{7.17}$$

Remark 7.10. When $\mathcal{D}$ is bounded, i.e., $\mathcal{D} \in \mathcal{L}(U;\mathcal{L}_2(V;O'))$, the condition $\mathcal{D}u_n \in L^2_{\mathbb{F}}(0,T;\mathcal{L}_2(V;O'))$ in Condition 7.1 holds automatically, and the inequality (7.17) in Condition 7.2 is equivalent to the following estimate:

$$\big|\mathcal{B}^*\xi\big|_{L^2_{\mathbb{F}}(0,\tau;U)} \leq \mathcal{C}\big|\xi_\tau\big|_{L^2_{\mathcal{F}_\tau}(\Omega;O)}. \tag{7.18}$$

Under Condition 7.2, the solutions to the system (7.8) are understood in the following sense:

Definition 7.11. *A process* $y \in C_{\mathbb{F}}([0,T]; L^2(\Omega;O'))$ *is called a transposition solution to the system (7.8) if for every* $\tau \in (0,T]$ *and* $\xi_\tau \in L^2_{\mathcal{F}_\tau}(\Omega;O)$, *it holds that*

$$\begin{aligned} &\mathbb{E}\big\langle y(\tau), \xi_\tau\big\rangle_{O',O} - \big\langle y_0, \xi(0)\big\rangle_{O',O} \\ &= \mathbb{E}\int_0^\tau \big\langle u(t), \mathcal{B}^*\xi(t) + \mathcal{D}^*\Xi(t)\big\rangle_U dt + \mathbb{E}\int_0^\tau \big\langle \mathbf{f}(t), \xi(t)\big\rangle_{O',O} dt, \end{aligned} \tag{7.19}$$

where (ξ,Ξ) *solves the equation (7.14).*

We have the following well-posedness result for the system (7.8).

Theorem 7.12. *Let Conditions 7.1 and 7.2 hold. For each $y_0 \in O'$ and $u \in \mathcal{U}_T$, the system (7.8) admits one and only one transposition solution $y \in C_{\mathbb{F}}([0,T];L^2(\Omega;O'))$. Moreover,*

$$|y|_{C_{\mathbb{F}}([0,T];L^2(\Omega;O'))} \leq \mathcal{C}\big(|y_0|_{O'} + |\mathbf{f}|_{L^2_{\mathbb{F}}(\Omega;L^1(0,T;O'))} + |u|_{\mathcal{U}_T}\big). \tag{7.20}$$

Proof: **Uniqueness of solutions.** Suppose that there are $y(\cdot)$ and $\hat{y}(\cdot)$ belong to $C_{\mathbb{F}}([0,T];L^2(\Omega;O'))$ such that (7.19) holds. Then,

$$\mathbb{E}\langle y(\tau),\xi_\tau\rangle_{O',O} = \mathbb{E}\langle \hat{y}(\tau),\xi_\tau\rangle_{O',O},$$

for all $\tau \in [0,T]$ and $\xi_\tau \in L^2_{\mathcal{F}_\tau}(\Omega;O)$. This concludes that $y(\cdot) = \hat{y}(\cdot)$.

Existence of solutions. Let $\{u_n\}_{n=1}^\infty \subset \mathcal{U}_T$ be the sequence given in Condition 7.1. Consider the following equation:

$$\begin{cases} dy_n(t) = \big(Ay_n(t) + \mathcal{F}(t)y_n(t) + \mathbf{f}(t) + \mathcal{B}u_n(t)\big)dt \\ \qquad\qquad +\big(\mathcal{G}(t)y_n(t) + \mathcal{D}u_n(t)\big)dW(t) & \text{in } (0,T], \\ y_n(0) = y_0. \end{cases} \tag{7.21}$$

By Theorem 3.14, we see that the equation (7.21) admits a unique mild solution $y_n \in C_{\mathbb{F}}([0,T];L^2(\Omega;O'))$. By Itô's formula, for each $n \in \mathbb{N}$, $\tau \in [0,T]$ and $\xi_\tau \in L^2_{\mathcal{F}_\tau}(\Omega;O)$, we obtain that

$$\begin{aligned} &\mathbb{E}\big\langle y_n(\tau),\xi_\tau\big\rangle_{O',O} - \big\langle y_0,\xi(0)\big\rangle_{O',O} \\ &= \mathbb{E}\int_0^\tau \big\langle \mathcal{B}u_n(t),\xi(t)\big\rangle_{O',O}dt + \mathbb{E}\int_0^\tau \big\langle \mathbf{f}(t),\xi(t)\big\rangle_{O',O}dt \\ &\quad +\mathbb{E}\int_0^\tau \big\langle \mathcal{D}u_n(t),\Xi(t)\big\rangle_{\mathcal{L}_2(V;O'),\mathcal{L}_2(V;O)}dt \\ &= \mathbb{E}\int_0^\tau \big\langle u_n(t),\mathcal{B}^*\xi(t) + \mathcal{D}^*\Xi(t)\big\rangle_U dt + \mathbb{E}\int_0^\tau \big\langle \mathbf{f}(t),\xi(t)\big\rangle_{O',O}dt, \end{aligned} \tag{7.22}$$

where (ξ,Ξ) solves (7.14). Therefore, for any $n,m \in \mathbb{N}$, it holds that

$$\mathbb{E}\big\langle y_n(\tau) - y_m(\tau),\xi_\tau\big\rangle_{O',O} = \mathbb{E}\int_0^\tau \big\langle u_n(t) - u_m(t),\mathcal{B}^*\xi(t) + \mathcal{D}^*\Xi(t)\big\rangle_U dt. \tag{7.23}$$

Let us choose $\xi_\tau \in L^2_{\mathcal{F}_\tau}(\Omega;O)$ such that

$$\begin{cases} |\xi_\tau|_{L^2_{\mathcal{F}_\tau}(\Omega;O)} = 1, \\ \mathbb{E}\big\langle y_n(\tau) - y_m(\tau),z_\tau\big\rangle_{O',O} \geq \dfrac{1}{2}|y_n(\tau) - y_m(\tau)|_{L^2_{\mathcal{F}_\tau}(\Omega;O')}. \end{cases}$$

These, together with (7.23) and Condition 7.2, imply that for any $\tau \in (0,T]$,

$$
\begin{aligned}
&\frac{1}{2}|y_n(\tau)-y_m(\tau)|_{L^2_{\mathcal{F}_\tau}(\Omega;O')}\\
&\le |u_n-u_m|_{L^2_{\mathbb{F}}(0,T;U)}|\mathcal{B}^*\xi(\cdot)+\mathcal{D}^*\Xi(\cdot)|_{\mathcal{U}_T}\\
&\le \mathcal{C}|\xi_\tau|_{L^2_{\mathcal{F}_\tau}(\Omega;O)}|u_n-u_m|_{\mathcal{U}_T}\le \mathcal{C}|u_n-u_m|_{\mathcal{U}_T},
\end{aligned}
$$

where $\mathcal{C}$ is independent of τ. Hence,

$$
|y_n-y_m|_{C_{\mathbb{F}}([0,T];L^2(\Omega;O'))}\le \mathcal{C}|u_n-u_m|_{\mathcal{U}_T}. \tag{7.24}
$$

By Condition 7.1, it is clear that $\{u_n\}_{n=1}^\infty$ is a Cauchy sequence in $\mathcal{U}_T$. Combining this with (7.24), we conclude that $\{y_n\}_{n=1}^\infty$ is a Cauchy sequence in $C_{\mathbb{F}}([0,T];L^2(\Omega;O'))$. Hence, there exists $y\in C_{\mathbb{F}}([0,T];L^2(\Omega;O'))$ such that $\lim_{n\to\infty}y_n=y$ in $C_{\mathbb{F}}([0,T];L^2(\Omega;O'))$. By letting $n\to+\infty$ in (7.22) and noting Condition 7.1, we see that for each $\tau\in[0,T]$ and $\xi_\tau\in L^2_{\mathcal{F}_\tau}(\Omega;O)$, the equality (7.19) holds. This indicates that y is a transposition solution to (7.21).

Choose a $\xi_\tau\in L^2_{\mathcal{F}_\tau}(\Omega;O)$ such that

$$
\begin{cases}
|\xi_\tau|_{L^2_{\mathcal{F}_\tau}(\Omega;O)}=1,\\
\mathbb{E}\big\langle y(\tau),\xi_\tau\big\rangle_{O',O}\ge \dfrac{1}{2}|y(\tau)|_{L^2_{\mathcal{F}_\tau}(\Omega;O')}.
\end{cases}
$$

These, together with (7.19), imply that for any $\tau\in(0,T]$,

$$
\begin{aligned}
&\frac{1}{2}|y(\tau)|_{L^2_{\mathcal{F}_\tau}(\Omega;O')}\\
&\le |y_0|_{O'}|\xi(0)|_O+|u|_{\mathcal{U}_T}|\mathcal{B}^*\xi(t)+\mathcal{D}^*\Xi(t)|_{L^2_{\mathbb{F}}(0,\tau;U)}\\
&\quad +|\mathbf{f}|_{L^2_{\mathbb{F}}(\Omega;L^1(0,T;O'))}|z|_{L^2_{\mathbb{F}}(\Omega;C([0,\tau];O))}\\
&\le \mathcal{C}|\xi_\tau|_{L^2_{\mathcal{F}_\tau}(\Omega;O')}\big(|y_0|_{O'}+|u|_{\mathcal{U}_T}+|\mathbf{f}|_{L^2_{\mathbb{F}}(\Omega;L^1(0,T;O'))}\big)\\
&\le \mathcal{C}\big(|y_0|_{O'}+|u|_{\mathcal{U}_T}+|\mathbf{f}|_{L^2_{\mathbb{F}}(\Omega;L^1(0,T;O'))}\big),
\end{aligned}
$$

where $\mathcal{C}$ is independent of τ. Hence, the desired estimate (7.20) holds. This completes the proof of Theorem 7.12. □

Similarly, in order to define transposition solutions to (7.9), for any $\tau\in[0,T]$, we introduce the following (forward stochastic) test equation:

$$
\begin{cases}
d\eta(t)=\big(A\eta(t)-\mathcal{J}(t)^*\eta(t)\big)dt+\big(-\mathcal{K}(t)^*\eta(t)+h(t)\big)dW(t) & \text{in }(\tau,T],\\
\eta(\tau)=\eta_\tau,
\end{cases}
\tag{7.25}
$$

where $\eta_\tau\in L^2_{\mathcal{F}_\tau}(\Omega;O)$ and $h\in L^2_{\mathbb{F}}(0,T;\mathcal{L}_2(V;O))$. By Theorem 3.14, the equation (7.25) admits a unique mild solution $\eta\in C_{\mathbb{F}}([\tau,T];L^2(\Omega;O))$ such that

$$|\eta|_{C_{\mathbb{F}}([\tau,T];L^2(\Omega;O))} \leq \mathcal{C}\big(|\eta_\tau|_{L^2_{\mathcal{F}_\tau}(\Omega;O)} + |h|_{L^2_{\mathbb{F}}(0,T;\mathcal{L}_2(V;O))}\big). \tag{7.26}$$

We need to introduce the following assumption:

Condition 7.3 *There exists a constant $\mathcal{C} > 0$ such that for any $\tau \in [0,T]$, $\eta_\tau \in L^2_{\mathcal{F}_\tau}(\Omega;O)$ and $h \in L^2_{\mathbb{F}}(0,T;\mathcal{L}_2(V;O))$, the corresponding mild solution η to (7.25) satisfies that*

$$|\mathcal{B}^*\eta|_{L^2_{\mathbb{F}}(0,\tau;U)} \leq \mathcal{C}\big(|\eta_\tau|_{L^2_{\mathcal{F}_\tau}(\Omega;O)} + |h|_{L^2_{\mathbb{F}}(0,T;\mathcal{L}_2(V;O))}\big). \tag{7.27}$$

Under Condition 7.3, the solutions to the system (7.9) are understood in the following sense:

Definition 7.13. *A pair of processes $(y,Y) \in C_{\mathbb{F}}([0,T];L^2(\Omega;O')) \times L^2_{\mathbb{F}}(0,T;\mathcal{L}_2(V;O'))$ is called a transposition solution to the system (7.9) if for every $\tau \in [0,T)$, $\eta_\tau \in L^2_{\mathcal{F}_\tau}(\Omega;O)$ and $h \in L^2_{\mathbb{F}}(0,T;\mathcal{L}_2(V;O))$, it holds that*

$$\begin{aligned}
&\mathbb{E}\big\langle y_T, \eta(T)\big\rangle_{O',O} - \mathbb{E}\big\langle y(\tau), \eta_\tau\big\rangle_{O',O}\\
&= \mathbb{E}\int_\tau^T \big\langle u(s), \mathcal{B}^*\eta(s)\big\rangle_U ds + \mathbb{E}\int_\tau^T \big\langle \mathbf{g}(s), \eta(s)\big\rangle_{O',O} ds\\
&\quad +\mathbb{E}\int_\tau^T \big\langle Y(s), h(s)\big\rangle_{\mathcal{L}_2(V;O'),\mathcal{L}_2(V;O)} ds,
\end{aligned} \tag{7.28}$$

where $\eta(\cdot)$ solves (7.25).

Similarly to Theorem 7.12, we have the following well-posedness result for the system (7.9).

Theorem 7.14. *Under Conditions 7.1 (with $\mathcal{D} = 0$) and 7.3, for each $y_T \in L^2_{\mathcal{F}_T}(\Omega;O')$ and $u \in \mathcal{U}_T$, the system (7.9) admits one and only one transposition solution $(y,Y) \in C_{\mathbb{F}}([0,T];L^2(\Omega;O')) \times L^2_{\mathbb{F}}(0,T;\mathcal{L}_2(V;O'))$. Moreover,*

$$\begin{aligned}
&|y|_{C_{\mathbb{F}}([0,T];L^2(\Omega;O'))} + |Y|_{L^2_{\mathbb{F}}(0,T;\mathcal{L}_2(V;O'))}\\
&\leq \mathcal{C}\big(|y_T|_{L^2_{\mathcal{F}_T}(\Omega;O')} + |\mathbf{g}|_{L^1_{\mathbb{F}}(0,T;L^2(\Omega;O'))} + |u|_{\mathcal{U}_T}\big).
\end{aligned} \tag{7.29}$$

Proof: The proof is very similarly to that of Theorem 7.12, hence we give below only a sketched proof of the existence of solutions. Let $\{u_n\}_{n=1}^\infty \subset \mathcal{U}_T$ be the sequence given in Condition 7.1 (with $\mathcal{D} = 0$). Consider the following equation:

$$\begin{cases}
dy_n(t) = -A^*y_n(t)dt + \big(\mathcal{J}(t)y_n(t) + \mathcal{K}(t)Y_n(t) + \mathbf{g}(t) + \mathcal{B}u_n(t)\big)dt\\
\qquad\qquad +Y_n(t)dW(t) \qquad \text{in } [0,T),\\
y_n(T) = y_T.
\end{cases} \tag{7.30}$$

Since $\mathbb{F}$ is assumed to be the natural filtration, by Theorem 4.10, the system (7.30) admits a unique mild solution $(y_n(\cdot), Y_n(\cdot)) \in L^2_{\mathbb{F}}(\Omega; C([0,T]; O')) \times L^2_{\mathbb{F}}(0,T;\mathcal{L}_2(V;O'))$. By Itô's formula, for each $n \in \mathbb{N}$, $\tau \in [0,T)$, $\eta_\tau \in L^2_{\mathcal{F}_\tau}(\Omega; O)$ and $h \in L^2_{\mathbb{F}}(0,T;\mathcal{L}_2(V;O))$, we obtain that

$$
\begin{aligned}
&\mathbb{E}\big\langle y_T, \eta(T)\big\rangle_{O',O} - \mathbb{E}\big\langle y_n(\tau), \eta_\tau\big\rangle_{O',O}\\
&= \mathbb{E}\int_\tau^T \big\langle u_n(s), \mathcal{B}^*\eta(s)\big\rangle_U ds + \mathbb{E}\int_\tau^T \big\langle \mathbf{g}(s), \eta(s)\big\rangle_{O',O} ds \\
&\quad +\mathbb{E}\int_\tau^T \big\langle Y_n(s), h(s)\big\rangle_{\mathcal{L}_2(V;O'),\mathcal{L}_2(V;O)} ds,
\end{aligned}
\tag{7.31}
$$

where $\eta(\cdot)$ solves (7.25). Therefore, for any $n, m \in \mathbb{N}$, it holds that

$$
\begin{aligned}
&\mathbb{E}\big\langle y_n(\tau) - y_m(\tau), \eta_\tau\big\rangle_{O',O} + \mathbb{E}\int_\tau^T \big\langle Y_n(s) - Y_m(s), h(s)\big\rangle_{\mathcal{L}_2(V;O'),\mathcal{L}_2(V;O)} ds\\
&= \mathbb{E}\int_\tau^T \big\langle u_m(s) - u_n(s), \mathcal{B}^*\eta(s)\big\rangle_U ds.
\end{aligned}
\tag{7.32}
$$

By the arbitrariness of τ, η_τ and h in (7.32), similarly to (7.24), by Condition 7.3, we can show that

$$
|(y_n - y_m, Y_n - Y_m)|_{C_{\mathbb{F}}([0,T];L^2(\Omega;O'))\times L^2_{\mathbb{F}}(0,T;\mathcal{L}_2(V;O'))} \le \mathcal{C}|u_n - u_m|_{\mathcal{U}_T}. \tag{7.33}
$$

By Condition 7.1 (with $\mathcal{D} = 0$), it is clear that $\{u_n\}_{n=1}^\infty$ is a Cauchy sequence in $\mathcal{U}_T$. Combining this with (7.33), we conclude that $\{(y_n, Y_n)\}_{n=1}^\infty$ is a Cauchy sequence in $C_{\mathbb{F}}([0,T]; L^2(\Omega;O')) \times L^2_{\mathbb{F}}(0,T;\mathcal{L}_2(V;O'))$. Hence, there exists $(y, Y) \in C_{\mathbb{F}}([0,T]; L^2(\Omega;O')) \times L^2_{\mathbb{F}}(0,T;\mathcal{L}_2(V;O'))$ such that

$$
\lim_{n\to\infty} (y_n, Y_n) = (y, Y) \;\text{ in }\; C_{\mathbb{F}}([0,T]; L^2(\Omega;O')) \times L^2_{\mathbb{F}}(0,T;\mathcal{L}_2(V;O')).
$$

By letting $n \to +\infty$ in (7.31) and noting Condition 7.1 (with $\mathcal{D} = 0$), we see that for each $\tau \in [0,T]$, $\eta_\tau \in L^2_{\mathcal{F}_\tau}(\Omega;O)$ and $h \in L^2_{\mathbb{F}}(0,T;\mathcal{L}_2(V;O))$, the equality (7.28) holds. This indicates that (y, Y) is a transposition solution to (7.9).

Finally, proceeding as that in the proof of Theorem 7.12, we can show that the desired estimate (7.29) holds. This completes the proof of Theorem 7.14. □

Remark 7.15. In view of Theorem 4.24, it seems that the regularity of the first component of the transposition solution $(y, Y) \in C_{\mathbb{F}}([0,T]; L^2(\Omega;O')) \times L^2_{\mathbb{F}}(0,T;\mathcal{L}_2(V;O'))$ to the system (7.9) can be improved as $y \in L^2_{\mathbb{F}}(\Omega; C([0,T]; O'))$ but this remains to be proved.

7.3 Reduction to the Observability of Dual Problems

In this section, we shall reduce the controllability problems presented in Section 7.1 to the observability of their dual problems.

To begin with, let us introduce respectively the following dual equations of (7.6) and (7.7):

$$\begin{cases} dz(t) = -\big(A^*z(t) + F(t)^*z(t) + G(t)^*Z(t)\big)dt + Z(t)dW(t) & \text{in } [0,T), \\ z(T) = z_T \end{cases} \tag{7.34}$$

and

$$\begin{cases} dx(t) = \big(Ax(t) - J(t)^*x(t)\big)dt - K(t)^*x(t)dW(t) & \text{in } (0,T], \\ x(0) = x_0, \end{cases} \tag{7.35}$$

where $z_T \in L^2_{\mathcal{F}_T}(\Omega;H)$ and $x_0 \in H$.

By Theorem 4.10, the backward stochastic evolution equation (7.34) admits a unique mild solution $(z,Z) \in L^2_{\mathbb{F}}(\Omega;C([0,T];H)) \times L^2_{\mathbb{F}}(0,T;\mathcal{L}_2^0)$; while by Theorem 3.14, the stochastic evolution equation (7.35) admits a unique mild solution $x(\cdot) \in C_{\mathbb{F}}([0,T];L^2(\Omega;H))$.

For the exact controllability of (7.6), we have the following result.

Theorem 7.16. *System (7.6) is exactly controllable at time T if and only if (7.34) is continuously finally observable on $[0,T]$ in the sense that all of its solutions satisfy the following (final time) observability estimate:*

$$|z_T|_{L^2_{\mathcal{F}_T}(\Omega;H)} \le \mathcal{C}|B^*z + D^*Z|_{\mathcal{U}_T}, \quad \forall\, z_T \in L^2_{\mathcal{F}_T}(\Omega;H). \tag{7.36}$$

Proof: Clearly, it suffices to consider the special case that the initial datum y_0 and the non-homogenous term f in the system (7.6) are equal to 0. Similarly to the proof of "1)$\Longleftrightarrow$2)" in Theorem 6.8, we define an operator $\mathcal{L} : \mathcal{U}_T \to L^2_{\mathcal{F}_T}(\Omega;H)$ by

$$\mathcal{L}(u(\cdot)) = y(T), \quad \forall\, u(\cdot) \in \mathcal{U}_T, \tag{7.37}$$

where $y(\cdot)$ is the corresponding solution to (7.6) with $y_0 = 0$ and $f = 0$. Then, applying Itô's formula to (7.6) (with $y_0 = 0$ and $f = 0$) and (7.34), we obtain that

$$\begin{aligned} \langle\, y(T), z_T \,\rangle_{L^2_{\mathcal{F}_T}(\Omega;H)} = \mathbb{E}\int_0^T \langle\, u(t), B^*z(t) + D^*Z(t) \,\rangle_U dt, \\ \forall\, z_T \in L^2_{\mathcal{F}_T}(\Omega;H), \end{aligned} \tag{7.38}$$

where $(z(\cdot),Z(\cdot)) \in L^2_{\mathbb{F}}(\Omega;C([0,T];H)) \times L^2_{\mathbb{F}}(0,T;\mathcal{L}_2^0)$ solves (7.34). By (7.37)–(7.38), it is clear that,

$$(\mathcal{L}^*z_T)(t) = B^*z(t) + D^*Z(t), \quad \text{a.e. } t \in [0,T]. \tag{7.39}$$

Now, the exact controllability of (7.6) is equivalent to that $\mathcal{R}(\mathcal{L}) = L^2_{\mathcal{F}_T}(\Omega; H)$; while the later, by (7.39) and using Theorem 1.7, is equivalent to the estimate (7.36). □

Next, for the null controllability of (7.6), we have the following result.

Theorem 7.17. *System* (7.6) *with* $f = 0$ *is null controllable at time* T *if and only if* (7.34) *is continuously initially observable on* $[0, T]$ *in the sense that all of its solutions satisfy the following (initial time) observability estimate:*

$$|z(0)|_H \leq \mathcal{C}|B^*z + D^*Z|_{\mathcal{U}_T}, \quad \forall\, z_T \in L^2_{\mathcal{F}_T}(\Omega; H). \tag{7.40}$$

Proof: Similarly to the proof of Theorem 7.16, we define an operator $\mathcal{L} : \mathcal{U}_T \to L^2_{\mathcal{F}_T}(\Omega; H)$ by (7.37) and an operator $\mathcal{M} : H \to L^2_{\mathcal{F}_T}(\Omega; H)$

$$\mathcal{M}y_0 = -y(T), \quad \forall\, y_0 \in H, \tag{7.41}$$

where $y(\cdot)$ is the corresponding solution to (7.6) with $u(\cdot) = 0$ and $f = 0$. Then, $\mathcal{L}^*$ can be expressed as that in (7.39). On the other hand, applying Itô's formula to (7.6) (with $u(\cdot) = 0$ and $f = 0$) and (7.34), we obtain that

$$\langle\, y(T), z_T \,\rangle_{L^2_{\mathcal{F}_T}(\Omega;H)} = \langle\, y_0, z(0) \,\rangle_H, \quad \forall\, z_T \in L^2_{\mathcal{F}_T}(\Omega; H), \tag{7.42}$$

where $(z(\cdot), Z(\cdot)) \in L^2_{\mathbb{F}}(\Omega; C([0,T]; H)) \times L^2_{\mathbb{F}}(0,T; \mathcal{L}^0_2)$ solves (7.34). By (7.41)–(7.42), it is clear that,

$$\mathcal{M}^* z_T = -z(0). \tag{7.43}$$

Now, the null controllability of (7.6) is equivalent to that $\mathcal{R}(\mathcal{M}) \subset \mathcal{R}(\mathcal{L})$; while the later, by (7.43) and (7.39), and in view of Theorem 1.7, is equivalent to the estimate (7.40). □

Remark 7.18. Compared with Theorem 7.16, we assume that $f = 0$ in Theorem 7.17. This condition is necessary. Indeed, Theorem 7.16 may be failed if $f \neq 0$. Such a phenomenon has already been found in the study of null controllability of deterministic heat equations.

Further, for the approximate controllability of (7.6), we have the following result.

Theorem 7.19. *System* (7.6) *is approximately controllable at time* T *if and only if solutions to* (7.34) *satisfy the following observability:*

$$B^*z + D^*Z \equiv 0 \textit{ in } (0,T), \text{a.s.} \Rightarrow z_T = 0. \tag{7.44}$$

Proof: Recall the operator $\mathcal{L} : \mathcal{U}_T \to L^2_{\mathcal{F}_T}(\Omega; H)$ defined by (7.37) (in the proof of Theorem 7.16). It is easy to show that the system (7.6) is approximately controllable at time T if and only if $\overline{\mathcal{R}(\mathcal{L})} = L^2_{\mathcal{F}_T}(\Omega; H)$. Note that $\mathcal{L}^*$ can be expressed as that in (7.39). Hence, by Theorem 1.12, $\overline{\mathcal{R}(\mathcal{L})} = L^2_{\mathcal{F}_T}(\Omega; H)$ if and only if solutions to (7.34) satisfy the property (7.44). □

In deterministic setting, by Theorem 7.1, if the control operator B is invertible, then it is easy to get controllability results (especially the null and approximate controllability of (7.1)). The same holds in the stochastic framework for this special case. Indeed, we have the following result.

Theorem 7.20. *If $B \in \mathcal{L}(U;H)$ is invertible, $D = 0$ and $f = 0$, then the system (7.6) is both null controllable and approximately controllable at any time $T > 0$.*

Proof: By Theorem 7.17, to prove that (7.6) (with $D = 0$ and $f = 0$) is null controllable, it suffices to show that solutions to (7.34) satisfy

$$|z(0)|_H^2 \leq \mathcal{C}\mathbb{E}\int_0^T |B^* z(t)|_U^2 dt, \quad \forall\, z_T \in L^2_{\mathcal{F}_T}(\Omega; H). \tag{7.45}$$

By Theorem 4.39, there is a constant $\mathcal{C} > 0$ such that

$$|z(0)|_H^2 \leq \mathcal{C}\mathbb{E}|z(t)|_H^2, \quad \forall\, t \in [0,T]. \tag{7.46}$$

Since B is invertible, we have that

$$|y|_H \leq \mathcal{C}|B^* y|_U, \quad \forall\, y \in H.$$

This, together with (7.46), implies the inequality (7.45).

By Theorem 7.19, in order to prove that (7.6) (with $D = 0$ and $f = 0$) is approximately controllable, it suffices to show that solutions to (7.34) satisfy

$$B^* z \equiv 0 \text{ in } (0,T), \text{a.s.} \Rightarrow z_T = 0. \tag{7.47}$$

Since B is invertible, we conclude that $z(\cdot) = 0$ in $L^2_{\mathbb{F}}(0,T;H)$. Noting that $z(\cdot) \in C_{\mathbb{F}}([0,T]; L^2(\Omega;H))$, we find that $z_T = 0$. □

By Proposition 7.6, it is easy to see that, generally speaking, the same result in Theorem 7.20 does NOT hold for the exact controllability of (7.6). Note that, this is because we choose the controls in (7.6) to be L^2-in time. The situation will be completely different if we choose L^1-in time controls. More precisely, similarly to Definition 6.12, we introduce the following notion:

Definition 7.21. *The system (7.6) with $D = 0$ is called exactly controllable at time T by means of controls in the space $L^1_{\mathbb{F}}(0,T;L^2(\Omega;U))$ if for any $y_0 \in H$ and $y_1 \in L^2_{\mathcal{F}_T}(\Omega;H)$, there exists a control $u(\cdot) \in L^1_{\mathbb{F}}(0,T;L^2(\Omega;U))$ so that the corresponding solution y to (7.6) with $D = 0$ verifies that $y(T) = y_1$, a.s.*

Proceeding as the proof of the "if" part in Theorem 6.15, one can show the following result (and hence we will not prove it).

Theorem 7.22. *If $B \in \mathcal{L}(U;H)$ is invertible and $D = 0$, then the system (7.6) is exactly controllable at time T by means of controls in the space $L^1_{\mathbb{F}}(0,T;L^2(\Omega;U))$.*

Similarly to Theorems 7.16, 7.17 and 7.19, we can prove the following controllability results for the system (7.7):

Theorem 7.23. 1) *System (7.7) is exactly controllable at time 0 if and only if (7.35) is continuously initially observable on $[0,T]$ in the sense that all of its solutions satisfy the following (initial time) observability estimate:*

$$|x_0|_H \le \mathcal{C}|B^*x|_{\mathcal{U}_T}, \quad \forall\, x_0 \in H. \tag{7.48}$$

2) *System (7.7) with $g=0$ is null controllable at time 0 if and only if (7.35) is continuously finally observable on $[0,T]$ in the sense that all of its solutions satisfy the following (finial time) observability estimate:*

$$|x(T)|_{L^2_{\mathcal{F}_T}(\Omega;H)} \le \mathcal{C}|B^*x|_{\mathcal{U}_T}, \quad \forall\, x_0 \in H. \tag{7.49}$$

3) *System (7.7) is approximately controllable at time 0 if and only if solutions to (7.35) satisfy the following observability:*

$$B^*x \equiv 0 \ in \ (0,T), \text{a.s.} \ \Rightarrow x_0 = 0. \tag{7.50}$$

Now, let us introduce respectively the following dual equations of (7.8) and (7.9):

$$\begin{cases} dz(t) = -\big(A^*z(t) + \mathcal{F}(t)^*z(t) + \mathcal{G}(t)^*Z(t)\big)dt + Z(t)dW(t) \ \text{ in } [0,T), \\ z(T) = z_T \end{cases} \tag{7.51}$$

and

$$\begin{cases} dx(t) = \big(Ax(t) - \mathcal{J}(t)^*x(t)\big)dt - \mathcal{K}(t)^*x(t)dW(t) \ \text{ in } (0,T], \\ x(0) = x_0, \end{cases} \tag{7.52}$$

where $z_T \in L^2_{\mathcal{F}_T}(\Omega;O)$ and $x_0 \in O$.

Similarly to Theorems 7.16, 7.17 and 7.19, we have the following controllability results for the system (7.8):

Theorem 7.24. *Under Conditions 7.1 and 7.2, the following assertions hold:*

1) *System (7.8) is exactly controllable at time T if and only if (7.51) is continuously finally observable on $[0,T]$ in the sense that all of its solutions satisfy the following (final time) observability estimate:*

$$|z_T|_{L^2_{\mathcal{F}_T}(\Omega;O)} \le \mathcal{C}|\mathcal{B}^*z + \mathcal{D}^*Z|_{\mathcal{U}_T}, \quad \forall\, z_T \in L^2_{\mathcal{F}_T}(\Omega;O). \tag{7.53}$$

2) *System (7.8) with $\mathbf{f}=0$ is null controllable at time T if and only if (7.51) is continuously initially observable on $[0,T]$ in the sense that all of its solutions satisfy the following (initial time) observability estimate:*

$$|z(0)|_O \le \mathcal{C}|\mathcal{B}^*z + \mathcal{D}^*Z|_{\mathcal{U}_T}, \quad \forall\, z_T \in L^2_{\mathcal{F}_T}(\Omega;O). \tag{7.54}$$

3) *System* (7.8) *is approximately controllable at time* T *if and only if solutions to* (7.51) *satisfy the following observability:*

$$\mathcal{B}^* z + \mathcal{D}^* Z \equiv 0 \text{ in } (0,T), \text{a.s.} \Rightarrow z_T = 0. \tag{7.55}$$

Proof: The proof of Theorem 7.24 is very similarly to that of Theorems 7.16–7.19 but it is even simpler. Indeed, the only difference is to use the equality (7.19) instead of the Itô formula to derive the duality relation. Hence, we only give below a proof of the conclusion 1) in Theorem 7.24. Similarly to the proof of Theorem 7.16, we define an operator $\mathcal{L} : \mathcal{U}_T \to L^2_{\mathcal{F}_T}(\Omega; O')$ by

$$\mathcal{L}(u(\cdot)) = y(T), \quad \forall\, u(\cdot) \in \mathcal{U}_T, \tag{7.56}$$

where $y(\cdot)$ is the corresponding transposition solution to (7.8) with $y_0 = 0$ and $\mathbf{f} = 0$. Then, by (7.19) (with $\tau = T$) in Definition 7.11, using (7.51) as a test equation for the transposition solution to (7.8) with $y_0 = 0$ and $\mathbf{f} = 0$, we obtain that

$$\begin{aligned} \mathbb{E}\langle\, y(T), z_T \,\rangle_{O',O} = \mathbb{E}\int_0^T \langle\, u(t), \mathcal{B}^* z(t) + \mathcal{D}^* Z(t) \,\rangle_U dt,& \\ \forall\, z_T \in L^2_{\mathcal{F}_T}(\Omega; O),& \end{aligned} \tag{7.57}$$

where $(z(\cdot), Z(\cdot)) \in L^2_{\mathbb{F}}(\Omega; C([0,T]; O)) \times L^2_{\mathbb{F}}(0,T; \mathcal{L}_2(V; O))$ solves (7.51). By (7.56)–(7.57), it is clear that,

$$(\mathcal{L}^* z_T)(t) = \mathcal{B}^* z(t) + \mathcal{D}^* Z(t), \quad \text{a.e. } t \in [0,T]. \tag{7.58}$$

Now, the exact controllability of (7.8) is equivalent to $\mathcal{R}(\mathcal{L}) = L^2_{\mathcal{F}_T}(\Omega; O')$; while the later, by (7.58) and using Theorem 1.7, is equivalent to the estimate (7.53). □

Similarly to Theorems 7.23 and 7.24, we can prove easily the following controllability results for the system (7.9) (Hence we will not prove it here):

Theorem 7.25. *Under Conditions 7.1 (with* $\mathcal{D} = 0$*) and 7.3, the following assertions hold:*

1) *System* (7.9) *is exactly controllable at time* 0 *if and only if* (7.52) *is continuously initially observable on* $[0,T]$ *in the sense that all of its solutions satisfy the following (initial time) observability estimate:*

$$|x_0|_O \le \mathcal{C}|\mathcal{B}^* x|_{\mathcal{U}_T}, \quad \forall\, x_0 \in O. \tag{7.59}$$

2) *System* (7.9) *with* $\mathbf{g} = 0$ *is null controllable at time* 0 *if and only if* (7.52) *is continuously finally observable on* $[0,T]$ *in the sense that all of its solutions satisfy the following (finial time) observability estimate:*

$$|x(T)|_{L^2_{\mathcal{F}_T}(\Omega;O)} \le \mathcal{C}|\mathcal{B}^* x|_{\mathcal{U}_T}, \quad \forall\, x_0 \in O. \tag{7.60}$$

3) *System* (7.9) *is approximately controllable at time* 0 *if and only if solutions to* (7.52) *satisfy the following observability:*

$$\mathcal{B}^* x \equiv 0 \text{ in } (0,T), \text{a.s.} \Rightarrow x_0 = 0. \tag{7.61}$$

7.4 Explicit Forms of Controls for the Controllability Problems

We have shown the existence of controls for exact/null/approxiate controllability of systems (7.6) and (7.7) (as well as systems (7.8) and (7.9)), provided that their dual systems satisfy suitable observability properties. These results allow concluding whether a system is controllable but do not give any information about the control which drives the state to the destination. In this section, stimulated by the simple control formula (1.5) in Theorem 1.3 for the exact controllability of linear ordinary differential equations and also [123] for a systematic development on similar problems but for deterministic partial differential equations, we shall present explicit forms of controls for the controllability problems of stochastic linear evolution equations. Such kind of results not only provide variational characterizations on the desired controls but also serve for numerical analysis for the controllability problems.

In the rest of this section, we fix any given $f, g \in L^2_{\mathbb{F}}(\Omega; L^1(0,T;H))$, $y_0 \in H$, $y_T \in L^2_{\mathcal{F}_T}(\Omega; H)$ and $\varepsilon \geq 0$. Define a functional on $L^2_{\mathcal{F}_T}(\Omega; H)$ as follows:

$$
\begin{aligned}
\mathfrak{J}_1(z_T) = \frac{1}{2}\mathbb{E}\int_0^T |B^*z(t) + D^*Z(t)|_U^2 dt - \mathbb{E}\langle y_T, z_T\rangle_H + \langle y_0, z(0)\rangle_H \\
+\mathbb{E}\int_0^T \langle f(t), z(t)\rangle_H dt + \varepsilon |z_T|_{L^2_{\mathcal{F}_T}(\Omega;H)}, \qquad \forall\, z_T \in L^2_{\mathcal{F}_T}(\Omega; H),
\end{aligned}
\tag{7.62}
$$

where (z, Z) solves (7.34). Also, we define a functional on H as follows:

$$
\begin{aligned}
\mathfrak{J}_2(x_0) = \frac{1}{2}\mathbb{E}\int_0^T |B^*x(t)|_U^2 dt - \mathbb{E}\langle x(T), y_T\rangle_H + \langle x_0, y_0\rangle_H \\
+\mathbb{E}\int_0^T \langle g(t), x(t)\rangle_H dt + \varepsilon |x_0|_H, \qquad \forall\, x_0 \in H,
\end{aligned}
\tag{7.63}
$$

where $x(\cdot)$ solves (7.35).

In what follows, we shall give the following explicit form of controls for exact/null/approxiate controllability of (7.6):

$$
u(t) = B^*z(t; \hat{z}_T) + D^*Z(t; \hat{z}_T), \tag{7.64}
$$

where $(z(\cdot; \hat{z}_T), Z(\cdot; \hat{z}_T))$ is the corresponding solution to (7.34) with the final datum z_T replaced by a suitable chosen $\hat{z}_T \in L^2_{\mathcal{F}_T}(\Omega; H)$, i.e., the minimizer of the functional $\mathfrak{J}_1(\cdot)$ under some assumptions.

First, for the exact controllability of (7.6), we have the following result:

Theorem 7.26. *If the equation* (7.34) *is continuously finally observable on* $[0,T]$, *then, among all controls transferring the state of* (7.6) *from* y_0 *to* y_T *at time* T, *the one given by* (7.64) *enjoys the minimal* $\mathcal{U}_T$*-norm, where* $\hat{z}_T \in L^2_{\mathcal{F}_T}(\Omega; H)$ *is the minimizer of the functional* $\mathfrak{J}_1(\cdot)$ *(defined by* (7.62)*) with* $\varepsilon = 0$.

Proof: Note that, in Theorem 7.26 the parameter ε in the functional $\mathfrak{J}_1(\cdot)$ is taken to be 0. Clearly, in this case the functional $\mathfrak{J}_1(\cdot) : L^2_{\mathcal{F}_T}(\Omega; H) \to \mathbb{R}$ is continuous and strictly convex. Since the equation (7.34) is continuously finally observable on $[0, T]$ (Recall Theorem 7.16 for the meaning of continuous final observability), we have

$$\begin{aligned}
&\mathfrak{J}_1(z_T)\\
&\geq \mathcal{C}\mathbb{E}|z_T|^2_H - |y_T|_{L^2_{\mathcal{F}_T}(\Omega;H)}|z_T|_{L^2_{\mathcal{F}_T}(\Omega;H)} - |y_0|_H|z(0)|_H\\
&\quad -|f|_{L^2_{\mathbb{F}}(\Omega;L^1(0,T;H))}|z|_{L^2_{\mathbb{F}}(\Omega;C([0,T];H))}\\
&\geq \mathcal{C}\mathbb{E}|z_T|^2_H - \mathcal{C}\big(|y_T|_{L^2_{\mathcal{F}_T}(\Omega;H)} + |y_0|_H + |f|_{L^2_{\mathbb{F}}(\Omega;L^1(0,T;H))}\big)|z_T|_{L^2_{\mathcal{F}_T}(\Omega;H)}.
\end{aligned}$$

This implies that $\mathfrak{J}_1(\cdot)$ is coercive. Hence, $\mathfrak{J}_1(\cdot)$ admits a unique minimizer $\hat{z}_T$. By computing the first order variation of $\mathfrak{J}_1(\cdot)$ (with $\varepsilon = 0$), we obtain that, for any $z_T \in L^2_{\mathcal{F}_T}(\Omega; H)$,

$$\begin{aligned}
&\mathbb{E}\int_0^T \big\langle B^*z(t;\hat{z}_T) + D^*Z(t;\hat{z}_T), B^*z(t) + D^*Z(t)\big\rangle_U dt\\
&= \mathbb{E}\langle y_T, z_T\rangle_H - \big\langle y_0, z(0)\big\rangle_H - \mathbb{E}\int_0^T \big\langle f(t), z(t)\big\rangle_H dt.
\end{aligned} \tag{7.65}$$

On the other hand, applying Itô's formula to (7.6) and (7.34), we obtain that

$$\begin{aligned}
&\mathbb{E}\langle y(T), z_T\rangle_H - \big\langle y_0, z(0)\big\rangle_H - \mathbb{E}\int_0^T \big\langle f(t), z(t)\big\rangle_H dt\\
&= \mathbb{E}\int_0^T \big\langle u(t), B^*z(t) + D^*Z(t)\big\rangle_U dt, \qquad \forall\, z_T \in L^2_{\mathcal{F}_T}(\Omega; H).
\end{aligned} \tag{7.66}$$

From (7.65) and (7.66), if we choose the control $u(\cdot)$ as (7.64), then

$$\mathbb{E}\langle y_T, z_T\rangle_H = \mathbb{E}\langle y(T), z_T\rangle_H, \quad \forall\, z_T \in L^2_{\mathcal{F}_T}(\Omega; H),$$

which implies that $y(T) = y_T$, a.s. Hence, the control $u(\cdot)$ transfers the state of (7.6) from y_0 to y_T at time T.

Next, let us prove that $u(\cdot)$ is the control with the minimal $\mathcal{U}_T$-norm, which transfers the state of (7.6) from y_0 to y_T at time T. Choosing $z_T = \hat{z}_T$ in (7.65), we obtain that

$$\begin{aligned}
&\mathbb{E}\int_0^T \big|B^*z(t;\hat{z}_T) + D^*Z(t;\hat{z}_T)\big|^2_U dt\\
&= \mathbb{E}\langle y_T, \hat{z}_T\rangle_H - \big\langle y_0, z(0;\hat{z}_T)\big\rangle_H - \mathbb{E}\int_0^T \big\langle f(t), z(t;\hat{z}_T)\big\rangle_H dt.
\end{aligned} \tag{7.67}$$

Let $u_1(\cdot)$ be another control which transfers the state of (7.6) from y_0 to y_T at time T. Then, it follows from (7.66) that

$$\begin{aligned}&\mathbb{E}\langle y_T,\hat z_T\rangle_H-\big\langle y_0,z(0;\hat z_T)\big\rangle_H-\mathbb{E}\int_0^T\big\langle f(t),z(t;\hat z_T)\big\rangle_H dt\\&=\mathbb{E}\int_0^T\big\langle u_1(t),B^*z(t;\hat z_T)+D^*Z(t;\hat z_T)\big\rangle_U dt.\end{aligned}\tag{7.68}$$

Combining (7.67) and (7.68), we arrive at

$$\begin{aligned}&\mathbb{E}\int_0^T\big|B^*z(t;\hat z_T)+D^*Z(t;\hat z_T)\big|_U^2dt\\&=\mathbb{E}\int_0^T\big\langle u_1(t),B^*z(t;\hat z_T)+D^*Z(t;\hat z_T)\big\rangle_U dt\\&\le\big|B^*z(t;\hat z_T)+D^*Z(t;\hat z_T)\big|_{\mathcal{U}_T}\big|u_1(\cdot)\big|_{\mathcal{U}_T},\end{aligned}$$

which implies that $|u|_{\mathcal{U}_T}\le|u_1|_{\mathcal{U}_T}$. Hence, we complete the proof of Theorem 7.26. □

Remark 7.27. Theorem 7.26 provides a way to compute the control u (with minimal $\mathcal{U}_T$-norm) which transfers the state of the system (7.6) from y_0 to y_T at time T, by means of the solution to its dual equation (7.34) with the final datum given by the minimizer of the functional $\mathfrak{J}_1(\cdot)$ (with $\varepsilon=0$) on $L^2_{\mathcal{F}_T}(\Omega;H)$. In the deterministic cases, people use numerical methods, such as implementing gradient like iterative algorithms, to compute such sort of controls (e.g., [395]). However, in the present stochastic setting, one will meet much more difficulties when employing similar methods to compute the desired controls, and it seems that a careful study of these problems remains to be done.

Next, similarly to Theorem 7.26, for the null controllability of the system (7.6), we have the following result:

Theorem 7.28. *If the system* (7.34) *is continuously initially observable on* $[0,T]$, *then, among all controls transferring the state of* (7.6) *with* $f=0$ *from* y_0 *to* 0 *at time* T, *the one given by* (7.64) *enjoys the minimal* $\mathcal{U}_T$*-norm, where* $\hat z_T\in L^2_{\mathcal{F}_T}(\Omega;H)$ *is the minimizer of the functional* $\mathfrak{J}_1(\cdot)$ *with* $\varepsilon=0$ *and* $f=0$.

Further, similarly to Theorem 7.26, for the approximate controllability of the system (7.6), we have the following result:

Theorem 7.29. *If the system* (7.34) *is finally observable on* $[0,T]$, *then for any given* $\varepsilon>0$, *among all controls transferring the state of* (7.6) *from* y_0 *to the* ε *neighborhood of* y_T *at time* T, *i.e.,* $|y(T)-y_T|_{L^2_{\mathcal{F}_T}(\Omega;H)}<\varepsilon$, *the one given by* (7.64) *enjoys the minimal* $\mathcal{U}_T$*-norm, where* $\hat z_T\in L^2_{\mathcal{F}_T}(\Omega;H)$ *is the minimizer of the functional* $\mathfrak{J}_1(\cdot)$.

Proof: It is easy to see that the functional $\mathfrak{J}_1(\cdot):L^2_{\mathcal{F}_T}(\Omega;H)\to\mathbb{R}$ is continuous and strictly convex. We claim that it is coercive under the assumption that the system (7.34) is finally observable on $[0,T]$. Indeed, choose any sequence $\{z_{T,k}\}_{k=1}^{\infty} \subset L^2_{\mathcal{F}_T}(\Omega;H)$ such that

$$|z_{T,k}|_{L^2_{\mathcal{F}_T}(\Omega;H)} \to +\infty \text{ as } k \to +\infty.$$

For each $k \in \mathbb{N}$, put

$$\tilde{z}_{T,k} = \frac{z_{T,k}}{|z_{T,k}|_{L^2_{\mathcal{F}_T}(\Omega;H)}}.$$

Denote by $(z(\cdot;\tilde{z}_{T,k}), Z(\cdot;\tilde{z}_{T,k}))$ the corresponding solution to (7.34) with the final datum z_T replaced by $\tilde{z}_{T,k}$. Then,

$$\begin{aligned}
&\frac{\mathfrak{J}_1(z_{T,k})}{|z_{T,k}|_{L^2_{\mathcal{F}_T}(\Omega;H)}}\\
&= \frac{|z_{T,k}|_{L^2_{\mathcal{F}_T}(\Omega;H)}}{2}\mathbb{E}\int_0^T |B^*z(t;\tilde{z}_{T,k}) + D^*Z(t;\tilde{z}_{T,k})|_U^2 dt\\
&\quad -\mathbb{E}\langle y_T, \tilde{z}_{T,k}\rangle_H + \langle y_0, z(0;\tilde{z}_{T,k})\rangle_H + \mathbb{E}\int_0^T \langle f(t), z(t;\tilde{z}_{T,k})\rangle_H dt + \varepsilon \qquad (7.69)\\
&\geq \frac{|z_{T,k}|_{L^2_{\mathcal{F}_T}(\Omega;H)}}{2}\mathbb{E}\int_0^T |B^*z(t;\tilde{z}_{T,k}) + D^*Z(t;\tilde{z}_{T,k})|_U^2 dt\\
&\quad -\mathcal{C}\big(|y_T|_{L^2_{\mathcal{F}_T}(\Omega;H)} + |y_0|_H + |f|_{L^2_{\mathbb{F}}(\Omega;L^1(0,T;H))}\big) + \varepsilon.
\end{aligned}$$

If

$$\varliminf_{k\to+\infty} \mathbb{E}\int_0^T |B^*z(t;\tilde{z}_{T,k}) + D^*Z(t;\tilde{z}_{T,k})|_U^2 dt > 0, \qquad (7.70)$$

then it follows from (7.69) that

$$\mathfrak{J}_1(z_{T,k}) \to +\infty \quad \text{as} \quad |z_{T,k}|_{L^2_{\mathcal{F}_T}(\Omega;H)} \to +\infty. \qquad (7.71)$$

Hence, we only need to consider the case that the left hand side of (7.70) equals 0. Since $|\tilde{z}_{T,k}|_{L^2_{\mathcal{F}_T}(\Omega;H)} = 1$ for all $k \in \mathbb{N}$, there is a subsequence $\{\tilde{z}_{T,k_j}\}_{j=1}^{\infty}$ of $\{\tilde{z}_{T,k}\}_{k=1}^{\infty}$ and a $\tilde{z}_T \in L^2_{\mathcal{F}_T}(\Omega;H)$ such that $\tilde{z}_{T,k_j}$ converges weakly to $\tilde{z}_T$ in $L^2_{\mathcal{F}_T}(\Omega;H)$. Denote by $(z(\cdot;\tilde{z}_T), Z(\cdot;\tilde{z}_T))$ the solution to (7.34) with the final datum z_T replaced by $\tilde{z}_T$. By the definition of the weak solution to (7.34), it follows that the corresponding solution $z(\cdot;\tilde{z}_{T,k_j})$ converges weak-star to $z(\cdot;\tilde{z}_T)$ in $L^2_{\mathbb{F}}(\Omega;L^\infty(0,T;H))$. Then,

$$\begin{aligned}
&\mathbb{E}\int_0^T |B^*z(t;\tilde{z}_T) + D^*Z(t;\tilde{z}_T)|_U^2 dt\\
&\leq \varliminf_{k\to+\infty} \mathbb{E}\int_0^T |B^*z(t;\tilde{z}_{T,k}) + D^*Z(t;\tilde{z}_{T,k})|_U^2 dt = 0.
\end{aligned}$$

Since the system (7.34) is finally observable on $[0,T]$, we conclude that $\tilde{z}_T=0$, a.s. Thus, $z(\cdot;\tilde{z}_T)=0$. It follows from the first equality in (7.69) that

$$\varliminf_{k\to+\infty}\frac{\mathfrak{J}_1(\tilde{z}_{T,k})}{|\tilde{z}_{T,k}|_{L^2_{\mathcal{F}_T}(\Omega;H)}}\geq\varepsilon,$$

which gives also (7.71). Hence, we prove the coercivity of $\mathfrak{J}_1(\cdot)$. Then, $\mathfrak{J}_1(\cdot)$ admits a unique minimizer $\hat{z}_T$.

By computing the first variation of $\mathfrak{J}_1(\cdot)$, we get that

$$\begin{aligned}&\mathbb{E}\int_0^T\big\langle B^*z(t;\hat{z}_T)+D^*Z(t;\hat{z}_T),B^*z(t)+D^*Z(t)\big\rangle_U dt\\&=\mathbb{E}\langle y_T,z_T\rangle_H-\big\langle y_0,z(0)\big\rangle_H-\mathbb{E}\int_0^T\langle f(t),z(t)\rangle_H dt-\frac{\varepsilon\mathbb{E}\langle z_T,\hat{z}_T\rangle_H}{|\hat{z}_T|_{L^2_{\mathcal{F}_T}(\Omega;H)}},\\&\qquad\forall\, z_T\in L^2_{\mathcal{F}_T}(\Omega;H).\end{aligned}\tag{7.72}$$

On the other hand, we still have (7.66). Hence, by (7.72) and (7.66), if we choose the control as (7.64), then

$$\mathbb{E}\langle y_T-y(T),z_T\rangle_H=\frac{\varepsilon\mathbb{E}\langle z_T,\hat{z}_T\rangle_H}{|\hat{z}_T|_{L^2_{\mathcal{F}_T}(\Omega;H)}}\leq\varepsilon|z_T|_{L^2_{\mathcal{F}_T}(\Omega;H)},\quad\forall\, z_T\in L^2_{\mathcal{F}_T}(\Omega;H).\tag{7.73}$$

By taking $z_T=y_T-y(T)$, we find from (7.73) that

$$|y_T-y(T)|_{L^2_{\mathcal{F}_T}(\Omega;H)}\leq\varepsilon.$$

Now we prove that $u(\cdot)$ given by (7.64) is the control with the minimal $\mathcal{U}_T$-norm, which transfers the state of the system (7.6) from y_0 to the ε neighborhood of y_T at time T. By choosing $z_T=\hat{z}_T$ in (7.72), we find that

$$\begin{aligned}&\mathbb{E}\int_0^T\big|B^*z(t;\hat{z}_T)+D^*Z(t;\hat{z}_T)\big|_U^2dt\\&=\mathbb{E}\langle y_T,\hat{z}_T\rangle_H-\big\langle y_0,z(0;\hat{z}_T)\big\rangle_H-\mathbb{E}\int_0^T\langle f(t),z(t;\hat{z}_T)\rangle_H dt-\varepsilon|\hat{z}_T|_{L^2_{\mathcal{F}_T}(\Omega;H)}.\end{aligned}\tag{7.74}$$

Let $u_1(\cdot)$ be another control which transfers the state of the system (7.6) from y_0 to the ε neighborhood of y_T at time T, i.e., the corresponding solution $y(\cdot;u_1)$ to (7.6) verifies

$$|y(T;u_1)-y_T|_{L^2_{\mathcal{F}_T}(\Omega;H)}<\varepsilon.\tag{7.75}$$

Then, similarly to (7.68), we have

$$\begin{aligned}&\mathbb{E}\langle y(T;u_1),\hat{z}_T\rangle_H-\big\langle y_0,z(0;\hat{z}_T)\big\rangle_H-\mathbb{E}\int_0^T\big\langle f(t),z(t;\hat{z})\big\rangle_H dt\\&=\mathbb{E}\int_0^T\big\langle u_1(t),B^*z(t;\hat{z}_T)+D^*Z(t;\hat{z}_T)\big\rangle_U dt.\end{aligned}\tag{7.76}$$

It follows from (7.74)–(7.76) that

$$\begin{aligned}
&\mathbb{E}\int_0^T \big|B^*z(t;\hat{z}_T)+D^*Z(t;\hat{z}_T)\big|_U^2 dt\\
&\le \mathbb{E}\int_0^T \big\langle u_1(t), B^*z(t;\hat{z}_T)+D^*Z(t;\hat{z}_T)\big\rangle_U dt\\
&\quad +\mathbb{E}\langle y_T-y(T;u_1),\hat{z}_T\rangle_H-\varepsilon|\hat{z}_T|_{L^2_{\mathcal{F}_T}(\Omega;H)}\\
&\le \mathbb{E}\int_0^T \big\langle u_1(t), B^*z(t;\hat{z}_T)+D^*Z(t;\hat{z}_T)\big\rangle_U dt\\
&\le \big|B^*z(t;\hat{z}_T)+D^*Z(t;\hat{z}_T)\big|_{\mathcal{U}_T}\big|u_1(\cdot)\big|_{\mathcal{U}_T},
\end{aligned}$$

which implies $|u|_{\mathcal{U}_T}\le |u_1|_{\mathcal{U}_T}$. Therefore, we complete the proof of Theorem 7.29. □

Similarly to (7.64), we shall give the following explicit form of controls for exact/null/approxiate controllability of (7.7):

$$u(t)=B^*x(t;\hat{x}_0), \tag{7.77}$$

where $x(\cdot;\hat{x}_0)$ is the corresponding solution to (7.35) with the initial datum x_0 replaced by a suitable chosen $\hat{x}_0$, i.e., the minimizer of the functional $\mathfrak{J}_2$ (defined by (7.63)) under some assumptions.

Proceeding exactly as that for Theorems 7.26, 7.28 and 7.29, one can prove the following result for exact/null/approxiate controllability of (7.7) (and hence we will not give the proof).

Theorem 7.30. 1) *If the equation (7.35) is continuously initially observable on $[0,T]$, then, among all controls transferring the state of (7.7) from $y_T\in L^2_{\mathcal{F}_T}(\Omega;H)$ to $y_0\in H$ at time 0, the one given by (7.77) enjoys the minimal $\mathcal{U}_T$-norm, where $\hat{x}_0\in H$ is the minimizer of the functional $\mathfrak{J}_2(\cdot)$ with $\varepsilon=0$.*

2) *If the system (7.35) is continuously finally observable on $[0,T]$, then, among all controls transferring the state of (7.7) with $g=0$ from y_T to 0 at time 0, the one given by (7.77) enjoys the minimal $\mathcal{U}_T$-norm, where $\hat{x}_0\in H$ is the minimizer of the functional $\mathfrak{J}_2(\cdot)$ with $\varepsilon=0$ and $g=0$.*

3) *If the system (7.35) is initially observable on $[0,T]$, then for any given $\varepsilon>0$, among all controls transferring the state of (7.7) from y_T to the ε neighborhood of y_0 at time 0, i.e., $|y(0)-y_0|_H<\varepsilon$, the one given by (7.77) enjoys the minimal $\mathcal{U}_T$-norm, where $\hat{x}_0\in H$ is the minimizer of the functional $\mathfrak{J}_2(\cdot)$.*

Finally, for the case of unbounded control operators, using similar arguments as that in Theorems 7.26, 7.28, 7.29 and 7.30, one may obtain the explicit forms of controls for the controllability problems of the systems (7.8) and (7.9) (Hence, we will not give the details here).

7.5 Relationship Between the Forward and the Backward Controllability

In this section, we shall present a relationship between the forward and the backward controllability, i.e., that for the controllability of the forward and the backward stochastic evolution equations.

7.5.1 The Case of Bounded Control Operators

In this subsection, we assume that A generates a C_0-group on H, $F(\cdot)$, $f(\cdot)$, B and $G(\cdot)$ are the same as that in the system (7.6). Let $\mathbf{U}$ be another Hilbert space which will serve as another control space, $D \in \mathcal{L}(\mathbf{U}; \mathcal{L}_2^0)$ be invertible, $B_1 \in \mathcal{L}(U; \mathcal{L}_2^0)$ and $D_1 \in \mathcal{L}(\mathbf{U}; H)$. We consider the following (forward) stochastic control system evolved on H:

$$\begin{cases} dy(t) = \big(Ay(t) + F(t)y(t) + f(t) + Bu(t) + D_1 v(t)\big)dt \\ \qquad\qquad +\big(G(t)y(t) + Dv(t) + B_1 u(t)\big)dW(t) & \text{in } (0,T], \\ y(0) = y_0, \end{cases} \tag{7.78}$$

and the following backward stochastic control system evolved on H:

$$\begin{cases} dy(t) = \big(Ay(t) + F(t)y(t) + f(t) + Bu(t) + D_1 D^{-1} Y(t)\big)dt \\ \qquad\qquad +\big(G(t)y(t) + Y(t) + B_1 u(t)\big)dW(t) & \text{in } [0,T), \\ y(T) = y_T. \end{cases} \tag{7.79}$$

Here, $y(\cdot)$ is the state variable, $u(\cdot)(\in \mathcal{U}_T)$ and $v(\cdot)(\in L^2_{\mathbb{F}}(0,T;\mathbf{U}))$ are the control variables, $y_0(\in H)$ and $y_T(\in L^2_{\mathcal{F}_T}(\Omega; H))$ are respectively the initial and the final states. Since A is assumed to generate a C_0-group on H, the system (7.78) admits a unique mild solution $y(\cdot) \in L^2_{\mathbb{F}}(\Omega; C([0,T]; H))$, while the system (7.79) admits a unique mild solution $(y(\cdot), Y(\cdot)) \in L^2_{\mathbb{F}}(\Omega; C([0,T]; H)) \times L^2_{\mathbb{F}}(0,T; \mathcal{L}_2^0)$. In (7.78), we emphasize that the control $u(\cdot)$ (*resp.* $v(\cdot)$) in the drift (*resp.* diffusion) term affects the diffusion (*resp.* drift) term (Nevertheless, if we regard $(u(\cdot), v(\cdot))$ as "one" control, then (7.78) is still in the form of (7.6)). Stimulated by [275], we shall show below that the controllability problems for the forward stochastic control system (7.78) (with "two" controls $u(\cdot)$ and $v(\cdot)$) can be reduced to that for the backward stochastic control system (7.79) (with only one control $u(\cdot)$).

For the exact controllability, we have the following result.

Theorem 7.31. *The system* (7.78) *is exactly controllable at time* T *if and only if the system* (7.79) *is exactly controllable at time* 0.

Proof: We fix any $y_0 \in H$ and $y_T \in L^2_{\mathcal{F}_T}(\Omega; H)$.

The **"only if" part**. Since (7.78) is exactly controllable at time T, one can find a pair of controls $(u, v) \in \mathcal{U}_T \times L^2_{\mathbb{F}}(0,T;\mathbf{U})$ such that the solution to (7.78) satisfies that $y(T) = y_T$. Let $Y = Dv$. Then (y, Y) is the solution to (7.79) corresponding to the final datum y_T and the control $u \in \mathcal{U}_T$, and $y(0) = y_0$. This implies that (7.79) is exactly controllable at time 0.

The **"if" part**. Fix any $y_0 \in H$ and $y_T \in L^2_{\mathcal{F}_T}(\Omega; H)$. Since (7.79) is exactly controllable at time 0, one can find a control $u \in \mathcal{U}_T$ such that the solution $(y(\cdot), Y(\cdot)) \in L^2_{\mathbb{F}}(\Omega; C([0,T]; H)) \times L^2_{\mathbb{F}}(0,T; \mathcal{L}_2^0)$ to (7.79) satisfies that $y(0) = y_0$. Let $v = D^{-1}Y$. Then y is the solution to (7.78) corresponding to the initial datum y_0 and the controls $(u, v) \in \mathcal{U}_T \times L^2_{\mathbb{F}}(0,T; \mathbf{U})$, and $y(T) = y_T$. This implies that (7.6) is exactly controllable at time T. □

Next, we give the relationship for the null controllability of (7.6) and exact controllability (7.79).

Theorem 7.32. *The system* (7.78) *is null controllable at time* T *if and only if the system* (7.79) *is exactly controllable at time* 0.

Proof: By Theorem 7.31, it suffices to prove the **"only if" part**. Fix any $y_0 \in H$ and $y_T \in L^2_{\mathcal{F}_T}(\Omega; H)$. Denote by $(\tilde{y}, \widetilde{Y})$ the solution to (7.79) with the final datum y_T and the control $u(\cdot) = 0$. Since (7.78) is null controllable, one can find a pair of controls $(u, v) \in \mathcal{U}_T \times L^2_{\mathbb{F}}(0,T; \mathbf{U})$ such that the corresponding solution $\hat{y}$ to (7.78) with initial datum y_0 replaced by $y_0 - \tilde{y}(0)$ satisfies $\hat{y}(T) = 0$. Then $(y, Y) = (\hat{y} + \tilde{y}, Dv + \widetilde{Y})$ is the solution to (7.79) corresponding to the final datum y_T and the control $u(\cdot) \in \mathcal{U}_T$ obtained above for (7.78), and satisfies $y(0) = y_0$. This implies that (7.79) is exactly controllable at time 0. □

Now, as an immediate consequence of Theorem 7.32, we have the following result:

Corollary 7.33. *The system* (7.78) *is null controllable at time* T *if and only if it is exactly controllable at time* T.

In view of Theorem 7.31, similarly to Theorem 7.23, the exact controllability problem of (7.78) can be reduced to a suitable observability estimate for the following stochastic evolution equation on H:

$$\begin{cases} dx = -\big(A^* + F^* - G^*(D^{-1})^* D_1^*\big)xdt - (D^{-1})^* D_1^* x dW(t) & \text{in } (0,T], \\ x(0) = x_0. \end{cases} \tag{7.80}$$

More precisely, we have the following result:

Theorem 7.34. *The system* (7.78) *is exactly controllable at time* T *if and only if all solutions to* (7.80) *satisfy the following observability estimate:*

$$|x_0|_H^2 \le C\mathbb{E}\int_0^T |(B - D_1 D^{-1} B_1)^* x(t)|_U^2 dt, \quad \forall\, x_0 \in H.$$

It follows from Theorem 7.34 that when $D_1 = 0$, the exact controllability of (7.78) can be reduced to the observability estimate for a random evolution equation. Indeed, let us consider the following random evolution equation on H:

$$\begin{cases} \dfrac{dx}{dt} = -(A^* + F^*)x \ \text{ in } (0, T], \\ x(0) = x_0. \end{cases} \tag{7.81}$$

As a special case of Theorem 7.34, we have the following result:

Theorem 7.35. *If $D_1 = 0$, then the following statements are equivalent:*

1) *The system (7.78) is exactly controllable at time T;*
2) *All solutions to (7.81) satisfy the following observability estimate:*

$$|x_0|_H^2 \leq C\mathbb{E} \int_0^T |B^* x(t)|_U^2 dt, \quad \forall\, x_0 \in H. \tag{7.82}$$

Remark 7.36. The equation (7.81) can be regarded as an evolution equation with a parameter ω. Since $F \in L^\infty_{\mathbb{F}}(0, T; \mathcal{L}(H))$, sometimes (7.82) can be obtained by the existing observability result for deterministic evolution equation immediately.

Remark 7.37. Since D is invertible, one may expect that the randomness of the system (7.78) can be eliminated simply by taking

$$v(\cdot) = -D^{-1}(G(\cdot)y(\cdot) + B_1 u(\cdot)),$$

and hence the system is reduced to the following controlled random evolution equation:

$$\begin{cases} \dfrac{dy(t)}{dt} = Ay(t) + \big(F(t) - D_1 D^{-1} G(\cdot)\big)y(t) + f(t) \\ \qquad\qquad +\big(B - D_1 D^{-1} B_1\big)u(t) \qquad \text{in } (0, T], \\ y(0) = y_0, \end{cases} \tag{7.83}$$

However, this does not help us to achieve our goal to find a control to transfer the state of the system (7.83) (or equivalently (7.78)) from a given y_0 to any given element in $L^2_{\mathcal{F}_T}(\Omega; H)$ at time T. Indeed, the dual equation of (7.83) is still a backward stochastic evolution equation given below:

$$\begin{cases} dz(t) = -\big[A^* z(t) + \big(F(t) - D_1 D^{-1} G(\cdot)\big)^* z(t)\big]dt + Z(t)dW(t) \ \text{ in } [0, T), \\ z(T) = z_T. \end{cases} \tag{7.84}$$

By Theorem 7.16, the system (7.83) (and hence (7.78)) is exactly controllable at time T if and only if solutions to (7.84) satisfy the following observability estimate:

$$|z_T|_{L^2_{\mathcal{F}_T}(\Omega;H)} \leq \mathcal{C}\big|\big(B - D_1 D^{-1} B_1\big)^* z\big|_{\mathcal{U}_T}, \quad \forall\, z_T \in L^2_{\mathcal{F}_T}(\Omega; H). \tag{7.85}$$

As far as we know, generally speaking, the observability estimate (7.85) cannot be proved directly by the existing controllability/observability results for deterministic evolution equations, even if in the special case that $D_1 = 0$.

Remark 7.38. We have proved that the system (7.78) is null controllable if and only if it is exactly controllable, provided that A generates a C_0-group and D is invertible. For deterministic systems, the condition that A generates a C_0-group is enough to establish this equivalence. However, according to Proposition 6.3, it is easy to see that, in the stochastic context the condition that D is invertible cannot be dropped.

Next, we give the relationship for the approximate controllability of (7.78) and (7.79).

Theorem 7.39. *The system* (7.78) *is approximately controllable at time* T *if and only if the system* (7.79) *is approximately controllable at time* 0.

Proof: We fix any $\varepsilon > 0$, $y_0 \in H$ and $y_T \in L^2_{\mathcal{F}_T}(\Omega; H)$.

The **"only if" part**. Since (7.78) is approximately controllable, for any $\varepsilon_1 > 0$, one can find a pair of controls $(u^1, v^1) \in \mathcal{U}_T \times L^2_{\mathbb{F}}(0,T;\mathbf{U})$ such that the corresponding mild solution $y^1(\cdot) \in L^2_{\mathbb{F}}(\Omega; C([0,T]; H))$ to (7.78) satisfies that $y^1(0) = y_0$ and $|y^1(T) - y_T|_{L^2_{\mathcal{F}_T}(\Omega;H)} < \varepsilon_1$. Denote by (y, Y) the corresponding solution to (7.79) with the control $u(\cdot)$ replaced by $u^1(\cdot)$. Then $(\hat{y}, \widehat{Y}) \overset{\Delta}{=} (y, Y) - (y^1, Dv^1)$ solves the following backward stochastic evolution equation:

$$\begin{cases} d\hat{y}(t) = \big(A\hat{y}(t) + F(t)\hat{y}(t) + D_1 D^{-1}\widehat{Y}(t)\big)dt \\ \qquad\qquad + \big(G(t)\hat{y}(t) + \widehat{Y}(t)\big)dW(t) & \text{in } [0,T), \\ \hat{y}(T) = y_T - y^1(T). \end{cases} \tag{7.86}$$

By Theorem 4.55, it follows that

$$|\hat{y}(0)|_H \leq \mathcal{C}|y^1(T) - y_T|_{L^2_{\mathcal{F}_T}(\Omega;H)} < \mathcal{C}\varepsilon_1.$$

Let us choose $\varepsilon_1 = \frac{\varepsilon}{\mathcal{C}}$. Then,

$$|y(0) - y_0|_H = |y(0) - y^1(0)|_H = |\hat{y}(0)|_H < \mathcal{C}\varepsilon_1 = \varepsilon.$$

This implies that (7.79) is approximately controllable.

The **"if" part**. Since (7.79) is approximately controllable, for any $\varepsilon^1 > 0$, one can find a control $u^1 \in \mathcal{U}_T$ such that the corresponding mild solution $(y^1(\cdot), Y^1(\cdot)) \in L^2_{\mathbb{F}}(\Omega; C([0,T]; H)) \times L^2_{\mathbb{F}}(0,T; \mathcal{L}_2^0)$ to (7.79) with the control $u(\cdot)$ replaced by $u^1(\cdot)$ satisfies $y^1(T) = y_T$ and $|y^1(0) - y_0|_H < \varepsilon_1$. Denote by

y the corresponding solution to (7.78) with the controls $(u(\cdot), v(\cdot))$ replaced by $(u^1(\cdot), D^{-1}Y^1(\cdot))$. Then $\hat{y} \overset{\Delta}{=} y - y^1$ solves the following equation:

$$\begin{cases} d\hat{y}(t) = \big(A\hat{y}(t) + F(t)\hat{y}(t)\big)dt + G(t)\hat{y}(t)dW(t) \ \text{ in } (0,T], \\ \hat{y}(0) = y_0 - y^1(0). \end{cases} \tag{7.87}$$

By Theorem 3.14, we have

$$|\hat{y}(T)|_{L^2_{\mathcal{F}_T}(\Omega;H)} \le \mathcal{C}|y^1(0) - y_0|_H < \mathcal{C}\varepsilon_1.$$

Let us choose $\varepsilon_1 = \frac{\varepsilon}{\mathcal{C}}$. Then,

$$|y(T) - y_T|_{L^2_{\mathcal{F}_T}(\Omega;H)} = |\hat{y}(T)|_{L^2_{\mathcal{F}_T}(\Omega;H)} < \mathcal{C}\varepsilon_1 = \varepsilon.$$

Therefore, y is the solution to (7.78) so that $|y(T) - y_T|_{L^2_{\mathcal{F}_T}(\Omega;H)} < \varepsilon$. This implies that (7.78) is approximately controllable. □

7.5.2 The Case of Unbounded Control Operators

In this subsection, O, O', $\mathcal{F}(\cdot)$, $\mathbf{f}(\cdot)$, $\mathcal{B}$ and $\mathcal{G}(\cdot)$ are the same as that for the system (7.8), and we assume that A generates a C_0-group on both O and O'. Let $\mathbf{U}$ be the same control space as that in the Subsection 7.5.1, $\mathcal{D} \in \mathcal{L}(\mathbf{U}; \mathcal{L}_2(V; O'))$ be invertible, $\mathcal{B}_1 \in \mathcal{L}(U; \mathcal{L}_2(V; O'))$ and $\mathcal{D}_1 \in \mathcal{L}(\mathbf{U}; O')$. We consider the following (forward) stochastic control system evolved on O':

$$\begin{cases} dy(t) = \big(Ay(t) + \mathcal{F}(t)y(t) + \mathbf{f}(t) + \mathcal{B}u(t) + \mathcal{D}_1 v(t)\big)dt \\ \qquad\qquad + \big(\mathcal{G}(t)y(t) + \mathcal{D}v(t) + \mathcal{B}_1 u(t)\big)dW(t) \qquad \text{in } (0,T], \\ y(0) = y_0, \end{cases} \tag{7.88}$$

and the following backward stochastic control system evolved on O':

$$\begin{cases} dy(t) = \big(Ay(t) + \mathcal{F}(t)y(t) + \mathbf{f}(t) + \mathcal{B}u(t) + \mathcal{D}_1\mathcal{D}^{-1}Y(t)\big)dt \\ \qquad\qquad + \big(\mathcal{G}(t)y(t) + Y(t) + \mathcal{B}_1 u(t)\big)dW(t) \qquad \text{in } [0,T), \\ y(T) = y_T. \end{cases} \tag{7.89}$$

Here, $y(\cdot)$ is the state variable, $u(\cdot)(\in \mathcal{U}_T)$ and $v(\cdot)(\in L^2_{\mathbb{F}}(0,T;\mathbf{U}))$ are the control variables, $y_0(\in O')$ and $y_T(\in L^2_{\mathcal{F}_T}(\Omega; O'))$ are respectively the initial and the final states.

Note that in both (7.88) and (7.89) there is only one unbounded control operator, i.e., $\mathcal{B}$. For this unbounded control operator, we need similar assumptions as that in Conditions 7.1, 7.2 and 7.3. More precisely, in this subsection we assume that

- There exists a sequence $\{u_n\}_{n=1}^{\infty} \subset \mathcal{U}_T$ such that $\mathcal{B}u_n \in L^2_{\mathbb{F}}(0,T;O')$ for each $n \in \mathbb{N}$ and (7.16) holds;

- There exists a constant $\mathcal{C} > 0$ such that for any $\tau \in [0,T]$, $\xi_\tau \in L^2_{\mathcal{F}_\tau}(\Omega;O)$, $\eta_\tau \in L^2_{\mathcal{F}_\tau}(\Omega;O)$ and $h \in L^2_{\mathbb{F}}(0,T;\mathcal{L}_2(V;O))$, the solution (ξ,Ξ) to (7.14) satisfies (7.18), and the solution η to the following equation

$$
\begin{cases}
d\eta(t) = -\big[A^*\eta(t) + \big(\mathcal{F}(t) - \mathcal{D}_1\mathcal{D}^{-1}\mathcal{G}(t)\big)^*\eta(t) + \mathcal{G}(t)^*h(t)\big]dt \\
\qquad\qquad +\big(-(\mathcal{D}^{-1})^*\mathcal{D}_1^*\eta(t) + h(t)\big)dW(t) \qquad \text{in } (\tau,T], \\
\eta(\tau) = \eta_\tau
\end{cases}
\tag{7.90}
$$

satisfies the inequality (7.27).

Under these assumptions, similarly to Definition 7.11, a process $y \in C_{\mathbb{F}}([0,T]; L^2(\Omega;O'))$ is called a transposition solution to the system (7.88) if for any $\tau \in [0,T]$ and $\xi_\tau \in L^2_{\mathcal{F}_\tau}(\Omega;O)$, it holds that

$$
\begin{aligned}
&\mathbb{E}\big\langle y(\tau),\xi_\tau\big\rangle_{O',O} - \big\langle y_0,\xi(0)\big\rangle_{O',O} \\
&= \mathbb{E}\int_0^\tau \big\langle u(t),\mathcal{B}^*\xi(t)\big\rangle_U dt + \mathbb{E}\int_0^\tau \big\langle \mathbf{f}(t) + \mathcal{D}_1 v(t),\xi(t)\big\rangle_{O',O}dt \\
&\quad +\mathbb{E}\int_0^\tau \big\langle \mathcal{D}v(t) + \mathcal{B}_1 u(t),\Xi(t)\big\rangle_{\mathcal{L}_2(V;O'),\mathcal{L}_2(V;O)}dt,
\end{aligned}
\tag{7.91}
$$

where (ξ,Ξ) solves (7.14). Also, similarly to Definition 7.13, a pair of processes $(y,Y) \in C_{\mathbb{F}}([0,T];L^2(\Omega;O')) \times L^2_{\mathbb{F}}(0,T;\ \mathcal{L}_2(V;O'))$ is called a transposition solution to the system (7.89) if for every $\tau \in [0,T)$, $\eta_\tau \in L^2_{\mathcal{F}_\tau}(\Omega;O)$ and $h \in L^2_{\mathbb{F}}(0,T;\mathcal{L}_2(V;O))$, it holds that

$$
\begin{aligned}
&\mathbb{E}\big\langle y_T,\eta(T)\big\rangle_{O',O} - \big\langle y(\tau),\eta_\tau\big\rangle_{O',O} \\
&= \mathbb{E}\int_\tau^T \big\langle u(t),\mathcal{B}^*\eta(t)\big\rangle_U dt + \mathbb{E}\int_\tau^T \big\langle \mathbf{f}(t) - \mathcal{D}_1\mathcal{D}^{-1}\mathcal{B}_1 u(t),\eta(t)\big\rangle_{O',O}dt \\
&\quad +\mathbb{E}\int_\tau^T \big\langle Y(t) + \mathcal{B}_1 u(t),h(t)\big\rangle_{\mathcal{L}_2(V;O'),\mathcal{L}_2(V;O)}dt,
\end{aligned}
\tag{7.92}
$$

where η solves (7.90). Proceeding exactly as that in Theorems 7.12 and 7.14, we can show that the systems (7.88) and (7.89) are well-posed in the sense of transposition solutions.

Similarly to Theorem 7.31, we have the following result.

Theorem 7.40. *The system* (7.88) *is exactly controllable at time* T *if and only if the system* (7.89) *is exactly controllable at time* 0.

Proof: We only prove the **"if" part** (The **"only if" part** can be proved similarly). Let $y_0 \in O'$ and $y_T \in L^2_{\mathcal{F}_T}(\Omega;O')$ be arbitrarily given. Since (7.89) is exactly controllable at time 0, one can find a control $u \in \mathcal{U}_T$ such that the transposition solution $(y,Y) \in C_{\mathbb{F}}([0,T];L^2(\Omega;O')) \times L^2_{\mathbb{F}}(0,T;\ \mathcal{L}_2(V;O'))$

to (7.89) satisfies that $y(0)=y_0$, and by (7.92), for any $\tau'\in[0,T]$, $\eta_{\tau'}\in L^2_{\mathcal{F}_{\tau'}}(\Omega;O)$ and $h\in L^2_{\mathbb{F}}(0,T;\mathcal{L}_2(V;O))$, it holds that

$$
\begin{aligned}
&\mathbb{E}\big\langle y_T,\eta(T)\big\rangle_{O',O}-\big\langle y(\tau'),\eta_{\tau'}\big\rangle_{O',O}\\
&=\mathbb{E}\int_{\tau'}^T\big\langle u(t),\mathcal{B}^*\eta(t)\big\rangle_U dt+\mathbb{E}\int_{\tau'}^T\big\langle \mathbf{f}(t)-\mathcal{D}_1\mathcal{D}^{-1}\mathcal{B}_1u(t),\eta(t)\big\rangle_{O',O}dt\\
&\quad+\mathbb{E}\int_{\tau'}^T\big\langle Y(t)+\mathcal{B}_1u(t),h(t)\big\rangle_{\mathcal{L}_2(V;O'),\mathcal{L}_2(V;O)}dt,
\end{aligned}
\tag{7.93}
$$

where η solves (7.90) with τ replaced by τ'.

We claim that the same y as above is the transposition solution to (7.88) with the above u and $v=D^{-1}Y$. Indeed, we only need to show that (7.91) holds for any $\tau\in[0,T]$ and $\xi_\tau\in L^2_{\mathcal{F}_\tau}(\Omega;O)$, and the corresponding mild solution (ξ,Ξ) to (7.14). For this purpose, first by (7.93) with $\tau'\in[0,T)$,

$$
\xi(t)=\eta(t),\quad \Xi(t)=-(\mathcal{D}^{-1})^*\mathcal{D}_1^*\eta(t)+h(t),\qquad \text{a.e. } t\in(\tau',T),
$$

where (ξ,Ξ) solves (7.14) with $\tau=T$ and $\xi_T\in L^2_{\mathcal{F}_T}(\Omega;O)$ given arbitrarily, we obtain that

$$
\begin{aligned}
&\mathbb{E}\big\langle y_T,\xi_T\big\rangle_{O',O}-\big\langle y(\tau'),\xi(\tau')\big\rangle_{O',O}\\
&=\mathbb{E}\int_{\tau'}^T\big\langle u(t),\mathcal{B}^*\xi(t)\big\rangle_U dt+\mathbb{E}\int_{\tau'}^T\big\langle \mathbf{f}(t)-\mathcal{D}_1\mathcal{D}^{-1}\mathcal{B}_1u(t),\xi(t)\big\rangle_{O',O}dt\\
&\quad+\mathbb{E}\int_{\tau'}^T\big\langle Y(t)+\mathcal{B}_1u(t),\Xi(t)+(\mathcal{D}^{-1})^*\mathcal{D}_1^*\xi(t)\big\rangle_{\mathcal{L}_2(V;O'),\mathcal{L}_2(V;O)}dt\\
&=\mathbb{E}\int_{\tau'}^T\big\langle u(t),\mathcal{B}^*\xi(t)\big\rangle_U dt+\mathbb{E}\int_{\tau'}^T\big\langle \mathbf{f}(t)+\mathcal{D}_1v(t),\xi(t)\big\rangle_{O',O}dt\\
&\quad+\mathbb{E}\int_{\tau'}^T\big\langle \mathcal{D}v(t)+\mathcal{B}_1u(t),\Xi(t)\big\rangle_{\mathcal{L}_2(V;O'),\mathcal{L}_2(V;O)}dt.
\end{aligned}
\tag{7.94}
$$

Particularly, by (7.94) and recalling that $y(0)=y_0$, we have

$$
\begin{aligned}
&\mathbb{E}\big\langle y_T,\xi_T\big\rangle_{O',O}-\big\langle y_0,\xi(0)\big\rangle_{O',O}\\
&=\mathbb{E}\int_0^T\big\langle u(t),\mathcal{B}^*\xi(t)\big\rangle_U dt+\mathbb{E}\int_0^T\big\langle \mathbf{f}(t)+\mathcal{D}_1v(t),\xi(t)\big\rangle_{O',O}dt\\
&\quad+\mathbb{E}\int_0^T\big\langle \mathcal{D}v(t)+\mathcal{B}_1u(t),\Xi(t)\big\rangle_{\mathcal{L}_2(V;O'),\mathcal{L}_2(V;O)}dt.
\end{aligned}
\tag{7.95}
$$

Combining (7.94) and (7.95), we get

$$
\begin{aligned}
&\mathbb{E}\big\langle y(\tau'),\xi(\tau')\big\rangle_{O',O}-\big\langle y_0,\xi(0)\big\rangle_{O',O}\\
&=\mathbb{E}\int_0^{\tau'}\big\langle u(t),\mathcal{B}^*\xi(t)\big\rangle_U dt+\mathbb{E}\int_0^{\tau'}\big\langle \mathbf{f}(t)+\mathcal{D}_1 v(t),\xi(t)\big\rangle_{O',O}dt \qquad (7.96)\\
&\quad+\mathbb{E}\int_0^{\tau'}\big\langle \mathcal{D}v(t)+\mathcal{B}_1 u(t),\Xi(t)\big\rangle_{\mathcal{L}_2(V;O'),\mathcal{L}_2(V;O)}dt.
\end{aligned}
$$

Since $\xi_T \in L^2_{\mathcal{F}_T}(\Omega;O)$ is given arbitrarily, by (7.96), we obtain (7.91).

Noting that $y(0)=y_0$ and $y(T)=y_T$, we conclude the exact controllability of (7.88) at time T. □

Similarly to Theorem 7.32, Corollary 7.33 and Theorem 7.39, we can prove following result.

Theorem 7.41. 1) *The system* (7.88) *is null controllable at time* T *if and only if the system* (7.89) *is exactly controllable at time* 0;

2) *The system* (7.88) *is null controllable at time* T *if and only if it is exactly controllable at time* T;

3) *The system* (7.88) *is approximately controllable at time* T *if and only if the system* (7.89) *is approximately controllable at time* 0.

Also, the exact controllability problem of (7.88) can be reduced to a suitable observability estimate for the following stochastic evolution equation on O:

$$
\begin{cases}
dx=-\big(A^*+\mathcal{F}^*-\mathcal{G}^*(\mathcal{D}^{-1})^*\mathcal{D}_1^*\big)xdt-(\mathcal{D}^{-1})^*\mathcal{D}_1^*xdW(t) & \text{in } (0,T],\\
x(0)=x_0.
\end{cases}
\qquad (7.97)
$$

More precisely, similarly to Theorem 7.34, we have the following result:

Theorem 7.42. *The system* (7.88) *is exactly controllable at time* T *if and only if all solutions to* (7.97) *satisfy the following observability estimate:*

$$
|x_0|_O^2\le C\mathbb{E}\int_0^T|(\mathcal{B}-\mathcal{D}_1\mathcal{D}^{-1}\mathcal{B}_1)^*x(t)|_U^2dt,\quad \forall\, x_0\in O.
$$

Remark 7.43. Clearly, similarly to Theorem 7.35, if $\mathcal{D}_1=0$, then, by Theorem 7.42, the exact controllability of (7.88) at time T is equivalent to that all solutions of the following random evolution equation on O:

$$
\begin{cases}
\dfrac{dx}{dt}=-(A^*+\mathcal{F}^*)x & \text{in } (0,T],\\
x(0)=x_0
\end{cases}
$$

satisfy the observability estimate:

$$
|x_0|_O^2\le C\mathbb{E}\int_0^T|\mathcal{B}^*x(t)|_U^2dt,\quad \forall\, x_0\in O.
$$

7.6 Notes and Comments

Since the seminal paper [163], controllability and observability became the basis of Control Theory. Controllability and observability problems for deterministic systems attracted many studies. It is impossible for us to list them comprehensively even though we restrict ourselves to systems governed by abstract evolution systems. Interested readers are referred to [8, 67, 325] and the rich references therein.

The relationships between controllability and observability for deterministic linear control system are well-known, which can be established by several differential methods (e.g. [123, 202, 207]). Similar relationship in the stochastic setting has been studied in [40, 125, 126, 303].

The main concern in this chapter is to reduce the controllability problems for both forward and backward stochastic evolution equations to the observability or observability estimates of their dual equations, particularly for the case of unbounded control operators. To do this, as in [40, 125, 126, 202, 303], we use the classical duality argument. Note that, for concrete stochastic partial differential equations, it is far from easy to establish the desired observability or observability estimates. Actually, in the next four chapters, the main task is to derive the needed observability estimates for stochastic transport equations, stochastic parabolic equations, stochastic wave equations and stochastic Schrödinger equations, respectively.

In deterministic setting, the variational characterization on the forms of controls for the controllability problems provides a useful way to compute the desired controls numerically (e.g. [123]). It would be quite interesting to extend these results to the stochastic setting. Nevertheless, as far as we know, there is no published result for the stochastic counterpart, especially on the numerical solutions to the stochastic controllability problems, even for that of stochastic evolution equations in finite dimensions.

There are several other concepts of controllability for stochastic evolution equations (e.g. [51, 75, 309, 310, 376]). Among them, we mention the so-called ε-controllability, which will be recalled below. For any $\varepsilon > 0$, $p \in [0,1]$ and $y_0 \in H$, put

$$S(\varepsilon, p; y_0) \stackrel{\Delta}{=} \big\{y_T \in H \bigm| \mathbb{P}(|y(T) - y_T|_H^2 > \varepsilon) \leq 1 - p \text{ for some } u \in \mathcal{U}_T\big\},$$

where $y(\cdot)$ solves (7.6). Following [309], we introduce the following notion of controllability:

Definition 7.44. *The system* (7.6) *is called completely ε-controllable in probability p (in the normed square sense), if $S(\varepsilon, p; y_0) = H$ for any $y_0 \in H$.*

Rough speaking, if the system (7.6) is completely ε-controllable, then it can transfer any given initial state y_0 into the ε-neighborhood of an arbitrary point in H with probability not less than p. This kind of controllability and some generalizations have been studied by several authors in the case that $D = 0$,

$G = 0$ and F is deterministic (in the system (7.6)) (e.g., [16, 17, 253, 254, 255]). In [17], a stronger controllability concept was introduced for the system (7.6), that is, not only $S(\varepsilon, p; y_0) = H$ for any $y_0 \in H$, but also $\mathcal{S}(\varepsilon, p; y_0) = H$ for any $y_0 \in H$, where

$$\mathcal{S}(\varepsilon, p; y_0) \stackrel{\Delta}{=} \big\{ y_T \in H \mid \mathbb{P}(|y(T) - y_T|_H^2 > \varepsilon) \le 1 - p \text{ and } \\ \mathbb{E}y(T) = y_T \text{ for some } u \in \mathcal{U}_T \big\},$$

and $y(\cdot)$ solves (7.6). The above assumptions on D, F and G are essential for the method used in [16, 17, 253, 254, 255], which reduces the ε-controllability problem of a stochastic control system to the exact/approximate controllability of a deterministic control system. To our best knowledge, there is no published works on the ε-controllability problem for general stochastic control systems.

Note that, unlike our definitions of controllability, when the stochastic control system reduces to a deterministic one, the above complete ε-controllability does not coincide with the usual notion of exact controllability. Because of this, it seems that our definitions are more physically meaningful.

In Section 4.3 of Chapter 4, we have introduced the notion of transposition solution for backward stochastic evolution equations with general filtration, to overcome the difficulty that there is no Martingale Representation Theorem. In Section 7.2, we also use the notion of transposition solution but the purpose is to overcome the difficulty introduced by the unbounded control operators. Note however that the idea to solve these two different problems is very similar, that is, to study the well-posedness of a "less-understood" equation by means of another "well-understood" one. To keep clarity, it seems that we would better use two different notions for them but we prefer keeping the present presentation because the readers should be able to detect the differences easily from the context. Note that, in this chapter, we assume that the filtration is natural. It would be interesting to consider the same problems (studied in this chapter) but in general filtration, for which people need to use the notion of transposition solutions to backward stochastic evolution equations (7.7), (7.9), (7.14), (7.34), (7.51), (7.79) and (7.89) (especially for (7.9) and (7.89) with unbounded control operators) but this remains to be done.

In many concrete control systems, it is very common that both the control and the observation operators are unbounded w.r.t. the state spaces. Typical examples are systems governed by partial differential equations in which the actuators and the sensors act on lower-dimensional hypersurfaces or on the boundary of a spatial domain. The unboundedness of the control and the observation operators leads to substantial technical difficulties even for the formulation of the state spaces. In the deterministic setting, to overcome this sort of difficulties, people introduced the notions of the admissible control operator and the admissible observation operator (e.g. [298]). On the other hand, people introduced the notion of well-posed linear system (with possibly unbounded control and observation operators), which satisfies, roughly speaking, that the map from the input space to the output one is bounded (e.g. [298, 305]). The well-posed linear systems form a very general class whose basic properties (such as that for feedback control, dynamic stabilization, tracking/disturbance rejection and so on) are rich enough to develop a parallel theory as that for control systems with bounded control and observation operators. The concept of well-posed linear system was generalized to the stochastic setting in [232] but it seems that there are many things to be done to develop a satisfactory theory for stochastic well-posed linear systems.

8

Exact Controllability for Stochastic Transport Equations

In this chapter, we are concerned with the exact boundary controllability for stochastic transport equations. By the duality argument, the controllability problem is reduced to a suitable observability estimate for backward stochastic transport equations, and we employ a stochastic version of global Carleman estimate to derive such an estimate.

8.1 Formulation of the Problem and the Main Result

In this chapter and the next three ones, we shall apply our general controllability results (for stochastic evolution equations) presented in the last chapter to some typical and concrete stochastic partial differential equations. As we shall see later, this is by no means an easy task. Indeed, general speaking, it is usually highly nontrivial to establish the needed observability estimates for the dual equations. In some sense, the transport equation is one of the simplest partial differential equations, and therefore in this chapter we shall focus on the exact controllability for stochastic transport equations.

Let $T > 0$ and $(\Omega, \mathcal{F}, \mathbf{F}, \mathbb{P})$ be a complete filtered probability space on which a one dimensional Brownian motion $W(\cdot)$ is defined and $\mathbf{F} = \{\mathcal{F}_t\}_{t\in[0,T]}$ is the natural filtration generated by $W(\cdot)$. Denote by $\mathbb{F}$ the progressive σ-field (in $[0,T] \times \Omega$) w.r.t. $\mathbf{F}$.

Let $G \subset \mathbb{R}^n$ ($n \in \mathbb{N}$) be a bounded domain with a C^1 boundary Γ. Put

$$Q \equiv (0,T) \times G, \quad \Sigma \equiv (0,T) \times \Gamma, \quad R = \max_{x\in\Gamma} |x|_{\mathbb{R}^n}.$$

Fix any $O \in \mathbb{R}^n$ satisfying $|O|_{\mathbb{R}^n} = 1$. Set

$$\begin{aligned} &\Gamma^- \overset{\Delta}{=} \big\{x \in \Gamma \bigm| O \cdot \nu(x) < 0\big\}, \qquad &&\Gamma^+ \overset{\Delta}{=} \Gamma \setminus \Gamma^-,\\ &\Sigma^- \overset{\Delta}{=} (0,T) \times \Gamma^-, &&\Sigma^+ \overset{\Delta}{=} (0,T) \times \Gamma^+. \end{aligned}$$

Q. Lü and X. Zhang, *Mathematical Control Theory for Stochastic Partial Differential Equations*, Probability Theory and Stochastic Modelling 101, https://doi.org/10.1007/978-3-030-82331-3_8

Define a Hilbert space $L^2_{\mathcal{O}}(\Gamma^-)$ as the completion of all $h \in C_0^\infty(\Gamma^-)$ w.r.t. the following norm

$$|h|_{L^2_{\mathcal{O}}(\Gamma^-)} \triangleq \Big(- \int_{\Gamma^-} \mathcal{O} \cdot \nu |h|^2 d\Gamma \Big)^{\frac{1}{2}}.$$

Clearly, the usual $L^2(\Gamma^-)$ is dense in $L^2_{\mathcal{O}}(\Gamma^-)$.

Let us recall first the controllability problem for the following deterministic transport equation:

$$\begin{cases} y_t + \mathcal{O} \cdot \nabla y = a_1^0 y & \text{in } Q, \\ y = u & \text{on } \Sigma^-, \\ y(0) = y_0 & \text{in } G, \end{cases} \tag{8.1}$$

where $a_1^0 \in L^\infty(0,T; L^\infty(G))$. In (8.1), y is the *state variable* and u is the *control variable*. The *state* and *control spaces* are chosen to be $L^2(G)$ and $L^2(0,T; L^2_{\mathcal{O}}(\Gamma^-))$, respectively. It is easy to show that, for any $y_0 \in L^2(G)$ and $u \in L^2(0,T; L^2_{\mathcal{O}}(\Gamma_-))$, (8.1) admits one and one (transposition) solution $y(\cdot) \in C([0,T]; L^2(G))$.

The system (8.1) is said to be exactly controllable at time T if for any given $y_0, y_1 \in L^2(G)$, one can find a control $u \in L^2(0,T; L^2_{\mathcal{O}}(\Gamma^-))$ such that the corresponding solution $y(\cdot) \in C([0,T]; L^2(G))$ to (8.1) satisfies $y(T) = y_1$.

In order to solve the above exact controllability of (8.1), people introduce the following dual equation:

$$\begin{cases} z_t + \mathcal{O} \cdot \nabla z = -a_1^0 z & \text{in } Q, \\ z = 0 & \text{on } \Sigma^+, \\ z(T) = z_T & \text{in } G. \end{cases} \tag{8.2}$$

By means of the classical duality argument (similarly to the conclusion 1) in Theorems 7.1 and 7.24), it is easy to show the following result.

Proposition 8.1. *The equation* (8.1) *is exactly controllable at time* T *if and only if solutions to the equation* (8.2) *satisfy the following observability estimate:*

$$|z_T|_{L^2(G)} \le \mathcal{C} |z|_{L^2_{\mathcal{O}}(\Sigma^-)}, \qquad \forall\, z_T \in L^2(G). \tag{8.3}$$

Controllability and observability problems for deterministic transposition equations are now well-understood. Indeed, one can use the global Carleman estimate to prove the observability inequality (8.3) for $T > 2R$ (c.f. [166]), and by which the exact controllability of (8.1) follows. In the rest of this chapter, we shall see that things are different in the stochastic setting.

Now, we consider the following controlled stochastic transport equation:

$$
\begin{cases}
dy + O \cdot \nabla y dt = \big(a_1 y + a_3 v\big)dt + \big(a_2 y + v\big)dW(t) & \text{in } Q, \\
y = u & \text{on } \Sigma^-, \\
y(0) = y_0 & \text{in } G.
\end{cases} \tag{8.4}
$$

Here $y_0 \in L^2(G)$, and $a_1, a_2, a_3 \in L^\infty_{\mathbb{F}}(0,T; L^\infty(G))$. In (8.4), y is the state, while $u \in L^2_{\mathbb{F}}(0,T; L^2_O(\Gamma^-))$ and $v \in L^2_{\mathbb{F}}(0,T; L^2(G))$ are two controls.

Remark 8.2. We introduce two controls into the system (8.4). Moreover, the control v acts on the whole domain, and also it affects the system through its drift term (in the form $a_3 v$). Compared with the deterministic transport equations, it seems that we choose too many controls. However, by Proposition 7.6, it is easy to see that two controls are really necessary for our exact controllability problem for (8.4) to be defined below.

Solutions to (8.4) are also understood in the transposition sense. For this, we introduce the following backward stochastic transport equation:

$$
\begin{cases}
dz + O \cdot \nabla z dt = -\big[(a_1 + a_2 a_3)z + a_2 Z\big]dt + (Z - a_3 z)dW(t) & \text{in } Q_\tau, \\
z = 0 & \text{on } \Sigma_\tau^+, \\
z(\tau) = z_\tau & \text{in } G,
\end{cases} \tag{8.5}
$$

where $\tau \in (0,T]$, $Q_\tau = (0,\tau) \times G$, $\Sigma_\tau^+ = (0,\tau) \times \Gamma^+$, $z_\tau \in L^2_{\mathcal{F}_\tau}(\Omega; L^2(G))$.

Define an unbounded operator A on $L^2(G)$ as follows:

$$
\begin{cases}
D(A) = \big\{h \in H^1(G) \;\big|\; h|_{\Gamma^-} = 0\big\}, \\
Ah = -O \cdot \nabla h, \quad \forall\, h \in D(A).
\end{cases} \tag{8.6}
$$

Clearly, $D(A)$ is dense in H and A is closed. Furthermore, for every $h \in D(A)$,

$$
\langle Ah, h\rangle_{L^2(G)} = -\int_G hO \cdot \nabla h dx = -\int_{\Gamma^+} O \cdot \nu |h|^2 d\Gamma^+ \le 0.
$$

One can easily check that the adjoint operator of A is

$$
\begin{cases}
D(A^*) = \big\{h \in H^1(G) \,|\, h = 0 \text{ on } \Gamma^+\big\}, \\
A^* h = O \cdot \nabla h, \quad \forall\, h \in D(A^*).
\end{cases} \tag{8.7}
$$

For every $h \in D(A^*)$, it holds that

$$
\langle A^* h, h\rangle_{L^2(G)} = \int_G hO \cdot \nabla h dx = \int_{\Gamma^-} O \cdot \nu |h|^2 d\Gamma^- \le 0.
$$

Hence, both A and A^* are dissipative operators. By the standard operator semigroup theory (e.g., [87, Page 84]), A generates a contractive C_0-semigroup on $L^2(G)$.

By Theorem 4.10, we have the following well-posedness result for the backward stochastic partial differential equation (8.5).

Proposition 8.3. *For any $\tau \in (0,T]$ and $z_\tau \in L^2_{\mathcal{F}_\tau}(\Omega;L^2(G))$, the equation* (8.5) *admits a unique mild solution $(z,Z) \in L^2_{\mathbb{F}}(\Omega;C([0,\tau];L^2(G))) \times L^2_{\mathbb{F}}(0,\tau;L^2(G))$ such that*

$$|z|_{L^2_{\mathbb{F}}(\Omega;C([0,\tau];L^2(G)))} + |Z|_{L^2_{\mathbb{F}}(0,\tau;L^2(G))} \leq \mathcal{C}|z_\tau|_{L^2_{\mathcal{F}_\tau}(\Omega;L^2(G))}, \tag{8.8}$$

where $\mathcal{C}$ is a constant, independent of τ and z_τ.

Note that, by Proposition 8.3, the first component z (of solution to (8.5)) belongs to $L^2_{\mathbb{F}}(\Omega;C([0,\tau];L^2(G)))$, hence, it is not obvious that $z|_{\Gamma^-} \in L^2_{\mathbb{F}}(0,\tau;L^2_{\mathcal{O}}(\Gamma^-))$. The latter is guaranteed by Proposition 8.6 in the next section. This fact will be used in the next notion:

Definition 8.4. *A stochastic process $y \in C_{\mathbb{F}}([0,T];L^2(\Omega;L^2(G)))$ is called a transposition solution to* (8.4) *if for any $\tau \in (0,T]$ and $z_\tau \in L^2_{\mathcal{F}_\tau}(\Omega;L^2(G))$, it holds that*

$$\begin{aligned}&\mathbb{E}\langle y(\tau), z_\tau\rangle_{L^2(G)} - \langle y_0, z(0)\rangle_{L^2(G)}\\&= \mathbb{E}\int_0^\tau \langle v, Z\rangle_{L^2(G)}dt - \mathbb{E}\int_0^\tau\int_{\Gamma^-} \mathcal{O}\cdot\nu uz d\Gamma^- dt.\end{aligned} \tag{8.9}$$

Here (z,Z) solves (8.5).

The main result in this chapter is the following result for the exact controllability of (8.4).

Theorem 8.5. *If $T > 2R$, then the system* (8.4) *is exactly controllable at time T, i.e., for any $y_0 \in L^2(G)$ and $y_T \in L^2_{\mathcal{F}_T}(\Omega;L^2(G))$, one can find a couple of controls $(u,v) \in L^2_{\mathbb{F}}(0,T;L^2_{\mathcal{O}}(\Gamma^-)) \times L^2_{\mathbb{F}}(0,T;L^2(G))$ such that the corresponding (transposition) solution y to* (8.4) *satisfies that $y(T) = y_T$ in G, a.s.*

Similarly to the conclusion 1) in Theorem 7.24 (in Section 7.3), in order to prove Theorem 8.5, it suffices to establish a suitable observability estimate for (8.5) with $\tau = T$. The latter will be done in Theorem 8.11 (in Section 8.3), by means of a stochastic version of global Carleman estimate.

8.2 Hidden Regularity and a Weighted Identity

This section is addressed to present some preliminary results. We first prove the following result:

Proposition 8.6. *Solutions to the equation* (8.5) *satisfy that*

$$|z|^2_{L^2_{\mathbb{F}}(0,\tau;L^2_{\mathcal{O}}(\Gamma^-))} \leq \mathcal{C}\mathbb{E}|z_\tau|^2_{L^2(G)}, \quad \forall\, \tau \in (0,T],\ z_\tau \in L^2_{\mathcal{F}_\tau}(\Omega;L^2(G)).$$

Proof: For simplicity, we only consider the case of $\tau = T$. By Theorem 4.10 and Proposition 4.7, one can show that if $z_T \in L^2_{\mathcal{F}_T}(\Omega; D(A^*))$, then the solution $(z, Z) \in \big(L^2_{\mathbb{F}}(\Omega; C([0,T]; L^2(G))) \cap L^2_{\mathbb{F}}(0,T; D(A^*))\big) \times L^2_{\mathbb{F}}(0,T; L^2(G))$. It follows from Itô's formula that

$$
\begin{aligned}
&\mathbb{E}|z_T|^2_{L^2(G)} - |z(0)|^2_{L^2(G)}\\
&= -2\mathbb{E}\int_0^T\int_G zO\cdot\nabla z dxdt\\
&\quad +\mathbb{E}\int_0^T\int_G \big\{-2z\big[(a_1 + a_2 a_3)z + a_2 Z\big] + |Z - a_3 z|^2\big\}dxdt.
\end{aligned}
$$

Therefore,

$$
\begin{aligned}
&-\mathbb{E}\int_0^T\int_{\Gamma^-} O\cdot\nu z^2 d\Gamma^- dt\\
&\le \mathbb{E}|z_T|^2_{L^2(G)} + 2\mathbb{E}\int_0^T\int_G z\big[(a_1 + a_2 a_3)z + a_2 Z\big]dxdt \le \mathcal{C}\mathbb{E}|z_T|^2_{L^2(G)}.
\end{aligned}
\tag{8.10}
$$

For any $z_T \in L^2_{\mathcal{F}_T}(\Omega; L^2(G))$, we can find a sequence $\{z_T^{(k)}\}_{k=1}^{\infty} \subset L^2_{\mathcal{F}_T}(\Omega; D(A^*))$ such that $\lim_{k\to\infty} z_T^{(k)} = z_T$ in $L^2_{\mathcal{F}_T}(\Omega; L^2(G))$. Hence, the inequality (8.10) also holds for $z_T \in L^2_{\mathcal{F}_T}(\Omega; L^2(G))$. □

Remark 8.7. The fact that $z|_{\Gamma^-} \in L^2_{\mathbb{F}}(0,\tau; L^2_O(\Gamma^-))$ (as shown in Proposition 8.6) is sometimes called a hidden regularity property, because it does not follow from the classical trace theorem of Sobolev spaces.

We have the following well-posedness result for (8.4) .

Proposition 8.8. *For any $y_0 \in L^2(G)$, $u \in L^2_{\mathbb{F}}(0,T; L^2_O(\Gamma^-))$ and $v \in L^2_{\mathbb{F}}(0,T; L^2(G))$, the equation* (8.4) *admits a unique (transposition) solution $y \in C_{\mathbb{F}}([0,T]; L^2(\Omega; L^2(G)))$ such that*

$$
\begin{aligned}
&|y|_{C_{\mathbb{F}}([0,T];L^2(\Omega;L^2(G)))}\\
&\le \mathcal{C}\big(|y_0|_{L^2(G)} + |u|_{L^2_{\mathbb{F}}(0,T;L^2_O(\Gamma^-))} + |v|_{L^2_{\mathbb{F}}(0,T;L^2(G))}\big).
\end{aligned}
\tag{8.11}
$$

Proof: Proposition 8.8 is actually a consequence of Theorem 7.12 (for abstract stochastic evolution equations). Here, for the reader's convenience, we give a "concrete" proof (without using the abstract result in Theorem 7.12). We only prove the existence. Since $u \in L^2_{\mathbb{F}}(0,T; L^2_O(\Gamma^-))$, there exists a sequence $\{u_m\}_{m=1}^{\infty} \subset C^2_{\mathbb{F}}([0,T];\ H_0^{3/2}(\Gamma^-))$ with $u_m(0) = 0$ for all $m \in \mathbb{N}$ such that

$$
\lim_{m\to\infty} u_m = u \quad \text{in } L^2_{\mathbb{F}}(0,T; L^2_O(\Gamma^-)). \tag{8.12}
$$

For each $m \in \mathbb{N}$, we can find a $\tilde{u}_m \in C^2_{\mathbb{F}}([0,T]; H^2(G))$ such that $\tilde{u}_m|_{\Gamma^-} = u_m$ and $\tilde{u}_m(0) = 0$.

Consider the following stochastic transport equation:

$$
\begin{cases}
d\tilde{y}_m + O\cdot\nabla\tilde{y}_m dt \\
= (a_1\tilde{y}_m + a_3 v + \zeta_m)dt + \big[a_2(\tilde{y}_m+\tilde{u}_m)+v\big]dW(t) & \text{in } Q,\\
\tilde{y}_m = 0 & \text{on } \Sigma,\\
\tilde{y}_m(0) = y_0 & \text{in } G,
\end{cases}
\tag{8.13}
$$

where $\zeta_m = -\tilde{u}_{m,t} - O\cdot\nabla\tilde{u}_m dt + a_1\tilde{u}_m$. By Theorem 3.13, the system (8.13) admits a unique mild solution $\tilde{y}_m \in C_{\mathbb{F}}([0,T];L^2(\Omega;L^2(G)))$.

Let $y_m = \tilde{y}_m + \tilde{u}_m$. For any $\tau\in(0,T]$ and $z_\tau\in L^2_{\mathcal{F}_\tau}(\Omega;L^2(G))$, by Itô's formula and using integration by parts, we have that

$$
\begin{aligned}
&\mathbb{E}\langle y_m(\tau), z_\tau\rangle_{L^2(G)} - \langle y_0, z(0)\rangle_{L^2(G)}\\
&= \mathbb{E}\int_0^\tau \langle v, Z\rangle_{L^2(G)}dt - \mathbb{E}\int_0^\tau\int_{\Gamma^-} O\cdot\nu u_m z d\Gamma^- dt,
\end{aligned}
\tag{8.14}
$$

where (z,Z) is the mild solution to (8.5). Consequently, for any $m_1,m_2\in\mathbb{N}$, it holds that

$$
\mathbb{E}\langle y_{m_1}(\tau) - y_{m_2}(\tau), z_\tau\rangle_{L^2(G)} = -\mathbb{E}\int_0^\tau\int_{\Gamma^-} O\cdot\nu(u_{m_1}-u_{m_2})z d\Gamma^- dt. \tag{8.15}
$$

Let us choose $z_\tau\in L^2_{\mathcal{F}_\tau}(\Omega;L^2(G))$ such that $|z_\tau|_{L^2_{\mathcal{F}_\tau}(\Omega;L^2(G))}=1$ and

$$
\mathbb{E}\langle y_{m_1}(\tau) - y_{m_2}(\tau), z_\tau\rangle_{L^2(G)} \geq \frac{1}{2}(|y_{m_1}(\tau)-y_{m_2}(\tau)|_{L^2_{\mathcal{F}_\tau}(\Omega;L^2(G))}. \tag{8.16}
$$

It follows from (8.15), (8.16) and Proposition 8.6 that

$$
\begin{aligned}
&|y_{m_1}(\tau)-y_{m_2}(\tau)|_{L^2_{\mathcal{F}_\tau}(\Omega;L^2(G))} \leq 2\Big|\mathbb{E}\int_0^\tau\int_{\Gamma^-} O\cdot\nu(u_{m_1}-u_{m_2})z d\Gamma^- ds\Big|\\
&\leq \mathcal{C}|u_{m_1}-u_{m_2}|_{L^2_{\mathbb{F}}(0,T;L^2_O(\Gamma^-))}|z_\tau|_{L^2_{\mathbb{F}}(0,T;L^2_O(\Gamma^-))}\\
&\leq \mathcal{C}|u_{m_1}-u_{m_2}|_{L^2_{\mathbb{F}}(0,T;L^2_O(\Gamma^-))},
\end{aligned}
$$

where the constant $\mathcal{C}$ is independent of τ. Consequently, it holds that

$$
|y_{m_1}-y_{m_2}|_{C_{\mathbb{F}}([0,T];L^2(\Omega;L^2(G)))} \leq \mathcal{C}|u_{m_1}-u_{m_2}|_{L^2_{\mathbb{F}}(0,T;L^2_O(\Gamma^-))}.
$$

Hence, $\{y_m\}_{m=1}^\infty$ is a Cauchy sequence in $C_{\mathbb{F}}([0,T];L^2(\Omega;L^2(G)))$. Denote by y the limit of $\{y_m\}_{m=1}^\infty$ in $C_{\mathbb{F}}([0,T];L^2(\Omega;L^2(G)))$. Letting $m\to\infty$ in (8.14), we see that y satisfies (8.9). Thus, y is a transposition solution to (8.4).

Let us choose $z_\tau\in L^2_{\mathcal{F}_\tau}(\Omega;L^2(G))$ such that $|z_\tau|_{L^2_{\mathcal{F}_\tau}(\Omega;L^2(G))}=1$ and

$$
\mathbb{E}\langle y(\tau), z_\tau\rangle_{L^2(G)} \geq \frac{1}{2}|y(\tau)|_{L^2_{\mathcal{F}_\tau}(\Omega;L^2(G))}. \tag{8.17}
$$

Combining (8.9), (8.17) and using Proposition 8.6 again, we obtain that

$$\begin{aligned}
&|y(\tau)|_{L^2_{\mathcal{F}_\tau}(\Omega;L^2(G))}\\
&\leq 2\Big(\big|\langle y_0, z(0)\rangle_{L^2(G)}\big| + \Big|\mathbb{E}\int_0^T \langle v, Z\rangle_{L^2(G)}dt\Big| + \Big|\mathbb{E}\int_0^T\int_{\Gamma^-} O\cdot\nu uzd\Gamma^- dt\Big|\Big)\\
&\leq \mathcal{C}\big(|y_0|_{L^2(G)} + |u|_{L^2_{\mathbb{F}}(0,T;L^2_O(\Gamma^-))} + |v|_{L^2_{\mathbb{F}}(0,T;L^2(G))}\big),
\end{aligned}$$

where the constant $\mathcal{C}$ is independent of τ. Therefore, we obtain the desired estimate (8.11). This completes the proof of Proposition 8.6. □

Next, we present a weighted identity for the stochastic transport operator $d + O\cdot\nabla dt$, which will play an important role in establishing the global Carleman estimate for (8.5).

Let $\lambda > 0$, and let $c \in (0,1)$ be such that $cT > 2R$. Put

$$\ell = \lambda\Big[|x|^2_{\mathbb{R}^n} - c\Big(t - \frac{T}{2}\Big)^2\Big] \quad \text{and} \quad \theta = e^{\ell}. \tag{8.18}$$

Similarly to (1.48) (at least in spirit), we have the following key weighted identity:

Proposition 8.9. *Let* $\mathbf{u}$ *be an* $H^1(\mathbb{R}^n)$*-valued Itô process, and* $\mathbf{v} = \theta\mathbf{u}$. *Then,*

$$\begin{aligned}
&-\theta\big(\ell_t + O\cdot\nabla\ell\big)\mathbf{v}\big(d\mathbf{u} + O\cdot\nabla\mathbf{u}dt\big)\\
&= -\frac{1}{2}d\big[\big(\ell_t + O\cdot\nabla\ell\big)\mathbf{v}^2\big] - \frac{1}{2}O\cdot\nabla\big[\big(\ell_t + O\cdot\nabla\ell\big)\mathbf{v}^2\big]dt + \frac{1}{2}\big[\ell_{tt} + O\cdot\nabla(O\cdot\nabla\ell)\\
&\quad +2O\cdot\nabla\ell_t\big]\mathbf{v}^2dt + \frac{1}{2}\big(\ell_t + O\cdot\nabla\ell\big)(d\mathbf{v})^2 + \big(\ell_t + O\cdot\nabla\ell\big)^2\mathbf{v}^2dt.
\end{aligned} \tag{8.19}$$

Proof: Clearly,

$$\begin{aligned}
\theta\big(d\mathbf{u} + O\cdot\nabla\mathbf{u}dt\big) &= \theta d\big(\theta^{-1}\mathbf{v}\big) + \theta O\cdot\nabla\big(\theta^{-1}\mathbf{v}\big)dt\\
&= d\mathbf{v} + O\cdot\nabla\mathbf{v}dt - \big(\ell_t + O\cdot\nabla\ell\big)\mathbf{v}dt.
\end{aligned}$$

Thus,

$$\begin{aligned}
&-\theta\big(\ell_t + O\cdot\nabla\ell\big)\mathbf{v}\big(d\mathbf{u} + O\cdot\nabla\mathbf{u}dt\big)\\
&= -\big(\ell_t + O\cdot\nabla\ell\big)\mathbf{v}\big[d\mathbf{v} + O\cdot\nabla\mathbf{v}dt - \big(\ell_t + O\cdot\nabla\ell\big)\mathbf{v}dt\big]\\
&= -\big(\ell_t + O\cdot\nabla\ell\big)\mathbf{v}\big(d\mathbf{v} + O\cdot\nabla\mathbf{v}dt\big) + \big(\ell_t + O\cdot\nabla\ell\big)^2\mathbf{v}^2dt.
\end{aligned} \tag{8.20}$$

It is easy to see that

$$\begin{cases} -\ell_t \mathbf{v} d\mathbf{v} = -\dfrac{1}{2} d(\ell_t \mathbf{v}^2) + \dfrac{1}{2}\ell_{tt}\mathbf{v}^2 dt + \dfrac{1}{2}\ell_t (d\mathbf{v})^2, \\ -O\cdot\nabla\ell\mathbf{v}d\mathbf{v} = -\dfrac{1}{2} d(O\cdot\nabla\ell\mathbf{v}^2) + \dfrac{1}{2}(O\cdot\nabla\ell)_t \mathbf{v}^2 dt + \dfrac{1}{2} O\cdot\nabla\ell(d\mathbf{v})^2, \\ -\ell_t\mathbf{v}O\cdot\nabla\mathbf{v}dt = -\dfrac{1}{2} O\cdot\nabla(\ell_t\mathbf{v}^2)dt + \dfrac{1}{2} O\cdot\nabla\ell_t\mathbf{v}^2 dt, \\ -O\cdot\nabla\ell\mathbf{v}O\cdot\nabla\mathbf{v}dt = -\dfrac{1}{2} O\cdot\nabla(O\cdot\nabla\ell\mathbf{v}^2)dt + \dfrac{1}{2} O\cdot\nabla(O\cdot\nabla\ell)\mathbf{v}^2 dt. \end{cases}$$

This, together with (8.20), implies (8.19). This completes the proof of Proposition 8.9. □

Remark 8.10. It is easy to see that, the spirit for proving the identity (8.19) in Proposition 8.9 is very close to that for the identity (1.48).

8.3 Observability Estimate for Backward Stochastic Transport Equations

In this subsection, we shall show the following observability estimate for the equation (8.5).

Theorem 8.11. *If $T > 2R$, then solutions to the equation* (8.5) *with $\tau = T$ satisfy that*

$$\begin{aligned} &|z_T|_{L^2_{\mathcal{F}_T}(\Omega;L^2(G))} \leq \mathcal{C}\big(|z|_{L^2_{\mathbb{F}}(0,T;L^2_O(\Gamma^-))} + |Z|_{L^2_{\mathbb{F}}(0,T;L^2(G))}\big), \\ &\qquad\qquad \forall\, z_T \in L^2_{\mathcal{F}_T}(\Omega; L^2(G)). \end{aligned} \tag{8.21}$$

Proof: Applying Proposition 8.9 to the equation (8.5) with $\mathbf{u} = z$, integrating (8.19) on Q, using integration by parts, noting (8.18) and taking expectation, we get that

$$\begin{aligned} &-2\mathbb{E}\int_Q \theta^2(\ell_t + O\cdot\nabla\ell)z(dz + O\cdot\nabla z dt)dx \\ &= \lambda\mathbb{E}\int_G (cT - 2O\cdot x)\theta^2(T)z^2(T)dx + \lambda\mathbb{E}\int_G (cT + 2O\cdot x)\theta^2(0)z^2(0)dx \\ &\quad +\lambda\mathbb{E}\int_{\Sigma_-} O\cdot\nu\big[c(T-2t) - 2O\cdot x\big]\theta^2 z^2 d\Sigma_- + 2(1-c)\lambda\mathbb{E}\int_Q \theta^2 z^2 dxdt \\ &\quad +\mathbb{E}\int_Q \theta^2(\ell_t + O\cdot\nabla\ell)(dz)^2 dx + 2\mathbb{E}\int_Q \theta^2(\ell_t + O\cdot\nabla\ell)^2 z^2 dxdt. \end{aligned} \tag{8.22}$$

By virtue of that (z, Z) solves (8.5) with $\tau = T$, we see that

$$
\begin{aligned}
&-2\mathbb{E}\int_Q \theta^2(\ell_t+O\cdot\nabla\ell)z(dz+O\cdot\nabla z dt)dx\\
&=2\mathbb{E}\int_Q \theta^2(\ell_t+O\cdot\nabla\ell)z\big[(a_1+a_2a_3)z+a_2Z\big]dxdt\\
&\le\frac{1}{2}\mathbb{E}\int_Q \theta^2(\ell_t+O\cdot\nabla\ell)^2z^2dxdt+C\mathbb{E}\int_Q\theta^2(z^2+Z^2)dxdt
\end{aligned}
\tag{8.23}
$$

and

$$
\begin{aligned}
&\mathbb{E}\int_Q \theta^2(\ell_t+O\cdot\nabla\ell)(dz)^2dx=\mathbb{E}\int_Q\theta^2(\ell_t+O\cdot\nabla\ell)|Z-a_3z|^2dxdt\\
&\le\frac{1}{2}\mathbb{E}\int_Q \theta^2(\ell_t+O\cdot\nabla\ell)^2z^2dxdt+C\mathbb{E}\int_Q\theta^2z^2dxdt\\
&\quad+4\mathbb{E}\int_Q\theta^2|\ell_t+O\cdot\nabla\ell|Z^2dxdt,
\end{aligned}
\tag{8.24}
$$

By (8.22)–(8.24) and $z(T)=z_T$ in G, it follows that

$$
\begin{aligned}
&\lambda\mathbb{E}\int_G(cT-2O\cdot x)\theta^2(T)z_T^2dx+\lambda\int_G(cT+2O\cdot x)\theta^2(0)z^2(0)dx\\
&\quad+2(1-c)\lambda\mathbb{E}\int_Q\theta^2z^2dxdt+\mathbb{E}\int_Q\theta^2(\ell_t+O\cdot\nabla\ell)^2z^2dxdt\\
&\le C\mathbb{E}\int_Q\theta^2(z^2+\lambda^2Z^2)dxdt-\lambda\mathbb{E}\int_{\Sigma_-}O\cdot\nu\big[c(T-2t)-2O\cdot x\big]\theta^2z^2d\Sigma_-,
\end{aligned}
\tag{8.25}
$$

where the constant $C>0$ is independent of λ.

Finally, recalling that $R=\max_{x\in\Gamma}|x|_{\mathbb{R}^n}$ and $c\in(0,1)$ are so that $cT>2R$, by choosing λ to be large enough in (8.25), we obtain the desired estimate (8.21). This completes the proof of Theorem 8.11. □

8.4 Notes and Comments

The main result in this chapter is a modification and generalization of that in [231].

Generally speaking, there are three methods to establish the exact controllability of deterministic transport equations. The first one is to utilize the explicit formula of their solutions. By this method, for some simple transport equations, one can explicitly give a control steering the system from a given initial state to another given final state, provided that the time is large enough. It seems that this method cannot be used to solve our stochastic problem s-ince generally we do not have the explicit formula for solutions to the system (8.4). The second one is the extension method. This method was introduced in [294] to prove the exact controllability of the classical wave equation. It is also effective to solve the exact controllability problem for many hyperbolic-type equations including the transport equations. However, it seems that it is only valid for time reversible systems, and therefore, it is not suitable for the stochastic problems in the framework of Itô's integral. The third method is based on the duality between controllability and observability, via which the exact controllability problem is reduced to suitable observability estimate for the dual equation, and the desired observability estimate is obtained by some global Carleman estimate (e.g., [43, 166]).

In the literature, in order to obtain the observability estimate, people usually combine the Carleman estimate and the usual energy estimate (e.g., [166, 384]). It deserves mentioning that the desired observability inequality (8.21) in this chapter is established directly by means of the global Carleman estimate (without using the energy estimate). Indeed, even for the observability estimate for deterministic transport equations, such a method provides a new proof which is simpler than that in [166].

As mentioned in Remark 8.2, in order to study the exact controllability of the system (8.4), people do need to introduce two controls u and v (and the control v acting on the whole domain in particular). Nevertheless, it seems unnecessary to introduce the extra control v if we are only concerned with null/approximate controllability of stochastic transport equations. It would be quite interesting to study such weak versions of controllability for (8.4) (with the control $v = 0$) but it seems that some new tool should be developed to obtain nontrivial results.

As far as we know, controllability problems for nonlinear stochastic transport equations are completely open.

9

Controllability and Observability of Stochastic Parabolic Systems

This chapter is devoted to studying the null/approximate controllability and observability of stochastic parabolic systems. The main results can be described as follows:

- When the coefficients of the underlying system are space-independent, based on the spectral method, we show the null/approximate controllability using only one control applied to the drift term;
- For a class of coupled stochastic parabolic system with only one control (applied to also only one equation), the null controllability is derived when the system is effectively coupled by the drift terms and un-coupled by the diffusion terms;
- By means of the global Carleman estimate, we establish the observability estimate for stochastic parabolic systems with considerably general coefficients;
- The null/approximate controllability of general stochastic parabolic systems with two controls are shown by means of duality argument.

In each of the above cases, we shall explain the main differences between the deterministic problem and its stochastic counterpart.

In this chapter, $T > 0$ is given, and $(\Omega, \mathcal{F}, \mathbf{F}, \mathbb{P})$ (with $\mathbf{F} \stackrel{\Delta}{=} \{\mathcal{F}_t\}_{t\in[0,T]}$) is a fixed filtered probability space, on which a one dimensional standard Brownian motion $W(\cdot)$ is defined, and $\mathbf{F}$ is the corresponding natural filtration. We denote by $\mathbb{F}$ the progressive σ-field w.r.t. $\mathbf{F}$.

9.1 Formulation of the Problems

Parabolic equations (and its special case, i.e., the heat equations) are one class of most typical partial differential equations. In the deterministic setting, the controllability theory for parabolic equations are studied most thoroughly. It is then very natural to see what will happen for its stochastic counterpart.

Q. Lü and X. Zhang, *Mathematical Control Theory for Stochastic Partial Differential Equations*, Probability Theory and Stochastic Modelling 101, https://doi.org/10.1007/978-3-030-82331-3_9

Throughout this chapter, $G \subset \mathbb{R}^n$ ($n \in \mathbb{N}$) is a given bounded domain with a C^∞ boundary Γ, and G_0 is a given nonempty open subset of G. Denote by χ_{G_0} the characteristic function of G_0. Put

$$Q \stackrel{\Delta}{=} (0,T) \times G, \quad \Sigma \stackrel{\Delta}{=} (0,T) \times \Gamma, \quad Q_0 \stackrel{\Delta}{=} (0,T) \times G_0.$$

Also, we assume that $a^{jk} : \overline{G} \to \mathbb{R}$ ($j,k = 1,2,\cdots,n$) satisfies $a^{jk} \in C^2(\overline{G})$, $a^{jk} = a^{kj}$, and for some $s_0 > 0$,

$$\sum_{j,k=1}^{n} a^{jk}(x)\xi_i\xi_j \geq s_0|\xi|^2, \quad \forall\, (x,\xi) \equiv (x_1,\cdots,x_n,\xi_1,\cdots,\xi_n) \in G \times \mathbb{R}^n. \tag{9.1}$$

We shall denote by $\mathcal{C}$ a generic positive constant depending only on G, G_0, T and $(a^{jk})_{n\times n}$ (unless other stated), which may change from one place to another.

We begin with the following controlled deterministic parabolic equation:

$$\begin{cases} y_t - \sum_{j,k=1}^{n} (a^{jk} y_{x_j})_{x_k} = a_1^0 \cdot \nabla y + a_2^0 y + \chi_{G_0} u & \text{in } Q, \\ y = 0 & \text{on } \Sigma, \\ y(0) = y_0 & \text{in } G, \end{cases} \tag{9.2}$$

where $a_1^0 \in L^\infty(0,T;W^{1,\infty}(G;\mathbb{R}^n))$ and $a_2^0 \in L^\infty(0,T;L^\infty(G))$. In (9.2), y and u are the *state variable* and *control variable*, the *state space* and *control space* are chosen to be $L^2(G)$ and $L^2(Q_0)$, respectively. The equation (9.2) is said to be null controllable (*resp.* approximately controllable) if for any given $y_0 \in L^2(G)$ (*resp.* for any given $\varepsilon > 0$, $y_0, y_1 \in L^2(G)$), one can find a control $u \in L^2(Q_0)$ such that the solution $y(\cdot) \in C([0,T];L^2(G)) \cap L^2(0,T;H_0^1(G))$ to (9.2) satisfies $y(T) = 0$ (*resp.* $|y(T) - y_1|_{L^2(G)} \leq \varepsilon$). Note that, due to the smoothing effect of solutions to the parabolic equation, the exact controllability for (9.2) is impossible, i.e., the above ε cannot be zero. The dual equation of (9.2) reads as follows

$$\begin{cases} z_t + \sum_{j,k=1}^{n} (a^{jk} z_{x_j})_{x_k} = -\nabla \cdot (a_1^0 z) + a_2^0 z & \text{in } Q, \\ z = 0 & \text{on } \Sigma, \\ z(T) = z_0 & \text{in } G. \end{cases} \tag{9.3}$$

Write $w(t,x) = z(T-t,x)$ and $w_0 = z_0$. By (9.3), it is easy to check that

$$\begin{cases} w_t - \sum_{j,k=1}^{n} (a^{jk} w_{x_j})_{x_k} = -\nabla \cdot (a_1^0 w) + a_2^0 w & \text{in } Q, \\ w = 0 & \text{on } \Sigma, \\ w(0) = w_0 & \text{in } G. \end{cases} \tag{9.4}$$

By means of the standard duality argument (similar to Theorem 7.17), it is easy to show the following result.

Proposition 9.1. i) *The equation (9.2) is null controllable if and only if solutions to the equation (9.4) satisfy the following observability estimate:*

$$|w(T)|_{L^2(G)} \leq C|w|_{L^2(Q_0)}, \qquad \forall\, w_0 \in L^2(G); \tag{9.5}$$

ii) *The equation (9.2) is approximately controllable if and only if any solution to the equation (9.4) satisfy the following unique continuation property:*

$$w = 0 \ \ in\ Q_0 \Longrightarrow w_0 = 0.$$

Controllability and observability of deterministic parabolic equations are now well-understood. Indeed, one can use the global Carleman estimate to prove the observability inequality (9.5) (c.f. [117, 357]), and by which the null controllability of (9.2) follows. On the other hand, by Proposition 8.1 and the backward uniqueness for solutions to the equation (9.4), it is easy to deduce the approximate controllability of (9.2) from the null controllability of the same equation. In the rest of this chapter, we shall see a quite different picture in the stochastic setting.

We now fix $m \in \mathbb{N}$ and consider the following controlled stochastic parabolic system:

$$\begin{cases} dy - \displaystyle\sum_{j,k=1}^{n} (a^{jk} y_{x_j})_{x_k} dt \\ \quad = \Big(\displaystyle\sum_{j=1}^{n} a_{1j} y_{x_j} + a_2 y + \chi_{G_0} u + a_4 v\Big) dt + (a_3 y + v)\, dW(t) & \text{in } Q, \\ y = 0 & \text{on } \Sigma, \\ y(0) = y_0 & \text{in } G, \end{cases} \tag{9.6}$$

where

$$\begin{cases} a_{1j} \in L^\infty_{\mathbb{F}}(0,T; W^{1,\infty}(G;\mathbb{R}^{m\times m})), & j = 1,2,\cdots,n, \\ a_k \in L^\infty_{\mathbb{F}}(0,T; L^\infty(G;\mathbb{R}^{m\times m})), & k = 2,3,4. \end{cases} \tag{9.7}$$

In the system (9.6), the initial state $y_0 \in L^2_{\mathcal{F}_0}(\Omega; L^2(G;\mathbb{R}^m))$, y is the state variable, and the control variable consists of the pair $(u,v) \in L^2_{\mathbb{F}}(0,T; L^2(G_0;\mathbb{R}^m)) \times L^2_{\mathbb{F}}(0,T; L^2(G;\mathbb{R}^m))$. The main goals in this chapter are to study the null and approximate controllability of (9.6).

Remark 9.2. Similar to the control system (8.4), the term $a_4 v$ reflects the effect of the control v in the diffusion term to the drift term. One can also consider the case that the control u in the drift term may influence the diffusion term. Nevertheless, since the control v in the diffusion term is acted on the whole domain, one can cancelation such effect by suitable choice of v.

From the proof of Theorem 3.24 and using Theorem 3.10, it is easy to show the following well-posedness result for the equation (9.6).

Proposition 9.3. *For any* $y_0 \in L^2_{\mathcal{F}_0}(\Omega; L^2(G;\mathbb{R}^m))$ *and* $(u,v) \in L^2_{\mathbb{F}}(0,T; L^2(G_0;\mathbb{R}^m)) \times L^2_{\mathbb{F}}(0,T; L^2(G;\mathbb{R}^m))$, *the system* (9.6) *admits a unique weak solution* $y \in L^2_{\mathbb{F}}(\Omega; C([0,T]; L^2(G;\mathbb{R}^m))) \cap L^2_{\mathbb{F}}(0,T; H^1_0(G;\mathbb{R}^m))$. *Moreover,*

$$\begin{aligned}&|y|_{L^2_{\mathbb{F}}(\Omega;C([0,T];L^2(G;\mathbb{R}^m)))\cap L^2_{\mathbb{F}}(0,T;H^1_0(G;\mathbb{R}^m))}\\ &\leq Ce^{C\kappa_1}\big(|y_0|_{L^2_{\mathcal{F}_0}(\Omega;L^2(G;\mathbb{R}^m))} + |(u,v)|_{L^2_{\mathbb{F}}(0,T;L^2(G_0;\mathbb{R}^m))\times L^2_{\mathbb{F}}(0,T;L^2(G;\mathbb{R}^m))}\big),\end{aligned} \tag{9.8}$$

where

$$\begin{aligned}\kappa_1 = &\sum_{j=1}^n |a_{1j}|_{L^\infty_{\mathbb{F}}(0,T;L^\infty(G;\mathbb{R}^{m\times m}))} + |a_2|_{L^\infty_{\mathbb{F}}(0,T;L^\infty(G;\mathbb{R}^{m\times m}))}\\ &+|a_3|^2_{L^\infty_{\mathbb{F}}(0,T;L^\infty(G;\mathbb{R}^{m\times m}))}.\end{aligned}$$

Note that we introduce two controls u and v in (9.6). In view of the controllability result for the deterministic parabolic equation (9.2), it is more natural to use only one control and consider the following controlled stochastic parabolic system (which is a special case of (9.6) with $v \equiv 0$):

$$\begin{cases} dy - \displaystyle\sum_{j,k=1}^n (a^{jk} y_{x_j})_{x_k} dt \\ \quad = \Big(\displaystyle\sum_{j=1}^n a_{1j} y_{x_j} + a_2 y + \chi_{G_0} u\Big)dt + a_3 y dW(t) & \text{in } Q, \\ y = 0 & \text{on } \Sigma, \\ y(0) = y_0 & \text{in } G. \end{cases} \tag{9.9}$$

It is easy to see that, the dual system of both (9.6) and (9.9) are respectively the following backward stochastic parabolic systems:

$$\begin{cases} dz + \displaystyle\sum_{j,k=1}^n (a^{jk} z_{x_j})_{x_k} dt \\ \quad = \Big[\displaystyle\sum_{j=1}^n (a_{1j}^\top z)_{x_j} - (a_2^\top - a_3^\top a_4^\top) z - a_3^\top Z\Big]dt + (-a_4^\top z + Z)dW(t) & \text{in } Q, \\ z = 0 & \text{on } \Sigma, \\ z(T) = z_T & \text{in } G. \end{cases} \tag{9.10}$$

$$
\begin{cases}
dz + \sum_{j,k=1}^{n} (a^{jk} z_{x_j})_{x_k} dt \\
\quad = \Big[\sum_{j=1}^{n} \big(a_{1j}^{\top} z\big)_{x_j} - a_2^{\top} z - a_3^{\top} Z \Big] dt + Z dW(t) & \text{in } Q, \\
z = 0 & \text{on } \Sigma, \\
z(T) = z_T & \text{in } G.
\end{cases}
\tag{9.11}
$$

Similar to the proof of Theorem 3.24 and using Theorem 4.11, one can show the following well-posedness result for the equations (9.10) and (9.11):

Proposition 9.4. *Under the condition* (9.7), *for any* $z_T \in L^2_{\mathcal{F}_T}(\Omega; L^2(G; \mathbb{R}^m))$, *the system* (9.10) *(resp.* (9.11)*) admits a unique weak solution* $(z, Z) \in \big(C_{\mathbb{F}}([0,T]; L^2(\Omega; L^2(G;\mathbb{R}^m))) \bigcap L^2_{\mathbb{F}}(0,T; H_0^1(G;\mathbb{R}^m))\big) \times L^2_{\mathbb{F}}(0,T; L^2(G;\mathbb{R}^m))$. *Moreover, for any* $t \in [0,T]$,

$$
\begin{aligned}
& |(z(\cdot), Z(\cdot))|_{\left(C_{\mathbb{F}}([0,t];L^2(\Omega;L^2(G;\mathbb{R}^m)))\cap L^2_{\mathbb{F}}(0,t;H_0^1(G;\mathbb{R}^m))\right)\times L^2_{\mathbb{F}}(0,t;L^2(G;\mathbb{R}^m))} \\
& \leq \mathcal{C} e^{\mathcal{C}\widetilde{\kappa}_1} |z(t)|_{L^2_{\mathcal{F}_t}(\Omega;L^2(G;\mathbb{R}^m))},
\end{aligned}
\tag{9.12}
$$

where

$$
\begin{aligned}
\widetilde{\kappa}_1 = & \sum_{j=1}^{n} |a_{1j}|_{L^\infty_{\mathbb{F}}(0,T;W^{1,\infty}(G;\mathbb{R}^{m\times m}))} + |a_2|_{L^\infty_{\mathbb{F}}(0,T;L^\infty(G;\mathbb{R}^{m\times m}))} \\
& + \sum_{k=3}^{4} |a_k|^2_{L^\infty_{\mathbb{F}}(0,T;L^\infty(G;\mathbb{R}^{m\times m}))}
\end{aligned}
\tag{9.13}
$$

$\Big($*resp.*

$$
\begin{aligned}
\widetilde{\kappa}_1 = & \sum_{j=1}^{n} |a_{1j}|_{L^\infty_{\mathbb{F}}(0,T;W^{1,\infty}(G;\mathbb{R}^{m\times m}))} + |a_2|_{L^\infty_{\mathbb{F}}(0,T;L^\infty(G;\mathbb{R}^{m\times m}))} \\
& + |a_3|^2_{L^\infty_{\mathbb{F}}(0,T;L^\infty(G;\mathbb{R}^{m\times m}))}. \Big)
\end{aligned}
$$

In order to obtain the null controllability of (9.9), by Theorem 7.17, we need to show that solutions to the system (9.11) satisfy the following observability estimate:

$$
\begin{aligned}
& |z(0)|_{L^2_{\mathcal{F}_0}(\Omega;L^2(G;\mathbb{R}^m))} \leq \mathcal{C} |\chi_{G_0} z|_{L^2_{\mathbb{F}}(0,T;L^2(G_0;\mathbb{R}^m))}, \\
& \hspace{15em} \forall\, z_T \in L^2_{\mathcal{F}_T}(\Omega; L^2(G;\mathbb{R}^m)).
\end{aligned}
\tag{9.14}
$$

Unfortunately, at this moment, we are not able to prove the observability estimate (9.14). Instead, we obtain a weak version of (9.14), i.e., a weak observability estimate (for the system (9.10)) in Theorem 9.37 (See Section 9.6). By duality, Theorem 9.37 implies the null controllability of (9.6).

There exist two main difficulties to establish (9.14). The first one is that, unlike the deterministic setting (See (9.3)–(9.4)), it is impossible to reduce the backward stochastic parabolic system (9.10) to a (forward) stochastic parabolic system. The second one is that, though the corrected term "Z" plays a "coercive" role for the well-posedness of (9.10), it seems to be a "bad" (non-homogeneous) term when one tries to prove (9.14) using the global Carleman estimate.

At least at this moment, generally speaking, technically we do need to introduce the second control v in (9.6). Indeed, it is unclear whether it is possible or not to use only one (**F**-adapted) control u such that the following random parabolic system

$$\begin{cases} y_t - \sum_{j,k=1}^{n} (a^{jk} y_{x_j})_{x_k} = \sum_{j=1}^{n} a_{1j} y_{x_j} + a_2 y + \chi_{G_0} u & \text{in } Q, \\ y = 0 & \text{on } \Sigma, \\ y(0) = y_0 & \text{in } G \end{cases} \tag{9.15}$$

is null controllable. The answer seems to be "yes" but we do not know how to prove it for the general case.

Of course, it is quite interesting to study the controllability of (9.9), in which only one control u is introduced. So far, this problem is much less understood. Indeed, positive results are available only for some special cases of (9.9) when the spectral method can be applied (See Sections 9.2–9.3). As we shall see later, even for these special cases, some new phenomenons will appear in the stochastic setting.

Although it does not imply any controllability of the same sort equations, the observability of stochastic parabolic equations is still an interesting control problem. Because of this, as another main goal in this chapter, we shall study observability estimates for the following stochastic parabolic equation:

$$\begin{cases} dz - \sum_{j,k=1}^{n} \big(a^{jk} z_{x_j}\big)_{x_k} dt = \Big(\sum_{j=1}^{n} b_{1j} z_{x_j} + b_2 z\Big) dt + b_3 z dW(t) & \text{in } Q, \\ z = 0 & \text{on } \Sigma, \\ z(0) = z_0 & \text{in } G, \end{cases} \tag{9.16}$$

where

$$\begin{cases} b_{1j} \in L^\infty_{\mathbb{F}}(0,T; L^\infty(G; \mathbb{R}^{m\times m})), \quad j = 1, 2, \cdots, n, \\ b_2, b_3 \in L^\infty_{\mathbb{F}}(0,T; L^\infty(G; \mathbb{R}^{m\times m})). \end{cases} \tag{9.17}$$

As a consequence of Proposition 9.3, it is easy to show that

Corollary 9.5. *For any $z_0 \in L^2(G; \mathbb{R}^m)$, the system (9.16) admits a unique weak solution $z \in L^2_{\mathbb{F}}(\Omega; C([0,T]; L^2(G; \mathbb{R}^m))) \cap L^2_{\mathbb{F}}(0,T; H^1_0(G; \mathbb{R}^m))$. Moreover, for any $t \in [0,T]$,*

$$|z(\cdot)|_{L^2_{\mathbb{F}}(\Omega;C([t,T];L^2(G;\mathbb{R}^m)))\cap L^2_{\mathbb{F}}(t,T;H^1_0(G;\mathbb{R}^m))} \leq \mathcal{C}e^{\mathcal{C}\kappa_2}|z(t)|_{L^2_{\mathcal{F}_t}(\Omega;L^2(G;\mathbb{R}^m))}, \tag{9.18}$$

where

$$\begin{aligned}\kappa_2 = &\sum_{j=1}^{n} |b_{1j}|_{L^\infty_{\mathbb{F}}(0,T;L^\infty(G;\mathbb{R}^{m\times m}))} + |b_2|_{L^\infty_{\mathbb{F}}(0,T;L^\infty(G;\mathbb{R}^{m\times m}))}\\ &+|b_3|^2_{L^\infty_{\mathbb{F}}(0,T;L^\infty(G;\mathbb{R}^{m\times m}))}.\end{aligned}$$

We shall give an observability estimate for (9.16) in Theorem 9.28 (of Section 9.5). Since we assume only that $b_3 \in L^\infty_{\mathbb{F}}(0,T;L^\infty(G;\mathbb{R}^m))$ (in (9.17)), one needs to overcome some difficulty to derive the observability estimate (9.93) for the system (9.16).

When $m = 1$ and $b_3 \in L^\infty_{\mathbb{F}}(0,T;W^{2,\infty}(G))$, (9.16) can be reduced to a random parabolic equation. To see this, we write $\vartheta = e^{-\int_0^t b_3(s)dW(s)}$, and introduce the following simple transformation:

$$\tilde{z} = \vartheta z. \tag{9.19}$$

Then, one can check that $\tilde{z}$ satisfies

$$\begin{cases} \tilde{z}_t - \sum_{j,k=1}^{n} \left(a^{jk}\tilde{z}_{x_j}\right)_{x_k} = \sum_{j=1}^{n} \tilde{b}_{1j}\tilde{z}_{x_j} + \tilde{b}_2\tilde{z} & \text{in } Q,\\ \tilde{z} = 0 & \text{on } \Sigma,\\ \tilde{z}(0) = z_0 & \text{in } G,\end{cases} \tag{9.20}$$

where

$$\tilde{b}_{1j} = b_{1j} - 2\sum_{k=1}^{n} a^{jk}\int_0^t b_{3,x_k}(s)dW(s), \quad j = 1,2,\cdots,n,$$

$$\begin{aligned}\tilde{b}_2 = b_2 - \frac{1}{2}b_3^2 - \sum_{j=1}^{n} b_{1j}\int_0^t b_{3,x_j}(s)dW(s) - \sum_{j,k=1}^{n}\left(a^{jk}\int_0^t b_{3,x_j}(s)dW(s)\right)_{x_k}\\ + \sum_{j,k=1}^{n} a^{jk}\int_0^t b_{3,x_j}(s)dW(s)\int_0^t b_{3,x_k}(s)dW(s).\end{aligned}$$

One may expect that the observability estimate (9.93) (in Theorem 9.28) for (9.16) follows from a similar observability estimate but for the random parabolic equation (9.20). But usually this is not the case. Indeed, it is easy to see that, in general neither $\tilde{b}_{1j}$ nor $\tilde{b}_2$ is guaranteed to be bounded (w.r.t. the sample point ω) unless b_3 is space-independent. Because of this, it seems that one cannot use Corollary 9.32 (which is a global Carleman estimate for random parabolic equations) directly to derive any useful observability estimate for (9.20).

Remark 9.6. When $m > 1$, general speaking, one cannot use the above technique (by introducing a suitable transformation like (9.19)) to reduce (9.16) to a random parabolic system even if $b_3 \in L^\infty_{\mathbb{F}}(0,T;W^{2,\infty}(G;\mathbb{R}^{m\times m}))$.

9.2 Controllability of a Class of Stochastic Parabolic Systems

This section is based on [225]. Throughout this section, we consider the following stochastic parabolic system:

$$\begin{cases} dy - \sum_{j,k=1}^{n} (a^{jk} y_{x_j})_{x_k} dt = [a(t)y + \chi_E(t)\chi_{G_0}(x)u]dt + b(t)ydW(t) & \text{in } Q, \\ y = 0 & \text{on } \Sigma, \\ y(0) = y_0 & \text{in } G, \end{cases} \tag{9.21}$$

where $a(\cdot) \in L^\infty_{\mathbb{F}}(0,T;\mathbb{R}^{m\times m})$ and $b(\cdot) \in L^\infty_{\mathbb{F}}(0,T;\mathbb{R}^{m\times m})$ are given, E is a fixed Lebesgue measurable subset in $(0,T)$ with a positive Lebesgue measure (i.e., $\mathbf{m}(E) > 0$). In (9.21), y is the state variable (valued in $L^2(G;\mathbb{R}^m)$), $y_0 \in L^2(G;\mathbb{R}^m)$ is the initial state, u is the control variable, and the control space is $L^\infty_{\mathbb{F}}(0,T;L^2(\Omega;L^2(G;\mathbb{R}^m)))$.

We have the following null controllability result for the system (9.21).

Theorem 9.7. *The system (9.21) is null controllable at time T.*

We shall give a proof of Theorem 9.7 in Subsection 9.2.2.

Remark 9.8. As we mentioned in Section 9.1 that, when $E = (0,T)$, one can use the global Carleman estimate to prove the corresponding null controllability result for the deterministic counterpart of (9.21). However, at least at this moment we do not know how to use a similar method to prove Theorem 9.7 even for the same case that $E = (0,T)$.

Next, we consider the approximate controllability for the system (9.21) under a stronger assumption on the controller $E \times G_0$ than that for the null controllability.

Theorem 9.9. *System (9.21) is approximately controllable at time T if and only if $\mathbf{m}((s,T) \cap E) > 0$ for any $s \in [0,T)$.*

We refer to Subsection 9.2.3 for a proof of Theorem 9.9.

At the first glance, it seems that Theorem 9.9 is unreasonable. If $b(\cdot) \equiv 0$, then the system (9.21) is like a deterministic parabolic equation with a random parameter. The readers may guess that one can obtain the approximate controllability by only assuming that $\mathbf{m}(E) > 0$. However, this is NOT the case. The reason for this comes from our definition of the approximate controllability for the system (9.21). We expect that any element belonging to $L^2_{\mathcal{F}_T}(\Omega; L^2(G))$ rather than $L^2_{\mathcal{F}_s}(\Omega; L^2(G))$ $(s < T)$ can be attached as close as one wants. Hence we need put the control u to be active until the time T.

In some sense, it is surprising that one needs a little more assumption in Theorem 9.9 for the approximate controllability of (9.21) than that in Theorem 9.7 for the null controllability. Indeed, it is well-known that in the deterministic setting, the null controllability is usually stronger than the approximate controllability. But this does NOT remain to be true in the stochastic case. Actually, from Theorem 9.9, we see that the additional condition (compared to the null controllability) that $\mathbf{m}((s,T)\bigcap E) > 0$ for any $s \in [0,T)$ is not only sufficient but also necessary for the approximate controllability of (9.21). Therefore, in the setting of stochastic distributed parameter systems, the null controllability does NOT imply the approximate controllability. This indicates that there exists some essential difference between the controllability theory between the deterministic parabolic equations and its stochastic counterpart (See also Remark 9.8).

9.2.1 Preliminaries

In this subsection, we collect some preliminary results that will be used subsequently.

To begin with, we recall the following known property about Lebesgue measurable sets.

Lemma 9.10. ([205, pp. 256–257]) *For a.e. $\tilde{t} \in E$, there exists a sequence of numbers $\{t_i\}_{i=1}^{\infty} \subset (0,T)$ such that*

$$t_1 < t_2 < \cdots < t_i < t_{i+1} < \cdots < \tilde{t}, \qquad t_i \to \tilde{t} \text{ as } i \to \infty, \tag{9.22}$$

$$\mathbf{m}(E \cap [t_i, t_{i+1}]) \geq \rho_1(t_{i+1} - t_i), \quad i = 1, 2, \cdots, \tag{9.23}$$

$$\frac{t_{i+1} - t_i}{t_{i+2} - t_{i+1}} \leq \rho_2, \quad i = 1, 2, \cdots, \tag{9.24}$$

where ρ_1 and ρ_2 are two positive constants which are independent of i.

Next, let us define an unbounded operator A on $L^2(G)$ as follows:

$$\begin{cases} D(A) = H^2(G) \cap H_0^1(G), \\ Ah = -\displaystyle\sum_{j,k=1}^{n} (a^{jk} h_{x_j})_{x_k}, \quad \forall\, h \in D(A). \end{cases} \tag{9.25}$$

Let $\{\lambda_i\}_{i=1}^{\infty}$ be the eigenvalues of A, and $\{e_i\}_{i=1}^{\infty}$ be the corresponding eigenfunctions satisfying $|e_i|_{L^2(G)} = 1$, $i = 1, 2, 3 \cdots$. It holds that $0 < \lambda_1 \leq \lambda_2 \leq \lambda_3 \leq \cdots \leq \lambda_k \leq \cdots \to \infty$. For any $r \geq \lambda_1$, write $\Lambda_r = \{i \in \mathbb{N} \mid \lambda_i \leq r\}$. We recall the following observability estimate (for partial sums of the eigenfunctions of A), established in [230, Theorem 1.2] (See also [192, Theorem 3] for a special case of this result).

Lemma 9.11. *There exist two positive constants $\mathcal{C}_1 \geq 1$ and $\mathcal{C}_2 \geq 1$ such that*

$$\sum_{i\in\Lambda_r} |a_i|^2 \leq \mathcal{C}_1 e^{\mathcal{C}_2\sqrt{r}} \int_{G_0} \Big| \sum_{i\in\Lambda_r} a_i e_i(x)\Big|^2 dx \tag{9.26}$$

holds for any $r \geq \lambda_1$ and $a_i \in \mathbb{C}$ with $i \in \Lambda_r$.

Further, for any s_1 and s_2 satisfying $0 \leq s_1 < s_2 \leq T$, we introduce the following backward stochastic parabolic system:

$$\begin{cases} dz + \displaystyle\sum_{j,k=1}^{n} (a^{jk} z_{x_j})_{x_k} dt = -[a(t)^\top z + b(t)^\top Z]dt + ZdW(t) & \text{in } (s_1, s_2) \times G, \\ z = 0 & \text{on } (s_1, s_2) \times \Gamma, \\ z(s_2) = \eta & \text{in } G, \end{cases} \tag{9.27}$$

where $\eta \in L^2_{\mathcal{F}_{s_2}}(\Omega; L^2(G;\mathbb{R}^m))$.

Put

$$r_0 = 2|a|_{L^\infty_{\mathbb{F}}(0,T;\mathbb{R}^{m\times m})} + |b|^2_{L^\infty_{\mathbb{F}}(0,T;\mathbb{R}^{m\times m})}.$$

For each $r \geq \lambda_1$, we set $H_r = \text{span}\,\{e_i \mid \lambda_i \leq r\}$ and denote by Π_r the orthogonal projection from $L^2(G)$ to H_r. Write

$$H_r^m = \overbrace{H_r \times H_r \times \cdots \times H_r}^{m \text{ times}}. \tag{9.28}$$

To simplify the notation, we also denote by Π_r the orthogonal projection from $L^2(G;\mathbb{R}^m)$ to H_r^m. We need the following observability result for (9.27) with the final data belonging to $L^2_{\mathcal{F}_{s_2}}(\Omega; H_r^m)$, a proper subspace of $L^2_{\mathcal{F}_{s_2}}(\Omega; L^2(G;\mathbb{R}^m))$.

Proposition 9.12. *For each $r \geq \lambda_1$, the solution to the system (9.27) with $\eta \in L^2_{\mathcal{F}_{s_2}}(\Omega; H_r^m)$ satisfies that*

$$\mathbb{E}\big|z(s_1)\big|^2_{L^2(G;\mathbb{R}^m)} \leq \frac{\mathcal{C}_1 e^{\mathcal{C}_2\sqrt{r}+r_0(s_2-s_1)}}{(\mathbf{m}(E\cap[s_1,s_2]))^2} \big|\chi_E \chi_{G_0} z\big|^2_{L^1_{\mathbb{F}}(s_1,s_2;L^2(\Omega;L^2(G;\mathbb{R}^m)))}, \tag{9.29}$$

whenever $\mathbf{m}(E\cap[s_1,s_2]) \neq 0$.

Proof: Each $\eta \in L^2_{\mathcal{F}_{s_2}}(\Omega; H_r^m)$ can be written as $\eta = \sum_{i\in\Lambda_r} \eta_i e_i(x)$ for some $\eta_i \in L^2_{\mathcal{F}_{s_2}}(\Omega;\mathbb{R}^m)$ with $i \in \Lambda_r$. The solution (z, Z) to (9.27) can be expressed as

$$z=\sum_{i\in\Lambda_r} z_i(t)e_i,\qquad Z=\sum_{i\in\Lambda_r} Z_i(t)e_i,$$

where $z_i(\cdot)\in C_{\mathbb{F}}([s_1,s_2];L^2(\Omega;\mathbb{R}^m))$ and $Z_i(\cdot)\in L^2_{\mathbb{F}}(s_1,s_2;\mathbb{R}^m)$, and satisfy the following equation

$$\begin{cases} dz_i-\lambda_i z_i dt=-[a(t)^\top z_i+b(t)^\top Z_i]dt+Z_i dW(t) & \text{in } [s_1,s_2],\\ z_i(T)=\eta_i. \end{cases}$$

By Lemma 9.11, for any $t\in[s_1,s_2]$, we have

$$\begin{aligned}\mathbb{E}\int_G |z(t)|^2_{\mathbb{R}^m}dx=\mathbb{E}\sum_{i\in\Lambda_r}|z_i(t)|^2_{\mathbb{R}^m}&\le \mathcal{C}_1 e^{\mathcal{C}_2\sqrt{r}}\mathbb{E}\int_{G_0}\Big|\sum_{i\in\Lambda_r} z_i(t)e_i\Big|^2_{\mathbb{R}^m}dx\\ &=\mathcal{C}_1 e^{\mathcal{C}_2\sqrt{r}}\mathbb{E}\int_{G_0}|z(t)|^2_{\mathbb{R}^m}dx.\end{aligned}\tag{9.30}$$

By Itô's formula, we find that

$$d(e^{r_0t}|z|^2_{\mathbb{R}^m})=r_0e^{r_0t}|z|^2_{\mathbb{R}^m}+e^{r_0t}\big(\langle dz,z\rangle_{\mathbb{R}^m}+\langle z,dz\rangle_{\mathbb{R}^m}\big)+e^{r_0t}|dz|^2_{\mathbb{R}^m}.$$

Hence,

$$\begin{aligned}&\mathbb{E}\Big(e^{r_0t}\int_G|z(t)|^2_{\mathbb{R}^m}dx\Big)-\mathbb{E}\Big(e^{r_0s_1}\int_G|z(s_1)|^2_{\mathbb{R}^m}dx\Big)\\ &=r_0\mathbb{E}\int_{s_1}^t\int_G e^{r_0s}|z(s)|^2_{\mathbb{R}^m}dxds+2\sum_{i\in\Lambda_r}\mathbb{E}\int_{s_1}^t e^{r_0s}\lambda_i|z_i(s)|^2_{\mathbb{R}^m}ds\\ &\quad+\mathbb{E}\int_{s_1}^t\int_G e^{r_0s}\big(-\langle a(s)^\top z(s)+b(s)^\top Z(s),z(s)\rangle_{\mathbb{R}^m}\\ &\quad-\langle z(s),a(s)^\top z(s)+b(s)^\top Z(s)\rangle_{\mathbb{R}^m}+|Z(s)|^2_{\mathbb{R}^m}\big)dxds\\ &\ge 2\sum_{i\in\Lambda_r}\mathbb{E}\int_{s_1}^t e^{r_0s}\lambda_i|z_i(s)|^2_{\mathbb{R}^m}ds\ge 0.\end{aligned}\tag{9.31}$$

From (9.30) and (9.31), we obtain that, for any $t\in[s_1,s_2]$,

$$\mathbb{E}\int_G|z(s_1,x)|^2_{\mathbb{R}^m}dx\le\mathcal{C}_1e^{\mathcal{C}_2\sqrt{r}+r_0(s_2-s_1)}\mathbb{E}\int_{G_0}|z(t,x)|^2_{\mathbb{R}^m}dx.\tag{9.32}$$

By (9.32), it follows that

$$\begin{aligned}&\int_{E\cap[s_1,s_2]}\Big[\mathbb{E}\int_G|z(s_1,x)|^2_{\mathbb{R}^m}dx\Big]^{\frac12}dt\\ &\le\big(\mathcal{C}_1e^{\mathcal{C}_2\sqrt{r}+r_0(s_2-s_1)}\big)^{\frac12}\int_{E\cap[s_1,s_2]}\Big[\mathbb{E}\int_{G_0}|z(t,x)|^2_{\mathbb{R}^m}dx\Big]^{\frac12}dt.\end{aligned}$$

Hence, when $\mathbf{m}(E\cap[s_1,s_2])\neq 0$, we obtain that for each $\eta\in L^2_{\mathcal{F}_{s_2}}(\Omega;H^m_r)$,

$$\begin{aligned}&\mathbb{E}\int_G|z(s_1,x)|^2_{\mathbb{R}^m}dx\\&\leq\frac{\mathcal{C}_1e^{\mathcal{C}_2\sqrt{r}+r_0(s_2-s_1)}}{(\mathbf{m}(E\cap[s_1,s_2]))^2}\Big\{\int_{s_1}^{s_2}\Big[\mathbb{E}\int_G|\chi_E(t)\chi_{G_0}(x)z(t,x)|^2_{\mathbb{R}^m}dx\Big]^{\frac{1}{2}}dt\Big\}^2\\&=\frac{\mathcal{C}_1e^{\mathcal{C}_2\sqrt{r}+r_0(s_2-s_1)}}{(\mathbf{m}(E\cap[s_1,s_2]))^2}|\chi_E\chi_{G_0}z|^2_{L^1_{\mathbb{F}}(s_1,s_2;L^2(\Omega;L^2(G;\mathbb{R}^m)))},\end{aligned}$$

which gives (9.29). □

By means of the usual duality argument, Proposition 9.12 yields a partial controllability result for the following controlled system:

$$\begin{cases}dy-\displaystyle\sum_{j,k=1}^n(a^{jk}y_{x_j})_{x_k}dt\\ \qquad=[a(t)y+\chi_E\chi_{G_0}u]dt+b(t)ydW(t) & \text{in }(s_1,s_2)\times G,\\ y=0 & \text{on }(s_1,s_2)\times\Gamma,\\ y(s_1)=y_{s_1} & \text{in }G,\end{cases}\tag{9.33}$$

where $y_{s_1}\in L^2_{\mathcal{F}_{s_1}}(\Omega;L^2(G;\mathbb{R}^m))$. That is, we have the following result.

Proposition 9.13. *If $\mathbf{m}(E\cap[s_1,s_2])\neq 0$, then for every $r\geq\lambda_1$ and $y_{s_1}\in L^2_{\mathcal{F}_{s_1}}(\Omega;L^2(G;\mathbb{R}^m))$, there exists a control $u_r\in L^\infty_{\mathbb{F}}(s_1,s_2;L^2(\Omega;L^2(G;\mathbb{R}^m)))$ such that the solution y to the system (9.33) with $u=u_r$ satisfies that $\Pi_r(y(s_2))=0$, a.s. Moreover, u_r verifies that*

$$|u_r|^2_{L^\infty_{\mathbb{F}}(s_1,s_2;L^2(\Omega;L^2(G;\mathbb{R}^m)))}\leq\frac{\mathcal{C}_1e^{\mathcal{C}_2\sqrt{r}+r_0(s_2-s_1)}}{(\mathbf{m}(E\cap[s_1,s_2]))^2}|y_{s_1}|^2_{L^2_{\mathcal{F}_{s_1}}(\Omega;L^2(G;\mathbb{R}^m))}.\tag{9.34}$$

Proof: Define a subspace H of $L^1_{\mathbb{F}}(s_1,s_2;L^2(\Omega;L^2(G;\mathbb{R}^m)))$:

$$H=\Big\{f=\chi_E\chi_{G_0}z\;\Big|\;(z,Z)\text{ solves (9.27) for some }\eta\in L^2_{\mathcal{F}_{s_2}}(\Omega;H^m_r)\Big\}$$

and a linear functional $\mathcal{L}$ on H:

$$\mathcal{L}(f)=-\mathbb{E}\int_G\langle y_{s_1},z(s_1)\rangle_{\mathbb{R}^m}dx.$$

By Proposition 9.12, it is easy to check that $\mathcal{L}$ is a bounded linear functional on H and

$$|\mathcal{L}|^2_{\mathcal{L}(H;\mathbb{R})}\leq\frac{\mathcal{C}_1e^{\mathcal{C}_2\sqrt{r}+r_0(s_2-s_1)}}{(\mathbf{m}(E\cap[s_1,s_2]))^2}|y_{s_1}|^2_{L^2_{\mathcal{F}_{s_1}}(\Omega;L^2(G;\mathbb{R}^m))}.$$

By the Hahn-Banach Theorem, $\mathcal{L}$ can be extended to a bounded linear functional $\widetilde{\mathcal{L}}$ (satisfying $|\widetilde{\mathcal{L}}|_{\mathcal{L}(L^1_{\mathbb{F}}(s_1,s_2;L^2(\Omega;L^2(G;\mathbb{R}^m)));\mathbb{R})} = |\mathcal{L}|_{\mathcal{L}(H;\mathbb{R})}$) on $L^1_{\mathbb{F}}(s_1, s_2; L^2(\Omega; L^2(G;\mathbb{R}^m)))$. By Theorem 2.73, there exists a control $u_r \in L^\infty_{\mathbb{F}}(s_1, s_2; L^2(\Omega; L^2(G;\mathbb{R}^m)))$ such that

$$\mathbb{E}\int_{s_1}^{s_2}\int_G \langle\, u_r, f\,\rangle_{\mathbb{R}^m} dxdt = \widetilde{\mathcal{L}}(f), \quad \forall\, f \in L^1_{\mathbb{F}}(s_1, s_2; L^2(\Omega; L^2(G;\mathbb{R}^m))).$$

Particularly, for any $\eta \in L^2_{\mathcal{F}_{s_2}}(\Omega; H^m_r)$, the corresponding solution (z, Z) to (9.27) satisfies

$$\mathbb{E}\int_{s_1}^{s_2}\int_G \langle\, u_r, \chi_E\chi_{G_0} z\,\rangle_{\mathbb{R}^m} dxdt = -\mathbb{E}\int_G \langle\, y_{s_1}, z(s_1)\,\rangle_{\mathbb{R}^m} dx. \tag{9.35}$$

Applying Itô's formula to $\langle\, y, z\,\rangle_{\mathbb{R}^m}$, where y solves the system (9.33) with $u = u_r$, we obtain that

$$\begin{aligned}&\mathbb{E}\int_G \langle\, y(s_2), \eta\,\rangle_{\mathbb{R}^m} dx - \mathbb{E}\int_G \langle\, y_{s_1}, z(s_1)\,\rangle_{\mathbb{R}^m} dx\\ &= \mathbb{E}\int_{s_1}^{s_2}\int_G \langle\, \chi_E\chi_{G_0} u_r, z\,\rangle_{\mathbb{R}^m} dxdt.\end{aligned} \tag{9.36}$$

Combining (9.35) and (9.36), we arrive at

$$\mathbb{E}\int_G \langle\, y(s_2), \eta\,\rangle_{\mathbb{R}^m} dx = 0, \qquad \forall\, \eta \in L^2_{\mathcal{F}_{s_2}}(\Omega; H^m_r),$$

which implies that $\Pi_r(y(s_2)) = 0$, a.s. Moreover,

$$|u_r|_{L^\infty_{\mathbb{F}}(s_1,s_2;L^2(\Omega;L^2(G;\mathbb{R}^m)))} = |\mathcal{L}|_{\mathcal{L}(H;\mathbb{R})},$$

which yields (9.34). $\square$

Finally, for any $s \in [0, T)$, we consider the following equation:

$$\begin{cases} dy - \displaystyle\sum_{j,k=1}^n (a^{jk} y_{x_j})_{x_k} dt = a(t)ydt + b(t)ydW(t) & \text{in } (s, T) \times G,\\ y = 0 & \text{on } (s, T) \times \Gamma,\\ y(s) = y_s & \text{in } G, \end{cases} \tag{9.37}$$

where $y_s \in L^2_{\mathcal{F}_s}(\Omega; L^2(G;\mathbb{R}^m))$. Let us show the following decay result for the system (9.37).

Proposition 9.14. *Let $r \geq \lambda_1$. Then, for any $y_s \in L^2_{\mathcal{F}_s}(\Omega; L^2(G;\mathbb{R}^m))$ with $\Pi_r(y_s) = 0$, a.s., the corresponding solution y to (9.37) satisfies that*

$$\mathbb{E}|y(t)|^2_{L^2(G;\mathbb{R}^m)} \leq e^{-(2r-r_0)(t-s)}|y_s|^2_{L^2_{\mathcal{F}_s}(\Omega;L^2(G;\mathbb{R}^m))}, \quad \forall\, t \in [s, T]. \tag{9.38}$$

Proof: Since $y_s \in L^2_{\mathcal{F}_s}(\Omega; L^2(G;\mathbb{R}^m))$ satisfying $\Pi_r(y_s) = 0$, we see that $y_s = \sum_{i\in\mathbb{N}\setminus\Lambda_r} y_s^i e_i$ for some $y_s^i \in L^2_{\mathcal{F}_s}(\Omega;\mathbb{R}^m)$ with $i \in \mathbb{N}\setminus\Lambda_r$. Clearly, the solution y to (9.37) can be expressed as $y = \sum_{i\in\mathbb{N}\setminus\Lambda_r} y^i(t)e_i$, where $y^i(\cdot) \in C_{\mathbb{F}}([s,T]; L^2(\Omega;\mathbb{R}^m))$ solves the following stochastic differential equation:

$$\begin{cases} dy^i + \lambda_i y^i dt = a(t)y^i dt + b(t)y^i dW(t) \ \text{ in } [s,T],\\ y^i(s) = y_s^i. \end{cases}$$

By Itô's formula, we have that

$$\begin{aligned} d\big(e^{(2r-r_0)(t-s)}|y|^2_{\mathbb{R}^m}\big) &= e^{(2r-r_0)(t-s)}\big(\langle\, dy, y\,\rangle_{\mathbb{R}^m} + \langle\, y, dy\,\rangle_{\mathbb{R}^m}\big)\\ &\quad + e^{(2r-r_0)(t-s)}|dy|^2_{\mathbb{R}^m} + (2r-r_0)e^{(2r-r_0)(t-s)}|y|^2_{\mathbb{R}^m}. \end{aligned}$$

Hence, by $\lambda_i > r$ for each $i \in \mathbb{N}\setminus\Lambda_r$ and recalling that $r_0 = 2|a|_{L^\infty_{\mathbb{F}}(0,T;\mathbb{R}^{m\times m})} + |b|^2_{L^\infty_{\mathbb{F}}(0,T;\mathbb{R}^{m\times m})}$, we arrive at

$$\begin{aligned} &\mathbb{E}\int_G e^{(2r-r_0)(t-s)}|y(t)|^2_{\mathbb{R}^m}dx - \mathbb{E}\int_G |y(s)|^2_{\mathbb{R}^m}dx\\ &= -2\sum_{i\in\mathbb{N}\setminus\Lambda_r}\lambda_i\mathbb{E}\int_s^t e^{(2r-r_0)(\sigma-s)}|y^i(\sigma)|^2_{\mathbb{R}^m}d\sigma\\ &\quad +\mathbb{E}\int_s^t\int_G e^{(2r-r_0)(\sigma-s)}\big(\langle\, ay, y\,\rangle_{\mathbb{R}^m} + \langle\, y, ay\,\rangle_{\mathbb{R}^m}\big)dxd\sigma\\ &\quad +\mathbb{E}\int_s^t\int_G e^{(2r-r_0)(\sigma-s)}|b(\sigma)y(\sigma)|^2_{\mathbb{R}^m}dxd\sigma\\ &\quad +(2r-r_0)\mathbb{E}\int_s^t e^{(2r-r_0)(\sigma-s)}|y(\sigma)|^2_{\mathbb{R}^m}dxd\sigma\\ &\le 0, \end{aligned}$$

which gives the desired estimate (9.38) immediately. □

9.2.2 Proof of the Null Controllability

This subsection is devoted to proving the null controllability result for the system (9.21), i.e., Theorem 9.7. For simplicity, we assume that $m = 1$. The proof for general m is similar.

By Lemma 9.10, we may take a number $\tilde{t} \in E$ with $\tilde{t} < T$ and a sequence $\{t_N\}_{N=1}^\infty \subset (0,T)$ such that (9.22)–(9.24) hold for some positive numbers ρ_1 and ρ_2.

Write $\tilde{y}_0 = \psi(t_1)$, where $\psi(\cdot)$ solves the following stochastic parabolic system:

$$\begin{cases} d\psi - \sum_{j,k=1}^{n} (a^{jk}\psi_{x_j})_{x_k} dt = a(t)\psi dt + b(t)\psi dW(t) & \text{in } (0,t_1)\times G, \\ \psi = 0 & \text{on } (0,t_1)\times \Gamma, \\ \psi(0) = y_0 & \text{in } G. \end{cases}$$

Let us consider the following controlled stochastic parabolic system:

$$\begin{cases} d\tilde{y} - \sum_{j,k=1}^{n} (a^{jk}\tilde{y}_{x_j})_{x_k} dt \\ \qquad = [a(t)\tilde{y} + \chi_E\chi_{G_0}\tilde{u}]dt + b(t)\tilde{y}dW(t) & \text{in } (t_1,\tilde{t}\,)\times G, \\ \tilde{y} = 0 & \text{on } (t_1,\tilde{t}\,)\times \Gamma, \\ \tilde{y}(t_1) = \tilde{y}_0 & \text{in } G. \end{cases} \tag{9.39}$$

It suffices to find a control $\tilde{u} \in L^\infty_{\mathbb{F}}(t_1,\tilde{t}; L^2(\Omega; L^2(G)))$ with

$$|\tilde{u}|^2_{L^\infty_{\mathbb{F}}(t_1,\tilde{t};L^2(\Omega;L^2(G)))} \le C\mathbb{E}|\tilde{y}_0|^2_{L^2(\Omega)}, \tag{9.40}$$

such that the solution $\tilde{y}$ to (9.39) satisfies $\tilde{y}(\tilde{t}\,) = 0$ in G, a.s.

Set $I_N = [t_{2N-1}, t_{2N}]$ and $J_N = [t_{2N}, t_{2N+1}]$ for $N \in \mathbb{N}$. Then $[t_1,\tilde{t}\,) = \bigcup_{N=1}^\infty (I_N \cup J_N)$. Clearly, $\mathbf{m}(E\cap I_N) > 0$ and $\mathbf{m}(E\cap J_N) > 0$. We will introduce a suitable control on each I_N and allow the system to evolve freely on every J_N. Also, we fix a suitable, strictly increasing sequence $\{r_N\}_{N=1}^\infty$ of positive integers (to be given later) satisfying that $\lambda_1 \le r_1 < r_2 < \cdots < r_N \to \infty$ as $N \to \infty$.

We consider first the controlled stochastic parabolic system on the interval $I_1 = [t_1, t_2]$ as follows:

$$\begin{cases} dy_1 - \sum_{j,k=1}^{n} (a^{jk}y_{1,x_j})_{x_k} dt \\ \qquad = [a(t)y_1 + \chi_E\chi_{G_0}u_1]dt + b(t)y_1 dW(t) & \text{in } (t_1,t_2)\times G, \\ y_1 = 0 & \text{on } (t_1,t_2)\times \Gamma, \\ y_1(t_1) = \tilde{y}_0 & \text{in } G. \end{cases} \tag{9.41}$$

By Proposition 9.13, there exists a control $u_1 \in L^\infty_{\mathbb{F}}(t_1,t_2; L^2(\Omega; L^2(G)))$ with the estimate:

$$|u_1|^2_{L^\infty_{\mathbb{F}}(t_1,t_2;L^2(\Omega;L^2(G)))} \le \frac{\mathcal{C}_1 e^{\mathcal{C}_2\sqrt{r_1}+r_0 T}}{(\mathbf{m}(E\cap[t_1,t_2]))^2}\mathbb{E}|\tilde{y}_0|^2_{L^2(G)},$$

such that $\Pi_{r_1}(y(t_2)) = 0$ in G, a.s. By (9.23), we see that

$$|u_1|^2_{L^\infty_{\mathbb{F}}(t_1,t_2;L^2(\Omega;L^2(G)))} \le \frac{\mathcal{C}_1 e^{\mathcal{C}_2\sqrt{r_1}+r_0 T}}{\rho_1^2(t_2-t_1)^2}\mathbb{E}|\tilde{y}_0|^2_{L^2(G)}. \tag{9.42}$$

Applying Itô's formula to $e^{-(r_0+1)t}|y_1(t)|^2_{L^2(G)}$, similar to the proof of (9.31) and Theorem 3.24, we obtain that

$$\begin{aligned}
&e^{-(r_0+1)t_2}\mathbb{E}|y_1(t_2)|^2_{L^2(G)}\\
&= e^{-(r_0+1)t_1}\mathbb{E}|y_1(t_1)|^2_{L^2(G)} - (r_0+1)\mathbb{E}\int_{t_1}^{t_2} e^{-(r_0+1)s}\int_G |y_1|^2 dxds\\
&\quad -2\sum_{j,k=1}^{n}\mathbb{E}\int_{t_1}^{t_2} e^{-(r_0+1)s}\int_G a^{jk}y_{1,x_j}y_{1,x_k}dxds\\
&\quad +\mathbb{E}\int_{t_1}^{t_2} e^{-(r_0+1)s}\int_G [2a(s)|y_1|^2+|b(s)y_1|^2]dxds\\
&\quad +2\mathbb{E}\int_{t_1}^{t_2} e^{-(r_0+1)s}\int_G \chi_E\chi_{G_0}u_1y_1dxds\\
&\le e^{-(r_0+1)t_1}\mathbb{E}|y_1(t_1)|^2_{L^2(G)} + \mathbb{E}\int_{t_1}^{t_2} e^{-(r_0+1)s}\int_G |u_1|^2dxds\\
&\le e^{-(r_0+1)t_1}\mathbb{E}|\tilde y_0|^2_{L^2(G)} + \frac{e^{-(r_0+1)t_1}-e^{-(r_0+1)t_2}}{r_0+1}|u_1|^2_{L^\infty_{\mathbb{F}}(t_1,t_2;L^2(\Omega;L^2(G)))}.
\end{aligned}$$

Hence, in view of (9.42),

$$\mathbb{E}|y_1(t_2)|^2_{L^2(G)} \le \frac{\mathcal{C}_3e^{\mathcal{C}_3\sqrt{r_1}}}{(t_2-t_1)^2}\mathbb{E}|\tilde y_0|^2_{L^2(G)}. \tag{9.43}$$

where $\mathcal{C}_3 = \max(2\rho_1^{-2}\mathcal{C}_1e^{(2r_0+1)T}, \mathcal{C}_2)$.

Then, on the interval $J_1 \equiv [t_2,t_3]$, we consider the following stochastic parabolic system without control:

$$\begin{cases}
dz_1 - \displaystyle\sum_{j,k=1}^{n}(a^{jk}z_{1,x_j})_{x_k}dt = a(t)z_1dt + b(t)z_1dW(t) & \text{in } (t_2,t_3)\times G,\\
z_1 = 0 & \text{on } (t_2,t_3)\times \Gamma,\\
z_1(t_2) = y_1(t_2) & \text{in } G.
\end{cases}$$

Since $\Pi_{r_1}(y_1(t_2)) = 0$, a.s., by Proposition 9.14, we have

$$\begin{aligned}
\mathbb{E}|z_1(t_3)|^2_{L^2(G)} &\le e^{(-2r_1+r_0)(t_3-t_2)}\mathbb{E}|y_1(t_2)|^2_{L^2(G)}\\
&\le \frac{\mathcal{C}_3e^{\mathcal{C}_3\sqrt{r_1}}}{(t_2-t_1)^2}e^{(-2r_1+r_0)(t_3-t_2)}\mathbb{E}|\tilde y_0|^2_{L^2(G)}.
\end{aligned} \tag{9.44}$$

Generally, on the interval I_N with $N\in\mathbb{N}\setminus\{1\}$, we consider a controlled stochastic parabolic system as follows:

$$\begin{cases} dy_N - \sum_{j,k=1}^{n} (a^{jk} y_{N,x_j})_{x_k} dt \\ \quad = [a(t)y_N + \chi_E \chi_{G_0} u_N]dt + b(t)y_N dW(t) & \text{in } (t_{2N-1}, t_{2N}) \times G, \\ y_N = 0 & \text{on } (t_{2N-1}, t_{2N}) \times \Gamma, \\ y_N(t_{2N-1}) = z_{N-1}(t_{2N-1}) & \text{in } G. \end{cases}$$

Similar to the above argument (See the proof of (9.42) and (9.43)), one can find a control $u_N \in L^\infty_{\mathbb{F}}(t_{2N-1}, t_{2N}; L^2(\Omega; L^2(G)))$ with the estimate:

$$|u_N|^2_{L^\infty_{\mathbb{F}}(t_{2N-1},t_{2N};L^2(\Omega;L^2(G)))} \leq \frac{C_1 e^{C_2\sqrt{r_N}+r_0 T}}{\rho_1^2 (t_{2N} - t_{2N-1})^2} \mathbb{E}|z_{N-1}(t_{2N-1})|^2_{L^2(G)}, \tag{9.45}$$

such that $\Pi_{r_N}(y_N(t_{2N})) = 0$ in G, a.s. Moreover,

$$\mathbb{E}|y_N(t_{2N})|^2_{L^2(G)} \leq \frac{C_3 e^{C_3\sqrt{r_N}}}{(t_{2N} - t_{2N-1})^2} \mathbb{E}|z_{N-1}(t_{2N-1})|^2_{L^2(G)}. \tag{9.46}$$

On the interval J_N, we consider the following stochastic parabolic system without control:

$$\begin{cases} dz_N - \sum_{j,k=1}^{n} (a^{jk} z_{N,x_j})_{x_k} dt \\ \quad = a(t)z_N dt + b(t)z_N dW(t) & \text{in } (t_{2N}, t_{2N+1}) \times G, \\ z_N = 0 & \text{on } (t_{2N}, t_{2N+1}) \times \Gamma, \\ z_N(t_{2N}) = y_N(t_{2N}) & \text{in } G. \end{cases}$$

Since $\Pi_{r_N}(y_N(t_{2N})) = 0$, a.s., by Proposition 9.14 and similar to (9.44), and recalling that $y_N(t_{2N-1}) = z_{N-1}(t_{2N-1})$ in G, we have

$$\begin{aligned} &\mathbb{E}|z_N(t_{2N+1})|^2_{L^2(G)} \\ &\leq \frac{C_3 e^{C_3\sqrt{r_N}}}{(t_{2N} - t_{2N-1})^2} e^{(-2r_N + r_0)(t_{2N+1} - t_{2N})} \mathbb{E}|y_N(t_{2N-1})|^2_{L^2(G)} \\ &\leq \frac{C_4 e^{C_4\sqrt{r_N}}}{(t_{2N} - t_{2N-1})^2} e^{-2r_N(t_{2N+1} - t_{2N})} \mathbb{E}|z_{N-1}(t_{2N-1})|^2_{L^2(G)}, \end{aligned} \tag{9.47}$$

where $C_4 = C_3 e^{r_0 T}$.

Inductively, by (9.24) and (9.47), we conclude that, for all $N \geq 1$,

$$
\begin{aligned}
&\mathbb{E}|z_N(t_{2N+1})|^2_{L^2(G)}\\
&\le \frac{\mathcal{C}_4^N e^{\mathcal{C}_4(\sqrt{r_N}+\sqrt{r_{N-1}}+\cdots+\sqrt{r_1})}}{(t_{2N}-t_{2N-1})^2(t_{2N-2}-t_{2N-3})^2\cdots(t_2-t_1)^2}\\
&\quad\times\exp\Big\{-2r_N(t_{2N+1}-t_{2N})-2r_{N-1}(t_{2N-1}-t_{2N-2})\\
&\quad-\cdots-2r_1(t_3-t_2)\Big\}\mathbb{E}|\tilde{y}_0|^2_{L^2(G)} \qquad (9.48)\\
&\le \frac{\mathcal{C}_4^N\exp\big\{\mathcal{C}_4 N\sqrt{r_N}-2r_N(t_{2N+1}-t_{2N})\big\}}{(t_{2N}-t_{2N-1})^2(t_{2N-2}-t_{2N-3})^2\cdots(t_2-t_1)^2}\mathbb{E}|\tilde{y}_0|^2_{L^2(G)}\\
&\le \frac{\mathcal{C}_4^N\rho_2^{2N(N-1)}\exp\big\{\mathcal{C}_4 N\sqrt{r_N}-2(t_2-t_1)\rho_2^{1-2N}r_N\big\}}{(t_2-t_1)^{2N}}\mathbb{E}|\tilde{y}_0|^2_{L^2(G)}.
\end{aligned}
$$

By (9.24), (9.45)–(9.46) and (9.48), we see that

$$
\begin{aligned}
&|u_N|^2_{L^\infty_{\mathbb{F}}(t_{2N-1},t_{2N};L^2(\Omega;L^2(G)))}\\
&\le \frac{\mathcal{C}_1\mathcal{C}_4^{N-1}\rho_2^{2N(N-1)}}{\rho_1^2(t_2-t_1)^{2N}}\exp\Big\{\mathcal{C}_2\sqrt{r_N}+r_0T+\mathcal{C}_4(N-1)\sqrt{r_{N-1}} \qquad (9.49)\\
&\quad-2(t_2-t_1)\rho_2^{3-2N}r_{N-1}\Big\}\mathbb{E}|\tilde{y}_0|^2_{L^2(G)}.
\end{aligned}
$$

and

$$
\begin{aligned}
&\mathbb{E}|y_N(t_{2N})|^2_{L^2(G)}\\
&\le \frac{\mathcal{C}_3\mathcal{C}_4^{N-1}\rho_2^{2N(N-1)}}{(t_2-t_1)^{2N}}\exp\Big\{\mathcal{C}_3\sqrt{r_N}+\mathcal{C}_4(N-1)\sqrt{r_{N-1}} \qquad (9.50)\\
&\quad-2(t_2-t_1)\rho_2^{3-2N}r_{N-1}\Big\}\mathbb{E}|\tilde{y}_0|^2_{L^2(G)}.
\end{aligned}
$$

We now choose $r_N=\max(2^{N^2},[\lambda_1]+1)$. From (9.49)–(9.50), it is easy to see that, whenever N is large enough,

$$
|u_N|^2_{L^\infty_{\mathbb{F}}(t_{2N-1},t_{2N};L^2(\Omega;L^2(G)))}\le\frac{1}{2^N}\mathbb{E}|\tilde{y}_0|^2_{L^2(G)} \qquad (9.51)
$$

and

$$
\mathbb{E}|y_N(t_{2N})|^2_{L^2(G)}\le\frac{1}{2^N}\mathbb{E}|\tilde{y}_0|^2_{L^2(G)}. \qquad (9.52)
$$

We now construct a control $\tilde{u}$ by setting

$$
\tilde{u}(t,x)=\begin{cases}u_N(t,x), & (t,x)\in I_N\times G,\quad N\ge 1,\\ 0, & (t,x)\in J_N\times G,\quad N\ge 1.\end{cases} \qquad (9.53)
$$

By (9.51), we see that $\tilde{u}\in L^\infty_{\mathbb{F}}(t_1,\tilde{t};L^2(\Omega;L^2(G)))$ satisfying (9.40). Let $\tilde{y}$ be the solution to the system (9.39) corresponding to the control constructed in (9.53). Then $\tilde{y}(\cdot)=y_N(\cdot)$ on $I_N\times G$. By (9.52) and recalling that $t_{2N}\to\tilde{t}$ as $N\to\infty$, we deduce that $\tilde{y}(\tilde{t})=0$, a.s. This completes the proof of Theorem 9.7. □

9.2.3 Proof of the Approximate Controllability

This subsection is devoted to giving a proof of the approximate controllability result for the system (9.21), i.e., Theorem 9.9.

To begin with, we show the following two preliminary results, which have some independent interests.

Proposition 9.15. *If* $\mathbf{m}((s,T)\bigcap E) > 0$ *for any* $s \in [0,T)$*, then for any given* $\eta \in L^2_{\mathcal{F}_T}(\Omega; L^2(G;\mathbb{R}^m))$*, the corresponding solution to* (9.27) *with* $s_1 = 0$ *and* $s_2 = T$ *satisfies*

$$|z(s)|_{L^2_{\mathcal{F}_s}(\Omega;L^2(G;\mathbb{R}^m))} \leq \mathcal{C}(s)|\chi_E\chi_{G_0}z|_{L^1_{\mathbb{F}}(s,T;L^2(\Omega;L^2(G;\mathbb{R}^m)))}. \tag{9.54}$$

Here and henceforth, $\mathcal{C}(s) > 0$ *is a generic constant depending on* s.

Proof: The proof of Proposition 9.15 is very similar to that of Theorem 7.17. We consider the following controlled stochastic parabolic system:

$$\begin{cases} dy - \displaystyle\sum_{j,k=1}^{n}(a^{jk}y_{x_j})_{x_k}dt \\ \qquad = [a(t)y + \chi_{(s,T)\cap E}\chi_{G_0}u]dt + b(t)ydW(t) & \text{in } (s,T)\times G, \\ y = 0 & \text{on } (s,T)\times \Gamma, \\ y(s) = y_s & \text{in } G, \end{cases} \tag{9.55}$$

where y is the state variable, u is the control variable, the initial state $y_s \in L^2_{\mathcal{F}_s}(\Omega; L^2(G;\mathbb{R}^m))$ and the control $u(\cdot) \in L^\infty_{\mathbb{F}}(s,T;L^2(\Omega;L^2(G;\mathbb{R}^m)))$. By the proof of Theorem 9.7, it is easy to show that the system (9.55) is null controllable, i.e., for any $y_s \in L^2_{\mathcal{F}_s}(\Omega;L^2(G;\mathbb{R}^m))$, there exists a control $u \in L^\infty_{\mathbb{F}}(s,T;L^2(\Omega;L^2(G;\mathbb{R}^m)))$ such that $y(T) = 0$ in G, a.s., and

$$|u|^2_{L^\infty_{\mathbb{F}}(s,T;L^2(\Omega;L^2(G;\mathbb{R}^m)))} \leq \mathcal{C}(s)|y_s|^2_{L^2_{\mathcal{F}_s}(\Omega;L^2(G;\mathbb{R}^m))}. \tag{9.56}$$

Applying Itô's formula to $\langle y,z\rangle_{\mathbb{R}^m}$, where y and (z,Z) solve respectively (9.55) and (9.27) with $s_1 = 0$ and $s_2 = T$, and noting that $y(T) = 0$ in G, a.s., we obtain that

$$-\mathbb{E}\int_G \langle y_s, z(s)\rangle_{\mathbb{R}^m}dx = \mathbb{E}\int_{(s,T)\cap E}\int_{G_0}\langle u,z\rangle_{\mathbb{R}^m}dxdt.$$

Choosing $y_s = -z(s)$ in (9.55), we then have

$$\begin{aligned} &\mathbb{E}\int_G |z(s)|^2_{\mathbb{R}^m}dx = \mathbb{E}\int_{(s,T)\cap E}\int_{G_0}\langle u,z\rangle_{\mathbb{R}^m}dxdt \\ &\leq |u|_{L^\infty_{\mathbb{F}}(s,T;L^2(\Omega;L^2(G;\mathbb{R}^m)))}|\chi_{(s,T)\cap E}\chi_{G_0}z|_{L^1_{\mathbb{F}}(s,T;L^2(\Omega;L^2(G;\mathbb{R}^m)))} \\ &\leq \mathcal{C}(s)\Big(\mathbb{E}\int_G |z(s)|^2_{\mathbb{R}^m}dx\Big)^{\frac{1}{2}}|\chi_{(s,T)\cap E}\chi_{G_0}z|_{L^1_{\mathbb{F}}(s,T;L^2(\Omega;L^2(G;\mathbb{R}^m)))}, \end{aligned}$$

which gives immediately the desired estimate (9.54). □

As an easy consequence of Proposition 9.15, we have the following unique continuation property for solutions to (9.27) with $s_1 = 0$ and $s_2 = T$.

Corollary 9.16. *If $\mathbf{m}((s,T)\cap E) > 0$ for any $s \in [0,T)$, then any solution (z,Z) to (9.27) with $s_1 = 0$ and $s_2 = T$ vanishes identically in Q, a.s. provided that $z = 0$ in $G_0 \times E$, a.s.*

Proof: Since $z = 0$ in $G_0 \times E$, a.s., by Proposition 9.15, we see that $z(s) = 0$ in G, a.s., for any $s \in [0,T)$. Therefore, $z \equiv 0$ in Q, a.s. □

Remark 9.17. If the condition $\mathbf{m}((s,T)\cap E) > 0$ for any $s \in [0,T)$ was not assumed, the conclusion in Corollary 9.16 might fail to be true. This can be shown by the following counterexample. Let E satisfy that $\mathbf{m}(E) > 0$ and $\mathbf{m}((s_0,T)\cap E) = 0$ for some $s_0 \in [0,T)$. Let $(z_1, Z_1) = 0$ in $(0,s_0)\times G$, a.s. and ξ_2 be a nonzero process in $L^2_{\mathbb{F}}(s_0,T;\mathbb{R}^m)$ (Then $Z_2 \equiv \xi_2 e_1$ is a nonzero process in $L^2_{\mathbb{F}}(s_0,T;L^2(G;\mathbb{R}^m))$). Solving the following forward stochastic differential equation:

$$\begin{cases} d\zeta_1 - \lambda_1\zeta_1 dt = -[a(t)^\top \zeta_1 + b(t)^\top \xi_2]dt + \xi_2 dW(t) \ \text{ in } [s_0,T],\\ \zeta_1(s_0) = 0, \end{cases}$$

we find a nonzero $\zeta_1 \in L^2_{\mathbb{F}}(\Omega;C([s_0,T];\mathbb{R}^m))$. In this way, we find a nonzero solution $(z_2, Z_2) \equiv (\zeta_1 e_1, \xi_2 e_1) \in L^2_{\mathbb{F}}(\Omega;C([s_0,T];L^2(G;\mathbb{R}^m))) \times L^2_{\mathbb{F}}(s_0,T;L^2(G;\mathbb{R}^m))$ to the following forward stochastic partial differential equation:

$$\begin{cases} dz_2 + \displaystyle\sum_{j,k=1}^{n} (a^{jk} z_{2,x_j})_{x_k} dt \\ = -[a(t)^\top z_2 + b(t)^\top Z_2]dt + Z_2 dW(t) & \text{in } (s_0,T)\times G,\\ z_2 = 0 & \text{on } (s_0,T)\times \Gamma,\\ z_2(s_0) = 0 & \text{in } G. \end{cases} \tag{9.57}$$

(Note however that one cannot solve the system (9.57) directly because this system is not well-posed). Put

$$(z,Z) = \begin{cases} (z_1, Z_1), & \text{in } (0,s_0)\times G,\\ (z_2, Z_2), & \text{in } (s_0,T)\times G. \end{cases}$$

Then, (z,Z) is a nonzero solution to (9.27) with $s_1 = 0$ and $s_2 = T$, and $z = 0$ in $G_0 \times E$, a.s. Note also that, the nonzero solution constructed for the system (9.57) indicates that, in general, the forward uniqueness does NOT hold for backward stochastic differential equations.

We are now in a position to prove Theorem 9.9.

Proof of Theorem 9.9: Similar to the proof of Theorem 7.19, the **"if" part** follows from Corollary 9.16. To prove the **"only if" part**, we use the contradiction argument. Assume that $\mathbf{m}((s_0,T)\cap E)=0$ for some $s_0\in[0,T)$. Since the system (9.21) is approximately controllable at time T, similar to the proof of Theorem 7.19 again, we deduce that any solution (z,Z) to (9.27) with $s_1=0$ and $s_2=T$ vanishes identically in Q provided that $z=0$ in $G_0\times E$, a.s. This contradicts the counterexample in Remark 9.17. □

9.3 Controllability of a Class of Stochastic Parabolic Systems by one Control

In this section, we consider the controllability problem for a class of coupled stochastic parabolic systems by using only one control. We shall see that, quite different from the deterministic setting, the controllability of this sort of stochastic partial differential equations is NOT robust with respect to the coefficients of lower order terms. This section is based on the paper [219].

We are concerned with the following controlled linear coupled stochastic parabolic system:

$$\begin{cases} dy-\displaystyle\sum_{j,k=1}^{n}(a^{jk}y_{x_j})_{x_k}dt \\ \quad=[a_1(t)y+a_2(t)z+\chi_{G_0}u]\,dt+[a_3(t)y+a_4(t)z]\,dW(t) & \text{in } Q,\\ dz-\displaystyle\sum_{j,k=1}^{n}(a^{ij}z_{x_j})_{x_k}dt \\ \quad=[a_5(t)y+a_6(t)z]\,dt+[a_7(t)z+a_8(t)y]\,dW(t) & \text{in } Q,\\ y=z=0 & \text{on } \Sigma,\\ y(0)=y_0,\ z(0)=z_0 & \text{in } G. \end{cases} \tag{9.58}$$

Here, u is the control variable, (y,z) is the state variable, $(y_0,z_0)\in L^2(G))\times L^2(G)$ is any given initial value, $a_i\in L^\infty_{\mathbb{F}}(0,T)(i=1,\cdots,8)$. In (9.58), the control u acts only on the first equation directly, and the control space is $L^2_{\mathbb{F}}(0,T;L^2(G_0))$; while the state variable y may be regarded as an indirect control entering the second equation through both the drift term a_5y and the diffusion term a_8y.

By means of Theorem 3.24, it is easy to check that for any $(y_0,z_0)\in L^2(G))\times L^2(G)$ and $u\in L^2_{\mathbb{F}}(0,T;L^2(G_0))$, (9.58) admits a unique weak solution $(y,z)\in L^2_{\mathbb{F}}(\Omega;C([0,T];L^2(G;\mathbb{R}^2)))\bigcap L^2_{\mathbb{F}}(0,T;H^1_0(G;\mathbb{R}^2))$.

We suppose that the coefficients a_5 and a_8 satisfy respectively the following conditions:

Condition 9.1 *There exist a constant $l_0 > 0$ and a nonempty interval $\mathbf{I} \subseteq [0,T]$ such that $a_5(\cdot) \geq l_0$ or $a_5(\cdot) \leq -l_0$ on $\mathbf{I}$,* a.e.

Condition 9.2 *There exists a nonempty interval $\tilde{\mathbf{I}} \subseteq \mathbf{I}$ such that $a_8(\cdot) = 0$ on $\tilde{\mathbf{I}}$,* a.e.

Write

$$r_1 \overset{\Delta}{=} \sum_{i=1}^{2} |a_i|_{L_{\mathbb{F}}^{\infty}(0,T)} + \sum_{i=3}^{4} |a_i|^2_{L_{\mathbb{F}}^{\infty}(0,T)} + \sum_{i=5}^{6} |a_i|_{L_{\mathbb{F}}^{\infty}(0,T)} + \sum_{i=7}^{8} |a_i|^2_{L_{\mathbb{F}}^{\infty}(0,T)}. \quad (9.59)$$

To simplify the presentation, we suppose that $\tilde{\mathbf{I}} = (t_1, t_2)$ for some t_1 and t_2 satisfying $0 \leq t_1 < t_2 \leq T$.

We have the following positive null controllability result for the system (9.58).

Theorem 9.18. *Under Conditions 9.1–9.2, the system* (9.58) *is null controllable at time T.*

On the other hand, we have the following negative result for the null controllability of (9.58).

Theorem 9.19. *The system* (9.58) *is not null controllable at time T provided that one of the following conditions is satisfied:*

1) $a_5(\cdot) = 0$ *in* $(0,T) \times \Omega$, a.e.;

2) $a_8(\cdot) \neq 0$ *in* $(0,T) \times \Omega$, a.e. *and* $\frac{a_5(\cdot)}{a_8(\cdot)} \in L_{\mathbb{F}}^{\infty}(0,T)$.

In some sense, Theorem 9.19 means that Conditions 9.1–9.2 are also necessary for the null controllability of (9.58) to hold.

Remark 9.20. Condition 9.1 is used to guarantee that the action of the control in the first equation of (9.58) can be transferred to the second one through the coupling in drift term. Theorem 9.19 indicates that the action transferred from the first equation to the second one should be good enough to control the second one. Indeed, the coupling in diffusion terms may destroy the controllability of (9.58).

To prove Theorem 9.18, we consider the following backward stochastic parabolic system:

$$\begin{cases} d\alpha + \sum_{j,k=1}^{n} (a^{jk}\alpha_{x_j})_{x_k} dt \\ \quad = \big(- a_6(t)\alpha - a_2(t)\beta - a_7(t)K - a_4(t)R\big)dt + KdW(t) & \text{in } Q, \\ d\beta + \sum_{j,k=1}^{n} (a^{jk}\beta_{x_j})_{x_k} dt \\ \quad = \big(- a_5(t)\alpha - a_1(t)\beta - a_3(t)R - a_8(t)K\big)dt + RdW(t) & \text{in } Q, \\ \alpha = \beta = 0 & \text{on } \Sigma, \\ \alpha(T) = \alpha_T, \ \beta(T) = \beta_T & \text{in } G, \end{cases} \tag{9.60}$$

where $(\alpha_T, \beta_T) \in L^2_{\mathcal{F}_T}(\Omega; L^2(G; \mathbb{R}^2))$. Clearly, (9.60) is the dual system of (9.58). By Theorem 4.11, the equation (9.60) admits a unique weak solution $(\alpha, \beta, K, R) \in \big(L^2_{\mathbb{F}}(\Omega; C([0,T]; L^2(G;\mathbb{R}^2))) \bigcap L^2_{\mathbb{F}}(0,T; H_0^1(G;\mathbb{R}^2))\big) \times L^2_{\mathbb{F}}(0,T; L^2(G;\mathbb{R}^2))$.

By Theorem 7.17, we have the following equivalence between the controllability of (9.58) and a suitable observability estimate for (9.60).

Proposition 9.21. *For any $T > 0$, the system* (9.58) *is null controllable at time T if and only if solutions of* (9.60) *satisfy the estimate:*

$$|\alpha(0)|^2_{L^2(G)} + |\beta(0)|^2_{L^2(G)} \leq \mathcal{C}\mathbb{E} \int_0^T \int_{G_0} \beta^2(t,x)dxdt, \tag{9.61}$$

for any $(\alpha_T, \beta_T) \in L^2_{\mathcal{F}_T}(\Omega; L^2(G; \mathbb{R}^2))$.

The corresponding controllability problem might be quite difficult when the coefficients $a_i(t)$ $(i = 1, \cdots, 8)$ in the equation (9.58) are replaced by $a_i(t,x)$ $(i = 1, \cdots, 8)$, respectively. The reason is, as we explained before, the controllability of (9.58) (needless to say the above general case) is not robust and therefore, it seems impossible to use the Carleman-type estimate to treat this problem. On the other hand, it is clear that the usual spectral method does not work for the general form of (9.58) (when its coefficients are both time and space-dependent), either.

9.3.1 Proof of the Null Controllability Result

This subsection is devoted to a proof of the null controllability result, i.e., Theorem 9.18. Similar to the proof of Theorem 9.7, it suffices to establish the analogue of Propositions 9.12–9.14 (for the system (9.58) or (9.60)).

In the sequel of this subsection, $\{\lambda_i\}_{i=1}^{\infty}$, $\{e_i\}_{i=1}^{\infty}$, Λ_r (with $r \geq \lambda_1$) and Π_r are the same as that in Subsection 9.2.1.

First, let $t_1 \leq s_1 < s_2 \leq t_2$ and $m \in \mathbb{N}$. Consider the following backward stochastic parabolic system:

$$\begin{cases} d\alpha+\sum\limits_{j,k=1}^{n}(a^{jk}\alpha_{x_j})_{x_k}dt \\ \quad =\big[-a_6(t)\alpha-a_2(t)\beta-a_7(t)K-a_4(t)R\big]dt+KdW(t) & \text{in } (s_1,s_2)\times G,\\ d\beta+\sum\limits_{j,k=1}^{n}(a^{jk}\beta_{x_j})_{x_k}dt \\ \quad =\big[-a_5(t)\alpha-a_1(t)\beta-a_3(t)R-a_8(t)K\big]dt+RdW(t) & \text{in } (s_1,s_2)\times G,\\ \alpha=\beta=0 & \text{on } (s_1,s_2)\times\Gamma,\\ \alpha(s_2)=\alpha_{s_2},\ \beta(s_2)=\beta_{s_2} & \text{in } G, \end{cases} \tag{9.62}$$

where $(\alpha_{s_2},\beta_{s_2})\in L^2_{\mathcal{F}_{s_2}}(\Omega;H_r^2)$ (Recall (9.28) for the definition of H_r^2).

Similar to Propositions 9.12, we have

Proposition 9.22. *Under Conditions 9.1–9.2, for each $r\geq\lambda_1$ and $(\alpha_{s_2},\beta_{s_2})\in L^2_{\mathcal{F}_{s_2}}(\Omega;H_r^2)$, the solution (α,β,K,R) to (9.62) satisfies that*

$$\mathbb{E}\,|\alpha(s_1)|^2_{L^2(G)}+\mathbb{E}\,|\beta(s_1)|^2_{L^2(G)}\leq\frac{\mathcal{C}e^{\mathcal{C}\sqrt{r}}}{(s_2-s_1)^5}\mathbb{E}\int_{s_1}^{s_2}\int_{G_0}\beta(t,x)^2dxdt. \tag{9.63}$$

Proof: We divide the proof into several steps.

Step 1. For any $(\alpha_{s_2},\beta_{s_2})\in L^2_{\mathcal{F}_{s_2}}(\Omega;H_r^2)$, write

$$\alpha_{s_2}=\sum_{i\in\Lambda_r}\alpha^i_{s_2}e_i,\quad \beta_{s_2}=\sum_{i\in\Lambda_r}\beta^i_{s_2}e_i,$$

where $\alpha^i_{s_2}$ and $\beta^i_{s_2}$ are suitable $\mathcal{F}_{s_2}$-measurable random variables. Then the solution (α,K,β,R) to (9.62) can be represented as follows:

$$\alpha=\sum_{i\in\Lambda_r}\alpha^i(t)e_i,\quad K=\sum_{i\in\Lambda_r}K^i(t)e_i,\quad \beta=\sum_{i\in\Lambda_r}\beta^i(t)e_i,\quad R=\sum_{i\in\Lambda_r}R^i(t)e_i,$$

where $(\alpha^i,\beta^i,K^i,R^i)\in L^2_{\mathbb{F}}(\Omega;C([s_1,s_2];\mathbb{R}^2))\times L^2_{\mathbb{F}}(s_1,s_2;\mathbb{R}^2)$ solves the following backward stochastic differential equation:

$$\begin{cases} d\alpha^i-\lambda_i\alpha^i dt \\ \quad =\big[-a_6(t)\alpha^i-a_2(t)\beta^i-a_7(t)K^i-a_4(t)R^i\big]dt+K^i dW(t) & \text{in } (s_1,s_2),\\ d\beta^i-\lambda_i\beta^i dt \\ \quad =\big[-a_5(t)\alpha^i-a_1(t)\beta^i-a_3(t)R^i-a_8(t)K^i\big]dt+R^i dW(t) & \text{in } (s_1,s_2),\\ \alpha^i(s_2)=\alpha^i_{s_2},\ \beta^i(s_2)=\beta^i_{s_2}. \end{cases}$$

From Lemma 9.11, it follows that, for a.e. $t\in(0,T)$ and $\omega\in\Omega$,

$$
\begin{aligned}
\int_G \beta(t,x)^2 dx &= \sum_{i\in\Lambda_r} |\beta^i(t)|^2 \le \mathcal{C}e^{\mathcal{C}\sqrt{r}} \int_{G_0} \Big|\sum_{i\in\Lambda_r} \beta^i(t)e_i(x)\Big|^2 dx \\
&= \mathcal{C}e^{\mathcal{C}\sqrt{r}} \int_{G_0} \beta(t,x)^2 dx.
\end{aligned}
\tag{9.64}
$$

On the other hand, note that (9.62) is a special case of (9.27). Hence, similarly to (9.31), recalling (9.59) for the definition of r_1, we see that, for any $t \in [s_1, s_2]$,

$$
\mathbb{E}\int_G [\alpha(s_1,x)^2 + \beta(s_1,x)^2]dx \le e^{r_1(s_2-s_1)}\mathbb{E}\int_G [\alpha(t,x)^2 + \beta(t,x)^2]dx. \tag{9.65}
$$

Step 2. We establish a local estimate for α (See (9.66) below). To this aim, we choose a function $\xi \in C^\infty[s_1, s_2]$ with $\xi(s_1) = \xi(s_2) = 0$ as follows

$$
\xi(t) = (t-s_1)^2(s_2-t)^2.
$$

By (9.62), we have that

$$
\begin{aligned}
& d\,(\xi\alpha\beta) \\
&= \xi_t\alpha\beta dt - \xi\alpha\Big[\sum_{j,k=1}^n (a^{jk}\beta_{x_j})_{x_k} + a_5(t)\alpha + a_1(t)\beta + a_3(t)R + a_8(t)K\Big]dt \\
&\quad -\xi\beta\Big[\sum_{j,k=1}^n (a^{jk}\alpha_{x_j})_{x_k} + a_6(t)\alpha + a_2(t)\beta + a_7(t)K + a_4(t)R\Big]dt \\
&\quad +\xi KRdt + \xi\alpha RdW(t) + \xi\beta KdW(t).
\end{aligned}
$$

Therefore, by Condition 9.2 (which leads to $a_8(\cdot) \equiv 0$ on (t_1,t_2)), for any (small) positive constant ε, we find that

$$
\begin{aligned}
& \Big|\mathbb{E}\int_{s_1}^{s_2}\int_G a_5(t)\xi\alpha^2 dxdt\Big| \\
&= \Big|\mathbb{E}\int_{s_1}^{s_2}\int_G \xi_t\alpha\beta dxdt + 2\mathbb{E}\int_{s_1}^{s_2} \xi\sum_{i\in\Lambda_r}\lambda_i\alpha^i(t)\beta^i(t)dt \\
&\quad +\mathbb{E}\int_{s_1}^{s_2}\int_G \xi\big(KR - a_1\alpha\beta - a_3\alpha R - a_6\alpha\beta - a_2\beta^2 - a_7\beta K - a_4\beta R\big)dxdt\Big| \\
&\le \frac{l_0}{2}\mathbb{E}\int_{s_1}^{s_2}\int_G \xi\alpha^2 dxdt + \varepsilon\mathbb{E}\int_{s_1}^{s_2}\int_G \xi^{\frac{3}{2}}\Big(K^2 + \sum_{i\in\Lambda_r}\lambda_i|\alpha^i(t)|^2\Big)dxdt \\
&\quad +\frac{\mathcal{C}}{\varepsilon}\mathbb{E}\int_{s_1}^{s_2}\int_G \beta^2 dxdt + \frac{\mathcal{C}}{\varepsilon}\mathbb{E}\int_{s_1}^{s_2}\int_G \xi^{\frac{1}{2}}\Big(R^2 + \sum_{i\in\Lambda_r}\lambda_i|\beta^i(t)|^2\Big)dxdt.
\end{aligned}
$$

This, together with Condition 9.1, implies that

$$
\begin{aligned}
&\frac{l_0}{2}\mathbb{E}\int_{s_1}^{s_2}\int_G \xi\alpha^2 dxdt \\
&\leq \varepsilon\mathbb{E}\int_{s_1}^{s_2}\int_G \xi^{\frac{3}{2}}\Big(K^2+\sum_{i\in\Lambda_r}\lambda_i|\alpha^i(t)|^2\Big)dxdt \\
&\quad +\frac{\mathcal{C}}{\varepsilon}\mathbb{E}\int_{s_1}^{s_2}\int_G \beta^2 dxdt+\frac{\mathcal{C}}{\varepsilon}\mathbb{E}\int_{s_1}^{s_2}\int_G \xi^{\frac{1}{2}}\Big(R^2+\sum_{i\in\Lambda_r}\lambda_i|\beta^i(t)|^2\Big)dxdt.
\end{aligned}
\tag{9.66}
$$

Step 3. In this step, we estimate the first and the third integrals in the right hand side of (9.66). Noticing that

$$
\begin{aligned}
d\big(\xi^{\frac{3}{2}}\alpha^2\big) &= \frac{3}{2}\xi^{\frac{1}{2}}\xi_t\alpha^2 dt+2\xi^{\frac{3}{2}}\alpha K dW(t)+\xi^{\frac{3}{2}}K^2 dt \\
&\quad -2\xi^{\frac{3}{2}}\alpha\Big[\sum_{j,k=1}^n (a^{jk}\alpha_{x_j})_{x_k}+a_6(t)\alpha+a_2(t)\beta+a_7(t)K+a_4(t)R\Big]dt,
\end{aligned}
$$

we obtain that

$$
\begin{aligned}
&\mathbb{E}\int_{s_1}^{s_2}\int_G \xi^{\frac{3}{2}}K^2 dxdt+2\mathbb{E}\int_{s_1}^{s_2}\xi^{\frac{3}{2}}\sum_{i\in\Lambda_r}\lambda_i|\alpha^i(t)|^2 dt \\
&= -\frac{3}{2}\mathbb{E}\int_{s_1}^{s_2}\int_G \frac{\xi_t}{\xi^{\frac{1}{2}}}\xi\alpha^2 dxdt+2\mathbb{E}\int_{s_1}^{s_2}\int_G \xi^{\frac{3}{2}}\alpha\big(a_6\alpha+a_2\beta+a_7K+a_4R\big)dxdt.
\end{aligned}
$$

Hence,

$$
\begin{aligned}
&\mathbb{E}\int_{s_1}^{s_2}\int_G \xi^{\frac{3}{2}}K^2 dxdt+\mathbb{E}\int_{s_1}^{s_2}\xi^{\frac{3}{2}}\sum_{i\in\Lambda_r}\lambda_i|\alpha^i(t)|^2 dt \\
&\leq \mathcal{C}\mathbb{E}\int_{s_1}^{s_2}\int_G (\xi\alpha^2+\beta^2+\xi^{\frac{1}{2}}R^2)dxdt.
\end{aligned}
\tag{9.67}
$$

On the other hand, by

$$
\begin{aligned}
d\big(\xi^{\frac{1}{2}}\beta^2\big) &= \frac{1}{2}\xi^{-\frac{1}{2}}\xi_t\beta^2 dt+2\xi^{\frac{1}{2}}\beta R dW(t)+\xi^{\frac{1}{2}}R^2 dt \\
&\quad -2\xi^{\frac{1}{2}}\beta\Big[\sum_{j,k=1}^n (a^{jk}\beta_{x_j})_{x_k}+a_5(t)\alpha+a_1(t)\beta+a_3(t)R+a_8(t)K\Big]dt,
\end{aligned}
$$

and using Condition 9.2 again, we obtain that

$$
\begin{aligned}
&\mathbb{E}\int_{s_1}^{s_2}\int_G \xi^{\frac{1}{2}}R^2 dxdt+2\mathbb{E}\int_{s_1}^{s_2}\xi^{\frac{1}{2}}\sum_{i\in\Lambda_r}\lambda_i|\beta^i(t)|^2 dt \\
&= -\frac{1}{2}\mathbb{E}\int_{s_1}^{s_2}\int_G \frac{\xi_t}{\xi^{\frac{1}{2}}}\beta^2 dxdt+2\mathbb{E}\int_{s_1}^{s_2}\int_G \xi^{\frac{1}{2}}\beta\big[a_5(t)\alpha+a_1(t)\beta+a_3(t)R\big]dxdt.
\end{aligned}
$$

Hence, for any (small) positive constant δ, it holds that

$$\begin{aligned}&\mathbb{E}\int_{s_1}^{s_2}\int_G \xi^{\frac{1}{2}} R^2 dxdt + \mathbb{E}\int_{s_1}^{s_2} \xi^{\frac{1}{2}} \sum_{i\in\Lambda_r} \lambda_i |\beta^i(t)|^2 dt \\ &\leq \delta\mathbb{E}\int_{s_1}^{s_2}\int_G \xi\alpha^2 dxdt + \frac{\mathcal{C}}{\delta}\mathbb{E}\int_{s_1}^{s_2}\int_G \beta^2 dxdt.\end{aligned} \tag{9.68}$$

Combining (9.66), (9.67) and (9.68), by suitably choosing ε and δ, we find that

$$\mathbb{E}\int_{s_1}^{s_2}\int_G \xi\alpha^2 dxdt \leq \mathcal{C}\mathbb{E}\int_{s_1}^{s_2}\int_G \beta^2 dxdt. \tag{9.69}$$

Choose $s_3 = s_1+(s_2-s_1)/4$ and $s_4 = s_2-(s_2-s_1)/4$. Then, $s_1 < s_3 < s_4 < s_2$. From (9.69), we see that

$$\mathbb{E}\int_{s_3}^{s_4}\int_G \alpha^2 dxdt \leq \frac{\mathcal{C}}{(s_2-s_1)^4}\mathbb{E}\int_{s_1}^{s_2}\int_G \beta^2 dxdt. \tag{9.70}$$

Step 4. Integrating (9.65) on (s_3, s_4) with respect to the variable t, we obtain that

$$\begin{aligned}&\mathbb{E}\int_G \alpha(s_1,x)^2 dx + \mathbb{E}\int_G \beta(s_1,x)^2 dx \\ &\leq \frac{\mathcal{C}}{s_2-s_1}\mathbb{E}\int_{s_3}^{s_4}\int_G [\alpha(t,x)^2 + \beta(t,x)^2] dxdt.\end{aligned} \tag{9.71}$$

By (9.70)–(9.71) and (9.64), we obtain that

$$\begin{aligned}&\mathbb{E}\int_G \alpha(s_1,x)^2 dx + \mathbb{E}\int_G \beta(s_1,x)^2 dx \\ &\leq \frac{\mathcal{C}}{(s_2-s_1)^5}\mathbb{E}\int_{s_1}^{s_2}\int_G \beta(t,x)^2 dxdt \\ &\leq \frac{\mathcal{C}e^{\mathcal{C}\sqrt{r}}}{(s_2-s_1)^5}\mathbb{E}\int_{s_1}^{s_2}\int_{G_0} \beta(t,x)^2 dxdt,\end{aligned}$$

which gives (9.63). This completes the proof of Proposition 9.22. □

Next, we consider the following control system:

$$
\begin{cases}
dy - \displaystyle\sum_{j,k=1}^{n} (a^{jk} y_{x_j})_{x_k} dt \\
\quad = \big(a_1(t)y + a_2(t)z + \chi_{G_0} u\big)dt + \big(a_3(t)y + a_4(t)z\big)dW(t) & \text{in } (s_1, s_2) \times G, \\
dz - \displaystyle\sum_{j,k=1}^{n} (a^{ij} z_{x_j})_{x_k} dt \\
\quad = \big(a_5(t)y + a_6(t)z\big)dt + \big(a_7(t)z + a_8(t)y\big)dW(t) & \text{in } (s_1, s_2) \times G, \\
y = z = 0 & \text{on } (s_1, s_2) \times \Gamma, \\
y(s_1) = y_{s_1}, \ z(s_1) = z_{s_1} & \text{in } G.
\end{cases}
\tag{9.72}
$$

Here $y_{s_1}, z_{s_1} \in L^2_{\mathcal{F}_{s_1}}(\Omega; H_r)$.

Proposition 9.22 implies the following partial controllability result for (9.72).

Proposition 9.23. *Let Conditions 9.1–9.2 hold. Then, for every $r \geq \lambda_1$ and $y_{s_1}, z_{s_1} \in L^2_{\mathcal{F}_{s_1}}(\Omega; H_r)$, there is a control $u_r \in L^2_{\mathbb{F}}(s_1, s_2; L^2(G_0))$, such that the corresponding solution (y, z) to (9.72) with $u = u_r$ satisfies*

$$
\Pi_r\,(y(s_2)) = \Pi_r\,(z(s_2)) = 0 \ \textit{in } G, \quad \textit{a.s.}
$$

Moreover, u_r verifies

$$
|u_r|^2_{L^2_{\mathbb{F}}(s_1,s_2;L^2(G_0))} \leq \frac{Ce^{C\sqrt{r}}}{(s_2 - s_1)^5} \mathbb{E}\left(|y(s_1)|^2_{L^2(G)} + |z(s_1)|^2_{L^2(G)} \right).
$$

The proof of Proposition 9.23 is very similar as that of Proposition 9.13 (follows from Theorem 7.17)), and hence we omit it here.

Further, as a special case of Proposition 9.14, we have the following decay result for (9.72).

Proposition 9.24. *Let $r \geq \lambda_1$. Then, for any $(y_{s_1}, z_{s_1}) \in L^2_{\mathcal{F}_{s_1}}(\Omega; L^2(G; \mathbb{R}^2))$ with $\Pi_r(y_{s_1}) = \Pi_r(z_{s_1}) = 0$, a.s., the corresponding solution (y, z) to (9.72) with $u \equiv 0$ satisfies that, for all $t \in [s_1, s_2]$,*

$$
\mathbb{E}\big(|y(t)|^2_{L^2(G)} + |z(t)|^2_{L^2(G)}\big) \leq e^{-(2r - r_1)(t - s_1)} \mathbb{E}\big(|y_{s_1}|^2_{L^2(G)} + |z_{s_1}|^2_{L^2(G)}\big).
$$

Based on Propositions 9.22–9.24, one can give a proof of Theorem 9.18 similar to that of Theorem 9.7, and therefore we omit the details here.

9.3.2 Proof of the Negative Null Controllability Result

In this subsection, we show the lack of null controllability for the system (9.58), presented in Theorem 9.19.

Proof of Theorem 9.19: First, we prove that the system (9.58) is not null controllable if $a_5(\cdot) = 0$ in $(0,T) \times \Omega$, a.e.

Without loss of generality, we may assume that the coefficient $a_6(\cdot)$ in the system (9.58) is equal to 0 (Otherwise, we introduce a simple transformation $\tilde{y} = y$, $\tilde{z}(t) = e^{-\int_0^t a_6(s)ds} z(t)$ and $\tilde{u} = u$ and consider the system for the new state variable $(\tilde{y}, \tilde{z})$ and the control variable $\tilde{u}$). Then, by the system (9.58), and noting that $a_5(\cdot) = a_6(\cdot) = 0$ in $(0,T) \times \Omega$, a.e., we find that $(\mathbb{E}y, \mathbb{E}z)$ solves

$$
\begin{cases}
(\mathbb{E}y)_t - \sum_{j,k=1}^{n} \big(a^{jk}(\mathbb{E}y)_{x_j}\big)_{x_k} = \mathbb{E}\big(a_1(t)y + a_2(t)z + \chi_{G_0}u\big) & \text{in } Q, \\
(\mathbb{E}z)_t - \sum_{j,k=1}^{n} \big(a^{ij}(\mathbb{E}z)_{x_j}\big)_{x_k} = 0 & \text{in } Q, \\
\mathbb{E}y = \mathbb{E}z = 0 & \text{on } \Sigma, \\
(\mathbb{E}y)(0) = y_0, \ \ (\mathbb{E}z)(0) = z_0 & \text{in } G.
\end{cases}
\tag{9.73}
$$

Since there is no control in the second equation of (9.73), $\mathbb{E}z$ cannot be driven to the rest for any time T if $z_0 \neq 0$ in G.

Next, we prove that the system (9.58) is not null controllable if $a_8(\cdot) \neq 0$ in $(0,T) \times \Omega$, a.e. and $\frac{a_5(\cdot)}{a_8(\cdot)} \in L^\infty_{\mathbb{F}}(0,T)$.

In the following, we construct a nontrivial solution (α, β, K, R) to (9.60) such that $a_5(t)\alpha + a_8(t)K = 0$ and $(\beta, R) = 0$ in Q, a.s. For this purpose, we consider the following linear stochastic differential equation:

$$
\begin{cases}
d\zeta - \lambda_1 \zeta dt = a_6(t)\zeta dt - \dfrac{a_5(t)}{a_8(t)} \zeta dW(t) \ \ \text{in } (0,T), \\
\zeta(0) = 1.
\end{cases}
\tag{9.74}
$$

Denote by $\zeta \in L^2_{\mathbb{F}}(\Omega; C([0,T]))$ the unique solution to (9.74). Write $q = \zeta(t)e_1$, then it is easy to check that $q \in L^2_{\mathbb{F}}(\Omega; C([0,T]; H^2(G) \cap H^1_0(G)))$ solves

$$
\begin{cases}
dq + \sum_{j,k=1}^{n} \big(a^{ij} q_{x_j}\big)_{x_k} dt = a_6(t) q dt - \dfrac{a_5(t)}{a_8(t)} q dW(t) & \text{in } Q, \\
q = 0 & \text{on } \Sigma, \\
q(0) = e_1 & \text{in } G.
\end{cases}
\tag{9.75}
$$

(Similarly to (9.57), one cannot solve the system (9.75) directly because it is not well-posed.) Furthermore, consider the following backward stochastic parabolic equation:

$$
\begin{cases}
dv + \sum_{j,k=1}^{n} \left(a^{ij} v_{x_j}\right)_{x_k} dt = a_6(t) v dt + V dW(t) & \text{in } Q, \\
v = 0 & \text{on } \Sigma, \\
v(T) = \zeta(T) e_1 & \text{in } G.
\end{cases}
\tag{9.76}
$$

By the well-posedness of the equation (9.76), it is easy to check that the solution (v, V) to (9.76) satisfies that $v = q$ and $V = -\frac{a_5(t)}{a_8(t)} q$ in Q, a.s.

It is easy to check that $(\alpha, \beta, K, R) = (v, 0, V, 0)$ solves the equation (9.60) with $(\alpha_T, \beta_T) = (\zeta(T)e_1, 0)$. However, $\mathbb{E}|\alpha(0)|^2_{L^2(G)} = 1$. This implies that (9.61) does not hold for the solution to (9.60) associated with the above (α_T, β_T). By means of Proposition 9.21, we deduce that, the system (9.58) is not null controllable. $\square$

Remark 9.25. Notice that by Theorem 9.18, when $a_8(\cdot) \equiv 0$ and $a_5(\cdot) \equiv 1$ on $(0, T)$, a.s., the system (9.58) is null controllable at any time. On the other hand, Theorem 9.19 shows that no matter how small the nonzero bounded function $a_8(\cdot)$ is, the corresponding system is not null controllable any more. This indicates that the controllability of the stochastic coupled system (9.58) is not robust with respect to the coupling coefficient in the diffusion term. Moreover, by Condition 9.2, we see that if the coefficient $a_8(\cdot)$ equals to zero on some time interval with an arbitrarily small (but still positive) measure, then (9.58) is null controllable again. Therefore, the controllability of stochastic coupled parabolic systems is influenced strongly by the coupling coefficient in diffusion terms. Also, it indicates that it is very hard (even impossible) to apply the Carleman estimate to prove Theorem 9.18.

9.4 Carleman Estimate for a Stochastic Parabolic-Like Operator

This section is addressed to deriving a Carleman estimate for a stochastic parabolic-like operator (to be given below), which will play key roles in the sequel. The main content of this section is taken from [315].

Throughout this section, we assume that

$$
b^{jk} = b^{kj} \in L^2_{\mathbb{F}}(\Omega; C^1([0, T]; W^{2,\infty}(G))), \qquad j, k = 1, 2, \cdots, n, \tag{9.77}
$$

$\ell \in C^{1,3}(Q)$ and $\Psi \in C^{1,2}(Q)$. Write

$$\begin{cases} \mathcal{A} = -\sum_{j,k=1}^{n} \big(b^{jk}\ell_{x_j}\ell_{x_k} - b^{jk}_{x_k}\ell_{x_j} - b^{jk}\ell_{x_jx_k}\big) - \Psi - \ell_t, \\ \mathcal{B} = 2\Big[\mathcal{A}\Psi - \sum_{j,k=1}^{n} \big(\mathcal{A}b^{jk}\ell_{x_j}\big)_{x_k}\Big] - \mathcal{A}_t - \sum_{j,k=1}^{n} (b^{jk}\Psi_{x_k})_{x_j}, \\ c^{jk} = \sum_{j',k'=1}^{n} \Big[2b^{jk'}\big(b^{j'k}\ell_{x_{j'}}\big)_{x_{k'}} - \big(b^{jk}b^{j'k'}\ell_{x_{j'}}\big)_{x_{k'}}\Big] - \frac{b^{jk}_t}{2} + \Psi b^{jk}. \end{cases} \tag{9.78}$$

First, we establish a fundamental weighted identity for the stochastic parabolic-like operator "$d - \sum_{j,k=1}^{n} \partial_{x_k}(b^{jk}\partial_{x_j})dt$"[1], in the spirit of the pointwise identities (1.48) and (8.19).

Theorem 9.26. *Let* $\mathbf{u}$ *be an* $H^2(G)$*-valued Itô process. Set* $\theta = e^{\ell}$ *and* $\mathbf{v} = \theta\mathbf{u}$*. Then, for any* $t \in [0,T]$ *and a.e.* $(x,\omega) \in G \times \Omega$*,*

$$\begin{aligned} &2\theta\Big[-\sum_{j,k=1}^{n} (b^{jk}\mathbf{v}_{x_j})_{x_k} + \mathcal{A}\mathbf{v}\Big]\Big[d\mathbf{u} - \sum_{j,k=1}^{n} (b^{jk}\mathbf{u}_{x_j})_{x_k}dt\Big] \\ &\quad +2\sum_{j,k=1}^{n} (b^{jk}\mathbf{v}_{x_j}d\mathbf{v})_{x_k} + 2\sum_{j,k=1}^{n}\Big[\sum_{j',k'=1}^{n} \big(2b^{jk}b^{j'k'}\ell_{x_{j'}}\mathbf{v}_{x_j}\mathbf{v}_{x_{k'}} \\ &\quad -b^{jk}b^{j'k'}\ell_{x_j}\mathbf{v}_{x_{j'}}\mathbf{v}_{x_{k'}}\big) + \Psi b^{jk}\mathbf{v}_{x_j}\mathbf{v} - b^{jk}\Big(\mathcal{A}\ell_{x_j} + \frac{\Psi_{x_j}}{2}\Big)\mathbf{v}^2\Big]_{x_k}dt \\ &= 2\sum_{j,k=1}^{n} c^{jk}\mathbf{v}_{x_j}\mathbf{v}_{x_k}dt + \mathcal{B}\mathbf{v}^2dt + d\Big(\sum_{j,k=1}^{n} b^{jk}\mathbf{v}_{x_j}\mathbf{v}_{x_k} + \mathcal{A}\mathbf{v}^2\Big) \\ &\quad +2\Big[-\sum_{j,k=1}^{n} \big(b^{jk}\mathbf{v}_{x_j}\big)_{x_k} + \mathcal{A}\mathbf{v}\Big]^2dt \\ &\quad -\theta^2\sum_{j,k=1}^{n} b^{jk}(d\mathbf{u}_{x_j} + \ell_{x_j}d\mathbf{u})(d\mathbf{u}_{x_k} + \ell_{x_k}d\mathbf{u}) - \theta^2\mathcal{A}(d\mathbf{u})^2. \end{aligned} \tag{9.79}$$

Proof: The proof is divided into four steps.

Step 1. Recalling $\theta = e^{\ell}$ and $\mathbf{v} = \theta\mathbf{u}$, one has $d\mathbf{u} = \theta^{-1}(d\mathbf{v} - \ell_t\mathbf{v}dt)$ and $\mathbf{u}_{x_j} = \theta^{-1}(\mathbf{v}_{x_j} - \ell_{x_j}\mathbf{v})$ for $i = 1, 2, \cdots, m$. By (9.77), it is easy to see that $\sum_{j,k=1}^{n} b^{jk}(\ell_{x_j}\mathbf{v}_{x_k} + \ell_{x_k}\mathbf{v}_{x_j}) = 2\sum_{j,k=1}^{n} b^{jk}\ell_{x_j}\mathbf{v}_{x_k}$. Hence,

---

[1] Since only the symmetry condition (9.77) is assumed for the coefficient matrix (b^{jk}), we call "$d - \sum_{j,k=1}^{n} \partial_{x_k}(b^{jk}\partial_{x_j})dt$" a stochastic parabolic-like operator.

$$
\begin{aligned}
\theta \sum_{j,k=1}^{n} (b^{jk}\mathbf{u}_{x_j})_{x_k} &= \theta \sum_{j,k=1}^{n} [\theta^{-1} b^{jk}(\mathbf{v}_{x_j} - \ell_{x_j}\mathbf{v})]_{x_k} \\
&= \sum_{j,k=1}^{n} [b^{jk}(\mathbf{v}_{x_j} - \ell_{x_j}\mathbf{v})]_{x_k} - \sum_{j,k=1}^{n} b^{jk}(\mathbf{v}_{x_j} - \ell_{x_j}\mathbf{v})\ell_{x_k} \\
&= \sum_{j,k=1}^{n} \Big[(b^{jk}\mathbf{v}_{x_j})_{x_k} - b^{jk}(\ell_{x_j}\mathbf{v}_{x_k} + \ell_{x_k}\mathbf{v}_{x_j}) \\
&\qquad + (b^{jk}\ell_{x_j}\ell_{x_k} - b^{jk}_{x_k}\ell_{x_j} - b^{jk}\ell_{x_j x_k})\mathbf{v}\Big] \\
&= \sum_{j,k=1}^{n} \Big[(b^{jk}\mathbf{v}_{x_j})_{x_k} - 2b^{jk}\ell_{x_j}\mathbf{v}_{x_k} + (b^{jk}\ell_{x_j}\ell_{x_k} - b^{jk}_{x_k}\ell_{x_j} - b^{jk}\ell_{x_j x_k})\mathbf{v}\Big].
\end{aligned}
\tag{9.80}
$$

Put

$$
\begin{cases}
I \stackrel{\Delta}{=} -\displaystyle\sum_{j,k=1}^{n} (b^{jk}\mathbf{v}_{x_j})_{x_k} + \mathcal{A}\mathbf{v}, \qquad I_1 \stackrel{\Delta}{=} \Big[-\displaystyle\sum_{j,k=1}^{n} (b^{jk}\mathbf{v}_{x_j})_{x_k} + \mathcal{A}\mathbf{v}\Big]dt, \\
I_2 \stackrel{\Delta}{=} d\mathbf{v} + 2\displaystyle\sum_{j,k=1}^{n} b^{jk}\ell_{x_j}\mathbf{v}_{x_k}dt, \qquad I_3 \stackrel{\Delta}{=} \Psi\mathbf{v}dt.
\end{cases}
\tag{9.81}
$$

By (9.80) and (9.81), it follows that

$$
\theta\Big[d\mathbf{u} - \sum_{j,k=1}^{n} (b^{jk}\mathbf{u}_{x_j})_{x_k}\, dt\Big] = I_1 + I_2 + I_3.
$$

Hence,

$$
2\theta\Big[-\sum_{j,k=1}^{n} (b^{jk}\mathbf{v}_{x_j})_{x_k} + \mathcal{A}\mathbf{v}\Big]\Big[d\mathbf{u} - \sum_{j,k=1}^{n} (b^{jk}\mathbf{u}_{x_j})_{x_k} dt\Big] = 2I(I_1 + I_2 + I_3).
\tag{9.82}
$$

Step 2. Let us compute $2II_2$. Utilizing (9.77) again, and noting that

$$
\sum_{j,k,j',k'=1}^{n} (b^{jk}b^{j'k'}\ell_{x_{j'}}\mathbf{v}_{x_j}\mathbf{v}_{x_k})_{x_{k'}} = \sum_{j,k,j',k'=1}^{n} (b^{jk}b^{j'k'}\ell_{x_j}\mathbf{v}_{x_{j'}}\mathbf{v}_{x_{k'}})_{x_k},
$$

we get

$$
\begin{aligned}
&2\sum_{j,k,j',k'=1}^{n} b^{jk}b^{j'k'}\ell_{x_{j'}}\mathbf{v}_{x_j}\mathbf{v}_{x_k x_{k'}} \\
&= \sum_{j,k,j',k'=1}^{n} b^{jk}b^{j'k'}\ell_{x_{j'}}(\mathbf{v}_{x_j}\mathbf{v}_{x_k x_{k'}} + \mathbf{v}_{x_k}\mathbf{v}_{x_j x_{k'}})
\end{aligned}
$$

$$\begin{aligned}
&= \sum_{j,k,j',k'=1}^{n} b^{jk}b^{j'k'}\ell_{x_{j'}}(\mathbf{v}_{x_j}\mathbf{v}_{x_k})_{x_{k'}} \qquad (9.83)\\
&= \sum_{j,k,j',k'=1}^{n} (b^{jk}b^{j'k'}\ell_{x_{j'}}\mathbf{v}_{x_j}\mathbf{v}_{x_k})_{x_{k'}} - \sum_{j,k,j',k'=1}^{n} (b^{jk}b^{j'k'}\ell_{x_{j'}})_{x_{k'}}\mathbf{v}_{x_j}\mathbf{v}_{x_k}\\
&= \sum_{j,k,j',k'=1}^{n} (b^{jk}b^{j'k'}\ell_{x_j}\mathbf{v}_{x_{j'}}\mathbf{v}_{x_{k'}})_{x_k} - \sum_{j,k,j',k'=1}^{n} (b^{jk}b^{j'k'}\ell_{x_{j'}})_{x_{k'}}\mathbf{v}_{x_j}\mathbf{v}_{x_k}.
\end{aligned}$$

Hence, by (9.83), and noting that

$$\sum_{j,k,j',k'=1}^{n} b^{jk}(b^{j'k'}\ell_{x_{j'}})_{x_k}\mathbf{v}_{x_j}\mathbf{v}_{x_{k'}} = \sum_{j,k,j',k'=1}^{n} b^{jk'}(b^{j'k}\ell_{x_{j'}})_{x_{k'}}\mathbf{v}_{x_j}\mathbf{v}_{x_k},$$

we obtain that

$$\begin{aligned}
&4\Big[-\sum_{j,k=1}^{n}(b^{jk}\mathbf{v}_{x_j})_{x_k} + \mathcal{A}\mathbf{v}\Big]\sum_{j,k=1}^{n} b^{jk}\ell_{x_j}\mathbf{v}_{x_k}\\
&= -4\sum_{j,k,j',k'=1}^{n} (b^{jk}b^{j'k'}\ell_{x_{j'}}\mathbf{v}_{x_j}\mathbf{v}_{x_{k'}})_{x_k} + 4\sum_{j,k,j',k'=1}^{n} b^{jk}(b^{j'k'}\ell_{x_{j'}})_{x_k}\mathbf{v}_{x_j}\mathbf{v}_{x_{k'}}\\
&\quad +4\sum_{j,k,j',k'=1}^{n} b^{jk}b^{j'k'}\ell_{x_{j'}}\mathbf{v}_{x_j}\mathbf{v}_{x_k x_{k'}} + 2\mathcal{A}\sum_{j,k=1}^{n} b^{jk}\ell_{x_j}(\mathbf{v}^2)_{x_k}\\
&= -2\sum_{j,k=1}^{n}\Big[\sum_{j',k'=1}^{n}(2b^{jk}b^{j'k'}\ell_{x_{j'}}\mathbf{v}_{x_j}\mathbf{v}_{x_{k'}} - b^{jk}b^{j'k'}\ell_{x_j}\mathbf{v}_{x_{j'}}\mathbf{v}_{x_{k'}}) \qquad (9.84)\\
&\quad -\mathcal{A}b^{jk}\ell_{x_j}\mathbf{v}^2\Big]_{x_k} + 2\sum_{j,k,j',k'=1}^{n}\Big[2b^{jk'}(b^{j'k}\ell_{x_{j'}})_{x_{k'}} - (b^{jk}b^{j'k'}\ell_{x_{j'}})_{x_{k'}}\Big]\mathbf{v}_{x_j}\mathbf{v}_{x_k}\\
&\quad -2\sum_{j,k=1}^{n}(\mathcal{A}b^{jk}\ell_{x_j})_{x_k}\mathbf{v}^2.
\end{aligned}$$

Using Itô's formula, we have

$$\begin{aligned}
&2\Big[-\sum_{j,k=1}^{n}(b^{jk}\mathbf{v}_{x_j})_{x_k} + \mathcal{A}\mathbf{v}\Big]d\mathbf{v}\\
&= -2\sum_{j,k=1}^{n}(b^{jk}\mathbf{v}_{x_j}d\mathbf{v})_{x_k} + 2\sum_{j,k=1}^{n} b^{jk}\mathbf{v}_{x_j}d\mathbf{v}_{x_k} + 2\mathcal{A}\mathbf{v}d\mathbf{v}\\
&= -2\sum_{j,k=1}^{n}(b^{jk}\mathbf{v}_{x_j}d\mathbf{v})_{x_k} + d\Big(\sum_{j,k=1}^{n} b^{jk}\mathbf{v}_{x_j}\mathbf{v}_{x_k} + \mathcal{A}\mathbf{v}^2\Big) \qquad (9.85)\\
&\quad -\sum_{j,k=1}^{n} b^{jk}_t\mathbf{v}_{x_j}\mathbf{v}_{x_k}dt - \mathcal{A}_t\mathbf{v}^2 dt - \sum_{j,k=1}^{n} b^{jk}d\mathbf{v}_{x_j}d\mathbf{v}_{x_k} - \mathcal{A}(d\mathbf{v})^2.
\end{aligned}$$

Now, from (9.81), (9.84) and (9.85), we arrive at

$$
\begin{aligned}
&2II_2\\
&= -2\sum_{j,k=1}^{n}\Big[\sum_{j',k'=1}^{n}\big(2b^{jk}b^{j'k'}\ell_{x_{j'}}\mathbf{v}_{x_j}\mathbf{v}_{x_{k'}} - b^{jk}b^{j'k'}\ell_{x_j}\mathbf{v}_{x_{j'}}\mathbf{v}_{x_{k'}}\big)\\
&\quad -\mathcal{A}b^{jk}\ell_{x_j}\mathbf{v}^2\Big]_{x_k}dt - 2\sum_{j,k=1}^{n}(b^{jk}\mathbf{v}_{x_j}d\mathbf{v})_{x_k} + d\Big(\sum_{j,k=1}^{n}b^{jk}\mathbf{v}_{x_j}\mathbf{v}_{x_k} + \mathcal{A}\mathbf{v}^2\Big)\\
&\quad +2\sum_{j,k=1}^{n}\Big[\sum_{j',k'=1}^{n}\Big(2b^{jk'}(b^{j'k}\ell_{x_{j'}})_{x_{k'}} - (b^{jk}b^{j'k'}\ell_{x_{j'}})_{x_{k'}}\Big) - \frac{b^{jk}_t}{2}\Big]\mathbf{v}_{x_j}\mathbf{v}_{x_k}dt\\
&\quad -\Big[\mathcal{A}_t + 2\sum_{j,k=1}^{n}(\mathcal{A}b^{jk}\ell_{x_j})_{x_k}\Big]\mathbf{v}^2dt - \sum_{j,k=1}^{n}b^{jk}d\mathbf{v}_{x_j}d\mathbf{v}_{x_k} - \mathcal{A}(d\mathbf{v})^2.
\end{aligned}
\tag{9.86}
$$

Step 3. Let us compute $2II_3$. By (9.81), we get

$$
\begin{aligned}
2II_3 &= 2\Big[-\sum_{j,k=1}^{n}(b^{jk}\mathbf{v}_{x_j})_{x_k} + \mathcal{A}\mathbf{v}\Big]\Psi\mathbf{v}dt\\
&= \Big[-2\sum_{j,k=1}^{n}\big(\Psi b^{jk}\mathbf{v}_{x_j}\mathbf{v}\big)_{x_k} + 2\Psi\sum_{j,k=1}^{n}b^{jk}\mathbf{v}_{x_j}\mathbf{v}_{x_k}\\
&\quad + \sum_{j,k=1}^{n}b^{jk}\Psi_{x_k}(\mathbf{v}^2)_{x_j} + 2\mathcal{A}\Psi\mathbf{v}^2\Big]dt\\
&= \Big\{-\sum_{j,k=1}^{n}\Big(2\Psi b^{jk}\mathbf{v}_{x_j}\mathbf{v} - b^{jk}\Psi_{x_j}\mathbf{v}^2\Big)_{x_k} + 2\Psi\sum_{j,k=1}^{n}b^{jk}\mathbf{v}_{x_j}\mathbf{v}_{x_k}\\
&\quad +\Big[-\sum_{j,k=1}^{n}(b^{jk}\Psi_{x_k})_{x_j} + 2\mathcal{A}\Psi\Big]\mathbf{v}^2\Big\}dt.
\end{aligned}
\tag{9.87}
$$

Step 4. Finally, combining the equalities (9.82), (9.86) and (9.87), and noting that

$$
\begin{aligned}
&\sum_{j,k=1}^{n}b^{jk}d\mathbf{v}_{x_j}d\mathbf{v}_{x_k} + \mathcal{A}(d\mathbf{v})^2\\
&= \theta^2\sum_{j,k=1}^{n}b^{jk}(d\mathbf{u}_{x_j} + \ell_{x_j}d\mathbf{u})(d\mathbf{u}_{x_k} + \ell_{x_k}d\mathbf{u}) + \theta^2\mathcal{A}(d\mathbf{u})^2,
\end{aligned}
$$

we conclude the desired equality (9.79) immediately. This completes the proof of Theorem 9.26. □

Next, we shall derive a Carleman estimate for the stochastic parabolic-like operator "$d - \sum_{j,k=1}^{n} \partial_{x_k}(b^{jk}\partial_{x_j})dt$".

For any fixed nonnegative and nonzero function $\psi \in C^4(\overline{G})$, and (large) parameters $\lambda > 1$ and $\mu > 1$, we choose

$$\theta = e^{\ell}, \quad \ell = \lambda\alpha, \quad \alpha(t,x) = \frac{e^{\mu\psi(x)} - e^{2\mu|\psi|_{C(\overline{G})}}}{t(T-t)}, \quad \varphi(t,x) = \frac{e^{\mu\psi(x)}}{t(T-t)}, \tag{9.88}$$

and

$$\Psi = 2\sum_{j,k=1}^{n} b^{jk}\ell_{x_j x_k}. \tag{9.89}$$

In what follows, for a positive integer r, we denote by $O(\mu^r)$ a function of order μ^r for large μ (which is independent of λ); by $O_\mu(\lambda^r)$ a function of order λ^r for fixed μ and for large λ. In a similar way, we use the notation $O(e^{\mu|\psi|_{C(\overline{G})}})$ and so on. For $j,k = 1,2,\cdots,n$, it is easy to check that

$$\ell_t = \lambda\alpha_t, \quad \ell_{x_j} = \lambda\mu\varphi\psi_{x_j}, \quad \ell_{x_j x_k} = \lambda\mu^2\varphi\psi_{x_j}\psi_{x_k} + \lambda\mu\varphi\psi_{x_j x_k} \tag{9.90}$$

and that

$$\alpha_t = \varphi^2 O(e^{2\mu|\psi|_{C(\overline{G})}}), \qquad \varphi_t = \varphi^2 O(e^{\mu|\psi|_{C(\overline{G})}}). \tag{9.91}$$

We have the following result.

Theorem 9.27. *Assume that either $(b^{jk})_{n\times n}$ or $-(b^{jk})_{n\times n}$ is a uniformly positive definite matrix, and its smallest eigenvalue is bigger than a constant $s_0 > 0$. Let $\mathbf{u}$ and $\mathbf{v} = \theta\mathbf{u}$ be that in Theorem 9.26 with θ being given in* (9.88). *Then, the equality* (9.79) *holds for any $t \in [0,T]$ and a.e. $(x,\omega) \in G \times \Omega$. Moreover, for $\mathcal{A}$, $\mathcal{B}$ and c^{jk} appeared in* (9.79) *(and given by* (9.78)*), when $|\nabla\psi(x)| > 0$, λ and μ are large enough, it holds that*

$$\begin{cases} \mathcal{A} = -\lambda^2\mu^2\varphi^2 \displaystyle\sum_{j,k=1}^{n} b^{jk}\psi_{x_j}\psi_{x_k} + \lambda\varphi^2 O(e^{2\mu|\psi|_{C(\overline{G})}}), \\ \mathcal{B} \geq 2s_0^2\lambda^3\mu^4\varphi^3|\nabla\psi|^4 + \lambda^3\varphi^3 O(\mu^3) \\ \qquad +\lambda^2\varphi^3 O(\mu^2 e^{2\mu|\psi|_{C(\overline{G})}}) + \lambda\varphi^3 O(e^{2\mu|\psi|_{C(\overline{G})}}), \\ \displaystyle\sum_{j,k=1}^{n} c^{jk}\mathbf{v}_{x_j}\mathbf{v}_{x_k} \geq [s_0^2\lambda\mu^2\varphi|\nabla\psi|^2 + \lambda\varphi O(\mu)]|\nabla\mathbf{u}|^2 \end{cases} \tag{9.92}$$

for any $t \in [0,T]$, a.s.

Proof: By Theorem 9.26, it remains to prove the estimates in (9.92).

Noting (9.89)–(9.90), from (9.78), we have $\ell_{x_j x_k} = \lambda\mu^2\varphi\psi_{x_j}\psi_{x_k} + \lambda\varphi O(\mu)$ and

$$\begin{aligned}
&\sum_{j,k=1}^{n} c^{jk}\mathbf{v}_{x_j}\mathbf{v}_{x_k}\\
&=\sum_{j,k=1}^{n}\Big\{\sum_{j',k'=1}^{n}\Big[2b^{jk'}b^{j'k}\ell_{x_{j'}x_{k'}}+b^{jk}b^{j'k'}\ell_{x_{j'}x_{k'}}+2b^{jk'}b^{j'k}_{x_{k'}}\ell_{x_{j'}}\\
&\qquad -(b^{jk}b^{j'k'})_{x_{k'}}\ell_{x_{j'}}\Big]-\frac{b^{jk}_t}{2}\Big\}\mathbf{v}_{x_j}\mathbf{v}_{x_k}\\
&=\sum_{j,k=1}^{n}\Big\{\sum_{j',k'=1}^{n}\Big[2\lambda\mu^2\varphi b^{jk'}b^{j'k}\psi_{x_{j'}}\psi_{x_{k'}}+\lambda\mu^2\varphi b^{jk}b^{j'k'}\psi_{x_{j'}}\psi_{x_{k'}}\Big]\\
&\qquad +\lambda\varphi O(\mu)\Big\}\mathbf{v}_{x_j}\mathbf{v}_{x_k}\\
&=2\lambda\mu^2\varphi\Big(\sum_{j,k=1}^{n}b^{jk}\psi_{x_j}\mathbf{v}_{x_k}\Big)^2+\lambda\mu^2\varphi\Big(\sum_{j,k=1}^{n}b^{jk}\psi_{x_j}\psi_{x_k}\Big)\Big(\sum_{j,k=1}^{n}b^{jk}\mathbf{v}_{x_j}\mathbf{v}_{x_k}\Big)\\
&\quad+\lambda\varphi|\nabla\mathbf{v}|^2O(\mu)\\
&\geq[s_0^2\lambda\mu^2\varphi|\nabla\psi|^2+\lambda\varphi O(\mu)]|\nabla\mathbf{v}|^2,
\end{aligned}$$

which gives the last inequality in (9.92).

Similarly, by the definition of $\mathcal{A}$ in (9.78), and noting (9.91), we see that

$$\begin{aligned}
\mathcal{A}&=-\sum_{j,k=1}^{n}\big(b^{jk}\ell_{x_j}\ell_{x_k}-b^{jk}_{x_k}\ell_{x_j}+b^{jk}\ell_{x_jx_k}\big)-\ell_t\\
&=-\lambda\mu\sum_{j,k=1}^{n}\Big[b^{jk}\lambda\mu\varphi^2\psi_{x_j}\psi_{x_k}-b^{jk}_{x_k}\varphi\psi_{x_j}+b^{jk}\big(\mu\varphi\psi_{x_j}\psi_{x_k}+\varphi\psi_{x_jx_k}\big)\Big]\\
&\quad+\lambda\varphi^2O(e^{2\mu|\psi|_{C(\overline{G})}})\\
&=-\lambda^2\mu^2\varphi^2\sum_{j,k=1}^{n}b^{jk}\psi_{x_j}\psi_{x_k}+\lambda\varphi^2O(e^{2\mu|\psi|_{C(\overline{G})}}).
\end{aligned}$$

Hence, we get the first estimate in (9.92).

Now, let us estimate $\mathcal{B}$ (recall (9.78) for the definition of $\mathcal{B}$). For this, by (9.90), and recalling the definitions of Ψ (in (9.89)), we see that

$$\begin{aligned}
\Psi&=2\lambda\mu\sum_{j,k=1}^{n}b^{jk}(\mu\varphi\psi_{x_j}\psi_{x_k}+\varphi\psi_{x_jx_k})\\
&=2\lambda\mu^2\varphi\sum_{j,k=1}^{n}b^{jk}\psi_{x_j}\psi_{x_k}+\lambda\varphi O(\mu);\\
\ell_{x_{j'}x_{k'}x_k}&=\lambda\mu^3\varphi\psi_{x_{j'}}\psi_{x_{k'}}\psi_{x_k}+\lambda\varphi O(\mu^2),
\end{aligned}$$

$$\begin{aligned}
\ell_{x_{j'}x_{k'}x_jx_k} &= \lambda\mu^4\varphi\psi_{x_{j'}}\psi_{x_{k'}}\psi_{x_j}\psi_{x_k} + \lambda\varphi O(\mu^3),\\
\Psi_{x_k} &= 2\sum_{j',k'=1}^{n}\big(b^{j'k'}\ell_{x_{j'}x_{k'}}\big)_{x_k} = 2\sum_{j',k'=1}^{n}\big(b^{j'k'}_{x_k}\ell_{x_{j'}x_{k'}} + b^{j'k'}\ell_{x_{j'}x_{k'}x_k}\big)\\
&= 2\lambda\mu^3\varphi\sum_{j',k'=1}^{n} b^{j'k'}\psi_{x_{j'}}\psi_{x_{k'}}\psi_{x_k} + \lambda\varphi O(\mu^2),\\
\Psi_{x_jx_k} &= 2\sum_{j',k'=1}^{n}\big(b^{j'k'}_{x_jx_k}\ell_{x_{j'}x_{k'}} + b^{j'k'}\ell_{x_{j'}x_{k'}x_jx_k} + 2b^{j'k'}_{x_k}\ell_{x_{j'}x_{k'}x_j}\big)\\
&= 2\lambda\mu^4\varphi\sum_{j',k'=1}^{n} b^{j'k'}\psi_{x_{j'}}\psi_{x_{k'}}\psi_{x_j}\psi_{x_k} + \lambda\varphi O(\mu^3),\\
-\sum_{j,k=1}^{n}\big(b^{jk}\Psi_{x_k}\big)_{x_j} &= -\sum_{j,k=1}^{n}\big(b^{jk}_{x_j}\Psi_{x_k} + b^{jk}\Psi_{x_jx_k}\big)\\
&= -2\lambda\mu^4\varphi\Big(\sum_{j,k=1}^{n} b^{jk}\psi_{x_j}\psi_{x_k}\Big)^2 + \lambda\varphi O(\mu^3).
\end{aligned}$$

Hence, recalling the definition of $\mathcal{A}$ (in (9.78)), and using (9.90) and (9.91), we have that

$$\begin{aligned}
\mathcal{A}\Psi &= -2\lambda^3\mu^4\varphi^3\Big(\sum_{j,k=1}^{n} b^{jk}\psi_{x_j}\psi_{x_k}\Big)^2 + \lambda^3\varphi^3 O(\mu^3) + \lambda^2\varphi^3 O(\mu^2 e^{2\mu|\psi|_{C(\overline{G})}}),\\
\mathcal{A}_{x_k} &= -\sum_{j',k'=1}^{n}\big(b^{j'k'}_{x_k}\ell_{x_{j'}}\ell_{x_{k'}} + 2b^{j'k'}\ell_{x_{j'}}\ell_{x_{k'}x_k} - b^{j'k'}_{x_{k'}x_k}\ell_{x_{j'}} - b^{j'k'}_{x_{k'}}\ell_{x_{j'}x_k}\\
&\qquad + b^{j'k'}_{x_k}\ell_{x_{j'}x_{k'}} + b^{j'k'}\ell_{x_{j'}x_{k'}x_k}\big) - \ell_{tx_k}\\
&= -\sum_{j',k'=1}^{n}\big(2b^{j'k'}\ell_{x_{j'}}\ell_{x_{k'}x_k} + b^{j'k'}\ell_{x_{j'}x_{k'}x_k}\big) - \ell_{tx_k} + \big(\lambda\varphi + \lambda^2\varphi^2\big)O(\mu^2)\\
&= -2\lambda^2\mu^3\varphi^2\sum_{j',k'=1}^{n} b^{j'k'}\psi_{x_{j'}}\psi_{x_{k'}}\psi_{x_k} + \lambda^2\varphi^2 O(\mu^2) + \lambda\varphi^2 O(\mu e^{2\mu|\psi|_{C(\overline{G})}}),\\
\sum_{j,k=1}^{n}\mathcal{A}_{x_k}b^{jk}\ell_{x_j} &= -2\lambda^3\mu^4\varphi^3\Big(\sum_{j,k=1}^{n} b^{jk}\psi_{x_j}\psi_{x_k}\Big)^2 + \lambda^3\varphi^3 O(\mu^3)\\
&\qquad + \lambda^2\varphi^3 O(\mu^2 e^{2\mu|\psi|_{C(\overline{G})}}),\\
\sum_{j,k=1}^{n}\big(\mathcal{A}b^{jk}\ell_{x_j}\big)_{x_k} &= \sum_{j,k=1}^{n}\mathcal{A}_{x_k}b^{jk}\ell_{x_j} + \mathcal{A}\sum_{j,k=1}^{n}\big(b^{jk}_{x_k}\ell_{x_j} + b^{jk}\ell_{x_jx_k}\big)\\
&= -3\lambda^3\mu^4\varphi^3\Big(\sum_{j,k=1}^{n} b^{jk}\psi_{x_j}\psi_{x_k}\Big)^2 + \lambda^3\varphi^3 O(\mu^3) + \lambda^2\varphi^3 O(\mu^2 e^{2\mu|\psi|_{C(\overline{G})}}),
\end{aligned}$$

and that

$$
\begin{aligned}
\mathcal{A}_t &= -\sum_{j,k=1}^{n}\Big(b^{jk}\ell_{x_j}\ell_{x_k} - b^{jk}_{x_k}\ell_{x_j} + b^{jk}\ell_{x_jx_k} - \ell_t\Big)_t\\
&= -\sum_{j,k=1}^{n}\Big[b^{jk}\big(\ell_{x_j}\ell_{x_k}\big)_t - b^{jk}_{x_k}\ell_{x_jt} + b^{jk}\ell_{x_jx_kt}\Big]\\
&\quad +\lambda^2\varphi^2 O(\mu^2) + \lambda\varphi^3 O(e^{2\mu|\psi|_{C(\overline{G})}})\\
&= \lambda^2\varphi^3 O(\mu^2 e^{2\mu|\psi|_{C(\overline{G})}}) + \lambda\varphi^3 O(e^{2\mu|\psi|_{C(\overline{G})}}).
\end{aligned}
$$

From the definition of $\mathcal{B}$ (see (9.78)), we have that

$$
\begin{aligned}
\mathcal{B} &= -4\lambda^3\mu^4\varphi^3\Big(\sum_{j,k=1}^{n} b^{jk}\psi_{x_j}\psi_{x_k}\Big)^2 + \lambda^3\varphi^3 O(\mu^3) + \lambda^2\varphi^3 O(\mu^2 e^{2\mu|\psi|_{C(\overline{G})}})\\
&\quad +6\lambda^3\mu^4\varphi^3\Big(\sum_{j,k=1}^{n} b^{jk}\psi_{x_j}\psi_{x_k}\Big)^2 + \lambda^3\varphi^3 O(\mu^3) + \lambda^2\varphi^3 O(\mu^2 e^{2\mu|\psi|_{C(\overline{G})}})\\
&\quad +\lambda^2\varphi^3 O(\mu^2 e^{2\mu|\psi|_{C(\overline{G})}}) + \lambda\varphi^3 O(e^{2\mu|\psi|_{C(\overline{G})}})\\
&\quad -2\lambda\mu^4\varphi\Big(\sum_{j,k=1}^{n} b^{jk}\psi_{x_j}\psi_{x_k}\Big)^2 + \lambda\varphi O(\mu^3)\\
&= 2\lambda^3\mu^4\varphi^3\Big(\sum_{j,k=1}^{n} b^{jk}\psi_{x_j}\psi_{x_k}\Big)^2 + \lambda^3\varphi^3 O(\mu^3)\\
&\quad +\lambda^2\varphi^3 O(\mu^2 e^{2\mu|\psi|_{C(\overline{G})}}) + \lambda\varphi^3 O(e^{2\mu|\psi|_{C(\overline{G})}}),
\end{aligned}
$$

which leads to the second estimate in (9.92). □

9.5 Observability Estimate for Stochastic Parabolic Equations

In this section, we shall establish an observability estimate for the stochastic parabolic equation (9.16). The main content of this section is taken from [218, 315].

We have the following result.

Theorem 9.28. *Let the condition* (9.17) *be satisfied. Then, solutions to* (9.16) *satisfy that, for any* $z_0 \in L^2_{\mathcal{F}_0}(\Omega; L^2(G;\mathbb{R}^m))$,

$$
|z(T)|_{L^2_{\mathcal{F}_T}(\Omega;L^2(G;\mathbb{R}^m))} \le \mathcal{C}e^{\mathcal{C}R_2^2}|z|_{L^2_{\mathbb{F}}(0,T;L^2(G_0;\mathbb{R}^m))}, \tag{9.93}
$$

where R_2 *is given in Corollary 9.5.*

We shall give a proof of Theorem 9.28 in Subsection 9.5.3. For simplicity, we consider only the case $m = 1$.

9.5.1 Global Carleman Estimate for Stochastic Parabolic Equations, I

In this subsection, as a preliminary to prove Theorem 9.28, we shall derive a global Carleman estimate for the following stochastic parabolic equation:

$$\begin{cases} dh - \sum_{j,k=1}^{n} (a^{jk} h_{x_j})_{x_k} dt = f dt + g dW(t), & \text{in } Q, \\ h = 0, & \text{on } \Sigma, \\ h(0) = h_0, & \text{in } G, \end{cases} \tag{9.94}$$

where $h_0 \in L^2_{\mathcal{F}_0}(\Omega; L^2(G))$, while f and g are suitable stochastic processes to be given later.

We begin with the following known technical result (See [117, p. 4, Lemma 1.1] and [337, Lemma 2.1] for its proof), which shows the existence of a nonnegative function with an arbitrarily given critical point location in G.

Lemma 9.29. *For any nonempty open subset G_1 of G, there is a $\psi \in C^\infty(\overline{G})$ such that $\psi > 0$ in G, $\psi = 0$ on Γ, and $|\nabla\psi(x)| > 0$ for all $x \in \overline{G \setminus G_1}$.*

In the rest of this section, we choose θ and ℓ as that in (9.88), and ψ given by Lemma 9.29 with G_1 being any fixed nonempty open subset of G such that $\overline{G_1} \subset G_0$. The desired Carleman estimate for (9.94) is stated as follows:

Theorem 9.30. *There is a constant $\mu_0 = \mu_0(G, G_0, (a^{jk})_{n\times n}, T) > 0$ such that for all $\mu \geq \mu_0$, one can find two constants $\mathcal{C} = \mathcal{C}(\mu) > 0$ and $\lambda_0 = \lambda_0(\mu) > 0$ such that for any $\lambda \geq \lambda_0$, $h_0 \in L^2_{\mathcal{F}_0}(\Omega; L^2(G))$, $f \in L^2_{\mathbb{F}}(0,T; L^2(G))$ and $g \in L^2_{\mathbb{F}}(0,T; H^1(G))$, the corresponding solution $h \in L^2_{\mathbb{F}}(\Omega; C([0,T]; L^2(G))) \bigcap L^2_{\mathbb{F}}(0,T; H^1_0(G))$ to (9.94) satisfies*

$$\begin{aligned} &\lambda^3\mu^4 \mathbb{E}\int_Q \theta^2\varphi^3 h^2 dxdt + \lambda\mu^2 \mathbb{E}\int_Q \theta^2\varphi|\nabla h|^2 dxdt \\ &\leq \mathcal{C}\mathbb{E}\Big[\int_Q \theta^2(f^2 + |\nabla g|^2 + \lambda^2\mu^2\varphi^2 g^2) dxdt + \lambda^3\mu^4\int_{Q_0} \theta^2\varphi^3 h^2 dxdt\Big]. \end{aligned} \tag{9.95}$$

Proof: We barrow some idea from [117]. We shall use Theorem 9.27 with b^{jk} being replaced by a^{jk} (and hence $\mathbf{u} = h$). The proof is divided into three steps.

Step 1. Integrating the equality (9.79) (with b^{jk} replaced by a^{jk}) on Q, taking mean value in both sides, and noting (9.92), we conclude that

$$2\mathbb{E}\int_Q \theta\Big[-\sum_{j,k=1}^{n}(a^{jk}\mathbf{v}_{x_j})_{x_k} + \mathcal{A}\mathbf{v}\Big]\Big[dh - \sum_{j,k=1}^{n}(a^{jk}h_{x_j})_{x_k}dt\Big]dx$$

$$
\begin{aligned}
&+2\mathbb{E}\int_Q \sum_{j,k=1}^n (a^{jk}\mathbf{v}_{x_j} d\mathbf{v})_{x_k} dx \\
&+2\mathbb{E}\int_Q \sum_{j,k=1}^n \Big[\sum_{j',k'=1}^n \Big(2a^{jk}a^{j'k'}\ell_{x_{j'}}\mathbf{v}_{x_j}\mathbf{v}_{x_{k'}} - a^{jk}a^{j'k'}\ell_{x_j}\mathbf{v}_{x_{j'}}\mathbf{v}_{x_{k'}}\Big) \\
&\qquad\qquad + \Psi a^{jk}\mathbf{v}_{x_j}\mathbf{v} - a^{jk}\Big(\mathcal{A}\ell_{x_j} + \frac{\Psi_{x_j}}{2}\Big)\mathbf{v}^2\Big]_{x_k} dxdt \qquad (9.96)\\
&\geq 2s_0^2\mathbb{E}\int_Q \Big[\varphi\big(\lambda\mu^2|\nabla\psi|^2 + \lambda O(\mu)\big)|\nabla\mathbf{v}|^2 + \varphi^3\Big(\lambda^3\mu^4|\nabla\psi|^4 + \lambda^3 O(\mu^3) \\
&\qquad\qquad + \lambda^2 O(\mu^2 e^{2\mu|\psi|_{C(\overline{G})}}) + \lambda O(e^{2\mu|\psi|_{C(\overline{G})}})\Big)\mathbf{v}^2\Big] dxdt \\
&+2\mathbb{E}\int_Q \Big| - \sum_{j,k=1}^n (a^{jk}\mathbf{v}_{x_j})_{x_k} + \mathcal{A}\mathbf{v}\Big|^2 dxdt \\
&-\mathbb{E}\int_Q \theta^2 \sum_{j,k=1}^n a^{jk}(dh_{x_j} + \ell_{x_j}dh)(dh_{x_k} + \ell_{x_k}dh)dx - \mathbb{E}\int_Q \theta^2\mathcal{A}(dh)^2 dx,
\end{aligned}
$$

where

$$
\mathcal{A} = -\sum_{j,k=1}^n (a^{jk}\ell_{x_j}\ell_{x_k} - a^{jk}_{x_k}\ell_{x_j} + a^{jk}\ell_{x_jx_k}) - \ell_t, \qquad \Psi = 2\sum_{j,k=1}^n a^{jk}\ell_{x_jx_k}.
$$

By (9.94), we find that

$$
\begin{aligned}
&2\mathbb{E}\int_Q \theta\Big[-\sum_{j,k=1}^n (a^{jk}\mathbf{v}_{x_j})_{x_k} + \mathcal{A}\mathbf{v}\Big]\Big[dh - \sum_{j,k=1}^n (a^{jk}h_{x_j})_{x_k}dt\Big]dx \\
&= 2\mathbb{E}\int_Q \theta\Big[-\sum_{j,k=1}^n (a^{jk}\mathbf{v}_{x_j})_{x_k} + \mathcal{A}\mathbf{v}\Big]\big(fdt + gdW(t)\big)dx \\
&= 2\mathbb{E}\int_Q \theta\Big[-\sum_{j,k=1}^n (a^{jk}\mathbf{v}_{x_j})_{x_k} + \mathcal{A}\mathbf{v}\Big]fdtdx \qquad (9.97)\\
&\leq \mathbb{E}\int_Q \Big| - \sum_{j,k=1}^n (a^{jk}\mathbf{v}_{x_j})_{x_k} + \mathcal{A}\mathbf{v}\Big|^2 dtdx + \mathbb{E}\int_Q \theta^2 f^2 dtdx.
\end{aligned}
$$

Since $h|_\Sigma = 0$, we have

$$
\mathbf{v}|_\Sigma = 0 \text{ and } \mathbf{v}_{x_j}|_\Sigma = \theta h_{x_j}|_\Sigma = \theta\frac{\partial h}{\partial\nu}\nu^j\Big|_\Sigma.
$$

Similarly, by Lemma 9.29, we get

$$
\ell_{x_j} = \lambda\mu\varphi\psi_{x_j} = \lambda\mu\varphi\frac{\partial\psi}{\partial\nu}\nu^j \quad \text{and} \quad \frac{\partial\psi}{\partial\nu} < 0 \quad \text{on } \Sigma.
$$

Therefore, utilizing integration by parts and noting $\mathbf{v}|_\Sigma = 0$, we obtain that

$$2\mathbb{E}\int_Q \sum_{j,k=1}^n (a^{jk}\mathbf{v}_{x_j} d\mathbf{v})_{x_k} dx = 2\mathbb{E}\int_\Sigma \sum_{j,k=1}^n a^{jk}\mathbf{v}_{x_j}\nu^k d\mathbf{v}dx = 0, \tag{9.98}$$

and that

$$\begin{aligned}
&2\mathbb{E}\int_Q \sum_{j,k=1}^n \Big[\sum_{j',k'=1}^n \Big(2a^{jk}a^{j'k'}\ell_{x_{j'}}\mathbf{v}_{x_j}\mathbf{v}_{x_{k'}} - a^{jk}a^{j'k'}\ell_{x_j}\mathbf{v}_{x_{j'}}\mathbf{v}_{x_{k'}}\Big)\\
&\qquad\qquad +\Psi a^{jk}\mathbf{v}_{x_j}\mathbf{v} - a^{jk}\Big(A\ell_{x_j} + \frac{\Psi_{x_j}}{2}\Big)\mathbf{v}^2\Big]_{x_k} dtdx\\
&= 2\mathbb{E}\int_\Sigma \sum_{j,k=1}^n \Big[\sum_{j',k'=1}^n \Big(2a^{jk}a^{j'k'}\ell_{x_{j'}}\mathbf{v}_{x_j}\mathbf{v}_{x_{k'}} - a^{jk}a^{j'k'}\ell_{x_j}\mathbf{v}_{x_{j'}}\mathbf{v}_{x_{k'}}\Big)\\
&\qquad\qquad +\Psi a^{jk}\mathbf{v}_{x_j}\mathbf{v} - a^{jk}\Big(A\ell_{x_j} + \frac{\Psi_{x_j}}{2}\Big)\mathbf{v}^2\Big]\nu^k d\Gamma dt\\
&= 2\mathbb{E}\int_\Sigma \sum_{j,k=1}^n \Big[\sum_{j',k'=1}^n \Big(2a^{jk}a^{j'k'}\ell_{x_{j'}}\mathbf{v}_{x_j}\mathbf{v}_{x_{k'}} - a^{jk}a^{j'k'}\ell_{x_j}\mathbf{v}_{x_{j'}}\mathbf{v}_{x_{k'}}\Big)\Big]\nu^k d\Gamma dt\\
&= 2\lambda\mu\mathbb{E}\int_\Sigma \theta^2\varphi\frac{\partial\psi}{\partial\nu}\Big(\frac{\partial h}{\partial\nu}\Big)^2 \sum_{j,k=1}^n \Big[\sum_{j',k'=1}^n \Big(2a^{jk}a^{j'k'}\nu^{j'}\nu^j\nu^{k'}\\
&\qquad\qquad -a^{jk}a^{j'k'}\nu^j\nu^{j'}\nu^{k'}\Big)\Big]\nu^k d\Gamma dt\\
&= 2\lambda\mu\mathbb{E}\int_\Sigma \theta^2\varphi\frac{\partial\psi}{\partial\nu}\Big(\frac{\partial h}{\partial\nu}\Big)^2\Big(\sum_{j,k=1}^n a^{jk}\nu^j\nu^k\Big)^2 d\Gamma dt \le 0.
\end{aligned}\tag{9.99}$$

By (9.94), we see that

$$\begin{aligned}
&\mathbb{E}\int_Q \theta^2 \sum_{j,k=1}^n a^{jk}(dh_{x_j} + \ell_{x_j}dh)(dh_{x_k} + \ell_{x_k}dh)dx + \mathbb{E}\int_Q \theta^2\mathcal{A}(dh)^2 dx\\
&= \mathbb{E}\int_Q \theta^2 \sum_{j,k=1}^n a^{jk}(g_{x_j} + \ell_{x_j}g)(g_{x_k} + \ell_{x_k}g)dxdt + \mathbb{E}\int_Q \theta^2\mathcal{A}g^2 dxdt.
\end{aligned}\tag{9.100}$$

Combining (9.96)–(9.100), we arrive at

$$\begin{aligned}
&2s_0^2\mathbb{E}\int_Q \Big[\varphi\big(\lambda\mu^2|\nabla\psi|^2 + \lambda O(\mu)\big)|\nabla\mathbf{v}|^2 + \varphi^3\Big(\lambda^3\mu^4|\nabla\psi|^4 + \lambda^3 O(\mu^3)\\
&\quad +\lambda^2 O(\mu^2 e^{2\mu|\psi|_{C(\overline{G})}}) + \lambda O(e^{2\mu|\psi|_{C(\overline{G})}})\Big)\mathbf{v}^2\Big]dxdt\\
&\le \mathbb{E}\int_Q \theta^2\Big[f^2 + \sum_{j,k=1}^n a^{jk}(g_{x_j} + \ell_{x_j}g)(g_{x_k} + \ell_{x_k}g) + \mathcal{A}g^2\Big]dxdt.
\end{aligned}\tag{9.101}$$

Step 2. By Lemma 9.29, $\min_{x\in G\setminus G_1}|\nabla\psi|>0$. Hence, there is a $\mu_0>0$ such that for all $\mu\geq\mu_0$, one can find a constant $\lambda_0=\lambda_0(\mu)$ so that for any $\lambda\geq\lambda_0$, it holds that

$$\begin{aligned}&2\mathbb{E}\int_0^T\int_{G\setminus G_1}\Big[\lambda\varphi\Big(\mu^2\min_{x\in G\setminus G_1}|\nabla\psi|^2+O(\mu)\Big)|\nabla\mathbf{v}|^2\\&\quad+\varphi^3\Big(\lambda^3\mu^4\min_{x\in G\setminus G_1}|\nabla\psi|^4+\lambda^3O(\mu^3)\\&\quad+\lambda^2O(\mu^2e^{2\mu|\psi|_{C(\overline{G})}})+\lambda O(e^{2\mu|\psi|_{C(\overline{G})}})\Big)\mathbf{v}^2\Big]dxdt\\&\geq\lambda\mu^2c_1\mathbb{E}\int_0^T\int_{G\setminus G_1}\varphi\Big(|\nabla\mathbf{v}|^2+\lambda^2\mu^2\varphi^2\mathbf{v}^2\Big)dxdt,\end{aligned}\tag{9.102}$$

where $c_1\overset{\Delta}{=}\min\big(\min_{x\in G\setminus G_1}|\nabla\psi|^2,\min_{x\in G\setminus G_1}|\nabla\psi|^4\big)$. By

$$h_{x_j}=\theta^{-1}(\mathbf{v}_{x_j}-\ell_{x_j}\mathbf{v})=\theta^{-1}(\mathbf{v}_{x_j}-\lambda\mu\varphi\psi_{x_j}\mathbf{v})$$

and

$$\mathbf{v}_{x_j}=\theta(h_{x_j}+\ell_{x_j}h)=\theta(h_{x_j}+\lambda\mu\varphi\psi_{x_j}h),$$

we obtain that

$$\frac{1}{\mathcal{C}}\theta^2\big(|\nabla h|^2+\lambda^2\varphi^2\mu^2h^2\big)\leq|\nabla\mathbf{v}|^2+\lambda^2\mu^2\varphi^2\mathbf{v}^2\leq\mathcal{C}\theta^2\big(|\nabla h|^2+\lambda^2\varphi^2\mu^2h^2\big).$$

Therefore, it follows from (9.102) that

$$\begin{aligned}&\lambda\mu^2\mathbb{E}\int_Q\theta^2\varphi(|\nabla h|^2+\lambda^2\varphi^2\mu^2h^2)dxdt\\&=\lambda\mu^2\mathbb{E}\int_0^T\Big(\int_{G\setminus G_1}+\int_{G_1}\Big)\theta^2\varphi(|\nabla h|^2+\lambda^2\varphi^2\mu^2h^2)dxdt\\&\leq\mathcal{C}\Big\{\mathbb{E}\int_Q\Big[\varphi(\lambda\mu^2|\nabla\psi|^2+\lambda O(\mu))|\nabla\mathbf{v}|^2+\varphi^3\Big(\lambda^3\mu^4|\nabla\psi|^4+\lambda^3O(\mu^3)\\&\qquad+\lambda^2O(\mu^2e^{2\mu|\psi|_{C(\overline{G})}})+\lambda O(e^{2\mu|\psi|_{C(\overline{G})}})\Big)\mathbf{v}^2\Big]dxdt\\&\quad+\lambda\mu^2\mathbb{E}\int_0^T\int_{G_1}\theta^2\varphi\big(|\nabla h|^2+\lambda^2\mu^2\varphi^2h^2\big)dxdt\Big\}.\end{aligned}\tag{9.103}$$

Now, combining (9.101) and (9.103), we end up with

$$\begin{aligned}&\lambda\mu^2\mathbb{E}\int_Q\theta^2\varphi(|\nabla h|^2+\lambda^2\varphi^2\mu^2h^2)dxdt\\&\leq\mathcal{C}\Big[\mathbb{E}\int_Q\theta^2\big(f^2+|\nabla g|^2+\lambda^2\mu^2\varphi^2g^2\big)dxdt\\&\quad+\lambda\mu^2\mathbb{E}\int_0^T\int_{G_1}\theta^2\varphi\big(|\nabla h|^2+\lambda^2\mu^2\varphi^2h^2\big)\,dxdt\Big].\end{aligned}\tag{9.104}$$

Step 3. We choose a cut-off function $\zeta \in C_0^\infty(G_0;[0,1])$ so that $\zeta \equiv 1$ in G_1. By $d(\theta^2\varphi h^2) = h^2(\theta^2\varphi)_t dt + 2\theta^2\varphi h dh + \theta^2\varphi(dh)^2$, recalling $\lim_{t\to 0^+}\varphi(t,\cdot) = \lim_{t\to T^-}\varphi(t,\cdot) \equiv 0$ and using (9.94), we find that

$$
\begin{aligned}
0 &= \mathbb{E}\int_{Q_0} \zeta^2[h^2(\theta^2\varphi)_t dt + 2\theta^2\varphi h dh + \theta^2\varphi(dh)^2]dx \\
&= \mathbb{E}\int_{Q_0} \theta^2\zeta^2\Big[h^2(\varphi_t + 2\lambda\varphi\alpha_t) + 2\varphi h\Big(\sum_{j,k=1}^n (a^{jk}h_{x_j})_{x_k} + f\Big) + \varphi g^2\Big]dxdt \\
&= \mathbb{E}\int_{Q_0} \theta^2\Big[\zeta^2h^2(\varphi_t + 2\lambda\varphi\alpha_t) - 2\zeta^2\varphi\sum_{j,k=1}^n a^{jk}h_{x_j}h_{x_k} \\
&\qquad -2\mu\zeta^2\varphi(1+2\lambda\varphi)h\sum_{j,k=1}^n a^{jk}h_{x_j}\psi_{x_k} \\
&\qquad -4\zeta\varphi h\sum_{j,k=1}^n a^{jk}h_{x_j}\zeta_{x_k} + 2\zeta^2\varphi fh + \zeta^2\varphi g^2\Big]dxdt.
\end{aligned}
\tag{9.105}
$$

From (9.105), we deduce that, for any $\varepsilon > 0$,

$$
\begin{aligned}
&2\mathbb{E}\int_{Q_0} \theta^2\zeta^2\varphi\sum_{j,k=1}^n a^{jk}h_{x_j}h_{x_k}dxdt \\
&= \mathbb{E}\int_{Q_0} \theta^2\Big[\zeta^2h^2(\varphi_t + 2\lambda\varphi\alpha_t) - 2\mu\zeta^2\varphi(1+2\lambda\varphi)h\sum_{j,k=1}^n a^{jk}h_{x_j}\psi_{x_k} \\
&\qquad -4\zeta\varphi h\sum_{j,k=1}^n a^{jk}h_{x_j}\zeta_{x_k} + 2\zeta^2\varphi fh + \zeta^2\varphi g^2\Big]dxdt \\
&\le \varepsilon\mathbb{E}\int_{Q_0} \theta^2\zeta^2\varphi|\nabla h|^2dxdt + \frac{\mathcal{C}}{\varepsilon}\mathbb{E}\int_{Q_0} \theta^2\Big(\frac{1}{\lambda^2\mu^2}f^2 + \lambda^2\mu^2\varphi^3h^2\Big)dxdt \\
&\quad +\mathbb{E}\int_{Q_0} \theta^2\varphi g^2dxdt.
\end{aligned}
\tag{9.106}
$$

From (9.106), we conclude that

$$
\begin{aligned}
&\mathbb{E}\int_0^T\int_{G_1} \theta^2\varphi|\nabla h|^2dxdt \\
&\le \mathcal{C}\Big[\mathbb{E}\int_Q \theta^2\Big(\frac{1}{\lambda^2\mu^2}f^2 + \varphi g^2\Big)dxdt + \lambda^2\mu^2\mathbb{E}\int_{Q_0} \theta^2\varphi^3h^2dxdt\Big].
\end{aligned}
\tag{9.107}
$$

Finally, combining (9.104) and (9.107), we obtain (9.95). □

Remark 9.31. In Theorem 9.30, it is assumed that $g \in L^2_{\mathbb{F}}(0,T;H^1(G))$. This assumption is not natural. In the next subsection, we shall relax this assumption to be $g \in L^2_{\mathbb{F}}(0,T;L^2(G))$.

As an easy consequence of Theorem 9.30, we have the following result.

Corollary 9.32. *For $\mu = \mu_0$ and any $\lambda \geq \lambda_0(= \lambda_0(\mu_0))$ given in Theorem 9.30, $h_0 \in L^2_{\mathcal{F}_0}(\Omega; L^2(G))$ and $f \in L^2_{\mathbb{F}}(0,T; L^2(G))$, the corresponding solution $h \in L^2_{\mathbb{F}}(\Omega; C([0,T];\, L^2(G))) \bigcap L^2_{\mathbb{F}}(0,T; H^1_0(G))$ to (9.94) with $g \equiv 0$ satisfies*

$$\begin{aligned} &\lambda^3 \mathbb{E} \int_Q \theta^2 \varphi^3 h^2 dxdt + \lambda \mathbb{E} \int_Q \theta^2 \varphi |\nabla h|^2 dxdt \\ &\leq C \mathbb{E} \Big(\int_Q \theta^2 f^2 dxdt + \lambda^3 \int_{Q_0} \theta^2 \varphi^3 h^2 dxdt \Big). \end{aligned}$$

Since the equation (9.94) with $g \equiv 0$ is a random parabolic equation, Corollary 9.32 follows also from the known global Carleman estimate for deterministic parabolic equations (e.g. [117]). Nevertheless, it seems that Theorem 9.30 has some independent interest.

9.5.2 Global Carleman Estimate for Stochastic Parabolic Equations, II

In this subsection, we derive an improved global Carleman estimate for the forward stochastic parabolic equation (9.94).

Throughout this subsection, $\mu = \mu_0$ and $\lambda \geq \lambda_0$ are given as that in Theorem 9.30, and θ and φ are the same as that in the last subsection.

For any fixed $f, g \in L^2_{\mathbb{F}}(0,T; L^2(G))$ and $h_0 \in L^2_{\mathcal{F}_0}(\Omega; L^2(G))$, let h denote the corresponding solution to the equation (9.94). Based on the Carleman estimate in Corollary 9.32, we give below a "partial" null controllability result for the following controlled backward stochastic parabolic equation:

$$\begin{cases} dr + \sum\limits_{j,k=1}^{n} \big(a^{jk} r_{x_j}\big)_{x_k} dt = (\lambda^3 \theta^2 \varphi^3 h + \chi_{G_1} u) dt + R dW(t) & \text{in } Q, \\ r = 0 & \text{on } \Sigma, \\ r(T) = 0 & \text{in } G, \end{cases} \tag{9.108}$$

where $u \in L^2_{\mathbb{F}}(0,T; L^2(G_1))$ is the control variable and (r, R) is the state variable.

Proposition 9.33. *There exists a control $\hat{u} \in L^2_{\mathbb{F}}(0,T; L^2(G_1))$ such that the corresponding solution $(\hat{r}, \widehat{R}) \in \big(L^2_{\mathbb{F}}(\Omega; C([0,T]; L^2(G))) \bigcap L^2_{\mathbb{F}}(0,T; H^1_0(G))\big) \times L^2_{\mathbb{F}}(0,T; L^2(G))$ to (9.108) with $u = \hat{u}$ satisfies $\hat{r}(0) = 0$ in G, a.s. Moreover,*

$$\mathbb{E} \int_Q \theta^{-2} \left(\hat{r}^2 + \lambda^{-3} \varphi^{-3} \hat{u}^2 + \lambda^{-2} \varphi^{-2} \widehat{R}^2 \right) dxdt \leq C \lambda^3 \mathbb{E} \int_Q \theta^2 \varphi^3 h^2 dxdt. \tag{9.109}$$

Proof: We divide the proof into several steps.

Step 1. Write

$$\mathcal{U}=\Big\{u\in L^2_{\mathbb{F}}(0,T;L^2(G_1))\;\Big|\;\mathbb{E}\int_0^T\int_{G_1}\theta^{-2}\varphi^{-3}u^2dxdt<\infty\Big\}.$$

For any $\varepsilon>0$, write $\alpha_\varepsilon\equiv\alpha_\varepsilon(t,x)=\dfrac{e^{\mu\psi(x)}-e^{2\mu|\psi|_{C(\overline{G})}}}{(t+\varepsilon)(T-t+\varepsilon)}$ and construct the following optimal control problem for the equation (9.108):

$$\inf\big\{\mathcal{J}(u)\;\big|\;u\in\mathcal{U}\big\},$$

where

$$\begin{aligned}\mathcal{J}(u)\triangleq\frac{1}{2}\mathbb{E}\Big(\int_Q e^{-2\lambda\alpha_\varepsilon}r^2dxdt+\lambda^{-3}\int_0^T\int_{G_1}\theta^{-2}\varphi^{-3}u^2dxdt\\+\frac{1}{\varepsilon}\int_G r^2(0)dx\Big).\end{aligned}$$

It is easy to check that for each $\varepsilon>0$, the above optimal control problem admits a unique optimal solution $(u_\varepsilon,r_\varepsilon,R_\varepsilon)\in\mathcal{U}\times L^2_{\mathbb{F}}(0,T;H^1_0(G))\times L^2_{\mathbb{F}}(0,T;L^2(G))$. Moreover,

$$u_\varepsilon=\chi_{G_1}\lambda^3\theta^2\varphi^3z_\varepsilon\qquad\text{in } Q,\quad \text{a.s.},\tag{9.110}$$

where z_ε satisfies

$$\begin{cases}z_{\varepsilon,t}-\sum\limits_{j,k=1}^n\big(a^{jk}z_{\varepsilon,x_j}\big)_{x_k}=e^{-2\lambda\alpha_\varepsilon}r_\varepsilon & \text{in } Q,\\ z_\varepsilon=0 & \text{on } \Sigma,\\ z_\varepsilon(0)=\dfrac{1}{\varepsilon}r_\varepsilon(0) & \text{in } G.\end{cases}\tag{9.111}$$

Step 2. We now establish a uniform estimate for the optimal solutions $(u_\varepsilon,r_\varepsilon,R_\varepsilon)$ w.r.t. $\varepsilon>0$. By (9.108), (9.111), Itô's formula and (9.110), it follows that

$$\begin{aligned}&-\mathbb{E}\int_G r_\varepsilon(0)z_\varepsilon(0)dx\\&=\mathbb{E}\int_Q\Big\{r_\varepsilon\Big[\sum_{j,k=1}^n\big(a^{jk}z_{\varepsilon,x_j}\big)_{x_k}+e^{-2\lambda\alpha_\varepsilon}r_\varepsilon\Big]\\&\qquad+z_\varepsilon\Big[-\sum_{j,k=1}^n\big(a^{jk}r_{\varepsilon,x_j}\big)_{x_k}+\lambda^3\theta^2\varphi^3h+\chi_{G_1}u_\varepsilon\Big]\Big\}dxdt\end{aligned}$$

$$= \mathbb{E}\int_Q \big(e^{-2\lambda\alpha_\varepsilon} r_\varepsilon^2 + \lambda^3\theta^2\varphi^3 z_\varepsilon h + \chi_{G_1}\lambda^3\theta^2\varphi^3 z_\varepsilon^2\big)\, dxdt.$$

This, together with the last equality of (9.111) and Corollary 9.32, indicates that for any $\rho > 0$,

$$\begin{aligned}
&\mathbb{E}\int_Q \big(e^{-2\lambda\alpha_\varepsilon} r_\varepsilon^2 + \chi_{G_1}\lambda^3\theta^2\varphi^3 z_\varepsilon^2\big)\, dxdt + \frac{1}{\varepsilon}\mathbb{E}\int_G r_\varepsilon^2(0)dx\\
&= -\lambda^3\mathbb{E}\int_Q \theta^2\varphi^3 z_\varepsilon h dxdt\\
&\le \rho\lambda^3\mathbb{E}\int_Q \theta^2\varphi^3 z_\varepsilon^2 dxdt + \frac{\lambda^3}{4\rho}\mathbb{E}\int_Q \theta^2\varphi^3 h^2 dxdt\\
&\le \rho\mathcal{C}\lambda^3\Big[\mathbb{E}\int_0^T\int_{G_1} \theta^2\varphi^3 z_\varepsilon^2 dxdt + \mathbb{E}\int_Q \theta^2(e^{-2\lambda\alpha_\varepsilon} r_\varepsilon)^2 dxdt\Big]\\
&\quad + \frac{\lambda^3}{4\rho}\mathbb{E}\int_Q \theta^2\varphi^3 h^2 dxdt.
\end{aligned} \tag{9.112}$$

Notice that $\theta^2 e^{-2\lambda\alpha_\varepsilon} \le 1$. Therefore, by (9.110) and (9.112), we see that

$$\begin{aligned}
&\mathbb{E}\int_Q \big(e^{-2\lambda\alpha_\varepsilon} r_\varepsilon^2 + \lambda^{-3}\theta^{-2}\varphi^{-3} u_\varepsilon^2\big)\, dxdt + \frac{1}{\varepsilon}\mathbb{E}\int_G r_\varepsilon^2(0)dx\\
&\le \mathcal{C}\lambda^3\mathbb{E}\int_Q \theta^2\varphi^3 h^2 dxdt.
\end{aligned} \tag{9.113}$$

On the other hand, by the first equation of (9.108), it follows that

$$\begin{aligned}
&d(e^{-2\lambda\alpha_\varepsilon}\varphi^{-2} r_\varepsilon^2)\\
&= (e^{-2\lambda\alpha_\varepsilon}\varphi^{-2})_t r_\varepsilon^2 dt + e^{-2\lambda\alpha_\varepsilon}\varphi^{-2}(dr_\varepsilon)^2\\
&\quad + 2e^{-2\lambda\alpha_\varepsilon}\varphi^{-2} r_\varepsilon\Big[-\sum_{j,k=1}^n (a^{jk} r_{\varepsilon,x_j})_{x_k} dt + \lambda^3\theta^2\varphi^3 h dt + \chi_{G_1} u_\varepsilon dt + R_\varepsilon dW(t)\Big].
\end{aligned}$$

This implies that

$$\begin{aligned}
&\mathbb{E}\int_Q e^{-2\lambda\alpha_\varepsilon}\varphi^{-2} R_\varepsilon^2 dxdt + 2\mathbb{E}\int_Q e^{-2\lambda\alpha_\varepsilon}\varphi^{-2}\sum_{j,k=1}^n a^{jk} r_{\varepsilon,x_j} r_{\varepsilon,x_k} dxdt\\
&= -\mathbb{E}\int_Q (e^{-2\lambda\alpha_\varepsilon}\varphi^{-2})_t r_\varepsilon^2 dxdt - 2\mathbb{E}\int_Q \sum_{j,k=1}^n a^{jk}(e^{-2\lambda\alpha_\varepsilon}\varphi^{-2})_{x_k} r_{\varepsilon,x_j} r_\varepsilon dxdt\\
&\quad - 2\lambda^3\mathbb{E}\int_Q \theta^2 e^{-2\lambda\alpha_\varepsilon}\varphi r_\varepsilon h dxdt - 2\mathbb{E}\int_Q \chi_{G_1} e^{-2\lambda\alpha_\varepsilon}\varphi^{-2} r_\varepsilon u_\varepsilon dxdt.
\end{aligned}$$

Notice that $e^{-2\lambda\alpha_\varepsilon} \le \theta^{-2}$. By a simple calculation, we find that

$$
\begin{aligned}
&\lambda^{-2}\mathbb{E}\int_Q e^{-2\lambda\alpha_\varepsilon}\varphi^{-2}R_\varepsilon^2 dxdt + 2\lambda^{-2}\mathbb{E}\int_Q e^{-2\lambda\alpha_\varepsilon}\varphi^{-2}|\nabla r_\varepsilon|^2 dxdt\\
&\le \mathcal{C}\Big(\lambda^{-1}\mathbb{E}\int_Q e^{-2\lambda\alpha_\varepsilon} r_\varepsilon^2 dxdt + \lambda^{-1}\mathbb{E}\int_Q e^{-2\lambda\alpha_\varepsilon}\varphi^{-1}|\nabla r_\varepsilon||r_\varepsilon| dxdt\\
&\qquad +\lambda\mathbb{E}\int_Q \theta e^{-\lambda\alpha_\varepsilon}\varphi|r_\varepsilon h| dxdt + \lambda^{-2}\mathbb{E}\int_Q e^{-2\lambda\alpha_\varepsilon}\varphi^{-2}|r_\varepsilon u_\varepsilon| dxdt\Big)\\
&\le \lambda^{-2}\mathbb{E}\int_Q e^{-2\lambda\alpha_\varepsilon}\varphi^{-2}|\nabla r_\varepsilon|^2 dxdt\\
&\quad +\mathcal{C}\mathbb{E}\int_Q \big(e^{-2\lambda\alpha_\varepsilon} r_\varepsilon^2 + \lambda^{-3}\theta^{-2}\varphi^{-3}u_\varepsilon^2 + \lambda^3\theta^2\varphi^3 h^2\big) dxdt.
\end{aligned}
\tag{9.114}
$$

Combining (9.114) with (9.113), we conclude that

$$
\begin{aligned}
&\mathbb{E}\int_Q \Big(e^{-2\lambda\alpha_\varepsilon} r_\varepsilon^2 + \lambda^{-3}\theta^{-2}\varphi^{-3}u_\varepsilon^2\Big) dxdt + \lambda^{-2}\mathbb{E}\int_Q e^{-2\lambda\alpha_\varepsilon}\varphi^{-2}|\nabla r_\varepsilon|^2 dxdt\\
&\quad +\frac{1}{\varepsilon}\mathbb{E}\int_G r_\varepsilon^2(0) dx + \lambda^{-2}\mathbb{E}\int_Q e^{-2\lambda\alpha_\varepsilon}\varphi^{-2}R_\varepsilon^2 dxdt\\
&\le \mathcal{C}\lambda^3\mathbb{E}\int_Q \theta^2\varphi^3 h^2 dxdt.
\end{aligned}
\tag{9.115}
$$

By (9.115), and noting (9.108), we deduce that there exists a triple $(\hat{u}, \hat{r}, \widehat{R}) \in L^2_{\mathbb{F}}(0,T;L^2(G_1)) \times L^2_{\mathbb{F}}(0,T;H^1_0(G)) \times L^2_{\mathbb{F}}(0,T;L^2(G))$ such that as $\varepsilon \to 0$,

$$
\begin{aligned}
u_\varepsilon &\to \hat{u} \quad \text{weakly in } L^2((0,T)\times\Omega;L^2(G_1));\\
r_\varepsilon &\to \hat{r} \quad \text{weakly in } L^2((0,T)\times\Omega;H^1_0(G));\\
R_\varepsilon &\to \widehat{R} \quad \text{weakly in } L^2((0,T)\times\Omega;L^2(G)).
\end{aligned}
\tag{9.116}
$$

Step 3. We conclude that $(\hat{r}, \widehat{R})$ is the solution to (9.108) with $u = \hat{u}$. In fact, by Theorem 4.11, the equation (9.108) with $g = \hat{g}$ admits one and only one solution $(\tilde{r}, \widetilde{R}) \in \big(L^2_{\mathbb{F}}(\Omega;C([0,T];L^2(G))) \bigcap L^2_{\mathbb{F}}(0,T;H^1_0(G))\big) \times L^2_{\mathbb{F}}(0,T;L^2(G))$. Then, for any $f_i \in L^2_{\mathbb{F}}(0,T;L^2(G))$ $(i = 1,2)$, consider the following forward stochastic parabolic equation:

$$
\begin{cases}
d\phi - \sum\limits_{j,k=1}^{n}\big(a^{jk}\phi_{x_j}\big)_{x_k} dt = f_1 dt + f_2 dW(t) & \text{in } Q,\\
\phi = 0 & \text{on } \Sigma,\\
\phi(0) = 0 & \text{in } G.
\end{cases}
\tag{9.117}
$$

By (9.108), (9.117) and Itô's formula, we see that

$$\mathbb{E}\int_Q (\lambda^3\theta^2\varphi^3 h + \chi_{G_1}\hat{u})\phi dxdt + \mathbb{E}\int_Q (\tilde{r}f_1 + \widetilde{R}f_2)dxdt = 0. \tag{9.118}$$

Likewise,

$$\mathbb{E}\int_Q (\lambda^3\theta^2\varphi^3 h + \chi_{G_1}u_\varepsilon)\phi dxdt + \mathbb{E}\int_Q (r_\varepsilon f_1 + R_\varepsilon f_2)dxdt = 0,$$

which, together with (9.116), indicates that

$$\mathbb{E}\int_Q (\lambda^3\theta^2\varphi^3 h + \chi_{G_1}\hat{u})\phi dxdt + \mathbb{E}\int_Q (\hat{r}f_1 + \widehat{R}f_2)dxdt = 0. \tag{9.119}$$

Combining (9.118) and (9.119), we see that $\tilde{r} = \hat{r}$ and $\widetilde{R} = \widehat{R}$ in Q, a.s.

Finally, by (9.115), we obtain that $\hat{r}(0) = 0$ in G, a.s. and the estimate (9.109) holds. □

Now, based on the above null controllability result for the backward parabolic equation (9.108), we show the following improved global Carleman estimate for (9.94) (compared to Theorem 9.30):

Theorem 9.34. *For $\mu = \mu_0$ and any $\lambda \geq \lambda_0$ given in Theorem 9.30, $h_0 \in L^2_{\mathcal{F}_0}(\Omega; L^2(G))$ and $f, g \in L^2_{\mathbb{F}}(0,T; L^2(G))$, the corresponding solution $h \in L^2_{\mathbb{F}}(\Omega; C([0,T]; L^2(G))) \bigcap L^2_{\mathbb{F}}(0,T; H^1_0(G))$ to (9.94) satisfies*

$$\begin{aligned} &\mathbb{E}\int_Q \theta^2\left(\lambda^3\varphi^3 h^2 + \lambda\varphi|\nabla h|^2\right)dxdt \\ &\leq \mathcal{C}\mathbb{E}\Big[\int_Q \theta^2(f^2 + \lambda^2\varphi^2 g^2)dxdt + \lambda^3\int_{Q_0}\theta^2\varphi^3 h^2 dxdt\Big]. \end{aligned} \tag{9.120}$$

Proof: For any $h_0 \in L^2_{\mathcal{F}_0}(\Omega; L^2(G))$ and $f, g \in L^2_{\mathbb{F}}(0,T; L^2(G))$, let $(\hat{r}, \widehat{R})$ be the solution to (9.108) with $u = \hat{u}$, which was given in Proposition 9.33. Then, by Itô's formula, we see that

$$\lambda^3\mathbb{E}\int_Q \theta^2\varphi^3 h^2 dxdt = -\mathbb{E}\int_Q (\chi_{G_1}\hat{u}h + \hat{r}f + \widehat{R}g)dxdt.$$

It follows that for any $\rho > 0$,

$$\begin{aligned} &\lambda^3\mathbb{E}\int_Q \theta^2\varphi^3 h^2 dxdt \\ &\leq \rho\mathbb{E}\int_Q \theta^{-2}\left(\lambda^{-3}\varphi^{-3}\hat{u}^2 + \hat{r}^2 + \lambda^{-2}\varphi^{-2}\widehat{R}^2\right)dxdt \\ &\quad + \frac{1}{4\rho}\Big[\lambda^3\mathbb{E}\int_0^T\int_{G_1}\theta^2\varphi^3 h^2 dxdt + \mathbb{E}\int_Q \theta^2\left(f^2 + \lambda^2\varphi^2 g^2\right)dxdt\Big]. \end{aligned}$$

By (9.109), this implies that

$$
\begin{aligned}
&\lambda^3\mathbb{E}\int_Q \theta^2\varphi^3 h^2 dxdt\\
&\le \mathcal{C}\Big[\lambda^3\mathbb{E}\int_0^T\int_{G_1}\theta^2\varphi^3h^2dxdt+\mathbb{E}\int_Q\theta^2\left(f^2+\lambda^2\varphi^2g^2\right)dxdt\Big].
\end{aligned}
\tag{9.121}
$$

On the other hand, by Itô's formula, we have that

$$
\begin{aligned}
d(\theta^2\varphi h^2) &= (\theta^2\varphi)_t h^2 dt + 2\theta^2\varphi h\Big(\sum_{j,k=1}^n (a^{jk}h_{x_j})_{x_k}dt + fdt + gdW(t)\Big)\\
&\quad +\theta^2\varphi(dh)^2.
\end{aligned}
$$

Hence, for any $\rho > 0$,

$$
\begin{aligned}
&2\mathbb{E}\int_Q \theta^2\varphi\sum_{j,k=1}^n a^{jk}h_{x_j}h_{x_k}dxdt\\
&= \mathbb{E}\int_Q\Big((\theta^2\varphi)_t h^2 - 2\sum_{j,k=1}^n a^{jk}(\theta^2\varphi)_{x_k}h_{x_j}h + 2\theta^2\varphi hf + \theta^2\varphi g^2\Big)dxdt\\
&\le \mathcal{C}\mathbb{E}\int_Q\left(\lambda\theta^2\varphi^3h^2+\lambda\theta^2\varphi^2|\nabla h||h| + \theta^2\varphi|hf| + \theta^2\varphi g^2\right)dxdt\\
&\le \rho\mathbb{E}\int_Q\theta^2\varphi|\nabla h|^2dxdt + \frac{\mathcal{C}}{\rho}\mathbb{E}\int_Q\theta^2\left(\lambda^2\varphi^3h^2+\lambda^{-1}f^2+\varphi g^2\right)dxdt.
\end{aligned}
$$

By (9.121), this implies that

$$
\begin{aligned}
&\lambda\mathbb{E}\int_Q \theta^2\varphi|\nabla h|^2 dxdt\\
&\le \mathcal{C}\Big[\lambda^3\mathbb{E}\int_0^T\int_{G_1}\theta^2\varphi^3h^2dxdt+\mathbb{E}\int_Q\theta^2\left(f^2+\lambda^2\varphi^2g^2\right)dxdt\Big].
\end{aligned}
\tag{9.122}
$$

Combining (9.121) and (9.122), we obtain the desired estimate (9.120). □

Remark 9.35. By a similar method used in the proof of Theorem 9.34, one can show the following global Carleman estimate for (9.94) in H^{-1}-space: For $\mu = \mu_0$ and any $\lambda \ge \lambda_0$ given in Theorem 9.30, $h_0 \in L^2_{\mathcal{F}_0}(\Omega; L^2(G))$, $f \in L^2_{\mathbb{F}}(0,T; H^{-1}(G))$ and $g \in L^2_{\mathbb{F}}(0,T;L^2(G))$, the corresponding solution $h \in L^2_{\mathbb{F}}(\Omega; C([0,T];L^2(G)))\bigcap L^2_{\mathbb{F}}(0,T;H^1_0(G))$ to (9.94) satisfies

$$
\begin{aligned}
&\mathbb{E}\int_Q\theta^2\left(\lambda\varphi^3h^2+\lambda^{-1}\varphi|\nabla h|^2\right)dxdt\\
&\le \mathcal{C}\Big(\lambda\mathbb{E}\int_0^T\int_{G_1}\theta^2\varphi^3h^2dxdt + \mathbb{E}\int_0^T\varphi^2|\theta f|^2_{H^{-1}(G)}dt + \mathbb{E}\int_Q\theta^2\varphi^2g^2dxdt\Big).
\end{aligned}
$$

9.5.3 Proof of the Observability Result

We are now in a position to prove Theorem 9.28.

Proof of Theorem 9.28: Applying Theorem 9.34 to the equation (9.16), recalling (9.90) and the definition of κ_2, noting also that we assume that $m=1$, we obtain that, for $\mu=\mu_0$ and any $\lambda \geq \lambda_0$ given in Theorem 9.30,

$$\begin{aligned}
&\lambda^3 \mathbb{E}\int_Q \theta^2\varphi^3 z^2 dxdt + \lambda\mathbb{E}\int_Q \theta^2\varphi|\nabla z|^2 dxdt \\
&\leq \mathcal{C}\mathbb{E}\Big\{\int_Q \theta^2\Big[\Big(\sum_{j=1}^n b_{1j}z_{x_j}+b_2 z\Big)^2+\lambda^2\varphi^2(b_3 z)^2\Big]dxdt + \lambda^3\mathbb{E}\int_{Q_0}\theta^2\varphi^3 z^2 dxdt\Big\} \\
&\leq \mathcal{C}\Big[\kappa_2^2\mathbb{E}\int_Q \theta^2\big(|\nabla z|^2+\lambda^2\varphi^3 z^2\big)\,dxdt + \lambda^3\mathbb{E}\int_{Q_0}\theta^2\varphi^3 z^2\,dxdt\Big].
\end{aligned} \tag{9.123}$$

Choosing $\lambda = \mathcal{C}(1+\kappa_2^2)$, from (9.123), we deduce that

$$\mathbb{E}\int_Q \theta^2\varphi^3 z^2 dxdt \leq \mathcal{C}e^{\mathcal{C}\kappa_2^2}\mathbb{E}\int_{Q_0}\theta^2\varphi^3 z^2 dxdt. \tag{9.124}$$

Noting that

$$\begin{aligned}
&\mathbb{E}\int_Q \theta^2\varphi^3 z^2 dxdt \geq \mathbb{E}\int_{T/4}^{3T/4}\int_G \theta^2\varphi^3 z^2 dxdt \\
&\geq \min_{x\in G}\Big(\theta^2(T/4,x)\varphi^3(T/2,x)\Big)\mathbb{E}\int_{T/4}^{3T/4}\int_G z^2 dxdt
\end{aligned}$$

and that

$$\mathbb{E}\int_{Q_0}\theta^2\varphi^3 z^2 dxdt \leq \max_{(t,x)\in Q}\Big(\theta^2(t,x)\varphi^3(t,x)\Big)\mathbb{E}\int_{Q_0} z^2 dxdt,$$

recalling (9.88) and $\theta = e^{\ell}$, we deduce that (9.124) implies that

$$\begin{aligned}
&\mathbb{E}\int_{T/4}^{3T/4}\int_G z^2 dxdt \\
&\leq \mathcal{C}e^{\mathcal{C}\kappa_2^2}\frac{\max_{(t,x)\in Q}\Big(\theta^2(t,x)\varphi^3(t,x)\Big)}{\min_{x\in G}\Big(\theta^2(T/4,x)\varphi^3(T/2,x)\Big)}\mathbb{E}\int_{Q_0} z^2 dxdt \\
&\leq \mathcal{C}e^{\mathcal{C}\kappa_2^2}\mathbb{E}\int_{Q_0} z^2 dxdt.
\end{aligned} \tag{9.125}$$

By Corollary 9.5, the inequality (9.125) implies the desired estimate (9.93). This completes the proof of Theorem 9.28. □

9.6 Null and Approximate Controllability of Stochastic Parabolic Equations

In this section, based on [315], we deal with the null and approximate controllability for (9.6). First of all, we have the following result.

Theorem 9.36. *Under the condition* (9.7)*, the system* (9.6) *is null controllable at any time* T.

In order to prove Theorem 9.36, in view of Theorem 7.17, we need the following observability result for (9.10) (Recall (9.13) for $\widetilde{\kappa}_1$):

Theorem 9.37. *Under the condition* (9.7)*, all solutions* $(z, Z) \in L^2_{\mathbb{F}}(\Omega; C([0,T]; L^2(G;\mathbb{R}^m))) \times L^2_{\mathbb{F}}(0,T; L^2(G;\mathbb{R}^m))$ *to the system* (9.10) *satisfy that*

$$
\begin{aligned}
&|z(0)|_{L^2_{\mathcal{F}_0}(\Omega;L^2(G;\mathbb{R}^m))}\\
&\leq \mathcal{C}e^{\mathcal{C}\widetilde{\kappa}_1^2}\big(|\chi_{G_0} z|_{L^2_{\mathbb{F}}(0,T;L^2(G;\mathbb{R}^m))} + |Z|_{L^2_{\mathbb{F}}(0,T;L^2(G;\mathbb{R}^m))}\big), \qquad (9.126)\\
&\qquad\qquad\qquad\qquad \forall\, z_T \in L^2_{\mathcal{F}_T}(\Omega; L^2(G;\mathbb{R}^m)).
\end{aligned}
$$

Remark 9.38. In Theorem 9.37, we assume that $a_{1j} \in L^\infty_{\mathbb{F}}(0,T; W^{1,\infty}(G; \mathbb{R}^{m\times m}))$ for $j = 1, 2, \cdots, n$ (See the condition (9.7)). It seems that this assumption can be weakened as $a_{1j} \in L^\infty_{\mathbb{F}}(0,T; L^\infty(G;\mathbb{R}^{m\times m}))$.

The rest of this section is mainly devoted to giving a proof of Theorem 9.37. For simplicity, we consider only the case $m = 1$.

9.6.1 Global Carleman Estimate for Backward Stochastic Parabolic Equations

As a key preliminary to prove Theorem 9.37, in this subsection we establish a global Carleman estimate for the following backward stochastic parabolic equation:

$$
\begin{cases}
dz + \displaystyle\sum_{j,k=1}^n (a^{jk} z_{x_j})_{x_k} dt = f dt + Z dW(t) & \text{in } Q,\\
z = 0 & \text{on } \Sigma,\\
z(T) = z_T & \text{in } G.
\end{cases}
\qquad (9.127)
$$

In the rest of this section, similar to that in Section 9.5, we choose θ and ℓ as that in (9.88), and ψ given by Lemma 9.29 with G_1 being any fixed nonempty open subset of G such that $\overline{G_1} \subset G_0$. The desired global Carleman estimate for (9.127) is stated as follows:

Theorem 9.39. *There is a constant $\mu_0 = \mu_0(G, G_0, (a^{jk})_{n\times n}, T) > 0$ such that for all $\mu \geq \mu_0$, one can find two constants $\mathcal{C} = \mathcal{C}(\mu) > 0$ and $\lambda_0 = \lambda_0(\mu) > 0$ such that for all $\lambda \geq \lambda_0$, $f \in L^2_{\mathbb{F}}(0,T;L^2(G))$ and $z_T \in L^2_{\mathcal{F}_T}(\Omega;L^2(G))$, the solution $(z,Z) \in [L^2_{\mathbb{F}}(\Omega;C([0,T];L^2(G)))\cap L^2_{\mathbb{F}}(0,T;H^1_0(G))]\times L^2_{\mathbb{F}}(0,T;L^2(G))$ to (9.127) satisfies that*

$$\begin{aligned}
&\lambda^3\mu^4\mathbb{E}\int_Q \theta^2\varphi^3 z^2 dxdt + \lambda\mu^2\mathbb{E}\int_Q \theta^2\varphi|\nabla z|^2 dxdt\\
&\leq \mathcal{C}\Big(\lambda^3\mu^4\mathbb{E}\int_{Q_0}\theta^2\varphi^3 z^2 dxdt + \mathbb{E}\int_Q \theta^2 f^2 dxdt + \lambda^2\mu^2\mathbb{E}\int_Q \theta^2\varphi^2 Z^2 dxdt\Big).
\end{aligned}\tag{9.128}$$

Proof: The proof is similar to that of Theorem 9.30. We shall use Theorem 9.27 with b^{jk} and h replaced respectively by $-a^{jk}$ and z (and hence $w = \theta z$).

Integrating the equality (9.79) (with b^{jk} replaced by $-a^{jk}$) on G, taking mean value in both sides, and noting (9.92), we conclude that

$$\begin{aligned}
&2\mathbb{E}\int_Q \theta\Big[\sum_{j,k=1}^n (a^{jk}w_{x_j})_{x_k} + \mathcal{A}w\Big]\Big[dz + \sum_{j,k=1}^n (a^{jk}z_{x_j})_{x_k}dt\Big]dx\\
&-2\mathbb{E}\int_Q \sum_{j,k=1}^n (a^{jk}w_{x_j}dw)_{x_k}dx\\
&+2\mathbb{E}\int_Q \sum_{j,k=1}^n\Big[\sum_{j',k'=1}^n\Big(2a^{jk}a^{j'k'}\ell_{x_{j'}}w_{x_j}w_{x_{k'}} - a^{jk}a^{j'k'}\ell_{x_j}w_{x_{j'}}w_{x_{k'}}\Big)\\
&\qquad\qquad -\Psi a^{jk}w_{x_j}w + a^{jk}\Big(\mathcal{A}\ell_{x_j} + \frac{\Psi_{x_j}}{2}\Big)w^2\Big]_{x_k}dxdt \qquad (9.129)\\
&\geq 2s_0^2\mathbb{E}\int_Q \Big[\varphi\big(\lambda\mu^2|\nabla\psi|^2 + \lambda O(\mu)\big)|\nabla w|^2 + \varphi^3\Big(\lambda^3\mu^4|\nabla\psi|^4 + \lambda^3 O(\mu^3)\\
&\qquad\qquad +\lambda^2 O(\mu^2 e^{2\mu|\psi|_{C(\overline{G})}}) + \lambda O(e^{2\mu|\psi|_{C(\overline{G})}})\Big)w^2\Big]dxdt\\
&+2\mathbb{E}\int_Q\Big|\sum_{j,k=1}^n (a^{jk}w_{x_j})_{x_k} + \mathcal{A}w\Big|^2 dxdt\\
&+\mathbb{E}\int_Q \theta^2\sum_{j,k=1}^n a^{jk}(dz_{x_j} + \ell_{x_j}dz)(dz_{x_k} + \ell_{x_k}dz)dx - \mathbb{E}\int_Q \theta^2\mathcal{A}(dz)^2dx,
\end{aligned}$$

where

$$\mathcal{A} = \sum_{j,k=1}^n (a^{jk}\ell_{x_j}\ell_{x_k} - a^{jk}_{x_k}\ell_{x_j} + a^{jk}\ell_{x_jx_k}) - \ell_t, \qquad \Psi = -2\sum_{j,k=1}^n a^{jk}\ell_{x_jx_k}.$$

By (9.127), it follows that

$$
\begin{aligned}
&2\mathbb{E}\int_Q \theta\Big[\sum_{j,k=1}^n (a^{jk}w_{x_j})_{x_k} + \mathcal{A}w\Big]\Big[dz + \sum_{j,k=1}^n (a^{jk}z_{x_j})_{x_k}dt\Big]dx\\
&= 2\mathbb{E}\int_Q \theta\Big[\sum_{j,k=1}^n (a^{jk}w_{x_j})_{x_k} + \mathcal{A}w\Big]\big(fdt + ZdW(t)\big)dx\\
&= 2\mathbb{E}\int_Q \theta\Big[-\sum_{j,k=1}^n (a^{jk}w_{x_j})_{x_k} + \mathcal{A}w\Big]fdtdx\\
&\le \mathbb{E}\int_Q \Big|\sum_{j,k=1}^n (a^{jk}w_{x_j})_{x_k} + \mathcal{A}w\Big|^2 dtdx + \mathbb{E}\int_Q \theta^2 f^2 dtdx.
\end{aligned}
\tag{9.130}
$$

It is clear that the term "$\mathbb{E}\int_Q \theta^2 \sum_{j,k=1}^n a^{jk}(dz_{x_j} + \ell_{x_j}dz)(dz_{x_k} + \ell_{x_k}dz)dx$" in (9.129) is nonnegative. Hence, by (9.129)–(9.130), similarly to the proof of (9.101), one can show that

$$
\begin{aligned}
&2s_0^2\mathbb{E}\int_Q \Big[\varphi(\lambda\mu^2|\nabla\psi|^2 + \lambda O(\mu))|\nabla w|^2 + \varphi^3\Big(\lambda^3\mu^4|\nabla\psi|^4 + \lambda^3 O(\mu^3)\\
&\qquad + \lambda^2 O(\mu^2 e^{2\mu|\psi|_{C(\overline{G})}}) + \lambda O(e^{2\mu|\psi|_{C(\overline{G})}})\Big)w^2\Big]dxdt\\
&\le \mathbb{E}\int_Q \theta^2(f^2 + \mathcal{A}Z^2)dxdt.
\end{aligned}
\tag{9.131}
$$

Similarly to (9.104), from (9.131), we conclude that there is a $\mu_0 > 0$ such that for all $\mu \ge \mu_0$, one can find a constant $\lambda_0 = \lambda_0(\mu)$ so that for any $\lambda \ge \lambda_0$, it holds that

$$
\begin{aligned}
&\lambda\mu^2\mathbb{E}\int_Q \theta^2\varphi\big(|\nabla z|^2 + \lambda^2\mu^2\varphi^2 z^2\big)dxdt\\
&\le \mathcal{C}\Big[\mathbb{E}\int_Q \theta^2(f^2 + \lambda^2\mu^2\varphi^2 Z^2)dxdt\\
&\qquad + \lambda\mu^2\mathbb{E}\int_0^T\int_{G_1} \theta^2\varphi\big(|\nabla z|^2 + \lambda^2\mu^2\varphi^2 z^2\big)dxdt\Big].
\end{aligned}
\tag{9.132}
$$

Choose a cut-off function $\zeta \in C_0^\infty(G_0;[0,1])$ so that $\zeta \equiv 1$ in G_1. Proceeding as in (9.105), from (9.127), we obtain that

$$
\begin{aligned}
0 = \mathbb{E}\int_{Q_0} \theta^2 \Big[& \zeta^2 z^2(\varphi_t + 2\lambda\varphi\eta_t) + 2\zeta^2\varphi \sum_{j,k=1}^{n} a^{jk} z_{x_j} z_{x_k} \\
& + 2\mu\zeta^2\varphi(1+2\lambda\varphi) z \sum_{j,k=1}^{n} a^{jk} z_{x_j} \psi_{x_k} \\
& + 4\zeta\varphi z \sum_{j,k=1}^{n} a^{jk} z_{x_j} \zeta_{x_k} + 2\zeta^2\varphi f z + \zeta^2\varphi Z^2 \Big] dxdt.
\end{aligned}
$$

Therefore, for any $\varepsilon > 0$, one has

$$
\begin{aligned}
& 2\mathbb{E}\int_{Q_0} \theta^2\zeta^2\varphi \sum_{j,k=1}^{n} a^{jk} z_{x_j} z_{x_k} dxdt + \mathbb{E}\int_{Q_0} \theta^2\zeta^2\varphi Z^2 dxdt \\
& \leq \varepsilon\mathbb{E}\int_{Q_0} \theta^2\zeta^2\varphi|\nabla z|^2 dxdt + \frac{\mathcal{C}}{\varepsilon}\mathbb{E}\int_{Q_0} \theta^2\Big(\frac{1}{\lambda^2\mu^2} f^2 + \lambda^2\mu^2\varphi^3 z^2\Big) dxdt.
\end{aligned}
\tag{9.133}
$$

Since the matrix $(a^{jk})_{1\leq i,j\leq n}$ is uniformly positive definite, we conclude from (9.133) that

$$
\mathbb{E}\int_0^T\int_{G_1} \theta^2\varphi|\nabla z|^2 dxdt \leq \mathcal{C}\mathbb{E}\int_{Q_0} \theta^2\Big(\frac{1}{\lambda^2\mu^2} f^2 + \lambda^2\mu^2\varphi^3 z^2\Big) dxdt. \tag{9.134}
$$

Combining (9.132) and (9.134), we obtain (9.128). This completes the proof of Theorem 9.39. □

9.6.2 Proof of the Observability Estimate for Backward Stochastic Parabolic Equations

We are now in a position to prove Theorem 9.37, which is the key observability estimate for the backward stochastic parabolic equation (9.10).

Proof of Theorem 9.37: Applying Theorem 9.39 to the equation (9.10), recalling (9.13) the definition of $\widetilde{\kappa}_1$, we deduce that, for all $\mu \geq \mu_0$ and $\lambda \geq \lambda_0(\mu)$,

$$
\begin{aligned}
& \lambda^3\mu^4\mathbb{E}\int_Q \theta^2\varphi^3 z^2 dxdt + \lambda\mu^2\mathbb{E}\int_Q \theta^2\varphi|\nabla z|^2 dxdt \\
& \leq \mathcal{C}\Big\{\lambda^3\mu^4\mathbb{E}\int_{Q_0} \theta^2\varphi^3 z^2 dxdt + \lambda^2\mu^2\mathbb{E}\int_Q \theta^2\varphi^2 Z^2 dxdt \\
& \quad + \mathbb{E}\int_Q \theta^2\Big[\sum_{j=1}^{n} \big(a_{1j} z\big)_{x_j} - (a_2 - a_3 a_4) z - a_3 Z\Big]^2 dxdt\Big\} \\
& \leq \mathcal{C}\Big[\lambda^3\mu^4\mathbb{E}\int_{Q_0} \theta^2\varphi^3 z^2 dxdt + \widetilde{\kappa}_1^2\mathbb{E}\int_Q \theta^2\big(|\nabla z|^2 + \lambda^2\mu^2\varphi^2 z^2 + Z^2\big) dxdt \\
& \quad + \lambda^2\mu^2\mathbb{E}\int_Q \theta^2\varphi^2 Z^2 dxdt\Big].
\end{aligned}
\tag{9.135}
$$

Choosing $\mu = \mu_0$ and $\lambda = \mathcal{C}(1+\widetilde{\kappa}_1^2)$, from (9.135), we obtain that

$$\begin{aligned}&\mathbb{E}\int_Q \theta^2\varphi^3 z^2 dxdt\\&\leq \mathcal{C}e^{\mathcal{C}\widetilde{\kappa}_1^2}\Big(\mathbb{E}\int_{Q_0}\theta^2\varphi^3|z|^2dxdt+\mathbb{E}\int_Q\theta^2\varphi^2|Z|^2dxdt\Big).\end{aligned}\tag{9.136}$$

Recalling (9.88), it follows from (9.136) that

$$\begin{aligned}&\mathbb{E}\int_{T/4}^{3T/4}\int_G z^2dxdt\\&\leq \mathcal{C}e^{\mathcal{C}\widetilde{\kappa}_1^2}\frac{\max\limits_{(t,x)\in Q}\Big(\theta^2(t,x)\varphi^3(t,x)+\theta^2(t,x)\varphi^2(t,x)\Big)}{\min\limits_{x\in G}\Big(\theta^2(T/4,x)\varphi^3(T/2,x)\Big)}\\&\quad\times\Big(\mathbb{E}\int_{Q_0}|z|^2dxdt+\mathbb{E}\int_Q|Z|^2dxdt\Big)\\&\leq \mathcal{C}e^{\mathcal{C}\widetilde{\kappa}_1^2}\Big(\mathbb{E}\int_{Q_0}|z|^2dxdt+\mathbb{E}\int_Q|Z|^2dxdt\Big).\end{aligned}\tag{9.137}$$

By (9.12) in Proposition 9.4, it follows that

$$\mathbb{E}\int_G z^2(0)dx\leq e^{\mathcal{C}\widetilde{\kappa}_1}\mathbb{E}\int_G z^2(t)dx,\qquad \forall\, t\in[0,T].\tag{9.138}$$

Finally, combining (9.137) and (9.138), we conclude that, the solution (z,Z) to the equation (9.10) satisfies (9.126). This completes the proof of Theorem 9.37. □

By modifying a little the proof of Theorem 9.37, we can prove easily the following result (hence we omit the proof):

Proposition 9.40. *Under the condition* (9.7), *any solution* $(z,Z)\in L^2_{\mathbb{F}}(\Omega; C([0,T];L^2(G;\mathbb{R}^m)))\times L^2_{\mathbb{F}}(0,T;L^2(G;\mathbb{R}^m))$ *to the system* (9.10) *vanishes identically provided that* $z=0$ *in* Q_0 *and* $Z=0$ *in* Q, *a.s.*

Remark 9.41. Since $Z = 0$ in Q, a.s., the system (9.10) becomes a (backward) random parabolic equation. Hence Proposition 9.40 follows also from the known observability estimate for deterministic parabolic equations (e.g. [117]).

By Theorem 7.19 and Proposition 9.40, one obtains immediately the following approximate controllability result for (9.6):

Corollary 9.42. *Under the condition* (9.7), *the system* (9.6) *is approximately controllable at any time* T.

9.7 Notes and Comments

There are numerous studies on the controllability theory of deterministic parabolic equations (See [77, 117, 154, 191, 336] and the references cited therein). Generally speaking, most of these results are based on either the global Carleman estimate introduced in [117] or the time iteration method introduced in [191]. Each method has advantages. In this chapter, we employ both of them to study the null controllability of stochastic parabolic equations. In order to establish the key global Carleman estimates for stochastic parabolic equations (or even other stochastic partial differential equations), so far there exist two ways in the literatures:

- The first way is that introduced in [315], which is based on a pointwise weighted identity for the stochastic parabolic-like operator, i.e., the identity (9.79) in Theorem 9.26.
- The second way is that introduced in [218], which uses surprisingly a known global Carleman estimate (See Corollary 9.32) for a random parabolic equation (which may follow from the existing results for deterministic parabolic equations) and a careful duality argument.

Each of the above two ways has its own merits. Indeed, each of them may need different conditions for different equations (See for example [218, 222, 315] for more details).

By virtue of some counterexamples in Section 9.3 (from [219]), it is found that the controllability of the coupled system (9.58) is not robust w.r.t. the coupling coefficient in the diffusion terms. Indeed, when this coefficient equals to zero on the whole time interval $[0, T]$, the system is null controllable (See Theorem 9.18). However, if it is a nonzero bounded function, no matter how small this function is, the corresponding system is uncontrollable any more. Hence, the Carleman-type estimates cannot be used to study the controllability of (9.58)). Notice that in [390] a similar phenomenon was found in studying the unique continuation for the characteristic Cauchy problem of (deterministic) linearized Benjamin-Bona-Mahony equations. In fact, the uniqueness of this equation depends strongly on the zero sets of its potentials. Hence, the desired unique continuation results in [390] were proved by means of spectral analysis and the eigenvector expansion of the solution rather than the Carleman-type estimates. Similarly, the controllability of the stochastic coupled system (9.58) was proved also by the spectral method.

One can find some other interesting works related to controllability, observability, unique continuation, stabilization, Hardy's uncertainty principle, insensitizing controls and so on for stochastic parabolic equations, say [11, 12, 54, 94, 142, 195, 196, 215, 237, 358, 359, 360, 361, 362, 365, 387].

There are some other interesting techniques to prove the controllability of deterministic parabolic equations, which should be extended to the stochastic setting, for example:

- In [93], the null controllability of the heat equation in one space dimension was obtained by solving a moment problem, based on the classical result on the linear independence of a suitably chosen families of real exponentials in $L^2(0,T)$. This method was generalized to obtain the controllability result for the heat equations in multi-dimensions (e.g., [7]). However, it seems that it is very hard to employ the same idea to prove the null controllability of stochastic partial differential equations. Indeed, so far it is unclear that how to reduce the latter stochastic null controllability problem to a suitable moment problem. For example, let us consider the following equation

$$
\begin{cases}
dy(t,x) - y_{xx}(t,x)dt \\
\quad = f(x)u(t)dt + a(x)y(t,x)dW(t) & \text{in } (0,T] \times (0,1), \\
y(t,0) = y(t,1) = 0 & \text{on } (0,T), \\
y(0,x) = y_0(x) & \text{in } (0,1).
\end{cases}
\tag{9.139}
$$

Here $y_0 \in L^2(0,1)$, $a \in L^\infty(0,1)$, $f \in L^2(0,1)$, y is the state variable and $u \in L^2_{\mathbb{F}}(0,T)$ is the control variable. One can see that it is not easy to reduce the null controllability problem for the system (9.139) to the usual moment problem (but, under some conditions it is still possible to reduce it to a stochastic moment problem, which remains to be done).
- In [293], it was shown that if the wave equation is exactly controllable for some $T > 0$ with controls supported in some subdomain G_0 of G, then the heat equation is null controllable for all $T > 0$ with controls supported in G_0. This result is not sharp when applied to the heat equation with smooth coefficients, since geometric restrictions are needed on the subset G_0 where the control applies. It seems that one can follow this idea to establish a connection between the null controllability of stochastic heat equations and stochastic wave equations. Nevertheless, at this moment the null controllability of stochastic wave equations is still open, which seems even more difficult than that of stochastic heat equations.
- In this chapter, we assume that the principle part coefficients $a^{jk} \in W^{2,\infty}(G)$. This assumption can be weaken. Indeed, one can easily see that the Carleman estimate approach works for Lipschitz continuous coefficients. Nevertheless, at this moment, we do not know what the minimal regularity condition is for the coefficients a^{jk}. A first controllability result for deterministic parabolic equations with piecewise constant principle part coefficients (by means of Carleman inequalities) has been established in [78] but imposing some monotonicity conditions for these coefficients at the interfaces. In [189], based on some modification of the tools from pseudo-differential operators, it has been proved that these monotonicity conditions can be dropped for the scalar coefficients. It seems that one can follow their methods to study the Carleman and observability estimates for stochastic parabolic equations with discontinuous principle part coefficients but this remains to be done.

There are many other unsolved problems related to the topic of this chapter. Here, we mention only two of them. One is to prove the null controllability of (9.6) with only one control, i.e., $v = 0$, for which we have explained the difficulty before. Another one is the null and approximate controllability for stochastic semi-linear parabolic equations. For deterministic semi-linear parabolic equation, people employ the Carleman estimate and some fixed point theorems to derive the null and approximate controllability (e.g., [77, 95, 117]). Since at this moment it is unclear whether these fixed point theorems can be generalized to the stochastic setting, we do not know how to use a similar method to solve the null and approximate controllability problems for stochastic semi-linear parabolic equations.

10

Exact Controllability for a Refined Stochastic Wave Equation

In this chapter, we shall prove that the usual stochastic wave equation, i.e., the classic wave equation perturbed by a term of Itô's integral, is not exactly controllable even if the controls are effective everywhere in both drift and diffusion terms, which means that some key feature is ignored in this model. Then, by means of a global Carleman estimate, we establish the exact controllability of a refined stochastic wave equation with three controls. Moreover, we give a result about the lack of exact controllability, which shows that the action of three controls is necessary. Our analysis indicates that, at least from the view point of Control Theory, the new stochastic wave equation introduced in our work is more reasonable than the one in the existing literature.

10.1 Formulation of the Problem

Similarly to parabolic equations, hyperbolic equations (and its special case, i.e., the wave equations) are another class of most typical partial differential equations. In the deterministic setting, the controllability theory for hyperbolic equations are also extensively studied. Hence, people hope to know what will happen for its stochastic counterpart. As we will see in this chapter, so far the stochastic situation is far from satisfactory.

For $T > 0$, $n \in \mathbb{N}$, a bounded domain $G \subset \mathbb{R}^n$ with a C^2 boundary Γ, a nonempty subset Γ_0 of Γ and a nonempty open subset G_0 of G, satisfying suitable assumptions to be given later, write

$$Q = (0,T) \times G, \quad \Sigma = (0,T) \times \Gamma,$$
$$\Sigma_0 = (0,T) \times \Gamma_0, \quad Q_0 = (0,T) \times G_0.$$

In the rest of this chapter, we assume $(a^{jk})_{1\le j,k\le n} \in C^3(\overline{G}; \mathbb{R}^{n\times n})$ satisfy that $a^{jk} = a^{kj}$ $(j,k = 1,2,\cdots,n)$ and for some constant $s_0 > 0$,

$$\sum_{j,k=1}^{n} a^{jk}\xi^j\xi^k \ge s_0|\xi|^2, \qquad \forall\,(x,\xi) \overset{\Delta}{=} (x,\xi^1,\cdots,\xi^n) \in G \times \mathbb{R}^n.$$

Q. Lü and X. Zhang, *Mathematical Control Theory for Stochastic Partial Differential Equations*, Probability Theory and Stochastic Modelling 101, https://doi.org/10.1007/978-3-030-82331-3_10

We begin with the following controlled (deterministic) wave equation:

$$\begin{cases} y_{tt} - \sum_{j,k=1}^{n} (a^{jk} y_{x_j})_{x_k} = a_1 \cdot \nabla y + a_2 y & \text{in } Q, \\ y = \chi_{\Sigma_0} h & \text{on } \Sigma, \\ y(0) = y_0, \quad y_t(0) = y_1 & \text{in } G. \end{cases} \tag{10.1}$$

Here $(y_0, y_1) \in L^2(G) \times H^{-1}(G)$, $a_1 \in L^\infty(0,T; W^{1,\infty}(G;\mathbb{R}^n))$, $a_2 \in L^\infty(Q)$, (y, y_t) is the state variable, and $h \in L^2(\Sigma_0)$ is the control variable. It is well-known that (e.g., [209]), the system (10.1) admits one and only one transposition solution $y \in C([0,T]; L^2(G)) \cap C^1([0,T]; H^{-1}(G))$.

Controllability (and also observability) for deterministic wave equations are now well-understood. Indeed, it is known that (e.g., [14, 80, 110, 207, 388, 396]), under some assumptions on (T, G, Γ_0) and the coefficients $(a^{jk})_{1\le j,k\le n}$, a_1 and a_2 (for example, for the case that $(a^{jk})_{1\le j,k\le n}$ is the $n \times n$ identity matrix, $a_1 = 0$, $a_2 = 0$, $\Gamma_0 = \{x \in \Gamma \mid (x - x_0) \cdot \nu(x) > 0\}$ for some $x_0 \in \mathbb{R}^n$, and $T > 2 \max_{x \in \overline{G}} |x - x_0|$), the system (10.1) is exactly controllable, i.e., for any given $(y_0, y_1), (z_0, z_1) \in L^2(G) \times H^{-1}(G)$, one can find a control $h \in L^2(\Sigma_0)$ such that the solution to (10.1) satisfying $y(T) = z_0$ and $y_t(T) = z_1$.

The main goal of this chapter is to study what happens when (10.1) is replaced by stochastic models. We shall see that, the corresponding stochastic controllability problems are much less understood.

As before, $(\Omega, \mathcal{F}, \mathbf{F}, \mathbb{P})$ (with $\mathbf{F} \overset{\Delta}{=} \{\mathcal{F}_t\}_{t\in[0,T]}$) is a fixed filtered probability space, on which a one dimensional standard Brownian motion $W(\cdot)$ is defined, and $\mathbf{F}$ is the corresponding natural filtration. We denote by $\mathbb{F}$ the progressive σ-field w.r.t. $\mathbf{F}$. Also, in the sequel we fix some stochastic coefficients $a_1 \in L^\infty_{\mathbb{F}}(0,T; W^{1,\infty}(G;\mathbb{R}^n))$, $a_2, a_3, a_4 \in L^\infty_{\mathbb{F}}(0,T; L^\infty(G))$ and $a_5 \in L^\infty_{\mathbb{F}}(0,T; W^{1,\infty}_0(G))$.

First, let us consider the following controlled stochastic wave equation:

$$\begin{cases} dy_t - \sum_{j,k=1}^{n} (a^{jk} y_{x_j})_{x_k} dt = (a_1 \cdot \nabla y + a_2 y + f)dt \\ \qquad\qquad\qquad\qquad\qquad +(a_3 y + g)dW(t) & \text{in } Q, \\ y = \chi_{\Sigma_0} h & \text{on } \Sigma, \\ y(0) = y_0, \quad y_t(0) = y_1 & \text{in } G, \end{cases} \tag{10.2}$$

where the initial datum $(y_0, y_1) \in L^2(G) \times H^{-1}(G)$, (y, y_t) is the state variable, and $f, g \in L^\infty_{\mathbb{F}}(0,T; H^{-1}(G))$ and $h \in L^2_{\mathbb{F}}(0,T; L^2(\Gamma_0))$ are three controls. As we shall see in Section 10.2, the equation (10.2) admits one and only one transposition solution $y \in C_{\mathbb{F}}([0,T]; L^2(\Omega; L^2(G))) \cap C^1_{\mathbb{F}}([0,T]; L^2(\Omega; H^{-1}(G)))$. In some sense, (10.2) looks quite natural because it is the classic (controlled) wave equation perturbed by a term of Itô's integral.

We introduce the following notion of exact controllability for (10.2).

Definition 10.1. *The system* (10.2) *is called exactly controllable at time* T *if for any* $(y_0, y_1) \in L^2(G) \times H^{-1}(G)$ *and* $(y_0', y_1') \in L^2_{\mathcal{F}_T}(\Omega; L^2(G)) \times L^2_{\mathcal{F}_T}(\Omega; H^{-1}(G))$, *one can find a triple of controls* $(f, g, h) \in L^2_{\mathbb{F}}(0, T; H^{-1}(G)) \times L^2_{\mathbb{F}}(0, T; H^{-1}(G)) \times L^2_{\mathbb{F}}(0, T; L^2(\Gamma_0))$ *such that the corresponding solution* y *to* (10.2) *satisfies that* $(y(T), y_t(T)) = (y_0', y_1')$.

Since three controls are introduced in (10.2), one may guess that the desired exact controllability should be trivially correct. Surprisingly, as we shall show below that, the exact controllability of (10.2) fails for any $T > 0$ and $\Gamma_0 \subset \Gamma$, even if the controls f and g are acted everywhere on the domain G.

Theorem 10.2. *The system* (10.2) *is not exactly controllable for any* $T > 0$ *and* $\Gamma_0 \subset \Gamma$.

Proof: We use the contradiction argument. Choose $\psi \in H_0^1(G)$ satisfying $|\psi|_{L^2(G)} = 1$ and let $\tilde{y}_0 = \xi\psi$, where ξ is the random variable given in Proposition 6.2. Assume that (10.2) was exactly controllable for some $T > 0$ and $\Gamma_0 \subset \Gamma$. Then, for any $y_0 \in L^2(G)$, we would find a triple of controls $(f, g, h) \in L^2_{\mathbb{F}}(0, T; H^{-1}(G)) \times L^2_{\mathbb{F}}(0, T; H^{-1}(G)) \times L^2_{\mathbb{F}}(0, T; L^2(\Gamma_0))$ such that the solution $y \in C_{\mathbb{F}}([0, T]; L^2(\Omega; L^2(G))) \cap C^1_{\mathbb{F}}([0, T]; L^2(\Omega; H^{-1}(G)))$ to the equation (10.2) satisfies that $y(T) = \tilde{y}_0$. Clearly,

$$\int_G \tilde{y}_0 \psi dx - \int_G y_0 \psi dx = \int_0^T \langle y_t, \psi \rangle_{H^{-1}(G), H_0^1(G)} dt,$$

which leads to

$$\xi = \int_G y_0 \psi dx + \int_0^T \langle y_t, \psi \rangle_{H^{-1}(G), H_0^1(G)} dt.$$

This contradicts Proposition 6.2. □

Note that, the above controls are the strongest possible ones that people can introduce into (10.2). Obviously, this differs significantly from the well-known controllability property (mentioned before) for deterministic wave equations.

Motivated by the above negative controllability result for (10.2), similarly to (5.26), in what follows, we consider the following refined version of controlled stochastic wave equation:

$$\begin{cases} dy = \hat{y}dt + (a_4 y + f)dW(t) & \text{in } Q, \\ d\hat{y} - \sum_{j,k=1}^{n} (a^{jk} y_{x_j})_{x_k} dt = (a_1 \cdot \nabla y + a_2 y + a_5 g)dt & \\ \qquad\qquad\qquad\qquad + (a_3 y + g)dW(t) & \text{in } Q, \\ y = \chi_{\Sigma_0} h & \text{on } \Sigma, \\ y(0) = y_0, \quad \hat{y}(0) = \hat{y}_0 & \text{in } G. \end{cases} \tag{10.3}$$

Here $(y_0, \hat{y}_0) \in L^2(G) \times H^{-1}(G)$, $(y, \hat{y})$ is the state variable, and $f \in L^2_{\mathbb{F}}(0, T; L^2(G))$, $g \in L^2_{\mathbb{F}}(0, T; H^{-1}(G))$ and $h \in L^2_{\mathbb{F}}(0, T; L^2(\Gamma_0))$ are three controls. As we shall see in Section 10.2, the equation (10.3) admits one and only one transposition solution $(y, \hat{y}) \in C_{\mathbb{F}}([0, T]; L^2(\Omega; L^2(G))) \times C_{\mathbb{F}}([0, T]; L^2(\Omega; H^{-1}(G)))$.

Remark 10.3. We put two controls f and g in the diffusion terms of (10.3). The first equation in this system can be regarded as a family of stochastic differential equations with a parameter $x \in G$; while the second one is a stochastic partial differential equation. Usually, if we put a control in the diffusion term, it may affect the drift term in one way or another. Here we assume that the effect is linear and in the form of "$a_5 g dt$" as that in the second equation of (10.3). One may consider more general cases, say, to add a term like "$a_6 f dt$" (in which $a_6 \in L^\infty_{\mathbb{F}}(0, T; L^\infty(G))$) into the first equation of (10.3). However, the corresponding controllability problem (as the one studied in this chapter) is still unsolved.

The exact controllability for (10.3) is defined as follows:

Definition 10.4. *The system* (10.3) *is called exactly controllable at time* T *if for any* $(y_0, \hat{y}_0) \in L^2(G) \times H^{-1}(G)$ *and* $(y_1, \hat{y}_1) \in L^2_{\mathcal{F}_T}(\Omega; L^2(G)) \times L^2_{\mathcal{F}_T}(\Omega; H^{-1}(G))$, *one can find a triple of controls* $(f, g, h) \in L^2_{\mathbb{F}}(0, T; L^2(G)) \times L^2_{\mathbb{F}}(0, T; H^{-1}(G)) \times L^2_{\mathbb{F}}(0, T; L^2(\Gamma_0))$ *such that the corresponding solution* $(y, \hat{y})$ *to* (10.3) *satisfies that* $(y(T), \hat{y}(T)) = (y_1, \hat{y}_1)$.

In this chapter, under some assumptions, we shall show that (10.3) is exactly controllable (See Theorem 10.12). Hence, from the viewpoint of controllability theory, the system (10.3) is a more reasonable model than (10.2). Noting that, we also introduce three controls into (10.3), which seems too many. However, we prove that none of these three controls can be ignored; moreover both f and g (the two internal controls) have to be effective everywhere in the domain G (See Theorem 10.9).

10.2 Well-Posedness of Stochastic Wave Equations With Boundary Controls

In order to define solutions to both (10.2) and (10.3) in the sense of transposition, we need to introduce the following backward stochastic wave equation:

$$\begin{cases} dz = \hat{z}dt + (b_5 z + Z)dW(t) & \text{in } Q_\tau, \\ d\hat{z} - \sum_{j,k=1}^{n} (a^{jk} z_{x_j})_{x_k} dt \\ = (b_1 \cdot \nabla z + b_2 z + b_3 Z + b_4 \widehat{Z})dt + \widehat{Z} dW(t) & \text{in } Q_\tau, \\ z = 0 & \text{on } \Sigma_\tau, \\ z(\tau) = z^\tau, \quad \hat{z}(\tau) = \hat{z}^\tau & \text{in } G, \end{cases} \tag{10.4}$$

where $\tau \in (0, T]$, $Q_\tau \stackrel{\Delta}{=} (0, \tau) \times G$, $\Sigma_\tau \stackrel{\Delta}{=} (0, \tau) \times \Gamma$, $(z^\tau, \hat{z}^\tau) \in L^2_{\mathcal{F}_\tau}(\Omega; H^1_0(G) \times L^2(G))$, $b_1 \in L^\infty_{\mathbb{F}}(0, T; W^{1,\infty}(G; \mathbb{R}^n))$, $b_i \in L^\infty_{\mathbb{F}}(0, T; L^\infty(G))$ $(i = 2, 3, 4)$ and $b_5 \in L^\infty_{\mathbb{F}}(0, T; W^{1,\infty}_0(G))$.

By Theorem 4.10, for any $(z^\tau, \hat{z}^\tau) \in L^2_{\mathcal{F}_\tau}(\Omega; H^1_0(G)) \times L^2_{\mathcal{F}_\tau}(\Omega; L^2(G))$, the equation (10.4) admits a unique mild solution $(z, \hat{z}, Z, \widehat{Z})$. Moreover,

$$
\begin{aligned}
&|z|_{C_{\mathbb{F}}([0,\tau]; H^1_0(G))} + |\hat{z}|_{C_{\mathbb{F}}([0,\tau]; L^2(G))} + |Z|_{L^2_{\mathbb{F}}(0,\tau; H^1_0(G))} + |\widehat{Z}|_{L^2_{\mathbb{F}}(0,\tau; L^2(G))}\\
&\le \mathcal{C} e^{\mathcal{C} r_1} \big(|z^\tau|_{L^2_{\mathcal{F}_\tau}(\Omega; H^1_0(G))} + |\hat{z}^\tau|_{L^2_{\mathcal{F}_\tau}(\Omega; L^2(G))} \big),
\end{aligned}
\tag{10.5}
$$

where

$$
r_1 \stackrel{\Delta}{=} |b_1|^2_{L^\infty_{\mathbb{F}}(0,T; W^{1,\infty}(G;\mathbb{R}^n))} + \sum_{i=2}^{4} |b_i|^2_{L^\infty_{\mathbb{F}}(0,T; L^\infty(G))} + |b_5|^2_{L^\infty_{\mathbb{F}}(0,T; W^{1,\infty}_0(G))}.
$$

Let us recall the following known result (See [207, p. 29] for its proof).

Proposition 10.5. *There exists a vector field* $\xi = (\xi^1, \cdots, \xi^n) \in C^1(\mathbb{R}^n; \mathbb{R}^n)$ *such that* $\xi = \nu$ *on* Γ.

The following regularity result for solutions to (10.4) will be needed later.

Proposition 10.6. *Let* $(z^\tau, \hat{z}^\tau) \in L^2_{\mathcal{F}_\tau}(\Omega; H^1_0(G)) \times L^2_{\mathcal{F}_\tau}(\Omega; L^2(G))$. *Then the (mild) solution* $(z, \hat{z}, Z, \widehat{Z})$ *to* (10.4) *satisfies* $\frac{\partial z}{\partial \nu}\big|_\Gamma \in L^2_{\mathbb{F}}(0, \tau; L^2(\Gamma))$. *Furthermore,*

$$
\Big| \frac{\partial z}{\partial \nu} \Big|_{L^2_{\mathbb{F}}(0,\tau; L^2(\Gamma))} \le \mathcal{C} e^{\mathcal{C} r_1} \big(|z^\tau|_{L^2_{\mathcal{F}_\tau}(\Omega; H^1_0(G))} + |\hat{z}^\tau|_{L^2_{\mathcal{F}_\tau}(\Omega; L^2(G))} \big), \tag{10.6}
$$

where the constant $\mathcal{C}$ *is independent of* τ.

Proof: For any $\eta \stackrel{\Delta}{=} (\eta^1, \cdots, \eta^n) \in C^1(\mathbb{R}_t \times \mathbb{R}^n_x; \mathbb{R}^n)$, by Itô's formula and the first equation of (10.4), we have

$$
\begin{aligned}
d(\hat{z}\eta \cdot \nabla z) &= d\hat{z}\eta \cdot \nabla z + \hat{z}\eta_t \cdot \nabla z dt + \hat{z}\eta \cdot \nabla dz + d\hat{z}\eta \cdot \nabla dz\\
&= d\hat{z}\eta \cdot \nabla z + \hat{z}\eta_t \cdot \nabla z dt + \hat{z}\eta \cdot \nabla \big(\hat{z} dt + (b_5 z + Z) dW(t) \big)\\
&\quad + d\hat{z}\eta \cdot \nabla dz\\
&= d\hat{z}\eta \cdot \nabla z + \hat{z}\eta_t \cdot \nabla z dt + \frac{1}{2} \Big[\mathrm{div}\, (\hat{z}^2 \eta) - (\mathrm{div}\, \eta) \hat{z}^2 \Big] dt\\
&\quad + \hat{z}\eta \cdot \nabla (b_5 z + Z) dW(t) + d\hat{z}\eta \cdot \nabla dz.
\end{aligned}
$$

It follows from a direct computation that

$$
\begin{aligned}
&\sum_{k=1}^n \Big(2(\eta \cdot \nabla z) \sum_{j=1}^n a^{jk} z_{x_j} - \eta^k \sum_{i,j=1}^n a^{ij} z_{x_i} z_{x_j} \Big)_{x_k}\\
&= 2 \Big[\sum_{j,k=1}^n (a^{jk} z_{x_j})_{x_k} \eta \cdot \nabla z + \sum_{i,j,k=1}^n a^{ij} z_{x_i} z_{x_k} \eta^k_{x_j} \Big] - \sum_{j,k=1}^n z_{x_j} z_{x_k} \mathrm{div}\, (a^{jk} \eta).
\end{aligned}
$$

Hence,

$$\begin{aligned}
&-\sum_{k=1}^{n}\Big[2(\eta\cdot\nabla z)\sum_{j=1}^{n}a^{jk}z_{x_j}+\eta^k\Big(\hat z^2-\sum_{i,j=1}^{n}a^{ij}z_{x_i}z_{x_j}\Big)\Big]_{x_k}dt\\
&=2\Big[-d(\hat z\eta\cdot\nabla z)+\Big(d\hat z-\sum_{j,k=1}^{n}(a^{jk}z_{x_j})_{x_k}dt\Big)\eta\cdot\nabla z+\hat z\eta_t\cdot\nabla z dt\\
&\quad-\sum_{i,j,k=1}^{n}a^{ij}z_{x_i}z_{x_k}\eta^k_{x_j}dt\Big]-(\mathrm{div}\,\eta)\hat z^2dt+\sum_{j,k=1}^{n}z_{x_j}z_{x_k}\mathrm{div}\,(a^{jk}\eta)dt\\
&\quad+2d\hat z\eta\cdot\nabla dz+2\hat z\eta\cdot\nabla(b_5z+Z)dW(t).
\end{aligned}\tag{10.7}$$

Let $\xi\in C^1(\mathbb{R}^n;\mathbb{R}^n)$ be given in Proposition 10.5. Setting $\eta=\xi$ in (10.7), integrating it in Q, and taking expectation on Ω, we obtain that

$$\begin{aligned}
&-\mathbb{E}\int_{\Sigma_\tau}\sum_{k=1}^{n}\Big[2(\eta\cdot\nabla z)\sum_{j=1}^{n}a^{jk}z_{x_j}+\eta^k\Big(\hat z^2-\sum_{i,j=1}^{n}a^{ij}z_{x_i}z_{x_j}\Big)\Big]\nu^k d\Gamma dt\\
&=-2\mathbb{E}\int_G\hat z^T\eta\cdot\nabla z^T dx+2\mathbb{E}\int_G\hat z(0)\eta\cdot\nabla z(0)dx\\
&\quad+2\int_{Q_\tau}\Big[\big(b_1\cdot\nabla z+b_2z+b_3Z+b_4\widehat Z\big)\eta\cdot\nabla z+\hat z\eta_t\cdot\nabla z\\
&\qquad-\sum_{i,j,k=1}^{n}a^{ij}z_{x_i}z_{x_k}\eta^k_{x_j}-(\mathrm{div}\,\eta)\hat z^2+\sum_{j,k=1}^{n}z_{x_k}z_{x_j}\mathrm{div}\,(a^{jk}\eta)\\
&\qquad+2\widehat Z\eta\cdot\nabla(b_5z+Z)\Big]dxdt.
\end{aligned}\tag{10.8}$$

Noting that $z=0$ on $(0,\tau)\times\Gamma$, we have

$$\begin{aligned}
&\mathbb{E}\int_{\Sigma_\tau}\sum_{k=1}^{n}\Big[2(\eta\cdot\nabla z)\sum_{j=1}^{n}a^{jk}z_{x_j}+\eta^k\Big(\hat z^2-\sum_{i,j=1}^{n}a^{ij}z_{x_i}z_{x_j}\Big)\Big]\nu^k d\Gamma dt\\
&=\mathbb{E}\int_{\Sigma_\tau}\Big[2\Big(\eta\cdot\nu\frac{\partial z}{\partial\nu}\Big)\sum_{j,k=1}^{n}a^{jk}\nu^k\nu^j\frac{\partial z}{\partial\nu}-\sum_{i,j,k=1}^{n}a^{ij}\nu^k\nu^i\eta^k\nu^j\Big|\frac{\partial z}{\partial\nu}\Big|^2\Big]d\Gamma dt\\
&=\mathbb{E}\int_{\Sigma_\tau}\sum_{j,k=1}^{n}a^{jk}\nu^k\nu^j\Big|\frac{\partial z}{\partial\nu}\Big|^2d\Gamma dt\geq s_0\mathbb{E}\int_{\Sigma_\tau}\Big|\frac{\partial z}{\partial\nu}\Big|^2d\Gamma dt.
\end{aligned}\tag{10.9}$$

It follows from the estimate (10.5) that

$$|\text{The right hand side of (10.8)}|\leq\mathcal{C}e^{\mathcal{C}r_1}\big(|z^T|_{L^2_{\mathcal{F}_\tau}(\Omega;H_0^1(G))}+|\hat z^T|_{L^2_{\mathcal{F}_\tau}(\Omega;L^2(G))}\big).$$

This, together with (10.8) and (10.9), yields the desired estimate (10.6). This complete the proof of Proposition 10.6. □

Remark 10.7. Proposition 10.6 shows that, solutions to (10.4) enjoy a better regularity on the boundary than the one provided by the classical trace theorem on Sobolev spaces. Such kind of results are called hidden regularities (for solutions to the equations under consideration). There are many studies on this topic for deterministic partial differential equations (e.g. [208]).

Systems (10.2) and (10.3) are nonhomogeneous boundary value problems. Their solutions are understood in the sense of transposition solution introduced in Section 7.2.

In view of Definition 7.11, *a stochastic process* $y \in C_{\mathbb{F}}([0,T];L^2(\Omega;L^2(G)))\cap C^1_{\mathbb{F}}([0,T];L^2(\Omega;H^{-1}(G)))$ *is called a transposition solution to* (10.2) *if for any* $\tau\in(0,T]$ *and* $(z^\tau,\hat z^\tau)\in L^2_{\mathcal{F}_\tau}(\Omega;H^1_0(G))\times L^2_{\mathcal{F}_\tau}(\Omega;L^2(G))$, *it holds that*

$$
\begin{aligned}
&\mathbb{E}\langle y_t(\tau),z^\tau\rangle_{H^{-1}(G),H^1_0(G)}-\mathbb{E}\langle y(\tau),\hat z^\tau\rangle_{L^2(G)}\\
&-\langle \hat y_0,z(0)\rangle_{H^{-1}(G),H^1_0(G)}+\langle y_0,\hat z(0)\rangle_{L^2(G)}\\
&=\mathbb{E}\int_0^\tau\langle f,z\rangle_{H^{-1}(G),H^1_0(G)}dt+\mathbb{E}\int_0^\tau\langle g,Z\rangle_{H^{-1}(G),H^1_0(G)}dt\\
&-\mathbb{E}\int_0^\tau\int_{\Gamma_0}h\frac{\partial z}{\partial\nu}d\Gamma ds,
\end{aligned}
\tag{10.10}
$$

where $(z,\hat z,Z,\widehat Z)$ *solves* (10.4) *with*

$$
b_1=-a_1,\quad b_2=-\operatorname{div}a_1+a_2,\quad b_3=a_3,\quad b_4=0,\quad b_5=0.
$$

Likewise, *a pair of stochastic processes* $(y,\hat y)\in C_{\mathbb{F}}([0,T];L^2(\Omega;L^2(G)))\times C_{\mathbb{F}}([0,T];\ L^2(\Omega;\ H^{-1}(G)))$ *is called a transposition solution to* (10.3) *if for any* $\tau\in(0,T]$ *and* $(z^\tau,\hat z^\tau)\in L^2_{\mathcal{F}_\tau}(\Omega;H^1_0(G))\times L^2_{\mathcal{F}_\tau}(\Omega;\ L^2(G))$, *it holds that*

$$
\begin{aligned}
&\mathbb{E}\langle \hat y(\tau),z^\tau\rangle_{H^{-1}(G),H^1_0(G)}-\mathbb{E}\langle y(\tau),\hat z^\tau\rangle_{L^2(G)}\\
&-\langle \hat y_0,z(0)\rangle_{H^{-1}(G),H^1_0(G)}+\langle y_0,\hat z(0)\rangle_{L^2(G)}\\
&=-\mathbb{E}\int_0^\tau\langle f,\widehat Z\rangle_{L^2(G)}dt+\mathbb{E}\int_0^\tau\langle g,Z\rangle_{H^{-1}(G),H^1_0(G)}dt-\mathbb{E}\int_0^\tau\int_{\Gamma_0}h\frac{\partial z}{\partial\nu}d\Gamma ds,
\end{aligned}
\tag{10.11}
$$

where $(z,\hat z,Z,\widehat Z)$ *solves* (10.4) *with*

$$
b_1=-a_1,\quad b_2=-\operatorname{div}a_1+a_2-a_3a_5,\quad b_3=a_3,\quad b_4=-a_4,\quad b_5=-a_5.
$$

Note that, by Proposition 10.6, the solution $(z,\hat z,Z,\widehat Z)$ to (10.4) satisfies $\frac{\partial z}{\partial\nu}\big|_\Gamma\in L^2_{\mathbb{F}}(0,\tau;L^2(\Gamma))$, hence the term "$\mathbb{E}\int_0^\tau\int_{\Gamma_0}h\frac{\partial z}{\partial\nu}d\Gamma ds$" in both (10.10) and (10.11) makes sense.

By Theorem 7.12, one can easily deduce the following well-posedness results for (10.2) and (10.3):

Proposition 10.8. 1) *For each* $(y_0, y_1) \in L^2(G) \times H^{-1}(G)$ *and* $(f, g, h) \in L^2_{\mathbb{F}}(0,T;\, H^{-1}(G)) \times L^2_{\mathbb{F}}(0,T; H^{-1}(G)) \times L^2_{\mathbb{F}}(0,T; L^2(\Gamma_0))$, *the equation* (10.2) *admits a unique transposition solution* $y \in C_{\mathbb{F}}([0,T]; L^2(\Omega; L^2(G))) \cap C^1_{\mathbb{F}}([0,T]; L^2(\Omega; H^{-1}(G)))$. *Furthermore,*

$$
\begin{aligned}
&|y|_{C_{\mathbb{F}}([0,T];L^2(\Omega;L^2(G)))\cap C^1_{\mathbb{F}}([0,T];L^2(\Omega;H^{-1}(G)))}\\
&\leq \mathcal{C}e^{\mathcal{C}r_3}\big(|y_0|_{L^2(G)} + |y_1|_{H^{-1}(G)} + |f|_{L^2_{\mathbb{F}}(0,T;H^{-1}(G))}\\
&\qquad + |g|_{L^2_{\mathbb{F}}(0,T;H^{-1}(G))} + |h|_{L^2_{\mathbb{F}}(0,T;L^2(\Gamma_0))}\big).
\end{aligned}
$$

Here

$$
r_3 = |a_1|^2_{L^\infty_{\mathbb{F}}(0,T;W^{1,\infty}(G;\mathbb{R}^n))} + \sum_{k=2}^{3} |a_2|^2_{L^\infty_{\mathbb{F}}(0,T;L^\infty(G))};
$$

2) *For each* $(y_0, \hat{y}_0) \in L^2(G) \times H^{-1}(G)$ *and* $(f, g, h) \in L^2_{\mathbb{F}}(0,T; L^2(G)) \times L^2_{\mathbb{F}}(0,T; H^{-1}(G)) \times L^2_{\mathbb{F}}(0,T; L^2(\Gamma_0))$, *the equation* (10.3) *admits one and only one transposition solution* $(y, \hat{y}) \in C_{\mathbb{F}}([0,T]; L^2(\Omega; L^2(G))) \times C_{\mathbb{F}}([0,T]; L^2(\Omega; H^{-1}(G)))$. *Moreover,*

$$
\begin{aligned}
&|(y,\hat{y})|_{C_{\mathbb{F}}([0,T];L^2(\Omega;L^2(G)))\times C_{\mathbb{F}}([0,T];L^2(\Omega;H^{-1}(G)))}\\
&\leq \mathcal{C}e^{\mathcal{C}r_2}\big(|y_0|_{L^2(G)} + |\hat{y}_0|_{H^{-1}(G)} + |f|_{L^2_{\mathbb{F}}(0,T;L^2(G))}\\
&\qquad + |g|_{L^2_{\mathbb{F}}(0,T;H^{-1}(G))} + |h|_{L^2_{\mathbb{F}}(0,T;L^2(\Gamma_0))}\big).
\end{aligned}
\tag{10.12}
$$

Here

$$
\begin{aligned}
r_2 \stackrel{\Delta}{=} &|a_1|^2_{L^\infty_{\mathbb{F}}(0,T;W^{1,\infty}(G;\mathbb{R}^n))} + \sum_{k=2}^{4} |a_k|^2_{L^\infty_{\mathbb{F}}(0,T;L^\infty(G))}\\
&+ |a_5|^2_{L^\infty_{\mathbb{F}}(0,T;W^{1,\infty}_0(G))}.
\end{aligned}
\tag{10.13}
$$

10.3 Main Controllability Results

We have introduced three controls (f, g and h) in the system (10.3). At a first glance, it seems this is unreasonable, especially for that the controls f and g in the diffusion term of (10.3) are acted on the whole domain G. One may ask whether localized controls are enough or the boundary control can be dropped. However, the answer is “no”. More precisely, we have the following negative controllability result for the system (10.3).

Theorem 10.9. *For any open subset* Γ_0 *of* Γ *and open subset* G_0 *of* G, *the system* (10.3) *is not exactly controllable at any time* $T > 0$, *provided that one of the following three conditions is satisfied:*

1) $a_4 \in C_{\mathbb{F}}([0,T]; L^\infty(G))$, $G \setminus \overline{G_0} \neq \emptyset$ *and* f *is supported in* G_0;
2) $a_3 \in C_{\mathbb{F}}([0,T]; L^\infty(G))$, $G \setminus \overline{G_0} \neq \emptyset$ *and* g *is supported in* G_0;
3) $h = 0$.

Proof: Let us employ the contradiction argument, and divide the proof into three cases.

Case 1) $a_4 \in C_{\mathbb{F}}([0,T]; L^\infty(G))$ **and** f **is supported in** G_0. Since $G_0 \subset G$ is an open subset and $G \setminus \overline{G_0} \neq \emptyset$, we can find $\rho \in C_0^\infty(G \setminus G_0)$ satisfying $|\rho|_{L^2(G)} = 1$.

Assume that (10.3) was exactly controllable. Then, for $(y_0, \hat{y}_0) = (0,0)$, one could find controls $(f,g,h) \in L^2_{\mathbb{F}}(0,T; L^2(G)) \times L^2_{\mathbb{F}}(0,T; H^{-1}(G)) \times L^2_{\mathbb{F}}(0,T; L^2(\Gamma_0))$ with $\operatorname{supp} f \subset G_0$, a.e. $(t,\omega) \in (0,T) \times \Omega$ such that the corresponding solution to (10.3) fulfills $(y(T), \hat{y}(T)) = (\rho\xi, 0)$, where ξ is given in Proposition 6.2. Thus,

$$\rho\xi = \int_0^T \hat{y}dt + \int_0^T (a_4 y + f)dW(t). \tag{10.14}$$

Multiplying both sides of (10.14) by ρ and integrating it in G, we obtain that

$$\xi = \int_0^T \langle \hat{y}, \rho \rangle_{H^{-1}(G), H_0^1(G)} dt + \int_0^T \langle a_4 y, \rho \rangle_{L^2(G)} dW(t). \tag{10.15}$$

Since the pair $(y, \hat{y}) \in C_{\mathbb{F}}([0,T]; L^2(\Omega; L^2(G))) \times C_{\mathbb{F}}([0,T]; L^2(\Omega; H^{-1}(G)))$ solves (10.3), then $\langle \hat{y}, \rho \rangle_{H^{-1}(G), H_0^1(G)} \in C_{\mathbb{F}}([0,T]; L^2(\Omega))$ and $\langle a_4 y, \rho \rangle_{L^2(G)} \in C_{\mathbb{F}}([0,T]; L^2(\Omega))$, which, together with (10.15), contradicts Proposition 6.2.

Case 2) $a_3 \in C_{\mathbb{F}}([0,T]; L^\infty(G))$ **and** g **is supported in** G_0. Choose ρ as in Case 1). If (10.3) was exactly controllable, then, for $(y_0, \hat{y}_0) = (0,0)$, one could find controls $(f,g,h) \in L^2_{\mathbb{F}}(0,T; L^2(G)) \times L^2_{\mathbb{F}}(0,T; H^{-1}(G)) \times L^2_{\mathbb{F}}(0,T; L^2(\Gamma_0))$ with $\operatorname{supp} g \subset G_0$, a.e. $(t,\omega) \in (0,T) \times \Omega$ such that the corresponding solution to (10.3) fulfills $(y(T), \hat{y}(T)) = (0, \xi)$.

It is clear that $(\phi, \hat{\psi}) = (\rho y, \rho \hat{y})$ solves the following equation:

$$\begin{cases} d\phi = \hat{\phi}dt + (a_4\phi + \rho f)dW(t) & \text{in } Q, \\ d\hat{\phi} - \displaystyle\sum_{j,k=1}^n (a^{jk}\phi_{x_j})_{x_k} dt = \zeta dt + a_3 \phi dW(t) & \text{in } Q, \\ \phi = 0 & \text{on } \Sigma, \\ \hat{\phi} = 0 & \text{on } \Sigma, \\ \phi(0) = 0, \quad \hat{\phi}(0) = 0 & \text{in } G, \end{cases} \tag{10.16}$$

where $\zeta = \displaystyle\sum_{j,k=1}^n [(a^{jk}\rho_{x_j} y)_{x_k} + a^{jk} y_{x_j} \rho_{x_k}] + \rho a_1 \cdot \nabla y + \rho a_2 y$. Further, we have $\phi(T) = 0$ and $\hat{\phi}(T) = \rho\xi$. Noting that $(\phi, \hat{\phi})$ is the weak solution to (10.16), we see that

$$
\begin{aligned}
&\big\langle \rho\xi, \rho \big\rangle_{H^{-2}(G),H_0^2(G)} \\
&= \int_0^T \Big[\Big\langle \sum_{j,k=1}^n (a^{jk}\phi_{x_j})_{x_k}, \rho \Big\rangle_{H^{-2}(G),H_0^2(G)} + \big\langle \zeta, \rho \big\rangle_{H^{-1}(G),H_0^1(G)} \Big] dt \\
&\quad + \int_0^T \big\langle a_3\phi, \rho \big\rangle_{L^2(G)} dW(t),
\end{aligned}
$$

which implies that

$$
\begin{aligned}
\xi = &\int_0^T \Big[\Big\langle \sum_{j,k=1}^n (a^{jk}\phi_{x_j})_{x_k}, \rho \Big\rangle_{H^{-2}(G),H_0^2(G)} + \big\langle \zeta, \rho \big\rangle_{H^{-1}(G),H_0^1(G)} \Big] dt \\
&+ \int_0^T \big\langle a_3\phi, \rho \big\rangle_{L^2(G)} dW(t).
\end{aligned}
\tag{10.17}
$$

Since $(\phi, \hat{\phi}) \in C_{\mathbb{F}}([0,T]; L^2(\Omega; L^2(G))) \times C_{\mathbb{F}}([0,T]; L^2(\Omega; H^{-1}(G)))$, then

$$
\Big\langle \sum_{j,k=1}^n (a^{jk}\phi_{x_j})_{x_k}, \rho \Big\rangle_{H^{-2}(G),H_0^2(G)} + \big\langle \zeta, \rho \big\rangle_{H^{-1}(G),H_0^1(G)} \in L^2_{\mathbb{F}}(0,T)
$$

and $\langle a_3\phi, \rho\rangle_{L^2(G)} \in C_{\mathbb{F}}([0,T]; L^2(\Omega))$. These, together with (10.17), contradict Proposition 6.2.

Case 3) $h = 0$. Assume that the system (10.3) was exactly controllable. Then, by Theorem 7.16, solutions $(z, \hat{z}, Z, \widehat{Z})$ to (10.4) (with $\tau = T$) would satisfy that, for all $(z^T, \hat{z}^T) \in L^2_{\mathcal{F}_T}(\Omega; H_0^1(G)) \times L^2_{\mathcal{F}_T}(\Omega; L^2(G))$,

$$
\begin{aligned}
&|(z^T, \hat{z}^T)|_{L^2_{\mathcal{F}_T}(\Omega; H_0^1(G)) \times L^2_{\mathcal{F}_T}(\Omega; L^2(G))} \\
&\leq \mathcal{C}\big(|Z|_{L^2_{\mathbb{F}}(0,T;H_0^1(G))} + |\widehat{Z}|_{L^2_{\mathbb{F}}(0,T;L^2(G))}\big).
\end{aligned}
\tag{10.18}
$$

For any nonzero $(\eta_0, \eta_1) \in H_0^1(G) \times L^2(G)$, let us consider the following random wave equation:

$$
\begin{cases}
\eta_{tt} - \displaystyle\sum_{j,k=1}^n (a^{jk}\eta_{x_j})_{x_k} dt = b_1 \cdot \nabla \eta + b_2 \eta & \text{in } (0,T) \times G, \\
z = 0 & \text{on } (0,T) \times \Gamma, \\
\eta(0) = \eta_0, \quad \eta_t(0) = \eta_1 & \text{in } G.
\end{cases}
\tag{10.19}
$$

Clearly, $(\eta, 0, \eta_t, 0)$ solves (10.4) with the final datum $(z^T, \hat{z}^T) = (\eta(T), \hat{\eta}(T))$, a contradiction to the inequality (10.18). □

Now, similarly to the deterministic setting, in order to give a positive controllability result for the system (10.3), we need the following additional assumptions on the coefficients $(a^{jk})_{1 \leq j,k \leq n}$:

Condition 10.1 *There exists a positive function* $\varphi(\cdot) \in C^2(\overline{G})$ *satisfying that:*

1) *For some constant* $\mu_0 > 0$,

$$\begin{aligned}&\sum_{j,k=1}^{n}\sum_{j',k'=1}^{n}\Big(2a^{jk'}(a^{j'k}\varphi_{x_{j'}})_{x_{k'}} - a^{jk}_{x_{k'}}a^{j'k'}\varphi_{x_{j'}}\Big)\xi^j\xi^k\\&\geq \mu_0\sum_{j,k=1}^{n} a^{jk}\xi^j\xi^k, \quad \forall\,(x,\xi^1,\cdots,\xi^n)\in\overline{G}\times\mathbb{R}^n.\end{aligned} \tag{10.20}$$

2) *The function* $\varphi(\cdot)$ *has no critical point in* $\overline{G}$, *i.e.,*

$$\min_{x\in\overline{G}}|\nabla\varphi(x)| > 0. \tag{10.21}$$

In the rest of this chapter, we shall choose the set Γ_0 as follows:

$$\Gamma_0 \stackrel{\Delta}{=} \Big\{x\in\Gamma \;\Big|\; \sum_{j,k=1}^{n} a^{jk}\varphi_{x_j}(x)\nu^k(x) > 0\Big\}. \tag{10.22}$$

Also, write

$$R_1 \stackrel{\Delta}{=} \sqrt{\max_{x\in\overline{G}}\varphi(x)}, \qquad R_0 \stackrel{\Delta}{=} \sqrt{\min_{x\in\overline{G}}\varphi(x)}.$$

It is easy to check that if $\varphi(\cdot)$ satisfies Condition 10.1, then for any given constants $\alpha \geq 1$ and $\beta\in\mathbb{R}$, $\tilde{\varphi} = \alpha\varphi + \beta$ still satisfies Condition 10.1 with μ_0 replaced by $\alpha\mu_0$. Therefore we may choose φ, μ_0, $c_0 > 0$, $c_1 > 0$ and T such that

Condition 10.2 *The following inequalities hold:*

1) $\displaystyle\frac{1}{4}\sum_{j,k=1}^{n} a^{jk}(x)\varphi_{x_j}(x)\varphi_{x_k}(x) \geq R_1^2, \quad \forall\, x\in\overline{G};$

2) $T > T_0 \stackrel{\Delta}{=} 2R_1;$

3) $\displaystyle\Big(\frac{2R_1}{T}\Big)^2 < c_1 < \frac{2R_1}{T}$ *and* $\displaystyle c_1 < \min\Big\{1, \frac{1}{16(1+|a_5|^2_{L^\infty_{\mathbb{F}}(0,T;L^\infty(G))})^2}\Big\};$

and

4) $\mu_0 - 4c_1 - c_0 > \sqrt{R_1}.$

Remark 10.10. Conditions 10.1-10.2 can be regarded as some modification of similar conditions introduced in [153] (See also [111, 109]). As we have explained, since $\sum_{j,k=1}^{n} a^{jk}\varphi_{x_j}\varphi_{x_k} > 0$, and one can choose μ_0 in Condition 10.1 large enough, Condition 10.2 could be satisfied obviously. We put it here merely to emphasize the relationship among $0 < c_0 < c_1 < 1$, μ_0 and T. In other words, once Condition 10.1 is fulfilled, Condition 10.2 can be always satisfied.

To be clearer, we give an example for the choice of φ when $(a^{jk})_{1\le j,k\le n}$ is the identity matrix. Let $x_0 \in \mathbb{R}^n \setminus \overline{G}$ such that $|x - x_0| \ge 1$ for all $x \in G$ and $\alpha_0 = \max_{x\in\overline{G}} |x - x_0|^2$. Then for all constant $\alpha \ge \max\{\alpha_0, 1\}$,

$$\alpha \ge \sqrt{\alpha} \max_{x\in\overline{G}} |x - x_0|.$$

Let $\varphi(x) = \alpha|x - x_0|^2$. Then the left hand side of (10.20) is specialized as

$$\sum_{j=1}^{n} \varphi_{x_j x_j} \xi_j^2 = 2\alpha|\xi|^2, \quad \forall\, (x, \xi^1, \cdots, \xi^n) \in \overline{G} \times \mathbb{R}^n. \tag{10.23}$$

Then, (10.20) holds with $\mu_0 = 2\alpha$. Further, it is obvious that (10.21) is true and

$$\frac{1}{4}\sum_{j,k=1}^{n} a^{jk}(x)\varphi_{x_j}(x)\varphi_{x_k}(x) = \frac{1}{4}\sum_{j=1}^{n} \varphi_{x_j}(x)^2 = \alpha^2|x - x_0|^2 \ge \max_{x\in\overline{G}} \varphi(x).$$

Hence, the first inequality in Condition 10.2 holds. Next, one can choose T large enough such that the second and third inequalities in Condition 10.2 hold and

$$c_1 < \min\Big\{1, \frac{1}{16(1 + |a_5|^2_{L^\infty_{\mathbb{F}}(0,T;L^\infty(G))})^2}\Big\}.$$

Let $\alpha \ge \{\alpha_0, 1, 4c_1 + c_0\}$. Then $\mu_0 - 4c_1 - c_0 > \sqrt{R_1}$.

Remark 10.11. To ensure that 3) in Condition 10.2 holds, the larger of the number $|a_5|_{L^\infty_{\mathbb{F}}(0,T;L^\infty(G))}$ is given, the smaller of c_1 and the longer of time T we should choose. In some sense, this seems reasonable because a_5 stands for the effect of the control in the diffusion term to the drift term. One needs time to get rid of such an effect.

Our exact controllability result for the system (10.3) is stated as follows:

Theorem 10.12. *Let Conditions 10.1 and 10.2 hold, and Γ_0 be given by* (10.22)*. Then, the system* (10.3) *is exactly controllable at time T.*

The proof of Theorem 10.12 is rather long and technical. Indeed, the main body in the rest of this chapter is devoted to proving this theorem.

Remark 10.13. Although it is necessary to put controls f and g on the whole domain, one may suspect that Theorem 10.12 is trivial. For instance, one may give a possible "proof" of Theorem 10.12 as follows:

Choosing $f = -a_4 y$ and $g = -a_3 y$, then the system (10.3) becomes

$$\begin{cases} dy = \hat{y}dt & \text{in } Q, \\ d\hat{y} - \displaystyle\sum_{j,k=1}^{n} (a^{jk} y_{x_j})_{x_k} dt = (a_1 \cdot \nabla y + a_2 y - a_5 a_3 y)dt & \text{in } Q, \\ y = \chi_{\Sigma_0} h & \text{on } \Sigma, \\ y(0) = y_0, \quad \hat{y}(0) = \hat{y}_0 & \text{in } G. \end{cases} \tag{10.24}$$

This is actually a wave equation with random coefficients. If one regards the sample point ω as a parameter, then for every given $\omega \in \Omega$, there is a control $u(\cdot,\cdot,\omega)$ such that the solution to (10.24) fulfills $(y(T,x,\omega),\hat{y}(T,x,\omega)) = (y_1(x,\omega),\hat{y}_1(x,\omega))$. It is easy to see that the control constructed in this way belongs to $L^2_{\mathcal{F}_T}(\Omega;L^2(0,T;L^2(\Gamma_0)))$. However, we do not know whether it is adapted to the filtration $\mathbb{F}$ or not. If it is not, then it means to determine the value of the control at present, one needs to use information in future, which is meaningless in the stochastic framework chosen in this book.

10.4 A Reduction of the Exact Controllability Problem

In this section, we shall reduce the exact controllability problem for the system (10.3) to the same problem but for the following controlled backward stochastic wave equation:

$$
\begin{cases}
d\mathbf{y} = \hat{\mathbf{y}}dt + (a_4\mathbf{y} + \mathbf{Y})dW(t) & \text{in } Q,\\
d\hat{\mathbf{y}} - \displaystyle\sum_{j,k=1}^{n}(a^{jk}\mathbf{y}_{x_j})_{x_k}dt = (a_1\cdot\nabla\mathbf{y} + a_2\mathbf{y} + a_5\widehat{\mathbf{Y}})dt & \\
\qquad\qquad\qquad\qquad\qquad\qquad +(a_3\mathbf{y} + \widehat{\mathbf{Y}})dW(t) & \text{in } Q,\\
\mathbf{y} = \chi_{\Sigma_0}\mathbf{h} & \text{on } \Sigma,\\
\hat{\mathbf{y}} = 0 & \text{on } \Sigma,\\
\mathbf{y}(T) = \mathbf{y}^T, \quad \hat{\mathbf{y}}(T) = \hat{\mathbf{y}}^T & \text{in } G.
\end{cases}
\tag{10.25}
$$

Here $(\mathbf{y}^T,\hat{\mathbf{y}}^T) \in L^2_{\mathcal{F}_T}(\Omega;L^2(G)) \times L^2_{\mathcal{F}_T}(\Omega;H^{-1}(G))$, $(\mathbf{y},\hat{\mathbf{y}},\mathbf{Y},\widehat{\mathbf{Y}})$ is the state variable and $\mathbf{h} \in L^2_{\mathbb{F}}(0,T;L^2(\Gamma_0))$ is the control variable. Note that there is only one control in the system (10.25).

Solutions to (10.25) are defined in the sense of transposition as well. For this, we need to introduce the following equation:

$$
\begin{cases}
d\mathbf{z} = \hat{\mathbf{z}}dt + (\mathbf{f} - a_5\mathbf{z})dW(t) & \text{in } Q^\tau,\\
d\hat{\mathbf{z}} - \displaystyle\sum_{j,k=1}^{n}(a^{jk}\mathbf{z}_{x_j})_{x_k}dt & \\
= [-a_1\cdot\nabla\mathbf{z} + (-\operatorname{div} a_1 + a_2 - a_3a_5)\mathbf{z} + a_3\mathbf{f} - a_4\hat{\mathbf{f}}]dt + \hat{\mathbf{f}}dW(t) & \text{in } Q^\tau,\\
\mathbf{z} = 0 & \text{on } \Sigma^\tau,\\
\mathbf{z}(\tau) = \mathbf{z}_\tau, \quad \hat{\mathbf{z}}(\tau) = \hat{\mathbf{z}}_\tau & \text{in } G.
\end{cases}
\tag{10.26}
$$

Here $Q^\tau \stackrel{\Delta}{=} (\tau,T)\times G$, $\Sigma^\tau \stackrel{\Delta}{=} (\tau,T)\times\Gamma$, $(\mathbf{z}_\tau,\hat{\mathbf{z}}_\tau) \in L^2_{\mathcal{F}_\tau}(\Omega;H^1_0(G)) \times L^2_{\mathcal{F}_\tau}(\Omega;L^2(G))$, $\mathbf{f} \in L^2_{\mathbb{F}}(0,T;H^1_0(G))$ and $\hat{\mathbf{f}} \in L^2_{\mathbb{F}}(0,T;L^2(G))$. Note that, unlike the backward stochastic wave equation (10.4), the system (10.26) is a forward stochastic wave equation. By means of Theorem 3.13, for any $(\mathbf{z}_\tau, \hat{\mathbf{z}}_\tau) \in L^2_{\mathcal{F}_\tau}(\Omega; H_0^1(G)) \times L^2_{\mathcal{F}_\tau}(\Omega; L^2(G))$, $\mathbf{f} \in L^2_{\mathbb{F}}(0,T; H_0^1(G))$ and $\hat{\mathbf{f}} \in L^2_{\mathbb{F}}(0,T; L^2(G))$, the system (10.26) admits a unique weak solution $(\mathbf{z}, \hat{\mathbf{z}}) \in C_{\mathbb{F}}([\tau,T]; H_0^1(G)) \times C_{\mathbb{F}}([\tau,T]; L^2(G))$. Moreover,

$$
\begin{aligned}
&|\mathbf{z}|_{C_{\mathbb{F}}([\tau,T];H_0^1(G))} + |\hat{\mathbf{z}}|_{C_{\mathbb{F}}([\tau,T];L^2(G))}\\
&\le \mathcal{C}e^{\mathcal{C}r_4}\big(|\mathbf{z}_\tau|_{L^2_{\mathcal{F}_\tau}(\Omega;H_0^1(G))} + |\hat{\mathbf{z}}_\tau|_{L^2_{\mathcal{F}_\tau}(\Omega;L^2(G))} + |\mathbf{f}|_{L^2_{\mathbb{F}}(0,T;H_0^1(G))}\\
&\qquad + |\hat{\mathbf{f}}|_{L^2_{\mathbb{F}}(0,T;L^2(G))}\big),
\end{aligned} \tag{10.27}
$$

where

$$
\begin{aligned}
r_4 \overset{\Delta}{=} & |a_1|^2_{L^\infty_{\mathbb{F}}(0,T;W^{1,\infty}(G;\mathbb{R}^n))} + \sum_{i=2}^{5} |a_i|^2_{L^\infty_{\mathbb{F}}(0,T;L^\infty(G))}\\
&+ \sum_{i=3}^{5} |a_i|^4_{L^\infty_{\mathbb{F}}(0,T;L^\infty(G))} + |a_5|^2_{L^\infty_{\mathbb{F}}(0,T;W_0^{1,\infty}(G))}
\end{aligned}
$$

and the constant $\mathcal{C}$ is independent of τ.

Similar to the proof of Proposition 10.6, one can show the following result:

Proposition 10.14. *The solution $(\mathbf{z}, \hat{\mathbf{z}})$ to (10.26) satisfies $\frac{\partial \mathbf{z}}{\partial \nu}\big|_\Gamma \in L^2_{\mathbb{F}}(\tau,T; L^2(\Gamma))$. Furthermore,*

$$
\begin{aligned}
\Big|\frac{\partial \mathbf{z}}{\partial \nu}\Big|_{L^2_{\mathbb{F}}(\tau,T;L^2(\Gamma))} &\le \mathcal{C}e^{\mathcal{C}r_4}\big(|\mathbf{z}_\tau|_{L^2_{\mathcal{F}_\tau}(\Omega;H_0^1(G))} + |\hat{\mathbf{z}}_\tau|_{L^2_{\mathcal{F}_\tau}(\Omega;L^2(G))}\\
&\qquad + |\mathbf{f}|_{L^2_{\mathbb{F}}(0,T;H_0^1(G))} + |\hat{\mathbf{f}}|_{L^2_{\mathbb{F}}(0,T;L^2(G))}\big),
\end{aligned} \tag{10.28}
$$

where the constant $\mathcal{C}$ is independent of τ.

In view of Definition 7.13, a quadruple of stochastic processes $(\mathbf{y}, \hat{\mathbf{y}}, \mathbf{Y}, \widehat{\mathbf{Y}}) \in C_{\mathbb{F}}([0,T]; L^2(\Omega; L^2(G))) \times C_{\mathbb{F}}([0,T]; L^2(\Omega; H^{-1}(G))) \times L^2_{\mathbb{F}}(0,T; L^2(G)) \times L^2_{\mathbb{F}}(0,T; H^{-1}(G))$ is called a transposition solution to (10.25), if for any $(\mathbf{z}_\tau, \hat{\mathbf{z}}_\tau) \in L^2_{\mathcal{F}_\tau}(\Omega; H_0^1(G)) \times L^2_{\mathcal{F}_\tau}(\Omega; L^2(G))$, $\mathbf{f} \in L^2_{\mathbb{F}}(0,T; H_0^1(G))$ and $\hat{\mathbf{f}} \in L^2_{\mathbb{F}}(0,T;\ L^2(G))$, it holds that

$$
\begin{aligned}
&\mathbb{E}\langle \hat{\mathbf{y}}^T, \mathbf{z}(T)\rangle_{H^{-1}(G),H_0^1(G)} - \mathbb{E}\langle \mathbf{y}^T, \hat{\mathbf{z}}(T)\rangle_{L^2(G)}\\
&\quad - \mathbb{E}\langle \hat{\mathbf{y}}(\tau), \mathbf{z}_\tau\rangle_{H^{-1}(G),H_0^1(G)} + \mathbb{E}\langle \mathbf{y}(\tau), \hat{\mathbf{z}}_\tau\rangle_{L^2(G)}\\
&= -\mathbb{E}\int_\tau^T \langle \mathbf{Y}, \hat{\mathbf{f}}\rangle_{L^2(G)} dt + \mathbb{E}\int_\tau^T \langle \widehat{\mathbf{Y}}, \mathbf{f}\rangle_{H^{-1}(G),H_0^1(G)} dt - \mathbb{E}\int_\tau^T\int_{\Gamma_0} \mathbf{h}\frac{\partial \mathbf{z}}{\partial \nu} d\Gamma ds.
\end{aligned} \tag{10.29}
$$

Here $(\mathbf{z}, \hat{\mathbf{z}})$ solves (10.26). Note that, by Proposition 10.14 the solution $(\mathbf{z}, \hat{\mathbf{z}})$ to (10.26) satisfies $\frac{\partial \mathbf{z}}{\partial \nu}\big|_\Gamma \in L^2_{\mathbb{F}}(\tau,T;\ L^2(\Gamma))$, hence the term "$\mathbb{E}\int_\tau^T\int_{\Gamma_0} \mathbf{h}\frac{\partial \mathbf{z}}{\partial \nu} d\Gamma ds$" in (10.29) makes sense.

By Theorem 7.14, for each $(\mathbf{y}^T, \hat{\mathbf{y}}^T) \in L^2_{\mathcal{F}_T}(\Omega; L^2(G)) \times L^2_{\mathcal{F}_T}(\Omega; H^{-1}(G))$ and $\mathbf{h} \in L^2_{\mathbb{F}}(0,T; L^2(\Gamma_0))$, the system (10.25) admits a unique transposition solution $(\mathbf{y}, \hat{\mathbf{y}}, \mathbf{Y}, \widehat{\mathbf{Y}})$. Moreover,

$$
\begin{aligned}
&|(\mathbf{y}, \hat{\mathbf{y}})|_{C_{\mathbb{F}}([0,T];L^2(\Omega;L^2(G)))\times C_{\mathbb{F}}([0,T];L^2(\Omega;H^{-1}(G)))}\\
&\quad +|(\mathbf{Y}, \widehat{\mathbf{Y}})|_{L^2_{\mathbb{F}}(0,T;L^2(G))\times L^2_{\mathbb{F}}(0,T;H^{-1}(G))}\\
&\le Ce^{Cr_2}\big(|\mathbf{y}^T|_{L^2_{\mathcal{F}_T}(\Omega;L^2(G))} + |\hat{\mathbf{y}}^T|_{L^2_{\mathcal{F}_T}(\Omega;H^{-1}(G))} + |\mathbf{h}|_{L^2_{\mathbb{F}}(0,T;L^2(\Gamma_0))}\big),
\end{aligned}
\tag{10.30}
$$

where r_2 is given by (10.13).

Similarly to that for (7.9), the system (10.25) is called exactly controllable at time T, if for any $(\mathbf{y}^T, \hat{\mathbf{y}}^T) \in L^2_{\mathcal{F}_T}(\Omega; L^2(G)) \times L^2_{\mathcal{F}_T}(\Omega; H^{-1}(G))$ and $(\mathbf{y}_0, \hat{\mathbf{y}}_0) \in L^2(G) \times H^{-1}(G)$, one can find a control $\mathbf{h} \in L^2_{\mathbb{F}}(0,T; L^2(\Gamma_0))$ such that the corresponding solution $(\mathbf{y}, \hat{\mathbf{y}})$ to (10.25) satisfies that $(\mathbf{y}(0), \hat{\mathbf{y}}(0)) = (\mathbf{y}_0, \hat{\mathbf{y}}_0)$.

Clearly, the following two results hold.

Proposition 10.15. *Let $\tau = T$ in (10.4) and $\tau = 0$ in (10.26). If $(z, \hat{z}, Z, \widehat{Z})$ is a solution to (10.4), then so is $(\mathbf{z}, \hat{\mathbf{z}}) = (z, \hat{z})$ to (10.26) with the initial data $(\mathbf{z}_0, \hat{\mathbf{z}}_0) = (z(0), \hat{z}(0))$ and the nonhomogeneous terms $(\mathbf{f}, \hat{\mathbf{f}}) = (Z, \widehat{Z})$. Conversely, if $(\mathbf{z}, \hat{\mathbf{z}})$ is a solution to (10.26), then so is $(z, \hat{z}, Z, \widehat{Z}) = (\mathbf{z}, \hat{\mathbf{z}}, \mathbf{f}, \hat{\mathbf{f}})$ to (10.4) with the final data $(z^T, \hat{z}^T) = (\mathbf{z}(T), \hat{\mathbf{z}}(T))$.*

Proposition 10.16. *If $(y, \hat{y})$ is a transposition solution to (10.3) for some $f \in L^2_{\mathbb{F}}(0,T; L^2(G))$, $g \in L^2_{\mathbb{F}}(0,T; H^{-1}(G))$ and $h \in L^2_{\mathbb{F}}(0,T; L^2(\Gamma_0))$, then so is $(\mathbf{y}, \hat{\mathbf{y}}, \mathbf{Y}, \widehat{\mathbf{Y}}) = (y, \hat{y}, f, g)$ to (10.25) with the final data $(\mathbf{y}^T, \hat{\mathbf{y}}^T) = (y(T), \hat{y}(T))$ and the nonhomogeneous term $\mathbf{h} = h$. Conversely, if $(\mathbf{y}, \hat{\mathbf{y}}, \mathbf{Y}, \widehat{\mathbf{Y}})$ is a transposition solution to (10.25) for some $\mathbf{h} \in L^2_{\mathbb{F}}(0,T; L^2(\Gamma_0))$, then so is $(y, \hat{y}) = (\mathbf{y}, \hat{\mathbf{y}})$ to (10.3) with the initial data $(y_0, \hat{y}_0) = (\mathbf{y}(0), \hat{\mathbf{y}}(0))$ and the nonhomogeneous terms $(f, g, h) = (\mathbf{Y}, \widehat{\mathbf{Y}}, \mathbf{h})$.*

By Propositions 10.15 and 10.16, and by borrowing some idea from [275], we obtain the following result:

Proposition 10.17. *The system (10.3) is exactly controllable at time T if and only if so is the system (10.25).*

Proof: **The "if" part.** Let $(y_0, \hat{y}_0) \in L^2(G) \times H^{-1}(G)$ and $(y^T, \hat{y}^T) \in L^2_{\mathcal{F}_T}(\Omega; L^2(G)) \times L^2_{\mathcal{F}_T}(\Omega; H^{-1}(G))$ be arbitrarily given. Since (10.25) is exactly controllable at time T, there exists $\mathbf{h} \in L^2_{\mathbb{F}}(0,T; L^2(\Gamma_0))$ such that the corresponding solution $(\mathbf{y}, \hat{\mathbf{y}}, \mathbf{Y}, \widehat{\mathbf{Y}})$ of (10.25) with $(\mathbf{y}^T, \hat{\mathbf{y}}^T) = (y^T, \hat{y}^T)$ satisfies that $(\mathbf{y}(0), \hat{\mathbf{y}}(0)) = (y_0, \hat{y}_0)$. Hence, $(y, \hat{y}) = (\mathbf{y}, \hat{\mathbf{y}})$ is a solution to (10.3) with a triple of controls $(f, g, h) = (\mathbf{Y}, \widehat{\mathbf{Y}}, \mathbf{h})$ such that $(y(T), \hat{y}(T)) = (y^T, \hat{y}^T)$. Hence, the system (10.3) is exactly controllable at time T.

The proof for **the "only if" part** is similar. □

By Proposition 10.17, the exact controllability of (10.3) is reduced to that of (10.25), which is easier to be handled, and therefore in the rest of this chapter we shall focus on the later.

10.5 A Fundamental Identity for Stochastic Hyperbolic-Like Operators

As a key preliminary to prove the exact controllability of (10.25), in this section we shall derive a fundamental identity for stochastic hyperbolic-like operator, which has some independent interest and may be applied in other places.

Throughout this section, we assume that $b^{jk} \in C^3((0,T) \times \mathbb{R}^n)$ satisfies $b^{jk} = b^{kj}$ for $j, k = 1, 2, \cdots, n$, and $\ell, \Psi \in C^2((0,T) \times \mathbb{R}^n)$. Write

$$\begin{cases} c^{jk} \stackrel{\Delta}{=} (b^{jk}\ell_t)_t + \displaystyle\sum_{j',k'=1}^{n} \Big(2b^{jk'}(b^{j'k}\ell_{x_{j'}})_{x_{k'}} - (b^{jk}b^{j'k'}\ell_{x_{j'}})_{x_{k'}}\Big) + \Psi b^{jk}, \\ \mathcal{A} \stackrel{\Delta}{=} (\ell_t^2 - \ell_{tt}) - \displaystyle\sum_{j,k=1}^{n} \Big(b^{jk}\ell_{x_j}\ell_{x_k} - (b^{jk}\ell_{x_j})_{x_k}\Big) - \Psi, \\ \mathcal{B} \stackrel{\Delta}{=} \mathcal{A}\Psi + (\mathcal{A}\ell_t)_t - \displaystyle\sum_{j,k=1}^{n} (\mathcal{A}b^{jk}\ell_{x_j})_{x_k} + \frac{1}{2}\Big(\Psi_{tt} - \displaystyle\sum_{j,k=1}^{n} (b^{jk}\Psi_{x_j})_{x_k}\Big). \end{cases} \tag{10.31}$$

Very similarly to the identities (1.48), (8.19) and (9.26), we have the following result.

Lemma 10.18. *Let z be an $H^2(\mathbb{R}^n)$-valued Itô process and $\hat{z}$ be an $L^2(\mathbb{R}^n)$-valued Itô process so that*

$$dz = \hat{z}dt + ZdW(t) \qquad \text{in } (0,T) \times \mathbb{R}^n \tag{10.32}$$

for some $Z \in L^2_{\mathbb{F}}(0,T; H^1(\mathbb{R}^n))$. Set $\theta = e^{\ell}$, $v = \theta z$ and $\hat{v} = \theta\hat{z} + \ell_t v$. Then, for a.e. $x \in \mathbb{R}^n$,

$$\begin{aligned} &\theta\Big(-2\ell_t\hat{v} + 2\sum_{j,k=1}^{n} b^{jk}\ell_{x_j}v_{x_k} + \Psi v\Big)\Big(d\hat{z} - \sum_{j,k=1}^{n} (b^{jk}z_{x_j})_{x_k}dt\Big) \\ &\quad + \sum_{j,k=1}^{n} \Big[\sum_{j',k'=1}^{n} \Big(2b^{jk}b^{j'k'}\ell_{x_{j'}}v_{x_j}v_{x_{k'}} - b^{jk}b^{j'k'}\ell_{x_j}v_{x_{j'}}v_{x_{k'}}\Big) \\ &\quad -2\ell_t b^{jk}v_{x_j}\hat{v} + b^{jk}\ell_{x_j}\hat{v}^2 + \Psi b^{jk}v_{x_j}v - \frac{\Psi_{x_j}}{2}b^{jk}v^2 - \mathcal{A}b^{jk}\ell_{x_j}v^2\Big]_{x_k} \\ &\quad + d\Big[\ell_t\sum_{j,k=1}^{n} b^{jk}v_{x_j}v_{x_k} + \ell_t\hat{v}^2 - 2\sum_{j,k=1}^{n} b^{jk}\ell_{x_j}v_{x_k}\hat{v} - \Psi v\hat{v} + \Big(\mathcal{A}\ell_t + \frac{\Psi_t}{2}\Big)v^2\Big] \\ &= \Big[\Big(\ell_{tt} + \sum_{j,k=1}^{n} (b^{jk}\ell_{x_j})_{x_k} - \Psi\Big)\hat{v}^2 + \sum_{j,k=1}^{n} c^{jk}v_{x_j}v_{x_k} + \mathcal{B}v^2 \\ &\quad -2\sum_{j,k=1}^{n} \Big((b^{jk}\ell_{x_k})_t + b^{jk}\ell_{tx_k}\Big)v_{x_j}\hat{v} + \Big(-2\ell_t\hat{v} + 2\sum_{j,k=1}^{n} b^{jk}\ell_{x_j}v_{x_k} + \Psi v\Big)^2\Big]dt \end{aligned} \tag{10.33}$$

$$
\begin{aligned}
&+\ell_t(d\hat v)^2-2\sum_{j,k=1}^n b^{jk}\ell_{x_j}dv_{x_k}d\hat v-\Psi dvd\hat v+\ell_t\sum_{j,k=1}^n b^{jk}(dv_{x_j})(dv_{x_k})\\
&+\mathcal{A}\ell_t(dv)^2-\Big[\theta\Big(-2\ell_t\hat v+2\sum_{j,k=1}^n b^{jk}\ell_{x_j}v_{x_k}+\Psi v\Big)\ell_t Z\\
&-\Big(2\sum_{j,k=1}^n b^{jk}(\theta Z)_{x_k}\ell_{x_j}\hat v-\theta\Psi_t vZ+\theta\Psi\hat vZ\Big)\\
&+2\Big(\sum_{j,k=1}^n b^{jk}v_{x_j}(\theta Z)_{x_k}+\theta\mathcal{A}vZ\Big)\ell_t\Big]dW(t),\quad a.s.,
\end{aligned}
$$

where $(dv)^2$ and $(d\hat v)^2$ denote the quadratic variation processes of v and $\hat v$, respectively.

Proof: By (10.32), and recalling $v=\theta z$ and $\hat v=\theta\hat z+\ell_t v$, we obtain that

$$
dv=d(\theta z)=\theta_t zdt+\theta dz=\ell_t\theta zdt+\theta\hat zdt+\theta ZdW(t)=\hat vdt+\theta ZdW(t).\tag{10.34}
$$

Hence,

$$
\begin{aligned}
d\hat z&=d[\theta^{-1}(\hat v-\ell_t v)]=\theta^{-1}[d\hat v-\ell_{tt}vdt-\ell_t dv-\ell_t(\hat v-\ell_t v)dt]\\
&=\theta^{-1}\Big[d\hat v-\big(2\ell_t\hat v+\ell_{tt}v-\ell_t^2 v\big)dt-\theta\ell_t ZdW(t)\Big].
\end{aligned}\tag{10.35}
$$

Similarly, by $b^{jk}=b^{kj}$ for $j,k=1,2,\cdots,n$, we have

$$
\begin{aligned}
&\sum_{j,k=1}^n (b^{jk}z_{x_j})_{x_k}\\
&=\theta^{-1}\sum_{j,k=1}^n\big[(b^{jk}v_{x_j})_{x_k}-2b^{jk}\ell_{x_j}v_{x_k}+(b^{jk}\ell_{x_j}\ell_{x_k}-b^{jk}_{x_k}\ell_{x_j}-b^{jk}\ell_{x_jx_k})v\big].
\end{aligned}\tag{10.36}
$$

Therefore, from (10.35)–(10.36) and the definition of $\mathcal{A}$ in (10.31), we obtain that

$$
\begin{aligned}
&\theta\Big(-2\ell_t\hat v+2\sum_{j,k=1}^n b^{jk}\ell_{x_j}v_{x_k}+\Psi v\Big)\Big(d\hat z-\sum_{j,k=1}^n (b^{jk}z_{x_j})_{x_k}dt\Big)\\
&=\Big(-2\ell_t\hat v+2\sum_{j,k=1}^n b^{jk}\ell_{x_j}v_{x_k}+\Psi v\Big)\Big[d\hat v-\sum_{j,k=1}^n (b^{jk}v_{x_j})_{x_k}dt+\mathcal{A}vdt\\
&\quad+\Big(-2\ell_t\hat v+2\sum_{j,k=1}^n b^{jk}\ell_{x_j}v_{x_k}+\Psi v\Big)dt-\theta\ell_t ZdW(t)\Big]\qquad(10.37)\\
&=\Big(-2\ell_t\hat v+2\sum_{j,k=1}^n b^{jk}\ell_{x_j}v_{x_k}+\Psi v\Big)d\hat v+\Big(-2\ell_t\hat v+2\sum_{j,k=1}^n b^{jk}\ell_{x_j}v_{x_k}+\Psi v\Big)^2 dt
\end{aligned}
$$

$$
\begin{aligned}
&+\Big(-2\ell_t\hat v+2\sum_{j,k=1}^n b^{jk}\ell_{x_j}v_{x_k}+\Psi v\Big)\Big(-\sum_{j,k=1}^n (b^{jk}v_{x_j})_{x_k}+\mathcal{A}v\Big)dt\\
&-\theta\Big(-2\ell_t\hat v+2\sum_{j,k=1}^n b^{jk}\ell_{x_j}v_{x_k}+\Psi v\Big)\ell_t Z dW(t).
\end{aligned}
$$

We now analyze the first and third terms in the right-hand side of (10.37). Using Itô's formula and noting (10.34), we have

$$
\begin{aligned}
&\Big(-2\ell_t\hat v+2\sum_{j,k=1}^n b^{jk}\ell_{x_j}v_{x_k}+\Psi v\Big)d\hat v\\
&=d\Big(-\ell_t\hat v^2+2\sum_{j,k=1}^n b^{jk}\ell_{x_j}v_{x_k}\hat v+\Psi v\hat v\Big)-2\sum_{j,k=1}^n b^{jk}\ell_{x_j}\hat v dv_{x_k}-\Psi\hat v dv\\
&\quad-\Big(-\ell_{tt}\hat v^2+2\sum_{j,k=1}^n \big(b^{jk}\ell_{x_j}\big)_t v_{x_k}\hat v+\Psi_t v\hat v\Big)dt+\ell_t(d\hat v)^2\\
&\quad-2\sum_{j,k=1}^n b^{jk}\ell_{x_j}dv_{x_k}d\hat v-\Psi dv d\hat v \qquad (10.38)\\
&=d\Big(-\ell_t\hat v^2+2\sum_{j,k=1}^n b^{jk}\ell_{x_j}v_{x_k}\hat v+\Psi v\hat v-\frac{\Psi_t}{2}v^2\Big)-\sum_{j,k=1}^n \big(b^{jk}\ell_{x_j}\hat v^2\big)_{x_k}dt\\
&\quad+\Big[\Big(\ell_{tt}+\sum_{j,k=1}^n (b^{jk}\ell_{x_j})_{x_k}-\Psi\Big)\hat v^2-2\sum_{j,k=1}^n (b^{jk}\ell_{x_k})_t v_{x_j}\hat v+\frac{\Psi_{tt}}{2}v^2\Big]dt\\
&\quad+\ell_t(d\hat v)^2-2\sum_{j,k=1}^n b^{jk}\ell_{x_j}dv_{x_k}d\hat v-\Psi dv d\hat v\\
&\quad-\Big(2\sum_{j,k=1}^n b^{jk}(\theta Z)_{x_k}\ell_{x_j}\hat v-\theta\Psi_t vZ+\theta\Psi\hat v Z\Big)dW(t).
\end{aligned}
$$

Next,

$$
\begin{aligned}
&-2\ell_t\hat v\Big(-\sum_{j,k=1}^n (b^{jk}v_{x_j})_{x_k}+\mathcal{A}v\Big)dt\\
&=2\sum_{j,k=1}^n (\ell_t b^{jk}v_{x_j}\hat v)_{x_k}dt-2\sum_{j,k=1}^n \ell_{tx_k}b^{jk}v_{x_j}\hat v dt-2\ell_t\sum_{j,k=1}^n b^{jk}v_{x_j}\hat v_{x_k}dt\\
&\quad-2\mathcal{A}\ell_t v\hat v dt \qquad (10.39)\\
&=2\sum_{j,k=1}^n (\ell_t b^{jk}v_{x_j}\hat v)_{x_k}dt-2\sum_{j,k=1}^n \ell_{tx_k}b^{jk}v_{x_j}\hat v dt-2\ell_t\sum_{j,k=1}^n b^{jk}v_{x_j}(dv-\theta Z dW(t))_{x_k}
\end{aligned}
$$

$$
\begin{aligned}
&-2\mathcal{A}\ell_t v(dv-\theta Z dW(t))\\
&=2\sum_{j,k=1}^{n}(\ell_t b^{jk}v_{x_j}\hat{v})_{x_k}dt-2\sum_{j,k=1}^{n}\ell_{tx_k}b^{jk}v_{x_j}\hat{v}dt-d\Big(\ell_t\sum_{j,k=1}^{n}b^{jk}v_{x_j}v_{x_k}+\mathcal{A}\ell_t v^2\Big)\\
&\quad+\sum_{j,k=1}^{n}\big(\ell_t b^{jk}\big)_t v_{x_j}v_{x_k}dt+(\mathcal{A}\ell_t)_t v^2 dt+\ell_t\sum_{j,k=1}^{n}b^{jk}(dv_{x_j})(dv_{x_k})\\
&\quad+\mathcal{A}\ell_t(dv)^2+2\Big(\sum_{j,k=1}^{n}b^{jk}v_{x_j}(\theta Z)_{x_k}+\theta\mathcal{A}vZ\Big)\ell_t dW(t).
\end{aligned}
$$

Further, by some direct computation, one may check that

$$
\begin{aligned}
&2\sum_{j,k=1}^{n}b^{jk}\ell_{x_j}v_{x_k}\Big(-\sum_{j,k=1}^{n}(b^{jk}v_{x_j})_{x_k}+\mathcal{A}v\Big)\\
&=-\sum_{j,k=1}^{n}\Big[\sum_{j',k'=1}^{n}\Big(2b^{jk}b^{j'k'}\ell_{x_{j'}}v_{x_j}v_{x_{k'}}-b^{jk}b^{j'k'}\ell_{x_j}v_{x_{j'}}v_{x_{k'}}\Big)-\mathcal{A}b^{jk}\ell_{x_j}v^2\Big]_{x_k}\\
&\quad+\sum_{j,k,j',k'=1}^{n}\Big(2b^{jk'}(b^{j'k}\ell_{x_{j'}})_{x_{k'}}-(b^{jk}b^{j'k'}\ell_{x_{j'}})_{x_{k'}}\Big)v_{x_j}v_{x_k}\\
&\quad-\sum_{j,k=1}^{n}(\mathcal{A}b^{jk}\ell_{x_j})_{x_k}v^2
\end{aligned}
\tag{10.40}
$$

and

$$
\begin{aligned}
&\Psi v\Big(-\sum_{j,k=1}^{n}(b^{jk}v_{x_j})_{x_k}+\mathcal{A}v\Big)\\
&=-\sum_{j,k=1}^{n}\Big(\Psi b^{jk}v_{x_j}v-\frac{\Psi_{x_j}}{2}b^{jk}v^2\Big)_{x_k}+\Psi\sum_{j,k=1}^{n}b^{jk}v_{x_j}v_{x_k}\\
&\quad+\Big(-\frac{1}{2}\sum_{j,k=1}^{n}(b^{jk}\Psi_{x_j})_{x_k}+\mathcal{A}\Psi\Big)v^2.
\end{aligned}
\tag{10.41}
$$

Finally, combining (10.37)–(10.41), we obtain the desired equality (10.33). This completes the proof of Lemma 10.18. □

10.6 Observability Estimate for the Stochastic Wave Equation

In order to prove Theorem 10.12, by Proposition 10.17 and Theorem 7.24, it suffices to establish the following observability estimate for the stochastic wave equation (10.26).

Theorem 10.19. *Under the assumptions of Theorem 10.12, all solutions to the equation* (10.26) *with* $\tau = 0$, $\mathbf{f} = 0$ *and* $\hat{\mathbf{f}} = 0$ *satisfy that*

$$|(\mathbf{z}_0, \hat{\mathbf{z}}_0)|_{H_0^1(G)\times L^2(G)} \le \mathcal{C}e^{\mathcal{C}r_4}\Big|\frac{\partial \mathbf{z}}{\partial \nu}\Big|_{L^2_{\mathbb{F}}(0,T;L^2(\Gamma_0))}, \qquad \forall\ (\mathbf{z}_0, \hat{\mathbf{z}}_0) \in H_0^1(G) \times L^2(G). \tag{10.42}$$

Proof: The proof is split into three steps.

Step 1. For each parameter $\lambda > 0$, let us choose

$$\ell(t,x) = \lambda\Big[\varphi(x) - c_1\Big(t - \frac{T}{2}\Big)^2\Big], \tag{10.43}$$

where $\varphi(\cdot)$ and c_1 are given in Conditions 10.1 and 10.2, respectively. Put

$$\Lambda_i \stackrel{\Delta}{=} \Big\{(t,x)\in Q \,\Big|\, \varphi(x) - c_1\Big(t-\frac{T}{2}\Big) > \frac{R_0^2}{2(i+2)}\Big\}, \qquad \text{for } i = 0,1,2. \tag{10.44}$$

Let

$$\begin{cases} T_i \stackrel{\Delta}{=} \dfrac{T}{2} - \varepsilon_i T, \qquad T_i' \stackrel{\Delta}{=} \dfrac{T}{2} + \varepsilon_i T, \\ Q_i \stackrel{\Delta}{=} (T_i, T_i') \times G, \end{cases} \qquad \text{for } i = 0,1, \tag{10.45}$$

where ε_0 and ε_1 are given below.

From Condition 10.2 and (10.43), it follows that

$$\ell(0,x) = \ell(T,x) \le \lambda\Big(R_1 - \frac{c_1 T^2}{4}\Big) < 0, \qquad \forall\ x \in G. \tag{10.46}$$

Hence, there exists $\varepsilon_1 \in (0, \frac{1}{2})$ such that

$$\Lambda_2 \subset Q_1 \tag{10.47}$$

and that

$$\ell(t,x) < 0, \qquad \forall\ (t,x) \in \big((0,T_1)\cup(T_1',T)\big)\times G. \tag{10.48}$$

Next, since $\{T/2\}\times G \subset \Lambda_0$, one can find $\varepsilon_0 \in (0, \varepsilon_1)$ such that

$$Q_0 \subset \Lambda_0. \tag{10.49}$$

Step 2. Recall Condition 10.2 for c_0. For each $(\mathbf{z}_0, \hat{\mathbf{z}}_0) \in H_0^1(G)\times L^2(G)$, let us apply Lemma 10.18 with $(b^{jk})_{1\le j,k\le n} = (a^{jk})_{1\le j,k\le n}$ and

$$\Psi = \ell_{tt} + \sum_{j,k=1}^{n} (a^{jk}\ell_{x_j})_{x_k} - c_0\lambda \tag{10.50}$$

to the corresponding solution $(\mathbf{z}, \hat{\mathbf{z}}) \in C_{\mathbb{F}}([0,T];H_0^1(G))\times C_{\mathbb{F}}([0,T];L^2(G))$ to (10.26) with $\tau = 0$, $\mathbf{f} = 0$ and $\hat{\mathbf{f}} = 0$, and then analyze the resulting terms in (10.33) one by one. In what follows, we use the notations $v = \theta \mathbf{z}$ and $\hat{v} = \theta \hat{\mathbf{z}} + \ell_t v$, where $\theta = e^{\ell}$ for ℓ given by (10.43).

Let us consider first the terms which stand for the "energy" of the solutions. To this end, we need to compute the orders of λ (when λ is sufficiently large) in the coefficients of $\hat{v}^2$, $|\nabla v|^2$ and v^2.

Clearly, the term for $\hat{v}^2$ reads

$$\Big(\ell_{tt} + \sum_{j,k=1}^{n} (a^{jk}\ell_{x_j})_{x_k} - \Psi\Big)\hat{v}^2 = c_0 \lambda \hat{v}^2. \tag{10.51}$$

Noting that $\ell_{tx_k} = \ell_{x_k t} = 0$, we get

$$2\sum_{j,k=1}^{n} \big((a^{jk}\ell_{x_k})_t + a^{jk}\ell_{tx_k}\big) v_{x_j}\hat{v} = 0. \tag{10.52}$$

From (10.50), we see that

$$\begin{aligned}
&(a^{jk}\ell_t)_t + \sum_{j',k'=1}^{n} \Big(2a^{jk'}\big(a^{j'k}\ell_{x_{j'}}\big)_{x_{k'}} - \big(a^{jk}a^{j'k'}\ell_{x_{j'}}\big)_{x_{k'}}\Big) + \Psi a^{jk} \\
&= a^{jk}\ell_{tt} + \sum_{j',k'=1}^{n} \Big(2a^{jk'}\big(a^{j'k}\ell_{x_{j'}}\big)_{x_{k'}} - \big(a^{jk}a^{j'k'}\ell_{x_{j'}}\big)_{x_{k'}}\Big) \\
&\quad + a^{jk}\Big(\ell_{tt} + \sum_{j',k'=1}^{\infty} \big(a^{j'k'}\ell_{x_{j'}}\big)_{x_{k'}} - c_0\lambda\Big) \\
&= 2a^{jk}\ell_{tt} + \sum_{j',k'=1}^{n} \Big(2a^{jk'}(a^{j'k}\ell_{x_{j'}})_{x_{k'}} - a^{jk}_{x_{k'}}a^{j'k'}\ell_{x_{j'}}\Big) - a^{jk}c_0\lambda \\
&= 2a^{jk}\ell_{tt} + \lambda \sum_{j',k'=1}^{n} \Big(2a^{jk'}(a^{j'k}\varphi_{x_{j'}})_{x_{k'}} - a^{jk}_{x_{k'}}a^{j'k'}\varphi_{x_{j'}}\Big) \\
&\quad + \lambda \sum_{j',k'=1}^{n} 2a^{jk'}a^{j'k}\varphi_{x_{j'}}\varphi_{x_{k'}} - a^{jk}c_0\lambda.
\end{aligned}$$

This, together with Condition 10.1, implies that

$$\begin{aligned}
&\sum_{j,k=1}^{n} c^{jk} v_{x_j} v_{x_k} \\
&= \sum_{j,k=1}^{n} \Big[(a^{jk}\ell_t)_t + \sum_{j',k'=1}^{n} \Big(2a^{jk'}(a^{j'k}\ell_{x_{j'}})_{x_{k'}} - (a^{jk}a^{j'k'}\ell_{x_{j'}})_{x_{k'}}\Big) + \\
&\qquad \Psi a^{jk}\Big] v_{x_j} v_{x_k} \\
&\geq \lambda(\mu_0 - 4c_1 - c_0) \sum_{j,k=1}^{n} a^{jk} v_{x_j} v_{x_k}.
\end{aligned} \tag{10.53}$$

Now we compute the coefficients of v^2. By (10.31), it is easy to obtain that

$$\begin{aligned}\mathcal{A} &= \ell_t^2 - \ell_{tt} - \sum_{j,k=1}^{n} \big(a^{jk}\ell_{x_j}\ell_{x_k} - (a^{jk}\ell_{x_j})_{x_k}\big) - \Psi \\ &= \lambda^2\Big[c_1^2(2t-T)^2 - \sum_{j,k=1}^{n} a^{jk}\varphi_{x_j}\varphi_{x_k}\Big] + 4c_1\lambda + c_0\lambda.\end{aligned} \tag{10.54}$$

By the definition of $\mathcal{B}$ in (10.31), we see that

$$\begin{aligned}\mathcal{B} &= \mathcal{A}\Psi + (\mathcal{A}\ell_t)_t - \sum_{j,k=1}^{n}(\mathcal{A}a^{jk}\ell_{x_j})_{x_k} + \frac{1}{2}\sum_{j,k=1}^{n}\big(\Psi_{tt} - (a^{jk}\Psi_{x_j})_{x_k}\big) \\ &= 2\mathcal{A}\ell_{tt} - \lambda c_0\mathcal{A} - \sum_{j,k=1}^{n} a^{jk}\ell_j\mathcal{A}_k + \mathcal{A}_t\ell_t \\ &\quad -\frac{1}{2}\sum_{j,k=1}^{n}\sum_{j',k'=1}^{n}\big[a^{jk}(a^{j'k'}\ell_{x_{j'}})_{x_{k'}x_j}\big]_{x_k} \\ &= 2\lambda^3\Big[-2c_1^3(2t-T)^2 + 2c_1\sum_{j,k=1}^{n} a^{jk}\varphi_{x_j}\varphi_{x_k}\Big] - \lambda^3 c_0 c_1^2(2t-T)^2 \\ &\quad +\lambda^3 c_0\sum_{j,k=1}^{n} a^{jk}\varphi_{x_j}\varphi_{x_k} + \lambda^3\sum_{j,k=1}^{n}\sum_{j',k'=1}^{n} a^{jk}\varphi_{x_j}(a^{j'k'}\varphi_{x_{j'}}\varphi_{x_{k'}})_{x_k} \\ &\quad -4\lambda^3 c_1^3(2t-T)^2 + O(\lambda^2) \\ &= (4c_1 + c_0)\lambda^3\sum_{j,k=1}^{n} a^{jk}\varphi_{x_j}\varphi_{x_k} + \lambda^3\sum_{j,k=1}^{n}\sum_{j',k'=1}^{n} a^{jk}\varphi_{x_j}(a^{j'k'}\varphi_{x_{j'}}\varphi_{x_{k'}})_{x_k} \\ &\quad -(8c_1^3 + c_0c_1^2)\lambda^3(2t-T)^2 + O(\lambda^2).\end{aligned} \tag{10.55}$$

Now we estimate $\sum_{j,k=1}^{n}\sum_{j',k'=1}^{n} a^{jk}\varphi_{x_j}(a^{j'k'}\varphi_{x_{j'}}\varphi_{x_{k'}})_{x_k}$. From Condition 10.1, we have

$$\begin{aligned}&\mu_0\sum_{j,k=1}^{n} a^{jk}\varphi_{x_j}\varphi_{x_k} \\ &\le \sum_{j,k=1}^{n}\sum_{j',k'=1}^{n}\Big(2a^{jk'}(a^{j'k}\varphi_{x_{j'}})_{x_{k'}} - a^{jk}_{x_{k'}}a^{j'k'}\varphi_{x_{j'}}\Big)\varphi_{x_j}\varphi_{x_k} \\ &= \sum_{j,k=1}^{n}\sum_{j',k'=1}^{n}\Big(2a^{jk'}a^{j'k}_{x_{k'}}\varphi_{x_{j'}} + 2a^{jk'}a^{j'k}\varphi_{x_{j'}x_{k'}} - a^{jk}_{x_{k'}}a^{j'k'}\varphi_{x_{j'}}\Big)\varphi_{x_j}\varphi_{x_k}\end{aligned}$$

$$
\begin{aligned}
&= \sum_{j,k=1}^{n} \sum_{j',k'=1}^{n} \Big(2a^{jk'} a^{j'k}_{x_{k'}} \varphi_{x_{j'}} \varphi_{x_j} \varphi_{x_k} + 2a^{jk'} a^{j'k} \varphi_{x_{j'}x_{k'}} \varphi_{x_j} \varphi_{x_k} \\
&\qquad\qquad - a^{jk}_{x_{k'}} a^{j'k'} \varphi_{x_{j'}} \varphi_{x_j} \varphi_{x_k} \Big) \\
&= \sum_{j,k=1}^{n} \sum_{j',k'=1}^{n} \Big(a^{j'k'} a^{jk}_{x_{k'}} \varphi_{x_{j'}} \varphi_{x_j} \varphi_{x_k} + a^{jk} a^{j'k'} \varphi_{x_{j'}x_k} \varphi_{x_j} \varphi_{x_{k'}} \\
&\qquad\qquad + a^{jk} a^{j'k'} \varphi_{x_{k'}x_k} \varphi_{x_j} \varphi_{x_{j'}} \Big) \\
&= \sum_{j,k=1}^{n} \sum_{j',k'=1}^{n} a^{jk} \varphi_{x_j} (a^{j'k'} \varphi_{x_{j'}} \varphi_{x_{k'}})_{x_k}.
\end{aligned}
\tag{10.56}
$$

From (10.55), (10.56) and Condition 10.2, we arrive at

$$
\begin{aligned}
\mathcal{B} &\geq \lambda^3 (4c_1 + c_0) \sum_{j,k=1}^{n} a^{jk} \varphi_{x_j} \varphi_{x_k} + \lambda^3 \mu_0 \sum_{j,k=1}^{n} a^{jk} \varphi_{x_j} \varphi_{x_k} \\
&\quad - (8c_1^3 + 2c_0 c_1^2) \lambda^3 (2t - T)^2 + O(\lambda^2) \\
&\geq (\mu_0 + 4c_1 + c_0) \lambda^3 \sum_{j,k=1}^{n} a^{jk} \varphi_{x_j} \varphi_{x_k} - 8c_1^2 (4c_1 + c_0) \Big(t - \frac{T}{2} \Big)^2 \lambda^3 + O(\lambda^2).
\end{aligned}
\tag{10.57}
$$

Since the diffusion term in the second equation of (10.26) (with $\tau = 0$, $\mathbf{f} = 0$ and $\hat{\mathbf{f}} = 0$) is zero, it follows that

$$
\mathbb{E}\big(\ell_t (d\hat{v})^2\big) = 0, \quad \mathbb{E}\left(\sum_{j,k=1}^{n} a^{jk} \ell_{x_j} dv_{x_k} d\hat{v} \right) = 0, \quad \mathbb{E}\left(\Psi dv d\hat{v}\right) = 0. \tag{10.58}
$$

From (10.43) and noting that $dv = \theta \ell_t \mathbf{z} dt + \theta d\mathbf{z} = \theta(\ell_t \mathbf{z} + \hat{\mathbf{z}})dt - a_5 v dW(t)$, we see that, for some constant $\mathcal{C}_0 > 0$,

$$
\begin{aligned}
&\mathbb{E}\left(\ell_t \sum_{j,k=1}^{n} a^{jk} (dv_{x_j})(dv_{x_k}) \right) \\
&= -2c_1 \lambda \Big(t - \frac{T}{2} \Big) \mathbb{E}\Big(\sum_{j,k=1}^{n} a^{jk} (a_5 v)_{x_j} (a_5 v)_{x_k} \Big) \\
&\geq -c_1 T \lambda \mathbb{E}\Big[(1 + |a_5|^2) \sum_{j,k=1}^{n} a^{jk} v_{x_j} v_{x_k} + \mathcal{C}_0 |v|^2 \Big].
\end{aligned}
\tag{10.59}
$$

Next, from (10.43) and (10.54), we find that

$$
\begin{aligned}
&\mathbb{E}\big[\mathcal{A}\ell_t(dv)^2\big]\\
&=-\lambda^3c_1(2t-T)\Big[c_1^2(2t-T)^2-\sum_{j,k=1}^n a^{jk}\varphi_{x_j}\varphi_{x_k}\Big]\mathbb{E}\big(|a_5|^2v^2\big)\\
&\quad-(4c_1+c_0)c_1(2t-T)\lambda^2\mathbb{E}\big(|a_5|^2v^2\big).
\end{aligned}
\tag{10.60}
$$

From (10.53), (10.57), (10.59) and (10.60), and noting the fourth inequality in Condition 10.2, we conclude that there is $c_2>0$ such that for all $(t,x)\in\Lambda_2$,

$$
\begin{aligned}
&\mathbb{E}\Big[\sum_{j,k=1}^n c^{jk}v_{x_j}v_{x_k}+\mathcal{B}v^2+\ell_t\sum_{j,k=1}^n a^{jk}(dv_{x_j})(dv_{x_k})+\mathcal{A}\ell_t(dv)^2\Big]\\
&=\mathbb{E}\Big\{\lambda\Big[\mu_0-4c_1-c_0-c_1T(1+|a_5|^2)\Big]\sum_{j,k=1}^n a^{jk}v_{x_j}v_{x_k}\\
&\qquad+(\mu_0+4c_1+c_0)\lambda^3\sum_{j,k=1}^n a^{jk}\varphi_{x_j}\varphi_{x_k}|v|^2-8c_1^2(4c_1+c_0)\lambda^3\Big(t-\frac{T}{2}\Big)^2|v|^2\\
&\qquad+4c_1\lambda^3\Big(t-\frac{T}{2}\Big)|a_5|^2\sum_{j,k=1}^n a^{jk}\varphi_{x_j}\varphi_{x_k}|v|^2-c_1^3\lambda^3(2t-T)^3|a_5|^2|v|^2\\
&\qquad+O(\lambda^2)|v|^2\Big\}\\
&\geq\mathbb{E}\big[c_2\lambda|\nabla v|^2+c_2\lambda^3|v|^2+O(\lambda^2)|v|^2\big].
\end{aligned}
$$

Thus, there exist $\lambda_1>0$ and $c_3>0$ such that for all $\lambda\geq\lambda_1$ and for every $(t,x)\in\Lambda_2$,

$$
\begin{aligned}
&\mathbb{E}\Big(\sum_{j,k=1}^n c^{jk}v_{x_j}v_{x_k}+\mathcal{B}v^2+\ell_t\sum_{j,k=1}^n a^{jk}(dv_{x_j})(dv_{x_k})+\mathcal{A}\ell_t(dv)^2\Big)\\
&\geq\mathbb{E}\big(c_3\lambda|\nabla v|^2+c_3\lambda^3|v|^2\big).
\end{aligned}
\tag{10.61}
$$

Step 3. For the boundary terms, by $v|_\Sigma=0$, we have the following equality:

$$
\begin{aligned}
&\mathbb{E}\int_\Sigma\sum_{j,k=1}^n\sum_{j',k'=1}^n\Big(2a^{jk}a^{j'k'}\ell_{x_{j'}}v_{x_j}v_{x_{k'}}-a^{jk}a^{j'k'}\ell_{x_j}v_{x_{j'}}v_{x_{k'}}\Big)\nu^kd\Sigma\\
&=\lambda\mathbb{E}\int_\Sigma\sum_{j,k=1}^n\sum_{j',k'=1}^n\Big(2a^{jk}a^{j'k'}\varphi_{x_{j'}}v_{x_j}v_{x_{k'}}-a^{jk}a^{j'k'}\varphi_{x_j}v_{x_{j'}}v_{x_{k'}}\Big)\nu^kd\Sigma\\
&=\lambda\mathbb{E}\int_\Sigma\sum_{j,k=1}^n\sum_{j',k'=1}^n\Big(2a^{jk}a^{j'k'}\varphi_{x_{j'}}\frac{\partial v}{\partial\nu}\nu^j\frac{\partial v}{\partial\nu}\nu^{k'}-a^{jk}a^{j'k'}\varphi_{x_j}\frac{\partial v}{\partial\nu}\nu^{j'}\frac{\partial v}{\partial\nu}\nu^{k'}\Big)\nu_kd\Sigma\\
&=\lambda\mathbb{E}\int_\Sigma\Big(\sum_{j,k=1}^n a^{jk}\nu^j\nu^k\Big)\Big(\sum_{j',k'=1}^n a^{j'k'}\varphi_{x_{j'}}\nu^{k'}\Big)\Big|\frac{\partial v}{\partial\nu}\Big|^2d\Sigma.
\end{aligned}
\tag{10.62}
$$

For any $\tau \in (0, T_1)$ and $\tau' \in (T_1', T)$, put

$$Q_\tau^{\tau'} \stackrel{\Delta}{=} (\tau, \tau') \times G. \tag{10.63}$$

Integrating (10.33) in $Q_\tau^{\tau'}$, taking expectation and by (10.51), (10.52), (10.58) and (10.61), we obtain that

$$\begin{aligned}
&\mathbb{E}\int_{Q_\tau^{\tau'}} \theta\Big(-2\ell_t\hat{v} + 2\sum_{i=1}^n a^{jk}\ell_{x_j}v_{x_k} + \Psi v\Big)\Big(d\hat{\mathbf{z}} - \sum_{j,k=1}^n (a^{jk}\mathbf{z}_{x_j})_{x_k}dt\Big)dx\\
&+\lambda\mathbb{E}\int_{\Sigma_0}\Big(\sum_{j,k=1}^n a^{jk}\nu^j\nu^k\Big)\Big(\sum_{j',k'=1}^n a^{j'k'}\varphi_{x_{j'}}\nu^{k'}\Big)\Big|\frac{\partial v}{\partial \nu}\Big|^2 d\Sigma\\
&+\mathbb{E}\int_{Q_\tau^{\tau'}} d\Big[\ell_t\sum_{j,k=1}^n a^{jk}v_{x_j}v_{x_k} + \ell_t\hat{v}^2 - 2\sum_{j,k=1}^n a^{jk}\ell_{x_j}v_{x_k}\hat{v} - \Psi v\hat{v}\\
&+\Big(\mathcal{A}\ell_t + \frac{\Psi_t}{2}\Big)v^2\Big]dx\\
&\geq c_0\lambda\mathbb{E}\int_{Q_\tau^{\tau'}}\hat{v}^2dxdt + \mathbb{E}\int_{Q_\tau^{\tau'}}\sum_{j,k=1}^n c^{jk}v_{x_j}v_{x_k}dxdt + \mathbb{E}\int_{Q_\tau^{\tau'}}\mathcal{B}v^2dxdt\\
&+\mathbb{E}\int_{Q_\tau^{\tau'}}\Big(\ell_t\sum_{j,k=1}^n a^{jk}(dv_{x_j})(dv_{x_k}) + \mathcal{A}\ell_t(dv)^2\Big)dx\\
&+\mathbb{E}\int_Q\Big(-2\ell_t\hat{v} + 2\sum_{j,k=1}^n a^{jk}\ell_{x_j}v_{x_k} + \Psi v\Big)^2dxdt.
\end{aligned} \tag{10.64}$$

Clearly,

$$\begin{aligned}
&\mathbb{E}\int_{Q_\tau^{\tau'}} d\Big[\ell_t\sum_{j,k=1}^n a^{jk}v_{x_j}v_{x_k} + \ell_t\hat{v}^2 - 2\sum_{j,k=1}^n a^{jk}\ell_{x_j}v_{x_k}\hat{v} - \Psi v\hat{v}\\
&\qquad+\Big(\mathcal{A}\ell_t + \frac{\Psi_t}{2}\Big)v^2\Big]dx\\
&=\mathbb{E}\int_G\Big[\ell_t\sum_{j,k=1}^n a^{jk}v_{x_j}(\tau')v_{x_k}(\tau') + \ell_t\hat{v}(\tau')^2 - 2\sum_{j,k=1}^n a^{jk}\ell_{x_j}v_{x_k}(\tau')\hat{v}(\tau')\\
&\qquad-\Psi v(\tau')\hat{v}(\tau') + \Big(\mathcal{A}\ell_t + \frac{\Psi_t}{2}\Big)v(\tau')^2\Big]dx\\
&\quad-\mathbb{E}\int_G\Big[\ell_t\sum_{j,k=1}^n a^{jk}v_{x_j}(\tau)v_{x_k}(\tau) + \ell_t\hat{v}(\tau)^2 - 2\sum_{j,k=1}^n a^{jk}\ell_{x_j}v_{x_k}(\tau)\hat{v}(\tau)\\
&\qquad-\Psi v(\tau)\hat{v}(\tau) + \Big(\mathcal{A}\ell_t + \frac{\Psi_t}{2}\Big)v(\tau)^2\Big]dx\\
&\leq \mathcal{C}\lambda^3\mathbb{E}\int_G\big[\big(\hat{v}(\tau)^2 + |\nabla v(\tau)|^2 + v(\tau)^2\big) + \big(\hat{v}(\tau')^2 + |\nabla v(\tau')|^2 + v(\tau')^2\big)\big]dx.
\end{aligned} \tag{10.65}$$

By $\theta = e^{\ell}$ and (10.48), there is a $\lambda_1 > 0$ such that for any $\lambda > \lambda_1$,

$$\lambda^3\theta(\tau) \le 1, \qquad \lambda^3\theta(\tau') \le 1. \tag{10.66}$$

Since $v = \theta\mathbf{z}$ and $\hat{v} = \theta\hat{\mathbf{z}}$, it follows from (10.66) that

$$\begin{aligned}&\lambda^3\mathbb{E}\int_G \big[\big(\hat{v}(\tau)^2 + |\nabla v(\tau)|^2 + v(\tau)^2\big) + \big(\hat{v}(\tau')^2 + |\nabla v(\tau')|^2 + v(\tau')^2\big)\big]dx\\ &\le \mathbb{E}\int_G \big[\big(\hat{\mathbf{z}}(\tau)^2 + |\nabla \mathbf{z}(\tau)|^2 + \mathbf{z}(\tau)^2\big) + \big(\hat{\mathbf{z}}(\tau')^2 + |\nabla \mathbf{z}(\tau')|^2 + \mathbf{z}(\tau')^2\big)\big]dx.\end{aligned} \tag{10.67}$$

From (10.44), (10.47) and (10.63), we obtain that $\Lambda_2 \subset Q_\tau^{\tau'}$. Thus,

$$\begin{aligned}&\lambda\mathbb{E}\int_{Q_\tau^{\tau'}\setminus\Lambda_2} \hat{v}^2 dxdt = \lambda\mathbb{E}\int_{Q_\tau^{\tau'}\setminus\Lambda_2} \big(\theta\hat{\mathbf{z}} + \ell_t\theta\mathbf{z}\big)^2 dxdt\\ &\le 2\lambda\mathbb{E}\int_{Q_\tau^{\tau'}\setminus\Lambda_2} \theta^2\hat{\mathbf{z}}^2 dxdt + 2\lambda^3\mathbb{E}\int_{Q_\tau^{\tau'}\setminus\Lambda_2} (2t-T)^2\theta^2\mathbf{z}^2 dxdt\end{aligned} \tag{10.68}$$

and

$$\begin{aligned}&\mathbb{E}\int_{Q_\tau^{\tau'}\setminus\Lambda_2} \sum_{j,k=1}^n c^{jk} v_{x_j} v_{x_k} dxdt = \mathbb{E}\int_{Q_\tau^{\tau'}\setminus\Lambda_2} \sum_{j,k=1}^n c^{jk}(\theta\mathbf{z}_{x_j})(\theta\mathbf{z}_{x_k})dxdt\\ &= \mathbb{E}\int_{Q_\tau^{\tau'}\setminus\Lambda_2} \sum_{j,k=1}^n c^{jk}\theta^2(\lambda\varphi_{x_j} + \mathbf{z}_{x_j})(\lambda\varphi_{x_k} + \mathbf{z}_{x_k})dxdt\\ &\le \mathcal{C}\lambda\mathbb{E}\int_{Q_\tau^{\tau'}\setminus\Lambda_2} \theta^2|\nabla\mathbf{z}|^2 dxdt + \mathcal{C}\lambda^3\mathbb{E}\int_{Q_\tau^{\tau'}\setminus\Lambda_2} \theta^2|\mathbf{z}|^2 dxdt.\end{aligned} \tag{10.69}$$

Furthermore, it follows from (10.55) that

$$\mathbb{E}\int_{Q_\tau^{\tau'}\setminus\Lambda_2} \mathcal{B}v^2 dxdt \le \mathcal{C}\lambda^3\mathbb{E}\int_{Q_\tau^{\tau'}\setminus\Lambda_2} \theta^2\mathbf{z}^2 dxdt. \tag{10.70}$$

Next, by (10.59) and (10.60), we obtain that

$$\begin{aligned}&\mathbb{E}\int_{Q_\tau^{\tau'}\setminus\Lambda_2} \Big[\ell_t \sum_{j,k=1}^n a^{jk}(dv_{x_j})(dv_{x_k}) + \mathcal{A}\ell_t(dv)^2\Big]dx\\ &\le \mathcal{C}\lambda\mathbb{E}\int_{Q_\tau^{\tau'}\setminus\Lambda_2} \theta^2|\nabla\mathbf{z}|^2 dxdt + \mathcal{C}\lambda^3\mathbb{E}\int_{Q_\tau^{\tau'}\setminus\Lambda_2} \theta^2|\mathbf{z}|^2 dxdt.\end{aligned} \tag{10.71}$$

From (10.44), we deduce that $\theta \le \exp(\lambda e^{R_0^2\mu/8})$ in $Q_\tau^{\tau'} \setminus \Lambda_2$. Consequently, there exists $\lambda_2 \ge \max\{\lambda_0, \lambda_1\}$ such that for all $\lambda \ge \lambda_2$,

$$\mathcal{C}\lambda \max_{(x,t)\in Q_\tau^{\tau'}\setminus\Lambda_2} \theta^2 \le e^{\lambda R_0^2/3}, \qquad \mathcal{C}\lambda^3 \max_{(x,t)\in Q_\tau^{\tau'}\setminus\Lambda_2} \theta^2 \le e^{\lambda R_0^2/3}. \tag{10.72}$$

It follows from (10.61) and (10.68)–(10.72) that

$$
\begin{aligned}
&\lambda\mathbb{E}\int_{Q_\tau^{\tau'}}\hat v^2dxdt+\mathbb{E}\int_{Q_\tau^{\tau'}}\sum_{j,k=1}^n c^{jk}v_{x_j}v_{x_k}dxdt+\mathbb{E}\int_{Q_\tau^{\tau'}}\mathcal{B}v^2dxdt\\
&+\mathbb{E}\int_{Q_\tau^{\tau'}}\Big(\ell_t\sum_{j,k=1}^n a^{jk}(dv_{x_j})(dv_{x_k})+\mathcal{A}\ell_t(dv)^2\Big)dx \qquad (10.73)\\
&\geq \lambda\mathbb{E}\int_{\Lambda_2}\hat v^2dxdt+c_2\lambda\mathbb{E}\int_{\Lambda_2}|\nabla v|^2dxdt+c_2\lambda^3\mathbb{E}\int_{\Lambda_2}|v|^2dxdt\\
&-e^{\lambda R_0^2/3}\mathbb{E}\int_Q(|\hat{\mathbf{z}}|^2+|\nabla\mathbf{z}|^2)dxdt.
\end{aligned}
$$

Noting that $(\mathbf{z},\hat{\mathbf{z}})$ solves the equation (10.26) with $\tau=0$, $\mathbf{f}=0$ and $\hat{\mathbf{f}}=0$, we deduce that

$$
\begin{aligned}
&\mathbb{E}\int_{Q_\tau^{\tau'}}\theta\Big(-2\ell_t\hat v+2\sum_{i=1}^n a^{jk}\ell_{x_j}v_{x_k}+\Psi v\Big)\Big(d\hat{\mathbf{z}}-\sum_{j,k=1}^n(a^{jk}\mathbf{z}_{x_j})_{x_k}dt\Big)dx\\
&=\mathbb{E}\int_{Q_\tau^{\tau'}}\theta\Big(-2\ell_t\hat v+2\sum_{i=1}^n a^{jk}\ell_{x_j}v_{x_k}+\Psi v\Big)\\
&\qquad\times\big[-a_1\cdot\nabla\mathbf{z}+\big(-\operatorname{div}a_1+a_2-a_3a_5\big)\mathbf{z}\big]dxdt\\
&\leq\mathbb{E}\int_{Q_\tau^{\tau'}}\theta\Big(-2\ell_t\hat v+2\sum_{i=1}^n a^{jk}\ell_{x_j}v_{x_k}+\Psi v\Big)^2dxdt\\
&+r_2\mathbb{E}\int_{Q_\tau^{\tau'}}(|\nabla\mathbf{z}|^2+\mathbf{z}^2)dxdt.
\end{aligned}
\qquad (10.74)
$$

Combing (10.64), (10.67), (10.73) and (10.74), we conclude that there exists $\lambda_3\geq\max\{\lambda_2, Cr_5+1\}$ such that for any $\lambda\geq\lambda_3$,

$$
\begin{aligned}
&\mathbb{E}\int_{\Lambda_1}\theta^2(|\hat{\mathbf{z}}|^2+|\nabla\mathbf{z}|^2)dxdt+\mathbb{E}\int_{\Lambda_1}\theta^2|\mathbf{z}|^2dxdt\\
&\leq\mathcal{C}\Big[e^{\lambda R_0^2/3}\mathbb{E}\int_Q(|\hat{\mathbf{z}}|^2+|\nabla\mathbf{z}|^2+|\mathbf{z}|^2)dxdt+e^{\lambda R_1^2}\mathbb{E}\int_{\Sigma_0}\Big|\frac{\partial\mathbf{z}}{\partial\nu}\Big|^2d\Sigma\\
&+\mathbb{E}\int_G\big(\hat{\mathbf{z}}(\tau)^2+|\nabla\mathbf{z}(\tau)|^2+\mathbf{z}(\tau)^2+\hat{\mathbf{z}}(\tau')^2+|\nabla\mathbf{z}(\tau')|^2+\mathbf{z}(\tau')^2\big)dx\Big].
\end{aligned}
\qquad (10.75)
$$

Integrating (10.75) with respect to τ and τ' on $[T_2,T_1]$ and $[T_1',T_2']$, respectively, we get that

$$
\begin{aligned}
&\mathbb{E}\int_{\Lambda_1}\theta^2(|\hat{\mathbf{z}}|^2+|\nabla\mathbf{z}|^2)dxdt+\mathbb{E}\int_{\Lambda_1}\theta^2|\mathbf{z}|^2dxdt\\
&\leq \mathcal{C}e^{\lambda R_0^2/3}\mathbb{E}\int_Q(|\hat{\mathbf{z}}|^2+|\nabla\mathbf{z}|^2)dxdt+\mathcal{C}e^{\lambda R_0^2/3}\mathbb{E}\int_Q|\mathbf{z}|^2dxdt\\
&\quad+\mathcal{C}e^{\lambda R_1^2}\mathbb{E}\int_{\Sigma_0}\Big|\frac{\partial\mathbf{z}}{\partial\nu}\Big|^2d\Sigma+\mathcal{C}\mathbb{E}\int_Q(\hat{\mathbf{z}}^2+|\nabla\mathbf{z}|^2+\mathbf{z}^2)dxdt.
\end{aligned}
\tag{10.76}
$$

From (10.44), we find that

$$
\begin{aligned}
&\mathbb{E}\int_{\Lambda_1}\theta^2(|\hat{\mathbf{z}}|^2+|\nabla\mathbf{z}|^2+|\mathbf{z}|^2)dxdt\\
&\geq \mathbb{E}\int_{\Lambda_0}\theta^2(|\hat{\mathbf{z}}|^2+|\nabla\mathbf{z}|^2+|\mathbf{z}|^2)dxdt\\
&\geq e^{\lambda R_0^2/2}\mathbb{E}\int_{Q_0}(|\hat{\mathbf{z}}|^2+|\nabla\mathbf{z}|^2+|\mathbf{z}|^2)dxdt.
\end{aligned}
\tag{10.77}
$$

Combing (10.76) and (10.77), we arrive at

$$
\begin{aligned}
&\mathbb{E}\int_{Q_0}(|\hat{\mathbf{z}}|^2+|\nabla\mathbf{z}|^2+|\mathbf{z}|^2)dxdt\\
&\leq \mathcal{C}\Big[e^{-\lambda R_0^2/6}\mathbb{E}\int_Q(|\hat{\mathbf{z}}|^2+|\nabla\mathbf{z}|^2+|\mathbf{z}|^2)dxdt+\lambda^2e^{-\lambda R_0^2/6}\mathbb{E}\int_Q|\mathbf{z}|^2dxdt\\
&\quad+e^{\lambda R_1^2}\mathbb{E}\int_{\Sigma_0}\Big|\frac{\partial\mathbf{z}}{\partial\nu}\Big|^2d\Sigma+\mathbb{E}e^{-\lambda R_0^2/2}\int_Q(\hat{\mathbf{z}}^2+|\nabla\mathbf{z}|^2+\mathbf{z}^2)dxdt\Big].
\end{aligned}
\tag{10.78}
$$

Applying the standard energy method to the equation (10.26) with $\tau=0$, $\mathbf{f}=0$ and $\hat{\mathbf{f}}=0$, we obtain that

$$
|(\mathbf{z}_0,\hat{\mathbf{z}}_0)|^2_{H_0^1(G)\times L^2(G)}\leq \mathcal{C}e^{\mathcal{C}r_4}\mathbb{E}\int_{Q_0}(|\hat{\mathbf{z}}|^2+|\nabla\mathbf{z}|^2+|\mathbf{z}|^2)dxdt \tag{10.79}
$$

and that

$$
|(\mathbf{z}_0,\hat{\mathbf{z}}_0)|^2_{H_0^1(G)\times L^2(G)}\geq \mathcal{C}e^{-\mathcal{C}r_4}\mathbb{E}\int_{Q}(|\hat{\mathbf{z}}|^2+|\nabla\mathbf{z}|^2+|\mathbf{z}|^2)dxdt. \tag{10.80}
$$

Now, by (10.78)–(10.80), we end up with

$$
\begin{aligned}
&|(\mathbf{z}_0,\hat{\mathbf{z}}_0)|^2_{H_0^1(G)\times L^2(G)}\\
&\leq \mathcal{C}e^{\mathcal{C}r_4}e^{-\lambda R_0^2/6}|(\mathbf{z}_0,\hat{\mathbf{z}}_0)|^2_{H_0^1(G)\times L^2(G)}+\mathcal{C}e^{\lambda R_1^2}\mathbb{E}\int_{\Sigma_0}\Big|\frac{\partial\mathbf{z}}{\partial\nu}\Big|^2d\Sigma.
\end{aligned}
\tag{10.81}
$$

Let us choose $\lambda_4\geq\lambda_3$ such that $\mathcal{C}e^{\mathcal{C}r_4}e^{-\lambda_4R_0^2/6}<1$. Then, from (10.81) with $\lambda\geq\lambda_4$, we obtain the desired inequality (10.42) immediately. This completes the proof of Theorem 10.19. □

10.7 Notes and Comments

There are numerous studies on the controllability and observability of deterministic hyperbolic equations (e.g., [7, 14, 63, 80, 88, 110, 123, 146, 153, 165, 199, 207, 295, 388, 395, 396] and the references cited therein).

Generally speaking, one can find the following five main methods to solve the exact controllability/observability problem of deterministic wave equations:

- The first one is the method of characteristic line ([199, 295]). This method only applies to wave equations in one space dimension.
- The second one is based on the Ingham type inequality ([7, 295]). This method works well for many partial differential equations involved in some special domains, i.e., intervals and rectangles. However, it seems that it is very hard to be applied to equations in general domains.
- The third one is the classical Rellich-type multiplier approach ([207]). Usually, it is used to treat several partial differential equations with time independent lower order terms.
- The forth one is the microlocal analysis approach ([14]). It is useful to solve controllability problems for many kinds of partial differential equations such as wave equations, Schrödinger equations and plate equations. Further, it can give sharp sufficient conditions for the exact controllability of wave equations. However, there exist lots of difficulties if one expects to adopt this approach to study stochastic control problems.
- The last one is the global Carleman estimate ([384]). This approach has the advantage of being more flexible and allowing to address variable coefficients. Further, it is robust with respect to the lower order terms and can be used to obtain explicit bounds on observability constants/control costs in terms of the involved coefficients.

So far, among the above mentioned five methods, only the one of global Carleman estimate was successfully extended to studying the controllability/observability problem of stochastic wave equations ([109, 224, 227, 238, 247, 386]). This chapter is based on our recent work [247].

There are many open problems related to the topic of this chapter. We shall list below some of them which, in our opinion, are particularly interesting:

1) Null and approximate controllability for stochastic wave equations. In Theorem 10.12, the exact controllability for (10.3) is presented. As an immediate consequence, we obtain the null and approximate controllability for the same system. Nevertheless, in order to show this later result, there seems no reason to introduce three controls. By Theorem 10.9, it is shown that only one control applied in the diffusion term is not enough. However, it seems that one boundary control is enough for the null and approximate controllability for both (10.2) and (10.3). Unfortunately, some essential difficulties appear when we try to prove such a result, following the method to prove Theorem 10.12. For example, for the null controllability, we should prove the following inequality for solutions to (10.4) with $\tau = T$:

$$|z(0)|^2_{H^1_0(G)} + |\hat{z}(0)|^2_{L^2(G)} \le \mathcal{C} \int_{\Sigma_0} \left|\frac{\partial z}{\partial \nu}\right|^2 d\Gamma dt,$$
$$\forall\, (z^T, \hat{z}^T) \in L^2_{\mathcal{F}_T}(\Omega; H^1_0(G) \times L^2(G)).$$

However, if we utilize the method developed in this chapter, we only obtain that, for all $(z^T, \hat{z}^T) \in L^2_{\mathcal{F}_T}(\Omega; H^1_0(G) \times L^2(G))$,

$$\begin{aligned} &|z(0)|^2_{H^1_0(G)} + |\hat{z}(0)|^2_{L^2(G)} \\ &\le \mathcal{C}\left(\int_{\Sigma_0} \left|\frac{\partial z}{\partial \nu}\right|^2 d\Gamma dt + \int_0^T |Z|^2_{H^1_0(G)} + \int_0^T |\widehat{Z}|^2_{L^2(G)} dt \right). \end{aligned}$$

There are two additional terms containing Z and $\widehat{Z}$ in the right hand side of the above inequality. These terms come from the fact that, in the Carleman estimate, we regard Z and $\widehat{Z}$ simply as nonhomogeneous terms rather than part of solutions. Therefore, we believe that one should introduce some new technique, for example, a Carleman estimate in which the fact that Z and $\widehat{Z}$ are part of solutions is essentially used, to get rid of the additional terms containing Z and $\widehat{Z}$. However, we do not know how to do it at this moment.

2) Exact controllability for stochastic wave equations with sharp conditions. In Theorem 10.12, we obtain the exact controllability of (10.3) for Γ_0 given by (10.22). It is well known that a sharp sufficient condition for exact controllability of deterministic wave equations with time-invariant coefficients is that the triple (G, Γ_0, T) satisfies the Geometric Control Condition introduced in [14]. It would be quite interesting and challenging to extend this result to the stochastic setting, and it seems that there are lots of things to be done before solving this problem. Indeed, the propagation of singularities for stochastic partial differential equations, at least, for stochastic hyperbolic equations, should be established. However, as far as we know, this topic is completely open.

3) Exact controllability for stochastic wave equations with more regular controls. In Theorem 10.12, we show the exact controllability of (10.3) by a triple of controls (f, g, h), where $g \in L^2_{\mathbb{F}}(0, T; H^{-1}(G))$, which is very irregular. It is very interesting to see whether (10.3) is exactly controllable with the control $g \in L^2_{\mathbb{F}}(0, T; L^2(G))$. By the standard duality argument, one can show that this is equivalent to the following observability estimate:

$$\begin{aligned} &|(z^T, \hat{z}^T)|_{L^2_{\mathcal{F}_T}(\Omega; H^1_0(G)\times L^2(G))} \\ &\le \mathcal{C}\left(\left|\frac{\partial z}{\partial \nu}\right|_{L^2_{\mathbb{F}}(0,T;L^2(\Gamma_0))} + |Z|_{L^2_{\mathbb{F}}(0,T;L^2(G))} + |\widehat{Z}|_{L^2_{\mathbb{F}}(0,T;L^2(G))} \right), \qquad (10.82) \\ &\qquad\qquad \forall\, (z^T, \hat{z}^T) \in L^2_{\mathcal{F}_T}(\Omega; H^1_0(G) \times L^2(G)), \end{aligned}$$

where $(z, \hat{z}, Z, \widehat{Z})$ is the solution to (10.4) with $\tau = T$. By Lemma 10.18, similarly to the proof of Theorem 10.19, one can prove that the inequality (10.82) holds if the term $|Z|_{L^2_{\mathbb{F}}(0,T;L^2(G))}$ is replaced by $|Z|_{L^2_{\mathbb{F}}(0,T;H^1_0(G))}$. However, at this moment we do not know whether (10.82) itself is true or not.

11

Exact Controllability for Stochastic Schrödinger Equations

This chapter is devoted to showing the exact controllability for stochastic Schrödinger equations by means of two controls, in which one is a boundary control and the other is an internal control acting everywhere in the diffusion term. Based on the duality argument, we solve this controllability problem by employing the global Carleman estimate to derive a suitable observability estimate for the dual equation.

11.1 Formulation of the Problem and the Main Result

In the previous three chapters, we studied the controllability/observability for stochastic transport, heat and wave equations. Another typical equation for which the controllability problems are extensively studied in the deterministic setting is the Schrödinger equation. It is notable that many properties of the Schrödinger equation are between that of the heat and the wave equations. In this chapter, we shall analyze the exact controllability for stochastic Schrödinger equations.

Let $T > 0$, $n \in \mathbb{N}$, $G \subset \mathbb{R}^n$ be a given bounded domain with a C^2 boundary Γ. Fix any $x_0 \in \mathbb{R}^n \setminus \overline{G}$ and put

$$\begin{cases} \Gamma_0 \overset{\Delta}{=} \big\{x \in \Gamma \ \big|\ (x - x_0) \cdot \nu(x) > 0\big\}, \\ Q \overset{\Delta}{=} (0,T) \times G, \quad \Sigma \overset{\Delta}{=} (0,T) \times \Gamma, \quad \Sigma_0 \overset{\Delta}{=} (0,T) \times \Gamma_0. \end{cases} \tag{11.1}$$

To begin with, let us recall the controllability problem for (deterministic) Schrödinger equation:

$$\begin{cases} iy_t + \Delta y = a_1 \cdot \nabla y + a_2 y & \text{in } Q, \\ y = \chi_{\Sigma_0} u & \text{on } \Sigma, \\ y(0) = y_0 & \text{in } G, \end{cases} \tag{11.2}$$

Q. Lü and X. Zhang, *Mathematical Control Theory for Stochastic Partial Differential Equations*, Probability Theory and Stochastic Modelling 101, https://doi.org/10.1007/978-3-030-82331-3_11

where the initial state $y_0 \in H^{-1}(G;\mathbb{C})$, the control $u \in L^2(\Sigma_0;\mathbb{C})$, and the coefficients a_k $(k=1,2)$ satisfy $a_1 \in L^\infty(0,T;W^{2,\infty}(G;\mathbb{R}^n) \cap W_0^{1,\infty}(G;\mathbb{R}^n))$ and $a_2 \in L^\infty(0,T;W^{1,\infty}(G;\mathbb{C}))$. One can show that, for any $y_0 \in H^{-1}(G;\mathbb{C})$ and $u \in L^2(\Sigma_0;\mathbb{C})$, (11.2) admits one and only one (transposition) solution $y(\cdot) \in C([0,T];H^{-1}(G;\mathbb{C}))$. The system (11.2) is called exactly controllable at time T if for any given $y_0, y_1 \in H^{-1}(G;\mathbb{C})$, one can find a control $u \in L^2(\Sigma_0;\mathbb{C})$ such that the corresponding solution $y(\cdot)$ to (11.2) satisfies $y(T) = y_1$.

Controllability/observability problems for deterministic Schrödinger equation are extensively studied (e.g., [4, 35, 18, 19, 41, 42, 187, 190, 252, 260, 280, 289, 317, 394]). It is well-known that under some assumptions on the controller Γ_0 (for example, Γ_0 is given in (11.1)) and the coefficients a_1 and a_2, the system (11.2) is exactly controllable at time T.

The main goal of this chapter is to study what happens when (11.2) is replaced by a stochastic model. Compared to the deterministic setting, there exist only very few results (in [224, 228, 229]) on the corresponding controllability problems for the stochastic Schrödinger equations.

As in the previous three chapters, let $(\Omega,\mathcal{F},\mathbf{F},\mathbb{P})$ (with $\mathbf{F} \stackrel{\Delta}{=} \{\mathcal{F}_t\}_{t\in[0,T]}$) be a filtered probability space on which a one dimensional Brownian motion $\{W(t)\}_{t\in[0,T]}$ is defined, and $\mathbf{F}$ be the natural filtration generated by $W(\cdot)$. Denote by $\mathbb{F}$ the progressive σ-field (in $[0,T]\times\Omega$) w.r.t. $\mathbf{F}$. Let us consider the following controlled stochastic Schrödinger equation:

$$\begin{cases} idy + \Delta y dt = \big(a_1\cdot\nabla y + a_2 y + a_4 v\big)dt + \big(a_3 y + v\big)dW(t) & \text{in } Q,\\ \chi_{\Sigma_0} y = u & \text{on } \Sigma,\\ y(0) = y_0 & \text{in } G. \end{cases} \tag{11.3}$$

Here, $y_0 \in H^{-1}(G;\mathbb{C})$, y is *the state variable*, both $u \in L^2_{\mathbb{F}}(0,T;L^2(\Gamma_0;\mathbb{C}))$ and $v \in L^2_{\mathbb{F}}(0,T;H^{-1}(G;\mathbb{C}))$ are *the control variables*, and the coefficients a_k $(k=1,2,3,4)$ satisfy $a_1 \in L^\infty_{\mathbb{F}}(0,T;W^{2,\infty}(G;\mathbb{R}^n)\cap W_0^{1,\infty}(G;\mathbb{R}^n))$, $a_2, a_3 \in L^\infty_{\mathbb{F}}(0,T;W^{1,\infty}(G;\mathbb{C}))$ and $a_4 \in L^\infty_{\mathbb{F}}(0,T;W_0^{1,\infty}(G;\mathbb{C}))$. It is easy to see that the control system (11.3) is a generalization of the system (5.33). As we shall see in Section 11.2, the system (11.3) admits one and only one transposition solution $y \in C_{\mathbb{F}}([0,T];L^2(\Omega;H^{-1}(G;\mathbb{C})))$.

Remark 11.1. Similarly to Remark 8.2, we introduce two controls into the system (11.3). Moreover, the control v acts on the whole domain, and also it affects the system through its drift term (in the form $a_4 v$). Compared with the deterministic Schrödinger equation, it seems that we choose too many controls. However, similarly to Proposition 7.6, one can show that two controls are really necessary for our exact controllability problem for (11.3) to be defined below.

We now introduce the following notion of exact controllability for (11.3).

Definition 11.2. *The system* (11.3) *is called exactly controllable at time* T *if for any* $y_0 \in H^{-1}(G;\mathbb{C})$ *and* $y_1 \in L^2_{\mathcal{F}_T}(\Omega; H^{-1}(G;\mathbb{C}))$, *one can find a pair of controls* $(u,v) \in L^2_{\mathbb{F}}(0,T;L^2(\Gamma_0;\mathbb{C})) \times L^2_{\mathbb{F}}(0,T;H^{-1}(G;\mathbb{C}))$ *such that the corresponding (transposition) solution* y *to* (11.3) *satisfies that* $y(T) = y_1$, *a.s.*

The main result in this chapter is stated as follows:

Theorem 11.3. *The system* (11.3) *is exactly controllable at (any given) time* $T > 0$.

One may follow the idea in Section 7.5 of Chapter 7 (or, more precisely, that in Chapter 10 for the exact controllability for stochastic wave equations) to prove Theorem 11.3 in the following two steps:

1. Reducing the exact controllability problem for the system (11.3) to another exact controllability problem for a suitable backward stochastic Schrödinger equation;
2. Solving the exact controllability problem for the above backward stochastic Schrödinger equation by establishing a suitable observability estimate for some forward stochastic Schrödinger equation.

Nevertheless, in this chapter we do not proceed in this way. Instead, in order to prove Theorem 11.3, we shall derive a key observability estimate for a suitable backward stochastic Schrödinger equation. This will be done in Theorem 11.14 by means of the global Carleman estimate.

11.2 Well-Posedness of the Control System

The control system (11.3) is a nonhomogeneous boundary value problem, and hence the notions of solutions introduced in Section 3.2 cannot be applied to it. Instead, as in the last chapter, we need to employ the notion of transposition solution presented in Section 7.2. For this, we first introduce the dual equation of (11.3). For any $\tau \in (0,T]$, let us consider the following backward stochastic Schrödinger equation:

$$\begin{cases} idz + \Delta z dt = \big(b_1 \cdot \nabla z + b_2 z + b_3 Z\big)dt + ZdW(t) & \text{in } (0,\tau) \times G, \\ z = 0 & \text{on } (0,\tau) \times \Gamma, \\ z(\tau) = z_\tau & \text{in } G, \end{cases} \tag{11.4}$$

where $z_\tau \in L^2_{\mathcal{F}_\tau}(\Omega; H^1_0(G;\mathbb{C}))$, the coefficients b_j $(j = 1,2,3)$ satisfy $b_1 \in L^\infty_{\mathbb{F}}(0,T;W^{1,\infty}_0(G;\mathbb{R}^n))$ and $b_2, b_3 \in L^\infty_{\mathbb{F}}(0,T;W^{1,\infty}(G;\mathbb{C}))$.

Remark 11.4. The regularity of the coefficients can be suitably relaxed. But we do not do this because we hope to present the key idea in a simple way.

Similarly to the proofs of Theorems 4.8 and 4.10, by means of the usual energy estimate and noting that the coefficient b_1 is $\mathbb{R}^n$-valued, it is easy to show the following well-posedness result for (11.4) (and hence we will not prove it here).

Lemma 11.5. *For any $z_\tau \in L^2_{\mathcal{F}_\tau}(\Omega; H^1_0(G;\mathbb{C}))$, the equation (11.4) admits a unique weak solution $(z, Z) \in L^2_{\mathbb{F}}(\Omega; C([0,\tau]; H^1_0(G;\mathbb{C}))) \times L^2_{\mathbb{F}}(0,\tau; H^1_0(G;\mathbb{C}))$. Moreover,*

$$|z|_{L^2_{\mathbb{F}}(\Omega;C([0,\tau];H^1_0(G;\mathbb{C})))} + |Z|_{L^2_{\mathbb{F}}(0,\tau;H^1_0(G;\mathbb{C}))} \le e^{Cr_1} |z_\tau|_{L^2_{\mathcal{F}_\tau}(\Omega;H^1_0(G;\mathbb{C}))}, \tag{11.5}$$

where

$$r_1 \stackrel{\Delta}{=} |b_1|^2_{L^\infty_{\mathbb{F}}(0,T;W^{1,\infty}_0(G;\mathbb{R}^n))} + \sum_{j=2}^{3} |b_j|^2_{L^\infty_{\mathbb{F}}(0,T;W^{1,\infty}(G;\mathbb{C}))} + 1.$$

Next, for the sake of completeness, we give an energy estimate for the equation (11.4).

Proposition 11.6. *For all solutions to the equation (11.4), it holds that*

$$\mathbb{E}|z(t)|^2_{H^1_0(G;\mathbb{C})} \le e^{C\sqrt{r_1}}\Big(\mathbb{E}|z(s)|^2_{H^1_0(G;\mathbb{C})} + |Z|^2_{L^2_{\mathbb{F}}(s,t;H^1_0(G;\mathbb{C}))}\Big), \tag{11.6}$$

for any $0 \le s \le t \le \tau$.

Proof : By Itô's formula, we have

$$\begin{aligned}
&\mathbb{E}|z(t)|^2_{L^2(G;\mathbb{C})} - \mathbb{E}|z(s)|^2_{L^2(G;\mathbb{C})}\\
&= \mathbb{E}\int_s^t \int_G \big(zd\bar z + \bar z dz + |Z|^2 d\sigma\big)dx\\
&= \mathbb{E}\int_s^t \int_G \Big\{2\mathrm{Re}\Big[i\bar z\big(\Delta z - b_1\cdot\nabla z - b_2 z - b_3 Z\big)\Big] + |Z|^2\Big\}dxd\sigma\\
&= \mathbb{E}\int_s^t \int_G \Big\{2\mathrm{Im}\Big[\mathrm{div}\,(\bar z\nabla z) - |\nabla z|^2 - \mathrm{div}\,(|z|^2 b_1) + (\mathrm{div}\, b_1)|z|^2 \\
&\qquad\qquad -b_2|z|^2 - b_3 i\bar z Z\Big] + |Z|^2\Big\}dxd\sigma \qquad\qquad (11.7)\\
&\le \mathbb{E}\int_s^t 2\Big[\big(|b_1|_{W^{1,\infty}(G;\mathbb{R}^n)} + |b_2|_{L^\infty(G;\mathbb{C})} + |b_3|_{L^\infty(G;\mathbb{C})} + 1\big)|z|^2_{L^2(G;\mathbb{C})}\\
&\qquad\qquad +|Z|^2_{L^2(G;\mathbb{C})}\Big]d\sigma
\end{aligned}$$

and

$$\mathbb{E}|\nabla z(t)|^2_{L^2(G;\mathbb{C})} - \mathbb{E}|\nabla z(s)|^2_{L^2(G;\mathbb{C})}$$

$$
\begin{aligned}
&= \mathbb{E}\int_s^t \int_G \Big(\nabla z\cdot d\nabla \bar z + \nabla \bar z\cdot d\nabla z + |\nabla Z|^2 d\sigma\Big)dx \qquad (11.8)\\
&= \mathbb{E}\int_s^t \int_G \Big[\mathrm{div}\,(\nabla z d\bar z) - \Delta z d\bar z + \mathrm{div}\,(\nabla \bar z dz) - \Delta \bar z dz + |\nabla Z|^2 d\sigma\Big]dx\\
&= \mathbb{E}\int_s^t \int_G \Big\{2\mathrm{Re}\Big[i\Delta z\big(\Delta \bar z - b_1\cdot\nabla\bar z - \overline{b_2 z} - \overline{b_3 Z}\big)\Big] + |\nabla Z|^2\Big\}dxd\sigma\\
&\leq 2\mathbb{E}\int_t^s \Big[\big(|b_1|_{W^{1,\infty}(G;\mathbb{R}^n)} + |b_3|_{W^{1,\infty}(G;\mathbb{C})} + 1\big)|\nabla z|^2_{L^2(G;\mathbb{C})}\\
&\qquad + \big(|b_2|_{W^{1,\infty}(G;\mathbb{C})} + |b_3|_{W^{1,\infty}(G;\mathbb{C})} + 1\big)|z|^2_{L^2(G;\mathbb{C})} + |Z|^2_{H_0^1(G;\mathbb{C})}\Big]d\sigma.
\end{aligned}
$$

It follows from (11.7) and (11.8) that

$$
\begin{aligned}
&\mathbb{E}|z(t)|^2_{H_0^1(G;\mathbb{C})} - \mathbb{E}|z(s)|^2_{H_0^1(G;\mathbb{C})}\\
&\leq 4\sqrt{r_1}\mathbb{E}\int_s^t |z(\sigma)|^2_{H_0^1(G;\mathbb{C})}d\sigma + 4\mathbb{E}\int_s^t |Z(\sigma)|^2_{H_0^1(G;\mathbb{C})}d\sigma.
\end{aligned}
$$

This, together with Gronwall's inequality, implies that

$$
\mathbb{E}|z(t)|^2_{H_0^1(G;\mathbb{C})} \leq e^{\mathcal{C}\sqrt{r_1}}\Big(\mathbb{E}|z(s)|^2_{H_0^1(G;\mathbb{C})} + \mathbb{E}\int_s^t |Z(\sigma)|^2_{H_0^1(G;\mathbb{C})}d\sigma\Big),
$$

which gives the desired inequality (11.6) immediately. The completes the proof of Proposition 11.6. □

Remark 11.7. The proof of Proposition 11.6 is almost standard. Indeed, if we regard z (in (11.4)) as a solution to a forward stochastic Schrödinger equation with a nonhomogeneous term Z, then the estimate as (11.6) follows from a standard energy estimate. Nevertheless, we still give a proof here for exhibiting how the constant in (11.6) depends on r_1.

Further, similarly to Proposition 8.6 in Chapter 8 and Proposition 10.6 in Chapter 10, we need the following hidden regularity result for solutions to (11.4) (Note that this sort of regularity result does NOT follow from Trace Theorem for Sobolev spaces).

Proposition 11.8. *For any $\tau \in (0,T]$ and $z_\tau \in L^2_{\mathcal{F}_\tau}(\Omega; H_0^1(G;\mathbb{C}))$, the unique weak solution $(z,Z) \in L^2_{\mathbb{F}}(\Omega; C([0,\tau]; H_0^1(G;\mathbb{C}))) \times L^2_{\mathbb{F}}(0,\tau; H_0^1(G;\mathbb{C}))$ to (11.4) satisfies $\displaystyle\frac{\partial z}{\partial\nu}\Big|_\Gamma \in L^2_{\mathbb{F}}(0,\tau; L^2(\Gamma;\mathbb{C}))$. Furthermore,*

$$
\Big|\frac{\partial z}{\partial \nu}\Big|_{L^2_{\mathbb{F}}(0,\tau;L^2(\Gamma;\mathbb{C}))} \leq e^{\mathcal{C}r_1}|z_\tau|_{L^2_{\mathcal{F}_\tau}(\Omega;H_0^1(G;\mathbb{C}))}, \qquad (11.9)
$$

where the constant $\mathcal{C}$ is independent of τ.

Proof: Since Γ is C^2, by Proposition 10.5, one can find a vector field $h = (h^1, \cdots, h^n) \in C^1(\overline{G}; \mathbb{R}^n)$ such that $h = \nu$ on Γ.

By a direct computation, we obtain that

$$\begin{aligned}
&\sum_{k=1}^{n}\sum_{j=1}^{n} h^k \bar{z}_{x_k} z_{x_j x_j} + \sum_{k=1}^{n}\sum_{j=1}^{n} h^k z_{x_k} \bar{z}_{x_j x_j} \\
&= \sum_{k=1}^{n}\sum_{j=1}^{n} \big[(h^k \bar{z}_{x_k} z_{x_j})_{x_j} + (h^k z_{x_k} \bar{z}_{x_j})_{x_j} + h^k_{x_k} |z_{x_j}|^2 \\
&\qquad -(h^k |z_{x_j}|^2)_{x_k} - 2h^k_{x_j} \bar{z}_{x_k} z_{x_j}\big]
\end{aligned} \tag{11.10}$$

and

$$\begin{aligned}
&i\sum_{k=1}^{n}(h^k \bar{z}_{x_k} dz - h^k z_{x_k} d\bar{z}) \\
&= i\sum_{k=1}^{n} \big[d(h^k \bar{z}_{x_k} z) - h^k z d\bar{z}_{x_k} - h^k d\bar{z}_{x_k} dz - (h^k z d\bar{z})_{x_k} \\
&\qquad + h^k z d\bar{z}_{x_k} + h^k_{x_k} z d\bar{z}\big] \\
&= i\sum_{k=1}^{n} \big[d(h^k \bar{z}_{x_k} z) - h^k d\bar{z}_{x_k} dz - (h^k z d\bar{z})_{x_k} + h^k_{x_k} z d\bar{z}\big].
\end{aligned} \tag{11.11}$$

Combining (11.10) and (11.11), we have

$$\begin{aligned}
&h\cdot\nabla\bar{z}(idz + \Delta z dt) + h\cdot\nabla z(-id\bar{z} + \Delta\bar{z} dt) \\
&= \nabla\cdot\Big[(h\cdot\nabla\bar{z})\nabla z + (h\cdot\nabla z)\nabla\bar{z} - i(zd\bar{z})h - |\nabla z|^2 h\Big]dt \\
&\quad + d(ih\cdot\nabla\bar{z}z) - 2\sum_{j,k=1}^{n} h^k_{x_k} z_j \bar{z}_k dt + (\nabla\cdot h)|\nabla z|^2 dt \\
&\quad + i(\nabla\cdot h) z d\bar{z} - i(h\cdot\nabla d\bar{z})dz.
\end{aligned} \tag{11.12}$$

Finally, integrating (11.12) in $(0,\tau)\times G$ and then taking the expectation in the both sides, noting that

$$\begin{aligned}
&\mathbb{E}\int_0^\tau\int_G \nabla\cdot\Big[(h\cdot\nabla\bar{z})\nabla z + (h\cdot\nabla z)\nabla\bar{z} - i(zd\bar{z})h - |\nabla z|^2 h\Big]dxdt \\
&= \mathbb{E}\int_0^\tau\int_\Gamma \Big[\frac{\partial\bar{z}}{\partial\nu}\frac{\partial z}{\partial\nu} + \frac{\partial\bar{z}}{\partial\nu}\frac{\partial z}{\partial\nu} - \Big|\frac{\partial z}{\partial\nu}\Big|^2\Big]d\Gamma dt = \mathbb{E}\int_0^\tau\int_\Gamma \Big|\frac{\partial z}{\partial\nu}\Big|^2 d\Gamma dt,
\end{aligned}$$

and using Lemma 11.5, we obtain (11.9). □

With the aid of Lemma 11.5 and Proposition 11.8, proceeding as that in Definition 7.13, we may define transposition solutions to (11.3) as follows:

Definition 11.9. *A process* $y \in C_{\mathbb{F}}([0,T]; L^2(\Omega; H^{-1}(G;\mathbb{C})))$ *is called a transposition solution to the system* (11.3) *if for every* $\tau \in [0,T]$ *and* $z_\tau \in L^2_{\mathcal{F}_\tau}(\Omega; H_0^1(G;\mathbb{C}))$, *it holds that*

$$
\begin{aligned}
&\mathbb{E}\langle y(\tau,\cdot), z_\tau(\cdot)\rangle_{H^{-1}(G;\mathbb{C}),H_0^1(G;\mathbb{C})} - \langle y_0(\cdot), z(0,\cdot)\rangle_{H^{-1}(G;\mathbb{C}),H_0^1(G;\mathbb{C})}\\
&= \mathbb{E}\int_0^\tau \int_{\Gamma_0} u\frac{\partial \overline{z}}{\partial \nu} dxds + \mathbb{E}\int_0^\tau \langle v, a_4 z + Z\rangle_{H^{-1}(G;\mathbb{C}),H_0^1(G;\mathbb{C})} ds.
\end{aligned}
\tag{11.13}
$$

Here (z,Z) *solves* (11.4) *with*

$$
b_1 = a_1, \quad b_2 = \operatorname{div} a_1 - a_2, \quad b_3 = -a_3.
$$

Now, similarly to the proof of Theorem 7.12, one can show the following well-posedness result for (11.3) (hence we omit the proof).

Proposition 11.10. *For each* $y_0 \in H^{-1}(G;\mathbb{C})$, *the system* (11.3) *admits a unique transposition solution* $y \in C_{\mathbb{F}}([0,T]; L^2(\Omega; H^{-1}(G;\mathbb{C})))$. *Moreover,*

$$
\begin{aligned}
&|y|_{C_{\mathbb{F}}([0,T];L^2(\Omega;H^{-1}(G;\mathbb{C})))}\\
&\leq e^{\mathcal{C}r_2}\big(|y_0|_{H^{-1}(G;\mathbb{C})} + |u|_{L^2_{\mathbb{F}}(0,T;L^2(\Gamma_0;\mathbb{C}))} + |v|_{L^2_{\mathbb{F}}(0,T;H^{-1}(G;\mathbb{C}))}\big).
\end{aligned}
\tag{11.14}
$$

Here

$$
r_2 \stackrel{\Delta}{=} |a_1|^2_{L^\infty_{\mathbb{F}}(0,T;W^{2,\infty}(G;\mathbb{R}^n))} + \sum_{k=2}^{4} |a_k|^2_{L^\infty_{\mathbb{F}}(0,T;W^{1,\infty}(G;\mathbb{C}))} + 1.
$$

Proceeding exactly as the proof of Theorem 7.24, one can show that the exact controllability of (11.3) can be reduced to an observability estimate for its dual equation, that is, the equation (11.4) with $\tau = T$ in the present case. The main task in the rest of this chapter is to derive such an estimate (See the inequality (11.56) in Theorem 11.14), via which the desired exact controllability result (for (11.3)) in Theorem 11.3 follows.

11.3 A Fundamental Identity for Stochastic Schrödinger-Like Operators

In this section, we shall establish a fundamental identity for a stochastic schrödinger-like operator, which is in spirit similarly to the identities (1.48), (8.19), (9.26) and (10.18). Besides the key role in proving observability estimates for backward stochastic Schrödinger equations, it can also be applied in the study of observability and unique continuation problems for forward stochastic Schrödinger equations.

Let $\beta(t,x) \in C^2(\mathbb{R}^{1+n})$. For $j,k = 1,2,\cdots,n$, let $b^{jk}(t,x) \in C^{1,2}(\mathbb{R}^{1+n})$ satisfy that

$$b^{jk} = b^{kj}. \tag{11.15}$$

Define a (formal) second order stochastic schrödinger-like operator $\mathcal{P}$ as

$$\mathcal{P}z \stackrel{\Delta}{=} i\beta(t,x)dz + \sum_{j,k=1}^{n} (b^{jk}(t,x)z_{x_j})_{x_k} dt. \tag{11.16}$$

We have the following result:

Theorem 11.11. *Let ℓ, $\Psi \in C^2(\mathbb{R}^{1+n})$. Assume that $\mathbf{u}$ is an $H^2(\mathbb{R}^n;\mathbb{C})$-valued Itô process. Put $\theta = e^{\ell}$ and $\mathbf{v} = \theta\mathbf{u}$. Then for a.e. $x \in \mathbb{R}^n$, it holds that*

$$\begin{aligned}
&\theta\big(\mathcal{P}\mathbf{u}\overline{I_1} + \overline{\mathcal{P}\mathbf{u}}I_1\big) + dM + \operatorname{div} V\\
&= 2|I_1|^2 dt + \sum_{j,k=1}^{n} c^{jk}\big(\mathbf{v}_{x_j}\bar{\mathbf{v}}_{x_k} + \bar{\mathbf{v}}_{x_j}\mathbf{v}_{x_k}\big)dt + \mathcal{D}|\mathbf{v}|^2 dt\\
&\quad + i\sum_{j,k=1}^{n}\Big[\big(\beta b^{jk}\ell_{x_j}\big)_t + b^{jk}\big(\beta\ell_t\big)_{x_j}\Big]\big(\bar{\mathbf{v}}_{x_k}\mathbf{v} - \mathbf{v}_{x_k}\bar{\mathbf{v}}\big)dt\\
&\quad + i\Big[\beta\Psi + \sum_{j,k=1}^{n}\big(\beta b^{jk}\ell_{x_j}\big)_{x_k}\Big]\big(\bar{\mathbf{v}}d\mathbf{v} - \mathbf{v}d\bar{\mathbf{v}}\big)\\
&\quad + \beta^2\ell_t\big|d\mathbf{v}\big|^2 + i\sum_{j,k=1}^{n}\beta b^{jk}\ell_{x_j}\big(d\mathbf{v}d\bar{\mathbf{v}}_{x_k} - d\mathbf{v}_{x_k}d\bar{\mathbf{v}}\big), \qquad a.s.
\end{aligned} \tag{11.17}$$

where

$$\left\{\begin{aligned}
&I_1 \stackrel{\Delta}{=} -i\beta\ell_t\mathbf{v} - 2\sum_{j,k=1}^{n} b^{jk}\ell_{x_j}\mathbf{v}_{x_k} + \Psi\mathbf{v},\\
&\mathcal{A} \stackrel{\Delta}{=} \sum_{j,k=1}^{n} b^{jk}\ell_{x_j}\ell_{x_k} - \sum_{j,k=1}^{n}(b^{jk}\ell_{x_j})_{x_k} - \Psi,\\
&M \stackrel{\Delta}{=} \beta^2\ell_t|\mathbf{v}|^2 + i\beta\sum_{j,k=1}^{n} b^{jk}\ell_{x_j}(\bar{\mathbf{v}}_{x_k}\mathbf{v} - \mathbf{v}_{x_k}\bar{\mathbf{v}}),
\end{aligned}\right. \tag{11.18}$$

$$\left\{\begin{aligned}
&V \stackrel{\Delta}{=} (V^1, \cdots, V^k, \cdots, V^n),\\
&V^k \stackrel{\Delta}{=} -i\beta\sum_{j=1}^{n}\Big[b^{jk}\ell_{x_j}(\mathbf{v}d\bar{\mathbf{v}} - \bar{\mathbf{v}}d\mathbf{v}) + b^{jk}\ell_t(\mathbf{v}_{x_j}\bar{\mathbf{v}} - \bar{\mathbf{v}}_{x_j}\mathbf{v})dt\Big]\\
&\qquad -\Psi\sum_{j=1}^{n} b^{jk}(\mathbf{v}_{x_j}\bar{\mathbf{v}} + \bar{\mathbf{v}}_{x_j}\mathbf{v})dt + \sum_{j=1}^{n} b^{jk}(2\mathcal{A}\ell_{x_j} + \Psi_{x_j})|\mathbf{v}|^2 dt\\
&\qquad + \sum_{j,j',k'=1}^{n}(2b^{jk'}b^{j'k} - b^{jk}b^{j'k'})\ell_{x_j}(\mathbf{v}_{x_{j'}}\bar{\mathbf{v}}_{x_{k'}} + \bar{\mathbf{v}}_{x_{j'}}\mathbf{v}_{x_{k'}})dt,
\end{aligned}\right. \tag{11.19}$$

and

$$\begin{cases} c^{jk} \stackrel{\Delta}{=} \sum_{j',k'=1}^{n} \Big[2(b^{j'k}\ell_{x_{j'}})_{x_{k'}} b^{jk'} - (b^{jk}b^{j'k'}\ell_{x_{j'}})_{x_{k'}} \Big] - b^{jk}\Psi, \\ \mathcal{D} \stackrel{\Delta}{=} (\beta^2 \ell_t)_t + \sum_{j,k=1}^{n} (b^{jk}\Psi_{x_j})_{x_k} + 2\Big[\sum_{j,k=1}^{n} (b^{jk}\ell_{x_j}\mathcal{A})_{x_k} + \mathcal{A}\Psi \Big]. \end{cases} \tag{11.20}$$

Proof: The proof is divided into three steps.

Step 1. By $\theta = e^{\ell}$ and $\mathbf{v} = \theta \mathbf{u}$, a straightforward computation shows that:

$$\begin{aligned} \theta \mathcal{P}\mathbf{u} &= i\beta d\mathbf{v} - i\beta \ell_t \mathbf{v} dt + \sum_{j,k=1}^{n} (b^{jk}\mathbf{v}_{x_j})_{x_k} dt + \sum_{j,k=1}^{n} b^{jk}\ell_{x_j}\ell_{x_k}\mathbf{v} dt \\ &\quad -2\sum_{j,k=1}^{n} b^{jk}\ell_{x_j}\mathbf{v}_{x_k} dt - \sum_{j,k=1}^{n} (b^{jk}\ell_{x_j})_{x_k}\mathbf{v} dt \\ &= I_1 dt + I_2, \end{aligned} \tag{11.21}$$

where I_1 is given in (11.18),

$$I_2 = i\beta d\mathbf{v} + \sum_{j,k=1}^{n} (b^{jk}\mathbf{v}_{x_j})_{x_k} dt + \mathcal{A}\mathbf{v} dt. \tag{11.22}$$

Hence,

$$\theta(P\mathbf{u}\overline{I_1} + \overline{P\mathbf{u}} I_1) = 2|I_1|^2 dt + (I_1\overline{I_2} + I_2\overline{I_1}). \tag{11.23}$$

Step 2. In this step, we compute $I_1\overline{I_2} + I_2\overline{I_1}$. Denote the three terms in I_1 and I_2 by I_1^j and I_2^j, $j = 1, 2, 3$, respectively. Then,

$$\begin{aligned} & I_1^1\overline{I_2^1} + I_2^1\overline{I_1^1} \\ &= -i\beta\ell_t \mathbf{v}\overline{(i\beta d\mathbf{v})} + i\beta d\mathbf{v}\overline{(-i\beta\ell_t\mathbf{v})} \\ &= -d(\beta^2\ell_t|\mathbf{v}|^2) + (\beta^2\ell_t)_t|\mathbf{v}|^2 dt + \beta^2\ell_t|d\mathbf{v}|^2. \end{aligned} \tag{11.24}$$

Noting that

$$\begin{cases} 2\mathbf{v}d\bar{\mathbf{v}} = d(|\mathbf{v}|^2) - (\bar{\mathbf{v}}d\mathbf{v} - \mathbf{v}d\bar{\mathbf{v}}) - |d\mathbf{v}|^2, \\ 2\mathbf{v}\bar{\mathbf{v}}_{x_k} = (|\mathbf{v}|^2)_{x_k} - (\bar{\mathbf{v}}\mathbf{v}_{x_k} - \mathbf{v}\bar{\mathbf{v}}_{x_k}), \end{cases} \tag{11.25}$$

we find first

$$
\begin{aligned}
&2i\sum_{j,k=1}^{n}(\beta b^{jk}\ell_{x_j}\mathbf{v}d\bar{\mathbf{v}})_{x_k}\\
&=i\sum_{j,k=1}^{n}\Big\{\beta b^{jk}\ell_{x_j}\big[d(|\mathbf{v}|^2)-(\bar{\mathbf{v}}d\mathbf{v}-\mathbf{v}d\bar{\mathbf{v}})-|d\mathbf{v}|^2\big]\Big\}_{x_k}\\
&=i\sum_{j,k=1}^{n}\Big\{\big(\beta b^{jk}\ell_{x_j}\big)_{x_k}d(|\mathbf{v}|^2)+\beta b^{jk}\ell_{x_j}\big[d(|\mathbf{v}|^2)\big]_{x_k}-\big[\beta b^{jk}\ell_{x_j}(\bar{\mathbf{v}}d\mathbf{v}-\mathbf{v}d\bar{\mathbf{v}})\big]_{x_k}\\
&\quad-\big(\beta b^{jk}\ell_{x_j}\big)_{x_k}|d\mathbf{v}|^2-\beta b^{jk}\ell_{x_j}d\mathbf{v}_{x_k}d\bar{\mathbf{v}}-\beta b^{jk}\ell_{x_j}d\bar{\mathbf{v}}d\mathbf{v}_{x_k}\Big\},
\end{aligned}
\tag{11.26}
$$

next

$$
\begin{aligned}
&-2i\sum_{j,k=1}^{n}\big(\beta b^{jk}\ell_{x_j}\big)_{x_k}\mathbf{v}d\bar{\mathbf{v}}\\
&=-i\sum_{j,k=1}^{n}\big(\beta b^{jk}\ell_{x_j}\big)_{x_k}\big[d(|\mathbf{v}|^2)-(\bar{\mathbf{v}}d\mathbf{v}-\mathbf{v}d\bar{\mathbf{v}})-|d\mathbf{v}|^2\big]\\
&=-i\sum_{j,k=1}^{n}\big[\big(\beta b^{jk}\ell_{x_j}\big)_{x_k}d(|\mathbf{v}|^2)-\big(\beta b^{jk}\ell_{x_j}\big)_{x_k}(\bar{\mathbf{v}}d\mathbf{v}-\mathbf{v}d\bar{\mathbf{v}})\\
&\quad-\big(\beta b^{jk}\ell_{x_j}\big)_{x_k}|d\mathbf{v}|^2\big],
\end{aligned}
\tag{11.27}
$$

then

$$
\begin{aligned}
&-2i\sum_{j,k=1}^{n}d\big(\beta b^{jk}\ell_{x_j}\mathbf{v}\bar{\mathbf{v}}_{x_k}\big)\\
&=-i\sum_{j,k=1}^{n}d\Big\{\beta b^{jk}\ell_{x_j}\big[(|\mathbf{v}|^2)_{x_k}-(\bar{\mathbf{v}}\mathbf{v}_{x_k}-\mathbf{v}\bar{\mathbf{v}}_{x_k})\big]\Big\}\\
&=-i\sum_{j,k=1}^{n}\Big\{\big(\beta b^{jk}\ell_{x_j}\big)_t(|\mathbf{v}|^2)_{x_k}dt+\beta b^{jk}\ell_{x_j}d\big[(|\mathbf{v}|^2)_{x_k}\big]\\
&\qquad\qquad-d\big[\beta b^{jk}\ell_{x_j}(\bar{\mathbf{v}}\mathbf{v}_{x_k}-\mathbf{v}\bar{\mathbf{v}}_{x_k})\big]\Big\},
\end{aligned}
\tag{11.28}
$$

and

$$
\begin{aligned}
&2i\sum_{j,k=1}^{n}\big(\beta b^{jk}\ell_{x_j}\big)_t\mathbf{v}\bar{\mathbf{v}}_{x_k}dt=i\sum_{j,k=1}^{n}d\big(\beta b^{jk}\ell_{x_j}\big)_t\big[(|\mathbf{v}|^2)_{x_k}-(\bar{\mathbf{v}}\mathbf{v}_{x_k}-\mathbf{v}\bar{\mathbf{v}}_{x_k})\big]dt\\
&=i\Big[\sum_{j,k=1}^{n}\big(\beta b^{jk}\ell_{x_j}\big)_t(|\mathbf{v}|^2)_{x_k}dt-\big(\beta b^{jk}\ell_{x_j}\big)_t(\bar{\mathbf{v}}\mathbf{v}_{x_k}-\mathbf{v}\bar{\mathbf{v}}_{x_k})dt\Big].
\end{aligned}
\tag{11.29}
$$

It follows from (11.26)–(11.29) that

$$\begin{aligned}
&(I_1^2+I_1^3)\overline{I_2^1}+I_2^1(\overline{I_1^2}+\overline{I_1^3})\\
&=\Big(-2\sum_{j,k=1}^n b^{jk}\ell_{x_j}\mathbf{v}_{x_k}+\Psi\mathbf{v}\Big)\overline{(i\beta d\mathbf{v})}+i\beta d\mathbf{v}\overline{\Big(-2\sum_{j,k=1}^n b^{jk}\ell_{x_j}\mathbf{v}_{x_k}+\Psi\mathbf{v}\Big)}\\
&=2i\sum_{j,k=1}^n \beta b^{jk}\ell_{x_j}(\mathbf{v}_{x_k}d\bar{\mathbf{v}}-\bar{\mathbf{v}}_{x_k}d\mathbf{v})+i\beta\Psi(\bar{\mathbf{v}}d\mathbf{v}-\mathbf{v}d\bar{\mathbf{v}})\\
&=2i\sum_{j,k=1}^n\Big[\big(\beta b^{jk}\ell_{x_j}\mathbf{v}d\bar{\mathbf{v}}\big)_{x_k}-\big(\beta b^{jk}\ell_{x_j}\big)_{x_k}\mathbf{v}d\bar{\mathbf{v}}-\beta b^{jk}\ell_{x_j}\mathbf{v}d\bar{\mathbf{v}}_{x_k}\Big]\\
&\quad-2i\sum_{j,k=1}^n\Big[d\big(\beta b^{jk}\ell_{x_j}\mathbf{v}\bar{\mathbf{v}}_{x_k}\big)-\big(\beta b^{jk}\ell_{x_j}\big)_t\mathbf{v}\bar{\mathbf{v}}_{x_k}dt-\beta b^{jk}\ell_{x_k}\mathbf{v}d\bar{\mathbf{v}}_{x_k}\Big]\\
&\quad+2i\sum_{j,k=1}^n\beta b^{jk}\ell_{x_j}d\mathbf{v}d\bar{\mathbf{v}}_{x_k}+i\beta\Psi(\bar{\mathbf{v}}d\mathbf{v}-\mathbf{v}d\bar{\mathbf{v}})\\
&=-i\sum_{j,k=1}^n\Big[\beta b^{jk}\ell_{x_j}(\bar{\mathbf{v}}d\mathbf{v}-\mathbf{v}d\bar{\mathbf{v}})\Big]_{x_k}dt-i\sum_{j,k=1}^n d\Big[\beta b^{jk}\ell_{x_j}(\mathbf{v}\bar{\mathbf{v}}_{x_k}-\bar{\mathbf{v}}\mathbf{v}_{x_k})\Big]\\
&\quad-i\sum_{j,k=1}^n(\beta b^{jk}\ell_{x_j})_t(\bar{\mathbf{v}}\mathbf{v}_{x_k}-\mathbf{v}\bar{\mathbf{v}}_{x_k})dt+i\Big[\beta\Psi+\sum_{j,k=1}^n(\beta b^{jk}\ell_{x_j})_{x_k}\Big](\bar{\mathbf{v}}d\mathbf{v}-\mathbf{v}d\bar{\mathbf{v}})\\
&\quad+i\sum_{j,k=1}^n\beta b^{jk}\ell_{x_j}\big(d\mathbf{v}d\bar{\mathbf{v}}_{x_k}-d\mathbf{v}_{x_k}d\bar{\mathbf{v}}\big).
\end{aligned}\tag{11.30}$$

Noting (11.15), we have that

$$\begin{aligned}
&I_1^1\overline{I_2^2}+I_2^2\overline{I_1^1}\\
&=-i\beta\ell_t\mathbf{v}\overline{\sum_{j,k=1}^n(b^{jk}\mathbf{v}_{x_j})_{x_k}dt}+\sum_{j,k=1}^n(b^{jk}\mathbf{v}_{x_j})_{x_k}\overline{(-i\beta\ell_t\mathbf{v})}\\
&=\sum_{j,k=1}^n\Big[i\beta b^{jk}\ell_t\big(\mathbf{v}_{x_j}\bar{\mathbf{v}}-\bar{\mathbf{v}}_{x_j}\mathbf{v}\big)\Big]_{x_k}dt+i\sum_{j,k=1}^n b^{jk}(\beta\ell_t)_{x_j}\big(\bar{\mathbf{v}}_{x_k}\mathbf{v}-\mathbf{v}_{x_k}\bar{\mathbf{v}}\big)dt.
\end{aligned}\tag{11.31}$$

Utilizing (11.15) again, we find that

$$\begin{aligned}
&\sum_{j,k,j',k'=1}^n b^{jk}b^{j'k'}\ell_{x_j}\big(\mathbf{v}_{x_{j'}}\bar{\mathbf{v}}_{x_kx_{k'}}+\bar{\mathbf{v}}_{x_{j'}}\mathbf{v}_{x_kx_{k'}}\big)\\
&=\sum_{j,k,j',k'=1}^n b^{jk}b^{j'k'}\ell_{x_j}\big(\mathbf{v}_{x_{j'}x_k}\bar{\mathbf{v}}_{x_{k'}}+\bar{\mathbf{v}}_{x_{j'}x_k}\mathbf{v}_{x_{k'}}\big).
\end{aligned}$$

Hence,

$$
\begin{aligned}
&2\sum_{j,k,j',k'=1}^{n} b^{jk}b^{j'k'}\ell_{x_j}\big(\mathbf{v}_{x_{j'}}\bar{\mathbf{v}}_{x_kx'_k}+\bar{\mathbf{v}}_{x_{j'}}\mathbf{v}_{x_kx'_k}\big)dt\\
&=\sum_{j,k,j',k'=1}^{n} b^{jk}b^{j'k'}\ell_{x_j}\big(\mathbf{v}_{x_{j'}}\bar{\mathbf{v}}_{x_kx'_k}+\bar{\mathbf{v}}_{x_{j'}}\mathbf{v}_{x_kx'_k}\big)dt\\
&\quad+\sum_{j,k,j',k'=1}^{n} b^{jk}b^{j'k'}\ell_{x_j}\big(\mathbf{v}_{x_{j'}x_k}\bar{\mathbf{v}}_{x_{k'}}+\bar{\mathbf{v}}_{x_{j'}x_k}\mathbf{v}_{x_{k'}}\big)dt\\
&=\sum_{j,k,j',k'=1}^{n} b^{jk}b^{j'k'}\ell_{x_j}\big(\mathbf{v}_{x_{j'}}\bar{\mathbf{v}}_{x_{k'}}+\bar{\mathbf{v}}_{x_{j'}}\mathbf{v}_{x_{k'}}\big)_{x_k}dt\\
&=\sum_{j,k,j',k'=1}^{n} \Big[b^{jk}b^{j'k'}\ell_{x_j}\big(\mathbf{v}_{x_{j'}}\bar{\mathbf{v}}_{x_{k'}}+\bar{\mathbf{v}}_{x_{j'}}\mathbf{v}_{x_{k'}}\big)\Big]_{x_k}dt\\
&\quad-\sum_{j,k,j',k'=1}^{n} \big(b^{jk}b^{j'k'}\ell_{x_j}\big)_{x_k}\big(\mathbf{v}_{x_{j'}}\bar{\mathbf{v}}_{x_{k'}}+\bar{\mathbf{v}}_{x_{j'}}\mathbf{v}_{x_{k'}}\big)dt.
\end{aligned}
\tag{11.32}
$$

By the equality (11.32), we get that

$$
\begin{aligned}
&I_1^2\overline{I_2^2}+I_2^2\overline{I_1^2}\\
&=-2\sum_{j,k=1}^{n} b^{jk}\ell_{x_j}\mathbf{v}_{x_k}\overline{\sum_{j,k=1}^{n}\big(b^{jk}\mathbf{v}_{x_j}\big)_{x_k}}dt-2\sum_{j,k=1}^{n}\big(b^{jk}\mathbf{v}_{x_j}\big)_{x_k}\overline{\sum_{j,k=1}^{n} b^{jk}\ell_{x_j}\mathbf{v}_{x_k}}dt\\
&=-2\sum_{j,k,j',k'=1}^{n}\Big[b^{jk}b^{j'k'}\ell_{x_j}\big(\mathbf{v}_{x_{j'}}\bar{\mathbf{v}}_{x_k}+\bar{\mathbf{v}}_{x_{j'}}\mathbf{v}_{x_k}\big)\Big]_{x_{k'}}dt\\
&\quad+2\sum_{j,k,j',k'=1}^{n} b^{j'k'}\big(b^{jk}\ell_{x_j}\big)_{x_{k'}}\big(\mathbf{v}_{x_{j'}}\bar{\mathbf{v}}_{x_k}+\bar{\mathbf{v}}_{x_{j'}}\mathbf{v}_{x_k}\big)dt\\
&\quad+2\sum_{j,k,j',k'=1}^{n} b^{jk}b^{j'k'}\ell_{x_j}\big(\mathbf{v}_{x_{j'}}\bar{\mathbf{v}}_{x_kx_{k'}}+\bar{\mathbf{v}}_{x_{j'}}\mathbf{v}_{x_kx_{k'}}\big)dt\\
&=-2\sum_{j,k,j',k'=1}^{n}\Big[b^{jk}b^{j'k'}\ell_{x_j}\big(\mathbf{v}_{x_{j'}}\bar{\mathbf{v}}_{x_k}+\bar{\mathbf{v}}_{x_{j'}}\mathbf{v}_{x_k}\big)\Big]_{x_{k'}}dt\\
&\quad+2\sum_{j,k,j',k'=1}^{n} b^{j'k'}\big(b^{jk}\ell_{x_j}\big)_{x_{k'}}\big(\mathbf{v}_{x_{j'}}\bar{\mathbf{v}}_{x_k}+\bar{\mathbf{v}}_{x_{j'}}\mathbf{v}_{x_k}\big)dt\\
&\quad+\sum_{j,k,j',k'=1}^{n}\Big[b^{jk}b^{j'k'}\ell_{x_j}\big(\mathbf{v}_{x_{j'}}\bar{\mathbf{v}}_{x_{k'}}+\bar{\mathbf{v}}_{x_{j'}}\mathbf{v}_{x_{k'}}\big)\Big]_{x_k}dt\\
&\quad-\sum_{j,k,j',k'=1}^{n}\big(b^{jk}b^{j'k'}\ell_{x_j}\big)_{x_k}\big(\mathbf{v}_{x_{j'}}\bar{\mathbf{v}}_{x_{k'}}+\bar{\mathbf{v}}_{x_{j'}}\mathbf{v}_{x_{k'}}\big)dt
\end{aligned}
\tag{11.33}
$$

$$
\begin{aligned}
&= -2 \sum_{j,k,j',k'=1}^{n} \Big[b^{jk'} b^{j'k} \ell_{x_j} \big(\mathbf{v}_{x_{j'}} \bar{\mathbf{v}}_{x_{k'}} + \bar{\mathbf{v}}_{x_{j'}} \mathbf{v}_{x_{k'}}\big)\Big]_{x_k} dt \\
&\quad +2 \sum_{j,k,j',k'=1}^{n} b^{jk'} \big(b^{j'k} \ell_{x_{j'}}\big)_{x_{k'}} \big(\mathbf{v}_{x_j} \bar{\mathbf{v}}_{x_k} + \bar{\mathbf{v}}_{x_j} \mathbf{v}_{x_k}\big) dt \\
&\quad + \sum_{j,k,j',k'=1}^{n} \Big[b^{jk} b^{j'k'} \ell_{x_j} \big(\mathbf{v}_{x_{j'}} \bar{\mathbf{v}}_{x_{k'}} + \bar{\mathbf{v}}_{x_{j'}} \mathbf{v}_{x_{k'}}\big)\Big]_{x_k} dt \\
&\quad - \sum_{j,k,j',k'=1}^{n} \big(b^{jk} b^{j'k'} \ell_{x_{j'}}\big)_{x_{k'}} \big(\mathbf{v}_{x_j} \bar{\mathbf{v}}_{x_k} + \bar{\mathbf{v}}_{x_j} \mathbf{v}_{x_k}\big) dt.
\end{aligned}
$$

Further, it holds that

$$
\begin{aligned}
& I_1^3 \overline{I_2^2} + I_2^2 \overline{I_1^3} \\
&= \Psi \mathbf{v} \overline{\sum_{j,k=1}^{n} (b^{jk} \mathbf{v}_{x_j})_{x_k} dt} + \sum_{j,k=1}^{n} (b^{jk} \mathbf{v}_{x_j})_{x_k} \overline{\Psi \mathbf{v}} dt \\
&= \sum_{j,k=1}^{n} \Big[\Psi b^{jk} (\mathbf{v}_{x_j} \bar{\mathbf{v}} + \bar{\mathbf{v}}_{x_j} \mathbf{v})\Big]_{x_k} dt - \sum_{j,k=1}^{n} \Psi b^{jk} (\mathbf{v}_{x_j} \bar{\mathbf{v}}_{x_k} + \bar{\mathbf{v}}_{x_j} \mathbf{v}_{x_k}) dt \\
&\quad - \sum_{j,k=1}^{n} \Psi_{x_j} b^{jk} (\mathbf{v}_{x_k} \bar{\mathbf{v}} + \bar{\mathbf{v}}_{x_k} \mathbf{v}) dt \qquad (11.34) \\
&= \sum_{j,k=1}^{n} \big[\Psi b^{jk} (\mathbf{v}_{x_j} \bar{\mathbf{v}} + \bar{\mathbf{v}}_{x_j} \mathbf{v})\big]_{x_k} dt - \sum_{j,k=1}^{n} \Psi b^{jk} (\mathbf{v}_{x_j} \bar{\mathbf{v}}_{x_k} + \bar{\mathbf{v}}_{x_j} \mathbf{v}_{x_k}) dt \\
&\quad - \sum_{j,k=1}^{n} \big(b^{jk} \Psi_{x_j} |\mathbf{v}|^2\big)_{x_k} dt + \sum_{j,k=1}^{n} (b^{jk} \Psi_{x_j})_{x_k} |\mathbf{v}|^2 dt.
\end{aligned}
$$

Finally,

$$
\begin{aligned}
& I_1 \overline{I_2^3} + I_2^3 \overline{I_1} \\
&= -i\beta \ell_t \mathbf{v} \overline{\mathcal{A} \mathbf{v}} dt + \mathcal{A} \mathbf{v} \overline{(-i\beta \ell_t \mathbf{v})} dt \\
&= -2 \sum_{j,k=1}^{n} (b^{jk} \ell_{x_j} \mathcal{A} |\mathbf{v}|^2)_{x_k} dt + 2 \Big[\sum_{j,k=1}^{n} (b^{jk} \ell_{x_j} \mathcal{A})_{x_k} + \mathcal{A} \Psi\Big] |\mathbf{v}|^2 dt.
\end{aligned}
\qquad (11.35)
$$

Step 3. Combining (11.23)–(11.35), we conclude the desired formula (11.17). This completes the proof of Theorem 11.11. □

Remark 11.12. Since we do not assume that the matrix $(b^{jk})_{1 \le j,k \le n}$ is positively definite, similar to [106] and based on the identity (11.17) in Theorem 11.11, we can deduce observability estimate not only for the backward stochastic Schrödinger equation, but also for deterministic hyperbolic, Schrödinger and plate equations, which had been derived via Carleman estimates (See [111], [187] and [385], respectively).

11.4 Observability Estimate for Backward Stochastic Schrödinger Equations

In this section, we shall prove the desired observability estimate for the backward stochastic Schrödinger equation (11.4) with $\tau = T$ (See Theorem 11.14 below). To this end, put

$$\psi(x) = |x - x_0|^2 + \sigma, \tag{11.36}$$

where σ is a positive constant such that

$$\min_{x\in\overline{G}} \psi(x) \geq \frac{5}{6}|\psi|_{L^\infty(G)}.$$

Let $\lambda > 0$ and $\mu > 0$. Put

$$\ell = \lambda\frac{e^{4\mu\psi} - e^{5\mu|\psi|_{L^\infty(G)}}}{t^2(T-t)^2}, \qquad \varphi = \frac{e^{4\mu\psi}}{t^2(T-t)^2}, \qquad \theta = e^{\ell}. \tag{11.37}$$

We first establish the following global Carleman estimate for backward stochastic Schrödinger equations:

Theorem 11.13. *There are $\lambda_1 > 0$ (depending on r_1) and $\mu_1 > 0$ such that for each $\lambda \geq \lambda_1$ and $\mu \geq \mu_1$, any solution to the equation (11.4) with $\tau = T$ satisfies that*

$$\begin{aligned}
&\mathbb{E}\int_Q \theta^2\Big(\lambda^3\mu^4\varphi^3|z|^2 + \lambda\mu\varphi|\nabla z|^2\Big)dxdt \\
&\leq \mathcal{C}\Big[\mathbb{E}\int_Q \theta^2(\lambda^2\mu^2\varphi^2|Z|^2 + |\nabla Z|^2)dxdt + \lambda\mu\mathbb{E}\int_{\Sigma_0}\theta^2\varphi\Big|\frac{\partial z}{\partial\nu}\Big|^2 d\Gamma dt\Big].
\end{aligned} \tag{11.38}$$

Proof: The proof is divided into three steps.

Step 1. Let $\beta = 1$ and $(b^{jk})_{1\leq j,k\leq n}$ be the $n\times n$ identity matrix. Put

$$\delta^{jk} = \begin{cases} 1, & \text{if } j = k,\\ 0, & \text{if } j \neq k.\end{cases}$$

For any solution (z, Z) to the equation (11.4) with $\tau = T$, applying Theorem 11.11 with θ given by (11.37), $\mathbf{u}$ replaced by z and $\mathbf{v} = \theta z$, we obtain that

$$\begin{aligned}
&\theta\mathcal{P}z\Big(i\beta\ell_t\bar{\mathbf{v}} - 2\sum_{j=1}^n \ell_{x_j}\bar{\mathbf{v}}_{x_j} + \Psi\bar{\mathbf{v}}\Big) + \theta\overline{\mathcal{P}z}\Big(-i\beta\ell_t\mathbf{v} - 2\sum_{j=1}^n \ell_{x_j}\mathbf{v}_{x_j} + \Psi\mathbf{v}\Big) \\
&\quad + dM + \operatorname{div} V \\
&= 2\Big| - i\beta\ell_t\mathbf{v} - 2\sum_{j=1}^n \ell_{x_j}\mathbf{v}_{x_j} + \Psi\mathbf{v}\Big|^2 dt
\end{aligned} \tag{11.39}$$

$$
\begin{aligned}
&+\sum_{j,k=1}^{n} c^{jk}(\mathbf{v}_{x_j}\bar{\mathbf{v}}_{x_k}+\bar{\mathbf{v}}_{x_j}\mathbf{v}_{x_k})dt+D|\mathbf{v}|^2dt\\
&+2i\sum_{j=1}^{n}(\ell_{x_jt}+\ell_{tx_j})(\bar{\mathbf{v}}_{x_j}\mathbf{v}-\mathbf{v}_{x_j}\bar{\mathbf{v}})dt+i(\Psi+\Delta\ell)(\bar{\mathbf{v}}d\mathbf{v}-\mathbf{v}d\bar{\mathbf{v}})+\ell_t|d\mathbf{v}|^2\\
&+i\sum_{j=1}^{n}\ell_{x_j}\big(d\bar{\mathbf{v}}_{x_j}d\mathbf{v}-d\mathbf{v}_{x_j}d\bar{\mathbf{v}}\big).
\end{aligned}
$$

Here

$$
\begin{cases}
M=\ell_t|\mathbf{v}|^2+i\displaystyle\sum_{j=1}^{n}\ell_{x_j}(\bar{\mathbf{v}}_{x_j}\mathbf{v}-\mathbf{v}_{x_j}\bar{\mathbf{v}}),\\
\mathcal{A}=\displaystyle\sum_{j=1}^{n}(\ell_{x_j}^2-\ell_{x_jx_j})-\Psi,\\
\mathcal{D}=\ell_{tt}+\displaystyle\sum_{j=1}^{n}\Psi_{x_jx_j}+2\sum_{j=1}^{n}(\ell_{x_j}\mathcal{A})_{x_j}+2\mathcal{A}\Psi,\\
c^{jk}=2\ell_{jk}-\delta^{jk}\Delta\ell-\delta^{jk}\Psi,
\end{cases}
$$

and

$$
\begin{aligned}
V_k=&-i\big[\ell_{x_k}(\mathbf{v}d\bar{\mathbf{v}}-\bar{\mathbf{v}}d\mathbf{v})+\ell_t(\mathbf{v}_{x_k}\bar{\mathbf{v}}-\bar{\mathbf{v}}_{x_k}\mathbf{v})dt\big]\\
&-\Psi(\mathbf{v}_{x_k}\bar{\mathbf{v}}+\bar{\mathbf{v}}_{x_k}\mathbf{v})dt+(2\mathcal{A}\ell_{x_k}+\Psi_{x_k})|\mathbf{v}|^2dt\\
&+2\sum_{j=1}^{n}\ell_{x_j}(\bar{\mathbf{v}}_{x_j}\mathbf{v}_{x_k}+\mathbf{v}_{x_j}\bar{\mathbf{v}}_{x_k})dt-2\sum_{j=1}^{n}\ell_{x_k}(\mathbf{v}_{x_j}\bar{\mathbf{v}}_{x_j})dt.
\end{aligned}
$$

Step 2. In this step, we estimate the terms in the right-hand side of the equality (11.39) one by one.

First, from the definitions of ℓ and φ in (11.37) and the choice of ψ in (11.36), we have that

$$
\begin{aligned}
|\ell_t|&=\Big|\lambda\frac{2(2t-T)}{t^3(T-t)^3}\big(e^{4\lambda\psi}-e^{5\mu|\psi|_{L^\infty(G)}}\big)\Big|\\
&\le\Big|\lambda\frac{2(2t-T)}{t^3(T-t)^3}e^{5\mu|\psi|_{L^\infty(G)}}\Big|\le\mathcal{C}\lambda\Big|\frac{1}{t^3(T-t)^3}e^{5\mu\psi}\Big|\le\mathcal{C}\lambda\varphi^{1+\frac{1}{2}},
\end{aligned}
\tag{11.40}
$$

and

$$
\begin{aligned}
|\ell_{tt}|&=\Big|\lambda\frac{20t^2-20tT+6T^2}{t^4(T-t)^4}\big(e^{4\mu\psi}-e^{5\mu|\psi|_{L^\infty(G)}}\big)\Big|\\
&\le\mathcal{C}\lambda\Big|\frac{1}{t^4(T-t)^4}e^{5\mu|\psi|_{L^\infty(G)}}\Big|\\
&\le\mathcal{C}\lambda\Big|\frac{1}{t^4(T-t)^4}e^{8\mu\psi}\Big|\le\mathcal{C}\lambda\varphi^2\le\mathcal{C}\lambda\varphi^3.
\end{aligned}
\tag{11.41}
$$

Let us choose $\Psi = -\Delta\ell$. Since

$$\mathcal{A} = \sum_{j=1}^{m} \ell_{x_j}^2 = \sum_{j=1}^{m} (4\lambda\mu\varphi\psi_{x_j})^2 = 16\lambda^2\mu^2\varphi^2|\nabla\psi|^2,$$

we find that

$$\begin{aligned}
\mathcal{D} &= \ell_{tt} + \sum_{j=1}^{n} \Psi_{x_jx_j} + 2\sum_{j=1}^{n}(\ell_{x_j}\mathcal{A})_{x_j} + 2\mathcal{A}\Psi \\
&= \ell_{tt} + \Delta(\Delta\ell) + 2\sum_{j=1}^{n} \big(4\lambda\mu\varphi\psi_{x_j} 16\lambda^2\mu^2\varphi^2|\nabla\psi|^2\big)_{x_j} \\
&\quad -32\lambda^2\mu^2\varphi^2|\nabla\psi|^2\Delta\ell \\
&= 384\lambda^3\mu^4\varphi^3|\nabla\psi|^4 - \mu^4\varphi O(\lambda) - \lambda^3\varphi^3 O(\mu^3) + \ell_{tt}.
\end{aligned} \tag{11.42}$$

Recalling that $x_0 \in \mathbb{R}^n \setminus \overline{G}$, we have $|\nabla\psi| > 0$ in $\overline{G}$. This, together with (11.41)–(11.42), implies that there exists a $\mu_0 > 0$ such that for all $\mu > \mu_0$, one can find a constant $\lambda_0 = \lambda_0(\mu_0)$ so that for any $\lambda > \lambda_0$, it holds that

$$\mathcal{D}|\mathbf{v}|^2 \geq \lambda^3\mu^4\varphi^3|\nabla\psi|^4|\mathbf{v}|^2. \tag{11.43}$$

Since

$$c^{jk} = 2\ell_{x_jx_k} - \delta^{jk}\Delta\ell - \delta^{jk}\Psi = 32\lambda\mu^2\varphi\psi_{x_j}\psi_{x_k} + 16\lambda\mu\varphi\psi_{x_jx_k},$$

we deduce that

$$\begin{aligned}
&\sum_{j,k=1}^{n} c^{jk}(\mathbf{v}_{x_j}\bar{\mathbf{v}}_{x_k} + v_{x_j}\bar{\mathbf{v}}_{x_k}) \\
&= 16\lambda\mu\varphi \sum_{j,k=1}^{n} (2\mu\psi_{x_j}\psi_{x_k} + \psi_{x_jx_k})(\mathbf{v}_{x_j}\bar{\mathbf{v}}_{x_k} + \mathbf{v}_{x_k}\bar{\mathbf{v}}_{x_j}) \\
&= 32\lambda\mu^2\varphi\Big[\sum_{j=1}^{n}(\psi_{x_j}\mathbf{v}_{x_j})\sum_{k=1}^{n}(\psi_{x_k}\bar{\mathbf{v}}_{x_k}) + \sum_{k=1}^{n}(\psi_{x_k}\mathbf{v}_{x_k})\sum_{j=1}^{n}(\psi_{x_j}\bar{\mathbf{v}}_{x_j})\Big] \\
&\quad +32\lambda\mu\varphi\sum_{j=1}^{n}(\mathbf{v}_{x_j}\bar{\mathbf{v}}_{x_j} + \bar{\mathbf{v}}_{x_j}\mathbf{v}_{x_j}) \\
&= 64\lambda\mu^2\varphi|\nabla\psi\cdot\nabla\mathbf{v}|^2 + 64s\mu\varphi|\nabla\mathbf{v}|^2 \\
&\geq 64\lambda\mu\varphi|\nabla\mathbf{v}|^2.
\end{aligned} \tag{11.44}$$

Now we estimate the other terms in the right-hand side of the equality (11.39). The first one reads

$$\begin{aligned}&\Big|2i\sum_{j=1}^{n}(\ell_{x_jt}+\ell_{tx_j})(\bar{\mathbf{v}}_{x_j}\mathbf{v}-\mathbf{v}_{x_j}\bar{\mathbf{v}})\Big|\\&=\Big|4i\sum_{j=1}^{n}\lambda\mu\psi_{x_j}\ell_t(\bar{\mathbf{v}}_{x_j}\mathbf{v}-\bar{\mathbf{v}}\mathbf{v}_{x_j})\Big|\le 2\lambda\varphi|\nabla\mathbf{v}|^2+2\lambda\mu^2\varphi^3|\nabla\psi|^2|\mathbf{v}^2|.\end{aligned}\tag{11.45}$$

The second one satisfies

$$i(\Psi+\Delta\ell)(\bar{\mathbf{v}}d\mathbf{v}-\mathbf{v}d\bar{\mathbf{v}})=0.\tag{11.46}$$

To estimate the third one, we take mean value and get that

$$\ell_t\mathbb{E}|d\mathbf{v}|^2=\ell_t\mathbb{E}|\theta\ell_tzdt+\theta dz|^2=\ell_t\theta^2\mathbb{E}|dz|^2\le 2\lambda\theta^2\varphi^{\frac{3}{2}}\mathbb{E}|Z|^2dt.\tag{11.47}$$

Here we utilize the inequality (11.40).

Further,

$$\begin{aligned}\mathbb{E}(d\bar{\mathbf{v}}_{x_k}d\mathbf{v})&=\mathbb{E}\big[\overline{(\theta\ell_tzdt+\theta dz)}_{x_k}(\theta\ell_tzdt+\theta dz)\big]=\mathbb{E}\big[\overline{(\theta dz)}_{x_k}\theta dz\big]\\&=\mathbb{E}\big[(\overline{\lambda\mu\varphi\psi_{x_k}\theta dz+\theta dz_{x_k}})(\theta dz)\big]\\&=\lambda\mu\varphi\psi_{x_k}\theta^2\mathbb{E}|dz|^2+\theta^2\mathbb{E}(d\bar{z}_{x_k}dz)\\&=\lambda\mu\varphi\psi_{x_k}\theta^2\mathbb{E}|Z|^2dt+\theta^2\mathbb{E}\big(\overline{Z}_{x_k}Z\big)dt.\end{aligned}$$

Similarly,

$$\mathbb{E}(d\mathbf{v}_{x_k}d\bar{\mathbf{v}})=\lambda\mu\varphi\psi_{x_k}\theta^2\mathbb{E}|Z|^2dt+\theta^2\mathbb{E}\big(\overline{Z}Z_{x_k}\big)dt.$$

Therefore, the fourth one satisfies that

$$\begin{aligned}&i\mathbb{E}\sum_{j=1}^{n}\ell_{x_j}\big(d\bar{\mathbf{v}}_{x_j}d\mathbf{v}-d\mathbf{v}_{x_j}d\bar{\mathbf{v}}\big)\\&=i\lambda\mu\varphi\sum_{j=1}^{n}\psi_{x_j}\big[\theta^2\mathbb{E}\big(\overline{Z}_{x_j}Z\big)-\theta^2\mathbb{E}\big(\overline{Z}_{x_j}Z\big)\big]dt.\end{aligned}\tag{11.48}$$

Step 3. Integrating the equality (11.39) in Q, taking mean value in both sides, and noting (11.40), (11.41), (11.43)–(11.48), we obtain that

$$\begin{aligned}&\mathbb{E}\int_Q\Big(\lambda^3\mu^4\varphi^3|\mathbf{v}|^2+\lambda\mu^2\varphi|\nabla\mathbf{v}|^2\Big)dxdt\\&\quad+2\mathbb{E}\int_Q\Big|-i\ell_t\mathbf{v}-2\sum_{j=1}^{n}\ell_{x_j}\mathbf{v}_{x_j}+\Psi\mathbf{v}\Big|^2dxdt\\&\le\mathbb{E}\int_Q\Big[\theta\mathcal{P}y\Big(i\ell_t\bar{\mathbf{v}}-2\sum_{j=1}^{n}\ell_{x_j}\bar{\mathbf{v}}_{x_j}+\Psi\bar{\mathbf{v}}\Big)\\&\quad+\theta\overline{\mathcal{P}y}\Big(-i\ell_t\mathbf{v}-2\sum_{j=1}^{n}\ell_{x_j}\mathbf{v}_{x_j}+\Psi\mathbf{v}\Big)\Big]dx\\&\quad+\mathcal{C}\mathbb{E}\int_Q\theta^2\big(s^2\mu^2\varphi^2|Z|^2+|\nabla Z|^2\big)dxdt+\mathbb{E}\int_Q dMdx+\mathbb{E}\int_Q\operatorname{div}V dx.\end{aligned}\tag{11.49}$$

Now we analyze the terms in the right-hand side of (11.49) one by one. The first one reads

$$\begin{aligned}
&\mathbb{E}\int_Q \Big[\theta\mathcal{P}z\Big(i\ell_t\bar{\mathbf{v}} - 2\sum_{j=1}^n \ell_{x_j}\bar{\mathbf{v}}_{x_j} + \Psi\bar{\mathbf{v}}\Big)\\
&\quad +\theta\overline{\mathcal{P}z}\Big(-i\ell_t\mathbf{v} - 2\sum_{j=1}^n \ell_{x_j}\mathbf{v}_{x_j} + \Psi\mathbf{v}\Big)\Big]dx\\
&= \mathbb{E}\int_Q \Big[\theta\big(b_1\cdot\nabla z + b_2 z + b_3 Z\big)\Big(i\ell_t\bar{\mathbf{v}} - 2\sum_{j=1}^n \ell_{x_j}\bar{\mathbf{v}}_{x_j} + \Psi\bar{\mathbf{v}}\Big)\\
&\quad +\theta\big(-b_1\cdot\nabla\bar{z} + b_2\bar{z} + b_3\overline{Z}\big)\Big(-i\ell_t\mathbf{v} - 2\sum_{j=1}^n \ell_{x_j}\mathbf{v}_{x_j} + \Psi\mathbf{v}\Big)\Big]dxdt\\
&\le \mathbb{E}\int_Q \Big(\theta^2\big|b_1\cdot\nabla z + b_2 z + b_3 Z\big|^2 + \Big| - i\beta\ell_t\mathbf{v} - 2\sum_{j=1}^n \ell_{x_j}\mathbf{v}_{x_j} + \Psi\mathbf{v}\Big|^2\Big)dxdt.
\end{aligned} \tag{11.50}$$

From the choice of θ, it is clear that $\mathbf{v}(0) = \mathbf{v}(T) = 0$. Hence, we obtain that

$$\int_Q dMdx = 0. \tag{11.51}$$

Further, noting that $\mathbf{v} = z = 0$ on Σ, we find

$$\begin{aligned}
&\mathbb{E}\int_Q \operatorname{div} V dx\\
&= \mathbb{E}\int_\Sigma 2\sum_{k=1}^n\sum_{j=1}^n \big[\ell_{x_j}(\bar{\mathbf{v}}_{x_j}\mathbf{v}_{x_k} + \mathbf{v}_{x_j}\bar{\mathbf{v}}_{x_k})\nu^k - \ell_{x_k}\nu^k\mathbf{v}_{x_j}\bar{\mathbf{v}}_{x_j}\big]d\Sigma\\
&= \mathbb{E}\int_\Sigma \Big(4\sum_{j=1}^n \ell_{x_j}\nu^j\Big|\frac{\partial\mathbf{v}}{\partial\nu}\Big|^2 - 2\sum_{j=1}^n \ell_{x_j}\nu^j\Big|\frac{\partial\mathbf{v}}{\partial\nu}\Big|^2\Big)d\Sigma\\
&= \mathbb{E}\int_\Sigma 2\sum_{j=1}^n \ell_{x_j}\nu^j\Big|\frac{\partial\mathbf{v}}{\partial\nu}\Big|^2 d\Sigma \le C\lambda\mu\mathbb{E}\int_{\Sigma_0}\theta^2\varphi\Big|\frac{\partial z}{\partial\nu}\Big|^2 d\Gamma dt.
\end{aligned} \tag{11.52}$$

It follows from (11.49)–(11.52) that

$$\begin{aligned}
&\mathbb{E}\int_Q \Big(\lambda^3\mu^4\varphi^3|\mathbf{v}|^2 + \lambda\mu\varphi|\nabla\mathbf{v}|^2\Big)dxdt\\
&\le C\mathbb{E}\int_Q \theta^2|b_1\cdot\nabla z + b_2 z + b_3 Z|^2 dxdt + C\lambda\mu\mathbb{E}\int_{\Sigma_0}\theta^2\varphi\Big|\frac{\partial z}{\partial\nu}\Big|^2 d\Gamma dt\\
&\quad + C\mathbb{E}\int_Q \theta^2\big(\lambda^2\mu^2\varphi^2|Z|^2 + |\nabla Z|^2\big)dxdt.
\end{aligned} \tag{11.53}$$

Noting that $z_{x_k} = \theta^{-1}(\mathbf{v}_{x_k} - \ell_{x_k}\mathbf{v}) = \theta^{-1}(\mathbf{v}_{x_k} - \lambda\mu\varphi\psi_{x_k}\mathbf{v})$, we get

$$\theta^2\big(|\nabla z|^2+\lambda^2\mu^2\varphi^2|z|^2\big)\leq \mathcal{C}\big(|\nabla \mathbf{v}|^2+\lambda^2\mu^2\varphi^2|\mathbf{v}|^2\big). \tag{11.54}$$

Therefore, it follows from (11.53) that

$$\begin{aligned}
&\mathbb{E}\int_Q \big(\lambda^3\mu^3\varphi^3|z|^2+\lambda\mu\varphi|\nabla z|^2\big)dxdt\\
&\leq \mathcal{C}\mathbb{E}\int_Q \theta^2\big(|b_1|^2|\nabla z|^2+b_2^2|z|^2+|b_3|^2|Z|^2\big)dxdt\\
&\quad+\mathcal{C}\lambda\mu\mathbb{E}\int_{\Sigma_0}\theta^2\varphi\Big|\frac{\partial z}{\partial\nu}\Big|^2 d\Gamma dt+\mathcal{C}\mathbb{E}\int_Q\theta^2\big(\lambda^2\mu^2\varphi^2|Z|^2+|\nabla Z|^2\big)dxdt.
\end{aligned} \tag{11.55}$$

Taking $\mu_1=\mu_0$ and $\lambda_1=\max(\lambda_0,\mathcal{C}r_1^2)$, and utilizing the inequality (11.55), we conclude the desired inequality (11.38). □

Now we are in a position to establish the following key observability estimate for the backward stochastic Schrödinger equation (11.4) with $\tau=T$.

Theorem 11.14. *All solutions to the backward stochastic Schrödinger equation* (11.4) *with* $\tau=T$ *satisfy that*

$$|z_T|_{L^2_{\mathcal{F}_T}(\Omega;H_0^1(G;\mathbb{C}))}\leq e^{\mathcal{C}r_1}\Big(\Big|\frac{\partial z}{\partial\nu}\Big|_{L^2_{\mathbb{F}}(0,T;L^2(\Gamma_0;\mathbb{C}))}+\big|a_4z+Z\big|_{L^2_{\mathbb{F}}(0,T;H_0^1(G;\mathbb{C}))}\Big). \tag{11.56}$$

Proof: Noting that

$$\begin{aligned}
&\int_Q\theta^2\Big(\lambda^2\mu^2\varphi^2|Z|^2+|\nabla Z|^2\Big)dxdt\\
&\leq 2\int_Q\theta^2\Big(\lambda^2\mu^2\varphi^2|Z+a_4z|^2+|\nabla(Z+a_4z)|^2\Big)dxdt\\
&\quad+2\int_Q\theta^2\Big(\lambda^2\mu^2\varphi^2|a_4z|^2+|\nabla(a_4z)|^2\Big)dxdt,
\end{aligned}$$

from (11.38), we obtain that

$$\begin{aligned}
&\mathbb{E}\int_Q\theta^2\Big(\lambda^3\mu^4\varphi^3|z|^2+\lambda\mu\varphi|\nabla z|^2\Big)dxdt\\
&\leq \mathcal{C}\mathbb{E}\Big[\int_Q\theta^2\Big(\lambda^2\mu^2\varphi^2|Z+a_4z|^2+|\nabla(Z+a_4z)|^2\Big)dxdt\\
&\quad+\int_Q\theta^2\Big(\lambda^2\mu^2\varphi^2|a_4z|^2+|\nabla(a_4z)|^2\Big)dxdt+\lambda\mu\int_{\Sigma_0}\theta^2\varphi\Big|\frac{\partial z}{\partial\nu}\Big|^2d\Gamma dt\Big].
\end{aligned} \tag{11.57}$$

Let $\lambda_2=\max\{\lambda_1,\mathcal{C}|a_4|_{L^\infty_{\mathbb{F}}(0,T;W_0^{1,\infty}(\Omega))}\}$. Then for all $\lambda\geq\lambda_2$, we deduce from (11.57) that

$$
\begin{aligned}
&\mathbb{E}\int_Q \theta^2\Big(\lambda^3\mu^4\varphi^3|z|^2+\lambda\mu\varphi|\nabla z|^2\Big)dxdt\\
&\le \mathcal{C}\mathbb{E}\Big[\int_Q \theta^2\Big(\lambda^2\mu^2\varphi^2|Z+a_4\hat z|^2+|\nabla(Z+a_4z)|^2\Big)dxdt\\
&\qquad+\lambda\mu\int_{\Sigma_0}\theta^2\varphi\Big|\frac{\partial z}{\partial\nu}\Big|^2 d\Gamma dt\Big].
\end{aligned}
\tag{11.58}
$$

It follows from the definition of ℓ and θ (in (11.37)) that

$$
\begin{aligned}
&\mathbb{E}\int_Q \theta^2\big(\varphi^3|z|^2+\varphi|\nabla z|^2\big)dxdt\\
&\ge \min_{x\in\overline{G}}\Big(\varphi\Big(\frac{T}{2},x\Big)\theta^2\Big(\frac{T}{4},x\Big)\Big)\mathbb{E}\int_{\frac{T}{4}}^{\frac{3T}{4}}\int_G\big(|z|^2+|\nabla z|^2\big)dxdt,
\end{aligned}
\tag{11.59}
$$

$$
\begin{aligned}
&\mathbb{E}\int_Q \theta^2\big(\varphi^2|Z+a_4z|^2+|\nabla(Z+a_4z)|^2\big)dxdt\\
&\le \max_{(x,t)\in\overline{Q}}\big(\varphi^2(t,x)\theta^2(t,x)\big)\mathbb{E}\int_Q\big(|Z+a_4z|^2+|\nabla(Z+a_4z)|^2\big)dxdt
\end{aligned}
\tag{11.60}
$$

and

$$
\mathbb{E}\int_{\Sigma_0}\theta^2\varphi\Big|\frac{\partial z}{\partial\nu}\Big|^2 d\Gamma dt\le \max_{(x,t)\in\overline{Q}}\big(\varphi(t,x)\theta^2(t,x)\big)\mathbb{E}\int_{\Sigma_0}\Big|\frac{\partial z}{\partial\nu}\Big|^2 d\Gamma dt.
\tag{11.61}
$$

Combining (11.38) and (11.59)–(11.61), we arrive at

$$
\begin{aligned}
&\mathbb{E}\int_{\frac{T}{4}}^{\frac{3T}{4}}\int_G\big(|z|^2+|\nabla z|^2\big)dxdt\\
&\le \mathcal{C}r_1\frac{\max_{(x,t)\in\overline{Q}}\Big(\varphi^2(t,x)\theta^2(t,x)\Big)}{\min_{x\in\overline{G}}\Big(\varphi(\frac{T}{2},x)\theta^2(\frac{T}{4},x)\Big)}\\
&\qquad\times\Big[\mathbb{E}\int_Q\big(|Z+a_4z|^2+|\nabla(Z+a_4z)|^2\big)dxdt+\mathbb{E}\int_{\Sigma_0}\Big|\frac{\partial z}{\partial\nu}\Big|^2 d\Gamma dt\Big]\\
&\le e^{\mathcal{C}r_1}\Big[\mathbb{E}\int_Q\big(|Z+a_4z|^2+|\nabla(Z+a_4z)|^2\big)dxdt+\mathbb{E}\int_{\Sigma_0}\Big|\frac{\partial z}{\partial\nu}\Big|^2 d\Gamma dt\Big].
\end{aligned}
\tag{11.62}
$$

Utilizing (11.62) and (11.6), we obtain that

$$
\begin{aligned}
&\mathbb{E}\int_G\big(|z_T|^2+|\nabla z_T|^2\big)dx\\
&\le e^{\mathcal{C}r_1}\Big[\mathbb{E}\int_Q\big(|Z+a_4z|^2+|\nabla(Z+a_4z)|^2\big)dxdt+\mathbb{E}\int_{\Sigma_0}\Big|\frac{\partial z}{\partial\nu}\Big|^2 d\Gamma dt\Big],
\end{aligned}
$$

which concludes Theorem 11.14 immediately. □

Remark 11.15. The constant $e^{Cr_1^2}$ in (11.56) can be improved. Nevertheless, as we explained before, since we want to present the key idea in a simple way, we do not pursue the sharpest result.

11.5 Notes and Comments

This chapter is an improved version of the results presented in [228, 229].

There are numerous studies on the controllability and observability of deterministic Schrödinger equations (e.g., [4, 35, 18, 19, 41, 42, 187, 190, 252, 260, 280, 289, 317, 394]).

Generally speaking, there are five useful methods to solve the exact controllability/observability problem for deterministic Schrödinger equations:

- The first one is based on the Ingham type inequality ([317]). It can be used to solve controllability/observability problem for Schrödinger equations involved in some special domains, i.e., intervals and rectangles. However, people do not know how to apply it to the equations evolved in general domains.
- The second one is the classical Rellich-type multiplier approach ([252]). It can be applied to Schrödinger equations with (small) time-invariant lower order terms.
- The third one is the microlocal analysis approach ([4, 190]), which gives sharp sufficient conditions for the exact controllability of Schrödinger equations with time independent lower order terms.
- The fourth one is the transmutation method ([280]). This method reduces the controllability problem for Schrödinger equations to the same problem but for parabolic equations.
- The last one is based on the global Carleman estimate ([187, 260]), which applies to Schrödinger equations with quite general lower order terms but the controller is not as sharp as those given by the first and the third methods (mentioned above).

Until now, as far as we know, only the last method was extended to studying the controllability/observability problem of stochastic Schrödinger equations (See [224, 228, 229]). Recently, in a very interesting paper [107], a significant weighted identity for quite general stochastic partial differential operators was established, via which one can obtain a unified treatment in establishing the weighted identities (including in particular (8.19), (9.26) and (11.17)) for some deterministic/stochastic partial differential operators. Hence, based on this identity, some known global Carleman estimates for stochastic transport equations, stochastic parabolic equations, stochastic Schrödinger equations and so on can be deduced.

It seems that people will face many essential difficulties to apply the other four methods mentioned above to study the controllability problem for stochastic Schrödinger equations. On the other hand, there are lots of open problems related to the topic of this chapter. In our opinion, the following are particularly interesting:

1) **Observability estimate and unique continuation for stochastic Schrödinger equations.** Consider the following stochastic Schrödinger equation:

$$\begin{cases} idy + \Delta ydt = \big(a_1 \cdot \nabla y + a_2 y\big)dt + a_3 ydW(t) & \text{in } Q, \\ y = 0 & \text{on } \Sigma, \\ y(0) = y_0 & \text{in } G. \end{cases} \tag{11.63}$$

As an easy consequence of Theorem 11.14, we can show that, the solution y to (11.63) equals zero in Q, a.s., provided that $y = 0$ in a neighborhood of Γ_0, a.e. $t \in (0, T)$. Compared with the classical unique continuation result for deterministic Schrödinger equations with time independent coefficients (e.g., [4, 190] for example), it seems too restrictive to assume that y vanishes in a neighborhood of Γ_0, a.e. $t \in (0, T)$. It would be quite interesting but may be challenging to prove whether the results in [4, 190] are true or not for stochastic Schrödinger equations. To do this, several new tools should be developed in the stochastic setting, such as the stochastic counterpart of microlocal defect measure and semiclassical defect measure.

2) **Null and approximate controllability for stochastic Schrödinger equations.** As immediate consequences of the main result of this chapter, we can obtain the null and approximate controllability for the same system. However, in order to get these two kinds of controllability, there are no reasons to employ two controls. By Proposition 7.8, it is possible to put only one control in the drift term to get the null controllability, at least for the system with deterministic coefficients. On the other hand, as suggested by the result in Section 9.2, we believe that one boundary control can guarantee the null and approximate controllability of (11.3). However, we will meet some essential difficulties to prove such a result. For example, to get the null controllability, we should prove that

$$\mathbb{E}|z(0)|^2_{H_0^1(G;\mathbb{C})} \leq \mathcal{C}\mathbb{E}\int_{\Sigma_0} \Big|\frac{\partial z}{\partial \nu}\Big|^2 d\Sigma_0. \tag{11.64}$$

If we utilize the method in this chapter, we only get that

$$\mathbb{E}|z(0)|^2_{H_0^1(G;\mathbb{C})} \leq \mathcal{C}\mathbb{E}\Big(\int_{\Sigma_0} \Big|\frac{\partial z}{\partial \nu}\Big|^2 d\Sigma_0 + \int_0^T |Z|^2_{H_0^1(G;\mathbb{C})} dt\Big).$$

An extra term Z appears in the right hand side of the above inequality. We believe that one should be able to introduce some new technique to get rid of this additional term. However, we do not know how to achieve this goal at this moment.

3) **Controllability for stochastic Schrödinger equations with bilinear controls**. In practice, it is more convenient to add controls in the coefficients of stochastic Schrödinger equations. This leads to the control system with bilinear controls. An example is as follows:

$$\begin{cases} idy + \Delta ydt = (a_1 + u)ydt + (a_1 + v)ydW(t) & \text{in } Q, \\ y = 0 & \text{on } \Sigma, \\ y(0) = y_0 & \text{in } G. \end{cases} \tag{11.65}$$

Here both u and v are controls belong to suitable function spaces. Bilinear control for deterministic Schrödinger equations are studied extensively in the literature (e.g., [19]). However, at this moment we have no idea on how to study the bilinear controllability problem for stochastic Schrödinger equations.

In this book, we consider only controllability/obseverbility problems for four kind of stochastic partial differential equations, i.e., stochastic transport equations, stochastic parabolic equations, stochastic wave equations and stochastic Schrödinger equations. In recent years, there are some interesting works on the same problems but for other stochastic partial differential equations, say [108] for stochastic complex Ginzburg-Landau equations, [118] for stochastic Korteweg-de Vries equations, [119] for stochastic Kawahara equations, [120] for stochastic fourth order Schrödinger equations, [121] for stochastic Kuramoto-Sivashinsky equations, [142] for stochastic coupled systems of fourth and second-order parabolic equations, [222] for stochastic degenerate parabolic equations, [358] for stochastic Grushin equations, and [391] for stochastic beam equations. As far as we know, there are no results on the controllability/obseverbility problems for stochastic plate equations, stochastic Maxwell equations, stochastic Benjamin-Bona-Mahony equations, stochastic elasticity/thermoelasticity equations, stochastic hyperbolic-parabolic coupled systems, stochastic Navier-Stokes equations and atmospheric equations in particular, etc. In our opinion, the direction of controllability/obseverbility for stochastic partial differential equations is full of open problems, and very likely people need to develop new tools to solve some of them.

12

Pontryagin-Type Stochastic Maximum Principle and Beyond

The main purpose of this chapter is to derive first order necessary conditions, i.e., Pontryagin-type maximum principle for optimal controls of general nonlinear stochastic evolution equations in infinite dimensions, in which both the drift and the diffusion terms may contain the control variables, and the control regions are allowed to be nonconvex. In order to do this, quite different from the deterministic infinite dimensional setting and the stochastic finite dimensional case, people have to introduce a suitable operator-valued backward stochastic evolution equation, served as the second order adjoint equation. It is very difficult to prove the existence of solutions to this equation for the general case. Indeed, in the infinite dimensional setting, there exists no such a satisfactory stochastic integration/evolution equation theory (in the previous literatures) that can be employed to establish the well-posedness of such an equation.

To overcome the above-mentioned difficulty, we employ our stochastic transposition method to introduce a concept of transposition solution/relaxed transposition solution to the desired operator-valued backward stochastic evolution equation, and develop carefully a way to study the corresponding well-posedness.

We shall also consider very quickly first order sufficient conditions and second order necessary conditions (for optimal controls), while the latter relies also essentially on the above well-posedness result, in particular on the characterization of the correction part in the above operator-valued backward stochastic evolution equation.

12.1 Formulation of the Optimal Control Problem

Throughout this chapter, $T > 0$, $(\Omega, \mathcal{F}, \mathbf{F}, \mathbb{P})$ (with $\mathbf{F} \stackrel{\Delta}{=} \{\mathcal{F}_t\}_{t\in[0,T]}$) is a fixed filtered probability space satisfying the usual condition, and we denote by $\mathbb{F}$ the progressive σ-field w.r.t. $\mathbf{F}$; H and V are two separable (complex) Hilbert

Q. Lü and X. Zhang, *Mathematical Control Theory for Stochastic Partial Differential Equations*, Probability Theory and Stochastic Modelling 101, https://doi.org/10.1007/978-3-030-82331-3_12

spaces (unless other stated); $\{h_j\}_{j=1}^{\infty}$ and $\{e_j\}_{j=1}^{\infty}$ are respectively orthonormal basis of H and V; we denote by I the identity operator on H, and write $\mathcal{L}_2^0 \stackrel{\Delta}{=} \mathcal{L}_2(V;H)$. Unless otherwise specified, $W(\cdot)$ is a V-valued, Q-Brownian motion or cylindrical Brownian motion but we only consider the case of cylindrical Brownian motion.

Let A be an unbounded linear operator (with domain $D(A)$ on H), which generates a C_0-semigroup $\{S(t)\}_{t\geq 0}$. Denote by A^* the dual operator of A. Clearly, $D(A)$ is a Hilbert space with the usual graph norm, and A^* is the infinitesimal generator of $\{S^*(t)\}_{t\geq 0}$, the dual C_0-semigroup of $\{S(t)\}_{t\geq 0}$. For any $\lambda \in \rho(A)$, the resolvent of A, denote by A_λ the Yosida approximation of A and by $\{S_\lambda(t)\}_{t\in\mathbb{R}}$ the C_0-group generated by A_λ. Let U be a separable metric space with a metric $d(\cdot,\cdot)$. Put

$$\mathcal{U}[0,T] \stackrel{\Delta}{=} \big\{u(\cdot) : [0,T]\times\Omega \to U \;\big|\; u(\cdot) \text{ is } \mathbf{F}\text{-adapted}\big\} \tag{12.1}$$

Throughout this chapter, $\mathcal{C}_L > 0$ is a generic constant, and we assume the following condition:

(S1) *Suppose that* $a(\cdot,\cdot,\cdot) : [0,T]\times\Omega\times H\times U \to H$ *and* $b(\cdot,\cdot,\cdot) : [0,T]\times\Omega\times H\times U \to \mathcal{L}_2^0$ *are two (vector-valued) functions satisfying: i) For any* $(x,u)\in H\times U$, *the functions* $a(\cdot,x,u) : [0,T]\times\Omega\to H$ *and* $b(\cdot,x,u) : [0,T]\times\Omega\to\mathcal{L}_2^0$ *are* $\mathbb{F}$*-measurable; ii) For any* $x\in H$ *and a.e.* $(t,\omega)\in(0,T)\times\Omega$, *the functions* $a(t,x,\cdot) : U\to H$ *and* $b(t,x,\cdot) : U\to\mathcal{L}_2^0$ *are continuous; and iii) For any* $(x_1,x_2,u)\in H\times H\times U$ *and a.e.* $(t,\omega)\in(0,T)\times\Omega$,

$$\begin{cases} |a(t,x_1,u)-a(t,x_2,u)|_H + |b(t,x_1,u)-b(t,x_2,u)|_{\mathcal{L}_2^0} \leq \mathcal{C}_L|x_1-x_2|_H,\\ |a(t,0,u)|_H + |b(t,0,u)|_{\mathcal{L}_2^0} \leq \mathcal{C}_L. \end{cases}$$

Let us consider the following controlled stochastic evolution equation:

$$\begin{cases} dx(t) = \big(Ax(t) + a(t,x(t),u(t))\big)dt \\ \qquad\qquad +b(t,x(t),u(t))dW(t) \qquad \text{in } (0,T], \\ x(0) = x_0, \end{cases} \tag{12.2}$$

where $u(\cdot)\in\mathcal{U}[0,T]$ is the control variable, $x(\cdot)$ is the state variable, and the (given) initial state $x_0\in L^{p_0}_{\mathcal{F}_0}(\Omega;H)$ for some given $p_0\geq 2$.

In the rest of this chapter, we shall denote by $\mathcal{C}$ a generic constant, depending on T, A, p_0 (or p to be introduced later) and $\mathcal{C}_L$ (or J and K to be introduced later), which may be different from one place to another. By Theorem 3.14, it is easy to show the following result:

Proposition 12.1. *Let the assumption (S1) hold. Then, for any* $x_0\in L^{p_0}_{\mathcal{F}_0}(\Omega;H)$ *and* $u(\cdot)\in\mathcal{U}[0,T]$, *the equation* (12.2) *admits a unique mild solution* $x(\cdot)\equiv x(\cdot\,;x_0,u)\in C_{\mathbb{F}}([0,T];L^{p_0}(\Omega;H))$. *Furthermore,*

$$|x(\cdot)|_{C_{\mathbb{F}}([0,T];L^{p_0}(\Omega;H))} \leq \mathcal{C}\big(1+|x_0|_{L^{p_0}_{\mathcal{F}_0}(\Omega;H)}\big).$$

Also, we need the following condition:

(S2) *Suppose that* $g(\cdot,\cdot,\cdot) : [0,T]\times\Omega\times H\times U \to \mathbb{R}$ *and* $h(\cdot) : \Omega\times H \to \mathbb{R}$ *are two functions satisfying: i) For any* $(x,u) \in H \times U$, *the function* $g(\cdot,x,u) : [0,T]\times\Omega \to \mathbb{R}$ *is* $\mathbb{F}$*-measurable, the function* $h(x) : \Omega \to \mathbb{R}$ *is* $\mathcal{F}_T$*-measurable; ii) For any* $x \in H$ *and a.e.* $(t,\omega) \in (0,T)\times\Omega$, *the function* $g(t,x,\cdot) : U \to \mathbb{R}$ *is continuous; and iii) For any* $(x_1,x_2,u) \in H\times H\times U$ *and a.e.* $(t,\omega) \in (0,T)\times\Omega$,

$$\begin{cases} |g(t,x_1,u)-g(t,x_2,u)|+|h(x_1)-h(x_2)| \le \mathcal{C}_L|x_1-x_2|_H, \\ |g(t,0,u)|+|h(0)| \le \mathcal{C}_L. \end{cases}$$

Define a cost functional $\mathcal{J}(\cdot)$ (for the controlled system (12.2)) as follows:

$$\mathcal{J}(u(\cdot)) \stackrel{\Delta}{=} \mathbb{E}\Big(\int_0^T g(t,x(t),u(t))dt + h(x(T))\Big), \quad \forall\, u(\cdot)\in\mathcal{U}[0,T], \tag{12.3}$$

where $x(\cdot)$ is the corresponding solution to (12.2). We consider the following optimal control problem for the controlled equation (12.2) with the cost functional (12.3):

Problem (OP) *Find a* $\bar{u}(\cdot) \in \mathcal{U}[0,T]$ *such that*

$$\mathcal{J}(\bar{u}(\cdot)) = \inf_{u(\cdot)\,\in\,\mathcal{U}[0,T]} \mathcal{J}(u(\cdot)). \tag{12.4}$$

Any $\bar{u}(\cdot) \in \mathcal{U}[0,T]$ satisfying (12.4) is called an *optimal control.* The corresponding state $\bar{x}(\cdot)$ (of (12.2)) is called an *optimal state*, and $(\bar{x}(\cdot),\bar{u}(\cdot))$ is called an *optimal pair* (of Problem (OP)).

The main goal of this chapter is to establish a first order necessary condition for optimal pairs of Problem (OP), in the spirit of the classical Pontryagin maximum principle for deterministic finite dimensional controlled systems ([281]). Also, we shall consider very quickly the related first order sufficient (optimality) condition and second order necessary (optimality) condition.

Stimulated by [273], for the case of general control regions, we need to assume the following further conditions[1]:

(S3) *For any* $u \in U$ *and a.e.* $(t,\omega) \in (0,T)\times\Omega$, *the functions* $a(t,\cdot,u) : H \to H$, $b(t,\cdot,u) : H \to \mathcal{L}_2^0$, $g(t,\cdot,u) : H \to \mathbb{R}$ *and* $h(\cdot) : H \to \mathbb{R}$ *are* C^2. *For any* $x \in H$ *and a.e.* $(t,\omega) \in (0,T)\times\Omega$, *the functions* $a_x(t,x,\cdot) : U \to \mathcal{L}(H)$, $b_x(t,x,\cdot) : U \to \mathcal{L}(H;\mathcal{L}_2^0)$, $g_x(t,x,\cdot) : U \to H$, $a_{xx}(t,x,\cdot) : U \to \mathcal{L}(H,H;H)$, $b_{xx}(t,x,\cdot) : U \to \mathcal{L}(H,H;\mathcal{L}_2^0)$ *and* $g_{xx}(t,x,\cdot) : U \to \mathcal{L}(H)$ *are continuous. Moreover, for any* $(x,u) \in H\times U$ *and a.e.* $(t,\omega) \in (0,T)\times\Omega$,

[1] See Subsection 2.11.1 for the notations $\mathcal{L}(H,H;H)$ and $\mathcal{L}(H,H;\mathcal{L}_2^0)$.

$$
\begin{cases}
|a_x(t,x,u)|_{\mathcal{L}(H)} + |b_x(t,x,u)|_{\mathcal{L}(H;\mathcal{L}_2^0)} + |g_x(t,x,u)|_H + |h_x(x)|_H \leq \mathcal{C}_L, \\
|a_{xx}(t,x,u)|_{\mathcal{L}(H,H;H)} + |b_{xx}(t,x,u)|_{\mathcal{L}(H,H;\mathcal{L}_2^0)} \\
\quad +|g_{xx}(t,x,u)|_{\mathcal{L}(H)} + |h_{xx}(x)|_{\mathcal{L}(H)} \\
\leq \mathcal{C}_L.
\end{cases}
$$

Problem (OP) is now well-understood for the case of finite dimensions (i.e., $\dim H < \infty$) and natural filtration. In this case, a Pontryagin-type maximum principle was obtained in [273] for general stochastic control systems with control-dependent diffusion coefficients and possibly non-convex control regions, and it was found that the corresponding result differs significantly from its deterministic counterpart.

At first glance, one might think that the study of Problem (OP) is simply a routine extension of that in [273]. However, the infinite dimensional setting leads to significantly new difficulties. To illustrate this, we first recall the main idea and result in [273]. Suppose that $(\bar{x}(\cdot), \bar{u}(\cdot))$ is a given optimal pair for the special case that $A = 0$, $H = \mathbb{R}^n$ (for some $n \in \mathbb{N}$), $V = \mathbb{R}$, $p_0 = 2$ and $\mathbf{F}$ is the natural filtration generated by $W(\cdot)$. First, similarly to the deterministic setting, one introduces the following first order adjoint equation (which is now however a backward stochastic differential equation in the stochastic case):

$$
\begin{cases}
dy(t) = -\Big(a_x(t,\bar{x}(t),\bar{u}(t))^\top y(t) + b_x(t,\bar{x}(t),\bar{u}(t))^\top Y(t) \\
\qquad\qquad -g_x(t,\bar{x}(t),\bar{u}(t))\Big)dt + Y(t)dW(t) \qquad \text{in } [0,T), \\
y(T) = -h_x(\bar{x}(T)).
\end{cases}
\tag{12.5}
$$

In (12.5), the unknown is a pair of $\mathbf{F}$-adapted processes $(y(\cdot), Y(\cdot)) \in L^2_{\mathbb{F}}(\Omega; C([0,T];\mathbb{R}^n)) \times L^2_{\mathbb{F}}(0,T;\mathbb{R}^n)$. Next, to establish the desired maximum principle for stochastic controlled systems with control-dependent diffusion and possibly nonconvex control regions, it was found in [273] that, except for the first order adjoint equation (12.5), one needs to introduce an additional second order adjoint equation as follows:

$$
\begin{cases}
dP(t) = -\Big(a_x(t,\bar{x}(t),\bar{u}(t))^\top P(t) + P(t)a_x(t,\bar{x}(t),\bar{u}(t)) \\
\qquad\qquad +b_x(t,\bar{x}(t),\bar{u}(t))^\top P(t) b_x(t,\bar{x}(t),\bar{u}(t)) \\
\qquad\qquad +b_x(t,\bar{x}(t),\bar{u}(t))^\top Q(t) + Q(t) b_x(t,\bar{x}(t),\bar{u}(t)) \\
\qquad\qquad +\mathbb{H}_{xx}(t,\bar{x}(t),\bar{u}(t),y(t),Y(t))\Big)dt + Q(t)dW(t) \ \ \text{in } [0,T] \\
P(T) = -h_{xx}(\bar{x}(T)).
\end{cases}
\tag{12.6}
$$

In (12.6), the *Hamiltonian* $\mathbb{H}(\cdot,\cdot,\cdot,\cdot,\cdot)$ is defined by

$$
\mathbb{H}(t,x,u,y_1,y_2) = \langle\, y_1, a(t,x,u)\,\rangle_{\mathbb{R}^n} + \langle\, y_2, b(t,x,u)\,\rangle_{\mathbb{R}^n} - g(t,x,u),
$$
$$
(t,\omega,x,u,y_1,y_2) \in [0,T] \times \Omega \times \mathbb{R}^n \times U \times \mathbb{R}^n \times \mathbb{R}^n.
$$

Clearly, the equation (12.6) is an $\mathbb{R}^{n\times n}$-valued backward stochastic differential equation in which the unknown is a pair of processes $(P(\cdot),Q(\cdot)) \in L^2_{\mathbb{F}}(\Omega;C([0,T];\mathbb{R}^{n\times n})) \times L^2_{\mathbb{F}}(0,T;\mathbb{R}^{n\times n})$. Then, associated with the 6-tuple $(\bar{x}(\cdot),\bar{u}(\cdot),y(\cdot),Y(\cdot),P(\cdot),Q(\cdot))$, define

$$\begin{aligned}\mathcal{H}(t,x,u) &\stackrel{\Delta}{=} \mathbb{H}(t,x,u,y(t),Y(t)) + \frac{1}{2}\langle P(t)b(t,x,u), b(t,x,u)\rangle_{\mathbb{R}^n}\\ &\quad -\langle P(t)b(t,\bar{x}(t),\bar{u}(t)), b(t,x,u)\rangle_{\mathbb{R}^n}.\end{aligned}$$

The main result in [273] asserts that any optimal pair $(\bar{x}(\cdot),\bar{u}(\cdot))$ satisfies the following maximum principle condition:

$$\mathcal{H}(t,\bar{x}(t),\bar{u}(t)) = \max_{u\in U}\mathcal{H}(t,\bar{x}(t),u), \quad \text{a.e. } t\in[0,T], \quad \text{a.s.}$$

It is easy to see that, in order to establish the Pontryagin-type necessary conditions for an optimal pair $(\bar{x}(\cdot),\bar{u}(\cdot))$ of Problem (OP), we need to introduce first the following H-valued backward stochastic evolution equation[2]:

$$\begin{cases} dy(t) = -A^*y(t)dt - \Big(a_x(t,\bar{x}(t),\bar{u}(t))^*y(t) + b_x(t,\bar{x}(t),\bar{u}(t))^*Y(t)\\ \qquad\qquad -g_x(t,\bar{x}(t),\bar{u}(t))\Big)dt + Y(t)dW(t) \quad \text{in } [0,T),\\ y(T) = -h_x\big(\bar{x}(T)\big), \end{cases} \tag{12.7}$$

which will serve as the first order adjoint equation. By Theorem 4.16, the equation (12.7) is well-posed in the sense of transposition solution (Note that we do not assume the filtration **F** is the natural one).

Next, inspired by [273], to deal with the case of possibly non-convex control region U, for any $p\in(1,2]$, we need to introduce the following formally $\mathcal{L}(H)$-valued backward stochastic evolution equation:

$$\begin{cases} dP = -(A^*+J^*)Pdt - P(A+J)dt - K^*PKdt\\ \qquad\quad -(K^*Q+QK)dt + Fdt + QdW(t) \qquad \text{in } [0,T),\\ P(T) = P_T, \end{cases} \tag{12.8}$$

where $(q = p/(p-1))$,

$$J \in L^{2q}_{\mathbb{F}}(0,T;L^\infty(\Omega;\mathcal{L}(H))), \quad K \in L^{2q}_{\mathbb{F}}(0,T;L^\infty(\Omega;\mathcal{L}(H;\mathcal{L}_2^0))), \tag{12.9}$$

$$F \in L^1_{\mathbb{F}}(0,T;L^p(\Omega;\mathcal{L}(H))), \quad P_T \in L^p_{\mathcal{F}_T}(\Omega;\mathcal{L}(H)). \tag{12.10}$$

For the special case when $H=\mathbb{R}^n$, similarly to (12.6), it is easy to see that (12.8) is an $\mathbb{R}^{n\times n}$ (matrix)-valued backward stochastic differential equation

[2] Throughout this book, for any operator-valued process (*resp.* random variable) R, we denote by R^* its pointwise dual operator-valued process (*resp.* random variable). For example, if $R \in L^1_{\mathbb{F}}(0,T;L^2(\Omega;\mathcal{L}(H)))$, then $R^* \in L^1_{\mathbb{F}}(0,T;L^2(\Omega;\mathcal{L}(H)))$, and $|R|_{L^1_{\mathbb{F}}(0,T;L^2(\Omega;\mathcal{L}(H)))} = |R^*|_{L^1_{\mathbb{F}}(0,T;L^2(\Omega;\mathcal{L}(H)))}$.

(which can be easily regarded as an $\mathbb{R}^{n^2}$(vector)-valued backward stochastic differential equation), and therefore, the desired well-posedness follows from the one for backward stochastic evolution equations valued in Hilbert spaces.

One has to face a real challenge in the study of (12.8) when $\dim H = \infty$, without further assumption on the data F and P_T. Indeed, as we mentioned in Remark 2.62, in the infinite dimensional setting, although $\mathcal{L}(H)$ is still a Banach space, it is neither reflexive (needless to say to be a Hilbert space) nor separable. To the best of the authors' knowledge, in the previous literatures there exists no such a stochastic integration/evolution equation theory in general Banach spaces that can be employed to treat the well-posedness of (12.8) in the usual sense, even if the filtration $\mathbf{F}$ is the natural one. For example, the existing results on stochastic integration/evolution equation in UMD Banach spaces (e.g., [328, 329]) do not fit the present case because, if a Banach space is UMD, then it is reflexive.

From the above analysis, it is clear that the key to establish the Pontryagin-type stochastic maximum principle for optimal pairs of Problem (OP) is to give a suitable sense of solutions to the equation (12.8) and prove the corresponding well-posedness. For this purpose, we shall adopt the stochastic transposition method developed in our previous work [241], which was addressed to the backward stochastic differential equations in $\mathbb{R}^n$ (See also Section 4.3 (in Chapter 4) for the well-posedness of H-valued backward stochastic evolution equations in the sense of transposition solutions).

The rest of this chapter is organized as follows. In Section 12.2, we review the Pontryagin maximum principle for optimal control problems of stochastic differential equations. In Section 12.3, a necessary condition for optimal controls for convex control domains is given. Section 12.4 is devoted to the study of operator-valued backward stochastic evolution equations. In Section 12.5, we establish the Pontryagin-type maximum principle for optimal controls for convex control domains. In Section 12.6, we give a sufficient condition for the optimal control. In Section 12.7, we give some integral-type second order necessary conditions for optimal controls. Finally, some remarks and open problems are given in Section 12.8.

12.2 The Case of Finite Dimensions

In this section, we consider the special case that $A = 0$, $H = \mathbb{R}^n$ and $V = \mathbb{R}$ (and hence $\{W(t)\}_{t\in[0,T]}$ is a one dimensional standard Brownian motion). As we mentioned before, for this case the well-posedness of the equation (12.8) follows from the one for $\mathbb{R}^{n^2}$(vector)-valued backward stochastic differential equations.

Let $(\bar{x}(\cdot), \bar{u}(\cdot))$ be an optimal pair for Problem (OP). Note that we do not assume $\mathbf{F}$ is the natural filtration in this section. Hence, the solution $(y(\cdot), Y(\cdot)) \in D_{\mathbb{F}}([0, T]; L^2(\Omega; \mathbb{R}^n)) \times L^2_{\mathbb{F}}(0, T; \mathbb{R}^n)$ to (12.5) is understood in the sense of transposition solution. By taking an inner product in $\mathbb{R}^{n\times n}$

as $\langle P_1, P_2\rangle_{\mathbb{R}^{n\times n}} = \mathrm{tr}\,(P_1 P_2^\top)$ for any $n\times n$ matrices P_1 and P_2, as a direct consequence of Theorem 4.16, we see that the equation (12.6) admits a unique transposition solution $(P(\cdot), Q(\cdot)) \in D_{\mathbb{F}}([0,T]; L^8(\Omega;\mathbb{R}^{n\times n})) \times L^8_{\mathbb{F}}(\Omega; L^2(0,T;\mathbb{R}^{n\times n}))$.

We have the following Pontryagin-type maximum principle for optimal pairs of Problem (OP).

Theorem 12.2. *Let (S1)–(S3) hold (for $H = \mathbb{R}^n$ and $V = \mathbb{R}$) and $x_0 \in L^8_{\mathcal{F}_0}(\Omega;\mathbb{R}^n)$. Then, for any optimal pair $(\bar{x}(\cdot), \bar{u}(\cdot))$ of Problem (OP), it holds that*

$$\begin{aligned}
&\mathbb{H}(t,\bar{x}(t),\bar{u}(t),y(t),Y(t)) - \mathbb{H}(t,\bar{x}(t),u,y(t),Y(t))\\
&\quad -\frac{1}{2}\big\langle P(t)\big(b(t,\bar{x}(t),\bar{u}(t)) - b(t,\bar{x}(t),u)\big), b(t,\bar{x}(t),\bar{u}(t)) - b(t,\bar{x}(t),u)\big\rangle_{\mathbb{R}^n}\\
&\geq 0, \qquad \forall\, u\in U, \quad \text{a.e. } t\in[0,T], \quad \text{a.s.}
\end{aligned} \tag{12.11}$$

Sketch of the proof of Theorem 12.2: The detailed proof of this theorem is very close to that of [273, Theorem 3] and [371, Theorem 3.2 in Chapter 3] (addressed to the stochastic maximum principle for controlled stochastic differential equations with the natural filtration generated by $W(\cdot)$). Hence, we only give below a sketch of proof for this theorem.

Fix any $u(\cdot)\in\mathcal{U}[0,T]$ and $\varepsilon > 0$. Similar to (1.44), let

$$u^\varepsilon(t) = \begin{cases} \bar{u}(t), & t\in[0,T]\setminus E_\varepsilon,\\ u(t), & t\in E_\varepsilon,\end{cases}$$

where $E_\varepsilon \subseteq [0,T]$ is a measurable set with the Lebesgue measure $\mathbf{m}(E_\varepsilon) = \varepsilon$. For $\psi = a, b, f$ and g, we write

$$\begin{cases}
\psi_1(t) = \psi_x(t,\bar{x}(t),\bar{u}(t)),\\
\psi_{11}(t) = \psi_{xx}(t,\bar{x}(t),\bar{u}(t)),\\
\delta\psi(t) = \psi(t,\bar{x}(t),u(t)) - \psi(t,\bar{x}(t),\bar{u}(t)),\\
\delta\psi_1(t) = \psi_x(t,\bar{x}(t),u(t)) - \psi_x(t,\bar{x}(t),\bar{u}(t)).
\end{cases}$$

Suppose that $x_1^\varepsilon(\cdot)$ and $x_2^\varepsilon(\cdot)$ solve respectively the following stochastic differential equations

$$\begin{cases}
dx_1^\varepsilon(t) = a_1(t)x_1^\varepsilon(t)dt + \big(b_1(t)x_1^\varepsilon(t) + \chi_{E_\varepsilon}(t)\delta b(t)\big)dW(t) \ \text{ in } (0,T],\\
x_1^\varepsilon(0) = 0,
\end{cases}$$

and

$$
\begin{cases}
dx_2^\varepsilon(t) = \Big(a_1(t)x_2^\varepsilon(t) + \chi_{E_\varepsilon}(t)\delta a(t) + \dfrac{1}{2}a_{11}(t)\big(x_1^\varepsilon(t), x_1^\varepsilon(t)\big)\Big)dt \\
\qquad\qquad + \Big(b_1(t)x_2^\varepsilon(t) + \chi_{E_\varepsilon}(t)\delta b_1(t)x_1^\varepsilon(t) + \dfrac{1}{2}b_{11}(t)\big(x_1^\varepsilon(t), x_1^\varepsilon(t)\big)\Big)dW(t) \\
\qquad\qquad\qquad\qquad\qquad\qquad\qquad\qquad\qquad\qquad\qquad\qquad \text{in } (0,T], \\
x_2^\varepsilon(0) = 0.
\end{cases}
$$

Then, when $\varepsilon \to 0$, by a very similar argument to the proof of [371, Theorem 4.4 in Chapter 3] (See [371, pp. 128–134]), one can obtain that

$$
\begin{aligned}
&\mathcal{J}(u^\varepsilon(\cdot)) - \mathcal{J}(\bar{u}(\cdot)) \\
&= \mathbb{E}\big\langle h_x(\bar{x}(T)), x_1^\varepsilon(T) + x_2^\varepsilon(T)\big\rangle_{\mathbb{R}^n} + \frac{1}{2}\mathbb{E}\big\langle h_{xx}(\bar{x}(T))x_1^\varepsilon(T), x_1^\varepsilon(T)\big\rangle_{\mathbb{R}^n} \\
&\quad + \mathbb{E}\int_0^T \Big(\big\langle g_1(t), x_1^\varepsilon(t) + x_2^\varepsilon(t)\big\rangle_{\mathbb{R}^n} + \frac{1}{2}\big\langle g_{11}(t)x_1^\varepsilon(t), x_1^\varepsilon(t)\big\rangle_{\mathbb{R}^n} \\
&\quad + \chi_{E_\varepsilon}(t)\delta g(t)\Big)dt + o(\varepsilon).
\end{aligned}
\tag{12.12}
$$

Since $(y(\cdot), Y(\cdot))$ is the transposition solution to the equation (12.5), we find that

$$
\begin{aligned}
&-\mathbb{E}\big\langle h_x(\bar{x}(T)), x_1^\varepsilon(T)\big\rangle_{\mathbb{R}^n} \\
&= \mathbb{E}\int_0^T \big(\big\langle g_1(t), x_1^\varepsilon(t)\big\rangle_{\mathbb{R}^n} + \chi_{E_\varepsilon}(t)\big\langle \delta b(t), Y(t)\big\rangle_{\mathbb{R}^n}\big)dt,
\end{aligned}
\tag{12.13}
$$

and

$$
\begin{aligned}
&-\mathbb{E}\langle h_x(\bar{x}(T)), x_2^\varepsilon(T)\rangle_{\mathbb{R}^n} \\
&= \mathbb{E}\int_0^T \Big[\big\langle g_1(t), x_2^\varepsilon(t)\big\rangle_{\mathbb{R}^n} + \frac{1}{2}\big(\big\langle y(t), a_{11}(t)\big(x_1^\varepsilon(t), x_1^\varepsilon(t)\big)\big\rangle_{\mathbb{R}^n} \\
&\qquad + \big\langle Y(t), b_{11}(t)\big(x_1^\varepsilon(t), x_1^\varepsilon(t)\big)\big\rangle_{\mathbb{R}^n}\big) \\
&\qquad + \chi_{E_\varepsilon}(t)\big(\big\langle y(t), \delta a(t)\big\rangle_{\mathbb{R}^n} + \big\langle Y(t), \delta b_1(t)x_1^\varepsilon(t)\big\rangle_{\mathbb{R}^n}\big)\Big]dt.
\end{aligned}
\tag{12.14}
$$

Further, put $x_3^\varepsilon(t) = x_1^\varepsilon(t)x_1^\varepsilon(t)^\top (\in \mathbb{R}^{n\times n})$. By Itô's formula, we get that $x_3^\varepsilon(\cdot)$ solves

$$
\begin{cases}
dx_3^\varepsilon(t) = \Big[a_1(t)x_3^\varepsilon(t) + x_3^\varepsilon(t)a_1(t)^\top + b_1(t)x_3^\varepsilon(t)b_1(t)^\top + \chi_{E_\varepsilon}(t)\delta b(t)\delta b(t)^\top \\
\qquad\qquad + \chi_{E_\varepsilon}(t)\Big(b_1(t)x_1^\varepsilon(t)\delta b(t)^\top + \delta b(t)x_1^\varepsilon(t)^\top b_1(t)^\top\Big)\Big]dt \\
\qquad\qquad + \Big[b_1(t)x_3^\varepsilon(t) + x_3^\varepsilon(t)b_1(t)^\top \\
\qquad\qquad + \chi_{E_\varepsilon}(t)\Big(\delta b(t)x_1^\varepsilon(t)^\top + x_1^\varepsilon(t)\delta b(t)^\top\Big)\Big]dW(t) \qquad \text{in } (0,T], \\
x_3^\varepsilon(0) = 0.
\end{cases}
\tag{12.15}
$$

Using the fact that $(P(\cdot), Q(\cdot))$ is the transposition solution to the equation (12.6) and noting that the inner product defined in $\mathbb{R}^{n\times n}$ is $\mathrm{tr}\,(P_1 P_2^\top)$ for $P_1, P_2 \in \mathbb{R}^{n\times n}$, we find that

$$
\begin{aligned}
&-\mathbb{E}\,\mathrm{tr}\,\big(h_{xx}(\bar{x}(T))x_3^\varepsilon(T)\big)\\
&= \mathbb{E}\int_0^T \mathrm{tr}\,\big(\chi_{E_\varepsilon}(t)\delta b(t)^\top P(t)\delta b(t) - \mathbb{H}_{xx}(t,\bar{x}(t),\bar{u}(t),y(t),Y(t))x_3^\varepsilon(t)\big)dt\\
&\quad +o(\varepsilon),
\end{aligned}
$$

which gives that

$$
\begin{aligned}
&-\mathbb{E}\big\langle h_{xx}(\bar{x}(T))x_1^\varepsilon(T), x_1^\varepsilon(T)\big\rangle_{\mathbb{R}^n}\\
&= \mathbb{E}\int_0^T \Big(\chi_{E_\varepsilon}(t)\big\langle P(t)\delta b(t), \delta b(t)\big\rangle_{\mathbb{R}^n} \\
&\qquad -\big\langle \mathbb{H}_{xx}(t,\bar{x}(t),\bar{u}(t),y(t),Y(t))x_1^\varepsilon(t), x_1^\varepsilon(t)\big\rangle_{\mathbb{R}^n}\Big)dt + o(\varepsilon).
\end{aligned}
\tag{12.16}
$$

From (12.12)–(12.16), we obtain that

$$
\begin{aligned}
&\mathcal{J}(u^\varepsilon(\cdot)) - \mathcal{J}(\bar{u}(\cdot))\\
&= \mathbb{E}\int_0^T \chi_{E_\varepsilon}(t)\Big(\mathbb{H}(t,\bar{x}(t),\bar{u}(t),y(t),Y(t)) - \mathbb{H}(t,\bar{x}(t),u(t),y(t),Y(t))\\
&\qquad -\frac{1}{2}\big\langle P(t)\delta b(t), \delta b(t)\big\rangle_{\mathbb{R}^n}\Big)dt + o(\varepsilon).
\end{aligned}
\tag{12.17}
$$

Since $\bar{u}(\cdot)$ is an optimal control, we have $\mathcal{J}(u^\varepsilon(\cdot)) - \mathcal{J}(\bar{u}(\cdot)) \geq 0$. This, together with (12.17), yields that

$$
\begin{aligned}
&\mathbb{E}\int_0^T \chi_{E_\varepsilon}(t)\Big(\mathbb{H}(t,\bar{x}(t),\bar{u}(t),y(t),Y(t)) - \mathbb{H}(t,\bar{x}(t),u(t),y(t),Y(t))\\
&\qquad -\frac{1}{2}\big\langle P(t)\delta b(t), \delta b(t)\big\rangle_{\mathbb{R}^n}\Big)dt\\
&\geq o(\varepsilon),
\end{aligned}
$$

which gives (12.11). □

12.3 Necessary Condition for Optimal Controls for Convex Control Regions

In this section, we shall present a necessary condition for optimal controls of Problem (OP) under the following condition:

(S4) *The control region U is a convex subset of a separable Hilbert space $\widetilde{H}$ and the metric of U is introduced by the norm of $\widetilde{H}$, i.e., $d(u_1,u_2)=|u_1-u_2|_{\widetilde{H}}$.*

We begin with the following preliminary result.

Lemma 12.3. *If $F(\cdot)\in L^2_{\mathbb{F}}(0,T;\widetilde{H})$ and $\bar u(\cdot)\in\mathcal{U}[0,T]$ such that*

$$\mathrm{Re}\,\mathbb{E}\int_0^T \big\langle F(t,\cdot),u(t,\cdot)-\bar u(t,\cdot)\big\rangle_{\widetilde{H}}dt\le 0, \tag{12.18}$$

holds for any $u(\cdot)\in\mathcal{U}[0,T]$ satisfying $u(\cdot)-\bar u(\cdot)\in L^2_{\mathbb{F}}(0,T;\widetilde{H}))$, then,

$$\mathrm{Re}\,\big\langle F(t,\omega),u-\bar u(t,\omega)\big\rangle_{\widetilde{H}}\le 0,\ \text{ a.e. } (t,\omega)\in[0,T]\times\Omega,\ \ \forall\, u\in U. \tag{12.19}$$

Proof: We use a contradiction argument. Suppose that the inequality (12.19) did not hold. Then, there would exist a $u_0\in U$ and an $\varepsilon>0$ such that

$$\alpha_\varepsilon\overset{\Delta}{=}\int_\Omega\int_0^T\chi_{\Lambda_\varepsilon}(t,\omega)dtd\mathbb{P}>0,$$

where

$$\Lambda_\varepsilon\overset{\Delta}{=}\big\{(t,\omega)\in[0,T]\times\Omega\ \big|\ \mathrm{Re}\,\big\langle F(t,\omega),u_0-\bar u(t,\omega)\big\rangle_{\widetilde{H}}\ge\varepsilon\big\}.$$

For any $m\in\mathbb{N}$, define

$$\Lambda_{\varepsilon,m}\overset{\Delta}{=}\Lambda_\varepsilon\cap\big\{(t,\omega)\in[0,T]\times\Omega\ \big|\ |\bar u(t,\omega)|_{\widetilde{H}}\le m\big\}.$$

It is clear that $\lim\limits_{m\to\infty}\Lambda_{\varepsilon,m}=\Lambda_\varepsilon$. Hence, there is an $m_\varepsilon\in\mathbb{N}$ such that

$$\int_\Omega\int_0^T\chi_{\Lambda_{\varepsilon,m}}(t,\omega)dtd\mathbb{P}>\frac{\alpha_\varepsilon}{2}>0,\qquad\forall\, m\ge m_\varepsilon.$$

Since $\big\langle F(\cdot),u_0-\bar u(\cdot)\big\rangle_{\widetilde{H}}$ is $\mathbb{F}$-adapted, so is the process $\chi_{\Lambda_{\varepsilon,m}}(\cdot)$. Define

$$\hat u_{\varepsilon,m}(t,\omega)=u_0\chi_{\Lambda_{\varepsilon,m}}(t,\omega)+\bar u(t,\omega)\chi_{\Lambda^c_{\varepsilon,m}}(t,\omega),\quad (t,\omega)\in[0,T]\times\Omega.$$

Noting that $|\bar u(\cdot)|_{\widetilde{H}}\le m$ on $\Lambda_{\varepsilon,m}$, we see that $\hat u_{\varepsilon,m}(\cdot)\in\mathcal{U}[0,T]$ satisfies $\hat u_{\varepsilon,m}(\cdot)-\bar u(\cdot)\in L^2_{\mathbb{F}}(0,T;\widetilde{H})$. Hence, for any $m\ge m_\varepsilon$, we obtain that

$$\begin{aligned}&\mathrm{Re}\,\mathbb{E}\int_0^T\big\langle F(t),\hat u_{\varepsilon,m}(t)-\bar u(t)\big\rangle_{\widetilde{H}}dt\\&=\int_\Omega\int_0^T\chi_{\Lambda_{\varepsilon,m}}(t,\omega)\mathrm{Re}\,\big\langle F(t,\omega),u_0-\bar u(t,\omega)\big\rangle_{\widetilde{H}}\,dt\mathbb{P}\\&\ge\varepsilon\int_\Omega\int_0^T\chi_{\Lambda_{\varepsilon,m}}(t,\omega)dt\mathbb{P}\ge\frac{\varepsilon\alpha_\varepsilon}{2}>0,\end{aligned}$$

which contradicts (12.18). This completes the proof of Lemma 12.3. □

In this section, we introduce the following further assumptions for $a(\cdot,\cdot,\cdot)$, $b(\cdot,\cdot,\cdot)$, $g(\cdot,\cdot,\cdot)$ and $h(\cdot)$.

(S5) *For a.e. $(t,\omega)\in(0,T)\times\Omega$, the functions $a(t,\cdot,\cdot):H\times U\to H$ and $b(t,\cdot,\cdot):H\times U\to\mathcal{L}_2^0$, $g(t,\cdot,\cdot):H\times U\to\mathbb{R}$ and $h(\cdot):H\to\mathbb{R}$ are C^1. Moreover, for any $(x,u)\in H\times U$ and a.e. $(t,\omega)\in(0,T)\times\Omega$,*

$$
\begin{cases}
|a_x(t,x,u)|_{\mathcal{L}(H)}+|b_x(t,x,u)|_{\mathcal{L}(H;\mathcal{L}_2^0)}+|g_x(t,x,u)|_H+|h_x(x)|_H\le \mathcal{C}_L,\\
|a_u(t,x,u)|_{\mathcal{L}(\widetilde H;H)}+|b_u(t,x,u)|_{\mathcal{L}(\widetilde H;\mathcal{L}_2^0)}+|g_u(t,x,u)|_{\widetilde H}\le \mathcal{C}_L.
\end{cases}
$$

We have the following necessary condition for optimal controls for Problem (OP) with convex control domains.

Theorem 12.4. *Assume that the assumptions (S1)–(S2) and (S4)–(S5) hold. Let $x_0\in L^2_{\mathcal{F}_0}(\Omega;H)$ and $(\bar x(\cdot),\bar u(\cdot))$ be an optimal pair for Problem (OP). Then,*

$$
\begin{array}{ll}
\mathrm{Re}\,\big\langle a_u(t,\bar x(t),\bar u(t))^*y(t)+b_u(t,\bar x(t),\bar u(t))^*Y(t)-g_u(t,\bar u(t),\bar x(t)),u-\bar u(t)\big\rangle_{\widetilde H}\\
\le 0,\qquad \text{a.e. } (t,\omega)\in[0,T]\times\Omega,\ \ \forall\, u\in U,
\end{array}
\tag{12.20}
$$

where $(y(\cdot),Y(\cdot))$ is the transposition solution to the equation (12.7).

Proof: We use the convex perturbation technique and divide the proof into several steps.

Step 1. For the optimal pair $(\bar x(\cdot),\bar u(\cdot))$, we fix arbitrarily a control $u(\cdot)\in\mathcal{U}[0,T]$ satisfying $u(\cdot)-\bar u(\cdot)\in L^2_{\mathbb{F}}(0,T;\widetilde H))$. Since U is convex, we see that

$$
u^\varepsilon(\cdot)=\bar u(\cdot)+\varepsilon(u(\cdot)-\bar u(\cdot))=(1-\varepsilon)\bar u(\cdot)+\varepsilon u(\cdot)\in\mathcal{U}[0,T],\quad\forall\,\varepsilon\in[0,1].
$$

Denote by $x^\varepsilon(\cdot)$ the state process of (12.2) corresponding to the control $u^\varepsilon(\cdot)$. By Proposition 12.1, it follows that

$$
|x^\varepsilon|_{C_{\mathbb{F}}([0,T];L^2(\Omega;H))}\le C\big(1+|x_0|_{L^2_{\mathcal{F}_0}(\Omega;H)}\big),\quad\forall\,\varepsilon\in[0,1].\tag{12.21}
$$

Write $x_1^\varepsilon(\cdot)=\dfrac{1}{\varepsilon}\big(x^\varepsilon(\cdot)-\bar x(\cdot)\big)$ and $\delta u(\cdot)=u(\cdot)-\bar u(\cdot)$. Since $(\bar x(\cdot),\bar u(\cdot))$ satisfies (12.2), it is easy to see that $x_1^\varepsilon(\cdot)$ satisfies the following stochastic evolution equation:

$$
\begin{cases}
dx_1^\varepsilon=\big(Ax_1^\varepsilon+a_1^\varepsilon x_1^\varepsilon+a_2^\varepsilon\delta u\big)dt+\big(b_1^\varepsilon x_1^\varepsilon+b_2^\varepsilon\delta u\big)dW(t)\ \text{ in } (0,T],\\
x_1^\varepsilon(0)=0,
\end{cases}\tag{12.22}
$$

where for $\psi=a,b$,

$$\begin{cases} \psi_1^\varepsilon(t) = \int_0^1 \psi_x(t, \bar{x}(t) + \sigma\varepsilon x_1^\varepsilon(t), u^\varepsilon(t))d\sigma, \\ \psi_2^\varepsilon(t) = \int_0^1 \psi_u(t, \bar{x}(t), \bar{u}(t) + \sigma\varepsilon\delta u(t))d\sigma. \end{cases} \tag{12.23}$$

Consider the following stochastic evolution equation:

$$\begin{cases} dx_2 = \big(Ax_2 + a_1(t)x_2 + a_2(t)\delta u\big)dt + \big(b_1(t)x_2 + b_2(t)\delta u\big)dW(t) \text{ in } (0,T], \\ x_2(0) = 0, \end{cases} \tag{12.24}$$

where for $\psi = a, b$,

$$\psi_1(t) = \psi_x(t, \bar{x}(t), \bar{u}(t)), \qquad \psi_2(t) = \psi_u(t, \bar{x}(t), \bar{u}(t)). \tag{12.25}$$

Step 2. In this step, we shall prove that

$$\lim_{\varepsilon\to 0+} |x_1^\varepsilon - x_2|_{L^\infty_{\mathbb{F}}(0,T;L^2(\Omega;H))} = 0. \tag{12.26}$$

First, applying Theorem 3.14 to (12.22) and by the assumption (S1), we find that

$$\begin{aligned} &\sup_{t\in[0,T]} \mathbb{E}|x_1^\varepsilon(t)|_H^2 \\ &\le C\mathbb{E}\Big[\Big(\int_0^T |a_2^\varepsilon(s)\delta u(s)|_H ds\Big)^2 + \int_0^T |b_2^\varepsilon(s)\delta u(s)|^2_{\mathcal{L}_2^0} ds\Big] \\ &\le C|\bar{u} - u|^2_{L^2_{\mathbb{F}}(0,T;\widetilde{H}))}. \end{aligned} \tag{12.27}$$

By a similar computation, we obtain that

$$\sup_{t\in[0,T]} \mathbb{E}|x_2(t)|_H^2 \le C|\bar{u} - u|^2_{L^2_{\mathbb{F}}(0,T;\widetilde{H}))}. \tag{12.28}$$

On the other hand, put $x_3^\varepsilon = x_1^\varepsilon - x_2$. Then, x_3^ε solves the following equation:

$$\begin{cases} dx_3^\varepsilon = \big[Ax_3^\varepsilon + a_1^\varepsilon(t)x_3^\varepsilon + \big(a_1^\varepsilon(t) - a_1(t)\big)x_2 + \big(a_2^\varepsilon(t) - a_2(t)\big)\delta u\big]dt \\ \qquad\quad + \big[b_1^\varepsilon(t)x_3^\varepsilon + \big(b_1^\varepsilon(t) - b_1(t)\big)x_2 + \big(b_2^\varepsilon(t) - b_2(t)\big)\delta u\big]dW(t) \text{ in } (0,T], \\ x_3^\varepsilon(0) = 0. \end{cases} \tag{12.29}$$

Applying Theorem 3.14 to (12.29), we obtain that

$$\begin{aligned} &\max_{t\in[0,T]} \mathbb{E}|x_3^\varepsilon(t)|_H^2 \\ &\le C\mathbb{E}\int_0^T \Big[\big(|a_1^\varepsilon(s) - a_1(s)|^2_{\mathcal{L}(H)} + |b_1^\varepsilon(s) - b_1(s)|^2_{\mathcal{L}(H;\mathcal{L}_2^0)}\big)|x_2(s)|_H^2 \\ &\quad + \big(|a_2^\varepsilon(s) - a_2(s)|^2_{\mathcal{L}(\widetilde{H};H)} + |b_2^\varepsilon(s) - b_2(s)|^2_{\mathcal{L}(\widetilde{H};\mathcal{L}_2^0)}\big)|u(s) - \bar{u}(s)|^2_{\widetilde{H}}\Big]ds. \end{aligned} \tag{12.30}$$

Note that (12.27) implies $x^\varepsilon(\cdot) \to \bar{x}(\cdot)$ (in H) in probability, as $\varepsilon \to 0$. Hence, by (12.23), (12.25) and the continuity of $a_x(t,\cdot,\cdot)$, $b_x(t,\cdot,\cdot)$, $a_u(t,\cdot,\cdot)$ and $b_u(t,\cdot,\cdot)$, and noting (12.28), we deduce that

$$\begin{aligned}&\mathbb{E}\int_0^T \Big[\big(|a_1^\varepsilon(s)-a_1(s)|^2_{\mathcal{L}(H)}+|b_1^\varepsilon(s)-b_1(s)|^2_{\mathcal{L}(H;\mathcal{L}_2^0)}\big)|x_2(s)|^2_H\\&\quad+\big(|a_2^\varepsilon(s)-a_2(s)|^2_{\mathcal{L}(\widetilde{H};H)}+|b_2^\varepsilon(s)-b_2(s)|^2_{\mathcal{L}(\widetilde{H};\mathcal{L}_2^0)}\big)|u(s)-\bar{u}(s)|^2_{\widetilde{H}}\Big]ds=0.\end{aligned}$$

This, combined with (12.30), gives (12.26).

Step 3. Since $(\bar{x}(\cdot),\bar{u}(\cdot))$ is an optimal pair for Problem (OP), from (12.26), we find that

$$\begin{aligned}0&\le \lim_{\varepsilon\to 0}\frac{\mathcal{J}(u^\varepsilon(\cdot))-\mathcal{J}(\bar{u}(\cdot))}{\varepsilon}\\&=\mathrm{Re}\Big[\mathbb{E}\int_0^T\Big(\big\langle g_1(t),x_2(t)\big\rangle_H+\big\langle g_2(t),\delta u(t)\big\rangle_{\widetilde{H}}\Big)dt\\&\qquad+\mathbb{E}\big\langle h_x(\bar{x}(T)),x_2(T)\big\rangle_H\Big],\end{aligned}\tag{12.31}$$

where $g_1(t)=g_x(t,\bar{x}(t),\bar{u}(t))$ and $g_2(t)=g_u(t,\bar{x}(t),\bar{u}(t))$.

By the definition of the transposition solution to (12.7), it follows that

$$\begin{aligned}&-\mathbb{E}\big\langle h_x(\bar{x}(T)),x_2(T)\big\rangle_H-\mathbb{E}\int_0^T\big\langle g_1(t),x_2(t)\big\rangle_H dt\\&=\mathbb{E}\int_0^T\Big(\big\langle a_2(t)\delta u(t),y(t)\big\rangle_H+\big\langle b_2(t)\delta u(t),Y(t)\big\rangle_{\mathcal{L}_2^0}\Big)dt.\end{aligned}\tag{12.32}$$

Combining (12.31) and (12.32), we find that

$$\mathrm{Re}\,\mathbb{E}\int_0^T\big\langle a_2(t)^*y(t)+b_2(t)^*Y(t)-g_2(t),u(t)-\bar{u}(t)\big\rangle_{\widetilde{H}}dt\le 0 \tag{12.33}$$

holds for any $u(\cdot)\in\mathcal{U}[0,T]$ satisfying $u(\cdot)-\bar{u}(\cdot)\in L^2_{\mathbb{F}}(0,T;\widetilde{H}))$. Hence, by means of Lemma 12.3, from (12.33), we conclude the desired inequality (12.20). This completes the proof of Theorem 12.4. □

12.4 Operator-Valued Backward Stochastic Evolution Equations

This section is addressed to the well-posedness and regularity of solutions to the operator-valued backward stochastic evolution equation (12.8).

12.4.1 Notions of Solutions

In this subsection, we shall define the solution to (12.8) in the transposition sense.

We begin with some simple observation. Any $Q \in \mathcal{L}(\mathcal{L}_2^0; H)$ induces a formal bilinear functional $\Psi(\cdot,\cdot)$ on $H \times \mathcal{L}_2^0$ as follows:

$$\Psi(h,v) \overset{\Delta}{=} \sum_{j=1}^{\infty} \langle Q(e_j \otimes h), ve_j \rangle_H, \qquad \forall\, h \in H, v \in \mathcal{L}_2^0. \tag{12.34}$$

Here, $e_j \otimes h$ is defined by (2.18) (Clearly, $e_j \otimes h \in \mathcal{L}_2^0$, and therefore $Q(e_j \otimes h) \in H$). Recall that $\mathcal{L}_2^0 = \mathcal{L}_2(V; H)$. It is easy to show that $\Psi(\cdot,\cdot)$ is a bounded bilinear functional on $H \times \mathcal{L}_2^0$ provided either $\dim V < \infty$ or $Q \in \mathcal{L}_2(\mathcal{L}_2^0; H)$. When $\Psi(\cdot,\cdot)$ is a bounded bilinear functional on $H \times \mathcal{L}_2^0$, it determines uniquely a bounded linear operator $\widetilde{Q} : H \to \mathcal{L}_2^0$ satisfying

$$\Psi(h,v) = \langle \widetilde{Q}h, v \rangle_{\mathcal{L}_2^0}, \qquad \forall\, h \in H, v \in \mathcal{L}_2^0.$$

In this case, we say that Q induces a bounded linear operator $\widetilde{Q} : H \to \mathcal{L}_2^0$, and by (12.34), we have

$$\langle \widetilde{Q}h, v \rangle_{\mathcal{L}_2^0} = \sum_{j=1}^{\infty} \langle Q(e_j \otimes h), ve_j \rangle_H, \qquad \forall\, h \in H, v \in \mathcal{L}_2^0. \tag{12.35}$$

On the other hand, any $\mathcal{Q} \in \mathcal{L}(H; \mathcal{L}_2^0)$ induces a formally linear operator $\widetilde{\mathcal{Q}} : \mathcal{L}_2^0 \to H$, defined by

$$\widetilde{\mathcal{Q}}\Big(\sum_{i,j=1}^{\infty} a_{ij} e_i \otimes h_j \Big) \overset{\Delta}{=} \sum_{i,j=1}^{\infty} a_{ij} (\mathcal{Q}h_j) e_i, \quad \forall\, a_{ij} \in \mathbb{C} \text{ with } \sum_{i,j=1}^{\infty} |a_{ij}|^2 < \infty. \tag{12.36}$$

It is easy to show that, if $\widetilde{\mathcal{Q}}$ is a bounded linear operator from $\mathcal{L}_2^0$ to H, then

$$\widetilde{\mathcal{Q}}(v \otimes h) = (\mathcal{Q}h)v, \qquad \forall\, v \in V,\ h \in H. \tag{12.37}$$

In this case, we say that $\mathcal{Q}$ induces a bounded linear operator $\widetilde{\mathcal{Q}} : \mathcal{L}_2^0 \to H$. It is easy to show that $\widetilde{\mathcal{Q}}$ is a bounded linear operator provided either $\dim V < \infty$ or $\mathcal{Q} \in \mathcal{L}_2(H; \mathcal{L}_2^0)$.

Now, suppose that $Q \in \mathcal{L}(\mathcal{L}_2^0; H)$ induces a bounded linear operator $\widetilde{Q} : H \to \mathcal{L}_2^0$ and $\widetilde{Q}$ induces a bounded linear operator $\overline{Q} : \mathcal{L}_2^0 \to H$. Then, $Q = \overline{Q}$. Conversely, suppose that $\mathcal{Q} \in \mathcal{L}(H; \mathcal{L}_2^0)$ induces a bounded linear operator $\widetilde{\mathcal{Q}} : \mathcal{L}_2^0 \to H$ and $\widetilde{\mathcal{Q}}$ induces a bounded linear operator $\overline{\mathcal{Q}} : H \to \mathcal{L}_2^0$. Then, $\mathcal{Q} = \overline{\mathcal{Q}}$.

We shall need the following result.

Proposition 12.5. *Any $\Lambda \in \mathcal{L}_2(V;\mathcal{L}_2(H))$ induces (uniquely) a bounded linear operator $Q \in \mathcal{L}_2(\mathcal{L}_2^0;H)$ and a bounded linear operator $\widetilde{Q} \in \mathcal{L}_2(H;\mathcal{L}_2^0)$ satisfying* (12.35) *and*

$$Q(v\otimes h)=(\Lambda v)h, \qquad \forall\, v\in V,\ h\in H. \tag{12.38}$$

Moreover,

$$|Q|_{\mathcal{L}_2(\mathcal{L}_2^0;H)}+|\widetilde{Q}|_{\mathcal{L}_2(H;\mathcal{L}_2^0)}\leq \mathcal{C}|\Lambda|_{\mathcal{L}_2(V;\mathcal{L}_2(H))}.$$

Proof: As in (12.36), we define a linear operator Q from $\mathcal{L}_2^0$ to H by

$$Q\Big(\sum_{i,j=1}^{\infty}a_{ij}e_i\otimes h_j\Big)\stackrel{\Delta}{=}\sum_{i,j=1}^{\infty}a_{ij}(\Lambda e_i)h_j,\quad \forall\, a_{ij}\in\mathbb{C}\text{ with }\sum_{i,j=1}^{\infty}|a_{ij}|^2<\infty.$$

By $\Lambda \in \mathcal{L}_2(V;\mathcal{L}_2(H))$, it is easy to check that $Q\in\mathcal{L}_2(\mathcal{L}_2^0;H)$ and

$$|Q|_{\mathcal{L}_2(\mathcal{L}_2^0;H)}\leq \mathcal{C}|\Lambda|_{\mathcal{L}_2(V;\mathcal{L}_2(H))}.$$

Then, Q induces a bounded linear operator $\widetilde{Q}\in\mathcal{L}_2(H;\mathcal{L}_2^0)$ satisfying (12.35) and

$$|\widetilde{Q}|_{\mathcal{L}_2(H;\mathcal{L}_2^0)}\leq \mathcal{C}|Q|_{\mathcal{L}_2(\mathcal{L}_2^0;H)}.$$

This completes the proof of Proposition 12.5. □

Let us introduce the following two (forward) stochastic evolution equations:

$$\begin{cases} dx_1=(A+J)x_1ds+u_1ds+Kx_1dW(s)+v_1dW(s) & \text{in } (t,T],\\ x_1(t)=\xi_1 \end{cases} \tag{12.39}$$

and

$$\begin{cases} dx_2=(A+J)x_2ds+u_2ds+Kx_2dW(s)+v_2dW(s) & \text{in } (t,T],\\ x_2(t)=\xi_2. \end{cases} \tag{12.40}$$

Here $\xi_1,\xi_2\in L^{2q}_{\mathcal{F}_t}(\Omega;H)$, $u_1,u_2\in L^2_{\mathbb{F}}(t,T;L^{2q}(\Omega;H))$, and $v_1,v_2\in L^2_{\mathbb{F}}(t,T;L^{2q}(\Omega;\mathcal{L}_2^0))$. Also, we need to introduce the solution space for (12.8). For this purpose, write

$$\begin{aligned} &D_{\mathbb{F},w}([0,T];L^p(\Omega;\mathcal{L}(H))\\ &\stackrel{\Delta}{=}\Big\{P(\cdot,\cdot)\ \Big|\ P(\cdot,\cdot)\in\mathcal{L}_{pd}\big(L^2_{\mathbb{F}}(0,T;L^{2q}(\Omega;H));\ L^2(0,T;L_{\mathbb{F}}^{\frac{2p}{p+1}}(\Omega;H))\big),\\ &\quad P(\cdot,\cdot)\xi\in D_{\mathbb{F}}([t,T];L^{\frac{2p}{p+1}}(\Omega;H))\ \text{ and }\ |P(\cdot,\cdot)\xi|_{D_{\mathbb{F}}([t,T];L^{\frac{2p}{p+1}}(\Omega;H))}\\ &\quad\leq C|\xi|_{L^{2q}_{\mathcal{F}_t}(\Omega;H)}\ \text{for every } t\in[0,T]\text{ and }\xi\in L^{2q}_{\mathcal{F}_t}(\Omega;H)\Big\}, \end{aligned} \tag{12.41}$$

and

$$
\begin{aligned}
&L^2_{\mathbb{F},w}(0,T;L^p(\Omega;\mathcal{L}(\mathcal{L}_2^0;H)))\\
&\stackrel{\Delta}{=}\Big\{Q(\cdot,\cdot)\ \Big|\ Q(\cdot,\cdot)\in\mathcal{L}_{pd}\big(L^2_{\mathbb{F}}(0,T;L^{2q}(\Omega;\mathcal{L}_2^0));\ L^1_{\mathbb{F}}(0,T;L^{\frac{2p}{p+1}}(\Omega;H))\big),\\
&\qquad\text{and } Q(t,\omega) \text{ induces a bounded linear operator } \widetilde{Q}(t,\omega):H\to\mathcal{L}_2^0\\
&\qquad\text{for a.e. } (t,\omega)\in[0,T]\times\Omega\Big\}.
\end{aligned}
\tag{12.42}
$$

We now introduce the notion of transposition solution to (12.8) as follows:

Definition 12.6. *We call* $(P(\cdot),Q(\cdot))\in D_{\mathbb{F},w}([0,T];L^p(\Omega;\mathcal{L}(H)))\times L^2_{\mathbb{F},w}(0,T;L^p(\Omega;\mathcal{L}(\mathcal{L}_2^0;H)))$ *a transposition solution to* (12.8) *if for any* $t\in[0,T]$, $\xi_1,\xi_2\in L^{2q}_{\mathcal{F}_t}(\Omega;H)$, $u_1(\cdot),u_2(\cdot)\in L^2_{\mathbb{F}}(t,T;L^{2q}(\Omega;H))$ *and* $v_1(\cdot),v_2(\cdot)\in L^2_{\mathbb{F}}(t,T;L^{2q}(\Omega;\mathcal{L}_2^0))$, *it holds that* [3]

$$
\begin{aligned}
&\mathbb{E}\big\langle P_T x_1(T),x_2(T)\big\rangle_H-\mathbb{E}\int_t^T\big\langle F(s)x_1(s),x_2(s)\big\rangle_H ds\\
&=\mathbb{E}\big\langle P(t)\xi_1,\xi_2\big\rangle_H+\mathbb{E}\int_t^T\big\langle P(s)u_1(s),x_2(s)\big\rangle_H ds\\
&\quad+\mathbb{E}\int_t^T\big\langle P(s)x_1(s),u_2(s)\big\rangle_H ds+\mathbb{E}\int_t^T\big\langle P(s)K(s)x_1(s),v_2(s)\big\rangle_{\mathcal{L}_2^0}ds\\
&\quad+\mathbb{E}\int_t^T\big\langle P(s)v_1(s),K(s)x_2(s)+v_2(s)\big\rangle_{\mathcal{L}_2^0}ds\\
&\quad+\mathbb{E}\int_t^T\big\langle Q(s)v_1(s),x_2(s)\big\rangle_H ds+\mathbb{E}\int_t^T\big\langle \widetilde{Q}(s)x_1(s),v_2(s)\big\rangle_{\mathcal{L}_2^0}ds.
\end{aligned}
\tag{12.43}
$$

Here, $x_1(\cdot)$ *and* $x_2(\cdot)$ *solve* (12.39) *and* (12.40), *respectively.*

We have the following uniqueness result for the transposition solution to (12.8).

Theorem 12.7. *If* $p\in(1,2]$, *and* J, K, F *and* P_T *satisfy* (12.9)–(12.10), *then the equation* (12.8) *admits at most one transposition solution* $(P(\cdot),Q(\cdot))\in D_{\mathbb{F},w}([0,T];L^p(\Omega;\mathcal{L}(H)))\times L^p_{\mathbb{F},w}(\Omega;L^2(0,T;\mathcal{L}(\mathcal{L}_2^0;H)))$.

We shall give a proof of Theorem 12.7 in Subsection 12.4.3.

Clearly, Definition 12.6 looks quite natural. However, the corresponding well-posedness of (12.8) is still unsolved. Because of this, we introduce below a weaker concept of solution to (12.8).

Write

$$
\mathcal{H}_t\stackrel{\Delta}{=}L^{2q}_{\mathcal{F}_t}(\Omega;H)\times L^2_{\mathbb{F}}(t,T;L^{2q}(\Omega;H))\times L^2_{\mathbb{F}}(t,T;L^{2q}(\Omega;\mathcal{L}_2^0)),\quad\forall\, t\in[0,T),
$$

---

[3] In (12.43), $\widetilde{Q}(\cdot)$ is the (pointwise defined) operator induced uniquely by $Q(\cdot)$.

and[4]

$$
\begin{aligned}
&\mathcal{Q}^p[0,T]\\
&\triangleq\Big\{\big(Q^{(\cdot)},\widehat{Q}^{(\cdot)}\big)\ \Big|\ Q^{(t)},\widehat{Q}^{(t)}\in\mathcal{L}(\mathcal{H}_t;\ L^2_{\mathbb{F}}(t,T;L^{\frac{2p}{p+1}}(\Omega;\mathcal{L}_2^0)))\\
&\qquad\text{and } Q^{(t)}(0,0,\cdot)^*=\widehat{Q}^{(t)}(0,0,\cdot)\text{ for any }t\in[0,T)\Big\}.
\end{aligned}
\tag{12.44}
$$

We now define the notion of relaxed transposition solution to (12.8) as follows:

Definition 12.8. *We call $\big(P(\cdot),\ Q^{(\cdot)},\ \widehat{Q}^{(\cdot)}\big)\in D_{\mathbb{F},w}([0,T];\ L^p(\Omega;\ \mathcal{L}(H)))\times\mathcal{Q}^p[0,T]$ a relaxed transposition solution to the equation* (12.8) *if for any $t\in[0,T]$, $\xi_1,\xi_2\in L^{2q}_{\mathcal{F}_t}(\Omega;H)$, $u_1(\cdot),u_2(\cdot)\in L^2_{\mathbb{F}}(t,T;L^{2q}(\Omega;H))$ and $v_1(\cdot),v_2(\cdot)\in L^2_{\mathbb{F}}(t,T;L^{2q}(\Omega;\mathcal{L}_2^0))$, it holds that*

$$
\begin{aligned}
&\mathbb{E}\big\langle P_Tx_1(T),x_2(T)\big\rangle_H-\mathbb{E}\int_t^T\big\langle F(s)x_1(s),x_2(s)\big\rangle_H ds\\
&=\mathbb{E}\big\langle P(t)\xi_1,\xi_2\big\rangle_H+\mathbb{E}\int_t^T\big\langle P(s)u_1(s),x_2(s)\big\rangle_H ds\\
&\quad+\mathbb{E}\int_t^T\big\langle P(s)x_1(s),u_2(s)\big\rangle_H ds+\mathbb{E}\int_t^T\big\langle P(s)K(s)x_1(s),v_2(s)\big\rangle_{\mathcal{L}_2^0}ds\\
&\quad+\mathbb{E}\int_t^T\big\langle P(s)v_1(s),K(s)x_2(s)+v_2(s)\big\rangle_{\mathcal{L}_2^0}ds\\
&\quad+\mathbb{E}\int_t^T\big\langle v_1(s),\widehat{Q}^{(t)}(\xi_2,u_2,v_2)(s)\big\rangle_{\mathcal{L}_2^0}ds\\
&\quad+\mathbb{E}\int_t^T\big\langle Q^{(t)}(\xi_1,u_1,v_1)(s),v_2(s)\big\rangle_{\mathcal{L}_2^0}ds,
\end{aligned}
\tag{12.45}
$$

where $x_1(\cdot)$ and $x_2(\cdot)$ solve respectively (12.39) *and* (12.40).

We refer to Remark 12.11 for the relationship between transposition solutions and relaxed transposition solutions.

We have the following well-posedness result for the equation (12.8) in the sense of relaxed transposition solution.

Theorem 12.9. *Assume that $p\in(1,2]$ and the Banach space $L^p_{\mathcal{F}_T}(\Omega;\mathbb{C})$ is separable. Then, for any J, K, F and P_T satisfying* (12.9)–(12.10)*, the equation* (12.8) *admits one and only one relaxed transposition solution $\big(P(\cdot),Q^{(\cdot)},\widehat{Q}^{(\cdot)}\big)\in D_{\mathbb{F},w}([0,T];L^p(\Omega;\mathcal{L}(H)))\times\mathcal{Q}^p[0,T]$. Furthermore,*

---

[4] By Theorem 2.73 and noting $\mathcal{L}_2^0$ is a Hilbert space, we see that $Q^{(t)}(0,0,\cdot)^*$ is a bounded linear operator from $L^2_{\mathbb{F}}(t,T;\ L^{\frac{2q}{2q-1}}(\Omega;\mathcal{L}_2^0))^*=L^2_{\mathbb{F}}(t,T;L^{2q}(\Omega;\mathcal{L}_2^0))$ to $L^2_{\mathbb{F}}(t,T;L^{2q}(\Omega;\mathcal{L}_2^0))^*=L^2_{\mathbb{F}}(t,T;\ L^{\frac{2q}{2q-1}}(\Omega;\mathcal{L}_2^0))$. Hence, $Q^{(t)}(0,0,\cdot)^*=\widehat{Q}^{(t)}(0,0,\cdot)$ makes sense.

$$\begin{aligned}
&|P|_{\mathcal{L}(L^2_{\mathbb{F}}(0,T;L^{2q}(\Omega;H));\ L^2(0,T;L^{2p/(p+1)}_{\mathbb{F}}(\Omega;H)))}\\
&\quad + \sup_{t\in[0,T)} \big|(Q^{(t)},\widehat{Q}^{(t)})\big|_{\mathcal{L}(\mathcal{H}_t;\ L^2_{\mathbb{F}}(t,T;L^{2p/(p+1)}(\Omega;\mathcal{L}_2^0)))^2} \qquad (12.46)\\
&\leq \mathcal{C}\big(|F|_{L^1_{\mathbb{F}}(\tau,T;\ L^p(\Omega;\mathcal{L}(H)))} + |P_T|_{L^p_{\mathcal{F}_T}(\Omega;\ \mathcal{L}(H))}\big).
\end{aligned}$$

The proof of Theorem 12.9 is quite long. We shall prove its "Uniqueness" part in Subsection 12.4.3, and its "Existence and Stability" part in Subsection 12.4.5.

12.4.2 Preliminaries

In this section, we present two preliminary results which will be used in the sequel.

Lemma 12.10. *Let $p \in (1,2]$, and J and K satisfy (12.9). Write*

$$M_{J,K,q}(\cdot) \stackrel{\Delta}{=} \big|J(\cdot)\big|^{2q}_{L^\infty(\Omega;\mathcal{L}(H))} + \big|K(\cdot)\big|^{2q}_{L^\infty(\Omega;\mathcal{L}(H;\mathcal{L}_2^0))}. \qquad (12.47)$$

Then, for each $t \in [0,T]$, the following three conclusions hold:

1) If $u_2 = v_2 = 0$ in the equation (12.40), then there exists an operator $U(\cdot,t) \in \mathcal{L}\big(L^{2q}_{\mathcal{F}_t}(\Omega;H);C_{\mathbb{F}}([t,T];L^{2q}(\Omega;H))\big)$ such that the solution to (12.40) can be represented as $x_2(\cdot) = \mathbf{U}(\cdot,t)\xi_2$. Further, for any $t \in [0,T)$, $\xi \in L^{2q}_{\mathcal{F}_t}(\Omega;H)$ and $\varepsilon > 0$, there is a $\delta \in (0,T-t)$ such that for any $s \in [t,t+\delta]$, it holds that

$$|\mathbf{U}(\cdot,t)\xi - \mathbf{U}(\cdot,s)\xi|_{L^\infty_{\mathbb{F}}(s,T;L^{2q}(\Omega;H))} < \varepsilon. \qquad (12.48)$$

2) If $\xi_2 = 0$ and $v_2 = 0$ in the equation (12.40), then there exists an operator $\mathbf{V}(\cdot,t) \in \mathcal{L}\big(L^{2q}_{\mathbb{F}}(\Omega;L^2(t,T;H));C_{\mathbb{F}}([t,T];L^{2q}(\Omega;H))\big)$ such that the solution to (12.40) can be represented as $x_2(\cdot) = \mathbf{V}(\cdot,t)u_2$.

3) If $\xi_2 = 0$ and $u_2 = 0$ in the equation (12.40), then there exists an operator $\Xi(\cdot,t) \in \mathcal{L}\big(L^{2q}_{\mathbb{F}}(\Omega;L^2(t,T;\mathcal{L}_2^0));C_{\mathbb{F}}([t,T];L^{2q}(\Omega;H))\big)$ such that the solution to (12.40) can be represented as $x_2(\cdot) = \Xi(\cdot,t)v_2$.

Proof: We prove only the first conclusion. Define $\mathbf{U}(\cdot,t)$ as follows:

$$\begin{cases} \mathbf{U}(\cdot,t): \ \ L^{2q}_{\mathcal{F}_t}(\Omega;H) \to C_{\mathbb{F}}([t,T];L^{2q}(\Omega;H)),\\ \mathbf{U}(s,t)\xi_2 = x_2(s), \qquad \forall\, s \in [t,T], \end{cases}$$

where $x_2(\cdot)$ is the mild solution to (12.40) with $u_2 = v_2 = 0$.

By Proposition 3.12 and Hölder's inequality, and noting the fact that $J \in L^{2q}_{\mathbb{F}}(0,T;L^\infty(\Omega;\mathcal{L}(H)))$ and $K \in L^{2q}_{\mathbb{F}}(0,T;L^\infty(\Omega;\mathcal{L}(H;\mathcal{L}_2^0)))$, we obtain that for any $s \in [t,T]$,

$$
\begin{aligned}
&\mathbb{E}|x_2(s)|_H^{2q} \\
&= \mathbb{E}\Big|S(s-t)\xi_2 + \int_t^s S(s-\sigma)J(\sigma)x_2(\sigma)d\sigma \\
&\qquad + \int_t^s S(s-\sigma)K(\sigma)x_2(\sigma)dW(\sigma)\Big|_H^{2q} \\
&\leq \mathcal{C}\Big[\mathbb{E}\big|S(s-t)\xi_2\big|_H^{2q} + \mathbb{E}\Big|\int_t^s S(s-\sigma)J(\sigma)x_2(\sigma)d\sigma\Big|_H^{2q} \\
&\qquad + \mathbb{E}\Big(\int_t^s \big|S(s-\sigma)K(\sigma)x_2(\sigma)\big|_{\mathcal{L}_2^0}^2 d\sigma\Big)^q\Big] \\
&\leq \mathcal{C}\Big(\mathbb{E}|\xi_2|_H^{2q} + \int_t^s M_{J,K,q}(\sigma)\mathbb{E}|x_2(\sigma)|_H^{2q}d\sigma\Big).
\end{aligned}
$$

This, together with Gronwall's inequality, implies that

$$
|x_2(\cdot)|_{C_{\mathbb{F}}([t,T];L^{2q}(\Omega;H))} \leq \mathcal{C}|\xi_2|_{L^{2q}_{\mathcal{F}_t}(\Omega;H)}.
$$

Hence, $\mathbf{U}(\cdot,t)$ is a bounded linear operator from $L^{2q}_{\mathcal{F}_t}(\Omega;H)$ to $C_{\mathbb{F}}([t,T];L^{2q}(\Omega;H))$ and $\mathbf{U}(\cdot,t)\xi_2$ solves the equation (12.40) with $u_2 = v_2 = 0$.

On the other hand, by the definition of $\mathbf{U}(\cdot,t)$ and $\mathbf{U}(\cdot,s)$, for each $s \in [t,T]$ and $r \in [s,T]$, we see that

$$
\begin{aligned}
\mathbf{U}(r,t)\xi = S(r-t)\xi &+ \int_t^r S(r-\tau)J(\tau)\mathbf{U}(\tau,t)\xi d\tau \\
&+ \int_t^r S(r-\tau)K(\tau)\mathbf{U}(\tau,t)\xi dW(\tau),
\end{aligned}
$$

and

$$
\begin{aligned}
\mathbf{U}(r,s)\xi = S(r-s)\xi &+ \int_s^r S(r-\tau)J(\tau)\mathbf{U}(\tau,s)\xi d\tau \\
&+ \int_s^r S(r-\tau)K(\tau)\mathbf{U}(\tau,s)\xi dW(\tau).
\end{aligned}
$$

Hence,

$$
\begin{aligned}
&\mathbb{E}|\mathbf{U}(r,s)\xi - \mathbf{U}(r,t)\xi|_H^{2q} \\
&\leq \mathcal{C}\mathbb{E}\Big[\Big|S(r-s)\xi - S(r-t)\xi\Big|_H^{2q} \\
&\quad + \Big|\int_s^r S(r-\tau)J(\tau)\big(\mathbf{U}(\tau,s)\xi - \mathbf{U}(\tau,t)\xi\big)d\tau\Big|_H^{2q} \\
&\quad + \Big|\int_s^r S(r-\tau)K(\tau)\big(\mathbf{U}(\tau,s)\xi - \mathbf{U}(\tau,t)\xi\big)dW(\tau)\Big|_H^{2q} \\
&\quad + \Big|\int_t^s S(r-\tau)J(\tau)\mathbf{U}(\tau,t)\xi d\tau\Big|_H^{2q} + \Big|\int_t^s S(s-\tau)K(\tau)\mathbf{U}(\tau,t)\xi dW(\tau)\Big|_H^{2q}\Big]
\end{aligned}
$$

$$
\begin{aligned}
&\leq \mathcal{C}\mathbb{E}\Big|S(r-s)\xi - S(r-t)\xi\Big|_H^{2q} + \mathcal{C}\int_s^r M_{J,K,q}(\tau)\mathbb{E}\Big|\mathbf{U}(\tau,s)\xi - \mathbf{U}(\tau,t)\xi\Big|_H^{2q} d\tau \\
&\quad + \mathcal{C}\int_t^s M_{J,K,q}(\tau)\mathbb{E}\Big|\mathbf{U}(\tau,t)\xi\Big|_H^{2q} d\tau \\
&\leq \mathcal{C}\mathbb{E}\Big|S(r-s)\xi - S(r-t)\xi\Big|_H^{2q} + \mathcal{C}\int_s^r M_{J,K,q}(\tau)\mathbb{E}\Big|\mathbf{U}(\tau,s)\xi - \mathbf{U}(\tau,t)\xi\Big|_H^{2q} d\tau \\
&\quad + \mathcal{C}\int_t^s M_{J,K,q}(\tau)d\tau\mathbb{E}|\xi|_H^{2q}.
\end{aligned}
$$

Then, by Gronwall's inequality, we find that

$$
\mathbb{E}\Big|\mathbf{U}(r,s)\xi - \mathbf{U}(r,t)\xi\Big|_H^{2q} \leq \mathcal{C}\Big(h(r,s,t) + \int_s^r h(\sigma,s,t)d\sigma\Big),
$$

where

$$
h(r,s,t) = \mathbb{E}\Big|S(r-s)\xi - S(r-t)\xi\Big|_H^{2q} + \int_t^s M_{J,K,q}(\tau)d\tau\mathbb{E}|\xi|_H^{2q}.
$$

Further, it is easy to see that

$$
|\xi - S(s-t)\xi|_H^{2q} \leq \mathcal{C}|\xi|_H^{2q}.
$$

By Lebesgue's dominated convergence theorem, we have

$$
\lim_{s\to t+0}\mathbb{E}|\xi - S(s-t)\xi|_H^{2q} = 0.
$$

Hence, there is a $\delta \in (0, T-t)$ such that (12.48) holds for any $s \in [t, t+\delta]$. This completes the proof of Lemma 12.10. □

Remark 12.11. It is easy to see that, if $\big(P(\cdot), Q(\cdot)\big)$ is a transposition solution to (12.8), then $\big(P(\cdot), Q^{(\cdot)}, \widehat{Q}^{(\cdot)}\big)$ is a relaxed transposition solution to the same equation, where (Recall Lemma 12.10 for $\mathbf{U}(\cdot,t)$, $\mathbf{V}(\cdot,t)$ and $\varXi(\cdot,t)$)

$$
\begin{cases}
Q^{(t)}(\xi,u,v) = \widetilde{Q}(\cdot)\mathbf{U}(\cdot,t)\xi + \widetilde{Q}(\cdot)\mathbf{V}(\cdot,t)u + \dfrac{\widetilde{Q}(\cdot)\varXi(\cdot,t) + \big(Q(\cdot)^*\varXi(\cdot,t)\big)^*}{2}v, \\
\widehat{Q}^{(t)}(\xi,u,v) = Q(\cdot)^*\mathbf{U}(\cdot,t)\xi + Q(\cdot)^*\mathbf{V}(\cdot,t)u + \dfrac{Q(\cdot)^*\varXi(\cdot,t) + \big(\widetilde{Q}(\cdot)\varXi(\cdot,t)\big)^*}{2}v,
\end{cases}
$$

for any $(\xi,u,v) \in \mathcal{H}_t$. However, it is unclear how to obtain a transposition solution $\big(P(\cdot), Q(\cdot)\big)$ to (12.8) by means of its relaxed transposition solution $\big(P(\cdot), Q^{(\cdot)}, \widehat{Q}^{(\cdot)}\big)$. It seems that this is possible (at least under some mild assumptions) but we cannot do it at this moment.

For any $t \in [0,T]$ and $\lambda \in \rho(A)$, consider the following two forward s-tochastic evolution equations:

$$\begin{cases} dx_1^\lambda = (A_\lambda + J)x_1^\lambda ds + u_1 ds + Kx_1^\lambda dW(s) + v_1 dW(s) \ \text{ in } (t,T], \\ x_1^\lambda(t) = \xi_1 \end{cases} \tag{12.49}$$

and

$$\begin{cases} dx_2^\lambda = (A_\lambda + J)x_2^\lambda ds + u_2 ds + Kx_2^\lambda dW(s) + v_2 dW(s) \ \text{ in } (t,T], \\ x_2^\lambda(t) = \xi_2. \end{cases} \tag{12.50}$$

Here (ξ_1, u_1, v_1) (*resp.* (ξ_2, u_2, v_2)) is the same as that in (12.39) (*resp.* (12.40)). We have the following result:

Lemma 12.12. *Under the assumptions in Lemma 12.10, the solutions of (12.49) and (12.50) satisfy*

$$\begin{cases} \lim\limits_{\lambda\to\infty} x_1^\lambda(\cdot) = x_1(\cdot) & \text{in } C_{\mathbb{F}}([t,T]; L^{2q}(\Omega; H)), \\ \lim\limits_{\lambda\to\infty} x_2^\lambda(\cdot) = x_2(\cdot) & \text{in } C_{\mathbb{F}}([t,T]; L^{2q}(\Omega; H)). \end{cases} \tag{12.51}$$

Here $x_1(\cdot)$ and $x_2(\cdot)$ are solutions of (12.39) *and* (12.40), *respectively.*

Proof: Clearly, for any $s \in [t,T]$, it holds that

$$\begin{aligned} &\mathbb{E}|x_1(s) - x_1^\lambda(s)|_H^{2q} \\ &= \mathbb{E}\Big|\big(S(s-t) - S_\lambda(s-t)\big)\xi_1 + \int_t^s \big(S(s-\sigma) - S_\lambda(s-\sigma)\big)u_1(\sigma)d\sigma \\ &\quad + \int_t^s \big(S(s-\sigma)J(\sigma)x_1(\sigma) - S_\lambda(s-\sigma)J(\sigma)x_1^\lambda(\sigma)\big)d\sigma \\ &\quad + \int_t^s \big(S(s-\sigma)K(\sigma)x_1(\sigma) - S_\lambda(s-\sigma)K(\sigma)x_1^\lambda(\sigma)\big)dW(\sigma) \\ &\quad + \int_t^s \big(S(s-\sigma) - S_\lambda(s-\sigma)\big)v_1(\sigma)dW(\sigma)\Big|_H^{2q}. \end{aligned}$$

Since A_λ is the Yosida approximation of A, one can find a positive constant $\mathcal{C} = \mathcal{C}(A,T)$, independent of λ, such that

$$|S_\lambda(\cdot)|_{L^\infty(0,T;\mathcal{L}(H))} \le \mathcal{C}. \tag{12.52}$$

Hence,

$$\begin{aligned} &\mathbb{E}\Big|\int_t^s \big(S(s-\sigma)J(\sigma)x_1(\sigma) - S_\lambda(s-\sigma)J(\sigma)x_1^\lambda(\sigma)\big)d\sigma\Big|_H^{2q} \\ &\le \mathcal{C}\mathbb{E}\int_t^s \Big|\big(S(s-\sigma) - S_\lambda(s-\sigma)\big)J(\sigma)x_1(\sigma)\Big|_H^{2q} d\sigma \\ &\quad + \mathcal{C}\mathbb{E}\int_t^s \Big|S_\lambda(s-\sigma)J(\sigma)\big[x_1(\sigma) - x_1^\lambda(\sigma)\big]\Big|_H^{2q} d\sigma \end{aligned}$$

$$
\begin{aligned}
&\le \mathcal{C}\mathbb{E}\Big|\int_t^s \big(S(s-\sigma)-S_\lambda(s-\sigma)\big)J(\sigma)x_1(\sigma)d\sigma\Big|_H^{2q}\\
&\quad +\mathcal{C}\mathbb{E}\int_t^s |J(\sigma)|^{2q}_{L^\infty(\Omega;\mathcal{L}(H))}|x_1(\sigma)-x_1^\lambda(\sigma)|_H^{2q}d\sigma.
\end{aligned}
$$

It follows from Proposition 3.12 that

$$
\begin{aligned}
&\mathbb{E}\Big|\int_t^s \big(S(s-\sigma)K(\sigma)x_1(\sigma)-S_\lambda(s-\sigma)K(\sigma)x_1^\lambda(\sigma)\big)dW(\sigma)\Big|_H^{2q}\\
&\le \mathcal{C}\mathbb{E}\int_t^s \Big|\big(S(s-\sigma)-S_\lambda(s-\sigma)\big)K(\sigma)x_1(\sigma)\Big|^{2q}_{\mathcal{L}_2^0}d\sigma\\
&\quad +\mathcal{C}\mathbb{E}\int_t^s \Big|S_\lambda(s-\sigma)K(\sigma)\big(x_1(\sigma)-x_1^\lambda(\sigma)\big)\Big|^{2q}_{\mathcal{L}_2^0}d\sigma\\
&\le \mathcal{C}\mathbb{E}\int_t^s \Big|\big(S(s-\sigma)-S_\lambda(s-\sigma)\big)K(\sigma)x_1(\sigma)\Big|^{2q}_{\mathcal{L}_2^0}d\sigma\\
&\quad +\mathcal{C}\mathbb{E}\int_t^s |K(\sigma)|^{2q}_{L^\infty(\Omega;\mathcal{L}(H;\mathcal{L}_2^0))}|x_1(\sigma)-x_1^\lambda(\sigma)|_H^{2q}d\sigma.
\end{aligned}
$$

Hence, for $t\le s\le T$ (Recall (12.47) for $M_{J,K,q}(\cdot)$),

$$
\mathbb{E}|x_1(s)-x_1^\lambda(s)|_H^{2q}\le \Lambda(\lambda,s)+\mathcal{C}\int_t^s M_{J,K,q}(\sigma)\mathbb{E}|x_1(\sigma)-x_1^\lambda(\sigma)|_H^{2q}d\sigma.
$$

Here

$$
\begin{aligned}
\Lambda(\lambda,s)&=\mathcal{C}\mathbb{E}\big|\big(S(s-t)-S_\lambda(s-t)\big)\xi_1\big|_H^{2q}\\
&\quad +\mathcal{C}\mathbb{E}\Big|\int_t^s \big(S(s-\sigma)-S_\lambda(s-\sigma)\big)u_1(\sigma)d\sigma\Big|_H^{2q}\\
&\quad +\mathcal{C}\mathbb{E}\int_t^s \big|\big(S(s-\sigma)-S_\lambda(s-\sigma)\big)v_1(\sigma)\big|_H^{2q}d\sigma\\
&\quad +\mathcal{C}\mathbb{E}\Big|\int_t^s \big(S(s-\sigma)-S_\lambda(s-\sigma)\big)J(\sigma)x_1(\sigma)d\sigma\Big|_H^{2q}\\
&\quad +\mathcal{C}\mathbb{E}\int_t^s \big|\big(S(s-\sigma)-S_\lambda(s-\sigma)\big)K(\sigma)x_1(\sigma)\big|_H^{2q}d\sigma.
\end{aligned}
$$

By Gronwall's inequality, it follows that

$$
\mathbb{E}|x_1(s)-x_1^\lambda(s)|_H^{2q}\le \Lambda(\lambda,s)+\mathcal{C}\int_t^s e^{\mathcal{C}(s-\tau)}\Lambda(\lambda,\tau)d\tau,\qquad t\le s\le T.
$$

Since A_λ is the Yosida approximation of A, we see that $\lim_{\lambda\to\infty}\Lambda(\lambda,s)=0$, which implies that

$$
\lim_{\lambda\to\infty}|x_1^\lambda(\cdot)-x_1(\cdot)|_{C_{\mathbb{F}}([t,T]:L^{2q}(\Omega;H))}=0.
$$

This leads to the first equality in (12.51). The second equality in (12.51) can be proved similarly. This completes the proof of Lemma 12.12. □

12.4.3 Proof of the Uniqueness Results

This subsection is devoted to proving the uniqueness of both transposition solution and relaxed transposition solution to the equation (12.8). To begin with, let us prove the following lemma.

Lemma 12.13. *The following set*

$$\mathfrak{R} \overset{\Delta}{=} \big\{ x_2(\cdot) \bigm| x_2(\cdot) \textit{ solves } (12.40) \textit{ with } t=0,\ \xi_2=0,\ v_2=0 \\ \textit{and some } u_2 \in L^2_{\mathbb{F}}(0,T;L^{2q}(\Omega;H)) \big\}$$

is dense in $L^{2q}_{\mathbb{F}}(\Omega;L^4(0,T;H))$.

Proof: Let us prove Lemma 12.13 by contradiction. If this is not the case, then we could find a nonzero $\rho \in L_{\mathbb{F}}^{\frac{2p}{1+p}}(\Omega;L^{\frac{4}{3}}(0,T;H))$ such that

$$\mathbb{E}\int_0^T \langle \rho, x_2\rangle_H ds = 0, \qquad \text{for any } x_2 \in \mathfrak{R}. \tag{12.53}$$

Let us consider the following H-valued backward stochastic evolution equation:

$$\begin{cases} dy = -A^*ydt + \big(\rho - J^*y - K^*Y\big)dt + YdW(t) & \text{in } [0,T), \\ y(T) = 0. \end{cases} \tag{12.54}$$

The solution to (12.54) is understood in the transposition sense. By Theorem 4.16, the equation (12.54) admits a unique transposition solution $(y(\cdot),Y(\cdot)) \in D_{\mathbb{F}}([0,T];L^{\frac{2p}{p+1}}(\Omega;H)) \times L^2_{\mathbb{F}}(0,T;L^{\frac{2p}{p+1}}(\Omega;\mathcal{L}_2^0))$. Hence, for any $\phi_1(\cdot) \in L^1_{\mathbb{F}}(0,T;L^{2q}(\Omega;H))$ and $\phi_2(\cdot) \in L^{2q}_{\mathbb{F}}(\Omega;L^2(0,T;\mathcal{L}_2^0))$, it holds that

$$\begin{aligned} &-\mathbb{E}\int_0^T \big\langle z(s), \rho(s) - J(s)^*y(s) - K(s)^*Y(s)\big\rangle_H ds \\ &= \mathbb{E}\int_0^T \big\langle \phi_1(s), y(s)\big\rangle_H ds + \mathbb{E}\int_0^T \big\langle \phi_2(s), Y(s)\big\rangle_{\mathcal{L}_2^0} ds, \end{aligned} \tag{12.55}$$

where $z(\cdot)$ solves

$$\begin{cases} dz = (Az+\phi_1)dt + \phi_2 dW(t), & \text{in } (0,T], \\ z(0) = 0. \end{cases} \tag{12.56}$$

In particular, for any $x_2(\cdot)$ solving (12.40) with $t=0$, $\xi_2=0$, $v_2=0$ and an arbitrarily given $u_2 \in L^2_{\mathbb{F}}(0,T;L^{2q}(\Omega;H))$, we choose $z=x_2$, $\phi_1 = Jx_2+u_2$ and $\phi_2 = Kx_2$. By (12.55), it follows that for all $u_2 \in L^2_{\mathbb{F}}(0,T;L^{2q}(\Omega;H))$,

$$-\,\mathbb{E}\int_0^T \big\langle x_2(s), \rho(s)\big\rangle_H ds = \mathbb{E}\int_0^T \big\langle u_2(s), y(s)\big\rangle_H ds. \tag{12.57}$$

By (12.57) and recalling (12.53), we conclude that $y(\cdot) = 0$. Hence, (12.55) is reduced to

$$-\mathbb{E}\int_0^T \big\langle z(s), \rho(s) - K(s)^*Y(s)\big\rangle_H ds = \mathbb{E}\int_0^T \big\langle \phi_2(s), Y(s)\big\rangle_{\mathcal{L}_2^0} ds. \tag{12.58}$$

Choosing $\phi_2(\cdot) = 0$ in (12.56) and (12.58), we obtain that for every $\phi_1(\cdot) \in L^1_{\mathbb{F}}(0,T; L^{2q}(\Omega; H))$,

$$\mathbb{E}\int_0^T \Big\langle \int_0^s S(s-\sigma)\phi_1(\sigma)d\sigma,\ \rho(s) - K(s)^*Y(s)\Big\rangle_H ds = 0. \tag{12.59}$$

Hence,

$$\int_\sigma^T S(s-\sigma)\big(\rho(s) - K(s)^*Y(s)\big)ds = 0, \qquad \forall\, \sigma \in [0,T]. \tag{12.60}$$

Then, for any given $\lambda_0 \in \rho(A)$ and $\sigma \in [0,T]$, we have

$$\begin{aligned} &\int_\sigma^T S(s-\sigma)(\lambda_0 - A)^{-1}\big(\rho(s) - K(s)^*Y(s)\big)ds \\ &= (\lambda_0 - A)^{-1}\int_\sigma^T S(s-\sigma)\big(\rho(s) - K(s)^*Y(s)\big)ds = 0. \end{aligned} \tag{12.61}$$

Differentiating the equality (12.61) with respect to σ, and noting (12.60), we see that

$$\begin{aligned} &(\lambda_0 - A)^{-1}\big(\rho(\sigma) - K(\sigma)^*Y(\sigma)\big) \\ &= -\int_\sigma^T S(s-\sigma)A(\lambda_0 - A)^{-1}\big(\rho(s) - K(s)^*Y(s)\big)ds \\ &= \int_\sigma^T S(s-\sigma)\big(\rho(s) - K(s)^*Y(s)\big)ds \\ &\quad -\lambda_0\int_\sigma^T S(s-\sigma)(\lambda_0 - A)^{-1}\big(\rho(s) - K(s)^*Y(s)\big)ds \\ &= 0, \qquad \forall\, \sigma \in [0,T]. \end{aligned}$$

Therefore,

$$\rho(\cdot) = K(\cdot)^*Y(\cdot). \tag{12.62}$$

By (12.62), the equation (12.54) is reduced to

$$\begin{cases} dy = -A^*ydt - J^*ydt + YdW(t) \ \text{ in } [0,T), \\ y(T) = 0. \end{cases} \tag{12.63}$$

It is clear that the unique transposition to (12.63) is $(y(\cdot), Y(\cdot)) = (0,0)$. Hence, by (12.62), we conclude that $\rho(\cdot) = 0$, which is a contradiction. Therefore, Ξ is dense in $L^{2q}_{\mathbb{F}}(\Omega; L^4(0,T;H))$. $\square$

Now we give the proof of Theorem 12.7.

Proof of Theorem 12.7: Assume that $(\overline{P}(\cdot),\overline{Q}(\cdot))$ is another transposition solution to the equation (12.8). Then, by Definition 12.6, it follows that, for any $t\in[0,T]$,

$$\begin{aligned}0=&\ \mathbb{E}\big\langle\big(\overline{P}(t)-P(t)\big)\xi_1,\xi_2\big\rangle_H+\mathbb{E}\int_t^T\big\langle\big(\overline{P}(s)-P(s)\big)u_1(s),x_2(s)\big\rangle_H ds\\&+\mathbb{E}\int_t^T\big\langle\big(\overline{P}(s)-P(s)\big)x_1(s),u_2(s)\big\rangle_H ds\\&+\mathbb{E}\int_t^T\big\langle\big(\overline{P}(s)-P(s)\big)K(s)x_1(s),v_2(s)\big\rangle_{\mathcal{L}_2^0}ds\\&+\mathbb{E}\int_t^T\big\langle\big(\overline{P}(s)-P(s)\big)v_1(s),K(s)x_2(s)+v_2(s)\big\rangle_{\mathcal{L}_2^0}ds\\&+\mathbb{E}\int_t^T\big\langle\big(\overline{Q}(s)-Q(s)\big)v_1(s),x_2(s)\big\rangle_H ds\\&+\mathbb{E}\int_t^T\big\langle\big(\widetilde{\overline{Q}}(s)-\widetilde{Q}(s)\big)x_1(s),v_2(s)\big\rangle_{\mathcal{L}_2^0}ds,\end{aligned}\tag{12.64}$$

where $\widetilde{\overline{Q}}(\cdot)$ and $\widetilde{Q}(\cdot)$ are respectively the (pointwise defined) operators induced uniquely by $\overline{Q}(\cdot)$ and $Q(\cdot)$.

Choosing $u_1=v_1=0$ and $u_2=v_2=0$ in the equations (12.39) and (12.40), respectively, by (12.64), we obtain that, for any $t\in[0,T]$,

$$0=\mathbb{E}\big\langle\big(\overline{P}(t)-P(t)\big)\xi_1,\xi_2\big\rangle_H,\qquad\forall\,\xi_1,\ \xi_2\in L^{2q}_{\mathcal{F}_t}(\Omega;H).$$

Hence, we find that $\overline{P}(\cdot)=P(\cdot)$. By this, it is easy to see that (12.64) becomes the following

$$\begin{aligned}0=&\ \mathbb{E}\int_t^T\big\langle\big(\overline{Q}(s)-Q(s)\big)v_1(s),x_2(s)\big\rangle_H ds\\&+\mathbb{E}\int_t^T\big\langle\big(\widetilde{\overline{Q}}(s)-\widetilde{Q}(s)\big)x_1(s),v_2(s)\big\rangle_{\mathcal{L}_2^0}ds,\quad\forall\,t\in[0,T].\end{aligned}\tag{12.65}$$

Choosing $t=0$, $\xi_2=0$ and $v_2=0$ in the equation (12.40), we see that (12.65) becomes

$$0=\mathbb{E}\int_t^T\big\langle\big(\overline{Q}(s)-Q(s)\big)v_1(s),x_2(s)\big\rangle_H ds.\tag{12.66}$$

Fix any $v_1(\cdot)\in L^2_{\mathbb{F}}(0,T;L^{2q}(\Omega;\mathcal{L}_2^0))$ and $N>0$. Write

$$v_1^N(t,\omega)=\chi_{\{|(\overline{Q}(\cdot)-Q(\cdot))v_1(\cdot)|_H\le N\}}(t,\omega)v_1(t,\omega).$$

Replacing v_1 in (12.66) by v_1^N and noting that both $\overline{Q}(\cdot)$ and $Q(\cdot)$ are pointwisely defined, we obtain that

$$0 = \mathbb{E}\int_t^T \big\langle \chi_{\{|(\overline{Q}(\cdot)-Q(\cdot))v_1(\cdot)|_H \le N\}}(s)\big(\overline{Q}(s) - Q(s)\big)v_1(s), x_2(s)\big\rangle_H ds.$$

This, together with Lemma 12.13, yields that, for any $v_1(\cdot) \in L^2_{\mathbb{F}}(0,T; L^{2q}(\Omega;\mathcal{L}_2^0))$ and $N > 0$,

$$\chi_{\{|(\overline{Q}(\cdot)-Q(\cdot))v_1(\cdot)|_H \le N\}}(\cdot)\big(\overline{Q}(\cdot) - Q(\cdot)\big)v_1(\cdot) = 0, \quad \text{a.e. } (t,\omega) \in (0,T)\times\Omega.$$

Hence $\overline{Q}(\cdot) = Q(\cdot)$. This completes the proof of Theorem 12.7. □

Next, we prove the uniqueness result in Theorem 12.9.

Proof of the "Uniqueness" Part in Theorem 12.9: Assume that both $\big(P(\cdot), Q^{(\cdot)}, \widehat{Q}^{(\cdot)}\big)$ and $\big(\overline{P}(\cdot), \overline{Q}^{(\cdot)}, \widehat{\overline{Q}}^{(\cdot)}\big)$ are two relaxed transposition solutions to the equation (12.8). Then, by (12.45), for any $t \in [0,T]$, it holds that

$$\begin{aligned}
0 &= \mathbb{E}\big\langle\big(\overline{P}(t) - P(t)\big)\xi_1, \xi_2\big\rangle_H + \mathbb{E}\int_t^T \big\langle\big(\overline{P}(s) - P(s)\big)u_1(s), x_2(s)\big\rangle_H ds \\
&\quad + \mathbb{E}\int_t^T \big\langle\big(\overline{P}(s) - P(s)\big)x_1(s), u_2(s)\big\rangle_H ds \\
&\quad + \mathbb{E}\int_t^T \big\langle\big(\overline{P}(s) - P(s)\big)K(s)x_1(s), v_2(s)\big\rangle_{\mathcal{L}_2^0} ds \\
&\quad + \mathbb{E}\int_t^T \big\langle v_1(s), \big(\overline{P}(s) - P(s)\big)\big(K(s)x_2(s) + v_2(s)\big)\big\rangle_{\mathcal{L}_2^0} ds \\
&\quad + \mathbb{E}\int_t^T \big\langle v_1(s), \big(\widehat{\overline{Q}}^{(t)} - \widehat{Q}^{(t)}\big)(\xi_2, u_2, v_2)(s)\big\rangle_{\mathcal{L}_2^0} ds \\
&\quad + \mathbb{E}\int_t^T \big\langle\big(\overline{Q}^{(t)} - Q^{(t)}\big)(\xi_1, u_1, v_1)(s), v_2(s)\big\rangle_{\mathcal{L}_2^0} ds.
\end{aligned} \tag{12.67}$$

Choosing $u_1 = u_2 = 0$ and $v_1 = v_2 = 0$ respectively in the test equations (12.39) and (12.40), by (12.67), we obtain that, for any $t \in [0,T]$,

$$0 = \mathbb{E}\big\langle\big(\overline{P}(t) - P(t)\big)\xi_1, \xi_2\big\rangle_H, \qquad \forall\, \xi_1,\ \xi_2 \in L^{2q}_{\mathcal{F}_t}(\Omega; H).$$

Hence, we find that $\overline{P}(\cdot) = P(\cdot)$. By this, it is easy to see that (12.67) becomes that

$$\begin{aligned}
0 &= \mathbb{E}\int_t^T \big\langle v_1(s), \big(\widehat{\overline{Q}}^{(t)} - \widehat{Q}^{(t)}\big)(\xi_2, u_2, v_2)(s)\big\rangle_{\mathcal{L}_2^0} ds \\
&\quad + \mathbb{E}\int_t^T \big\langle\big(\overline{Q}^{(t)} - Q^{(t)}\big)(\xi_1, u_1, v_1)(s), v_2(s)\big\rangle_{\mathcal{L}_2^0} ds, \quad \forall\, t \in [0,T].
\end{aligned} \tag{12.68}$$

Choosing $v_2 = 0$ in the test equation (12.40), we see that (12.68) becomes

$$0 = \mathbb{E}\int_t^T \big\langle v_1(s), \big(\widehat{\overline{Q}}^{(t)} - \widehat{Q}^{(t)}\big)(\xi_2, u_2, 0)(s)\big\rangle_H ds. \tag{12.69}$$

Noting that v_1 is arbitrary in $L^2_{\mathbb{F}}(0,T;L^{2q}(\Omega;\mathcal{L}_2^0))$, we conclude from (12.69) that $\widehat{\overline{Q}}^{(t)}(\cdot,\cdot,0)=\widehat{Q}^{(t)}(\cdot,\cdot,0)$. Similarly, $\overline{Q}^{(t)}(\cdot,\cdot,0)=Q^{(t)}(\cdot,\cdot,0)$. Hence,

$$
\begin{aligned}
0=\ &\mathbb{E}\int_t^T \big\langle v_1(s),\big(\widehat{\overline{Q}}^{(t)}-\widehat{Q}^{(t)}\big)(0,0,v_2)(s)\big\rangle_{\mathcal{L}_2^0}ds\\
&+\mathbb{E}\int_t^T \big\langle\big(\overline{Q}^{(t)}-Q^{(t)}\big)(0,0,v_1)(s),v_2(s)\big\rangle_{\mathcal{L}_2^0}ds.
\end{aligned}
\tag{12.70}
$$

Since $\overline{Q}^{(t)}(0,0,\cdot)^*=\widehat{\overline{Q}}^{(t)}(0,0,\cdot)$ and $Q^{(t)}(0,0,\cdot)^*=\widehat{Q}^{(t)}(0,0,\cdot)$, from (12.70), we find that

$$
0=2\mathbb{E}\int_t^T \big\langle v_1(s),\big(\widehat{\overline{Q}}^{(t)}-\widehat{Q}^{(t)}\big)(0,0,v_2)(s)\big\rangle_{\mathcal{L}_2^0}ds,
\tag{12.71}
$$

which implies that $\overline{Q}^{(t)}(0,0,\cdot)=Q^{(t)}(0,0,\cdot)$ and $\widehat{\overline{Q}}^{(t)}(0,0,\cdot)=\widehat{Q}^{(t)}(0,0,\cdot)$. Hence $\overline{Q}^{(t)}(\cdot,\cdot,\cdot)=Q^{(t)}(\cdot,\cdot,\cdot)$ and $\widehat{\overline{Q}}^{(t)}(\cdot,\cdot,\cdot)=\widehat{Q}^{(t)}(\cdot,\cdot,\cdot)$. This completes the proof of the "Uniqueness" part in Theorem 12.9. □

12.4.4 Well-Posedness Result for a Special Case

This section is addressed to proving the following well-posedness result for the transposition solutions to the operator-valued backward stochastic evolution equation (12.8) with some special data P_T and F.

Theorem 12.14. *If $p\in(1,2]$, $F\in L^1_{\mathbb{F}}(0,T;L^p(\Omega;\mathcal{L}_2(H)))$, $P_T\in L^p_{\mathcal{F}_T}(\Omega;\mathcal{L}_2(H))$ and J,K satisfy (12.9), then the equation (12.8) admits a unique transposition solution $\big(P(\cdot),Q(\cdot)\big)$ with the following regularity:*

$$
\big(P(\cdot),Q(\cdot)\big)\in D_{\mathbb{F}}([0,T];L^p(\Omega;\mathcal{L}_2(H)))\times L^2_{\mathbb{F}}(0,T;L^p(\Omega;\mathcal{L}_2(\mathcal{L}_2^0;H)).
$$

Furthermore,

$$
\begin{aligned}
&|(P,Q)|_{D_{\mathbb{F}}([0,T];L^p(\Omega;\mathcal{L}_2(H)))\times L^2_{\mathbb{F}}(0,T;L^p(\Omega;\mathcal{L}_2(\mathcal{L}_2^0;H)))}\\
&\le \mathcal{C}\big(|F|_{L^1_{\mathbb{F}}(0,T;L^p(\Omega;\mathcal{L}_2(H)))}+|P_T|_{L^p_{\mathcal{F}_T}(\Omega;\mathcal{L}_2(H))}\big).
\end{aligned}
\tag{12.72}
$$

Proof: We consider only the case that H is a real Hilbert space (The case of complex Hilbert spaces can be treated similarly). We divide the proof into several steps.

Step 1. Define a family of operators $\{\mathcal{T}(t)\}_{t\ge 0}$ on $\mathcal{L}_2(H)$ as follows:

$$
\mathcal{T}(t)O=S(t)OS^*(t),\quad \forall\, O\in\mathcal{L}_2(H).
$$

We claim that $\{\mathcal{T}(t)\}_{t\ge0}$ is a C_0-semigroup on $\mathcal{L}_2(H)$. Indeed, for any $O\in\mathcal{L}_2(H)$ and nonnegative s and t, we have

$$\mathcal{T}(t+s)O = S(t+s)OS^*(t+s) = S(t)S(s)OS^*(s)S^*(t) = \mathcal{T}(t)\mathcal{T}(s)O.$$

Hence, $\{\mathcal{T}(t)\}_{t\geq 0}$ is a semigroup on $\mathcal{L}_2(H)$. Next, we choose an orthonormal basis $\{h_i\}_{i=1}^{\infty}$ of H. For any $O \in \mathcal{L}_2(H)$ and $t \in [0,\infty)$,

$$\begin{aligned}
&\lim_{s\to t^+} \big|\mathcal{T}(s)O - \mathcal{T}(t)O\big|^2_{\mathcal{L}_2(H)} \\
&\leq |S(t)|^2_{\mathcal{L}(H)} \lim_{s\to t^+} \big|S(s-t)OS^*(s-t) - O\big|^2_{\mathcal{L}_2(H)} |S^*(t)|^2_{\mathcal{L}(H)} \\
&\leq |S(t)|^4_{\mathcal{L}(H)} \lim_{s\to t^+} \sum_{i=1}^{\infty} \big|S(s-t)OS^*(s-t)h_i - Oh_i\big|^2_H \\
&\leq 2|S(t)|^4_{\mathcal{L}(H)} \lim_{s\to t^+} \sum_{i=1}^{\infty} \Big(\big|S(s-t)OS^*(s-t)h_i - S(s-t)Oh_i\big|^2_H \\
&\qquad\qquad + \big|S(s-t)Oh_i - Oh_i\big|^2_H\Big).
\end{aligned} \tag{12.73}$$

For the first series in the right hand side of (12.73), we have

$$\begin{aligned}
&\sum_{i=1}^{\infty} \big|S(s-t)OS^*(s-t)h_i - S(s-t)Oh_i\big|^2_H \\
&\leq \mathcal{C}\sum_{i=1}^{\infty} \big|OS^*(s-t)h_i - Oh_i\big|^2_H = \mathcal{C}\big|OS^*(s-t) - O\big|^2_{\mathcal{L}_2(H)} \\
&= \mathcal{C}\big|\big(OS^*(s-t) - O\big)^*\big|^2_{\mathcal{L}_2(H)} = \mathcal{C}\sum_{i=1}^{\infty} \big|\big(S(s-t)O^* - O^*\big)h_i\big|^2_H.
\end{aligned}$$

For each $i \in \mathbb{N}$,

$$\big|\big(S(s-t)O^* - O^*\big)h_i\big|^2_H \leq 2\big|S(s-t)O^*h_i\big|^2_H + \big|O^*h_i\big|^2_H \leq \mathcal{C}\big|O^*h_i\big|^2_H.$$

It is clear that

$$\sum_{i=1}^{\infty} \big|O^*h_i\big|^2_H = |O^*|^2_{\mathcal{L}_2(H)} = |O|^2_{\mathcal{L}_2(H)}.$$

Hence,

$$\begin{aligned}
&\lim_{s\to t^+} \sum_{i=1}^{\infty} \big|S(s-t)OS^*(s-t)h_i - S(s-t)Oh_i\big|^2_H \\
&\leq \mathcal{C} \lim_{s\to t^+} \sum_{i=1}^{\infty} \big|OS^*(s-t)h_i - Oh_i\big|^2_H \\
&= \mathcal{C} \sum_{i=1}^{\infty} \lim_{s\to t^+} \big|OS^*(s-t)h_i - Oh_i\big|^2_H = 0.
\end{aligned} \tag{12.74}$$

Similarly, it follows that

$$\lim_{s\to t^+}\sum_{i=1}^{\infty}\left|S(s-t)Oh_i-Oh_i\right|_H^2=0. \tag{12.75}$$

From (12.73)–(12.75), we find that

$$\lim_{s\to t^+}\left|\mathcal{T}(s)O-\mathcal{T}(t)O\right|_{\mathcal{L}_2(H)}^2=0,\quad \forall\, t\in[0,\infty) \text{ and } O\in\mathcal{L}_2(H).$$

Similarly,

$$\lim_{s\to t^-}\left|\mathcal{T}(s)O-\mathcal{T}(t)O\right|_{\mathcal{L}_2(H)}^2=0,\quad \forall\, t\in(0,\infty) \text{ and } O\in\mathcal{L}_2(H).$$

Hence, $\{\mathcal{T}(t)\}_{t\geq 0}$ is a C_0-semigroup on $\mathcal{L}_2(H)$.

Step 2. Denote by $\mathcal{A}$ the infinitesimal generater of $\{\mathcal{T}(t)\}_{t\geq 0}$. Then, $\mathcal{A}^*$ generates a C_0-semigroup on $\mathcal{L}_2(H)$. We consider the following $\mathcal{L}_2(H)$-valued backward stochastic evolution equation[5]:

$$\begin{cases} dP=-\mathcal{A}^*Pdt+f(t,P,\Lambda)dt+\Lambda dW & \text{in } [0,T),\\ P(T)=P_T, \end{cases} \tag{12.76}$$

where

$$f(t,P,\Lambda)=-J^*P-PJ-K^*PK-K^*\Lambda-\Lambda K+F. \tag{12.77}$$

Note that $\mathcal{L}_2(H)$ is a Hilbert space, and the Hilbert spaces $\mathcal{L}_2(V;\mathcal{L}_2(H))$ and $\mathcal{L}_2(\mathcal{L}_2^0;H)(=\mathcal{L}_2(\mathcal{L}_2(V;H);H))$ are isomorphism. Hence, the composition operator ΛK in (12.77) makes sense, and the function $f(\cdot,\cdot,\cdot)$ satisfies Condition 4.1, in which the Hilbert space H is replaced by $\mathcal{L}_2(H)$. By Theorem 4.19, the equation (12.76) admits one and only one transposition solution $(P,\Lambda)\in D_{\mathbb{F}}([0,T];L^p(\Omega;\mathcal{L}_2(H)))\times L^2_{\mathbb{F}}(0,T;L^p(\Omega;\mathcal{L}_2(V;\mathcal{L}_2(H)))$ (in the sense of Definition 4.13). Further, (P,Λ) satisfies

$$\begin{aligned}&|(P,\Lambda)|_{D_{\mathbb{F}}([0,T];L^p(\Omega;\mathcal{L}_2(H)))\times L^2_{\mathbb{F}}(0,T;L^p(\Omega;\mathcal{L}_2(V;\mathcal{L}_2(H)))}\\ &\leq \mathcal{C}\big(|F|_{L^1_{\mathbb{F}}(0,T;L^p(\Omega;\mathcal{L}_2(H)))}+|P_T|_{L^p_{\mathcal{F}_T}(\Omega;\mathcal{L}_2(H))}\big).\end{aligned} \tag{12.78}$$

Denote by $\mathbf{T}(\cdot)$ the tensor product of $x_1(\cdot)$ and $x_2(\cdot)$, i.e., $\mathbf{T}(\cdot)=x_1(\cdot)\otimes x_2(\cdot)$, where x_1 and x_2 solve respectively (12.39) and (12.40). Clearly, $\mathbf{T}(t,\omega)\in\mathcal{L}_2(H)$. We shall derive below the stochastic evolution equation (valued in $\mathcal{L}_2(H)$) such that $\mathbf{T}(\cdot)$ is the mild solution of this stochastic evolution equation.

For any $\lambda\in\rho(A)$, define a family of operators $\{\mathcal{T}_\lambda(t)\}_{t\geq 0}$ on $\mathcal{L}_2(H)$ as follows:

[5] Generally, in the equation (12.76) we do NOT have $\mathcal{A}^*P=AP+PA^*$ (Actually, we do not need this equality for the definition of mild solution to (12.76)). Nevertheless, this equality does hold when $A\in\mathcal{L}(H)$.

$$\mathcal{T}_\lambda(t)O = S_\lambda(t)OS_\lambda^*(t), \quad \forall\, O \in \mathcal{L}_2(H).$$

By the result proved in Step 1, it follows that $\{\mathcal{T}_\lambda(t)\}_{t\geq 0}$ is a C_0-semigroup on $\mathcal{L}_2(H)$. Further, for any $O \in \mathcal{L}_2(H)$, we have

$$\begin{aligned}
&\lim_{t\to 0^+} \frac{\mathcal{T}_\lambda(t)O - O}{t} = \lim_{t\to 0^+} \frac{S_\lambda(t)OS_\lambda^*(t) - O}{t}\\
&= \lim_{t\to 0^+} \frac{S_\lambda(t)OS_\lambda^*(t) - OS_\lambda^*(t) + OS_\lambda^*(t) - O}{t} = A_\lambda O + OA_\lambda^*.
\end{aligned}$$

Hence, the infinitesimal generater $\mathcal{A}_\lambda$ of $\{\mathcal{T}_\lambda(t)\}_{t\geq 0}$ is given as follows:

$$\mathcal{A}_\lambda O = A_\lambda O + OA_\lambda^*, \qquad \forall\, O \in \mathcal{L}_2(H). \tag{12.79}$$

Now, for any $O \in \mathcal{L}_2(H)$, it holds that

$$\begin{aligned}
&\lim_{\lambda\to\infty} \big|\mathcal{T}(t)O - \mathcal{T}_\lambda(t)O\big|_{\mathcal{L}_2(H)}\\
&= \lim_{\lambda\to\infty} \big|S(t)OS^*(t) - S_\lambda(t)OS_\lambda^*(t)\big|_{\mathcal{L}_2(H)}\\
&\leq \lim_{\lambda\to\infty} \big|S(t)OS^*(t) - S(t)OS_\lambda^*(t)\big|_{\mathcal{L}_2(H)}\\
&\quad + \lim_{\lambda\to\infty} \big|S(t)OS_\lambda^*(t) - S_\lambda(t)OS_\lambda^*(t)\big|_{\mathcal{L}_2(H)}.
\end{aligned}$$

Let us compute each term in the right hand side of the above inequality. First,

$$\begin{aligned}
&\big|S(t)OS^*(t) - S(t)OS_\lambda^*(t)\big|^2_{\mathcal{L}_2(H)} \leq \mathcal{C}\big|OS^*(t) - OS_\lambda^*(t)\big|^2_{\mathcal{L}_2(H)}\\
&= \mathcal{C}\big|S(t)O^* - S_\lambda(t)O^*\big|^2_{\mathcal{L}_2(H)} = \mathcal{C}\sum_{i=1}^{\infty}\big|\big(S(t) - S_\lambda(t)\big)O^*h_i\big|^2_H.
\end{aligned} \tag{12.80}$$

Since

$$\big|\big(S(t) - S_\lambda(t)\big)O^*h_i\big|^2_H \leq \mathcal{C}|O^*h_i|^2_H$$

and

$$\sum_{i=1}^{\infty}\big|O^*h_i\big|^2_H = |O|^2_{\mathcal{L}_2(H)} < \infty,$$

by means of Lebesgue's dominated theorem and (12.80), we find that

$$\lim_{\lambda\to\infty} \big|S(t)OS^*(t) - S(t)OS_\lambda^*(t)\big|^2_{\mathcal{L}_2(H)} = 0.$$

Similarly, we get that

$$\lim_{\lambda\to\infty} \big|S(t)OS_\lambda^*(t) - S_\lambda(t)OS_\lambda^*(t)\big|^2_{\mathcal{L}_2(H)} = 0.$$

Hence,

$$\lim_{\lambda\to\infty} \big|\mathcal{T}(t)O - \mathcal{T}_\lambda(t)O\big|_{\mathcal{L}_2(H)} = 0, \qquad \text{for any } t \geq 0. \tag{12.81}$$

Write $\mathbf{T}^\lambda = x_1^\lambda \otimes x_2^\lambda$, where x_1^λ and x_2^λ solve accordingly (12.49) and (12.50). Let $\{e_n\}_{n=1}^\infty$ be an orthonormal basis of V. For any $f, g \in \mathcal{L}_2^0$, denote by $f \otimes_{1,1} g$ the bounded bilinear functional on H defined by

$$(f \otimes_{1,1} g)(x, y) = \sum_{j=1}^{\infty} \langle x, fe_j \rangle_H \langle ge_j, y \rangle_H, \qquad \forall\, x, y \in H.$$

Also, for any $h \in H$, denote respectively by $f \otimes_{1,0} h$ and $h \otimes_{0,1} g$ the bounded linear operators from V to $\mathcal{L}(H)$, defined by

$$(f \otimes_{1,0} h)v = (fv) \otimes h \quad \text{ and } \quad (h \otimes_{0,1} g)v = h \otimes (gv), \qquad \forall\, v \in V.$$

Recall that any bounded linear operator on H can be viewed as a bounded bilinear functional on H, and *vice versa*. In what follows, we will also view $x_1^\lambda \otimes x_2^\lambda$, $f \otimes_{1,1} g$ and so on as bounded bilinear functionals on H. By Itô's formula, we have

$$\begin{aligned}
d\mathbf{T}^\lambda &= d(x_1^\lambda \otimes x_2^\lambda)\\
&= \big(dx_1^\lambda\big) \otimes x_2^\lambda + x_1^\lambda \otimes d\big(x_2^\lambda\big) + \big(dx_1^\lambda\big) \otimes d\big(x_2^\lambda\big)\\
&= \Big[(A_\lambda + J)x_1^\lambda\Big] \otimes x_2^\lambda ds + x_1^\lambda \otimes \Big[(A_\lambda + J)x_2^\lambda\Big] ds\\
&\quad + \Big[u_1 \otimes x_2^\lambda + x_1^\lambda \otimes u_2 + \big(Kx_1^\lambda\big) \otimes_{1,1} \big(Kx_2^\lambda\big) + \big(Kx_1^\lambda\big) \otimes_{1,1} v_2\\
&\quad + v_1 \otimes_{1,1} \big(Kx_2^\lambda\big) + v_1 \otimes_{1,1} v_2\Big] ds + \Big[\big(Kx_1^\lambda\big) \otimes_{1,0} x_2^\lambda\\
&\quad + x_1^\lambda \otimes_{0,1} \big(Kx_2^\lambda\big) + v_1 \otimes_{1,0} x_2^\lambda + x_1^\lambda \otimes_{0,1} v_2\Big] dW(s),
\end{aligned} \tag{12.82}$$

On the other hand, for any $h \in H$, we find

$$((A_\lambda x_1^\lambda) \otimes x_2^\lambda)(h) = \langle h, A_\lambda x_1^\lambda \rangle_H x_2^\lambda = \langle A_\lambda^* h, x_1^\lambda \rangle_H x_2^\lambda = \big(x_1^\lambda \otimes x_2^\lambda\big) A_\lambda^* h.$$

Thus,

$$\big(A_\lambda x_1^\lambda\big) \otimes x_2^\lambda = \mathbf{T}^\lambda A_\lambda^*. \tag{12.83}$$

Similarly, we have the following equalities:

$$\begin{cases}
x_1^\lambda \otimes \big(A_\lambda x_2^\lambda\big) = A_\lambda \mathbf{T}^\lambda,\\
\big(Jx_1^\lambda\big) \otimes x_2^\lambda + x_1^\lambda \otimes \big(Jx_2^\lambda\big) = \mathbf{T}^\lambda J^* + J\mathbf{T}^\lambda,\\
\big(Kx_1^\lambda\big) \otimes_{1,1} \big(Kx_2^\lambda\big) = K\mathbf{T}^\lambda K^*,\\
\big(Kx_1^\lambda\big) \otimes_{1,0} x_2^\lambda + x_1^\lambda \otimes_{0,1} \big(Kx_2^\lambda\big) = \mathbf{T}^\lambda K^* + K\mathbf{T}^\lambda.
\end{cases} \tag{12.84}$$

By (12.79), (12.82)–(12.84), we see that $\mathbf{T}^\lambda$ solves the following equation:

$$\begin{cases} d\mathbf{T}^\lambda = \mathcal{A}_\lambda \mathbf{T}^\lambda ds + \alpha^\lambda ds + \beta^\lambda dW(s) \ \text{ in } (t,T], \\ \mathbf{T}^\lambda(t) = \xi_1 \otimes \xi_2, \end{cases} \tag{12.85}$$

where

$$\begin{cases} \alpha^\lambda = J\mathbf{T}^\lambda + \mathbf{T}^\lambda J^* + u_1 \otimes x_2^\lambda + x_1^\lambda \otimes u_2 + K\mathbf{T}^\lambda K^* \\ \qquad +\big(Kx_1^\lambda\big) \otimes_{1,1} v_2 + v_1 \otimes_{1,1} \big(Kx_2^\lambda\big) + v_1 \otimes_{1,1} v_2, \\ \beta^\lambda = K\mathbf{T}^\lambda + \mathbf{T}^\lambda K^* + v_1 \otimes_{1,0} x_2^\lambda + x_1^\lambda \otimes_{0,1} v_2. \end{cases} \tag{12.86}$$

Hence, for any $s \in [t,T]$,

$$\begin{aligned} \mathbf{T}^\lambda(s) = \mathcal{T}_\lambda(s-t)(\xi_1 \otimes \xi_2) + \int_t^s \mathcal{T}_\lambda(\tau - t)\alpha^\lambda(\tau)d\tau \\ + \int_t^s \mathcal{T}_\lambda(\tau - t)\beta^\lambda(\tau)dW(\tau). \end{aligned} \tag{12.87}$$

We claim that

$$\lim_{\lambda\to\infty} |\mathbf{T}^\lambda(\cdot) - \mathbf{T}(\cdot)|_{C_{\mathbb{F}}([t,T];L^q(\Omega;\mathcal{L}_2(H)))} = 0, \qquad \forall\, t \in [0,T]. \tag{12.88}$$

Indeed, for any $s \in [t,T]$, we have

$$\begin{aligned} &|\mathbf{T}^\lambda(s) - \mathbf{T}(s)|^2_{\mathcal{L}_2(H)} = \sum_{i=1}^\infty \big|\mathbf{T}^\lambda(s)h_i - \mathbf{T}(s)h_i\big|_H^2 \\ &= \sum_{i=1}^\infty \big|\langle h_i, x_1^\lambda(s)\rangle_H x_2^\lambda(s) - \langle h_i, x_1(s)\rangle_H x_2(s)\big|_H^2 \\ &\le 2\sum_{i=1}^\infty \big|\langle h_i, x_1^\lambda(s)\rangle_H x_2^\lambda(s) - \langle h_i, x_1^\lambda(s)\rangle_H x_2(s)\big|_H^2 \\ &\quad +2\sum_{i=1}^\infty \big|\langle h_i, x_1^\lambda(s)\rangle_H x_2(s) - \langle h_i, x_1(s)\rangle_H x_2(s)\big|_H^2 \\ &\le 2\big|x_2^\lambda(s) - x_2(s)\big|_H^2 \sum_{i=1}^\infty \big|\langle h_i, x_1^\lambda(s)\rangle_H\big|^2 \\ &\quad +2\big|x_2(s)\big|_H^2 \sum_{i=1}^\infty \big|\langle h_i, x_1^\lambda(s) - x_1(s)\rangle_H\big|^2 \\ &= 2\big|x_1(s)\big|_H^2\big|x_2^\lambda(s) - x_2(s)\big|_H^2 + 2\big|x_2(s)\big|_H^2\big|x_1^\lambda(s) - x_1(s)\big|_H^2. \end{aligned}$$

This, together Lemma 12.12, implies that (12.88) holds. Write

$$\begin{cases} \alpha = J\mathbf{T} + \mathbf{T}J^* + u_1 \otimes x_2 + x_1 \otimes u_2 + K\mathbf{T}K^* \\ \qquad +\big(Kx_1\big) \otimes_{1,1} v_2 + v_1 \otimes_{1,1} \big(Kx_2\big) + v_1 \otimes_{1,1} v_2, \\ \beta = K\mathbf{T} + \mathbf{T}K^* + v_1 \otimes_{1,0} x_2 + x_1 \otimes_{0,1} v_2. \end{cases} \tag{12.89}$$

Similar to the proof of (12.88), we have

$$
\begin{aligned}
&\lim_{\lambda\to\infty} |\alpha^\lambda(\cdot)-\alpha(\cdot)|_{L^2_{\mathbb{F}}(t,T;L^q(\Omega;\mathcal{L}_2(H)))}=0,\\
&\lim_{\lambda\to\infty} |\beta^\lambda(\cdot)-\beta(\cdot)|_{L^2_{\mathbb{F}}(t,T;L^q(\Omega;\mathcal{L}_2(V;\mathcal{L}_2(H))))}=0.
\end{aligned}
\tag{12.90}
$$

By (12.81), using (12.88), (12.90) and Proposition 3.12, one can show that, for any $t\in[0,T]$, it holds that

$$
\begin{cases}
\displaystyle\int_t^{\cdot} \mathcal{T}_\lambda(\tau-t)\alpha^\lambda(\tau)d\tau \to \int_t^{\cdot} \mathcal{T}(\tau-t)\alpha(\tau)d\tau,\\
\displaystyle\int_t^{\cdot} \mathcal{T}_\lambda(\tau-t)\beta^\lambda(\tau)dW(\tau) \to \int_t^{\cdot} \mathcal{T}(\tau-t)\beta(\tau)dW(\tau)\\
\qquad\qquad\qquad\qquad\qquad\qquad \text{in } C_{\mathbb{F}}([t,T];L^q(\Omega;\mathcal{L}_2(H))),
\end{cases}
\tag{12.91}
$$

as $\lambda\to\infty$. Hence, $\mathbf{T}(\cdot)$ is the (unique) mild solution to the following stochastic evolution equation (valued in $\mathcal{L}_2(H)$):

$$
\begin{cases}
d\mathbf{T}(s)=\mathcal{A}\mathbf{T}(s)ds+\alpha ds+\beta dW(s) \ \text{ in } (t,T],\\
\mathbf{T}(t)=\xi_1\otimes\xi_2.
\end{cases}
\tag{12.92}
$$

Step 3. Since (P,Λ) solves (12.76) in the transposition sense and by (12.92), it follows that

$$
\begin{aligned}
&\mathbb{E}\big\langle \mathbf{T}(T),P_T\big\rangle_{\mathcal{L}_2(H)}-\mathbb{E}\int_t^T \big\langle \mathbf{T}(s),f(s,P(s),\Lambda(s))\big\rangle_{\mathcal{L}_2(H)}ds\\
&=\mathbb{E}\big\langle \xi_1\otimes\xi_2,P(t)\big\rangle_{\mathcal{L}_2(H)}+\mathbb{E}\int_t^T \big\langle \alpha(s),P(s)\big\rangle_{\mathcal{L}_2(H)}ds\\
&\quad+\mathbb{E}\int_t^T \big\langle \beta(s),\Lambda(s)\big\rangle_{\mathcal{L}_2(V;\mathcal{L}_2(H))}ds.
\end{aligned}
\tag{12.93}
$$

By (12.77) and recalling that $\mathbf{T}(\cdot)=x_1(\cdot)\otimes x_2(\cdot)$, we find that

$$
\begin{aligned}
&\mathbb{E}\int_t^T \big\langle \mathbf{T}(s),f(s,P(s),\Lambda(s))\big\rangle_{\mathcal{L}_2(H)}ds\\
&=\mathbb{E}\int_t^T \big\langle\big(-J(s)^*P(s)-P(s)J(s)-K^*(s)P(s)K(s)\\
&\qquad -K^*(s)\Lambda(s)-\Lambda(s)K(s)+F(s)\big)x_1(s),x_2(s)\big\rangle_H ds.
\end{aligned}
\tag{12.94}
$$

Further, by the first equality in (12.89), we have

$$
\mathbb{E}\int_t^T \big\langle \alpha(s),P(s)\big\rangle_{\mathcal{L}_2(H)}ds
$$

$$
\begin{aligned}
&= \mathbb{E}\int_t^T \big\langle P(s)x_1(s), J(s)x_2(s)\big\rangle_H ds + \mathbb{E}\int_t^T \big\langle P(s)J(s)x_1(s), x_2(s)\big\rangle_H ds \\
&\quad +\mathbb{E}\int_t^T \big\langle P(s)u_1(s), x_2(s)\big\rangle_H ds + \mathbb{E}\int_t^T \big\langle P(s)x_1(s), u_2(s)\big\rangle_H ds \qquad (12.95)\\
&\quad +\mathbb{E}\int_t^T \big\langle K^*(s)P(s)K(s)x_1(s), x_2(s)\big\rangle_H ds \\
&\quad +\mathbb{E}\int_t^T \big\langle P(s)K(s)x_1(s), v_2(s)\big\rangle_{\mathcal{L}_2^0} ds \\
&\quad +\mathbb{E}\int_t^T \big\langle P(s)v_1(s), K(s)x_2(s)\big\rangle_{\mathcal{L}_2^0} ds + \mathbb{E}\int_t^T \big\langle P(s)v_1(s), v_2(s)\big\rangle_{\mathcal{L}_2^0} ds.
\end{aligned}
$$

Further, by Proposition 12.5, $\Lambda(\cdot)$ induces a bounded linear operator $Q(\cdot) \in \mathcal{L}_2(\mathcal{L}_2^0; H)$ and a bounded linear operator $\widetilde{Q}(\cdot) \in \mathcal{L}_2(H; \mathcal{L}_2^0)$ satisfying (12.35) and 12.38 pointwise. Hence,

$$
\begin{aligned}
&\big\langle v_1(\cdot) \otimes_{1,0} x_2(\cdot), \Lambda(\cdot)\big\rangle_{\mathcal{L}_2(V;\mathcal{L}_2(H))} = \sum_{j=1}^{\infty} \big\langle \big(v_1(\cdot) \otimes_{1,0} x_2(\cdot)\big)e_j, \Lambda(\cdot)e_j\big\rangle_{\mathcal{L}_2(H)} \\
&= \sum_{j=1}^{\infty} \big\langle \big(v_1(\cdot)e_j\big) \otimes x_2(\cdot), \Lambda(\cdot)e_j\big\rangle_{\mathcal{L}_2(H)} = \sum_{j=1}^{\infty} \big\langle \big(\Lambda(\cdot)e_j\big)\big(v_1(\cdot)e_j\big), x_2(\cdot)\big\rangle_H \\
&= \sum_{j=1}^{\infty} \big\langle Q(\cdot)\big(e_j \otimes \big(v_1(\cdot)e_j\big)\big), x_2(\cdot)\big\rangle_H = \sum_{j=1}^{\infty} \big\langle e_j \otimes \big(v_1(\cdot)e_j\big), Q(\cdot)^* x_2(\cdot)\big\rangle_{\mathcal{L}_2^0} \\
&= \sum_{j=1}^{\infty} \big\langle \big(Q(\cdot)^* x_2(\cdot)\big)e_j, v_1(\cdot)e_j\big\rangle_H = \big\langle Q(\cdot)^* x_2, v_1(\cdot)\big\rangle_{\mathcal{L}_2^0} = \big\langle Q(\cdot)v_1(\cdot), x_2(\cdot)\big\rangle_H,
\end{aligned}
$$

and

$$
\begin{aligned}
&\big\langle x_1(\cdot) \otimes_{0,1} v_2(\cdot), \Lambda(\cdot)\big\rangle_{\mathcal{L}_2(V;\mathcal{L}_2(H))} = \sum_{j=1}^{\infty} \big\langle \big(x_1(\cdot) \otimes_{0,1} v_2(\cdot)\big)e_j, \Lambda(\cdot)e_j\big\rangle_{\mathcal{L}_2(H)} \\
&= \sum_{j=1}^{\infty} \big\langle x_1(\cdot) \otimes \big(v_2(\cdot)e_j\big), \Lambda(\cdot)e_j\big\rangle_{\mathcal{L}_2(H)} = \sum_{j=1}^{\infty} \big\langle \big(\Lambda(\cdot)e_j\big)x_1(\cdot), v_2(\cdot)e_j\big\rangle_H \\
&= \sum_{j=1}^{\infty} \big\langle Q(\cdot)\big(e_j \otimes x_1(\cdot)\big), v_2(\cdot)e_j\big\rangle_H = \big\langle \widetilde{Q}(\cdot)x_1(\cdot), v_2(\cdot)\big\rangle_{\mathcal{L}_2^0}.
\end{aligned}
$$

Thus, by the second equality in (12.89), we have

$$
\begin{aligned}
&\mathbb{E}\int_t^T \big\langle \beta(s), \Lambda(s)\big\rangle_{\mathcal{L}_2(V;\mathcal{L}_2(H))} ds \\
&= \mathbb{E}\int_t^T \big\langle K^*(s)\Lambda(s)x_1(s), x_2(s)\big\rangle_H ds + \mathbb{E}\int_t^T \big\langle \Lambda(s)K(s)x_1(s), x_2(s)\big\rangle_H ds
\end{aligned}
$$

$$+\mathbb{E}\int_t^T \big\langle Q(s)v_1(s), x_2(s)\big\rangle_H ds + \mathbb{E}\int_t^T \big\langle \widetilde{Q}(s)x_1(s), v_2(s)\big\rangle_{\mathcal{L}_2^0} ds. \qquad (12.96)$$

By (12.93)–(12.96), it follows that $(P(\cdot), Q(\cdot))$ satisfies (12.43) and thus is a transposition solution of (12.8) due to Definition 12.6. The uniqueness of which follows theorem 12.7. As a result, the proof for Theorem 12.14 is completed . □

12.4.5 Proof of the Existence and Stability for the General Case

This subsection is devoted to proving the existence and stability result (for the equation (12.8) in the sense of relaxed transposition solution) in Theorem 12.9.

Proof of the "Existence and Stability" part in Theorem 12.9: We consider only the case that H is a real Hilbert space (The case of complex Hilbert spaces can be treated similarly). The proof is divided into several steps.

Step 1. In this step, we introduce a suitable approximation to the equation (12.8).

Let $\{h_n\}_{n=1}^{\infty}$ be an orthonormal basis of H and $\{\Gamma_n\}_{n=1}^{\infty}$ be the standard projection operator from H onto its subspace span $\{h_1, h_2, \cdots, h_n\}$, that is, $\Gamma_n x = \sum_{i=1}^{n} x_i h_i$ for any $x = \sum_{i=1}^{\infty} x_i h_i \in H$. Write

$$F^n = \Gamma_n F \Gamma_n, \quad P_T^n = \Gamma_n P_T \Gamma_n. \qquad (12.97)$$

Obviously, $F^n \in L^1_{\mathbb{F}}(0,T; L^p(\Omega; \mathcal{L}_2(H)))$, $P_T^n \in L^p_{\mathcal{F}_T}(\Omega; \mathcal{L}_2(H))$. Furthermore,

$$\begin{aligned}
&|F^n|_{L^1_{\mathbb{F}}(0,T;L^p(\Omega;\mathcal{L}(H)))} + |P_T^n|_{L^p_{\mathcal{F}_T}(\Omega;\mathcal{L}(H))}\\
&\leq |F^n|_{L^1_{\mathbb{F}}(0,T;L^p(\Omega;\mathcal{L}_2(H)))} + |P_T^n|_{L^p_{\mathcal{F}_T}(\Omega;\mathcal{L}_2(H))} \qquad (12.98)\\
&\leq \mathcal{C}\big(|F|_{L^1_{\mathbb{F}}(0,T;L^p(\Omega;\mathcal{L}(H)))} + |P_T|_{L^p_{\mathcal{F}_T}(\Omega;\mathcal{L}(H))}\big).
\end{aligned}$$

Here and henceforth $\mathcal{C}$ denotes a generic constant, independent of n.

Let us consider the following $\mathcal{L}_2(H)$-valued backward stochastic evolution equation:

$$\begin{cases} dP^n = -(A^* + J^*)P^n dt - P^n(A+J)dt - K^*P^nKdt \\ \qquad -(K^*Q^n + Q^nK)dt + F^n dt + Q^n dW(t) \quad \text{in } [0,T), \\ P^n(T) = P_T^n. \end{cases} \qquad (12.99)$$

Clearly, for each $n \in \mathbb{N}$, (12.99) can be regarded as an approximation of the equation (12.8). In the rest of the proof, we shall construct the desired solution to the equation (12.8) by means of the solution to (12.99).

According to Theorem 12.14, the equation (12.99) admits one and only one transposition solution

$$\begin{aligned}&\big(P^n(\cdot),Q^n(\cdot)\big)\\&\in D_{\mathbb{F}}([0,T];L^p(\Omega;\mathcal{L}_2(H)))\times L^2_{\mathbb{F}}(0,T;L^p(\Omega;\mathcal{L}_2(V;\mathcal{L}_2(H)))),\end{aligned}\tag{12.100}$$

in the sense of Definition 12.6. Hence, for any $t\in[0,T]$, $\xi_1,\xi_2\in L^{2q}_{\mathcal{F}_t}(\Omega;H)$, $u_1(\cdot),u_2(\cdot)\in L^2_{\mathbb{F}}(t,T;L^{2q}(\Omega;H))$ and $v_1(\cdot),v_2(\cdot)\in L^2_{\mathbb{F}}(t,T;L^{2q}(\Omega;\mathcal{L}_2^0))$, it holds that

$$\begin{aligned}&\mathbb{E}\big\langle P^n_T x_1(T),x_2(T)\big\rangle_H-\mathbb{E}\int_t^T\big\langle F^n(s)x_1(s),x_2(s)\big\rangle_H ds\\&=\mathbb{E}\big\langle P^n(t)\xi_1,\xi_2\big\rangle_H+\mathbb{E}\int_t^T\big\langle P^n(s)u_1(s),x_2(s)\big\rangle_H ds\\&\quad+\mathbb{E}\int_t^T\big\langle P^n(s)x_1(s),u_2(s)\big\rangle_H ds+\mathbb{E}\int_t^T\big\langle P^n(s)K(s)x_1(s),v_2(s)\big\rangle_{\mathcal{L}_2^0}ds\\&\quad+\mathbb{E}\int_t^T\big\langle P^n(s)v_1(s),K(s)x_2(s)+v_2(s)\big\rangle_{\mathcal{L}_2^0}ds\\&\quad+\mathbb{E}\int_t^T\big\langle Q^n(s)v_1(s),x_2(s)\big\rangle_H ds+\mathbb{E}\int_t^T\big\langle \widetilde{Q}^n(s)x_1(s),v_2(s)\big\rangle_{\mathcal{L}_2^0}ds.\end{aligned}\tag{12.101}$$

Here, $\widetilde{Q}^n(s)$ is the operator induced by $Q^n(s)$, and $x_1(\cdot)$(*resp.* $x_2(\cdot)$) solves (12.39)(*resp.* (12.40)). As we shall see below, the variational equality (12.101) can be viewed as an approximation of (12.45).

Step 2. In this step, we take $n\to\infty$ in (12.101). For this purpose, we need to establish some *a priori* estimates for $P^n(\cdot)$ an $Q^n(\cdot)$ (in suitable sense).

Let $u_1=v_1=0$ in (12.39) and $u_2=v_2=0$ in (12.40). By (12.101), we obtain that, for any $t\in[0,T]$ (Recall Lemma 12.10 for $\mathbf{U}(\cdot,\cdot)$),

$$\begin{aligned}&\mathbb{E}\big\langle P^n_T\mathbf{U}(T,t)\xi_1,\mathbf{U}(T,t)\xi_2\big\rangle_H-\mathbb{E}\int_t^T\big\langle F^n(s)\mathbf{U}(s,t)\xi_1,\mathbf{U}(s,t)\xi_2\big\rangle_H ds\\&=\mathbb{E}\big\langle P^n(t)\xi_1,\xi_2\big\rangle_H,\qquad\forall\,\xi_1,\xi_2\in L^{2q}_{\mathcal{F}_t}(\Omega;H).\end{aligned}\tag{12.102}$$

Hence,

$$\begin{aligned}&\mathbb{E}\Big\langle\mathbf{U}^*(T,t)P^n_T\mathbf{U}(T,t)\xi_1-\int_t^T\mathbf{U}^*(s,t)F^n(s)\mathbf{U}(s,t)\xi_1 ds,\xi_2\Big\rangle_H\\&=\mathbb{E}\big\langle P^n(t)\xi_1,\xi_2\big\rangle_H,\qquad\forall\,\xi_1,\xi_2\in L^{2q}_{\mathcal{F}_t}(\Omega;H).\end{aligned}$$

This leads to

$$\begin{aligned}&\mathbb{E}\Big(\mathbf{U}^*(T,t)P^n_T\mathbf{U}(T,t)\xi_1-\int_t^T\mathbf{U}^*(s,t)F^n(s)\mathbf{U}(s,t)\xi_1 ds\;\Big|\;\mathcal{F}_t\Big)\\&=P^n(t)\xi_1,\quad\text{a.s., }\forall\,t\in[0,T],\ \xi_1\in L^{2q}_{\mathcal{F}_t}(\Omega;H).\end{aligned}\tag{12.103}$$

By (12.98) and (12.102), it follows that

$$\begin{aligned}&\big|\mathbb{E}\langle P^n(t)\xi_1,\xi_2\rangle_H\big|\\&\leq \mathcal{C}\big(|P_T|_{L^p_{\mathcal{F}_T}(\Omega;\mathcal{L}(H))}+|F|_{L^1_{\mathbb{F}}(0,T;\,L^p(\Omega;\mathcal{L}(H)))}\big)\\&\quad\times|\xi_1|_{L^{2q}_{\mathcal{F}_t}(\Omega;H)}|\xi_2|_{L^{2q}_{\mathcal{F}_t}(\Omega;H)}.\end{aligned}\tag{12.104}$$

Here and henceforth $\mathcal{C}$ denotes a generic constant, independent of n and t. For $P^n(t)$, we can find a $\xi_{1,n}\in L^{2q}_{\mathcal{F}_t}(\Omega;H)$ with $|\xi_{1,n}|_{L^{2q}_{\mathcal{F}_t}(\Omega;H)}=1$ such that

$$\big|P^n(t)\xi_{1,n}\big|_{L^{\frac{2q}{2q-1}}_{\mathcal{F}_t}(\Omega;H)}\geq\frac{1}{2}\big|P^n(t)\big|_{L^p_{\mathcal{F}_t}(\Omega;\mathcal{L}(H))}.\tag{12.105}$$

Moreover, we can find a $\xi_{2,n}\in L^{2q}_{\mathcal{F}_t}(\Omega;H)$ with $|\xi_{2,n}|_{L^{2q}_{\mathcal{F}_t}(\Omega;H)}=1$ such that

$$\mathbb{E}\big\langle P^n(t)\xi_{1,n},\xi_{2,n}\big\rangle_H\geq\frac{1}{2}\big|P^n(t)\xi_{1,n}\big|_{L^{\frac{2q}{2q-1}}_{\mathcal{F}_t}(\Omega;H)}.\tag{12.106}$$

From (12.104)–(12.106), we obtain that, for all $n\in\mathbb{N}$,

$$\begin{aligned}&|P^n|_{L^\infty_{\mathbb{F}}(0,T;L^p(\Omega;\mathcal{L}(H)))}\\&\leq\mathcal{C}\big(|P_T|_{L^p_{\mathcal{F}_T}(\Omega;\mathcal{L}(H))}+|F|_{L^1_{\mathbb{F}}(0,T;L^p(\Omega;\mathcal{L}(H)))}\big).\end{aligned}\tag{12.107}$$

By Theorem 2.77, one can find $P\in\mathcal{L}_{pd}\big(L^2_{\mathbb{F}}(0,T;L^{2q}(\Omega;H));\ L^2_{\mathbb{F}}(0,T;L^{\frac{2q}{2q-1}}(\Omega;H))\big)$ such that

$$\begin{aligned}&|P|_{\mathcal{L}(L^2_{\mathbb{F}}(0,T;L^{2q}(\Omega;H));L^2_{\mathbb{F}}(0,T;L^{\frac{2q}{2q-1}}(\Omega;H)))}\\&\leq\mathcal{C}\big(|P_T|_{L^p_{\mathcal{F}_T}(\Omega;\mathcal{L}(H))}+|F|_{L^1_{\mathbb{F}}(0,T;L^p(\Omega;\mathcal{L}(H)))}\big),\end{aligned}\tag{12.108}$$

and a subsequence $\{n_k^{(1)}\}_{k=1}^\infty\subset\{n\}_{n=1}^\infty$ so that, for any $u_1\in L^2_{\mathbb{F}}(0,T;L^{2q}(\Omega;H))$,

$$(\mathrm{w})\text{-}\lim_{k\to\infty}P^{n_k^{(1)}}u_1=Pu_1\quad\text{in }L^2_{\mathbb{F}}(0,T;L^{\frac{2q}{2q-1}}(\Omega;H)).\tag{12.109}$$

Further, by (12.107) and Theorem 2.79, for each fixed $t\in[0,T)$, there exist an $R^{(t)}\in\mathcal{L}_{pd}\big(L^{2q}_{\mathcal{F}_t}(\Omega;H);L^{\frac{2q}{2q-1}}_{\mathcal{F}_t}(\Omega;H)\big)$ and a subsequence $\{n_k^{(2)}\}_{k=1}^\infty\subset\{n_k^{(1)}\}_{k=1}^\infty$ (Generally speaking, each $n_k^{(2)}$ may depend on t) such that, for all $\xi\in L^{2q}_{\mathcal{F}_t}(\Omega;H)$,

$$(\mathrm{w})\text{-}\lim_{k\to\infty}P^{n_k^{(2)}}(t)\xi=R^{(t)}\xi\quad\text{in }L^{\frac{2q}{2q-1}}_{\mathcal{F}_t}(\Omega;H).\tag{12.110}$$

Next, let $u_1=v_1=0$ and $\xi_2=0$, $u_2=0$ in (12.39) and (12.40), respectively. From (12.101), we find that

$$
\begin{aligned}
&\mathbb{E}\int_t^T \big\langle \widetilde{Q}^n(s)\mathbf{U}(s,t)\xi_1, v_2(s)\big\rangle_{\mathcal{L}_2^0} ds\\
&= \mathbb{E}\big\langle P_T^n x_1(T), x_2(T)\big\rangle_H - \mathbb{E}\int_t^T \big\langle F^n(s)x_1(s), x_2(s)\big\rangle_H ds\\
&\quad -\mathbb{E}\int_t^T \big\langle P^n(s)K(s)x_1(s), v_2(s)\big\rangle_{\mathcal{L}_2^0} ds.
\end{aligned}
$$

This implies that

$$
\begin{aligned}
&\mathbb{E}\int_t^T \big\langle \widetilde{Q}^n(s)\mathbf{U}(s,t)\xi_1, v_2(s)\big\rangle_{\mathcal{L}_2^0} ds\\
&\le \mathcal{C}\big(|P_T|_{L^p_{\mathcal{F}_T}(\Omega;\mathcal{L}(H))} + |F|_{L^1_{\mathbb{F}}(0,T;L^p(\Omega;\mathcal{L}(H)))}\big)\\
&\quad \times |\xi_1|_{L^{2q}_{\mathcal{F}_t}(\Omega;H)} |v_2|_{L^2_{\mathbb{F}}(t,T;L^{2q}(\Omega;\mathcal{L}_2^0))},\\
&\qquad \forall\, \xi_1 \in L^{2q}_{\mathcal{F}_t}(\Omega;H),\ v_2 \in L^2_{\mathbb{F}}(t,T;L^{2q}(\Omega;\mathcal{L}_2^0)).
\end{aligned}
\tag{12.111}
$$

We define two operators $Q_1^{n,t}$ and $\widehat{Q}_1^{n,t}$ from $L^{2q}_{\mathcal{F}_t}(\Omega;\ H)$ to $L^2_{\mathbb{F}}(t,T;L^{\frac{2q}{2q-1}}(\Omega;\ \mathcal{L}_2^0))$ as follows (Recall Lemma 12.10 for $\mathbf{U}(\cdot,\cdot)$):

$$
\begin{cases}
Q_1^{n,t}\xi = \widetilde{Q}^n(\cdot)\mathbf{U}(\cdot,t)\xi, \\
\widehat{Q}_1^{n,t}\xi = Q^n(\cdot)^*\mathbf{U}(\cdot,t)\xi,
\end{cases}
\qquad \forall\, \xi \in L^{2q}_{\mathcal{F}_t}(\Omega;H).
$$

It is easy to see that $Q_1^{n,t}, \widehat{Q}_1^{n,t} \in \mathcal{L}(L^{2q}_{\mathcal{F}_t}(\Omega;H); L^2_{\mathbb{F}}(t,T;L^{\frac{2q}{2q-1}}(\Omega;\mathcal{L}_2^0)))$. By (12.111), it follows that

$$
\begin{aligned}
&\big|Q_1^{n,t}\big|_{\mathcal{L}(L^{2q}_{\mathcal{F}_t}(\Omega;H);L^2_{\mathbb{F}}(t,T;L^{\frac{2q}{2q-1}}(\Omega;\mathcal{L}_2^0)))}\\
&\le \mathcal{C}\big(|P_T|_{L^p_{\mathcal{F}_T}(\Omega;\mathcal{L}(H))} + |F|_{L^1_{\mathbb{F}}(0,T;L^p(\Omega;\mathcal{L}(H)))}\big).
\end{aligned}
\tag{12.112}
$$

Similarly,

$$
\begin{aligned}
&\big|\widehat{Q}_1^{n,t}\big|_{\mathcal{L}(L^{2q}_{\mathcal{F}_t}(\Omega;H);L^2_{\mathbb{F}}(t,T;L^{\frac{2q}{2q-1}}(\Omega;\mathcal{L}_2^0)))}\\
&\le \mathcal{C}\big(|P_T|_{L^p_{\mathcal{F}_T}(\Omega;\mathcal{L}(H))} + |F|_{L^1_{\mathbb{F}}(0,T;L^p(\Omega;\mathcal{L}(H)))}\big).
\end{aligned}
\tag{12.113}
$$

By Lemma 2.61, for each t, there exist two bounded linear operators Q_1^t and $\widehat{Q}_1^t$, from $L^{2q}_{\mathcal{F}_t}(\Omega;H)$ to $L_{\mathbb{F}}^{\frac{2q}{2q-1}}(t,T;L^{\frac{2q}{2q-1}}(\Omega;\mathcal{L}_2^0))$, and a subsequence $\{n_k^{(3)}\}_{k=1}^{\infty} \subset \{n_k^{(2)}\}_{n=1}^{\infty}$ such that, for any $\xi \in L^{2q}_{\mathcal{F}_t}(\Omega;H)$,

$$
\begin{cases}
\text{(w)-}\lim\limits_{k\to\infty} Q_1^{n_k^{(3)},t}\xi = Q_1^t\xi & \text{in } L^2_{\mathbb{F}}(t,T;L^{\frac{2q}{2q-1}}(\Omega;\mathcal{L}_2^0)),\\
\text{(w)-}\lim\limits_{k\to\infty} \widehat{Q}_1^{n_k^{(3)},t}\xi = \widehat{Q}_1^t\xi & \text{in } L^2_{\mathbb{F}}(t,T;L^{\frac{2q}{2q-1}}(\Omega;\mathcal{L}_2^0)).
\end{cases}
\tag{12.114}
$$

In view of (12.112)–(12.113), we have

$$
\begin{aligned}
&\big|Q_1^t\big|_{\mathcal{L}(L^{2q}_{\mathcal{F}_t}(\Omega;H);L^2_{\mathbb{F}}(t,T;L^{\frac{2q}{2q-1}}(\Omega;\mathcal{L}_2^0)))}\\
&+\big|\widehat{Q}_1^t\big|_{\mathcal{L}(L^{2q}_{\mathcal{F}_t}(\Omega;H);L^2_{\mathbb{F}}(t,T;L^{\frac{2q}{2q-1}}(\Omega;\mathcal{L}_2^0)))}\\
&\leq \mathcal{C}\big(|P_T|_{L^p_{\mathcal{F}_T}(\Omega;\mathcal{L}(H))}+|F|_{L^1_{\mathbb{F}}(0,T;L^p(\Omega;\mathcal{L}(H)))}\big).
\end{aligned}
\tag{12.115}
$$

Further, we choose $\xi_1 = 0$, $v_1 = 0$ in (12.39) and $\xi_2 = 0$, $u_2 = 0$ in (12.40). From (12.101), we obtain that

$$
\begin{aligned}
&\mathbb{E}\big\langle P_T^n x_1(T), x_2(T)\big\rangle_H - \mathbb{E}\int_t^T \big\langle F^n(s)x_1(s), x_2(s)\big\rangle_H ds\\
&= \mathbb{E}\int_t^T \big\langle P^n(s)u_1(s), x_2(s)\big\rangle_H ds + \mathbb{E}\int_t^T \big\langle P^n(s)K(s)x_1(s), v_2(s)\big\rangle_{\mathcal{L}_2^0} ds\\
&\quad +\mathbb{E}\int_t^T \big\langle \widetilde{Q}^n(s)x_1(s), v_2(s)\big\rangle_{\mathcal{L}_2^0} ds.
\end{aligned}
\tag{12.116}
$$

Define an operator $Q_2^{n,t}$ from $L^2_{\mathbb{F}}(t,T;L^{2q}(\Omega;H))$ to $L^2_{\mathbb{F}}(t,T;L^{\frac{2q}{2q-1}}(\Omega;\mathcal{L}_2^0))$ as follows (Recall Lemma 12.10 for $\mathbf{V}(\cdot,\cdot)$):

$$
\big(Q_2^{n,t}u_2\big)(\cdot) = \widetilde{Q}^n(\cdot)\mathbf{V}(\cdot,t)u_2, \quad \forall\, u_2 \in L^2_{\mathbb{F}}(t,T;L^{2q}(\Omega;H)).
$$

From (12.116), we get that

$$
\begin{aligned}
&\mathbb{E}\int_t^T \big\langle \big(Q_2^{n,t}u_1\big)(s), v_2(s)\big\rangle_{\mathcal{L}_2^0} ds\\
&= \mathbb{E}\big\langle P_T^n x_1(T), x_2(T)\big\rangle_H - \mathbb{E}\int_t^T \big\langle F^n(s)x_1(s), x_2(s)\big\rangle_H ds\\
&\quad -\mathbb{E}\int_t^T \big\langle P^n(s)u_1(s), x_2(s)\big\rangle_H ds - \mathbb{E}\int_t^T \big\langle P^n(s)K(s)x_1(s), v_2(s)\big\rangle_{\mathcal{L}_2^0} ds\\
&\leq \mathcal{C}\big(|P_T|_{L^p_{\mathcal{F}_T}(\Omega;\mathcal{L}(H))}\\
&\qquad +|F|_{L^1_{\mathbb{F}}(0,T;L^p(\Omega;\mathcal{L}(H)))}\big)|u_1|_{L^2_{\mathbb{F}}(t,T;L^{2q}(\Omega;H))}|v_2|_{L^2_{\mathbb{F}}(t,T;L^{2q}(\Omega;\mathcal{L}_2^0))},\\
&\qquad\quad \forall\, u_1 \in L^2_{\mathbb{F}}(t,T;L^{2q}(\Omega;H)),\ v_2 \in L^2_{\mathbb{F}}(t,T;L^{2q}(\Omega;\mathcal{L}_2^0)).
\end{aligned}
\tag{12.117}
$$

From (12.117), we see that

$$
\begin{aligned}
&\big|Q_2^{n,t}\big|_{\mathcal{L}(L^2_{\mathbb{F}}(t,T;L^{2q}(\Omega;H));L^2_{\mathbb{F}}(t,T;L^{\frac{2q}{2q-1}}(\Omega;\mathcal{L}_2^0)))}\\
&\leq \mathcal{C}\big(|P_T|_{L^p_{\mathcal{F}_T}(\Omega;\mathcal{L}(H))}+|F|_{L^1_{\mathbb{F}}(0,T;L^p(\Omega;\mathcal{L}(H)))}\big).
\end{aligned}
\tag{12.118}
$$

Also, we define a linear operator $\widehat{Q}_2^{n,t}$ from $L^2_{\mathbb{F}}(t,T;L^{2q}(\Omega;H))$ to $L^2_{\mathbb{F}}(t,T;$ $L^{\frac{2q}{2q-1}}(\Omega;\mathcal{L}_2^0))$ by

$$\big(\widehat{Q}_2^{n,t}u_2\big)(\cdot) = Q^n(\cdot)^*\mathbf{V}(\cdot,t)u_2, \quad \forall\, u_2 \in L^2_{\mathbb{F}}(t,T;L^{2q}(\Omega;H)).$$

By a similar argument to derive the inequality (12.118), we find that

$$\begin{aligned}&\big|\widehat{Q}_2^{n,t}\big|_{\mathcal{L}(L^2_{\mathbb{F}}(t,T;L^{2q}(\Omega;H));L^2_{\mathbb{F}}(t,T;L^{\frac{2q}{2q-1}}(\Omega;\mathcal{L}_2^0)))}\\&\le \mathcal{C}\big(|P_T|_{L^p_{\mathcal{F}_T}(\Omega;\mathcal{L}(H))}+|F|_{L^1_{\mathbb{F}}(0,T;L^p(\Omega;\mathcal{L}(H)))}\big).\end{aligned}\tag{12.119}$$

By Lemma 2.61, for each t, there exist two bounded linear operators Q_2^t and $\widehat{Q}_2^t$ from $L^2_{\mathbb{F}}(t,T;L^{2q}(\Omega;H))$ to $L^2_{\mathbb{F}}(t,T;L^{\frac{2q}{2q-1}}(\Omega;\mathcal{L}_2^0))$ and a subsequence $\{n_k^{(4)}\}_{k=1}^\infty \subset \{n_k^{(3)}\}_{n=1}^\infty$ such that, for all $u_2 \in L^2_{\mathbb{F}}(t,T;L^{2q}(\Omega;H))$,

$$\begin{cases}\text{(w)-}\lim\limits_{k\to\infty} Q_2^{n_k^{(4)},t}u_2 = Q_2^t u_2 & \text{in } L^2_{\mathbb{F}}(t,T;L^{\frac{2q}{2q-1}}(\Omega;\mathcal{L}_2^0)),\\[2mm] \text{(w)-}\lim\limits_{k\to\infty} \widehat{Q}_2^{n_k^{(4)},t}u_2 = \widehat{Q}_2^t u_2 & \text{in } L^2_{\mathbb{F}}(t,T;L^{\frac{2q}{2q-1}}(\Omega;\mathcal{L}_2^0)).\end{cases}\tag{12.120}$$

In terms of (12.118) and (12.119), we obtain that

$$\begin{aligned}&\big|Q_2^t\big|_{\mathcal{L}(L^2_{\mathbb{F}}(t,T;L^{2q}(\Omega;H));L^2_{\mathbb{F}}(t,T;L^{\frac{2q}{2q-1}}(\Omega;\mathcal{L}_2^0)))}\\&\quad+\big|\widehat{Q}_2^t\big|_{\mathcal{L}(L^2_{\mathbb{F}}(t,T;L^{2q}(\Omega;H));L^2_{\mathbb{F}}(t,T;L^{\frac{2q}{2q-1}}(\Omega;\mathcal{L}_2^0)))}\\&\le \mathcal{C}\big(|P_T|_{L^p_{\mathcal{F}_T}(\Omega;\mathcal{L}(H))}+|F|_{L^1_{\mathbb{F}}(0,T;L^p(\Omega;\mathcal{L}(H)))}\big).\end{aligned}\tag{12.121}$$

Now, we choose $\xi_1 = 0$ and $u_1 = 0$ in (12.39), and $\xi_2 = 0$ and $u_2 = 0$ in (12.40). From (12.101), we obtain that

$$\begin{aligned}&\mathbb{E}\big\langle P_T^n x_1(T), x_2(T)\big\rangle_H - \mathbb{E}\int_t^T \big\langle F^n(s)x_1(s), x_2(s)\big\rangle_H ds\\&= \mathbb{E}\int_t^T \big\langle P^n(s)K(s)x_1(s), v_2(s)\big\rangle_{\mathcal{L}_2^0} ds\\&\quad+\mathbb{E}\int_t^T \big\langle P^n(s)v_1(s), K(s)x_2(s)+v_2(s)\big\rangle_{\mathcal{L}_2^0} ds\\&\quad+\mathbb{E}\int_t^T \big\langle Q^n(s)v_1(s), x_2(s)\big\rangle_H ds + \mathbb{E}\int_t^T \big\langle \widetilde{Q}^n(s)x_1(s), v_2(s)\big\rangle_{\mathcal{L}_2^0} ds.\end{aligned}\tag{12.122}$$

We define a bilinear functional $\mathcal{B}_{n,t}(\cdot,\cdot)$ on $L^2_{\mathbb{F}}(t,T;L^{2q}(\Omega;\mathcal{L}_2^0)) \times L^2_{\mathbb{F}}(t,T;$ $L^{2q}(\Omega;\mathcal{L}_2^0))$ as follows:

$$\begin{aligned}\mathcal{B}_{n,t}(v_1,v_2) &= \mathbb{E}\int_t^T \big\langle Q^n(s)v_1(s), x_2(s)\big\rangle_H ds\\&\quad+\mathbb{E}\int_t^T \big\langle \widetilde{Q}^n(s)x_1(s), v_2(s)\big\rangle_{\mathcal{L}_2^0} ds,\\&\qquad \forall\, v_1, v_2 \in L^2_{\mathbb{F}}(t,T;L^{2q}(\Omega;\mathcal{L}_2^0)).\end{aligned}\tag{12.123}$$

It is easy to check that $\mathcal{B}_{n,t}(\cdot,\cdot)$ is a bounded bilinear functional. From (12.122), it follows that

$$\begin{aligned}
&\mathcal{B}_{n_k^{(1)},t}(v_1,v_2)\\
&=\mathbb{E}\Big\langle P_T^{n_k^{(1)}}x_1(T),x_2(T)\Big\rangle_H-\mathbb{E}\int_t^T\Big\langle F^{n_k^{(1)}}(s)x_1(s),x_2(s)\Big\rangle_H ds\\
&\quad-\mathbb{E}\int_t^T\Big\langle P^{n_k^{(1)}}(s)K(s)x_1(s),v_2(s)\Big\rangle_{\mathcal{L}_2^0}ds\\
&\quad-\mathbb{E}\int_t^T\Big\langle P^{n_k^{(1)}}(s)v_1(s),K(s)x_2(s)+v_2(s)\Big\rangle_{\mathcal{L}_2^0}ds.
\end{aligned}\tag{12.124}$$

It is easy to show that

$$\begin{cases}
\displaystyle\lim_{k\to\infty}\mathbb{E}\big\langle P_T^{n_k^{(1)}}x_1(T),x_2(T)\big\rangle_H=\mathbb{E}\big\langle P_Tx_1(T),x_2(T)\big\rangle_H,\\
\displaystyle\lim_{k\to\infty}\mathbb{E}\int_t^T\big\langle F^{n_k^{(1)}}(s)x_1(s),x_2(s)\big\rangle_H ds=\mathbb{E}\int_t^T\big\langle F(s)x_1(s),x_2(s)\big\rangle_H ds,\\
\displaystyle\lim_{k\to\infty}\mathbb{E}\int_t^T\big\langle P^{n_k^{(1)}}(s)K(s)x_1(s),v_2(s)\big\rangle_{\mathcal{L}_2^0}ds\\
\displaystyle\quad=\mathbb{E}\int_t^T\big\langle P(s)K(s)x_1(s),v_2(s)\big\rangle_{\mathcal{L}_2^0}ds,\\
\displaystyle\lim_{k\to\infty}\mathbb{E}\int_t^T\big\langle P^{n_k^{(1)}}(s)v_1(s),K(s)x_2(s)+v_2(s)\big\rangle_{\mathcal{L}_2^0}ds\\
\displaystyle\quad=\mathbb{E}\int_t^T\big\langle P(s)v_1(s),K(s)x_2(s)+v_2(s)\big\rangle_{\mathcal{L}_2^0}ds,
\end{cases}$$

where x_1 (*resp.* x_2) solves the equation (12.39) (*resp.* (12.40)) with $\xi_1=0$ and $u_1=0$ (*resp.* $\xi_2=0$ and $u_2=0$). This, together with (12.124), implies that

$$\begin{aligned}
\mathcal{B}^t(v_1,v_2)&\stackrel{\Delta}{=}\lim_{k\to\infty}\mathcal{B}_{n_k^{(1)},t}(v_1,v_2)\\
&=\mathbb{E}\big\langle P_Tx_1(T),x_2(T)\big\rangle_H-\mathbb{E}\int_t^T\big\langle F(s)x_1(s),x_2(s)\big\rangle_H ds\\
&\quad-\mathbb{E}\int_t^T\big\langle P(s)K(s)x_1(s),v_2(s)\big\rangle_{\mathcal{L}_2^0}ds\\
&\quad-\mathbb{E}\int_t^T\big\langle P(s)v_1(s),K(s)x_2(s)+v_2(s)\big\rangle_{\mathcal{L}_2^0}ds.
\end{aligned}\tag{12.125}$$

Noting that the solution of (12.39) (with $\xi_1=0$ and $u_1=0$) satisfies

$$\begin{aligned} x_1(s) = & \int_t^s S(s-\tau)J(\tau)x_1(\tau)d\tau + \int_t^s S(s-\tau)K(\tau)x_1(\tau)d\tau \\ & + \int_t^s S(s-\tau)v_1(\tau)dW(\tau), \end{aligned}$$

by means of Proposition 3.12 and Gronwall's inequality, we conclude that

$$|x_1|_{L^\infty_{\mathbb{F}}(t,T;L^{2q}(\Omega;H))} \leq \mathcal{C}|v_1|_{L^2_{\mathbb{F}}(t,T;L^{2q}(\Omega;\mathcal{L}_2^0))}. \tag{12.126}$$

Similarly,

$$|x_2|_{L^\infty_{\mathbb{F}}(t,T;L^{2q}(\Omega;H))} \leq \mathcal{C}|v_2|_{L^2_{\mathbb{F}}(t,T;L^{2q}(\Omega;\mathcal{L}_2^0))}. \tag{12.127}$$

Combining (12.125), (12.126), (12.127) and (12.108), we obtain that

$$\begin{aligned} |\mathcal{B}^t(v_1,v_2)| \leq & \mathcal{C}\big(|P_T|_{L^p_{\mathcal{F}_T}(\Omega;\mathcal{L}(H))} + |F|_{L^1_{\mathbb{F}}(0,T;L^p(\Omega;\mathcal{L}(H)))}\big) \\ & \times |v_1|_{L^2_{\mathbb{F}}(t,T;L^{2q}(\Omega;\mathcal{L}_2^0))}|v_2|_{L^2_{\mathbb{F}}(t,T;L^{2q}(\Omega;\mathcal{L}_2^0))}. \end{aligned}$$

Hence, $\mathcal{B}^t(\cdot,\cdot)$ is a bounded bilinear functional on $L^2_{\mathbb{F}}(t,T;L^{2q}(\Omega;\mathcal{L}_2^0))\times L^2_{\mathbb{F}}(t,T;L^{2q}(\Omega;\mathcal{L}_2^0))$. Now, for any fixed $v_2 \in L^2_{\mathbb{F}}(t,T;L^{2q}(\Omega;\mathcal{L}_2^0))$, it is easy to see that $\mathcal{B}^t(\cdot,v_2)$ is a bounded linear functional on $L^2_{\mathbb{F}}(t,T;L^{2q}(\Omega;\mathcal{L}_2^0))$. Therefore, by Theorem 2.73, we can find a unique $\tilde{v}_1 \in L^2_{\mathbb{F}}(t,T;L^{\frac{2q}{2q-1}}(\Omega;\mathcal{L}_2^0))$ such that, for all $v_2 \in L^2_{\mathbb{F}}(t,T;L^{2q}(\Omega;\mathcal{L}_2^0))$,

$$\mathcal{B}^t(v_1,v_2) = \big\langle \tilde{v}_1, v_2\big\rangle_{L^2_{\mathbb{F}}(t,T;L^{\frac{2q}{2q-1}}(\Omega;H)),\ L^2_{\mathbb{F}}(t,T;L^{2q}(\Omega;\mathcal{L}_2^0))}.$$

Define an operator $\widehat{Q}_3^t$ from $L^2_{\mathbb{F}}(t,T;L^{2q}(\Omega;\mathcal{L}_2^0))$ to $L^2_{\mathbb{F}}(t,T;L^{\frac{2q}{2q-1}}(\Omega;\mathcal{L}_2^0))$ as follows:

$$\widehat{Q}_3^t\, v_1 = \tilde{v}_1.$$

From the uniqueness of $\tilde{v}_1$, it is clear that $\widehat{Q}_3^t$ is well-defined. Further,

$$\begin{aligned} & |\widehat{Q}_3^t v_1|_{L^2_{\mathbb{F}}(t,T;L^{\frac{2q}{2q-1}}(\Omega;\mathcal{L}_2^0))} = |\tilde{v}_1|_{L^2_{\mathbb{F}}(t,T;L^{\frac{2q}{2q-1}}(\Omega;\mathcal{L}_2^0))} \\ & \leq \mathcal{C}\big(|P_T|_{L^p_{\mathcal{F}_T}(\Omega;\mathcal{L}(H))} + |F|_{L^1_{\mathbb{F}}(0,T;L^p(\Omega;\mathcal{L}(H)))}\big)|v_1|_{L^2_{\mathbb{F}}(t,T;L^{2q}(\Omega;\mathcal{L}_2^0))}. \end{aligned} \tag{12.128}$$

This shows that $\widehat{Q}_3^t$ is a bounded operator. For any $\alpha,\beta\in\mathbb{R}$ and $v_2,v_3,v_4 \in L^2_{\mathbb{F}}(t,T;L^{2q}(\Omega;\mathcal{L}_2^0))$,

$$\begin{aligned} & \big\langle \widehat{Q}_3^t(\alpha v_3+\beta v_4), v_2\big\rangle_{L^2_{\mathbb{F}}(t,T;L^{\frac{2q}{2q-1}}(\Omega;\mathcal{L}_2^0)),\ L^2_{\mathbb{F}}(t,T;L^{2q}(\Omega;\mathcal{L}_2^0))} \\ & = \mathcal{B}^t(\alpha v_3+\beta v_4, v_2) = \alpha\mathcal{B}^t(v_3,v_2) + \beta\mathcal{B}^t(v_4,v_2), \end{aligned}$$

which indicates that

$$\widehat{Q}_3^t(\alpha v_3+\beta v_4) = \alpha\widehat{Q}_3^t v_3 + \beta\widehat{Q}_3^t v_4.$$

Hence, $\widehat{Q}_3^t$ is a bounded linear operator from $L^2_{\mathbb{F}}(t,T;L^{2q}(\Omega;\mathcal{L}_2^0))$ to $L^2_{\mathbb{F}}(t,T;L^{\frac{2q}{2q-1}}(\Omega;\mathcal{L}_2^0))$. Put $Q_3^t = \frac{1}{2}\widehat{Q}_3^t$. Then, for any $v_1, v_2 \in L^2_{\mathbb{F}}(t,T;L^{2q}(\Omega;\mathcal{L}_2^0))$, it holds that

$$
\begin{aligned}
\mathcal{B}^t(v_1,v_2) &= \big\langle Q_3^t v_1, v_2\big\rangle_{L^2_{\mathbb{F}}(t,T;L^{\frac{2q}{2q-1}}(\Omega;\mathcal{L}_2^0)),\ L^2_{\mathbb{F}}(t,T;L^{2q}(\Omega;\mathcal{L}_2^0))}\\
&\quad+\big\langle v_1, \big(Q_3^t\big)^* v_2\big\rangle_{L^2_{\mathbb{F}}(t,T;L^{2q}(\Omega;\mathcal{L}_2^0)),\ L^2_{\mathbb{F}}(t,T;L^{\frac{2q}{2q-1}}(\Omega;\mathcal{L}_2^0))}.
\end{aligned}
\tag{12.129}
$$

Here, by Theorem 2.73, $\big(Q_3^t\big)^*$ is a bounded linear operator from $L^2_{\mathbb{F}}(t,T;L^{\frac{2q}{2q-1}}(\Omega;\mathcal{L}_2^0))^* = L^2_{\mathbb{F}}(t,T;L^{2q}(\Omega;\mathcal{L}_2^0))$ to $L^2_{\mathbb{F}}(t,T;L^{2q}(\Omega;\mathcal{L}_2^0))^* = L^2_{\mathbb{F}}(t,T;L^{\frac{2q}{2q-1}}(\Omega;\mathcal{L}_2^0))$. It follows from (12.128) that

$$
\begin{aligned}
&|Q_3^t|_{\mathcal{L}(L^2_{\mathbb{F}}(t,T;L^{2q}(\Omega;\mathcal{L}_2^0));\ L^2_{\mathbb{F}}(t,T;L^{\frac{2q}{2q-1}}(\Omega;\mathcal{L}_2^0)))}\\
&\leq \mathcal{C}\big(|P_T|_{L^p_{\mathcal{F}_T}(\Omega;\mathcal{L}(H))} + |F|_{L^1_{\mathbb{F}}(0,T;L^p(\Omega;\mathcal{L}(H)))}\big).
\end{aligned}
\tag{12.130}
$$

For any $t\in[0,T]$, we define two operators $Q^{(t)}$ and $\widehat{Q}^{(t)}$ on $L^{2q}_{\mathcal{F}_t}(\Omega;H)\times L^2_{\mathbb{F}}(t,T;L^{2q}(\Omega;H))\times L^2_{\mathbb{F}}(t,T;L^{2q}(\Omega;\mathcal{L}_2^0))$ as follows:

$$
\begin{cases}
Q^{(t)}(\xi,u,v) = Q_1^t\xi + Q_2^t u + Q_3^t v,\\
\widehat{Q}^{(t)}(\xi,u,v) = \widehat{Q}_1^t\xi + \widehat{Q}_2^t u + \big(Q_3^t\big)^* v,\\
\forall\,(\xi,u,v)\in L^{2q}_{\mathcal{F}_t}(\Omega;H)\times L^2_{\mathbb{F}}(t,T;L^{2q}(\Omega;H))\times L^2_{\mathbb{F}}(t,T;L^{2q}(\Omega;\mathcal{L}_2^0)).
\end{cases}
\tag{12.131}
$$

Thanks to the definition of Q_1^t, Q_2^t and Q_3^t (*resp.* $\widehat{Q}_1^t$, $\widehat{Q}_2^t$ and $\big(Q_3^t\big)^*$), we find that $Q^{(t)}(\cdot,\cdot,\cdot)$ (*resp.* $\widehat{Q}^{(t)}(\cdot,\cdot,\cdot)$) is a bounded linear operator from $L^{2q}_{\mathcal{F}_t}(\Omega;H)\times L^2_{\mathbb{F}}(t,T;\ L^{2q}(\Omega;H))\times L^2_{\mathbb{F}}(t,T;L^{2q}(\Omega;\mathcal{L}_2^0))$ to $L^2_{\mathbb{F}}(t,T;L^{\frac{2q}{2q-1}}(\Omega;\mathcal{L}_2^0))$ and $Q^{(t)}(0,0,\cdot)^* = \widehat{Q}^{(t)}(0,0,\cdot)$.

For any $t\in[0,T]$, by means of (12.101), (12.109), (12.110), (12.114), (12.120), (12.123), (12.125), (12.129) and (12.131), we see that, for all $(\xi_1,u_1,v_1),(\xi_2,u_2,v_2)\in L^{2q}_{\mathcal{F}_t}(\Omega;H)\times L^2_{\mathbb{F}}(t,T;L^{2q}(\Omega;H))\times L^2_{\mathbb{F}}(t,T;L^{2q}(\Omega;\mathcal{L}_2^0))$,

$$
\begin{aligned}
&\mathbb{E}\big\langle P_T x_1(T), x_2(T)\big\rangle_H - \mathbb{E}\int_t^T \big\langle F(s)x_1(s), x_2(s)\big\rangle_H ds\\
&= \mathbb{E}\big\langle R^{(t)}\xi_1,\xi_2\big\rangle_H + \mathbb{E}\int_t^T \big\langle P(s)u_1(s), x_2(s)\big\rangle_H ds\\
&\quad+\mathbb{E}\int_t^T \big\langle P(s)x_1(s), u_2(s)\big\rangle_H ds + \mathbb{E}\int_t^T \big\langle P(s)K(s)x_1(s), v_2(s)\big\rangle_{\mathcal{L}_2^0} ds\\
&\quad+\mathbb{E}\int_t^T \big\langle P(s)v_1(s), K(s)x_2(s)+v_2(s)\big\rangle_{\mathcal{L}_2^0} ds\\
&\quad+\mathbb{E}\int_t^T \big\langle v_1(s), \widehat{Q}^{(t)}(\xi_2,u_2,v_2)(s)\big\rangle_{\mathcal{L}_2^0} ds
\end{aligned}
\tag{12.132}
$$

$$+\mathbb{E}\int_t^T \big\langle Q^{(t)}(\xi_1,u_1,v_1)(s),v_2(s)\big\rangle_{\mathcal{L}_2^0}ds.$$

Step 3. In this step, we shall show that $P(\cdot)\in D_{\mathbb{F},w}([0,T];L^p(\Omega;\mathcal{L}(H)))$ (recall (12.41) for the definition of this space) and

$$P(t)=R^{(t)},\quad \text{a.e. } t\in[0,T]. \tag{12.133}$$

Similar to the proof of (12.103), by (12.132), one can show that

$$\begin{aligned}&\mathbb{E}\Big(\mathbf{U}^*(T,t)P_T\mathbf{U}(T,t)\xi_1-\int_t^T \mathbf{U}^*(s,t)F(s)\mathbf{U}(s,t)\xi_1 ds\;\Big|\;\mathcal{F}_t\Big)\\&=R^{(t)}\xi_1,\quad \text{a.s., }\forall\, t\in[0,T],\ \xi_1\in L^{2q}_{\mathcal{F}_t}(\Omega;H).\end{aligned}\tag{12.134}$$

Now we show that $R^{(\cdot)}\xi\in D_{\mathbb{F}}([t,T];L^{\frac{2p}{p+1}}(\Omega;H))$ for any $\xi\in L^{2q}_{\mathcal{F}_t}(\Omega;H)$. By (12.100), it remains to show that

$$\lim_{n\to\infty}\big|R^{(\cdot)}\xi-P^n(\cdot)\xi\big|_{L^\infty_{\mathbb{F}}(t,T;L^{\frac{2q}{2q-1}}(\Omega;H))}=0. \tag{12.135}$$

For this purpose, by (12.103) and (12.134), for any $\tau\in[t,T]$, we see that

$$\begin{aligned}&\mathbb{E}\big|R^{(\tau)}\xi-P^n(\tau)\xi\big|_H^{\frac{2q}{2q-1}}\\&\le C\mathbb{E}\Big|\int_\tau^T \mathbf{U}^*(s,\tau)\big(F(s)-F_n(s)\big)\mathbf{U}(s,\tau)\xi ds\Big|_H^{\frac{2q}{2q-1}}\\&\quad+C\mathbb{E}\Big|\mathbf{U}^*(T,\tau)\big(P_T-P_T^n\big)\mathbf{U}(T,\tau)\xi\Big|_H^{\frac{2q}{2q-1}}.\end{aligned}$$

By the first conclusion in Lemma 12.10, we deduce that for any $\varepsilon_1>0$, there is a $\delta_1>0$ so that for all $\tau\in[t,T]$ and $\tau\le\sigma\le\tau+\delta_1$,

$$\mathbb{E}\big|\mathbf{U}(r,\tau)\xi-\mathbf{U}(r,\sigma)\xi\big|_H^{\frac{2q}{2q-1}}<\varepsilon_1,\qquad \forall\, r\in[\sigma,T]. \tag{12.136}$$

Now, we choose a monotonically increasing sequence $\{\tau_i\}_{i=1}^{N_1}\subset[0,T]$ for N_1 being sufficiently large such that $\tau_{i+1}-\tau_i\le\delta_1$ with $\tau_1=t$ and $\tau_{N_1}=T$, and that

$$\Big(\int_{\tau_i}^{\tau_{i+1}}\big(\mathbb{E}|F(s)|^p_{\mathcal{L}(H)}\big)^{\frac{1}{p}}ds\Big)^{\frac{2q}{2q-1}}<\varepsilon_1,\ \text{ for all } i=1,\cdots,N_1-1. \tag{12.137}$$

For any $\tau(\tau_i,\tau_{i+1}]$, recalling $F^n=\Gamma_nF\Gamma_n$, we conclude that

$$\begin{aligned}&\mathbb{E}\Big|\int_\tau^T \mathbf{U}^*(s,\tau)\big(F(s)-F^n(s)\big)\mathbf{U}(s,\tau)\xi ds\Big|_H^{\frac{2q}{2q-1}}\\&\le C\mathbb{E}\Big|\int_{\tau_i}^T \mathbf{U}^*(s,\tau)\big(F(s)-F^n(s)\big)\mathbf{U}(s,\tau_i)\xi ds\Big|_H^{\frac{2q}{2q-1}}\end{aligned}$$

$$\begin{aligned}&+\mathcal{C}\mathbb{E}\Big|\int_\tau^{\tau_i}\mathbf{U}^*(s,\tau)\big(F(s)-F^n(s)\big)\mathbf{U}(s,\tau_i)\xi ds\Big|_H^{\frac{2q}{2q-1}} \qquad (12.138)\\&+\mathcal{C}\mathbb{E}\Big|\int_\tau^{T}\mathbf{U}^*(s,\tau)\big(F(s)-F^n(s)\big)\big(\mathbf{U}(s,\tau_i)-\mathbf{U}(s,\tau)\big)\xi ds\Big|_H^{\frac{2q}{2q-1}}\\&\leq\mathcal{C}\int_{\tau_i}^T\mathbb{E}\Big|\big(F(s)-F^n(s)\big)\mathbf{U}(s,\tau_i)\xi\Big|_H^{\frac{2q}{2q-1}}ds\\&+\mathcal{C}\Big(\int_{\tau_i}^{\tau_{i+1}}\big(\mathbb{E}|F(s)|^p_{\mathcal{L}(H)}\big)^{\frac{1}{p}}ds\Big)^{\frac{2q}{2q-1}}\\&+\mathcal{C}\max_{s\in[\tau,T]}\mathbb{E}\Big|\big(\mathbf{U}(s,\tau_i)-\mathbf{U}(s,\tau)\big)\xi\Big|_H^{\frac{2q}{2q-1}}ds.\end{aligned}$$

By the choice of F^n, there is an integer $N_2(\varepsilon_1)>0$ so that for all $n>N_2$ and $i=1,\cdots,N_1-1$,

$$\int_{\tau_i}^T\mathbb{E}\Big|\big(F(s)-F^n(s)\big)\mathbf{U}(s,\tau_i)\xi\Big|_H^{\frac{2q}{2q-1}}ds\leq\varepsilon_1. \qquad (12.139)$$

By (12.136)–(12.139), we conclude that for all $n>N_2$ and $\tau\in[t,T]$,

$$\mathbb{E}\Big|\int_\tau^T\mathbf{U}^*(s,\tau)\big(F(s)-F^n(s)\big)\mathbf{U}(s,\tau)\xi ds\Big|_H^{\frac{2q}{2q-1}}\leq\mathcal{C}_1\varepsilon_1. \qquad (12.140)$$

Here the constant $\mathcal{C}_1$ is independent of ε_1, n and τ. Similarly, there is an integer $N_3(\varepsilon_1)>0$ such that for every $n>N_3$,

$$\mathbb{E}\Big|\mathbf{U}^*(T,\tau)\big(P_T-P_T^n\big)\mathbf{U}(T,\tau)\xi\Big|_H^{\frac{2q}{2q-1}}\leq\mathcal{C}_2\varepsilon_1, \qquad (12.141)$$

for a constant $\mathcal{C}_2$ which is independent of ε_1, n and τ. Now for any $\varepsilon>0$, let us choose $\varepsilon_1=\dfrac{\varepsilon}{\mathcal{C}_1+\mathcal{C}_2}$. Then,

$$\mathbb{E}\big|R^{(\cdot)}\xi-P^n(\cdot)\xi\big|_H^{\frac{2q}{2q-1}}<\varepsilon \qquad \forall\, n>\max\{N_2(\varepsilon_1),\,N_3(\varepsilon_1)\},\forall\,\tau\in[t,T].$$

Therefore, we obtain the desired result (12.135).

To show (12.133), for any $0\leq t_1<t_2<T$ and $\eta_1,\eta_2\in L^{2q}_{\mathcal{F}_{t_1}}(\Omega;H)$, we choose $\xi_1=\eta_1$ and $u_1=v_1=0$ in the equation (12.39), and $\xi_2=0$, $u_2(\cdot)=\frac{\chi_{[t_1,t_2]}}{t_2-t_1}\eta_2$ and $v_2=0$ in the equation (12.40). By (12.132) and recalling the definition of the evolution operator $\mathbf{U}(\cdot,\cdot)$ (in Lemma 12.10), we see that

$$\begin{aligned}&\frac{1}{t_2-t_1}\mathbb{E}\int_{t_1}^{t_2}\big\langle P(s)\mathbf{U}(s,t_1)\eta_1,\eta_2\big\rangle_H ds \qquad (12.142)\\&=\mathbb{E}\big\langle P_T\mathbf{U}(T,t_1)\eta_1,x_{2,t_2}(T)\big\rangle_H-\mathbb{E}\int_{t_1}^T\big\langle F(s)\mathbf{U}(s,t_1)\eta_1,x_{2,t_2}(s)\big\rangle_H ds,\end{aligned}$$

where $x_{2,t_2}(\cdot)$ stands for the solution to the equation (12.40) with the above choice of ξ_2, u_2 and v_2. It is clear that

$$x_{2,t_2}(s)=\begin{cases}\displaystyle\int_{t_1}^{s} S(s-\tau)J(\tau)x_{2,t_2}(\tau)d\tau+\frac{1}{t_2-t_1}\int_{t_1}^{s} S(s-\tau)\eta_2 d\tau \\ \displaystyle\quad+\int_{t_1}^{s} S(s-\tau)K(\tau)x_{2,t_2}(\tau)dW(\tau), \qquad s\in[t_1,t_2], \\ \mathbf{U}(s,t_2)x_{2,t_2}(t_2), \qquad s\in[t_2,T].\end{cases} \tag{12.143}$$

Then, by Proposition 3.12, we see that for all $s\in[t_1,t_2]$ (Recall (12.47) for $M_{J,K,q}(\cdot)$),

$$\mathbb{E}\big|x_{2,t_2}(s)\big|_H^{2q}\leq \mathcal{C}\Big(\int_{t_1}^{s} M_{J,K,q}(\tau)\mathbb{E}\big|x_{2,t_2}(\tau)\big|_H^{2q}d\tau+\mathbb{E}|\eta_2|_H^{2q}\Big).$$

By Gronwall's inequality, it follows that

$$\big|x_{2,t_2}(\cdot)\big|_{L_{\mathbb{F}}^{\infty}(t_1,t_2;L^{2q}(\Omega;H))}\leq \mathcal{C}|\eta_2|_{L^{2q}_{\mathcal{F}_{t_1}}(\Omega;H)}, \tag{12.144}$$

where the constant $\mathcal{C}$ is independent of t_2. On the other hand, by (12.143), we have

$$\begin{aligned}&\mathbb{E}\big|x_{2,t_2}(t_2)-\eta_2\big|_H^{2q}\\&\leq \mathcal{C}\Big(\int_{t_1}^{s} M_{J,K,q}(\tau)\mathbb{E}\big|x_{2,t_2}(\tau)\big|_H^{2q}d\tau+\mathbb{E}\Big|\frac{1}{t_2-t_1}\int_{t_1}^{t_2} S(t_2-\tau)\eta_2 d\tau-\eta_2\Big|_H^{2q}\Big).\end{aligned}$$

This, together with (12.144), implies that

$$\lim_{t_2\to t_1+0}\mathbb{E}\big|x_{2,t_2}(t_2)-\eta_2\big|_H^{2q}\leq \mathcal{C}\lim_{t_2\to t_1+0}\mathbb{E}\Big|\frac{1}{t_2-t_1}\int_{t_1}^{t_2} S(t_2-\tau)\eta_2 d\tau-\eta_2\Big|_H^{2q}=0.$$

Therefore, for any $s\in[t_2,T]$,

$$\begin{aligned}&\lim_{t_2\to t_1+0}\mathbb{E}\big|\mathbf{U}(s,t_2)x_{2,t_2}(t_2)-\mathbf{U}(s,t_1)\eta_2\big|_H^{2q}\\&\leq \mathcal{C}\lim_{t_2\to t_1+0}\Big(\mathbb{E}\big|\mathbf{U}(s,t_2)x_{2,t_2}(t_2)-\mathbf{U}(s,t_2)\eta_2\big|_H^{2q}\\&\qquad+\mathbb{E}\big|\mathbf{U}(s,t_2)\eta_2-\mathbf{U}(s,t_1)\eta_2\big|_H^{2q}\Big)\\&\leq \mathcal{C}\lim_{t_2\to t_1+0}\Big(\mathbb{E}\big|x_{2,t_2}(t_2)-\eta_2\big|_H^{2q}+\mathbb{E}\big|\mathbf{U}(s,t_2)\eta_2-\mathbf{U}(s,t_1)\eta_2\big|_H^{2q}\Big)\\&=0.\end{aligned}$$

Hence,

$$\lim_{t_2\to t_1+0}x_{2,t_2}(s)=\mathbf{U}(s,t_1)\eta_2\ \text{ in } L^{2q}_{\mathcal{F}_s}(\Omega;H), \qquad \forall\, s\in[t_2,T]. \tag{12.145}$$

By (12.144) and (12.145), we conclude that

$$\begin{aligned}&\lim_{t_2\to t_1+0}\Big(\mathbb{E}\big\langle P_T\mathbf{U}(T,t_1)\eta_1,x_{2,t_2}(T)\big\rangle_H\\&\qquad-\mathbb{E}\int_{t_1}^T\big\langle F(s)\mathbf{U}(s,t_1)\eta_1,x_{2,t_2}(s)\big\rangle_H ds\Big)\qquad(12.146)\\&=\mathbb{E}\big\langle P_T\mathbf{U}(T,t_1)\eta_1,\mathbf{U}(T,t_1)\eta_2\big\rangle_H-\mathbb{E}\int_{t_1}^T\big\langle F(s)\mathbf{U}(s,t_1)\eta_1,\mathbf{U}(s,t_1)\eta_2\big\rangle_H ds.\end{aligned}$$

Choosing $\xi_1=\eta_1$ and $u_1=v_1=0$ in (12.39), and $\xi_2=\eta_2$ and $u_2=v_2=0$ in (12.40), by (12.132), we find that

$$\begin{aligned}\mathbb{E}\big\langle R^{(t_1)}\eta_1,\eta_2\big\rangle_H&=\mathbb{E}\big\langle P_T\mathbf{U}(T,t_1)\eta_1,\mathbf{U}(T,t_1)\eta_2\big\rangle_H\\&\quad-\mathbb{E}\int_{t_1}^T\big\langle F(s)\mathbf{U}(s,t_1)\eta_1,\mathbf{U}(s,t_1)\eta_2\big\rangle_H ds.\end{aligned}\qquad(12.147)$$

Combining (12.142), (12.146) and (12.147), we obtain that

$$\lim_{t_2\to t_1+0}\frac{1}{t_2-t_1}\mathbb{E}\int_{t_1}^{t_2}\big\langle P(s)\mathbf{U}(s,t_1)\eta_1,\eta_2\big\rangle_H ds=\mathbb{E}\big\langle R^{(t_1)}\eta_1,\eta_2\big\rangle_H.\qquad(12.148)$$

On the other hand, by Lemma 4.14, there is a monotonically decreasing sequence $\{t_2^{(n)}\}_{n=1}^\infty$ with $t_2^{(n)}>t_1$ for every n, such that for a.e. $t_1\in[0,T)$,

$$\lim_{t_2^{(n)}\to t_1+0}\frac{1}{t_2^{(n)}-t_1}\mathbb{E}\int_{t_1}^{t_2^{(n)}}\big\langle P(s)\mathbf{U}(s,t_1)\eta_1,\eta_2\big\rangle_H ds=\mathbb{E}\big\langle P(t_1)\eta_1,\eta_2\big\rangle_H.$$

This, together with (12.148), implies that

$$\mathbb{E}\big\langle R^{(t_1)}\eta_1,\eta_2\big\rangle_H=\mathbb{E}\big\langle P(t_1)\eta_1,\eta_2\big\rangle_H,\quad\text{for a.e. } t_1\in[0,T).$$

Since η_1 and η_2 are arbitrary elements in $L^{2q}_{\mathcal{F}_{t_1}}(\Omega;H)$, we conclude (12.133).

By (12.132) and (12.133), we see that $\big(P(\cdot),Q^{(\cdot)},\widehat{Q}^{(\cdot)}\big)$ satisfies (12.45). Hence, $\big(P(\cdot),Q^{(\cdot)},\widehat{Q}^{(\cdot)}\big)$ is a relaxed transposition solution to (12.8), and by (12.108), (12.115), (12.121), (12.130) and (12.131), it satisfies the estimate (12.46). This completes the proof of Theorem 12.9. □

12.4.6 A Regularity Result

In this subsection, we shall derive a regularity result for relaxed transposition solutions to the equation (12.8), which will play a key role in the proof of the general Pontryagin-type stochastic maximum principle in Section 12.5.

Let $\{\Delta_m\}_{m=1}^\infty$ be a sequence of partitions of $[0,T]$, that is,

$$\Delta_m \stackrel{\Delta}{=} \Big\{ t_i^m \;\Big|\; i = 0, 1, \cdots, m, \text{ and } 0 = t_0^m < t_1^m < \cdots < t_m^m = T \Big\}$$

such that $\Delta_m \subset \Delta_{m+1}$ and $|\Delta_m| \stackrel{\Delta}{=} \max_{0 \le i \le m-1}(t_{i+1}^m - t_i^m) \to 0$ as $m \to \infty$. Recall that $\{e_k\}_{k=1}^\infty$ is an orthonormal basis of V. For $m, n \in \mathbb{N}$, we introduce the following subspaces of $L^2_{\mathbb{F}}(0,T; L^{2q}(\Omega; \mathcal{L}_2^0))$:

$$\begin{aligned}\mathcal{H}_{m,n} = \Big\{ \sum_{i=0}^{m-1} \sum_{k=1}^{n} \chi_{[t_i^m, t_{i+1}^m)}(\cdot) a_{ki} e_k \otimes \big(\mathbf{U}(\cdot, t_i^m) f_i\big) \;\Big|\; a_{ki} \in \mathbb{C}, \\ f_i \in L^{2q}_{\mathcal{F}_{t_i^m}}(\Omega; H) \Big\}.\end{aligned} \tag{12.149}$$

Here $\mathbf{U}(\cdot, \cdot)$ is the operator introduced in Lemma 12.10. In the sequel, we shall use the following norm for the linear space $\mathcal{H}_{m,n}$,

$$\Big| \sum_{i=0}^{m-1} \sum_{k=1}^{n} \chi_{[t_i^m, t_{i+1}^m)}(\cdot) a_{ki} e_k \otimes f_i \Big|_{L^2_{\mathbb{F}}(0,T; L^{2q}(\Omega; \mathcal{L}_2^0))},$$

for any $v = \sum_{i=0}^{m-1} \sum_{k=1}^{n} \chi_{[t_i^m, t_{i+1}^m)}(\cdot) a_{ki} e_k \otimes \big(\mathbf{U}(\cdot, t_i^m) f_i\big) (\in \mathcal{H}_{m,n})$ with $a_{ki} \in \mathbb{C}$ and $f_i \in L^{2q}_{\mathcal{F}_{t_i^m}}(\Omega; H)$.

We have the following result.

Proposition 12.15. *The set* $\bigcup_{m,n=1}^{\infty} \mathcal{H}_{m,n}$ *is dense in* $L^2_{\mathbb{F}}(0,T; L^{2q}(\Omega; \mathcal{L}_2^0))$.

Proof: We introduce the following subspace of $L^2_{\mathbb{F}}(0,T; L^{2q}(\Omega; \mathcal{L}_2^0))$:

$$\widetilde{\mathcal{H}}_{m,n} = \Big\{ \sum_{i=0}^{m-1} \sum_{k=1}^{n} \chi_{[t_i^m, t_{i+1}^m)}(\cdot) a_{ki} e_k \otimes f_i \;\Big|\; a_{ki} \in \mathbb{C},\ f_i \in L^{2q}_{\mathcal{F}_{t_i^m}}(\Omega; H) \Big\}.$$

It is clear that $\bigcup_{m,n=1}^{\infty} \widetilde{\mathcal{H}}_{m,n}$ is dense in $L^2_{\mathbb{F}}(0,T; L^{2q}(\Omega; \mathcal{L}_2^0))$.

For any $m, n \in \mathbb{N}$, $f_i \in L^{2q}_{\mathcal{F}_{t_i^m}}(\Omega; H)$ and $a_{ki} \in \mathbb{C}$ ($i \in \{0, 1, \cdots, m-1\}$, $k \in \{1, 2, \cdots, n\}$), put

$$\tilde{v}_{m,n} \stackrel{\Delta}{=} \sum_{i=0}^{m-1} \sum_{k=1}^{n} \chi_{[t_i^m, t_{i+1}^m)}(\cdot) a_{ki} e_k \otimes f_i.$$

Clearly, $\tilde{v}_{m,n} \in \widetilde{H}_{m,n}$. We claim that for any $\varepsilon > 0$, there exist an $m_\varepsilon \in \mathbb{N}$ and a $v_{m_\varepsilon,n} \in \mathcal{H}_{m_\varepsilon,n}$ such that

$$\big| \tilde{v}_{m,n} - v_{m_\varepsilon,n} \big|_{L^2_{\mathbb{F}}(0,T; L^{2q}(\Omega; \mathcal{L}_2^0))} < \varepsilon. \tag{12.150}$$

Indeed, by Lemma 12.10, for each f_i, there is a $\delta_i^m > 0$ such that for all $t \in [t_i^m, T - \delta_i^m)$ and $s \in [t, t + \delta_i^m]$, it holds that

$$\begin{aligned}&\big|\mathbf{U}(s,t)f_i - f_i\big|_{L^{2q}_{\mathcal{F}_T}(\Omega;H)}\\&< \frac{\varepsilon}{\sqrt{T}n\max\big\{|a_{ki}|+1 \bigm| 0\le i\le m-1, 1\le k\le n\big\}}.\end{aligned} \tag{12.151}$$

Now we choose an $m_\varepsilon \in \mathbb{N}$ and a partition Δ_{m_ε} of $[0,T]$ such that $\Delta_m \subset \Delta_{m_\varepsilon}$ and $|\Delta_{m_\varepsilon}| \le \min_{0\le i\le m-1}\{\delta_i^m\}$. Let

$$v_{m_\varepsilon,d} = \sum_{j=0}^{m_\varepsilon-1}\sum_{k=1}^{n}\chi_{[t_j^{m_\varepsilon}, t_{j+1}^{m_\varepsilon})}(\cdot)\tilde a_{kj}e_k \otimes \big(\mathbf{U}(\cdot, t_j^{m_\varepsilon})\tilde f_j\big),$$

where $\tilde a_{kj} = a_{ki}$ and $\tilde f_j = f_i$ whenever $[t_j^{m_\varepsilon}, t_{j+1}^{m_\varepsilon}) \subset [t_i^m, t_{i+1}^m)$. From (12.151), we find that

$$\begin{aligned}&\big|\tilde v_{m,n} - v_{m_\varepsilon,n}\big|_{L^2_{\mathbb{F}}(0,T;L^{2q}(\Omega;\mathcal{L}_2^0))}\\&= \Big|\sum_{i=0}^{m-1}\sum_{k=1}^{n}\chi_{[t_i^m, t_{i+1}^m)}(\cdot)a_{ki}e_k\otimes f_i\\&\quad - \sum_{j=0}^{m_\varepsilon-1}\sum_{k=1}^{n}\chi_{[t_j^{m_\varepsilon}, t_{j+1}^{m_\varepsilon})}(\cdot)\tilde a_{kj}e_k \otimes \big(\mathbf{U}(\cdot, t_j^{m_\varepsilon})\tilde f_j\big)\Big|_{L^2_{\mathbb{F}}(0,T;L^{2q}(\Omega;\mathcal{L}_2^0))}\\&= \Big|\sum_{j=0}^{m_\varepsilon-1}\sum_{k=1}^{n}\chi_{[t_j^{m_\varepsilon}, t_{j+1}^{m_\varepsilon})}(\cdot)\tilde a_{kj}e_k \otimes \big(\mathbf{U}(\cdot, t_j^{m_\varepsilon})\tilde f_j - \tilde f_j\big)\Big|_{L^2_{\mathbb{F}}(0,T;L^{2q}(\Omega;\mathcal{L}_2^0))} < \varepsilon.\end{aligned}$$

This proves (12.150). Hence, $\displaystyle\bigcup_{m,d=1}^{\infty}\mathcal{H}_{m,n}$ is dense in $L^2_{\mathbb{F}}(0,T;L^{2q}(\Omega;\mathcal{L}_2^0))$.

□

Our regularity result for solutions to (12.8) can be stated as follows.

Theorem 12.16. *Suppose that the assumptions in Theorem 12.9 hold and let* $(P(\cdot), Q^{(\cdot)}, \widehat Q^{(\cdot)})$ *be the relaxed transposition solution to the equation (12.8). Then, for each* $m, n \in \mathbb{N}$, *there exist two pointwise defined, bounded linear operators* $Q_{m,n}$ *and* $\widehat Q_{m,n}$, *both of which are from* $\mathcal{H}_{m,n}$ *to* $L^2_{\mathbb{F}}(0,T;L^{\frac{2p}{p+1}}(\Omega;H))$, *such that, for any* $\xi_1, \xi_2 \in L^{2q}_{\mathcal{F}_0}(\Omega;H)$, $u_1(\cdot), u_2(\cdot) \in L^{2q}_{\mathbb{F}}(\Omega;L^2(0,T;H))$ *and* $v_1(\cdot), v_2(\cdot) \in \mathcal{H}_{m,n}$, *it holds that*

$$\begin{aligned}&\mathbb{E}\int_0^T \big\langle v_1(s), \widehat Q^{(0)}(\xi_2,u_2,v_2)(s)\big\rangle_{\mathcal{L}_2^0}ds\\&+\mathbb{E}\int_0^T \big\langle Q^{(0)}(\xi_1,u_1,v_1)(s), v_2(s)\big\rangle_{\mathcal{L}_2^0}ds\\&= \mathbb{E}\int_0^T \big(\big\langle (Q_{m,n}v_1)(s), x_2(s)\big\rangle_H + \big\langle x_1(s), (\widehat Q_{m,n}v_2)(s)\big\rangle_H\big)ds,\end{aligned} \tag{12.152}$$

where, $x_1(\cdot)$ and $x_2(\cdot)$ solve accordingly (12.39) *and* (12.40) *with* $t = 0$.

Proof: From the proof of the "Existence and Stability" part in Theorem 12.9, for each $\ell \in \mathbb{N}$, one can find a pair of $(P^\ell(\cdot), Q^\ell(\cdot)) \in L^p_{\mathbb{F}}(\Omega; D([0,T]; \mathcal{L}_2(H))) \times L^p_{\mathbb{F}}(\Omega; L^2(0,T; \mathcal{L}_2(\mathcal{L}_2^0; H)))$ so that the variational equality (12.101) holds.

For any $m \in \mathbb{N}$, $i \in \{0,1,\cdots,m-1\}$, $\xi_1 \in L^{2q}_{\mathcal{F}_{t_i^m}}(\Omega; H)$ and $v_2 \in L^2_{\mathbb{F}}(t_i^m, T; L^{2q}(\Omega; \mathcal{L}_2^0))$, letting $u_1 = 0$ and $v_1 = 0$ in the equation (12.39), and letting $\xi_2 = 0$ and $u_2 = 0$ in the equation (12.40), by (12.101) with $t = t_i^m$, we find that

$$\begin{aligned} &\mathbb{E}\big\langle P_T^\ell x_1(T), x_2(T)\big\rangle_H - \mathbb{E}\int_{t_i^m}^T \big\langle F^\ell(s)x_1(s), x_2(s)\big\rangle_H ds \\ &= \mathbb{E}\int_{t_i^m}^T \big\langle P^\ell(s)K(s)x_1(s), v_2(s)\big\rangle_{\mathcal{L}_2^0} ds + \mathbb{E}\int_{t_i^m}^T \big\langle \widetilde{Q^\ell}(s)U(s,t_i^m)\xi_1, v_2(s)\big\rangle_{\mathcal{L}_2^0} ds, \end{aligned} \tag{12.153}$$

where F^ℓ and P_T^ℓ are given in (12.97). For these data ξ_1, u_1, v_1, ξ_2, u_2 and v_2, from the variational equality (12.101) with $t = t_{i+1}^m$, we obtain that

$$\begin{aligned} &\mathbb{E}\big\langle P_T^\ell x_1(T), x_2(T)\big\rangle_H - \mathbb{E}\int_{t_{i+1}^m}^T \big\langle F^\ell(s)x_1(s), x_2(s)\big\rangle_H ds \\ &= \mathbb{E}\big\langle P^\ell(t_{i+1}^m)x_1(t_{i+1}^m), x_2(t_{i+1}^m)\big\rangle_H + \mathbb{E}\int_{t_{i+1}^m}^T \big\langle P^\ell(s)K(s)x_1(s), v_2(s)\big\rangle_{\mathcal{L}_2^0} ds \\ &\quad +\mathbb{E}\int_{t_{i+1}^m}^T \big\langle \widetilde{Q^\ell}(s)\mathbf{U}(s,t_i^m)\xi_1, v_2(s)\big\rangle_{\mathcal{L}_2^0} ds. \end{aligned} \tag{12.154}$$

From (12.153) and (12.154), it follows that

$$\begin{aligned} &\mathbb{E}\big\langle P^\ell(t_{i+1}^m)\xi_1, x_2(t_{i+1}^m)\big\rangle_H - \mathbb{E}\int_{t_i^m}^{t_{i+1}^m} \big\langle F^\ell(s)x_1(s), x_2(s)\big\rangle_H ds \\ &= \mathbb{E}\int_{t_i^m}^{t_{i+1}^m} \big\langle P^\ell(s)K(s)x_1(s), v_2(s)\big\rangle_{\mathcal{L}_2^0} ds \\ &\quad +\mathbb{E}\int_{t_i^m}^{t_{i+1}^m} \big\langle \widetilde{Q^\ell}(s)U(s,t_i^m)\xi_1, v_2(s)\big\rangle_{\mathcal{L}_2^0} ds, \end{aligned} \tag{12.155}$$

holds for any $i \in \{0,1,\cdots,m-1\}$, $\xi_1 \in L^{2q}_{\mathcal{F}_{t_i^m}}(\Omega; H)$ and $v_2 \in L^2_{\mathbb{F}}(t_i^m, t_{i+1}^m; L^{2q}(\Omega; \mathcal{L}_2^0))$. We choose $\xi_1 \in L^{2q}_{\mathcal{F}_{t_i^m}}(\Omega; H)$ with $|\xi_1|_{L^{2q}_{\mathcal{F}_{t_i^m}}(\Omega;H)} = 1$ such that

$$\begin{aligned} &\big|\widetilde{Q^\ell}(\cdot)\mathbf{U}(\cdot,t_i^m)\xi_1\big|_{L^2_{\mathbb{F}}(t_i^m,t_{i+1}^m;L^{\frac{2p}{p+1}}(\Omega;\mathcal{L}_2^0))} \\ &\geq \frac{1}{2}\big|\widetilde{Q^\ell}(\cdot)\mathbf{U}(\cdot,t_i^m)\big|_{\mathcal{L}(L^{2q}_{\mathcal{F}_{t_i^m}}(\Omega;H);\ L^2_{\mathbb{F}}(t_i^m,t_{i+1}^m;L^{\frac{2p}{p+1}}(\Omega;\mathcal{L}_2^0)))}. \end{aligned}$$

We then find a $v_2 \in L^2_{\mathbb{F}}(t_i^m, t_{i+1}^m; L^{2q}(\Omega; \mathcal{L}_2^0))$ with $|v_2|_{L^2_{\mathbb{F}}(t_i^m, t_{i+1}^m; L^{2q}(\Omega;\mathcal{L}_2^0))} = 1$ so that

$$\begin{aligned}&\mathbb{E}\int_{t_i^m}^{t_{i+1}^m} \big\langle \widetilde{Q^\ell}(s)\mathbf{U}(s,t_i^m)\xi_1, v_2(s)\big\rangle_{\mathcal{L}_2^0} ds\\&\geq \frac{1}{2}\big|\widetilde{Q^\ell}(\cdot)\mathbf{U}(\cdot,t_i^m)\xi_1\big|_{L^2_{\mathbb{F}}(t_i^m,t_{i+1}^m;L^{\frac{2p}{p+1}}(\Omega;\mathcal{L}_2^0))}.\end{aligned}$$

Hence,

$$\begin{aligned}&\mathbb{E}\int_{t_i^m}^{t_{i+1}^m} \big\langle \widetilde{Q^\ell}(s)\mathbf{U}(s,t_i^m)\xi_1, v_2(s)\big\rangle_{\mathcal{L}_2^0} ds\\&\geq \frac{1}{4}\big|\widetilde{Q^\ell}(\cdot)\mathbf{U}(\cdot,t_i^m)\big|_{\mathcal{L}(L^{2q}_{\mathcal{F}_{t_i^m}}(\Omega;H);\ L^2_{\mathbb{F}}(t_i^m,t_{i+1}^m;L^{\frac{2p}{p+1}}(\Omega;\mathcal{L}_2^0))}.\end{aligned} \tag{12.156}$$

Also, it is easy to see that

$$\begin{aligned}&\Big|P^\ell(t_{i+1}^m)\xi_1, x_2(t_{i+1}^m)\big\rangle_H - \mathbb{E}\int_{t_i^m}^{t_{i+1}^m} \big\langle F^\ell(s)x_1(s), x_2(s)\big\rangle_H ds\\&-\mathbb{E}\int_{t_i^m}^{t_{i+1}^m} \big\langle P^\ell(s)K(s)x_1(s), v_2(s)\big\rangle_{\mathcal{L}_2^0} ds\Big|\\&\leq \mathcal{C}\big(|P_T|_{L^p_{\mathcal{F}_T}(\Omega;\mathcal{L}(H))} + |F|_{L^1_{\mathbb{F}}(0,T;L^p(\Omega;\mathcal{L}(H)))}\big).\end{aligned} \tag{12.157}$$

Combining (12.153), (12.156) and (12.157), we find that

$$\begin{aligned}&\big|\widetilde{Q^\ell}(\cdot)\mathbf{U}(\cdot,t_i^m)\big|_{\mathcal{L}(L^{2q}_{\mathcal{F}_{t_i^m}}(\Omega;H);\ L^2_{\mathbb{F}}(t_i^m,t_{i+1}^m;L^{\frac{2p}{p+1}}(\Omega;\mathcal{L}_2^0)))}\\&\leq \mathcal{C}\big(|P_T|_{L^p_{\mathcal{F}_T}(\Omega;\mathcal{L}(H))} + |F|_{L^1_{\mathbb{F}}(0,T;L^p(\Omega;\mathcal{L}(H)))}\big).\end{aligned} \tag{12.158}$$

By (12.158) and Lemma 2.78, there exist a bounded, pointwisely defined linear operator $\mathcal{Q}_{t_i^m}$ from $L^{2q}_{\mathcal{F}_{t_i^m}}(\Omega;H)$ to $L^2_{\mathbb{F}}(t_i^m, t_{i+1}^m; L^{\frac{2p}{p+1}}(\Omega;\mathcal{L}_2^0))$, and a subsequence $\{\ell_{1,k}\}_{k=1}^\infty$ of $\{\ell\}_{\ell=1}^\infty$ such that, for all $\xi \in L^{2q}_{\mathcal{F}_{t_i^m}}(\Omega;H)$,

$$\text{(w)-}\lim_{k\to\infty} \widetilde{Q^{\ell_{1,k}}}(\cdot)\mathbf{U}(\cdot,t_i^m)\xi = \mathcal{Q}_{t_i^m}(\cdot)\xi \quad \text{in } L^2_{\mathbb{F}}(t_i^m, t_{i+1}^m; L^{\frac{2p}{p+1}}(\Omega;\mathcal{L}_2^0)). \tag{12.159}$$

Since $\mathcal{Q}_{t_i^m}$ is pointwisely defined, for a.e. $(t,\omega) \in (t_i^m, t_{i+1}^m) \times \Omega$, there is a $q_{t_i^m}(t,\omega) \in \mathcal{L}(H;\mathcal{L}_2^0)$ such that, for any $\xi \in L^{2q}_{\mathcal{F}_{t_i^m}}(\Omega;H)$,

$$\big(\mathcal{Q}_{t_i^m}\xi\big)(t,\omega) = q_{t_i^m}(t,\omega)\xi(\omega).$$

Now, for any $m,n \in \mathbb{N}$, let us define an operator $Q_{m,n}$ from $\mathcal{H}_{m,n}$ to $L^2_{\mathbb{F}}(0,T;L^{\frac{2p}{p+1}}(\Omega;H))$ as follows:

$$
\begin{aligned}
&\big(Q_{m,n}v\big)(t,\omega)\\
&=\sum_{i=0}^{m-1}\sum_{k=1}^{n}\chi_{[t_i^m,t_{i+1}^m)}(t)a_{ki}\big(q_{t_i^m}(t,\omega)f_i\big)e_k, \qquad \text{a.e. } (t,\omega)\in(0,T)\times\Omega,
\end{aligned}
$$

where $v=\sum_{i=0}^{m-1}\sum_{k=1}^{n}\chi_{[t_i^m,t_{i+1}^m)}(\cdot)a_{ki}e_k\otimes\big(\mathbf{U}(\cdot,t_i^m)f_i\big)\in\mathcal{H}_{m,n}$ with $a_{ki}\in\mathbb{C}$ and $f_i\in L^{2q}_{\mathcal{F}_{t_i^m}}(\Omega;H)$. It is easy to check that $Q_{m,n}v\in L^2_{\mathbb{F}}(0,T;L^{\frac{2p}{p+1}}(\Omega;H))$, $Q_{m,n}$ is a pointwisely defined linear operator from $\mathcal{H}_{m,n}$ to $L^2_{\mathbb{F}}(0,T;L^{\frac{2p}{p+1}}(\Omega;H))$ and

$$
\big|Q_{m,n}v\big|_{L^2_{\mathbb{F}}(0,T;L^{\frac{2p}{p+1}}(\Omega;H))}\le\mathcal{C}|v|_{\mathcal{H}_{m,n}}.
$$

For the above v, we have

$$
\begin{aligned}
Q^{\ell_{1,k}}(s)v(s)&=\sum_{i=0}^{m-1}\sum_{k=1}^{n}\chi_{[t_i^m,t_{i+1}^m)}(\cdot)a_{ki}Q^{\ell_{1,k}}(s)\big(e_k\otimes\big(\mathbf{U}(\cdot,t_i^m)f_i\big)\big)\\
&=\sum_{i=0}^{m-1}\sum_{k=1}^{n}\chi_{[t_i^m,t_{i+1}^m)}(\cdot)a_{ki}\widetilde{Q^{\ell_{1,k}}}(s)\big(\big(\mathbf{U}(\cdot,t_i^m)f_i\big)e_k.
\end{aligned}
$$

Hence,

$$
\begin{aligned}
&Q^{\ell_{1,k}}(\cdot)v(\cdot)-\big(Q_{m,n}v\big)(\cdot)\\
&=\sum_{i=0}^{m-1}\sum_{k=1}^{n}\chi_{[t_i^m,t_{i+1}^m)}(\cdot)a_{ki}\Big(\widetilde{Q^{\ell_{1,k}}}(s)\big(\big(\mathbf{U}(\cdot,t_i^m)f_i\big)e_k-\big(q_{t_i^m}(t,\omega)f_i\big)e_k\Big).
\end{aligned}
$$

This gives that

$$
\text{(w)-}\lim_{k\to\infty}Q^{\ell_{1,k}}(\cdot)v(\cdot)=Q_{m,n}v \quad \text{in } L^2_{\mathbb{F}}(0,T;L^{\frac{2p}{p+1}}(\Omega;H)),\quad \forall\, v\in\mathcal{H}_{m,n}. \tag{12.160}
$$

Similarly, one can find a subsequence $\{\ell_{2,k}\}_{k=1}^{\infty}\subset\{\ell_{1,k}\}_{k=1}^{\infty}$ and a pointwise defined linear operator $\widehat{Q}_{m,n}$ from $\mathcal{H}_{m,n}$ to $L^2_{\mathbb{F}}(0,T;L^{\frac{2p}{p+1}}(\Omega;H))$ such that

$$
\text{(w)-}\lim_{k\to\infty}\widetilde{Q^{\ell_{2,k}}}^{*}(\cdot)v(\cdot)=\widehat{Q}_{m,n}v \quad \text{in } L^2_{\mathbb{F}}(0,T;L^{\frac{2p}{p+1}}(\Omega;H)),\quad \forall\, v\in\mathcal{H}_{m,n}. \tag{12.161}
$$

For any $\xi_1,\xi_2\in L^{2q}_{\mathcal{F}_0}(\Omega;H)$, $u_1(\cdot),u_2(\cdot)\in L^2_{\mathbb{F}}(0,T;L^{2q}(\Omega;H))$ and $v_1(\cdot)$, $v_2(\cdot)\in\mathcal{H}_{m,n}$, by (12.160)–(12.161), it is easy to see that

$$\begin{aligned}&\lim_{k\to\infty}\mathbb{E}\int_0^T\Big(\big\langle Q^{\ell_{2,k}}(s)v_1(s),x_2(s)\big\rangle_H+\big\langle \widetilde{Q^{\ell_{2,k}}}(s)x_1(s),v_2(s)\big\rangle_{\mathcal{L}_2^0}\Big)ds\\&=\lim_{k\to\infty}\mathbb{E}\int_0^T\Big(\big\langle Q^{\ell_{2,k}}(s)v_1(s),x_2(s)\big\rangle_H+\big\langle x_1(s),\widetilde{Q^{\ell_{2,k}}}^*(s)v_2(s)\big\rangle_H\Big)ds\\&=\mathbb{E}\int_0^T\Big(\big\langle\big(Q_{m,n}v_1\big)(s),x_2(s)\big\rangle_H+\big\langle x_1(s),\big(\widehat{Q}_{m,n}v_2\big)(s)\big\rangle_H\Big)ds.\end{aligned}\tag{12.162}$$

On the other hand, from the proof of the "Existence and Stability" part in Theorem 12.9, one can show that there exists a subsequence $\{\ell_{3,k}\}_{k=1}^\infty$ of $\{\ell_{2,k}\}_{k=1}^\infty$ such that

$$\begin{aligned}&\lim_{k\to\infty}\mathbb{E}\int_0^T\Big(\big\langle Q^{\ell_{3,k}}(s)v_1(s),x_2(s)\big\rangle_H+\big\langle \widetilde{Q^{\ell_{3,k}}}(s)x_1(s),v_2(s)\big\rangle_{\mathcal{L}_2^0}\Big)ds\\&=\mathbb{E}\int_0^T\big[\big\langle v_1(s),\widehat{Q}^{(0)}(\xi_2,u_2,v_2)(s)\big\rangle_{\mathcal{L}_2^0}+\big\langle Q^{(0)}(\xi_1,u_1,v_1)(s),v_2(s)\big\rangle_{\mathcal{L}_2^0}\big]ds.\end{aligned}\tag{12.163}$$

Combining (12.162) and (12.163), we obtain (12.152). This completes the proof of Theorem 12.16. □

12.5 Pontryagin-Type Maximum Principle

In this section, we shall present our result on Pontryagin-type maximum principle for Problem (OP). Let

$$\begin{aligned}\mathbb{H}(t,x,u,k_1,k_2)&\overset{\Delta}{=}\big\langle k_1,a(t,x,u)\big\rangle_H+\big\langle k_2,b(t,x,u)\big\rangle_{\mathcal{L}_2^0}-g(t,x,u),\\&\qquad (t,x,u,k_1,k_2)\in[0,T]\times H\times U\times H\times\mathcal{L}_2^0.\end{aligned}\tag{12.164}$$

We have the following result.

Theorem 12.17. *Suppose that H is a separable Hilbert space, $L^p_{\mathcal{F}_T}(\Omega;\mathbb{C})$ ($1\le p<\infty$) is a separable Banach space, U is a separable metric space, and $x_0\in L^8_{\mathcal{F}_0}(\Omega;H)$. Let the assumptions (S1)–(S3) hold, and let $(\bar x(\cdot),\bar u(\cdot))$ be an optimal pair of Problem (OP). Let $\big(y(\cdot),Y(\cdot)\big)$ be the transposition solution to* (12.7)*. Assume that $(P(\cdot),Q^{(\cdot)},\widehat{Q}^{(\cdot)})$ is the relaxed transposition solution to the equation* (12.8) *in which $F(\cdot)$, $J(\cdot)$, $K(\cdot)$ and P_T are given by*

$$\begin{cases}F(t)=-\mathbb{H}_{xx}\big(t,\bar x(t),\bar u(t),y(t),Y(t)\big), & J(t)=a_x(t,\bar x(t),\bar u(t)),\\ K(t)=b_x(t,\bar x(t),\bar u(t)), & P_T=-h_{xx}\big(\bar x(T)\big).\end{cases}\tag{12.165}$$

Then,

$$
\begin{aligned}
&\operatorname{Re}\mathbb{H}\big(t,\bar{x}(t),\bar{u}(t),y(t),Y(t)\big)-\operatorname{Re}\mathbb{H}\big(t,\bar{x}(t),u,y(t),Y(t)\big)\\
&\quad-\frac{1}{2}\big\langle P(t)\big(b\big(t,\bar{x}(t),\bar{u}(t)\big)-b\big(t,\bar{x}(t),u\big)\big),b\big(t,\bar{x}(t),\bar{u}(t)\big)-b\big(t,\bar{x}(t),u\big)\big\rangle_{\mathcal{L}_2^0}\\
&\geq 0,\qquad \text{a.e. } [0,T]\times\Omega,\quad \forall\, u\in U.
\end{aligned}
\tag{12.166}
$$

Proof: We divide the proof into several steps.

Step 1. For each $\varepsilon\in(0,T)$, let $E_\varepsilon\subset[0,T]$ be a measurable set with measure ε. Similar to (1.44), put

$$
u^\varepsilon(t)=\begin{cases}\bar{u}(t), & t\in[0,T]\setminus E_\varepsilon,\\ u(t), & t\in E_\varepsilon.\end{cases}
\tag{12.167}
$$

where $u(\cdot)$ is an arbitrarily given element in $\mathcal{U}[0,T]$.

We introduce some notations which will be used later. For $\psi=a,b,g$,

$$
\begin{cases}
\psi_1(t)=\psi_x(t,\bar{x}(t),\bar{u}(t)), \qquad \psi_{11}(t)=\psi_{xx}(t,\bar{x}(t),\bar{u}(t)),\\
\psi_1^\varepsilon(t)=\displaystyle\int_0^1\psi_x\big(t,\bar{x}(t)+\sigma(x^\varepsilon(t)-\bar{x}(t)),u^\varepsilon(t)\big)d\sigma,\\
\psi_{11}^\varepsilon(t)=2\displaystyle\int_0^1(1-\sigma)\psi_{xx}\big(t,\bar{x}(t)+\sigma(x^\varepsilon(t)-\bar{x}(t)),u^\varepsilon(t)\big)d\sigma,
\end{cases}
\tag{12.168}
$$

and

$$
\begin{cases}
\delta\psi(t)=\psi(t,\bar{x}(t),u(t))-\psi(t,\bar{x}(t),\bar{u}(t)),\\
\delta\psi_1(t)=\psi_x(t,\bar{x}(t),u(t))-\psi_x(t,\bar{x}(t),\bar{u}(t)),\\
\delta\psi_{11}(t)=a_{xx}(t,\bar{x}(t),u(t))-\psi_{xx}(t,\bar{x}(t),\bar{u}(t)).
\end{cases}
\tag{12.169}
$$

Let $x^\varepsilon(\cdot)$ be the state process of the system (12.2) corresponding to the control $u^\varepsilon(\cdot)$, that is, $x^\varepsilon(\cdot)$ solves

$$
\begin{cases}
dx^\varepsilon=\big(Ax^\varepsilon+a(t,x^\varepsilon,u^\varepsilon)\big)dt+b(t,x^\varepsilon,u^\varepsilon)dW(t) \ \text{ in } (0,T],\\
x^\varepsilon(0)=x_0.
\end{cases}
\tag{12.170}
$$

From Proposition 12.1, it follows that

$$
|x^\varepsilon|_{C_{\mathbb{F}}([0,T];L^8(\Omega;H))}\leq \mathcal{C}\big(1+|x_0|_{L^8_{\mathcal{F}_0}(\Omega;H)}\big),\quad \forall\,\varepsilon\in(0,T).
\tag{12.171}
$$

Let $x_1^\varepsilon(\cdot)=x^\varepsilon(\cdot)-\bar{x}(\cdot)$. Then, by (12.171) and noting that the optimal pair $(\bar{x}(\cdot),\bar{u}(\cdot))$ solves the equation (12.2), we see that $x_1^\varepsilon(\cdot)$ satisfies the following stochastic evolution equation:

$$
\begin{cases}
dx_1^\varepsilon=\big(Ax_1^\varepsilon+a_1^\varepsilon(t)x_1^\varepsilon+\chi_{E_\varepsilon}(t)\delta a(t)\big)dt\\
\qquad\quad+\big(b_1^\varepsilon(t)x_1^\varepsilon+\chi_{E_\varepsilon}(t)\delta b(t)\big)dW(t) \quad \text{in } (0,T],\\
x_1^\varepsilon(0)=0.
\end{cases}
\tag{12.172}
$$

Consider the following two stochastic evolution equations:

$$\begin{cases} dx_2^\varepsilon = \big(Ax_2^\varepsilon + a_1(t)x_2^\varepsilon\big)dt + \big(b_1(t)x_2^\varepsilon + \chi_{E_\varepsilon}(t)\delta b(t)\big)dW(t) & \text{in } (0,T], \\ x_2^\varepsilon(0) = 0 \end{cases} \tag{12.173}$$

and[6]

$$\begin{cases} dx_3^\varepsilon = \Big(Ax_3^\varepsilon + a_1(t)x_3^\varepsilon + \chi_{E_\varepsilon}(t)\delta a(t) + \dfrac{1}{2}a_{11}(t)\big(x_2^\varepsilon, x_2^\varepsilon\big)\Big)dt \\ \qquad\quad + \Big(b_1(t)x_3^\varepsilon + \chi_{E_\varepsilon}(t)\delta b_1(t)x_2^\varepsilon + \dfrac{1}{2}b_{11}(t)\big(x_2^\varepsilon, x_2^\varepsilon\big)\Big)dW(t) \quad \text{in } (0,T], \\ x_3^\varepsilon(0) = 0. \end{cases} \tag{12.174}$$

In the following **Steps 2–4**, we shall prove that

$$|x_1^\varepsilon - x_2^\varepsilon - x_3^\varepsilon|_{C_{\mathbb{F}}([0,T];L^2(\Omega;H))} = o(\varepsilon), \qquad \text{as } \varepsilon \to 0. \tag{12.175}$$

Step 2. In this step, we provide some estimates on x_i^ε $(i = 1,2,3)$.

First of all, applying Theorem 3.14 to (12.172), by the conditions (S1) and (S3), we find that

$$\begin{aligned} &\sup_{t\in[0,T]} \mathbb{E}|x_1^\varepsilon(t)|_H^8 \\ &\leq \mathcal{C}\Big[\mathbb{E}\Big(\int_0^T \chi_{E_\varepsilon}(s)|\delta a(s)|_H ds\Big)^8 + \mathbb{E}\Big(\int_0^T \chi_{E_\varepsilon}(s)|\delta b(s)|_{\mathcal{L}_2^0}^2 ds\Big)^4\Big]. \end{aligned} \tag{12.176}$$

By the condition in (S3), it follows that

$$\begin{aligned} &|\delta a(s)|_H = \big|a\big(s,\bar x(s),u(s)\big) - a\big(s,\bar x(s),\bar u(s)\big)\big|_H \\ &\leq \big|a\big(s,\bar x(s),u(s)\big) - a\big(s,0,u(s)\big)\big|_H + \big|a\big(s,0,\bar u(s)\big) - a\big(s,\bar x(s),\bar u(s)\big)\big|_H \\ &\quad + \big|a\big(s,0,u(s)\big) - a\big(s,0,\bar u(s)\big)\big|_H \\ &\leq \Big|\int_0^1 a_x\big(s,\sigma\bar x(s),u(s)\big)\bar x(s)d\sigma\Big|_H + \Big|\int_0^1 a_x\big(s,\sigma\bar x(s),\bar u(s)\big)\bar x(s)d\sigma\Big|_H + \mathcal{C}_L \\ &\leq \mathcal{C}\big(|\bar x(s)|_H + 1\big), \qquad \text{a.e. } s \in [0,T]. \end{aligned} \tag{12.177}$$

Hence,

[6] Recall that, for any Banach spaces X and Y, C^2-function $f(\cdot) : X \to Y$ and $x_0 \in X$, $f_{xx}(x_0) \in \mathcal{L}(X,X;Y)$. This means that, for any $x_1, x_2 \in X$, $f_{xx}(x_0)(x_1,x_2) \in Y$. Hence, by (12.168), $a_{11}(t)\big(x_2^\varepsilon, x_2^\varepsilon\big)$ (in (12.174)) stands for $a_{xx}(t,\bar x(t),\bar u(t))\big(x_2^\varepsilon(t), x_2^\varepsilon(t)\big)$. One has a similar meaning for $b_{11}(t)\big(x_2^\varepsilon, x_2^\varepsilon\big)$ and so on.

$$
\begin{aligned}
&\mathbb{E}\Big(\int_0^T \chi_{E_\varepsilon}(s)\big|\delta a(s)\big|_H ds\Big)^8\\
&\leq \mathcal{C}\mathbb{E}\Big[\int_0^T \chi_{E_\varepsilon}(s)\big(|\bar{x}(s)|_H+1\big)ds\Big]^8\\
&\leq \mathcal{C}\mathbb{E}\Big\{\Big(\int_0^T \chi_{E_\varepsilon}(s)ds\Big)^{7/8}\Big[\int_0^T \chi_{E_\varepsilon}(s)\big(|\bar{x}(s)|_H^8+1\big)ds\Big]^{1/8}\Big\}^8 \qquad (12.178)\\
&\leq \mathcal{C}\varepsilon^7\int_0^T \chi_{E_\varepsilon}(s)\big(\mathbb{E}|\bar{x}(s)|_H^8+1\big)ds\\
&\leq \mathcal{C}\big(|x(\cdot)|^8_{C_{\mathbb{F}}([0,T];L^8(\Omega;H))}+1\big)\varepsilon^7\int_0^T \chi_{E_\varepsilon}(s)ds \leq \mathcal{C}(x_0)\varepsilon^8.
\end{aligned}
$$

Here and henceforth, $\mathcal{C}(x_0)$ is a generic constant (depending on x_0, T, A and $\mathcal{C}_L$), which may be different from line to line. As (12.177), we have

$$
|\delta b(s)|_{\mathcal{L}_2^0} \leq \mathcal{C}\big(|\bar{x}(s)|_H+1\big), \qquad \text{a.e. } s\in[0,T].
$$

Hence, similar to (12.178), one has

$$
\begin{aligned}
&\Big(\mathbb{E}\int_0^T \chi_{E_\varepsilon}(s)|\delta b(s)|^2_{\mathcal{L}_2^0}ds\Big)^4\\
&\leq \mathcal{C}\mathbb{E}\Big[\int_0^T \chi_{E_\varepsilon}(s)\big(|\bar{x}(s)|_H^2+1\big)ds\Big]^4\\
&\leq \mathcal{C}\mathbb{E}\Big\{\Big(\int_0^T \chi_{E_\varepsilon}(s)ds\Big)^{3/4}\Big[\int_0^T \chi_{E_\varepsilon}(s)\big(|\bar{x}(s)|_H^8+1\big)ds\Big]^{1/4}\Big\}^4 \qquad (12.179)\\
&\leq \mathcal{C}\varepsilon^3\int_0^T \chi_{E_\varepsilon}(s)\big(\mathbb{E}|\bar{x}(s)|_H^8+1\big)ds\\
&\leq \mathcal{C}\big(|x(\cdot)|^8_{C_{\mathbb{F}}([0,T];L^8(\Omega;H))}+1\big)\varepsilon^3\int_0^T \chi_{E_\varepsilon}(s)ds \leq \mathcal{C}(x_0)\varepsilon^4.
\end{aligned}
$$

Therefore, combining (12.176), (12.178) and (12.179), we end up with

$$
|x_1^\varepsilon(\cdot)|^8_{C_{\mathbb{F}}([0,T];L^8(\Omega;H))} \leq \mathcal{C}(x_0)\varepsilon^4. \tag{12.180}
$$

From the inequality (12.180) and Hölder's inequality, we find that

$$
\begin{cases}
|x_1^\varepsilon(\cdot)|^4_{C_{\mathbb{F}}([0,T];L^4(\Omega;H))} \leq \mathcal{C}(x_0)\varepsilon^2,\\
|x_1^\varepsilon(\cdot)|^2_{C_{\mathbb{F}}([0,T];L^2(\Omega;H))} \leq \mathcal{C}(x_0)\varepsilon.
\end{cases} \tag{12.181}
$$

By a similar computation, for $x_2^\varepsilon(\cdot)$, by (12.173), we have

$$
\begin{cases}
|x_2^\varepsilon(\cdot)|^8_{C_{\mathbb{F}}([0,T];L^8(\Omega;H))} \leq \mathcal{C}(x_0)\varepsilon^4,\\
|x_2^\varepsilon(\cdot)|^4_{C_{\mathbb{F}}([0,T];L^4(\Omega;H))} \leq \mathcal{C}(x_0)\varepsilon^2,\\
|x_2^\varepsilon(\cdot)|^2_{C_{\mathbb{F}}([0,T];L^2(\Omega;H))} \leq \mathcal{C}(x_0)\varepsilon.
\end{cases} \tag{12.182}
$$

Similar to (12.178), we have

$$\mathbb{E}\Big(\int_0^T \chi_{E_\varepsilon}(s)|\delta a(s)|_H ds\Big)^4 \leq \mathcal{C}(x_0)\varepsilon^4.$$

Hence, applying Theorem 3.14 to (12.174), it follows from (12.182) that

$$\begin{aligned}
&\sup_{t\in[0,T]} \mathbb{E}|x_3^\varepsilon(t)|_H^4\\
&\leq \mathcal{C}\Big\{\mathbb{E}\Big[\int_0^T \big(\chi_{E_\varepsilon}(s)|\delta a(s)|_H + \big|a_{11}(s)\big(x_2^\varepsilon(s), x_2^\varepsilon(s)\big)\big|_H\big)ds\Big]^4\\
&\quad +\mathbb{E}\Big[\int_0^T \big(\chi_{E_\varepsilon}(s)|\delta b_1(s)x_2^\varepsilon(s)|_{\mathcal{L}_2^0}^2 + \big|b_{11}(s)\big(x_2^\varepsilon(s), x_2^\varepsilon(s)\big)\big|_{\mathcal{L}_2^0}^2\big)ds\Big]^2\Big\}\\
&\leq \mathcal{C}\mathbb{E}\Big[\Big(\int_0^T \chi_{E_\varepsilon}(s)|\delta a(s)|_H ds\Big)^4 + \int_0^T |x_2^\varepsilon(t)|_H^8 ds + \Big|\int_0^T |\chi_{E_\varepsilon} x_2^\varepsilon(t)|_H^2 dt\Big|^2\Big]\\
&\leq \mathcal{C}\Big[\varepsilon^4 + \mathbb{E}\Big(\Big|\int_0^T \chi_{E_\varepsilon}(s)ds\Big|^{1/2}\Big|\int_0^T \chi_{E_\varepsilon}(s)|x_2^\varepsilon(s)|_H^4 ds\Big|^{1/2}\Big)^2\Big]\\
&\leq \mathcal{C}\Big(\varepsilon^4 + \varepsilon\int_0^T \chi_{E_\varepsilon}(s)\mathbb{E}|x_2^\varepsilon(s)|_H^4 ds\Big)\\
&\leq \mathcal{C}(x_0)\varepsilon^4.
\end{aligned}$$

Then, by Hölder's inequality, we conclude that

$$|x_3^\varepsilon(\cdot)|_{C_{\mathbb{F}}([0,T];L^2(\Omega;H))}^2 \leq \mathcal{C}(x_0)\varepsilon^2. \tag{12.183}$$

Step 3. We now estimate $x_4^\varepsilon \overset{\Delta}{=} x_1^\varepsilon - x_2^\varepsilon$. Clearly, x_4^ε solves the following equation:

$$\begin{cases}
dx_4^\varepsilon = \big[Ax_4^\varepsilon + a_1^\varepsilon(t)x_4^\varepsilon + \big(a_1^\varepsilon(t) - a_1(t)\big)x_2^\varepsilon + \chi_{E_\varepsilon}(t)\delta a(t)\big]dt\\
\qquad\quad +\big[b_1^\varepsilon(t)x_4^\varepsilon + \big(b_1^\varepsilon(t) - b_1(t)\big)x_2^\varepsilon\big]dW(t) \qquad \text{in } (0,T],\\
x_4^\varepsilon(0) = 0.
\end{cases} \tag{12.184}$$

Hence, applying Theorem 3.14 to (12.184), by the condition (S3), we have

$$\begin{aligned}
&\sup_{t\in[0,T]} \mathbb{E}|x_4^\varepsilon(t)|_H^2\\
&\leq \mathcal{C}\Big\{\mathbb{E}\Big|\int_0^T \big[\big|\big(a_1^\varepsilon(s) - a_1(s)\big)x_2^\varepsilon(s)\big|_H + \chi_{E_\varepsilon}(s)|\delta a(s)|_H\big]ds\Big|^2 \qquad (12.185)\\
&\quad +\mathbb{E}\int_0^T \big|\big(b_1^\varepsilon(s) - b_1(s)\big)x_2^\varepsilon(s)\big|_{\mathcal{L}_2^0}^2 ds\Big\}.
\end{aligned}$$

By the condition (S3) again, we have

$$
\begin{aligned}
&\big|a_1^\varepsilon(s)-a_1(s)\big|_{\mathcal{L}(H)}\\
&=\Big|\int_0^1\big(a_x(s,\bar{x}(s)+\sigma x_1^\varepsilon(s),u^\varepsilon(s))-a_x(s,\bar{x}(s),\bar{u}(s))\big)d\sigma\Big|_{\mathcal{L}(H)}\\
&=\Big|\int_0^1\big(a_x(s,\bar{x}(s)+\sigma x_1^\varepsilon(s),u^\varepsilon(s))-a_x(s,\bar{x}(s),u^\varepsilon(s))\\
&\quad+a_x(s,\bar{x}(s),u^\varepsilon(s))-a_x(s,\bar{x}(s),\bar{u}(s))\big)d\sigma\Big|_{\mathcal{L}(H)}\\
&=\Big|\int_0^1\Big(\sigma\int_0^1 a_{xx}(s,\bar{x}(s)+\eta\sigma x_1^\varepsilon(s),u^\varepsilon(s))x_1^\varepsilon(s)d\eta+\chi_{E_\varepsilon}(s)\delta a_1(s)\Big)d\sigma\Big|_{\mathcal{L}(H)}\\
&\leq\mathcal{C}\left(\big|x_1^\varepsilon(s)\big|_H+\chi_{E_\varepsilon}(s)\right),\qquad\text{a.e. } s\in[0,T].
\end{aligned}
\tag{12.186}
$$

Hence, it follows from (12.182) and (12.186) that

$$
\begin{aligned}
&\mathbb{E}\Big[\int_0^T\big|\big(a_1^\varepsilon(s)-a_1(s)\big)x_2^\varepsilon(s)\big|_H ds\Big]^2\\
&\leq\mathcal{C}|x_2^\varepsilon(\cdot)|^2_{C_{\mathbb{F}}([0,T];L^4(\Omega;H))}\int_0^T\Big(\mathbb{E}\big|a_1^\varepsilon(s)-a_1(s)\big|^4_{\mathcal{L}(H)}\Big)^{1/2}ds\\
&\leq\mathcal{C}(x_0)\varepsilon\int_0^T\Big(\chi_{E_\varepsilon}(t)+\mathbb{E}\big|x_1^\varepsilon(t)\big|_H^4\Big)^{1/2}dt\\
&\leq\mathcal{C}(x_0)\varepsilon\int_0^T\Big(\chi_{E_\varepsilon}(t)+\big(\mathbb{E}\big|x_1^\varepsilon(t)\big|_H^4\big)^{1/2}\Big)dt\leq\mathcal{C}(x_0)\varepsilon^2.
\end{aligned}
\tag{12.187}
$$

Similar to (12.178), we have

$$
\mathbb{E}\Big(\int_0^T\chi_{E_\varepsilon}(s)|\delta a(s)|_H ds\Big)^2\leq\mathcal{C}(x_0)\varepsilon^2. \tag{12.188}
$$

As (12.186), we have

$$
\big|b_1^\varepsilon(s)-b_1(s)\big|_{\mathcal{L}(H;\mathcal{L}_2^0)}\leq\mathcal{C}\left(\big|x_1^\varepsilon(s)\big|_H+\chi_{E_\varepsilon}(s)\right),\qquad\text{a.e. } s\in[0,T].
$$

Hence, similar to (12.187), one obtains that

$$
\begin{aligned}
&\mathbb{E}\int_0^T\big|\big(b_1^\varepsilon(s)-b_1(s)\big)x_2^\varepsilon(s)\big|^2_{\mathcal{L}_2^0}ds\\
&\leq\mathcal{C}\mathbb{E}\int_0^T\big|b_1^\varepsilon(s)-b_1(s)\big|^2_{\mathcal{L}(H;\mathcal{L}_2^0)}\big|x_2^\varepsilon(s)\big|_H^2 ds\leq\mathcal{C}(x_0)\varepsilon^2.
\end{aligned}
\tag{12.189}
$$

Combining (12.185), (12.187), (12.188) and (12.189), we obtain that

$$
|x_4^\varepsilon(\cdot)|_{C_{\mathbb{F}}([0,T];L^2(\Omega;H))}\leq\mathcal{C}(x_0)\varepsilon. \tag{12.190}
$$

Step 4. We are now in a position to estimate $\mathbb{E}|x_1^\varepsilon(t) - x_2^\varepsilon(t) - x_3^\varepsilon(t)|_H^2 = \mathbb{E}|x_4^\varepsilon(t) - x_3^\varepsilon(t)|_H^2$.

Let $x_5^\varepsilon(\cdot) = x_4^\varepsilon(\cdot) - x_3^\varepsilon(\cdot)$. It is clear that

$$x_5^\varepsilon(\cdot) = x_1^\varepsilon(t) - x_2^\varepsilon(t) - x_3^\varepsilon(t) = x^\varepsilon(\cdot) - \bar{x}(\cdot) - x_2^\varepsilon(t) - x_3^\varepsilon(t).$$

We claim that $x_5^\varepsilon(\cdot)$ solves the following equation (recall (12.168)–(12.169) for the notations):

$$\begin{cases} dx_5^\varepsilon = \Big[Ax_5^\varepsilon + a_1(t)x_5^\varepsilon + \chi_{E_\varepsilon}(t)\delta a_1(t)x_1^\varepsilon + \frac{1}{2}\big(a_{11}^\varepsilon(t) - a_{xx}(t,\bar{x},u^\varepsilon)\big)\big(x_1^\varepsilon, x_1^\varepsilon\big) \\ \qquad +\frac{1}{2}\chi_{E_\varepsilon}(t)\delta a_{11}(t)\big(x_1^\varepsilon, x_1^\varepsilon\big) + \frac{1}{2}a_{11}(t)\big(x_1^\varepsilon, x_1^\varepsilon\big) - \frac{1}{2}a_{11}(t)\big(x_2^\varepsilon, x_2^\varepsilon\big)\Big]dt \\ \qquad +\Big[b_1(t)x_5^\varepsilon + \chi_{E_\varepsilon}(t)\delta b_1(t)x_4^\varepsilon + \frac{1}{2}\big(b_{11}^\varepsilon(t) - b_{xx}(t,\bar{x},u^\varepsilon)\big)\big(x_1^\varepsilon, x_1^\varepsilon\big) \\ \qquad +\frac{1}{2}\chi_{E_\varepsilon}(t)\delta b_{11}(t)\big(x_1^\varepsilon, x_1^\varepsilon\big) + \frac{1}{2}b_{11}(t)\big(x_1^\varepsilon, x_1^\varepsilon\big) \\ \qquad -\frac{1}{2}b_{11}(t)\big(x_2^\varepsilon, x_2^\varepsilon\big)\Big]dW(t) \quad \text{in } (0,T], \\ x_5^\varepsilon(0) = 0. \end{cases} \tag{12.191}$$

Indeed, by (12.170), (12.172), (12.173) and (12.174), it is easy to see that the drift term for the equation solved by $x_5^\varepsilon(\cdot)$ is

$$\begin{aligned} &Ax^\varepsilon + a(t,x^\varepsilon,u^\varepsilon) - A\bar{x} - a(t,\bar{x},\bar{u}) - Ax_2^\varepsilon - a_1(t)x_2^\varepsilon - Ax_3^\varepsilon \\ &\quad -a_1(t)x_3^\varepsilon - \chi_{E_\varepsilon}(t)\delta a(t) - \frac{1}{2}a_{11}(t)\big(x_2^\varepsilon, x_2^\varepsilon\big) \\ &= Ax_5^\varepsilon + a(t,x^\varepsilon,u^\varepsilon) - a(t,\bar{x},u^\varepsilon) - a_1(t)(x_2^\varepsilon + x_3^\varepsilon) - \frac{1}{2}a_{11}(t)\big(x_2^\varepsilon, x_2^\varepsilon\big). \end{aligned} \tag{12.192}$$

For $\sigma \in [0,1]$, write $f(\sigma) = a(t, \bar{x} + \sigma x_1^\varepsilon, u^\varepsilon)$. Then, by Taylor's formula with the integral type remainder, we see that

$$f(1) - f(0) = f'(0) + \int_0^1 (1-\sigma)f''(\sigma)d\sigma.$$

Since $f'(\sigma) = a_x(t, \bar{x} + \sigma x_1^\varepsilon, u^\varepsilon)x_1^\varepsilon$ and $f''(\sigma) = a_{xx}(t, \bar{x} + \sigma x_1^\varepsilon, u^\varepsilon)(x_1^\varepsilon, x_1^\varepsilon)$, we obtain that

$$\begin{aligned} &a(t,x^\varepsilon,u^\varepsilon) - a(t,\bar{x},u^\varepsilon) \\ &= a_x(t,\bar{x},u^\varepsilon)x_1^\varepsilon + \int_0^1 (1-\sigma)a_{xx}(t,\bar{x}+\sigma x_1^\varepsilon,u^\varepsilon)(x_1^\varepsilon,x_1^\varepsilon)d\sigma \\ &= a_x(t,\bar{x},u^\varepsilon)x_1^\varepsilon + \frac{1}{2}a_{11}^\varepsilon(t)(x_1^\varepsilon,x_1^\varepsilon). \end{aligned} \tag{12.193}$$

Next,

$$\begin{aligned}&a_x(t,\bar{x},u^\varepsilon)x_1^\varepsilon - a_1(t)(x_2^\varepsilon + x_3^\varepsilon)\\&= a_x(t,\bar{x},u^\varepsilon)x_1^\varepsilon - a_x(t,\bar{x},\bar{u})x_1^\varepsilon + a_1(t)(x_1^\varepsilon - x_2^\varepsilon - x_3^\varepsilon)\\&= \chi_{E_\varepsilon}\big[a_x(t,\bar{x},u) - a_x(t,\bar{x},\bar{u})\big]x_1^\varepsilon + a_1(t)x_5^\varepsilon\\&= \chi_{E_\varepsilon}\delta a_1(t)x_1^\varepsilon + a_1(t)x_5^\varepsilon.\end{aligned}\tag{12.194}$$

Further,

$$\begin{aligned}&\frac{1}{2}a_{11}^\varepsilon(t)(x_1^\varepsilon,x_1^\varepsilon) - \frac{1}{2}a_{11}(t)(x_2^\varepsilon,x_2^\varepsilon)\\&= \frac{1}{2}a_{11}^\varepsilon(x_1^\varepsilon,x_1^\varepsilon) - \frac{1}{2}a_{xx}(t,\bar{x},u^\varepsilon)(x_1^\varepsilon,x_1^\varepsilon) + \frac{1}{2}a_{xx}(t,\bar{x},u^\varepsilon)(x_1^\varepsilon,x_1^\varepsilon)\\&\quad-\frac{1}{2}a_{11}(t)(x_1^\varepsilon,x_1^\varepsilon) + \frac{1}{2}a_{11}(t)(x_1^\varepsilon,x_1^\varepsilon) - \frac{1}{2}a_{11}(t)(x_2^\varepsilon,x_2^\varepsilon)\\&= \frac{1}{2}\Big(a_{11}^\varepsilon(t) - a_{xx}(t,\bar{x},u^\varepsilon)\Big)(x_1^\varepsilon,x_1^\varepsilon) + \frac{1}{2}\chi_{E_\varepsilon}\delta a_{11}(t)(x_1^\varepsilon,x_1^\varepsilon)\\&\quad+\frac{1}{2}a_{11}(t)(x_1^\varepsilon,x_1^\varepsilon) - \frac{1}{2}a_{11}(t)(x_2^\varepsilon,x_2^\varepsilon).\end{aligned}\tag{12.195}$$

By (12.192)–(12.195), we conclude that

$$\begin{aligned}&Ax^\varepsilon + a(t,x^\varepsilon,u^\varepsilon) - A\bar{x} - a(t,\bar{x},\bar{u}) - Ax_2^\varepsilon - a_1(t)x_2^\varepsilon - Ax_3^\varepsilon\\&\quad -a_1(t)x_3^\varepsilon - \chi_{E_\varepsilon}(t)\delta a(t) - \frac{1}{2}a_{11}(t)\big(x_2^\varepsilon,x_2^\varepsilon\big) - \chi_{E_\varepsilon}(t)\delta b_1(t)x_2^\varepsilon\\&= Ax_5^\varepsilon + a_1(t)x_5^\varepsilon + \chi_{E_\varepsilon}(t)\delta a_1(t)x_1^\varepsilon + \frac{1}{2}\big(a_{11}^\varepsilon(t) - a_{xx}(t,\bar{x},u^\varepsilon)\big)\big(x_1^\varepsilon,x_1^\varepsilon\big)\\&\quad+\frac{1}{2}\chi_{E_\varepsilon}(t)\delta a_{11}(t)\big(x_1^\varepsilon,x_1^\varepsilon\big) + \frac{1}{2}a_{11}(t)\big(x_1^\varepsilon,x_1^\varepsilon\big) - \frac{1}{2}a_{11}(t)\big(x_2^\varepsilon,x_2^\varepsilon\big).\end{aligned}$$

Similarly, the diffusion term (for the equation solved by $x_5^\varepsilon(\cdot)$) is

$$\begin{aligned}&b(t,x^\varepsilon,u^\varepsilon) - b(t,\bar{x},\bar{u}) - b_1(t)x_2^\varepsilon - b_1(t)x_3^\varepsilon - \chi_{E_\varepsilon}(t)\delta b(t)x_2^\varepsilon - \frac{1}{2}b_{11}(t)\big(x_2^\varepsilon,x_2^\varepsilon\big)\\&= b_1(t)x_5^\varepsilon + \chi_{E_\varepsilon}(t)\delta b_1(t)x_4^\varepsilon + \frac{1}{2}\big(b_{11}^\varepsilon(t) - b_{xx}(t,\bar{x},u^\varepsilon)\big)\big(x_1^\varepsilon,x_1^\varepsilon\big)\\&\quad+\frac{1}{2}\chi_{E_\varepsilon}(t)\delta b_{11}(t)\big(x_1^\varepsilon,x_1^\varepsilon\big) + \frac{1}{2}b_{11}(t)\big(x_1^\varepsilon,x_1^\varepsilon\big) - \frac{1}{2}b_{11}(t)\big(x_2^\varepsilon,x_2^\varepsilon\big).\end{aligned}$$

This verifies that $x_5^\varepsilon(\cdot)$ satisfies the equation (12.191).

Applying Theorem 3.14 to (12.191), we see that

$$
\begin{aligned}
&\sup_{t\in[0,T]} \mathbb{E}\big|x_5^\varepsilon(t)\big|_H^2\\
&\leq \mathcal{C}\mathbb{E}\Big\{\Big|\int_0^T \Big[\chi_{E_\varepsilon}(s)\Big(|\delta a_1(s)x_1^\varepsilon(s)|_H + \big|\delta a_{11}(s)\big(x_1^\varepsilon(s),x_1^\varepsilon(s)\big)\big|_H\Big)\\
&\qquad +\big|\big(a_{11}^\varepsilon(s)-a_{xx}(s,\bar{x}(s),u^\varepsilon(s))\big)\big(x_1^\varepsilon(s),x_1^\varepsilon(s)\big)\big|_H\\
&\qquad +\big|a_{11}(s)\big(x_1^\varepsilon(s),x_1^\varepsilon(s)\big)-a_{11}(s)\big(x_2^\varepsilon(s),x_2^\varepsilon(s)\big)\big|_H\Big]ds\Big|^2 \qquad (12.196)\\
&\qquad +\int_0^T \Big[\chi_{E_\varepsilon}(s)\Big(|\delta b_1(s)x_4^\varepsilon(s)|_{\mathcal{L}_2^0}^2 + \big|\delta b_{11}(s)\big(x_1^\varepsilon(s),x_1^\varepsilon(s)\big)\big|_{\mathcal{L}_2^0}^2\Big)\\
&\qquad +\big|\big(b_{11}^\varepsilon(s)-b_{xx}(s,\bar{x}(s),u^\varepsilon(s))\big)\big(x_1^\varepsilon(s),x_1^\varepsilon(s)\big)\big|_{\mathcal{L}_2^0}^2\\
&\qquad +\big|b_{11}(s)\big(x_1^\varepsilon(s),x_1^\varepsilon(s)\big)-b_{11}(s)\big(x_2^\varepsilon(s),x_2^\varepsilon(s)\big)\big|_{\mathcal{L}_2^0}^2\Big]ds\Big\}.
\end{aligned}
$$

We now estimate the "drift" terms in the right hand side of (12.196). By (12.181), and using the condition (S3), we have the following estimate:

$$
\begin{aligned}
&\mathbb{E}\Big|\int_0^T \chi_{E_\varepsilon}(s)\Big(|\delta a_1(s)x_1^\varepsilon(s)|_H + \big|\delta a_{11}(s)\big(x_1^\varepsilon(s),x_1^\varepsilon(s)\big)\big|_H\Big)ds\Big|^2\\
&\leq \mathcal{C}\mathbb{E}\Big[\int_0^T \chi_{E_\varepsilon}(s)ds\int_0^T \chi_{E_\varepsilon}(s)\big(|x_1^\varepsilon(s)|_H^2+|x_1^\varepsilon(s)|_H^4\big)ds\Big]\\
&\leq \mathcal{C}\varepsilon\int_0^T \chi_{E_\varepsilon}(s)\big(\mathbb{E}|x_1^\varepsilon(s)|_H^2+\mathbb{E}|x_1^\varepsilon(s)|_H^4\big)ds \qquad (12.197)\\
&\leq \mathcal{C}\varepsilon\big(|x_1^\varepsilon(\cdot)|^2_{C_{\mathbb{F}}([0,T];L^2(\Omega;H))}+|x_1^\varepsilon(\cdot)|^4_{C_{\mathbb{F}}([0,T];L^4(\Omega;H))}\big)\int_0^T\chi_{E_\varepsilon}(s)ds\\
&\leq \mathcal{C}(x_0)\varepsilon^3.
\end{aligned}
$$

By (12.168) and recalling that $x_1^\varepsilon(\cdot)=x^\varepsilon(\cdot)-\bar{x}(\cdot)$, we see that, for a.e. $s\in[0,T]$,

$$
\begin{aligned}
&\big|a_{11}^\varepsilon(s)-a_{xx}(s,\bar{x}(s),u^\varepsilon(s))\big|_{\mathcal{L}(H,H;H)}\\
&=\Big|2\int_0^1(1-\sigma)a_{xx}\big(s,\bar{x}(s)+\sigma x_1^\varepsilon(s),u^\varepsilon(s)\big)d\sigma - a_{xx}\big(s,\bar{x}(s),u^\varepsilon(s)\big)\Big|_{\mathcal{L}(H,H;H)}\\
&=\Big|2\int_0^1(1-\sigma)\Big(a_{xx}\big(s,\bar{x}(s)+\sigma x_1^\varepsilon(s),\bar{u}(s)\big)-a_{xx}\big(s,\bar{x}(s),\bar{u}(s)\big)\Big)d\sigma\\
&\quad +2\int_0^1(1-\sigma)\chi_{E_\varepsilon}(s)\Big(a_{xx}\big(s,\bar{x}(s)+\sigma x_1^\varepsilon(s),u(s)\big) \qquad (12.198)\\
&\qquad\qquad -a_{xx}\big(s,\bar{x}(s)+\sigma x_1^\varepsilon(s),\bar{u}(s)\big)\Big)d\sigma\\
&\quad -\chi_{E_\varepsilon}(s)\big(a_{xx}(s,\bar{x}(s),u(s))-a_{xx}(s,\bar{x}(s),\bar{u}(s))\big)\Big|_{\mathcal{L}(H,H;H)}
\end{aligned}
$$

$$
\begin{aligned}
&\le \mathcal{C}\Big(\int_0^1 \big| a_{xx}\big(s,\bar{x}(s)+\sigma x_1^\varepsilon(s),\bar{u}(s)\big) - a_{xx}\big(s,\bar{x}(s),\bar{u}(s)\big)\big|_{\mathcal{L}(H,H;H)} d\sigma \\
&\qquad +\chi_{E_\varepsilon}(s)\Big).
\end{aligned}
$$

Hence, by (12.180) and noting the continuity of $a_{xx}(t,x,u)$ with respect to x, we have

$$
\begin{aligned}
&\mathbb{E}\Big[\int_0^T \big| \big(a_{11}^\varepsilon(s) - a_{xx}(s,\bar{x}(s),u^\varepsilon(s))\big)\big(x_1^\varepsilon(s),x_1^\varepsilon(s)\big)\big|_H ds\Big]^2 \\
&\le \mathcal{C}\mathbb{E}\int_0^T \big|a_{11}^\varepsilon(s) - a_{xx}(s,\bar{x}(s),u^\varepsilon(s))\big|^2_{\mathcal{L}(H,H;H)} |x_1^\varepsilon(s)|_H^4 dt \qquad (12.199)\\
&\le \mathcal{C}|x_1^\varepsilon(\cdot)|^4_{C_{\mathbb{F}}([0,T];L^8(\Omega;H))} \int_0^T \Big(\mathbb{E}\big|a_{11}^\varepsilon(s) - a_{xx}(s,\bar{x}(s),u^\varepsilon(s))\big|^4_{\mathcal{L}(H,H;H)}\Big)^{1/2} ds \\
&\le \mathcal{C}(x_0)\varepsilon^2 \int_0^T \Big(\mathbb{E}\int_0^1 \big|a_{xx}\big(s,\bar{x}(s)+\sigma x_1^\varepsilon(s),\bar{u}(s)\big) \\
&\qquad\qquad -a_{xx}\big(s,\bar{x}(s),\bar{u}(s)\big)\big|^4_{\mathcal{L}(H,H;H)} d\sigma + \chi_{E_\varepsilon}(s)\Big)^{1/2} ds \\
&= o(\varepsilon^2), \qquad \text{as } \varepsilon\to 0.
\end{aligned}
$$

By means of (12.181), (12.182) and (12.190), and noting that $x_4^\varepsilon = x_1^\varepsilon - x_2^\varepsilon$, we obtain that

$$
\begin{aligned}
&\mathbb{E}\Big(\int_0^T \big| a_{11}(s)\big(x_1^\varepsilon(s),x_1^\varepsilon(s)\big) - a_{11}(s)\big(x_2^\varepsilon(s),x_2^\varepsilon(s)\big)\big|_H ds\Big)^2 \\
&= \mathbb{E}\Big(\int_0^T \big| a_{11}(s)\big(x_4^\varepsilon(s),x_1^\varepsilon(s)\big) + a_{11}(s)\big(x_2^\varepsilon(s),x_4^\varepsilon(s)\big)\big|_H ds\Big)^2 \qquad (12.200)\\
&\le \mathcal{C}\big(|x_1^\varepsilon(\cdot)|^2_{C_{\mathbb{F}}([0,T];L^2(\Omega;H))} + |x_2^\varepsilon(\cdot)|^2_{C_{\mathbb{F}}([0,T];L^2(\Omega;H))}\big)|x_4^\varepsilon(\cdot)|^2_{C_{\mathbb{F}}([0,T];L^2(\Omega;H))} \\
&\le \mathcal{C}(x_0)\varepsilon^3.
\end{aligned}
$$

Next, we estimate the "diffusion" terms in the right hand side of (12.196). Similar to (12.197) and noting (12.190), we obtain that

$$
\begin{aligned}
&\mathbb{E}\int_0^T \chi_{E_\varepsilon}(s)\Big(|\delta b_1(s)x_4^\varepsilon(s)|^2_{\mathcal{L}_2^0} + \big|\delta b_{11}(s)\big(x_1^\varepsilon(s),x_1^\varepsilon(s)\big)\big|^2_{\mathcal{L}_2^0}\Big) ds \\
&\le \mathcal{C}\int_0^T \chi_{E_\varepsilon}(s)\big(\mathbb{E}|x_4^\varepsilon(s)|_H^2 + \mathbb{E}|x_1^\varepsilon(s)|_H^4\big) ds \le \mathcal{C}(x_0)\varepsilon^3.
\end{aligned}
\qquad (12.201)
$$

By virtue of (12.180) again, similar to (12.199), we find that

$$
\begin{aligned}
&\mathbb{E}\int_0^T \big| \big(b_{11}^\varepsilon(s) - b_{xx}(s,\bar{x}(s),u^\varepsilon(s))\big)\big(x_1^\varepsilon(s),x_1^\varepsilon(s)\big)\big|^2_{\mathcal{L}_2^0} ds \\
&\le \mathcal{C}|x_1^\varepsilon(\cdot)|^4_{C_{\mathbb{F}}([0,T];L^8(\Omega;H))} \int_0^T \Big(\mathbb{E}\big|b_{11}^\varepsilon(s) - b_{xx}(s,\bar{x}(s),u^\varepsilon(s))\big|^4_{\mathcal{L}(H,H;\mathcal{L}_2^0)}\Big)^{1/2} ds
\end{aligned}
$$

$$\leq \mathcal{C}(x_0)\varepsilon^2 \int_0^T \Big(\mathbb{E} \int_0^1 \big|b_{xx}\big(s, \bar{x}(s) + \sigma x_1^\varepsilon(s), \bar{u}(s)\big) \\ - b_{xx}(s, \bar{x}(s), \bar{u}(s))\big|^4_{\mathcal{L}(H,H;\mathcal{L}_2^0)} d\sigma + \chi_{E_\varepsilon}(s)\Big)^{1/2} ds \\ = o(\varepsilon^2), \qquad \text{as } \varepsilon \to 0. \tag{12.202}$$

Similar to (12.200), it holds that

$$\mathbb{E}\int_0^T \big|b_{11}(s)\big(x_1^\varepsilon(s), x_1^\varepsilon(s)\big) - b_{11}(s)\big(x_2^\varepsilon(s), x_2^\varepsilon(s)\big)\big|^2_{\mathcal{L}_2^0} ds \leq \mathcal{C}(x_0)\varepsilon^3. \tag{12.203}$$

From (12.196)–(12.197) and (12.199)–(12.203), we conclude that

$$|x_5^\varepsilon(\cdot)|^2_{C_{\mathbb{F}}([0,T];L^2(\Omega;H))} = o(\varepsilon^2), \qquad \text{as } t \to 0. \tag{12.204}$$

This gives (12.175).

Step 5. We now compute $\mathcal{J}(u^\varepsilon(\cdot)) - \mathcal{J}(\bar{u}(\cdot))$.

By the definition of $\mathcal{J}(\cdot)$ in (12.3), using Taylor's formula with the integral type remainder as that in (12.193), we obtain that

$$\begin{aligned}
&\mathcal{J}(u^\varepsilon(\cdot)) - \mathcal{J}(\bar{u}(\cdot)) \\
&= \mathbb{E}\int_0^T \big(g(t, x^\varepsilon(t), u^\varepsilon(t)) - g(t, \bar{x}(t), \bar{u}(t))\big)dt + \mathbb{E}h\big(x^\varepsilon(T)\big) - \mathbb{E}h\big(\bar{x}(T)\big) \\
&= \mathrm{Re}\,\mathbb{E}\int_0^T \Big(\chi_{E_\varepsilon}(t)\delta g(t) + \big\langle g_x(t, \bar{x}(t), u^\varepsilon(t)), x_1^\varepsilon(t)\big\rangle_H \\
&\qquad + \int_0^1 \big\langle (1-\sigma) g_{xx}\big(t, \bar{x}(t) + \sigma x_1^\varepsilon(t), u^\varepsilon(t)\big) x_1^\varepsilon(t), x_1^\varepsilon(t)\big\rangle_H d\sigma\Big)dt \\
&\quad + \mathrm{Re}\,\mathbb{E}\big\langle h_x(\bar{x}(T)), x_1^\varepsilon(T)\big\rangle_H \\
&\quad + \mathrm{Re}\,\mathbb{E}\int_0^1 \big\langle (1-\sigma) h_{xx}\big(\bar{x}(T) + \sigma x_1^\varepsilon(T)\big) x_1^\varepsilon(T), x_1^\varepsilon(T)\big\rangle_H d\sigma.
\end{aligned}$$

This, together with the definition of $x_1^\varepsilon(\cdot)$, $x_2^\varepsilon(\cdot)$, $x_3^\varepsilon(\cdot)$, $x_4^\varepsilon(\cdot)$ and $x_5^\varepsilon(\cdot)$, yields that

$$\begin{aligned}
&\mathcal{J}(u^\varepsilon(\cdot)) - \mathcal{J}(\bar{u}(\cdot)) \\
&= \mathrm{Re}\,\mathbb{E}\int_0^T \Big[\chi_{E_\varepsilon}(t)\delta g(t) + \big\langle \delta g_1(t), x_1^\varepsilon(t)\big\rangle_H \chi_{E_\varepsilon}(t) + \big\langle g_1(t), x_2^\varepsilon(t) + x_3^\varepsilon(t)\big\rangle_H \\
&\qquad + \big\langle g_1(t), x_5^\varepsilon(t)\big\rangle_H + \int_0^1 \big\langle (1-\sigma)\big(g_{xx}\big(t, \bar{x}(t) + \sigma x_1^\varepsilon(t), u^\varepsilon(t)\big) \\
&\qquad\qquad - g_{xx}\big(t, \bar{x}(t), u^\varepsilon(t)\big)\big) x_1^\varepsilon(t), x_1^\varepsilon(t)\big\rangle_H d\sigma \\
&\qquad + \frac{1}{2}\big\langle \delta g_{11}(t) x_1^\varepsilon(t), x_1^\varepsilon(t)\big\rangle_H \chi_{E_\varepsilon}(t) + \frac{1}{2}\big\langle g_{11}(t) x_2^\varepsilon(t), x_2^\varepsilon(t)\big\rangle_H
\end{aligned}$$

$$
\begin{aligned}
&+\frac{1}{2}\langle g_{11}(t)x_4^\varepsilon(t), x_1^\varepsilon(t)+x_2^\varepsilon(t)\rangle_H\Big]dt \\
&+\mathrm{Re}\,\mathbb{E}\Big[\big\langle h_x\big(\bar{x}(T)\big), x_2^\varepsilon(t)+x_3^\varepsilon(t)\big\rangle_H+\big\langle h_x\big(\bar{x}(T)\big), x_5^\varepsilon(t)\big\rangle_H \\
&+\frac{1}{2}\big\langle h_{xx}\big(\bar{x}(T)\big)x_2^\varepsilon(t), x_2^\varepsilon(t)\big\rangle_H+\frac{1}{2}\big\langle h_{xx}\big(\bar{x}(T)\big)x_4^\varepsilon(T), x_1^\varepsilon(T)+x_2^\varepsilon(T)\big\rangle_H \\
&+\int_0^1\big\langle(1-\sigma)\big(h_{xx}\big(\bar{x}(T)+\sigma x_1^\varepsilon(T)\big)-h_{xx}\big(\bar{x}(T)\big)\big)x_1^\varepsilon(T), x_1^\varepsilon(T)\big\rangle_H d\sigma\Big].
\end{aligned}
\tag{12.205}
$$

Similar to (12.198), we find that, for a.e. $t\in[0,T]$,

$$
\begin{aligned}
&\Big|\int_0^1(1-\sigma)\big(g_{xx}\big(t,\bar{x}(t)+\sigma x_1^\varepsilon(t),u^\varepsilon(t)\big)-g_{xx}\big(t,\bar{x}(t),u^\varepsilon(t)\big)\big)d\sigma\Big|_{\mathcal{L}(H)} \\
&\leq\mathcal{C}\Big(\int_0^1\big|g_{xx}\big(t,\bar{x}(t)+\sigma x_1^\varepsilon(t),\bar{u}(t)\big)-g_{xx}\big(t,\bar{x}(t),\bar{u}(t)\big)\big|_{\mathcal{L}(H)}d\sigma+\chi_{E_\varepsilon}(t)\Big),
\end{aligned}
\tag{12.206}
$$

and

$$
\begin{aligned}
&\Big|\int_0^1(1-\sigma)\big(h_{xx}\big(\bar{x}(T)+\sigma x_1^\varepsilon(T)\big)-h_{xx}\big(\bar{x}(T)\big)\big)d\sigma\Big|_{\mathcal{L}(H)} \\
&\leq\mathcal{C}\Big(\int_0^1\big|h_{xx}(\bar{x}(T)+\sigma x_1^\varepsilon(T))-h_{xx}(\bar{x}(T))\big|_{\mathcal{L}(H)}d\sigma+\chi_{E_\varepsilon}(t)\Big).
\end{aligned}
\tag{12.207}
$$

By means of (12.205), noting (12.180), (12.182), (12.5), (12.190), (12.204), (12.206) and (12.207), and using the continuity of both $h_{xx}(x)$ and $g_{xx}(x)$ with respect to x, we end up with

$$
\begin{aligned}
&\mathcal{J}(u^\varepsilon(\cdot))-\mathcal{J}(\bar{u}(\cdot)) \\
&=\mathrm{Re}\,\mathbb{E}\int_0^T\Big(\big\langle g_1(t), x_2^\varepsilon(t)+x_3^\varepsilon(t)\big\rangle_H+\frac{1}{2}\big\langle g_{11}(t)x_2^\varepsilon(t), x_2^\varepsilon(t)\big\rangle_H \\
&\qquad+\chi_{E_\varepsilon}(t)\delta g(t)\Big)dt+\mathrm{Re}\,\mathbb{E}\big\langle h_x\big(\bar{x}(T)\big), x_2^\varepsilon(T)+x_3^\varepsilon(T)\big\rangle_H \\
&\quad+\frac{1}{2}\mathrm{Re}\,\mathbb{E}\big\langle h_{xx}\big(\bar{x}(T)\big)x_2^\varepsilon(t), x_2^\varepsilon(t)\big\rangle_H+o(\varepsilon),\qquad\text{as }\varepsilon\to 0.
\end{aligned}
\tag{12.208}
$$

We now get rid of $x_2^\varepsilon(\cdot)$ and $x_3^\varepsilon(\cdot)$ in (12.208) by solutions to the equations (12.7) and (12.8). By the definition of transposition solution to the equation (12.7), and recalling the notations in (12.168), we obtain that

$$
\begin{aligned}
&-\mathbb{E}\big\langle h_x(\bar{x}(T))), x_2^\varepsilon(T)\big\rangle_H-\mathbb{E}\int_0^T\big\langle g_1(t), x_2^\varepsilon(t)\big\rangle_H dt \\
&=\mathbb{E}\int_0^T\big\langle Y(t), \delta b(t)\big\rangle_{\mathcal{L}_2^0}\chi_{E_\varepsilon}(t)dt
\end{aligned}
\tag{12.209}
$$

and

$$
\begin{aligned}
&-\mathbb{E}\big\langle h_x(\bar{x}(T))), x_3^\varepsilon(T)\big\rangle_H - \mathbb{E}\int_0^T \big\langle g_1(t), x_3^\varepsilon(t)\big\rangle_H dt\\
&= \mathbb{E}\int_0^T \Big[\frac{1}{2}\Big(\big\langle y(t), a_{11}(t)\big(x_2^\varepsilon(t), x_2^\varepsilon(t)\big)\big\rangle_H + \big\langle Y(t), b_{11}(t)\big(x_2^\varepsilon(t), x_2^\varepsilon(t)\big)\big\rangle_{\mathcal{L}_2^0}\Big)\\
&\qquad + \chi_{E_\varepsilon}(t)\Big(\big\langle y(t), \delta a(t)\big\rangle_H + \big\langle Y, \delta b_1(t) x_2^\varepsilon(t)\big\rangle_{\mathcal{L}_2^0}\Big)\Big] dt. \qquad (12.210)
\end{aligned}
$$

According to (12.208)–(12.210), we conclude that, when $\varepsilon \to 0$,

$$
\begin{aligned}
&\mathcal{J}(u^\varepsilon(\cdot)) - \mathcal{J}(\bar{u}(\cdot))\\
&= \frac{1}{2}\mathrm{Re}\,\mathbb{E}\int_0^T \Big(\big\langle g_{11}(t)x_2^\varepsilon(t), x_2^\varepsilon(t)\big\rangle_H - \big\langle y(t), a_{11}(t)\big(x_2^\varepsilon(t), x_2^\varepsilon(t)\big)\big\rangle_H\\
&\qquad - \big\langle Y, b_{11}(t)\big(x_2^\varepsilon(t), x_2^\varepsilon(t)\big)\big\rangle_{\mathcal{L}_2^0}\Big) dt \qquad (12.211)\\
&\quad + \mathrm{Re}\,\mathbb{E}\int_0^T \chi_{E_\varepsilon}(t)\Big(\delta g(t) - \big\langle y(t), \delta a(t)\big\rangle_H - \big\langle Y(t), \delta b(t)\big\rangle_{\mathcal{L}_2^0}\Big) dt\\
&\quad + \frac{1}{2}\mathrm{Re}\,\mathbb{E}\big\langle h_{xx}\big(\bar{x}(T)\big)x_2^\varepsilon(T), x_2^\varepsilon(T)\big\rangle_H + o(\varepsilon).
\end{aligned}
$$

By the definition of the relaxed transposition solution to the equation (12.8) (with $F(\cdot)$, $J(\cdot)$, $K(\cdot)$ and P_T given by (12.165)), we obtain that

$$
\begin{aligned}
&-\mathbb{E}\big\langle h_{xx}\big(\bar{x}(T)\big)x_2^\varepsilon(T), x_2^\varepsilon(T)\big\rangle_H\\
&\quad + \mathbb{E}\int_0^T \big\langle \mathbb{H}_{xx}\big(t, \bar{x}(t), \bar{u}(t), y(t), Y(t)\big)x_2^\varepsilon(t), x_2^\varepsilon(t)\big\rangle_H dt\\
&= \mathbb{E}\int_0^T \chi_{E_\varepsilon}(t)\big\langle P(t)b_1(t)x_2^\varepsilon(t), \delta b(t)\big\rangle_{\mathcal{L}_2^0} dt\\
&\quad + \mathbb{E}\int_0^T \chi_{E_\varepsilon}(t)\big\langle P(t)\delta b(t), b_1(t)x_2^\varepsilon(t)\big\rangle_{\mathcal{L}_2^0} dt \qquad (12.212)\\
&\quad + \mathbb{E}\int_0^T \chi_{E_\varepsilon}(t)\big\langle P(t)\delta b(t), \delta b(t)\big\rangle_{\mathcal{L}_2^0} dt\\
&\quad + \mathbb{E}\int_0^T \big\langle \chi_{E_\varepsilon}(t)\delta b(t), \widehat{Q}^{(0)}(0, 0, \chi_{E_\varepsilon}\delta b)(t)\big\rangle_{\mathcal{L}_2^0} dt\\
&\quad + \mathbb{E}\int_0^T \big\langle Q^{(0)}(0, 0, \chi_{E_\varepsilon}\delta b)(t), \chi_{E_\varepsilon}(t)\delta b(t)\big\rangle_{\mathcal{L}_2^0} dt.
\end{aligned}
$$

By (12.182), we have

$$
\begin{aligned}
&\Big|\mathbb{E}\int_0^T \chi_{E_\varepsilon}(t)\big\langle P(t)b_1(t)x_2^\varepsilon(t), \delta b(t)\big\rangle_{\mathcal{L}_2^0} dt\\
&\quad + \mathbb{E}\int_0^T \chi_{E_\varepsilon}(t)\big\langle P(t)\delta b(t), b_1(t)x_2^\varepsilon(t)\big\rangle_{\mathcal{L}_2^0} dt\Big| = o(\varepsilon). \qquad (12.213)
\end{aligned}
$$

By (12.164) and (12.211)–(12.213), we arrive at

$$\begin{aligned}
&\mathcal{J}(u^\varepsilon(\cdot)) - \mathcal{J}(\bar u(\cdot))\\
&= \mathrm{Re}\,\mathbb{E}\int_0^T \chi_{E_\varepsilon}(t)\Big(\delta g(t) - \big\langle y(t), \delta a(t)\big\rangle_H - \big\langle Y(t), \delta b(t)\big\rangle_{\mathcal{L}_2^0}\Big)dt\\
&\quad -\mathrm{Re}\,\mathbb{E}\int_0^T \big\langle \chi_{E_\varepsilon}(t)\delta b(t), \widehat Q^{(0)}(0,0,\chi_{E_\varepsilon}\delta b)(t)\big\rangle_{\mathcal{L}_2^0} dt\\
&\quad -\mathrm{Re}\,\mathbb{E}\int_0^T \big\langle Q^{(0)}(0,0,\chi_{E_\varepsilon}\delta b)(t), \chi_{E_\varepsilon}(t)\delta b(t)\big\rangle_{\mathcal{L}_2^0} dt + o(\varepsilon).
\end{aligned}\tag{12.214}$$

Step 6. In this step, we shall get rid of the terms containing $Q^{(0)}$ and $\widehat Q^{(0)}$ in (12.214).

In what follows, for any $\tau \in [0,T)$, we choose $E_\varepsilon = [\tau, \tau+\varepsilon] \subset [0,T]$.

By Proposition 12.15, we can find a sequence $\{\beta_{m_n,\ell_n}\}_{n=1}^\infty$ such that $\beta_{m_n,\ell_n} \in \mathcal{H}_{m_n,\ell_n}$ (Recall (12.149) for the definition of $\mathcal{H}_{m_n,\ell_n}$) and

$$\lim_{n\to\infty} \beta_{m_n,\ell_n} = \delta b \quad \text{in } L^2_{\mathbb{F}}(0,T; L^4(\Omega; \mathcal{L}_2^0)).$$

Hence, for some positive constant $\mathcal{C}(x_0)$ (depending on x_0),

$$|\beta_{m_n,\ell_n}|_{L^2_{\mathbb{F}}(0,T;L^4(\Omega;\mathcal{L}_2^0))} \le \mathcal{C}(x_0) < \infty, \qquad \forall\, n \in \mathbb{N}, \tag{12.215}$$

and there is a subsequence $\{n_k\}_{k=1}^\infty \subset \{n\}_{n=1}^\infty$ such that

$$\lim_{k\to\infty} |\beta_{m_{n_k},\ell_{n_k}}(t) - \delta b(t)|_{L^4_{\mathcal{F}_t}(\Omega;\mathcal{L}_2^0)} = 0 \quad \text{for a.e. } t \in [0,T]. \tag{12.216}$$

Denote by $Q_{m_{n_k},\ell_{n_k}}$ and $\widehat Q_{m_{n_k},\ell_{n_k}}$ the corresponding pointwise defined linear operators from $\mathcal{H}_{m_{n_k},\ell_{n_k}}$ to $L^2_{\mathbb{F}}(0,T; L^{\frac43}(\Omega; H))$, given in Theorem 12.16. To simplify notations, we write $\beta_k \stackrel{\Delta}{=} \beta_{m_{n_k},\ell_{n_k}}$, $Q_k \stackrel{\Delta}{=} Q_{m_{n_k},\ell_{n_k}}$ and $\widehat Q_k \stackrel{\Delta}{=} \widehat Q_{m_{n_k},\ell_{n_k}}$.

Consider the following equation:

$$\begin{cases} dx_{2,n_k}^\varepsilon = \big(Ax_{2,n_k}^\varepsilon + a_1(t)x_{2,n_k}^\varepsilon\big)dt\\ \qquad\qquad +\big(b_1(t)x_{2,n_k}^\varepsilon + \chi_{E_\varepsilon}(t)\beta_k(t)\big)dW(t) \quad \text{in } (0,T],\\ x_{2,n_k}^\varepsilon(0) = 0. \end{cases}\tag{12.217}$$

Applying Theorem 3.14 to (12.217) and by (12.215), we obtain that

$$\begin{aligned}
|x_{2,n_k}^\varepsilon(\cdot)|^4_{C_{\mathbb{F}}([0,T];L^4(\Omega;H))} &\le \mathcal{C}\mathbb{E}\Big(\int_0^T \chi_{E_\varepsilon}(s)|\beta_k(s)|^2_{\mathcal{L}_2^0} ds\Big)^2\\
&\le \mathcal{C}\varepsilon \int_{E_\varepsilon} \mathbb{E}|\beta_k(s)|^4_{\mathcal{L}_2^0} ds \le \mathcal{C}(x_0,k)\varepsilon^2.
\end{aligned}\tag{12.218}$$

Here and henceforth, $\mathcal{C}(x_0,k)$ is a generic constant (depending on x_0, k, T, A and $\mathcal{C}_L$), which may be different from line to line. For any fixed $k\in\mathbb{N}$, since $Q_k\beta_k\in L^2_{\mathbb{F}}(0,T;L^{\frac{4}{3}}(\Omega;H))$, by (12.218), we find that

$$
\begin{aligned}
&\Big|\mathbb{E}\int_0^T \chi_{E_\varepsilon}(t)\big\langle (Q_k\beta_k)(t), x^\varepsilon_{2,n_k}(t)\big\rangle_H dt\Big|\\
&\le |x^\varepsilon_{2,n_k}(\cdot)|_{C_{\mathbb{F}}([0,T];L^4(\Omega;H))}\int_{E_\varepsilon}\big|(Q_k\beta_k)(t)\big|_{L^{\frac{4}{3}}_{\mathcal{F}_t}(\Omega;H)}dt \qquad (12.219)\\
&\le \mathcal{C}(x_0,k)\sqrt{\varepsilon}\int_{E_\varepsilon}\big|(Q_k\beta_k)(t)\big|_{L^{\frac{4}{3}}_{\mathcal{F}_t}(\Omega;H)}dt=o(\varepsilon),\qquad \text{as } \varepsilon\to 0.
\end{aligned}
$$

Similarly,

$$
\Big|\mathbb{E}\int_0^T \chi_{E_\varepsilon}(t)\big\langle x^\varepsilon_{2,n_k}(t), (\widehat{Q}_k\beta_k)(t)\big\rangle_H dt\Big| = o(\varepsilon),\qquad \text{as } \varepsilon\to 0. \qquad (12.220)
$$

From (12.152) in Theorem 12.16, and noting that both Q_k and $\widehat{Q}_k$ are pointwise defined, we arrive at the following equality:

$$
\begin{aligned}
&\mathbb{E}\int_0^T \big\langle \chi_{E_\varepsilon}(t)\beta_k(t), \widehat{Q}^{(0)}(0,0,\chi_{E_\varepsilon}\beta_k)(t)\big\rangle_{\mathcal{L}_2^0}dt\\
&\quad+\mathbb{E}\int_0^T \big\langle Q^{(0)}(0,0,\chi_{E_\varepsilon}\beta_k)(t), \chi_{E_\varepsilon}(t)\beta_k(t)\big\rangle_{\mathcal{L}_2^0}dt \qquad (12.221)\\
&=\mathbb{E}\int_0^T \chi_{E_\varepsilon}(t)\Big(\big\langle (Q_k\beta_k)(t), x^\varepsilon_{2,n_k}(t)\big\rangle_H+\big\langle x^\varepsilon_{2,n_k}(t),(\widehat{Q}_k\beta_k)(t)\big\rangle_H\Big)dt.
\end{aligned}
$$

Hence,

$$
\begin{aligned}
&\mathbb{E}\int_0^T \big\langle \chi_{E_\varepsilon}(t)\delta b(t), \widehat{Q}^{(0)}(0,0,\chi_{E_\varepsilon}\delta b)(t)\big\rangle_{\mathcal{L}_2^0}dt\\
&\quad+\mathbb{E}\int_0^T \big\langle Q^{(0)}(0,0,\chi_{E_\varepsilon}\delta b)(t), \chi_{E_\varepsilon}(t)\delta b(t)\big\rangle_{\mathcal{L}_2^0}dt\\
&\quad-\mathbb{E}\int_0^T \chi_{E_\varepsilon}(t)\Big(\big\langle (Q_k\beta_k)(t), x^\varepsilon_{2,n_k}(t)\big\rangle_H+\big\langle x^\varepsilon_{2,n_k}(t),(\widehat{Q}_k\beta_k)(t)\big\rangle_H\Big)dt\\
&=\mathbb{E}\int_0^T \big\langle \chi_{E_\varepsilon}(t)\delta b(t), \widehat{Q}^{(0)}(0,0,\chi_{E_\varepsilon}\delta b)(t)\big\rangle_{\mathcal{L}_2^0}dt \qquad (12.222)\\
&\quad+\mathbb{E}\int_0^T \big\langle Q^{(0)}(0,0,\chi_{E_\varepsilon}\delta b)(t), \chi_{E_\varepsilon}(t)\delta b(t)\big\rangle_{\mathcal{L}_2^0}dt\\
&\quad-\mathbb{E}\int_0^T \big\langle \chi_{E_\varepsilon}(t)\beta_k(t), \widehat{Q}^{(0)}(0,0,\chi_{E_\varepsilon}\beta_k)(t)\big\rangle_{\mathcal{L}_2^0}dt\\
&\quad-\mathbb{E}\int_0^T \big\langle Q^{(0)}(0,0,\chi_{E_\varepsilon}\beta_k)(t), \chi_{E_\varepsilon}(t)\beta_k(t)\big\rangle_{\mathcal{L}_2^0}dt.
\end{aligned}
$$

It is easy to see that

$$
\begin{aligned}
&\Big|\mathbb{E}\int_0^T \big\langle \chi_{E_\varepsilon}(t)\delta b(t), \widehat{Q}^{(0)}(0,0,\chi_{E_\varepsilon}\delta b)(t)\big\rangle_{\mathcal{L}_2^0}dt \\
&\quad -\mathbb{E}\int_0^T \big\langle \chi_{E_\varepsilon}(t)\beta_k(t), \widehat{Q}^{(0)}(0,0,\chi_{E_\varepsilon}\beta_k)(t)\big\rangle_{\mathcal{L}_2^0}dt\Big| \\
&\leq \Big|\mathbb{E}\int_0^T \big\langle \chi_{E_\varepsilon}(t)\delta b(t), \widehat{Q}^{(0)}(0,0,\chi_{E_\varepsilon}\delta b)(t)\big\rangle_{\mathcal{L}_2^0}dt \\
&\quad -\mathbb{E}\int_0^T \big\langle \chi_{E_\varepsilon}(t)\delta b(t), \widehat{Q}^{(0)}(0,0,\chi_{E_\varepsilon}\beta_k)(t)\big\rangle_{\mathcal{L}_2^0}dt\Big| \\
&\quad +\Big|\mathbb{E}\int_0^T \big\langle \chi_{E_\varepsilon}(t)\delta b(t), \widehat{Q}^{(0)}(0,0,\chi_{E_\varepsilon}\beta_k)(t)\big\rangle_{\mathcal{L}_2^0}dt \\
&\quad -\mathbb{E}\int_0^T \big\langle \chi_{E_\varepsilon}(t)\beta_k(t), \widehat{Q}^{(0)}(0,0,\chi_{E_\varepsilon}\beta_k)(t)\big\rangle_{\mathcal{L}_2^0}dt\Big|.
\end{aligned}
\tag{12.223}
$$

From (12.216) and the density of the Lebesgue points, we find that for a.e. $\tau \in [0,T)$, it holds that

$$
\begin{aligned}
&\lim_{k\to\infty}\lim_{\varepsilon\to 0}\frac{1}{\varepsilon}\Big|\mathbb{E}\int_0^T \big\langle \chi_{E_\varepsilon}(t)\delta b(t), \widehat{Q}^{(0)}(0,0,\chi_{E_\varepsilon}\delta b)(t)\big\rangle_{\mathcal{L}_2^0}dt \\
&\qquad -\mathbb{E}\int_0^T \big\langle \chi_{E_\varepsilon}(t)\delta b(t), \widehat{Q}^{(0)}(0,0,\chi_{E_\varepsilon}\beta_k)(t)\big\rangle_{\mathcal{L}_2^0}dt\Big| \\
&\leq \lim_{k\to\infty}\lim_{\varepsilon\to 0}\frac{1}{\varepsilon}\Big[\int_0^T \chi_{E_\varepsilon}(t)\Big(\mathbb{E}|\delta b(t)|_{\mathcal{L}_2^0}^4\Big)^{\frac{1}{2}}dt\Big]^{\frac{1}{2}} \\
&\qquad \times|\widehat{Q}^{(0)}(0,0,\chi_{E_\varepsilon}(\delta b-\beta_k))|_{L^2_{\mathbb{F}}(0,T;L^{\frac{4}{3}}(\Omega;\mathcal{L}_2^0))} \\
&\leq \mathcal{C}\lim_{k\to\infty}\lim_{\varepsilon\to 0}\frac{1}{\varepsilon}\Big[\int_0^T \chi_{E_\varepsilon}(t)\Big(\mathbb{E}|\delta b(t)|_{\mathcal{L}_2^0}^4\Big)^{\frac{1}{2}}dt\Big]^{\frac{1}{2}}|\chi_{E_\varepsilon}(\delta b-\beta_k)|_{L^2_{\mathbb{F}}(0,T;L^4(\Omega;\mathcal{L}_2^0))} \\
&\leq \mathcal{C}\lim_{k\to\infty}\lim_{\varepsilon\to 0}\frac{|\delta b(\tau)|_{L^4_{\mathcal{F}_\tau}(\Omega;\mathcal{L}_2^0)}}{\sqrt{\varepsilon}}\Big[\int_0^T \chi_{E_\varepsilon}(t)\Big(\mathbb{E}|\delta b(t)-\beta_k(t)|_{\mathcal{L}_2^0}^4\Big)^{\frac{1}{2}}dt\Big]^{\frac{1}{2}} \\
&= \mathcal{C}\lim_{k\to\infty}\lim_{\varepsilon\to 0}|\delta b(\tau)|_{L^4_{\mathcal{F}_\tau}(\Omega;\mathcal{L}_2^0)}\Big[\frac{1}{\varepsilon}\int_\tau^{\tau+\varepsilon}|\delta b(t)-\beta_k(t)|^2_{L^4_{\mathcal{F}_t}(\Omega;\mathcal{L}_2^0)}dt\Big]^{\frac{1}{2}} \\
&= \mathcal{C}\lim_{k\to\infty}|\delta b(\tau)|_{L^4_{\mathcal{F}_\tau}(\Omega;\mathcal{L}_2^0)}|\delta b(\tau)-\beta_k(\tau)|_{L^4_{\mathcal{F}_\tau}(\Omega;\mathcal{L}_2^0)} \\
&= 0.
\end{aligned}
\tag{12.224}
$$

Similarly,

$$
\begin{aligned}
&\lim_{k\to\infty}\lim_{\varepsilon\to 0}\frac{1}{\varepsilon}\Big|\mathbb{E}\int_0^T \big\langle \chi_{E_\varepsilon}(t)\delta b(t), \widehat{Q}^{(0)}(0,0,\chi_{E_\varepsilon}\beta_k)(t)\big\rangle_{\mathcal{L}_2^0}dt \\
&\qquad -\mathbb{E}\int_0^T \big\langle \chi_{E_\varepsilon}(t)\beta_k(t), \widehat{Q}^{(0)}(0,0,\chi_{E_\varepsilon}\beta_k)(t)\big\rangle_{\mathcal{L}_2^0}dt\Big|
\end{aligned}
$$

$$
\begin{aligned}
&\leq \lim_{k\to\infty}\lim_{\varepsilon\to 0}\frac{1}{\varepsilon}\big|\widehat{Q}^{(0)}(0,0,\chi_{E_\varepsilon}\beta_k)\big|_{L^2_{\mathbb{F}}(0,T;L^{\frac{4}{3}}(\Omega;\mathcal{L}_2^0))}\\
&\qquad\qquad \times\Big[\int_0^T\chi_{E_\varepsilon}(t)\Big(\mathbb{E}|\delta b(t)-\beta_k(t)|^4_{\mathcal{L}_2^0}\Big)^{\frac{1}{2}}dt\Big]^{\frac{1}{2}}\\
&\leq \mathcal{C}\lim_{k\to\infty}\lim_{\varepsilon\to 0}\frac{1}{\varepsilon}\big|\chi_{E_\varepsilon}\beta_k\big|_{L^2_{\mathbb{F}}(0,T;L^4(\Omega;\mathcal{L}_2^0))}\\
&\qquad\qquad \times\Big[\int_0^T\chi_{E_\varepsilon}(t)\Big(\mathbb{E}|\delta b(t)-\beta_k(t)|^4_{\mathcal{L}_2^0}\Big)^{\frac{1}{2}}dt\Big]^{\frac{1}{2}} \qquad (12.225)\\
&\leq \mathcal{C}\lim_{k\to\infty}\lim_{\varepsilon\to 0}\frac{1}{\varepsilon}\Big\{\big|\chi_{E_\varepsilon}\delta b\big|_{L^2_{\mathbb{F}}(0,T;L^4(\Omega;\mathcal{L}_2^0))}\\
&\qquad\qquad \times\Big[\int_0^T\chi_{E_\varepsilon}(t)\Big(\mathbb{E}|\delta b(t)-\beta_k(t)|^4_{\mathcal{L}_2^0}\Big)^{\frac{1}{2}}dt\Big]^{\frac{1}{2}}\\
&\qquad\qquad\qquad +\int_0^T\chi_{E_\varepsilon}(t)\Big(\mathbb{E}|\delta b(t)-\beta_k(t)|^4_{\mathcal{L}_2^0}\Big)^{\frac{1}{2}}dt\Big\}\\
&\leq \mathcal{C}\lim_{k\to\infty}\lim_{\varepsilon\to 0}\Big\{\frac{|\delta b(\tau)|_{L^4_{\mathcal{F}_\tau}(\Omega;\mathcal{L}_2^0)}}{\sqrt{\varepsilon}}\Big[\int_0^T\chi_{E_\varepsilon}(t)\Big(\mathbb{E}|\delta b(t)-\beta_k(t)|^4_{\mathcal{L}_2^0}\Big)^{\frac{1}{2}}dt\Big]^{\frac{1}{2}}\\
&\qquad\qquad +\frac{1}{\varepsilon}\int_0^T\chi_{E_\varepsilon}(t)\Big(\mathbb{E}|\delta b(t)-\beta_k(t)|^4_{\mathcal{L}_2^0}\Big)^{\frac{1}{2}}dt\Big\}\\
&= \mathcal{C}\lim_{k\to\infty}\lim_{\varepsilon\to 0}\Big[|\delta b(\tau)|_{L^4_{\mathcal{F}_\tau}(\Omega;\mathcal{L}_2^0)}\Big(\frac{1}{\varepsilon}\int_\tau^{\tau+\varepsilon}|\delta b(t)-\beta_k(t)|^2_{L^4_{\mathcal{F}_t}(\Omega;\mathcal{L}_2^0)}dt\Big)^{\frac{1}{2}}\\
&\qquad\qquad +\frac{1}{\varepsilon}\int_\tau^{\tau+\varepsilon}|\delta b(t)-\beta_k(t)|^2_{L^4_{\mathcal{F}_t}(\Omega;\mathcal{L}_2^0)}dt\Big]\\
&= \mathcal{C}\lim_{k\to\infty}\Big(|\delta b(\tau)|_{L^4_{\mathcal{F}_\tau}(\Omega;\mathcal{L}_2^0)}|\delta b(\tau)-\beta_k(\tau)|_{L^4_{\mathcal{F}_\tau}(\Omega;\mathcal{L}_2^0)}\\
&\qquad +|\delta b(\tau)-\beta_k(\tau)|^2_{L^4_{\mathcal{F}_\tau}(\Omega;\mathcal{L}_2^0)}\Big)\\
&= 0.
\end{aligned}
$$

From (12.223)–(12.225), we find that

$$
\begin{aligned}
&\lim_{k\to\infty}\lim_{\varepsilon\to 0}\frac{1}{\varepsilon}\Big|\mathbb{E}\int_0^T\big\langle\chi_{E_\varepsilon}(t)\delta b(t),\widehat{Q}^{(0)}(0,0,\chi_{E_\varepsilon}\delta b)(t)\big\rangle_{\mathcal{L}_2^0}dt\\
&\qquad\qquad -\mathbb{E}\int_0^T\big\langle\chi_{E_\varepsilon}(t)\beta_k(t),\widehat{Q}^{(0)}(0,0,\chi_{E_\varepsilon}\beta_k)(t)\big\rangle_{\mathcal{L}_2^0}dt\Big| \qquad (12.226)\\
&= 0.
\end{aligned}
$$

By a similar argument, we obtain that

$$
\begin{aligned}
&\lim_{k\to\infty}\lim_{\varepsilon\to 0}\frac{1}{\varepsilon}\Big|\mathbb{E}\int_0^T \big\langle Q^{(0)}(0,0,\chi_{E_\varepsilon}\delta b)(t),\chi_{E_\varepsilon}(t)\delta b(t)\big\rangle_{\mathcal{L}_2^0}dt\\
&\qquad\qquad -\mathbb{E}\int_0^T \big\langle Q^{(0)}(0,0,\chi_{E_\varepsilon}\beta_k)(t),\chi_{E_\varepsilon}(t)\beta_k(t)\big\rangle_{\mathcal{L}_2^0}dt\Big| \qquad (12.227)\\
&=0.
\end{aligned}
$$

From (12.219)–(12.222) and (12.226)–(12.227), we obtain that

$$
\begin{aligned}
&\Big|\mathbb{E}\int_0^T \chi_{E_\varepsilon}(t)\big\langle \delta b(t),\widehat{Q}^{(0)}(0,0,\chi_{E_\varepsilon}\delta b)(t)\big\rangle_{\mathcal{L}_2^0}dt\\
&\quad+\mathbb{E}\int_0^T \chi_{E_\varepsilon}(t)\big\langle Q^{(0)}(0,0,\delta b)(t),\delta b(t)\big\rangle_{\mathcal{L}_2^0}dt\Big| \qquad (12.228)\\
&=o(\varepsilon),\qquad \text{as } \varepsilon\to 0.
\end{aligned}
$$

Combining (12.211), (12.212), (12.213) and (12.228), we end up with

$$
\begin{aligned}
&\mathcal{J}(u^\varepsilon(\cdot))-\mathcal{J}(\bar u(\cdot))\\
&=\mathrm{Re}\,\mathbb{E}\int_0^T \Big(\delta g(t)-\big\langle y(t),\delta a(t)\big\rangle_H-\big\langle Y(t),\delta b(t)\big\rangle_{\mathcal{L}_2^0}\\
&\quad -\frac{1}{2}\big\langle P(t)\delta b(t),\delta b(t)\big\rangle_{\mathcal{L}_2^0}\Big)\chi_{E_\varepsilon}(t)dt+o(\varepsilon).
\end{aligned}
$$

Since $\bar u(\cdot)$ is the optimal control, $\mathcal{J}(u^\varepsilon(\cdot))-\mathcal{J}(\bar u(\cdot))\geq 0$. Thus,

$$
\begin{aligned}
&\mathrm{Re}\,\mathbb{E}\int_0^T \chi_{E_\varepsilon}(t)\Big(\big\langle y(t),\delta a(t)\big\rangle_H+\big\langle Y(t),\delta b(t)\big\rangle_{\mathcal{L}_2^0}\\
&\qquad\qquad -\delta g(t)+\frac{1}{2}\big\langle P(t)\delta b(t),\delta b(t)\big\rangle_{\mathcal{L}_2^0}\Big)dt\leq o(\varepsilon), \qquad (12.229)
\end{aligned}
$$

as $\varepsilon\to 0$.

Step 7. We are now ready to complete the proof, i.e., we shall deduce (12.166) from (12.229). This is more or less standard (e.g., [378]).

Since $L^2_{\mathcal{F}_T}(\Omega;\mathbb{C})$ is separable, for any $t\in[0,T]$, $\mathcal{F}_t$ is countably generated by a sequence $\{M_k\}_{k=1}^\infty\subset\mathcal{F}_t$, that is, for any $M\subset\mathcal{F}_t$, there exists a subsequence $\{M_{k_j}\}_{j=1}^\infty\subset\{M_k\}_{k=1}^\infty$ such that

$$
\lim_{j\to\infty}\mathbb{P}\big((M\setminus M_{k_j})\cup(M_{k_j}\setminus M)\big)=0.
$$

Denote by $\{t_i\}_{i=1}^\infty$ the sequence of rational numbers in $[0,T)$, and by $\{v_i\}_{i=1}^\infty$ a dense subset of U. For each $i\in\mathbb{N}$, we choose $\{M_{ij}\}_{j=1}^\infty\subset\mathcal{F}_{t_i}$ to be a sequence which generates $\mathcal{F}_{t_i}$.

Fix $i,j,k\in\mathbb{N}$ arbitrarily. For any $\tau\in[t_i,T)$ and $\theta\in(0,T-\tau)$, write $E_\theta^i=[\tau,\tau+\theta)$. Put

$$u_{ij}^{k,\theta} = \begin{cases} v^k, & \text{if } (t,\omega) \in E_\theta^i \times M_{ij}, \\ \bar{u}(t,\omega), & \text{if } (t,\omega) \in ([0,T] \times \Omega) \setminus (E_\theta^i \times M_{ij}). \end{cases}$$

Clearly, $u_{ij}^{k,\theta} \in \mathcal{U}[0,T]$ and

$$u_{ij}^{k,\theta}(t,\omega) - \bar{u}(t,\omega) = (v^k - \bar{u}(t,\omega))\chi_{M_{ij}}(\omega)\chi_{E_\theta^i}(t), \quad (t,\omega) \in [0,T] \times \Omega.$$

From (12.229), it follows that

$$\begin{aligned} &\operatorname{Re}\mathbb{E}\int_\theta^{\theta+\tau} \Big(\mathbb{H}\big(t,\bar{x}(t),\bar{u}(t),y(t),Y(t)\big) - \mathbb{H}\big(t,\bar{x}(t),u_{ij}^{k,\theta}(t),y(t),Y(t)\big) \\ &\qquad\qquad -\frac{1}{2}\big\langle P(t)\big(b\big(t,\bar{x}(t),\bar{u}(t)\big) - b\big(t,\bar{x}(t),u_{ij}^{k,\theta}(t)\big)\big), \\ &\qquad\qquad\qquad b\big(t,\bar{x}(t),\bar{u}(t)\big) - b\big(t,\bar{x}(t),u_{ij}^{k,\theta}(t)\big)\big\rangle_{\mathcal{L}_2^0}\Big)dt \\ &\le o(\varepsilon). \end{aligned} \tag{12.230}$$

Divide both sides of (12.230) by τ and let $\tau \to 0^+$, by the property of Lebesgue point, we deduce that for any $i,j,k \in \mathbb{N}$, there exists a Lebesgue measurable set $E_{i,j}^k \subset [t_i,T)$ with Lebesgue measure $\mathbf{m}(E_{i,j}^k) = 0$ such that

$$\begin{aligned} &\operatorname{Re}\mathbb{E}\Big[\Big(\mathbb{H}\big(t,\bar{x}(t),\bar{u}(t),y(t),Y(t)\big) - \mathbb{H}\big(t,\bar{x}(t),v^k(t),y(t),Y(t)\big) \\ &\quad -\frac{1}{2}\big\langle P(t)\big(b\big(t,\bar{x}(t),\bar{u}(t)\big) - b\big(t,\bar{x}(t),v^k(t)\big)\big), \\ &\quad\quad b\big(t,\bar{x}(t),\bar{u}(t)\big) - b\big(t,\bar{x}(t),v^k(t)\big)\big\rangle_{\mathcal{L}_2^0}\Big)\chi_{M_{ij}}\Big] \le 0, \ \ \forall\, t \in [t_i,T) \setminus E_{i,j}^k. \end{aligned}$$

Let $E_0 = \bigcup\limits_{i,j,k\in\mathbb{N}} E_{i,j}^k$. Then its Lebesgue measure $\mathbf{m}(E_0) = 0$, and for any $i,j,k \in \mathbb{N}$ and $t \in [t_i,T) \setminus E_0$,

$$\begin{aligned} &\operatorname{Re}\mathbb{E}\Big[\Big(\mathbb{H}\big(t,\bar{x}(t),\bar{u}(t),y(t),Y(t)\big) - \mathbb{H}\big(t,\bar{x}(t),v^k(t),y(t),Y(t)\big) \\ &\quad -\frac{1}{2}\big\langle P(t)\big(b\big(t,\bar{x}(t),\bar{u}(t)\big) - b\big(t,\bar{x}(t),v^k(t)\big)\big), \\ &\quad\quad b\big(t,\bar{x}(t),\bar{u}(t)\big) - b\big(t,\bar{x}(t),v^k(t)\big)\big\rangle_{\mathcal{L}_2^0}\Big)\chi_{M_{ij}}\Big] \le 0. \end{aligned} \tag{12.231}$$

By the construction of $\{M_{ij}\}_{j=1}^\infty$, the right continuity of the filter $\mathbb{F}$ and the density of $\{v_k\}_{k=1}^\infty$, we conclude from (12.231) that (12.166) holds. This completes the proof of Theorem 12.17. □

12.6 Sufficient Condition for Optimal Controls

Theorem 12.17 gives a necessary condition for the optimal control. In this section, we shall present a sufficient optimality condition. For simplicity, we assume that H is a real separable Hilbert space, the control region U is a convex subset of the separable Hilbert space $\widetilde{H}$ appeared in (S4), and $L^p_{\mathcal{F}_T}(\Omega)$ $(p \geq 1)$ is separable.

12.6.1 Clarke's Generalized Gradient

In this subsection, we recall the definition and some basic properties of Clarke's generalized gradient. A detailed introduction to this topic can be found in [59].

Let O be a domain in H and $\varphi : O \to \mathbb{R}$ be a locally Lipschitz continuous function. For any $x \in O$, the Clarke's generalized gradient $\partial\varphi(x)$ of φ at x is defined as follows:

$$\partial\varphi(x) \overset{\Delta}{=} \Big\{ \xi \in H \;\Big|\; \langle \xi, y \rangle_H \leq \overline{\lim_{\substack{z\to x, z\in O\\ t\downarrow 0}}} \frac{\varphi(z+ty)-\varphi(z)}{t}, \ \forall\, y \in H \Big\}. \qquad (12.232)$$

We need the following lemma.

Lemma 12.18. *For the above function φ, the following assertions hold:*

1) *$\partial\varphi(x)$ is a nonempty, bounded, convex subset of H.*

2) $\partial(-\varphi)(x) = -\partial\varphi(x)$.

3) *$0 \in \partial\varphi(x)$ if φ attains a local minimum or maximum at x.*

Proof: 1) The convexity and boundedness of $\partial\varphi(x)$ follow immediately from (12.232). We only need to show the non-emptyness.

Since $\varphi : O \to \mathbb{R}$ is locally Lipschitz continuous, for any $x \in O$ and a small $\delta > 0$ such that $B_\delta(x) \overset{\Delta}{=} \big\{ z \in H \;\big|\; |z-x| \leq \delta \big\} \subseteq O$, there exists a constant $\mathcal{C}_L > 0$ so that

$$|\varphi(z) - \varphi(\widehat{z})| \leq \mathcal{C}_L |z - \widehat{z}|_H, \qquad \forall\, z, \widehat{z} \in B_\delta(x).$$

Thus, for any fixed $y \in H$ and $t > 0$ is small enough,

$$\frac{|\varphi(z+ty)-\varphi(z)|}{t} \leq \mathcal{C}_L |y|_H, \qquad \forall\, z \in B_{\delta/2}(x).$$

Consequently, the following functional is well-defined:

$$\varphi^0(x; y) \overset{\Delta}{=} \overline{\lim_{\substack{z\to x, z\in O\\ t\downarrow 0}}} \frac{\varphi(z+ty)-\varphi(z)}{t}. \qquad (12.233)$$

It is an easy matter to see that

$$\varphi^0(x; \lambda y) = \lambda \varphi^0(x; y), \quad \forall\, y \in H,\ \lambda \geq 0 \qquad (12.234)$$

and

$$\varphi^0(x; y+z) \leq \varphi^0(x; y) + \varphi^0(x; z), \quad \forall\, y, z \in H. \tag{12.235}$$

Thus, the map $y \mapsto \varphi^0(x; y)$ is convex. Further, the inequality (12.235) implies that

$$-\varphi^0(x; y) \leq \varphi^0(x; -y), \quad \forall\, y \in H. \tag{12.236}$$

Next, we fix a $z \in H$ and define $F : \{\lambda z \mid \lambda \in \mathbb{R}\} \to \mathbb{R}$ by

$$F(\lambda z) = \lambda \varphi^0(x; z), \quad \forall \lambda \in \mathbb{R}.$$

Then for $\lambda \geq 0$,

$$\begin{cases} F(\lambda z) \equiv \lambda \varphi^0(x; z) = \varphi^0(x; \lambda z), \\ F(-\lambda z) \equiv -\lambda \varphi^0(x; z) = -\varphi^0(x; \lambda z) \leq \varphi^0(x, -\lambda z), \end{cases}$$

which yields

$$F(\lambda z) \leq \varphi^0(x; \lambda z), \quad \forall \lambda \in \mathbb{R}. \tag{12.237}$$

Therefore, F is a linear functional defined on the linear space spanned by z, and it is dominated by the convex function $\varphi^0(x; \cdot)$. By the Hahn-Banach theorem, F can be extended to be a bounded linear functional on H. Then, by the classical Riesz representation theorem, there exists $\xi \in H$, such that

$$\begin{cases} \langle \xi, \lambda z \rangle = F(\lambda z) \equiv \lambda \varphi^0(x; z), \quad \forall \lambda \in \mathbb{R}, \\ \langle \xi, y \rangle \leq \varphi^0(x; y), \quad \forall\, y \in H. \end{cases} \tag{12.238}$$

This implies $\xi \in \partial\varphi(x)$. Consequently, $\partial\varphi(x)$ is nonempty.

2) It follows from (12.233) that

$$\begin{aligned} (-\varphi)^0(x; y) &= \overline{\lim_{\substack{z \to x, z \in O \\ t \downarrow 0}}} \frac{-\varphi(z+ty) + \varphi(z)}{t} \\ &= \overline{\lim_{\substack{z' \to x, z' \in O \\ t \downarrow 0}}} \frac{-\varphi(z') + \varphi(z' - ty)}{t} = \varphi^0(x; -y). \end{aligned}$$

Thus, $\xi \in \partial(-\varphi)(x)$ if and only if

$$\langle -\xi, y \rangle_H = \langle \xi, -y \rangle_H \leq (-\varphi)^0(x; -y) = \varphi^0(x; y), \quad \forall y \in H,$$

which is equivalent to $-\xi \in \partial\varphi(x)$.

3) Suppose φ attains a local minimum at x. Then

$$\begin{aligned} \varphi^0(x; y) &= \overline{\lim_{\substack{z \to x, z \in O \\ t \downarrow 0}}} \frac{\varphi(z+ty) - \varphi(z)}{t} \\ &\geq \overline{\lim_{t \to 0^+}} \frac{\varphi(x+ty) - \varphi(x)}{t} \geq 0 = \langle 0, y \rangle_H, \quad \forall y \in H, \end{aligned}$$

which implies $0 \in \partial \varphi(x)$.

If φ attains a local maximum at x, then the conclusion follows from the fact that $-\varphi$ attains a local minimum at x. □

By fixing some arguments in a function, one may naturally define its *partial generalized gradient*. For example, if $\psi(\cdot,\cdot) : H \times \widetilde{H} \to \mathbb{R}$ is locally Lipschitz, by $\partial_x \psi(x,u)$ (*resp.* $\partial_u \psi(x,u)$), we mean the partial generalized gradient of ψ in x (*resp.* in u) at $(x,u) \in H \times U$.

In the rest of this subsection, we give two technical lemmas.

Lemma 12.19. *Let ψ be a convex or concave function on $H \times \widetilde{H}$. Assume that $\psi(x,u)$ is differentiable in x and $\psi_x(x,u)$ is continuous in (x,u). Then*

$$\big\{(\psi_x(\hat{x},\hat{u}),r) \;\big|\; r \in \partial_u \psi(\hat{x},\hat{u})\big\} \subseteq \partial_{x,u}\psi(\hat{x},\hat{u}), \quad \forall\, (\hat{x},\hat{u}) \in H \times U. \tag{12.239}$$

Proof: Once (12.239) is proved for convex ψ, it is also true for concave ψ by noting that $-\psi$ is convex and the assertion 2) in Lemma 12.18. Hence, we only handle the case that ψ is convex.

For any $\xi \in H$ and $u \in \widetilde{H}$, we choose a sequence $\{(x_j,\delta_j)\}_{j=1}^{\infty} \subset H \times \mathbb{R}$ as follows:

$$\begin{aligned}&(x_j,\hat{u}) \in H \times U, \quad (x_j+\delta_j\xi, \hat{u}+\delta_j u) \in H \times U,\\ &\delta_j \to 0^+, \text{ as } j \to \infty, \text{ and } |x_j - \hat{x}|_H \le \delta_j^2.\end{aligned}$$

By the convexity of ψ, we have

$$\begin{aligned}&\lim_{j\to\infty} \frac{\psi(x_j+\delta_j\xi,\hat{u}+\delta_j u) - \psi(\hat{x},\hat{u}+\delta_j u)}{\delta_j}\\ &\ge \lim_{j\to\infty} \frac{\langle \psi_x(\hat{x},\hat{u}+\delta_j u), x_i - \hat{x} + \delta_j \xi\rangle_H}{\delta_j} = \langle \psi_x(\hat{x},\hat{u}),\xi\rangle_H.\end{aligned} \tag{12.240}$$

Similarly,

$$\lim_{j\to\infty} \frac{\psi(\hat{x},\hat{u}+\delta_j u) - \psi(\hat{x},\hat{u})}{\delta_j} \ge \langle r,u\rangle_H. \tag{12.241}$$

Also,

$$\lim_{j\to\infty} \frac{\psi(\hat{x},\hat{u}) - \psi(x_j,\hat{u})}{\delta_j} \ge \lim_{j\to\infty} \frac{\langle \psi_x(x_j,\hat{u}), \hat{x} - x_j\rangle_H}{\delta_j} = 0. \tag{12.242}$$

It follows from (12.240)–(12.242) that

$$\lim_{j\to\infty} \frac{\psi(x_j+\delta_j\xi,\hat{u}+\delta_j u) - \psi(x_j,\hat{u})}{\delta_j} \ge \langle \psi_x(\hat{x},\hat{u}),\xi\rangle_H + \langle r,u\rangle_H.$$

This, together with the definition of the generalized gradient, implies that $(\psi_x(\hat{x},\hat{u}),r) \in \partial_{x,u}\psi(\hat{x},\hat{u})$. □

12.6.2 A Sufficient Condition for Optimal Controls

In this subsection, following [371, 393] (for stochastic controls in finite dimensions), we shall derive a sufficient condition for optimal controls for Problem (OP), i.e., for a given admissible control $\bar{u}(\cdot) \in \mathcal{U}[0,T]$ and the corresponding state $\bar{x}(\cdot)$, we hope to find a suitable condition to guarantee (12.4) holds. For this purpose, for the above admissible pair $(\bar{x}(\cdot), \bar{u}(\cdot))$, denote by $\big(y(\cdot), Y(\cdot)\big)$ be the corresponding transposition solution to (12.7), and by $(P(\cdot), Q^{(\cdot)}, \widehat{Q}^{(\cdot)})$ the corresponding relaxed transposition solution to the equation (12.8) in which $F(\cdot)$, $J(\cdot)$, $K(\cdot)$ and P_T are given by (12.165)[7]. Recall (12.164) for the definition of $\mathbb{H}(\cdot,\cdot,\cdot,\cdot,\cdot)$, and write

$$\begin{aligned}\mathcal{H}(t,x,u) &\overset{\Delta}{=} \mathbb{H}\big(t,x,u,y(t),Y(t)\big) + \frac{1}{2}\big\langle P(t)b\big(t,x,u\big), b\big(t,x,u\big)\big\rangle_{\mathcal{L}_2^0}\\ &\quad -\big\langle P(t)b\big(t,\bar{x}(t),\bar{u}(t)\big), b\big(t,x,u\big)\big\rangle_{\mathcal{L}_2^0}, \quad \forall\, (t,x,u) \in [0,T]\times H \times U.\end{aligned}$$

We need the following intermediate result:

Lemma 12.20. *Let (S1)–(S4) hold. Then, for a.e.* $(t,\omega) \in [0,T]\times\Omega$,

$$\partial_u \mathbb{H}(t,\bar{x}(t),\bar{u}(t),y(t),Y(t)) = \partial_u \mathcal{H}(t,\bar{x}(t),\bar{u}(t)). \tag{12.243}$$

Proof: Fix a $t\in[0,T]$. Denote

$$\begin{cases}\mathbb{H}(u) \overset{\Delta}{=} \mathbb{H}\big(t,\bar{x}(t),u,y(t),Y(t)\big), \qquad \mathcal{H}(u) \overset{\Delta}{=} \mathcal{H}\big(t,\bar{x}(t),u\big),\\ b(u) \overset{\Delta}{=} b\big(t,\bar{x}(t),u\big), \qquad \psi(u) \overset{\Delta}{=} \dfrac{1}{2}\big\langle P(t)b(u), b(u)\big\rangle_{\mathcal{L}_2^0} - \big\langle P(t)b(\bar{u}(t)), b(u)\big\rangle_{\mathcal{L}_2^0}.\end{cases}$$

Then,

$$\mathcal{H}(u) = \mathbb{H}(u) + \psi(u).$$

Note that for any $r\to 0^+$, $u, v\in U$, with $u\to\bar{u}(t)$,

$$\begin{aligned}&\psi(u+rv) - \psi(u)\\ &= \frac{1}{2}\big\langle P(t)\big[b(u+rv) + b(u) - 2b(u(t))\big], b(u+rv) - b(u)\big\rangle_{\mathcal{L}_2^0} = o(r).\end{aligned}$$

Thus,

$$\varlimsup_{\substack{u\to\bar{u}(t),u\in U\\ r\downarrow 0}} \frac{\mathcal{H}(u+rv) - \mathcal{H}(u)}{r} = \varlimsup_{\substack{u\to\bar{u}(t),u\in U\\ r\downarrow 0}} \frac{\mathbb{H}(u+rv) - \mathbb{H}(u)}{r}.$$

Consequently, by (12.232), the desired result (12.243) follows. □

Now, let us prove the following sufficient optimality condition for Problem (OP).

[7] Clearly, the data in both (12.7) and (12.8) can be well-defined for any given admissible pair $(\bar{x}(\cdot), \bar{u}(\cdot))$, not only for the optimal ones.

Theorem 12.21. *Let (S1)–(S4) hold. Suppose that the function* $h(\cdot)$ *is convex,* $\mathbb{H}(t,\cdot,\cdot,y(t),Y(t))$ *is concave for a.e.* $t\in[0,T]$, *a.s., and*

$$\mathcal{H}(t,\bar x(t),\bar u(t))=\max_{u\in U}\mathcal{H}(t,\bar x(t),u),\quad \text{a.e. } (t,\omega)\in[0,T]\times\Omega. \tag{12.244}$$

Then $(\bar x(\cdot),\bar u(\cdot))$ *is an optimal pair for Problem (OP).*

Proof: By the maximum condition (12.244), Lemma 12.20 and the assertion 3) in Lemma 12.18, we have

$$0\in\partial_u\mathcal{H}(t,\bar x(t),\bar u(t))=\partial_u\mathbb{H}(t,\bar x(t),\bar u(t),y(t),Y(t)).$$

Hence, by Lemma 12.19, it follows that

$$\Big(\mathbb{H}_x(t,\bar x(t),\bar u(t),y(t),Y(t)),0\Big)\in\partial_{x,u}\mathbb{H}(t,\bar x(t),\bar u(t),y(t),Y(t)).$$

This, combined with the concavity of $\mathbb{H}(t,\cdot,\cdot,y(t),Y(t))$, yields that

$$\begin{aligned}&\int_0^T\big(\mathbb{H}(t,x(t),u(t),y(t),Y(t))-\mathbb{H}(t,\bar x(t),\bar u(t),y(t),Y(t))\big)dt\\&\le\int_0^T\big\langle\mathbb{H}_x(t,\bar x(t),\bar u(t),y(t),Y(t)),x(t)-\bar x(t)\big\rangle_H dt,\end{aligned}\tag{12.245}$$

for any admissible pair $(x(\cdot),u(\cdot))$. Let $\xi(t)\overset{\Delta}{=}x(t)-\bar x(t)$. Then $\xi(t)$ satisfies the following equation:

$$\begin{cases}d\xi(t)=\big(A\xi(t)+a_x(t,x(t),u(t))\xi(t)+\alpha(t)\big)dt\\\qquad\qquad+\big(b_x(t,x(t),u(t))\xi(t)+\beta(t)\big)dW(t)\quad\text{in }(0,T],\\\xi(0)=0,\end{cases}\tag{12.246}$$

where

$$\begin{cases}\alpha(t)\overset{\Delta}{=}-a_x\big(t,\bar x(t),\bar u(t)\big)\xi(t)+a\big(t,x(t),u(t)\big)-a\big(t,\bar x(t),\bar u(t)\big),\\\beta(t)\overset{\Delta}{=}-b_x\big(t,\bar x(t),\bar u(t)\big)\xi(t)+b\big(t,x(t),u(t)\big)-b\big(t,\bar x(t),\bar u(t)\big).\end{cases}$$

By (12.245)–(12.246), it follows from the definition of the transposition solution to (12.7) that

$$\begin{aligned}
&\mathbb{E}\big\langle h_x(\bar{x}(T)),\xi(T)\big\rangle_H = -\mathbb{E}\big\langle y(T),\xi(T)\big\rangle_H + \mathbb{E}\big\langle y(0),\xi(0)\big\rangle_H\\
&= -\mathbb{E}\int_0^T \Big[\big\langle g_x(t,\bar{x}(t),\bar{u}(t)),\xi(t)\big\rangle_H + \big\langle y(t),\alpha(t)\big\rangle_H + \big\langle Y(t),\beta(t)\big\rangle_H\Big]dt\\
&= \mathbb{E}\int_0^T \big\langle \mathbb{H}_x(t,\bar{x}(t),\bar{u}(t),y(t),Y(t)),\xi(t)\big\rangle_H dt\\
&\quad -\mathbb{E}\int_0^T \Big(\big\langle y(t), a(t,x(t),u(t)) - a(t,\bar{x}(t),\bar{u}(t))\big\rangle_H\\
&\qquad\qquad + \big\langle Y(t), b(t,x(t),u(t)) - b(t,\bar{x}(t),\bar{u}(t))\big\rangle_H\Big)dt\\
&\geq \mathbb{E}\int_0^T \Big(\mathbb{H}(t,x(t),u(t),y(t),Y(t)) - \mathbb{H}(t,\bar{x}(t),\bar{u}(t),y(t),Y(t))\Big)dt\\
&\quad -\mathbb{E}\int_0^T \Big(\big\langle y(t), a(t,x(t),u(t)) - a(t,\bar{x}(t),\bar{u}(t))\big\rangle_H\\
&\qquad\qquad + \big\langle Y(t), b(t,x(t),u(t)) - b(t,\bar{x}(t),\bar{u}(t))\big\rangle_H\Big)dt\\
&= -\mathbb{E}\int_0^T \Big(g(t,x(t),u(t)) - g(t,\bar{x}(t),\bar{u}(t))\Big)dt.
\end{aligned}$$

On the other hand, the convexity of h implies that

$$\mathbb{E}\big\langle h_x(\bar{x}(T)),\xi(T)\big\rangle_H \leq \mathbb{E}h(x(T)) - \mathbb{E}h(\bar{x}(T)).$$

Combining the above two, we arrive at

$$\mathcal{J}(\bar{u}(\cdot)) \leq \mathcal{J}(u(\cdot)). \tag{12.247}$$

Since $u(\cdot) \in \mathcal{U}[0,T]$ is arbitrary, (12.247) means that $\bar{u}(\cdot)$ is an optimal control for Problem (OP). This completes the proof of Theorem 12.21. □

12.7 Second Order Necessary Condition for Optimal Controls

In Section 12.5, we have given in Theorem 12.17 a Pontryagin-type maximum for Problem (OP), which is a first order necessary for optimal controls. Similarly to Calculus of Variations (or even the elementary Calculus), in addition to first-order necessary conditions, some second order necessary conditions are needed to distinguish optimal controls from the candidates which satisfy the first order necessary conditions, especially when the controls are singular, i.e., when these controls satisfy the first order necessary conditions trivially. In this section, we shall give an integral-type second order necessary condition for optimal controls.

For simplicity, as in Section 12.6, we assume that H is a real separable Hilbert space, the control region U is a convex subset of the separable Hilbert space $\widetilde{H}$ appeared in (S4), and $L^p_{\mathcal{F}_T}(\Omega)$ $(p \geq 1)$ is separable.

We need to introduce the following assumption[8].

(S6) *For a.e.* $(t,\omega) \in (0,T)\times\Omega$, *the functions* $a(t,\cdot,\cdot) : H\times U \to H$, $b(t,\cdot,\cdot) : H \times U \to \mathcal{L}_2^0$, *and* $g(t,\cdot,\cdot) : H \times U \to \mathbb{R}$ *are* C^2. *Moreover, for a.e.* $(t,\omega) \in (0,T) \times \Omega$ *and any* $(x,u) \in H \times U$,

$$\begin{cases} |a_{xu}(t,x,u)|_{\mathcal{L}(H,\widetilde{H};H)} + |b_{xu}(t,x,u)|_{\mathcal{L}(H,\widetilde{H};\mathcal{L}_2^0)} \\ \quad +|a_{uu}(t,x,u)|_{\mathcal{L}(\widetilde{H},\widetilde{H};H)} + |b_{uu}(t,x,u)|_{\mathcal{L}(\widetilde{H},\widetilde{H};\mathcal{L}_2^0)} \\ \leq \mathcal{C}_L, \\ |g_{xu}(t,x,u)|_{\mathcal{L}(H;\widetilde{H})} + |g_{uu}(t,x,u)|_{\mathcal{L}(\widetilde{H})} \leq \mathcal{C}_L. \end{cases}$$

Let $(\bar{x}(\cdot),\bar{u}(\cdot))$ be an optimal pair for Problem (OP). Let us introduce some notations. For $\psi = a, b, g$, put

$$\begin{cases} \psi_1(t) = \psi_x(t,\bar{x}(t),\bar{u}(t)), & \psi_2(t) = \psi_u(t,\bar{x}(t),\bar{u}(t)), \\ \psi_{11}(t) = \psi_{xx}(t,\bar{x}(t),\bar{u}(t)), & \psi_{22}(t) = \psi_{uu}(t,\bar{x}(t),\bar{u}(t)), \\ \psi_{12}(t) = \psi_{xu}(t,\bar{x}(t),\bar{u}(t)). \end{cases} \tag{12.248}$$

Let $(y(\cdot),Y(\cdot))$ be the transposition solution of the equation (12.7). Put

$$\begin{cases} \mathbb{H}_1(t) = \mathbb{H}_x(t,\bar{x}(t),\bar{u}(t),y(t),Y(t)), \\ \mathbb{H}_2(t) = \mathbb{H}_u(t,\bar{x}(t),\bar{u}(t),y(t),Y(t)), \\ \mathbb{H}_{11}(t) = \mathbb{H}_{xx}(t,\bar{x}(t),\bar{u}(t),y(t),Y(t)), \\ \mathbb{H}_{12}(t) = \mathbb{H}_{xu}(t,\bar{x}(t),\bar{u}(t),y(t),Y(t)), \\ \mathbb{H}_{22}(t) = \mathbb{H}_{uu}(t,\bar{x}(t),\bar{u}(t),y(t),Y(t)). \end{cases}$$

Let $(P(\cdot),Q^{(\cdot)},\widehat{Q}^{(\cdot)})$ be the relaxed transposition solution to the equation (12.8) in which $F(\cdot)$, $J(\cdot)$, $K(\cdot)$ and P_T are given by (12.165), i.e.,

$$\begin{cases} P_T = -h_{xx}(\bar{x}(T)), & J(t) = a_1(t), \\ K(t) = b_1(t), & F(t) = -\mathbb{H}_{11}(t). \end{cases}$$

For any admissible control $u(\cdot) \in \mathcal{U}[0,T]$, let $x_1(\cdot)$ be the corresponding solution to the following equation:

$$\begin{cases} dx_1(t) = \big(Ax_1(t) + a_1(t)x_1(t) + a_2(t)\delta u(t)\big)dt \\ \qquad\qquad +\big(b_1(t)x_1(t) + b_2(t)\delta u(t)\big)dW(t) \quad \text{in } (0,T], \\ x_1(0) = 0, \end{cases} \tag{12.249}$$

[8] See Subsection 2.11.1 for the notations $\mathcal{L}(H,\widetilde{H};H)$, $\mathcal{L}(H,\widetilde{H};\mathcal{L}_2^0)$, $\mathcal{L}(\widetilde{H},\widetilde{H};H)$ and $\mathcal{L}(\widetilde{H},\widetilde{H};\mathcal{L}_2^0)$.

where $\delta u(\cdot) = u(\cdot) - \bar{u}(\cdot)$.

The desired integral-type second order necessary condition for optimal controls for Problem (OP) is stated as follows.

Theorem 12.22. *Let (S1)–(S6) hold, and let $\bar{u}(\cdot) \in L^4_{\mathbb{F}}(0,T;\widetilde{H}) \cap \mathcal{U}[0,T]$ be an optimal control for Problem (OP). Then, for any $u(\cdot) \in L^4_{\mathbb{F}}(0,T;\widetilde{H}) \cap \mathcal{U}[0,T]$ with*

$$\mathbb{E}\int_0^T \big\langle \mathbb{H}_2(t), u(t) - \bar{u}(t)\big\rangle_{\widetilde{H}} dt = 0, \tag{12.250}$$

the following second order necessary condition holds:

$$\begin{aligned}
&\mathbb{E}\int_0^T \Big[\big\langle \mathbb{H}_{22}(t)\big(u(t)-\bar{u}(t)\big), u(t)-\bar{u}(t)\big\rangle_{\widetilde{H}} \\
&\qquad + \big\langle b_2(t)^* P(t) b_2(t)\big(u(t)-\bar{u}(t)\big), u(t)-\bar{u}(t)\big\rangle_{\widetilde{H}}\Big] dt \\
&\quad + 2\mathbb{E}\int_0^T \big\langle \big(\mathbb{H}_{12}(t) + a_2(t)^* P(t) + b_2(t)^* P(t) b_1(t)\big) x_1(t), u(t)-\bar{u}(t)\big\rangle_{\widetilde{H}} dt \\
&\quad + \mathbb{E}\int_0^T \big\langle \big(\widehat{Q}^{(0)} + Q^{(0)}\big)\big(0, a_2(t)\big(u(t)-\bar{u}(t)\big), b_2(t)\big(u(t)-\bar{u}(t)\big)\big), \\
&\qquad\qquad b_2(t)\big(u(t)-\bar{u}(t)\big)\big)\big\rangle_{\mathcal{L}_2^0} dt \\
&\le 0.
\end{aligned} \tag{12.251}$$

Proof: The proof will be divided into four steps.

Step 1. In this step, we introduce some notations.

Obviously, $\delta u(\cdot) = u(\cdot) - \bar{u}(\cdot) \in L^4_{\mathbb{F}}(0,T;\widetilde{H})$. Let $\delta x(\cdot) = x(\cdot) - \bar{x}(\cdot)$. Since U is convex, we see that, for any $\varepsilon \in [0,1]$,

$$u^\varepsilon(\cdot) = \bar{u}(\cdot) + \varepsilon\delta u(\cdot) = (1-\varepsilon)\bar{u}(\cdot) + \varepsilon u(\cdot) \in L^4_{\mathbb{F}}(0,T;\widetilde{H}) \cap \mathcal{U}[0,T].$$

Denote by $x^\varepsilon(\cdot)$ the corresponding solution to (12.2) with $u(\cdot)$ replaced by $u^\varepsilon(\cdot)$. Let $\delta x^\varepsilon(\cdot) = x^\varepsilon(\cdot) - \bar{x}(\cdot)$ and for $\psi = a, b, g$, put

$$\left\{\begin{aligned}
\tilde{\psi}^\varepsilon_{11}(t) &\triangleq \int_0^1 (1-\theta)\psi_{xx}(t, \bar{x}(t) + \theta\delta x^\varepsilon(t), \bar{u}(t) + \theta\varepsilon\delta u(t))d\theta, \\
\tilde{\psi}^\varepsilon_{12}(t) &\triangleq \int_0^1 (1-\theta)\psi_{xu}(t, \bar{x}(t) + \theta\delta x^\varepsilon(t), \bar{u}(t) + \theta\varepsilon\delta u(t))d\theta, \\
\tilde{\psi}^\varepsilon_{22}(t) &\triangleq \int_0^1 (1-\theta)\psi_{uu}(t, \bar{x}(t) + \theta\delta x^\varepsilon(t), \bar{u}(t) + \theta\varepsilon\delta u(t))d\theta.
\end{aligned}\right.$$

Also, we define

$$\tilde{h}_{xx}^{\varepsilon}(T) \stackrel{\Delta}{=} \int_0^1 (1-\theta) h_{xx}(\bar{x}(T) + \theta \delta x^{\varepsilon}(T)) d\theta.$$

Obviously, δx^{ε} solves the following stochastic evolution equation:

$$\begin{cases} d\delta x^{\varepsilon} = \Big[A\delta x^{\varepsilon} + a_1(t)\delta x^{\varepsilon} + \varepsilon a_2(t)\delta u + \tilde{a}_{11}^{\varepsilon}(t)\big(\delta x^{\varepsilon}, \delta x^{\varepsilon}\big) \\ \qquad +2\varepsilon \tilde{a}_{12}^{\varepsilon}(t)\big(\delta x^{\varepsilon}, \delta u\big) + \varepsilon^2 \tilde{a}_{22}^{\varepsilon}(t)\big(\delta u, \delta u\big)\Big] dt \\ \qquad + \Big[b_1(t)\delta x^{\varepsilon} + \varepsilon b_2(t)\delta u + \tilde{b}_{11}^{\varepsilon}(t)\big(\delta x^{\varepsilon}, \delta x^{\varepsilon}\big) \\ \qquad +2\varepsilon \tilde{b}_{12}^{\varepsilon}(t)\big(\delta x^{\varepsilon}, \delta u\big) + \varepsilon^2 \tilde{b}_{22}^{\varepsilon}(t)\big(\delta u, \delta u\big)\Big] dW(t) \quad \text{in } (0,T], \\ \delta x^{\varepsilon}(0) = 0. \end{cases} \tag{12.252}$$

Consider the following linearized stochastic evolution equation:

$$\begin{cases} dx_2 = \big[Ax_2 + a_1(t)x_2 + a_{11}(t)(x_1, x_1) + 2a_{12}(t)(x_1, \delta u) \\ \qquad +a_{22}(t)(\delta u, \delta u)\big] dt + \big[b_1(t)x_2 + b_{11}(t)(x_1, x_1) \\ \qquad +2b_{12}(t)(x_1, \delta u) + b_{22}(t)(\delta u, \delta u)\big] dW(t) \qquad \text{in } (0,T], \\ x_2(0) = 0. \end{cases} \tag{12.253}$$

Similar to Step 2 in the proof of Theorem 12.4, we can prove the following estimates (for some constant $\mathcal{C}$, independent of ε):

$$\begin{cases} |\delta x^{\varepsilon}|_{C_{\mathbb{F}}([0,T];L^2(\Omega;H))} \le \mathcal{C}\varepsilon |\delta u|_{L^2_{\mathbb{F}}(0,T;\widetilde{H})}, \\ |x_1|_{C_{\mathbb{F}}([0,T];L^2(\Omega;H))} \le \mathcal{C} |\delta u|_{L^2_{\mathbb{F}}(0,T;\widetilde{H})}, \\ |x_2|_{C_{\mathbb{F}}([0,T];L^2(\Omega;H))} \le \mathcal{C} |\delta u|^2_{L^4_{\mathbb{F}}(0,T;\widetilde{H})}, \\ |\delta x - x_1|_{C_{\mathbb{F}}([0,T];L^2(\Omega;H))} \le \mathcal{C} |\delta u|^2_{L^4_{\mathbb{F}}(0,T;\widetilde{H})}. \end{cases} \tag{12.254}$$

Step 2. We claim that there exists a subsequence $\{\varepsilon_n\}_{n=1}^{\infty} \subset (0,1]$ such that $\lim_{n\to\infty} \varepsilon_n = 0$ and

$$\Big|\delta x^{\varepsilon_n} - \varepsilon_n x_1 - \frac{\varepsilon_n^2}{2} x_2\Big|_{C_{\mathbb{F}}([0,T];L^2(\Omega;H))} = o(\varepsilon_n^2), \qquad \text{as } n \to \infty. \tag{12.255}$$

To show this, we write $r_2^{\varepsilon}(\cdot) = \varepsilon^{-2}\Big(\delta x^{\varepsilon}(\cdot) - \varepsilon x_1(\cdot) - \dfrac{\varepsilon^2}{2} x_2(\cdot)\Big)$. Then, $r_2^{\varepsilon}(\cdot)$ fulfills

$$
\begin{cases}
dr_2^\varepsilon = \Big\{ Ar_2^\varepsilon + a_1(t)r_2^\varepsilon + \Big[a_{11}^\varepsilon(t)\Big(\frac{\delta x^\varepsilon}{\varepsilon}, \frac{\delta x^\varepsilon}{\varepsilon}\Big) - \frac{1}{2}a_{11}(t)(x_1, x_1)\Big] \\
\qquad + \Big[2\tilde a_{12}^\varepsilon(t)\Big(\frac{\delta x^\varepsilon}{\varepsilon}, \delta u\Big) - a_{12}(t)(x_1, \delta u)\Big] \\
\qquad + \Big(\tilde a_{22}^\varepsilon(t) - \frac{1}{2}a_{22}(t)\Big)(\delta u, \delta u)\Big\}dt \\
\qquad + \Big\{ b_1(t)r_2^\varepsilon + \Big[\tilde b_{11}^\varepsilon(t)\Big(\frac{\delta x^\varepsilon}{\varepsilon}, \frac{\delta x^\varepsilon}{\varepsilon}\Big) - \frac{1}{2}b_{11}(t)(x_1, x_1)\Big] \\
\qquad + \Big[2\tilde b_{12}^\varepsilon(t)\Big(\frac{\delta x^\varepsilon}{\varepsilon}, \delta u\Big) - b_{12}(t)(x_1, \delta u)\Big] \\
\qquad + \Big(\tilde b_{22}^\varepsilon(t) - \frac{1}{2}b_{22}(t)\Big)(\delta u, \delta u)\Big\}dW(t) \quad \text{in } (0, T], \\
r_2^\varepsilon(0) = 0.
\end{cases}
\tag{12.256}
$$

Put

$$
\begin{aligned}
\Psi_{1,\varepsilon}(t) = & \Big[\tilde a_{11}^\varepsilon(t)\Big(\frac{\delta x^\varepsilon(t)}{\varepsilon}, \frac{\delta x^\varepsilon(t)}{\varepsilon}\Big) - \frac{1}{2}a_{11}(t)(x_1(t), x_1(t))\Big] \\
& + \Big[2\tilde a_{12}^\varepsilon(t)\Big(\frac{\delta x^\varepsilon(t)}{\varepsilon}, \delta u(t)\Big) - a_{12}(t)(x_1(t), \delta u(t))\Big] \\
& + \Big(\tilde a_{22}^\varepsilon(t) - \frac{1}{2}a_{22}(t)\Big)(\delta u(t), \delta u(t))
\end{aligned}
$$

and

$$
\begin{aligned}
\Psi_{2,\varepsilon}(t) = & \Big[\tilde b_{11}^\varepsilon(t)\Big(\frac{\delta x^\varepsilon(t)}{\varepsilon}, \frac{\delta x^\varepsilon(t)}{\varepsilon}\Big) - \frac{1}{2}b_{11}(t)(x_1(t), x_1(t))\Big] \\
& + \Big[2\tilde b_{12}^\varepsilon(t)\Big(\frac{\delta x^\varepsilon(t)}{\varepsilon}, \delta u(t)\Big) - b_{12}(t)(x_1(t), \delta u(t))\Big] \\
& + \Big(\tilde b_{22}^\varepsilon(t) - \frac{1}{2}b_{22}(t)\Big)(\delta u(t), \delta u(t)).
\end{aligned}
$$

We have that

$$
\begin{aligned}
\mathbb{E}\big|r_2^\varepsilon(t)\big|_H^2 = & \mathbb{E}\Big|\int_0^t S(t-s)a_1(s)r_2^\varepsilon(s)ds + \int_0^t S(t-s)b_1(s)r_2^\varepsilon(s)dW(s) \\
& + \int_0^t S(t-s)\Psi_{1,\varepsilon}(s)ds + \int_0^t S(t-s)\Psi_{2,\varepsilon}(s)dW(s)\Big|_H^2 \\
\le & \,\mathcal{C}\Big(\mathbb{E}\int_0^t |r_2^\varepsilon(s)|_H^2 ds + \mathbb{E}\int_0^t |\Psi_{1,\varepsilon}(s)|_H^2 ds + \mathbb{E}\int_0^t |\Psi_{2,\varepsilon}(s)|_{\mathcal{L}_2^0}^2 ds\Big).
\end{aligned}
\tag{12.257}
$$

By (12.254), there exists a subsequence $\{\varepsilon_n\}_{n=1}^\infty \subset (0, T]$ such that $\varepsilon_n \to 0$ and $x^{\varepsilon_n}(\cdot) \to \bar x(\cdot)$ (in H) a.e. in $[0, T] \times \Omega$, as $n \to \infty$. Then, by (12.254), the condition (S4) and Lebesgue's dominated convergence theorem, we deduce that

$$
\begin{aligned}
&\lim_{n\to\infty}\mathbb{E}\int_0^t |\Psi_{1,\varepsilon_n}(t)|_H^2 dt\\
&\le \lim_{n\to\infty}\mathbb{E}\int_0^T \Big|\Big[\tilde a_{11}^{\varepsilon_n}(t)\Big(\frac{\delta x^{\varepsilon_n}(t)}{\varepsilon_n},\frac{\delta x^{\varepsilon_n}(t)}{\varepsilon_n}\Big)-\frac{1}{2}a_{11}(t)(x_1(t),x_1(t))\Big]\\
&\qquad +\Big[2\tilde a_{12}^{\varepsilon_n}(t)\Big(\frac{\delta x^{\varepsilon_n}(t)}{\varepsilon_n},\delta u(t)\Big)-a_{12}(t)(x_1(t),\delta u(t))\Big]\\
&\qquad +\Big(\tilde a_{22}^{\varepsilon_n}(t)-\frac{1}{2}a_{22}(t)\Big)(\delta u(t),\delta u(t))\Big|_H^2 dt\\
&\le \mathcal{C}\lim_{n\to\infty}\mathbb{E}\int_0^T\Big[\Big|\tilde a_{11}^{\varepsilon_n}(t)\Big(\frac{\delta x_n^{\varepsilon}(t)}{\varepsilon_n},\frac{\delta x_n^{\varepsilon}(t)}{\varepsilon_n}\Big)-\tilde a_{11}^{\varepsilon_n}(t)(x_1(t),x_1(t))\Big|_H^2 \qquad (12.258)\\
&\qquad +\Big|\tilde a_{11}^{\varepsilon_n}(t)-\frac{1}{2}a_{11}(t)\Big|^2_{\mathcal{L}(H,H;H)}|x_1(t)|_H^4\\
&\qquad +2\Big|\tilde a_{12}^{\varepsilon_n}(t)\Big(\frac{\delta x^{\varepsilon_n}(t)}{\varepsilon_n},\delta u(t)\Big)-\tilde a_{12}^{\varepsilon_n}(t)(x_1(t),\delta u(t))\Big|_H^2\\
&\qquad +\big|2\tilde a_{12}^{\varepsilon_n}(t)-a_{12}(t)\big|^2_{\mathcal{L}(H,\widetilde H;H)}|x_1(t)|_H^2|\delta u(t)|_{\widetilde H}^2\\
&\qquad +\Big|\tilde a_{22}^{\varepsilon_n}(t)-\frac{1}{2}a_{22}(t)\Big|^2_{\mathcal{L}(\widetilde H,\widetilde H;H)}|\delta u(t)|_{\widetilde H}^4\Big]dt\\
&=0.
\end{aligned}
$$

Similarly,

$$
\lim_{n\to\infty}\mathbb{E}\int_0^t |\Psi_{2,\varepsilon_n}(t)|^2_{\mathcal{L}_2^0}dt=0. \qquad (12.259)
$$

Combining (12.257), (12.258) with (12.259) and using Gronwall's inequality, we obtain (12.255).

Step 3. By Taylor's formula, we see that

$$
\begin{aligned}
&g(t,x^\varepsilon(t),u^\varepsilon(t))-g(t,\bar x(t),\bar u(t))\\
&=\big\langle g_1(t),\delta x^\varepsilon(t)\big\rangle_H+\varepsilon\big\langle g_2(t),\delta u(t)\big\rangle_{\widetilde H}+\big\langle \tilde g_{11}^\varepsilon(t)\delta x^\varepsilon(t),\delta x^\varepsilon(t)\big\rangle_H \qquad (12.260)\\
&\quad +2\varepsilon\big\langle \tilde g_{12}^\varepsilon(t)\delta x^\varepsilon(t),\delta u(t)\big\rangle_{\widetilde H}+\varepsilon^2\big\langle \tilde g_{22}^\varepsilon(t)\delta u(t),\delta u(t)\big\rangle_{\widetilde H}
\end{aligned}
$$

and

$$
\begin{aligned}
&h(x^\varepsilon(T))-h(\bar x(T))\\
&=\big\langle h_x(\bar x(T)),\delta x^\varepsilon(T)\big\rangle_H+\big\langle \tilde h_{xx}^\varepsilon(T)\delta x^\varepsilon(T),\delta x^\varepsilon(T)\big\rangle_H.
\end{aligned} \qquad (12.261)
$$

Using a similar argument as the proof of (12.255), we can show that, for the subsequence $\{\varepsilon_n\}_{n=1}^\infty$ such that $x^{\varepsilon_n}(\cdot)\to\bar x(\cdot)$ (in H) a.e. in $[0,T]\times\Omega$ (as $n\to\infty$), the following equalities hold:

$$
\lim_{n\to\infty}\frac{1}{\varepsilon_n^2}\mathbb{E}\int_0^T\Big(\big\langle \tilde g_{11}^{\varepsilon_n}(t)\delta x^{\varepsilon_n}(t),\delta x^{\varepsilon_n}(t)\big\rangle_H-\frac{\varepsilon_n^2}{2}\big\langle g_{11}(t)x_1(t),x_1(t)\big\rangle_H\Big)dt=0,
$$

$$\lim_{n\to\infty}\frac{1}{\varepsilon_n^2}\mathbb{E}\int_0^T\Big(2\big\langle \tilde g_{12}^{\varepsilon_n}(t)\delta x^{\varepsilon_n}(t),\varepsilon_n\delta u(t)\big\rangle_{\widetilde H}-\varepsilon_n^2\big\langle g_{12}(t)x_1(t),\delta u(t)\big\rangle_{\widetilde H}\Big)dt=0,$$

$$\lim_{n\to\infty}\mathbb{E}\int_0^T\Big(\big\langle \tilde g_{22}^{\varepsilon_n}(t)\delta u(t),\delta u(t)\big\rangle_{\widetilde H}-\frac{1}{2}\big\langle g_{22}(t)\delta u(t),\delta u(t)\big\rangle_{\widetilde H}\Big)dt=0$$

and

$$\lim_{n\to\infty}\frac{1}{\varepsilon_n^2}\mathbb{E}\Big(\big\langle \tilde h_{xx}^{\varepsilon_n}(\bar x(T))\delta x^{\varepsilon_n}(T),\delta x^{\varepsilon_n}(T)\big\rangle_H-\frac{\varepsilon_n^2}{2}\big\langle h_{xx}(\bar x(T))x_1(T),x_1(T)\big\rangle_H\Big)$$
$$=0.$$

These, together with (12.255), imply that

$$\begin{aligned}
&\mathcal{J}(u^{\varepsilon_n})-\mathcal{J}(\bar u)\\
&=\mathbb{E}\int_0^T\Big[\varepsilon_n\big\langle g_1(t),x_1(t)\big\rangle_H+\frac{\varepsilon_n^2}{2}\big\langle g_1(t),x_2(t)\big\rangle_H+\varepsilon_n\big\langle g_2(t),\delta u(t)\big\rangle_{\widetilde H}\\
&\qquad+\frac{\varepsilon_n^2}{2}\Big(\big\langle g_{11}(t)x_1(t),x_1(t)\big\rangle_H+2\big\langle g_{12}(t)x_1(t),\delta u(t)\big\rangle_{\widetilde H}\\
&\qquad+\big\langle g_{22}(t)\delta u(t),\delta u(t)\big\rangle_{\widetilde H}\Big)\Big]dt\\
&\quad+\mathbb{E}\Big(\varepsilon_n\big\langle h_x(\bar x(T)),x_1(T)\big\rangle_H+\frac{\varepsilon_n^2}{2}\big\langle h_x(\bar x(T)),x_2(T)\big\rangle_H\\
&\qquad+\frac{\varepsilon_n^2}{2}\big\langle h_{xx}(\bar x(T))x_1(T),x_1(T)\big\rangle_H\Big)+o(\varepsilon_n^2),\quad\text{as } n\to\infty.
\end{aligned}\tag{12.262}$$

Step 4. By the definition of transposition solution to (10.4), we have that

$$\begin{aligned}
&\mathbb{E}\big\langle h_x(\bar x(T)),y(T)\big\rangle_H\\
&=-\mathbb{E}\int_0^T\Big(\big\langle y(t),a_2(t)\delta u(t)\big\rangle_H+\big\langle Y(t),b_2(t)\delta u(t)\big\rangle_{\mathcal{L}_2^0}\\
&\qquad+\big\langle g_1(t),x_1(t)\big\rangle_H\Big)dt
\end{aligned}\tag{12.263}$$

and

$$\begin{aligned}
&\mathbb{E}\big\langle h_x(\bar x(T)),x_2(T)\big\rangle_H\\
&=-\mathbb{E}\int_0^T\Big(\big\langle y(t),a_{11}(t)(x_1(t),x_1(t))\big\rangle_H+2\big\langle y(t),a_{12}(t)(x_1(t),\delta u(t))\big\rangle_H\\
&\qquad+\big\langle y(t),a_{22}(t)(\delta u(t),\delta u(t))\big\rangle_H+\big\langle Y(t),b_{11}(t)(x_1(t),x_1(t))\big\rangle_{\mathcal{L}_2^0}\\
&\qquad+2\big\langle Y(t),b_{12}(t)(x_1(t),\delta u(t))\big\rangle_{\mathcal{L}_2^0}+\big\langle Y(t),b_{22}(t)(\delta u(t),\delta u(t))\big\rangle_{\mathcal{L}_2^0}\\
&\qquad+\big\langle g_1(t),x_2(t)\big\rangle_H\Big)dt.
\end{aligned}\tag{12.264}$$

On the other hand, by the definition of relaxed transposition solution to (12.8), we obtain that

$$
\begin{aligned}
&\mathbb{E}\big\langle h_{xx}(\bar{x}(T))x_1(T), x_1(T)\big\rangle_H\\
&= -\mathbb{E}\int_0^T \Big(\big\langle P(t)x_1(t), a_2(t)\delta u(t)\big\rangle_H + \big\langle x_1(t), P(t)a_2(t)\delta u(t)\big\rangle_H\\
&\quad +\big\langle P(t)b_1(t)x_1(t), b_2(t)\delta u(t)\big\rangle_{\mathcal{L}_2^0} + \big\langle b_1(t)x_1(t), P(t)b_2(t)\delta u(t)\big\rangle_{\mathcal{L}_2^0}\\
&\quad +\big\langle P(t)b_2(t)\delta u(t), b_2(t)\delta u(t)\big\rangle_{\mathcal{L}_2^0} + \big\langle \widehat{Q}^{(0)}(0, a_2(t)\delta u, b_2(t)\delta u)(t), b_2(t)\delta u(t)\big\rangle_{\mathcal{L}_2^0}\\
&\quad +\big\langle Q^{(0)}(0, a_2(t)\delta u, b_2(t)\delta u)(t), b_2(t)\delta u(t)\big\rangle_{\mathcal{L}_2^0} - \big\langle \mathbb{H}_{11}(t)x_1(t), x_1(t)\big\rangle_H\Big)dt.
\end{aligned}
\tag{12.265}
$$

Combining (12.262)–(12.265) with (12.250), we obtain that, for sufficiently large n,

$$
\begin{aligned}
0 &\le \frac{\mathcal{J}(u^{\varepsilon_n}) - \mathcal{J}(\bar{u})}{\varepsilon_n^2}\\
&= -\mathbb{E}\int_0^T \Big[\frac{1}{\varepsilon_n}\Big(\big\langle y(t), a_2(t)\delta u(t)\big\rangle_H + \big\langle Y(t), b_2(t)\delta u(t)\big\rangle_{\mathcal{L}_2^0} - \big\langle g_2(t), \delta u(t)\big\rangle_{\widetilde{H}}\Big)\\
&\qquad +\frac{1}{2}\Big(\big\langle y(t), a_{22}(t)\big(\delta u(t), \delta u(t)\big)\big\rangle_H + \big\langle Y(t), b_{22}(t)\big(\delta u(t), \delta u(t)\big)\big\rangle_{\mathcal{L}_2^0}\\
&\qquad -\big\langle g_{22}(t)\delta u(t), \delta u(t)\big\rangle_{\widetilde{H}}\Big) + \frac{1}{2}\big\langle P(t)b_2(t)\delta u(t), b_2(t)\delta u(t)\big\rangle_{\mathcal{L}_2^0}\\
&\qquad +\Big(-\big\langle g_{12}(t)x_1(t), \delta u(t)\big\rangle_{\widetilde{H}} + \big\langle y(t), a_{12}(t)(x_1, \delta u)\big\rangle_H\\
&\qquad +\big\langle Y(t), b_{12}(t)(x_1, \delta u)\big\rangle_{\mathcal{L}_2^0} + \big\langle a_2(t)^* P(t)x_1(t), \delta u(t)\big\rangle_{\widetilde{H}}\\
&\qquad +\big\langle b_2(t)^* P(t)b_1(t)x_1(t), \delta u(t)\big\rangle_{\widetilde{H}}\\
&\qquad +\frac{1}{2}\big\langle \widehat{Q}^{(0)}(0, a_2(t)\delta u, b_2(t)\delta u)(t) + Q^{(0)}(0, a_2(t)\delta u, b_2(t)\delta u)(t),\\
&\qquad\qquad b_2(t)\delta u(t)\big\rangle_{\mathcal{L}_2^0}\Big)\Big]dt + o(1)\\
&= -\mathbb{E}\int_0^T \Big[\frac{1}{\varepsilon_n}\big\langle \mathbb{H}_2(t), \delta u(t)\big\rangle_{\widetilde{H}} + \frac{1}{2}\big\langle \mathbb{H}_{22}(t)\delta u(t), \delta u(t)\big\rangle_{\widetilde{H}}\\
&\qquad +\frac{1}{2}\big\langle b_2(t)^* P(t)b_2(t)\delta u(t), \delta u(t)\big\rangle_{\widetilde{H}}\Big]dt\\
&\quad -\mathbb{E}\int_0^T \big\langle \big[\mathbb{H}_{12}(t) + a_2(t)^* P(t) + b_2(t)^* P(t)b_1(t)\big]x_1(t), \delta u(t)\big\rangle_{\widetilde{H}}dt\\
&\quad -\frac{1}{2}\mathbb{E}\int_0^T \big\langle \widehat{Q}^{(0)}(0, a_2(t)\delta u, b_2(t)\delta u)(t) + Q^{(0)}(0, a_2(t)\delta u, b_2(t)\delta u)(t),\\
&\qquad\qquad b_2(t)\delta u(t)\big\rangle_{\mathcal{L}_2^0}dt + o(1)\\
&= -\mathbb{E}\int_0^T \Big(\frac{1}{2}\big\langle \mathbb{H}_{22}(t)\delta u(t), \delta u(t)\big\rangle_{\widetilde{H}} + \frac{1}{2}\big\langle b_2(t)^* P(t)b_2(t)\delta u(t), \delta u(t)\big\rangle_{\widetilde{H}}\Big)dt
\end{aligned}
$$

$$
\begin{aligned}
&-\mathbb{E}\int_0^T \big\langle \big(\mathbb{H}_{12}(t) + a_2(t)^* P(t) + b_2(t)^* P(t) b_1(t)\big) x_1(t), \delta u(t)\big\rangle_{\widetilde{H}} dt \\
&-\frac{1}{2}\mathbb{E}\int_0^T \big\langle \widehat{Q}^{(0)}(0, a_2(t)\delta u, b_2(t)\delta u)(t) + Q^{(0)}(0, a_2(t)\delta u, b_2(t)\delta u)(t), \\
&\qquad\qquad b_2(t)\delta u(t)\big\rangle_{\mathcal{L}_2^0} dt + o(1).
\end{aligned}
$$

Finally, letting $n \to \infty$ in the above inequality, we then obtain the desired second order necessary condition (12.251). This completes the proof of Theorem 12.22. □

Remark 12.23. It is easy to see that, the correction part $(Q^{(\cdot)}, \widehat{Q}^{(\cdot)})$ of the relaxed transposition solution $(P(\cdot), (Q^{(\cdot)}, \widehat{Q}^{(\cdot)})$ to the equation (12.8) appears explicitly in the second order necessary condition (12.251) in Theorem 12.22.

12.8 Notes and Comments

The main body of this chapter is an improved version of the results in [242, 244], except that Sections 12.2 and 12.7 are based respectively on [241] and [240], while the sufficient optimality condition in Section 12.6 can be regarded as an infinite version of the similar result (for stochastic controls in finite dimensions) in [371, 393].

Pontryagin's maximum principle for deterministic optimal control problems in finite dimensions is one of the three milestones of modern optimal control theory. Since the classical works in [34], the theory of maximum principle for controlled ordinary differential equations has been studied extensively and many different versions of maximum principle were established for more complex systems in various aspects, such as control systems governed by ordinary differential equations on manifolds, partial differential equations, stochastic differential equations in particular. We refer the readers to [1, 202, 371] for Pontryagin-type maximum principle for these three kinds of control systems, respectively.

Soon after the work [34], the maximum principle has been extended to the controlled stochastic differential equations. To our best knowledge, the first paper concerning the stochastic maximum principle is [180]. At that early time, people usually assumed that the diffusion term is nondegenerate, and hence one can apply the Girsanov transformation to obtain the desired necessary conditions for stochastic optimal controls (e.g., [139]). A breakthrough was made in [31], in which the backward stochastic differential equation was introduced to study the maximum principle for stochastic control systems, and via the nondegenerate condition on the diffusion terms was dropped. Further development in this respect can be found in [25, 33], etc.

Until 1988, all the works on maximum principle for controlled stochastic differential equations were obtained under one of the following assumptions:

- The diffusion term is independent of the controls (e.g., [25, 31]);
- The coefficients of the diffusion term is nondegenerate (e.g., [139]);
- The control region is convex (e.g., [25]).

In [273], a new maximum principle was obtained for general controlled stochastic differential equations without the above mentioned assumptions, and it was found that the corresponding result in the general case differs essentially from its deterministic counterpart.

Naturally, one expects to establish the Pontryagin-type maximum for control systems governed by stochastic evolution equations. A pioneer work along this line is [26]. Further progresses are available in the literature [148, 314, 324, 392] and so on. Similar to the study of controlled stochastic differential equations, all of the existing published works before 2012 on the necessary conditions for optimal controls of infinite dimensional stochastic evolution equations addressed only the case that one of the following conditions holds:

- The diffusion term does NOT depend on the control variable, or the control region is convex;
- For a.e. $(t,\omega)\in(0,T)\times\Omega$ and any $(x,u,k_1,k_2)\in H\times U\times H\times\mathcal{L}_2^0$, the data $\langle k_1,a_{xx}(t,x,u)\rangle_H$, $\langle k_2,b_{xx}(t,x,u)\rangle_{\mathcal{L}_2^0}$, $g_{xx}(t,x,u)$ and $h_{xx}(x)$ are Hilbert-Schmidt valued operators.

The main difficulty in the infinite dimensional setting is how to handle the well-posedness of (12.8). There were several works ([79, 112, 242], of which the arXiv versions were all posted in 2012) addressed to overcome the above difficulty. The main idea is to solve the equation (12.8) in a weak sense. In [79, 112], a partial well-posedness result for (12.8) was established, that is, only P was obtained by Riesz Representation Theorem (See [113, 114] for further related works). In [242], a complete well-posedness result for (12.8) was derived by means of the stochastic transposition method. Results in [242] have been further improved in [244, 245]. Although the correction part Q (in the equation (12.8)) is not need in establishing the Pontryagin-type maximum principle for controlled stochastic evolution equations, as mentioned in Remark 12.23 (for a weak version of Q), it plays an important role in the study of the second order necessary conditions for optimal controls (See [103, 104, 240] for more details).

As far as we know, the sufficiency of the Pontryagin-type maximum principle was first studied in [33] for controlled stochastic differential equations. In [393], it was proved that the maximum condition in [273] is also sufficient, provided that some convexity conditions on the control regions and the cost functionals are imposed. Some further development in this respect can be found in [269] and the references cited therein. To our best knowledge, there exist only very few published works on the sufficiency of the Pontryagin-type maximum principle for controlled stochastic evolution equations in infinite dimensions (e.g., [194, 268]).

There are many open problems related to the topic of this chapter. We shall list below some of them which, in our opinion, are particularly interesting and/or important:

1) Well-posedness of (12.8) in the sense of transposition solution

In Theorem 12.9, we established the well-posedness of the equation (12.8) in the sense of relaxed transposition solution. It would be quite interesting (and also very important for some problems) to prove that this equation is also well-posed in the sense of transposition solution (See Definition 12.6). Nevertheless, so far the well-posedness of (12.8) in the sense of transposition solution is know only for some very special case (Se Theorem 12.14). As far as we know, it is a challenging (unsolved) problem to prove the existence of transposition solution to (12.8) (See Theorem 12.7 for the uniqueness result), even for the following special case:

$$\begin{cases} dP = -A^*Pdt - PAdt + Fdt + QdW(t) & \text{in } [0,T), \\ P(T) = P_T, \end{cases} \tag{12.266}$$

where $F \in L^1_{\mathbb{F}}(0,T;L^2(\Omega;\mathcal{L}(H)))$, $P_T \in L^2_{\mathcal{F}_T}(\Omega;\mathcal{L}(H))$. The same can be said for (12.266) even for concrete problems, say when $A = -\Delta$, the Laplacian with the usual homogenous Dirichlet boundary condition (even for the case of one space dimension!).

2) Optimal control problems with endpoint/state constraints

In this chapter, we have not considered the endpoint/state constraints to the control systems since this topic is quite difficult even for the case $H = \mathbb{R}^n$. For some special constraints, such as $\mathbb{E}x(T) > 0$, one can use the Ekeland principle to establish a similar Pontryagin type maximum principle with nontrivial Lagrange multipliers. However, for the general case, one does need some further condition to obtain nontrivial results in this repect, as shown in [202] (even for the deterministic optimal control problems in infinite dimensions). In the study of the aforementioned deterministic optimal control problems, people introduce the so-called finite codimension condition to guarantee the nontriviality of the Lagrange multiplier (e.g., [202, 220]). There is some attempt to generalize this condition to the stochastic framework (e.g., [221]) but so far the results are still not so satisfactory. Another way is to use some tools from the set-valued analysis, as developed in the recent paper [103].

3) Higher order necessary conditions for optimal controls

As mentioned at the very beginning of Section 12.7, similarly to Calculus of Variations or even the elementary Calculus, in addition to the first-order necessary conditions, sometimes higher order necessary conditions should be established to distinguish optimal controls from the candidates which satisfy the first order necessary conditions trivially. In this chapter, we only gave one result of integral-type second order necessary condition for optimal controls of Problem (OP). Some further results in this direction can be found in [103, 104,

240]. However, the theory on higher order necessary conditions for stochastic optimal control problems is far from satisfactory. Indeed, as far as we know, so far only second order necessary conditions for stochastic optimal controls were under consideration, and pointwise second order necessary conditions (which are more useful in applications than the integral ones) were obtained only under very strong assumptions, even for the stochastic control problems in finite dimensions.

4) Necessary conditions for optimal controls of partially observed control systems

In this chapter, we actually assumed that the system state is completely observed. For many practical problems, it often happens that only a part of the state can be observed, and also there might exist noises in the observation. A typical example is as follows:

Let W_1 and W_2 be two independent cylindrical Brownian motions on the filtered probability space $(\Omega, \mathcal{F}, \mathbb{F}, \mathbb{P})$, with values in two Hilbert spaces V_1 and V_2, respectively. For a given control $u(\cdot) \in \mathcal{U}[0,T]$ (See (12.1) for the definition of $\mathcal{U}[0,T]$), let us consider the following stochastic evolution equation:

$$\begin{cases} dx(t) = \big(Ax(t) + a(t,x(t),u(t))\big)dt \\ \qquad\qquad + \displaystyle\sum_{j=1}^{2} b_j(t,x(t),u(t))dW_j(t) \qquad \text{in } (0,T], \\ x(0) = x_0, \end{cases} \tag{12.267}$$

and the following (partial) observation equation:

$$\begin{cases} dy(t) = c(t,x(t),y(t),u(t))dt + f(t,x(t),y(t),u(t))dW_2(t) \ \text{ in } [0,T], \\ y(0) = y_0, \end{cases} \tag{12.268}$$

with suitably given initial values x_0 and y_0, and functions a, b_1, b_2, c and f.

For any $t \in [0,T]$, write

$$\mathcal{Y}_t \overset{\Delta}{=} \sigma\big\{y(s) \bigm| s \in [0,t]\big\}.$$

We call the above control $u(\cdot)$ *an admissible control for the above partially observed system* if it happens to be $\{\mathcal{Y}_t\}_{t\in[0,T]}$-adapted. Denote by $\mathcal{V}[0,T]$ the class of all admissible controls for the above partially observed system.

People then hope to find a control $\bar{u}(\cdot) \in \mathcal{V}[0,T]$, which minimizes the following cost functional:

$$\mathcal{J}(u(\cdot)) = \mathbb{E}\Big[\int_0^T g(t,x(t),y(t),u(t))dt + h(x(T),y(T))\Big], \quad u(\cdot) \in \mathcal{V}[0,T],$$

in which the functions g and h are suitably given.

Motivated by the works for the same problems but in finite dimensions (e.g., [27]), one may follow the following two main steps to study the above optimal control problems of partially observable systems governed by stochastic evolution equations:

- Step 1. *Computing the filtering of the state*;
- Step 2. *Solving a complete information optimal control problem driven by the filtering and obtaining the optimal control.*

Nevertheless, compared with the partially observed systems in finite dimensions (e.g., [27, 339, 363]), there exist some essential difficulties to achieve the above two steps. For example, in Step 1, one has to solve a stochastic partial differential equation with infinitely many variables to compute the filtering of the state. Unfortunately, at this moment the well-posedness of such an equation is far from well-understood.

13

Linear Quadratic Optimal Control Problems

In this chapter, we are concerned with linear quadratic optimal control problems (LQ problems for short) for stochastic evolution equations, in which the diffusion terms depend on the control variables and the coefficients are stochastic. In such a general setting, one has to introduce suitable operator-valued backward stochastic evolution equations (to characterize the optimal controls in the form of Pontryagin-type maximum principle or in the feedback forms), served as the second order adjoint equations or the Riccati type equations. As in the previous chapter, it is very difficult to show the existence of solutions to these equations. We shall use the stochastic transposition method to overcome this difficulty.

Compared with the general optimal control problems, the main novelty of LQ problems is that the optimal controls can be represented in the feedback forms, which keep the corresponding control strategies robust w.r.t. (small) perturbation/disturbance. This is particularly important in many practical applications for which perturbation/disturbance are usually unavoidable. Nevertheless, it is actually very difficult to find feedback controls for general control problems. So far, the most successful attempt in this respect is that for various LQ problems in the deterministic setting.

13.1 Formulation of the Problem

Throughout this chapter, $T > 0$, $(\Omega, \mathcal{F}, \mathbf{F}, \mathbb{P})$ (with $\mathbf{F} \stackrel{\Delta}{=} \{\mathcal{F}_t\}_{t\in[0,T]}$) is a fixed filtered probability space satisfying the usual condition, and we denote by $\mathbb{F}$ the progressive σ-field w.r.t. $\mathbf{F}$; H, V and U are three separable Hilbert spaces. We denote by I the identity operator on H and by I_n the identity matrix on $\mathbb{R}^n$ ($n \in \mathbb{R}$), and $\mathcal{L}_2^0 \stackrel{\Delta}{=} \mathcal{L}_2(V; H)$. Unless otherwise specified, $W(\cdot)$ is a V-valued, Q-Brownian motion or cylindrical Brownian motion but we only consider the case of cylindrical Brownian motion. Let A be an unbounded linear operator (with domain $D(A)$ on H), which generates a C_0-semigroup $\{S(t)\}_{t\geq 0}$. Denote by A^* the dual operator of A.

Q. Lü and X. Zhang, *Mathematical Control Theory for Stochastic Partial Differential Equations*, Probability Theory and Stochastic Modelling 101, https://doi.org/10.1007/978-3-030-82331-3_13

Denote by $\mathcal{S}(H)$ the Banach space of all self-adjoint (linear bounded) operators on H. For $M, N \in \mathcal{S}(H)$, we use the notation $M \geq N$ (*resp.* $M > N$) to indicate that $M - N$ is positive semi-definite (*resp.* positive definite). For any $\mathcal{S}(H)$-valued stochastic process F on $[0,T]$, we write $F \geq 0$ (*resp.* $F > 0$, $F \gg 0$) if $F(t,\omega) \geq 0$ (*resp.* $F(t,\omega) > 0$, $F(t,\omega) \geq \delta I$ for some $\delta > 0$) for a.e. $(t,\omega) \in [0,T] \times \Omega$. One can define $F \ll 0$ and so on in a similar way.

For any given $\eta \in H$, we consider a control system governed by the following linear stochastic evolution equation:

$$\begin{cases} dx(t) = \big(Ax(t) + B(t)u(t)\big)dt \\ \qquad\qquad\quad +\big(C(t)x(t) + D(t)u(t)\big)dW(t) \ \text{ in } (0,T], \\ x(0) = \eta, \end{cases} \tag{13.1}$$

where $B(\cdot) \in L^\infty_{\mathbb{F}}(0,T;\mathcal{L}(U;H))$, $C(\cdot) \in L^\infty_{\mathbb{F}}(0,T;\mathcal{L}(H;\mathcal{L}_2^0))$ and $D(\cdot) \in L^\infty_{\mathbb{F}}(0,T;\mathcal{L}(U;\mathcal{L}_2^0))$. In (13.1), $u(\cdot)(\in L^2_{\mathbb{F}}(0,T;U))$ is the control variable, $x(\cdot)$ is the state variable. In view of Theorem 3.13, for any $u \in L^2_{\mathbb{F}}(0,T;U)$, the system (13.1) admits a unique (mild) solution $x(\cdot) \equiv x(\cdot;\eta,u) \in C_{\mathbb{F}}([0,T];L^2(\Omega;H))$ such that

$$|x(\cdot)|_{C_{\mathbb{F}}([0,T];L^2(\Omega;H))} \leq \mathcal{C}\big(|\eta|_H + |u(\cdot)|_{L^2_{\mathbb{F}}(0,T;U)}\big). \tag{13.2}$$

Associated with the system (13.1), we consider the following quadratic cost functional

$$\begin{aligned} \mathcal{J}(\eta;u(\cdot)) = \frac{1}{2}\mathbb{E}\Big[\int_0^T \big(\langle M(t)x(t),x(t)\rangle_H + \langle R(t)u(t),u(t)\rangle_U\big)dt \\ +\langle Gx(T),x(T)\rangle_H\Big], \end{aligned} \tag{13.3}$$

where $M(\cdot) \in L^\infty_{\mathbb{F}}(0,T;\mathcal{S}(H))$, $R(\cdot) \in L^\infty_{\mathbb{F}}(0,T;\mathcal{S}(U))$ and $G \in L^\infty_{\mathcal{F}_T}(\Omega;\mathcal{S}(H))$. In what follows, to simplify the notations, the time variable t will be suppressed in $B(\cdot)$, $C(\cdot)$, $D(\cdot)$, $M(\cdot)$ and $R(\cdot)$, and therefore we shall simply write them as B, C, D, M and R, respectively (if there is no confusion).

Remark 13.1. In this chapter, we assume that $B \in L^\infty_{\mathbb{F}}(0,T;\mathcal{L}(U;H))$ and $D \in L^\infty_{\mathbb{F}}(0,T;\mathcal{L}(U;\mathcal{L}_2^0))$. Thus, our results can only be applied to controlled stochastic partial differential equations with internal controls. To study systems with boundary/pointwise controls, one needs to introduce some further assumptions, such as the semigroup $\{S(t)\}_{t\in[0,T]}$ has some smoothing effect. Under these conditions, many results of this chapter can be generalized to systems with unbounded control operators. More details will be discussed in Section 13.10.

The main concern of this chapter is to study the following *stochastic linear quadratic optimal control problem*:

Problem (SLQ). For each $\eta \in H$, find a $\bar{u}(\cdot) \in L^2_{\mathbb{F}}(0,T;U)$ such that

$$\mathcal{J}(\eta;\bar{u}(\cdot)) = \inf_{u(\cdot)\,\in\, L^2_{\mathbb{F}}(0,T;U)} \mathcal{J}(\eta;u(\cdot)). \tag{13.4}$$

Remark 13.2. Unlike Chapter 12, we denote by $\mathcal{J}(\eta; u)$ rather than $\mathcal{J}(u)$ for the cost functional. The reason is that we want to emphasize the relationship among the initial data, the optimal controls and the optimal feedback operators.

Let us begin with the following notions.

Definition 13.3. 1) *Problem (SLQ) is called a standard LQ problem if*

$$M(\cdot) \geq 0, \quad R \gg 0, \quad G \geq 0;$$

2) *Problem (SLQ) is said to be finite at $\eta \in H$ if the right hand side of (13.4) is finite;*

3) *Problem (SLQ) is said to be (uniquely) solvable at $\eta \in H$ if there exists a (unique) control $\bar{u}(\cdot) \in L^2_{\mathbb{F}}(0,T;U)$ satisfying (13.4). In this case, $\bar{u}(\cdot)$ is called an (the) optimal control, the corresponding $\bar{x}(\cdot)$ and $(\bar{x}(\cdot), \bar{u}(\cdot))$ are called an (the) optimal state and an (the) optimal pair, respectively;*

4) *Problem (SLQ) is said to be finite (resp. (uniquely) solvable) if for any $\eta \in H$, it is finite (resp. (uniquely) solvable) at η.*

Let us briefly review the deterministic LQ problems in finite dimensions (See [193, 370] for more materials).

For any $m, n \in \mathbb{N}$ and $\eta \in \mathbb{R}^n$, we consider the following control system:

$$\begin{cases} \dot{x}(t) = A(t)x(t) + B(t)u(t) \ \text{ in } [0,T], \\ x(0) = \eta, \end{cases} \tag{13.5}$$

where $A(\cdot) \in L^\infty(0,T;\mathbb{R}^{n\times n})$, $B(\cdot) \in L^\infty(0,T;\mathbb{R}^{n\times m})$ and the control $u(\cdot) \in L^2(0,T;\mathbb{R}^m)$. The cost functional is

$$\begin{aligned} \mathcal{J}(\eta; u(\cdot)) = \frac{1}{2}\Big[\int_0^T \big(\langle M(t)x(t), x(t)\rangle_{\mathbb{R}^n} + \langle R(t)u(t), u(t)\rangle_{\mathbb{R}^m}\big)dt \\ + \langle Gx(T), x(T)\rangle_{\mathbb{R}^n}\Big], \end{aligned} \tag{13.6}$$

with $M(\cdot) \in L^\infty(0,T;\mathcal{S}(\mathbb{R}^n))$, $R(\cdot) \in L^\infty(0,T;\mathcal{S}(\mathbb{R}^m))$ and $G \in \mathcal{S}(\mathbb{R}^n)$. Similarly to Problem (SLQ), we consider the following (deterministic) *linear quadratic optimal control problem:*

Problem (DLQ). For each $\eta \in \mathbb{R}^n$, find a control $\bar{u}(\cdot) \in L^2(0,T;\mathbb{R}^m)$, such that

$$\mathcal{J}(\eta; \bar{u}(\cdot)) = \inf_{u(\cdot) \in L^2(0,T;\mathbb{R}^m)} \mathcal{J}(\eta; u(\cdot)).$$

A special case of the above problem has already been introduced in Section 1.3. Similarly to that for Problem (SLQ), one can define the so-called standard Problem (DLQ), and the finiteness and (unique) solvability of Problem (DLQ) (and hence we omit the details).

The following result (e.g., [370, p. 227]) shows that $R \geq 0$ is a necessary condition for Problem (DLQ) to be finite.

Proposition 13.4. *If Problem (DLQ) is finite, then* $R \geq 0$.

Remark 13.5. Surprisingly, it was found in [55] that in the stochastic setting $R \geq 0$ is NOT a necessary condition anymore for Problem (SLQ) to be finite even for the case of finite dimensions (See also Example 13.16).

The following example shows that Problem (DLQ) may not be finite (needless to say solvable) even if $R \gg 0$ (but Q and G are indefinite).

Example 13.6. For $m = n = 1$, let us consider the control system

$$\begin{cases} \dfrac{dx}{dt} = u & \text{in } [0, T], \\ x(0) = \eta \in \mathbb{R}, \end{cases} \tag{13.7}$$

with the cost functional

$$\mathcal{J}(\eta; u(\cdot)) = \frac{1}{2}\Big[\int_0^T u(t)^2 dt - \frac{2}{T}x(T)^2\Big]. \tag{13.8}$$

Let $\lambda \in \mathbb{R}$ and take $u_\lambda(t) \equiv \lambda$ for $t \in [0, T]$. Then

$$\mathcal{J}(\eta; u_\lambda(\cdot)) = \frac{1}{2}\lambda^2 T - \frac{1}{T}\big(\lambda T + \eta\big)^2 \to -\infty \text{ as } \lambda \to \infty.$$

Thus, the corresponding LQ problem is not finite.

Because of Example 13.6, in order that Problem (DLQ) is solvable, one needs more assumptions besides $R \gg 0$. For any $u(\cdot) \in L^2(0, T; \mathbb{R}^m)$, define two operators $L : L^2(0, T; \mathbb{R}^m) \to L^2(0, T; \mathbb{R}^n)$ and $L_T : L^2(0, T; \mathbb{R}^m) \to \mathbb{R}^n$ as follows:

$$\big(Lu(\cdot)\big)(t) \equiv x(t; 0, u(\cdot)), \quad L_T u \equiv x(T; 0, u(\cdot)),$$

where $x(\cdot; 0, u(\cdot))$ solves the equation (13.5) with $\eta = 0$. Further, define

$$N = L^* M L + L_T^* G L_T + R.$$

The following result (e.g., [370, p. 229]) characterizes the unique solvability of Problem (DLQ) under suitable assumptions:

Theorem 13.7. Suppose that $N \geq 0$ and $R(\cdot) \gg 0$. Then, Problem (DLQ) is uniquely solvable if and only if the following *coupled* forward-backward system

$$\begin{cases} \dot{\bar{x}}(t) = A(t)\bar{x}(t) - B(t)R^{-1}(t)B^\top(t)\bar{y}(t), & t \in [0, T], \\ \dot{\bar{y}}(t) = -A(t)^\top \bar{y}(t) + M(t)\bar{x}(t), \ t \in [0, T], \\ \bar{x}(0) = \eta, \qquad \bar{y}(T) = -G\bar{x}(T) \end{cases} \tag{13.9}$$

admits a unique solution $(\bar{x}(\cdot), \bar{y}(\cdot))$. In this case,

$$\bar{u}(t) = R^{-1}(t)B^\top(t)\bar{y}(t), \qquad t \in [0, T], \tag{13.10}$$

is the optimal control.

One task of this chapter is to establish a characterization on the unique solvability of Problem (SLQ) similarly to that in Theorem 13.7 for Problem (DLQ). This will be done in Corollary 13.33 (in Section 13.5).

Note that the optimal control given by (13.10) is not in a feedback form. In Control Theory, another fundamental issue is to find feedback controls, which are particularly important in practical applications. Indeed, the main advantage of feedback controls is that they keep the corresponding control strategy robust w.r.t. small perturbation/disturbance, which are usually unavoidable in realistic background.

In order to find a feedback form of $\bar{u}(\cdot)$, formally, we assume that, for some $\mathbb{R}^{n\times n}$-valued function $P(\cdot)$ (to be determined later),

$$\bar{y}(t) = -P(t)\bar{x}(t), \qquad t \in [0,T]. \tag{13.11}$$

Combining this with (13.10), we obtain an optimal feedback control as follows:

$$\bar{u}(t) = -R^{-1}B^\top P(t)\bar{x}(t), \qquad t \in [0,T]. \tag{13.12}$$

In order to find the above $P(\cdot)$, proceeding as that in [370, pp. 230–231], differentiating (13.11) w.r.t. t, using (13.9) and (13.12), we obtain that

$$\begin{aligned} A(t)^\top P\bar{x} + M\bar{x} &= \frac{d\bar{y}}{dt} = -\frac{dP}{dt}\bar{x} - P\big(A(t)\bar{x} + B\bar{u}\big) \\ &= -\frac{dP}{dt}\bar{x} - P\big(A(t)\bar{x} - BR^{-1}B^\top P\bar{x}\big), \end{aligned}$$

which yields

$$0 = \Big(\frac{dP}{dt} + PA(t) + A(t)^\top P - PBR^{-1}B^\top P + M\Big)\bar{x}.$$

Thus, if we choose $P(\cdot)$ to solve the following *Riccati equation* (associated with Problem (DLQ)):

$$\begin{cases} \dfrac{dP}{dt} + PA(t) + A(t)^\top P - PBR^{-1}B^\top P + M = 0 \ \text{ in } [0,T], \\ P(T) = G, \end{cases} \tag{13.13}$$

then (13.11) and (13.12) hold. Clearly, (13.13) is a matrix-valued nonlinear ordinary differential equation.

The following result (e.g., [370, p. 234]) shows the equivalence between the unique solvability of Problem (DLQ) and the global solvability of the equation (13.13) under suitable assumptions:

Theorem 13.8. *Suppose that $R \gg 0$. Then, for any $\eta \in \mathbb{R}^n$, Problem (DLQ) is uniquely solvable if and only if the Riccati equation* (13.13) *admits a global solution $P(\cdot) \in C([0,T];\mathcal{S}(\mathbb{R}^n))$. In this case, the optimal feedback control is given by* (13.12), *and*

$$\inf_{u(\cdot)\,\in\, L^2(0,T;\mathbb{R}^m)} \mathcal{J}(\eta; u(\cdot)) = \frac{1}{2}\langle P(0)\eta, \eta\rangle_{\mathbb{R}^n}. \tag{13.14}$$

It is easy to show that any standard LQ problem (even in the stochastic setting) is uniquely solvable. Hence, thanks to Theorem 13.8, one can find a unique optimal feedback control for any standard Problem (DLQ). Nevertheless, things are completely different in the stochastic situation. Indeed, as shown in [236, Example 6.2] (See also Example 13.22 in Section 13.3), a standard Problem (SLQ) (even in finite dimensions, i.e., both H and U are Euclidean spaces) does NOT need to have feedback controls. This is a significant difference between stochastic LQ problems and their deterministic counterparts. Indeed, Theorem 13.8 means that, under the assumption $R \gg 0$, one can always find the desired feedback control through the corresponding Riccati equation whenever Problem (DLQ) is uniquely solvable.

Because of the difference mentioned above, in the present stochastic setting it is quite natural to ask such a question: *Is it possible to study directly the existence of optimal feedback controls (rather than the solvability) for Problem (SLQ)?* Clearly, from the viewpoint of applications, it is more desirable to study the existence of feedback controls for Problem (SLQ) than its solvability.

Another task of this chapter is to give an affirmative answer to the above question under sharp assumptions on the coefficients appearing in (13.1) and (13.3). For this purpose, let us introduce below more notions[1].

Definition 13.9. *Any $\Theta(\cdot) \in \Upsilon_2(H;U)$ is called an admissible feedback operator for Problem (SLQ).*

Remark 13.10. In order to keep the feedback controls to be robust w.r.t. small perturbations, we should choose the feedback operators so that they are bounded linear operators from $L^2_{\mathbb{F}}(\Omega; C([0,T];H))$ to $L^2_{\mathbb{F}}(0,T;H)$. This is the reason to introduce $\Upsilon_2(H;U)$.

Now, motivated by [236, 307, 308], we introduced the following notion of optimal feedback operator for Problem (SLQ):

Definition 13.11. *An admissible feedback operator $\Theta(\cdot) \in \Upsilon_2(H;U)$ is called an optimal feedback operator for Problem (SLQ) if*

$$\mathcal{J}(\eta; \Theta(\cdot)\bar{x}(\cdot)) \le \mathcal{J}(\eta; u(\cdot)), \qquad \forall\, (\eta, u(\cdot)) \in H \times L^2_{\mathbb{F}}(0,T;U), \tag{13.15}$$

where $\bar{x}(\cdot) = \bar{x}(\cdot\,; \eta, \Theta(\cdot)\bar{x}(\cdot))$ solves

$$\begin{cases} d\bar{x} = \big(A\bar{x} + B\Theta\bar{x}\big)dt + \big(C\bar{x} + D\Theta\bar{x}\big)dW(t) \quad \text{in } (0,T], \\ \bar{x}(0) = \eta. \end{cases} \tag{13.16}$$

Clearly, for a fixed $\eta \in H$, the inequality (13.15) implies that the control $\bar{u}(\cdot) \equiv \Theta(\cdot)\bar{x}(\cdot) \in L^2_{\mathbb{F}}(0,T;U)$ is optimal for Problem (SLQ). Therefore, for Problem (SLQ), the existence of an optimal feedback operator implies the existence of an optimal control for any initial state $\eta \in H$, but not vice versa.

[1] Recall (3.50) for the definition of $\Upsilon_2(H;U)$.

In the study of Problem (DLQ), people introduced the Riccati equation (13.13) to construct the desired feedback controls. Stimulated by this and especially by the pioneer work [32] (for stochastic LQ problems in finite dimensions), in order to study the optimal feedback operator for Problem (SLQ), we introduce the following *operator-valued backward stochastic Riccati equation*:

$$\begin{cases} dP = -\big(PA + A^*P + \Lambda C + C^*\Lambda + C^*PC + M \\ \qquad\quad -L^*K^{-1}L\big)dt + \Lambda dW(t) & \text{in } [0,T), \\ P(T) = G, \end{cases} \tag{13.17}$$

where

$$K = R + D^*PD, \qquad L = B^*P + D^*(PC + \Lambda). \tag{13.18}$$

Note that, the presence of the control variable in the diffusion term of (13.1) and the infinite dimensional setting make Problem (SLQ) (and in particular the study of (13.17)) significantly different from LQ problems for ordinary differential equations or stochastic differential equations. To illustrate this, we present in the next two sections some basic results for LQ problems for the latter two cases, respectively. Then, we go back to the study of Problem (SLQ). One can see the difference caused by the infinite dimensional setting.

The rest of this chapter is organized as follows. In Sections 13.2 and 13.3, we revisit some important aspects in the LQ problems for ordinary differential equations and stochastic differential equations, respectively. In Section 13.4, we give the finiteness and solvability of LQ problems for stochastic evolution equations. In Section 13.5, we establish Pontryagin-type maximum principle for optimal controls. In Section 13.6 optimal feedback operator and the stochastic Riccati equation are introduced. In Section 13.7, the optimal feedback operator is represented through the solution to the stochastic Riccati equation. Section 13.8 is devoted to the solvability of the stochastic Riccati equation. In Section 13.9, we present some examples for LQ problems of stochastic partial differential equations which satisfies the assumptions in this chapter. Finally, some remarks and open problems are given in Section 13.10.

13.2 Optimal Feedback for Deterministic LQ Problem in Finite Dimensions

In this section, we focus on the relationship between the existence of optimal feedback operator for Problem (DLQ) (introduced in the last section) and the global solvability of the Riccati equation (13.13). All contents in this section are classical. We refer the readers to [193, 371] for a comprehensive introduction on deterministic LQ problems in finite dimensions.

The following concept is actually a special case of that in Definition 13.11.

Definition 13.12. *We call $\Theta(\cdot) \in L^2(0,T;\mathbb{R}^{m\times n})$ an optimal feedback operator for Problem (DLQ) if*

$$\mathcal{J}(\eta;\Theta(\cdot)\bar{x}(\cdot)) \leq \mathcal{J}(\eta;u(\cdot)), \quad \forall\,(\eta,u(\cdot)) \in \mathbb{R}^n \times L^2(0,T;\mathbb{R}^m), \tag{13.19}$$

where $\bar{x}(\cdot)$ solves the following equation:

$$\begin{cases} \dfrac{d\bar{x}}{dt} = (A(t) + B\Theta)\bar{x} \quad in\ [0,T], \\ \bar{x}(0) = \eta. \end{cases} \tag{13.20}$$

We have the following result:

Theorem 13.13. *Assume $R \gg 0$. If the Riccati equation* (13.13) *admits a global solution $P(\cdot) \in C([0,T];\mathcal{S}(\mathbb{R}^n))$, then Problem (DLQ) has an optimal feedback operator in the following form*

$$\Theta(\cdot) = -R^{-1}(\cdot)B^\top(\cdot)P(\cdot), \tag{13.21}$$

and (13.14) *holds.*

Proof: Let $\bar{x}(\cdot)$ satisfy

$$\begin{cases} \dfrac{d\bar{x}}{dt} = \big(A(t) - BR^{-1}B^\top P\big)\bar{x} \ \text{ in } [0,T], \\ \bar{x}(0) = \eta. \end{cases} \tag{13.22}$$

Define $\bar{u}(\cdot)$ by (13.12). Then $\bar{x}(\cdot)$ is the corresponding state.

For any $u(\cdot) \in L^2(0,T;\mathbb{R}^m)$, let $x(\cdot) = x(\cdot;\eta,u(\cdot))$. By (13.13), we have

$$\begin{aligned} &\langle P(T)x(T),x(T)\rangle_{\mathbb{R}^n} - \langle P(0)\eta,\eta\rangle_{\mathbb{R}^n} \\ &= \int_0^T \big[\langle (PBR^{-1}B^\top P - M)x(t),x(t)\rangle_{\mathbb{R}^n} + 2\,\langle B^\top P(t)x(t),u(t)\rangle_{\mathbb{R}^m}\big]dt. \end{aligned}$$

Consequently,

$$\begin{aligned} &\mathcal{J}(\eta;u(\cdot)) - \frac{1}{2}\langle P(0)\eta,\eta\rangle_{\mathbb{R}^n} \\ &= \frac{1}{2}\int_0^T \Big(\langle Ru,u\rangle_{\mathbb{R}^m} + \langle PBR^{-1}B^\top Px,x\rangle_{\mathbb{R}^n} + 2\,\langle B^\top Px,u\rangle_{\mathbb{R}^m}\Big)dt \qquad (13.23) \\ &= \frac{1}{2}\int_0^T \langle R\big(u + R^{-1}B^\top Px\big), u + R^{-1}B^\top Px\rangle^2_{\mathbb{R}^m}\,dt. \end{aligned}$$

This implies

$$\mathcal{J}(\eta,\bar{u}(\cdot)) = \frac{1}{2}\langle P(0)\eta,\eta\rangle_{\mathbb{R}^n} \leq \mathcal{J}(\eta;u(\cdot)),$$

which concludes that $\bar{u}(\cdot)$ is an optimal control, $\Theta(\cdot)$ given by (13.21) is an optimal feedback operator, and (13.14) holds. This completes the proof of Theorem 13.13. $\square$

Since (13.13) is locally Lipschitz in the unknown $P(\cdot)$, it is locally solvable, that is, there exists $s < T$ such that (13.13) admits a solution on $[s, T]$. Nevertheless, due to the quadratic term, the global well-posedness is not guaranteed, unless there are some additional conditions. The following result, together with Theorem 13.13, gives an equivalence between the solvability of the Riccati equation (13.13) and the existence of an optimal feedback operator for Problem (DLQ).

Theorem 13.14. *Assume $R \gg 0$. If there exists an optimal feedback operator for Problem (DLQ), then the Riccati equation* (13.13) *admits a unique, global solution $P(\cdot) \in C([0,T]; \mathcal{S}(\mathbb{R}^n))$.*

Proof: The uniqueness of the solution follows from the boundedness of the coefficients along with the Gronwall's inequality. Consequently, we only need to show the existence of the solution.

Let Θ be an optimal feedback operator for Problem (DLQ). By (1.41) and (1.43), the following *de-coupled* forward-backward (ordinary differential) system:

$$\begin{cases} \dfrac{d\bar{x}}{dt} = A(t)\bar{x} + B\Theta\bar{x} & \text{in } [0,T], \\ \dfrac{dy}{dt} = -A(t)^\top y + M\bar{x} & \text{in } [0,T], \\ \bar{x}(0) = \eta, \quad y(T) = -G\bar{x}(T), \end{cases} \tag{13.24}$$

admits a (unique) solution $(\bar{x}(\cdot), y(\cdot))^\top \in C([0,T]; \mathbb{R}^{2n})$, and

$$\bar{u}(t) = R^{-1}B^\top y(t) = \Theta\bar{x}(t), \qquad t \in [0,T] \tag{13.25}$$

gives an optimal control, where $\bar{x}(\cdot)$ is the corresponding optimal state.

Define $\mathbb{R}^{n\times n}$-valued functions $X(\cdot)$ and $Y(\cdot)$ by

$$\begin{cases} X(t)\eta \overset{\Delta}{=} \bar{x}(t;\eta), \\ Y(t)\eta \overset{\Delta}{=} y(t;\eta), \end{cases} \qquad \forall\ (t,\eta) \in [0,T] \times \mathbb{R}^n. \tag{13.26}$$

Clearly, $(X(\cdot), Y(\cdot))$ solves the following matrix-valued forward-backward ordinary differential equation

$$\begin{cases} \dfrac{dX}{dt} = A(t)X - B\Theta X & \text{in } [0,T], \\ \dfrac{dY}{dt} = -A(t)^\top Y + MX & \text{in } [0,T], \\ X(0) = I_n, \quad Y(T) = -GX(T). \end{cases} \tag{13.27}$$

Consider the following ordinary differential equation:

$$\begin{cases} \frac{d\widetilde{X}}{dt} = \big(- A(t) - B\Theta \big)^\top \widetilde{X} \text{ in } [0,T], \\ \widetilde{X}(0) = I_n. \end{cases} \tag{13.28}$$

Then,

$$\frac{d\Big(X(t)\widetilde{X}(t)^\top\Big)}{dt} = 0, \qquad \forall t \in [0,T].$$

This implies that $X(t)^{-1} = \widetilde{X}(t)^\top$ for all $t \in [0,T]$.

From (13.25), it follows that

$$R^{-1}B^\top Y(t) = \Theta X(t), \qquad t \in [0,T]. \tag{13.29}$$

Let

$$P(t) \stackrel{\Delta}{=} -Y(t)X(t)^{-1}, \qquad t \in [0,T]. \tag{13.30}$$

From (13.27), (13.28) and (13.29), we have

$$\begin{aligned} \frac{dP}{dt} &= -\big(MX - A(t)^\top Y\big)X^{-1} + Y\big(X^{-1}A(t) - X^{-1}BR^{-1}B^\top YX^{-1}\big) \\ &= -M - A(t)^\top P - PA(t) + PBR^{-1}B^\top P. \end{aligned}$$

This, together with $P(T) = G$, proves that $P(\cdot)$ is the solution to (13.13). This completes the proof of Theorem 13.14. □

Remark 13.15. Compared with the coupled forward-backward system (13.9), due to the assumption on the existence of optimal feedback operator for Problem (DLQ), we obtain the de-coupled forward-backward system (13.24), which plays a key role in the proof of Theorem 13.14. As we shall see in the sequel, this point is also the key for us to characterize the optimal feedback operator for Problem (SLQ).

13.3 Optimal Feedback for Stochastic LQ Problem in Finite Dimensions

In this section, we recall the main results on the optimal feedback for stochastic LQ problems in finite dimensions. To this end, we set $H = \mathbb{R}^n$, $V = \mathbb{R}$ and $U = \mathbb{R}^m$. The main content of this section are taken from [236, 371].

13.3.1 Differences Between Deterministic and Stochastic LQ Problems in Finite Dimensions

For any $\eta \in \mathbb{R}^n$, we consider the following control system

$$\begin{cases} dx(t) = \big(A(t)x + Bu\big)dt + \big(Cx + Du\big)dW(t) \ \text{ in } [0,T], \\ x(0) = \eta, \end{cases} \tag{13.31}$$

with the quadratic cost functional

$$\mathcal{J}(\eta; u(\cdot)) = \frac{1}{2}\mathbb{E}\Big[\int_0^T \big(\langle Mx, x\rangle_{\mathbb{R}^n} + \langle Ru, u\rangle_{\mathbb{R}^m}\big)dt + \langle Gx(T), x(T)\rangle_{\mathbb{R}^n}\Big], \tag{13.32}$$

where $u(\cdot) \in L^2_{\mathbb{F}}(0,T;\mathbb{R}^m)$ is the control variable, and

$$A(\cdot), C(\cdot) \in L^\infty_{\mathbb{F}}(0,T;\mathbb{R}^{n\times n}), \qquad B(\cdot), D(\cdot) \in L^\infty_{\mathbb{F}}(0,T;\mathbb{R}^{n\times m}),$$

$$M(\cdot) \in L^\infty_{\mathbb{F}}(0,T;\mathcal{S}(\mathbb{R}^n)), \quad R(\cdot) \in L^\infty_{\mathbb{F}}(0,T;\mathcal{S}(\mathbb{R}^m)), \quad G \in L^\infty_{\mathcal{F}_T}(\Omega;\mathcal{S}(\mathbb{R}^n)).$$

It is easy to see that, the system (13.31) admits a unique solution $x(\cdot) \in L^2_{\mathbb{F}}(\Omega; C([0,T];\mathbb{R}^n))$ and hence (13.32) is well-defined.

The following optimal control problem can be viewed as a finite dimensional version of Problem (SLQ):

Problem (FSLQ). For each $\eta \in \mathbb{R}^n$, find a $\bar{u}(\cdot) \in L^2_{\mathbb{F}}(0,T;\mathbb{R}^m)$ such that

$$\mathcal{J}\big(\eta; \bar{u}(\cdot)\big) = \inf_{u(\cdot)\in L^2_{\mathbb{F}}(0,T;\mathbb{R}^m)} \mathcal{J}\big(\eta; u(\cdot)\big).$$

Similarly to that of Problem (SLQ), one can define the so-called standard Problem (FSLQ), and the finiteness and (unique) solvability of Problem (FSLQ) (and hence we omit the details).

Let us recall that, Proposition 13.4 shows the necessity of the condition $R \gg 0$ for Problem (DLQ) to be finite. The following example, however, reveals a completely different phenomenon in the stochastic setting.

Example 13.16. For $m = n = 1$, consider the stochastic control system

$$\begin{cases} dx = udW(t) \ \text{ in } [0,T], \\ x(0) = \eta, \end{cases} \tag{13.33}$$

with the cost functional

$$\mathcal{J}(\eta; u(\cdot)) = \frac{1}{2}\mathbb{E}\Big(-\int_0^T u^2 dt + 2x(T)^2\Big). \tag{13.34}$$

Then,

$$\mathcal{J}(\eta; u(\cdot)) = -\frac{1}{2}\mathbb{E}\int_0^T u^2 dt + \mathbb{E}\int_0^T u^2 dt + \eta^2 = \frac{1}{2}\mathbb{E}\int_0^T u^2 dt + \eta^2 \geq 0.$$

Hence, this LQ problem is finite though, in this example, the corresponding $R = -1$!

A necessary condition for the optimal control is given below.

Theorem 13.17. *Let Problem (FSLQ) be solvable at $\eta \in \mathbb{R}^n$ with $(\bar{x}(\cdot), \bar{u}(\cdot))$ being an optimal pair. Then there exists a transposition solution $(\bar{y}(\cdot), \overline{Y}(\cdot))$ to*

$$\begin{cases} d\bar{y} = -\big(A(t)^\top \bar{y} + C^\top \overline{Y} - M\bar{x}\big)dt + \overline{Y}dW(t) & \text{in } [0,T], \\ \bar{y}(T) = -G\bar{x}(T), \end{cases} \tag{13.35}$$

and a transposition solution $(P(\cdot), Q(\cdot))$ to

$$\begin{cases} dP = -\big(A(t)^\top P + PA(t) + C^\top PC + QC + C^\top Q - M\big)dt \\ \qquad\quad +QdW(t) \quad \text{in } [0,T], \\ P(T) = -G, \end{cases} \tag{13.36}$$

such that

$$R(t)\bar{u}(t) - B(t)^\top \bar{y}(t) - D(t)^\top \overline{Y}(t) = 0, \quad \text{a.e. } (t,\omega) \in [0,T] \times \Omega. \tag{13.37}$$

and

$$R(t) - D(t)^\top P(t) D(t) \geq 0, \quad \text{a.e. } (t,\omega) \in [0,T] \times \Omega. \tag{13.38}$$

Theorem 13.17 can be viewed as a special case of Theorem 13.30, and hence we will not prove it here.

Remark 13.18. Compared with the deterministic case (Proposition 13.4), we see that, by (13.38), R may be negative. Namely, the necessary condition might still be satisfied even if R is negative, provided that the term $-D^\top PD$ is sufficiently positive definite.

The following concept can be viewed as a special case of that in Definition 13.11.

Definition 13.19. *Any $\Theta(\cdot) \in L^\infty_{\mathbb{F}}(\Omega; L^2(0,T;\mathbb{R}^{m\times n}))$ is called an admissible feedback operator for Problem (FSLQ). Further, it is called an optimal feedback operator for Problem (FSLQ) if*

$$\mathcal{J}(\eta; \Theta(\cdot)\bar{x}(\cdot)) \leq \mathcal{J}(\eta; u(\cdot)), \qquad \forall\, (\eta, u(\cdot)) \in \mathbb{R}^n \times \mathcal{U}[0,T], \tag{13.39}$$

where $\bar{x}(\cdot) = \bar{x}(\cdot\,; \eta, \Theta(\cdot)\bar{x}(\cdot))$ solves

$$\begin{cases} d\bar{x}(t) = \big(A(t) + B\Theta\big)\bar{x}dt + \big(C + D\Theta\big)\bar{x}dW(t) & \text{in } [0,T], \\ \bar{x}(0) = \eta. \end{cases} \tag{13.40}$$

Remark 13.20. In the above, we choose $\Theta \in L^\infty_{\mathbb{F}}(\Omega; L^2(0,T;\, \mathbb{R}^{m\times n}))$ rather than $\Theta \in L^p_{\mathbb{F}}(\Omega;\, L^2(0,T;\mathbb{R}^{n\times n}))$ for any $p \in [1,\infty)$. This is because we hope to keep the control strategy to be robust w.r.t. small perturbations. Indeed, under the latter assumption on Θ (i.e., $\Theta \in L^p_{\mathbb{F}}(\Omega;\, L^2(0,T;\mathbb{R}^{m\times n}))$ for any $p \in [1,\infty)$), if there is a measurement error $\delta x \in L^2_{\mathbb{F}}(\Omega; C([0,T];\mathbb{R}^n))$ with $|\delta x|_{L^2_{\mathbb{F}}(\Omega;C([0,T];\mathbb{R}^n))} = \varepsilon > 0$ for ε being small enough in the observation of the state, then one cannot conclude that $\Theta(\bar{x} + \delta x)$ is an admissible control.

As in the deterministic case, we shall construct the optimal feedback operator by means of a suitable Riccati equation. Let us first derive such an equation formally. For simplicity, in the rest of this section, we assume that $\mathbf{F}$ is the natural filtration generated by $W(\cdot)$.

Suppose that Problem (FSLQ) is uniquely solvable. For any $\eta \in \mathbb{R}^n$, let $(\bar{x}(\cdot), \bar{u}(\cdot))$ be the optimal pair and $(\bar{y}(\cdot), \overline{Y}(\cdot)$ the classical adapted solution to (13.35). It follows from (13.37) that

$$R\bar{u}(t) - B^\top \bar{y}(t) - D^\top \overline{Y}(t) = 0, \qquad \text{a.e. } (t,\omega) \in [0,T] \times \Omega. \tag{13.41}$$

Assume that there is an Itô process

$$dP = Fdt + \Lambda dW(t) \tag{13.42}$$

with suitable F and Λ such that

$$\bar{y}(t) = -P(t)\bar{x}(t), \qquad t \in [0,T]. \tag{13.43}$$

Then, by Itô's formula,

$$\begin{aligned} &d(P\bar{x}) \\ &= (dP)\bar{x} + P\big(A(t)\bar{x} + B\bar{u}\big)dt + P\big(C\bar{x} + D\bar{u}\big)dW(t) + \Lambda\big(C\bar{x} + D\bar{u}\big)dt \\ &= -d\bar{y} = \big(A(t)^\top \bar{y} + C^\top \overline{Y} - M\bar{x}\big)dt - \overline{Y}dW(t). \end{aligned} \tag{13.44}$$

Consequently,

$$\begin{aligned} (dP)\bar{x} = \big(&-PA(t)\bar{x} - PB\bar{u} - M\bar{x} + A(t)^\top \bar{y} + C^\top \overline{Y}\big)dt - \Lambda\big(C\bar{x} + D\bar{u}\big)dt \\ &-P\big(C\bar{x} + D\bar{u}\big)dW(t) - \overline{Y}dW(t). \end{aligned} \tag{13.45}$$

From (13.42) and (13.45), we see that

$$\Lambda\bar{x} = -P(C\bar{x} + D\bar{u}) - \overline{Y}.$$

This, together with (13.41) and (13.43), implies that

$$\begin{cases} \bar{u} = -(R + D^\top PD)^{-1}(B^\top P + D^\top PC + D^\top \Lambda)\bar{x}, \\ \overline{Y} = -\Lambda\bar{x} - PC\bar{x} + (R + D^\top PD)^{-1}(B^\top P + D^\top PC + D^\top \Lambda)\bar{x}. \end{cases} \tag{13.46}$$

By (13.45) and (13.46), we have

$$\begin{cases} dP = -\big(PA(t) + A(t)^\top P + \Lambda C + C^\top \Lambda + C^\top PC \\ \qquad\quad +M - L^\top K^{-1} L\big)dt + \Lambda dW(t) \qquad \text{in } [0,T], \\ P(T) = G, \end{cases} \tag{13.47}$$

where

$$K = R + D^\top PD, \qquad L = B^\top P + D^\top (PC + \Lambda).$$

By (13.46), we see then that, formally, the optimal feedback operator is

$$\Theta = -K^{-1}L. \tag{13.48}$$

We call (13.47) the *matrix-valued backward stochastic Riccati equation* associated with Problem (FSLQ). It is a nonlinear backward stochastic differential equation and its well-posedness is highly nontrivial.

Compared with Problem (DLQ), clearly Problem (FSLQ) is much more complicated. Both the well-posedness of (13.47) and the relationship between the solution (P, Λ) and the solvability of Problem (FSLQ) are much different from the deterministic problem, for example

1) The uniqueness of the solution to the Riccati equation (13.47) is not guaranteed;
2) The Θ given by (13.48) may not be an admissible feedback operator, i.e. it may not belong to $L^\infty_{\mathbb{F}}(\Omega; L^2(0,T;\mathbb{R}^{m\times n}))$.

Let us illustrate the above two differences by the following two examples, respectively.

Example 13.21. For $m = n = 1$, consider the control system

$$\begin{cases} dx = (A(t)x + Bu)dt + udW(t) \text{ in } [0,1], \\ x(0) = \eta, \end{cases}$$

with the cost functional

$$\mathcal{J}(\eta; u(\cdot)) = \frac{1}{2}\mathbb{E}\Big[- x(1)^2 + \int_0^1 (Mx^2 + Ru^2)dt \Big],$$

where

$$\begin{aligned} R(t) &= \Big(t - \frac{3}{2}\Big)^2 + \frac{3}{4} > 0, \qquad M(t) = -\frac{1}{R(t)}, \\ A(t) &= \frac{1}{2}\Big[\frac{(R(t)-1)^2}{R(t)^2} - 1\Big], \quad B(t) = \frac{R(t)-1}{R(t)}. \end{aligned} \tag{13.49}$$

In this case, the corresponding stochastic Riccati equation becomes an ordinary differential equation:

$$\begin{cases} \dfrac{dP}{dt} + 2A(t)P + M - \dfrac{B^2P^2}{R+P} = 0 \text{ in } [0,1], \\ P(1) = -1. \end{cases} \tag{13.50}$$

It follows from (13.49) that

$$\frac{B^2P^2}{R+P}-2A(t)P-M=\frac{(B^2-2A(t))P^2-(M+2A(t)R)P-MR}{R+P}$$
$$=\frac{P^2+2P+1}{R+P}.$$

This, together with (13.50), implies that

$$\begin{cases}\dfrac{dP}{dt}=\dfrac{P^2+2P+1}{R+P} \quad \text{in } [0,1],\\ P(1)=-1.\end{cases}\tag{13.51}$$

Clearly,

$$P_1(t)=-1, \qquad t\in[0,1],$$

and

$$P_2(t)=t-2, \qquad t\in[0,1]$$

are solutions to (13.51).

Example 13.22. Define one-dimensional stochastic processes $M(\cdot)$, $\zeta(\cdot)$ and a stopping time τ as follows:

$$\begin{cases}M(t)\overset{\Delta}{=}\displaystyle\int_0^t\frac{1}{\sqrt{T-s}}dW(s), \qquad t\in[0,T),\\ \tau\overset{\Delta}{=}\inf\big\{t\in[0,T)\big|\ |M(t)|>1\big\}\wedge T,\\ \zeta(t)\overset{\Delta}{=}\dfrac{\pi}{2\sqrt{2}\sqrt{T-t}}\chi_{[0,\tau]}(t), \qquad t\in[0,T).\end{cases}\tag{13.52}$$

By some direct computations, we have

$$\begin{aligned}\Big|\int_0^T\zeta(s)dW(s)\Big|&=\frac{\pi}{2\sqrt{2}}\Big|\int_0^\tau\frac{1}{\sqrt{T-t}}dW(t)\Big|\\&=\frac{\pi}{2\sqrt{2}}\big|M(\tau)\big|\leq\frac{\pi}{2\sqrt{2}},\end{aligned}\tag{13.53}$$

and

$$\mathbb{E}\Big[\exp\Big(\int_0^T|\zeta(t)|^2dt\Big)\Big]=\infty.\tag{13.54}$$

Consider the following backward stochastic differential equation:

$$z(t)=\int_0^T\zeta(s)dW(s)+\frac{\pi}{2\sqrt{2}}+1-\int_t^TZ(s)dW(s), \qquad t\in[0,T].$$

This equation admits a unique solution

$$z(t)=\int_0^t\zeta(s)dW(s)+\frac{\pi}{2\sqrt{2}}+1, \quad Z(t)=\zeta(t), \qquad \text{a.e. } t\in[0,T].$$

From (13.52)–(13.54), it is easy to see that

$$\begin{cases} 1 \le z(\cdot) \le \dfrac{\pi}{\sqrt{2}} + 1, \\ Z(\cdot) \notin L^\infty_{\mathbb{F}}(\Omega; L^2(0,T)). \end{cases} \tag{13.55}$$

Consider a *standard* Problem (FSLQ) with the following setting:

$$\begin{cases} m = k = 1, \quad A(\cdot) = B = C = 0, \quad D = 1, \\ M = 0, \quad R = \dfrac{1}{4} > 0, \quad G = z(T)^{-1} - \dfrac{1}{4} > 0. \end{cases} \tag{13.56}$$

For this problem, the corresponding Riccati equation reads

$$\begin{cases} dP = (R + P)^{-1} \Lambda^2 dt + \Lambda dW(t) \ \text{ in } [0,T], \\ P(T) = G, \end{cases} \tag{13.57}$$

and $\Theta(\cdot) = -(R + P(\cdot))^{-1}\Lambda(\cdot)$.

Put

$$\widetilde{P}(\cdot) \stackrel{\Delta}{=} P(\cdot) + R, \qquad \widetilde{\Lambda} \stackrel{\Delta}{=} \Lambda.$$

It follows from (13.57) that

$$\begin{cases} d\widetilde{P} = \widetilde{P}^{-1}\widetilde{\Lambda}^2 ds + \widetilde{\Lambda} dW(s) \ \text{ in } [0,T], \\ \widetilde{P}(T) = z(T)^{-1}. \end{cases} \tag{13.58}$$

Applying Itô's formula to $z(\cdot)^{-1}$, we deduce that $(\widetilde{P}(\cdot), \widetilde{\Lambda}(\cdot)) = (z(\cdot)^{-1}, -z(\cdot)^{-2}Z(\cdot))$ is the unique solution to (13.58). As a result,

$$(P(\cdot), \Lambda(\cdot)) \stackrel{\Delta}{=} (z(\cdot)^{-1} - R, -z(\cdot)^{-2}Z(\cdot))$$

is the unique solution to the Riccati equation (13.57). Moreover, $\Theta(\cdot) = -z(\cdot)^{-1}Z(\cdot)$. By (13.55), we see that $\Theta(\cdot)$ does not belong to $L^\infty_{\mathbb{F}}(\Omega; L^2(0,T))$. Hence, it is not an admissible feedback operator.

13.3.2 Characterization of Optimal Feedbacks for Stochastic LQ Problems in Finite Dimensions

From Examples 13.21 and 13.22, we see that neither the well-posedness of the matrix-valued backward stochastic Riccati equation (13.47) nor the existence of the optimal feedback operator is true even for a standard Problem (FSLQ). Nevertheless, we have the following result which gives the equivalence between the well-posedness of the equation (13.47) and the existence of optimal feedback operator for Problem (FSLQ).

Theorem 13.23. *Assume that Problem (FSLQ) is standard. Then, this problem admits an optimal feedback operator* $\Theta(\cdot) \in L^{\infty}_{\mathbb{F}}(\Omega; L^2(0,T;\mathbb{R}^{m\times n}))$ *if and only if the matrix-valued backward stochastic Riccati equation* (13.47) *admits a solution* $\big(P(\cdot), \Lambda(\cdot)\big) \in L^{\infty}_{\mathbb{F}}(\Omega; C([0,T];\mathcal{S}(\mathbb{R}^n))) \times L^2_{\mathbb{F}}(0,T;\mathcal{S}(\mathbb{R}^n))$ *such that*

$$K \gg 0 \tag{13.59}$$

and

$$K(\cdot)^{-1}L(\cdot) \in L^{\infty}_{\mathbb{F}}(\Omega; L^2(0,T;\mathbb{R}^{m\times n})). \tag{13.60}$$

In this case, the optimal feedback operator $\Theta(\cdot)$ *is given as*

$$\Theta(\cdot) = -K(\cdot)^{-1}L(\cdot). \tag{13.61}$$

Furthermore,

$$\inf_{u(\cdot)\in L^2_{\mathbb{F}}(0,T;\mathbb{R}^m)} \mathcal{J}(\eta;u) = \frac{1}{2}\langle P(0)\eta,\eta\rangle_{\mathbb{R}^n}. \tag{13.62}$$

Proof: The "**if**" part. Assume that the equation (13.47) admits a solution $\big(P(\cdot), \Lambda(\cdot)\big) \in L^{\infty}_{\mathbb{F}}(\Omega; C([0,T];\mathcal{S}(\mathbb{R}^n))) \times L^2_{\mathbb{F}}(0,T;\mathcal{S}(\mathbb{R}^n))$ such that (13.60) holds. Then, it is clear that the function $\Theta(\cdot)$ given by (13.61) belongs to $L^{\infty}_{\mathbb{F}}(\Omega; L^2(0,T;\mathbb{R}^{m\times n}))$. For any $\eta \in \mathbb{R}^n$ and $u(\cdot) \in L^2_{\mathbb{F}}(0,T;\mathbb{R}^m)$, let $x(\cdot) \equiv x(\cdot\,;\eta,u(\cdot))$ be the corresponding state process for (13.31). By Itô's formula, and using (13.31) and (13.47), we obtain that

$$\begin{aligned}
& d\big\langle Px,x\big\rangle_{\mathbb{R}^n} \\
&= \big\langle dPx,x\big\rangle_{\mathbb{R}^n} + \big\langle Pdx,x\big\rangle_{\mathbb{R}^n} + \big\langle Px,dx\big\rangle_{\mathbb{R}^n} \\
&\quad +\big\langle dPdx,x\big\rangle_{\mathbb{R}^n} + \big\langle dPx,dx\big\rangle_{\mathbb{R}^n} + \big\langle Pdx,dx\big\rangle_{\mathbb{R}^n} \\
&= \big\langle -\big(PA(t)+A(t)^\top P+\Lambda C+C^\top\Lambda+C^\top PC+M-L^\top K^{-1}L\big)x,x\big\rangle_{\mathbb{R}^n}dr \\
&\quad +\big\langle P(A(t)x+Bu),x\big\rangle_{\mathbb{R}^n}dr + \big\langle P(Cx+Du),x\big\rangle_{\mathbb{R}^n}dW(r) \qquad (13.63)\\
&\quad +\big\langle Px,A(t)x+Bu\big\rangle_{\mathbb{R}^n}dr + \big\langle Px,Cx+Du\big\rangle_{\mathbb{R}^n}dW(r) \\
&\quad +\big\langle \Lambda(Cx+Du),x\big\rangle_{\mathbb{R}^n}dr + \big\langle \Lambda x,Cx+Du\big\rangle_{\mathbb{R}^n}dr \\
&\quad +\big\langle P(Cx+Du),Cx+Du\big\rangle_{\mathbb{R}^n}dr + \big\langle \Lambda x,x\big\rangle_{\mathbb{R}^n}dW(r) \\
&= -\big\langle (M-L^\top K^{-1}L)x,x\big\rangle_{\mathbb{R}^n}dr + \big\langle PBu,x\big\rangle_{\mathbb{R}^n}dr \\
&\quad +\big\langle Px,Bu\big\rangle_{\mathbb{R}^n}dr + \big\langle PCx,Du\big\rangle_{\mathbb{R}^n}dr + \big\langle PDu,Cx+Du\big\rangle_{\mathbb{R}^n}dr \\
&\quad +\big\langle Du,\Lambda x\big\rangle_{\mathbb{R}^n}dr + \big\langle \Lambda x,Du\big\rangle_{\mathbb{R}^n}dr + \langle P(Cx+Du),x\rangle_{\mathbb{R}^n}dW(r) \\
&\quad +\langle Px,Cx+Du\rangle_{\mathbb{R}^n}dW(r) + \big\langle \Lambda x,x\big\rangle_{\mathbb{R}^n}dW(r) \\
&= -\big\langle (M-L^\top K^{-1}L)x,x\big\rangle_{\mathbb{R}^n}dr + 2\langle L^\top u,x\rangle_{\mathbb{R}^n}dr + \langle D^\top PDu,u\rangle_{\mathbb{R}^m}dr \\
&\quad +\big[2\langle P(Cx+Du),x\rangle_{\mathbb{R}^n} + \langle \Lambda x,x\rangle_{\mathbb{R}^n}\big]dW(r).
\end{aligned}$$

From the definition of Θ in (13.138), we derive that,

$$\Theta^\top K\Theta = -\Theta^\top KK^{-1}L = L^\top K^{-1}L.$$

This, together with (13.63), implies that

$$\begin{aligned} d\big\langle Px,x\big\rangle_{\mathbb{R}^n} &= -\big\langle (M-\Theta^\top K\Theta)x,x\big\rangle_{\mathbb{R}^n}dr + 2\langle L^\top u,x\rangle_{\mathbb{R}^n}dr \\ &\quad +\langle D^\top PDu,u\rangle_{\mathbb{R}^m}dr + \big[2\langle P(Cx+Du),x\rangle_{\mathbb{R}^n} \\ &\quad +\langle \Lambda x,x\rangle_{\mathbb{R}^n}\big]dW(r). \end{aligned} \tag{13.64}$$

To handle the stochastic integral above, we introduce a sequence of stopping times $\{\tau_j\}_{j=1}^\infty$ as

$$\tau_j \stackrel{\Delta}{=} \inf\Big\{t\geq 0 \;\Big|\; \int_0^t |\Lambda(r)|^2 dr \geq j\Big\}\wedge T.$$

It is easy to see that $\tau_j\to T$, a.s., as $j\to\infty$. Using (13.64), we obtain that,

$$\begin{aligned} &\mathbb{E}\langle P(\tau_j)x(\tau_j),x(\tau_j)\rangle_{\mathbb{R}^n} + \mathbb{E}\int_0^T \chi_{[0,\tau_j]}\big[\langle Mx,x\rangle_{\mathbb{R}^n}+\langle Ru,u\rangle_{\mathbb{R}^m}\big]dr \\ &= \mathbb{E}\langle P(0)\eta,\eta\rangle_{\mathbb{R}^n} + \mathbb{E}\int_0^T \chi_{[0,\tau_j]}\big[\langle \Theta^\top K\Theta x,x\rangle_{\mathbb{R}^n} + 2\langle L^\top u,x\rangle_{\mathbb{R}^n}\big]dr \\ &\quad +\mathbb{E}\int_0^T \chi_{[s,\tau_j]}\langle (R+D^\top PD)u,u\rangle_{\mathbb{R}^m}dr. \end{aligned}$$

Clearly,

$$|\langle P(\tau_j)x(\tau_j),x(\tau_j)\rangle_{\mathbb{R}^n}| \leq |P|_{L^\infty_{\mathbb{F}}(0,T;\mathbb{R}^{n\times n})}|x|^2_{L^2_{\mathbb{F}}(\Omega;C([0,T];\mathbb{R}^n)}.$$

By the Dominated Convergence Theorem, we obtain that

$$\lim_{j\to\infty}\langle P(\tau_j)x(\tau_j),x(\tau_j)\rangle_{\mathbb{R}^n} = \langle P(T)x(T),x(T)\rangle_{\mathbb{R}^n}. \tag{13.65}$$

Furthermore,

$$\begin{aligned} &\big|\chi_{[0,\tau_j]}\big[\langle Mx,x\rangle_{\mathbb{R}^n}+\langle Ru,u\rangle_{\mathbb{R}^m}\big]\big| \\ &\leq \big|\big[\langle Mx,x\rangle_{\mathbb{R}^n}+\langle Ru,u\rangle_{\mathbb{R}^m}\big]\big| \in L^1_{\mathbb{F}}(0,T;\mathbb{R}). \end{aligned}$$

By the Dominated Convergence Theorem again, we get that

$$\begin{aligned} &\lim_{j\to\infty}\mathbb{E}\int_0^T \chi_{[0,\tau_j]}\big[\langle Mx,x\rangle_{\mathbb{R}^n}+\langle Ru,u\rangle_{\mathbb{R}^m}\big]dr \\ &= \mathbb{E}\int_0^T \big[\langle Mx,x\rangle_{\mathbb{R}^n}+\langle Ru,u\rangle_{\mathbb{R}^m}\big]dr. \end{aligned} \tag{13.66}$$

Similarly, we can obtain that

$$
\begin{aligned}
&\lim_{j\to\infty}\mathbb{E}\int_0^T \chi_{[0,\tau_j]}\big[\langle\Theta^\top K\Theta x,x\rangle_{\mathbb{R}^n}+2\langle L^\top u,x\rangle_{\mathbb{R}^n}\big]dr\\
&\quad+\lim_{j\to\infty}\mathbb{E}\int_0^T \chi_{[0,\tau_j]}\langle(R+D^\top PD)u,u\rangle_{\mathbb{R}^m}dr\\
&=\mathbb{E}\int_0^T \big[\langle\Theta^\top K\Theta x,x\rangle_{\mathbb{R}^n}+2\langle L^\top u,x\rangle_{\mathbb{R}^n}\big]dr\\
&\quad+\mathbb{E}\int_0^T \langle(R+D^\top PD)u,u\rangle_{\mathbb{R}^m}dr.
\end{aligned}
\tag{13.67}
$$

It follows from (13.65)–(13.67) that

$$
\begin{aligned}
&2\mathcal{J}(\eta;u(\cdot))\\
&=\mathbb{E}\langle Gx(T),x(T)\rangle_{\mathbb{R}^n}+\mathbb{E}\int_0^T \big[\langle Mx,x\rangle_{\mathbb{R}^n}+\langle Ru,u\rangle_{\mathbb{R}^m}\big]dr\\
&=\mathbb{E}\langle P(0)\eta,\eta\rangle_{\mathbb{R}^n}+\mathbb{E}\int_0^T \big[\langle\Theta^\top K\Theta x,x\rangle_{\mathbb{R}^n}+2\langle L^\top u,x\rangle_{\mathbb{R}^n}+\langle Ku,u\rangle_{\mathbb{R}^m}\big]dr\\
&=\mathbb{E}\Big[\big\langle P(0)\eta,\eta\big\rangle_{\mathbb{R}^n}+\int_0^T\Big(\big\langle K\Theta x,\Theta x\big\rangle_{\mathbb{R}^m}-2\big\langle K\Theta x,u\big\rangle_{\mathbb{R}^m}+\langle Ku,u\rangle_{\mathbb{R}^m}\Big)dr\Big]\\
&=2\mathcal{J}(\eta;\Theta\bar{x})+\mathbb{E}\int_0^T\big\langle K(u-\Theta x),u-\Theta x\big\rangle_{\mathbb{R}^m}dr,
\end{aligned}
$$

where we have used the fact that $L^\top=-\Theta^\top K$. Hence, by $K\gg 0$, we get

$$
\mathcal{J}(\eta;\Theta\bar{x})\le\mathcal{J}(\eta;u),\quad \forall\, u(\cdot)\in L^2_{\mathbb{F}}(0,T;\mathbb{R}^m).
$$

Thus, the function $\Theta(\cdot)$ given by (13.138) is an optimal feedback operator for Problem (FSLQ). This completes the proof of the “if” part of Theorem 13.23.

The “**only if**” part. Let us divide the proof into three steps.

Step 1. Let $\Theta(\cdot)\in L^\infty_{\mathbb{F}}(\Omega;L^2(0,T;\mathbb{R}^{m\times n}))$ be an optimal feedback operator for Problem (FSLQ). Then, by Theorem 13.17, for any $\zeta\in\mathbb{R}^n$, the following forward-backward stochastic differential equation

$$
\begin{cases}
dx=(A(t)+B\Theta)xdt+(C+D\Theta)xdW(t) & \text{in } [0,T],\\
dy=-\big(A(t)^\top y+C^\top Y+Mx\big)dt+YdW(t) & \text{in } [0,T],\\
x(0)=\zeta,\qquad y(T)=Gx(T)
\end{cases}
\tag{13.68}
$$

admits a unique solution $(x(\cdot),y(\cdot),Y(\cdot))\in L^2_{\mathbb{F}}(\Omega;C([0,T];\mathbb{R}^n))\times L^2_{\mathbb{F}}(\Omega;C([0,T];\mathbb{R}^n))\times L^2_{\mathbb{F}}(0,T;\mathbb{R}^n)$ such that

$$
R\Theta x+B^\top y+D^\top Y=0,\quad \text{a.e. } (t,\omega)\in(0,T)\times\Omega.
\tag{13.69}
$$

Also, consider the following stochastic differential equation:

$$\begin{cases} d\tilde{x} = \big[- A(t) - B\Theta + \big(C + D\Theta\big)^2\big]^\top \tilde{x}dt - \big(C + D\Theta\big)^\top \tilde{x}dW(t) & \text{in } [0,T], \\ \tilde{x}(0) = \zeta, \end{cases} \tag{13.70}$$

which has a unique solution $\tilde{x} \in L^2_{\mathbb{F}}(\Omega; C([0,T];\mathbb{R}^n))$.

Further, consider the following $\mathbb{R}^{n\times n}$-valued forward-backward stochastic differential equation:

$$\begin{cases} d\mathbb{X} = (A(t) + B\Theta)\mathbb{X}dt + (C + D\Theta)\mathbb{X}dW(t) & \text{in } [0,T], \\ d\mathbb{Y} = -\big(A(t)^\top \mathbb{Y} + C^\top \mathbf{Y} + M\mathbb{X}\big)dt + \mathbf{Y}dW(t) & \text{in } [0,T], \\ \mathbb{X}(0) = I_n, \quad \mathbb{Y}(T) = G\mathbb{X}(T) \end{cases} \tag{13.71}$$

and $\mathbb{R}^{n\times n}$-valued stochastic differential equation:

$$\begin{cases} d\widetilde{\mathbb{X}} = \big[- A(t) - B\Theta + \big(C + D\Theta\big)^2\big]^\top \widetilde{\mathbb{X}}dt \\ \qquad\quad -(C + D\Theta)^\top \widetilde{\mathbb{X}}dW(t) & \text{in } [0,T], \\ \widetilde{\mathbb{X}}(0) = I_n. \end{cases} \tag{13.72}$$

Equations (13.71) and (13.72) admit, respectively, unique solutions $(\mathbb{X}, \mathbb{Y}, \mathbf{Y}) \in L^2_{\mathbb{F}}(\Omega; C([0,T];\mathbb{R}^{n\times n})) \times L^2_{\mathbb{F}}(\Omega; C([0,T];\mathbb{R}^{n\times n})) \times L^2_{\mathbb{F}}(0,T;\mathbb{R}^{n\times n})$ and $\widetilde{\mathbb{X}} \in L^2_{\mathbb{F}}(\Omega; C([0,T];\mathbb{R}^{n\times n}))$.

It follows from (13.68) to (13.72) that, for any $\zeta \in \mathbb{R}^n$,

$$\begin{array}{ll} x(t;\zeta) = \mathbb{X}(t)\zeta, \;\; y(t;\zeta) = \mathbb{Y}(t)\zeta, \;\; \tilde{x}(t;\zeta) = \widetilde{\mathbb{X}}(t)\zeta, & \forall\, t \in [0,T], \\ z(t;\zeta) = \mathbf{Y}(t)\zeta, & \text{a.e. } t \in [0,T]. \end{array} \tag{13.73}$$

By (13.69) and (13.73), we find that

$$R\Theta\mathbb{X} + B^\top \mathbb{Y} + D^\top \mathbf{Y} = 0, \quad \text{a.e. } (t,\omega) \in [0,T] \times \Omega. \tag{13.74}$$

For any $\zeta, \rho \in \mathbb{R}^n$ and $t \in [0,T]$, by Itô's formula, we have

$$\begin{aligned} &\big\langle x(t;\zeta), \tilde{x}(t;\rho)\big\rangle_{\mathbb{R}^n} - \big\langle \zeta, \rho\big\rangle_{\mathbb{R}^n} \\ &= \int_0^t \big\langle (A(t) + B\Theta)x(r;\zeta), \tilde{x}(r;\rho)\big\rangle_{\mathbb{R}^n} dr \\ &\quad + \int_0^t \big\langle (C + D\Theta)x(r;\zeta), \tilde{x}(r;\rho)\big\rangle_{\mathbb{R}^n} dW(r) \\ &\quad + \int_0^t \big\langle x(r;\zeta), \big[- A(t) - B\Theta + \big(C + D\Theta\big)^2\big]^\top \tilde{x}(r;\rho)\big\rangle_{\mathbb{R}^n} dr \\ &\quad - \int_0^t \big\langle x(r;\zeta), \big(C + D\Theta\big)^\top \tilde{x}(r;\rho)\big\rangle_{\mathbb{R}^n} dW(r) \\ &\quad - \int_0^t \big\langle (C + D\Theta)x(r;\zeta), \big(C + D\Theta\big)^\top \tilde{x}(r;\rho)\big\rangle_{\mathbb{R}^n} dr \\ &= 0. \end{aligned}$$

Consequently,

$$\big\langle \mathbb{X}(t)\zeta, \widetilde{\mathbb{X}}(t)\rho\big\rangle_{\mathbb{R}^n} = \big\langle x(t;\zeta), \tilde{x}(t;\rho)\big\rangle_{\mathbb{R}^n} = \big\langle \zeta, \rho\big\rangle_{\mathbb{R}^n}, \quad \text{a.s.}$$

This implies that $\mathbb{X}(t)\widetilde{\mathbb{X}}(t)^\top = I_n$ for all $t \in [0,T]$, a.s., that is, $\widetilde{\mathbb{X}}(t)^\top = \mathbb{X}(t)^{-1}$ for all $t \in [0,T]$, a.s.

Step 2. Put

$$P(t,\omega) \overset{\Delta}{=} \mathbb{Y}(t,\omega)\mathbb{X}(t,\omega)^{-1}, \quad \Pi(t,\omega) \overset{\Delta}{=} \mathbf{Y}(t,\omega)\mathbb{X}(t,\omega)^{-1}, \quad t \in [0,T]. \quad (13.75)$$

By Itô's formula,

$$\begin{aligned} dP &= \Big\{ - \big(A(t)^\top \mathbb{Y} + C^\top \mathbf{Y} + M\mathbb{X}\big)\mathbb{X}^{-1} + \mathbb{Y}\mathbb{X}^{-1}\big[(C+D\Theta)^2 - A(t) - B\Theta\big] \\ &\quad - \mathbf{Y}\mathbb{X}^{-1}(C+D\Theta)\Big\}dt + \Big[\mathbf{Y}\mathbb{X}^{-1} - \mathbb{Y}\mathbb{X}^{-1}(C+D\Theta)\Big]dW(t) \\ &= \Big\{ - A(t)^\top P - C^\top \Pi - M + P\big[(C+D\Theta)^2 - A(t) - B\Theta\big] \\ &\quad - \Pi(C+D\Theta)\Big\}dt + \big[\Pi - P(C+D\Theta)\big]dW(t). \end{aligned}$$

Let

$$\Lambda \overset{\Delta}{=} \Pi - P(C+D\Theta). \quad (13.76)$$

Then, $(P(\cdot), \Lambda(\cdot))$ solves the following $\mathbb{R}^{n\times n}$-valued backward stochastic differential equation:

$$\begin{cases} dP = -\big[PA(t) + A(t)^\top P + \Lambda C + C^\top \Lambda + C^\top PC \\ \qquad\quad +(PB + C^\top PD + \Lambda D)\Theta + M\big]dt + \Lambda dW(t) \ \text{ in } [0,T], \\ P(T) = G. \end{cases} \quad (13.77)$$

By Theorem 4.2, we conclude that $(P, \Lambda) \in L^\infty_{\mathbb{F}}(\Omega; C([0,T]; \mathbb{R}^{n\times n})) \times L^2_{\mathbb{F}}(0,T; \mathbb{R}^{n\times n})$.

For any $t \in [0,T)$ and $\eta \in L^2_{\mathcal{F}_t}(\Omega; \mathbb{R}^n)$, let us consider the following forward-backward stochastic differential equation:

$$\begin{cases} dx^t(r) = \big(A(t) + B\Theta\big)x^t dr + \big(C + D\Theta\big)x^t dW(r) & \text{in } [t,T], \\ dy^t(r) = -\big(A(t)^\top y^t + C^\top Y^t + Mx^t\big)dr + Y^t dW(r) & \text{in } [t,T], \\ x^t(t) = \eta, \quad y^t(T) = Gx^t(T). \end{cases} \quad (13.78)$$

Clearly, the equation (13.78) admits a unique solution $\big(x^t(\cdot), y^t(\cdot), z^t(\cdot)\big) \in L^2_{\mathbb{F}}(\Omega; C([t,T]; \mathbb{R}^n)) \times L^2_{\mathbb{F}}(\Omega; C([t,T]; \mathbb{R}^n)) \times L^2_{\mathbb{F}}(t,T; \mathbb{R}^n)$. Also, we consider the following forward-backward stochastic differential equation:

$$\begin{cases} d\mathbb{X}^t(r) = \big(A(t) + B\Theta\big)\mathbb{X}^t dr + \big(C + D\Theta\big)\mathbb{X}^t dW(r) & \text{in } [t,T], \\ d\mathbb{Y}^t(r) = -\big(A(t)^\top \mathbb{Y}^t + C^\top \mathbf{Y}^t + M\mathbb{X}^t\big)dr + \mathbf{Y}^t dW(r) & \text{in } [t,T], \\ \mathbb{X}^t(t) = I_n, \quad \mathbb{Y}^t(T) = G\mathbb{X}^t(T). \end{cases} \quad (13.79)$$

Likewise, the equation (13.79) admits a unique solution $\big(\mathbb{X}^t(\cdot), \mathbb{Y}^t(\cdot), \mathbf{Y}^t(\cdot)\big) \in L^2_{\mathbb{F}}(\Omega; C([t,T]; \mathbb{R}^{n\times n})) \times L^2_{\mathbb{F}}(\Omega; C([t,T]; \mathbb{R}^{n\times n})) \times L^2_{\mathbb{F}}(t,T; \mathbb{R}^{n\times n})$. It follows from (13.78) and (13.79) that, for any $\eta \in L^2_{\mathcal{F}_t}(\Omega; \mathbb{R}^n)$,

$$\begin{aligned} & x^t(r) = \mathbb{X}^t(r)\eta, \quad y^t(r) = \mathbb{Y}^t(r)\eta, \qquad \forall\, r \in [t,T]. \\ & Y^t(r) = \mathbf{Y}^t(r)\eta, \qquad\qquad\qquad\qquad \text{a.e. } r \in [t,T]. \end{aligned} \tag{13.80}$$

By the uniqueness of the solution to (13.68), for any $\zeta \in \mathbb{R}^n$ and $t \in [0,T]$, we have that

$$\mathbb{X}^t(r)\mathbb{X}(t)\zeta = x^t(r; \mathbb{X}(t)\zeta) = x(r;\zeta), \quad \text{a.s.}$$

Thus,

$$\mathbb{Y}^t(t)\mathbb{X}(t)\zeta = y^t(t; \mathbb{X}(t)\zeta) = \mathbb{Y}(t)\zeta, \quad \text{a.s.}$$

This implies that for all $t \in [0,T]$,

$$\mathbb{Y}^t(t) = \mathbb{Y}(t)\mathbb{X}(t)^{-1} = P(t), \quad \text{a.s.} \tag{13.81}$$

Let $\eta, \xi \in L^2_{\mathcal{F}_t}(\Omega; \mathbb{R}^n)$. Since $\mathbb{Y}^t(r)\eta = y^t(r;\eta)$ and $\mathbb{X}^t(r)\xi = x^t(r;\xi)$, applying Itô's formula to $\langle x^t(\cdot), y^t(\cdot)\rangle_{\mathbb{R}^n}$, we get that

$$\begin{aligned} & \mathbb{E}\langle \xi, P(t)\eta\rangle_{\mathbb{R}^n} \\ & = \mathbb{E}\langle G\mathbb{X}^t(T)\eta, \mathbb{X}^t(T)\xi\rangle_{\mathbb{R}^n} + \mathbb{E}\int_t^T \langle M\mathbb{X}^t\eta, \mathbb{X}^t\xi\rangle_{\mathbb{R}^n} dr \\ & -\mathbb{E}\int_t^T \langle B\Theta\mathbb{X}^t\xi, \mathbb{Y}^t\eta\rangle_{\mathbb{R}^n} dr - \mathbb{E}\int_t^T \langle D\Theta\mathbb{X}^t\xi, \mathbf{Y}^t\eta\rangle_{\mathbb{R}^n} dr. \end{aligned} \tag{13.82}$$

This implies that

$$\begin{aligned} \mathbb{E}\langle P(t)\eta, \xi\rangle_{\mathbb{R}^n} = \mathbb{E}\langle G\mathbb{X}^t(T)\eta, \mathbb{X}^t(T)\xi\rangle_{\mathbb{R}^n} + \mathbb{E}\int_t^T \big(\langle M\mathbb{X}^t\eta, \mathbb{X}^t\xi\rangle_{\mathbb{R}^n} \\ +\langle R\Theta\mathbb{X}^t\eta, \Theta\mathbb{X}^t\xi\rangle_{\mathbb{R}^n}\big) dr. \end{aligned}$$

Therefore,

$$\begin{aligned} & \mathbb{E}\langle P(t)\eta, \xi\rangle_{\mathbb{R}^n} \\ & = \mathbb{E}\Big\langle \xi, \mathbb{X}^t(T)^\top G\mathbb{X}^t(T)\eta + \mathbb{E}\int_t^T \big[(\mathbb{X}^t)^\top M\mathbb{X}^t\eta + (\mathbb{X}^t)^\top\Theta^\top R\Theta\mathbb{X}^t\eta\big] dr \Big\rangle_{\mathbb{R}^n}, \end{aligned}$$

which concludes that

$$P(t) = \mathbb{E}\Big(\mathbb{X}^t(T)^\top G\mathbb{X}^t(T) + \mathbb{E}\int_t^T \big[(\mathbb{X}^t)^\top Q\mathbb{X}^t + (\mathbb{X}^t)^\top\Theta^\top R\Theta\mathbb{X}^t\big] dr \Big| \mathcal{F}_t\Big).$$

This, together with the assumption that Problem (FSLQ) is standard, proves that

$$P(t) \geq 0, \quad \text{a.s.}, \quad \forall\, t \in [0, T]. \tag{13.83}$$

Clearly, $(P^\top, \Lambda^\top)$ satisfies that

$$\begin{cases} dP^\top = -\big[P^\top A(t) + A(t)^\top P^\top + \Lambda^\top C + C^\top \Lambda^\top + C^\top P^\top C \\ \qquad\qquad +\Theta^\top (PB + C^\top PD + \Lambda D)^\top + M\big]dt + \Lambda^\top dW(t) \ \text{ in } [0, T], \\ P(T)^\top = G. \end{cases} \tag{13.84}$$

According to (13.77) and (13.84), and noting that $P(\cdot)$ is symmetric, we find that for any $t \in [0, T]$,

$$\begin{aligned} 0 = -\int_0^t \Big\{&\big[\Lambda C + C^\top \Lambda + (PB + C^\top PD + \Lambda D)\Theta\big] \\ &-\big[\Lambda C + C^\top \Lambda + (PB + C^\top PD + \Lambda D)\Theta\big]^\top\Big\}d\tau \\ &+\int_0^t (\Lambda - \Lambda^\top)dW(\tau). \end{aligned} \tag{13.85}$$

The first term in the right hand side of (13.85) is absolutely continuous with respect to t, a.s. Hence, $\int_0^t (\Lambda - \Lambda^\top)dW(\tau)$ is absolutely continuous with respect to t, a.s. Consequently, its quadratic variation is 0, a.s. Namely,

$$\int_0^T |\Lambda - \Lambda^\top|^2_{\mathbb{R}^{n\times n}} ds = 0, \quad \text{a.s.}$$

This concludes that

$$\Lambda(t, \omega) = \Lambda(t, \omega)^\top, \quad \text{a.e. } (t, \omega) \in (0, T) \times \Omega. \tag{13.86}$$

Step 3. In this step, we show that (P, Λ) is a pair of stochastic processes satisfying (13.47), (13.59) and (13.60).

It follows from $R \gg 0$ and (13.83) that (13.59) holds. By (13.74), we get that

$$B^\top P + D^\top \Pi + R\Theta = 0, \quad \text{a.e. } (t, \omega) \in [0, T] \times \Omega. \tag{13.87}$$

This implies that

$$\begin{aligned} 0 &= B^\top P + D^\top\big[\Lambda + P(C + D\Theta)\big] + R\Theta \\ &= B^\top P + D^\top PC + D^\top \Lambda + K\Theta = L + K\Theta. \end{aligned} \tag{13.88}$$

From (13.59) and (13.88), we obtain

$$\Theta = -K^{-1}L.$$

This, together with the fact that $\Theta \in L^\infty_{\mathbb{F}}(\Omega; L^2(0, T; \mathbb{R}^{m\times n}))$, leads to (13.60). Therefore, it follows from (13.86) and (13.88) that

$$(PB + C^\top PD + \Lambda D)\Theta = L^\top \Theta = -\Theta^\top K \Theta = -L^\top K^{-1} L. \tag{13.89}$$

Hence, by (13.77), we conclude that (P, Λ) is a solution to (13.47). This completes the proof of the necessity in Theorem 13.23. □

Remark 13.24. By Example 13.22, we see that sometimes it is impossible to construct the desired optimal feedback operator by solving the Riccati equation (13.47). Because of this, we need to introduce the condition (13.60). Clearly, this is quite different from the deterministic case.

Remark 13.25. Actually, the condition that Problem (FSLQ) is standard can be dropped (e.g., [236]). Nevertheless, in order to present the key idea in the simplest way, we do not pursue the full generality in this section.

Remark 13.26. One can see that the main ideas in the proof of Theorem 13.23 are similarly to the ones in the proof of Theorems 13.13 and 13.14. Nevertheless, the control dependent diffusion term $(Cx + Du)dW(t)$ leads to some new difficulty.

13.4 Finiteness and Solvability of Problem (SLQ)

In this section, similar to [371], we study the finiteness and solvability of Problem (SLQ).

Since the state equation (13.1) is linear, the state process can be explicitly expressed in terms of the initial state and the control in a linear form. Substituting this formula into the cost functional (which is quadratic in the state and control), one can obtain a quadratic functional w.r.t. the initial state and the control. Thus, Problem (SLQ) can be transformed to a quadratic optimization problem in the Hilbert space $L^2_{\mathbb{F}}(0, T; U)$. This leads to some necessary and sufficient conditions for the finiteness and solvability of Problem (SLQ). Let us explain the details below. To this end, we define the following operators:

For any $\eta \in H$ and $u(\cdot) \in L^2_{\mathbb{F}}(0, T; U)$,

$$\begin{cases} F_1 : L^2_{\mathbb{F}}(0, T; U) \to L^2_{\mathbb{F}}(0, T; H), \\ (F_1 u(\cdot))(\cdot) \stackrel{\Delta}{=} x(\cdot\,; 0, u), \quad \forall\, u(\cdot) \in L^2_{\mathbb{F}}(0, T; U); \end{cases}$$

$$\begin{cases} \widehat{F}_1 : L^2_{\mathbb{F}}(0, T; U) \to L^2_{\mathcal{F}_T}(\Omega; H), \\ \widehat{F}_1 u(\cdot) \stackrel{\Delta}{=} x(T; 0, u), \quad \forall\, u(\cdot) \in L^2_{\mathbb{F}}(0, T; U); \end{cases}$$

$$\begin{cases} F_2 : H \to L^2_{\mathbb{F}}(0, T; H), \\ (F_2 \eta)(\cdot) \stackrel{\Delta}{=} x(\cdot\,; \eta, 0), \quad \forall\, \eta \in H; \end{cases}$$

$$\begin{cases} \widehat{F}_2 : H \to L^2_{\mathcal{F}_T}(\Omega; H), \\ \widehat{F}_2\eta \overset{\Delta}{=} x(T; \eta, 0), \quad \forall\, \eta \in H, \end{cases}$$

where $x(\cdot) \equiv x(\cdot; \eta, u(\cdot))$ solves the equation (13.1). Then, for any $\eta \in H$ and $u(\cdot) \in L^2_{\mathbb{F}}(0, T; U)$, the corresponding state process $x(\cdot)$ and its terminal value $x(T)$ (for the equation (13.1)) are given respectively by

$$\begin{cases} x(\cdot) = (F_2\eta)(\cdot) + (F_1 u)(\cdot), \\ x(T) = \widehat{F}_2\eta + \widehat{F}_1 u. \end{cases} \tag{13.90}$$

Next, we compute the adjoint operators of F_1, $\widehat{F}_1$, F_2 and $\widehat{F}_2$:

$$\begin{cases} F_1^* : L^2_{\mathbb{F}}(0, T; H) \to L^2_{\mathbb{F}}(0, T; U), \\ \widehat{F}_1^* : L^2_{\mathcal{F}_T}(\Omega; H) \to L^2_{\mathbb{F}}(0, T; U), \\ F_2^* : L^2_{\mathbb{F}}(0, T; H) \to H, \\ \widehat{F}_2^* : L^2_{\mathcal{F}_T}(\Omega; H) \to H. \end{cases}$$

For this purpose, we introduce the following backward stochastic evolution equation:

$$\begin{cases} dy = -\big(A^*y + C^*Y + \xi\big)dt + YdW(t) \ \text{ in } (0, T], \\ y(T) = y_T, \end{cases} \tag{13.91}$$

where $y_T \in L^2_{\mathcal{F}_T}(\Omega; H)$ and $\xi(\cdot) \in L^2_{\mathbb{F}}(0, T; H)$.

Proposition 13.27. *For any $\xi(\cdot) \in L^2_{\mathbb{F}}(0, T; H)$, let $(y_0(\cdot), Y_0(\cdot))$ be the transposition solution to (13.91) with $y_T = 0$. Then*

$$\begin{cases} (F_1^*\xi)(t) = B^*y_0(t) + D^*Y_0(t), \qquad \text{a.e. } t \in [0, T], \\ F_2^*\xi = y_0(0). \end{cases} \tag{13.92}$$

On the other hand, for any $y_T \in L^2_{\mathcal{F}_T}(\Omega; H)$, let $(y_1(\cdot), Y_1(\cdot))$ be the transposition solution to (13.91) with $\xi(\cdot) = 0$. Then

$$\begin{cases} (\widehat{F}_1^* y_T)(t) = B^*y_1(t) + D^*Y_1(t), \qquad \text{a.e. } t \in [0, T], \\ \widehat{F}_2^* y_T = y_1(0). \end{cases} \tag{13.93}$$

Proof: For any $y_T \in L^2_{\mathcal{F}_T}(\Omega; H)$ and $\xi(\cdot) \in L^2_{\mathbb{F}}(0, T; H)$, there exists a unique transposition solution $(y(\cdot), Y(\cdot)) \in D_{\mathbb{F}}([0, T]; L^2(\Omega; H)) \times L^2_{\mathbb{F}}(0, T; \mathcal{L}_2^0)$ to (13.91), which satisfies that

$$\sup_{0 \le t \le T} \mathbb{E}|y(t)|_H^2 + \mathbb{E}\int_0^T |Y(t)|^2_{\mathcal{L}_2^0} dt \le \mathcal{C}\mathbb{E}\Big(|y_T|_H^2 + \int_0^T |\xi(t)|_H^2 dt\Big),$$

for some constant $\mathcal{C} > 0$. Therefore, we know that all the operators defined in (13.92) and (13.93) are bounded.

Next, for any $\eta \in H$ and $u(\cdot) \in L^2_{\mathbb{F}}(0,T;U)$, let $x(\cdot)$ be the solution of (13.1). From the definition of the transposition solution to the equation (13.91), we find

$$\begin{aligned}&\mathbb{E}\langle x(T), y_T\rangle_H - \mathbb{E}\langle \eta, y(0)\rangle_H\\&= \mathbb{E}\int_0^T \big(\langle u(t), B^*y(t) + D^*Y(t)\rangle_U - \langle x(t), \xi(t)\rangle_H\big)dt,\end{aligned}$$

which implies that

$$\begin{aligned}&\mathbb{E}\big(\langle \widehat{F}_2\eta + \widehat{F}_1 u, y_T\rangle_H - \langle \eta, y(0)\rangle_H\big)\\&= \mathbb{E}\int_0^T \big(\langle u(t), B^*y(t) + D^*Y(t)\rangle_U - \langle (F_2\eta)(t) + (F_1u)(t), \xi(t)\rangle_H\big)dt.\end{aligned} \tag{13.94}$$

Let $y_T = 0$ and $\eta = 0$ in (13.94). We get

$$\mathbb{E}\int_0^T \langle (F_1u)(t), \xi(t)\rangle_H dt = \mathbb{E}\int_0^T \langle u(t), B^*y_0(t) + D^*Y_0(t)\rangle_U dt.$$

This proves the first equality in (13.92).

Letting $u(\cdot) = 0$ and $y_T = 0$ in (13.94), we obtain

$$\mathbb{E}\langle \eta, F_2^*\xi\rangle_H = \mathbb{E}\int_0^T \langle (F_2\eta)(t), \xi(t)\rangle_H dt = \mathbb{E}\langle \eta, y_0(0)\rangle_H.$$

This gives the second equality in (13.92).

Let $\eta = 0$ and $\xi(\cdot) = 0$ in (13.94). We find that

$$\begin{aligned}&\mathbb{E}\int_0^T \mathbb{E}\langle u(t), (\widehat{F}_1^* y_T)(t)\rangle_H dt = \mathbb{E}\langle \widehat{F}_1 u, y_T\rangle_H\\&= \mathbb{E}\int_0^T \langle u(t), B^*y_1(t) + D^*Y_1(t)\rangle_H dt.\end{aligned}$$

This demonstrates the first equality in (13.93).

At last, let $u(\cdot) = 0$ and $\xi(\cdot) = 0$ in (13.94). We see that

$$\mathbb{E}\langle \eta, \widehat{F}_2^* y_T\rangle_H = \mathbb{E}\langle \widehat{F}_2\eta, y_T\rangle_H = \mathbb{E}\langle \eta, y_1(0)\rangle_H.$$

This verifies the second equality in (13.93). □

From Proposition 13.27, we immediately obtain the following result, which is a representation for the cost functional (13.3).

Proposition 13.28. *The cost functional* (13.3) *can be represented as*

$$\mathcal{J}(\eta; u(\cdot)) = \frac{1}{2}\Big[\mathbb{E}\int_0^T \big(\langle \mathcal{N} u, u\rangle_U + 2\langle \mathcal{H}(\eta), u\rangle_U\big)dt + \mathcal{M}(\eta)\Big]. \qquad (13.95)$$

Here

$$\begin{cases} \mathcal{N} = R + F_1^* M F_1 + \widehat{F}_1^* G \widehat{F}_1, \\ \mathcal{H}(\eta) = \big(F_1^* M F_2 \eta\big)(\cdot) + \big(\widehat{F}_1^* G \widehat{F}_2 \eta\big)(\cdot), \\ \mathcal{M}(\eta) = \big\langle M F_2 \eta, F_2 \eta\big\rangle_{L^2_{\mathbb{F}}(0,T;H)} + \big\langle G\widehat{F}_2 \eta, \widehat{F}_2 \eta\big\rangle_{L^2_{\mathcal{F}_T}(\Omega;H)}. \end{cases} \qquad (13.96)$$

As a corollary of Proposition 13.28, we have the following result for the finiteness and solvability of Problem (SLQ) .

Theorem 13.29. 1) *If Problem (SLQ) is finite at some $\eta \in H$, then*

$$\mathcal{N} \geq 0. \qquad (13.97)$$

2) *Problem (SLQ) is solvable at $\eta \in H$ if and only if $\mathcal{N} \geq 0$ and there exists $\bar{u}(\cdot) \in L^2_{\mathbb{F}}(0,T;U)$, such that*

$$\mathcal{N}\bar{u}(\cdot) + \mathcal{H}(\eta) = 0. \qquad (13.98)$$

In this case, $\bar{u}(\cdot)$ is an optimal control.

3) *If $\mathcal{N} \gg 0$, then for any $\eta \in H$, $\mathcal{J}(\eta; \cdot)$ admits a unique minimizer*

$$\bar{u}(\cdot) = -\mathcal{N}^{-1}\mathcal{H}(\eta). \qquad (13.99)$$

In this case, it holds

$$\begin{aligned} &\inf_{u(\cdot) \in L^2_{\mathbb{F}}(0,T;U)} \mathcal{J}(\eta; u(\cdot)) = \mathcal{J}(\eta; \bar{u}(\cdot)) \\ &= \frac{1}{2}\Big(\mathcal{M}(\eta) - \langle \mathcal{N}^{-1}\mathcal{H}(\eta), \mathcal{H}(\eta)\rangle_{L^2_{\mathbb{F}}(0,T;U)}\Big), \qquad \forall\, \eta \in H. \end{aligned} \qquad (13.100)$$

Proof: "Proof of the assertion 1)". Suppose that (13.97) does not hold. Then there is $u_0 \in L^2_{\mathbb{F}}(0,T;U)$ such that

$$\mathbb{E}\int_0^T \langle \mathcal{N} u_0(\cdot), u_0(\cdot)\rangle_H ds < 0.$$

Define a sequence $\{u_k\}_{k=1}^{\infty} \subset L^2_{\mathbb{F}}(0,T;U)$ as follows:

$$u_k(\cdot) = k u_0(\cdot) \text{ in } [0,T].$$

Then

$$\mathcal{J}(\eta; u_k(\cdot)) \to -\infty \text{ as } k \to \infty,$$

which contradicts that Problem (SLQ) is finite at η.

"Proof of the assertion 2)". The **"if"** part. Let $\bar{u}(\cdot) \in L^2_{\mathbb{F}}(0,T;U)$ be an optimal control of Problem (SLQ) for $\eta \in H$. By the optimality of $\bar{u}(\cdot)$, for any $u(\cdot) \in L^2_{\mathbb{F}}(0,T;U)$, it holds that

$$\begin{aligned} 0 &\leq \varliminf_{\lambda \to 0} \frac{1}{\lambda}\Big(\mathcal{J}(\eta;\bar{u}(\cdot)+\lambda u(\cdot)) - \mathcal{J}(\eta;\bar{u}(\cdot))\Big) \\ &= \mathbb{E}\int_0^T \langle \mathcal{N}\bar{u}(\cdot) + \mathcal{H}(\eta), u(\cdot)\rangle_U dt. \end{aligned}$$

Consequently,

$$\mathcal{N}\bar{u}(\cdot) + \mathcal{H}(\eta) = 0.$$

The **"only if"** part. Let $(\eta, \bar{u}(\cdot)) \in H \times L^2_{\mathbb{F}}(0,T;U)$ satisfy (13.98). For any $u \in L^2_{\mathbb{F}}(0,T;U)$, from (13.97), we see

$$\begin{aligned} &\mathcal{J}(\eta;u(\cdot)) - \mathcal{J}(\eta;\bar{u}(\cdot)) = \mathcal{J}\big(\eta;\bar{u}(\cdot)+u(\cdot)-\bar{u}(\cdot)\big) - \mathcal{J}(\eta;\bar{u}(\cdot)) \\ &= \mathbb{E}\int_0^T \big\langle \mathcal{N}\bar{u}(\cdot)+\mathcal{H}(\eta), u(\cdot)-\bar{u}(\cdot)\big\rangle_U dt \\ &\quad +\frac{1}{2}\mathbb{E}\int_0^T \big\langle \mathcal{N}\big(u(\cdot)-\bar{u}(\cdot)\big), u(\cdot)-\bar{u}(\cdot)\big\rangle_U dt \\ &= \frac{1}{2}\mathbb{E}\int_0^T \big\langle \mathcal{N}\big(u(\cdot)-\bar{u}(\cdot)\big), u(\cdot)-\bar{u}(\cdot)\big\rangle_U dt \geq 0. \end{aligned}$$

This concludes that $\bar{u}(\cdot)$ is an optimal control.

"Proof of the assertion 3)". Since all optimal controls should satisfy (13.98) and $\mathcal{N}$ is invertible, we get assertion 3) immediately. □

Clearly, if Problem (SLQ) is standard, then $\mathcal{N} \gg 0$, and by Theorem 13.29, it is uniquely solvable and the optimal control is given by (13.99).

There is a main drawback of the formula (13.99), that is, it is very difficult to compute the inverse of the operator $\mathcal{N}$.

13.5 Pontryagin-Type Maximum Principle for Problem (SLQ)

In this section, we shall derive the following Pontryagin type maximum principle for Problem (SLQ).

Theorem 13.30. *Let Problem (SLQ) be solvable at $\eta \in H$ with $(\bar{x}(\cdot), \bar{u}(\cdot))$ being an optimal pair. Then, for the transposition solution $(y(\cdot), Y(\cdot))$ to*

$$\begin{cases} dy = -\big(A^*y + C^*Y - M\bar{x}\big)dt + YdW(t) & \textit{in } [0,T), \\ y(T) = -G\bar{x}(T), \end{cases} \tag{13.101}$$

and the relaxed transposition solution $(P(\cdot), Q^{(\cdot)}, \widehat{Q}^{(\cdot)})$ *to*

$$\begin{cases} dP = -\big(A^*P + PA + C^*PC + QC + C^*Q - M\big)dt \\ \qquad\quad +QdW(t) \qquad \textit{in } [0,T), \\ P(T) = -G, \end{cases} \tag{13.102}$$

it holds that

$$R(t)\bar{u}(t) - B(t)^*y(t) - D(t)^*Y(t) = 0, \quad \text{a.e. } (t,\omega) \in [0,T] \times \Omega \tag{13.103}$$

and

$$R(t) - D(t)^*P(t)D(t) \geq 0, \quad \text{a.e. } (t,\omega) \in [0,T] \times \Omega. \tag{13.104}$$

Proof: Since the assumption **(S2)** in Chapter 12 is not satisfied, we cannot apply Theorem 12.17 to Problem (SLQ) directly. The proof is divided into two steps.

Step 1. In this step, we prove (13.103) by the convex perturbation technique.

For the optimal pair $(\bar{x}(\cdot), \bar{u}(\cdot))$ and a control $u(\cdot) \in L^2_{\mathbb{F}}(0,T;U)$, we have that

$$u^\varepsilon(\cdot) = \bar{u}(\cdot) + \varepsilon(u(\cdot) - \bar{u}(\cdot)) = (1-\varepsilon)\bar{u}(\cdot) + \varepsilon u(\cdot) \in L^2_{\mathbb{F}}(0,T;U), \quad \forall\, \varepsilon \in [0,1].$$

Denote by $x^\varepsilon(\cdot)$ the solution of (13.1) corresponding to the control $u^\varepsilon(\cdot)$. It is easy to get that

$$|x^\varepsilon|_{C_{\mathbb{F}}([0,T];L^2(\Omega;H))} \leq \mathcal{C}\big(1 + |\eta|_H\big), \quad \forall\, \varepsilon \in [0,1]. \tag{13.105}$$

Write $x_1^\varepsilon(\cdot) = \dfrac{1}{\varepsilon}\big(x^\varepsilon(\cdot) - \bar{x}(\cdot)\big)$ and $\delta u(\cdot) = u(\cdot) - \bar{u}(\cdot)$. Then $x_1^\varepsilon(\cdot)$ solves the following stochastic evolution equation:

$$\begin{cases} dx_1^\varepsilon = \big(Ax_1^\varepsilon + B\delta u\big)dt + \big(Cx_1^\varepsilon + D\delta u\big)dW(t) & \text{in } (0,T], \\ x_1^\varepsilon(s) = 0. \end{cases} \tag{13.106}$$

Since $(\bar{x}(\cdot), \bar{u}(\cdot))$ is an optimal pair of Problem (SLQ) , we have

$$\begin{aligned} 0 &\leq \lim_{\varepsilon \to 0} \frac{\mathcal{J}(\eta; u^\varepsilon(\cdot)) - \mathcal{J}(\eta; \bar{u}(\cdot))}{\varepsilon} \\ &= \mathbb{E}\int_0^T \big(\langle M\bar{x}, x_1^\varepsilon\rangle_H + \langle R\bar{u}, \delta u\rangle_U\big)dt + \mathbb{E}\big\langle G\bar{x}(T), x_1^\varepsilon(T)\big\rangle_H. \end{aligned} \tag{13.107}$$

It follows from the definition of the transposition solution to (13.101) that

$$
\begin{aligned}
&-\mathbb{E}\big\langle G\bar{x}(T), x_1^\varepsilon(T)\big\rangle_H + \mathbb{E}\int_0^T \big\langle M\bar{x}, x_1^\varepsilon\big\rangle_H dt \\
&= \mathbb{E}\int_0^T \big(\big\langle B\delta u, y\big\rangle_H + \big\langle D\delta u, Y\big\rangle_{\mathcal{L}_2^0}\big)dt.
\end{aligned}
\tag{13.108}
$$

Combining (13.107) and (13.108), we find that

$$
\mathbb{E}\int_0^T \big\langle R\bar{u} - B^*y - D^*Y, u - \bar{u}\big\rangle_U dt \geq 0, \quad \forall\, u(\cdot) \in L^2_{\mathbb{F}}(0,T;U).
$$

Hence, by Lemma 12.3, we conclude that

$$
\begin{aligned}
&\big\langle R(t)\bar{u}(t) - B(t)^*y(t) - D(t)^*Y(t), u - \bar{u}(t)\big\rangle_U \geq 0, \\
&\qquad\qquad\qquad\qquad \text{a.e. } [0,T]\times\Omega,\ \forall\, u \in U.
\end{aligned}
\tag{13.109}
$$

This implies (13.103).

Step 2. In this step, we prove (13.104) by the spike variation method.

For each $\varepsilon > 0$ and $\tau \in [0, T-\varepsilon)$, let $E_\varepsilon = [\tau, \tau+\varepsilon]$. Let $u \in L^2_{\mathbb{F}}(0,T;U)$ be such that $|\bar{u} - u|_{L^\infty_{\mathbb{F}}(0,T;U)} < +\infty$. Put

$$
u^\varepsilon(t) = \begin{cases} \bar{u}(t), & \text{if } t \in [0,T]\setminus E_\varepsilon, \\ u(t), & \text{if } t \in E_\varepsilon. \end{cases}
\tag{13.110}
$$

Let $x^\varepsilon(\cdot)$ be the solution to (13.1) corresponding to the control $u^\varepsilon(\cdot)$. Consider the following two stochastic evolution equations:

$$
\begin{cases} dx_2^\varepsilon = Ax_2^\varepsilon dt + \big(Cx_2^\varepsilon + \chi_{E_\varepsilon} D\delta u\big)dW(t) & \text{in } (0,T], \\ x_2^\varepsilon(0) = 0 \end{cases}
$$

and

$$
\begin{cases} dx_3^\varepsilon = \big(Ax_3^\varepsilon + \chi_{E_\varepsilon} B\delta u\big)dt + Cx_3^\varepsilon dW(t) & \text{in } (0,T], \\ x_3^\varepsilon(0) = 0. \end{cases}
$$

Clearly, $x^\varepsilon - \bar{x} = x_2^\varepsilon + x_3^\varepsilon$. Similar to the proof of (12.181), we can obtain that

$$
\begin{cases} |x_2^\varepsilon(\cdot)|_{L^\infty_{\mathbb{F}}(0,T;L^2(\Omega;H))} \leq \mathcal{C}\sqrt{\varepsilon}, \\ |x_3^\varepsilon(\cdot)|_{L^\infty_{\mathbb{F}}(0,T;L^2(\Omega;H))} \leq \mathcal{C}\varepsilon. \end{cases}
\tag{13.111}
$$

Similar to the proof of (12.208), we can get that

$$
\begin{aligned}
&\mathcal{J}(\eta; u^\varepsilon(\cdot)) - \mathcal{J}(\eta; \bar{u}(\cdot)) \\
&= \mathbb{E}\int_0^T \Big[\big\langle M\bar{x}, x_2^\varepsilon + x_3^\varepsilon\big\rangle_H + \frac{1}{2}\big\langle Mx_2^\varepsilon, x_2^\varepsilon\big\rangle_H \\
&\qquad + \chi_{E_\varepsilon}\big(\big\langle R\delta u, \delta u\big\rangle_U + \frac{1}{2}\big\langle R\delta u, \delta u\big\rangle_U\big)\Big]dt \\
&\quad + \mathbb{E}\big\langle G\bar{x}(T), x_2^\varepsilon(T) + x_3^\varepsilon(T)\big\rangle_H + \frac{1}{2}\mathbb{E}\big\langle Gx_2^\varepsilon(T), x_2^\varepsilon(T)\big\rangle_H + o(\varepsilon).
\end{aligned}
\tag{13.112}
$$

It follows from the definition of the transposition solution to (13.101) that

$$
\begin{aligned}
&-\mathbb{E}\big\langle G\bar{x}(T), x_2^\varepsilon(T)+x_3^\varepsilon(T)\big\rangle_H+\mathbb{E}\int_0^T \langle M\bar{x}, x_2^\varepsilon+x_3^\varepsilon\rangle_H dt \\
&=\mathbb{E}\int_0^T \big(\langle B\delta u, y\rangle_H+\langle D\delta u, Y\rangle_{\mathcal{L}_2^0}\big)dt.
\end{aligned}
\tag{13.113}
$$

By the definition of the relaxed transposition solution to (13.102), we obtain that

$$
\begin{aligned}
&-\mathbb{E}\big\langle Gx_2^\varepsilon(T), x_2^\varepsilon(T)\big\rangle_H+\mathbb{E}\int_0^T \langle Mx_2^\varepsilon, x_2^\varepsilon\rangle_H dt \\
&=\mathbb{E}\int_0^T \chi_{E_\varepsilon}\langle Cx_2^\varepsilon, P^*D\delta u\rangle_{\mathcal{L}_2^0}dt+\mathbb{E}\int_0^T \chi_{E_\varepsilon}\langle PD\delta u, Cx_2^\varepsilon\rangle_{\mathcal{L}_2^0}dt \\
&\quad+\mathbb{E}\int_0^T \chi_{E_\varepsilon}\langle PD\delta u, D\delta u\rangle_{\mathcal{L}_2^0}dt \\
&\quad+\mathbb{E}\int_0^T \chi_{E_\varepsilon}\big\langle D\delta u, \widehat{Q}^{(0)}(0,0,\chi_{E_\varepsilon}D\delta u)\big\rangle_{\mathcal{L}_2^0}dt \\
&\quad+\mathbb{E}\int_0^T \chi_{E_\varepsilon}\langle Q^{(0)}(0,0,\chi_{E_\varepsilon}D\delta u), D\delta u\rangle_{\mathcal{L}_2^0}dt.
\end{aligned}
\tag{13.114}
$$

From (13.111), we have that

$$
\begin{aligned}
&\Big|\mathbb{E}\int_0^T \chi_{E_\varepsilon}\langle Cx_2^\varepsilon, P^*\chi_{E_\varepsilon}D\delta u\rangle_{\mathcal{L}_2^0}dt+\mathbb{E}\int_0^T \chi_{E_\varepsilon}\langle P\chi_{E_\varepsilon}D\delta u, Cx_2^\varepsilon\rangle_{\mathcal{L}_2^0}dt\Big| \\
&=o(\varepsilon).
\end{aligned}
\tag{13.115}
$$

Similarly to the proof of (12.227), one can show that

$$
\begin{aligned}
&\Big|\mathbb{E}\int_0^T \chi_{E_\varepsilon}\big\langle D\delta u, \widehat{Q}^{(0)}(0,0,\chi_{E_\varepsilon}D\delta u)\big\rangle_{\mathcal{L}_2^0}dt \\
&\quad+\mathbb{E}\int_0^T \chi_{E_\varepsilon}\langle Q^{(0)}(0,0,\chi_{E_\varepsilon}D\delta u), D\delta u\rangle_{\mathcal{L}_2^0}dt\Big| \\
&=o(\varepsilon), \quad \text{as } \varepsilon\to 0.
\end{aligned}
\tag{13.116}
$$

By (13.112)–(13.116), we obtain

$$
\begin{aligned}
&\mathcal{J}(\eta;u^\varepsilon(\cdot))-\mathcal{J}(\eta;\bar{u}(\cdot)) \\
&=\mathbb{E}\int_0^T \chi_{E_\varepsilon}(t)\langle R\bar{u}-B^*y-D^*Y, \delta u\rangle_U dt \\
&\quad+\frac{1}{2}\mathbb{E}\int_0^T \chi_{E_\varepsilon}\Big(\langle R\delta u, \delta u\rangle_U-\langle PD\delta u, D\delta u\rangle_{\mathcal{L}_2^0}\Big)dt+o(\varepsilon) \\
&=\frac{1}{2}\mathbb{E}\int_0^T \chi_{E_\varepsilon}(t)\big\langle (R-D^*PD)\delta u, \delta u\big\rangle_U dt+o(\varepsilon).
\end{aligned}
$$

Since $\bar{u}(\cdot)$ is an optimal control, $\mathcal{J}(\eta; u^{\varepsilon}(\cdot)) - \mathcal{J}(\eta; \bar{u}(\cdot)) \geq 0$. Thus,

$$\frac{1}{2}\mathbb{E}\int_0^T \chi_{E_\varepsilon}\big\langle \big(R - D^*PD\big)\delta u, \delta u\big\rangle_U dt \geq o(\varepsilon), \tag{13.117}$$

as $\varepsilon \to 0$. By (13.117), similarly to Step 7 in the proof of Theorem 12.17, we can show that for all $u \in U$,

$$\big\langle \big(R - D^*PD\big)u, u\big\rangle_U \geq 0, \qquad \text{a.e. } (t, \omega) \in [0, T] \times \Omega,$$

which gives (13.104). This completes the proof of Theorem 13.30. □

Next, we introduce the following decoupled *forward-backward stochastic evolution equation*:

$$\begin{cases} dx = \big(Ax + Bu\big)dt + \big(Cx + Du\big)dW(t) & \text{in } (0, T], \\ dy = -\big(A^*y - Mx + C^*Y\big)dt + YdW(t) & \text{in } [0, T), \\ x(0) = \eta, \qquad y(T) = -Gx(T). \end{cases} \tag{13.118}$$

We call $(x(\cdot), y(\cdot), Y(\cdot))$ a transposition solution to the equation (13.118) if $x(\cdot)$ is the mild solution to the forward stochastic evolution equation and $(y(\cdot), Y(\cdot))$ is the transposition solution of the backward one.

Since the equation (13.118) is decoupled, its well-posedness is easy to be obtained. Given $\eta \in H$ and $u(\cdot) \in L^2_{\mathbb{F}}(0, T; U)$, one can first solve the forward one to get $x(\cdot)$, and then solve the backward one. Consequently, the equation (13.118) admits a unique transposition solution $(x(\cdot), y(\cdot), Y(\cdot))$ corresponding to η and $u(\cdot)$. The following result is a consequence of Proposition 13.27.

Proposition 13.31. *For any* $(\eta, u(\cdot)) \in H \times L^2_{\mathbb{F}}(0, T; U)$, *let* $(x(\cdot), y(\cdot), Y(\cdot))$ *be the transposition solution to* (13.118). *Then*

$$\begin{aligned} \big(\mathcal{N}u + \mathcal{H}(\eta)\big)(t) = Ru(t) - B^*y(t) - D^*Y(t), \\ \text{a.e. } (t, \omega) \in [0, T] \times \Omega. \end{aligned} \tag{13.119}$$

In particular, if $\big(x_0(\cdot), y_0(\cdot), Y_0(\cdot)\big)$ *is the transposition solution to* (13.118) *with* $\eta = 0$, *then*

$$(\mathcal{N}u)(t) = R(t)u(t) - B(t)^*y_0(t) - D(t)^*Y_0(t), \quad \text{a.e. } (t, \omega) \in [0, T] \times \Omega. \tag{13.120}$$

Proof: Let $(x(\cdot), y(\cdot), Y(\cdot))$ be the transposition solution of (13.118). From (13.96), we obtain that

$$\begin{aligned} &\big(\mathcal{N}u + \mathcal{H}(\eta)\big)(\cdot) \\ &= \big[(R + F_1^*MF_1 + \widehat{F}_1^*G\widehat{F}_1)u\big](\cdot) + (F_1^*MF_2\eta)(\cdot) + (\widehat{F}_1^*G\widehat{F}_2\eta)(\cdot) \\ &= (Ru)(\cdot) + F_1^*M\big[(F_2\eta)(\cdot) + F_1u(\cdot)\big] + \widehat{F}_1^*G(\widehat{F}_2\eta + \widehat{F}_1u) \\ &= (Ru)(\cdot) + (F_1^*Mx)(\cdot) + \big(\widehat{F}_1^*Gx(T)\big)(\cdot). \end{aligned} \tag{13.121}$$

By (13.92) and (13.93), we know that

$$(F_1^* Mx)(\cdot) + \big(\widehat{F}_1^* Gx(T)\big)(\cdot) = -B(t)^* y(\cdot) - D(t)^* Y(\cdot).$$

This, together with (13.121), implies (13.119) immediately. At last, (13.120) follows from (13.119) as a special case. □

The following result (for which the conditions are given in terms of the transposition solution to the forward-backward stochastic evolution equation (13.118)) is an immediate corollary of Theorem 13.29.

Theorem 13.32. *Problem (SLQ) is solvable at $\eta \in H$ with an optimal pair $(\bar{x}(\cdot), \bar{u}(\cdot))$ if and only if the following two conditions hold:*

1) *The unique transposition solution $(x(\cdot), y(\cdot), Y(\cdot))$ to (13.118) with $u(\cdot) = \bar{u}(\cdot)$, verifies that $x(\cdot) = \bar{x}(\cdot)$ and*

$$R(t)\bar{u}(t) - B(t)^* y(t) - D(t)^* Y(t) = 0, \quad \text{a.e. } (t,\omega) \in [0,T] \times \Omega. \tag{13.122}$$

2) *For any $u(\cdot) \in L^2_{\mathbb{F}}(0,T;U)$, the unique transposition solution $(x_0(\cdot), y_0(\cdot), Y_0(\cdot))$ to (13.118) with $\eta = 0$ satisfies*

$$\mathbb{E}\int_0^T \big\langle Ru - B^* y_0 - D^* Y_0, u\big\rangle_U dt \geq 0. \tag{13.123}$$

Proof: The **"only if"** part. The equality (13.122) follows from Theorem 13.30 while (13.123) follows from Theorem 13.29 and (13.120).

The **"if"** part. By Proposition 13.31, the inequality (13.123) is equivalent to $\mathcal{N} \geq 0$. Now, let $(\bar{x}(\cdot), y(\cdot), Y(\cdot))$ be a transposition solution to the equation (13.118) such that (13.122) holds. Then, by Proposition 13.31, we see that (13.122) is the same as (13.98). Hence by Theorem 13.29, Problem (SLQ) is solvable. □

Theorem 13.32 is nothing but a restatement of Theorem 13.29. However, when $R(t)$ is invertible for all t and

$$R(\cdot)^{-1} \in L^\infty_{\mathbb{F}}(0,T;\mathcal{L}(U)), \tag{13.124}$$

it gives us a way to find the optimal control by solving the following coupled forward-backward stochastic evolution equation:

$$\begin{cases} d\bar{x} = \big(A\bar{x} + BR^{-1}B^* y + BR^{-1}D^* Y\big)dt \\ \qquad\qquad + \big(C\bar{x} + DR^{-1}B^* y + DR^{-1}D^* Y\big)dW(t) & \text{in } (0,T], \\ dy(t) = -\big(A^* y + M\bar{x} - C^* Y\big)dt + Y dW(t) & \text{in } [0,T), \\ \bar{x}(0) = \eta, \qquad y(T) = -G\bar{x}(T). \end{cases} \tag{13.125}$$

As a direct consequence of Theorem 13.32, we have the following result.

Corollary 13.33. *Let* (13.124) *hold and* $\mathcal{N} \geq 0$. *Then Problem (SLQ) is uniquely solvable at* $\eta \in H$ *if and only if the forward-backward stochastic evolution equation* (13.125) *admits a unique transposition solution* $(\bar{x}(\cdot), y(\cdot), Y(\cdot))$. *In this case, the optimal control is given by*

$$\bar{u}(t) = R(t)^{-1}\big[B(t)^* y(t) + D(t)^* Y(t)\big], \quad \text{a.e. } (t,\omega) \in [0,T] \times \Omega. \tag{13.126}$$

When Problem (SLQ) is a standard SLQ problem, (13.124) holds and $\mathcal{N} \gg 0$. Consequently, we have the following result.

Corollary 13.34. *If Problem (SLQ) is standard, then the equation* (13.125) *admits a unique transposition solution* $(\bar{x}(\cdot), y(\cdot), Y(\cdot))$ *and Problem (SLQ) is uniquely solvable with the optimal control given by* (13.126).

13.6 Transposition Solutions to Operator-Valued Backward Stochastic Riccati Equations

From this section to Section 13.8, we shall study the relationship between the existence of optimal feedback controls for Problem (SLQ) and the global solvability of the operator-valued backward stochastic Riccati equation (13.17).

Though the equations (13.17) and (13.47) are in the same form, there exists an essential difference between them. Indeed, similarly to the relationship between the equations (12.8) and (12.6) (in Chapter 12), (13.47) is an $\mathbb{R}^{n\times n}$ (matrix)-valued backward stochastic differential equation (which can be easily regarded as an $\mathbb{R}^{n^2}$-valued backward stochastic differential equation), and therefore, the desired well-posedness follows from the one for backward stochastic differential equations valued in $\mathbb{R}^{n^2}$. Also, as mentioned before, there exists no such a stochastic integration/evolution equation theory in general Banach spaces that can be employed to treat the well-posedness of (13.17) in the usual sense.

On the other hand, compared with the linear operator-valued backward stochastic evolution equation (12.8), it is clear that (13.17) is a nonlinear operator-valued backward stochastic evolution equation with quadratical nonlinearities. Generally speaking, in order to study the difficult (deterministic or stochastic) nonlinear partial differential equations, people need to introduce suitable new concept of solutions, such as viscosity solutions for Hamilton-Jacobi equations ([64]) and fully nonlinear second-order equations ([213]), and renormalized solutions for the KPZ equation ([136]). In order to overcome the difficulties mentioned above, similarly to Chapter 12, we employ the transposition method to study (13.17). More precisely, we need to introduce another type of solution, i.e., transposition solution to this equation. To this end, let us introduce the following assumptions:

(**AS1**) *The eigenvectors* $\{e_j\}_{j=1}^{\infty}$ *of* A *such that* $|e_j|_H = 1$ *for all* $j \in \mathbb{N}$ *constitute an orthonormal basis of* H.

Let $\{\mu_j\}_{j=1}^{\infty}$ (corresponding to $\{e_j\}_{j=1}^{\infty}$) be the eigenvalues of A. Let $\{\lambda_j\}_{j=1}^{\infty} \in \ell^2$ be an arbitrarily given sequence of positive real numbers. Define a norm $|\cdot|_{H_\lambda}$ on H as follows:

$$|h|_{H_\lambda} = \sqrt{\sum_{j=1}^{\infty} \lambda_j^2 |e_j|_{D(A)}^{-2} h_j^2}, \qquad \forall\, h = \sum_{j=1}^{\infty} h_j e_j \in H.$$

Denote by H_λ the completion of H with respect to this norm. Clearly, H_λ is a Hilbert space, $H \subset H_\lambda$ and $\{\lambda_j^{-1}|e_j|_{D(A)} e_j\}_{j=1}^{\infty}$ is an orthonormal basis of H_λ. Write $\mathcal{V}_H$ for the set of all such kind of Hilbert spaces.

Write H'_λ for the dual space of H_λ with respect to the pivot space $H \equiv H'$. For any $H_\lambda \in \mathcal{V}_H$, from the definition of H_λ, it is easy to see that $\{\lambda_j |e_j|_{D(A)}^{-1} e_j\}_{j=1}^{\infty} \subset H'_\lambda$ is an orthonormal basis of H'_λ and the norm on H'_λ is given by

$$|\xi|_{H'_\lambda} = \sqrt{\sum_{j=1}^{\infty} \xi_j^2 |e_j|_{D(A)}^2 \lambda_j^{-2}}, \qquad \forall\, \xi \in H'_\lambda,$$

where $\xi_j = \langle \xi, e_j \rangle_H$.

We also need the following technical condition:

(**AS2**) *There exists $H_\lambda \in \mathcal{V}_H$ such that $C \in L^\infty_{\mathbb{F}}(0,T;\mathcal{L}(H_\lambda;\mathcal{L}_2(V;H_\lambda)))$, $M \in L^\infty_{\mathbb{F}}(0,T;\mathcal{L}(H_\lambda))$, $G \in L^\infty_{\mathcal{F}_T}(\Omega;\mathcal{L}(H_\lambda))$ and $C \in L^\infty_{\mathbb{F}}(0,T;\mathcal{L}(H'_\lambda;\mathcal{L}_2(V;H'_\lambda)))$, $M \in L^\infty_{\mathbb{F}}(0,T;\mathcal{L}(H'_\lambda))$, $G \in L^\infty_{\mathcal{F}_T}(\Omega;\mathcal{L}(H'_\lambda))$.*

Remark 13.35. In (**AS2**), for $C \in L^\infty_{\mathbb{F}}(0,T;\mathcal{L}(H_\lambda;\mathcal{L}_2(V;H_\lambda)))$, we mean that for a.e. $(t,\omega) \in [0,T] \times \Omega$, $C(t,\omega)$ can be extended to be a bounded linear operator $C_{H_\lambda}(t,\omega)$ from H_λ to $\mathcal{L}_2(V;H_\lambda)$ and after the extension, $C_{H_\lambda} \in L^\infty_{\mathbb{F}}(0,T;\mathcal{L}(H_\lambda;\mathcal{L}_2(V;H_\lambda)))$. For simplicity of notations, we still denote the extension by C if there is no confusion. On the other hand, $C \in L^\infty_{\mathbb{F}}(0,T;\mathcal{L}(H'_\lambda;\,\mathcal{L}_2(V;H'_\lambda)))$ means for a.e. $(t,\omega) \in [0,T] \times \Omega$, the restriction of $C(t,\omega)$ on H'_λ (denoted by $C_{H'_\lambda}(t,\omega)$) belongs to $\mathcal{L}(H'_\lambda;\mathcal{L}_2(V;H'_\lambda))$ and $C_{H'_\lambda} \in L^\infty_{\mathbb{F}}(0,T;\mathcal{L}(H'_\lambda;\mathcal{L}_2(V;H'_\lambda)))$. The other notations in (**AS2**) can be understood in a similar way.

Lemma 13.36. *Let $H_\lambda \in \mathcal{V}_H$ and **(AS2)** hold. If $\{S(t)\}_{t\in\mathbb{R}}$ is a C_0-group (resp. C_0-semigroup) on H, then it is a C_0-group on H'_λ (resp. C_0-semigroup), and it can be uniquely extended to a C_0-group (resp. C_0-semigroup) (also denoted by itself) on H_λ.*

Proof: We only prove that $\{S(t)\}_{t\geq 0}$ is a C_0-group on H'_λ. Proofs of the other conclusions are similar.

Let $\xi = \sum_{j=1}^{\infty} \xi_j \lambda_j |e_j|_{D(A)}^{-1} e_j \in H'_\lambda$ with $\{\xi_j\}_{j=1}^{\infty} \in \ell^2$. Then, $\tilde{\xi} = \sum_{j=1}^{\infty} \xi_j e_j \in H$ and $|\tilde{\xi}|_H = |\xi|_{H'_\lambda}$. Clearly,

$$S(t)\xi = \sum_{j=1}^{\infty} \xi_j \lambda_j |e_j|_{D(A)}^{-1} e^{\mu_j t} e_j.$$

For any $t_1, t_2 \in \mathbb{R}$,

$$\begin{aligned} S(t_2)S(t_1)\xi &= \sum_{j=1}^{\infty} \xi_j \lambda_j |e_j|_{D(A)}^{-1} S(t_2)S(t_1)e_j \\ &= \sum_{j=1}^{\infty} \xi_j \lambda_j |e_j|_{D(A)}^{-1} e^{\mu_j(t_1+t_2)} e_j = S(t_2+t_1)\xi. \end{aligned} \tag{13.127}$$

This indicates that $\{S(t)\}_{t\geq 0}$ is a group on H'_λ.

For any $t_2 > t_1 > 0$,

$$\begin{aligned} &\left|(S(t_2) - S(t_1))\xi\right|_{H'_\lambda} = \Big|\sum_{j=1}^{\infty} \xi_j \lambda_j |e_j|_{D(A)}^{-1} (S(t_2) - S(t_1))e_j\Big|_{H'_\lambda} \\ &= \Big|\sum_{j=1}^{\infty} \xi_j \lambda_j |e_j|_{D(A)}^{-1} \big(e^{\mu_j t_2} - e^{\mu_j t_1}\big) e_j\Big|_{H'_\lambda} \\ &= \Big[\sum_{j=1}^{\infty} \xi_j^2 \big(e^{\mu_j t_2} - e^{\mu_j t_1}\big)^2\Big]^{\frac{1}{2}} = \left|(S(t_2) - S(t_1))\tilde{\xi}\right|_H. \end{aligned}$$

This, together with that $\{S(t)\}_{t\geq 0}$ is strongly continuous on H, implies that $\{S(t)\}_{t\geq 0}$ is strongly continuous on H'_λ. □

Next, similarly to Subsection 12.4.1, we need to introduce some more concepts/notations. Beginners may skip this and only consider the case that $V = \mathbb{R}$ in the rest of this chapter.

Let $\{\tilde{e}_j\}_{j=1}^{\infty}$ be an orthonormal basis of V. Any $\Lambda \in \mathcal{L}(\mathcal{L}_2(V; H'_\lambda); H_\lambda)$ induces a bounded bilinear functional $\Psi(\cdot,\cdot)$ on $H'_\lambda \times \mathcal{L}_2(V; H'_\lambda)$ as follows:

$$\Psi(h, v) \overset{\Delta}{=} \sum_{j=1}^{\infty} \langle \Lambda(\tilde{e}_j \otimes h), v\tilde{e}_j \rangle_{H_\lambda, H'_\lambda}, \quad \forall\, h \in H'_\lambda,\ v \in \mathcal{L}_2(V; H'_\lambda). \tag{13.128}$$

Here, $\tilde{e}_j \otimes h$ is defined by (2.18) (Clearly, $\tilde{e}_j \otimes h \in \mathcal{L}_2(V; H'_\lambda)$, and therefore $\Lambda(\tilde{e}_j \otimes h) \in H_\lambda$).

Define a bounded linear operator $\widetilde{\Lambda} : H'_\lambda \to \mathcal{L}_2(V; H_\lambda)$ as follows:

$$\Psi(h, v) = \langle \widetilde{\Lambda} h, v \rangle_{\mathcal{L}_2(V;H_\lambda), \mathcal{L}_2(V;H'_\lambda)}, \qquad \forall\, h \in H'_\lambda,\ v \in \mathcal{L}_2(V; H'_\lambda).$$

It follows from (12.34) that

$$\begin{aligned} \langle \widetilde{\Lambda} h, v \rangle_{\mathcal{L}_2(V;H_\lambda), \mathcal{L}_2(V;H'_\lambda)} &= \sum_{j=1}^{\infty} \langle \Lambda(\tilde{e}_j \otimes h), v\tilde{e}_j \rangle_{H_\lambda, H'_\lambda}, \\ &\qquad \forall\, h \in H'_\lambda, v \in \mathcal{L}_2(V; H'_\lambda). \end{aligned} \tag{13.129}$$

On the other hand, any $\Lambda \in \mathcal{L}(H'_\lambda; \mathcal{L}_2(V; H_\lambda))$ induces a bounded linear operator $\widetilde{\Lambda} : \mathcal{L}_2(V; H'_\lambda) \to H_\lambda$ as follows:

$$\widetilde{\Lambda}\Big(\sum_{i,j=1}^{\infty} a_{ij}\tilde{e}_i \otimes h_j\Big) \stackrel{\Delta}{=} \sum_{i,j=1}^{\infty} a_{ij}(\Lambda h_j)\tilde{e}_i, \quad \forall\, a_{ij} \in \mathbb{C} \text{ with } \sum_{i,j=1}^{\infty} |a_{ij}|^2 < \infty. \tag{13.130}$$

Clearly, if $\widetilde{\Lambda} \in \mathcal{L}(\mathcal{L}_2(V; H'_\lambda); H_\lambda)$, then

$$\widetilde{\Lambda}(v \otimes h) = (\Lambda h)v, \qquad \forall\, (v, h) \in V \times H'_\lambda. \tag{13.131}$$

In this case, we say that Λ induces an operator $\widetilde{\Lambda} \in \mathcal{L}(\mathcal{L}_2(V; H'_\lambda); H_\lambda)$.

Now, suppose that $\Lambda \in \mathcal{L}(\mathcal{L}_2(V; H'_\lambda); H_\lambda)$ induces an operator $\widetilde{\Lambda} \in \mathcal{L}(H'_\lambda; \mathcal{L}_2(V; H_\lambda))$ and $\widetilde{\Lambda}$ induces an operator $\overline{\Lambda} \in \mathcal{L}(\mathcal{L}_2(V; H'_\lambda); H_\lambda)$. Then, $\Lambda = \overline{\Lambda}$. Conversely, suppose that $\Lambda \in \mathcal{L}(H'_\lambda; \mathcal{L}_2(V; H_\lambda))$ induces an operator $\widetilde{\Lambda} \in \mathcal{L}(\mathcal{L}_2(V; H'_\lambda); H_\lambda)$ and $\widetilde{\Lambda}$ induces an operator $\overline{\Lambda} \in \mathcal{L}(H'_\lambda; \mathcal{L}_2(V; H_\lambda))$. Then, $\Lambda = \overline{\Lambda}$.

We shall need the following result.

Proposition 13.37. *Any $\Xi \in \mathcal{L}_2(V; \mathcal{L}_2(H'_\lambda; H_\lambda))$ induces (uniquely) an operator $\Lambda \in \mathcal{L}_2(\mathcal{L}_2(V; H'_\lambda); H_\lambda)$ and an operator $\widetilde{\Lambda} \in \mathcal{L}_2(H'_\lambda; \mathcal{L}_2(V; H_\lambda))$ satisfying* (13.130) *and*

$$\Lambda(v \otimes h) = (\Xi v)h, \qquad \forall\, (v, h) \in H_\lambda \times H. \tag{13.132}$$

Moreover,

$$|\Lambda|_{\mathcal{L}_2(\mathcal{L}_2(V; H'_\lambda); H_\lambda)} + |\widetilde{\Lambda}|_{\mathcal{L}_2(H'_\lambda; \mathcal{L}_2(V; H_\lambda))} \leq \mathcal{C}|\Xi|_{\mathcal{L}_2(V; \mathcal{L}_2(H'_\lambda; H_\lambda))}.$$

Proof: As in (12.36), we define a linear operator Λ from $\mathcal{L}_2(V; H'_\lambda)$ to H_λ by

$$\Lambda\Big(\sum_{i,j=1}^{\infty} a_{ij}\tilde{e}_i \otimes h_j\Big) \stackrel{\Delta}{=} \sum_{i,j=1}^{\infty} a_{ij}(\Xi\tilde{e}_i)h_j, \quad \forall\, a_{ij} \in \mathbb{C} \text{ with } \sum_{i,j=1}^{\infty} |a_{ij}|^2 < \infty.$$

By $\Xi \in \mathcal{L}_2(V; \mathcal{L}_2(H'_\lambda; H_\lambda))$, it is easy to check that $\Lambda \in \mathcal{L}_2(\mathcal{L}_2(V; H'_\lambda); H_\lambda)$ and

$$|\Lambda|_{\mathcal{L}_2(\mathcal{L}_2(V; H'_\lambda); H_\lambda)} \leq \mathcal{C}|\Xi|_{\mathcal{L}_2(V; \mathcal{L}_2(H'_\lambda; H_\lambda))}.$$

Then, Λ induces a bounded linear operator $\widetilde{\Lambda} \in \mathcal{L}_2(H'_\lambda; \mathcal{L}_2(V; H_\lambda))$ satisfying (13.130) and

$$|\widetilde{\Lambda}|_{\mathcal{L}_2(H'_\lambda; \mathcal{L}_2(V; H_\lambda))} \leq \mathcal{C}|\Lambda|_{\mathcal{L}_2(\mathcal{L}_2(V; H'_\lambda); H_\lambda)}.$$

This completes the proof of Proposition 13.37. □

Now, let us consider the following two (forward) stochastic evolution e-quations:

$$\begin{cases} dx_1 = \big(Ax_1 + u_1\big)d\tau + \big(Cx_1 + v_1\big)dW(\tau) & \text{in } (t,T], \\ x_1(t) = \xi_1 \end{cases} \tag{13.133}$$

and

$$\begin{cases} dx_2 = \big(Ax_2 + u_2\big)d\tau + \big(Cx_2 + v_2\big)dW(\tau) & \text{in } (t,T], \\ x_2(t) = \xi_2. \end{cases} \tag{13.134}$$

Here $t \in [0,T)$, ξ_1, ξ_2 are suitable random variables and u_1, u_2, v_1, v_2 are suitable stochastic processes.

By Lemma 13.36 and Theorem 3.20, we obtain the following result immediately.

Corollary 13.38. *Let* ***(AS1)–(AS2)*** *hold. Then, for* $j = 1, 2$, *for any* $\xi_j \in L^4_{\mathcal{F}_t}(\Omega; H'_\lambda)$, $u_j(\cdot) \in L^4_{\mathbb{F}}(\Omega; L^2(t,T; H'_\lambda))$ *and* $v_j(\cdot) \in L^4_{\mathbb{F}}(\Omega; L^2(t,T; \mathcal{L}_2(V; H'_\lambda)))$, *the mild solution* $x_1(\cdot)$ *(resp.* $x_2(\cdot)$*) to* (13.133) *(resp.* (13.134)*) belongs to* $C_{\mathbb{F}}([t,T]; L^4(\Omega; H'_\lambda))$.

Put

$$\begin{aligned} & C_{\mathbb{F},w}([0,T]; L^\infty(\Omega; \mathcal{L}(H))) \\ & \triangleq \Big\{ P \in \Upsilon_2(H) \;\Big|\; P(t,\omega) \in \mathcal{S}(H), \text{ a.e. } (t,\omega) \in [0,T] \times \Omega, \ |P(\cdot)|_{\mathcal{L}(H)} \in L^\infty_{\mathbb{F}}(0,T), \\ & \qquad\qquad \text{and } P(\cdot)\zeta \in C_{\mathbb{F}}([0,T]; L^\infty(\Omega; H)), \ \forall\, \zeta \in H \Big\} \end{aligned}$$

and

$$\begin{aligned} & L^2_{\mathbb{F},w}(0,T; \mathcal{L}(H)) \\ & \triangleq \Big\{ \Lambda \in L^2_{\mathbb{F}}(0,T; \mathcal{L}_2(\mathcal{L}_2(V; H'_\lambda); H_\lambda)) \Big| D^*\Lambda \in \Upsilon_2(U; H), \text{ a.e. } (t,\omega) \in [0,T] \times \Omega \Big\}. \end{aligned}$$

Now, we introduce the notion of *transposition solution* to (13.17):

Definition 13.39. *A pair of operator-valued stochastic processes* $\big(P(\cdot), \Lambda(\cdot)\big) \in C_{\mathbb{F},w}([0,T]; L^\infty(\Omega; \mathcal{L}(H))) \times L^2_{\mathbb{F},w}(0,T; \mathcal{L}(H))$ *is called a transposition solution to* (13.17) *if the following three conditions hold:*

1) $K(t,\omega)\big(\equiv R(t,\omega) + D(t,\omega)^* P(t,\omega) D(t,\omega)\big) > 0$ *and its left inverse* $K(t,\omega)^{-1}$ *is a densely defined closed operator for a.e.* $(t,\omega) \in [0,T] \times \Omega$;

2) *For any* $t \in [0,T]$, $\xi_1, \xi_2 \in L^4_{\mathcal{F}_t}(\Omega; H'_\lambda)$, $u_1(\cdot), u_2(\cdot) \in L^4_{\mathbb{F}}(\Omega; L^2(t,T; H'_\lambda))$ *and* $v_1(\cdot), v_2(\cdot) \in L^4_{\mathbb{F}}(\Omega; L^2(t,T; \mathcal{L}_2(V; H'_\lambda)))$, *it holds that*

$$\begin{aligned} & \mathbb{E}\langle Gx_1(T), x_2(T)\rangle_H + \mathbb{E}\int_t^T \big\langle M(\tau)x_1(\tau), x_2(\tau)\big\rangle_H d\tau \\ & -\mathbb{E}\int_t^T \big\langle K(\tau)^{-1}L(\tau)x_1(\tau), L(\tau)x_2(\tau)\big\rangle_H d\tau \end{aligned}$$

$$
\begin{aligned}
&= \mathbb{E}\big\langle P(t)\xi_1, \xi_2\big\rangle_H + \mathbb{E}\int_t^T \big\langle P(\tau)u_1(\tau), x_2(\tau)\big\rangle_H d\tau \\
&\quad + \mathbb{E}\int_t^T \big\langle P(\tau)x_1(\tau), u_2(\tau)\big\rangle_H d\tau + \mathbb{E}\int_t^T \big\langle P(\tau)C(\tau)x_1(\tau), v_2(\tau)\big\rangle_{\mathcal{L}_2^0} d\tau \\
&\quad + \mathbb{E}\int_t^T \big\langle P(\tau)v_1(\tau), C(\tau)x_2(\tau) + v_2(\tau)\big\rangle_{\mathcal{L}_2^0} d\tau \\
&\quad + \mathbb{E}\int_t^T \big\langle \Lambda(\tau)v_1(\tau), x_2(\tau)\big\rangle_{H_\lambda, H_\lambda'} d\tau \\
&\quad + \mathbb{E}\int_t^T \big\langle \widetilde{\Lambda}(\tau)x_1(\tau), v_2(\tau)\big\rangle_{\mathcal{L}_2(V;H_\lambda), \mathcal{L}_2(V;H_\lambda')} d\tau,
\end{aligned}
\tag{13.135}
$$

where $\widetilde{\Lambda}(\cdot)$ is the operator induced by $\Lambda(\cdot)$, and $x_1(\cdot)$(resp. $x_2(\cdot)$) solves (13.133)*(resp.* (13.134)*);* [2] *and*

3) *For any $t \in [0,T]$, $\xi_1, \xi_2 \in L^2_{\mathcal{F}_t}(\Omega; H)$, $u_1(\cdot), u_2(\cdot) \in L^2_{\mathbb{F}}(t,T;H)$ and $v_1(\cdot), v_2(\cdot) \in L^2_{\mathbb{F}}(t,T;U)$, it holds that*

$$
\begin{aligned}
&\mathbb{E}\langle Gx_1(T), x_2(T)\rangle_H + \mathbb{E}\int_t^T \big\langle M(\tau)x_1(\tau), x_2(\tau)\big\rangle_H d\tau \\
&- \mathbb{E}\int_t^T \big\langle K(\tau)^{-1}L(\tau)x_1(\tau), L(\tau)x_2(\tau)\big\rangle_H d\tau \\
&= \mathbb{E}\big\langle P(t)\xi_1, \xi_2\big\rangle_H + \mathbb{E}\int_t^T \big\langle P(\tau)u_1(\tau), x_2(\tau)\big\rangle_H d\tau \\
&\quad + \mathbb{E}\int_t^T \big\langle P(\tau)x_1(\tau), u_2(\tau)\big\rangle_H d\tau + \mathbb{E}\int_t^T \big\langle P(\tau)C(\tau)x_1(\tau), D(\tau)v_2(\tau)\big\rangle_{\mathcal{L}_2^0} d\tau \\
&\quad + \mathbb{E}\int_t^T \big\langle P(\tau)D(\tau)v_1(\tau), C(\tau)x_2(\tau) + D(\tau)v_2(\tau)\big\rangle_{\mathcal{L}_2^0} d\tau \\
&\quad + \mathbb{E}\int_t^T \big\langle \Lambda(\tau)D(\tau)v_1(\tau), x_2(\tau)\big\rangle_H d\tau + \mathbb{E}\int_t^T \big\langle D(\tau)^*\widetilde{\Lambda}(\tau)x_1(\tau), v_2(\tau)\big\rangle_U d\tau.
\end{aligned}
\tag{13.136}
$$

Here, $x_1(\cdot)$ and $x_2(\cdot)$ solve (13.133) *and* (13.134) *with v_1 and v_2 replaced by Dv_1 and Dv_2, respectively.*

13.7 Existence of Optimal Feedback Operator for Problem (SLQ)

In this section, we shall prove the existence of optimal feedback operator for Problem (SLQ), provided that the operator-valued backward stochastic Riccati equation (13.17) admits a transposition solution. To this end, we should introduce the following assumption.

[2] By Theorem 3.20, one has $x_1(\cdot), x_2(\cdot) \in L^4_{\mathbb{F}}(\Omega; C([0,T]; H_\lambda'))$.

(**AS3**) *Let* $\{\varphi_j\}_{j=1}^{\infty}$ *be an orthonormal basis of* U. *There is a* $\widetilde{U} \subset U$ *such that* $\widetilde{U}$ *is dense in* U, $\{\varphi_j\}_{j=1}^{\infty} \subset \widetilde{U}$, $R \in L_{\mathbb{F}}^{\infty}(0,T;\mathcal{L}(\widetilde{U}))$, $B \in L_{\mathbb{F}}^{\infty}(0,T;\mathcal{L}(\widetilde{U};H_{\lambda}'))$ *and* $D \in L_{\mathbb{F}}^{\infty}(0,T;\mathcal{L}(\widetilde{U};\mathcal{L}_2(V;H_{\lambda}')))$, *where* H_{λ} *is given in Assumption* (**AS2**).

Theorem 13.40. *Let (AS1)–(AS3) hold. If the operator-valued backward stochastic Riccati equation* (13.17) *admits a transposition solution* $\big(P(\cdot),\Lambda(\cdot)\big) \in C_{\mathbb{F},w}([0,T];L^{\infty}(\Omega;\mathcal{L}(H))) \times L_{\mathbb{F},w}^{2}(0,T;\mathcal{L}(H))$ *such that*

$$K(\cdot)^{-1}\big[B(\cdot)^*P(\cdot)+D(\cdot)^*P(\cdot)C(\cdot)+D(\cdot)^*\widetilde{\Lambda}(\cdot)\big] \in \Upsilon_2(H;U)\cap\Upsilon_2(H_{\lambda}';\widetilde{U}), \tag{13.137}$$

then Problem (SLQ) is uniquely solvable and admits an optimal feedback operator $\Theta(\cdot) \in \Upsilon_2(H;U)\cap\Upsilon_2(H_{\lambda}';\widetilde{U})$. *In this case, the optimal feedback operator* $\Theta(\cdot)$ *is given by*

$$\Theta(\cdot) = -K(\cdot)^{-1}[B(\cdot)^*P(\cdot)+D(\cdot)^*P(\cdot)C(\cdot)+D(\cdot)^*\widetilde{\Lambda}(\cdot)]. \tag{13.138}$$

Furthermore,

$$\inf_{u(\cdot)\,\in\, L_{\mathbb{F}}^2(0,T;U)} \mathcal{J}(\eta;u) = \frac{1}{2}\langle P(0)\eta,\eta\rangle_H. \tag{13.139}$$

Proof: Let us assume that the equation (13.17) admits a transposition solution $\big(P(\cdot),\Lambda(\cdot)\big) \in C_{\mathbb{F},w}([0,T];L^{\infty}(\Omega;\mathcal{L}(H)))\times L_{\mathbb{F},w}^{2}(0,T;\mathcal{L}(H))$ such that (13.137) holds. Then,

$$\Theta \overset{\Delta}{=} -K^{-1}(B^*P+D^*PC+D^*\widetilde{\Lambda}) \in \Upsilon_2(H;U)\cap\Upsilon_2(H_{\lambda}';\widetilde{U}). \tag{13.140}$$

For any $t\in[0,T)$, $\eta\in L_{\mathcal{F}_t}^2(\Omega;H)$ and $u(\cdot)\in L_{\mathbb{F}}^2(0,T;U)$, choose $\xi_1=\xi_2=\eta$, $u_1=u_2=Bu$ and $v_1=v_2=Du$ in (13.133)–(13.134). From (13.18), (13.136) and the pointwise self-adjointness of $K(\cdot)$, we obtain that

$$\begin{aligned}
&\mathbb{E}\langle Gx(T),x(T)\rangle_H+\mathbb{E}\int_t^T\big\langle M(r)x(r),x(r)\big\rangle_H dr\\
&-\mathbb{E}\int_t^T\big\langle \Theta(r)^*K(r)\Theta(r)x(r),x(r)\big\rangle_H dr\\
&=\mathbb{E}\big\langle P(t)\eta,\eta\big\rangle_H+\mathbb{E}\int_t^T\big\langle P(r)B(r)u(r),x(r)\big\rangle_H dr\\
&+\mathbb{E}\int_t^T\big\langle P(r)x(r),B(r)u(r)\big\rangle_H dr+\mathbb{E}\int_t^T\big\langle P(r)C(r)x(r),D(r)u(r)\big\rangle_{\mathcal{L}_2^0} dr\\
&+\mathbb{E}\int_t^T\big\langle P(r)D(r)u(r),C(r)x(r)+D(r)u(r)\big\rangle_{\mathcal{L}_2^0} dr\\
&+\mathbb{E}\int_t^T\big\langle \Lambda(r)D(r)u(r),x(r)\big\rangle_H dr+\mathbb{E}\int_t^T\big\langle D(r)^*\widetilde{\Lambda}(r)x(r),u(r)\big\rangle_U dr.
\end{aligned} \tag{13.141}$$

Then, by (13.3) and (13.141), recalling the definition of $L(\cdot)$ and $K(\cdot)$, we arrive at

$$\begin{aligned}
&\mathbb{E}\Big[\int_t^T \Big(\big\langle Mx(r), x(r)\big\rangle_H + \big\langle Ru(r), u(r)\big\rangle_U\Big)dr + \big\langle Gx(T), x(T)\big\rangle_H\Big]\\
&= \mathbb{E}\big\langle P(t)\eta, \eta\big\rangle_H + \mathbb{E}\int_t^T \big\langle PBu(r), x(r)\big\rangle_H dr + \mathbb{E}\int_t^T \big\langle Px(r), Bu(r)\big\rangle_H dr\\
&\quad +\mathbb{E}\int_t^T \big\langle PCx(r), Du(r)\big\rangle_{\mathcal{L}_2^0} dr + \mathbb{E}\int_t^T \big\langle PDu(r), Cx(r) + Du(r)\big\rangle_{\mathcal{L}_2^0} dr\\
&\quad +\mathbb{E}\int_t^T \big\langle \Lambda Du(r), x(r)\big\rangle_H dr + \mathbb{E}\int_t^T \big\langle D^*\widetilde{\Lambda}x(r), u(r)\big\rangle_U dr\\
&\quad +\mathbb{E}\int_t^T \big\langle \Theta^* K\Theta x(r), x(r)\big\rangle_H dr + \mathbb{E}\int_t^T \big\langle Ru(r), u(r)\big\rangle_U dr\\
&= \mathbb{E}\Big[\big\langle P(t)\eta, \eta\big\rangle_H + \int_t^T \Big(\big\langle \Theta^* K\Theta x(r), x(r)\big\rangle_H + 2\big\langle Lx(r), u(r)\big\rangle_U\\
&\qquad\qquad +\big\langle Ku(r), u(r)\big\rangle_U\Big)dr\Big].
\end{aligned}$$

This, together with (13.138), implies that

$$\begin{aligned}
&\mathbb{E}\Big[\int_t^T \Big(\big\langle Mx(r), x(r)\big\rangle_H + \big\langle Ru(r), u(r)\big\rangle_U\Big)dr + \big\langle Gx(T), x(T)\big\rangle_H\Big]\\
&= \frac{1}{2}\mathbb{E}\Big[\big\langle P(t)\eta, \eta\big\rangle_H + \int_t^T \big(\big\langle K\Theta x, \Theta x\big\rangle_U - 2\big\langle K\Theta x, u\big\rangle_U + \langle Ku, u\rangle_U\big)dr\Big]\\
&= \frac{1}{2}\mathbb{E}\Big(\big\langle P(t)\eta, \eta\big\rangle_H + \int_t^T \big\langle K(u - \Theta x), u - \Theta x\big\rangle_U dr\Big).
\end{aligned} \tag{13.142}$$

By taking $t = 0$ in (13.142), we get that

$$\begin{aligned}
&\mathcal{J}(\eta; u(\cdot))\\
&= \frac{1}{2}\mathbb{E}\Big[\big\langle P(0)\eta, \eta\big\rangle_H + \int_0^T \big(\big\langle K\Theta x, \Theta x\big\rangle_U - 2\big\langle K\Theta x, u\big\rangle_U + \langle Ku, u\rangle_U\big)dr\Big]\\
&= \frac{1}{2}\mathbb{E}\Big(\big\langle P(0)\eta, \eta\big\rangle_H + \int_0^T \big\langle K(u - \Theta x), u - \Theta x\big\rangle_U dr\Big)\\
&= \mathcal{J}(\eta; \Theta\bar{x}) + \frac{1}{2}\mathbb{E}\int_0^T \big\langle K(u - \Theta x), u - \Theta x\big\rangle_U dr.
\end{aligned} \tag{13.143}$$

Hence,

$$\mathcal{J}(\eta; \Theta\bar{x}) \leq \mathcal{J}(\eta; u), \quad \forall\, u(\cdot) \in L^2_{\mathbb{F}}(0, T; U).$$

Consequently, $\Theta(\cdot)$ is an optimal feedback operator for Problem (SLQ), and (13.139) holds. Further, since $K > 0$, we know that the optimal control is unique. This completes the proof of Theorem 13.40. □

13.8 Global Solvability of Operator-Valued Backward Stochastic Riccati Equations

Theorem 13.40 in the last section concludes the existence of optimal feedback operator for Problem (SLQ) under the assumption that the operator-valued backward stochastic Riccati equation (13.17) admits a transposition solution satisfying (13.137). In this section, we shall show that the converse of this theorem is also true under some more conditions.

The main result of this section (together with Theorem 13.40), which reveals the relationship between the existence of optimal feedback operator for Problem (SLQ) and the global solvability of (13.17) in the sense of transposition solution (See Definition 13.39), is stated as follows:

Theorem 13.41. *Let* ***(AS1)–(AS3)*** *hold and A generate a C_0-group on H, and let $\mathbb{F}$ be the natural filtration generated by $W(\cdot)$. If Problem (SLQ) is uniquely solvable and admits an optimal feedback operator $\Theta(\cdot) \in \Upsilon_2(H;U) \cap \Upsilon_2(H'_\lambda;\widetilde{U})$, then the equation* (13.17) *admits a unique transposition solution $\big(P(\cdot),\Lambda(\cdot)\big) \in C_{\mathbb{F},w}([0,T];L^\infty(\Omega;\mathcal{L}(H))) \times L^2_{\mathbb{F},w}(0,T;\mathcal{L}(H))$ satisfying* (13.137) *and the optimal feedback operator $\Theta(\cdot)$ is given by* (13.138). *Furthermore,* (13.139) *holds.*

The proof of Theorem 13.41 is quite long, and will be given in the Subsection 13.8.2, after some careful preliminaries presented in the Subsection 13.8.1.

Several remarks are in order.

Remark 13.42. In Theorem 13.41, we only conclude that $K(t,\omega)$ has left inverse for a.e. $(t,\omega) \in (0,T)\times\Omega$, and therefore $K(t,\omega)^{-1}$ may be unbounded. Nevertheless, this result cannot be improved. Let us show this by the following example.

Let $\mathcal{O} \subset \mathbb{R}^k$ (for some $k \in \mathbb{N}$) be a bounded domain with a smooth boundary $\partial\mathcal{O}$. Let $H = H_0^1(\mathcal{O})\times L^2(\mathcal{O})$, $V=\mathbb{R}$, $U = L^2(\mathcal{O})$ and $A = \begin{pmatrix} 0 & I \\ \Delta & 0 \end{pmatrix}$, where Δ is the Laplacian on $\mathcal{O}$ with the usual homogeneous Dirichlet boundary condition. Let $B = \begin{pmatrix} 0 \\ I \end{pmatrix}$, $C = \begin{pmatrix} I \\ 0 \end{pmatrix}$, $D=0$, $M=0$, $R=(-\Delta)^{-1}$ and $G=0$. Then (13.1) is specialized as

$$\begin{cases} dx = \big(Ax+Bu\big)dt + CxdW(t) & \text{in } (0,T], \\ x(0)=\eta. \end{cases} \tag{13.144}$$

The cost functional reads

$$\mathcal{J}(\eta;u(\cdot)) = \frac{1}{2}\mathbb{E}\int_0^T \big\langle (-\Delta)^{-1}u(t), u(t)\big\rangle_{L^2(\mathcal{O})}dt. \tag{13.145}$$

Clearly, for any $\eta \in H_0^1(\mathcal{O}) \times L^2(\mathcal{O})$, there is a unique optimal control $u \equiv 0$. For the present case, it is easy to check that $(P(\cdot), \Lambda(\cdot)) = (0, 0)$ is the unique transposition solution to (13.17). However, $K = (-\Delta)^{-1}$ is not surjective and K^{-1} is unbounded.

Remark 13.43. In Theorem 13.41, we assume that A generates a C_0-group on H and $\mathbb{F}$ is the natural filtration generated by $W(\cdot)$. These assumptions are used to guarantee the well-posednss of the stochastic evolution equation (13.219) in the sequel. Actually, the C_0-group condition can be dropped (See [235] for the details).

Remark 13.44. By Theorem 3.20, if $\xi_j \in L^2_{\mathcal{F}_t}(\Omega; H'_\lambda)$, $u_j(\cdot) \in L^2_{\mathbb{F}}(t, T; H'_\lambda)$ and $v_j(\cdot) \in L^2_{\mathbb{F}}(t, T; \mathcal{L}_2(V; H'_\lambda))$, then the solutions x_j $(j = 1, 2)$ to (13.133)–(13.134) belong to $L^2_{\mathbb{F}}(\Omega; C([t, T]; H'_\lambda))$. This plays a key role in Step 5 in the proof of Theorem 13.41. We believe that this assumption can be dropped. However, we do not know how to do it at this moment.

Remark 13.45. In Theorem 13.41, the most natural choice of optimal feedback operator should be an element in $\Upsilon_2(H; U)$ rather than an element in $\Upsilon_2(H; U) \cap \Upsilon_2(H'_\lambda; \widetilde{U})$. Nevertheless, at this moment, in the proof of Theorem 13.41, we do need to suppose that $\Theta(\cdot) \in \Upsilon_2(H; U) \cap \Upsilon_2(H'_\lambda; \widetilde{U})$.

In this rest of this section, *unless otherwise stated*, we assume the assumptions in Theorem 13.41.

13.8.1 Some Preliminary Results

In this subsection, we present some results for the approximation of stochastic evolution equations and backward stochastic evolution equations by stochastic differential equations and backward stochastic differential equations, respectively, which will be useful in this section. Besides proving Theorem 13.41, we believe that these results have their own interest.

First, for any $\eta \in H$, consider the following stochastic evolution equation:

$$\begin{cases} dx = [(A + \mathcal{A})x + f]dt + (\mathcal{B}x + g)dW(t) \ \text{ in } (0, T], \\ x(0) = \eta. \end{cases} \tag{13.146}$$

Here $\mathcal{A} \in \Upsilon_1(H)$, $\mathcal{B} \in \Upsilon_2(H; \mathcal{L}_2^0)$, $\eta \in H$, $f \in L^2_{\mathbb{F}}(\Omega; L^1(0, T; H))$ and $g \in L^2_{\mathbb{F}}(0, T; \mathcal{L}_2^0)$. By Theorem 3.20, we know that the equation (13.146) admits a unique mild solution $x(\cdot) \in L^2_{\mathbb{F}}(\Omega; C([0, T]; H))$, and

$$\begin{aligned} &|x(\cdot)|_{L^2_{\mathbb{F}}(\Omega; C([0,T]; H))} \\ &\leq \mathcal{C}\big(|\eta|_H + |f|_{L^2_{\mathbb{F}}(\Omega; L^1(0,T;H))} + |g|_{L^2_{\mathbb{F}}(0,T;\mathcal{L}_2^0)}\big). \end{aligned} \tag{13.147}$$

Next, we consider the following backward stochastic evolution equation:

$$\begin{cases} dy = -\big(A^*y + \mathcal{D}Y + h\big)dt + YdW(t) & \text{in } [0,T), \\ y(T) = \xi. \end{cases} \tag{13.148}$$

Here $\xi \in L^2_{\mathcal{F}_T}(\Omega; H)$, $\mathcal{D} \in L^\infty_{\mathbb{F}}(0,T;\mathcal{L}(\mathcal{L}_2^0;H))$ and $h \in L^2_{\mathbb{F}}(0,T;H)$. By Theorem 4.10, we know that the equation (13.148) admits a unique mild solution $(y(\cdot),Y(\cdot)) \in L^2_{\mathbb{F}}(\Omega;C([0,T];H)) \times L^2_{\mathbb{F}}(0,T;\mathcal{L}_2^0)$, and

$$|(y(\cdot),Y(\cdot))|_{L^2_{\mathbb{F}}(\Omega;C([0,T];H))\times L^2_{\mathbb{F}}(0,T;\mathcal{L}_2^0)} \le \mathcal{C}\big(|\xi|_{L^2_{\mathcal{F}_T}(\Omega;H)} + |h|_{L^2_{\mathbb{F}}(0,T;H)}\big).$$

For each $n \in \mathbb{N}$, denote by Γ_n the projection operator from H to the finite dimensional space $H_n \overset{\Delta}{=} \text{span}_{1\le j\le n}\{e_j\}$ (recall that $\{e_j\}_{j=1}^\infty$ is an orthonormal basis of H). Let

$$\begin{array}{llll} A_n = \Gamma_n A \Gamma_n, & \mathcal{A}_n = \Gamma_n \mathcal{A}\Gamma_n, & \mathcal{B}_n = \Gamma_n \mathcal{B}\Gamma_n, & \mathcal{D}_n = \Gamma_n \mathcal{D}\Gamma_n, \\ G_n = \Gamma_n G \Gamma_n, & f_n = \Gamma_n f, & g_n = \Gamma_n g, & h_n = \Gamma_n h. \end{array} \tag{13.149}$$

It is easy to show that

$$\lim_{n\to\infty} \mathcal{A}_n\zeta = \mathcal{A}\zeta, \ \lim_{n\to\infty} \mathcal{B}_n\zeta = \mathcal{B}\zeta, \quad \text{for all } \zeta \in H \text{ and a.e. } (t,\omega) \in [0,T]\times\Omega, \tag{13.150}$$

$$\lim_{n\to\infty} \mathcal{D}_n\zeta = \mathcal{D}\zeta, \quad \text{for all } \zeta \in \mathcal{L}_2^0 \text{ and a.e. } (t,\omega) \in [0,T]\times\Omega, \tag{13.151}$$

$$\lim_{n\to\infty} G_n\zeta = G\zeta, \quad \text{for all } \zeta \in H \text{ and a.e. } \omega \in \Omega, \tag{13.152}$$

$$\begin{cases} \lim_{n\to\infty} f_n = f, \quad \lim_{n\to\infty} h_n = h \quad \text{in } L^2_{\mathbb{F}}(0,T;H), \\ \lim_{n\to\infty} g_n = g \quad \text{in } L^2_{\mathbb{F}}(0,T;\mathcal{L}_2^0). \end{cases} \tag{13.153}$$

For any $\xi \in D(A)$,

$$\begin{aligned} &\lim_{n\to\infty} |A_n\xi - A\xi|_H = \lim_{n\to\infty} |\Gamma_n A\Gamma_n\xi - A\xi|_H \\ &\le \lim_{n\to\infty} |\Gamma_n(A\Gamma_n\xi - A\xi)|_H + \lim_{n\to\infty} |(\Gamma_n - I)A\xi|_H \\ &\le \lim_{n\to\infty} |(A\Gamma_n\xi - A\xi)|_H + \lim_{n\to\infty} |(\Gamma_n - I)A\xi|_H \\ &\le \lim_{n\to\infty} |A|_{\mathcal{L}(D(A);H)}|(\Gamma_n\xi - \xi)|_{D(A)} + \lim_{n\to\infty} |(\Gamma_n - I)A\xi|_H = 0. \end{aligned}$$

By the Trotter-Kato approximation theorem (e.g. [87, page 209]), we have that, for any $\zeta \in H$,

$$\lim_{n\to\infty} e^{A_n t}\zeta = S(t)\zeta \ \text{ in } H, \quad \text{uniformly for } t \in [0,T]. \tag{13.154}$$

Similarly, we have that for any $\zeta \in H_\lambda$,

$$\lim_{n\to\infty} e^{A_n t}\zeta = S(t)\zeta \ \text{ in } H_\lambda, \quad \text{uniformly for } t \in [0,T]. \tag{13.155}$$

Lemma 13.46. *Let **(AS1)** hold, and $H_\lambda \in \mathcal{V}_H$. Then $A \in \mathcal{L}_2(H; H_\lambda)$ and $\lim_{n\to\infty} |A_n - A|_{\mathcal{L}_2(H;H_\lambda)} = 0$.*

Proof: Since $\{e_j\}_{j=1}^\infty$ are eigenvectors of A, it holds that

$$\sum_{j=1}^\infty |Ae_j|_{H_\lambda}^2 = \sum_{j=1}^\infty |\mu_j e_j|_{H_\lambda}^2 = \sum_{j=1}^\infty \lambda_j^2\mu_j^2 |e_j|_{D(A)}^{-2} \le \sum_{j=1}^\infty \lambda_j^2 < \infty.$$

Hence, $A \in \mathcal{L}_2(H; H_\lambda)$.

Next,

$$\begin{aligned}\lim_{n\to\infty} |A_n - A|_{\mathcal{L}_2(H;H_\lambda)}^2 &= \lim_{n\to\infty} \sum_{j=1}^\infty |(A_n - A)e_j|_{H_\lambda}^2\\ &= \lim_{n\to\infty} \sum_{j=n+1}^\infty \lambda_j^2\mu_j^2 |e_j|_{D(A)}^{-2} \le \lim_{n\to\infty} \sum_{j=n+1}^\infty \lambda_j^2 = 0.\end{aligned}$$

This completes the proof of Lemma 13.46. □

Now we introduce sequences of stochastic differential equations and backward stochastic differential equations, which approximate the stochastic evolution equation (13.146) and backward stochastic evolution equation (13.148), respectively.

For any $\eta \in H$, $\xi \in L^2_{\mathcal{F}_T}(\Omega; H)$ and $n \in \mathbb{N}$, consider the following two equations:

$$\begin{cases} dx_n = \big[(A_n + \mathcal{A}_n)x_n + f_n\big]dt + (\mathcal{B}_n x_n + g_n)dW(t) \ \text{ in } [0,T],\\ x_n(0) = \Gamma_n \eta, \end{cases} \tag{13.156}$$

and

$$\begin{cases} dy_n = -\big(A_n^* y_n + \mathcal{D}_n Y_n + h_n\big)dt + Y_n dW(t) \ \text{ in } [0,T],\\ y_n(T) = \Gamma_n \xi. \end{cases} \tag{13.157}$$

Lemma 13.47. *For any $\eta \in H$ and $\xi \in L^2_{\mathcal{F}_T}(\Omega; H)$, it holds that*

$$\begin{cases} \lim_{n\to\infty} \mathbb{E}\big(\sup_{t\in[0,T]} |x_n(t) - x(t)|_H^2\big) = 0,\\ \lim_{n\to\infty} \mathbb{E}\big(\sup_{t\in[0,T]} |y_n(t) - y(t)|_H^2\big) = 0,\\ \lim_{n\to\infty} |Y_n(\cdot) - Y(\cdot)|_{L^2_{\mathbb{F}}(0,T;\mathcal{L}_2^0)} = 0. \end{cases} \tag{13.158}$$

Proof: **Step 1**. In this step, we prove the first equality in (13.158).

Put

$$\mathcal{M} \stackrel{\Delta}{=} \sup_{t\in[0,T]} |S(t)|_{\mathcal{L}(H)}\big(|\mathcal{A}|_{\mathcal{L}(H)} + |\mathcal{B}|_{\mathcal{L}(H;\mathcal{L}_2^0)}\big)$$

and let $t_0 \in [0,T]$ satisfying that

$$t_0 \stackrel{\Delta}{=} \max\Big\{t\in[0,T]\Big|\ \max\{t,t^2\} \leq \frac{1}{4\mathcal{M}^2}\Big\}. \tag{13.159}$$

From (13.146) and (13.156), we have that

$$\begin{aligned}
&\mathbb{E}\big(\sup_{t\in[0,t_0]} |x_n(t)-x(t)|_H^2\big)\\
&\leq \mathcal{C}\mathbb{E}\sup_{t\in[0,t_0]}\Big(\Big|S(t)\eta - e^{A_n t}\Gamma_n\eta\Big|_H^2\\
&\quad+\Big|\int_0^t S(t-r)\mathcal{A}x(r)dr - \int_0^t e^{A_n(t-r)}\mathcal{A}_n x_n(r)dr\Big|_H^2\\
&\quad+\Big|\int_0^t S(t-r)f(r)dr - \int_0^t e^{A_n(t-r)}f_n(r)dr\Big|_H^2\\
&\quad+\Big|\int_0^t S(t-r)\mathcal{B}x(r)dW(r) - \int_0^t e^{A_n(t-r)}\mathcal{B}_n x_n(r)dW(r)\Big|_H^2\\
&\quad+\Big|\int_0^t S(t-r)g(r)dr - \int_0^t e^{A_n(t-r)}g_n(r)dW(r)\Big|_H^2\Big).
\end{aligned} \tag{13.160}$$

Let us estimate the terms in the right hand side of (13.160) one by one.

First,

$$\begin{aligned}
&\sup_{t\in[0,t_0]}\big|S(t)\eta - e^{A_n t}\Gamma_n\eta\big|_H^2\\
&\leq 2\sup_{t\in[0,t_0]}\big|S(t)\eta - e^{A_n t}\eta\big|_H^2 + 2\sup_{t\in[0,t_0]}\big|e^{A_n t}\eta - e^{A_n t}\Gamma_n\eta\big|_H^2\\
&\leq 2\sup_{t\in[0,t_0]}\big|S(t)\eta - e^{A_n t}\eta\big|_H^2 + \mathcal{C}\mathbb{E}\big|\eta - \Gamma_n\eta\big|_H^2.
\end{aligned} \tag{13.161}$$

Next, from the definition of A_n and $\mathcal{A}_n$, we know that

$$\big(S(t) - e^{A_n t}\big)\mathcal{A}_n = 0, \qquad \forall t\in[0,T]. \tag{13.162}$$

Thus, we have

$$\begin{aligned}
&\mathbb{E}\sup_{t\in[0,t_0]}\Big|\int_0^t S(t-r)\mathcal{A}x(r)dr - \int_0^t e^{A_n(t-r)}\mathcal{A}_n x_n(r)dr\Big|_H^2\\
&\leq 2\mathbb{E}\sup_{t\in[0,t_0]}\Big|\int_0^t S(t-r)\mathcal{A}x(r)dr - \int_0^t e^{A_n(t-r)}\mathcal{A}_n x(r)dr\Big|_H^2\\
&\quad+2\mathbb{E}\sup_{t\in[0,t_0]}\Big|\int_0^t e^{A_n(t-r)}\mathcal{A}_n x(r)dr - \int_0^t e^{A_n(t-r)}\mathcal{A}_n x_n(r)dr\Big|_H^2
\end{aligned} \tag{13.163}$$

$$
\begin{aligned}
&\le 2\mathbb{E}\int_0^{t_0} \big|\big[\big(S(t-r)-e^{A_n(t-r)}\big)\mathcal{A}+\big(S(t-r)-e^{A_n(t-r)}\big)\mathcal{A}_n\big]x(r)\big|_H dr\\
&\quad +\mathcal{M}^2t_0^2\mathbb{E}\sup_{t\in[0,t_0]}|x(t)-x_n(t)|_H^2\\
&\le 2\mathbb{E}\int_0^{t_0} \big|\big(S(t-r)-e^{A_n(t-r)}\big)\mathcal{A}x(r)\big|_H dr\\
&\quad +\mathcal{M}^2t_0^2\mathbb{E}\sup_{t\in[0,t_0]}|x(t)-x_n(t)|_H^2.
\end{aligned}
$$

Using (13.162) again, we get that

$$
\begin{aligned}
&\mathbb{E}\sup_{t\in[0,t_0]}\Big|\int_0^t S(t-r)f(r)dr-\int_0^t e^{A_n(t-r)}f_n(r)dr\Big|_H^2\\
&\le 2\mathbb{E}\sup_{t\in[0,t_0]}\Big|\int_0^t S(t-r)f(r)dr-\int_0^t e^{A_n(t-r)}f(r)dr\Big|_H^2\\
&\quad +2\mathbb{E}\sup_{t\in[0,t_0]}\Big|\int_0^t e^{A_n(t-r)}f(r)dr-\int_0^t e^{A_n(t-r)}f_n(r)dr\Big|_H^2 \qquad (13.164)\\
&\le 2\mathbb{E}\int_0^{t_0}\big|\big(S(t-r)-e^{A_n(t-r)}\big)f(r)\big|_H^2dr+\mathcal{C}\mathbb{E}\int_0^{t_0}|f(r)-f_n(r)|_H^2dr.
\end{aligned}
$$

By Theorem 3.18 (Burkholder-Davis-Gundy inequality) and (13.162), we have that

$$
\begin{aligned}
&\mathbb{E}\sup_{t\in[0,t_0]}\Big|\int_0^t S(t-r)\mathcal{B}x(r)dW(r)-\int_0^t e^{A_n(t-r)}\mathcal{B}_nx_n(r)dW(r)\Big|_H^2\\
&\le 2\mathbb{E}\sup_{t\in[0,t_0]}\Big|\int_0^t S(t-r)\mathcal{B}x(r)dW(r)-\int_0^t e^{A_n(t-r)}\mathcal{B}_nx(r)dW(r)\Big|_H^2 \qquad (13.165)\\
&\quad +2\mathbb{E}\sup_{t\in[0,t_0]}\Big|\int_0^t e^{A_n(t-r)}\mathcal{B}_nx(r)dW(r)-\int_0^t e^{A_n(t-r)}\mathcal{B}_nx_n(r)dW(r)\Big|_H^2\\
&\le 2\mathbb{E}\int_0^{t_0}\big|\big(S(t-r)-e^{A_n(t-r)}\big)\mathcal{B}x(r)\big|_{\mathcal{L}_2^0}^2dr+\mathcal{M}^2t_0\mathbb{E}\sup_{t\in[0,t_0]}|x(t)-x_n(t)|_H^2
\end{aligned}
$$

and

$$
\begin{aligned}
&\mathbb{E}\sup_{t\in[0,t_0]}\Big|\int_0^t S(t-r)g(r)dr-\int_0^t e^{A_n(t-r)}g_n(r)dW(r)\Big|_H^2\\
&\le 2\mathbb{E}\sup_{t\in[0,t_0]}\Big|\int_0^t S(t-r)g(r)dW(r)-\int_0^t e^{A_n(t-r)}g(r)dW(r)\Big|_H^2 \qquad (13.166)\\
&\quad +2\mathbb{E}\sup_{t\in[0,t_0]}\Big|\int_0^t e^{A_n(t-r)}g(r)dW(r)-\int_0^t e^{A_n(t-r)}g_n(r)dW(r)\Big|_H^2\\
&\le 2\mathbb{E}\int_0^{t_0}\big|\big(S(t-r)-e^{A_n(t-r)}\big)g(r)\big|_{\mathcal{L}_2^0}^2dr+\mathcal{C}\mathbb{E}\int_0^{t_0}|g(r)-g_n(r)|_{\mathcal{L}_2^0}^2dr.
\end{aligned}
$$

From (13.159) to (13.166), we find that

$$
\begin{aligned}
&\mathbb{E}\big(\sup_{t\in[0,t_0]}|x_n(t)-x(t)|_H^2\big)\\
&\leq \mathcal{C}\Big(\mathbb{E}\sup_{t\in[0,t_0]}\big|S(t-s)\eta-e^{A_n(t-s)}\eta\big|_H^2+\mathbb{E}\big|\eta-\Gamma_n\eta\big|_H^2\\
&\quad+\mathbb{E}\int_0^{t_0}\big|\big(S(t_0-r)-e^{A_n(t_0-r)}\big)\mathcal{A}x(r)\big|_H dr \qquad (13.167)\\
&\quad+\mathbb{E}\int_0^{t_0}\big|\big(S(t_0-r)-e^{A_n(t_0-r)}\big)f(r)\big|_H^2 dr+\mathbb{E}\int_0^{t_0}|f(r)-f_n(r)|_H^2 dr\\
&\quad+\mathbb{E}\int_0^{t_0}\big|\big(S(t_0-r)-e^{A_n(t_0-r)}\big)\mathcal{B}x(r)\big|_{\mathcal{L}_2^0}^2 dr\\
&\quad+\mathbb{E}\int_0^{t_0}\big|\big(S(t_0-r)-e^{A_n(t_0-r)}\big)g(r)\big|_{\mathcal{L}_2^0}^2 dr+\mathbb{E}\int_0^{t_0}|g(r)-g_n(r)|_{\mathcal{L}_2^0}^2 dr\Big).
\end{aligned}
$$

For all $n\in\mathbb{N}$,

$$
\sup_{t\in[0,T]}\big|S(t)\eta-e^{A_nt}\eta\big|_H^2\leq\mathcal{C}|\eta|_H^2.
$$

This, together with Lebesgue's dominated convergence theorem and (13.150), implies that

$$
\begin{aligned}
&\lim_{n\to\infty}\mathbb{E}\sup_{t\in[0,T]}\big|S(t)\eta-e^{A_nt}\eta\big|_H^2\\
&=\mathbb{E}\lim_{n\to\infty}\sup_{t\in[0,T]}\big|S(t)\eta-e^{A_nt}\eta\big|_H^2=0.
\end{aligned}
\qquad (13.168)
$$

Similarly, we can prove that

$$
\lim_{n\to\infty}\mathbb{E}\big|\eta-\Gamma_n\eta\big|_H^2=0. \qquad (13.169)
$$

Next, noting that for all $n\in\mathbb{N}$,

$$
\big|\big(S(t-r)-e^{A_n(t-r)}\big)\mathcal{A}x(r)\big|_H^2\leq\mathcal{C}\big|x(r)\big|_H^2,\qquad \text{a.e. } (t,\omega)\in[0,t_0]\times\Omega,
$$

it follows from Lebesgue's dominated convergence theorem and (13.150) that

$$
\lim_{n\to\infty}\mathbb{E}\int_0^{t_0}\big|\big(S(t_0-r)-e^{A_n(t_0-r)}\big)\mathcal{A}x(r)\big|_H^2 dr=0. \qquad (13.170)
$$

Similar to the above arguments, we can show that all the terms in the right hand side of (13.167) tend to zero as n tends to ∞. Consequently, we obtain that

$$
\lim_{n\to\infty}\mathbb{E}\big(\sup_{t\in[0,t_0]}|x_n(t)-x(t)|_H^2\big)=0.
$$

If $t_0=T$, then we complete our proof. Otherwise, let

$$t_1 \overset{\Delta}{=} \max\Big\{t \in [t_0, T]\Big|\ \max\{t - t_0, (t - t_0)^2\} \leq \frac{1}{4\mathcal{M}^2}\Big\}. \tag{13.171}$$

Repeating the above argument, we get that

$$\lim_{n\to\infty} \mathbb{E}\big(\sup_{t\in[0,t_1]} |x_n(t) - x(t)|_H^2\big) = 0.$$

By an induction argument, we can obtain that

$$\lim_{n\to\infty} \mathbb{E}\big(\sup_{t\in[0,T]} |x_n(t) - x(t)|_H^2\big) = 0.$$

Step 2. In this step, we prove that the second and third equalities in (13.158) hold.

Denote by $\sigma(A^*)$ the spectrum of A^*. Since A^* generates a C_0-semigroup on H, there exists $\mathcal{C}_0 > 0$ such that

$$\sup_{\mu\in\sigma(A^*)} \{\mathrm{Re}\, \mu\} \leq \mathcal{C}_0. \tag{13.172}$$

Recall that the eigenvectors $\{e_k\}_{k=1}^\infty$ of A constitutes an orthonormal basis of H. For any $\xi = \sum_{k=1}^\infty \xi_k e_k \in D(A)$, we have

$$\mathrm{Re}\, \langle A\xi, \xi\rangle_H = \mathrm{Re}\Big\langle \sum_{k=1}^\infty \mu_k\xi_k e_k, \sum_{k=1}^\infty \xi_k e_k\Big\rangle_H \leq \mathcal{C}_0|\xi|_H^2. \tag{13.173}$$

For each $\mu \in \rho(A^*)$, let $R(\mu) = \mu(\mu I - A^*)^{-1}$. Introduce respectively the approximating equations of (13.148) and (13.157) as follows:

$$\begin{cases} dy_\mu = -A^* y_\mu dt - R(\mu)\big(\mathcal{D}Y + h\big)dt + Y_\mu dW(t) & \text{in } [0, T),\\ y_\mu(T) = R(\mu)\xi, \end{cases} \tag{13.174}$$

$$\begin{cases} dy_{n,\mu} = -A_n^* y_{n,\mu} - \Gamma_n R(\mu)\big(\mathcal{D}_n Y_n + h_n\big)dt + Y_{n,\mu} dW(t) \ \text{ in } [0, T],\\ y_{n,\mu}(T) = \Gamma_n R(\mu)\Gamma_n\xi. \end{cases} \tag{13.175}$$

By Theorem 4.12, we obtain that

$$\lim_{\mu\to\infty} \big(|y_{n,\mu} - y_n|_{L^2_{\mathbb{F}}(\Omega;C([0,T];H))} + |Y_{n,\mu} - Y_n|_{L^2_{\mathbb{F}}(0,T;\mathcal{L}_2^0)}\big) = 0 \tag{13.176}$$

and

$$\lim_{\mu\to\infty} \big(|y_\mu - y|_{L^2_{\mathbb{F}}(\Omega;C([0,T];H))} + |Y_\mu - Y|_{L^2_{\mathbb{F}}(0,T;\mathcal{L}_2^0)}\big) = 0. \tag{13.177}$$

By Itô's formula, (13.173) and noting $Ay_{n,\mu} = A_n y_{n,\mu}$ as well as (13.172), we have

$$
\begin{aligned}
&|y_\mu(t) - y_{n,\mu}(t)|_H^2 + \int_t^T |Y_\mu(r) - Y_{n,\mu}(r)|_{\mathcal{L}_2^0} dr \\
&= |R(\mu)\xi - \Gamma_n R(\mu)\Gamma_n \xi|_H^2 + \int_t^T \langle A^*(y_\mu(r) - y_{n,\mu}(r)), y_\mu(r) - y_{n,\mu}(r)\rangle_H dr \\
&\quad + \int_t^T \langle R(\mu)\mathcal{D}Y(r) - \Gamma_n R(\mu)\mathcal{D}_n Y_n(r), y_\mu(r) - y_{n,\mu}(r)\rangle_H dr \\
&\quad + \int_t^T \langle R(\mu)h(r) - \Gamma_n R(\mu)h_n(r), y_\mu(r) - y_{n,\mu}(r)\rangle_H dr \qquad (13.178)\\
&\quad + \int_t^T \langle (Y_\mu(r) - Y_{n,\mu}(r))dW(r), y_\mu(r) - y_{n,\mu}(r)\rangle_H \\
&\leq |R(\mu)\xi - \Gamma_n R(\mu)\Gamma_n \xi|_H^2 + \mathcal{C}_0 \int_t^T |y_\mu(r) - y_{n,\mu}(r)|_H^2 dr \\
&\quad + \int_t^T \langle R(\mu)\mathcal{D}Y(r) - \Gamma_n R(\mu)\mathcal{D}_n Y_n(r), y_\mu(r) - y_{n,\mu}(r)\rangle_H dr \\
&\quad + \int_t^T \langle R(\mu)h(r) - \Gamma_n R(\mu)h_n(r), y_\mu(r) - y_{n,\mu}(r)\rangle_H dr \\
&\quad + \int_t^T \langle (Y_\mu(r) - Y_{n,\mu}(r))dW(r), y_\mu(r) - y_{n,\mu}(r)\rangle_H .
\end{aligned}
$$

Letting $\mu \to \infty$ in (13.178), from (13.176) and (13.177), using Burkholder-Davis-Gundy inequality, we have

$$
\begin{aligned}
&\mathbb{E}|y(t) - y_n(t)|_H^2 + \mathbb{E}\int_t^T |Y - Y_n|_{\mathcal{L}_2^0} dr \\
&\leq \mathbb{E} \sup_{t\in[0,T]} |y(t) - y_n(t)|_H^2 + \mathbb{E}\int_t^T |Y - Y_n|_{\mathcal{L}_2^0} dr \\
&\leq \mathbb{E}|\xi - \Gamma_n \xi|_H^2 + C\mathbb{E}\int_t^T |y - y_n|_H^2 dr + \frac{1}{2}\mathbb{E}\int_t^T |Y - Y_n|_{\mathcal{L}_2^0}^2 dr \qquad (13.179)\\
&\quad + \int_t^T |h - h_n|_H^2 dr.
\end{aligned}
$$

This, together with Gronwall's inequality, implies that

$$
\begin{aligned}
&\mathbb{E}|y(t) - y_n(t)|_H^2 + \mathbb{E}\int_t^T |Y - Y_n|_{\mathcal{L}_2^0} dr \\
&\leq \mathcal{C}\Big(\mathbb{E}|\xi - \Gamma_n \xi|_H^2 + \mathbb{E}\int_t^T |h - h_n|_H^2 dr\Big). \qquad (13.180)
\end{aligned}
$$

From (13.180), we see that

$$\lim_{n\to\infty}\Big(\sup_{t\in[0,T]}\mathbb{E}|y(t)-y_n(t)|_H^2+\mathbb{E}\int_0^T|Y-Y_n|_{\mathcal{L}_2^0}dr\Big)=0. \tag{13.181}$$

By the second and third line of (13.179), and using (13.181), we obtain the second and third equalities in (13.158). □

Consider the following stochastic evolution equation:

$$\begin{cases} dx_{1,n}=\big(Ax_{1,n}+\Gamma_n u_1\big)d\tau+\big(\Gamma_n Cx_{1,n}+\Gamma_n v_1\big)dW(\tau) \text{ in } (0,T],\\ x_{1,n}(s)=\Gamma_n\xi_1. \end{cases} \tag{13.182}$$

Similar to Lemma 13.47, we can establish the following result.

Lemma 13.48. *For any $\xi_1\in L^4_{\mathcal{F}_t}(\Omega;H'_\lambda)$, $u_1(\cdot)\in L^4_{\mathbb{F}}(\Omega;L^2(0,T;H'_\lambda))$ and $v_1(\cdot)\in L^4_{\mathbb{F}}(\Omega;L^2(0,T;\mathcal{L}_2(V;H'_\lambda)))$, the solution $x_{1,n}(\cdot)\in L^4_{\mathbb{F}}(\Omega;C([0,T];H'_\lambda))$ to (13.182) satisfies*

$$\lim_{n\to\infty}|x_{1,n}(\cdot)-x_1(\cdot)|_{L^4_{\mathbb{F}}(\Omega;C([0,T];H'_\lambda))}=0, \tag{13.183}$$

where $x_1(\cdot)$ is the solution to (13.133).

For a.e. $\tau\in[0,T]$, define six operators Φ, Φ_n, Ψ, Ψ_n, Ξ and Ξ_n as follows:

$$\begin{cases}\Phi: H\to L^2_{\mathbb{F}}(\Omega;C([0,T];H)),\\ (\Phi\eta)(\tau)=x(\tau), \qquad \forall\,\eta\in H,\end{cases}$$

$$\begin{cases}\Phi_n: H\to L^2_{\mathbb{F}}(\Omega;C([0,T];H)),\\ (\Phi_n\eta)(\tau)=x_n(\tau), \qquad \forall\,\eta\in H,\end{cases}$$

$$\begin{cases}\Psi: H\to L^2_{\mathbb{F}}(\Omega;C([0,T];H)),\\ (\Psi\eta)(\tau)=y(\tau), \qquad \forall\,\eta\in H,\end{cases}$$

$$\begin{cases}\Psi_n: H\to L^2_{\mathbb{F}}(\Omega;C([0,T];H)),\\ (\Psi_n\eta)(\tau)=y_n(\tau), \qquad \forall\,\eta\in H,\end{cases}$$

$$\begin{cases}\Xi: H\to L^2_{\mathbb{F}}(0,T;\mathcal{L}_2^0),\\ (\Xi\eta)(\tau)=Y(\tau), \qquad \forall\,\eta\in H,\end{cases}$$

and

$$\begin{cases}\Xi_n: H\to L^2_{\mathbb{F}}(0,T;\mathcal{L}_2^0),\\ (\Xi_n\eta)(\tau)=Y_n(\tau), \qquad \forall\eta\in H.\end{cases}$$

Here $x(\cdot)$ (*resp.* $x_n(\cdot)$) is the solution to (13.146) (*resp.* (13.156)) with $f=g=0$, $(y(\cdot),Y(\cdot))$ (*resp.* $(y_n(\cdot),Y_n(\cdot))$) is the solution to (13.148) (*resp.* (13.157)) with h and ξ replaced by $\mathcal{K}x$ for some $\mathcal{K} \in L^\infty_{\mathbb{F}}(0,T;\mathcal{L}(H))$ and $Gx(T)$ (*resp.* h_n and ξ_n replaced by $\mathcal{K}_n x_n$ with $\mathcal{K}_n = \Gamma_n \mathcal{K} \Gamma_n$ and $G_n x_n(T)$), respectively.

Denote by I_{HH_λ} the embedding operator from H to H_λ. We have the following result.

Lemma 13.49. *Suppose* $\mathcal{A} \in \Upsilon_1(H_\lambda)$, $\mathcal{B} \in \Upsilon_2(H_\lambda;\mathcal{L}_2(V;H_\lambda))$, $\mathcal{D} \in L^\infty_{\mathbb{F}}(0,T;\mathcal{L}(\mathcal{L}_2(V;H_\lambda);H_\lambda))$ *and* $\mathcal{K} \in L^\infty_{\mathbb{F}}(0,T;\mathcal{L}(H_\lambda))$. *Then,*

$$\lim_{n\to\infty} |I_{HH_\lambda}\Phi_n - I_{HH_\lambda}\Phi|_{L^4_{\mathbb{F}}(\Omega;C([0,T];\mathcal{L}_2(H;H_\lambda)))} = 0, \tag{13.184}$$

$$\begin{cases} \lim\limits_{n\to\infty} |I_{HH_\lambda}\Psi_n - I_{HH_\lambda}\Psi|_{L^4_{\mathbb{F}}(\Omega;C([0,T];\mathcal{L}_2(H;H_\lambda)))} = 0, \\ \lim\limits_{n\to\infty} |I_{HH_\lambda}\Xi_n - I_{HH_\lambda}\Xi|_{L^4_{\mathbb{F}}(\Omega;L^2(0,T;\mathcal{L}_2(H;\mathcal{L}_2(V;H_\lambda))))} = 0. \end{cases} \tag{13.185}$$

Proof: We first prove (13.184). It is easy to show that, for any $\varrho \in L^\infty_{\mathbb{F}}(0,T;L^2(\Omega;H_\lambda))$,

$$\begin{cases} \lim\limits_{n\to\infty} |\mathcal{A}_n\varrho - \mathcal{A}\varrho|_{L^4_{\mathbb{F}}(\Omega;L^1(0,T;H_\lambda))} = 0, \\ \lim\limits_{n\to\infty} |\mathcal{B}_n\varrho - \mathcal{B}\varrho|_{L^4_{\mathbb{F}}(\Omega;L^2(0,T;\mathcal{L}_2(H_\lambda;\mathcal{L}_2(V;H_\lambda)))} = 0. \end{cases} \tag{13.186}$$

From the definitions of Φ and Φ_n, we see that, for any $\eta \in H$ and $t \in [0,T]$,

$$\begin{aligned} \Phi(t)\eta = S(t)\eta &+ \int_0^t S(t-r)\mathcal{A}(r)\Phi(r)\eta dr \\ &+ \int_0^t S(t-r)\mathcal{B}(r)\Phi(r)\eta dW(r) \quad \text{in } H, \quad \text{a.s.}, \end{aligned}$$

and

$$\begin{aligned} \Phi_n(t)\eta = e^{A_n t}\Gamma_n\eta &+ \int_0^t e^{A_n(t-r)}\mathcal{A}_n(r)\Phi_n(r)\eta dr \\ &+ \int_0^t e^{A_n(t-r)}\mathcal{B}_n(r)\Phi_n(r)\eta dW(r) \quad \text{in } H, \quad \text{a.s.} \end{aligned}$$

Noting $e^{A_n t}\Gamma_n\eta = e^{A_n t}\eta$, we have

$$\begin{aligned} I_{HH_\lambda}\Phi(t)\eta = I_{HH_\lambda}S(t)\eta &+ \int_0^t I_{HH_\lambda}S(t-r)\mathcal{A}(r)\Phi(r)\eta dr \\ &+ \int_0^t I_{HH_\lambda}S(t-r)\mathcal{B}(r)\Phi(r)\eta dW(r) \quad \text{in } H_\lambda, \quad \text{a.s.} \end{aligned} \tag{13.187}$$

and

$$\begin{aligned} I_{HH_\lambda}\Phi_n(t)\eta = I_{HH_\lambda}e^{A_n t}\eta &+ \int_0^t I_{HH_\lambda}e^{A_n(t-r)}\mathcal{A}_n(r)\Phi_n(r)\eta dr \\ &+ \int_0^t I_{HH_\lambda}e^{A_n(t-r)}\mathcal{B}_n(r)\Phi_n(r)\eta dW(r) \quad \text{in } H_\lambda, \quad \text{a.s.} \end{aligned}$$

Since $I_{HH_\lambda}e_j = e_j$ for all $j \in \mathbb{N}$, if $O \in \mathcal{L}(H)$ can be extended to a bounded linear operator on H_λ, then

$$\begin{aligned}\big|I_{HH_\lambda}O - OI_{HH_\lambda}\big|^2_{\mathcal{L}_2(H;H_\lambda)} &= \sum_{j=1}^{\infty}\big|I_{HH_\lambda}Oe_j - OI_{HH_\lambda}e_j\big|^2_{H_\lambda}\\ &= \sum_{j=1}^{\infty}\big|Oe_j - Oe_j\big|^2_{H_\lambda} = 0.\end{aligned}$$

Consequently, $I_{HH_\lambda}O = OI_{HH_\lambda}$. This, together with (13.187), implies that

$$\begin{aligned}I_{HH_\lambda}\Phi(t)\eta &= I_{HH_\lambda}S(t)\eta + \int_0^t S(t-r)\mathcal{A}(r)I_{HH_\lambda}\Phi(r)\eta dr\\ &\quad + \int_0^t S(t-r)\mathcal{B}(r)I_{HH_\lambda}\Phi(r)\eta dW(r) \quad \text{in } H_\lambda, \quad \text{a.s.}\end{aligned}$$

Since $\mathcal{L}_2(H;H_\lambda)$ is a Hilbert space, for any $t \in [0,T]$ and a.s.,

$$\begin{aligned}I_{HH_\lambda}\Phi(t) &= I_{HH_\lambda}S(t) + \int_0^t S(t-r)\mathcal{A}(r)I_{HH_\lambda}\Phi(r)dr\\ &\quad + \int_0^t S(t-r)\mathcal{B}(r)I_{HH_\lambda}\Phi(r)dW(r) \quad \text{in } \mathcal{L}_2(H;H_\lambda).\end{aligned} \tag{13.188}$$

Similarly, we can prove that for any $t \in [0,T]$ and a.s.,

$$\begin{aligned}I_{HH_\lambda}\Phi_n(t) &= I_{HH_\lambda}e^{A_n t} + \int_0^t e^{A_n(t-r)}\mathcal{A}_n(r)I_{HH_\lambda}\Phi_n(r)dr\\ &\quad + \int_0^t e^{A_n(t-r)}\mathcal{B}_n(r)I_{HH_\lambda}\Phi_n(r)dW(r) \quad \text{in } \mathcal{L}_2(H;H_\lambda).\end{aligned} \tag{13.189}$$

In what follows, to simplify notations, we omit the operator I_{HH_λ} if there is no confusion.

It follows from (13.188) and (13.189) that for any stopping time τ_0 with $\tau_0(\omega) \in (0,T]$, a.s.,

$$\begin{aligned}&\mathbb{E}\sup_{r\in[0,\tau_0]}\big|\Phi(r) - \Phi_n(r)\big|^4_{\mathcal{L}_2(H;H_\lambda)}\\ &\le \mathcal{C}\mathbb{E}\sup_{r\in[0,\tau_0]}\Big[\big|S(r) - e^{A_n r}\big|^4_{\mathcal{L}_2(H;H_\lambda)}\\ &\quad + \Big|\int_0^r \big(S(r-\tau)\mathcal{A}(\tau)\Phi(\tau) - e^{A_n(r-\tau)}\mathcal{A}_n(\tau)\Phi_n(\tau)\big)d\tau\Big|^4_{\mathcal{L}_2(H;H_\lambda)}\\ &\quad + \Big|\int_0^r \big(S(r-\tau)\mathcal{B}(\tau)\Phi(\tau) - e^{A_n(r-\tau)}\mathcal{B}_n(\tau)\Phi_n(\tau)\big)dW(\tau)\Big|^4_{\mathcal{L}_2(H;H_\lambda)}\Big].\end{aligned} \tag{13.190}$$

Hence, by Burkholder-Davis-Gundy inequality, and noting that $|\mathcal{A}_n(\cdot)|_{\mathcal{L}(H_\lambda)} \leq |\mathcal{A}(\cdot)|_{\mathcal{L}(H_\lambda)}$ as well as $|\mathcal{B}_n(\cdot)|_{\mathcal{L}(H_\lambda;\mathcal{L}_2(V;H_\lambda))} \leq |\mathcal{B}(\cdot)|_{\mathcal{L}(H_\lambda;\mathcal{L}_2(V;H_\lambda))}$, we deduce that

$$
\begin{aligned}
&\mathbb{E} \sup_{r\in[0,\tau_0]} \big|\Phi(r)-\Phi_n(r)\big|^4_{\mathcal{L}_2(H;H_\lambda)}\\
&\leq \mathcal{C}\Big[\mathbb{E}\Big(\sup_{r\in[0,\tau_0]} \big|S(r)-e^{A_n r}\big|^4_{\mathcal{L}_2(H;H_\lambda)}\Big)\\
&\quad +\mathbb{E}\Big(\int_0^{\tau_0} \Big|\big(S(\tau_0-\tau)\mathcal{A}(\tau)-e^{A_n(\tau_0-\tau)}\mathcal{A}_n(\tau)\big)\Phi(\tau)\\
&\qquad\qquad +e^{A_n(\tau_0-\tau)}\mathcal{A}_n(\tau)\big(\Phi(\tau)-\Phi_n(\tau)\big)\Big|_{\mathcal{L}_2(H;H_\lambda)}d\tau\Big)^4\\
&\quad +\mathbb{E}\Big(\int_0^{\tau_0} \Big|\big(S(\tau_0-\tau)\mathcal{B}(\tau)-e^{A_n(\tau_0-\tau)}\mathcal{B}_n(\tau)\big)\Phi(\tau)\\
&\qquad\qquad +e^{A_n(\tau_0-\tau)}\mathcal{B}_n(\tau)\big(\Phi(\tau)-\Phi_n(\tau)\big)\Big|^2_{\mathcal{L}_2(H;\mathcal{L}_2(V;H_\lambda))}d\tau\Big)^2\Big]\\
&\leq \mathcal{C}\Big[\mathbb{E}\Big(\sup_{r\in[0,\tau_0]} |S(r)-e^{A_n r}|^4_{\mathcal{L}_2(H;H_\lambda)}\Big) \qquad (13.191)\\
&\quad +\mathbb{E}\Big(\int_0^{T} \Big|\big(S(T-\tau)\mathcal{A}(\tau)-e^{A_n(T-\tau)}\mathcal{A}_n(\tau)\big)\Phi(\tau)\Big|_{\mathcal{L}_2(H;H_\lambda)}d\tau\Big)^4\\
&\quad +\Big(\big||\mathcal{A}(\cdot)|_{\mathcal{L}(H_\lambda)}\big|^4_{L^\infty_{\mathbb{F}}(\Omega;L^1(0;\tau_0))}+\big||\mathcal{B}(\cdot)|_{\mathcal{L}(H_\lambda;\mathcal{L}_2(V;H_\lambda))}\big|^4_{L^\infty_{\mathbb{F}}(\Omega;L^2(0;\tau_0))}\Big)\\
&\qquad \times\mathbb{E}\Big(\sup_{r\in[0,\tau_0]} |\Phi(r)-\Phi_n(r)|^4_{\mathcal{L}_2(H;H_\lambda)}\Big)\\
&\quad +\mathbb{E}\Big(\int_0^{T} \big|\big(S(T-\tau)\mathcal{B}(\tau)-e^{A_n(T-\tau)}\mathcal{B}_n(\tau)\big)\Phi(\tau)\big|^2_{\mathcal{L}_2(H;\mathcal{L}_2(V;H_\lambda))}d\tau\Big)^2\Big].
\end{aligned}
$$

Noting that $\mathcal{A}\in\Upsilon_1(H_\lambda)$ and $\mathcal{B}\in\Upsilon_2(H_\lambda;\mathcal{L}_2(V;H_\lambda))$, we conclude that there is a stopping time $\tau_0\in(0,T]$, a.s., such that

$$
\mathcal{C}\Big(\Big|\int_0^{\tau_0} |\mathcal{A}(\tau)|_{\mathcal{L}(H_\lambda)}d\tau\Big|^4_{L^\infty(\Omega)}+\Big|\int_0^{\tau_0} |\mathcal{B}(\tau)|^2_{\mathcal{L}(H_\lambda;\mathcal{L}_2(V;H_\lambda))}d\tau\Big|^2_{L^\infty(\Omega)}\Big)\leq\frac{1}{2}.
$$

For this τ_0, it follows from (13.190) that

$$
\begin{aligned}
&\mathbb{E}\Big(\sup_{r\in[0,\tau_0]} |\Phi(r)-\Phi_n(r)|^4_{\mathcal{L}_2(H;H_\lambda)}\Big)\\
&\leq \mathcal{C}\Big[\mathbb{E}\Big(\sup_{r\in[0,\tau_0]} |S(r)-e^{A_n r}|^4_{\mathcal{L}_2(H;H_\lambda)}\Big)\\
&\quad +\mathbb{E}\Big|\int_0^{T} \big(S(T-\tau)\mathcal{A}(\tau)-e^{A_n(T-\tau)}\mathcal{A}_n(\tau)\big)\Phi(\tau)d\tau\Big|^4_{\mathcal{L}_2(H;H_\lambda)} \qquad (13.192)\\
&\quad +\mathbb{E}\Big(\int_0^{T} \big|\big(S(T-\tau)\mathcal{B}(\tau)-e^{A_n(T-\tau)}\mathcal{B}_n(\tau)\big)\Phi(\tau)\big|^2_{\mathcal{L}_2(H;\mathcal{L}_2(V;H_\lambda))}d\tau\Big)^2\Big].
\end{aligned}
$$

From (13.154), we have

$$\lim_{n\to\infty} \mathbb{E}\Big(\sup_{r\in[0,\tau_0]} |S(r) - e^{A_n r}|^4_{\mathcal{L}_2(H;H_\lambda)}\Big) = 0.$$

By (13.154) and (13.186), we get that

$$\lim_{n\to\infty} \mathbb{E}\Big| \int_0^T \big(S(T-\tau)\mathcal{A}(\tau) - e^{A_n(T-\tau)}\mathcal{A}_n(\tau)\big)\varPhi(\tau)d\tau\Big|^4_{\mathcal{L}_2(H;H_\lambda)} = 0$$

and

$$\lim_{n\to\infty} \mathbb{E}\Big(\int_0^T \big|\big(S(T-\tau)\mathcal{B}(\tau) - e^{A_n(T-\tau)}\mathcal{B}_n(\tau)\big)\varPhi(\tau)\big|^2_{\mathcal{L}_2(H;\mathcal{L}_2(V;H_\lambda))}d\tau\Big)^2 = 0.$$

These, together with (13.192), imply that

$$\lim_{n\to\infty} |\varPhi_n - \varPhi|_{L^4_{\mathbb{F}}(\Omega;L^\infty(0,T;\mathcal{L}_2(H;H_\lambda)))} = 0. \tag{13.193}$$

Repeating the above argument gives (13.184).

Next, we prove (13.185). Clearly, for any $\varrho \in L^\infty_{\mathbb{F}}(0,T;L^2(\Omega;H_\lambda))$ and $\vartheta \in L^\infty_{\mathbb{F}}(0,T;L^2(\Omega;\mathcal{L}_2(V;H_\lambda)))$,

$$\begin{cases} \lim\limits_{n\to\infty} |\mathcal{K}_n\varrho - \mathcal{K}\varrho|_{L^\infty_{\mathbb{F}}(0,T;L^2(\Omega;H_\lambda))} = 0, \\ \lim\limits_{n\to\infty} |\mathcal{D}_n\vartheta - \mathcal{D}\vartheta|_{L^\infty_{\mathbb{F}}(0,T;L^2(\Omega;H_\lambda))} = 0. \end{cases} \tag{13.194}$$

Similar to the proof of (13.188), we obtain that for any $t \in [0,T]$ and a.s.,

$$\begin{aligned} \varPsi(t) = S(T-t)^*G\varPhi(T) &+ \int_t^T S(r-t)^*\big(\mathcal{K}(r)\varPsi(r) + \mathcal{D}(r)\varXi(r)\big)dr \\ &- \int_t^T S(r-t)^*\varXi(r)dW(r) \quad \text{in } \mathcal{L}_2(H;H_\lambda) \end{aligned} \tag{13.195}$$

and

$$\begin{aligned} \varPsi_n(t) = e^{A_n^*(T-t)}G_n\varPhi_n(T) &+ \int_t^T e^{A_n^*(r-t)}\big(\mathcal{K}_n(r)\varPsi_n(r) + \mathcal{D}_n(r)\varXi_n(r)\big)dr \\ &- \int_t^T e^{A_n^*(r-t)}\varXi_n(r)dW(r) \quad \text{in } \mathcal{L}_2(H;H_\lambda). \end{aligned} \tag{13.196}$$

Since $\mathcal{L}_2(H;H_\lambda)$ is a Hilbert space, by (13.195)–(13.196), it is easy to see that $(\varPsi,\varXi)$ and $(\varPsi_n,\varXi_n)$ are respectively weak solutions of the following $\mathcal{L}_2(H;H_\lambda)$-valued backward stochastic evolution equations

$$\begin{cases} d\varPsi = -(A^*\varPsi + \mathcal{K}\varPsi + \mathcal{D}\varXi)dt + \varXi dW(t) \quad \text{in } [0,T), \\ \varPsi(T) = G\varPhi(T) \end{cases} \tag{13.197}$$

and

$$\begin{cases} d\Psi_n = -(A_n^*\Psi_n + \mathcal{K}_n\Psi_n + \mathcal{D}_n\Xi_n)dt + \Xi_n dW(t) & \text{in } [0,T), \\ \Psi_n(T) = G_n\Phi_n(T). \end{cases} \tag{13.198}$$

Then, for any $t \in (0,T]$, by Itô's formula and noting that $(A^* - A_n^*)\Psi_n = 0$ (by our assumption **(AS1)**),

$$\begin{aligned} & |\Psi(t) - \Psi_n(t)|^2_{\mathcal{L}_2(H;H_\lambda)} + \int_t^T |\Xi(r) - \Xi_n(r)|^2_{\mathcal{L}_2(H;\mathcal{L}_2(V;H_\lambda))}dr \\ & = |G\Phi(T) - G_n\Phi_n(T)|^2_{\mathcal{L}_2(H;H_\lambda)} + 2\int_t^T \big[\big\langle A^*(\Psi - \Psi_n), \Psi - \Psi_n\big\rangle_{\mathcal{L}_2(H;H_\lambda)} \\ & \quad +\big\langle(\mathcal{K} - \mathcal{K}_n)\Psi, \Psi - \Psi_n\big\rangle_{\mathcal{L}_2(H;H_\lambda)} + \big\langle\mathcal{K}_n(\Psi - \Psi_n), \Psi - \Psi_n\big\rangle_{\mathcal{L}_2(H;H_\lambda)} \\ & \quad +\big\langle(\mathcal{D} - \mathcal{D}_n)\Xi, \Psi - \Psi_n\big\rangle_{\mathcal{L}_2(H;H_\lambda)} + \big\langle\mathcal{D}_n(\Xi - \Xi_n), \Psi - \Psi_n\big\rangle_{\mathcal{L}_2(H;H_\lambda)}\big]d\tau \\ & \quad -2\int_t^T \big\langle(\Xi - \Xi_n)dW(\tau), \Psi - \Psi_n\big\rangle_{\mathcal{L}_2(H;H_\lambda)}. \end{aligned} \tag{13.199}$$

Since A^* generates a C_0-group on H_λ, we have that for any $\varrho \in \mathcal{L}_2(H;H_\lambda)$,

$$\langle A^*\varrho, \varrho\rangle_{\mathcal{L}_2(H;H_\lambda)} = \sum_{k=1}^{\infty}\langle A^*(\varrho e_k), \varrho e_k\rangle_{H_\lambda} \le \mathcal{C}\sum_{k=1}^{\infty}|\varrho e_k|^2_{H_\lambda} = \mathcal{C}|\varrho|^2_{\mathcal{L}_2(H;H_\lambda)}.$$

Thus,

$$\int_t^T \big\langle A^*(\Psi - \Psi_n), \Psi - \Psi_n\big\rangle_{\mathcal{L}_2(H;H_\lambda)}d\tau \le \mathcal{C}\int_t^T |\Psi - \Psi_n|^2_{\mathcal{L}_2(H;H_\lambda)}d\tau. \tag{13.200}$$

Clearly,

$$\begin{aligned} & \int_t^T \big\langle\mathcal{K}_n(\Psi - \Psi_n), \Psi - \Psi_n\big\rangle_{\mathcal{L}_2(H;H_\lambda)}d\tau \\ & \quad + \int_t^T \big\langle\mathcal{D}_n(\Xi - \Xi_n), \Psi - \Psi_n\big\rangle_{\mathcal{L}_2(H;H_\lambda)}d\tau \\ & \le |\mathcal{K}_n|_{L^\infty_{\mathbb{F}}(0,T;H_\lambda)}\int_t^T \big|\Psi - \Psi_n\big|^2_{\mathcal{L}_2(H;H_\lambda)}d\tau \\ & \quad +|\mathcal{D}_n|_{L^\infty_{\mathbb{F}}(0,T;H_\lambda)}\int_t^T \big|\Psi - \Psi_n\big|_{\mathcal{L}_2(H;H_\lambda)}\big|\Xi - \Xi_n\big|_{\mathcal{L}_2(H;\mathcal{L}_2(V;H_\lambda))}d\tau \\ & \le \mathcal{C}\int_t^T \big|\Psi - \Psi_n\big|^2_{\mathcal{L}_2(H;H_\lambda)}d\tau + \frac{1}{2}\int_t^T \big|\Xi - \Xi_n\big|^2_{\mathcal{L}_2(H;\mathcal{L}_2(V;H_\lambda))}d\tau. \end{aligned} \tag{13.201}$$

From (13.199)–(13.201), we find that

$$\mathbb{E}|\Psi(t) - \Psi_n(t)|^4_{\mathcal{L}_2(H;H_\lambda)} + \mathbb{E}\Big(\int_t^T |\Xi(r) - \Xi_n(r)|^2_{\mathcal{L}_2(H;\mathcal{L}_2(V;H_\lambda))}dr\Big)^2$$

$$\leq \mathcal{C}\mathbb{E}\Big[|G\Phi(T) - G_n\Phi_n(T)|^4_{\mathcal{L}_2(H;H_\lambda)} + \Big(\int_t^T |\Psi - \Psi_n|^2_{\mathcal{L}_2(H;H_\lambda)} d\tau\Big)^2 \tag{13.202}$$
$$+\Big(\int_t^T \big|(\mathcal{K} - \mathcal{K}_n)\Psi\big|^2_{\mathcal{L}_2(H;H_\lambda)} d\tau\Big)^2 + \Big(\int_t^T \big|(\mathcal{D} - \mathcal{D}_n)\Xi\big|^2_{\mathcal{L}_2(H;H_\lambda)} d\tau\Big)^2\Big].$$

Since

$$\lim_{n\to\infty} \mathbb{E}\Big[\Big(\int_0^T \big|(\mathcal{K} - \mathcal{K}_n)\Psi\big|^2_{\mathcal{L}_2(H;H_\lambda)} d\tau\Big)^2 + \Big(\int_0^T \big|(\mathcal{D} - \mathcal{D}_n)\Xi\big|^2_{\mathcal{L}_2(H;H_\lambda)} d\tau\Big)^2\Big] = 0,$$

the estimate (13.202) implies that

$$\lim_{n\to\infty} \Big[\sup_{t\in[0,T]} \mathbb{E}|\Psi(t) - \Psi_n(t)|^4_{\mathcal{L}_2(H;H_\lambda)} + \mathbb{E}\Big(\int_0^T |\Xi(r) - \Xi_n(r)|^2_{\mathcal{L}_2(H;\mathcal{L}_2(V;H_\lambda))} dr\Big)^2\Big] = 0. \tag{13.203}$$

This gives the second equality in (13.185).

It remains to prove the first equality in (13.185). From (13.195) and (13.196), it follows that, for any $s_0 \in [0, T)$,

$$\begin{aligned}
&\mathbb{E} \sup_{r\in[s_0,T]} |\Psi(r) - \Psi_n(r)|^4_{\mathcal{L}_2(H;H_\lambda)} \\
&\leq \mathcal{C}\mathbb{E} \sup_{r\in[s_0,T]} \Big[|S(T-r)^* G\Phi(T) - e^{A_n^*(T-r)} G_n\Phi_n(T)|^4_{\mathcal{L}_2(H;H_\lambda)} \\
&\quad + \Big|\int_r^T \big(S(\tau - r)^*\mathcal{K}(\tau)\Psi(\tau) - e^{A_n^*(\tau-r)}\mathcal{K}_n(\tau)\Psi_n(\tau)\big) d\tau\Big|^4_{\mathcal{L}_2(H;H_\lambda)} \\
&\quad + \Big|\int_r^T \big(S(\tau - r)^*\mathcal{D}(\tau)\Xi(\tau) - e^{A_n^*(\tau-r)}\mathcal{D}_n(\tau)\Xi_n(\tau)\big) d\tau\Big|^4_{\mathcal{L}_2(H;H_\lambda)} \\
&\quad + \Big|\int_r^T \big(S(\tau - r)^*\Xi(\tau) - e^{A_n^*(\tau-r)}\Xi_n(\tau)\big) dW(\tau)\Big|^4_{\mathcal{L}_2(H;H_\lambda)}\Big].
\end{aligned} \tag{13.204}$$

Therefore, by Burkholder-Davis-Gundy inequality, similar to (13.191), we obtain that

$$\begin{aligned}
&\mathbb{E} \sup_{r\in[s_0,T]} |\Psi(r) - \Psi_n(r)|^4_{\mathcal{L}_2(H;H_\lambda)} \\
&\leq \mathcal{C}\Big[\mathbb{E} \sup_{r\in[0,T]} |S(T-r)^* G\Phi(T) - e^{A_n^*(T-r)} G_n\Phi_n(T)|^4_{\mathcal{L}_2(H;H_\lambda)} \\
&\quad + \mathbb{E}\Big(\int_0^T \Big|\big(S(\tau)^*\mathcal{K}(\tau) - e^{A_n^*\tau}\mathcal{K}_n(\tau)\big)\Psi(\tau)\Big|_{\mathcal{L}_2(H;H_\lambda)} d\tau\Big)^4 \\
&\quad + (T - s_0)^3 |\mathcal{K}|^4_{L^\infty_{\mathbb{F}}(0,T;\mathcal{L}(H_\lambda))} \mathbb{E} \sup_{r\in[s_0,T]} |\Psi(r) - \Psi_n(r)|^4_{\mathcal{L}_2(H;H_\lambda)}
\end{aligned} \tag{13.205}$$

$$+\mathbb{E}\Big(\int_0^T \big|\big(S(\tau)^*\mathcal{D}(\tau)-e^{A_n^*\tau}\mathcal{D}_n(\tau)\big)\Xi(\tau)\big|^2_{\mathcal{L}_2(H;H_\lambda)}d\tau\Big)^2$$
$$+\big(|\mathcal{D}|^4_{L^\infty_{\mathbb{F}}(0,T;\mathcal{L}(\mathcal{L}_2(V;H_\lambda);H_\lambda))}+1\big)|\Xi-\Xi_n|^4_{L^2_{\mathbb{F}}(0,T;\mathcal{L}_2(H;\mathcal{L}_2(V;H_\lambda)))}\Big)\Big].$$

Let $s_0=T-(2\mathcal{C}|\mathcal{K}|^4_{L^\infty_{\mathbb{F}}(0,T;\mathcal{L}(H_\lambda))})^{-\frac{1}{3}}$. Then $\mathcal{C}(T-s_0)^3|\mathcal{K}|^4_{L^\infty_{\mathbb{F}}(0,T;\mathcal{L}(H_\lambda))}=\frac{1}{2}$. By (13.153), (13.184) and (13.205), we conclude that

$$\lim_{n\to\infty}|\Psi_n-\Psi|_{L^4_{\mathbb{F}}(\Omega;L^\infty(s_0,T;\mathcal{L}_2(H;H_\lambda)))}=0.$$

Repeating this argument gives the first equality in (13.185). □

To end this subsection, we provide below a controllability result concerning the trajectories of solutions to (13.134), which is a variant of Lemma 12.13 and will play an important role in the proof of the uniqueness of transposition solutions to (13.17).

Lemma 13.50. *The set*

$$\mathfrak{R}\overset{\Delta}{=}\big\{x_2(\cdot)\bigm| x_2(\cdot)\ \textit{solves}\ (13.134)\ \textit{with}\ t=0,\ \xi_2=0,\ v_2=0\ \textit{and}\ u_2\in L^4_{\mathbb{F}}(\Omega;L^2(0,T;H'_\lambda))\big\}$$

is dense in $L^2_{\mathbb{F}}(0,T;H'_\lambda)$.

Proof: The proof of Lemma 13.50 is very similar to the proof of Lemma 12.13. We give it here for the sake of completeness.

If Lemma 13.50 were not true, then there would be a nonzero $\rho\in L^2_{\mathbb{F}}(0,T;H_\lambda)$ such that

$$\mathbb{E}\int_0^T\langle\rho,x_2\rangle_{H_\lambda,H'_\lambda}ds=0,\qquad \text{for every } x_2\in\mathfrak{R}. \tag{13.206}$$

Consider the following backward stochastic evolution equation:

$$\begin{cases} dy=-A^*ydt+\big(\rho-C^*Y\big)dt+YdW(t) & \text{in } [0,T),\\ y(T)=0, \end{cases} \tag{13.207}$$

which admits a unique solution

$$(y(\cdot),Y(\cdot))\in L^2_{\mathbb{F}}(\Omega;C([0,T];H_\lambda))\times L^2_{\mathbb{F}}(0,T;H_\lambda).$$

Hence, for any $\phi_1(\cdot)\in L^1_{\mathbb{F}}(0,T;L^4(\Omega;H'_\lambda))$ and $\phi_2(\cdot)\in L^2_{\mathbb{F}}(0,T;L^4(\Omega;H'_\lambda))$, we have

$$\begin{aligned} &-\mathbb{E}\int_0^T\big\langle z(s),\rho(s)-C^*Y(s)\big\rangle_{H'_\lambda,H_\lambda}ds\\ &=\mathbb{E}\int_0^T\big\langle\phi_1(s),y(s)\big\rangle_{H'_\lambda,H_\lambda}ds+\mathbb{E}\int_0^T\big\langle\phi_2(s),Y(s)\big\rangle_{H'_\lambda,H_\lambda}ds, \end{aligned} \tag{13.208}$$

where $z(\cdot)$ solves

$$\begin{cases} dz = (Az + \phi_1)dt + \phi_2 dW(t) \ \text{ in } (0,T], \\ z(0) = 0. \end{cases} \tag{13.209}$$

In particular, for any $x_2(\cdot)$ solving (12.40) with $t = 0$, $\xi_2 = 0$, $v_2 = 0$ and an arbitrarily given $u_2 \in L^4_{\mathbb{F}}(0,T;H'_\lambda)$, we choose $z = x_2$, $\phi_1 = u_2$ and $\phi_2 = Cx_2$. It follows from (13.208) that

$$\begin{aligned} -\mathbb{E}\int_0^T \big\langle x_2(s), \rho(s)\big\rangle_{H'_\lambda,H_\lambda} ds = \mathbb{E}\int_0^T \big\langle u_2(s), y(s)\big\rangle_{H'_\lambda,H_\lambda} ds, \\ \forall\, u_2 \in L^4_{\mathbb{F}}(0,T;H'_\lambda). \end{aligned} \tag{13.210}$$

By (13.210) and recalling (13.206), we get that $y(\cdot) = 0$. Consequently, (13.208) is reduced to

$$-\,\mathbb{E}\int_0^T \big\langle z(s), \rho(s) - C^*Y(s)\big\rangle_{H'_\lambda,H_\lambda} ds = \mathbb{E}\int_0^T \big\langle \phi_2(s), Y(s)\big\rangle_{H'_\lambda,H_\lambda} ds. \tag{13.211}$$

Choosing $\phi_2(\cdot) = 0$ in (13.209), by (13.211), we obtain that

$$\begin{aligned} \mathbb{E}\int_0^T \Big\langle \int_0^s S(s-\sigma)\phi_1(\sigma)d\sigma, \rho(s) - C^*Y(s)\Big\rangle_{H'_\lambda,H_\lambda} ds = 0, \\ \forall\, \phi_1(\cdot) \in L^1_{\mathbb{F}}(0,T;L^4(\Omega;H'_\lambda)). \end{aligned} \tag{13.212}$$

Hence,

$$\int_\sigma^T S(s-\sigma)\big[\rho(s) - C^*Y(s)\big]ds = 0, \qquad \forall\, \sigma \in [0,T]. \tag{13.213}$$

Then, for any given $\lambda_0 \in \rho(A)$ and $\sigma \in [0,T]$, it holds that

$$\begin{aligned} &\int_\sigma^T S(s-\sigma)(\lambda_0 - A)^{-1}\big(\rho(s) - C^*Y(s)\big)ds \\ &= (\lambda_0 - A)^{-1}\int_\sigma^T S(s-\sigma)\big(\rho(s) - C^*Y(s)\big)ds = 0. \end{aligned} \tag{13.214}$$

Differentiating the equality (13.214) with respect to σ, and noting (13.213), we obtain that

$$\begin{aligned}
&(\lambda_0 - A)^{-1}\big(\rho(\sigma) - C^*Y(\sigma)\big)\\
&= -\int_\sigma^T S(s-\sigma)A(\lambda_0 - A)^{-1}\big(\rho(s) - C^*Y(s)\big)ds\\
&= \int_\sigma^T S(s-\sigma)\big[\rho(s) - C^*Y(s)\big]ds\\
&\quad -\lambda_0 \int_\sigma^T S(s-\sigma)(\lambda_0 - A)^{-1}\big(\rho(s) - C^*Y(s)\big)ds\\
&= 0, \qquad \forall\, \sigma \in [0,T].
\end{aligned}$$

Therefore,

$$\rho(\cdot) = C(\cdot)^*Y(\cdot). \tag{13.215}$$

By (13.215), the equation (13.207) is reduced to

$$\begin{cases} dy = -A^*ydt + YdW(t) & \text{in } [0,T),\\ y(T) = 0. \end{cases} \tag{13.216}$$

It is clear that the unique solution of (13.216) is $(y(\cdot), Y(\cdot)) = (0,0)$. Hence, by (13.215), we conclude that $\rho(\cdot) = 0$, a contradiction. This completes the proof of Lemma 13.50. □

13.8.2 Proof of the Main Solvability Result

In this subsection, we shall prove Theorem 13.41. To avoid the beginners being trapped in the technical details, we only handle the case that $V = \mathbb{R}$, that is $W(\cdot)$ is a standard one dimensional Brownian motion (Note that, in this case, $\mathcal{L}_2^0 = \mathbb{R} \otimes H$ is isometric isomorphism to H. Hence, in the rest of this chapter, we simply regard $\mathcal{L}_2^0$ as H). With the preliminary results in Subsection 13.8.1, interested readers can follow the proof in this subsection to deal with the general case. The proof (even for the case $V = \mathbb{R}$) is so long that we have to divide it into several steps.

Step 1. In this step, we introduce some operators $\mathbb{X}(\cdot)$ [3], $\mathbb{Y}(\cdot)$, $\mathbf{Y}(\cdot)$ and $\widetilde{\mathbb{X}}(\cdot)$.

Let $\Theta(\cdot) \in \Upsilon_2(H;U) \cap \Upsilon_2(H'_\lambda; \widetilde{U})$ be an optimal feedback operator of Problem (SLQ). Then, by Theorems 3.20 and 4.10, for any $\zeta \in H$, the following forward-backward stochastic evolution equation

$$\begin{cases} d\hat{x} = \big(A + B\Theta\big)\hat{x}dt + (C + D\Theta)\hat{x}dW(t) & \text{in } (0,T],\\ dy = -\big(A^*y + C^*Y + M\hat{x}\big)dt + YdW(t) & \text{in } [0,T),\\ \hat{x}(0) = \zeta, \quad y(T) = G\hat{x}(T) \end{cases} \tag{13.217}$$

[3] In the sequel, we shall interchangeably use $\mathbb{X}(t)$, $\mathbb{X}(t,\cdot)$, or even $\mathbb{X}$ to denote the operator $\mathbb{X}(\cdot)$. The same can be said for $\mathbb{Y}(\cdot)$, $\mathbf{Y}(\cdot)$ and $\widetilde{\mathbb{X}}(\cdot)$.

admits a unique mild solution $(\hat{x}(\cdot), y(\cdot), Y(\cdot)) \in L^2_{\mathbb{F}}(\Omega; C([0,T]; H)) \times L^2_{\mathbb{F}}(\Omega; C([0,T]; H)) \times L^2_{\mathbb{F}}(0,T; H)$ such that

$$R\Theta\hat{x} + B^*y + D^*Y = 0, \quad \text{a.e. } (t,\omega) \in (0,T) \times \Omega. \tag{13.218}$$

Further, consider the following stochastic evolution equation:

$$\begin{cases} d\tilde{x} = \big[-A - B\Theta + \big(C + D\Theta\big)^2\big]^* \tilde{x}dt \\ \qquad\quad -\big(C + D\Theta\big)^* \tilde{x}dW(t) & \text{in } (0,T], \\ \tilde{x}(0) = \zeta. \end{cases} \tag{13.219}$$

Note that A generates a C_0-group, and hence, so does $-A^*$. By Theorem 3.20, the equation (13.219) admits a unique mild solution $\tilde{x}(\cdot) \in L^2_{\mathbb{F}}(\Omega; C([0,T]; H))$.

For each $n \in \mathbb{N}$, denote by $\widetilde{\Gamma}_n$ the projection operator from U to $U_n \stackrel{\Delta}{=} \text{span}_{1 \le j \le n}\{\varphi_j\}$ (Recall that $\{\varphi_j\}_{j=1}^{\infty}$ is an orthonormal basis of U). Write

$$B_n = \Gamma_n B \widetilde{\Gamma}_n, \quad C_n = \Gamma_n C \Gamma_n, \quad D_n = \Gamma_n D \widetilde{\Gamma}_n,$$

$$M_n = \Gamma_n M \Gamma_n, \quad R_n = \widetilde{\Gamma}_n R \widetilde{\Gamma}_n, \quad \Theta_n = \widetilde{\Gamma}_n \Theta \Gamma_n.$$

Clearly,

$$\begin{cases} \lim\limits_{n\to+\infty} C_n\zeta = C\zeta & \text{in } H, \\ \lim\limits_{n\to+\infty} M_n\zeta = M\zeta & \text{in } H, \\ \lim\limits_{n\to+\infty} \Theta_n\zeta = \Theta\zeta & \text{in } U, \end{cases} \tag{13.220}$$
$$\text{for all } \zeta \in H \text{ and a.e. } (t,\omega) \in [0,T] \times \Omega,$$

and

$$\begin{cases} \lim\limits_{n\to+\infty} B_n\varsigma = B\varsigma & \text{in } H, \\ \lim\limits_{n\to+\infty} D_n\varsigma = D\varsigma & \text{in } H, \\ \lim\limits_{n\to+\infty} R_n\varsigma = R\varsigma & \text{in } U, \end{cases} \tag{13.221}$$
$$\text{for all } \varsigma \in U \text{ and a.e. } (t,\omega) \in [0,T] \times \Omega.$$

Consider the following forward-backward stochastic differential equation:

$$\begin{cases} d\hat{x}_n = (A_n + B_n\Theta_n)\hat{x}_n dt + (C_n + D_n\Theta_n)\hat{x}_n dW(t) & \text{in } [0,T], \\ dy_n = -\big(A_n^* y_n + C_n^* Y_n + M_n\hat{x}_n\big)dt + Y_n dW(t) & \text{in } [0,T], \\ \hat{x}_n(0) = \Gamma_n\zeta, \quad y_n(T) = G_n\hat{x}_n(T) \end{cases} \tag{13.222}$$

and the following stochastic differential equation:

$$\begin{cases} d\tilde{x}_n = \big[-A_n - B_n\Theta_n + \big(C_n + D_n\Theta_n\big)^2\big]^\top \tilde{x}_n dt \\ \qquad\quad -(C_n + D_n\Theta_n)^\top \tilde{x}_n dW(t) & \text{in } [0,T], \\ \tilde{x}_n(0) = \Gamma_n\zeta, \end{cases} \tag{13.223}$$

where A_n and G_n are given in (13.149). For each $t \in [0,T]$, define three operators $\mathbb{X}_{n,t}$, $\mathbb{Y}_{n,t}$ and $\widetilde{\mathbb{X}}_{n,t}$ on H_n as follows:

$$\begin{cases} \mathbb{X}_{n,t}\Gamma_n\zeta \overset{\Delta}{=} \hat{x}_n(t;\Gamma_n\zeta), \\ \mathbb{Y}_{n,t}\Gamma_n\zeta \overset{\Delta}{=} y_n(t;\Gamma_n\zeta), & \forall\, \zeta \in H. \\ \widetilde{\mathbb{X}}_{n,t}\Gamma_n\zeta \overset{\Delta}{=} \tilde{x}_n(t;\Gamma_n\zeta), \end{cases} \tag{13.224}$$

For a.e. $t \in [0,T]$, define an operator $\mathbf{Y}_{n,t}$ on H_n by

$$\mathbf{Y}_{n,t}\Gamma_n\zeta \overset{\Delta}{=} Y_n(t;\Gamma_n\zeta), \qquad \forall\, \zeta \in H. \tag{13.225}$$

By the well-posedness results for the equations (13.222) and (13.223), and the fact that both A and $-A^*$ generate C_0-semigroups on H (because A generates a C_0-group on H), we see that

$$\begin{cases} |\mathbb{X}_{n,t}\Gamma_n\zeta|_{L^2_{\mathcal{F}_t}(\Omega;H)} \le \mathcal{C}|\zeta|_H, \\ |\mathbb{Y}_{n,t}\Gamma_n\zeta|_{L^2_{\mathcal{F}_t}(\Omega;H)} \le \mathcal{C}|\zeta|_H, \\ |\mathbf{Y}_{n,\cdot}\Gamma_n\zeta|_{L^2_{\mathbb{F}}(0,T;H)} \le \mathcal{C}|\zeta|_H, \\ |\widetilde{\mathbb{X}}_{n,t}\Gamma_n\zeta|_{L^2_{\mathcal{F}_t}(\Omega;H)} \le \mathcal{C}|\zeta|_H, \end{cases}$$

where the constant $\mathcal{C}$ is independent of n. This implies that

$$\begin{cases} |\mathbb{X}_{n,t}\Gamma_n|_{\mathcal{L}(H;L^2_{\mathcal{F}_t}(\Omega;H))} \le \mathcal{C}, \\ |\mathbb{Y}_{n,t}\Gamma_n|_{\mathcal{L}(H;L^2_{\mathcal{F}_t}(\Omega;H))} \le \mathcal{C}, \\ |\mathbf{Y}_{n,\cdot}\Gamma_n|_{\mathcal{L}(H;L^2_{\mathbb{F}}(0,T;H))} \le \mathcal{C}, \\ |\widetilde{\mathbb{X}}_{n,t}\Gamma_n|_{\mathcal{L}(H;L^2_{\mathcal{F}_t}(\Omega;H))} \le \mathcal{C}. \end{cases} \tag{13.226}$$

Consider the following equations:

$$\begin{cases} d\mathbb{X}_n = (A_n + B_n\Theta_n)\mathbb{X}_n dt + (C_n + D_n\Theta_n)\mathbb{X}_n dW(t) & \text{in } [0,T], \\ d\mathbb{Y}_n = -\big(A_n^*\mathbb{Y}_n + C_n^*\mathbf{Y}_n + M_n\mathbb{X}_n\big)dt + \mathbf{Y}_n dW(t) & \text{in } [0,T], \\ \mathbb{X}_n(0) = I_n, \quad \mathbb{Y}_n(T) = G_n\mathbb{X}_n(T) \end{cases} \tag{13.227}$$

and

$$\begin{cases} d\widetilde{\mathbb{X}}_n = \big[-A_n - B_n\Theta_n + \big(C_n + D_n\Theta_n\big)^2\big]^\top \widetilde{\mathbb{X}}_n dt \\ \qquad\qquad -(C_n + D_n\Theta_n)^\top \widetilde{\mathbb{X}}_n dW(t) \quad \text{in } [0,T], \\ \widetilde{\mathbb{X}}_n(0) = I_n. \end{cases} \tag{13.228}$$

Clearly, both (13.227) and (13.228) can be viewed as $\mathbb{R}^{n\times n} \equiv \mathbb{R}^{n^2}$-valued equations. By Theorems 3.20 and 4.10, the equations (13.227) and (13.228) admit solutions $(\mathbb{X}_n, \mathbb{Y}_n, \mathbf{Y}_n) \in L^2_{\mathbb{F}}(\Omega; C([0,T]; \mathbb{R}^{n\times n})) \times L^2_{\mathbb{F}}(\Omega; C([0,T]; \mathbb{R}^{n\times n})) \times L^2_{\mathbb{F}}(0,T; \mathbb{R}^{n\times n})$ and $\widetilde{\mathbb{X}}_n \in L^2_{\mathbb{F}}(\Omega; C([0,T]; \mathbb{R}^{n\times n}))$, respectively. It follows from (13.224)–(13.228) that, for a.e. $t \in [0,T]$,

$$\begin{cases} \mathbb{X}_{n,t}\Gamma_n\zeta = \mathbb{X}_n(t)\Gamma_n\zeta, \\ \mathbb{Y}_{n,t}\Gamma_n\zeta = \mathbb{Y}_n(t)\Gamma_n\zeta, \\ \mathbf{Y}_{n,t}\Gamma_n\zeta = \mathbf{Y}_n(t)\Gamma_n\zeta, \\ \widetilde{\mathbb{X}}_{n,t}\Gamma_n\zeta = \widetilde{\mathbb{X}}_n(t)\Gamma_n\zeta, \end{cases} \qquad \forall\, \zeta \in H. \tag{13.229}$$

Thus,

$$\begin{cases} \mathbb{X}_{n,t}\Gamma_n \in L^2_{\mathcal{F}_t}(\Omega; \mathcal{L}(H)), \\ \mathbb{Y}_{n,t}\Gamma_n \in L^2_{\mathcal{F}_t}(\Omega; \mathcal{L}(H)), \\ \widetilde{\mathbb{X}}_{n,t}\Gamma_n \in L^2_{\mathcal{F}_t}(\Omega; \mathcal{L}(H)), \\ \mathbf{Y}_{n,\cdot}\Gamma_n \in L^2_{\mathbb{F}}(0,T; \mathcal{L}(H)). \end{cases}$$

Clearly,

$$\begin{cases} \mathbb{X}_{n,t}\Gamma_n \in \mathcal{L}_{pd}(H; L^2_{\mathcal{F}_t}(\Omega; H)) \;\; \text{for a.e. } t \in [0,T], \\ \mathbb{Y}_{n,t}\Gamma_n \in \mathcal{L}_{pd}(H; L^2_{\mathcal{F}_t}(\Omega; H)) \;\; \text{for a.e. } t \in [0,T], \\ \widetilde{\mathbb{X}}_{n,t}\Gamma_n \in \mathcal{L}_{pd}(H; L^2_{\mathcal{F}_t}(\Omega; H)) \;\; \text{for a.e. } t \in [0,T], \\ \mathbf{Y}_{n,\cdot}\Gamma_n \in \mathcal{L}_{pd}(H; L^2_{\mathbb{F}}(0,T; H)). \end{cases}$$

By (13.226) and using Theorem 2.63, we deduce that, there exist suitable subsequences $\{\mathbb{X}_{n_k,t}\}_{k=1}^\infty \subset \{\mathbb{X}_{n,t}\}_{n=1}^\infty$, $\{\mathbb{Y}_{n_k,t}\}_{k=1}^\infty \subset \{\mathbb{Y}_{n,t}\}_{n=1}^\infty$, $\{\mathbf{Y}_{n_k,t}\}_{k=1}^\infty \subset \{\mathbf{Y}_{n,t}\}_{n=1}^\infty$ and $\{\widetilde{\mathbb{X}}_{n_k,t}\}_{k=1}^\infty \subset \{\widetilde{\mathbb{X}}_{n,t}\}_{n=1}^\infty$ (these sequences may depend on t), and (pointwise defined) operators $\mathbb{X}(t,\cdot), \mathbb{Y}(t,\cdot), \widetilde{\mathbb{X}}(t,\cdot) \in \mathcal{L}_{pd}(H; L^2_{\mathcal{F}_t}(\Omega; H))$ (for each $t \in [0,T]$) and $\mathbf{Y}(\cdot,\cdot) \in \mathcal{L}_{pd}(H; L^2_{\mathbb{F}}(0,T; H))$ such that

$$\begin{cases} \lim\limits_{k\to+\infty} \mathbb{X}_{n_k,t}\Gamma_{n_k}\zeta = \mathbb{X}(t,\cdot)\zeta \;\; \text{weakly in } L^2_{\mathcal{F}_t}(\Omega; H), \\ \lim\limits_{k\to+\infty} \mathbb{Y}_{n_k,t}\Gamma_{n_k}\zeta = \mathbb{Y}(t,\cdot)\zeta \;\; \text{weakly in } L^2_{\mathcal{F}_t}(\Omega; H), \\ \lim\limits_{k\to+\infty} \mathbf{Y}_{n_k,t}\Gamma_{n_k}\zeta = \mathbf{Y}(\cdot,\cdot)\zeta \;\; \text{weakly in } L^2_{\mathbb{F}}(0,T; H), \\ \lim\limits_{k\to+\infty} \widetilde{\mathbb{X}}_{n_k,t}\Gamma_{n_k}\zeta = \widetilde{\mathbb{X}}(t,\cdot)\zeta \;\; \text{weakly in } L^2_{\mathcal{F}_t}(\Omega; H), \end{cases} \tag{13.230}$$

and that

$$\begin{cases} |\mathbb{X}(t,\cdot)\zeta|_{L^2_{\mathcal{F}_t}(\Omega;H)} \leq \mathcal{C}|\zeta|_H, \\ |\mathbb{Y}(t,\cdot)\zeta|_{L^2_{\mathcal{F}_t}(\Omega;H)} \leq \mathcal{C}|\zeta|_H, \\ |\mathbf{Y}(\cdot,\cdot)\zeta|_{L^2_{\mathbb{F}}(0,T;H)} \leq \mathcal{C}|\zeta|_H, \\ |\widetilde{\mathbb{X}}(t,\cdot)\zeta|_{L^2_{\mathcal{F}_t}(\Omega;H)} \leq \mathcal{C}|\zeta|_H. \end{cases} \tag{13.231}$$

On the other hand, from the definition of $(\hat{x}_n(\cdot;\Gamma_n\zeta), y_n(\cdot;\Gamma_n\zeta), Y_n(\cdot;\Gamma_n\zeta))$ and $\tilde{x}_n(\cdot;\Gamma_n\zeta)$, by Lemma 13.47, we have that

$$\begin{cases} \lim\limits_{n\to\infty} \hat{x}_n(\cdot;\Gamma_n\zeta) = \hat{x}(\cdot;\zeta) \quad \text{in } L^2_{\mathbb{F}}(\Omega;C([0,T];H)), \\ \lim\limits_{n\to\infty} y_n(\cdot;\Gamma_n\zeta) = y(\cdot;\zeta) \quad \text{in } L^2_{\mathbb{F}}(\Omega;C([0,T];H)), \\ \lim\limits_{n\to\infty} Y_n(\cdot;\Gamma_n\zeta) = Y(\cdot;\zeta) \;\; \text{in } L^2_{\mathbb{F}}(0,T;H), \\ \lim\limits_{n\to\infty} \tilde{x}_n(\cdot;\Gamma_n\zeta) = \tilde{x}(\cdot;\zeta) \quad \text{in } L^2_{\mathbb{F}}(\Omega;C([0,T];H)). \end{cases} \tag{13.232}$$

Hence, in view of (13.224), we find that

$$\begin{cases} \lim\limits_{n\to\infty} \mathbb{X}_{n,t}\Gamma_n\zeta = \hat{x}(t;\zeta) \quad \text{strongly in } L^2_{\mathcal{F}_t}(\Omega;H), \\ \lim\limits_{n\to\infty} \mathbb{Y}_{n,t}\Gamma_n\zeta = y(t;\zeta) \quad \text{strongly in } L^2_{\mathcal{F}_t}(\Omega;H), \\ \lim\limits_{n\to\infty} \mathbf{Y}_{n,t}\Gamma_n\zeta = Y(t;\zeta) \;\; \text{strongly in } L^2_{\mathbb{F}}(0,T;H), \\ \lim\limits_{n\to\infty} \widetilde{\mathbb{X}}_{n,t}\Gamma_n\zeta = \tilde{x}(t;\zeta) \quad \text{strongly in } L^2_{\mathcal{F}_t}(\Omega;H). \end{cases} \tag{13.233}$$

According to (13.230) and (13.233), we obtain that

$$\begin{aligned} &\mathbb{X}(t,\cdot)\zeta = \hat{x}(t;\zeta), \quad \mathbb{Y}(t,\cdot)\zeta = y(t;\zeta), \\ &\mathbf{Y}(t,\cdot)\zeta = Y(t;\zeta), \quad \widetilde{\mathbb{X}}(t,\cdot)\zeta = \tilde{x}(t;\zeta). \end{aligned} \tag{13.234}$$

Also, from the equality (13.218) and noting (13.234), we find that

$$R\Theta\mathbb{X} + B^*\mathbb{Y} + D^*\mathbf{Y} = 0, \quad \text{for a.e. } (t,\omega) \in [0,T]\times\Omega. \tag{13.235}$$

Combining (13.232) and (13.234), we get that

$$\begin{cases} \lim\limits_{n\to\infty} \mathbb{X}_{n,t}\Gamma_n\zeta = \mathbb{X}(t,\cdot)\zeta \quad \text{strongly in } L^2_{\mathcal{F}_t}(\Omega;H), \\ \lim\limits_{n\to\infty} \mathbb{Y}_{n,t}\Gamma_n\zeta = \mathbb{Y}(t,\cdot)\zeta \quad \text{strongly in } L^2_{\mathcal{F}_t}(\Omega;H), \\ \lim\limits_{n\to\infty} \mathbf{Y}_{n,t}\Gamma_n\zeta = \mathbf{Y}(\cdot,\cdot)\zeta \;\; \text{strongly in } L^2_{\mathbb{F}}(0,T;H), \\ \lim\limits_{n\to\infty} \widetilde{\mathbb{X}}_{n,t}\Gamma_n\zeta = \widetilde{\mathbb{X}}(t,\cdot)\zeta \quad \text{strongly in } L^2_{\mathcal{F}_t}(\Omega;H). \end{cases} \tag{13.236}$$

Moreover, from Lemma 13.49, it follows that

$$
\begin{aligned}
&\mathbb{X}, \mathbb{Y}, \widetilde{\mathbb{X}} \in L^4_{\mathbb{F}}(\Omega; C([0,T]; \mathcal{L}_2(H; H_\lambda))),\\
&\mathbf{Y} \in L^4_{\mathbb{F}}(\Omega; L^2(0,T; \mathcal{L}_2(H; H_\lambda)))
\end{aligned}
\tag{13.237}
$$

and

$$
\begin{cases}
\lim\limits_{n\to\infty} \mathbb{X}_n = \mathbb{X} & \text{in } L^4_{\mathbb{F}}(\Omega; C([0,T]; \mathcal{L}_2(H; H_\lambda))),\\
\lim\limits_{n\to\infty} \mathbb{Y}_n = \mathbb{Y} & \text{in } L^4_{\mathbb{F}}(\Omega; C([0,T]; \mathcal{L}_2(H; H_\lambda))),\\
\lim\limits_{n\to\infty} \mathbf{Y}_n = \mathbf{Y} & \text{in } L^4_{\mathbb{F}}(\Omega; L^2(0,T; \mathcal{L}_2(H; H_\lambda))),\\
\lim\limits_{n\to\infty} \widetilde{\mathbb{X}}_n = \widetilde{\mathbb{X}} & \text{in } L^4_{\mathbb{F}}(\Omega; C([0,T]; \mathcal{L}_2(H; H_\lambda))).
\end{cases}
\tag{13.238}
$$

Step 2. In this step, we shall prove that

$$
\mathbb{X}(t,\cdot)\widetilde{\mathbb{X}}(t,\cdot)^* = I \quad \forall t \in [0,T], \text{ a.s.} \tag{13.239}
$$

For any $\zeta, \rho \in H$ and $t \in [0,T]$, by Itô's formula, we have

$$
\begin{aligned}
&\big\langle \hat{x}_n(t; \Gamma_n\zeta), \tilde{x}_n(t; \Gamma_n\rho)\big\rangle_{H_n} - \big\langle \Gamma_n\zeta, \Gamma_n\rho\big\rangle_{H_n}\\
&= \int_0^t \big\langle (A_n + B_n\Theta_n)\hat{x}_n(r; \Gamma_n\zeta), \tilde{x}_n(r; \Gamma_n\rho)\big\rangle_{H_n} d\tau\\
&\quad + \int_0^t \big\langle (C_n + D_n\Theta_n)\hat{x}_n(r; \Gamma_n\zeta), \tilde{x}_n(r; \Gamma_n\rho)\big\rangle_{H_n} dW(\tau)\\
&\quad + \int_0^t \big\langle \hat{x}_n(r; \Gamma_n\zeta), \big[-A_n - B_n\Theta_n + (C_n + D_n\Theta_n)^2\big]^\top \tilde{x}_n(r; \Gamma_n\rho)\big\rangle_{H_n} d\tau\\
&\quad - \int_0^t \big\langle \hat{x}_n(r; \Gamma_n\zeta), (C_n + D_n\Theta_n)^\top \tilde{x}_n(r; \Gamma_n\rho)\big\rangle_{H_n} dW(\tau)\\
&\quad - \int_0^t \big\langle (C_n + D_n\Theta_n)\hat{x}_n(r; \Gamma_n\zeta), (C_n + D_n\Theta_n)^\top \tilde{x}_n(r; \Gamma_n\rho)\big\rangle_{H_n} d\tau = 0.
\end{aligned}
$$

Hence, for every $t \in [0,T]$,

$$
\begin{aligned}
\big\langle \mathbb{X}_{n,t}\Gamma_n\zeta, \widetilde{\mathbb{X}}_{n,t}\Gamma_n\rho\big\rangle_{H_n} &= \big\langle \mathbb{X}_n(t)\Gamma_n\zeta, \widetilde{\mathbb{X}}_n(t)\Gamma_n\rho\big\rangle_{H_n}\\
&= \big\langle \hat{x}_n(t; \Gamma_n\zeta), \tilde{x}_n(t; \Gamma_n\rho)\big\rangle_{H_n} = \big\langle \Gamma_n\zeta, \Gamma_n\rho\big\rangle_{H_n}, \quad \text{a.s.}
\end{aligned}
$$

This implies that $\mathbb{X}_n(t)\widetilde{\mathbb{X}}_n(t)^\top = I_n$, a.s. Namely, for all $t \in [0,T]$,

$$
\widetilde{\mathbb{X}}_n(t)^\top = \mathbb{X}_n(t)^{-1}, \qquad \text{a.s.}
$$

By (13.229), (13.232) and (13.234), for any $\zeta \in H$ and $t \in [0,T]$,

$$\mathbb{X}(t,\cdot)\widetilde{\mathbb{X}}(t,\cdot)^*\zeta = \zeta \quad \text{in } H_\lambda, \quad \text{a.s.}$$

Furthermore, it follows from $\mathbb{X}(t,\cdot), \widetilde{\mathbb{X}}(t,\cdot) \in \mathcal{L}(H)$, a.s. that

$$\mathbb{X}(t,\cdot)\widetilde{\mathbb{X}}(t,\cdot)^*\zeta = \zeta \quad \text{in } H, \quad \text{a.s.},$$

which implies (13.239).

Step 3. Put

$$\begin{cases} P(\cdot) = \mathbb{Y}(\cdot)\widetilde{\mathbb{X}}(\cdot)^*, \\ \Pi(\cdot) = \mathbf{Y}(\cdot)\widetilde{\mathbb{X}}(\cdot)^*, \\ \Lambda(\cdot) = \Pi(\cdot) - P(\cdot)\big(C(\cdot) + D(\cdot)\Theta(\cdot)\big). \end{cases} \tag{13.240}$$

From (13.237), it follows that $\widetilde{\mathbb{X}}(\cdot)^* \in L^4_{\mathbb{F}}(\Omega; C([0,T]; \mathcal{L}_2(H'_\lambda; H)))$. This, together with (13.240), implies that

$$\begin{cases} P(\cdot) \in L^2_{\mathbb{F}}(\Omega; C([0,T]; \mathcal{L}_2(H'_\lambda; H_\lambda))), \\ \Lambda(\cdot) \in L^2_{\mathbb{F}}(0,T; \mathcal{L}_2(H'_\lambda; H_\lambda)). \end{cases} \tag{13.241}$$

Put

$$\begin{cases} P_n(\cdot) = \mathbb{Y}_n(\cdot)\widetilde{\mathbb{X}}_n(\cdot)^\top, \quad \Pi_n(\cdot) = \mathbf{Y}_n(\cdot)\widetilde{\mathbb{X}}_n(\cdot)^\top, \\ \Lambda_n(\cdot) = \Pi_n(\cdot) - P_n(\cdot)(C_n(\cdot) + D_n(\cdot)\Theta_n(\cdot)). \end{cases} \tag{13.242}$$

Similar to the proof of (13.83) and (13.86), we can show that

$$\begin{cases} P_n(t) = P_n(t)^\top, \quad \text{a.s.,} \quad \forall t \in [0,T], \\ \Lambda_n(t,\omega) = \Lambda_n(t,\omega)^\top, \quad \text{a.e. } (t,\omega) \in (0,T) \times \Omega. \end{cases} \tag{13.243}$$

It follows from Lemma 13.49 that

$$\begin{aligned} &\lim_{n\to\infty} |\widetilde{\mathbb{X}}_n^\top - \widetilde{\mathbb{X}}^*|_{L^4_{\mathbb{F}}(\Omega; C([0,T]; \mathcal{L}_2(H'_\lambda; H)))} \\ &= \lim_{n\to\infty} |\widetilde{\mathbb{X}}_n - \widetilde{\mathbb{X}}|_{L^4_{\mathbb{F}}(\Omega; C([0,T]; \mathcal{L}_2(H; H_\lambda)))} = 0, \end{aligned} \tag{13.244}$$

$$\lim_{n\to\infty} |\mathbb{Y}_n - \mathbb{Y}|_{L^4_{\mathbb{F}}(\Omega; C([0,T]; \mathcal{L}_2(H; H_\lambda)))} = 0, \tag{13.245}$$

and

$$\lim_{n\to\infty} |\mathbf{Y}_n - \mathbf{Y}|_{L^4_{\mathbb{F}}(\Omega; L^2(0,T; \mathcal{L}_2(H; H_\lambda)))} = 0. \tag{13.246}$$

Now, by (13.240), (13.242) and (13.244)–(13.246), we deduce that

$$\lim_{n\to\infty} |P_n - P|_{L^2_{\mathbb{F}}(\Omega; C([0,T]; \mathcal{L}_2(H'_\lambda; H_\lambda)))} = 0, \tag{13.247}$$

and

$$\lim_{n\to\infty} |\Lambda_n - \Lambda|_{L^2_{\mathbb{F}}(0,T;\mathcal{L}_2(H'_\lambda;H_\lambda))} = 0. \tag{13.248}$$

Combining (13.247), (13.248) and (13.243), we obtain that

$$\begin{cases} P(t,\omega) = P(t,\omega)^*, \\ \Lambda(t,\omega) = \Lambda(t,\omega)^*, \end{cases} \quad \text{a.e. } (t,\omega) \in (0,T) \times \Omega. \tag{13.249}$$

By Itô's formula, and noting (13.227)–(13.228), we obtain that

$$\begin{aligned} dP_n &= \Big\{ - \big[A_n^\top \mathbb{Y}_n + C_n^\top \mathbf{Y}_n + M_n \mathbb{X}_n\big]\mathbb{X}_n^{-1} \\ &\quad + \mathbb{Y}_n \mathbb{X}_n^{-1}\big[(C_n + D_n\Theta_n)^2 - A_n - B_n\Theta_n\big] - \mathbf{Y}_n\mathbb{X}_n^{-1}(C_n + D_n\Theta_n)\Big\}dt \\ &\quad + \Big[\mathbf{Y}_n\mathbb{X}_n^{-1} - \mathbb{Y}_n\mathbb{X}_n^{-1}(C_n + D_n\Theta_n)\Big]dW(t) \\ &= \Big\{ - A_n^\top P_n - C_n^\top \Pi_n - M_n + P_n\big[(C_n + D_n\Theta_n)^2 - A_n - B_n\Theta_n\big] \\ &\quad - \Pi_n(C_n + D_n\Theta_n)\Big\}dt + \big[\Pi_n - P_n(C_n + D_n\Theta_n)\big]dW(t). \end{aligned}$$

Hence, by (13.242), $(P_n(\cdot), \Lambda_n(\cdot))$ solves the following $\mathbb{R}^{n\times n}$-valued backward stochastic differential equation:

$$\begin{cases} dP_n = -\big[P_nA_n + A_n^\top P_n + \Lambda_nC_n + C_n^\top\Lambda_n + C_n^\top P_nC_n \\ \qquad\quad + (P_nB_n + C_n^\top P_nD_n + \Lambda_nD_n)\Theta_n + M_n\big]dt \\ \qquad\quad + \Lambda_n dW(t) \qquad\qquad \text{in } [0,T], \\ P_n(T) = G_n. \end{cases} \tag{13.250}$$

For $\xi_k \in L^4_{\mathcal{F}_t}(\Omega; H'_\lambda)$, $u_k \in L^4_{\mathbb{F}}(\Omega; L^2(t,T;H'_\lambda))$ and $v_k \in L^4_{\mathbb{F}}(\Omega; L^2(t,T; H'_\lambda))$ $(k = 1,2)$, denote by $x_1(\cdot)$ and $x_2(\cdot)$ respectively the mild solutions to the equations (13.133) and (13.134). For $k = 1,2$, let us introduce the following stochastic differential equations:

$$\begin{cases} dx_{k,n} = \big(A_nx_{k,n} + u_{k,n}\big)dr + \big(C_nx_{k,n} + v_{k,n}\big)dW(r) \ \text{ in } [t,T], \\ x_{k,n}(t) = \xi_{k,n}, \end{cases} \tag{13.251}$$

where $u_{k,n} = \Gamma_n u_k$, $v_{k,n} = \Gamma_n v_k$ and $\xi_{k,n} = \Gamma_n\xi_k$. Clearly,

$$\begin{cases} \lim\limits_{n\to\infty} \xi_{k,n} = \xi_k & \text{in } L^4_{\mathcal{F}_t}(\Omega; H'_\lambda), \\ \lim\limits_{n\to\infty} u_{k,n} = u_k & \text{in } L^4_{\mathbb{F}}(\Omega; L^2(t,T;H'_\lambda)), \\ \lim\limits_{n\to\infty} v_{k,n} = v_k & \text{in } L^4_{\mathbb{F}}(\Omega; L^2(t,T;H'_\lambda)). \end{cases} \tag{13.252}$$

From Lemma 13.48 and (13.252), for $k=1,2$, we get that

$$\lim_{n\to\infty} x_{k,n}(\cdot) = x_k(\cdot) \text{ in } L^4_{\mathbb{F}}(\Omega; C([t,T]; H'_\lambda)). \tag{13.253}$$

By Itô's formula, and using (13.250)–(13.251), we arrive at

$$\begin{aligned}
& d\big\langle P_n x_{1,n}, x_{2,n}\big\rangle_{H_n}\\
&= \big\langle dP_n x_{1,n}, x_{2,n}\big\rangle_{H_n} + \big\langle P_n dx_{1,n}, x_{2,n}\big\rangle_{H_n} + \big\langle P_n x_{1,n}, dx_{2,n}\big\rangle_{H_n}\\
&\quad +\big\langle dP_n dx_{1,n}, x_{2,n}\big\rangle_{H_n} + \big\langle dP_n x_{1,n}, dx_{2,n}\big\rangle_{H_n} + \big\langle P_n dx_{1,n}, dx_{2,n}\big\rangle_{H_n}\\
&= \big\langle -\big[P_n A_n + A_n^\top P_n + \Lambda_n C_n + C_n^\top \Lambda_n + C_n^\top P_n C_n\\
&\quad +(P_n B_n + C_n^\top P_n D_n + \Lambda_n D_n)\Theta_n + M_n\big] x_{1,n}, x_{2,n}\big\rangle_{H_n} dr\\
&\quad +\big\langle \Lambda_n x_{1,n}, x_{2,n}\big\rangle_{H_n} dW(r) + \big\langle P_n(A_n x_{1,n} + u_{1,n}), x_{2,n}\big\rangle_{H_n} dr\\
&\quad +\big\langle P_n(C_n x_{1,n} + v_{1,n}), x_{2,n}\big\rangle_{H_n} dW(r) + \big\langle P_n x_{1,n}, A_n x_{2,n} + u_{2,n}\big\rangle_{H_n} dr\\
&\quad +\big\langle P_n x_{1,n}, (C_n x_{2,n} + v_{2,n})\big\rangle_{H_n} dW(r) + \big\langle \Lambda_n(C_n x_{1,n} + v_{1,n}), x_{2,n}\big\rangle_{H_n} dr\\
&\quad +\big\langle \Lambda_n x_{1,n}, C_n x_{2,n} + v_{2,n}\big\rangle_{H_n} dr\\
&\quad +\big\langle P_n(C_n x_{1,n} + v_{1,n}), C_n x_{2,n} + v_{2,n}\big\rangle_{H_n} dr\\
&= \big\langle -\big[(P_n B_n + C_n^\top P_n D_n + \Lambda_n D_n)\Theta_n + M_n\big] x_{1,n}, x_{2,n}\big\rangle_{H_n} dr\\
&\quad +\big\langle P_n u_{1,n}, x_{2,n}\big\rangle_{H_n} dr + \big\langle P_n x_{1,n}, u_{2,n}\big\rangle_{H_n} dr\\
&\quad +\big\langle P_n C_n x_{1,n}, v_{2,n}\big\rangle_{H_n} dr + \big\langle P_n v_{1,n}, C_n x_{2,n} + v_{2,n}\big\rangle_{H_n} dr\\
&\quad +\big\langle \Lambda_n v_{1,n}, x_{2,n}\big\rangle_{H_n} dr + \big\langle \Lambda_n x_{1,n}, v_{2,n}\big\rangle_{H_n} dr\\
&\quad +\big\langle \Lambda_n x_{1,n}, x_{2,n}\big\rangle_{H_n} dW(r) + \langle P_n(C_n x_{1,n} + v_{1,n}), x_{2,n}\rangle_{H_n} dW(r)\\
&\quad +\langle P_n x_{1,n}, (C_n x_{2,n} + v_{2,n})\rangle_{H_n} dW(r).
\end{aligned}$$

This implies that, for any $t\in[0,T]$,

$$\begin{aligned}
& \mathbb{E}\langle G_n x_{1,n}(T), x_{2,n}(T)\rangle_{H_n} + \mathbb{E}\int_t^T \big\langle M_n(\tau) x_{1,n}(\tau), x_{2,n}(\tau)\big\rangle_{H_n} d\tau\\
&\quad +\mathbb{E}\int_t^T \big\langle \big[P_n(\tau)B_n(\tau) + C_n(\tau)^\top P_n(\tau) D_n(\tau) + \Lambda_n(\tau) D_n(\tau)\big]\\
&\qquad\qquad \times\Theta_n(\tau) x_{1,n}(\tau), x_{2,n}(\tau)\big\rangle_{H_n} d\tau\\
&= \mathbb{E}\big\langle P_n(t)\xi_{1,n}, \xi_{2,n}\big\rangle_{H_n} + \mathbb{E}\int_t^T \big\langle P_n(\tau) u_{1,n}(\tau), x_{2,n}(\tau)\big\rangle_{H_n} d\tau\\
&\quad +\mathbb{E}\int_t^T \big\langle P_n(\tau) x_{1,n}(\tau), u_{2,n}(\tau)\big\rangle_{H_n} d\tau\\
&\quad +\mathbb{E}\int_t^T \big\langle P_n(\tau) C_n(\tau) x_{1,n}(\tau), v_{2,n}(\tau)\big\rangle_{H_n} d\tau\\
&\quad +\mathbb{E}\int_t^T \big\langle P_n(\tau) v_{1,n}(\tau), C_n(\tau) x_{2,n}(\tau) + v_{2,n}(\tau)\big\rangle_{H_n} d\tau
\end{aligned} \tag{13.254}$$

$$+\mathbb{E}\int_t^T \big\langle \Lambda_n(\tau)v_{1,n}(\tau), x_{2,n}(\tau)\big\rangle_{H_n} d\tau + \mathbb{E}\int_t^T \big\langle \Lambda_n(\tau)x_{1,n}(\tau), v_{2,n}(\tau)\big\rangle_{H_n} d\tau.$$

Step 4. In this step, we derive some properties of P and Λ.

Let $t \in [0,T)$ and $\eta \in L^2_{\mathcal{F}_t}(\Omega;H)$. Consider the following forward-backward stochastic evolution equation:

$$\begin{cases} dx^t(r) = \big(A + B\Theta\big)x^t dr + \big(C + D\Theta\big)x^t dW(r) & \text{in } (t,T], \\ dy^t(r) = -\big(A^*y^t + C^*Y^t + Mx^t\big)dr + Y^t dW(r) & \text{in } [t,T), \\ x^t(t) = \eta, \quad y^t(T) = Gx^t(T). \end{cases} \tag{13.255}$$

By Theorems 3.20 and 4.10, we deduce that (13.255) admits a unique mild solution $\big(x^t(\cdot), y^t(\cdot), Y^t(\cdot)\big)\big(\equiv \big(x^t(\cdot;\eta), y^t(\cdot;\eta), z^t(\cdot;\eta)\big)\big) \in L^2_{\mathbb{F}}(\Omega;C([t,T];H)) \times L^2_{\mathbb{F}}(\Omega;C([t,T];H)) \times L^2_{\mathbb{F}}(t,T;H)$ such that

$$\big|\big(x^t(\cdot), y^t(\cdot), Y^t(\cdot)\big)\big|_{L^2_{\mathbb{F}}(\Omega;C([t,T];H))\times L^2_{\mathbb{F}}(\Omega;C([t,T];H))\times L^2_{\mathbb{F}}(t,T;H)} \leq \mathcal{C}|\eta|_{L^2_{\mathcal{F}_t}(\Omega;H)}. \tag{13.256}$$

For every $r \in [t,T]$, define two families of operators $\mathbb{X}^t_r$ and $\mathbb{Y}^t_r$ on $L^2_{\mathcal{F}_t}(\Omega;H)$ as follows:

$$\mathbb{X}^t_r\eta \overset{\Delta}{=} x^t(r;\eta), \quad \mathbb{Y}^t_r\eta \overset{\Delta}{=} y^t(r;\eta).$$

For a.e. $r \in [t,T]$, define a family of operators $\mathbf{Y}^t(r)$ on $L^2_{\mathcal{F}_t}(\Omega;H)$ by

$$\mathbf{Y}^t(r)\eta \overset{\Delta}{=} Y^t(r;\eta).$$

It follows from (13.256) that for any $r \in [t,T]$ and $\eta \in L^2_{\mathcal{F}_t}(\Omega;H)$,

$$\begin{cases} |\mathbb{X}^t_r\eta|_{L^2_{\mathcal{F}_r}(\Omega;H)} \leq \mathcal{C}|\eta|_{L^2_{\mathcal{F}_t}(\Omega;H)}, \\ |\mathbb{Y}^t_r\eta|_{L^2_{\mathcal{F}_r}(\Omega;H)} \leq \mathcal{C}|\eta|_{L^2_{\mathcal{F}_t}(\Omega;H)}, \\ |\mathbf{Y}^t(\cdot)\eta|_{L^2_{\mathbb{F}}(t,T;H)} \leq \mathcal{C}|\eta|_{L^2_{\mathcal{F}_t}(\Omega;H)}. \end{cases} \tag{13.257}$$

This indicates that for every $r \in [t,T]$,

$$\mathbb{X}^t_r \in \mathcal{L}(L^2_{\mathcal{F}_t}(\Omega;H); L^2_{\mathcal{F}_r}(\Omega;H)), \quad \mathbb{Y}^t_r \in \mathcal{L}(L^2_{\mathcal{F}_t}(\Omega;H); L^2_{\mathcal{F}_r}(\Omega;H))$$

and

$$\mathbf{Y}^t(\cdot) \in \mathcal{L}(L^2_{\mathcal{F}_t}(\Omega;H); L^2_{\mathbb{F}}(t,T;H)).$$

By (13.217) and (13.255), it is easy to see that, for any $\zeta \in H$,

$$\mathbb{X}^t_r\mathbb{X}(t)\zeta = x^t(r;\mathbb{X}(t)\zeta) = \hat{x}(r;\zeta).$$

Thus,

$$\mathbb{Y}^t_t\mathbb{X}(t)\zeta = y^t(t;\mathbb{X}(t)\zeta) = \mathbb{Y}(t)\zeta$$

and

$$\mathbf{Y}^t(\tau)\mathbb{X}(t)\zeta = Y^t(\tau;\mathbb{X}(t)\zeta) = \mathbf{Y}(\tau)\zeta, \qquad \text{for a.e. } \tau \in [t,T].$$

This implies that

$$\mathbb{Y}_t^t = \mathbb{Y}(t)\widetilde{\mathbb{X}}(t)^* \quad \text{for all } t \in [0,T], \ \text{ a.s.,} \tag{13.258}$$

and

$$\mathbf{Y}^t(\tau) = \mathbf{Y}(\tau)\widetilde{\mathbb{X}}(t)^* \quad \text{for a.e. } t \in [0,T], \ \tau \in [t,T], \ \text{a.s.} \tag{13.259}$$

Further, the inequality (13.257) implies that for all $t \in [0,T]$ and $\eta \in L^2_{\mathcal{F}_t}(\Omega;H)$,

$$\mathbb{E}|\mathbb{Y}_t^t\eta|_H^2 \le \mathcal{C}\mathbb{E}|\eta|_H^2, \tag{13.260}$$

where $\mathcal{C}$ is independent of $t \in [0,T]$. According to (13.260), we find that

$$|\mathbb{Y}(t)\widetilde{\mathbb{X}}(t)^*|_{\mathcal{L}(L^2_{\mathcal{F}_t}(\Omega;H);\ L^2_{\mathcal{F}_t}(\Omega;H))} \le \mathcal{C}. \tag{13.261}$$

Thus, from (13.240), (13.258) and (13.261), it follows that, for some positive constant $\mathcal{C}_0$,

$$|P(t)|_{\mathcal{L}(L^2_{\mathcal{F}_t}(\Omega;H);\ L^2_{\mathcal{F}_t}(\Omega;H))} \le \mathcal{C}_0, \qquad \forall\, t \in [0,T]. \tag{13.262}$$

We claim that,

$$|P(t)|_{\mathcal{L}(H)} \le \mathcal{C}_0, \qquad \forall\, t \in [0,T], \quad \text{a.s.} \tag{13.263}$$

Otherwise, there would exist $\varepsilon_0 > 0$ and $\widetilde{\Omega} \in \mathcal{F}_t$ with $\mathbb{P}(\widetilde{\Omega}) > 0$ such that

$$|P(t,\omega)|_{\mathcal{L}(H)} > \mathcal{C}_0 + \varepsilon_0, \qquad \text{for a.e. } \omega \in \widetilde{\Omega}.$$

Let $\{\eta_k\}_{k=1}^\infty$ be a dense subset of the unit sphere of H. Then, for a.e. $\omega \in \widetilde{\Omega}$, there is an $\eta_\omega \in \{\eta_k\}_{k=1}^\infty$ such that

$$|P(t,\omega)\eta_\omega|_H \ge |P(t,\omega)|_{\mathcal{L}(H)} - \frac{\varepsilon_0}{2} > \mathcal{C}_0 + \frac{\varepsilon_0}{2}.$$

Write

$$\begin{cases} \Omega_1 = \Big\{\omega \in \widetilde{\Omega} \;\Big|\; |P(t,\omega)\eta_1|_H > |P(t,\omega)|_{\mathcal{L}(H)} - \dfrac{\varepsilon_0}{2}\Big\}, \\[2mm] \Omega_j = \Big\{\omega \in \widetilde{\Omega} \;\Big|\; |P(t,\omega)\eta_j|_H > |P(t,\omega)|_{\mathcal{L}(H)} - \dfrac{\varepsilon_0}{2}\Big\} \setminus \Big(\displaystyle\bigcup_{k=1}^{j-1}\Omega_k\Big), \\ \hfill \text{for } j = 2,3,\cdots \end{cases}$$

Since $P(t,\omega)\eta_j \in L^2_{\mathcal{F}_t}(\Omega;H)$ for all $j \in \mathbb{N}$, we see that $\{\Omega_j\}_{j=1}^\infty \subset \mathcal{F}_t$ and $\mathbb{P}(\widetilde{\Omega}) = \displaystyle\sum_{j=1}^\infty \mathbb{P}(\Omega_j)$. Hence,

$$\Big|P(t,\omega)\sum_{j=1}^{\infty}\chi_{\Omega_j}\eta_j\Big|_H \geq \mathcal{C}_0+\frac{\varepsilon_0}{2}, \qquad \text{for a.e. } \omega\in\widetilde{\Omega}.$$

Therefore,

$$\mathbb{E}\Big|P(t,\omega)\sum_{j=1}^{\infty}\chi_{\Omega_j}\eta_j\Big|_H^2 \geq \Big(\mathcal{C}_0+\frac{\varepsilon_0}{2}\Big)^2\mathbb{P}(\widetilde{\Omega}).$$

On the other hand, it follows from (13.262) that

$$\begin{aligned}\mathbb{E}\Big|P(t,\omega)\sum_{j=1}^{\infty}\chi_{\Omega_j}\eta_j\Big|_H^2 &\leq \mathcal{C}_0^2\mathbb{E}\Big|\sum_{j=1}^{\infty}\chi_{\Omega_j}\eta_j\Big|_H^2 = \mathcal{C}_0^2\sum_{j=1}^{\infty}\mathbb{P}(\Omega_j)\\ &= \mathcal{C}_0^2\mathbb{P}(\widetilde{\Omega}).\end{aligned}$$

The above two inequalities contradict each other. Hence, (13.263) holds.

Since the constant $\mathcal{C}_0$ (in (13.262)) is independent of $t\in[0,T]$, it holds that

$$|P(t,\omega)|_{\mathcal{L}(H)}\leq \mathcal{C}_0, \qquad \text{for a.e. } (t,\omega)\in[0,T]\times\Omega. \tag{13.264}$$

Similar to the proof of (13.262), we can show that for any $\eta\in L^2_{\mathcal{F}_t}(\Omega;H'_\lambda)$,

$$|P(t)|_{\mathcal{L}(L^2_{\mathcal{F}_t}(\Omega;H'_\lambda);\,L^2_{\mathcal{F}_t}(\Omega;H'_\lambda))}\leq \mathcal{C}.$$

Then, similar to the proof of (13.264), we obtain that

$$|P(t,\omega)|_{\mathcal{L}(H'_\lambda)}\leq \mathcal{C}, \qquad \text{for a.e. } (t,\omega)\in[0,T]\times\Omega. \tag{13.265}$$

From (13.235), it follows that

$$R\Theta+B^*P+D^*\Pi=0, \quad \text{a.e. } (t,\omega)\in[0,T]\times\Omega. \tag{13.266}$$

This, together with (13.262), (13.265) and (**AS3**), implies that

$$D^*\Pi\in\Upsilon_2(H;U)\cap\Upsilon_2(H'_\lambda;\widetilde{U}). \tag{13.267}$$

According to (13.240), (13.267) and (**AS3**), it holds that

$$D^*\Lambda=D^*\Pi-D^*P(C+D\Theta)\in\Upsilon_2(H;U)\cap\Upsilon_2(H'_\lambda;\widetilde{U}). \tag{13.268}$$

From (13.249) and (13.268), we see that

$$\Lambda D\in\Upsilon_2(U;H)\cap\Upsilon_2(\widetilde{U};H'_\lambda). \tag{13.269}$$

Step 5. In this step, we prove that a variant of (13.135) holds. We shall do this by taking $n\to\infty$ in (13.254).

Denote by $\widetilde{U}'$ the dual space of $\widetilde{U}$ with respect to the pivot space U. From (**AS3**), (13.247), (13.243) (13.248) and (13.253), we obtain that

$$
\begin{aligned}
&\lim_{n\to\infty}\mathbb{E}\int_t^T\big\langle\big[P_n(\tau)B_n(\tau)+C_n(\tau)^\top P_n(\tau)D_n(\tau)\\
&\qquad\qquad +\Lambda_n(\tau)D_n(\tau)\big]\Theta_n(\tau)x_{1,n}(\tau),x_{2,n}(\tau)\big\rangle_{H_n}d\tau\\
&=\mathbb{E}\int_t^T\big\langle\big[P(\tau)B(\tau)+C(\tau)^*P(\tau)D(\tau)\\
&\qquad\qquad +\Lambda(\tau)D(\tau)\big]\Theta(\tau)x_1(\tau),x_2(\tau)\big\rangle_{H_\lambda,H_\lambda'}d\tau.
\end{aligned}
\tag{13.270}
$$

By (13.248) and (13.253), we see that, for $k=1,2$,

$$
\begin{aligned}
&\lim_{n\to\infty}\big|\Lambda_n(\cdot)x_{k,n}(\cdot)-\Lambda(\cdot)x_k(\cdot)\big|_{L_{\mathbb{F}}^{\frac{4}{3}}(\Omega;L^2(0,T;H_\lambda))}\\
&\le\lim_{n\to\infty}\Big[\big|\Lambda_n(\cdot)\big(x_{k,n}(\cdot)-x_k(\cdot)\big)\big|_{L_{\mathbb{F}}^{\frac{4}{3}}(\Omega;L^2(0,T;H_\lambda))}\\
&\qquad\qquad +\big|\big(\Lambda_n(\cdot)-\Lambda(\cdot)\big)x_k(\cdot)\big|_{L_{\mathbb{F}}^{\frac{4}{3}}(\Omega;L^2(0,T;H_\lambda))}\Big]\\
&\le\lim_{n\to\infty}\big|\Lambda(\cdot)\big|_{L_{\mathbb{F}}^2(0,T;\mathcal{L}_2(H_\lambda';H_\lambda))}\big|x_{k,n}(\cdot)-x_i(\cdot)\big|_{L_{\mathbb{F}}^4(\Omega;L^\infty(0,T;H_\lambda'))}\\
&\quad+\lim_{n\to\infty}\big|\Lambda_n(\cdot)-\Lambda(\cdot)\big|_{L_{\mathbb{F}}^2(0,T;\mathcal{L}_2(H_\lambda';H_\lambda))}\big|x_k(\cdot)\big|_{L_{\mathbb{F}}^4(\Omega;L^\infty(0,T;H_\lambda'))}=0.
\end{aligned}
\tag{13.271}
$$

By (13.270)–(13.271), using a similar argument for other terms in (13.254), we can take $n\to\infty$ on both sides of this equality to get that

$$
\begin{aligned}
&\mathbb{E}\langle Gx_1(T),x_2(T)\rangle_H+\mathbb{E}\int_t^T\big\langle M(\tau)x_1(\tau),x_2(\tau)\big\rangle_H d\tau\\
&\quad+\mathbb{E}\int_t^T\big\langle\big[P(\tau)B(\tau)+C(\tau)^*P(\tau)D(\tau)+\Lambda(\tau)^*D(\tau)\big]\Theta(\tau)x_1(\tau),x_2(\tau)\big\rangle_{H_\lambda,H_\lambda'}d\tau\\
&=\mathbb{E}\big\langle P(t)\xi_1,\xi_2\big\rangle_{H_\lambda,H_\lambda'}+\mathbb{E}\int_t^T\big\langle u_1(\tau),P(\tau)^*x_2(\tau)\big\rangle_{H_\lambda',H_\lambda}d\tau\\
&\quad+\mathbb{E}\int_t^T\big\langle P(\tau)x_1(\tau),u_2(\tau)\big\rangle_{H_\lambda,H_\lambda'}d\tau+\mathbb{E}\int_t^T\big\langle P(\tau)C(\tau)x_1(\tau),v_2(\tau)\big\rangle_{H_\lambda,H_\lambda'}d\tau\\
&\quad+\mathbb{E}\int_t^T\big\langle P(\tau)v_1(\tau),C(\tau)x_2(\tau)+v_2(\tau)\big\rangle_{H_\lambda,H_\lambda'}d\tau\\
&\quad+\mathbb{E}\int_t^T\big\langle\Lambda(\tau)x_2(\tau),v_1(\tau)\big\rangle_{H_\lambda,H_\lambda'}d\tau+\mathbb{E}\int_t^T\big\langle\Lambda(\tau)x_1(\tau),v_2(\tau)\big\rangle_{H_\lambda,H_\lambda'}d\tau.
\end{aligned}
\tag{13.272}
$$

Noting that $P(t)\in\mathcal{L}(L_{\mathcal{F}_t}^2(\Omega;H);L_{\mathcal{F}_t}^2(\Omega;H))$, we have

$$
\begin{aligned}
&\mathbb{E}\big\langle P(t)\xi_1,\xi_2\big\rangle_{H_\lambda,H_\lambda'}+\mathbb{E}\int_t^T\big\langle u_1(\tau),P(\tau)^*x_2(\tau)\big\rangle_{H_\lambda',H_\lambda}d\tau\\
&\quad+\mathbb{E}\int_t^T\big\langle P(\tau)x_1(\tau),u_2(\tau)\big\rangle_{H_\lambda,H_\lambda'}d\tau+\mathbb{E}\int_t^T\big\langle P(\tau)C(\tau)x_1(\tau),v_2(\tau)\big\rangle_{H_\lambda,H_\lambda'}d\tau\\
&\quad+\mathbb{E}\int_t^T\big\langle P(\tau)v_1(\tau),C(\tau)x_2(\tau)+v_2(\tau)\big\rangle_{H_\lambda,H_\lambda'}d\tau
\end{aligned}
\tag{13.273}
$$

$$
\begin{aligned}
&= \mathbb{E}\big\langle P(t)\xi_1, \xi_2\big\rangle_H + \mathbb{E}\int_t^T \big\langle u_1(\tau), P(\tau)^* x_2(\tau)\big\rangle_H d\tau \\
&\quad + \mathbb{E}\int_t^T \big\langle P(\tau)x_1(\tau), u_2(\tau)\big\rangle_{H_\lambda, H'_\lambda} d\tau + \mathbb{E}\int_t^T \big\langle P(\tau)C(\tau)x_1(\tau), v_2(\tau)\big\rangle_H d\tau \\
&\quad + \mathbb{E}\int_t^T \big\langle P(\tau)v_1(\tau), C(\tau)x_2(\tau) + v_2(\tau)\big\rangle_H d\tau.
\end{aligned}
$$

From (13.269) and noting again $P(t) \in \mathcal{L}(L^2_{\mathcal{F}_t}(\Omega; H); L^2_{\mathcal{F}_t}(\Omega; H))$, we obtain that

$$
\begin{aligned}
&\mathbb{E}\int_t^T \big\langle \big[P(\tau)B(\tau) + C(\tau)^* P(\tau)D(\tau) + \Lambda(\tau)D(\tau)\big]\Theta(\tau)x_1(\tau), x_2(\tau)\big\rangle_{H_\lambda, H'_\lambda} d\tau \\
&= \mathbb{E}\int_t^T \big\langle \big[P(\tau)B(\tau) + C(\tau)^* P(\tau)D(\tau) + \Lambda(\tau)D(\tau)\big]\Theta(\tau)x_1(\tau), x_2(\tau)\big\rangle_H d\tau.
\end{aligned} \tag{13.274}
$$

Combing (13.272)–(13.274), we conclude that

$$
\begin{aligned}
&\mathbb{E}\langle Gx_1(T), x_2(T)\rangle_H + \mathbb{E}\int_t^T \big\langle M(\tau)x_1(\tau), x_2(\tau)\big\rangle_H d\tau \\
&\quad + \mathbb{E}\int_t^T \big\langle \big[P(\tau)B(\tau) + C(\tau)^* P(\tau)D(\tau) + \Lambda(\tau)^* D(\tau)\big]\Theta(\tau)x_1(\tau), x_2(\tau)\big\rangle_H d\tau \\
&= \mathbb{E}\big\langle P(t)\xi_1, \xi_2\big\rangle_H + \mathbb{E}\int_t^T \big\langle P(\tau)u_1(\tau), x_2(\tau)\big\rangle_H d\tau \qquad (13.275) \\
&\quad + \mathbb{E}\int_t^T \big\langle P(\tau)x_1(\tau), u_2(\tau)\big\rangle_H d\tau + \mathbb{E}\int_t^T \big\langle P(\tau)C(\tau)x_1(\tau), v_2(\tau)\big\rangle_H d\tau \\
&\quad + \mathbb{E}\int_t^T \big\langle P(\tau)v_1(\tau), C(\tau)x_2(\tau) + v_2(\tau)\big\rangle_H d\tau \\
&\quad + \mathbb{E}\int_t^T \big\langle \Lambda(\tau)v_1(\tau), x_2(\tau)\big\rangle_{H_\lambda, H'_\lambda} d\tau + \mathbb{E}\int_t^T \big\langle \Lambda(\tau)x_1(\tau), v_2(\tau)\big\rangle_{H_\lambda, H'_\lambda} d\tau.
\end{aligned}
$$

Step 6. In this step, we prove that a variant of (13.136) holds.

For any $\xi_k \in L^2_{\mathcal{F}_t}(\Omega; H)$, $u_k \in L^2_{\mathbb{F}}(t, T; H)$ and $v_k \in L^2_{\mathbb{F}}(t, T; U)$ $(k = 1, 2)$, denote by $x_1(\cdot)$ and $x_2(\cdot)$ respectively the mild solutions to the equations (13.133) and (13.134). We can find six sequences $\{\xi_k^j\}_{j=1}^\infty \in L^4_{\mathcal{F}_t}(\Omega; H'_\lambda)$, $\{u_k^j\}_{j=1}^\infty \subset L^4_{\mathbb{F}}(\Omega; L^2(t, T; H'_\lambda))$ and $\{v_k^j\}_{j=1}^\infty \subset L^4_{\mathbb{F}}(\Omega; L^2(t, T; \widetilde{U}))$, such that

$$
\begin{cases}
\lim\limits_{j\to\infty} \xi_k^j = \xi_k & \text{in } L^2_{\mathcal{F}_t}(\Omega; H), \\
\lim\limits_{j\to\infty} u_k^j = u_k & \text{in } L^2_{\mathbb{F}}(t, T; H), \\
\lim\limits_{j\to\infty} v_k^j = v_k & \text{in } L^2_{\mathbb{F}}(t, T; U).
\end{cases} \tag{13.276}
$$

Denote by $x_1^j(\cdot)$ (*resp.* $x_2^j(\cdot)$) the mild solution to the equation (13.133) (*resp.* (13.134)) with ξ_1, u_1 and v_1 (*resp.* ξ_2, u_2 and v_2) replaced by ξ_1^j, u_1^j and v_1^j (*resp.* ξ_2^j, u_2^j and v_2^j), and by $x_{k,n}^j$ the solution to (13.251) with $\xi_{k,n}$, $u_{k,n}$ and $v_{k,n}$ replaced respectively by $\Gamma_n\xi_k^j$, $\Gamma_n u_k^j$ and $D_n v_k^j$. It follows from (13.254) that

$$\begin{aligned}
&\mathbb{E}\langle G_n x_{1,n}^j(T), x_{2,n}^j(T)\rangle_{H_n} + \mathbb{E}\int_t^T \big\langle M_n(\tau)x_{1,n}^j(\tau), x_{2,n}^j(\tau)\big\rangle_{H_n} d\tau \\
&\quad +\mathbb{E}\int_t^T \big\langle \big[P_n(\tau)B_n(\tau) + C_n(\tau)^\top P_n(\tau)D_n(\tau) + \Lambda_n(\tau)D_n(\tau)\big] \\
&\qquad\qquad \times\Theta_n(\tau)x_{1,n}^j(\tau), x_{2,n}^j(\tau)\big\rangle_{H_n} d\tau \qquad\qquad (13.277)\\
&= \mathbb{E}\big\langle P_n(t)\xi_1^j, \xi_2^j\big\rangle_{H_n} + \mathbb{E}\int_t^T \Big[\big\langle P_n(\tau)C_n(\tau)x_{1,n}^j(\tau), D_n(\tau)v_2^j(\tau)\big\rangle_{H_n} \\
&\quad +\big\langle P_n(\tau)u_1^j(\tau), x_{2,n}^j(\tau)\big\rangle_{H_n} + \big\langle P_n(\tau)x_{1,n}^j(\tau), u_2^j(\tau)\big\rangle_{H_n} \\
&\quad +\big\langle P_n(\tau)D_n(\tau)v_1^j(\tau), C_n(\tau)x_{2,n}^j(\tau) + D_n(\tau)v_2^j(\tau)\big\rangle_{H_n} \\
&\quad +\big\langle D_n(\tau)v_1^j(\tau), \Lambda_n(\tau)x_{2,n}^j(\tau)\big\rangle_{H_n} + \big\langle \Lambda_n(\tau)x_{1,n}^j(\tau), D_n(\tau)v_2^j(\tau)\big\rangle_{H_n}\Big] d\tau.
\end{aligned}$$

From Assumption (**AS3**) and (13.271), for $k = 1, 2$ and $j \in \mathbb{N}$, we have that

$$\lim_{n\to\infty} D(\cdot,\cdot)^*\Lambda_n(\cdot,\cdot)x_{k,n}^j(\cdot,\cdot) = D(\cdot,\cdot)^*\Lambda(\cdot,\cdot)x_k^j(\cdot,\cdot) \ \text{ in } \ L_{\mathbb{F}}^{\frac{4}{3}}(\Omega; L^2(t,T;\widetilde{U}')).$$

Consequently, we get that

$$\begin{aligned}
&\lim_{n\to\infty} \mathbb{E}\int_t^T \big\langle D_n(\tau)v_1^j(\tau), \Lambda_n(\tau)x_{2,n}^j(\tau)\big\rangle_{H_n} d\tau \\
&= \mathbb{E}\int_t^T \langle v_1^j(\tau), D(\tau)^*\Lambda(\tau)x_2^j(\tau)\rangle_{\widetilde{U},\widetilde{U}'} d\tau.
\end{aligned} \tag{13.278}$$

By (13.268), we see that $D^*\Lambda x_2^j \in L_{\mathbb{F}}^2(t,T;U)$. Hence,

$$\begin{aligned}
&\lim_{j\to\infty} \mathbb{E}\int_t^T \langle v_1^j(\tau), D(\tau)^*\Lambda(\tau)x_2^j(\tau)\rangle_{\widetilde{U},\widetilde{U}'} d\tau \\
&= \mathbb{E}\int_t^T \big\langle v_1(\tau), D(\tau)^*\Lambda(\tau)x_2(\tau)\big\rangle_U d\tau.
\end{aligned} \tag{13.279}$$

Similarly,

$$\begin{aligned}
&\lim_{j\to\infty}\lim_{n\to\infty} \mathbb{E}\int_t^T \big\langle \Lambda_n(\tau)x_{1,n}^j(\tau), D_n(\tau)v_{2,n}^j(\tau)\big\rangle_{H_n} d\tau \\
&= \mathbb{E}\int_t^T \big\langle D(\tau)^*\Lambda(\tau)x_1(\tau), v_2(\tau)\big\rangle_U d\tau.
\end{aligned} \tag{13.280}$$

Noting that $P(t) \in \mathcal{L}(L^2_{\mathcal{F}_t}(\Omega; H); L^2_{\mathcal{F}_t}(\Omega; H))$, we have

$$\lim_{j\to\infty} P(t)\xi_1^j = P(t)\xi_1 \quad \text{in} \quad L^2_{\mathcal{F}_t}(\Omega; H).$$

Thus, we get that

$$\lim_{j\to\infty} \mathbb{E}\big\langle P(t)\xi_1^j, \xi_2^j\big\rangle_{H_\lambda, H'_\lambda} = \lim_{j\to\infty} \mathbb{E}\big\langle P(t)\xi_1^j, \xi_2^j\big\rangle_H = \mathbb{E}\big\langle P(t)\xi_1, \xi_2\big\rangle_H.$$

Since $P(\cdot) \in \Upsilon_2(H)$, we see that

$$\lim_{j\to\infty} P(\cdot)x_2^j = P(\cdot)x_2 \quad \text{in} \quad L^2_{\mathbb{F}}(t, T; H).$$

Consequently,

$$\begin{aligned}
&\lim_{j\to\infty} \mathbb{E}\int_t^T \big\langle u_1^j(\tau), P(\tau)x_2^j(\tau)\big\rangle_{H'_\lambda, H_\lambda} d\tau = \lim_{j\to\infty} \mathbb{E}\int_t^T \big\langle u_1^j(\tau), P(\tau)x_2^j(\tau)\big\rangle_H d\tau\\
&= \mathbb{E}\int_t^T \big\langle u_1(\tau), P(\tau)x_2(\tau)\big\rangle_H d\tau = \mathbb{E}\int_t^T \big\langle P(\tau)u_1(\tau), x_2(\tau)\big\rangle_H d\tau.
\end{aligned}$$

By (13.279)–(13.280) and a similar proof of (13.135), we can get that

$$\begin{aligned}
&\mathbb{E}\langle Gx_1(T), x_2(T)\rangle_H + \mathbb{E}\int_t^T \big\langle M(\tau)x_1(\tau), x_2(\tau)\big\rangle_H d\tau\\
&\quad + \mathbb{E}\int_t^T \big\langle \big[P(\tau)B(\tau) + C(\tau)^*P(\tau)D(\tau) + \Lambda(\tau)D(\tau)\big]\Theta(\tau)x_1(\tau), x_2(\tau)\big\rangle_H d\tau\\
&= \mathbb{E}\big\langle P(t)\xi_1, \xi_2\big\rangle_H + \mathbb{E}\int_t^T \big\langle P(\tau)u_1(\tau), x_2(\tau)\big\rangle_H d\tau \qquad (13.281)\\
&\quad + \mathbb{E}\int_t^T \big\langle P(\tau)x_1(\tau), u_2(\tau)\big\rangle_H d\tau + \mathbb{E}\int_t^T \big\langle P(\tau)C(\tau)x_1(\tau), D(\tau)v_2(\tau)\big\rangle_H d\tau\\
&\quad + \mathbb{E}\int_t^T \big\langle P(\tau)D(\tau)v_1(\tau), C(\tau)x_2(\tau) + D(\tau)v_2(\tau)\big\rangle_H d\tau\\
&\quad + \mathbb{E}\int_t^T \big\langle v_1(\tau), D(\tau)^*\Lambda(\tau)x_2(\tau)\big\rangle_U d\tau + \mathbb{E}\int_t^T \big\langle D(\tau)^*\Lambda(\tau)x_1(\tau), v_2(\tau)\big\rangle_U d\tau.
\end{aligned}$$

Step 7. In this step, we prove that the assertion 1) in Definition 13.39 holds. We first show that

$$K \geq 0, \quad \text{a.e. } (t, \omega) \in [0, T] \times \Omega.$$

Let us replace the v_1 in (13.133) and v_2 in (13.134) by Dv_1 and Dv_2, respectively.

From (13.266), we see that

$$\begin{aligned}
0 &= B^*P + D^*\big[\Lambda + P(C + D\Theta)\big] + R\Theta\\
&= B^*P + D^*PC + D^*\Lambda + K\Theta.
\end{aligned} \qquad (13.282)$$

Thus,

$$PB + C^*PD + \Lambda D = -\Theta^* K^*. \tag{13.283}$$

Thanks to (13.281) and (13.283), and noting $K(\cdot)^* = K(\cdot)$, we obtain that

$$\begin{aligned}
&\mathbb{E}\langle Gx_1(T), x_2(T)\rangle_H + \mathbb{E}\int_t^T \big\langle M(\tau)x_1(\tau), x_2(\tau)\big\rangle_H d\tau \\
&-\mathbb{E}\int_t^T \big\langle \Theta(\tau)^* K(\tau)\Theta(\tau)x_1(\tau), x_2(\tau)\big\rangle_H d\tau \\
&= \mathbb{E}\big\langle P(t)\xi_1, \xi_2\big\rangle_H + \mathbb{E}\int_t^T \big\langle P(\tau)u_1(\tau), x_2(\tau)\big\rangle_H d\tau \\
&+\mathbb{E}\int_t^T \big\langle P(\tau)x_1(\tau), u_2(\tau)\big\rangle_H d\tau + \mathbb{E}\int_t^T \big\langle P(\tau)C(\tau)x_1(\tau), D(\tau)v_2(\tau)\big\rangle_H d\tau \\
&+\mathbb{E}\int_t^T \big\langle P(\tau)D(\tau)v_1(\tau), C(\tau)x_2(\tau)+D(\tau)v_2(\tau)\big\rangle_H d\tau \\
&+\mathbb{E}\int_t^T \big\langle v_1(\tau), D(\tau)^*\Lambda(\tau)x_2(\tau)\big\rangle_U d\tau + \mathbb{E}\int_t^T \big\langle D(\tau)^*\Lambda(\tau)x_1(\tau), v_2(\tau)\big\rangle_U d\tau.
\end{aligned} \tag{13.284}$$

For any $t \in [0,T)$, $\eta \in L^2_{\mathcal{F}_t}(\Omega; H)$ and $u(\cdot) \in L^2_{\mathbb{F}}(0,T;U)$, choose $\xi_1 = \xi_2 = \eta$, $u_1 = u_2 = Bu$ and $v_1 = v_2 = Du$ in (13.133)–(13.134). Similar to the proof of (13.143), thanks to (13.284), and noting (13.18) and (13.283), we can show that

$$\begin{aligned}
&\mathcal{J}(\eta; u(\cdot)) \\
&= \frac{1}{2}\mathbb{E}\Big[\big\langle P(0)\eta, \eta\big\rangle_H + \int_0^T \big(\big\langle K\Theta x, \Theta x\big\rangle_U - 2\big\langle K\Theta x, u\big\rangle_U + \langle Ku, u\rangle_U\big)dr\Big] \\
&= \frac{1}{2}\mathbb{E}\Big(\big\langle P(0)\eta, \eta\big\rangle_H + \int_0^T \big\langle K(u-\Theta x), u-\Theta x\big\rangle_U dr\Big).
\end{aligned} \tag{13.285}$$

Hence,

$$\frac{1}{2}\mathbb{E}\langle P(0)\eta, \eta\rangle_H = \mathcal{J}(\eta; \Theta(\cdot)\bar{x}(\cdot)) \le \mathcal{J}(\eta; u(\cdot)), \ \forall\, u(\cdot) \in L^2_{\mathbb{F}}(0,T;U), \tag{13.286}$$

if and only if

$$K \ge 0, \quad \text{a.e. } (t,\omega) \in [0,T] \times \Omega.$$

Put

$$\mathfrak{U}_1 \overset{\Delta}{=} \big\{(t,\omega) \in (0,T) \times \Omega \;\big|\; K(t,\omega)h = 0 \text{ for some nonzero } h \in U\big\}$$

and

$$\mathfrak{U}_2 \overset{\Delta}{=} \big\{(t,\omega) \in (0,T) \times \Omega \;\big|\; |K(t,\omega)h|_U > 0 \text{ for all } h \in \mathcal{B}_U\big\},$$

where $\mathcal{B}_U \overset{\Delta}{=} \big\{h \in U \;\big|\; |h|_U = 1\big\}$.

Clearly, $\mathfrak{U}_1 \cap \mathfrak{U}_2 = \emptyset$ and $\mathfrak{U}_1 \cup \mathfrak{U}_2 = (0, T) \times \Omega$. By the definition of $\mathfrak{U}_2$, we see that

$$\mathfrak{U}_2 = \bigcup_{m=1}^{\infty} \Big\{ (t, \omega) \in (0, T) \times \Omega \ \Big|\ |K(t, \omega)h|_U > \frac{1}{m} \text{ for all } h \in \mathcal{B}_U \Big\}.$$

Let $\mathcal{B}_U^0$ be a countable dense subset of $\mathcal{B}_U$. Then

$$\begin{aligned}\mathfrak{U}_2 &= \bigcup_{m=1}^{\infty} \Big\{ (t, \omega) \in (0, T) \times \Omega \ \Big|\ |K(t, \omega)h|_U > \frac{1}{m} \text{ for all } h \in \mathcal{B}_U^0 \Big\} \\ &= \bigcup_{m=1}^{\infty} \bigcap_{h \in \mathcal{B}_U^0} \Big\{ (t, \omega) \in (0, T) \times \Omega \ \Big|\ |K(t, \omega)h|_U > \frac{1}{m} \Big\}.\end{aligned} \tag{13.287}$$

Since $K(\cdot, \cdot)h \in L^2_{\mathbb{F}}(0, T; U)$, we get that, for any $h \in U$,

$$\Big\{ (t, \omega) \in (0, T) \times \Omega \ \Big|\ |K(t, \omega)h|_U > \frac{1}{m} \Big\} \in \mathbb{F}.$$

This, together with (13.287), implies that $\mathfrak{U}_2 \in \mathbb{F}$. So does $\mathfrak{U}_1$.

We now show that $K > 0$ for a.e. $(t, \omega) \in [0, T] \times \Omega$. Let us use the contradiction argument and assume that this were untrue. Then the measure (given by the product measure of the Lebesgue measure on $[0, T]$ and the probability measure $\mathbb{P}$) of $\mathfrak{U}_1$ would be positive.

For a.e. $(t, \omega) \in \mathfrak{U}_1$, put

$$\Upsilon(t, \omega) \stackrel{\Delta}{=} \big\{ h \in \mathcal{B}_U \ \big|\ K(t, \omega)h = 0 \big\}.$$

Clearly, $\Upsilon(t, \omega)$ is closed in U. Define a map $F : (0, T) \times \Omega \to 2^U$ as follows:

$$F(t, \omega) = \begin{cases} \Upsilon(t, \omega), & \text{if } (t, \omega) \in \mathfrak{U}_1 \\ 0, & \text{if } (t, \omega) \in \mathfrak{U}_2. \end{cases}$$

Then, $F(t, \omega)$ is closed for a.e. $(t, \omega) \in (0, T) \times \Omega$.

We now prove that F is $\mathbb{F}$-measurable. Let O be a closed subset of U and $O_1 = O \cap B_U$. Put

$$\begin{aligned}\Sigma_1 &\stackrel{\Delta}{=} \big\{ (t, \omega) \in (0, T) \times \Omega \ \big|\ F(t, \omega) \cap O \neq \emptyset \big\}, \\ \Sigma_2 &\stackrel{\Delta}{=} \big\{ (t, \omega) \in (0, T) \times \Omega \ \big|\ F(t, \omega) \cap O_1 \neq \emptyset \big\}.\end{aligned} \tag{13.288}$$

Clearly, $\Sigma_1 \supset \Sigma_2$. Moreover,

$$\Sigma_1 = \begin{cases} \Sigma_2 \cup \mathfrak{U}_2, & \text{if } 0 \in O, \\ \Sigma_2, & \text{if } 0 \notin O. \end{cases}$$

Write

$$\Sigma_3 \stackrel{\Delta}{=} \{(t,\omega) \in (0,T) \times \Omega \mid |K(t,\omega)h|_U > 0 \text{ for all } h \in O_1\}.$$

Clearly, $\Sigma_2 \cap \Sigma_3 = \emptyset$ and $(0,T) \times \Omega = \Sigma_2 \cup \Sigma_3$. Similar to the above (for the proof of $\mathfrak{U}_2 \in \mathbb{F}$), we can show that $\Sigma_3 \in \mathbb{F}$. Hence, $\Sigma_2 \in \mathbb{F}$, and therefore so does Σ_1.

Before continuing, let us recall the following known measurable selection result (e.g. [334]).

Lemma 13.51. *Let $(\widetilde{\Omega}, \widetilde{\mathcal{F}})$ be a measurable space. Let $F : (\widetilde{\Omega}, \widetilde{\mathcal{F}}) \to 2^H$ be a closed-valued set mapping such that $F(\tilde{\omega}) \neq \emptyset$ for every $\tilde{\omega} \in \widetilde{\Omega}$, and for each open set $O \subset H$,*

$$F^{-1}(O) \stackrel{\Delta}{=} \{\tilde{\omega} \in \widetilde{\Omega} \mid F(\tilde{\omega}) \cap O \neq \emptyset\} \in \widetilde{\mathcal{F}}.$$

Then F has a measurable selection $f : \widetilde{\Omega} \to H$, i.e., there is an H-valued, $\widetilde{\mathcal{F}}$-measurable function f such that $f(\tilde{\omega}) \in F(\tilde{\omega})$ for every $\tilde{\omega} \in \widetilde{\Omega}$.

Now we apply Lemma 13.51 to $F(\cdot,\cdot)$ with $(\widetilde{\Omega}, \widetilde{\mathcal{F}}) = ((0,T) \times \Omega, \mathbb{F})$ to find an $\mathbf{F}$-adapted process f such that

$$Kf = 0 \quad \text{for a.e. } (t,\omega) \in (0,T) \times \Omega.$$

Noting that

$$|f(t,\omega)|_U \leq 1 \quad \text{for a.e. } (t,\omega) \in (0,T) \times \Omega,$$

we find that $f \in L^2_{\mathbb{F}}(0,T;U)$. Furthermore, we have

$$|f(t,\omega)|_U = 1 \quad \text{for a.e. } (t,\omega) \in \mathfrak{U}_1,$$

which concludes that $|f|_{L^2_{\mathbb{F}}(0,T;U)} > 0$.

By (13.285), we see that $\Theta\bar{x} + f$ is also an optimal control. This contradicts the uniqueness of the optimal control. Hence,

$$K(t,\omega) \text{ is invertible for a.e. } (t,\omega) \in [0,T] \times \Omega,$$

(but $K(t,\omega)^{-1}$ does not need to be bounded).

Further, we show that the domain of $K(t,\omega)^{-1}$ is dense in U for a.e. $(t,\omega) \in [0,T] \times \Omega$. Denote by $\mathcal{R}(K(t,\omega))$ the range of $K(t,\omega)$. Clearly, $\mathcal{R}(K(t,\omega)) \subset U$.

Put

$$\widetilde{\mathfrak{U}}_1 \stackrel{\Delta}{=} \{(t,\omega) \in (0,T) \times \Omega \mid \mathcal{R}(K(t,\omega))^{\perp} \neq \{0\}\}$$

and

$$\widetilde{\mathfrak{U}}_2 \stackrel{\Delta}{=} \{(t,\omega) \in (0,T) \times \Omega \mid \mathcal{R}(K(t,\omega))^{\perp} = \{0\}\}.$$

Clearly, $\widetilde{\mathfrak{U}}_1 \cup \widetilde{\mathfrak{U}}_2 = (0,T) \times \Omega$. By the definition of $\widetilde{\mathfrak{U}}_2$, we see that

$$\widetilde{\mathfrak{U}}_2 = \bigcup_{m=1}^{\infty} \Big\{(t,\omega) \in (0,T) \times \Omega \;\Big|\; \forall\, \tilde{h} \in \mathcal{B}_U^0, \text{ there is an } h \in \mathcal{B}_U^0 \text{ such that } \big|\langle K(t,\omega)h, \tilde{h}\rangle_U\big| > \frac{1}{m}\Big\}.$$

Then

$$\widetilde{\mathfrak{U}}_2 = \bigcap_{\tilde{h}\in\mathcal{B}_U^0} \bigcup_{h\in\mathcal{B}_U^0} \bigcup_{m=1}^{\infty} \Big\{(t,\omega) \in (0,T) \times \Omega \;\Big|\; \big|\langle K(t,\omega)h, \tilde{h}\rangle_U\big| > \frac{1}{m}\Big\}. \quad (13.289)$$

Since $K(\cdot,\cdot)h \in L^2_{\mathbb{F}}(0,T;U)$, it follows that, for any $h, \tilde{h} \in U$,

$$\Big\{(t,\omega) \in (0,T) \times \Omega \;\Big|\; \big|\langle K(t,\omega)h, \tilde{h}\rangle_U\big| > \frac{1}{m}\Big\} \in \mathbb{F}. \quad (13.290)$$

From (13.289) and (13.290), we see that $\widetilde{\mathfrak{U}}_2 \in \mathbb{F}$. Hence, $\widetilde{\mathfrak{U}}_1 \in \mathbb{F}$.

It suffices to prove that $\mathcal{R}(K(t,\omega))$ is dense in U for a.e. $(t,\omega) \in (0,T)\times\Omega$. To show this, we use the contradiction argument. If $\mathcal{R}(K(t,\omega))$ were not dense in U for a.e. $(t,\omega) \in (0,T) \times \Omega$., then the measure of $\widetilde{\mathfrak{U}}_1$ would be positive.

For a.e. $(t,\omega) \in \widetilde{\mathfrak{U}}_1$, put

$$\widetilde{\Upsilon}(t,\omega) \overset{\Delta}{=} \big\{\tilde{h} \in \mathcal{B}_U \mid \langle K(t,\omega)h, \tilde{h}\rangle_U = 0,\ \forall\, h \in U\big\}.$$

Clearly, $\widetilde{\Upsilon}(t,\omega)$ is closed in U.

Define a map $\widetilde{F} : (0,T) \times \Omega \to 2^U$ as follows:

$$\widetilde{F}(t,\omega) = \begin{cases} \widetilde{\Upsilon}(t,\omega), & \text{if } (t,\omega) \in \widetilde{\mathfrak{U}}_1 \\ 0, & \text{if } (t,\omega) \in \widetilde{\mathfrak{U}}_2. \end{cases}$$

Then, $\widetilde{F}(t,\omega)$ is closed for a.e. $(t,\omega) \in (0,T) \times \Omega$.

We now prove that $\widetilde{F}$ is $\mathbb{F}$-measurable. Similar to (13.288), put

$$\begin{aligned} \widetilde{\Sigma}_1 &\overset{\Delta}{=} \big\{(t,\omega) \in (0,T) \times \Omega \mid \widetilde{F}(t,\omega) \cap O \neq \emptyset\big\}, \\ \widetilde{\Sigma}_2 &\overset{\Delta}{=} \big\{(t,\omega) \in (0,T) \times \Omega \mid \widetilde{F}(t,\omega) \cap O_1 \neq \emptyset\big\}. \end{aligned}$$

If $0 \in O$, then $\widetilde{\Sigma}_1 = \widetilde{\Sigma}_2 \cup \widetilde{\mathfrak{U}}_2$. If $0 \notin O$, then $\widetilde{\Sigma}_1 = \widetilde{\Sigma}_2$. Hence, we only need to show that $\widetilde{\Sigma}_2 \in \mathbb{F}$. Write

$$\widetilde{\Sigma}_3 \overset{\Delta}{=} \Big\{(t,\omega) \in (0,T) \times \Omega \;\Big|\; \forall\, \tilde{h} \in O_1, \text{ there is an } h \in \mathcal{B}_U^0 \text{ so that } \langle K(t,\omega)h, \tilde{h}\rangle_U > 0\Big\}.$$

Then, $(0,T) \times \Omega = \widetilde{\Sigma}_2 \cup \widetilde{\Sigma}_3$ and $\widetilde{\Sigma}_2 \cap \widetilde{\Sigma}_3 = \emptyset$. Hence, it suffices to show that $\widetilde{\Sigma}_3 \in \mathbb{F}$. Let O_0 be a countable dense subset of O_1. Clearly,

$$
\begin{aligned}
\widetilde{\Sigma}_3 &= \bigcup_{m=1}^{\infty} \Big\{ (t,\omega) \in (0,T)\times\Omega \;\Big|\; \forall\, \tilde h \in O_1, \text{ there is an } h \in \mathcal{B}_U^0 \text{ such that} \\
&\qquad\qquad\qquad\qquad \langle K(t,\omega)h, \tilde h\rangle_U > \frac{1}{m} \Big\} \qquad (13.291)\\
&= \bigcup_{m=1}^{\infty} \Big\{ (t,\omega) \in (0,T)\times\Omega \;\Big|\; \forall\, \tilde h \in O_0, \text{ there is an } h \in \mathcal{B}_U^0 \text{ such that} \\
&\qquad\qquad\qquad\qquad \langle K(t,\omega)h, \tilde h\rangle_U > \frac{1}{m} \Big\} \\
&= \bigcup_{m=1}^{\infty} \bigcap_{\tilde h \in O_0} \bigcup_{h\in\mathcal{B}_U^0} \Big\{ (t,\omega) \in (0,T)\times\Omega \;\Big|\; \langle K(t,\omega)h, \tilde h\rangle_U > \frac{1}{m} \Big\}.
\end{aligned}
$$

For any $m \in \mathbb{N}$, $\tilde h \in O_0$ and $h \in \mathcal{B}_U^0$, noting that $K(\cdot,\cdot)h \in L^2_{\mathbb{F}}(0,T;U)$, we deduce that

$$
\Big\{ (t,\omega) \in (0,T)\times\Omega \;\Big|\; \langle K(t,\omega)h, \tilde h\rangle_U > \frac{1}{m} \Big\} \in \mathbb{F}. \tag{13.292}
$$

From (13.291) and (13.292), it follows that $\widetilde{\Sigma}_3 \in \mathbb{F}$. Hence, $\widetilde{\Sigma}_2 \in \mathbb{F}$.

Now we apply Lemma 13.51 to $\widetilde{F}(\cdot,\cdot)$ with $(\widetilde{\Omega},\widetilde{\mathcal{F}}) = ((0,T)\times\Omega,\mathbb{F})$ to find an $\mathbf{F}$-adapted process $\tilde f$ such that

$$
\langle K(t,\omega)h, \tilde f(t,\omega)\rangle_U = 0, \quad \forall\, h \in U, \text{ a.e. } (t,\omega) \in (0,T)\times\Omega.
$$

Since

$$
|\tilde f(t,\omega)|_U \le 1 \text{ for a.e. } (t,\omega) \in (0,T)\times\Omega,
$$

it holds that $\tilde f \in L^2_{\mathbb{F}}(0,T;U)$. Furthermore, we have

$$
|\tilde f(t,\omega)|_U = 1 \text{ for a.e. } (t,\omega) \in \widetilde{\mathfrak{U}}_1,
$$

which implies that $|\tilde f|_{L^2_{\mathbb{F}}(0,T;U)} > 0$.

We claim that $\Theta\bar x + \tilde f$ is also an optimal control. Indeed, by the choice of $\tilde f$, it holds that

$$
\mathbb{E}\int_0^T \big\langle K(u - \Theta\bar x - \tilde f), \tilde f\big\rangle_U dr = 0
$$

and

$$
\mathbb{E}\int_0^T \big\langle K\tilde f, u - \Theta\bar x - \tilde f\big\rangle_U dr = \mathbb{E}\int_s^T \big\langle \tilde f, K(u - \Theta x - \tilde f)\big\rangle_U dr = 0.
$$

Therefore, for any $u(\cdot) \in L^2_{\mathbb{F}}(0,T;U)$,

$$
\begin{aligned}
&\mathbb{E}\int_0^T \big\langle K(u - \Theta\bar x - \tilde f), u - \Theta\bar x - \tilde f\big\rangle_U dr \\
&= \mathbb{E}\int_0^T \big\langle K(u - \Theta\bar x), u - \Theta\bar x\big\rangle_U dr.
\end{aligned} \tag{13.293}
$$

According to (13.285) and (13.293), we obtain that

$$\mathcal{J}(\eta;\Theta(\cdot)\bar{x}(\cdot)+\tilde{f})\leq\mathcal{J}(\eta;u),\quad\forall u(\cdot)\in L^2_{\mathbb{F}}(0,T;U),$$

which indicates that $\Theta\bar{x}+\tilde{f}$ is also an optimal control. This leads to a contradiction to the uniqueness of the optimal controls. Hence, $K(t,\omega)^{-1}$ is densely defined in U for a.e. $(t,\omega)\in(0,T)\times\Omega$.

Now, let us prove that $K(t,\omega)^{-1}$ is a closed operator for a.e. $(t,\omega)\in(0,T)\times\Omega$. Let $\{\rho_j\}_{j=1}^{\infty}\subset\mathcal{D}(K(t,\omega)^{-1})$, $\rho\in U$ and $\hat{\rho}\in U$ satisfy

$$\lim_{j\to\infty}\rho_j=\rho\text{ in }U \tag{13.294}$$

and

$$\lim_{j\to\infty}K(t,\omega)^{-1}\rho_j=\hat{\rho}\text{ in }U. \tag{13.295}$$

From (13.295), we obtain that

$$\lim_{j\to\infty}\rho_j=K(t,\omega)\hat{\rho}\quad\text{ in }U.$$

This, together with (13.294), implies that $\rho=K(t,\omega)\hat{\rho}$. Consequently, $\rho\in\mathcal{D}(K(t,\omega)^{-1})$ and $K(t,\omega)^{-1}\rho=\hat{\rho}$. This indicates that the operator $K(t,\omega)^{-1}$ is closed. Therefore, the assertion 1) in Definition 13.39 holds.

By (13.282), we find that

$$-K^{-1}(B^*P+D^*\Lambda+D^*PC)=\Theta. \tag{13.296}$$

This implies (13.137). Moreover, from (13.275), (13.284) and (13.296), we see that the assertions 2) and 3) in Definition 13.39 holds.

Step 8. In this step, we prove the uniqueness of the transposition solution to (13.17).

Assume that $(P_1(\cdot),\Lambda_1(\cdot)),(P_2(\cdot),\Lambda_2(\cdot))\in C_{\mathbb{F},w}([0,T];L^\infty(\Omega;\mathcal{L}(H)))\times L^2_{\mathbb{F},w}(0,T;\mathcal{L}(H))$ are two transposition solutions to (13.17). Similar to the proof of (13.142), we can obtain that

$$\begin{aligned}&\mathbb{E}\Big[\int_t^T\Big(\big\langle Mx(r),x(r)\big\rangle_H+\big\langle Ru(r),u(r)\big\rangle_U\Big)dr+\big\langle Gx(T),x(T)\big\rangle_H\Big]\\&=\frac{1}{2}\mathbb{E}\Big(\big\langle P_1(t)\eta,\eta\big\rangle_H+\int_t^T\big\langle K_1(u-\Theta x),u-\Theta x\big\rangle_U dr\Big)\\&=\frac{1}{2}\mathbb{E}\Big(\big\langle P_2(t)\eta,\eta\big\rangle_H+\int_t^T\big\langle K_2(u-\Theta x),u-\Theta x\big\rangle_U dr\Big),\end{aligned} \tag{13.297}$$

where $K_j=R+DP_jD$ for $j=1,2$.

By taking $u=\Theta x$ in (13.297), we get that for any $t\in[0,T)$ and $\eta\in L^2_{\mathcal{F}_t}(\Omega;H)$,

$$\mathbb{E}\langle P_1(t)\eta,\eta\rangle_H=\mathbb{E}\langle P_2(t)\eta,\eta\rangle_H. \tag{13.298}$$

Thus, for any $\xi, \eta \in L^2_{\mathcal{F}_t}(\Omega; H)$, we have that

$$\mathbb{E}\langle P_1(t)(\eta + \xi), \eta + \xi \rangle_H = \mathbb{E}\langle P_2(t)(\eta + \xi), \eta + \xi \rangle_H,$$

and

$$\mathbb{E}\langle P_1(t)(\eta - \xi), \eta - \xi \rangle_H = \mathbb{E}\langle P_2(t)(\eta - \xi), \eta - \xi \rangle_H.$$

These, together with $P_1(\cdot) = P_1(\cdot)^*$ and $P_2(\cdot) = P_2(\cdot)^*$, imply that

$$\mathbb{E}\langle P_1(t)\eta, \xi \rangle_H = \mathbb{E}\langle P_2(t)\eta, \xi \rangle_H, \qquad \forall\, \xi, \eta \in L^2_{\mathcal{F}_t}(\Omega; H). \tag{13.299}$$

Hence, $P_1(t) = P_2(t)$ for any $t \in [0, T]$, a.s.

Let $\xi_1 = \xi_2 = 0$, $u_1 = 0$ and $v_2 = 0$. By (13.135), we see that for any $u_2 \in L^4_{\mathbb{F}}(\Omega; L^2(0, T; H'_\lambda))$ and $v_1 \in L^4_{\mathbb{F}}(\Omega;\, L^2(0, T; H'_\lambda))$,

$$0 = \mathbb{E} \int_0^T \big\langle v_1(r), \big(\Lambda_1(r) - \Lambda_2(r)\big) x_2(r) \big\rangle_{H'_\lambda, H_\lambda} dr,$$

which implies that

$$\big(\Lambda_1 - \Lambda_2\big) x_2 = 0 \ \text{ in } \ L^{\frac{4}{3}}_{\mathbb{F}}(\Omega; L^2(0, T; H_\lambda)).$$

This, together with Lemma 13.50, concludes that $\Lambda_1 = \Lambda_2$ in $L^2_{\mathbb{F}}(0, T; \mathcal{L}_2(H'_\lambda; H_\lambda))$. Consequently, the desired uniqueness follows. □

13.9 Some Examples

In this section, we shall give two illuminating examples for LQ problems governed respectively by stochastic wave and Schrödinger equations. One can also consider LQ problems for other controlled stochastic partial differential equations. We leave it to interested readers.

Throughout this section, $n \in \mathbb{N}$ and $\mathcal{O} \subset \mathbb{R}^n$ is a bounded domain with a C^∞ boundary $\partial\mathcal{O}$.

13.9.1 LQ Problems for Stochastic Wave Equations

Consider the following controlled stochastic wave equation:

$$\begin{cases} dy_t - \Delta y dt = b_1 u dt + (ay + b_2 u) dW(t) & \text{in } \mathcal{O} \times (0, T), \\ y = 0 & \text{on } \partial\mathcal{O} \times (0, T), \\ y(0) = y_0, \quad y_t(0) = y_1 & \text{in } \mathcal{O}, \end{cases} \tag{13.300}$$

with the cost functional

$$\mathcal{J}(y_0,y_1;u)\stackrel{\Delta}{=}\mathbb{E}\int_0^T\int_{\mathcal{O}}(q|y|^2+r|u|^2)dxdt+\mathbb{E}\int_{\mathcal{O}}g|y(T)|^2dx.$$

Here $(y_0,y_1)\in H_0^1(\mathcal{O})\cap L^2(\mathcal{O})$, $u\in L^2_{\mathbb{F}}(0,T;L^2(\mathcal{O}))$,

$$\begin{cases}a,b_1,b_2,q,r\in L^\infty_{\mathbb{F}}(0,T;C^{2n}(\overline{\mathcal{O}})),\\ g\in L^\infty_{\mathcal{F}_T}(\Omega;C^{2n}(\overline{\mathcal{O}})),\qquad q\geq 0,\ \ r\geq 1,\ \ g\geq 0.\end{cases}\tag{13.301}$$

Our optimal control problem is as follows:

Problem (wSLQ). For each $(y_0,y_1)\in H_0^1(\mathcal{O})\times L^2(\mathcal{O})$, find $\bar u(\cdot)\in L^2_{\mathbb{F}}(0,T;L^2(\mathcal{O}))$ such that

$$\mathcal{J}\big(y_0,y_1;\bar u(\cdot)\big)=\inf_{u(\cdot)\in L^2_{\mathbb{F}}(0,T;L^2(\mathcal{O}))}\mathcal{J}\big(y_0,y_1;u(\cdot)\big).\tag{13.302}$$

Problem (wSLQ) is a concrete example of Problem (SLQ) in the following setting:

- $H=H_0^1(\mathcal{O})\times L^2(\mathcal{O})$ and $U=L^2(\mathcal{O})$;
- The operator A is defined as follows:

$$\begin{cases}D(A)=[H^2(\mathcal{O})\cap H_0^1(\mathcal{O})]\times H_0^1(\mathcal{O}),\\ A\begin{pmatrix}\varphi_1\\ \varphi_2\end{pmatrix}=\begin{pmatrix}\varphi_2\\ \Delta\varphi_1\end{pmatrix},\quad\forall\begin{pmatrix}\varphi_1\\ \varphi_2\end{pmatrix}\in D(A);\end{cases}$$

- $Bu=(0,b_1u)^\top$, $C(y,y_t)^\top=(0,ay)^\top$ and $Du=(0,b_2u)^\top$;
- The operators M, R and G are given by

$$\begin{cases}\langle M(y,y_t),(y,y_t)^\top\rangle_H=\displaystyle\int_{\mathcal{O}}q(|\nabla y|^2+|y_t|^2)dx,\\ \langle Ru,u\rangle_U=\displaystyle\int_{\mathcal{O}}r|u|^2dx,\\ \langle G(y(T),y_t(T))^\top,(y(T),y_t(T))^\top\rangle_H=\displaystyle\int_{\mathcal{O}}g(|\nabla y(T)|^2+|y_t(T)|^2)dx.\end{cases}$$

Since $q\geq 0$, $r\geq 1$ and $g\geq 0$, we have that Problem (wSLQ) is uniquely solvable.

Define an operator $\widehat A$ as follows:

$$\begin{cases}D(\widehat A)=H^2(\mathcal{O})\cap H_0^1(\mathcal{O}),\\ \widehat A\varphi=-\Delta\varphi,\quad\forall\varphi\in D(\widehat A).\end{cases}$$

Denote by $\{\hat{\lambda}_j\}_{j=1}^{\infty}$ the eigenvalues of $\widehat{A}$ and $\{\hat{e}_j\}_{j=1}^{\infty}$ the corresponding eigenvectors with $|\hat{e}_j|_{L^2(\mathcal{O})} = 1$ for $j \in \mathbb{N}$. Clearly, $\{\pm i\hat{\lambda}_j^{\frac{1}{2}}\}_{j=1}^{\infty}$ are the eigenvalues of A and $\Big\{\Big(\pm \frac{1}{i\hat{\lambda}_j^{\frac{1}{2}}}\hat{e}_j, \hat{e}_j\Big)\Big\}_{j=1}^{\infty}$ are the corresponding eigenvectors. It is well known that $\Big\{\Big(\pm \frac{1}{i\hat{\lambda}_j^{\frac{1}{2}}}\hat{e}_j, \hat{e}_j\Big)\Big\}_{j=1}^{\infty}$ constitutes an orthonormal basis of $H_0^1(\mathcal{O}) \times L^2(\mathcal{O})$. Hence, (**AS1**) holds.

Denote by H_λ the completion of the Hilbert space $H_0^1(\mathcal{O}) \times L^2(\mathcal{O})$ with the norm

$$
|(f_1, f_2)|_{H_\lambda}^2 = \sum_{j=1}^{\infty} |\hat{\lambda}_j|^{-n}\big(|f_{1,j}|^2 + |f_{2,j}|^2\big),
$$

$$
\text{for all } f_1 = \sum_{j=1}^{\infty} f_{1,j}\hat{\lambda}_j^{-\frac{1}{2}} e_j \in H_0^1(\mathcal{O}) \ \text{ and } \ f_2 = \sum_{j=1}^{\infty} f_{2,j} e_j \in L^2(\mathcal{O}).
$$

By the asymptotic distribution of the eigenvalues of $\widehat{A}$ (e.g., [297, Chapter 1, Theorem 1.2.1]), we see that

$$
\begin{aligned}
|I|_{\mathcal{L}_2(H;H_\lambda)}^2 &= \sum_{j=1}^{\infty} \Big\langle \Big(\pm i\hat{\lambda}_j^{-\frac{1}{2}} e_j, e_j\Big)^\top, \Big(\pm i\hat{\lambda}_j^{-\frac{1}{2}} e_j, e_j\Big)^\top \Big\rangle_{H_\lambda} \\
&\le \mathcal{C} \sum_{j=1}^{\infty} |\hat{\lambda}_j|^{-n} < \infty.
\end{aligned}
$$

Hence, the embedding from $H_0^1(\mathcal{O}) \times L^2(\mathcal{O})$ to H_λ is Hilbert-Schmidt. From the definition of H_λ, it follows that $H_\lambda' = D(A^n)$.

Let $\alpha \in C^{2n}(\overline{\mathcal{O}})$. For any $f \in H_\lambda'$, one has

$$
\begin{aligned}
|\alpha f|_{H_\lambda'} &= |\alpha f|_{D(A^n)} = |A^n(\alpha f)|_{H_0^1(\mathcal{O}) \times L^2(\mathcal{O})} \\
&\le |\alpha|_{C^{2n}(\overline{\mathcal{O}})} |f|_{D(A^n)} = |\alpha|_{C^{2n}(\overline{\mathcal{O}})} |f|_{H_\lambda'}.
\end{aligned}
$$

From this, we conclude that $C \in L_{\mathbb{F}}^\infty(0,T;\mathcal{L}(H_\lambda'))$, $G \in L_{\mathcal{F}_T}^\infty(\Omega;\mathcal{L}(H_\lambda'))$ and $M \in L_{\mathbb{F}}^\infty(0,T;\mathcal{L}(H_\lambda'))$. Thus, (**AS2**) holds.

Let $\widetilde{U} = D(\widehat{A}^n)$. Clearly, $\widetilde{U}$ is dense in $L^2(\mathcal{O})$. From the definition of B, D and R, we find that $R \in L_{\mathbb{F}}^\infty(0,T;\mathcal{L}(\widetilde{U}))$ and $B, D \in L_{\mathbb{F}}^2(0,T;\mathcal{L}(\widetilde{U};H_\lambda'))$. Therefore, (**AS3**) holds.

13.9.2 LQ problems for Stochastic Schrödinger Equations

In this section, we consider the LQ problem for the following controlled stochastic Schrödinger equations:

$$\begin{cases} dy - i\Delta y dt = b_1 u dt + (ay + b_2 u) dW(t) & \text{in } \mathcal{O} \times (0,T), \\ y = 0 & \text{on } \partial\mathcal{O} \times (0,T), \\ y(0) = y_0 & \text{in } \mathcal{O}, \end{cases} \qquad (13.303)$$

with the following cost functional

$$\mathcal{J}(y_0; u) \stackrel{\Delta}{=} \mathbb{E}\int_0^T \int_{\mathcal{O}} (q|y|^2 + r|u|^2) dx dt + \mathbb{E}\int_{\mathcal{O}} g|y(T)|^2 dx.$$

Here $y_0 \in L^2(\mathcal{O})$, $u \in L^2_{\mathbb{F}}(0,T; L^2(\mathcal{O}))$, and the coefficients fulfill (13.301).

Our optimal control problem is as follows:

Problem (sSLQ). For each $y_0 \in L^2(\mathcal{O})$, find a $\bar{u}(\cdot) \in L^2_{\mathbb{F}}(0,T; L^2(\mathcal{O}))$ such that

$$\mathcal{J}\big(y_0; \bar{u}(\cdot)\big) = \inf_{u(\cdot) \in L^2_{\mathbb{F}}(0,T;L^2(\mathcal{O}))} \mathcal{J}\big(y_0; u(\cdot)\big). \qquad (13.304)$$

Problem (sSLQ) is a concrete example of Problem (SLQ) in the following setting:

- $H = U = L^2(\mathcal{O})$;
- The operator A is defined as follows:

$$\begin{cases} D(A) = H^2(\mathcal{O}) \cap H_0^1(\mathcal{O}), \\ A\varphi = i\Delta\varphi, \quad \forall \varphi \in D(A); \end{cases}$$

- $Bu = b_1 u$, $Cy = ay$ and $Du = b_2 u$;
- The operators M, R and G are given by

$$\begin{cases} \langle My, y\rangle_H = \displaystyle\int_{\mathcal{O}} q|y|^2 dx, \qquad \langle Ru, u\rangle_U = \int_{\mathcal{O}} r|u|^2 dx, \\ \langle Gy(T), y(T)\rangle_H = \displaystyle\int_{\mathcal{O}} g|y(T)|^2 dx. \end{cases}$$

Since $q \geq 0$, $r \geq 1$ and $g \geq 0$, Problem (sSLQ) is uniquely solvable.

Write $\{\mu_j\}_{j=1}^{\infty}$ for the eigenvalues of A and $\{e_j\}_{j=1}^{\infty}$ the corresponding eigenvectors with $|e_j|_{L^2(\mathcal{O})} = 1$ for $j \in \mathbb{N}$. It is well known that $\{e_j\}_{j=1}^{\infty}$ constitutes an orthonormal basis of $L^2(\mathcal{O})$. Hence, **(AS1)** holds.

Denote by H_λ the completion of the Hilbert space $L^2(\mathcal{O})$ with the norm

$$|f|^2_{H_\lambda} = \sum_{j=1}^{\infty} |\mu_j|^{-n} |f_j|^2, \qquad \text{for all } f = \sum_{j=1}^{\infty} f_j e_j \in L^2(\mathcal{O}).$$

By the asymptotic distribution of the eigenvalues of A (e.g., [297, Chapter 1, Theorem 1.2.1]), we see that

$$|I|^2_{\mathcal{L}_2(H;H_\lambda)} = \sum_{j=1}^{\infty} \langle e_j, e_j \rangle_{H_\lambda} = 4\sum_{j=1}^{\infty} |\mu_j|^{-n} < \infty.$$

Hence, the embedding from $L^2(\mathcal{O})$ to H_λ is Hilbert-Schmidt. From the definition of H_λ, it follows that $H'_\lambda = D((iA)^{n/2})$.

Let $\alpha \in C^{2n}(\overline{\mathcal{O}})$. For any $f \in H'_\lambda$, one has

$$\begin{aligned}|\alpha f|_{H'_\lambda} &= |\alpha f|_{D((iA)^{n/2})} = |(iA)^{n/2}(\alpha f)|_{L^2(\mathcal{O})}\\ &\le |\alpha|_{C^{2n}(\overline{\mathcal{O}})}|f|_{D((iA)^{n/2})} = |\alpha|_{C^{2n}(\overline{\mathcal{O}})}|f|_{H'_\lambda}.\end{aligned}$$

From this, we conclude that

$$C \in L^\infty_{\mathbb{F}}(0,T;\mathcal{L}(H'_\lambda)), \quad G \in L^\infty_{\mathcal{F}_T}(\Omega;\mathcal{L}(H'_\lambda)), \quad M \in L^\infty_{\mathbb{F}}(0,T;\mathcal{L}(H'_\lambda)).$$

Thus, (**AS2**) holds.

Let $\widetilde{U} = D((iA)^n)$. Clearly, $\widetilde{U}$ is dense in $L^2(\mathcal{O})$. From the definitions of B, D and R, we find that

$$R \in L^\infty_{\mathbb{F}}(0,T;\mathcal{L}(\widetilde{U})), \quad B, D \in L^2_{\mathbb{F}}(0,T;\mathcal{L}(\widetilde{U};H'_\lambda)).$$

Therefore, (**AS3**) holds.

13.10 Notes and Comments

Unless other stated, the material of this chapter is mainly based on [248]. Other works that are related to this chapter can be found in [23, 76, 150, 235], etc.

LQ problem for (deterministic) finite dimensional systems is one of the three milestones in modern control theory and has been extensively studied for different kinds of control systems. It is fundamentally important due to the following reasons:

- It can be used to model many problems in applications;
- Many optimal control problems for nonlinear control systems can be reasonably approximated by LQ problems.

To our best knowledge, the deterministic LQ problem was first studied in [22]. In [163], the optimal feedback control was found via the matrix-valued Riccati equation, and hence a systematic LQ control theory was established. This theory was extended to infinite dimensions as well (e.g., [185, 202]).

LQ problems for stochastic finite dimensional systems were first studied in [177] by means of the dynamic programming method. In [351, 352], Riccati equation was first introduced to study LQ problems for stochastic differential equations with deterministic coefficients. The LQ problems for stochastic differential equations with random coefficients were first studied in [32], in which a matrix-valued nonlinear backward stochastic differential equation in the form of (13.47) was derived. These works have inspired many new progresses and applications on LQ problems for stochastic differential equations (e.g., [55, 236, 285, 307, 308, 313] and the rich references cited therein).

Compared with LQ problems for stochastic differential equations, there exist only a quite limited number of works dwelling on some special cases of LQ problems for stochastic distributed parameter systems (e.g., [2, 76, 130, 131, 134, 152, 233, 318]). We list below some of these typical works:

- In [152, 318], Problem (SLQ) was studied under the condition that the diffusion term in (13.1) is specialized as $Cx(t)dW_1(t)+Du(t)dW_2(t)$, where $W_1(\cdot)$ and $W_2(\cdot)$ are mutually independent Brownian motions. With this assumption, the corresponding Riccati equation reads

$$\begin{cases} dP = -\big[P(A+A_1)+(A+A_1)^*P+C^*PC+Q \\ \qquad\qquad -PBK^{-1}B^*P\big]dt & \text{in } [0,T), \\ P(T)=G, \end{cases} \tag{13.305}$$

  which is a random operator-valued Riccati equation (rather than operator-valued backward stochastic Riccati equation).
- In [2], under the assumption that the diffusion term in (13.1) is $\sigma dW(t)$ with σ being a suitable $\mathbf{F}$-adapted H-valued process independent of both the state and the control, Problem (SLQ) was studied and the optimal feedback control was obtained by solving a random operator-valued Riccati equation (similarly to (13.305)) and a backward stochastic evolution equation.
- In [130], Problem (SLQ) was considered for the case that $R = I_U$, the identity operator on U, and $D = 0$. In this case, the equation (13.17) is specialized as

$$\begin{cases} dP = -\big(PA+A^*P+\Lambda C+C^*\Lambda+C^*PC+Q \\ \qquad\qquad -PBB^*P\big)dt+\Lambda dW(t) & \text{in } [0,T), \\ P(T)=G. \end{cases} \tag{13.306}$$

  Although (13.306) looks much simpler than (13.17), it is also an operator-valued backward stochastic evolution equation (because the "bad" term "$\Lambda dW(t)$" is still in (13.306)). Nevertheless, [130] considered a kind of generalized solution to (13.306), which was a "weak limit" of solutions to some suitable finite dimensional approximations of (13.306). A little more precisely, it was shown in [130] that the finite dimensional approximations P_n of P are convergent in some weak sense, and via which P may be obtained as a suitable generalized solution to the equation (13.306) although nothing can be said about Λ (to the same equation). This is enough for the case that $D = 0$ since the corresponding optimal feedback operator in (13.138) is then specialized as

$$\Theta(\cdot) = -K(\cdot)^{-1}B(\cdot)^*P(\cdot),$$

which is independent of Λ.

- In [131], the well-posedness of (13.306) was further studied when A is a self-adjoint operator on H and there exists an orthonormal basis $\{e_j\}_{j=1}^{\infty}$ in H and an increasing sequence of positive numbers $\{\mu_j\}_{j=1}^{\infty}$ so that $Ae_j = -\mu_j e_j$ for $j \in \mathbb{N}$ and $\sum_{j=1}^{\infty} \mu_j^{-r} < \infty$ for some $r \in (\frac{1}{4}, \frac{1}{2})$. Clearly, this assumption is not satisfied by many controlled stochastic partial differential equations, such as stochastic wave equations, stochastic Schrödinger equations, stochastic KdV equations, stochastic beam equations, etc. It is even not fulfilled by the classical m-dimensional stochastic heat equation for $m \geq 2$. The well-posedness result (in [131]) for (13.306) was then applied to Problem (SLQ) in the case that $R = I_U$ and $D = 0$.

In this chapter, we do not assume that $R = I_U$ and $D = 0$. If $D = 0$, then the nonlinear term $L^*K^{-1}L$ is specialized as PBB^*P, which enjoys a "good" sign in the energy estimate, and therefore, it is not very hard to obtain the well-posedness of the linearized equation of (13.17), at least for some special situation. Once the well-posedness of this linearized equation is established, the well-posedness of (13.17) follows from a fixed point argument (e.g. [130, 131]). However, if $D \neq 0$, it is not an easy task even to derive the well-posedness of the linearized version of (13.17).

In order to study the well-posedness of (13.17), we introduce a sequence of finite dimensional approximate equations, and prove that the limit of the solutions for these approximate equations is the transposition solution to (13.17). To this end, we need to assume that A generates a C_0 group on H. As pointed out in Remark 13.43, this assumption can be dropped (See [235]). However, we keep the use of the present method due to the following two reasons:

- This method is rather elementary and provides an intrinsic connection between the well-posedness of (13.17) and the forward-backward stochastic evolution equation (13.217). Consequently, it shows the relationship between the Pontryagin-type maximum principle (Theorem 13.32 and Corollary 13.33) and the optimal feedback operator.
- It provides a numerical scheme to compute the solution to (13.17). Of course, since the convergence is very weak, it is not easy to compute the numerical solution to (13.17) by this method. However, as far as we know, so far it is the only existing convergent numerical scheme.

There are many open problems related to the main topic of this chapter. We shall list below some of them which, in our opinion, are particularly interesting:

1) Control systems with unbounded control operators.

As we have pointed out in Remark 13.1, In this paper, we assume that both $B(\cdot)$ and $D(\cdot)$ are bounded operator-valued process. As a result, the system (13.1) does not cover controlled stochastic partial differential equations with boundary/pointwise controls. To drop this restriction, one needs to make some further assumptions, such as the semigroup $\{S(t)\}_{t\geq 0}$ has some smoothing effect. When $D = 0$, some results along this line are obtained (e.g. [134]). On the other hand, when $D \neq 0$, there is no published result in the literature. We believe that one can combine the transposition methods introduced in Section 7.2 and this chapter to deal with that case. But several technical difficulties should be overcome.

2) LQ problems with incomplete state information.

In this chapter, we only consider LQ problems with complete state information. When the control system is only partially observed, that is, what we know is only $z = Ox + \xi$, where O is a suitable operator and ξ is the observation noise. LQ problems for partially observed stochastic differential equations are extensively studied (e.g. [27, 339]). However, as far as we know, except for some very particular case, such as the coefficients are deterministic and the noise is additive, there exist no results for LQ problems of stochastic distributed parameter systems with incomplete state information.

References

1. A. A. Agrachev and Y. L. Sachkov. *Control theory, from the geometric viewpoint.* Encyclopaedia of Mathematical Sciences, **87**. Control Theory and Optimization, II. Springer-Verlag, Berlin, 2004.
2. N. U. Ahmed. *Stochastic control on Hilbert space for linear evolution equations with random operator-valued coefficients. SIAM J. Control Optim.* **19** (1981), 401–430.
3. A. Al-Hussein. *Strong, mild and weak solutions of backward stochastic evolution equations. Random Oper. Stochastic Equations.* **13** (2005), 129–138.
4. N. Anantharaman, M. Léautaud and F. Macià. *Wigner measures and observability for the Schrödinger equation on the disk. Invent. Math.* **206** (2016), 485–599.
5. A. Arapostathis, V. S. Borkar and M. K. Ghosh. *Ergodic control of diffusion processes.* Encyclopedia of Mathematics and its Applications, **143**. Cambridge University Press, Cambridge, 2012.
6. K. J. Åström. *Introduction to stochastic control theory.* Mathematics in Science and Engineering, **70**. Academic Press, New York-London 1970.
7. S. A. Avdonin and S. A. Ivanov. *Families of exponentials. The method of moments in controllability problems for distributed parameter systems.* Translated from the Russian and revised by the authors. Cambridge University Press, Cambridge, 1995.
8. K. Balachandran and J. P. Dauer. *Controllability of nonlinear systems in Banach spaces: a survey. J. Optim. Theory Appl.* **115** (2002), 7–28.
9. A. V. Balakrishnan. *Optimal Control Problems in Banach Spaces. J. Soc. Indust. Appl. Math. Ser. A Control.* **3** (1965), 152–180.
10. H. T. Banks and K. Kunisch. *Estimation techniques for distributed parameter systems.* Systems & Control: Foundations & Applications, **1**. Birkhäuser Boston, Inc., Boston, MA, 1989.
11. V. Barbu. *Note on the internal stabilization of stochastic parabolic equations with linearly multiplicative Gaussian noise. ESAIM Control Optim. Calc. Var.* **19** (2013), 1055–1063.
12. V. Barbu, A. Răşcanu and G. Tessitore. *Carleman estimate and cotrollability of linear stochastic heat equatons. Appl. Math. Optim.* **47** (2003), 97–120.
13. V. Barbu and L. Tubaro. *Exact controllability of stochastic differential equations with multiplicative noise. Systems Control Lett.* **122** (2018), 19–23.

Q. Lü and X. Zhang, *Mathematical Control Theory for Stochastic Partial Differential Equations*, Probability Theory and Stochastic Modelling 101, https://doi.org/10.1007/978-3-030-82331-3

14. C. Bardos, G. Lebeau and J. Rauch. *Sharp sufficient conditions for the observation, control and stabilizion of waves from the boundary. SIAM J. Control Optim.* **30** (1992), 1024–1065.
15. A. Bashirov. *Partially observable linear systems under dependent noises.* Birkhäuser Verlag, Basel-Boston-Berlin, 2003.
16. A. Bashirov and K. R. Kerimov. *On controllability conception for stochastic systems. SIAM J. Control Optim.* **35** (1997), 384–398.
17. A. Bashirov and N. I. Mahmudov. *On concepts of controllability for deterministic and stochastic systems. SIAM J. Control Optim.* **37** (1999), 1808–1821.
18. K. Beauchard, Y. Chitour, D. Kateb and R. Long. *Spectral controllability for 2D and 3D linear Schrödinger equations. J. Funct. Anal.* **256** (2009), 3916–3976.
19. K. Beauchard and C. Laurent. *Local controllability of 1D linear and nonlinear Schrödinger equations with bilinear control. J. Math. Pures Appl.* **94** (2010), 520–554.
20. R. Bellman. *Dynamic Programming.* Princeton Univ. Press, Princeton, New Jersey, 1957.
21. R. Bellman. *Dynamic programming and stochastic control process. Inform. & Control.* **1** (1958), 228–239.
22. R. Bellman, I. Glicksberg and O. A. Gross. *Some aspects of the mathematical theory of control processes.* Rand Corporation, Santa Monica, Calif., Rep. No. R-313, 1958.
23. P. Benner and C. Trautwein. *A linear quadratic control problem for the stochastic heat equation driven by Q-Wiener processes. J. Math. Anal. Appl.*bf 457 (2018), 776–802.
24. A. Bensoussan. *Contrôle optimal stochastique de système gouvernés par des équations aux dérivées partielles de type parabolique. Rend. Mat.* **2** (1969), 135–173.
25. A. Bensoussan. *Lecture on Stochastic Control.* In: *Nonlinear Filtering and Stochastic Control.* Edited by S. K. Mitter and A. Moro. Lecture Notes in Mathematics, **972**. Springer-Verlag, Berlin, 1982, 1–62.
26. A. Bensoussan. *Stochastic maximum principle for distributed parameter systems. J. Franklin Inst.* **315** (1983), 387–406.
27. A. Bensoussan. *Stochastic control of partially observable systems.* Cambridge University Press, Cambridge, 1992.
28. L. D. Berkovitz. *Optimal control theory.* Springer-Verlag, New York, 1974.
29. D. P. Bertsekas. *Dynamic programming and stochastic control.* Mathematics in Science and Engineering, **125**. Academic Press [Harcourt Brace Jovanovich, Publishers], New York-London, 1976.
30. M. Kh. Bikchantayev. *Optimal control of systems with distributed parameters and stochastic signals. Engrg. Cybernetics.* **1969** (1969), 147–157.
31. J.-M. Bismut. *Analyse convexe et probabilitiés.* Ph D Thesis, Faculté des Sciences de Paris, Paris, France, 1973.
32. J.-M. Bismut. *Linear quadratic optimal stochastic control with random coefficients. SIAM J. Control Optim.* **14** (1976), 419–444.
33. J.-M. Bismut. *An introductory approach to duality in optimal stochastic control. SIAM Rev.* **20** (1978), 62–78.
34. V. G. Boltyanski, R. V. Gamkrelidze and L. S. Pontryagin. *The theory of optimal processes, I: The maximum principle. Izv. Akad. Nauk SSSR Ser. Mat.*,

24 (1960), 3–42 (in Russian); English transl. in *Amer. Math. Soc. Transl.* **18** (1961), 341–382.

35. U. Boscain, M. Caponigro, T. Chambrion and M. Sigalotti. *A weak spectral condition for the controllability of the bilinear Schrödinger equation with application to the control of a rotating planar molecule. Comm. Math. Phys.* **311** (2012), 423–455.
36. R. Bouldin. *A counterexample in the factorization of Banach space operators. Proc. Amer. Math. Soc.* **68** (1978), 327–327.
37. A. Bressan and B. Piccoli. *Introduction to the mathematical theory of control.* AIMS Series on Applied Mathematics, **2**. American Institute of Mathematical Sciences (AIMS), Springfield, MO, 2007.
38. W. Bryc. *The normal distribution. Characterizations with applications.* Lecture Notes in Statistics, **100**. Springer-Verlag, New York, 1995.
39. R. Buckdahn and J. Li. *Stochastic differential games and viscosity solutions of Hamilton-Jacobi-Bellman-Isaacs equations. SIAM J. Control Optim.* **47** (2008), 444–475.
40. R. Buckdahn, M. Quincampoix and G. Tessitore. *A Characterization of approximately controllable linear stochastic differential equations. Stochastic partial differential equations and applicationsVII.* Lect. Notes Pure Appl. Math., **245**, Chapman & Hall/CRC, Boca Raton, FL, 2006, 53–60.
41. J. Bourgain, N. Burq and M. Zworski. *Control for Schrödinger equations on 2-tori: rough potentials. J. Eur. Math. Soc.* **15** (2013), 1597–1628.
42. N. Burq and M. Zworski. *Geometric control in the presence of a black box. J. Amer. Math. Soc.* **17** (2004), 443–471.
43. A. L. Bukhgeĭm and M. V. Klibanov. *Uniqueness in the large of a class of multidimensional inverse problems. Dokl. Akad. Nauk SSSR.* **260** (1981), 269–272.
44. A. G. Butkovskiĭ. *Distributed Control Systems.* Translated from the Russian by S. Technica. Inc. Translation Editor: George M. Kranc. Modern Analytic and Computational Methods in Science and Mathematics, No. **11**. American Elsevier Publishing Co., Inc., New York, 1969.
45. P. Cannarsa and C. Sinestrari. *Semiconcave functions, Hamilton-Jacobi equations, and optimal control.* Birkhäuser, Boston, 2004.
46. P. Cannarsa, P. Martinez and J. Vancostenoble. *Global Carleman estimates for degenerate parabolic operators with applications. Mem. Amer. Math. Soc.* **239** (2016), no. 1133, ix+209 pp.
47. T. Carleman. *Sur un problème d'unicité pour les systèmes d'équations aux dérivées partielles à deux variables indépendantes. Ark. Mat. Astr. Fys. 26 B.* **17** (1939), 1–9.
48. R. Carmona. *Lectures on BSDEs, stochastic control, and stochastic differential games with financial applications.* Financial Mathematics, **1**. Society for Industrial and Applied Mathematics (SIAM), Philadelphia, PA, 2016.
49. E. Casas and F. Tröltzsch. *Recent advances in the analysis of pointwise state-constrained elliptic optimal control problems. ESAIM Control Optim. Calc. Var.* **16** (2010), 581–600.
50. F. W. Chaves-Silva, X. Zhang and E. Zuazua. *Controllability of evolution equations with memory. SIAM J. Control Optim.* **55** (2017), 2437–2459.
51. H. F. Chen. *On stochastic observability and controllability. Automatica.* **16** (1980), 179–190.

52. H. F. Chen. *Stochastic approximation and its applications.* Nonconvex Optimization and its Applications, **64**. Kluwer Academic Publishers, Dordrecht, 2002.
53. H. F. Chen and L. Guo. *Identification and stochastic adaptive control.* Systems & Control: Foundations & Applications. Birkhäuser Boston, Inc., Boston, MA, 1991.
54. M. Chen. *Null controllability with constraints on the state for stochastic heat equation. J. Dyn. Control Syst.* **24** (2018), 39–50.
55. S. Chen, X. Li and X. Zhou. *Stochastic linear quadratic regulators with indefinite control weight costs. SIAM J. Control Optim.* **36** (1998), 1685–1702.
56. Z. Chen and L. Epstein. *Ambiguity, risk, and asset returns in continuous time. Econometrica.* **70** (2002), 1403–1443.
57. K. Chung. *A course in probability theory. Second edition.* Probability and Mathematical Statistics, **21**. Academic Press, New York-London, 1974.
58. K. Chung and R. J. Williams. *Introduction to stochastic integration.* Modern Birkhäuser Classics. Birkhäuser/Springer, New York, 2014.
59. F. Clarke. *Functional analysis, calculus of variations and optimal control.* Graduate Texts in Mathematics, **264**. Springer, London, 2013.
60. J. B. Conway. *A Course in Functional Analysis. Second edition.* Springer-Verlag, New York, 1994.
61. J.-M. Coron. *Control and nonlinearity.* Mathematical Surveys and Monographs, 136. American Mathematical Society, Providence, RI, 2007.
62. J.-M. Coron. *On the controllability of nonlinear partial differential equations. Proceedings of the International Congress of Mathematicians. Vol. I,* 238–264, Hindustan Book Agency, New Delhi, 2010.
63. J.-M. Coron, O. Glass and Z. Wang. *Exact boundary controllability for 1-D quasilinear hyperbolic systems with a vanishing characteristic speed. SIAM J. Control Optim.* **48** (2009/10), 3105–3122.
64. M. G. Crandall and P.-L. Lions. *Viscosity solutions of Hamilton-Jacobi equations. Trans. Amer. Math. Soc.* **277** (1983), 1–42.
65. R. Curtain and K. Morris. *Transfer functions of distributed parameter systems: a tutorial. Automatica J. IFAC.* **45** (2009), 1101–1116.
66. R. F. Curtain and A. J. Pritchard. *Infinite dimensional linear systems theory.* Lecture Notes in Control and Information Sciences, **8**. Springer-Verlag, Berlin-New York, 1978.
67. R. F. Curtain and H. Zwart. *An introduction to infinite-dimensional linear systems theory.* Texts in Applied Mathematics, **21**. Springer-Verlag, New York, 1995.
68. G. Da Prato and J. Zabczyk. *Stochastic equations in infinite dimensions. Second edition.* Encyclopedia of Mathematics and Its Applications, **152**. Cambridge University Press, Cambridge, 2014.
69. L. Dai, Y. Zhang and J. Zou. *Numerical schemes for forward-backward stochastic differential equations using transposition solutions.* Preprint.
70. R. Dalang, D. Khoshnevisan, C. Mueller, D. Nualart and Y. Xiao. *A minicourse on stochastic partial differential equations.* Lecture Notes in Mathematics, **1962**. Springer-Verlag, Berlin, 2009.
71. D. A. Dawson. *Stochastic evolution equations. Math. Biosci.* **15** (1972), 287–316.
72. A. de Bouard and A. Debussche. *A stochastic nonlinear Schrödinger equation with multiplicative noise. Comm. Math. Phys.* **205** (1999), 161–181.

73. F. Delbaen. *BSDE and risk measures. Proceedings of the International Congress of Mathematicians, Vol. IV.* Hyderabad, India, 2010, 3054–3060.
74. J. Diestel and J. J. Uhl, Jr. *Vector Measures,* Math. Surveys, vol. 15, American Mathematical Society, Providence, R.I., 1977.
75. F. Dou and Q. Lü. *Partial approximate controllability for linear stochastic control systems. SIAM J. Control Optim.* **57** (2019), 1209–1229.
76. F. Dou and Q. Lü. *Time-inconsistent linear quadratic optimal control problems for stochastic evolution equations. SIAM J. Control Optim.* **58** (2020), 485–509.
77. A. Doubova, E. Fernández-Cara, M. González-Burgos and E. Zuazua. *On the controllability of parabolic systems with a nonlinear term involving the state and the gradient. SIAM J. Control Optim.* **41** (2002), 798–819.
78. A. Doubova, A. Osses and J.-P. Puel. *Exact controllability to trajectories for semilinear heat equations with discontinuous diffusion coefficients. ESAIM Control Optim. Calc. Var.* **8** (2002), 621–661.
79. K. Du and Q. Meng. *A maximum principle for optimal control of stochastic evolution equations. SIAM J. Control Optim.* **51** (2013), 4343–4362.
80. T. Duyckaerts, X. Zhang and E. Zuazua. *On the optimality of the observability inequalities for parabolic and hyperbolic systems with potentials. Ann. Inst. H. Poincaré Anal. Non Linéaire.* **25** (2008), 1–41.
81. D. Duffie and L. G. Epstein. *Stochastic differential utility. Econometrica.* **60** (1992), 353–394.
82. T. E. Duncan. *Some topics in stochastic control. Discrete Contin. Dyn. Syst. Ser. B.* **14** (2010), 1361–1373.
83. T. E. Duncan, B. Maslowski and B. Pasik-Duncan. *Adaptive boundary and point control of linear stochastic distributed parameter system. SIAM J. Control Optim.* **32** (1994), 648–672.
84. T. Dunst and A. Prohl. *The forward-backward stochastic heat equation: numerical analysis and simulation. SIAM J. Sci. Comput.* **38** (2016) A2725–A2755.
85. Y. V. Egorov. *Some problems in the theory of optimal control. Ž. Vyčisl Mat. i Mat. Fiz.* **3** (1963), 887–904. (In Russian)
86. N. El Karoui, S. Peng and M. C. Quenez. *Backward stochastic differential equations in finance. Math. Finance.* 7 (1997), 1–71.
87. K. J. Engel and R. Nagel. *One-parameter semigroups for linear evolution equations.* Graduate Texts in Mathematics, 194. Springer-Verlag, New York, 2000.
88. S. Ervedoza and E. Zuazua. *Numerical approximation of exact controls for waves.* Springer Briefs in Mathematics. Springer, New York, 2013.
89. G. Fabbri, F Gozzi and A. Święch. *Stochastic optimal control in infinite dimension, Dynamic programming and HJB equations.* With a contribution by M. Fuhrman and G. Tessitore. Probability Theory and Stochastic Modelling, **82**. Springer, Cham, 2017.
90. P. L. Falb. *Infinite-dimensional filtering: The Kalman-Bucy filter in Hilbert space. Information and Control.* **11** (1967), 102–137.
91. H. O. Fattorini. *Control in finite time of differential equations in Banach space. Comm. Pure Appl. Math.* **19** (1966), 17–34.
92. H. O. Fattorini. *Infinite-dimensional optimization and control theory.* Encyclopedia of Mathematics and its Applications, **62**. Cambridge University Press, Cambridge, 1999.

93. H. O. Fattorini and D. L. Russell. *Exact controllability theorems for linear parabolic equations in one space dimension*. Arch. Rational Mech. Anal. **43** (1971), 272–292.
94. A. Fernández-Bertolin and J. Zhong. *Hardy's uncertainty principle and unique continuation property for stochastic heat equations*. ESAIM Control Optim. Calc. Var. **26** (2020), no. 9.
95. E. Fernández-Cara. *The control of PDEs: some basic concepts, recent results and open problems*. Topics in mathematical modeling and analysis, 49–107, Jindřich Nečas Cent. Math. Model. Lect. Notes, **7**. Matfyzpress, Prague, 2012.
96. W. H. Fleming. *Distributed parameter stochastic systems in population biology*. Control theory, numerical methods and computer systems modelling (Internat. Sympos., IRIA LABORIA, Rocquencourt, 1974). Lecture Notes in Econom. and Math. Systems, **107**, Springer, Berlin, 1975, 179–191.
97. W. H. Fleming and M. Nisio. *On the existence of optimal stochastic controls*. J. Math. Mech. **15** (1966), 777–794.
98. W. H. Fleming and H. M. Soner. *Controlled Markov processes and viscosity solutions. Second edition*. Stochastic Modelling and Applied Probability, **25**. Springer, New York, 2006.
99. W. H. Fleming and P. E. Souganidis. *On the existence of value functions of two-player, zero-sum stochastic differential games*. Indiana Univ. Math. J. **38** (1989), 293–314.
100. J. J. Florentin. *Optimal control of continuous time, Markov, stochastic systems*. J. Electronics Control. **10** (1961), 473–488.
101. C. Foias, H. Özbay and A. Tannenbaum. *Robust control of infinite-dimensional systems. Frequency domain methods*. Lecture Notes in Control and Information Sciences, **209**. Springer-Verlag London, Ltd., London, 1996.
102. H. Frankowska. *Optimal control under state constraints*. Proceedings of the International Congress of Mathematicians, Vol. IV. Hyderabad, India, 2010, 2915–2943.
103. H. Frankowska and Q. Lü. *First and second order necessary optimality conditions for controlled stochastic evolution equations with control and state constraints*. J. Differential Equations. **268** (2020), 2949–3015.
104. H. Frankowska and X. Zhang. *Necessary conditions for stochastic optimal control problems in infinite dimensions*. Stochastic Process. Appl., **130** (2020), 4081–4103.
105. P. K. Friz and M. Hairer, *A course on rough paths. With an introduction to regularity structures*. Universitext. Springer, Cham, 2014.
106. X. Fu. *A weighted identity for partial differential operators of second order and its applications*. C. R. Math. Acad. Sci. Paris. **342** (2006), 579–584.
107. X. Fu, X. Liu. *A weighted identity for stochastic partial differential operators and its applications*. J. Differential Equations **262** (2017), 3551–3582.
108. X. Fu, X. Liu. *Controllability and observability of some stochastic complex Ginzburg-Landau equations*. SIAM J. Control Optim. **55** (2017), 1102–1127.
109. X. Fu, X. Liu, Q. Lü and X. Zhang. *An internal observability estimate for stochastic hyperbolic equations*. ESAIM Control Optim. Calc. Var. , **22** (2016), 1382–1411.
110. X. Fu, Q. Lü and X. Zhang. *Carleman Estimates for Second Order Partial Differential Operators and Applications, a Unified Approach*. Springer, Cham, 2019.

111. X. Fu, J. Yong and X. Zhang. *Exact controllability for the multidimensional semilinear hyperbolic equations. SIAM J. Control Optim.* **46** (2007), 1578–1614.
112. M. Fuhrman, Y. Hu and G. Tessitore. *Stochastic maximum principle for optimal control of SPDEs.* Appl. Math. Optim. **68** (2013), 181–217.
113. M. Fuhrman, Y. Hu and G. Tessitore. *Stochastic maximum principle for optimal control of partial differential equations driven by white noise. Stoch. Partial Differ. Equ. Anal. Comput.* **6** (2018), 255–285.
114. M. Fuhrman and C. Orrier. *Stochastic maximum principle for optimal control of a class of nonlinear SPDEs with dissipative drift. SIAM J. Control Optim.* **54** (2016), 341–371.
115. T. Funaki. *Random motion of strings and related stochastic evolution equations. Nagoya Math. J.* **89** (1983), 129–193.
116. A. V. Fursikov. *Optimal control of distributed systems. Theory and applications.* Translations of Mathematical Monographs, 187. American Mathematical Society, Providence, RI, 2000.
117. A. V. Fursikov and O. Yu. Imanuvilov. *Controllability of Evolution Equations.* Lecture Notes Series **34**, Research Institute of Mathematics, Seoul National University, Seoul, Korea, 1994.
118. P. Gao. *Carleman estimate and unique continuation property for the linear stochastic Korteweg-de Vries equation. Bull. Aust. Math. Soc.* **90** (2014), 283–294.
119. P. Gao. *Global Carleman estimates for linear stochastic Kawahara equation and their applications. Math. Control Signals Systems.* **28** (2016), Art. 21, 22 pp.
120. P. Gao. *Carleman estimates for forward and backward stochastic fourth order Schrödinger equations and their applications. Evol. Equ. Control Theory.* **7** (2018), 465–499.
121. P. Gao, M. Chen and Y. Li. *Observability estimates and null controllability for forward and backward linear stochastic Kuramoto-Sivashinsky equations. SIAM J. Control Optim.* **53** (2015), 475–500.
122. J. Glimm. *Nonlinear and stochastic phenomena: the grand challenge for partial differential equations. SIAM Rev.* **33** (1991), 626–643.
123. R. Glowinski, J.-L. Lions and J. He. *Exact and approximate controllability for distributed parameter systems. A numerical approach.* Encyclopedia of Mathematics and its Applications, 117. Cambridge University Press, Cambridge, 2008.
124. Ĭ. Ī. Gīhman and A. V. Skorohod. *Controlled stochastic processes.* Springer-Verlag, New York-Heidelberg, 1979.
125. D. Goreac. *A Kalman-type condition for stochastic approximate controllability. C. R. Math. Acad. Sci. Paris.* **346** (2008), 183–188.
126. D. Goreac. *Approximate controllability for linear stochastic differential equations in infinite dimensions. Appl. Math. Optim.* **60** (2009), 105–132.
127. D. Goreac. *Approximately reachable directions for piecewise linear switched systems. Math. Control Signals Systems.* **31** (2019), 333–362.
128. F. Gozzi, S. S. Sritharan, A. Święch. *Bellman equations associated to the optimal feedback control of stochastic Navier-Stokes equations. Comm. Pure Appl. Math.* **58** (2005), 671–700.
129. W. Grecksch and C. Tudor. *Stochastic evolution equations. A Hilbert space approach.* Mathematical Research, 85. Akademie-Verlag, Berlin, 1995.

130. G. Guatteri and G. Tessitore. *On the backward stochastic Riccati equation in infinite dimensions. SIAM J. Control Optim.* **44** (2005), 159–194.
131. G. Guatteri and G. Tessitore. *Well posedness of operator valued backward stochastic Riccati equations in infinite dimensional spaces. SIAM J. Control Optim.* **52** (2014), 3776–3806.
132. M. Gunzburger and J. Ming. *Optimal control of stochastic flow over a backward-facing step using reduced-order modeling. SIAM J. Sci. Comput.* **33** (2011), 2641–2663.
133. B. Guo and J. Wang. *Control of wave and beam PDEs. The Riesz basis approach.* Communications and Control Engineering Series. Springer, Cham, 2019.
134. C. Hafizoglu, I. Lasiecka, T. Levajković, H. Mena and A. Tuffaha. *The stochastic linear quadratic control problem with singular estimates. SIAM J. Control Optim.* **55** (2017), 595–626.
135. P. R. Halmos. *A Hilbert space problem book.* Springer-Verlag, New York-Berlin, 1982.
136. M. Hairer. *Solving the KPZ equation. Ann. of Math.* **178** (2013), 559–664.
137. M. Hairer. *A theory of regularity structures. Invent. Math.* **198** (2014), 269–504.
138. E. Hausenblas and J. Seidler. *A note on maximal inequality for stochastic convolutions. Czechoslovak Math. J.* **51** (2001), 785–790.
139. U. G. Haussmann. *General necessary conditions for optimal control of stochastic system. Math. Prog. Study.* **6** (1976), 34–48.
140. S. He, J. Wang and J. Yan. *Semimartingale theory and stochastic calculus.* Kexue Chubanshe (Science Press), Beijing; CRC Press, Boca Raton, FL, 1992.
141. Y. He. *Switching Controls for Linear Stochastic Differential Systems. Math. Control Relat. Fields.* Published online, doi:10.3934/mcrf.2020005.
142. V. Hernández-Santamaría and L. Peralta. *Controllability results for stochastic coupled systems of fourth and second-order parabolic equations.* arXiv:2003.01334v1.
143. E. Hille and R. S. Phillips. *Functional analysis and semi-groups. Revised edition.* American Mathematical Society Colloquium Publications, **31**. American Mathematical Society, Providence, R. I., 1957.
144. M. Hinze and A. Rösch. *Discretization of optimal control problems. Constrained optimization and optimal control for partial differential equations,* 391–430, Internat. Ser. Numer. Math., 160, Birkhäuser/Springer Basel AG, Basel, 2012.
145. H. Holden, B. Øksendal, J. Ubøe and T. Zhang. *Stochastic partial differential equations. A modeling, white noise functional approach. Second edition.* Universitext. Springer, New York, 2010.
146. L. Hu. *Sharp time estimates for exact boundary controllability of quasilinear hyperbolic systems. SIAM J. Control Optim.* **53** (2015), 3383–3410.
147. M. Hu, S. Ji, S. Peng and Y. Song. *Backward stochastic differential equations driven by G-Brownian motion. Stochastic Process. Appl.* **124** (2014), 759–784.
148. Y. Hu and S. Peng. *Maximum principle for semilinear stochastic evolution control systems. Stoch. & Stoch. Rep.* **33** (1990), 159–180.
149. Y. Hu and S. Peng. *Adapted solution of a backward semilinear stochastic evolution equation. Stochastic Anal. Appl.* **9** (1991), 445–459.
150. Y. Hu and S. Tang. *Nonlinear backward stochastic evolutionary equations driven by a space-time white noise. Math. Control Relat. Fields.* **8** (2018), 739–751.

151. T. Hytönen, J. van Neerven, M. Veraar and L. Weis. *Analysis in Banach spaces. Vol. I. Martingales and Littlewood-Paley theory.* Ergebnisse der Mathematik und ihrer Grenzgebiete. 3. Folge. A Series of Modern Surveys in Mathematics, **63**. Springer, Cham, 2016.
152. A. Ichikawa. *Dynamic programming approach to stochastic evolution equations. SIAM J. Control Optim.* **17** (1979), 152–174.
153. O. Yu. Imanuvilov. *On Carlerman estimates for hyperbolic equations. Asymptot. Anal.* **32** (2002), 185–220.
154. O. Yu. Imanuvilov and M. Yamamoto. *Carleman inequalities for parabolic equations in a Sobolev spaces of negative order and exact controllability for semilinear parabolic equations. Publ. RIMS Kyoto Univ.* **39** (2003), 227–274.
155. K. Itô. *Stochastic integral. Proc. Imp. Acad. Tokyo.* **20** (1944), 519–524.
156. K. Itô. *On a stochastic integral equation. Proc. Japan Acad.* **22** (1946). 32–35.
157. K. Itô. *Foundations of stochastic differential equations in infinite-dimensional spaces.* CBMS-NSF Regional Conference Series in Applied Mathematics, **47**. Society for Industrial and Applied Mathematics, Philadelphia, PA, 1984.
158. K. Itô and K. Kunisch. *Novel concepts for nonsmooth optimization and their impact on science and technology. Proceedings of the International Congress of Mathematicians.* Volume IV, 3061–3090, Hindustan Book Agency, New Delhi, 2010.
159. K. Itô, Y. Zhang and J. Zou. *Fully discrete schemes and their analyses for forward-backward stochastic differential equations.* arXiv:1804.10944v1.
160. J. Jacod, S. Méléard and P. Protter. *Explicit form and robustness of martingale representations. Ann. Probab.* **28** (2000), 1747–1780.
161. G. Kallianpur and J. Xiong. *Stochastic differential equations in infinite dimensional spaces.* Institute of Mathematial Statistics, Lecture Notes-Monograph Series, **26**, Institute of Mathematical Statistics, United States, 1995.
162. R. E. Kalman. *A new approach to linear filtering and prediction problems. J. Fluids Eng.* 82 (1960), 35–45.
163. R. E. Kalman. *On the general theory of control systems. Proceedings of the First IFAC Congress. Moscow, 1960;* Butterworth, London, 1961, vol.1, 481–492.
164. I. Karatzas and S. E. Shreve. *Brownian motion and stochastic calculus.* Springer-Verlag, New York, 1988.
165. M. A. Kazemi and M. V. Klibanov. *Stability estimates for ill-posed Cauchy problems involving hyperbolic equations and inequalities. Appl. Anal.* **50** (1993), 93–102.
166. M. V. Klibanov and M. Yamamoto. *Exact controllability for the time dependent transport equation. SIAM J. Control Optim.* **46** (2007), 2071–2195.
167. P. S. Knopov and O. N. Deriyeva. *Estimation and control problems for stochastic partial differential equation.* Springer, New York, Heidelberg, Dordrecht, London, 2013.
168. P. I. Kogut and G. R. Leugering. *Optimal control problems for partial differential equations on reticulated domains. Approximation and asymptotic analysis.* Systems & Control: Foundations & Applications. Birkhäuser/Springer, New York, 2011.
169. V. N. Kolokoltsov. *Semiclassical analysis for diffusions and stochastic processes.* Lecture Notes in Mathematics, **1724**, Springer-Verlag, Berlin, 2000.
170. V. Komornik and P. Loreti. *Fourier series in control theory.* Springer Monographs in Mathematics. Springer-Verlag, New York, 2005.

171. P. Kotelenez. *Stochastic ordinary and stochastic partial differential equations. Transition from microscopic to macroscopic equations.* Stochastic Modelling and Applied Probability, **58**. Springer, New York, 2008.
172. M. Krstic and A. Smyshlyaev. *Boundary control of PDEs. A course on backstepping designs.* Advances in Design and Control, **16**. Society for Industrial and Applied Mathematics (SIAM), Philadelphia, PA, 2008.
173. N. V. Krylov. *Controlled diffusion processes. Reprint of the 1980 edition.* Stochastic Modelling and Applied Probability, 14. Springer-Verlag, Berlin, 2009.
174. H. Kunita. *On backward stochastic differential equations. Stochastics.* **6** (1982), 293–313.
175. H. Kunita and S. Watanabe. *On square integrable martingales. Nagoya Math. J.* **30** (1967), 209–245.
176. H.-H. Kuo. *Introduction to stochastic integration.* Universitext. Springer, New York, 2006.
177. H. J. Kushner. *Optimal stochastic control. IRE Trans. Auto. Control.* **7** (1962), 120–122.
178. H. J. Kushner. *On the optimal control of a system governed by a linear parabolic equation with white noise inputs. SIAM J. Control.* **6** (1968), 596–614.
179. H. J. Kushner. *Introduction to stochastic control.* Holt, Rinehart and Winston, Inc., New York-Montreal, Que.-London, 1971.
180. H. J. Kushner and F. C. Schweppe. *A maximum principle for stochastic control systems. J. Math. Anal. Appl.* **8** (1964), 287–302.
181. H. J. Kushner and G. Yin. *Stochastic approximation and recursive algorithms and applications. Second edition.* Applications of Mathematics (New York), **35**. Stochastic Modelling and Applied Probability. Springer-Verlag, New York, 2003.
182. D. F. Kuznetsov. *Multiple Itô and Stratonovich stochastic integrals: Fourier-Legendge and trigonometric expansions, approximations, formulas. Differ. Uravn. Protsessy Upr.* 2017, no. 1, 385 pp.
183. C. C. Kwan and K. N. Wang. Sur la stabilisation de la vibration élastique. *Sci. Sinica.* **17** (1974), 446–467.
184. L. D. Landau and E. M. Lifshitz. *Quantum mechanics: non-relativistic theory. Course of Theoretical Physics, Vol. 3.* Addison-Wesley Series in Advanced Physics. Pergamon Press Ltd., London-Paris; for U.S.A. and Canada: Addison-Wesley Publishing Co., Inc., Reading, Mass; 1958.
185. I. Lasiecka and R. Triggiani. *Differential and algebraic Riccati equations with application to boundary/point control problems: continuous theory and approximation theory.* Lecture Notes in Control and Information Sciences, 164. Springer-Verlag, Berlin, 1991.
186. I. Lasiecka and R. Triggiani. *Control Theory for Partial Differential Equations: Continuous and Approximation Theories. I. Abstract Parabolic Systems.* Encyclopedia of Mathematics and its Applications, 74. Cambridge University Press, Cambridge, 2000.
187. I. Lasiecka, R. Triggiani and X. Zhang. *Global uniqueness, observability and stabilization of nonconservative Schrödinger equations via pointwise Carleman estimates: Part I. H^1-estimates. J. Inv. Ill-posed Problems.* **11**(2004), 43–123.
188. M. M. Lavrent'ev, V. G. Romanov and S. P. Shishat·skiĭ. *Ill-posed problems of mathematical physics and analysis.* Translations of Mathematical Monographs, vol. 64. American Mathematical Society, Providence, RI, 1986.

189. J. Le Rousseau and L. Robbiano, *Local and global Carleman estimates for parabolic operators with coefficients with jumps at interfaces.* Invent. Math. **183** (2011), 245–336.
190. G. Lebeau. *Contrôle de léquation de Schrödinger. J. Math. Pures Appl.* **71** (1992), 267–291.
191. G. Lebeau and L. Robbiano. *Contrôle exact de l'équation de la chaleur. Comm. Partial Differential Equations.* **20** (1995), 335–356.
192. G. Lebeau and E. Zuazua. *Null controllability of a system of linear thermoelasticity. Arch. Rational Mech. Anal.* **141** (1998), 297–329.
193. E. B. Lee and L. Markus. *Foundations of optimal control theory.* John Wiley, New York, 1967.
194. S. Lenhart, J. Xiong and J. Yong. *Optimal controls for stochastic partial differential equations with an application in population modeling. SIAM J. Control Optim.* **54** (2016), 495–535.
195. H. Li and Q. Lü. *Null controllability for some systems of two backward stochastic heat equations with one control force. Chin. Ann. Math. Ser. B.* **33** (2012), 909–918.
196. H. Li and Q. Lü. *A quantitative boundary unique continuation for stochastic parabolic equations. J. Math. Anal. Appl.* **402** (2013), 518–526.
197. J. Li and Z. Zhang. *Topological classification of linear control systemsAn elementary analytic approach. J. Math. Anal. Appl.* **402** (2013), 84–102.
198. J. Li, H. Liu and S. Ma. *Determining a random Schrödinger equation with unknown source and potential. SIAM J. Math. Anal.* **51** (2019), 3465–3491.
199. T.-T. Li. *Controllability and observability for quasilinear hyperbolic systems.* AIMS Series on Applied Mathematics, vol. 3. American Institute of Mathematical Sciences, Springfield, MO, 2010.
200. T.-T. Li and B. Rao. *Boundary synchronization for hyperbolic systems.* Progress in Nonlinear Differential Equations and their Applications, **94**. Subseries in Control. Birkhäuser/Springer, Cham, 2019.
201. W. Li and X. Zhang. *Controllability of parabolic and hyperbolic equations: towards a unified theory.* in *Control Theory of Partial Differential Equations*, Lect. Notes Pure Appl. Math., 242, Chapman & Hall/CRC, Boca Raton, FL, 2005, 157–174.
202. X. Li and J. Yong. *Optimal control theory for infinite dimensional systems.* Systems & Control: Foundations & Applications. Birkhäuser Boston, Inc., Boston, MA, 1995.
203. G. Liang, T. Lyons and Z. Qian, *Backward stochastic dynamics on a filtered probability space. Ann. Probab.* **39** (2011), 1422–1448.
204. A. Lindquist and G. Picci. *Linear stochastic systems. A geometric approach to modeling, estimation and identification.* Series in Contemporary Mathematics, **1**. Springer, Heidelberg, 2015.
205. J.-L. Lions. *Optimal control of systems governed by partial differential equations.* Springer-Verlag, Berlin, Heidelberg, New York, 1971.
206. J.-L. Lions. *Sur la théorie du contrôle. Actes du Congrès International des Mathématiciens 1974.* Vol. 1, 139–154, Vancouver, Canada, Canadian Mathematical Congress, 1975.
207. J.-L. Lions. *Contrôlabilité exacte, perturbations et stabilisation de systèmes distribués,* Tome 1, *Contrôlabilité exacte*, Rech. Math. Appl. 8, Masson, Paris, 1988.

208. J.-L. Lions and E. Magenes. *Non-homogeneous boundary value problems and applications. Vol. I.* Die Grundlehren der mathematischen Wissenschaften, Band 181. Springer-Verlag, New York-Heidelberg, 1972.
209. J.-L. Lions and E. Magenes. *Non-homogeneous boundary value problems and applications. Vol. II.* Die Grundlehren der mathematischen Wissenschaften, Band 181. Springer-Verlag, New York-Heidelberg, 1972.
210. P.-L. Lions. *Optimal control of diffusion processes and Hamilton-Jacobi-Bellman equations. I. The dynamic programming principle and applications.* Comm. Partial Differential Equations. **8** (1983), 1101–1174.
211. P.-L. Lions. *Optimal control of diffusion processes and Hamilton-Jacobi-Bellman equations. II. Viscosity solutions and uniqueness.* Comm. Partial Differential Equations. **8** (1983), 1229–1276.
212. P.-L. Lions. *Optimal control of diffusion processes and Hamilton-Jacobi-Bellman equations. III. Regularity of the optimal cost function.* Nonlinear partial differential equations and their applications. Collège de France seminar, Vol. V (Paris, 1981/1982), 95–205, Res. Notes in Math., 93, Pitman, Boston, MA, 1983.
213. P.-L. Lions. *Viscosity solutions of fully nonlinear second-order equations and optimal stochastic control in infinite dimensions. I. The case of bounded stochastic evolutions.* Acta Math. **161** (1988), 243–278.
214. P.-L. Lions. *Viscosity solutions of fully nonlinear second-order equations and optimal stochastic control in infinite dimensions. III. Uniqueness of viscosity solutions for general second-order equations.* J. Funct. Anal. **86** (1989), 1–18.
215. L. Liu and X. Liu. *Controllability and observability of some coupled stochastic parabolic systems.* Math. Control Relat. Fields. **8** (2018), 829–854.
216. K. Liu. *Stability of Infinite Dimensional Stochastic Differential Equations with Applications.* Pitman Monographs and Surveys in Pure and Applied Mathematics, vol. **135**. Chapman & Hall/CRC, 2006.
217. W. Liu and M. Röckner. *Stochastic partial differential equations: an introduction.* Universitext. Springer, Cham, 2015.
218. X. Liu. *Global Carleman estimate for stochastic parabolic equations, and its application.* ESAIM Control Optim. Calc. Var. **20** (2014), 823–839.
219. X. Liu. *Controllability of some coupled stochastic parabolic systems with fractional order spatial differential operators by one control in the drift.* SIAM J. Control Optim. **52** (2014), 836–860.
220. X. Liu, Q. Lü and X. Zhang. *Finite codimensional controllability and optimal control problems with endpoint state constraints.* J. Math. Pures Appl. **138** (2020), 164–203.
221. X. Liu, Q. Lü, H. Zhang and X. Zhang. *Finite codimensionality technique in optimization and optimal control problems.* Preprint.
222. X. Liu and Y. Yu. *Carleman estimates of some stochastic degenerate parabolic equations and application.* SIAM J. Control Optim. **57** (2019), 3527–3552.
223. S. V. Lototsky and B. L. Rozovsky. *Stochastic Partial Differential Equations.* Universitext. Springer, Cham, 2017.
224. Q. Lü. *Control and observation of stochastic partial differential equations.* Ph D Thesis, Sichuan University, 2010.
225. Q. Lü. *Some results on the controllability of forward stochastic parabolic equations with control on the drift.* J. Funct. Anal. **260** (2011), 832–851.
226. Q. Lü. *Carleman estimate for stochastic parabolic equations and inverse stochastic parabolic problems.* Inverse Problems **28** (2012), 045008.

227. Q. Lü. *Observability estimate and state observation problems for stochastic hyperbolic equations.* Inverse Problems. 29 (2013), 095011.
228. Q. Lü. *Exact controllability for stochastic Schrödinger equations.* J. Differential Equations. **255** (2013), 2484–2504.
229. Q. Lü. *Observability estimate for stochastic Schrödinger equations and its applications.* SIAM J. Control Optim. **51** (2013), 121–144.
230. Q. Lü. *A lower bound on local energy of partial sum of eigenfunctions for Laplace-Beltrami operators.* ESAIM Control Optim. Calc. Var. **19** (2013), 255–273.
231. Q. Lü. *Exact controllability for stochastic transport equations.* SIAM J. Control Optim. **52** (2014), 397–419.
232. Q. Lü. *Stochastic well-posed systems and well-posedness of some stochastic partial differential equations with boundary control and observation.* SIAM J. Control Optim. **53** (2015), 3457–3482.
233. Q. Lü. *Well-posedness of stochastic Riccati equations and closed-loop solvability for stochastic linear quadratic optimal control problems.* J. Differential Equations. **267** (2019), 180–227.
234. Q. Lü and J. van Neerven. *Conditional expectations in $L^p(\mu; L^q(\nu; X))$.* Positivity. **23** (2019), 11–19.
235. Q. Lü and T. Wang. *Characterization of optimal feedback for linear quadratic control problems of stochastic evolution equations.* Preprint.
236. Q. Lü, T. Wang and X. Zhang. *Characterization of optimal feedback for stochastic linear quadratic control problems.* Probab. Uncertain. Quant. Risk **2** (2017), Paper No. 11, 20 pp.
237. Q. Lü and Z. Yin. *Unique continuation for stochastic heat equations.* ESAIM Control Optim. Calc. Var. **21** (2015), 378–398.
238. Q. Lü and Z. Yin. *Local state observation for stochastic hyperbolic equations.* ESAIM Control Optim. Calc. Var. Published online, doi: 10.1051/cocv/2019049.
239. Q. Lü, J. Yong and X. Zhang. *Representation of Itô integrals by Lebesgue/ Bochner integrals.* J. Eur. Math. Soc. **14** (2012), 1795–1823. (Erratum: J. Eur. Math. Soc. **20** (2018), 259–260).
240. Q. Lü, H. Zhang and X. Zhang. *Second order necessary conditions for optimal control problems of stochastic evolution equations.* arXiv:1811.07337.
241. Q. Lü and X. Zhang. *Well-posedness of backward stochastic differential equations with general filtration.* J. Differential Equations. **254** (2013), 3200–3227.
242. Q. Lü and X. Zhang. *General Pontryagin-type stochastic maximum principle and backward stochastic evolution equations in infinite dimensions.* Springer Briefs in Mathematics. Springer, Cham, 2014.
243. Q. Lü and X. Zhang. *Global uniqueness for an inverse stochastic hyperbolic problem with three unknowns,* Comm. Pure Appl. Math., **68** (2015), 948–963.
244. Q. Lü and X. Zhang. *Transposition method for backward stochastic evolution equations revisited, and its application.* Math. Control Relat. Fields. **5** (2015), 529–555.
245. Q. Lü and X. Zhang. *Operator-valued backward stochastic Lyapunov equations in infinite dimensions, and its application.* Math. Control Relat. Fields. **8** (2018), 337–381.
246. Q. Lü and X. Zhang. *A mini-course on stochastic control.* Control and Inverse Problems for Partial Differential Equations. Edited by G. Bao, J.-M. Coron and T.-T. Li. Higher Education Press, Beijing, 2019, 171–254.

247. Q. Lü and X. Zhang. *Exact controllability for a refined stochastic wave equation.* arXiv:1901.06074.
248. Q. Lü and X. Zhang. *Optimal feedback for stochastic linear quadratic control and backward stochastic Riccati equations in infinite dimensions.* arXiv:1901.00978.
249. T. J. Lyons, M. Caruana and T. Lévy. *Differential equations driven by rough paths.* Lecture Notes in Mathematics, **1908**. Springer, Berlin, 2007.
250. T. J. Lyons and Z. Qian. *System control and rough paths.* Oxford Mathematical Monographs. Oxford Science Publications. Oxford University Press, Oxford, 2002.
251. J. Ma and J. Yong. *Forward-backward differential equations and their applications.* Lecture Notes in Mathematics, vol. 1702. Berlin, Springer, 1999.
252. E. Machtyngier. *Exact controllability for the Schrödinger equation. SIAM J. Control Optim.* **32** (1994), 24–34.
253. N. I. Mahmudov. *Controllability of linear stochastic systems. IEEE Trans. Automat. Control* **46** (2001), 724–731.
254. N. I. Mahmudov. *Controllability of linear stochastic systems in Hilbert spaces. J. Math. Anal. Appl.* **259** (2001), 64–82.
255. N. I. Mahmudov. *Controllability of semilinear stochastic systems in Hilbert spaces. J. Math. Anal. Appl.* **288** (2003), 197–211.
256. N. I. Mahmudov and M. A. McKibben. *On backward stochastic evolution equations in Hilbert spaces and optimal control. Nonlinear Anal.* **67** (2007), 1260–1274.
257. A. J. Majda, I. Timofeyev and E. Vanden Eijnden. *A mathematical framework for stochastic climate models. Comm. Pure Appl. Math.* **54** (2001), 891–974.
258. A. M. Márquez-Durán and J. Real. *Some results on nonlinear backward stochastic evolution equations. Stochastic Anal. Appl.* **22** (2004), 1273–1293.
259. J. Martínez-Frutos and F. Periago Esparza. *Optimal control of PDEs under uncertainty. An introduction with application to optimal shape design of structures.* SpringerBriefs in Mathematics. BCAM SpringerBriefs. Springer, Cham, 2018.
260. A. Mercado, A. Osses and L. Rosier. *Inverse problems for the Schrödinger equation via Carleman inequalities with degenerate weights. Inverse Problems.* **24** (2008), 015017.
261. A. P. Meyer. *Probability and potentials.* Blaisdell Publishing Co. Ginn and Co., Waltham, Mass.-Toronto, Ont.-London, 1966.
262. A. S. Monin and A. M. Yaglom. *Statistical fluid mechanics: mechanics of turbulence. Vol. I.* Translated from the 1965 Russian original. Edited and with a preface by John L. Lumley. English edition updated, augmented and revised by the authors. Reprinted from the 1971 edition. Dover Publications, Inc., Mineola, NY, 2007.
263. R. M. Murray *et al. Control in an information rich world. Report of the Panel on Future Directions in Control, Dynamics, and Systems.* Papers from the meeting held in College Park, MD, July 16-17, 2000. Society for Industrial and Applied Mathematics, Philadelphia, PA, 2003.
264. J. Nedoma. *Note on the generalized random variables. Trans. of the first Prague conference on information theory, etc.* 1957, 139–141.
265. E. Nelson. *Dynamical Theories of Brownian Motion.* Princeton University Press, Princeton, N.J. 1967.

266. D. Ocone and É. Pardoux. *A generalized Itô-Ventzell formula. Application to a class of anticipating stochastic differential equations.* Ann. Inst. H. Poincaré Probab. Statist. **25** (1989), 39–71.
267. B. Øksendal. *Stochastic differential equations. An introduction with applications. Sixth edition.* Universitext. Springer-Verlag, Berlin, 2003.
268. B. Øksendal. *Optimal control of stochastic partial differential equations. Stoch. Anal. Appl.* **23** (2005), 165–179.
269. B. Øksendal and A. Sulem. *Maximum principles for optimal control of forward-backward stochastic differential equations with jumps. SIAM J. Control Optim.* **48** (2009/10), 2945–2976.
270. S. Omatu and J. H. Seinfeld. *Distributed parameter systems. Theory and applications.* Oxford Mathematical Monographs. Clarendon Press, Oxford, 1989.
271. É. Pardoux and S. Peng. *Adapted solution of backward stochastic equation. Systems Control Lett.* **14** (1990), 55–61.
272. E. Pardoux and A. Răşcanu. *Stochastic differential equations, backward SDEs, partial differential equations.* Stochastic Modelling and Applied Probability, **69**. Springer, Cham, 2014.
273. S. Peng. *A general stochastic maximum principle for optimal control problems. SIAM J. Control Optim.* **28** (1990), 966–979.
274. S. Peng. *Probabilistic interpretation for systems of quasilinear parabolic partial differential equations. Stoch. Stoch. Rep.* **37** (1991) 61–74.
275. S. Peng. *Backward stochastic differential equation and exact controllability of stochastic control systems. Progr. Natur. Sci. (English Ed.).* **4** (1994), 274–284.
276. S. Peng. *G-expectation, G-Brownian motion and related stochastic calculus of Itô type. Stochastic analysis and applications.* 541–567, Abel Symp., **2**, Springer, Berlin, 2007.
277. S. Peng. *Nonlinear expectations and stochastic calculus under uncertainty. With robust CLT and G-Brownian motion.* Probability Theory and Stochastic Modelling, **95**. Springer, Berlin, 2019.
278. S. Peng and Y. Shi. *A type of time-symmetric forward-backward stochastic differential equations. C. R. Math. Acad. Sci. Paris.* **336** (2003), 773–778.
279. H. Pham. *Continuous-time stochastic control and optimization with financial applications.* Stochastic Modelling and Applied Probability, **61**. Springer-Verlag, Berlin, 2009.
280. K. D. Phung. *Observability and control of Schrödinger equations. SIAM J. Control Optim.* **40** (2001), 211–230.
281. L. S. Pontryagin, V. G. Boltyanskii, R. V. Gamkrelidze and E. F. Mischenko. *Mathematical theory of optimal processes.* Wiley, New York, 1962.
282. C. Prévôt and M. Röckner. *A concise course on stochastic partial differential equations.* Lecture Notes in Mathematics, **1905**. Springer, Berlin, 2007.
283. A. Prohl and Y. Wang. *Strong rates of convergence for space-time discretization of the backward stochastic heat equation, and of a linear-quadratic control problem for the stochastic heat equation.* In submission.
284. E. P. Protter. *Stochastic integration and differential equations. Second edition. Version 2.1. Corrected third printing.* Stochastic Modelling and Applied Probability, **21**. Springer-Verlag, Berlin, 2005.
285. Z. Qian and X. Y. Zhou. *Existence of solutions to a class of indefinite stochastic Riccati equations. SIAM J. Control Optim.* 51 (2013), 221–229.
286. J. Ren, M. Röckner and F. Wang. *Stochastic generalized porous media and fast diffusion equations. J. Differential Equations.* **238** (2007), 118–152.

287. D. Revuz and M. Yor. *Continuous martingales and Brownian motion. Second edition.* Grundlehren der Mathematischen Wissenschaften, **293**. Springer-Verlag, Berlin, 1994.
288. L. Robbiano. *Carleman estimates, results on control and stabilization for partial differential equations.* International Congress of Mathematicians. Vol. IV, 897-920, Kyung Moon Sa Co. Ltd., Seoul, Korea, 2014.
289. L. Rosier and B. Zhang. *Exact boundary controllability of the nonlinear Schrödinger equation.* J. Differential Equations. **246** (2009), 4129–4153.
290. P. Rouchon. *Models and feedback stabilization of open quantum systems.* International Congress of Mathematicians. Vol. IV, 921-946, Kyung Moon Sa Co. Ltd., Seoul, Korea, 2014.
291. W. Rudin. *Functional Analysis.* McGraw-Hill Series in Higher Mathematics, McGraw-Hill Book Co., New York-Düsseldorf-Johannesburg, 1973.
292. D. L. Russell. *Nonharmonic fourier series in the control theory of distributed parameter systems.* J. Math. Anal. Appl. **18** (1967), 542–560.
293. D. L. Russell. *A unified boundary controllability theory for hyperbolic and parabolic partial differential equations.* Studies in Appl. Math. **52** (1973), 189–211.
294. D. L. Russell. *Exact boundary value controllability theorems for wave and heat processes in star-complemented regions.* Differential games and control theory. Proc. NSF–CBMS Regional Res. Conf., Univ. Rhode Island, Kingston, R.I., Dekker, New York, 1974, 291–319.
295. D. L. Russell. *Controllability and stabilizability theory for linear partial differential equations: recent progress and open problems.* SIAM Rev. **20** (1978), 639–739.
296. K. K. Sabelfeld. *Random fields and stochastic Lagrangian models. Analysis and applications in turbulence and porous media.* Walter de Gruyter & Co., Berlin, 2013.
297. Yu. Safarov and D. Vassiliev. *The Asymptotic Distribution of Eigenvalues of Partial Differential Operators.* American Mathematical Society, Providence, RI, 1997.
298. D. Salamon. *Infinite-dimensional linear systems with unbounded control and observation: A functional analytic approach.* Trans. Amer. Math. Soc. **300** (1987), 383–431.
299. R. Schatten. *Norm ideals of completely continuous operators. Second printing.* Ergebnisse der Mathematik und ihrer Grenzgebiete, Band 27 Springer-Verlag, Berlin-New York, 1970.
300. T. Schlick. *Modeling superhelical DNA—recent analytical and dynamical approaches.* Cuur. Opin. Struct. Biol. **5** (1995), 245–262.
301. A. Shirikyan. *Control and mixing for 2D Navier-Stokes equations with space-time localised noise.* Ann. Sci. Éc. Norm. Supér. **48** (2015), 253–280.
302. T. K. Sirazetdinov. *Optimal control of stochastic processes with distributed parameters.* Trudy Kazan. Aviacion. Inst. Vyp. **89** (1965), 88–101. (In Russian)
303. M. Sîrbu and G. Tessitore. *Null controllability of an infinite dimensional SDE with state- and control-dependent noise.* Systems Control Lett. **44** (2001), 385–394.
304. S. S. Sritharan. *Stochastic Navier-Stokes equations: solvability, control, and filtering.* Stochastic partial differential equations and applications–VII, 273–280, Lect. Notes Pure Appl. Math., 245, Chapman & Hall/CRC, Boca Raton, FL, 2006.

305. O. Staffans. *Well-posed linear systems.* Encyclopedia of Mathematics and its Applications, **103**. Cambridge University Press, Cambridge, 2005.
306. K. Stowe. *An introduction to thermodynamics and statistical mechanics. Second edition.* Cambridge University Press, Cambridge, 2007.
307. J. Sun and J. Yong. *Linear quadratic stochastic differential games: open-loop and closed-loop saddle points. SIAM J. Control Optim.* **52** (2014), 4082–4121.
308. J. Sun, X. Li and J. Yong. *Open-loop and closed-loop solvabilities for stochastic linear quadratic optimal control problems. SIAM J. Control Optim.* **54** (2016), 2274–2308.
309. Y. Sunahara, T. Kabeuchi, Y. Asada, S. Aihara and K. Kishino. *On stochastic controllability for nonlinear systems. IEEE Trans. Automatic Control.* **19** (1974), 49–54.
310. Y. Sunahara, S. Aihara, K. Kamejima and K. Kishino. *On stochastic observability and controllability for nonlinear distributed parameter systems. Information and Control* **34** (1977), 348–371.
311. H. J. Sussmann. *Orbits of families of vector fields and integrability of distributions. Trans. Amer. Math. Soc.* **180** (1973), 171–188.
312. B. Sz.-Nagy, C. Foias, H. Bercovici and L. Kérchy. *Harmonic analysis of operators on Hilbert space. Revised and enlarged edition.* Springer, New York, 2010.
313. S. Tang. *General linear quadratic optimal stochastic control problems with random coefficients: linear stochastic Hamilton systems and backward stochastic Riccati equations. SIAM J. Control Optim.* **42** (2003), 53–75.
314. S. Tang and X. Li. *Maximum principle for optimal control of distributed parameter stochastic systems with random jumps.* In: *Differential Equations, Dynamical Systems, and Control Science.* Lecture Notes in Pure and Appl. Math. vol. **152**. Dekker, New York, 1994, 867–890.
315. S. Tang and X. Zhang. *Null controllability for forward and backward stochastic parabolic equations. SIAM J. Control Optim.* **48** (2009), 2191–2216.
316. R. Temam. *Équations aux dérivées partielles stochastiques. Problems in nonlinear analysis (Centro Internaz. Mat. Estivo (C.I.M.E.),* IV Ciclo, Varenna, 1970), 431–462. Edizioni Cremonese, Rome, 1971.
317. G. Tenenbaum and M. Tucsnak. *Fast and strongly localized observation for the Schrödinger equation. Trans. Amer. Math. Soc.* **361** (2009), 951–977.
318. G. Tessitore. *Some remarks on the Riccati equation arising in an optimal control problem with state- and control-dependent noise. SIAM J. Control Optim.* **30** (1992), 717–744.
319. G. Tessitore. *Existence, uniqueness and space regularity of the adapted solutions of a backward SPDE. Stochastic Anal. Appl.* **14** (1996), 461–486.
320. E. Trélat. *Optimal shape and location of sensors or actuators in PDE models. Proceedings of the International Congress of Mathematicians-Rio de Janeiro 2018. Vol. IV. Invited lectures,* World Sci. Publ., Hackensack, NJ, 2018, 3843–3863.
321. N. Touzi. *Second order backward SDEs, fully nonlinear PDEs, and applications in finance. Proceedings of the International Congress of Mathematicians, Vol. IV.* Hyderabad, India, 2010, 3232–3150.
322. N. Touzi. *Optimal stochastic control, stochastic target problems, and backward SDE.* Fields Institute Monographs, **29**. Springer, New York; Fields Institute for Research in Mathematical Sciences, Toronto, ON, 2013.

323. F. Tröltzsch. *Optimal control of partial differential equations. Theory, methods and applications.* Graduate Studies in Mathematics, **112**. American Mathematical Society, Providence, RI, 2010.
324. C. Tudor. *Optimal control for semilinear stochastic evolution equations. Appl. Math. Optim.* **20** (1989), 319–331.
325. M. Tucsnak and G. Weiss. *Observation and control for operator semigroups.* Birkhäuser Advanced Texts: Basler Lehrbücher. Birkhäuser Verlag, Basel, 2009.
326. G. Tzafestas and J. M. Nightingale. *Optimal control of a class of linear stochastic distributed-parameter systems. Proc. Inst. Elec. Engrs.* **115** (1968), 1213–1220.
327. J. van Neerven and M. C. Veraar. *On the stochastic Fubini theorem in infinite dimensions. Stochastic partial differential equations and applications–VII*, 323–336, *Lect. Notes Pure Appl. Math.*, **245**, Chapman & Hall/CRC, Boca Raton, FL, 2006.
328. J. M. A. M. van Neerven, M. C. Veraar and L. W. Weis. *Stochastic integration in UMD Banach spaces. Ann. Probab.* **35** (2007), 1438–1478.
329. J. M. A. M. van Neerven, M. C. Veraar and L. W. Weis. *Stochastic evolution equations in UMD Banach spaces. J. Funct. Anal.* **255** (2008), 940–993.
330. A. Vieru. *On null controllability of linear systems in Banach spaces. Systems Control Lett.* **54** (2005), 331–337.
331. M. I. Vishik, A. V. Fursikov. *Mathematical problems in statistical hydromechanics.* Kluwer, Dordrecht, 1988.
332. J. vom Scheidt. *Stochastic equations of mathematical physics.* Mathematische Lehrbücher und Monographien, II. Abteilung: Mathematische Monographien, **76**. Akademie-Verlag, Berlin, 1990.
333. G. Wachsmuth and D. Wachsmuth. *Convergence and regularization results for optimal control problems with sparsity functional. ESAIM Control Optim. Calc. Var.* **17** (2011), 858–886.
334. D. H. Wagner. *Survey of measurable selection theorems. SIAM J. Control Optim.* **15** (1977), 859–903.
335. J. B. Walsh. *An introduction to stochastic partial differential equations.* École d'été de probabilités de Saint-Flour, XIV1984, 265-C439, Lecture Notes in Math., **1180**, Springer, Berlin, 1986.
336. G. Wang. *L^∞-null controllability for the heat equation and its consequences for the time optimal control problem. SIAM J. Control Optim.* **47** (2008), 1701–1720.
337. G. Wang and L. Wang. *The Carleman inequality and its application to periodic optimal control governed by semilinear parabolic differential equations. J. Optim. Theory Appl.* **118** (2003), 249–461.
338. G. Wang, L. Wang, Y. Xu and Y. Zhang. *Time optimal control of evolution equation.* Progress in Nonlinear Differential Equations and Their Applications: Subseries in Control, **92**, Birkhäuser, Springer International Publishing AG, part of Springer Nature, 2018.
339. G. Wang, Z. Wu and J. Xiong. *An introduction to optimal control of FBSDE with incomplete information.* SpringerBriefs in Mathematics. Springer, Cham, 2018.
340. L. Y. Wang, G. Yin, J.-F. Zhang and Y. Zhao. *System identification with quantized observations.* Systems & Control: Foundations & Applications. Birkhäuser Boston, Ltd., Boston, MA, 2010.

341. P. Wang, Y. Wang and X. Zhang. *Numerical analysis on backward stochastic differential equations by finite transposition method.* Preprint.
342. P. Wang and X. Zhang. *Numerical solutions of backward stochastic differential equations: a finite transposition method. C. R. Math. Acad. Sci. Paris.* **349** (2011), 901–903.
343. P. K. C. Wang. *Control of distributed parameter systems.* Advances in Control Systems, Theory and Applications, Vol. 1, Edited by C. T. Leondes, 75–172. Academic Press, New York, 1964.
344. Y. Wang. *Transposition Solutions of Backward Stochastic Differential Equations and Numerical Schemes.* Ph.D. thesis, Academy of Mathematics and Systems Science, Chinese Academy of Sciences, 2013.
345. Y. Wang. *A semidiscrete Galerkin scheme for backward stochastic parabolic differential equations. Math. Control Relat. Fields.* **6** (2016), 489–515.
346. Y. Wang. *L^2-regularity of solutions to linear backward stochastic heat equations, and a numerical application. J. Math. Anal. Appl.* **486** (2020), 123870, 18 pp.
347. Y. Wang, D. Yang, J. Yong and Z. Yu. *Exact controllability of linear stochastic differential equations and related problems. Math. Control Relat. Fields.* **7** (2017), 305–345.
348. Y. Wang and Z. Yu. *On the partial controllability of SDEs and the exact controllability of FBSDEs. ESAIM Control Optim. Calc. Var.* In press.
349. N. Wiener. *Cybernetics or control and communication in the animal and the machine.* The MIT Press, Cambridge, Massachusetts, 1948.
350. J.C. Willems. *Topological classification and structural stability of linear systems. J. Differential Equations.* **35** (1980), 306–318.
351. W. M. Wonham. *On the separation theorem of stochastic control. SIAM J. Control.* **6** (1968), 312–326.
352. W. M. Wonham. *On a matrix Riccati equation of stochastic control. SIAM J. Control.* **6** (1968), 681–697.
353. B. Wu, Q. Chen and Z. Wang. *Carleman estimates for a stochastic degenerate parabolic equation and applications to null controllability and an inverse random source problem. Inverse Problems.* Published online, DOI: 10.1088/1361-6420/ab89c3.
354. P. Wu and G. Wang. *Representation theorems for generators of BSDEs and the extended g-expectations in probability spaces with general filtration. J. Math. Anal. Appl.* **487** (2020), 124010.
355. D. Xia, S. Li, S. Yan, W. Shu. *Theory of Real Functions and Functional Analysis, Vol. 2 (Second version).* Gaodeng Jiaoyu Chubanshe (Higher Education Press), Beijing, 1985 (In Chinese).
356. J. Xiong. *An introduction to stochastic filtering theory.* Oxford Graduate Texts in Mathematics, **18**. Oxford University Press, Oxford, 2008.
357. M. Yamamoto. *Carleman estimates for parabolic equations and applications. Inverse Problems.* **25** (2009), 123013.
358. L. Yan, B. Wu, S. Lu and Y. Wang. *Null controllability and inverse source problem for stochastic Grushin equation with boundary degeneracy and singularity.* arXiv:2001.01877.
359. Y. Yan. *Carleman estimates for stochastic parabolic equations with Neumann boundary conditions and applications. J. Math. Anal. Appl.* **457** (2018), 248–272.
360. Y. Yan and F. Sun. *Insensitizing controls for a forward stochastic heat equation. J. Math. Anal. Appl.* **384** (2011), 138–150.

361. D. Yang and J. Zhong. *Observability inequality of backward stochastic heat equations for measurable sets and its applications. SIAM J. Control Optim.* **54** (2016), 1157–1175.
362. D. Yang and J. Zhong. *Optimal actuator location of the minimum norm controls for stochastic heat equations. Math. Control Relat. Fields.* **8** (2018), 1081–1095.
363. Y. Yavin. *Feedback strategies for partially observable stochastic systems.* Lecture Notes in Control and Information Sciences, **48**. Springer-Verlag, Berlin, 1983.
364. G. Yin and Q. Zhang. *Continuous-time Markov chains and applications. A singular perturbation approach.* Applications of Mathematics (New York), **37**. Springer-Verlag, New York, 1998.
365. Z. Yin. *A quantitative internal unique continuation for stochastic parabolic equations. Math. Control Relat. Fields.* **5** (2015), 165–176.
366. Z. Yin. *Lipschitz stability for a semi-linear inverse stochastic transport problem. J. Inverse Ill-Posed Probl.* **28** (2020), 185–193.
367. J. Yong. *Backward stochastic Volterra integral equations and some related problems. Stoch. Proc. Appl.* **116** (2006), 779–795.
368. J. Yong. *Well-posedness and regularity of backward stochastic Volterra integral equations. Prob. Theory Rel. Fields.* **142** (2008), 21–77.
369. J. Yong, *Time-inconsistent optimal control problems. Proceedings of the International Congress of Mathematicians 2014, Vol. IV.* Seoul, Korea, 2014, 947–969.
370. J. Yong and H. Lou. *A Concise Course on Optimal Control Theory.* Higher Education Press, Beijing, 2006. (In Chinese)
371. J. Yong and X. Y. Zhou. *Stochastic controls: Hamiltonian systems and HJB equations.* Springer-Verlag, New York, 1999.
372. K. Yosida. *Functional analysis.* Classics in Mathematics. Springer-Verlag, Berlin, 1995.
373. G. Yuan. *Determination of two kinds of sources simultaneously for a stochastic wave equation. Inverse Problems.* **31** (2015), 085003.
374. G. Yuan. *Conditional stability in determination of initial data for stochastic parabolic equations. Inverse Problems.* **33** (2017), 035014.
375. G. Yuan. *Determination of two unknowns simultaneously for stochastic Euler-Bernoulli beam equations. J. Math. Anal. Appl.* **450** (2017), 137–151.
376. J. Zabczyk. *Controllability of stochastic linear systems. Systems Control Lett.* **1** (1981/82), 25–31.
377. J. Zabczyk. *Mathematical control theory. An introduction. Reprint of the 1995 edition.* Modern Birkhäuser Classics. Birkhäuser Boston, Inc., Boston, MA, 2008.
378. H. Zhang. *Second-order necessary conditions for stochastic optimal controls.* Ph D thesis, Sichuan University, 2014.
379. H. Zhang and X. Zhang. *Pointwise second-order necessary conditions for stochastic optimal controls, Part I: The case of convex control constraint. SIAM J. Control Optim.* **53** (2015), 2267–2296.
380. H. Zhang and X. Zhang. *Second-order necessary conditions for stochastic optimal control problems. SIAM Rev.*, **60** (2018), 139–178.
381. J. Zhang. *Backward stochastic differential equations. From linear to fully nonlinear theory.* Probability Theory and Stochastic Modelling, **86**. Springer, New York, 2017.

382. R. Zhang and L. Guo. *Controllability of stochastic game-based control systems. SIAM J. Control Optim.* **57** (2019), 3799–3826.
383. W. Zhang, L. Xie and B-S. Chen. *Stochastic H_2/H_∞ control. A Nash game approach.* CRC Press, Boca Raton, FL, 2017.
384. X. Zhang. *Explicit observability estimate for the wave equation with potential and its application. R. Soc. Lond. Proc. Ser. A Math. Phys. Eng. Sci.* **456** (2000), 1101–1115.
385. X. Zhang. *Exact controllability of the semilinear plate equations. Asymptot. Anal.* **27** (2001), 95–125.
386. X. Zhang. *Carleman and observability estimates for stochastic wave equations. SIAM J. Math. Anal.* **40** (2008), 851–868.
387. X. Zhang. *Unique continuation for stochastic parabolic equations. Differential Integral Equations.* **21** (2008), 81–93.
388. X. Zhang. *A unified controllability/observability theory for some stochastic and deterministic partial differential equations. Proceedings of the International Congress of Mathematicians.* Volume IV, 3008–3034, Hindustan Book Agency, New Delhi, 2010.
389. X. Zhang. *A brief talk on mathematical control theory. Lectures at Institute of Mathematics 2014.* Edited by N. Xi, Q. Feng, X. Zhang and B. Fu. Science Press, Beijing, 2017, 1–38. (In Chinese)
390. X. Zhang and E. Zuazua. *Unique continuation for the linearized Benjamin-Boma-Mahony equation with space dependent potential. Math. Ann.* **325** (2003), 543–582.
391. M. Zhen, J. He, H. Xu and M. Yang. *Carleman and observability estimates for stochastic beam equation.* arXiv:1710.02910v4.
392. X. Y. Zhou. *On the necessary conditions of optimal controls for stochastic partial differential equations. SIAM J. Control Optim.* **31** (1993), 1462–1478.
393. X. Y. Zhou. *Sufficient conditions of optimality for stochastic systems with controllable diffusions. IEEE Trans. Automat. Control.* **41** (1996), 1176–1179.
394. E. Zuazua. *Remarks on the controllability of the Schrödinger equation. Quantum control: mathematical and numerical challenges,* 193–211, CRM Proc. Lecture Notes, 33, Amer. Math. Soc., Providence, RI, 2003.
395. E. Zuazua. *Propagation, observation, and control of waves approximated by finite difference methods. SIAM Rev.* **47** (2005), 197–243.
396. E. Zuazua. *Controllability and observability of partial differential equations: some results and open problems.* in *Handbook of Differential Equations: Evolutionary Equations, vol. 3,* Elsevier Science, 2006, 527–621.

Index

Q. Lü and X. Zhang, *Mathematical Control Theory for Stochastic Partial Differential Equations*, Probability Theory and Stochastic Modelling 101, https://doi.org/10.1007/978-3-030-82331-3

Printed by Printforce, the Netherlands